Advances in Oil and Gas Exploration & Production

Series editor

Rudy Swennen, Department of Earth and Environmental Sciences,
K.U. Leuven, Heverlee, Belgium

The book series Advances in Oil and Gas Exploration & Production publishes scientific monographs on a broad range of topics concerning geophysical and geological research on conventional and unconventional oil and gas systems, and approaching those topics from both an exploration and a production standpoint. The series is intended to form a diverse library of reference works by describing the current state of research on selected themes, such as certain techniques used in the petroleum geoscience business or regional aspects. All books in the series are written and edited by leading experts actively engaged in the respective field.

The Advances in Oil and Gas Exploration & Production series includes both single and multi-authored books, as well as edited volumes. The Series Editor, Dr. Rudy Swennen (KU Leuven, Belgium), is currently accepting proposals and a proposal form can be obtained from our representative at Springer, Dr. Alexis Vizcaino (Alexis.Vizcaino@springer.com).

More information about this series at http://www.springer.com/series/15228

Xinghe Yu · Shengli Li
Shunli Li

Clastic Hydrocarbon Reservoir Sedimentology

Springer

Xinghe Yu
China University
of Geosciences
Beijing
China

Shengli Li
China University
of Geosciences
Beijing
China

Shunli Li
China University of Geosciences
Beijing
China

ISSN 2509-372X ISSN 2509-3738 (electronic)
Advances in Oil and Gas Exploration & Production
ISBN 978-3-030-09947-3 ISBN 978-3-319-70335-0 (eBook)
https://doi.org/10.1007/978-3-319-70335-0

Translated from Xinghe Yu. 2006. Clastic Hydrocarbon Reservoir Sedimentology (Second Edition). Beijing: Petroleum Industry Press. ISBN 978-7-5021-6399-0

Softcover re-print of the Hardcover 1st edition 2018

Printed on acid-free paper

This Springer imprint is published by the registered company Springer International Publishing AG part of Springer Nature
The registered company address is: Gewerbestrasse 11, 6330 Cham, Switzerland

Preface

It has been over five years since the first publication of *Clastic Petroleum Reservoir Sedimentology* in September 2002, and since this time, it has been used as a textbook for graduates majoring in geology in universities and colleges where recognition and good reputation have been attributed to teachers and students, particularly at the China University of Geosciences (CUG) (Beijing). The book has also been used as a text in over 20 domestic professional training courses related to petroleum reservoir and sedimentology, particularly at three giant petroleum companies (PetroChina, Sinopec, and CNOOC), which reflects the considerable attention accorded to it by professionals in industrial and academic circles. With the development of hydrocarbon exploration, a larger number of petroleum geological scientists consider that this book not only carries the function of theoretical direction for scientific research, but also has significant practical value in the practice of hydrocarbon exploration. Hence, there is a great expectation for the second edition.

For the first edition, some readers pointed out the shortcomings therein and proposed certain amendments, which were valued by me. Owing to new academic accumulation in reservoir sedimentology and the advancement of scientific research in recent years, my intention was to make systematical modification and supplementation to the book, and thus had the idea of publishing the second edition. For the above-mentioned reasons, I completed the second edition in 200 days by further systematizing the original manuscript, referring to lecture notes, scientific achievements, published articles, and new academic points of view over recent years, and by making word-by-word modifications of the first edition and adding new chapters. This second edition is now about to be published, and I would thus like to cite the same famous Chinese proverb, as in the foreword to the first edition, "Human life is limited, but knowledge is limitless." Since I am restricted in view and shallow in understanding, this book will thus be used in the way of throwing out a minnow to catch a whale.

There are thirteen chapters in this book, covering many aspects of clastic petroleum reservoir sedimentology. This book, in addition to the latest trends in domestic and overseas reservoir studies, and along with basic knowledge, such as essential reservoir characteristics, diagenesis, and reservoir heterogeneity, also covers depositional features and reservoir characteristics of all types of depositional systems (alluvial fans, rivers, lakes, delta, shorelands, and deepwater) for clastic rock. Furthermore, this book presents detailed discussions on the classification of each system, geologic features, sedimentary sequences, identification marks, facies models, and corresponding reservoir characteristics. Moreover, in consideration of the fast development of sequence stratigraphy in China, Chapter 4, "Theory and Methods for Studying Clastic Sequence Stratigraphy," has been added.

In the process of re-publication, I would like to thank the people who participated in the compilation and modification, as follows: Assoc. Prof. Li Shengli (Chaps. 4 and 11), Dr. Zheng Xiujuan (Chap. 1), Master Bai Zhenhua (Chaps. 4 and 5), and Master Fu Ju (Chap. 6); in addition, people who participated in error checks, supplementation, and data compilation include Zou Dejiang, Yang Fan, Ren Xiaojun, Chang Shuyun, Zhan Lufeng, Zhang Shuping, Li Mei, Sun Xiangcan, Yuan Kun, and Gao Dongchen; in addition, Dr. Zheng Xiujuan, who fully proofread the text from the first edition. I would like to express my gratitude to Profs. Qiu Yi'nan, Wang Defa, Zheng Junmao, Tian Shicheng, and other experts and scholars who proposed excellent suggestions for amendment. It should be noted that the second edition of this book is generously subsidized by the Graduate School of CUG (Beijing) and the Specialized Research Fund for the Doctoral Program of Higher Education of China (20050491001). I would also like to mention my wife Miss Hu Yihong and family members who have been involved in the compilation of this book. Many thanks for their great patience, passion, and complete support during the long period of compilation and modification.

Beijing, China
December 2016

Xinghe Yu
Shengli Li
Shunli Li

Contents

1 Introduction

With continuing global hydrocarbon exploration and development, improved descriptions of hydrocarbon reservoirs, and associated studies, the appearance (spatial distribution) and internal properties (physical property) of petroleum reservoirs have become the focus of hydrocarbon exploration and developmental research. However, the external shape (configuration) and distribution law of the internal properties within petroleum reservoirs under different geological conditions are related to their formation environment and associated conditions, and sedimentary petroleum reservoirs that are formed by particular depositional systems have differing distribution laws and heterogeneity characteristics. It is thus necessary to analyze the geological processes and sedimentary environment of the various types of reservoirs. Diagenesis also imposes a remarkable influence on the internal properties of reservoirs. Therefore, the overall characteristics of a petroleum reservoir depend on three factors affecting heterogeneity, the "stage of tectonic evolution, diversity of the sedimentary environment, and the complexity of diagenesis," and gaining a basic knowledge of these areas and associated theory gives a solid foundation in reservoir sedimentology.

As a key branch of sedimentology, reservoir sedimentology is categorized as applied sedimentology. It is therefore necessary to conduct an overall review of the formation and development of sedimentology, as doing so provides a basis for the analysis of historical studies and associated developmental trends of reservoir sedimentology. Sedimentary environment and sedimentary facies are key components of sedimentology, and are the basis for lithofacies restoration and palaeogeography, research on the distribution of depositional systems, explanations of seismic facies, the building of isochronous sequence stratigraphic frameworks, basin analysis, and prediction of the distribution of favorable reservoirs. Therefore, to enable study in these areas, it is firstly necessary to define the concept and connotations of a sedimentary environment and facies, and to gain an understanding of the role of the facies model, as well as similarities and differences between reservoir models.

1.1 Formation and Development of Modern Sedimentology

The definition of sedimentology was first put forward by H. A. Wadell in 1932 as, "A science for sediments research." In 1978, Friedman and Sanders then gave a complete definition of the sedimentology research field in the *Principles of Sedimentology* as, "A science for research on sediments, sedimentation process, and sedimentary rock and sedimentary environment." Furthermore, in 1980 the *Glossary of Geology* defined sedimentology as, "A science of research on sediment resources, description and classification of sedimentary rocks and research on the formation process of sediments."

X. Yu et al., *Clastic Hydrocarbon Reservoir Sedimentology*, Advances in Oil and Gas Exploration & Production, https://doi.org/10.1007/978-3-319-70335-0_1

From the initial proposal of its concept to the present day, sedimentology has been constantly developed, enriched, and improved through the tireless efforts of global sedimentologists. In the beginning, actualism (i.e., uniformitarianism) contributed to the development of the subject, when, the formally published *Principles of Geology* (1837), Lyell proposed that, "Ongoing geologic processes in modern times once occurred with the essentially same intensities or ways during the whole geological period, and ancient geological events may be explained by phenomena and roles observed today." This definition was then encapsulated by A. Geikie in 1905, with the phrase, "The Present is the Key to the Past." Since then, the intersection and infiltration of different subjects, such as interinfiltrations between sedimentology and oceanography, physics, chemistry, aerology, hydrology, plate tectonics, petroleum geology, paleontology, geophysical logging, and seismic stratigraphy have played a considerable role in its development. Lately, the introduction of new technologies, such as the invention, use, and popularization of the polarizing microscope, X diffractometer, scanning electron microscope, energy disperse spectroscope (EDS), mass spectrometer, computed tomography (CT), nanometer (NM), and computers, have accelerated the development of sedimentology.

It can be said that sedimentology has mainly transitioned through two historical development periods, early modern times and modern times, and its development in early modern times can be further divided into three phases, as follows.

1.1.1 Emerging Phase (1830–1894)

Most of the research conducted during this phase focused on paleobiological strata, coal, and bioherms. In 1837, Lyell published an epoch-making monograph, the *Principles of Geology*, in which he put forward the principle and method of actualism, and stated that, "The Present is the Key to the Past." This book offered guidelines for later geological scientific research. Then, in 1857, Sorby first used the polarizing microscope in rock research, and determined that, "Using all polarization means is an absolute necessity," which lifted the veil for the microscopic study of rocks. In 1884, Murray and Renard collaborated to write *Deep-sea Deposits*, which could be used in the classification and description of deep-sea sediments. Furthermore, in 1894, Walther wrote the *Introduction to Geology as a Historical Science*, and put forward the concept of a "facies sequence," namely, Walther's law (law of correlation of facies), which determined, "The principle that facies that occur in conformable vertical successions of strata also occur in laterally adjacent environments. [Only laterally adjacent and dependent facies can be vertically superimposed in the absence of a hiatus]." A limited number of monographs emerged during this period, and their enormous influence and considered academic viewpoint on sedimentology cannot be ignored, even today (Table 1.1).

1.1.2 Initial Formation Phase (1894–1931)

Due to the high demand for minerals, geological exploration evolved in line with social development. Sedimentary studies (sedimentary rocks) were historically encouraged, data accumulated, and a collection of monographs and articles (Table 1.1) were published. As early as 1881, sedimentologists began studying heavy minerals using a microscope, and thus analyze the source direction and properties of sediments. In 1913, F. H. Hatch and R. H. Rastall co-authored the book, *Petrology of the Sedimentary Rocks*, and in 1914, G. K. Gilbert conducted the first flume experiments with sand of different grain sizes and different flow intensities to solve problems in the use of hydraulic power to mine gold, which was an important step in hydrodynamic experiments. Following this, H. B. Milner published *Sedimentary Petrography* in 1922, and W. H. Twenhorel edited *Treatise on Sedimentation* in 1926. Also in 1922, Wentworth first determined the power of 2 as a grain size boundary for dividing clastic sediment particles, and 2 mm as the upper limit grain size of sand, thereby leading the

Table 1.1 Representative works in the development of sedimentology before 1950 and associated contribution

Phase	Year	Author	Book or Article Title	Main Viewpoint or Functions, and Significance
Emergence	1837	C. Lyell	*Principles of Geology*	Proposing the principle of actualism, that "The Present is the Key to the Past"
	1884	J. Murray et al.	*Deep-sea Deposits*	Beginning of classification and description of deep-sea sediments
	1894	J. Walther	*Introduction to Geology as a Historical Science* (in three volumes)	Proposing Walther's law, and first use of the phrase, "comparative petrology"
Initial formation	1905	van't Hoff	*Evaporite*	Proposing the viewpoint of chemical precipitation
	1913	F. H. Hatch et al.	*Petrology of the Sedimentary Rocks*	Sedimentary petrology starts to serve as an independent branch of geoscience separated from stratigraphy. Later, research begins to be conducted on flume experiments; the division of grain size tends to become scientifically quantitative, and more attention is paid to research on sedimentation
	1922	H. B. Milner	*Sedimentary Petrography*	
	1922	C. K. Wentworth	*A Scale of Grade and Class Terms for Clastic Sediments*	
	1925	Twenhorel	*Sedimentation*	
	1926	W. H. Twenhorel	*Treatise on Sedimentation*	
	1929	Lucien Cayeux	*Sedimentary Rocks of France*	
	1931	Society for Sedimentary Geology (SEPM) in United States	First publication of *Journal of Sedimentary Petrology*	Marking the emergence of sedimentology as an independent discipline
Professional studies	1932	Nalivkin (Д. В. Наивки)	*Study of Facies*	Sedimentary petrology trends veer from qualitative research to become semi-quantitative, and greater importance is attached to research on sedimentation and sediment formation mechanisms
	1932	M. C. Svecovs. (M. C. Швецов)	*Petrology of the Sedimentary Rocks*	
	1939	P. D. Trask	*Recent Marine Sediments*	
	1939	Twenhorel	*Principles of Sedimentation*	
	1940	M- C- Pustovalov. (M- C- вечов)	*Petrology of the Sedimentary Rocks*	
	1949	F. J. Pettijohn	*Sedimentary Rocks* (3rd edition, revised and published in 1975)	A significant milestone in the maturity of sedimentary petrology
	1950	P. D. Trask	*Applied Sedimentation*	Proposing the concept and quantification method of sorting coefficients

classification of sandstones and conglomerates to the field of science quantization. Subsequently, in 1929, Lucien Cayeux from France published the first volume of *Sedimentary Rocks of France*, focusing on siliceous rocks. Thereafter in 1931, SEPM began publication of the *Journal of Sedimentary Petrology*. The abovementioned works show the development of sedimentary petrology into an independent discipline that is separate from stratigraphy.

1.1.3 Professional Study Phase (1932–1950)

With the application of new technologies in the field of sedimentology, including differential thermal analysis and X-ray diffraction, research on sedimentary rocks tended to become semi-quantitative. A series of monographs and articles (Table 1.1) on sedimentary petrology were successively published and released during this period. In 1932, Nalivkin (Д. В. Наивкин) published the *Study of Facies*; and in 1932 and 1940, M. C. Svecovs. (M.C. Швецов) and M- C-Pustovalov (M- C- вечов) respectively published the *Petrology of the Sedimentary Rocks*. P. In 1934, Krumbein conducted quantitative research on a sedimentary environment, applied the roundness and shapes of clasts, and determined three main factors–boundary conditions, grains, and energy within a depositional system–and took the lead in applying statistics to sedimentology. D. Trask published *Recent Marine Sediments* in 1939 and *Applied Sedimentation* in 1950. In 1940, M. T. Halbouty researched the provenance, marine transgression and regression, change of an ancient shoreline, and the stratigraphic pinch out and petroleum prospect of the Gulf of Mexico. Furthermore, in 1949, Pettijohn wrote *Sedimentary Rocks*, which marked the mature growth stage of sedimentary petrology. Consequently, the development of early modern sedimentology focused mainly on the discussion of the formation mechanism of various conventional sedimentary rocks from the angle of a single sedimentary petrology discipline.

1.2 Development and Features of Modern Sedimentology

Since the end of World War II and the subsequent start of a global cold war, all countries have attached great importance to, and increased investment in, the development of science and technology. Due to the above stimulation and the influence of new demands on the scientific community, accelerated development has occurred within all disciplines. The division of labor of science itself tends to be elaborate, without exception of sedimentology development. It is precisely because of the rapid development of modern science that leaps have been made in the progress of sedimentology every 10 years, and as a subject it experienced four obvious development stages between the mid-20th century to the early 21st century. These are presented as follows:

1.2.1 Basic Maturity Phase (1950s–1960s)

After World War II, countries globally accelerated construction of their economies (particularly European countries and the USA), and the global industrial revolution entered a new period. Owing to demands for mineral resources in accordance with industrial development, sedimentology gradually transformed from a basic pure scientific theory into an applied science, and became increasingly connected with production practice, particularly in oil and gas exploration. After 1950, there was significant development in terms of research scale and direction with respect to modern sediments, and a number of monographs and literature (Table 1.2) were published and released. Representative works include *From Sedimentary Petrology to Sedimentology*, published by D. J. Doeglas in 1951; *Petrology of Sedimentary Rocks,* released by A. Vatan in 1954; *Primary Structures in Some Recent Sediments*, released by E. D. Mckee in 1957; *Paleocurrents and Basin Analysis*, co-authored by F. J. Pettijohn and P. E. Potter in 1963, *Atlas and Glossary of Primary Sedimentary Structures*, published in 1964; and Use of Vertical Profile in Environmental Reconstruction, proposed by Visher in 1965. During the same period, H. N. Fisk investigated the modern Mississippi Delta in detail, and the study of a Bahamas beach by L. V. Illing represented the first detailed research on modern carbonate deposition, which

Table 1.2 Representative works during the basic maturity phase and associated contribution (1951–1969)

Year	Author	Book or Article Title	Main Viewpoints or Functions, and Significance
1950	P. H. Kuenen et al.	*Turbidity Currents as a Cause of Graded Bedding*	A new chapter of research on turbidity current was opened
1951	D. J. Doeglas	*From Sedimentary Petrology to Sedimentology*	Sedimentology was at a new stage during the development of sedimentary petrology and a new field of geoscience was developed
1955	Ye Lianjun	*Sedimentary Conditions of Manganese Deposit in China*	A huge impact was made in relation to research on sedimentary deposits in China
1955	Liu Hongyun	*Paleogeographic Map of China*	Compilation of the first complete palaeogeographic atlas in China
1957	E. D. Mckee	*Primary Structures in Some Recent Sediments*	Emphasis on primary structure features formed under different sedimentary environments
1957	F. J. Pettijohn	*Sedimentary Rocks*	A significant mark in the maturation of sedimentary petrology, and an historical summary of sedimentary petrology over the last half century
1959	R. L. Folk	*Practical Petrographic Classification of Limestones*	Marking a new stage in research of carbonate rock
1961	A. Fi. Bouma et al.	*Turbidite*	Proposition of the famous Bouma Sequence
1961	Chiefly edited by Zeng Yunfu and Liu Baojun	*Sedimentary Facies and Paleogeography*	Ending the primary use of work of international scholars in domestic teaching colleges in China, and an attempt to use textbooks in sedimentary petrology written by domestic scholars
1962	Dai Donglin	*Petrology of the Sedimentary Rocks*	
1962	Chiefly edited by Liu Baojun	*Research Method of Sedimentary Rocks*	
1962	Chiefly edited by Wu Chongjun	*Petrology of the Sedimentary Rocks*	
1963	P. E. Potter et al.	*Paleocurrents and Basin Analysis*	A new lead is taken in conducting research on sedimentology by considering the basin as a whole
1964	Ye Zhizheng, Meng Xianghua and He Qixiang	*Texture-generic Classification of Carbonate Rock*	Marking the beginning of the modern concept of carbonate rock research in China
1967	W. L. Fisher et al.	*Depositional systems in Wilcox Group (Eocene) of Texas and Their Relationship to Occurrence of Oil and Gas on Gulf Shores*	Proposition of the concept and connotations of a depositional system

thereby played a significant role in guiding scientists to recognize carbonate rock from a new perspective. The progress made during this phase covers modeling applications, and genetic interpretation and graphic methods, particularly in relation to the subjects listed below; thus the development of sedimentology was of epoch-making significance.

1.2.1.1 Knowledge of Turbidity Current and Tractional Currents

It was initially believed that tractional currents were the main hydrodynamic process involved in the formation mechanism of clastic rocks. However, a new chapter of research on turbidity currents developed following the publication of *Turbidity Currents as a Cause of Graded Bedding* by Kuenen and Migiorini in 1950. After this publication, research on the formation and distribution of turbidity currents earned widespread attention in the field of geoscience, which resulted in a revolution in the field of sedimentology. Thereafter, under the guidance of Kuenen, A. H. Bouma conducted research on turbidity currents and flysch formation, proposed features of turbidity sediments, and built the famous "Bouma Sequence" in 1961. He then edited *Turbidite* together with A. Bronwer, forming two characteristic theories between the 1950s and 1960s: The Biochemical Origin of Carbonate Rock and The "Bouma Sequence" of a Turbidity Current.

1.2.1.2 Role of Flume Experiments in Sedimentary Structure Interpretation

In the early 1960s, American geologists applied the concept of hydraulics in a flume experiment, and from their experimental results determined the hydraulic conditions for the formation of bedding and ripples (Simons and Richardson 1962). Their experiments determined that the bedform changing sequence of a nonviscous bed when flow velocity is increased and depth reduced is that of no movement → lower flat bed → sand ripples → sand waves → dune → transition area → upper flat bed → antidunes → scour pit and chute. The change in sandy bedform was well explained for various sedimentary structures, where the mechanism and hydrodynamic conditions were formed by tabular, trough cross bedding, and parallel bedding.

1.2.1.3 Grain Size Analysis for the Interpretation of Hydrodynamic Conditions

R. Passega and G. S. Visher put forward a C-M diagram and cumulative probability graph explained by a tractional current in 1964 and 1969, respectively, which analyze and explain a sedimentary environment based on hydrodynamic conditions by virtue of grain size quantification; thus, genetic analysis became more scientific and feasible.

1.2.1.4 New Phase for Research on Carbonate Rocks

In 1959, R. L. Folk introduced a genetic point of view for clastic rocks in relation to the classification of carbonate rocks. He also classified and explained clastic rocks, and revealed the identity of carbonate rocks and terrigenous clastic rocks in terms of their formation processes and mechanisms. This marked a new phase in research methods and associated knowledge of carbonate rocks.

1.2.1.5 Construction of Vertical Sedimentary Sequence or Facies Model

In 1965, Visher constructed 13 facies models from the perspective of vertical sedimentary sequences. These included four river models, four regression models, two transgression models, one delta model, one deep-sea model, and one lake facies model. The construction of a vertical model formed the foundation for the recognition of sedimentary facies, especially for the specific operation of analyzing sedimentary facies from field and downhole data.

1.2.2 Summary and Improvement Phase (1970s)

In the early 1970s, A. V. Carozzi translated the book *Carbonate Rocks* (written in French by Caye) into English for wider circulation, and in 1971, R. G. C. Bathurst published *Carbonate Sedimentary and their Diagenesis.* These achievements represented the mature phase in the study of carbonate rocks. During this period, a large number of summary monographs on sedimentology emerged (Table 1.3), for example Morgan edited *Deltaic sedimentation: Modern and Ancient* in 1970, R. C. Selley completed *Ancient Sedimentary Environments*, J. R. Allen wrote *Physical Process of Sedimentation*, and R. M. Garrels and F. T. Mackenzie coauthored *Evolution of Sedimentary Rocks.* In 1972, Blatt and Middleton co-authored *Origin of Sedimentary Rocks*, which summarized various mechanisms for the formation of sedimentary rocks, and quoted research achievements from hydromechanics. In the same year, *Sand and Sandstone,* coauthored by F. J. Pettijohn, P. E. Potter, and R. Siever, was formally published; this book summarizes important achievements made during research on clastic rocks previously determined at a seminar in Canada in 1964, and holds that sandstone can play an important role in tracing stratum history. In 1973, Reineck from Germany and Singh from Indian co-authored and published *Depositional Sedimentary Environments with Reference to Terrigenous Clastics* from the perspective of sedimentary structure research. In 1976, R. G. Walker from Canada compiled *Facies Models*, and R. C. Selley rewrote *An Introduction to Sedimentology* (2nd Edition) (Selley 1976). In the late 1970s, the most noteworthy achievement is that of *Sedimentary Environments and Facies*, which was chiefly edited by H. G. Reading in 1978, and *Principles of Sedimentology* published by G. M. Friedman and J. L. Sanders, which systematically summarizes geologic features and the formation mechanisms of various sedimentary environments, thereby reflecting the highest level of sedimentology research at that time.

The above works comprehensively upgraded and summarized the theory of sedimentology to an authoritative stage, particularly in relation to clastic rocks, thereby laying a firm foundation for further developments in sedimentology. At this point in time, Chinese scholars mainly learned and applied new theories and technologies obtained from overseas, and universities began writing trial textbooks on the *Petrology of Sedimentary Rocks* and *Sedimentary Environments and Facies*, thus guiding college students in China within a discipline originally consisting of scattered knowledge towards one with systematic study. The author began to learn the science of sedimentology under the guidance of these works. It is worth mentioning that some universities in China (for example, Beijing Graduate School of Wuhan Geological College, and Chengdu College of Geology), together with production departments, conducted research on the sedimentary environment of sedimentary minerals and petroleum resources, as well as the genesis of minerals, and published a number of papers and monographs on continental deposition in our country, thereby opening a new page in the development of sedimentology in China.

To conclude, the general features during this period are as follows: ① extensive research on identifying marks of sedimentary facies was conducted, and facies models were constructed for various sedimentary environments to enhance the operability of sedimentology research; ② research on sedimentary facies had an obvious evolutionary viewpoint; namely, analyzing the changes in depositional facies from the perspective of evolution; ③ facies analysis within global sedimentary basins was initiated; ④ knowledge of gravity flow was incorporated in the interpretation and classification of a particle supported mechanism; and ⑤ the theory of sedimentology was applied in hydrocarbon exploration and development.

In the 1960s–1970s, when the petroleum industry faced a large transition, reservoir sedimentology as a new science emerged at the right moment to better explain the formation and evolution of petroleum reservoirs. It used the

Table 1.3 Representative works during the summary and improvement phases associated contribution (1970–1980)

Year	Author	Book or Article title	Main viewpoints or functions, and significance	Remarks
1970	S. J. Pieson	*Geologic Well Log Analysis*	Applied well-logging to research on oildom sedimentology for the first time	
1970	R. C. Selley	*Ancient Sedimentary Environments*	Defined the relationship between environment and facies, and systematically introduced the geologic features of various sedimentary environments	2nd edition in 1976; 3rd edition in 1985
1970	J. R. Allen	*Physical Process of Sedimentation*	Put forward the control of sedimentary processes and sediment distribution, and a mutual relation between "deposition rate" and "settlement rate"	
1971	R. G. C. Bathurst	*Carbonate Sediments and Their Diagenesis*	Incorporated diagenesis as the main research content in relation to carbonate rock, thereby bringing associated research to maturity	
1972	H. Blatt et al.	*Origin of Sedimentary Rocks*	Summarized various formation mechanisms of sedimentary rocks, and quoted achievements in hydromechanics	2nd edition in 1980
1972	F. J. Pettijohn et al.	*Sand and Sandstone*	Proposed the viewpoint that "sandstone is able to play an important role in retrospecting stratum history"	2nd edition in 1987
1973	H. E. Reineck et al.	*Depositional Sedimentary Environments with Reference to Terrigenous Clastics*	Elucidated the geologic features of various terrigenous clastic sedimentary environments based on sedimentary structure, in order to lay a theoretical foundation for recognition of sedimentary facies	2nd edition in 1980
1974	K. J. Hsu et al.	*Pelagic Sediments: On Land and Under the Sea*	Provided a new understanding of deep-sea sediments	
1976	R. G. Walker	*Facies Models*	Expounded facies models formed by various sedimentary environments to give research on sedimentology had an analogous example	2nd edition in 1986; 3rd edition in 1992
1977	P. R. Vail	*Seismic Stratigraphy*	First to combine seismic data and sedimentary facies analysis	
1976	R. C. Selley	*An Introduction to Sedimentology*	Systematically expounded hydrodynamic conditions and methods of sediment transport and sedimentation mechanisms under different environments	2nd edition in 1982
1978	H. G. Reading	*Sedimentary Environments and Facies*	These two monumental works gave an overall summary of the theory of sedimentology, and reflected the highest, most current, level of research on sedimentology. They are thus classic works on sedimentology	2nd edition in 1986; 3rd edition in 1996
1978	G. M. Friedman et al.	*Principles of Sedimentology*		
1973	Sun Shu	*Studies on The Phosphate Rocks in Western Szechuan*	First to apply the theory of sedimentology in analysis of the distribution of sedimentary minerals. Published reports and monographs satisfied sedimentary features of continental facies in China, which opened a new page for the development of sedimentology in China	
1978	Li Jiliang et al.	*On the Features of Turbidite Sequences in Some Regions of China*		
1978	He Qixiang	*Sedimentary Rocks and Depositional Ore Deposits*		
1980	Liu Baoli	*Petrology of the Sedimentary Rocks*		

basic theory and method of sedimentology, and involved the prediction of macroscopic and microscopic characteristics of reservoirs in the search for hydrocarbon reservoirs.

1.2.3 Theory Sublimation Phase (1980s)

This was a comprehensive, fast, and vigorous development period for global sedimentology, and also a rapid developmental stage for sedimentology in China (Table 1.4). During this period, many new viewpoints and theories were expounded, and knowledge was gained particularly in relation to the genesis of special sediments, mainly in the following areas.

1.2.3.1 Storm Deposition

In 1975, J. C. Harms discovered hummocky cross bedding, which is iconic bedding from a storm event. Although hummocky cross bedding is found extensively at a continental shelf near the shore (Duke 1985), it was also discovered in estuaries, tidal flats, and delta-marginal environments (J. Bourgeois 1980), and even as deep-water turbidite. It was therefore evident that hummocky cross bedding and other various representative signs of storm events needed to be comprehensively distinguished. Also during the 1980s, R. H. Dott (1988) put forward the concept of episodic sedimentation, and pointed out that there may be an average status or balanced state in some environments, and also a deviation from the average status. Using the deposition of a

Table 1.4 Main Representative works during theory sublimation phase and associated contribution (1981–1990)

Year	Author	Book or Article title	Main viewpoints or functions, and significance
1983	W. E. Galloway	*Terrigenous Clastic Depositional Systems*	Gradually led theory and methods used in depositional system analysis to a systematic stage, and to practical applications
1983	R. A. Davis	*Depositional Systems*	
1984	A. D. Miall	*Principles of Sedimentary Basin Analysis*	Presented basin analysis as an integration of a number of studies such as stratigraphy, structural geology, and sedimentology
1985	A. D. Miail	*Architectural-Elements Analysis: A new Method of Facies Analysis Applied to Fluvial Deposits*	Proposed new concepts for architectural element and bounding surface hierarchy, divided rivers into 12 categories, and conducted quantitative research on the heterogeneity of fluvial reservoirs at different levels, with the aim of making research on sedimentology more operable
1988	A. D. Miall	*Reservoir Heterogeneities in Fluvial Sandstones*	
1988	W. Nemec et al.	*Fan Deltas*	Put forward a genetic classification method for a delta structure
1988	P. R. Vail	*Handbook of Sequence Stratigraphy*	Marks the birth of sequence stratigraphy, highlights an isochronal stratigraphic framework, and determines four effects that affect the sequence (i.e., tectonic movement, sea level eustacy, sediment supply, and climate change)
1988	J. B. Sagree	*Basics of Sequence Stratigraphy*	
1988	J. C. Wagoner	*SEPM Special Publication in Sequence Stratigraphy*	

(continued)

Table 1.4 (continued)

Year	Author	Book or Article title	Main viewpoints or functions, and significance
1981	Institute of Geology and Geophysics (CAS)	*Petrology of the Sedimentary Rocks*	Based on high start, sedimentology in China applied advanced overseas theories to systematically summarize and upgrade domestic geologic features, thereby forming a theory of sedimentology using Chinese characteristics
1981	Lithofacies Palaeogeography Committee of China	Initial issue of *Lithofacies Palaeogeography*	
1981–1982	Sun Shu et al.	*Evolution of Henan-Shaanxi Sedimentary Basin of the Middle And Late Proterozoic Age*	
1982	Wang Yinghua et al.	*Early Paleozoic Carbonate Petrology in Northern China Platform*	
1983	Sedimentology Society of China	*Initial issue of Acta Sedimentologica Sinica*	
1983	Ye Lianjun	*Sedimentary Associations of the Northern China Platform*	
1985	Wang Hongzhen	*Paleogeographic Atlas of China*	
1986	Wu Chongjun	*Sedimentation Types of Lake Sand Body*	
1986	Zeng Yunfu, Xia Wenjie	*Petrology of the Sedimentary Rocks*	
1986	Sun Yongchuan, Li Huisheng	*Clastic Sedimentary Facies and Sedimentary Environment*	
1986	He Qixiang et al.	*Reef facies Deposition in Xisha Islands*	
1987	Sun Shu et al.	*Sedimentation of Extensional Basins in Platform Regions of China*	
1987	He Jingyu, Meng Xianghua	*Sedimentary Rock, Sedimentary Facies Model and Construction*	
1987	Yu Suyu, He Jingyu	*Petrology of the Sedimentary Rocks*	
1987	Ye Lianjun	*Current Status and Development Trends of Sedimentology*	
1988	Ye Lianjun, Sun Shu et al.	*Advances and Prospects of Sedimentology in China*	
1989	Sha Qing'an	*Deposition of Holocene Beach in Dafu Bay, Pingtan Island, Fujian and Diagenesis Thereof*	
1989	Zhu Xia	*Sedimentary Basins in China*	The first volume of *World Sedimentary Basins*
1989	Zheng Junmao, Pang Ming	*Research on Diagenesis of Clastic Reservoirs*	Systematically introduced methods for studying diagenesis of clastic reservoirs

nearshore wind-wave zone as an example, positive deviation may cause storm deposition, while negative deviation may cause non-deposition or a hard bottom.

1.2.3.2 Flysch Formation

As early as 1978, Friedman and Sanders in *Principles of Sedimentology* defined flysch as a product specific to marine strata. Prior to this, it had been considered that flysch was only produced in deep-sea troughs, and that steep slopes and turbidity currents were thought to be necessary and sufficient conditions for its formation. In the 1980s, it became acknowledged that flysch may be produced in other environments (such as in basins on continental slopes or continental margins).

1.2.3.3 Basin Analysis and Research on Sedimentary System

In accordance with a geotectonic environment, Sun Shu, a Chinese scholar, conducted systematic research on the clastic sedimentary environment and lithofacies paleogeography of sedimentary basins in Henan and Shaanxi in 1981, and summarized their formation and evolutionary history to discuss the zonation of sedimentary iron deposits.

In 1984, A. D. Midi determined basin analysis to be an integration of stratigraphy, structural geology, and sedimentology. He also answered paleogeographic evolutionary problems relating to basins, and published the *Principles of Sedimentary Basin Analysis*. Furthermore, Galloway and Hobday published *Terrigenous Clastic Depositional Systems* in 1983, and R. A. Davis published *Depositional Systems* in the same year (Davis 1983). In 1982, the editor-in-chief, Scholle and Spearing, compiled the publication, *Sandstone Depositional Environments*, which contained a large number of colorful pictures. In addition, the theory of plate tectonics became one of the theoretical foundations of sedimentation.

Studying the sedimentation and evolutionary laws of different types of basins is a major topic in the understanding of relations between tectonics and sedimentation. Construction of a depositional basin model involves analysis of the depositional system of different types of basins, and integrated research into the sedimentary filling pattern and basin evolutionary process. At present, theory and methods for the analysis of depositional systems have moved into a systematic stage.

1.2.3.4 Impact of Diagenesis on Porosity

Following further progress and developments made in hydrocarbon exploration, and the constant introduction of new technologies and new methods, there was a breakthrough in research into the diagenesis of clastic rocks in the late 1970s. Among such breakthroughs, was the discovery that a large amount of secondary porosity is formed by diagenesis in sandstone. Representative articles include, *Diagenesis in Sandstones* by Blatt (1979), *Diagenesis and Pore Evolution of Sandstones (Sand Layers)* by Zhu Guohua (1982), *The Chemistry of Secondary Porosity* by Surdam et al. (1984), *Application of Organic/Inorganic Diagenesis to Porosity Prediction* by L. J. Crossey (1986), and *Diagenesis of Clastic Reservoirs* by Zheng Junmao (1989). All these mark the mature stage of diagenesis research and theory.

1.2.3.5 Hierarchical Analysis and Formation of Configuration Concept

At the Second International Symposium on Rivers in 1985, Miall put forward the concept of "Configuration or Architecture Elements," and "Bounding Surface Hierarchy" for the hierarchical classification of rivers, and divided river sediments into 8 basic architectural elements. He also proposed 12 river depositional models, which guided research on the depositional system into a new historical period. Three features or ideas in this respect were as follows: highlighting the hierarchical concept; embodying the three-dimensional (3D) structure of sediments and internal lithofacies; and obtaining research on rivers using a simple morphological classification stage. These are of historical significance in the formation and development of reservoir sedimentology, and on improvements in sedimentology, and have been of evident guidance and reference to date. In particular, these concepts have been extensively applied in high water-cut oilfields when conducting research on intercalation during efficient development.

1.2.3.6 Further Developments in Geotectonic Sedimentology

Along with the systematization of classifying sedimentary basins according to plate tectonics (Dickinson 1979; Ye Lianjun 1983; Xu Jinghua 1985; Klein 1987; Zhu Xia 1989), scholars around the world have conducted systematic research on the sedimentation of different sedimentary basins, and discussed the formation, development, and evolution of sedimentary basins in various geotectonic backgrounds based on the fundamental theories of continental dynamics, plate tectonics, and sedimentology. Furthermore, from the perspective of construction and evolution, scholars have focused on discussing the distribution law of the sedimentary system within the basin, on the basis of sedimentary control, in order to predict the distribution of sedimentary minerals.

1.2.3.7 Delta Structure—Genetic Classification

In the publication, *Fan Deltas*, which was chiefly edited by Nemec and Steel in 1988, Orton and Mcpherson (1988) further divided a megaclast delta into a fan delta and a braided river delta, and applied a structure in the form of a genetic classification method to enable the scientific partition of a delta depositional system. In this respect, great leaps were made in delta research.

1.2.3.8 Rise of Sequence Stratigraphy

Vail et al. published an essay on seismic stratigraphy with the American Association of Petroleum Geologists (AAPG) in 1977. The essay proposed two major viewpoints, the first of which was that a sequence consists of mutually integrated and genesis-related strata, of which the top and bottom margins are the plane of unconformity and corresponding conformity surface. The second viewpoint was that the genesis of a sequence is basically or completely controlled by changes in global sea level. Following this, in 1988, as editor-in-chief, C. K. Wilgas compiled a special issue called *Integrated Research Method for Sea-level Changes–Sequence Stratigraphy*. In addition, P. R. Vail and J. B. Sagree chiefly edited the *Handbook of Sequence Stratigraphy* and *Basics of Sequence Stratigraphy*, where it was determined that *a* stratigraphic sequence is produced by the interaction of geological factors such as tectonic movement, global sea level eustasy, sedimentation, and climate change. A new science, known as sequence stratigraphy, was born with the advent of these works.

In relation to the knowledge contained in the above texts, a large number of famous international scholars revised their monographs and textbooks, and published second or third editions (Table 1.3), such as *Sedimentary Environments and Facies* (2nd edition in 1986) chiefly edited by Reading, *Facies Models* (2nd edition in 1986) edited by R. G. Walker, *Depositional Sedimentary Environments with Reference to Terrigenous Clastics* (2nd edition in 1980) edited by H. E. Reineck, *An Introduction to Sedimentology* (2nd edition in 1982), and *Ancient Sedimentary Environments* (2nd edition in 1978, and 3rd edition in 1985) edited by Selley, *Sandstone Depositional Model* (2nd edition in 1978 and 3rd edition in 1985) edited by Klein, *Sand and Sandstone* (2nd edition in 1987) edited by Pettijohn, and *Origin of Sedimentary Rocks* (2nd edition in 1980) edited by H. Blatt. The extra sections in these re-publications mostly include basin analysis, the principle of sequence stratigraphy, and research on depositional systems, which makes these works more systematic and theoretical.

Geoscientific research in China at the time was mainly "Following Thought, Imitable Research," and was focused on studying related knowledge from developed countries within Europe and the USA. Therefore, scientists were able to single out principal contradictions, and as a result a number of texts were compiled by systematically summarizing and upgrading geologic features by applying advanced theories from overseas. For example, Chongjun (1986) conducted systematic research on a Mesozoic-Cenozoic lake basin sand body in China, and determined that the distribution and characteristics of the lake basin's sand body were controlled by tectonic activity, terrain, provenance, and climatic condition. In addition, he supported a link between the determination of sand body types and the zonal division of depositional facies within a lake basin, and advocated that a sandy lake basin could be divided into five categories: a turbidity sand body, delta sand body, fan delta sand body, submarine alluvial fan sand body, and a beach bar sand body; the distribution, formation and characteristics of which were then systematically summarized. During this period, a large number of representative monographs and textbooks (Table 1.4) were published: *Petrology of the Sedimentary Rocks* (1980), chiefly edited by Baoban; *Paleogeographic Atlas of China* (1985), chiefly edited by Wang Hongzhen; *Lithofacies Paleogeography Basis and Working Method* (1986), chiefly edited by Liu Baoban and Zeng Yunfu; *Petrology of the Sedimentary Rocks*, chiefly edited by Zeng Yunfu and Xia Wenjie; *Clastic Sedimentary Facies and Sedimentary Environment* (1985), chiefly edited by Yong-chuan, Huisheng; and *Petrology of Sedimentary Rocks*, chiefly (1987) edited by Yu Suyu and He Jingyu.

1.2.4 Discipline Permeation Phase (1990s)

Both the depth and width of research on sedimentology expanded in line with the introduction of new technologies and approaches. Rapid developments in seismic technology and integrated electric well-logging interpretation technology opened a new and wider approach towards research into sedimentology. Previous studies had mostly focused on the analysis of one-dimensional or two-dimensional sections, and it was therefore necessary to introduce a new concept of 3D space to apply the new approaches. Between the late 1980s and early 1990s, research on depositional systems entered a true 3D space era through the use of 3D seismic data, and promoted the rapid development of the graphic workstation.

New technologies and approaches were also reflected in the development of new instruments, equipment, and tools, such as the isotope assay, cathode luminescence, X-ray diffraction, scanning electron microscope (SEM), electronic probe, paleomagnetism, paramagnetic resonance, automated imager, and remote sensing technology, all of which have played an important role in research to the present day.

China began conducting research on continental reservoir sedimentology in the early 1970s. By the late 1980s, reservoir sedimentology had entered a new stage by virtue of implementation of "China Petroleum Reservoir Research" by the former CNPC, which was headed by Qiu Yinan, and a series of research reports and monographs were published in relation to this project. With almost 30 years of practice, the future of reservoir sedimentology based on Chinese characteristics and research methods was summarized by Qiu Yinan (1992) as, "The main task of reservoir sedimentology is to acquire abundant quantitative geological knowledge to set up a continental reservoir geologic model using an outcrop survey." In the 1990s, reservoir sedimentology had already become a key subdiscipline of sedimentology.

The science of sedimentology had penetrated a number of other disciplines and embarked on a journey at a higher level. To be specific, a large number of classic monographs, and those concerning sequence stratigraphy, were revised and re-published (Table 1.5). The most representative works include *Sea-level Changes: An Integrated Approach*, edited by C. K. Wilgas from the USA in 1991; *Facies Models: Response to Sea Level Change*, re-published by Roger G. Walker and Noel P. James from Canada in 1992; and *Depositional System* (2nd edition), republished by R. A. Davis from the USA in 1992. In addition, Einsele from Germany edited and published *Sedimentary Basins: Evolution, Facies and Sediments Budget*, and later revised it as a

Table 1.5 Main representative works at discipline permeation and integration phase and associated contribution (1991–present)

Year	Author	Monograph or Article title	Main content or function, and significance
1992	Roger G. Walker et al.	*Facies Models: Response to Sea Level Change* (3rd edition)	These works were not only the cornerstone of various monographs with respect to applied sedimentology after the 1990s, but also presented the theoretical basis for research and development of present-day sedimentology
1992	G. Einsele	*Sedimentary Basins: Evolution, Facies and Sediments Budget*	
1992	R. A. Davis	*Depositional Systems* (2nd edition)	
1996	H. G. Reading	*Sedimentary Environments: Processes, Facies and Stratigraphy* (3rd edition)	
1991	Song Tianjing et al.	*Proterozoic Sedimentary Rocks in Northern China*	These works represent China entering its heyday in relation to developments in sedimentology
1992	Wu Chongjun, Xue Shuhao	*Sedimentology of Petroliferous Basins in China*	
1992	Li Sitian et al.	*Sequence Stratigraphy and Depositional System Analysis of the Northeastern Ordos Basin*	
1992	Qiu Yinan et al.	*Developments of Continental Clastic Reservoir Sedimentology in China*	
1993	Feng Zengzhao	*Petrology of the Sedimentary Rocks*	
1994	Wang Defa et al.	*Mesozoic-Cenozoic Fault Basin Depositional System and Pattern of Filling in China*	
1996	Tian Zaiyi	*Sedimentology of Petroliferous Basins in China*	
1996	Wang Liangzhen et al.	*Sedimentary Environments and Sedimentary Facies*	
1997	Qiu Yi'nan, Xue Shuhao et al.	*Continental Oil and Gas Reservoirs in China*	
1997	Xu Xiaosong, Liu Baojun et al.	*Analysis of Large-scale Basins in West China and Their Geodynamic Characteristics*	
1999	Ma Yongsheng	*Carbonate Reservoir Sedimentology*	
2001	Zhao Chenglin	*Principles of Sedimentology*	
2002	Xue Shuhao, Liu Wenlin et al.	*Sedimentary Geology of Lakes and Basins and Hydrocarbon Exploration*	
2002	Yu Xinghe	*Clastic Petroleum Reservoir Sedimentology*	

2nd edition in 2000. This book is considered to be one of the most classic textbooks of the end of 20th century. It features an intuitive and systematic description of various types of basins and depositional systems using stereograms. Furthermore, in 1993, H. G. Reading from the UK revised *Sedimentary Environments and Facies* and republished *Sedimentary Environments: Processes, Facies and Stratigraphy* (3rd edition). All these great works act as cornerstones for various monographs published after the 1990s with respect to applied sedimentology, as well as being the theoretical basis for research and development of present-day sedimentology.

Nowadays, sedimentology in China has entered its heyday and Chinese scholars are approaching a period of "Comprehensible Thought, Creative Research." A large number of excellent works have now been published, including many monographs, with respect to professional sedimentology and continental sequence stratigraphy (Table 1.5). Monographs that represent the research achievements of the older generation of sedimentologists include *Sedimentology of Petroliferous Basins in China* (Chongjun and Shuhao 1993), *Sedimentology in China* (Zengzhao et al. 1994), *Continental Oil and Gas Reservoirs in China* (Qiu Yinan et al. 1997), and *Petroleum Reservoir Evaluation Technique* (Qiu Yi'nan and Xue Shuhao 1997).

To summarize, the development of sedimentology in the second half of the 20th century had the following characteristics: a change in concepts, the application of new technologies and new approaches, gradual improvements in theory, and penetration of and intersection with other disciplines. The most remarkable characteristic is that sedimentology changed from a theoretical research field to a practical research field, and also from a local scale to a global scale, due to the massive emergence of interdisciplines with respect to sedimentology, such as sequence stratigraphy, reservoir sedimentology, environmental sedimentology, geotectonic sedimentology, event sedimentology, global cyclic stratigraphy, geodynamics sedimentology and experimental sedimentology.

1.3 Related Concepts, Connotations, and Development of Sedimentology

1.3.1 Sedimentary Environment

The concept of the environment is derived from geography. Geographers divide the Earth's surface into different geographic landscape units, e.g., mountains, rivers, lakes, deserts and seas, all of which are collectively referred to as the physicogeographical environment. Sedimentology is mainly related to the physicogeographical environment in relation to the way in which sediments are formed, namely the sedimentary or depositional environment (Fig. 1.1).

In 1976, Selley determined that the "…sedimentary environment is a surface that is physically, chemically, and biologically different from adjoining areas." His work emphasized that the physiographic landscape and geomorphic features are determined by a mixture of three disciplines: ① physics—in the speed, direction, and change of wind, waves, and running water, and

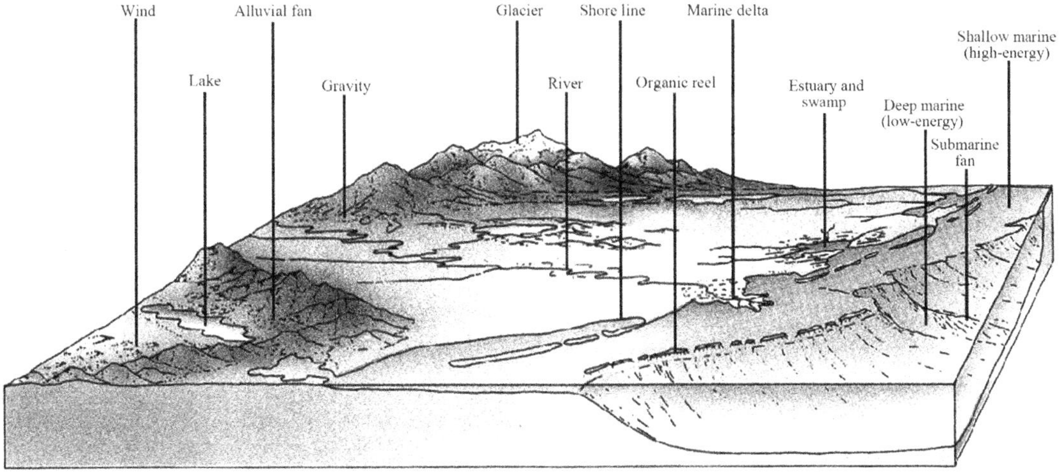

Fig. 1.1 Various sedimentary environments existing on the earth surface (by Laing 1991)

in relation to climate, weathering, and variations in temperature; ② chemistry—the constituents of water are related to the sedimentary environment and the lithogeochemistry properties within the collection area; and ③ biology—including the effects of animals and plants. The environment may therefore be studied using either one, two, or three of these disciplines, and divided accordingly.

1.3.2 Formation and Development of Concept of Depositional Facies and Depositional System

1.3.2.1 Depositional Facies

The environment is a concept that encompasses the expression of surface features. The modern environment refers to the modern Earth's surface and associated characteristics, whereas the palaeoenvironment is the Earth's surface in ancient times. The sedimentary rocks (sediments) examined by sedimentologists are traces or products that have been deposited in an ancient sedimentary environment, and can be directly observed. It was thus necessary to develop a concept to determine a unit representing a sedimentary product, and therefore the concept of "facies" or "lithofacies" was established.

"Facies" or "depositional facies" is a basic term in sedimentology, but has been a long-debated concept. As early as 1669, the word "facies" was introduced in geological literature by N. Steno, a Danish scholar. However, it was the Swedish geologist, A. Gressly, who endowed sedimentology with neoteric meanings during his work in the Alps in 1838. While researching a Jurassic system in northwest Switzerland, he discovered considerable differences in the stratum between lithological characters and palaeontological features, and thus adopted the term "facies" or "aspect" to express a rock unit. In addition, he considered that rock units with the same petrologic and paleontological characteristics could be considered as one "facies."

However, from that time onwards, the term "facies" has been used differently by geologists. It is currently used in four different ways: ① to indicate the composition of sediments as "lithofacies," such as "sandstone facies" and "limestone facies"; ② to indicate the genesis of sediments as "genetic facies," such as "turbidite facies" and "biohermal facies"; ③ to indcate the environment of sediment formation as "depositional facies," such as "fluvial facies" and "littoral facies"; and ④ to indicate the tectonic background of sedimentation as "tectonic facies," such as "molasse facies" and "flysch facies." In 1986, H. G. Reading considered that, "...as long as "facies" is clearly defined, it will be feasible for various uses." Under normal circumstances, facies means one, or a set of, characteristic rock(s) formed under a certain sedimentary condition. This sedimentary condition reflects a specific sedimentary process and environment. Chinese scholars divide facies into subfacies and microfacies in accordance with the hierarchy of geoscience.

Currently, scholars globally tend to understand "facies" as meaning "palaeoenvironmental products" (Selley 1976), i.e., exclusively referring to the "material expression of the sedimentary environment." Some environments have specific material expressions, where, for example, "facies" refers to bio-facies in relation to a biologic comprehensive expression, and lithofacies is the comprehensive expression of lithology and sedimentary structure. In certain sedimentary environments, particular sedimentation arises and certain sedimentary associations form. Various characteristics of the sedimentary environment and sedimentation inevitably leave a record within sedimentary products, and these mainly involve discrepancies among rock components, geometrical morphology, rock structure, sedimentary structure, and biological fossils. As a result, "facies" represents the synthesized law of lithologic characters and paleotological features relating to a sedimentary condition (A- B. Rukhin (Л- Б. Рухин) 1953 and 1959; Reineck and Singh 1980). Based on this definition, the concept of "facies" is different from that of "environment." "Environment" refers to conditions and causes while "facies" involves the products and results of all the effects within an

Table 1.6 Relationship between environment and facies

Cause			Process	Result
Condition	Background	Effect	Time	Product
Physical		Erosion		
Chemical	Environment	Equalization	→	Depositional facies
Biologic		Sedimentation		

Source Selley 1976; later revised in 1985

environment. Selley (first proposed in 1976, but later revised in 1985) gave a concise explanation of the causal relationship (Table 1.6) between "environment" and "facies."

From the above, it can be inferred that "depositional facies" is an explanatory term that has genetic meanings in relation to recovering the palaeoenvironment. Therefore, during practical work, the various pieces of information collected (such as rocks, biological features, chemical features, thickness, form, and contact relations) that reflect the sedimentary environment need to be comprehensively analyzed. The petroleum region (oilfield) also involves the use of logging and seismic data, and thus "depositional facies" can be used after a comprehensive judgment has been made in relation to the formation of the sedimentary environment.

1.3.2.2 Depositional System

The depositional system is comprised of a combination of 3D lithofacies (with respect to genetic relations in a sedimentary environment) and sedimentation (W. L. Fisher and J. H. Mcgowe 1967). A depositional system therefore consists of a three-dimensional depositional facies body and associated spatial relations, and may be described using acquired underground data. Whenever possible, it is necessary to accurately describe sections of core or outcrops. This method of facies analysis is highly dependent on recovering the basin shape and bedding structure, determining overall lithology, making a quantitative description of skeleton sandstone geometry, and recognizing vertical and lateral sequences as well as combinations of general facies. Overseas, the depositional system is generally divided into an alluvial fan system, fluvial system, delta system, shore zone system, lacustrine system, and eolian system, whereas in China it is identical to the sedimentary facies, and is further divided into subfacies and microfacies, i.e., any depositional facies may be divided into several subfacies, and the subfacies may then also be divided into multiple microfacies, which is easier for use and for operations within the production department (such as at an oilfield). However, attention always needs to be paid to universal concepts and meanings when communicating internationally.

1.3.3 Depositional Model

In English, the term "model" is used to describe both a pattern and a model, and this is obviously different in Chinese. For example, the term "model" in relation to the first meaning refers to an overall summary of the study object, and highlights appearance with the purpose of explaining the genetic mechanism, whereas the second meaning is a specific depiction of the study object and emphasizes the internal properties, with the purpose of quantitatively representing characteristics and change. The former definition involves a qualitative description, which can be used to guide follow-up studies and make references for copies, whereas the latter refers to quantitative research, with the effect of designing a plan for implementation, which can be used for direct and selective use.

The term "depositional model" is also commonly used in sedimentology. This model in general involves an ideal and simplified model that assists in delivering knowledge about complicated natural phenomena and processes, and the term has a specific meaning. In 1963, F. J. Pettijohn and P. E. Potter stated that the "... depositional model is essentially a feature for describing and reproducing sedimentation," and in 1978 Roger G. Walker held that "the depositional model is a generalization of sedimentary characteristics, or rather, a generalization of comprehensive geological characters, evolution and its spatial combination." This generalization, comprises two aspects: one is the summary of sedimentary characteristics; and the other is generalization of the formation mechanism. Therefore, "depositional model" is an integrated interpretation of sedimentary products formed in the sedimentary environment, with an explanatory genetic significance.

The role of the depositional model is to assist in the discovery of inherent relations among various environments, and to depict geological process and products that are expected to be discovered in these environments. In 1984, taking the turbidity current model as an example, Walker determined that the facies model may: ① play a standard role during comparison; ② play the role of giving an outline and guidance during observation; ③ play the role of prediction in a new region; and ④ play a fundamental role during hydrodynamic interpretation. Consequently, the establishment and mastering of depositional models in different environments not only contributes to explaining the genesis of various ancient sediments, but also has a guiding role in predicting the mineral distribution law during the exploration and development of hydrocarbons and other sedimentary minerals.

References

Baojun, Liu (ed.). 1980. Petrology of the sedimentary rocks. Beijing: Geological Publishing House.

Blatt, H. 1979. Diagenetic processes in sandstone. *SEPM (Society of Sedimentary Geology)* 26: 141–157.

Blatt, H., G.V. Middleton, and R. Murray. 1972. *Origin of sedimentary rocks*. Englewood Cliffs, New Jersey: 2nd Prentice-Hall.

Chongjun, Wu. 1986. Sedimentation types of lake sand body. *Acta Sedimentologica Sinica* 4.

Chongjun, Wu, and Xue Shuhao. 1993. *Sedimentology of petroliferous basins in China*. Beijing: Petroleum Industry Press.

Crossey, L.J., R.C. Surdam, and R. Lahann. 1986. Application of organic/inorganic diagenesis to porosity prediction. *SEPM* 147–155.

Davis, R.A.Jr. 1983. *Depositional systems*. Englewood Cliffs, New Jersey: Prentice-Hall Inc.

Dickinson, W.R., and C.A. Suczek. 1979. Plate tectonics and sandstone compositions. *AAPG* 63: 189–194.

Duke, W.L. 1985. Hummody cross stratification, tropical hurricanes, and intense winter storms. *Sedimentology* 32: 167–194.

Einsele, G. 2000. *Sedimentary basins-evolution, facies, and sediment budget*, 2nd ed, 1–792. London: Springer.

Friedman, G.M., and J.E. Sanders. 1978. *Principles of sedimentology*, 729. New York: Wiley.

Galloway, W.E., and D.K. Hobday. 1983. *Terrigenous clastic depositional systems*. New York: Springer.

Guohua, Zhu. 1982. Evolution of diagenesis and sandstone (layer) pore. *Acta Sedimentologica Sinica* 1: 45–49.

Junmao, Zheng, and Pang Ming. 1989. *Research on diagenesis of clastic reservoirs*. Wuhan: China University of Geosciences Press.

Klein, G.D. 1985. Sandstone depositional models for exploration for fossil fuels, 3rd ed. D. Reidel publishing company.

Klein, G.V. 1987. Current aspects of basin analysis. *Sedimentary Geology* 50: 95–118.

Krumbein, W.C. 1934. Size frequency distribution. *Journal Sedimentary Petrology* 4: 65–77.

Kuenen, P.H., and C.I. Migiorini. 1950. Turbidity currents as a cause of graded bedding. *Journal Geology* 58: 41–127.

Laing, David. 1991. *The earth system: An introduction to earth science*. USA: Wm. C. Brown Publishers.

Lyell, C. 1837. *Principles of geology*. London: John Murry.

Mcpherson, J.G., G. Shanmugam, and R.J. Moiola. 1988. Pan deltas and braid deltas: Conceptual problems. In *Fan deltas: Sedimentology and tectonic setting*, ed. Nemec W, and Steel R.J., 14–22. London: Blackie and Son.

Miall, A.D. 1985. Architectural element analysis: A new method of facies analysis applied to fluvial deposits. *Earth Science Reviews* 22: 261–308.

Morgan, J.P. 1970. Deltas—A resume. *Journal of Geological Education* 18: 107–117.

Murrary, J., and A.F. Renard. 1884. On the nomenclature origin, and distribution of deep—Sea deposit. *Royal Society of Edinburgh, Proceed* 12: 495–529.

Nemec, W., and R.J. Steel. 1988. Fan delta. Blackie and Son Ltd.

Pettijohn, F.J. 1949. *Sedimentary Rocks*. New York: Harper & Row.

Pettijohn, F.J. 1987. *Sand and sandstones*, 2nd ed. Berlin, Heidelberg, New York: Springer.

Reading, H.G. 1986. Sedimentary environments and facies, 2nd ed. Blackwell Scientific Publications.

Reineck, H.E., and I.B. Singh. 1973. *Depositional sedimentary environments with reference to terrigenous elastics*, 439. Berlin: Springer.

Reinek, H.E., and I.B. Singh. 1980. Depositional sedimentary environments with reference to terrigenous clastic, 2nd ed. Berlin, Heidelberg, New York: Springer.

Scholle, P.A., and D.R. Spearing. 1982. Sandstone depositional environments, Mem. No. 31. O-klahoma: AAPG Tulas.

Selley, R.C. 1976. *An introduction to sedimentology*, 408. London: Academic Press.

Selly, R.C. 1985. *Ancient sedimentary environments*, 3rd ed. London: Chapman & Hall.

Simons, D.B., and E.Y. Richardson. 1962. The effect of bed roughness on depth-discharge relations in alluvial channels. US Geological Survey, Water-Supply Paper, 1498.

Sorby, H.C. 1857. On the origin of the Cleveland Hill ironstone. *Proceedings of the Yorkshire Geological Society* 3: 457–461.

Surdam, R.C., S.W. Boese, and L.J. Crossey. 1984. The chemistry of secondary porosity (in Clastic diagenesis). *AAPG Memoir* 37: 127–149.

Vail, P.R., R.M. Mitchum Jr, R.G. Todd, et al. 1977. Seismic stratigraphy and globe changes of sea lever. In Payton CE (ed) Seismic stratigraphy application to hydrocarbon exploration. Am Assoc Petrol Geol Mem 26: 49–212.

Visher, G.S. 1965. Use of vertical profile in environment reconstruction. *AAPG* 49: 41–61.

Walker, R.G. 1984. Facies models, 2nd ed. Ainsworth press Limited.

Walther, J. 1894. Einleitung in die geologia historische Wissenschaft. In *Beobachtungen Uber Die Bildung Der Gesteine Und ihrer organischen Einschlusse*. Jena: Gustav Fischer.

Wentworth, C.K. 1922. A scale of grade and class terms for clastic sediments. *Journal Geology* 30: 377–392.

Yinan, Qiu, and Xue Shuhao. 1997. *Evaluation technique of petroleum reservoir*. Beijing: Petroleum Industry Press.

Yinan, Qiu, Xue Shuhao, and Ying Fengxiang. 1997. *Continental oil and gas reservoirs in China*. Beijing: Petroleum Industry Press.

Yongchuan, Sun, and Li Huisheng. 1985. *Clastic sedimentary facies and sedimentary environment*. Beijing: Geological Publishing House.

Zengzhao, Feng, Wang Yinghua, Liu Huanjie, Sha Qingan, Wang Defa, et al. 1994. *Sedimentology in China*. Beijing: Petroleum Industry Press.

2 Formation, Development, and Trends in Reservoir Sedimentology

All disciplines have specific research purposes, missions, and characteristics, and thus defining their purposes and missions is the starting point and cornerstone in scientific research. This is certainly no exception in petroleum reservoir sedimentology, which, from the perspective of sedimentology, describes, evaluates, and predicts the main features (geometry, spatial distribution, and physical properties) of a petroleum reservoir, i.e., the appearance and internal properties, in order to improve the effects of hydrocarbon exploration and development.

2.1 Concept, Nature, and Task of Reservoir Sedimentology

Since the beginning of the 20th century, oil and gas resources as the most important energy sources for global industries and national economic development, have gained considerable global attention. However, over 90% of the world's hydrocarbon reserves are located in sedimentary rocks, particularly in China. Therefore, research on reservoir sedimentology is necessary for the discovery of more hydrocarbon resources, and the problems faced in exploration and development and the extensive use of hydrocarbon resources are the driving forces promoting the rapid development of reservoir sedimentology.

2.1.1 Rise of Reservoir Sedimentology

The purpose of reservoir sedimentology is to describe the sedimentary characteristics and heterogeneity of a (hydrocarbon) reservoir body formed under various environments, using the theory of applied sedimentology and facies analytical methods and means.[1] Reservoir sedimentology as an important branch of sedimentology, and reservoir geology, lie within the scope of applied sedimentology. During the 1960s, when large numbers of large oil and gas fields were discovered all over the world, explorationists and reservoir engineers needed to make relatively correct evaluations and predictions of the characteristics and distribution of each petroleum reservoir using minimal drilling data, in order to achieve good economic effects. This means that the spatial distribution of hydrocarbon reservoirs and the transformation law of their internal properties, in particular, needed to be described and predicted in a scientific way, and this need was hastened by actual production requirements. Therefore, petroleum geologists and reservoir

[1] Qiu Yi'nan's internal text of his speech, 2000.

X. Yu et al., *Clastic Hydrocarbon Reservoir Sedimentology*, Advances in Oil and Gas Exploration & Production, https://doi.org/10.1007/978-3-319-70335-0_2

engineers were immediately brought to the forefront, and the field of reservoir sedimentology was ultimately generated. In the late 1970s, with the rapid development of the petroleum industry and the emergence of various testing means, reservoir sedimentology gradually matured as a science.

To date, a number of successful applications have been made in the field of hydrocarbon exploration and development. In the autumn of 1976, two teams were organized during the annual meeting of SPE in the US, which was themed "Synergy," to discuss issues relating to depositional facies in petroleum reservoir sedimentology, reservoir continuity, and heterogeneity. Papers were published in a special issue of the *Journal of Petroleum Technology* in 1977, which most scholars (such as Qiu, Yinan) considered to be a new "milestone." In 1987, R. W. Tillman compiled a text entitled *Reservoir Sedimentology*. At the end of the 20th century, Richardson, a famous petroleum engineer from the USA, listed reservoir sedimentology as one of five technologies used to improve the petroleum recovery ratio in 2000.

It is of note that the book *Reservoir Sedimentology* was translated by the Institute of Scientific Information of the former CNPC in 1990, and it covers geological issues with respect to improving the petroleum recovery ratio, from the viewpoint of sedimentology. In the middle of the 1980s, under the advocacy of Mr. Qiu, Yinan, the former CNPC determined a major subject named "Research on Petroleum Reservoirs in China," which not only discusses how to improve the recovery ratio in hydrocarbon development, but also involves many issues relevant to sedimentology in reservoir research for hydrocarbon exploration and development. This marked the point at which China's petroleum reservoir sedimentology matured as a science and entered into a new prosperous era. Today, reservoir sedimentology is applied in the comprehensive evaluation of reservoirs at each phase of hydrocarbon exploration and development, and has expanded from use in the evaluation and prediction of petroleum fields, to the nuclear industry, and to other mineral products.

2.1.2 Definitions and Related Concepts

2.1.2.1 Reservoir Sedimentology

Reservoir sedimentology is a synthetic discipline that applies various data in the study and explanation of the sedimentary environment, diagenesis, and formation mechanisms of hydrocarbon reservoir, in order to analyze and determine associated geological information and improve the effect of hydrocarbon exploration and development. It is also aimed at improving oilfield exploration and development, and focuses on the analysis and prediction of petroleum reservoir heterogeneity at different levels by the comprehensive utilization of geological, seismic, logging, and well testing data. Furthermore, it uses results from various reservoir testing methods, guidance from the principles of sedimentology, and study objectives of petroleum reservoir.

Reservoir sedimentology is a core content in petroliferous basin analysis and reservoir description, because petroleum reservoirs are direct targets for oil-gas exploration and development. Hence, research on reservoir sedimentology is significant in gaining a gradual grasp on the macroscopic distribution characteristics (anisotropy) of petroleum reservoirs, and internal architecture characteristics of reservoir space and their property distribution laws (heterogeneity).

2.1.2.2 Reservoir Description

The concept of reservoir description was firstly applied by Schlumberger Limited in the 1970s, when they launched a reservoir description service mainly for use in well logging. Since the 1980s, petroleum companies and researchers have expanded this concept so that it pertains to hydrocarbon reservoir research and involves

Table 2.1 Objectives and research content of reservoir description throughout various phases

Phase	Exploration phase		Development phase	
	Early phase of exploration	Middle and later phases of exploration	Early phase of development	Middle and later phases of development
Data status	One discovery well	Evaluation well as basis	First batch of development well pattern	Production performance
Objectives	Ascertaining hydrocarbon reservoir	Evaluating hydrocarbon reservoir	Optimizing development program to improve development efficiency	Adjusting development program, infill drilling, and enhancement of oil recovery
Research contents	Researching type of hydrocarbon reservoir, scale of reservoir body, and distribution of hydrocarbon reservoir with the aid of seismic information	Describing the three-dimensional distribution of structures and parameters of hydrocarbon reservoirs, while taking advantage of multi-well comprehensive evaluations	Utilizing existing development of well pattern for fine division and comparison of oil reservoir and research on sedimentary microfacies; conducting comprehensive electric well-logging interpretation and evaluation; and analyzing seepage geological features of reservoir	Combining dynamic data to carryout inter-well reservoir characterization, analysis of the transformation law of fluid attribute parameters, and predicting distribution of remaining oil
Final results	Submitting controlled reserves and proposing well location evaluation, and optimizing exploration deployment	Establishing conceptual model for an oil reservoir and submitting proven reserves	Establishing static geological model	Establishing predictable reservoir model, and submitting the well and reservoir location of infill drilling

multidisciplinary angles such as seismology, well logging, and geology. In this respect, an integrated research method for hydrocarbon reservoirs has now been formed.

Reservoir description refers to a comprehensive research method system that focuses on the qualitative analysis and quantitative description of characteristics of reservoirs in a three-dimensional space, under the guidance of the theories of sedimentology, structural geology, and petroleum geology, and using geological, seismic, logging, and computer applications. The contents of reservoir description in relation to reservoir sedimentology and geology cover petroleum reservoir type, internal architecture, and external geometry of the reservoir, the scale of sedimentary bodies and petroleum reservoirs, and parametric variation and fluid distribution of a reservoir. Reservoir description is aimed at making a quantitative description and prediction using the various characteristics of a reservoir (trap, reservoir, fluid) in a three-dimensional way. Due to the obvious stage features in hydrocarbon reservoir exploration and development, reservoir description involves different research contents and objectives during its various phases (Table 2.1).

2.1.2.3 Reservoir Characterization

L. W. Lake defined Reservoir Characterization in the *Collected Works of the First International Conference on Reservoir Characterization* (1986) as "…an uncertainty process of quantitative determination of reservoir property and identification of geological information and space variation." The term "geological information" involves two concepts (see Table 2.2): ① the reservoir property, which refers to the nonuniformity or heterogeneity of physical properties in a particular reservoir body (porosity, permeability, oil saturation, and electrical properties); and ② the geometry of the reservoir, which refers to the spatial characteristics of the reservoir (i.e., features of change, or the expansion range of lithology in a three-dimensional space), and mainly refers to anisotropy during reservoir modeling. In relation to the former concept, it is necessary to analyze the internal properties, particularly the distribution features of porosity, permeability, and oil saturation in the interior reservoir, within the interlayer, and at the surface. For the latter concept, it is mainly necessary to research the sedimentation microfacies of the reservoir and the associated spatial distribution. The essential attributes, along with the geometrical characteristics of the reservoir, mainly refers to the reservoir configuration or architecture. Furthermore, one of the two key characteristics controlling and affecting a reservoir is the sedimentation formed by the reservoir, while the other is diagenesis, or the diversity of the sedimentary framework or sedimentation and complexity of diagenesis. The author holds the opinion that the "uncertainty process" in the definition above refers to the establishment of multi-models instead of one, as the differences between multi-models can deliver uncertainty. In this respect, optimization of multi-models should be performed so that either one is selected or more certain models are developed.

Table 2.2 Connotations of the meaning "Geological Information" relating to petroleum reservoirs

Reservoir geological information	① Property i.e., heterogeneity
	② Geometry i.e., anisotropy

Hence, the purpose of reservoir characterization is to provide a reservoir configuration framework, i.e., a reservoir geological model that has the above two features: heterogeneity and anisotropy. In a sense, the key is to establish a predictable model, because the reservoir geological model is the main element determining oil reservoir or reservoir simulation results.

Specifically, reservoir characterization according to sedimentation or sedimentary genesis, requires dividing the reservoir into several different levels or the use of a unit scale in relation to the sand layer, or as an architectural element unit or genetic unit. Following this, all the property values of the units (four physical properties: lithology, physical properties, and oil-bearing and electric properties) need to be determined before establishing a reservoir geological model. Therefore, the scope of research includes almost all the aspects of reservoir characteristics, such as geologic features, geometry, and the distribution laws of a sand body, the frequency and size of muddy intercalation, variability in the spatial distribution of properties, and determination of both simulation parameters and fluid movement.

However, the overall aim is to determine the spatial variation of two major reservoir characteristics, i.e., anisotropy and heterogeneity, to determine the appearance and internal properties of a reservoir, and to ultimately maximize exploration efficiency and the final recovery ratio of a petroleum reservoir.

2.1.2.4 Similarities and Differences

In terms of definition, there is a significant difference between reservoir description and characterization, but in practice these terms are used synonymously. Most scientists consider that the focus of reservoir description is to make a detailed description of the various characteristics of an actual petroleum reservoir using the disciplines of geology and geophysics. In addition to the reservoir itself, description also includes fluid characteristics and reservoir types. However, the focus of reservoir characterization is qualitative research and quantitative characterization of the geological features of the reservoir itself.

Therefore, the former topic principally describes the geological characteristics of a petroleum reservoir and emphasizes static study and its applied technique, while the latter attaches greater importance to the research and genetic interpretation of the major characteristics of a reservoir, and stresses quantification and the spatial change process. Hence, the former term mainly uses deterministic modeling methods and has the establishment of a static model of a reservoir as its objective, while the latter applies the method of geostatistics to establish a predictable reservoir model.

Reservoir sedimentology is at the core of both of these topics. It takes research on the sedimentary characteristics as its center, and utilizes and interprets geological and geophysical data to conduct comprehensive analysis and research on various methods and techniques, with the purpose of solving issues of anisotropy and heterogeneity within the reservoir, and with the ultimate aim of establishing various types of reservoir geological models (concept, static, and predictable models).

2.1.3 Tasks, Objectives, and Research Content

The main task is to study the macroscopic distribution law of the petroleum reservoir (anisotropy), the internal structural characteristics of the reservoir space and distribution characteristics of permeability (heterogeneity), and to interpret genesis from the perspective of sedimentology to establish a corresponding depositional and geological model. The objective from the perspective of sedimentology is to make a description, characterization, evaluation, and prediction of the two major features of a petroleum reservoir, in order to improve the effect of oil and gas field exploration and development (Table 2.3).

2.2 Dynamics, Trends, and Reservoir Sedimentology Research Methods

2.2.1 Status and Challenges Involved in Reservoir Research

2.2.1.1 Perspectives from Technical Development History of Petroleum Industry

The global petroleum industry has experienced three large technological revolutions. The first was the improvement of drilling technology, which enhanced drilling speed and depth to increase opportunities of finding new oil and gas fields, and the second was the innovation of seismic exploration technology, which evolved

Table 2.3 Tasks, objectives, and research content of reservoir sedimentology

Task	Objectives		Content	Discipline basis
	Wider objectives	Narrower objectives		
Establishing corresponding depositional model and various geological models	① Predict favorable reservoir facies belt ② Enhancing oil recovery ③ Serving reserves calculation	① Determining distribution range and property changes of reservoir ② Arranging infill welling ③ Determining epitaxial well	① Physical property or internal property—heterogeneity ② Geometry or appearance—anisotropy	① Sedimentology ② Petroleum geology ③ Oil Reservoir Physics ④ Geophysics ⑤ Mathematical geology ⑥ Computer technology

from light spot recording in the 1960s to analog magnetic recording in the 1970s, and to digital magnetic recording in 1980s. Since the 1990s, the emergence of high-tech interpretation technologies, such as three-dimensional seismology and man-machine interaction, have not only enlarged exploration domains from high exploration areas to low exploration areas and from shallow layers to deep layers, but from outcrop areas to coverage areas, and from finding structural traps to finding stratigraphic and lithologic subtle traps. In addition, direct use of seismic data has been used to predict the internal architecture of reservoir and reservoir performance, e.g., the interwell seismic technique of CGG. Undoubtedly, these two technological revolutions have contributed to the rapid development of the petroleum industry. The third technological revolution is related to the various technological innovations pertaining to reservoir research, i.e., petroleum reservoir characterization and stochastic modeling.

In recent years, there has been a need to study reservoirs for exploration and development in new areas and fields (e.g., deep layers), in order to take full advantage of geological, seismic, logging, and well testing data to predict the geometry and spatial distribution of a sand body (reservoir body) and to discover subtle strata and lithologic hydrocarbon reservoirs. Moreover, with the increased density of development of well patterns in old oilfields, and other advancements at a development level, water saturation has become more predominant. There is therefore a need to make full use of static data and development performance data, such as geophysical prospecting, well logging, and geological data, to carry out comprehensive research and establish various reservoir geological models, with the aim of predicting the distribution of the remaining oil by stochastic modeling and/or numerical simulation. Such models would serve the formulation, modification, and implementation of a development scheme for tertiary oil recovery and to find potential hydrocarbon resources. As the exploration and development of conventional reservoirs continue to forge ahead, new strategic backup reserves need to be established, and thus greater attention needs to be paid to research on extra-low permeability reservoirs and special complex reservoirs by scientists in the area of petroleum geology and development.

2.2.1.2 Perspectives on the Current Status of Petroleum and Natural Gas Exploration

Large global oilfields were discovered in the 1960s into the early 1970s. However, only a few oilfields, the size of the North Sea Oilfield and Alaskan Oilfield, have been found, even though extensive and deep prospecting work has been conducted throughout the world (J. R. Michael 1993)[2]. Although a considerable amount of hydrocarbon reserves have been discovered, they are restricted to a few large oilfields with recoverable reserves of above 5×10^8 bbl. Therefore, with further development of these oil and gas fields over the past 20 years, most of the petroliferous basins have approached, or are reaching, the middle or elderly stage.

There are two methods that can be used to assist with this problem. The first of these is exploration and the discovery of new hydrocarbon domains, positions, and thus oil and gas fields; and the second is development, which involves enhancing the recovery ratio of the developed hydrocarbon reservoirs and predicting remaining oil distribution. The main task of the former is to find new methods of searching for petroleum and establishing a new oil theory, while the main task for the latter is to develop and use other new methods. The former will increase geological reserves, while the latter will increase reserves (recoverable) and production. In terms of objective, the former aims to find targets and make appraisals of wells, and the latter determines the heterogeneity of reservoirs and the distribution of remaining oil, and arranges expansion for infill drilling and oilfields. The biggest challenge for the former is to discover large oil and gas fields, and for the latter it is to enhance the recovery ratio of oil and gas fields. Petroleum geologists have now begun to turn from research on exploration methods to innovation and intensive application of oilfield exploration and developmental technology, which in turn has now progressed from a simple to a progressive type. Over the past 20 years, more and more theories and methods of reservoir geology have appeared. Aside from new methods, such as the CT technique, reservoir geology has bonded together petroleum geophysical well logging, seismic, well testing, and reservoir protection knowledge with conventional petroleum geology theory. However, the potential for research as a single discipline lags behind the times. Reservoir geology has developed from the original reservoir description to a prediction of the distribution of reservoir characteristics. In order to enhance oil recovery and infill drilling arrangements, and to determine the extensional well and interpret high moisture content, quantitative methods and techniques (i.e., petroleum geological statistics) have been introduced to this area, new concepts of reservoir characterization are being proposed, and a new science is being formed.

2.2.1.3 Challenges Involved in Exploring New Reserves

World oil prices have slumped twice, in 1986 and 1998, which stimulated cost reductions throughout the entire petroleum industry. The most prosperous petroleum basins with low exploration maturity are located in remote and border areas where high exploration expenditure is needed. Therefore, major global oil producers focus on maturing exploration areas and increasing reserves and production in developed oilfields, with the aim of making improvements in exploration and development efficiency. At present, oil producing countries are paying increasing attention to reservoir research. According to an estimation by the American Bureau of Economic Geology, which was made at the end of the 20th century, in addition to Alaska the supplemented recoverable petroleum reserves in developed onshore reserves in America (though infill drilling, expansion, or recompletion of oil wells) amount to about 800×10^8 bbl (114×10^8 t), and natural gas reaches 5.1×10^{12} m^3.

However, there are two new challenges involved in excavating such reserves. Firstly, the two main features of a reservoir must be accurately determined in order to describe the continuity, superimposed relations, and spatial distribution of physical properties and internal microscopic characteristics of the sand body. The second challenge is to improve and enhance the methods of understanding reservoirs, including dynamic and static methods. To solve these problems it is thus necessary to conduct reservoir characterization research and stochastic simulation.

All of the major global petroleum companies are facing the same problems. For instance, in the ten years from 1981 to 1990, with the exception of North America, there was an increase in the recoverable reserves of oilfields operated by the Shell Oil Company to 36×10^8 standard barrels, and after further oilfield extension this could increase by 10×10^8 standard barrels (J. R. Michael 1993).[2] However, the ability to do this depends on the following factors:

(1) Reservoir characterization (50%);
(2) Correction of data for geology, seismology, and petrophysics (30%); and
(3) Revaluation and development of drilling results (20%).

Currently, over 20% of movable oil reserves cannot be exploited globally due to various heterogeneity obstructions of reservoirs in vertical and planar directions. Furthermore, with greater advancement in China's hydrocarbon exploration and development, two urgent problems stand in the way. Firstly, there are desperately inadequate oil and gas backup reserves, and secondly, over 80% of the oilfields in the East have entered a high water cut exploitation period. The distribution of underground oil-gas-water is very complicated and the remaining oil is distributed dispersedly due to the various heterogeneity obstructions. The main method used to solve this problem is to conduct a subtle quantitative reservoir description or reservoir

[2]Internal data of the Shell Oil Company.

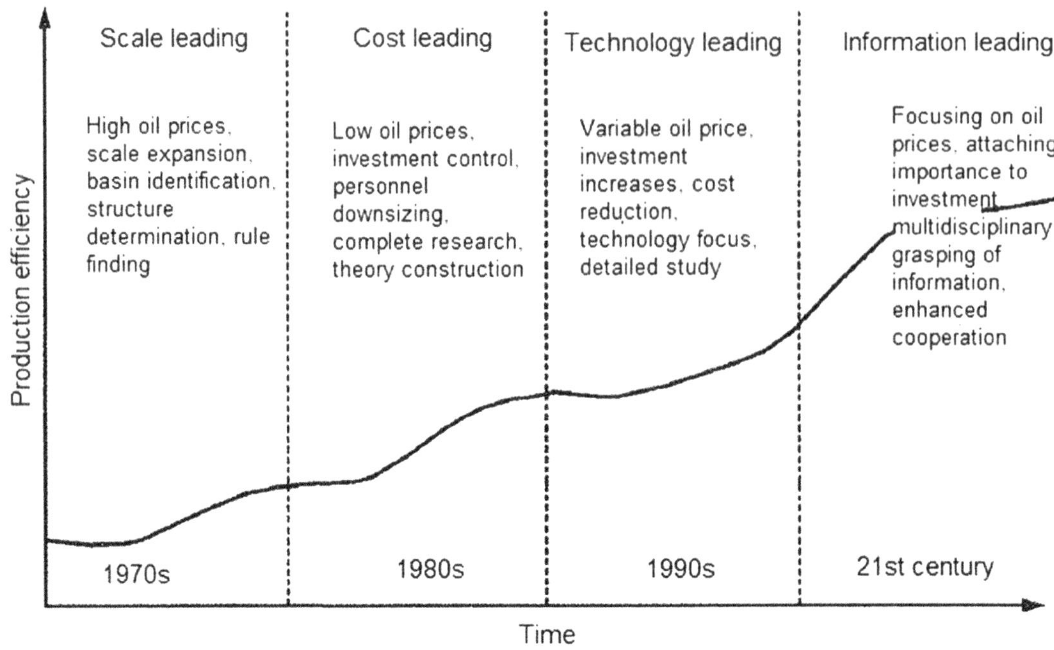

Fig. 2.1 In the development process of nearly 20 years, changes of development strategy of transnational corporations

characterization study. Therefore, the establishment of accurate geologic reservoir models (concept, static, and predictable) is a significant challenge in current reservoir research.

2.2.1.4 The Development Strategy of World-Class Petroleum Companies

Multinational petroleum companies have experienced a development course spanning 40 years, during which time innovation in exploration technology and methods, solutions for theoretical problems and changes in global oil prices have played restrictive roles in relation to macroscopic strategies (Fig. 2.1). From the above, we can see that in the 21 century of the information era, the main future strategy will be advantage integration, resource sharing, industry-university-research cooperation, multidisciplinary joint research and determination of what is possible and what is not.

2.2.1.5 From the Perspective of Discipline Development

At present, research on international and domestic petroleum geological exploration and development is progressing rapidly, and new methods and new theories are continuously emerging, which pushes this discipline from that of singular development to one of multidisciplinary coordinated development. From the perspective of hydrocarbon exploration, domestic and overseas petroleum geology theories can be divided into three parts: basin formation, hydrocarbon generation, and reservoir formation (Table 2.4), which are all related but have respective characteristics. There are three aspects to the study content of petroleum geology and applied theoretical methods, basin analysis/simulation, seismic/sequence stratigraphy, and reservoir characterization/reservoir description, and these three aspects have now become three

Table 2.4 Comparison between levels of research and focus of current domestic and overseas hydrocarbon exploration systems

Item	Objectives	Overseas	Domestic	Comparison
Basin formation system	Basin resources evaluation	Analyzing the formation time of oil-bearing basins from the evolution of global plate tectonics, and evaluating the potential of hydrocarbon resources of the basin based on the palaeolatitude position (positioning)	Taking the basin as the basic geological unit for petroleum generation, migration, and accumulation; research is limited to construction and transformation within the basin over years	Difficult to be theoretically innovative due to a lack of perspectives on global plate evolution
Hydrocarbon generation system	Timing, positioning, and quantification of source rock	Mature theory for marine facies of petroleum, comprehensive analytical test and advanced technology	Continental facies-dominated, mature theory, incomplete comprehensive analytical tests, and advanced technology	Following the herd but with certain innovations
Reservoir forming system	Timing, positioning, and quantification of oil reservoir	Focusing on fluid and hydrocarbon accumulation mechanism; mature research on petroleum system, seal compartments, and abnormal pressure belt	Focusing on reservoir-forming elements analysis, with less mechanism analysis, and attaching importance to research on the petroleum system, seal compartment, and abnormal pressure belt	Following the herd but having or needing to have some innovations on copied research

major research fields, or subdisciplines of petroleum geology, and are described below.

1. Basin analysis/simulation

Basin analysis mainly refers to the analysis of three basin formation histories, i.e., the history of tectonic evolution, thermal history, and sedimentary burial history; while simulation mainly refers to the simulation of five major processes of petroleum (generation, migration, accumulation, expulsion, and dispersion). Basin analysis has three objectives: ① the prediction of favorable exploration zones; ② the selection of exploration targets; and ③ undertaking a comprehensive analysis of a petroleum system. The former two form the basis of simulation and the latter represents the specific content.

2. Seismic/sequence stratigraphy studies

Seismic stratigraphy acts as a research basis, while sequence stratigraphy is a specific research point. There are four objectives: ① to determine an isochronic strata framework; ② to analyze the temporal and spatial distribution law and arrangement of the sedimentary system tract; ③ to predict the spatial distribution of sedimentary bodies; and ④ to serve as basin simulation and reservoir modeling.

3. Reservoir characterization/reservoir description

Objectives: ① to understand and conduct quantitative descriptions of the anisotropy and heterogeneity of reservoirs; ② to serve in the enhancement of oil recovery and formulation of development plans; and ③ to determine locations for infill drilling and epitaxial wells.

In relation to the three major research fields in Chinese petroleum geology listed above, considerable progress has been made following the balancing of relations between introduction and absorption. In addition, these three aspects are keys to petroleum geology research, and are currently hot topics.

2.2.1.6 China's Petroleum Industry Strategy

Due to China's urgent need for gas and oil in industrial development and economic construction, great changes are now taking place in China's oil and gas industry strategic policies in comparison with the previous period (in the early 1980s). The general principles of these policies are: ① "developing the western region while stabilizing the eastern region"; ② "centering on economic effectiveness and implementing oil and gas exploration and development"; ③ setting research priorities so that they "promote exploration and development simultaneously," "enable more intensive study," "move from the shallows to the deep sea," and "become international." This shows that China's petroleum geology research centers will, in the next 10–20 years, be focused on the following popular areas of research: ① increasing reserves (mainly as recoverable reserves) and production; ② energetically developing secondary or tertiary oil recovery; ③ searching for deep or hidden hydrocarbon reservoirs; and ④ building a predictable model and a quantitative static model with petroleum reservoir characterization as the core.

2.2.2 History and Prospects of Petroleum Reservoir Research

2.2.2.1 Practice and Status of China's Petroleum Reservoir Research

For more than half a century, it has been shown through petroleum industry exploration in China that there are many types of petroleum reservoirs within the country, ranging from marine facies carbonate rock to continental facies clastic rocks, and from volcanic rock to metamorphic rock, among which continental facies clastic petroleum reservoirs are the most important and account for above 90% of total proven reserves. It is thus evident that China's petroleum reservoirs have rich variety and complexity as well as variability, and can be determined as having multiple layers, small thicknesses, quick horizontal changes, poor connectivities, and relatively serious heterogeneities. In the 1950s, China's petroleum geological scientists began to study the characteristics of reservoirs, in addition to studying and understanding features of the petroleum source bed in China. However, there have been many new understandings since the 1970s or 1980s, in terms of depositional facies, diagenesis, and predictions of a secondary porosity development zone. At this time, teams were formed according to majors or research directions and research on reservoir sedimentology became increasingly popular; therefore research was conducted on depositional facies in oildom and great achievements were made. In addition, depositional facies models of various conventional sedimentary bodies and reservoirs in different basins were established, which have provided a basis in sedimentology for understanding the hydrocarbon distribution laws of different basins, and in particular for finding various hidden reservoirs in mature exploration areas. In addition, research on the microfacies of various depositional systems and reservoir heterogeneity has been conducted in accordance with the progressive exploration and development of oilfields, and heterogeneity models have been established for different depositional systems, which have provided a scientific basis for reference in the exploration and development of similar oilfields.

2.2.2.2 General Characters of China's Major Oildoms (Output Above 1000 × 10^4 T/a)

At the end of 2001, there were four oil-producing regions, or oilfields, in China with annual outputs above ten-million-tons (Daqing, Shengli, Liaohe, and the Eastern South China Sea), and their common characteristics are: ① they are all located in Eastern China; ② they are all located in tensional fault basins (rift valleys); ③ their reservoirs are dominated by sandstone, featuring river-delta sedimentation; and ④ their production layers are all Mesozoic and Cenozoic, which all belong to the Paleogene and Neogene except Daqing; and ⑤ they are all located near heavily industrial cities, which is favorable for sustainable development.

However, by the end of 2005, there were 8 oil-producing regions or oilfields in China with annual outputs above ten-million-ton: Daqing Oilfield, Shengli Oilfield, Changqing Oilfield, CNOOC Tianjin Branch, Tarim Oilfield, Karamay Oilfield, Liaohe Oilfield, and CNOOC Zhanjiang Branch. Their common characteristics are: ① they are all located in large sedimentary basins; ② their reservoirs are dominated by sandstone, featuring river-delta sedimentation; ③ the production layers are all Mesozoic and Cenozoic except Changqing Gasfield which is of Paleozoic erathem; and④ they are all located near heavy industrial cities, which is favorable for sustainable development.

2.2.2.3 Focus of China's Current Petroleum Research

There is an increasingly higher requirement for reservoir evaluation and prediction in the current process of China's hydrocarbon exploration and development, such as ① gaining an accurate prediction of favorable reservoir facies belts; ② making interwell predictions and extrapolating reservoirs; ③ promoting reservoir protection in the process of drilling and oil recovery operations and water injection; ④ transforming the low-permeability layer; ⑤ using dynamic monitoring techniques in the process of development; ⑥ predicting remaining oil distribution; and ⑦ selecting appropriate oil recovery enhancement methods. To solve these problems, the reservoir is the object of study, reservoir assessment is the basis of study, and the objective is to make favorable reservoir predictions.

2.2.2.4 History of Petroleum Reservoir Research Over the Past 20 Years

After the concept of reservoir characterization was first proposed, the United States Department of Energy held four international conferences between 1985 and 1997 (Table 2.5), and the number of participants, mainly European and American, increased each time. The features of these conferences were as follows: ① the theme of the conference varied based on ideological and geological problems facing oilfield development; ② the content covered a wide range, such as new methods and techniques; and ③ their priorities were a combination of fundamental geology, new technology, and discipline. In addition, since the beginning of the 21st century, at the American AAPG annual conference the themes have always been related to reservoir characterization and geological modeling, and the main content has involved determining the heterogeneity (anisotropy) of reservoir spatial distribution, which maximizes services for oil recovery enhancement. The main task and goal of such research is to establish a predictable reservoir model through research on reservoir characterization, to serve the arrangement of infill wells and epitaxial wells, the enhancement of oil recovery, and calculate simulated reservoir parameters.

In June 1989, the second international reservoir characterization seminar was held in Dallas, during which 49 papers were published. These papers were related to petroleum geology, reservoir sedimentology, oil reservoir physics, seepage mechanics, well logging technology, seismic technology, petroleum reservoir engineering, production testing technology, and horizontal well technology. The seminar was divided into four groups according to the research scale to enable discussion and communication of all papers, which related specifically to microscopic heterogeneity, macroscopic heterogeneity, megascopic heterogeneity, and gigascopic heterogeneity. In conclusion, this seminar focused on a series of issues relating to

Table 2.5 Previous international petroleum reservoir characterization conferences

Year	Session No.	Place and attendance	Topic
April 1985	First session	175 persons, Galleria in Texas	Four topics: ① Geological and petrophysical basis; ② space variance; ③ determination of simulation parameters; and ④ fluid flow in heterogeneous medium
June 1989	Second session	233 persons, Dallas in Texas	This session was divided into four technical parallel branches for the first time, based on scales of microscopic heterogeneity, macroscopic heterogeneity, megascopic heterogeneity, and gigascopic heterogeneity
November 1991	Third session	247 persons, Tulsa in Oklahoma	① Heterogeneity and anisotropy; ② oilfield study and data requirements; ③ reservoir modeling and interwell description; and ④ optimal reservoir management
March 1997	Fourth session	256 persons, Houston in Texas	The objective of this conference was to discuss the latest developments of reservoir characterization technology that had occurred since the first conference in 1985. The main topics included reservoir description, improvement, and enhancement of oil recovery, associated methods and techniques, fracture analysis, simulation and modeling
May 2009	First session	115 persons, NapaValley, California	This conference focused on key factors controlling hydrocarbon migration and pressure distribution in the petroleum system. This session covered a wide range of problems from rock properties in sandstone reservoirs and cata-clastic deformation zones to properties of shale sequences

oilfield development. It is also worth mentioning that the research by scholars from various countries tended to be unified with respect to the research level of reservoir heterogeneity.

To bridge the gap between China and advanced countries, and to enhance the quality of current hydrocarbon exploration and development with the particular aim of serving secondary or tertiary oil recovery, the former CNPC (under the advocacy of Qiu, Yinan) established significant research with extensively related contents. This was called "China Petroleum Reservoir Research," and it involved investment funds of up to several million yuan. Over 20 units from national petroleum system and colleges and universities, and over 400 scientific and technical personnel actively participated in this research project.

It is considered that the years before 1990 were a preparatory stage, and in order to exchange research achievements in a better way it was necessary to understand the status of domestic and overseas reservoir research to undertake useful research in China. To make necessary preparations for tackling hard-nut problems in science and technology during the 8th Five-Year Plan, the lead and coordinating group for this project edited and published the collected papers of *Research Progress on Reservoir Evaluation.* This represented the beginning of comprehensive and large-scale national petroleum reservoir research. The collected papers include the preliminary summary of China's reservoir research and an evaluation of domestic and overseas research methods and trends, which reflected, to some extent, the situation and level of China's petroleum reservoir evaluation and research at that time. At the time, this project (No. 85–103) was included in the national 8th Five-Year Plan. To understand the situation and trends in overseas reservoir study, the leading coordination group organized the compilation and translation of two collected papers: *Foreign Reservoir Modeling Technology* and *Foreign Reservoir Description Technology.* The system of Ministry of Geology and Mineral Resources also established many sub-projects related to petroleum reservoir research during the

8th Five-Year Plan, and organized the compilation and translation of a number of informative publications and collected papers relevant to reservoir modeling and description. This shows the significance of reservoir research in relation to petroleum geology and development geology in China.

Then, in June 1991, the former CNPC carried out an achievement report and evaluation of the "7th Five-Year Plan" and "China Petroleum Reservoir Research" in Hangzhou. One of the features of this work was the energetic development of research relating to the reservoir concept geological model and the accumulation of a geological database. It was thus evident that research had developed from conventional qualitative research to quantitative or semi-quantitative research, and although there were large differences between applied methods of the past, significant progress had been made in terms of a systematic summary of China's different basin reservoir characteristics and a unification of concepts. Upon completion of this project, a series of monographs were published representing the level of China's petroleum reservoir research, such as *Evaluation Techniques of Petroleum Reservoirs* and *Petroleum Reservoir Research Atlas of China* (1997) by Qiu, Y. N. and Xue, and *China Continental Facies Petroleum Reservoir* (1997) by Qiu, Y. N., Xue, S. H., Ying, F.X., et al.

Since the 1990s, the lithologic hydrocarbon reservoir has become the main reservoir type for oil and gas exploration and development of today. In particular, the deep development of water-injection technology (secondary oil recovery) during oilfield development, and the rise of experimental investigations for various technologies to enhance oil recovery are representative of new reservoir research, and they have promoted the rapid development of reservoir sedimentology, or reservoir geology theory, and relevant technology.

In the mid and late 1990s, with the constant emergence of various stochastic modeling software and the application of various mathematical statistics methods in reservoir modeling and simulation, certain new software emerged. Although a number of articles and monographs (such as Zhang and Wang 1995; Yu 1996; and Wu 1999) related to stochastic modeling were published in China at the time, Chinese systematic software had not been established due to insufficient funding and the associated small scale. Therefore, a large gap still existed with respect to software, systems, and methods involved in stochastic modeling. This is attributed to the fact that geological researchers did not have adequate computer or programming skills, and that computer experts did not have an adequate understanding of geologic features. Therefore, the use of computer skills to achieve scientific geological modeling had not yet been realized in China, and therefore most scientists continued to use overseas software.

2.2.2.5 Outlook on Reservoir Sedimentology

Recent progress in sedimentology has ensured that reservoir sedimentology is a prosperous research field in relation to petroleum production. This popularity is related to the establishment of various model concepts for depositional facies based on accumulated sedimentation data and outcrop surveys, and insights into formation mechanisms of various sedimentation phenomena derived from basic research theory dominated by flume experiments. Therefore, reservoir sedimentology research is now focused on increasing the reservoir description required in petroleum production and associated developments both macroscopically and microscopically. In addition, reservoir sedimentology scientists are striving to describe and predict reservoirs from a qualitative to semi-quantitative or quantitative perspective.

Recent essential practice in petroleum development and management has enhanced the efficiency of data acquisition systems, made correct descriptions of geologic features of petroleum reservoirs, clarified the oil-displacement mechanism, established a quantitative petroleum reservoir model with complete parameters, predicted the production performance using a numerical simulation technique, and determined a development strategy and concrete measures according to prediction outcomes to obtain the most superior development effect. Two key technologies are involved–reservoir modeling and numerical simulation. The former is the foundation, as simulation results can only be validated using an accurate reservoir model. Although it was estimated during the 11th World Petroleum Congress (WPC) that outstanding mathematics and physics issues in numerical simulation technology would soon be satisfactorily resolved, the ability to make correct reservoir models requires a considerable amount of time.

Hydrocarbon exploration and development requires knowledge of various geological characteristics to enable detailed reservoir characterization. Such knowledge has developed since the early stages of focusing on the geometry of macroscopic sand bodies, continuity and change of holes, and the permeability plane, and now focuses on microphenomena such as in-layer permeability change, discontinuous barrier bed distribution, bedding structure, and pore structure. These phenomena, however, are difficult to control using the current development well pattern (with a well spacing of hundreds of meters). Therefore, the ability to correctly predict and describe reservoir characterization under a scattered well pattern is the current challenge for reservoir sedimentology researchers.

2.2.3 Thoughts and Methods for Reservoir Sedimentology Research

2.2.3.1 Uniformitarianism

At present, the most common method is to make analogies between reservoirs and characteristics using similar ancient depositional environments via a depositional model established using modern sedimentary surveys. Since the investigation by Fisk on modern Mississippi river delta deposits in 1944, there has been considerable growth in the use of depositional models, and they are now used in various environments and have assisted with the development of modern sedimentology. After its initial application in predicting reservoir distribution in petroleum exploration, the depositional model was established through the use of modern sedimentary surveys. The establishment of this model means that practice must be guided by ideas related to observations. By the late 1970s, it was considered (for instance, Leblanc et. al. 1991) that all clastic rocks under various environments should be observed and studied directly in modern sedimentation, except those of harbor type delta and turbidities. Over the years, depositional models have also been used in oilfield development to understand ancient reservoirs, and are effective for use in estimating features such as the continuity of a sand body and in-layer heterogeneous rhythmicity.

2.2.3.2 Reasoning Method for Outcrop Studies

A prototype is needed as a guide to establishing a geological model of a subsurface petroleum reservoir, and the establishment of such a prototype needs firstly to be based on the study of a field outcrop (and to thus establish the model based on this outcrop) before spreading underground. Therefore, if an outcrop exists near a studied reservoir, the method is relatively accurate and reliable, and many predictions of reservoir characteristics for oilfields in North America have applied this method. In addition, although most of China's oilfields do not have this prerequisite, outcrop research on the depositional system plays a positive and instructional role in predicting the characteristics of a subsurface petroleum reservoir, particularly with respect to macroscopic features. There is a limitation however, in obtaining a complete observation of the three-dimensional space.

2.2.3.3 Evolution of Sedimentary Facies and Change of Base Level

Depositional environments evolve in various ways in line with the evolution of a basin's structure, and environmental changes bring about alterations in the distribution of sedimentary facies, which are also subject to change at a base level. In the process of cycle change at the base level, the volume allocation effect and facies differentiation effect of sediments in the same sedimentary system tract or facies region, due to the change in ratio of accommodation to sediment supply (A/S), results in changes in factors such as the preservation degree of sediments, stratigraphic accumulation style, facies sequence, facies type, and rock structure, thereby leading to differences in the spatial superimposed form of sedimentary sand bodies. The core to this approach is the use of genesis dynamic analysis to analyze the formation mechanism of a reservoir from the perspective of evolution.

2.2.3.4 Sedimentation-Heterogeneity Response Relation

Different types of sedimentation, or depositional modes, form varying types of sedimentary sand bodies, and regular changes occur in the internal properties of these structures, as indicated by the various features of heterogeneity. There are many types of sand bodies controlled by microfacies, and it is thus difficult to establish one "model" as this requires finding inherent laws from the relation of "sedimentation–heterogeneity response," (i.e., sublimate sedimentology theory), to analyze the formation mechanisms (processes) of all types of sedimentary bodies in order to grasp all features of heterogeneity.

2.2.3.5 Application of Empirical Equations and Parameters

There has been profound development in comparative sedimentology with respect to making estimations of the spatial distribution range of subsurface reservoirs of the same type; for instance the application of the width/thickness ratio in a large number of different sedimentary microfacies. This has been achieved by obtaining quantitative and semi-quantitative data from modern sedimentary surveys and ancient outcrop research, which has enabled the establishment of quantitative geological databases. There are currently a large number of empirical equations and quantitative geological databases that were built in the process of researching particular areas. They can thus be used as references, and make reservoir sedimentology quantitative research possible. Their core features involve dialectic relations between generality and specialty-laws and typical cases.

The main aim of petroleum reservoir sedimentology research is to be able to infer or predict reservoir distribution in a three-dimensional space using fewer drilling and seismic data, particularly the estimation of three-dimensional distribution, according to data gained from a one-dimensional section. Unremitting efforts have been made by both domestic and overseas scholars in this respect. However, evident achievements have not been made due to the complexity of geological processes and the difficulty of quantification, and such a lack of progress agrees with the point of view put forward by Mr. Qiu, Yinan, i.e., the necessity of establishing quantitative geological databases will be the focus of reservoir sedimentology research in future. Therefore, the method of studying the spatial distribution range of sedimentary bodies (sand bodies) and determining the variogram eigenvalue of reservoir modeling is a significant practice.

2.2.3.6 Analysis on Hydrodynamic Conditions—Flume Experiments

Flume experiments were firstly used to study hydrodynamic conditions in the early 20th century, and have since solved many problems relating to the genetic mechanism of sand bodies, such as the formation mechanism of sandy bedform, influential factors in differing types of formations, and the generation and deposition mechanisms of turbidity currents. A recent trend has evolved in outcrop research and underground data research, whereby certain easily-defined palaeoenvironment parameters are determined and simulations using flume experiments are conducted to make these parameters identical with those of already established sedimentary characteristics. This is achieved by undertaking repeated tests on uncertain parameters, and finally conducting a comprehensive simulation of all the sedimentary features of the sand body using these parameters. This trend further illustrates that flume experiments remain one of the most significant basic experiments used in sedimentology.

2.2.3.7 Isochronous Sub-layer Correlation of Sedimentary Microfacies Guide

The method of stratigraphic correlation in sequence stratigraphy has become a point of focus in present sedimentary basins analysis and reservoir modeling. However, sub-layer correlation has always been a controversial issue in exploration geology. This new approach involves a detailed correlation of the sub-layer sand body using distribution features of sedimentary microfacies. Owing to the obvious hierarchy of depositional facies and depositional sequences, the correlation of oil groups, sand groups, and sand layers (sub-layer) can be controlled by a level hierarchy to realize the hierarchical control, facies sequence guidance, and layer-by-layer comparison, thereby reflecting the evolution and superimposition of the spatial sand body and ensuring isochronism of the facies control.

2.2.3.8 Analysis of Seismic Facies and Electrofacies

Seismic and log data are the key information used in petroleum reservoir research, and therefore seismic facies are not only the basis for use in sequence stratigraphy analysis, but also the cornerstone for depositional facies or depositional system research in oildoms, in which the depositional facies model and its spatial concept guide analysis. Research on electrofacies uses logging feature analysis as its approach, establishment of a vertical sequence model as its method, and the geological response of a well logging series as its theory. Its resolution ratio can therefore reach the scale of a single sand body, and can thus be the basis for subfacies analysis and sub-layer correlation, both of which involve analogy analysis and research of morphological characteristics (which requires establishing a series of models, the establishment of which needs detailed understanding of actual data and a theoretical model). Close integration of the two can more effectively play the role of reservoir sedimentology. In recent years, the development of well logging and seismic technology have enabled greater reservoir description, and the cores of these lie in the relation between the concept of geologic facies and geophysical responses. However, there are current limitations in the closeness of well and seismic data, in terms of the resolution ratio.

2.3 Trends in Domestic and International Petroleum Reservoir Research

2.3.1 Evolution from Macroscopic to Microscopic Research in Reservoir Sedimentology

(1) With the further development of clastic petroleum reservoirs and exploration of subtle hydrocarbon reservoirs, petroleum geologists are now required to master spatial distribution laws and geometric features of sand bodies formed under a variety of different sedimentary environments (Ravenne et al. 1989; Dreyer 1993). In order to improve hydrocarbon exploration and development, it is necessary to fully comprehend the geometric features and continuity of a single sand body in reservoir research, i.e., macroscopic heterogeneity. For this reason, more attention is being paid to research on reservoir sedimentology and geological modeling. However, the microscopic pore structure of reservoirs and the clay pore matrix and authigenic clay minerals not only have a significant impact on oil displacement efficiency, but damage the reservoir to varying degrees. It is therefore necessary to study the microscopic heterogeneity of a reservoir and to provide a reliable geological basis for reservoir protection.

(2) Through continuous innovation and further advancement of various water-flooding development technologies, it has been determined that heterogeneity within a reservoir

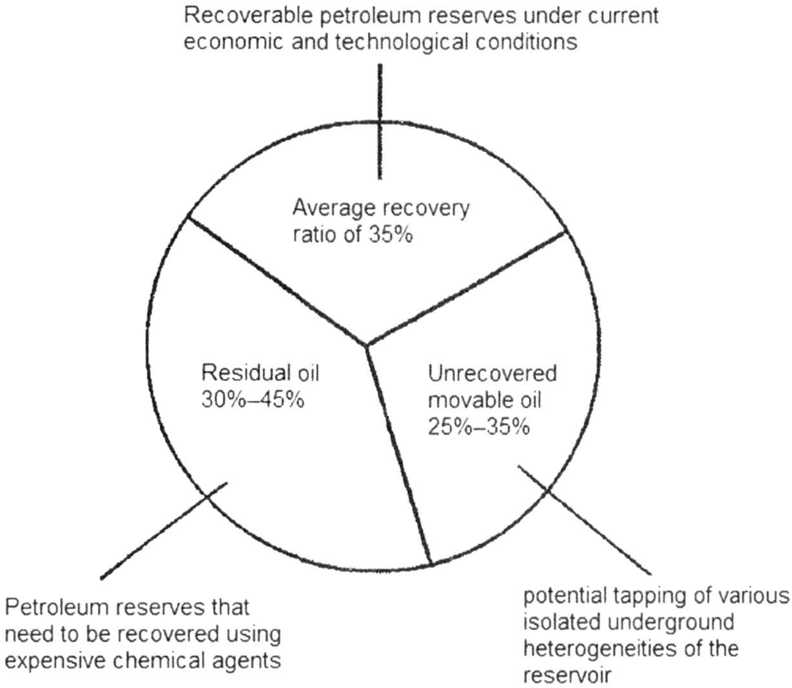

Fig. 2.2 Oil and gas production will change under different technical conditions (by Qiu et al. 1993)

greatly affects the volumetric sweep efficiency and internal oil displacement efficiency. In addition, the internal wave and coefficient are mainly affected by differential degrees of reservoir permeability, the rhythm type of permeability, and the distribution of discontinuous muddy intercalation within the reservoir. Therefore, petroleum geologists are expressing considerable interest and emphasis on outcrop and mature oilfield research as the basis for reliable and accurate heterogeneity geological modeling of a reservoir.

(3) At present, the distribution law of secondary porosity of different sedimentary basins has become the main focus for oil and gas field exploration, which is both exciting in the search for hydrocarbon reserves and promotes studies on diagenesis. In particular, research on secondary porosity is now enjoying a rise in popularity and status within reservoir evaluation. New insights have been gained and models established on the formation mechanism of various secondary porosities (Meshri 1986; Scherer et al. 1987; Surdam et al. 1989; Guohua 1992; Zheng and Pang 1989; Yu et al. 1999), which provides a basis for the prediction and simulation of reservoir properties. However, quantitative research on petroleum reservoirs and secondary porosity in relation to the differing causes of basin creation and associated predictions is still in the exploratory stage.

(4) A reservoir is subdivided into a plurality of flow units in the excavation of movable oil not involved in secondary oil recovery (Fig. 2.2). In 1984, Hearn et al. first proposed the concept of the petrophysical flow unit when studying the Shannon Sandstone Reservoir in the Hartzog Draw Oilfield in Wyoming, USA. They then defined the flow unit as a horizontal and vertical continuous reservoir belt, and stated that the petrophysical properties affecting fluid flow are similar to the rock characteristics. In subsequent articles, this was further explained as a "… mobile unit is a space where the fluid continuously flows in lateral and vertical direction in a genetic unit."

According to statistics published by the United States, conventional ultimate recovery is 35%, residual oil accounts for 30–45%, and unrecovered movable oil is around 16%, on average. Movable oil is mainly blocked by various heterogeneities within a reservoir, and can be extracted by infill drilling, horizontal drilling, or a change in the injection-production system. However, it is essential to understand the controlling factor in a reservoir to determine the distribution of this remaining movable oil. The current challenge in this respect, therefore, is to predict the distribution of this remaining movable oil. However, there are many geological factors controlling it, such as the heterogeneity of different levels and the degree of well pattern control, due to barriers such as faults. Currently, a description of macroscopic heterogeneity is not considered important, and the main difficulty lies in gaining an in-depth study on a finer scale, in-depth knowledge of micro-level heterogeneity within the reservoir, and a three-dimensional description of certain geological phenomena; which are known collectively as interwell prediction.

In general, reservoir geological determinants for defining flow units mainly include: ① the architecture or configuration of a reservoir sand body; ② the connectivity and isolation of geological interfaces at all levels; ③ continuous or discontinuous muddy interlayers and other barrier beds; ④ continuity of the genetic unit; ⑤ characteristics of pore structure within the reservoir; and ⑥ faults and fractures.

2.3.2 Qualitative to Quantitative Evolution of Reservoir Description and Prediction

(1) A considerable amount of research has been conducted to calculate and predict the porosity and permeability of a subsurface reservoir sandbody, and meet the actual needs for petroleum production (Scherer 1987; Dutton and Scholle 1992), and empirical formulae or mathematical models have been proposed. However, the parameters in such formulae are hard to determine in practical applications, and this has restricted popularization and validation of the formulae or quantitative models, which center on the constant scope of empirical formula and adaptive conditions thereof.

(2) In reservoir numerical simulation technology, it has been necessary in recent years to require quantitative descriptions of the three-dimensional spatial distribution of physical parameters of a reservoir. The number of field experiments conducted for tertiary recovery has been far less ideal than those for indoor tests, or have even failed. The main problem and reason for this is the large number of differences between reservoir geological models used in indoor simulation study and real models, and their associated inaccuracies. The current research trends are as follows: firstly, to establish a reservoir geological database and geological model prototype through modern deposition investigations and outcrop studies, apply geostatistical methods to deduce a variety of empirical formula; use actual parameters obtained from one-dimensional profiles to predict the distribution of a sand body in a two-dimensional or three-dimensional space, and apply valid empirical formula and methods obtained from outcrop and mature field studies after verification of a new area and underground. Thus far, and despite some successful results, this is still in the research stage. The second trend is to utilize high-resolution seismic technology to track reservoirs in a lateral form and the characteristics of constrained inversion, to predict the spatial distribution of a sand body.

(3) With continuous developments in oilfield exploration and development, quantitative and accurate predictions of the physical properties of subsurface clastic rock reservoirs, such as porosity, permeability, and oil saturation are expected. Hence, reservoir modeling (especially stochastic modeling) and simulation has become the main method used in the prediction of reservoir property and heterogeneity.

It has been recognized that the current methods of predicting these reservoir properties consider restrictions related to major geological conditions, such as the formation of the sedimentary environment, tectonic evolution of the basin, and the diagenetic mechanism. Realization is most likely to be achieved only after considering first-hand research data relating to systematic and detailed quantitative research, and determining the formation mechanism using detailed mathematical statistical analysis.

(4) Realizing the quantitative description and prediction of a sand body and the continuity or spatial distribution characteristics in a lateral direction (i.e., stochastic modeling or simulation study of reservoir) has become the focus of research for reservoir geologists in recent years. This has been undertaken to meet the urgent need for oilfield development, infill drilling, and oilfield expansion. Reservoir sedimentology is the basis of such studies, and computer technology is important in achieving these aims. The key to stochastic modeling is to develop geostatistical techniques and various conditional and non-conditional simulation techniques, and to obtain the heterogeneous appearance of the spatial distribution of reservoir parameters on the basis of geological constraints (such as facies control), and limitations of property statistical laws; thereby determining the geological reality.

2.3.3 Progress from Theoretical Sedimentology to Applied Sedimentology and Construction of Reservoir Characterization Technology

(1) International Sedimentology Congresses have focused on "Reservoir Sedimentology and Geological Modeling," which has been a technical topic of discussion since the 1990s. This illustrates that applied sedimentology is now valued over theoretical sedimentology. In real applications, reservoir sedimentology is top of the list in direct discussions of technical issues involved in geological modeling. Geological modeling is classified as research in petroleum development geology, and is a current hot field and focus in obtaining "reservoir descriptions" and "reservoir characterizations."

(2) Special discussion groups for reservoir sedimentology were formed in many World Petroleum Congress (WPC) meetings in the 1990s, and they always had the same focus: how to establish a quantitative geological model. Besides, this issue is also listed as the key topic for discussion by the world's highest-level academic conference from two different academic areas. In particular, the Sedimentological Congress, which always focuses on the theoretical development of sedimentology itself, also raises sedimentology to a special status, which proves the importance of petroleum reservoir characterization or the modeling technique.

(3) Both the World Petroleum Congress and the World Geological Congress have been convened in recent years, and both have placed research on reservoir sedimentology in a prominent position. In addition, the number of relevant published papers and monographs on the subject has increased annually, and this is associated with the speedy recovery and rise of reservoir geology and developmental geology in the late 1980s. Even the American Association of Petroleum Geologists Bulletin (AAPG), which mainly reports geological exploration-related contents, published a special issue with the slogan, "Returning reservoir geology to its rightful place."

(4) Since 1985, there have been four international conferences on "Reservoir Characterization" sponsored by the Research Institute

of United States Department of Energy, and four collected papers have been published. Reservoir characterization and stochastic modeling techniques have been formed on the basis of reservoir geology. In the 21st century, the AAPG discusses the topic of reservoir geology annually, and this is the best example of its transformation from theory cognition to real application.

2.3.4 Synergetic Research and Development of Reservoir Characterization from Single Discipline to Multi-discipline

Current research on reservoir geology and reservoir sedimentology is no longer only centered on petrophysical properties and the sedimentary environment of a reservoir, but concentrates on various characteristics from a multidisciplinary research perspective (such as geology, geophysical and seismic logging, mathematical statistics, and computer studies), which, as a result, have contributed to the rapid development of other geological disciplines. According to current international research on reservoirs, there are three main research perspectives that have the focus of carrying out comprehensive analysis and research on physical and geometric properties.

2.3.4.1 Outcrop Reservoir and Underground Geological Research

Outcrop reservoir and underground geological research for reservoir geological modeling has been newly incorporated into reservoir geology research. It intends to describe reservoir properties using the theory of reservoir sedimentology, focusing on applied sedimentology and heterogeneity responses, in combination with the evolution of diagenesis. However, the ability to make a quantitative description of geometric and spatial variation is still limited as a prototype and its regularity or mathematical degree doesn't achieve the ideal level.

In addition to the above, reservoir geology has made considerable breakthroughs in the past decade, especially in relation to the introduction of the level or sedimentary scale concept. A large amount of useful knowledge has been gained through reservoir research by the implementation of genetic unit divisions and bounding surface hierarchies. However, problems exist in relation to the identification of surface hierarchy in underground or electric well-logging curves, and this is a key issue currently being tackled. Presently, large overseas petroleum companies and research institutions are committing to outcrop research, because geological models and distribution features of reservoir heterogeneity thus obtained can be used to conduct, or guide, interwell prediction. Thus, building geological outcrop model prototypes with various types of sand bodies can be used to guide and restrain interwell stochastic modeling, and used to inspect various technical application conditions and effects on the outcrop.

Intensive sampling is firstly carried out on the field outcrop to measure petrophysical parameters such as porosity and permeability, and the property change within the sand body is then faithfully represented for the depositional system being researched. At a later date, various other geostatistical methods are used to simulate and rarefy control points, and some mathematical statistics methods used to simulate parameters between control points and approximate the actual geology, for application in actual subsurface geological work. This type of method, which is used to simulate reservoir property changes with various probability statistical models, is known as Conditional Simulation, and is a statistical technique for producing property parameter distribution between known points.

2.3.4.2 Digital Processing Technology for Log Data

The development of digital processing technology for log data has become the foundation for reservoir description and simulation, in addition to validation. Log analysis is required in hydrocarbon exploration and development. It has expanded from a comprehensive analysis and evaluation of a borehole to that of a zone and become an emerging integrated technology for reservoir description. Geological, drilling, seismic, and log data are integrated together for application to a large-scale software package system to undertake analysis of the reservoir lithology, property parameters, planar and spatial distribution laws of oil-gas-water, and determine structural configuration. Reservoir description is thus becoming an increasingly digitized and automated science.

In addition, the emergence of new logging and testing methods has played a great role in promoting reservoir research. For example, in recent years, new imaging technology has been developed overseas, including microelectrode scanning (FMS), array induction logging (five induction curves with 0.25–2.25 m of detection range), and three-dimensional acoustic borehole imaging technology. All these technologies enjoy intuitive features, and their use enables direct observation of essential characteristics of the downhole geological body, making geological interpretation in reservoir research, or reservoir description, more accurate. In addition, more than 10 chemical elements can be detected by induced natural gamma-ray spectral logging (IGRS) through geochemical logging, thereby determining mineral, clay type, and petrophysical properties. Furthermore, nuclear magnetic resonance (NMR) logging, enables direct evaluation of the porosity, permeability, and saturation of the bound water and residual oil within the reservoir, thereby improving the quantitative reliability of reservoir physical parameters. Finally, the innovation of measuring while drilling (MWD) has greatly improved the quality and accuracy of various log data, thus providing superior first-hand information for digital processing and interpretation of high-resolution logging information.

2.3.4.3 High-Resolution 3D Seismic Acquisition Processing and Interpretation Technique

Owing to the rapid development of computer technology, reservoir seismic exploration and modeling prediction technology have become further developed and innovative. With the appearance of 3D seismic exploration, and greater acquisition processing and interpretation techniques for use in enhancing seismic resolution, seismic exploration data and high-resolution seismic contrast data are being produced and used in reservoir description and lateral sandbody prediction. For this purpose, development geophysics, reservoir geophysics, and other new techniques are being shaped. However, this technique is used to combine logging technology at a high vertical resolution and seismic technology at a large lateral coverage, to ultimately make a three-dimensional quantitative reservoir description (particularly a reservoir characterization/reservoir description) on an inter-well scale. Although core and log data enjoy good vertical resolution, the spatial resolution is insufficient due to the sparse well distribution. This creates an evident technical "gap" between conventional seismic and logging methods, namely, the lack of geophysical exploration technology to enable accurate imaging within the interwell range. The interwell seismic and/or reservoir seismic modeling prediction technique, by combining the advantages of seismic and acoustic logging technology, fills in this "gap," such as by interwell seismic and multichannel seismic waves inversion modeling (ROVIM) by CGG, seismic lithologic modeling (SLIM) by Western Geophysical, and multi-parameter constrained inversion modeling (LCI) by Geophysical Service Incorporated (GSI for short). Results indicate that such seismic modeling prediction techniques are effective in solving geological problems during hydrocarbon exploration and development, particularly for quantitative prediction on interwell net pay. It is thus envisaged that seismic modeling prediction and/or interwell seismic technology will be used as key components or primary research focuses in the development of seismic technology and geological development, and widely used in reservoir characteristics description, dynamic production monitoring, and the prediction of effective reservoirs. In conclusion, all disciplines and specialties need to be integrated in reservoir characterization research and/or fine reservoir description in order to service the middle and later stages of oilfield development as well as the oil recovery enhancing stage. In real practice, the research objectives and key issues that need to be solved should be used as the basis for carrying out work using certain technology.

2.3.5 Emergence of Various Simulation Methods and Software

The emergence of various simulation methods and software has facilitated further computerization of reservoir studies. Due to the rapid development of computer technology since the 1990s, particularly the emergence of graphic workstations and the expansion of computer capacity, various reservoir simulation methods (or modeling) and software for reservoir prediction and oilfield development have also emerged. These can also be used for interwell sand body prediction and determination of the sand body scale (which is an urgent problem).

2.3.5.1 Common Reservoir Simulation or Stochastic Modeling Methods

① Turning Bands Method; ② Boolen Simulation; ③ Marked Point Process Simulation; ④ Enhance Truncated Gauss; ⑤ Sequential Gauss Simulation; ⑥ Sequential Indicator Simulation; ⑦ Fractal Geometry; and ⑧ Simulated Annealing.

2.3.5.2 Main Reservoir Modeling Software

There are many types of simulation software, and the followings have excellent functions:

(1) The TUBA software, developed by the New Mexico Institute of Mining and Technology, is used for a vertical and horizontal comparison of reservoirs and was designed combining the turning bands method and the indicator Kriging method. Its mathematics is based on the Bessel function and indicator-related function.
(2) SGM is a geological model computer system software mainly operated on an SGI graphic workstation and was developed by the Strata Model Company, USA. This software uses a slice method to present the distribution characteristics of porosity, permeability, and a sand body on a continuous section or slice, and has stochastic simulation as its mathematical basis.
(3) The "Monarch" software was launched by the Royal Dutch/Shell Group, and was designed based on the conditional probability method, which is mainly used to simulate three-dimensional connectedness and configuration in a sandstone reservoir.
(4) StatMod from Jason Company in the Netherlands is a software using stochastic simulation on the basis of geostatistics and seismic characteristic inversion, and was developed with the help of technology from British Petroleum (BP). It focuses on analysis of the spatial variation of reservoir properties (heterogeneity) and spatial distribution characteristics of reservoir lithologic characteristics (anisotropy), to calculate the eigenvalues, build a mathematical model, and then simulate different variable types using the three-dimensional Kriging and CoKriging methods, three-dimensional sequential Gauss conditional simulation, and three-dimensional sequential indicator (e.g., continuous variables of porosity and permeability and discrete variables of lithology or sedimentary facies). Stochastic modeling and inversion are conducted to provide a reservoir spatial model following not only geostatistical variable characteristic value distribution and a geological mode, but also seismic reflection characteristics.
(5) The Gridstat software is a unique and advanced reservoir description software working under the principle of geostatistics. This software, which originated in the Technical Dept. of Texaco, Inc. from the USA, is a necessary tool for production

personnel and researchers. This software can integrate log data and seismic data for three-dimensional stochastic modeling.

(6) STORM software: This software was developed by Smedvig Technologies from Norway, and uses the initials of stochastic reservoir modeling. It is a software that closely integrates the stochastic simulation method with geological conditions, and its main features are as follows. It can generate a variety of equiprobable three-dimensional reservoir heterogeneity models, which can be used as input for dynamic reservoir simulation, process and identify a variety of complex fault systems and edit while displaying, evaluate the uncertainty of petroleum reserves, evaluate the uncertainty of a fluid producing level, and is able to provide multiple stochastic simulation algorithms. Its modeling process roughly contains five stages: building the strata framework model; selecting different stochastic simulation methods for different depositional environments; establishing a sedimentary facies model; simulating the spatial distribution of reservoir property parameters within simulation of a sedimentary unit after considering the control action of the depositional facies; upscaling the established model to match with the numerical reservoir simulation; and conducting sequencing and selection in a random field. In the early days, this software only ran on a workstation, but it has now been translated for use on a PC, where it has been renamed, RMS.

The RMS geological modeling software is derived from the ROXAR Company. It combines the advantages of STORM in terms of sedimentary facies simulation, integrates seismic, logging, reservoir performance, geological databases and other information, and applies geostatistics, sequence stratigraphy, modern sedimentology, and stochastic theory for a comprehensive study of reservoir geology and performance. Its main function includes data integration, stratum modeling, fault modeling, sedimentary facies modeling, petrophysical parameters modeling, volume calculation, model upscaling, streamline analysis, graphic display, and well site design, centering on the reservoir sedimentary facies stochastic simulation method based on the theory of the reservoir depositional system and the genetic unit of a depositional facies.

(7) The RC^2 software is a set of integrated business software packages used for stochastic modeling. This software includes 10 modules: database preparation, generation and quality control (ResPrep); spatial data analysis (ResMod); geostatistical reservoir model (ResMod); Interactive block modeling (ReFrame); fractured reservoir interpretation and modeling (ResFrac); upscaling technology (ResScale); waveform evaluation and seismic inversion (ResSeis); streamline fluid flow simulation (RealFlow); model operating, well design and uncertainty analysis (ResCalc); and interactive model visualization (ResScape).

(8) The GOCAD software was developed by the T-SURF Company from the USA, and can run on multiple operating system platforms such as SUN, WIN95, and NT. GOCAD has the following characteristics: it has a flexible and convenient user interface, a powerful fault modeling capability, complex three-dimensional mesh generation, reservoir lithofacies and petrophysical properties modeling, excellent openness (users can interact with the system at five levels), and wonderful visualization. This software consists of four modules: structure modeling, velocity modeling, reservoir modeling, and reservoir risk assessment.

(9) The Petrel TM software is from Schlumberger Limited, and is a Windows-based three-dimensional visualization modeling software currently dominating the international market. It integrates seismic interpretation, structure modeling, facies modeling, reservoir property modeling, and reservoir numerical simulation display and virtual reality into one whole, thus providing a shared information platform for geologists,

geophysicists, petrophysicists, and reservoir engineering staff. In addition, Petrel TM applies various advanced technologies such as a powerful structure modeling technique, high-precision three-dimensional grid technology, deterministic and stochastic depositional facies modeling technique, scientific petrophysical modeling technique, advanced three-dimensional computer visualization, and a virtual reality technique.

From the functions and effects of the above software it is possible to observe their individual strengths. However, their common outstanding feature is that they combine the spatial distribution characteristics of reservoir lithology and physical properties (porosity and permeability) to realize stereoscopic display and/or an arbitrary slice in a three-dimensional space. They require a large memory capacity and high speed due to the large amount of reservoir simulation data. Since the high-resolution color graphic display requires graphic card configuration, these software were initially only used with a graphic workstation. However, with the acceleration of computer graphics card developments and computing speed, most of the software has now been ported to the PC.

These software are able to simulate marine facies stratum with slight heterogeneity, but in China, where there are serious heterogeneity reservoirs mainly within petroliferous continental fault basins, it will be necessary to develop reservoir simulation software suitable for China's unique situation; this would require combining such studies with scientific research. At present, China has created a small amount of software, but they are still not systematic or complete. A considerable gap remains between China and overseas countries in terms of stochastic modeling, and this is therefore the primary mission of the new generation of reservoir geological scientists.

2.3.6 Current Hot Issues in Petroleum Reservoir Research

In addition to the further advancement of domestic and international hydrocarbon exploration and development, conventional methods for petroleum reservoir and testing technology have gradually matured. New research ideas and technology have appeared in relation to various reservoir characteristics and problems in production, all of which have enabled a consensual view on the geological characteristics of a petroleum reservoir to some extent. However, a number of insurmountable problems remain, and these are mainly as follows:

2.3.6.1 Analysis of Internal Architecture or Configuration of Sand Body

Geologists globally have attached great importance in determining the three-dimensional attributes of geological bodies (reservoir bodies), since A. D. Miall first proposed the concept of configuration in 1985. This is mainly because the geological characteristics of a three-dimensional reservoir are the main content required when building a reservoir geological model, and the development of modern computer technology makes it easier to reproduce the changes in three-dimensional space from different angles, especially visualization technology. Determining the three-dimensional attributes mainly involves the following aspects in the area of geological research.

1. **Determination of fluid flow unit**

After oil and gas field development began, and particularly after a period of development, research began into predicting the amount of remaining oil. This now needed to be classified and researched in relation to the flow unit of subsurface reservoirs. The concept of the flow unit was proposed by C. L. Hearn et al. in 1984. It is currently agreed that, "it is continuous in horizontal and vertical directions, and has a reservoir belt with similar permeability, porosity and flow characteristics." However, determining the flow unit depends not only on its geological characteristics and location in vertical sequence, but also on its petrophysical characteristic, especially porosity and permeability. For this reason, division of the flow unit became the focus of later geological research after hydrocarbon development. At present, four division methods are mainly used: ① division based on the spatial distribution of sedimentary microfacies; ② classification based on distribution and characteristics of intercalation; ③ quantitative distinction and research (e.g., FZI) according to pore structure characteristic parameters of reservoir; and ④ determination according to characteristics and pressure condition of fluid. For a sedimentary reservoir it can be said that the flow unit is a fundamental element of the internal architecture of a sand body.

2. **Analysis of the distribution and law of discontinuous (muddy) barriers**

Research on the spatial distribution of discontinuous barriers involves a detailed analysis of the interlayer and intercalation, and particularly involves an evaluation of intercalation with relative low permeability. As such, this is one of the focuses in determining reservoir heterogeneity research. Qualitative analysis and evaluation on the discontinuous barrier relates directly to an understanding and comprehensive evaluation of petroleum analysis in a subsurface reservoir. However, one of core issues lies in sub-layer correlation standards and methods, and how to reflect facies sequence and its genesis, which requires the researcher to use a variety of research modes. Hence, the core task is to carry out scientific analysis on genesis prior to distribution research; only when genesis is fully understood can the spatial distribution law be correctly determined.

In addition, emphasis during research on interlayer and intercalation with relative low permeability is placed on defining the upper limit of permeability. However, the reservoir characteristics of different basins with different depositional systems and different ages are not the same, and thus the petroleum properties of the reservoir are also different. Scientific quantitative research can only be carried out based on specific circumstances, and the spatial distribution law is mainly subject to sedimentary microfacies and diagenesis.

3. **Bounding surface hierarchy**

Hierarchical division is one of the main features of geology, and specific methods thereof have been the constant focus of geological research (for example the emergence and development of sequence stratigraphy over the past 20 years). A. D. Miall proposed detailed breakdown principles and methods for the division and identification of the surface at all levels, while putting forward the concept of architecture or configuration. In the meantime, the key to isochronous contrast of stratum or the sedimentary layer is to ensure correctness of the boundary surface division.

4. **Laminae identification**

Laminae can be said to be the smallest unit of a sedimentary structure of a sedimentary body and can only be identified by core, imaging logging, and outcrop analysis. Laminae identification not only contributes to genesis discussion, but is also of great significance in the analysis of reservoir configuration. Its principal benefit is to reflect the characteristics of various sedimentary structures, the formation process of the sandy bedform, and the energy unit–particularly the lithofacies unit. Lithofacies analysis assists in recognizing the formation and spatial distribution laws of a reservoir.

2.3.6.2 Interwell Prediction on Reservoir Property

After the establishment of well patterns in an oil and gas field, and after a period of development, the petroleum reservoir engineer will begin to pay close attention to the characteristics of the interwell reservoir, and the variation law of the reservoir properties (porosity, permeability, and saturation). Different methods are usually applied to evaluate the characteristics change of different reservoir characteristics and the well pattern. At present, the most common method is deterministic modeling interpolation. In case of a sparse well, the difficulty lies in determining the characteristic values of the variogram during stochastic modeling.

2.3.6.3 Determination of Sand Body or Reservoir Continuity

Geologists hope to be able to determine the continuity and effective range of a reservoir, during both the exploration and development stages, especially when limited information is available. However, there is large variety in the reservoir continuity of different basins and in varying depositional facies. The establishment of a quantitative geological database proposed by Qiu, Yinan (1995) would solve this issue and the main content of this would involve setting the width-to-thickness ratio data of different sedimentary sand bodies.

1. **Geometry & size determination of genetic unit**

This involves two main methods: ① using available information to conduct the division of a genetic unit, and a comparative study of the same, determining the spatial distribution and compiling geometrical statistics on this basis, and developing empirical formulae for prediction or calculation; ② establishing a quantitative database based on research on an outcrop, such as pertaining to the width-to-thickness ratio, and carrying out analogy and calculation according to the size.

2. **Analysis on the connectedness of a sand body**

In addition to the methods described above, it is also possible to analyze the spatial continuity of a sand body using production performance analysis, the drilling-encounter ratio, RFT, pressure tests, and tracer tracking. However, the primary issues in this respect are to ensure the reliability and isochronism of the sand body (sub-layer) contrast. To ensure the isochronism of facies control, the key is to realize hierarchical control, layer-by-layer contrast, and reflection of spatial evolution and superposition.

2.3.6.4 Prediction of Favorable Porosity and Permeability Distribution

It is well acknowledged that porosity features have an obvious horizontal and vertical zonation. However, difficulties currently exist in determining how to quantitatively evaluate the characteristics and laws of these two aspects.

1. **Porosity zonation**

In relation to the vertical zonation of porosity, regularity can usually be obtained from research on diagenesis, but it is difficult to evaluate several different pore zones within an area. Therefore, one major difficulty is to apply indicators other than porosity for quantitative research and evaluation. This is especially the case when secondary porosity is rather developed and therefore requires systematic analysis on its genesis.

2. **Prediction on low-permeability zone**

Prediction on the low-permeability zone is always difficult in the field of petroleum geological research, especially for a gas reservoir or gas field, as there are no very effective available methods. Therefore, genesis analysis and log interpretation are currently the major research tools.

3. **Development mechanism of deep secondary porosity**

Deep hydrocarbon exploration is a very important part of hydrocarbon exploration in the 21st century. Although basic theory and methods for research on diagenesis are already mature, the emergence of testing methods is promoting an understanding of the diagenesis evolution laws of a reservoir. The current main trend is to use the concept of lithofacies and apply diagenetic indicators for quantification, thus advancing from qualitative to quantitative prediction. However, this is presently in an exploratory stage.

2.3.6.5 Reservoir Damage and Protection

Process measures and engineering fluid applied in the development of oil and gas fields have direct implications for reservoir damage and protection. Different reservoir properties, in particular lithology, are directly related to the drilling fluid and process, and also directly influence hydrocarbon exploration in the later stages, especially for deep and low-permeability reservoirs.

1. **Sensitivity analysis**

Sensitivity analysis focuses on four parts: water sensitivity, acid sensitivity, velocity sensitivity, and salt sensitivity; the first three are usually more critical. Sensitivity analysis is used as a scientific basis for proposing a specific dosing scheme and process, and has experimental analysis at its core.

2. **Reservoir fluid interaction**

Different reservoir fluids and lithologies have different effects and chemical reactions. Under certain temperature and pressure conditions, interactions may also be different between different fluids, which will directly and indirectly affect the ultimate recovery ratio. Computer simulation and laboratory test analysis are the main research methods in this respect. The invasion of foreign particles and interactions between external fluid and rock can damage the reservoir; the engineering fluid is the main source of foreign particle invasion.

2.3.6.6 Fracture and In Situ Stress Analysis

Quantitative evaluation and comprehensive research on fractured reservoirs is posing major difficulties for current petroleum reservoir study. The principal concern is related to the restraint of fracture formation by in situ stress. Petroleum reservoirs experience several tectonic movements during geological formation, and there are many difficulties involved in recovering the ancient stress field itself. The degree of fracture development varies and the direction changes in areas with different lithologies and structures; it thus becomes more difficult to study and determine the size differences. In addition, identification of a subsurface reservoir mostly depends on the use of coring and downhole television, which also suffers from serious limitations.

1. **Fracture stratigraphy**

At present, the use of fracture stratigraphy, which is mainly directed at determining the characteristics of a fracture within a stratum, is becoming popular overseas. Using the main data from drilling, it combines the use of stress field recovery and the fracture development status in research and statistical analysis, thereby enabling the study of the distribution of reservoir fractures.

2. **Fracture spacing**

The law of fracture variation in a downhole reservoir can be speculated using measurements and statistics of the actual fracture spacing. The central issue is now how to predict interwell reservoir fracture development and distribution. The main research procedures and methods in this respect include system statistics, stress analysis, Mechanical stratigraphy, and stochastic modeling.

References

Dutton, S.P., and H.S. Hamlin. 1992. Interaction of Burial History and Diagenesis of the Upper Cretaceous Frontier Formation, Moxa Arch, Green River Basin, Wyoming [J]. 37–50. Wyoming Geological Association.

Fisk, H.N. 1944. *Geological investigation of the alluvial valley of the lower Mississippi River*, 78p. Vicksburg: Mississippi River commission.

Guohua, Zhu, 1992. The formation, evolution and prediction of porosity in reservoir detrital rocks [J]. *Acta Sedimentologica Sinica*, 10(3): 114–123.

Hearn, C.L., W.J. Ebanks, and R.S. Tye, et al. 1984. Geological Factors Influencing Reservoir Performance of the Hartzog Draw Field, Wyoming. *Journal of Petroleum Technology*, 36(8): 1335–1344.

Lake, L.W., Jr, H.B. Carroll. 1986. WORKSHOP SESSION REPORT, In *Reservoir Characterization*, 643–650.

Leblanc, L.R., S.G. Schock, S. Panda. 1991. Pulse and Aperture Design Cons Ideations For A Marine Sediment of Classification Chirp Sonar. In *Oceans'91. Ocean Technologies and Opportunities in the Pacific for the 90's. Proceedings*, 1027–1027. IEEE.

Meshri, I.D. 1986. On the reactivity of carbonic and organic acids and generation of secondary porosity [M]. *Roles of Organic Matter in Sediment Diagenesis*, 36–50.

Miall, A.D. 1985. Architectural-element analysis: A new method of facies analysis applied to fluvial deposits [J]. *Earth Science Reviews*, 22(4): 261–308.

Qiu, Y.N., and S.H. Xue. 1997. *Evaluation technique of petroleum reservoir*. Beijing: Petroleum Industry Press.

Qiu, Y.N., et al. 1993. *Evaluation technique of petroleum reservoir*. Beijing: Petroleum Industry Press.

Ravenne, C., et al. 1989. Heterogeneity and geometry of sedimentary bodies in fluvial-deltaic reservoir. *SPEFE*, 239–246.

Scherer, M. 1987. Parameters influencing porosity in sandstones: A model for sandstone porosity prediction. *AAPG* 71: 5.

Scherer, G., A. Moissant, G. Claus, et al. 1987. Valorisation des brais durs de pétrole par combustion [J]. *Oil & Gas Science & Technology*, 42(5): 318–320.

Surdam, R.C., L.J. Crossey, E.S. Hagen, et al. 1989. Organic-inorganic and sandstone diagenesis. *AAPG Bulletin* 73: 1–23.

Tillman, R.W., K.J. Weber, 1987. Reservoir sedimentology [M]. *Society of Economic Paleontologists and Mineralogists*.

Wu, S.H. 1999. *Reservoir modelling*. Petroleum Industry Press.

Yu, X.H. 1996. *Reservoir characterization and stochastic simulation of clastic reservoirs*. Geological Publishing House.

Yu, X.H., D.J. Zhang, S.L. Li, et al. 1999. The diagenetic evolutive law of deep burial strata at Shahejie Formation in eastern and western depressions of Liaohe Oil Field [J]. *Experimental Petroleum Geology*, 12(4): 290–296.

Zhang, T.F., and J.H. Wang. 1995. The principle of stochastic modelling and stochastic simulation of reservoir. *Well Logging Technology*, 19(6): 391–397.

Zheng, J.M., and M. Pang. 1989. Study on Diagenesis of Reservoir Detrital Rocks [M]. *Wuhan: China University of Geosciences Publishing House*, 15–85.

3 Basic Features of Clastic Reservoirs

Oil and gas reservoirs worldwide are dominated by sedimentogenic clastic rock and carbonate rock strata. Therefore, it is necessary to study the relationships among the various sedimentary environments of petroleum reservoirs, paleo-geographic conditions, spatial distribution features of sedimentary systems, and various sedimentary facies belts to build a reservoir depositional model and geological model. This would help to comprehensively and accurately evaluate and predict the spatial distribution, morphological characteristics, and vertical and horizontal variation laws of physical properties and thus satisfy the reservoir range (determination of epitaxial well) and inter-well characteristics (property and geometry) required for hydrocarbon exploration and development.

Compared with carbonate rock and other special rock reservoirs, clastic reservoirs have the following advantages: ① oriented intergranular pores (carbonate rocks mostly feature intragranular pores); ② sedimentation is strongly controllable; ③ grain size uniformly influences porosity and permeability; and ④ compaction process is relatively clear and quantitative analysis is easy.

3.1 Petrologic Features of Clastic Reservoirs

Clastic petroleum reservoirs vary considerably from carbonate reservoirs and other rock reservoirs. They have unique sedimentary genesis, structures, architectures, and property formation mechanisms.

3.1.1 Rock Type

A clastic rock can be classified into conglomerate, sandstone, siltstone, and mudstone by grain size. Rock consisting of volcanic clastic materials is called volcaniclastic rock (abnormal clastic rock).

In 1922, Wentworth suggested taking the power of 2 as a grade scale (Table 3.1), which is also known as the Wentworth standard in foreign countries, for dividing clastic sediment particles, and 2 mm as the upper limit of grain size for sand, which leads the classification of clastic rocks toward scientific quantization. Siltstone is categorized as mudstone abroad, mainly because siltstone cannot reflect the underlying sedimentary environment or pattern. However, siltstone may serve as a reservoir. Hence, siltstone is considered as sandstone in most classifications in China.

Conglomerate is mainly composed of coarse clastic particles with grain size greater than 2 mm (less than -1Φ). Coarse clastic particles are divided into granules, pebbles, cobbles, and boulders by roundnessdiameter. Correspondingly, conglomerate (or breccia) is classified as boulder conglomerate (breccia), cobble conglomerate (breccia), pebble conglomerate (breccia), and granule conglomerate (breccia) by

X. Yu et al., *Clastic Hydrocarbon Reservoir Sedimentology*, Advances in Oil and Gas Exploration & Production, https://doi.org/10.1007/978-3-319-70335-0_3

Table 3.1 Granularity division standard for clastic sediments (according to Wentworth 1922)

Description	Description	Particle diameter		Settling rate (cm/s)
		(mm)	Φ Value	
Gravel	Boulder	>256	<−8	>4.29×10^6
	Cobble	64–256	−6 to −8	2.68×10^5–4.29×10^6
	Pebble	4–64	−2 to −6	1.05×10^3–2.68×10^5
	Granule	2–4	−1 to −2	2.62×10^2–1.05×10^3
Sand	Very coarse	1–2	0 to −1	65.5–262
	Coarse	0.5–1	1–0	16.4–65.5
	Medium	0.25–0.5	2–1	4.09–16.4
	Fine	0.125–0.25	3–2	1.02–4.09
	Very fine	0.0625–0.125	4–3	0.256–1.02
Mud	Silt	0.0039–0.0625	8–4	9.96×10^{-4}–0.256
	Clay	<0.0039	>8	<9.96×10^{-4}

Note $\Phi = -\log_2 D$ (D refers to particle diameter)

gravel (rubble) size. Granules, pebbles, cobbles, and boulders, which are composed mainly of detritus, are classified into monomict conglomerate and polymictic conglomerate based on the complexity of clastic constituents. Although there are not many conglomerate reservoirs, they are abundant in type and complex in pore structure, and in particular, complex in terms of gravel components and support types.

Rock fragment is dominated by terrigenous clast, as well as intrabasinal clast. Clastic constituents mainly include quartz (Q), feldspar (F), and rock detritus (R). Quartz is a stable component, whereas feldspar and detritus are unstable. In general, the compositional maturity of rock is represented by its relative content (i.e., Q/(F + R)) of stable and unstable components. Sandstone refers to rock with grain sizes of 2–0.0625 mm (−1 to 4Φ). It can be divided into very coarse sandstone, coarse sandstone, medium sandstone, and fine sandstone by grain size or into arenite and greywacke (matrix content greater than 15%) by matrix content. According to clastic components, sandstone can be categorized into quartz sandstone (Q > 50%, F < 25%, R < 25%), arkose (F > 25%, Q < 75%, R < 25%), and lithic sandstone (R > 25%, Q < 75%, F < 25%). Sandstone reservoirs feature a wider distribution and good oil storage property, and over 80% of petroleum reservoirs in China are composed of sandstones.

Siltstone, with grain sizes ranging from 0.0625 to 0.0039 mm (4–8Φ), is mainly composed of silt-sized clastics, and it may be divided into coarse and fine by grain size. Coarse siltstone can form a good petroleum reservoir. Siltstone and fine sandstone reservoirs have relatively poor properties owing to particle fineness such as in the main Palaeogene hydrocarbon reservoirs of the Dongpu Depression in the Zhongyuan Oilfield.

Mudstone refers to rock composed mainly of clay minerals with a grain size less than 0.0039 mm (greater than 8Φ). Transition types between mudstone and siltstone are diverse, such as argillaceous siltstone and silty mudstone. In general, mudstones do not form conventional reservoirs, but they may form fractured reservoirs when fractures are developed.

Volcaniclastic rock is a type of special clastic rock that is composed mainly of volcanic clastic materials (detritus, crystal fragments, and vitric fragments), and combines the characteristics of both volcanic and clastic rocks. Major rock types include agglomerate, volcanic breccia, tuff, ignimbrite, and sedimentary tuff.

3.1.2 Fabric Features

Clastic rock comprises three basic components, namely, clastic particles, interstitial materials, and pores, among which clastic particles account for more than 50% of its total composition.

Clastic interstitial materials include matrix and cement. Matrix generally refers to non-chemically precipitated fine silt and clay with a grain size less than 0.0315 mm (greater than 5Φ). Cement refers to chemical precipitates from intergranular solutions, which comprise mainly carbonate minerals (calcite, dolomite, ferrocalcite, ankerite, and siderite), silica minerals (quartz, chalcedony, and opal), clay minerals (kaolinite, smectite, illite, chlorite, and mixed-layer illite/smectite), sulfate minerals (gypsum, anhydrite, celestite, and barite), zeolite minerals (analcite, laumontite, epistilbite, thomsonite, mordenite, and geolyte), and iron minerals (hematite, limonite, and pyrite).

The texture of clastic rocks involves features (grain size, shape, sphericity, roundness, and surface features) intrinsic to clastic particles, sorting of particles, features of cement, and relationships (cementing type) between clast and interstitial material. Texture maturity refers to the degree to which detrital materials are close to the ultimate structural features under weathering, transport, and sedimentation, and is represented mainly by matrix content, sorting, and roundness.

3.1.3 Support Form

As is well known, clastic rocks are generally supported by two forms, namely, particle and matrix. Clastic rock may be grouped into five categories (Fig. 3.1) according to grain size, distribution characteristics, and matrix content.

(1) Peer particle supported: Refers to a rock particle–supported framework composed basically of gravels or sands of the same grain size.
(2) Multi-level particle supported: Refers to rock-supported framework structured with particles of different grain sizes in sequence by level. "In sequence by level" refers to the fact that in a support framework void of particles in the upper level, multi-level particle support is formed sequentially by a secondary support framework composed of lower-level particles. Its quantitative characterization can be expressed as a granularity histogram, which is characterized mainly by bimodality or multimodality.
(3) Local matrix supported: Refers to rock partly supported by a fine matrix and partly by clastic particles.
(4) Matrix supported: Refers to rock in which particles are distributed freely in the matrix.
(5) Hybrid supported: Refers to rock in which the support framework is formed by the combination of peer particle support, multi-level particle support, and little matrix support.

3.2 Reservoir Properties

The properties of petroleum reservoirs mainly include basic features of porosity, permeability, and saturation, which are not only basic objects of reservoir research but also core contents for reservoir evaluation and prediction, as well as the most fundamental parameters for quantitative reservoir research.

3.2.1 Porosity of Reservoir Rock

Broadly, rock pores refer to the spaces not filled with solid substances within rocks, which are also known as reservoir spaces or voids; this includes intergranular pores, intragranular pores, fractures, and karst caves. However, in a narrow sense, pore refers to intergranular and intragranular voids, as well as voids within the interstitial material in rocks.

3.2.1.1 Pore Classification

According to different research studies and purposes, pores may be classified based on pore

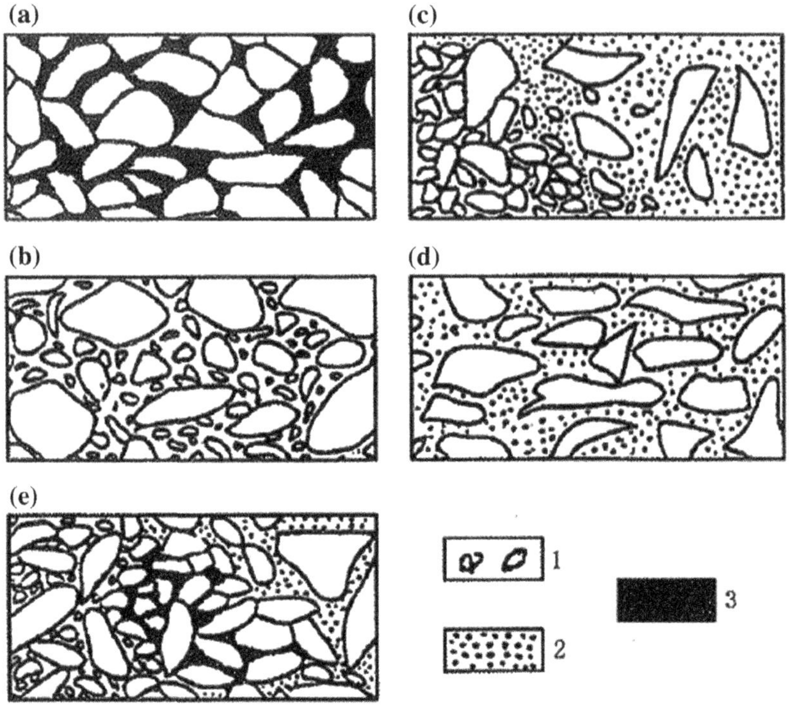

Fig. 3.1 The clastic rocks are mainly supported by five ways (by Qide and Youliang 1996) **a** peer particle supported; **b** multi-level particle supported; **c** local matrix supported; **d** matrix supported; and **e** hybrid supported, 1—coarse matrix and pore; 2—fine matrix; and 3—hybrid (coarse and fine) matrix and pore

Table 3.2 Common pore classification for clastic rock

Method	Standard	Classification Result
By contact relations between pore and particle	Pore position in rock	Intergranular pore
		Intragranular pore
		Pore within interstitial material
By pore genesis	Before or during diagenesis	Primary pore
	After diagenesis	Secondary pore
By grain size and fracture width	Aperture > 0.5 mm	Supercapillary pore
	0.0002 mm < Aperture < 0.5 mm	Capillary pore
	Aperture < 0.0002 mm	Microcapillary pore
By fluid seepage of pore	Pore communication	Effective pore
	Pore isolation	Ineffective pore

genesis, pore size, and contact relations with particles. Therefore, classification results vary according to the basis of classification (Table 3.2).

1. **Genetic classification**

Common types of pores in clastic rocks include primary intergranular pores, micropores within the matrix, dissolved pores, intracrystalline pores, and fracture pores. These types can be divided into two categories (Table 3.3) based on pore genesis: ① primary porosity, which refers to pores formed after sediment deposition, and before or during diagenesis; and ② secondary porosity, which refers to pores produced due to dissolution, recrystallization, and dolomitization upon diagenesis. Strictly speaking, all intercrustal rocks have pores, however they vary in size, structure, and quantity.

1.1 Primary pores

Primary pores in clastic rocks mainly comprise primary intergranular pores and micropores within the matrix.

(1) Primary intergranular pores

Rock is particle supported or matrix supported, which includes a small amount of cement. Intergranular pores are named after their location, that is, pore surrounded by particles. This is the most common pore type in sandstone. The size and shape of intergranular pores depend on grain size, sorting, sphericity, contact direction, filling mode, and degree of compaction of particles. The distribution of this pore directly relates to the sedimentary environment, and it is subject to change due to later diagenesis. Sandstone reservoirs, which mainly have intergranular pores, are characterized by large porosity, coarse throat, good connectedness, and excellent reservoir permeability.

(2) Micropores within a matrix

Micropores within a matrix mainly relate to the pores formed by contracting matrix sediments during weathering, as well as recrystallized intercrystal pores of clay mineral. It can be found in kaolin, chlorite, mica, illite, and carbonate argillaceous matrix.

This micropore is very tiny, generally less than 0.2 μm, and it can be identified clearly only under a high-power microscope. Its total quantity is vast, sometimes even accounting for over 50% of rock pores, but its permeability is extremely poor. Almost all sandstones have this micropore.

1.2 Secondary pore

Secondary pores mainly refer to solution enhanced pores, crystal secondary intergranular pores, and fractures.

(1) Dissolved pores

Dissolved pores originate from the dissolution of carbonate, sulfate, feldspar, or other soluble components in rock. By location, it can be classified into intergranular dissolved pores and intragranular dissolved pores.

① Intergranular dissolved pores are not limited by particle boundary, and they feature embayed edges, irregular shapes, and larger sizes, sometimes, even much larger than adjacent particles.
② Mold pores are formed by the complete dissolution of soluble mineral particles.
③ Intragranular dissolved pores and dissolved pores within cement are formed by partially dissolving rock particles and partially dissolving cement, respectively.

(2) Crystal secondary intergranular pore

This pore type mainly refers to the pores remaining after the primary pores are filled with crystalline quartz particles upon secondary enlargement. When quartz is subject to secondary enlargement, pore space is reduced significantly, which narrows the pore and throat thus resulting in decreased rock permeability.

(3) Fracture

Tiny, flaky microfractures formed within a sandstone reservoir due to ground stress features bend the joint surface, bypass particle boundaries, and are subject to ground stress in terms of

Table 3.3 Type, genesis, and reservoir permeability of sandstone pores

Type		Genesis	Reservoir permeability feature	Histological feature		
Primary pore	Primary intergranular pore	Sedimentation	Large and high porosity; good reservoir permeability	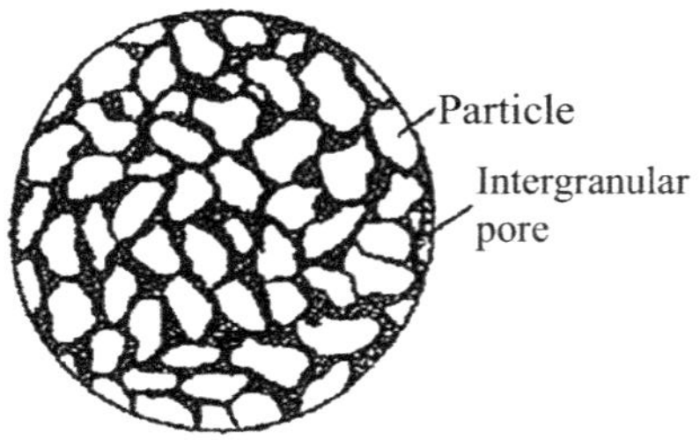		
	Micropore within matrix	Complex-genesis	Small and low porosity; relatively poor reservoir permeability	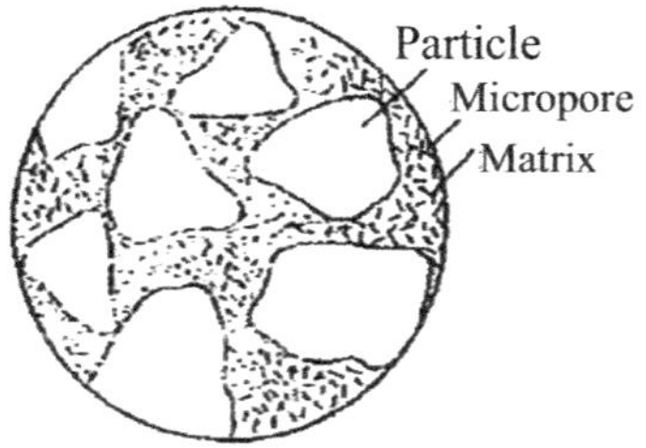		
Secondary pore	Dissolved pore	Dissolution	Small and low porosity; good reservoir permeability	Cement	A. Dissolved pore in cement	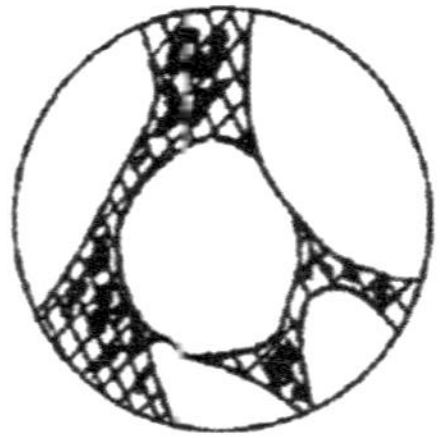

(continued)

Table 3.3 (continued)

Type		Genesis	Reservoir permeability feature	Histological feature		
				Intragranular	B. Intragranular dissolved pore	
				Intergranular	C. Intergranular dissolved pore	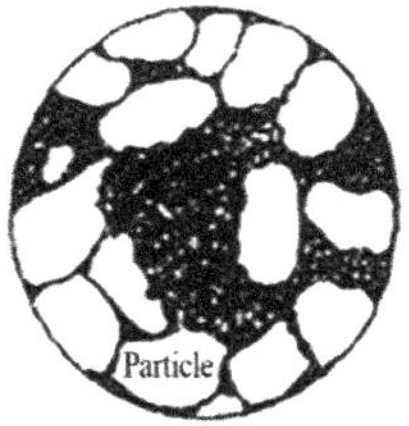
					D. Mold pore within sandstone	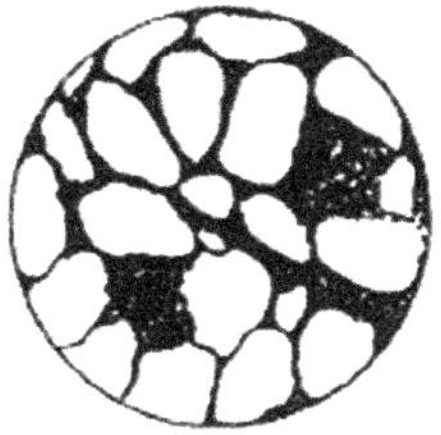

(continued)

Table 3.3 (continued)

Type		Genesis	Reservoir permeability feature	Histological feature
	Crystal secondary intergranular pore	Pressure solution	Small but high porosity; poor reservoir permeability	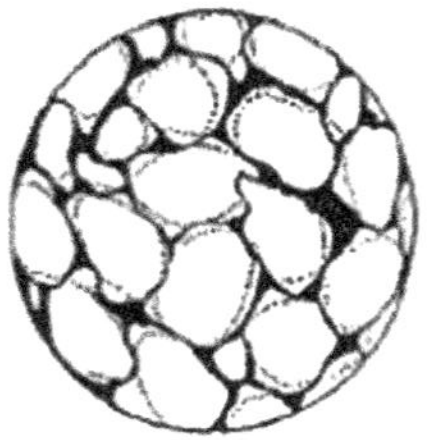
	Fracture	Ground stress Effect	Small and low porosity; good reservoir permeability	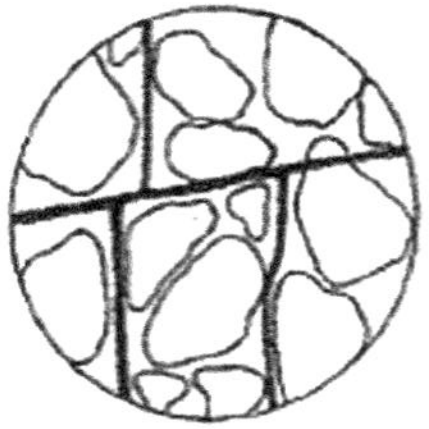

arrangement. Fracture width is generally parallel to the direction of the minimum ground stress.

In sandstone reservoirs, fracture width ranges from less than 1 μm to dozens of microns. Albeit with a small amount of total porosity, less than 5%, fractures can greatly improve rock permeability. Another feature is that fractures change as ground stress undergoes remarkable changes. For example, fracture closure occurs when ground stress increases along the vertical direction of a fracture, which results in a sharp decrease in stratum permeability.

2. **Pore size classification**

Based on pore diameter, fracture or fracture width, and the impact on fluid, pores can be grouped into three types:

(1) Supercapillary pores: Pore size is greater than 0.5 mm or fracture width is greater than 0.25 mm. Under natural conditions, fluid may flow freely under the action of gravity, and cemented loose sand body mostly contains supercapillary pores. Fluid flow through the pore follows the general rule of hydrostatics.
(2) Capillary pores: Pore diameter is 0.5–0.0002 mm, and fracture width is 0.25–0.0001 mm. This pore is always under molecular attraction, either between fluid particles or between fluid and pore walls. Owing to the action of capillary force, fluid cannot flow freely; it can flow only when any external force is greater than the intrinsic capillary force. In general, most sandstone pores are capillary pores.
(3) Microcapillary pores: Pore diameter is less than 0.0002 mm, and fracture width is less than 0.0001 mm. Intermolecular attraction is huge in this type of pore, so a very high pressure gradient is required for fluid flow in the pore. Consequently, fluid flow under normal stratigraphic conditions is difficult, which is why a pore radius threshold of 0.1 μm is taken as a boundary for fluid flow. This pore is usually developed in clay rock and dense shale.

Pores can also be classified according to fluid seepage as follows:

(1) Effective pores: Refer to interconnected pore spaces that allow fluid flow under natural conditions.
(2) Ineffective pores: Refer to isolated and disconnected pores, as well as microscopic capillary pores within rocks.

3.2.1.2 Porosity

As an important physical parameter controlling oil and gas reserves and energy storage, porosity is a typical scalar and an indispensable object of study in reservoir research, evaluation and prediction. It is the fundamental scalar for reservoir research. Generally, it can be classified into absolute porosity and effective porosity according to pore size and connectivity.

1. **Absolute porosity**

This refers to the ratio of the total space volume of all pores to the total volume of the rock sample, as expressed by the formula below:

$$\Phi_a = \frac{\Sigma V_{ap}}{V_r} \times 100\%$$

where

Φ_a absolute porosity, %;
ΣV_{ap} total volume of all pores within a rock sample, cm^3; and
V_r volume of rock sample, cm^3

2. **Effective porosity**

This refers to the ratio of total volume (namely, volume of effective pores) of pores that allows fluid flow under certain differential pressures (greater than ordinary pressure) to the bulk volume of the rock, and it is expressed as follows:

$$\Phi_e = \frac{\Sigma V_{ep}}{V_r} \times 100\%$$

Table 3.4 General standards for clastic reservoir porosity in eastern China

Level	Effective porosity (%)	Evaluation
I	>25	Very good
II	25–20	Good
III	20–15	Better
IV	15–10	Medium
V	10–5	Relatively poor
VI	<5	Worthless

where

Φ_e effective porosity, %; and
ΣV_{ep} total volume of effective pores, cm^3

Obviously, effective porosity is less than absolute porosity in the same rock sample, and the effective porosity of a reservoir is generally 5–30%, and most commonly 10–25%. Reservoir properties may be evaluated roughly based on the effective porosity of a reservoir (Table 3.4).

In recent years, the concepts of connected porosity and flowing porosity have been proposed. Owing to the uneven distribution of pores within rocks, a few pores are of the supercapillary type, through which fluid can flow, while others are of the microcapillary and stagnant types, through which fluid cannot flow.

Connected porosity is a percentage obtained by deducting the content of stagnant pores from the total porosity, while flowing porosity is connected porosity minus microcapillary porosity. Therefore, flowing porosity is less than connected porosity, which is less than total porosity. As for loose sandstone, its connected porosity is close to its total porosity. On the contrary, these two values differ considerably in the case of cemented dense reservoirs. Often, the porosity values mentioned in scientific literature are total porosity values.

3.2.1.3 Determination of Porosity

Rock porosity can generally be determined by two methods: directly (slices, mercury intrusion, and experimental testing) and indirectly (seismic and electric well-logging interpretation and calculation methods). The method of obtaining surface porosity by microscopic statistics with slices is a direct method. For porosity determination and prediction of underground reservoirs in an oilfield, especially stratum without a core, indirect geophysical methods are mostly applied, which include logging and seismic methods. The well test method is applicable as well. Method selection depends mainly on the data of the study area, research purpose, and the researcher's field of work. Generally, accuracy of the results obtained using the above test methods differs considerably, where the accuracy of the rock sample test is the highest and that of seismic method is the lowest. Nevertheless, logging calculation is the most widely used method for oilfield development, and its accuracy relies on template or core calibration. Therefore, when a low-accuracy method is applied for porosity prediction and evaluation, the best results can be obtained by performing a certain calibration with high-accuracy methods to improve research reliability.

3.2.1.4 Factors Influencing Porosity

In general, because general clastic rock is formed by the crushing, transportation, cementing, and compacting of parent rocks, the porosity of clastic rock is mainly influenced by the type of clast, quantity, and post-diagenetic compaction.

1. **Mineral composition of rock**

Provided that other conditions are the same, the oil reservoir property of quartz sandstone is generally better than that of arkose because the lipophilicity and hydrophilicity of feldspar are superior to that of quartz. When wetted with oil and water, the liquid membrane formed on the surface of arkose is usually immobile, which reduces the flow section of pores and reservoir volume to some extent.

Moreover, differences among mineral particles considerably influence porosity and permeability owing to the expansion of clay minerals in the presence of water, in addition to the effect of form on porosity (for example, quartz is of a granular shape, while mica is of a flaky shape).

2. **Particle arrangement and sorting**

The form and size of pore spaces are largely impacted by particle arrangement. Graton and H. J. Fraser (1935), among others, successively studied isometric spherical particles to determine the loosest and tightest arrangements (Fig. 3.2). They believed that the porosity of ideal soil is independent of the grain size of its particles, and depends only on the mode of their arrangement (namely, the θ angle). When $\theta = 90°$, $\Phi = 47.6\%$; and when $\theta = 60°$, $\Phi = 25.9\%$. The two aforementioned values represent two end member types of maximum porosity and minimum porosity, respectively. However, the conclusion that porosity is independent of particle diameter is not applicable to real rocks. By applying the laws of statistics to thousands of sandstones, it was found that porosity is related to particle diameter. Porosity decreases as particle diameter increases because fine clasts lead to better porosity than relatively coarse sand given that fine clasts have poor roundness and angular shape, which result in a relatively loose state during particle support (Gengsheng 1994).

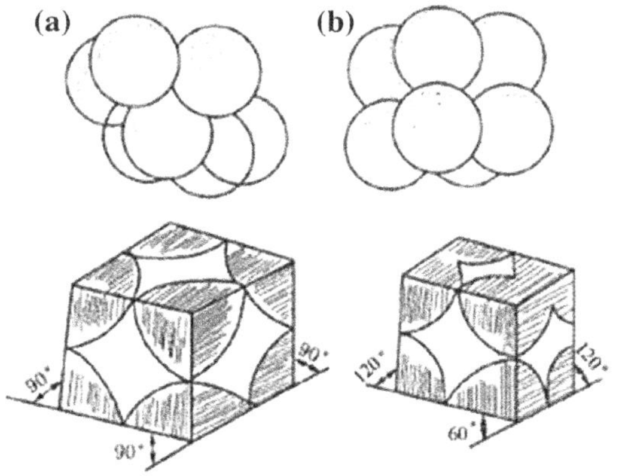

Fig. 3.2 Typical ordered porous medium structure and corresponding porosity (by Graton and H. J. Fraser 1935) **a** Cubic arrangement of isometric balls, $\Phi = 47.64\%$ and **b** Rhombohedral arrangement of isometric balls, $\Phi = 25.96\%$

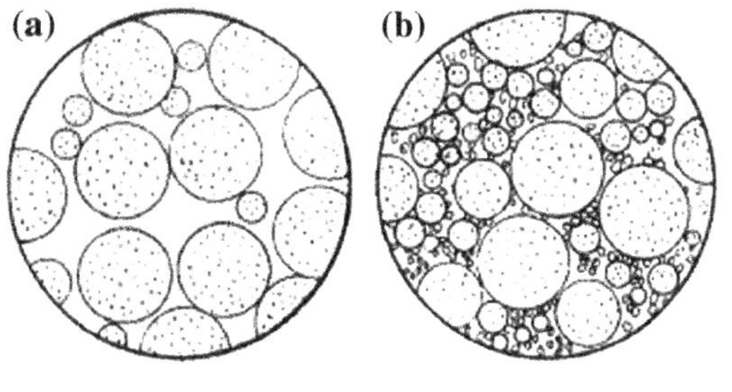

Fig. 3.3 Different degrees of sorting have different influences on the porosity; the better the degree of sorting, the larger the porosity (by Graton and H. J. Fraser 1935) **a** Well-sorted substances, $\Phi = 32\%$; and **b** Poorly sorted substances, $\Phi = 17\%$

In addition to particle diameter and mode of arrangement, the degree of sorting greatly influences porosity. When rock sorting is poor, small granulated clasts are filled with interparticle pores and throats, which reduce porosity and permeability (Fig. 3.3).

Assuming that the stacking manner of clastic particles in a natural sand layer has a random form, Scherer (1987) deduced the sorting coefficient (S_o) of granularity using Baerd and Weyl's (1973) data. The relationship between sorting coefficient and primary porosity can be expressed by the following formula, using which the porosity of a sand body in a wet surface environment can be calculated.

$$\Phi_{primary} = 20.91 + (22.9/S_o)$$

where

- S_o trask sorting coefficient (Trask $= \sqrt{Q_1/Q_3}$);
- Q_1 first quartile, equivalent to particle diameter value at 25%; and
- Q_3 third quartile, equivalent to particle diameter value at 75%

3. **Buried depth**

As the thickness and buried depth of the overlying strata of sedimentary rocks increase, static pressure and temperature of the stratum increase, and leads to a tighter arrangement, inelastic and irreversible interparticle movement, and rapid decrease in porosity (Fig. 3.4). When the tightness of arrangement peaks, any further increase of the overlying formation pressure facilitates partial particle dissolution at the contact point. Later, dissolved minerals (such as quartz) will form new crystals or minerals in pore spaces, further lowering porosity or even resulting in the disappearance of pores, thus leading to impermeability. Therefore, porosity, especially primary porosity, decreases with an increase in buried depth under normal circumstances.

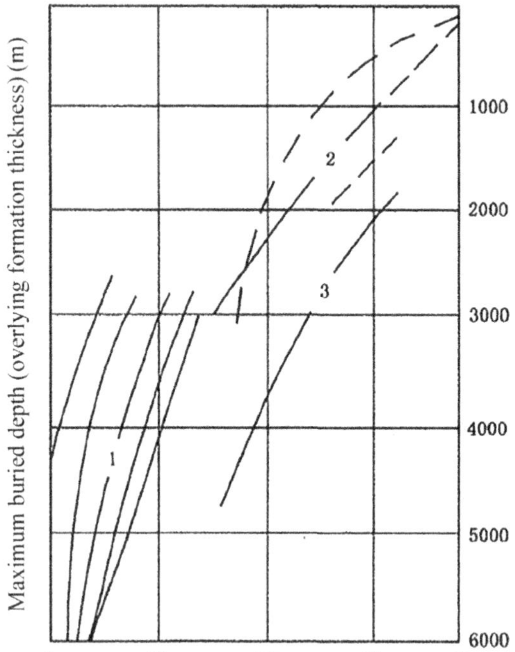

Fig. 3.4 The porosity and the maximum buried depth have a certain relationship (by Jackson et al. 1978), 1-argillaceous sandstone (containing mica); 2-jurassic-cretaceous quartz sandstone; and 3-paleogene-neogene quartz sandstone

4. **Diagenesis**

Owing to the presence of certain aqueous solutions underground, different aqueous solutions may selectively dissolve rock minerals at certain temperatures and pressures to form secondary pores. In general, organic aqueous acid solutions easily dissolve silicate minerals other than carbonate minerals, while inorganic aqueous acid solutions do just the opposite. For further details, please see the contents hereunder.

3.2.2 Permeability of Reservoir Rock

Permeability of reservoir rock refers to the extent to which the rock allows fluid to flow through it under a certain differential pressure. Similar to porosity, it is one of the most important parameters in reservoir research. It not only affects hydrocarbon storage but also, and more importantly, controls hydrocarbon productivity. Permeability performance is usually represented by permeability, which serves as a vector due to its obvious directivity, unlike porosity. In other words, permeability varies significantly in different directions, and it is usually divided into horizontal permeability (K_h) and vertical permeability (K_v).

3.2.2.1 Absolute Permeability

This refers to the fluid penetration capacity of rock, as measured in a laboratory. For liquids, the calculation formula is

$$K = \frac{Q\mu L}{(P_1 - P_2)Ft}$$

where

K absolute permeability of rock sample, μm^2;
Q volume of fluid that flows through a rock sample within time t, cm^3;
P_1 front-end pressure of rock sample, 10^5 Pa;
P_2 back-end pressure of rock sample, 10^5 Pa;
F sectional area of rock sample, cm^2;
L length of rock sample, cm;
μ liquid viscosity, cP; and

t duration of fluid flow through a rock sample, s

Permeability is expressed in Darcy (D) units. As specified, when 1 cm^3 of fluid with a viscosity of 1 cP flows through a porous medium with a cross section of 1 cm^2, the permeability of the porous medium will be 1D because the pressure difference is 1×10^5 Pa and the flow distance is 1 cm within 1 s. In practical applications, millidarcy (mD) is often applied because Darcy (D) is too large (1D = 1000 mD). In standardized measurements, permeability is expressed in μm^2 or 10^{-3} μm^2.

$$1\,\mu m^2 = 1.013\,D;\ \text{and}\ 1 \times 10^{-3}\,\mu m^2 = 1.013\,mD$$

In practical work, gas is often used to determine absolute permeability, which is also called air permeability. Gas flows passing through each point are different owing to different pressures at each point in the rock sample. Thus, volumetric flow in the Darcy formula should be indicated by average gas flow. Accordingly, the permeability formula can be expressed as follows:

$$K = \frac{\overline{Q}\mu L}{(P_1 - P_2)Ft}$$

where

$\overline{Q}$ volumetric flow of gas that passes through the middle part of rock sample within time t, cm^3; and

μ_s liquid viscosity, cP

Absolute permeability is a physical parameter and it is independent of fluid property but dependent on the pore structure of rock. Currently, air permeability (referred to as permeability in geological literature) is often considered absolute permeability for current production, and its evaluation indicators can be generally divided into five levels (Table 3.5) for continental clastic oilfields in eastern China.

3.2.2.2 Effective Permeability

When two or more fluids are present in rock, effective permeability can be obtained by measuring the permeability of one of the fluids. Effective permeability refers to the capacity of rock to conduct a particular fluid in the presence of other fluids. It depends on both pore structure and fluid saturation. Generally, the symbols K_o, K_g, and K_w are used to express the effective permeability of oil, gas, and water, respectively.

3.2.2.3 Relative Permeability

Relative permeability refers to the ratio of the effective permeability of various fluids to the absolute permeability of rock. It is a direct indicator of the capability of a particular fluid to pass through rocks. The relative permeabilities of oil, gas, and water are expressed by the symbols $\frac{K_o}{K}$, $\frac{K_g}{K}$, and $\frac{K_w}{K}$, respectively.

Many practices and lab experiments have proven that effective permeability and relative permeability are dependent on rock property, fluid property, and saturation. With an increase in facies saturation, effective permeability increases until all facies are saturated by a particular monophasic fluid, at which point effective permeability is equal to absolute permeability (Fig. 3.5).

Table 3.5 General standards for clastic reservoir permeability in eastern China

Level	Scope ($\times 10^{-3}$ μm^2)	Evaluation
I	$K \geq 2000$	Extra-high permeability
II	$2000 > K \geq 500$	High permeability
III	$500 > K \geq 100$	Medium permeability
IV	$100 > K \geq 10$	Low permeability
V	$K < 10$	Extra-low permeability

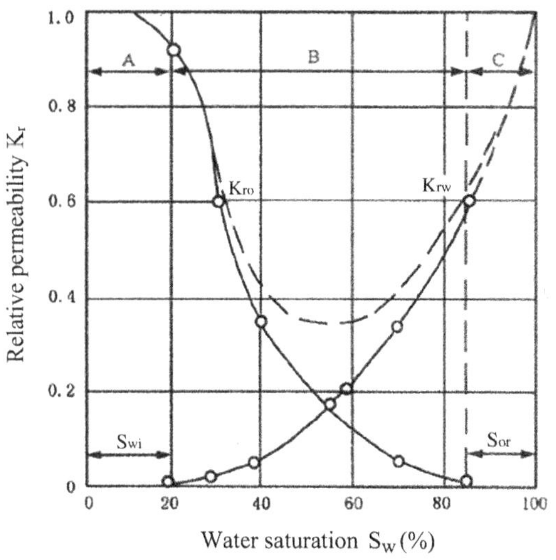

Fig. 3.5 Oil-water relative permeability curve (by Geng-sheng 1994)

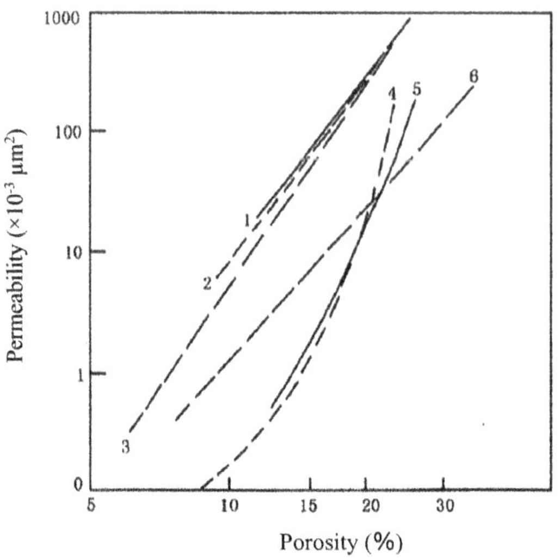

Fig. 3.6 Relations between porosity and permeability (by Timmerman 1982) 1-clean sand; 2-sorted fine sand; 3-extremely well-sorted fine sand; 4-well-sorted fine sand; 5-moderately sorted fine sand; and 6-poorly sorted fine sand

3.2.2.4 Factors Influencing Permeability

According to the concepts of absolute permeability and relative permeability, absolute permeability depends only on the pore structure of a rock and not on fluid property, while relative permeability is closely related to both factors.

1. **Factors influencing absolute permeability**

Absolute permeability can be affected by many factors, among which the following three are major ones.

(1) Rock characteristics

This refers mainly to the granularity, sorting, cement, and bedding, all of which influence permeability. In sedimentary reservoirs with positive rhythm, granularity tapers off upward and permeability decreases accordingly, which lead to premature flooding at the bottom of an oil reservoir during water injection.

(2) Pore structure

In general, absolute permeability is not only related to porosity but also mainly to pore structure (Fig. 3.6). All factors influencing the pore structure of rock can also affect its permeability. In addition, connectedness, tortuosity, and inner surface roughness are influencing factors.

(3) Temperature and pressure

At constant temperature, permeability decreases with an increase in directionless pressure; when the pressure exceeds a certain value, permeability declines sharply. The permeability of argillaceous sandstone decreases faster than that of sandstone.

At high temperatures, the influence of pressure on permeability decreases, especially under low pressures. This is because the increase in temperature leads to expansion of the rock matrix and the fluid in pores, which hinders compaction, and leads to a sharp decrease in absolute permeability with an increase in pressure.

2. **Factors influencing relative permeability**

(1) Wettability

Regarding the influence of rock wettability on relative permeability, as the core changes from strong hydrophilic to strong oleophylic, the relative permeability of oil tends to decrease. As for hydrophilic rock, water is usually distributed in tiny pores, stagnant pores, or particle surfaces. Consequently, it has little impact on oil permeability. For oleophylic rock, under the same saturation, water does not stay stagnant in pores as a water film, but in the form of droplets and streaming flows, which hamper oil seepage. Moreover, oil itself is attached onto the particle surface or in small pores as oil film; thus, the relative permeability of oil decreases in the case of saturation with the same oil.

(2) Pore structure

Because fluid saturation distribution and fluid flow channels are directly related to pore size and distribution, the relative permeability curve, which reflects the flow resistance of each facies of rock, is inevitably affected. The scope of the two-phase seepage zone of high–porosity, high-permeability sandstone is large, and irreducible water saturation is low; the reverse is true for low–porosity, low-permeability sandstone, because the seepage channels of large pores are larger than those of fine pores.

(3) Temperature

The influence of temperature on the relative permeability of oil and water is of great importance in research on the seepage and displacement process of thermal oil recovery. The key feature lies in the fact that irreducible water saturation increases with an increase in temperature. Moreover, temperature increase can lead to thermal expansion in rock, thus leading to a change in pore structure, which causes permeability to change accordingly.

In addition to these factors, there are many other influencing factors, such as fluid viscosity. If the viscosity of the non-wetting phase is considerably higher than that of the wetting phase, the relative permeability of the non-wetting phase increases as the viscosity ratio of the two phases increases, whereas the relative permeability of the wetting phase is not influenced by the viscosity ratio.

3.2.3 Fluid Saturation

Generally, pores in petroleum reservoirs are saturated with oil, gas, and water. However, they are saturated with oil and water in a reservoir where the pressure is higher than the saturation pressure. The percentage of saturated oil, gas, and water contents in the total pore volume is called the saturation of oil, gas, and water.

If oil, gas, and water exist in a reservoir, then

Oil saturation

$$S_o = \frac{V_o}{V_p} = \frac{V_o}{\Phi \cdot V_f} \times 100\%$$

Gas saturation

$$S_g = \frac{V_g}{V_p} = \frac{V_g}{\Phi \cdot V_f} \times 100\%$$

Water saturation

$$S_w = \frac{V_w}{V_p} = \frac{V_w}{\Phi \cdot V_f} \times 100\%$$

$$S_o + S_g + S_w = 1$$

where V_o, V_g, and V_w are the volume of oil, gas, and water, respectively, in a reservoir pore;

V_o pore volume;

Φ rock porosity; and

V_f rock volume

As is well known, most reservoirs are composed of sedimentary rocks, which are initially completely saturated with water. Gas and oil migrate and gather in such reservoirs from the side or the bottom only in later periods, during which upward migration of oil gradually expulses the water originally saturated in pores. This process is subject to the oil and water pore system. Water discharge during upward movement of the oil column and water expulsion depends on the nature of the oil and water, pore size and distribution in the rock, and formation pressure.

Because oil saturation is a key parameter for hydrocarbon exploration and development, only when initial oil saturation is determined can reserves be calculated accurately. Determination of oil saturation in the mid and later periods can provide an understanding of the developed performance of an oilfield, help conduct dynamic monitoring, and help calculate the remaining reserves and their distribution. Therefore, fluid saturation has always been a critical parameter in oilfield studies. It is different from porosity and permeability because it is neither a scalar nor a vector but a time-varying variable that is difficult to calculate accurately.

Notably, rock water saturation in an oil reservoir is related to many factors such as agglomeration of petroleum in the original aquifer, viscosity of petroleum, surface tension at the oil–water interface, particle distribution in rock, degree of closeness between oil–water contact and coring position, and clay content in rock, in particular rock pore size and distribution. Hence, permeability cannot be determined based solely on water saturation of the oil reservoir.

3.2.4 Reservoir Concept and Classification

3.2.4.1 Concept

All rocks available for oil and gas storage and fluid infiltration are called reservoir rocks. Fluid reserve is mainly determined by rock porosity, whereas fluid infiltration is determined by rock permeability and both factors are indispensable. In other words, reservoir rocks must have two basic elements: porosity and permeability. The former is the key determinant of energy storage, while the latter acts as the major influencing factor of productivity. As a result, the stratum formed by reservoir rocks is called a reservoir. If oil and gas flow with industrial value exists in a reservoir, it is usually referred to as an oil reservoir, a gas reservoir, or hydrocarbon reservoir.

Oil and gas exploration practices indicate that oilfields are found in three major rock groups, but more than 95% of oil and gas reserves are accumulated in sedimentary rocks, in which clastic rocks and carbonate rock account for a large proportion. In China, more than 90% of oil and gas reserves are distributed in clastic strata.

3.2.4.2 Reservoir Classification

To better evaluate the basic characteristics of various reservoirs in order to reflect reservoir quality and characteristics, the first step is to classify reservoirs (Table 3.6). In addition to the larger difference owing to different research purposes, the classification standard should reflect reservoir characteristics and the differences between various types of reservoirs. Moreover, the standard should be representative, especially of the overall sedimentation unit (genetic unit or energy unit) to meet the requirements of numerical reservoir simulation and all types of reservoir modeling.

Most oilfields are classified based on porosity and permeability. The common classification scheme in terms of porosity (Φ) and permeability (K) employed for continental oilfields in eastern China is summarized in Table 3.7.

Both rock porosity and permeability are subject to pore structure, but there is no strict functional relationship between the porosity and the permeability of reservoir rock. In general, the larger the effective porosity, the higher is absolute permeability. In the case of no fractures and karst cave in rock, the permeability of rock with a smaller pore throat is lower than that of rock with a larger pore throat, and the permeability of rock with a complex pore configuration is lower than that of rock with a simple pore configuration. Different configuration relationships between pores and throats can result in different reservoir properties. Briefly, rock permeability is proportional to the square of pore throat size, and it is related to throat tortuosity.

Table 3.6 Common reservoir classification scheme

Standard		Classification result	Standard	Classification result
Petrophysics		Clastic reservoir	Reservoir space	Porous reservoir
		Carbonate reservoir		Cave reservoir
		Other rock reservoir		Fractured reservoir
Physical property	Porosity	High-porosity reservoir		Porous-type reservoir
		Medium-porosity reservoir		Fractured-vuggy reservoir
		Low-porosity reservoir		
	Permeability	High-permeability reservoir	Oil and gas properties	Heavy oil reservoir
		Medium-permeability reservoir		Conventional oil reservoir
		Low-permeability reservoir		Coal-formed oil reservoir

Table 3.7 Common reservoir classification scheme for continental oilfields in central and eastern China

Type	Porosity (%)	Permeability ($\times 10^{-3}\ \mu m^2$)
High-porosity, high-permeability reservoir	>30	>500
Medium-porosity, medium-permeability reservoir	30–20	500–100
Medium-porosity, low-permeability reservoir	20–10	100–10
Low-porosity, low-permeability reservoir	15–10	10–1.0
Tight reservoir	10–5	1.0–0.02
Super-tight reservoir	<5	<0.02

In recent years, according to the requirements of reservoir evaluation, researchers have started applying the comprehensive configuration relationships among pore throat radius, porosity, and permeability for classification (Table 3.8). This is because throat radius and expulsion pressure are well related quantitatively. This method not only overcomes the shortcomings of other classification schemes, but also better reflects the correspondence between the physical characteristics of pore structure, and porosity and permeability.

However, the classification range of reservoir properties is of great importance for evaluating and predicting reservoir property changes. The determination should follow a distribution law based on statistical histograms rather than following the law of normal distribution. Different property ranges directly impact the threshold and step length in the comprehensive evaluation and stochastic modeling of petroleum reservoirs.

3.3 Reservoir Architecture

For a long time, the geometric features of geological bodies have been one of the main objects of geological description. Owing to the difference between geological formations and researchers image views, description languages and nouns are diversified, and quantification is difficult to impossible. To some extent, this phenomenon restricts the research, understanding, and sublimation of sedimentology.

Petroleum reservoir geometry is one of the key components in the research and prediction of two features of petroleum reservoirs, particularly in research on the geometry of clastic petroleum

Table 3.8 Classification standards for porosity, permeability and throat reservoir of a Chinese gas field (Yu Xinghe 1996)

Porosity (%)	<9	9–15	>15
	Low porosity	Medium porosity	High porosity
Permeability ($\times$ 10^{-3} μm^2)	<10	10–20	>20
	Low permeability	Medium permeability	High permeability
Throat radius (μm)	<1.2	1.2–5.8	>5.8
	Fine throat	Medium throat	Coarse throat

reservoirs. It not only plays a vital role in oil and gas field exploration and development but also provides an important geological basis for determining the connectedness of oil reservoirs and the oil–water interface. Moreover, spatial geometric features of individual oil reservoirs or genetic units are critical for determining depositional environments and predicting the distribution laws of sand bodies.

One of the key factors for judging sedimentary environments is sand body architecture because the formation, size, and distribution of sand bodies usually reflect the formation environment and sedimentation, which are related directly to the sedimentary environment. Hence, sand body geometry serves as one of the key bases for seismic facies identification. It covers shape and size, where shape includes both the plane form and profile form. The concept of size is only relative and qualitative, thus it is hard to give a specific quantitative standard or scope. In addition, sand body geometry is one of the main evidence and principles for the establishment of reservoir geological models, particularly skeleton models of sand bodies. From the viewpoint of stochastic simulation, sand body geometry acts as the criterion for determining the main structural parameters of the variation function.

3.3.1 Sectional Geometric Feature of Sand Bodies

Research on the sectional geometric feature of sand bodies can not only help with the identification of sedimentary environments but also help with the correlation of sand bodies between wells, particularly when analyzing the superimposition of a sand body with different geometrical features, which is the principle to be followed and research focus for researchers. Consequently, research on the sectional geometric feature of a single sand body of a reservoir usually focuses on the following five aspects: ① principle for well-to-well correlation; ② guarantee of sand body connectedness, i.e., superimposition relationship; ③ speculation of sand body extensions; ④ analysis of spatial distribution of sedimentary facies (microfacies); and ⑤ establishment of reservoir skeleton models.

3.3.1.1 Lens with Flat Top and Concave Bottom

This profile morphology, which is the result of various filling sedimentations, is usually the major characteristic of a cross section of various channel sedimentary sand bodies. It is mostly manifested as a fining-upward structure in the vertical direction, usually with a scour surface at the bottom. On the electric well-logging curve, it is mainly bell-shaped or low-amplitude serrated bell–shaped, but occurs as a cylinder (braided river channel) in particular cases. The width/thickness ratio depends on the nature of different channels or rivers.

3.3.1.2 Lens with Flat Bottom and Convex Top

A sand body of this shape is mainly a sedimentary body formed by the progradation or winnowing accretion of waves, for instance, a distributary mouth bar deposit in a delta. Moreover, profile shapes of longshore bars, distal bars, and barrier islands fall into this category. It is manifested as a coarsening-upward structure in the vertical direction. On the electric well-logging curve, it is mostly funnel shaped. The width/thickness ratio depends mainly on the source supply and the topographic slope.

3.3.1.3 Lens with Convex Top and Concave Bottom

A single sand body with this shape is rare. The shape is usually related to some small sedimentary sand bodies such as delta-front finger bars. The width/thickness ratio of this sand body is usually small. Moreover, occasionally, various bars in a channel can also show the sectional structure of this form.

3.3.1.4 Wedge

The longitudinal profiles of various fans parallel to the flow direction (longitudinal direction), such as alluvial fans or basin floor fan sand bodies, are usually wedge shaped, with a decreasing sand body toward the basin.

3.3.1.5 Plane

This profile shape is usually not associated with specific sedimentary sand body types. In general, the longitudinal section of a channel sand body and the cross section of beach sand deposition could be of this shape, which is characterized by no apparent grain size change along the vertical plane. In most cases, it has a cylindrical shape on the electric well-logging curve, with a width/thickness ratio usually higher than 3:1 but lower than 20:1.

3.3.1.6 Stripe

This profile shape, with a width/thickness ratio higher than 20:1, is usually the product of delta-front sheet sand and riverbank sand sedimentation. This sand body usually features low thickness and fine granularity, without obvious interior rhythmic variations.

3.3.2 Planar Geometry of Sand Bodies

The planar geometry of a sand body is usually expressed by a sandstone thickness contour map. To better reflect the planar modality of a sand body, it is best to prepare the map considering the overall law of the sand content contour line under the premise of scientific comparison in order to reflect the true form and sedimentary structure of the sand body.

3.3.2.1 Classification

Geologic description of the planar geometry of sand bodies is usually classified according to the length/width ratio.

1. **Sheet**

   The flat area is large, L/W $\approx$ 1:1 (L refers to length, and W refers to width, similarly hereinafter). It has an equiaxed shape on the plane, and is characterized by thin, but stable thickness. For instance, shelf sand or beach sand is of this structure.

2. **Fan**

   This can be divided into fan shape, lobe shape, leaf shape, and bird's foot shape (Fig. 3.7), L/W $\leq$ 3:1. Sand body thickness increases toward the basin and scatters in a fan shape, for example, alluvial fan, submarine fan, or fan delta sand body. Usually, an alluvial fan is of a thick fan shape, a turbidite fan has a lamellar lobe, and fan delta has a progradational leafy shape, but this depends on the specific depositional conditions and background. For example,

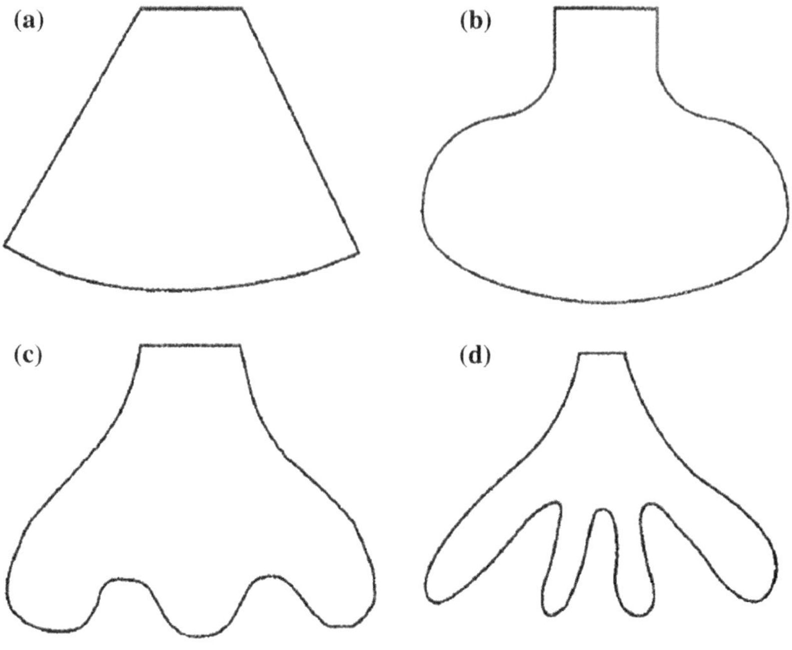

Fig. 3.7 Planar features of fan-shaped sand body **a** fan shape; **b** lobe shape; **c** leaf shape; and **d** bird's foot shape

steep deltas are mostly lobe shapes, while macroaxis fluvial-dominated deltas on fault lacustrine basins are mainly bird foot–shaped.

3. **Elongate**

This shape (3:1<L/W $\leq$ 20:1) is characterized by unstable thickness, and it can be further divided into ① stripe (L/W > 3:1), which could be 20:1 or higher sometimes; ② arborization, relatively bent with branching or bifurcation; ③ belt, formed by lateral movement of stripe sand body to combine with arborization sand body, for example, longshore bars, barrier islands, rivers, deltas and tidal channels can form a microscler sand body; and ④ shoelace shape (L/W > 20:1), which is similar to a highly tortuous meandering river.

4. **Lens**

This is also called the pod shape or henhouse shape, and it covers particularly small distribution areas (L/W < 3:1).

To better reflect the morphological changes and distribution features of sand bodies in three-dimensional space, the length/width/thickness ratio is applied for description (Berg, 1986). Length, width, and thickness are denoted by L, W, and T, respectively.

(1) Sheet: L = W > 100T;
(2) Lobe shape (leaf): L = W > 100T;
(3) Ellipticity: L > W > 100T;
(4) Linear: L > 10 W > 300T;
(5) Finger shape; L > 10 W > 100T (Fig. 3.8).

The major difference between the sheet and lobe shapes lies in that the former is presented as equiaxed on the plane, while the latter scatters in a certain direction but converges toward another direction. The main differences between the finger shape and the linear shape are in their distribution directions and variation trends. The finger shape becomes thinner and wedges out in a single direction, whereas the linear shape thins

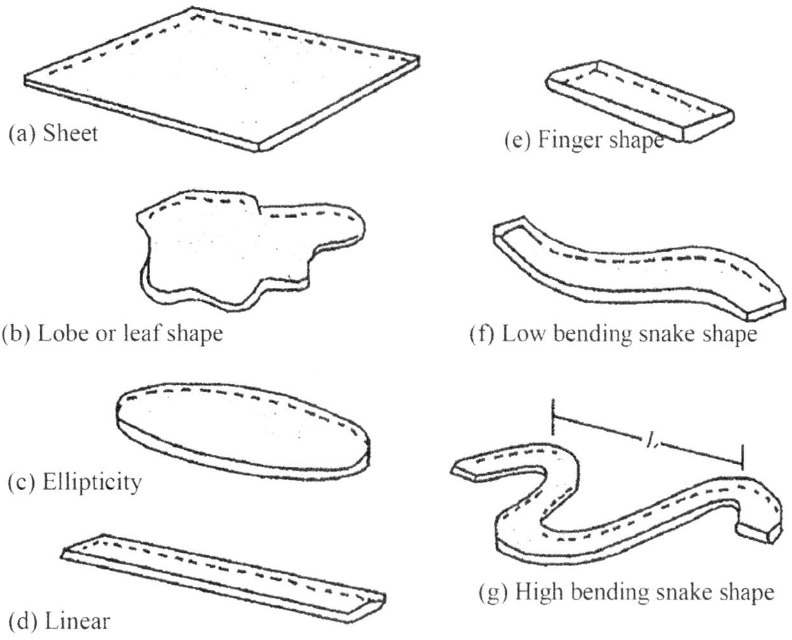

Fig. 3.8 Spatial form type of sandy sediment combination (by Berg, 1986)

gradually, wedges out, or becomes parallel toward two ends.

3.3.3 Mechanism for Controlling Plane Modality of Depositional System

In terms of sequence stratigraphy, the plane distribution of a depositional system relates mainly to the ratio of accommodation (A) to sediment supply (S). Accommodation reflects the capability of piling sediments formed by the ascent and descent of the stratum base level, which relates to tectonic movement, sediment compaction, differentiation, and changes in water level. In case the earth's surface moves downward relative to the base level, that is, the base level rises, sediment storage potential increases or accommodation increases. In case the earth's surface moves above the stratum base level, denudation potential increases. Sediment supply controls the action, product, and redistribution of sedimentation; therefore, the earth's surface can move upward or downward, near or far away from the stratum with an increase or decrease in sediments, respectively. The main factors restricting sediment supply include climate, topographical gradient, terrain elevation, vegetation growth, rock type in source region, nutrient supply, biological enrichment and productivity, weathering denudation rate, and hydrodynamic energy.

Trends of increase or decrease in the of A/S value are in line with upward and downward migration of the relevant sedimentary environment in the side direction along the slope. Considering the changes in different coastal deltaic types, for example, different types of deltaic deposit bodies are formed as the A/S value changes (Fig. 3.9). As the base level descends, accommodation in the direction of land decreases, as does sediment storage or accumulation capacity at higher positions in the basin. Most of the sediments bypassing coastal plains are carried onto the shoreface continental shelf for deposition. In the case of a rise in the base level, accommodation in the direction of land increases, as does sediment storage capacity at higher positions in the basin. Most sediments are deposited, and only a few are carried to the shoreface (Cotton 1918; Cross et al. 1993; Hongwen et al. 2002).

Regarding the estuary role of deltas, friction factors dominate to form the bird's foot shape in

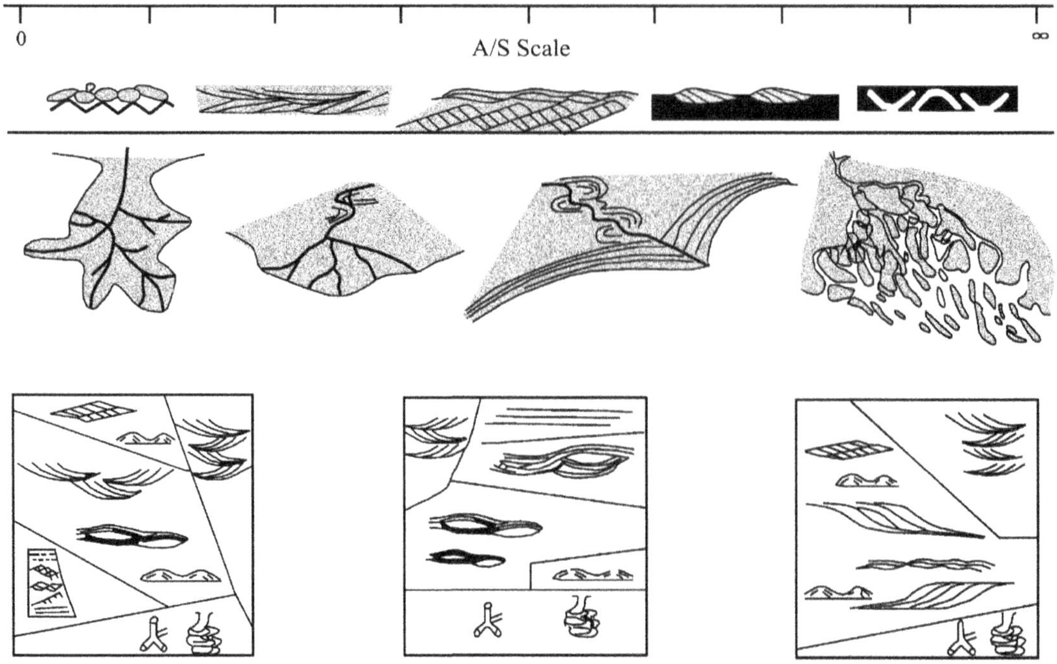

Fig. 3.9 With changes of A/S value, the delta type and the phase composition also change (by Cross and Gardner, 1997). A is accommodation, S is sediment supply

the case of flat terrain, inertia factors dominate to form the fan shape and buoyancy factors dominate to form the lobe shape in the case of steep terrain. They are subject to the combined influence of topographic slope, water flow, velocity, and density.

3.3.4 Geometrical Superimposition and Genesis of Sand Bodies

Sand body connectivity is mainly impacted by sedimentation. Taking a river as an example, its connectivity is usually of the following forms (Fig. 3.10): unilateral or multilateral (mostly interconnected in the side direction), multi-layer (also known as superimposed type, mostly interconnected in the vertical direction), and isolated (not connected to any other sand body). The form is determined mainly by the depositional environment, water flow energy, terrain, tectonic setting, and provenance supply amount.

Similarly, the superimposition mode of sand bodies is also impacted by the A/S value or change in the base level. In the case of a low A/S value, the degree of channel sand body superimposition in the vertical direction increases obviously to produce sedimentation, which is vertical accretion-oriented and supplemented by lateral accretion. This reduces the diversity of sediment microfacies, leads to multi-layer- or multilateral-type dominant sand body, and increases sand body connectedness. In the case of a high A/S value, enhancement of the degree of preservation of original landscape elements results in a channel sand body dominated by lateral accretion and aggradation. Hence, the diversity of sediment microfacies increases, and the resulting sand body is of the unilateral-isolated form (Fig. 3.11).

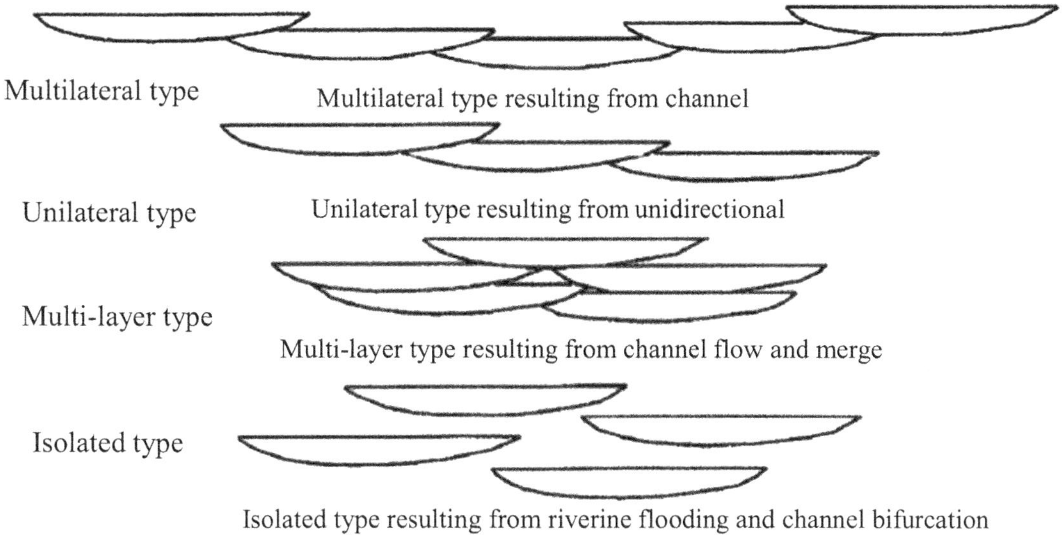

Fig. 3.10 Change of vertical superimposition or communication mode of sand body reflects the corresponding sedimentation (by Xinghe 2002)

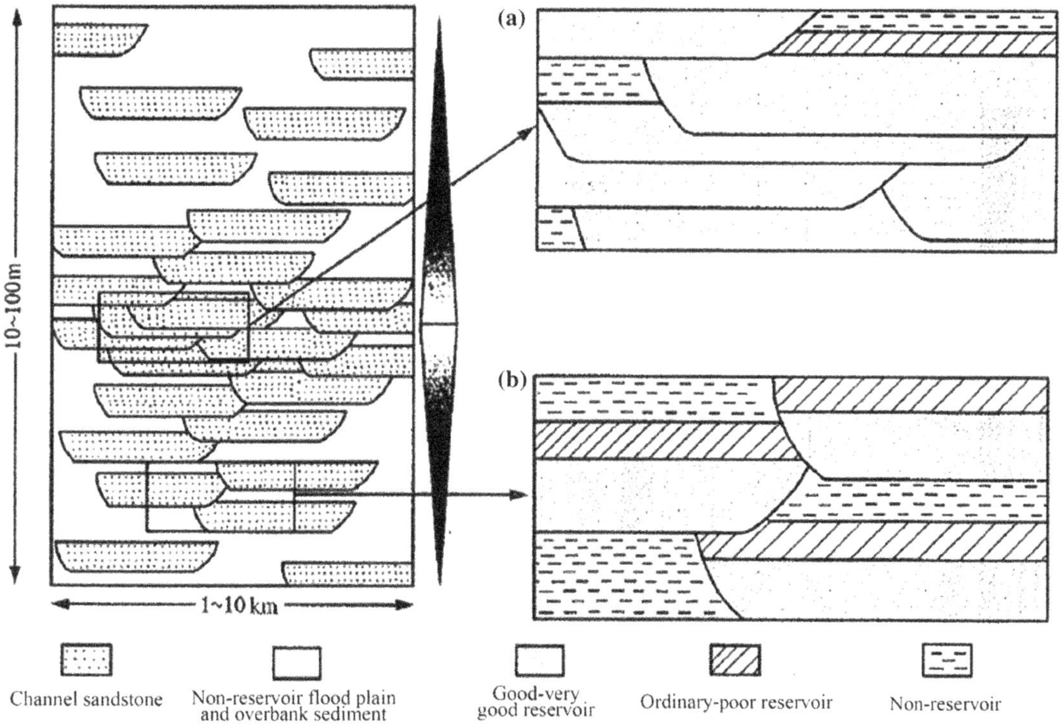

Fig. 3.11 With changes of A/S ratio and base level cycle, the vertical superimposition mode and communication mode of channel sand body are different. When A/S ratio is lower (a), the channel sand body is dominated by vertical accretion in the vertical direction; when A/S ratio is higher (b), the channel sand body is dominated by lateral accretion and aggradation (by Cross and Gardner, 1997)

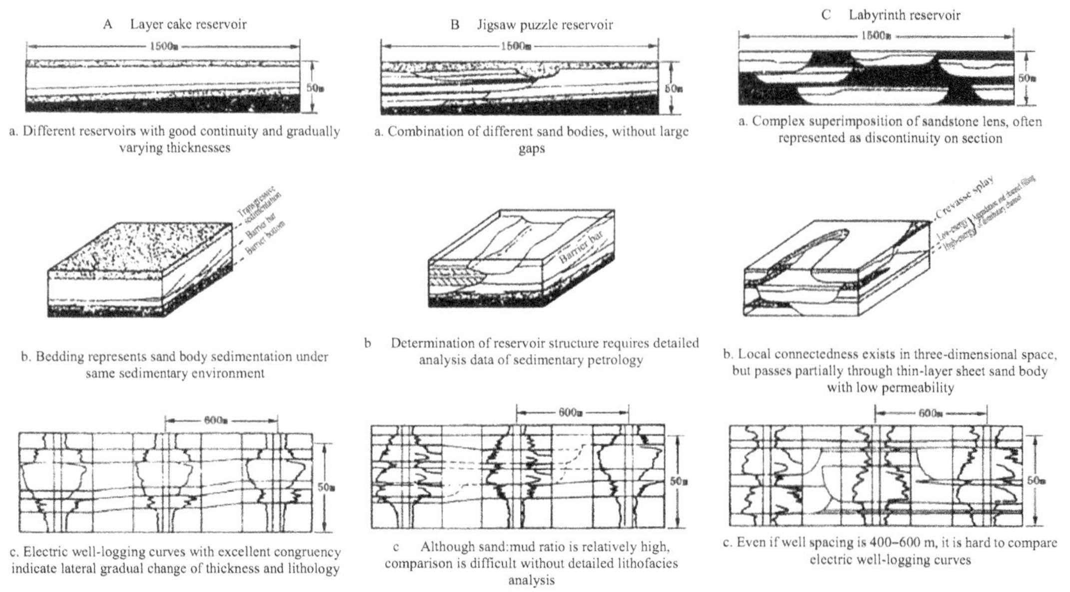

Average well density data required for deterministic contrast of main sand body unit

Well pattern	Well spacing (m)	(Well number/km^2)
Rectangular	1000	1
Triangular	2000	0.8
Random		1.3

Well pattern	Well spacing (m)	(Well number/km^2)
Rectangular		3
Triangular	800	2
Random		

Well pattern	Well spacing (m)	(Well number/km^2)
Rectangular	200	25
Triangular	300	13
Random		32

Fig. 3.12 With changes of reservoir type and drilling density, the reservoir has three different structural forms (by Webber and Geuns Van 1990)

3.3.5 Reservoir Structure or Model

Webber, K.J., and van Geuns, L.C.V. (1990) from the Shell Oil Company hold that all reservoir models can actually be considered superimposed combinations of multiple homospheres in space. Thus, to better describe and simplify sand body geometry, the spatial superimposition mode and distribution characteristics are integrated into spatial structure forms including ① jigsaw puzzle; ② labyrinth; and ③ layer cake (Fig. 3.12). The above three forms are the main spatial structure forms of marine clastic rock reservoirs. Considering the sedimentary features of continental clastic rocks, a fourth form, namely, ④ Stuffing Pie, was proposed (Yu, 1996) (Table 2.9).

Detailed determination and comparison are impossible if well spacing is too large or reservoir structure is too complicated. Consequently, a statistical approach is required to obtain a possible outline of reservoir structure. Current research mainly focuses on conceptual modeling technology based on the quantification of sand body geometry, description of sand body thickness, superimposition mode, pure thickness to total ratio, and positioning.

Most permeability data originate indirectly from log data. Rock types can be distinguished by integrating log data, and the relationships between porosity and permeability can be established with sufficient well numbers. In some situations, the sedimentary structure of sedimentary sand bodies must be considered to obtain the ideal stratified vertical and horizontal permeability.

Reservoirs classified as having the layer cake structure include single or multiple barrier bars,

Table 3.9 Characteristics of four spatial structure types of continental clastic reservoirs and performance under different sedimentary environments

Type	Sand combination mode	Characteristics between sand bodies	Petrophysics	Features of single sand body	Continental facies	Littoral facies	Marine facies	Schematic diagram (■ mud and □ sand)
Jigsaw puzzle structure	Multi-layer superimposition	No macro-pores among sands, with occasionally impermeable barrier layer	Large petrophysics change between sand bodies	Good continuity, large and stable thicknesses	Braided river sediments, meandering river point bar, lake/alluviation mixed sedimentation, and aeolian/dry valley mixed sedimentation	Sedimentary facies complex, such as barrier bar and tidal channel filling complex, distributary channel and distributary mouth bar complex with high net/gross ratio	Complex of storm sand lens, turbidity channel of middle fan, and turbidity sand	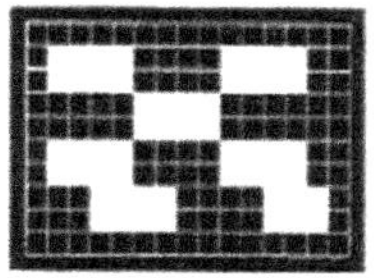
Labyrinth structure	Isolated combination of multiple sandstone lenses	Sand body connected by thin-layer sheet sandstone with low permeability	Large petrophysics change between sand bodies	Smallness and poor continuity	Outwash deposits with low net/gross ratio, and low-tortuosity channel	Deposits filled in the distributary channel of delta with low tortuosity	Sand slump rocks in turbidity channel of inner fan, and storm deposits with low net/gross ratio	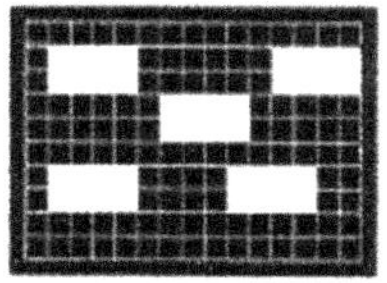

(continued)

Table 3.9 (continued)

Type	Sand combination mode	Characteristics between sand bodies	Petrophysics	Features of single sand body	Continental facies	Littoral facies	Marine facies	Schematic diagram (■ mud and □ sand)
Layer cake structure	Unilateral or multilateral sand body superimposition	Sand body boundary consistent with property change or flow resistance boundary	Stable and continuous horizontal permeability, and gradually changing vertical permeability	Good horizontal continuity and gradually varied thicknesses	Lake sheet sand, delta-front sheet sand, and wind dune	Barrier bar, chenier deposits, and transgressive sands	Neritic sheet sand, offshore bar, and turbidity sand of outer fan	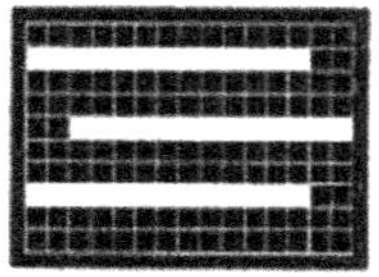
Stuffing pie Structure	Isolated sand superimposition	Poor connectedness	Great change	Medium but relatively poor continuity, and greatly changed thickness	Anastomosing river filled deposits, high-sinuosity meandering river sand body, and crevasse splay sand body	Longshore bar, delta-front distal bar, delta-front crevasse sand body	Offshore bar or spit, outer fan turbidity sand lobe, and offshore sandy ridge	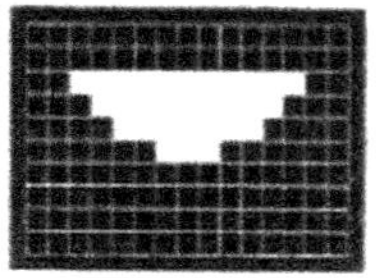

neritic sheet sands, transgressive sands, offshore bars, turbidite, braided channel sands, and aeolian sands. However, a few sand bodies are combined in the form of bricklaying or splicing, such as braided river deposits, meandering river point bars, or complexes of intermediate channel of turbidite fan and turbidity sand. Some barrier bars show tidal channel filling intercalation or a combination of alluvial channels and lake sheet sand. In contrast, some reservoirs are in a non-continuous labyrinth form, which is composed of narrow channel-filling sand bodies. Consequently, four clastic rock reservoir structure models (Table 3.9) can be concluded as follows.

3.3.5.1 Layer Cake Structure

The layer cake reservoir structure model is constituted of the combination of very wide sand bodies or by the superimposition of unilateral or multilateral sand bodies, and feature better sand body continuity. Each sand body has its own petrophysics available for graphical representation without great discontinuity or change in horizontal permeability along the lateral direction. The thickness of a single-layer sand body is not always consistent, and it is subject to gradual change. Boundaries between single layers are consistent with property change or a flow-resistant boundary. In addition, the vertical permeability of single layers varies gradually.

3.3.5.2 Jigsaw Puzzle Structure

The jigsaw puzzle structure model, which comprises a series of spliced sand bodies, is a reservoir body superimposed by multi-layer sand bodies, without large gaps between units. Occasionally, reservoirs are mixed with low-permeability or impermeable sand bodies. Impermeable barrier beds are also visible between some superimposed sand bodies. In this structure, the sand body has relatively good continuity, and the thickness of a single-layer sand body is usually large and stable. Great changes in petrophysics may occur between sand layers. The internal properties of a few sand layers are highly uneven, and they can be quantified by simulation. Using a detailed fence diagram, the structural characteristics of the reservoir can be well determined.

3.3.5.3 Labyrinth Structure

The labyrinth reservoir model comprises isolated combinations of multiple sandstone lenses. In general, a single sand body is small and poor in continuity, and is often presented on the section part. A detailed comparison of connection parts, which are linked by thin-layer sheet sandstones with low permeability, may be made at places with smaller well spacing. The results of such comparisons show that sand continuity is often directional. Regarding this structure, an accurate three-dimensional reservoir model is hard to build, but a probability model is realizable.

3.3.5.4 Stuffing Pie Structure

This structure is mainly superimposed by isolated sand bodies, and is characterized by medium but relatively poor sand body continuity, medium thickness change, and poor connectedness between sand bodies. It is mostly presented on cross sections developed by river and distributary mouth bars, and slump gravity flow developed at prodeltas, which show the large-bit-balling sand structure in the cross section.

Taking a coarse delta for example, the multi-layer jigsaw puzzle and layer cake structures are usually formed on the upper delta plain due to the obvious superimposition of gravel sandbank and plane jet action of flood. The transition from multilateral jigsaw puzzle to labyrinth takes place on the lower delta plain due to channel distributary and migration. Owing to the formation of a distributary mouth bar and development of sheet sand, the isolated stuffing pie and layer cake structures can be found on the delta front. The stuffing pie structure is found mostly in the prodelta, where the sand bodies may have formed because of slump or wavy transformations.

Based on the above reservoir structure models, a flow unit model, i.e., reservoir model gridding, may be further established. The gird boundary may represent the boundary of each sand body. In the case of obvious stratification in the genetic unit, a secondary partition may be performed. Representative grids may be 100 m wide and 1–2 m high, which is generally large. A full-field 3D simulation model consists of rectangular grids–measuring 100 m × 100 m–300 m × 300 m.

References

Beard, D.C., and Weyl, P.K. 1973. Influence of Texture on Porosity and Permeability of Unconsolidated Sand [J]. *AAPG Bulletin*, 57(2): 349–369.

Berg, R.R. 1986. Reservoir sandstones [M]. Prentice-Hall.

Cotton, C.A. 1918. Conditions of deposition on the continental shelf and slope. *Journal of Geology* 26: 135–160.

Cross, T.A. et al. 1993. Applications of high-resolution sequence stratigraphy to reservoir analysis. In *Subsurface reservoir characterization from outcrop observations,* ed. R. Es-chard, and B. Doligez. *Proceedings of the 7th IFF Exploration and Production Research Conference*, 11–33. Paris: Technip.

Cross, T.A., and Gardner, M.H. 1997. Base-Level Concepts and Sequence Stratigraphy. Research Program in Finance Working Papers.

Deng, H.W., Wang, H.L., Zhu, H.J., et al. 2002. *High resolution sequence stratigraphy-theory and application*. Beijing: Geological Publishing House.

Dai,f Q.D., and Ji, Y.L. 1996. *Petroleum reservoir geology*. Dongying: China University of Petroleum Press.

Graton, L.C., and Fraser, H.J. 1935. Systematic Packing of Spheres: With Particular Relation to Porosity and Permeability [J]. *Journal of Geology*, 43(8, Part 1): 785–909.

He, G.S. 1994. *Oil reservoir physics*. Beijing: Petroleum Industry Press.

Jackson, P.D. Smith, D.T., and Stanford, P.N. 1978. Resistivity-porosity-particle shape relationships for marine sands [J]. *Geophysics*, 43(6): 1250–1268.

Scherer, M. 1987. Parameters influencing porosity in sandstones: A model for sandstone porosity prediction. *AAPG* 71 (5).

Timmerman, E.H. 1982. *Practical reservoir engineering*. Pennwell Pub.co.

Webber, K.J., and Geuns, L.C.V. 1990. Framework for Constructing Clastic Reservoir Simulation Models [J]. *Journal of Petroleum Technology*, 42(10): 1248–1297.

Wentworth, C.K. 1922. A scale of grade and class terms for clastic sediments. *Journal Geology* 30: 377–392.

Yu, X.H. 1996. Reservoir characterization and stochastic simulation of clastic reservoirs. *Geological Publishing House*.

Yu, X.H. 2002. *Clastic petroleum reservoir sedimentology*. Beijing: Petroleum Industry Press.

4 Theory and Methods for Studying Clastic Sequence Stratigraphy

Sequence stratigraphy is the scientific study of rock relationships using a repetitive and genetically geochronologic sequence stratigraphic framework, which is demarcated by an erosion surface, a non-sedimentation surface, or a correlative conformity surface. It is a new approach for dividing, contrasting, and analyzing sedimentary strata. When combined with biostratigraphy and tectonic subsidence analysis, it enhances the feasibility of chronostratigraphic correlation, paleogeographic reconstruction, and prediction of petroleum reservoirs, source rock, and seal bed before drilling. The application of sequence stratigraphy to sedimentary strata can potentially provide a complete and unified stratigraphic concept, similar to the complete and unified structure concept for plate tectonics. Moreover, sequence stratigraphy has changed the fundamental principle for stratigraphic record analysis and started a new stage in understanding the history of the Earth. As a result, it triggered a revolution in geology in the 1980s.

4.1 Formation and Development of Sequence Stratigraphy

The theoretically guided study of sequence stratigraphy, which has existed since the 1980s, has greatly changed our recognition of stratigraphic formations and the control of the basin-fill mechanism. It has contributed to geoscience by providing an easy isochronostratigraphic correlation method for geologists and paved the way for the emerging field of lithostratigraphic hydrocarbon reservoir exploration (Zhou et al. 2006). Moreover, its concept and meaning are also constantly developing through theoretical research and extensive practice. The development of sequence stratigraphy can be roughly divided into three stages.

4.1.1 Emergence and Establishment of the Sequence Concept (1948–1976)

In 1948, Sloss, at a symposium on "Sedimentary Facies in Geological History," proposed the earliest concept that "a sequence is an unconformity-bounded stratigraphic unit," which marked the emergence of the "sequence." Siclen (1958) provided a stratigraphic response diagram for showing the changes in a continental margin relative to the sea level and changes in sediment supply; this diagram is very similar to the current sequence model. In addition, Wheeler (1959) proposed the concept of depositional episodes bounded by regional unconformity. Based on the boundary of regional unconformity, Sloss (1963) divided the strata between the late Precambrian of North American craton and the Holocene epoch into six stratigraphic units, namely, Sauk, Tippecanoe, Kaskaskia, Absaroka, Zuni, and Tejas from the bottom up (Fig. 4.1). He called these strata "sequence" and defined sequence as

X. Yu et al., *Clastic Hydrocarbon Reservoir Sedimentology*, Advances in Oil and Gas Exploration & Production, https://doi.org/10.1007/978-3-319-70335-0_4

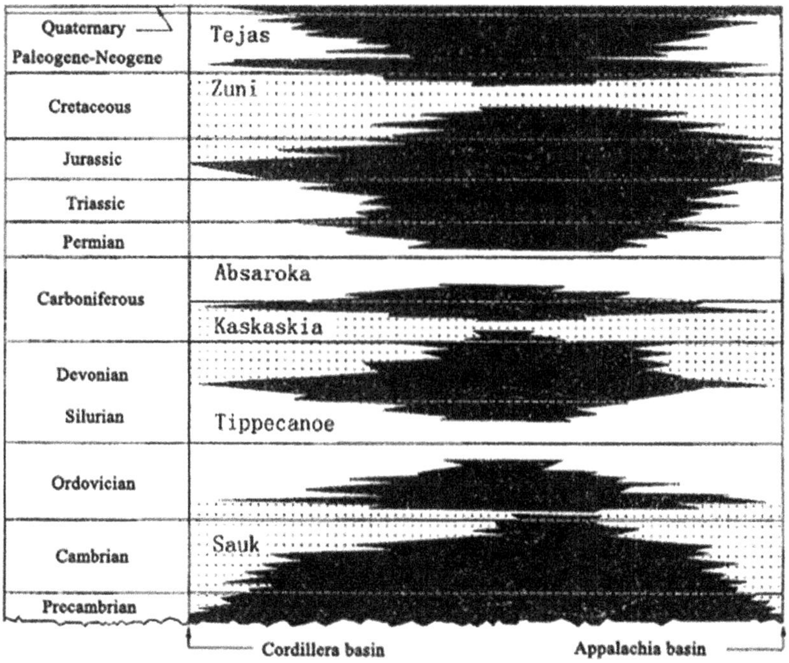

Fig. 4.1 Sloss identified the relations between Phanerozoic Craton sequence and the orogenic episode on Cratonic margin in 1963, and this marked the germination of this concept "sequence" (Sloss 1963)

"a lithostratigraphic unit that is higher than the level of ordinary group, large group, and super-group; bounded by regional unconformity; and traceable over a large area." Moreover, as a practical geological unit, sequence is used for mapping. He holds that some of these sequences are of chronostratigraphic significance. However, because the time spans represented by these sequence units are too large to meet the requirements for detailed layered correlation, the concept of "sequence" proposed by Sloss et al. was not accepted widely before the 1970s. However, it did lay the foundation for the development of current sequence stratigraphy.

4.1.2 Formation and Development of Seismic Stratigraphy (1977–1986)

Seismic Stratigraphy: Application to Hydrocarbon Exploration, by P. R. Vail (1977) and his co-workers based on their experience working in Carter and Exxon petroleum companies, is a key milestone in the development of sequence stratigraphy, as well as the most cited work in sequence stratigraphy literature. In this book, they, in the real sense, proposed the concept of sedimentary sequence—a genetically related stratigraphic unit—and emphasized the termination method and identification method of seismic reflection (Fig. 4.2). Moreover, they proposed the concept of seismic sequence stratigraphy and exported the sea level change curve from the seismic cross section. The outstanding progress made in this stage is shown below.

(1) The development of high-resolution seismic exploration and processing technologies is the basis for confirming the parallel formation level of seismic reflection, rather than a single lithologic boundary, to impart chronostratigraphic significance to the seismic reflection boundary and to perform tracking and correlation within a basin.
(2) Relative consistency of global sea-level change is defined clearly to establish a global sea-level change curve as a reference for chronostratigraphic correlation.
(3) Vail et al. defined "sequence" as "a relatively conformable succession of genetically related strata at their upper surface and base by

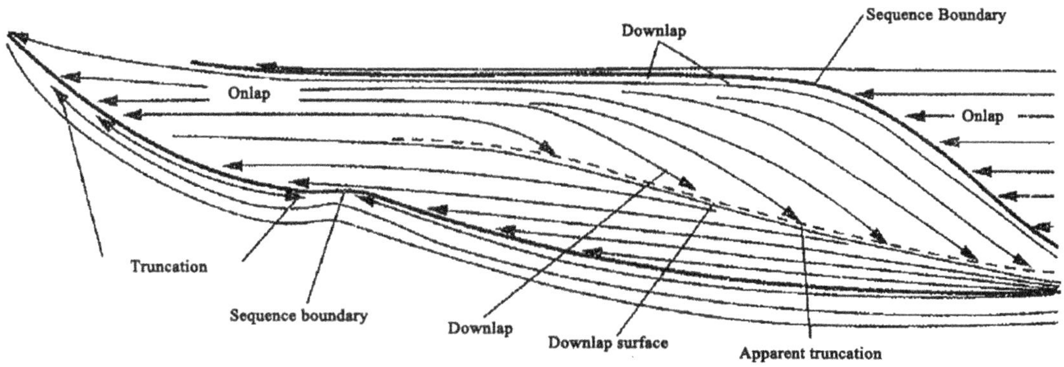

Fig. 4.2 Concept of seismic stratigraphy and termination method of seismic reflection (by P. R. Vail 1977)

unconformities and their correlative conformities." This concept of "sequence" is much smaller in timescale than Sloss' Sequence, and it can be applied to seismic data and drilling and well logging data to predict and determine stratigraphic textures, depositional facies types, and distribution at the basin scale or sequence at three levels or higher.

(4) Regarding the genesis interpretation of sequences, Vail et al. took sea level change as a leading factor for sequence formation and evolution. In basin analysis, seismic stratigraphy has shown its effectiveness in terms of a detailed definition of complicated basin stratigraphic framework, recognition over a large scale, unconformity correlation, and recovery of the filling evolution history of a basin, with a focus on the sea level change process represented by the stratum termination mode in sequence units and various other termination modes.

4.1.3 Formation and Development of Sequence Stratigraphy (from 1987 to Date)

Although the theory of seismic stratigraphy solves the sequence formation problem, it fails to clearly expound the stratigraphic correlation within a sequence. Moreover, mainly seismic data obtained within a basin is used for sequence stratigraphy. With deeper research, Vail et al. applied a greater number of outcrops and more well logging data to further refine the primary theories and concepts. In the *Atlas of Seismic Stratigraphy*, compiled in Bally by Vail and Wagoner et al. in 1987, "sequence stratigraphy" was clearly applied for the first time. Based on the sequence concept, the inner part of each sequence was divided into three system tracts based on the sea level change tendency, and the inner part of each system tract was further divided into a secondary unit-parasequence. Moreover, the sedimentation of typical sequences in a passive continental margin, stratigraphic distribution models, and a series of relevant new concepts and terminologies was proposed to set up the stratigraphic distribution law and genetic relationship within a sequence and provide an important means and basis for stratigraphic division.

In 1988, special issue No. 42 of SEPM (*Sea-level Changes: an Integrated Approach*), which was chiefly compiled by C. K. Wilgus, marked the birth of a new science—sequence stratigraphy. Later, more papers on sequence stratigraphy were published, heralding the start of a mature and comprehensive development stage. Xu et al. (1991), a Chinese scholar, translated the aforementioned book into Chinese. The translation was renamed *Principle of Sequence Stratigraphy* and **published in 1993**, and P. R. Vail wrote a preface to it. This is the earliest textbook that systematically introduced sequence stratigraphy in China. In the 1990s, with the rapid development of sequence stratigraphy, many branches have been presented successively, such as continental sequence stratigraphy, carbonate sequence stratigraphy, high-resolution sequence stratigraphy, tectonic sequence stratigraphy, and diagenetic sequence stratigraphy. These branch subjects have not only enriched and developed the theory of sequence stratigraphy but have also extended its applications extensively.

4.1.4 Different Schools of Sequence Stratigraphy

Since the emergence of sequence stratigraphy, three disciplines have been formed based on different boundary classification schemes: ① The unconformity sequence boundary proposed by Vail et al. Vail and his partners from Exxon believe that a plane of unconformity and its corresponding conformity surface are sequence boundaries, and the strata between two planes of unconformity (or corresponding conformity surfaces) are regarded as a sequence. In the 1990s, based on this school of thought, Professor T. A. Cross from the Colorado School of Mines proposed the idea of high-resolution sequence stratigraphy by applying the concepts of base level, facies differentiation, accommodation, and volume distribution principle of sediments. ② The maximum flooding surface sequence boundary proposed by Galloway, who upholds that the maximum flooding surface is the best marker bed horizon for contrast, can serve as the sequence boundary, and the strata therebetween constitute a genetic stratigraphic sequence. ③ Transgressive sequence boundary proposed by Johnson. He suggested that a transgressive-regressive stratigraphic unit is a transgressive-regressive sequence (T-R cycle) (Fig. 4.3).

Sequence stratigraphic study should indicate a sequence boundary division scheme or allow for sequence division in combination with one or two schemes. In recent studies on continental sequence stratigraphy, most researchers in China have applied the typical sequence stratigraphy technique of taking a plane of unconformity as a sequence boundary, and the high-resolution sequence stratigraphy proposed by Vail and Cross. The author believes that the sequence stratigraphy concepts proposed by Vail and Galloway are more applicable to large stratigraphic correlations in hydrocarbon exploration research, while the high-resolution sequence stratigraphy concept proposed by Cross in combination with the T-R cycle (may be extended to the lacustrine transgressive-regressive cycle for a continental stratum) proposed by Johnson are relatively applicable to thin layers in development geology research, especially for small sub-layer correlations.

4.1.5 Terrestrial Clastic Sequence Stratigraphy

Because marine sequence stratigraphy is applied extensively in the petroleum exploration and development and because petroleum exploration and development in continental strata led to the gradual development of sequence stratigraphy, continental sequence stratigraphy emerged at the right moment. In foreign countries, research on sequence stratigraphy has been conducted on main types of continental deposits such as rivers, lakes, and eolian deposits, however most of this research employs theories and methods of marine sequence stratigraphy, with no further research on terrestrial sequence stratigraphy. In China, since most petroliferous basins are continental deposits, research on continental sequence stratigraphy has been developed vigorously and recognized gradually by the broad mass of petroleum geologists. Research in this regard is focused on the influence on sequence-filling resulting from lake level changes, sediment supply, tectonic movement, and climate change, with the division method of high-resolution sequence stratigraphy proposed by Cross as main the research approach. Its fundamental principle is to emphasize the influence of accommodation, base level change, stacking pattern, volumetric partitioning, and facies differentiation on sedimentary sequence filling.

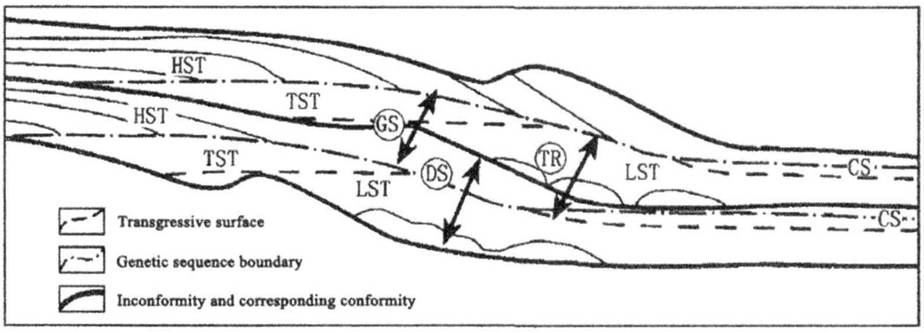

Fig. 4.3 Since establishment, the sequence stratigraphy has formed three major schools according to the division scheme of different interfaces, and these are three different sequence definitions (by Armentrout et al. 1993). DS-sequence bounded by stratigraphic unconformity or correlative conformity surface, proposed by the school represented by Vail in Exxon; GS-maximum flooding surface applied as sequence boundary, proposed by Galloway as a representative of Fraizer. TR transgressive-regressive cycle (T-R cycle) of sequence by taking transgressive surface as sequence boundary, as emphasized by Johnson, LST lowstand systems tract; TST transgressive systems tract, HST highstand systems tract, and CS condensed section

Recently, three theory systems and five approach systems have been formulated in sequence stratigraphy. Specifically, these theory systems involve marine sequence stratigraphy, continental sequence stratigraphy, and high-resolution sequence stratigraphy; and the five approach systems are seismic sequence stratigraphy analysis, outcrop sequence stratigraphy analysis, logging sequence stratigraphy analysis, sequence stratigraphy simulation analysis, and biological sequence stratigraphy analysis.

4.2 Theory System of Marine Clastic Sequence Stratigraphy

Sequence stratigraphy analysis depends on the recognition of stratigraphic units at all levels, bounded by etched planes of unconformity, sedimentary discontinuities, and the corresponding bedding surfaces. Stratigraphic units of different sizes are defined by boundaries of different levels, and the interpretation of a stratigraphic unit relies on the analysis of genesis control factors. There are three important concept systems (Table 4.1) in sequence stratigraphy research, namely, sequence architecture system, sequence boundary system, and sequence genesis control system.

4.2.1 Sequence and Its Formation System

4.2.1.1 Sequence

A sequence is a relatively conformable succession of genetically related strata at their upper surface and base by unconformities and their correlative conformities (Vail et al. 1977). It is composed of a succession of genetically linked deposition systems thought to be deposited between eustatic-fall inflection points (Posamentier et al. 1988). A complete vertical sequence is composed of the following elements from the bottom to the top: bottom stratum boundary (BSB), lowstand systems tract (LST), initial flooding surface (IFS), transgressive systems tract (TST), maximum flooding surface (MFS), highstand systems tract (HST), and top sequence boundary (TSB) (Fig. 4.4).

Sequences boundary I and II (Fig. 4.5) can be identified in continental margin basins. When the sea level decreases faster than the basin structure subsidence rate at the depositional shoreline break, sequence I is formed. By this time, the sea level descends relatively and falls below the shelf slope-break, namely, the shelf is exposed above the sea level or is subject to large-scale denudation, resulting in a plane of unconformity with a relatively wide scope. In contrast, when the sea level decreases slower than the basin structure

Table 4.1 Conceptual systems of sequence stratigraphy

Conceptual system	Sequence (sedimentary unit) formation system	Sequence boundary system	Sequence genetic interpretation system	Geomorphologic system for sequence control
Concept	① Sequence	① Sequence boundary	① Accommodation	① River equalization surface
	② Systems tract	② Maximum water flooding surface	② Stratigraphic base level	② Slope-break belt
	③ Parasequence set	③ Initial water flooding surface	③ Relative sea level change	③ Incised valley
	④ Parasequence	④ Flooding surface	④ Absolute sea level	
	⑤ Condensed section	⑤ Main flooding surface		

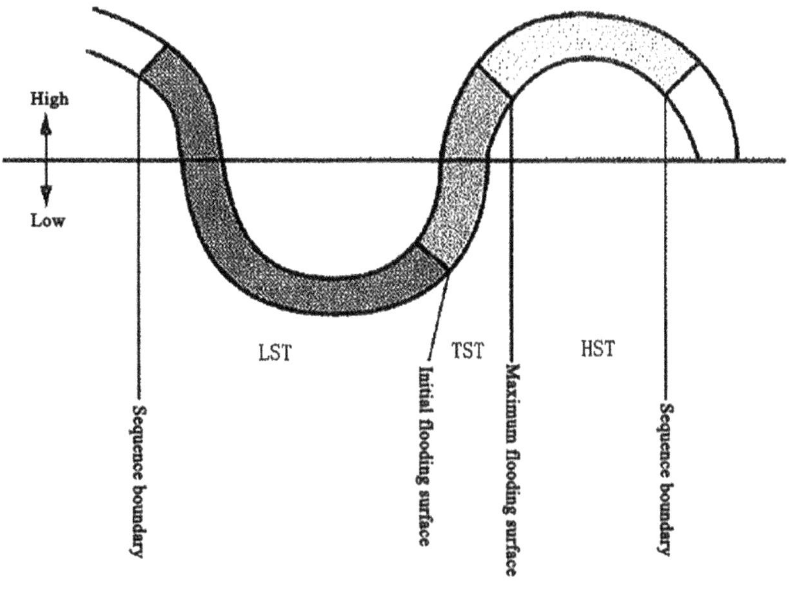

Fig. 4.4 The sea level variation, the sequence interface and the inside constitute a certain corresponding relationship (by Posamentier et al. 1988)

subsidence rate at the depositional shoreline break, sequence II is formed. At that time, obvious relative decline of the sea level fails to occur on the shoreline slope-break, namely, the shelf is exposed above the sea level or subject to small-scale denudation, resulting in a plane of unconformity with a relatively narrow scope.

4.2.1.2 Parasequence and Parasequence Set

A parasequence is defined as a relatively conformable succession of genetically related beds or bedsets bounded by marine flooding surfaces and their correlated surfaces. There are three types of parasequences: upward coarsening, upward fining, and uniform (Fig. 4.6). They have the same basic characteristic wherein water shallows gradually on the vertical combination and causes one marine flooding event. In other words, a parasequence is formed between two small marine flooding events. A parasequence boundary is used for separating deep-water rocks (such as shelf mudstone) thereabove from shallow-water rocks (such as shore sandstone) there below.

The flooding surface is flat, and it often has only very small topographic relief ranging from several centimeters to dozens of meters, most

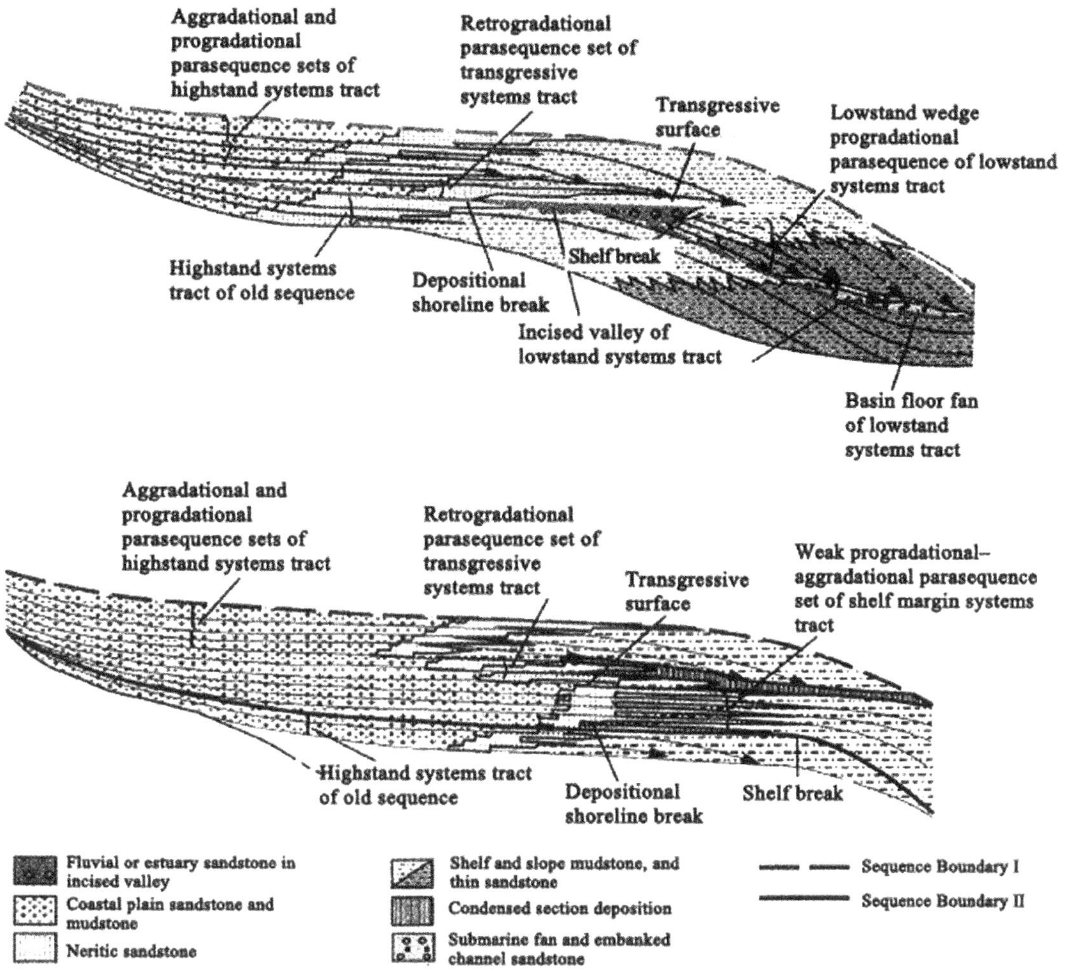

Fig. 4.5 Two sequences can be identified in the continental margin basin, and two main sequence types and the characteristics are shown as the figure (by Van Wangoner 1988)

commonly several meters. A flooding surface generally has a correlative surface in the coastal plain and on the shelf. The correlative surface in the coastal plain does not show obvious subaerial erosion owing to rejuvenation of the waves, downward transfer of coastal onlap, lithofacies transfer toward the basin, and overlying strata onlap. The symbolic characteristics of this surface include local erosion arising out of fluvial action and small onshore exposure. Analysis of the stratigraphic lithofacies of correlative surfaces does not generally indicate a significant change in water depth. The existence of a correlative surface on a coastal plain or shelf can be proven only by contrasting the contra-dip or downdip of marine flooding surfaces.

A parasequence set is a specific vertical overlay formed by a series of genetically related parasequences, with the main marine flooding surface or correlative surface as its boundary. According to the overlaying style of all the parasequences in a parasequence set, parasequence sets can be divided into progradation, retrogradation, and aggradation types (Fig. 4.7) depending on the ratio of deposition rate: increment rate of accommodation. A progradational parasequence is formed when the deposition rate is greater than the increment rate of

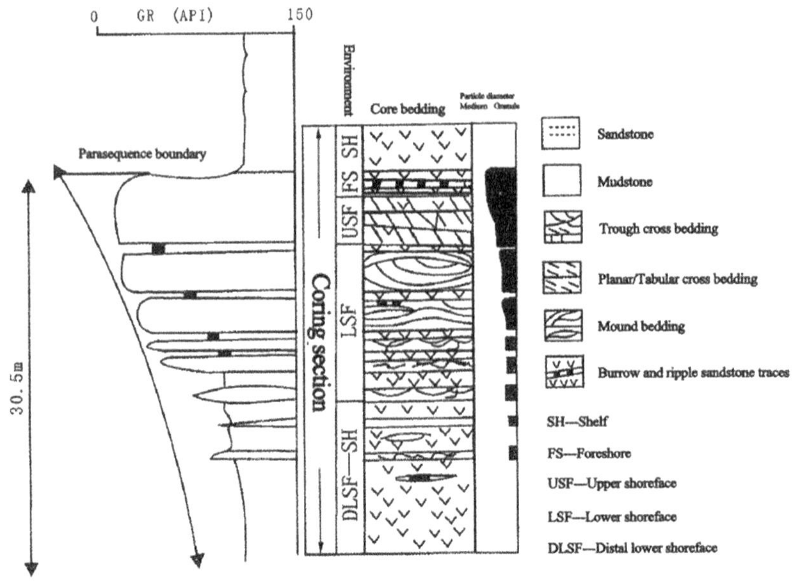

(a) Characteristics of coarsening-upward parasequence in coastal environment (deposition rate of accommodation is larger than its increment rate; formed in sandy, wavy, or fluvial-dominated beach environments).

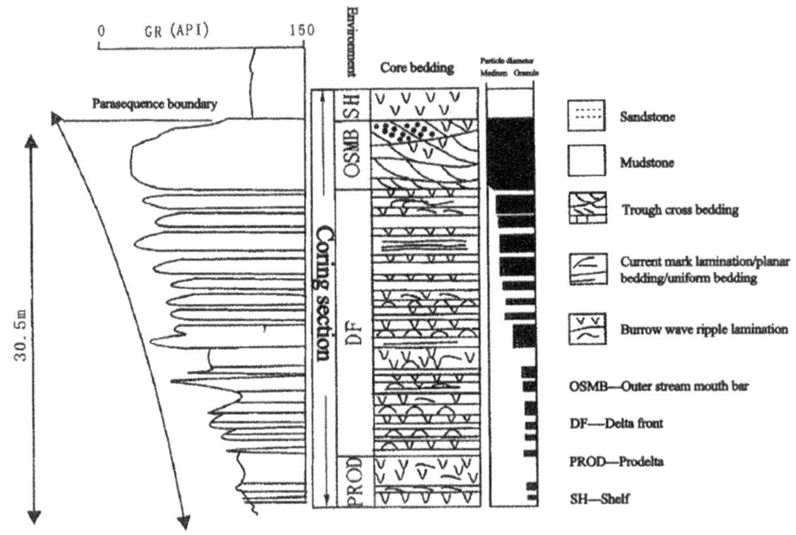

(b) Characteristics of coarsening-upward parasequence in delta (formed in sandy, wavy, or fluvial-dominated deltas).

Fig. 4.6 The shoreland, delta and tidal flat environments express different parasequence characteristics (by Van Wagoner 1990)

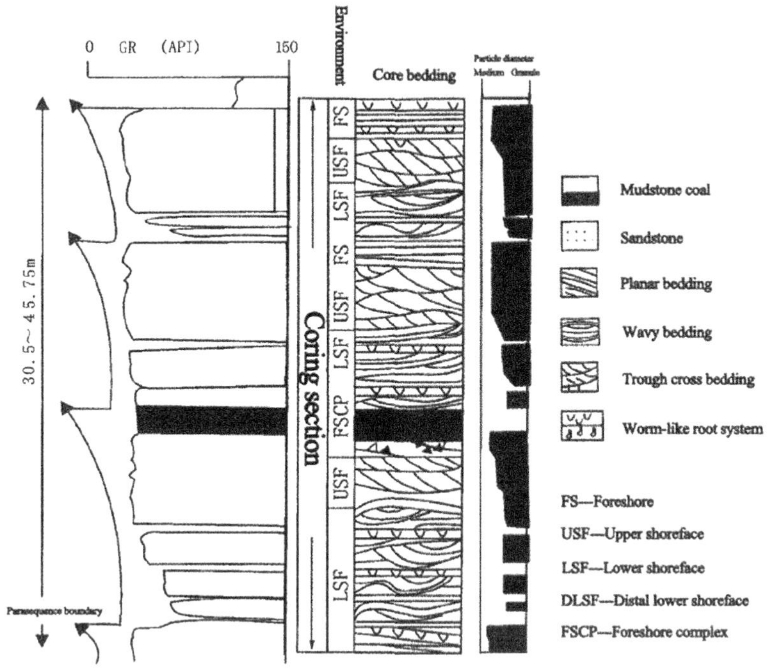

(c) Characteristics of coarsening-upward parasequence when deposition rate is equal to subsidence rate (formed in sandy, wavy, or fluvial-dominated beach environments).

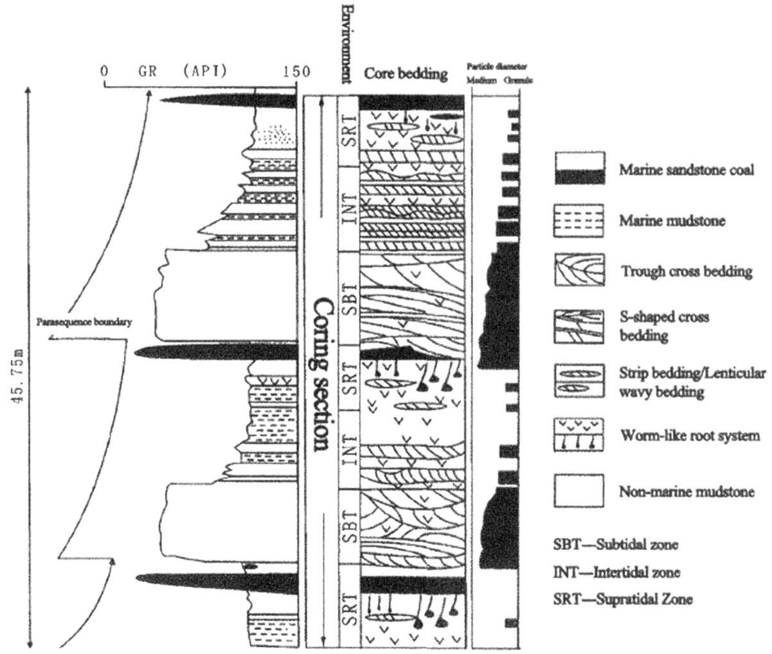

(d) Characteristics of fining-upward overlaying parasequence (formed in tidal shoal-subtidal environment of sandy, tide-dominated coast or fluvial-dominated coasts)

Fig. 4.6 (continued)

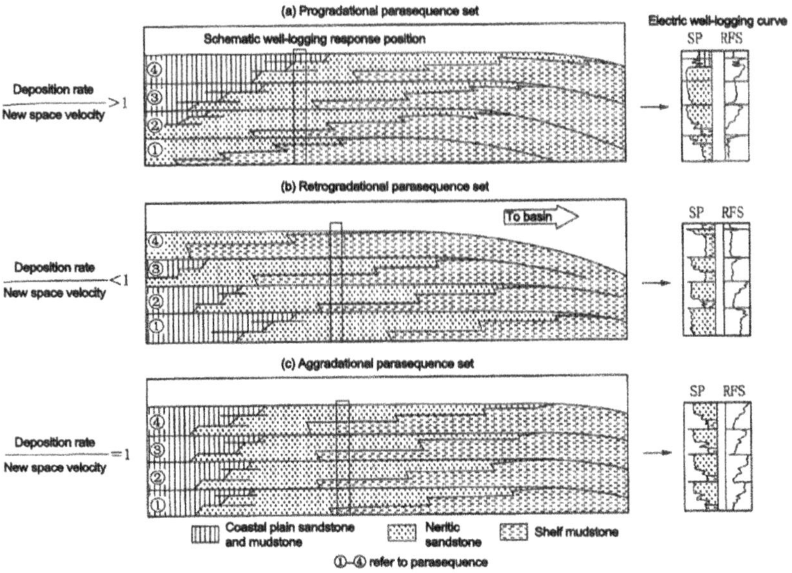

Fig. 4.7 When the ratio of the deposition rate to the accommodation changes, the type of parasequence set and the superposition mode will also change (by Van Wagoner 1990)

accommodation. Younger parasequences migrate sequentially toward the basin to form a parasequence overlay pattern with the thickness of single sandstone increasing upward and that of musdstone thinning, increasing sand-mud ratio, and gradually decreasing water depth. A retrogradational parasequence is formed when the deposition rate is less than the increment rate of accommodation. Younger parasequences return landward sequentially, resulting in the following characteristics: thinning single sandstone upward, thickening musdstone, and decreasing sand-mud ratio. An aggradational parasequence features younger parasequences overlaid sequentially upward, without large changes in water depth, sand-mudstone thickness, and sand-mud ratio from the bottom up.

4.2.1.3 Systems Tract and Development

Systems tract refers to the combination of a series of depositional systems formed during the same period. According to the relative sea level position and the trend of change in its fluctuation, the interior part of a sequence can be divided into three components, namely, lowstand systems tract (or shelf margin systems tract), transgressive systems tract, and highstand systems tract (Fig. 4.8).

The existence of various systems tracts and their degrees of development are affected greatly by sea level eustasy (Fig. 4.9). Moreover, the sedimentary characteristics and developments (Fig. 4.10) of these three systems tracts are different in different types of basins. Lowstand systems tracts are distributed mainly on basin slopes and their lower parts. Thus, shelf exposure and river incision are present and form the plane of unconformity when the sea level decreases faster than the basin-margin tectonic subsidence rate. Deposits carried with the river and substances eroded from the shelf and slope are carried toward the basin direction and deposited on the part below the shelf margin, in the form of basin floor fans, slope fans, and lowstand (progradational) wedges, to form a lowstand systems tract (Fig. 4.9b, c). In this systems tract, the lowstand basin floor fan, in the form of point sources, is deposited by deep-incised valleys and slope valleys. It shows the characteristics of gravity flow sediments and coarser sediments. Slope fans are distributed as the skirt type along the shelf margin, resulting in the formation of movable bank channels and overbank sheet sand. The progradational wedge is composed of progradational and aggradational parasequences, which are distributed mainly on the seaward side of the shelf break. The proximal part of the

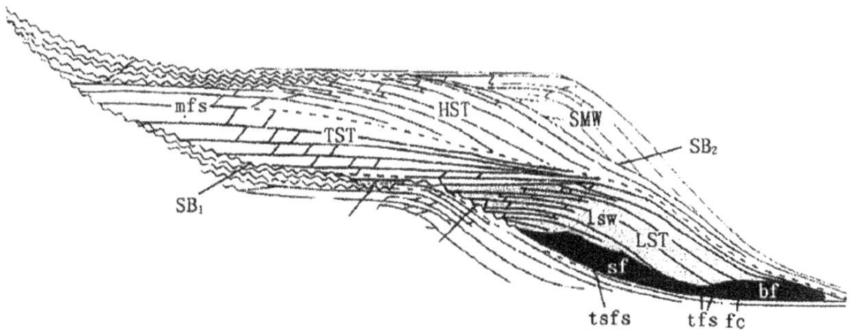

Fig. 4.8 The standard sequence has three system tracts: LST, TST and HST. LST lowstand systems tract, TST transgressive systems tract, HST highstand systems tract, MFS maximum flooding surface, SMW shelf margin systems tract, LSW lowstand wedge, sf low slope fan, BF lowstand basin floor fan; TSFS top slope fan surface, TFS top fan surface, FC fan channel, SB_1 sequence boundary 1; and SB_2 sequence boundary 2

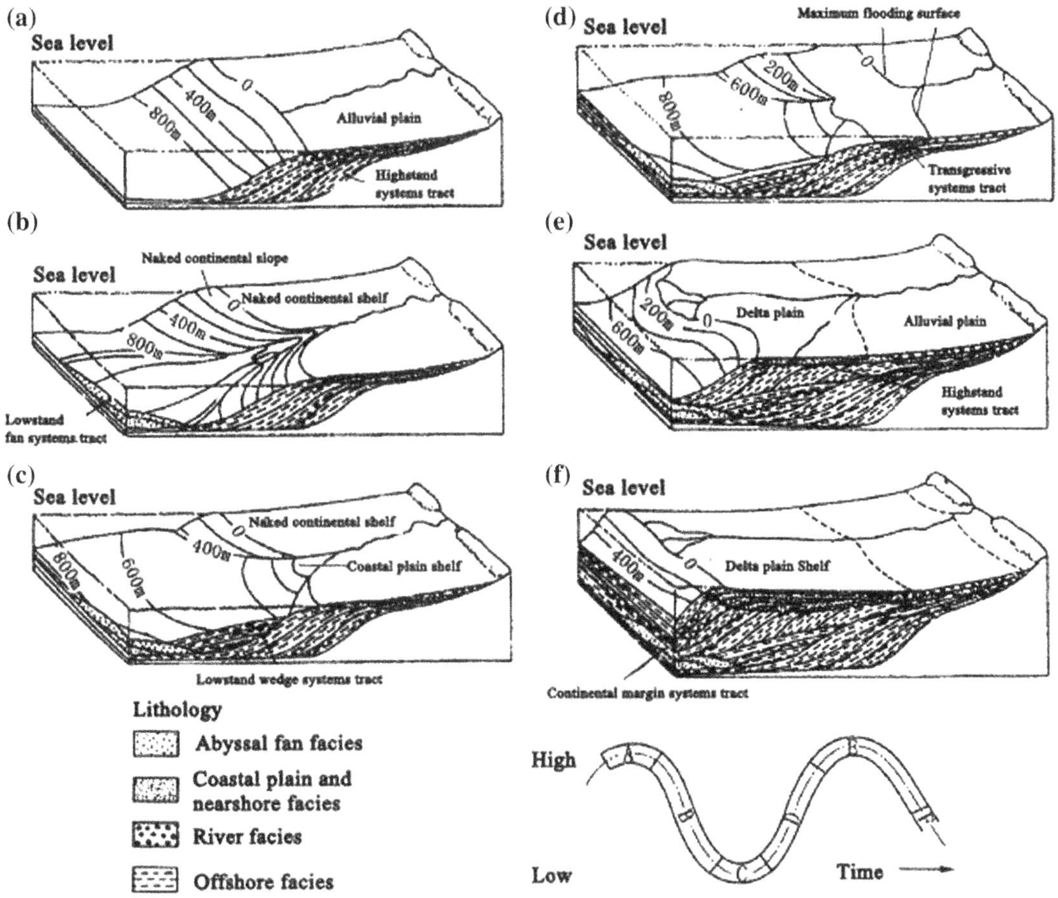

Fig. 4.9 Different stratigraphic sedimentary patterns have been developed during sea level rise and fall (by Posamentier et al. 1988)

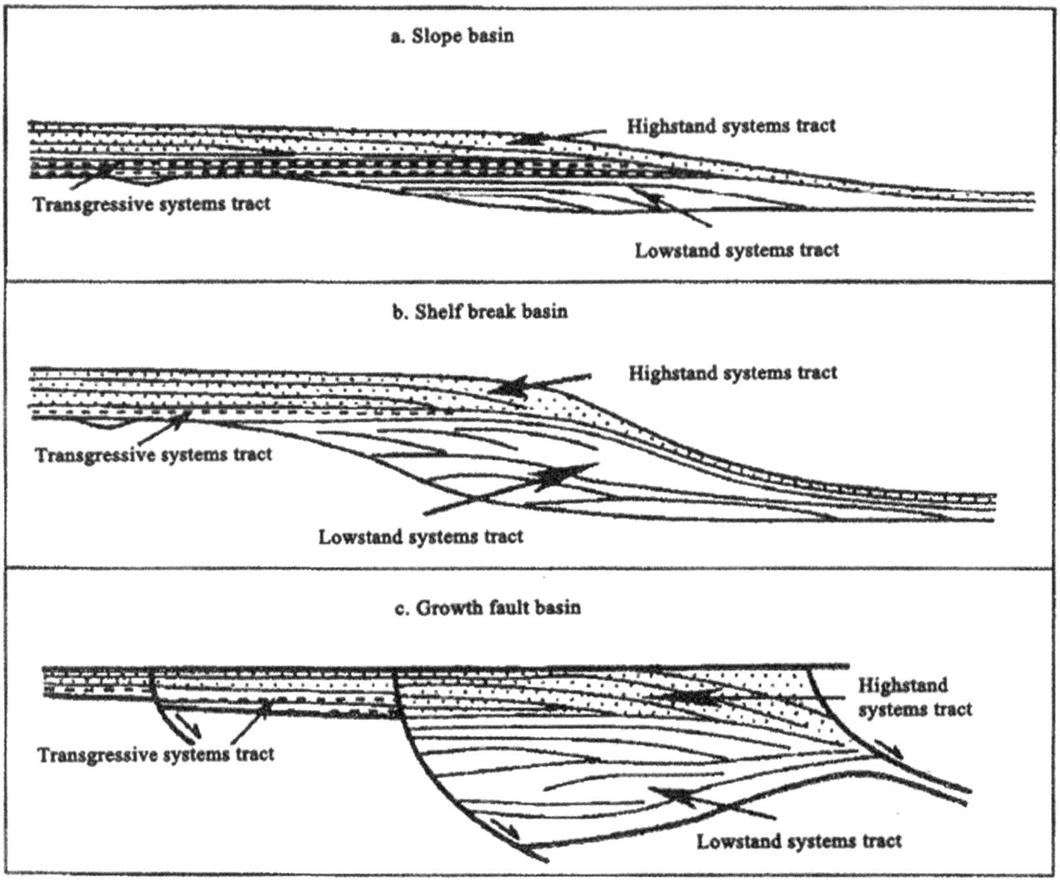

Fig. 4.10 In slope basin, continental shelf break basin and growth fault basin, the developed sequence stratigraphic models are different (by Vail et al. 1991)

wedge has deep-incised valley-filled deposits, while the distal part is composed of sand-mud interbedded wedge-shaped progradational units.

A transgressive systems tract (Fig. 4.9d), which is characterized mainly by retrogradation, is formed by a gradual increase in sea level, leading to gradual submerging of the shelf and the transfer of the depocenter landward. A transgressive systems tract is constituted of a series of relatively thin parasequence sets that retreats continuously toward the direction of land in a ladder shape. The depositional system mainly includes shelf deltas, near-shore plains, coal-rich interactive sea-land deposits, lagoons, and lacustrine deposit. When transgression peaks, a condensed section (or intensive section) characterized by fine grains under slow deposition is formed due to the insufficient supply of deposits in the middle and external parts of the shelf and slope deposits. This constitutes a major source of the source bed section or seal bed section in a sequence.

As the highstand sea level stops rising rapidly, littoral and neritic facies deposits start to move toward the sea to form a highstand systems tract (Fig. 4.9e), mainly featuring progradation. This is mainly composed of one or more parasequences from aggradation to progradation, characterized by delta and channel sand development. At the end of the highstand systems tract, the sea level descends again to form a plane of unconformity at the top.

In a highstand systems tract, actually, the sedimentary course is from slow rise to slow

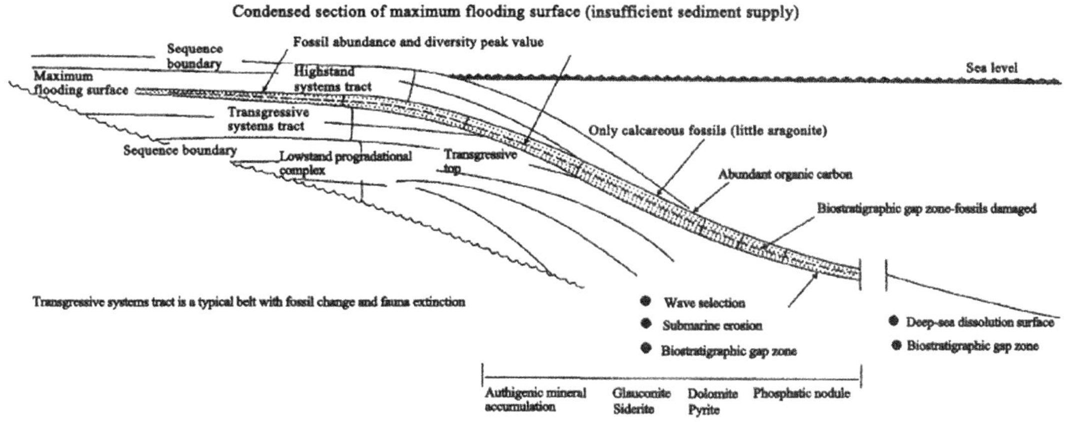

Fig. 4.11 Relations between maximum marine flooding surface and condensed layer (Loutit et al. 1988)

decline of the relative seal level. The slow rise is based on aggradation, whereas the slow decline is based on progradation. Posamentier emphasized the differences between the two in later research, according to which the aggradational parasequence sets formed during relative sea level rise are called highstand systems tracts, while the progradational parasequence sets formed during relative sea level decline are called regressive systems tracts.

4.2.1.4 Condensed Section

The condensed section is a thinner marine stratigraphic section (deposition rate <1–10 mm/ka) featuring an extremely low sedimentation rate (according to Vail et al. 1984). It is developed in the lower part of highstand deposition and the upper part of transgressive deposition (Fig. 4.11). Owing to various progenesis and diagenesis processes resulting from long-term contact between sediments and sea water, in the condensed section of a siliceous clastic set, shali shale, abundant marine microfossils or super microfossils, authigenic glauconite, siderite, apatite, and protogenetic dolomite are often found. Low resistance and gamma ray values frequently correspond to a large mudstone section in the logging response.

4.2.2 Boundary System in Sequence Stratigraphy

4.2.2.1 Sequence Boundary and Type

A sequence boundary is a surface characterized by a plane of unconformity or correlative conformity surface that is distributed transversely, continuously, and extensively, at least over the entire basin. This surface acts as a boundary separating an old stratum from a new stratum. Along the boundary, there is sub-aerial erosion truncation or onshore exposure, indicating a significant depositional break. No evidence for onshore or submarine erosion and obvious lacuna is shown along the boundary.

In the Exxon sequence model, the plane of unconformity is divided into two types (Figs. 4.12 and 4.13) according to the scope of terrestrial erosion and seaward transfer volume of the facies belt.

1. **Plane of unconformity I**

This is formed in a rapid tectonic subsidence period when the sea level falls quickly. With deep incisions by shelf-incised valleys and submarine canyons, the land surface is subject to wide erosion due to shifting of the coastline to

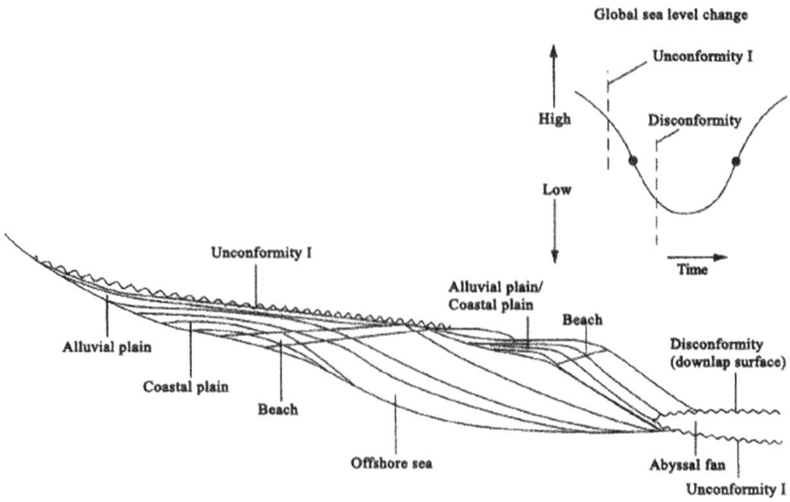

Fig. 4.12 The sedimentary facies belt migrates towards the basin rapidly in the type I unconformity; the earlier highstand system tract under the unconformity suffers from the wide erosion effect (by Posamentier et al. 1988)

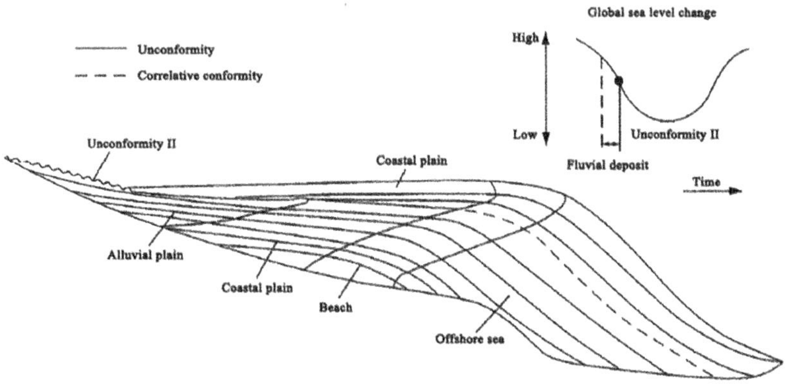

Fig. 4.13 Only a small amount of land exposures and erosion effects can be founded in the type II unconformity (by Posamentier et al. 1988)

the shelf margin or below. A wide lowstand systems tract is formed by the transport of clastic rock blocks to the bottom of the slope along the canyon system. In the plane of unconformity I, the depositional facies belt shifts quickly toward the basin direction. Early (or old sequence) highstand systems tracts are subject to wide erosion under unconformity. Sequence I (Fig. 4.12) is formed by the sequence bounded by plane of unconformity I.

2. **Plane of unconformity II**

This is formed during the period when relative sea level falls slowly, causing the facies regions to migrate toward the sea gradually, accompanied by a small amount of onshore exposure and erosion. A shelf margin systems tract is developed in correspondence with the lowstand systems tract in sequence I, while a sequence with plane of unconformity II as its bottom boundary is called sequence II (Fig. 4.13).

There are differences (Table 4.2) between the formation and distribution characteristics of planes of unconformity I and II. As a whole, type I unconformity surface is more apparent, is easy to identify, and type II unconformity surface is not obvious.

4.2.2.2 Initial Flooding Surface and Maximum Flooding Surface

A sequence is formed within a sea level change cycle. When the sea level rises upwards from the low-stand state and initially crosses the shelf break, an initial flooding surface is formed in the

Table 4.2 Characteristics of planes of unconformity I and II

Sequence boundary type	Sea level change	Coastal plain erosion	Sequence type	Distribution range
Plane of unconformity I	Rapid and obvious rate of decline	Obvious and relatively large erosion, with incised valleys mostly formed	Sequence I, LST, TST, and HST	Wide distribution, extending to below the coastal plain
Plane of unconformity II	Slow and less obvious rate of decline	Less obvious or smaller erosion, without incised valleys	Sequence boundary II, SMST, TST, and HST	Narrow distribution; mainly on the coastal plain

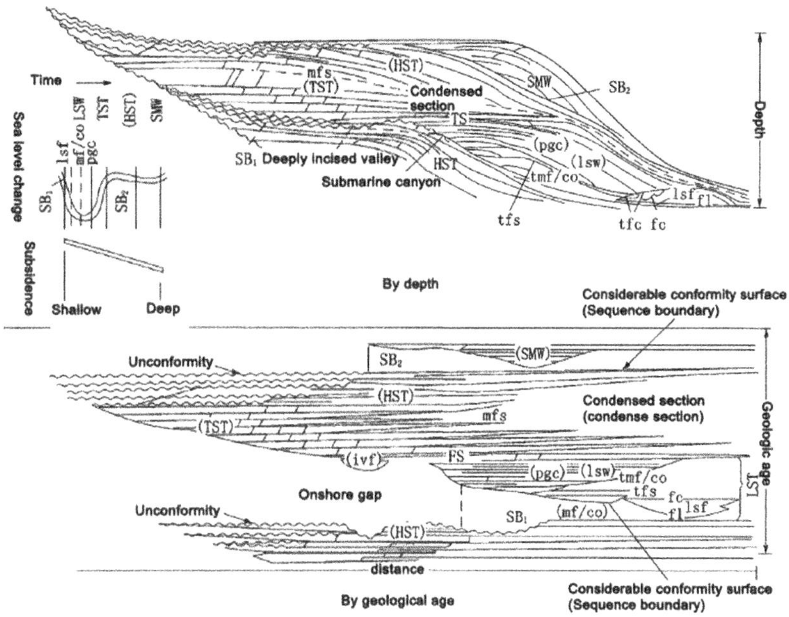

Fig. 4.14 Depositional systems tract and key interface of the sequence stratigraphy

vast shelf area. When the sea level continues to rise to reach the maximum transgressive period in a sequence, the maximum flooding surface is formed with the characteristics of deepest sedimentary water and widest transgressive scope. The initial flooding surface is a physical boundary between a lowstand systems tract and a transgressive systems tract, while the maximum flooding surface is a boundary between a transgressive systems tract and a highstand systems tract. A condensed section (Fig. 4.14) is formed between the upper and lower sections of the maximum flooding surface.

4.2.2.3 Flooding Surface and Main Flooding Surface

The flooding surface is a boundary separating a new stratum from an old stratum, with the characteristic that water deepens suddenly after the boundary is passed through, which is usually accompanied by underwater erosion and depositional break on a small scale without sedimentation. There are different scales of flooding surfaces, where the smaller ones are boundaries for defining a parasequence, but the main one is the boundary for defining a parasequence set.

4.2.3 Concept System of Sequence Genesis Control

4.2.3.1 Relative Sea Level Change and Absolute Sea Level Change

Exxon, represented by Vail and colleagues, has focused on tectonic subsidence, climate conditions, and source supply rate during research on sequence genesis control, however sea level change, which determines sequence formation, is the main point, and even the core of sequence stratigraphy. Sea level changes are of two basic types, namely, global sea level (absolute sea level) change and relative sea level change. Global level change is a unit function in which the sea level moves corresponding to some fixed reference point (such as the Earth's core), while relative sea level change is a dual function (Fig. 4.15) of sea level movement and thalassogenic movement.

Relative sea level change is a fundamental cause for the formation of a sedimentary sequence bounded by a plane of unconformity and correlative conformity surface. Relative sea level change is irrelevant to sediment accumulation above the seabed, and it is not equal to sedimentary depth change. Water depth is the distance between sediment surface and sea level, and it is subject to various factors such as global sea level change, tectonic subsidence, and sediment supply. When the sediment accumulation rate is higher than the speed of sea level rise, water depth decreases even if the sea level rises.

4.2.3.2 Accommodation

Accommodation refers to the potential space for sediment accumulation, namely, the space between the sedimentary base level and the sediment surface. It is a function of global sea level change, sediment supply, and tectonic subsidence. The relative fluctuation change characteristics of sea level and structure decide the existence of accommodation capable of accumulating sediments.

Based on changes in the accommodation growth rate to sediment supply rate (i.e., A/S) ratio, accommodation migrates to different space parts, together with facies differentiation accompanied by volumetric partitioning. Changes in a reservoir's spatial distribution and superimposition lead to changes in reservoir properties. When the A/S value is relatively high, a marine-transgressive terrestrial-regressive sequence is formed, and the properties of marine sandstone improve because the injection of more terrigenous debris, especially fine grained terrigenous debris, ceases. When the A/S value is relatively low, a marine-regressive terrestrial-transgressive sequence is formed, and the properties of a marine reservoir worsen owing to the adequate supply of terrigenous clasts.

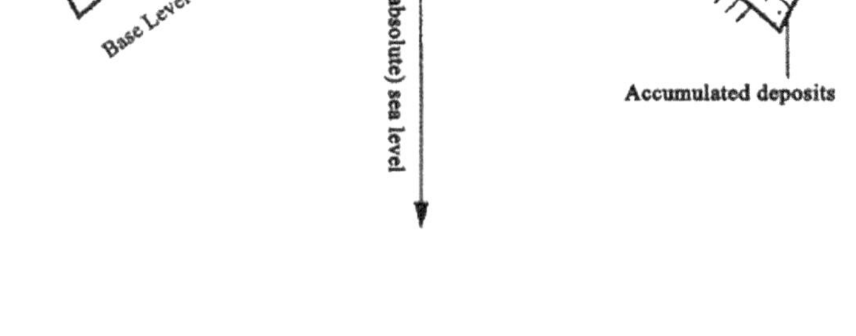

Fig. 4.15 Definitions of global sea level, relative sea level and water depth (by Jervey 1988)

4.2.3.3 Stratigraphic Base Level

The base level is another important concept in the interpretation of sedimentary sequence genesis. From the perspective of stratigraphic preservation, Wheeler (1964) believes that the base level is neither the sea level nor a horizontal plane on which the sea level extends landward, but an abstract surface (non-physical surface) that fluctuates in a wavy shape corresponding to the Earth's surface and descends continuously toward the basin direction. This surface, whose position, movement direction, and fluctuation vary with time, rises and falls corresponding to the Earth's surface (Deng et al. 2002). Two energy components, of which one is used for carrying and the other is used for saving, act on the sediments. The base level is a state in which the two aforementioned energy components are balanced.

T. A. Cross (1991) considers the stratigraphic base level as a potential energy surface that reflects the degree of imbalance between the Earth's surface and the Earth's surface process striving for a good balance. To achieve this balance, the Earth's surface must move toward the direction close to the base level by changing its form through sedimentation or erosion. Therefore, this surface describes a type of energy that forces the Earth's surface to move to some position where terrain gradient, sediment supply, and accommodation are balanced (Fig. 4.16). The stratigraphic base level can be simply regarded as the top boundary of accommodation. It may be located above or below the Earth's surface or intersect with the Earth's surface.

Among the various factors controlling the development and evolution of sequence strata, periodic change in global sea level and tectonic subsidence are deemed the important factors. With the combined action of periodic fluctuation and basement settlement in geological history, the global sea level determines the relative sea level change in some areas of the basin, thus determining the potential space (accommodation) for sediment accumulation. Different control factors have different effects on stratum development: tectonic subsidence provides space for sediment deposition, global sea level changes determine strata and lithofacies distribution models, sediment supply determines sedimentary filling mode and palaeobathymetry, and climate determines sediment type.

4.3 Theoretical System of Terrestrial Clastic Sequence Stratigraphy

Continental sequence stratigraphy, as a new sub discipline of sequence stratigraphy, has attracted increasing attention from researchers and triggered academic debates on different viewpoints.

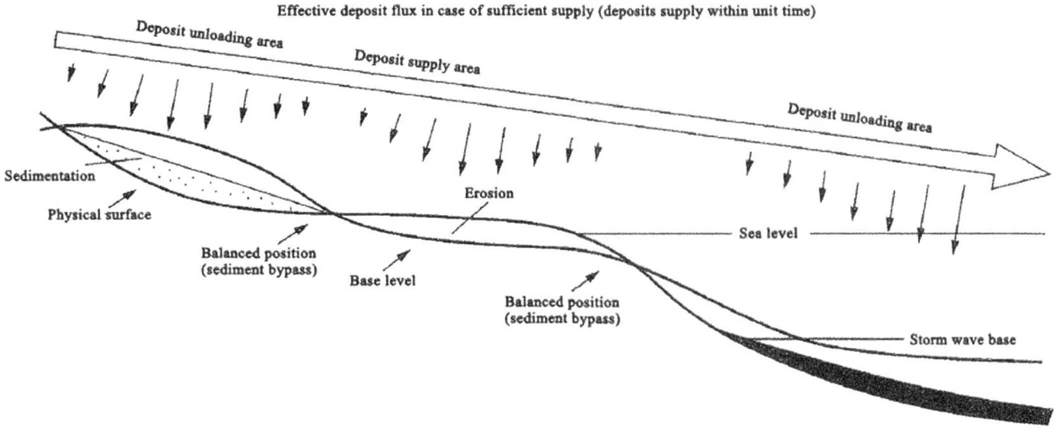

Fig. 4.16 The stratigraphic base level can be considered as a potential energy surface, on which topographic gradient, sediment supply and accommodation space are balanced (by Cross 1991; Deng et al. 2002)

Table 4.3 Comparison between factors influencing the study of marine and continental basin sequence stratigraphy (modified from Gu 1995)

Type	Marine basin	Continental basin
Determinants	Mainly affected by sea level change, basin subsidence, deposition rate and climate	Mainly affected by tectonic subsidence, sediment supply, climate, and paleotopography
Sedimentary hydrodynamic conditions	Main geological agents include wave, tide, ocean current, storm, turbidity current, contour current, and submarine volcano	Mainly affected by continental water (river), wave, and tide
Deposition area and range	Tens of meters to hundreds of kilometers of transverse range across coastal zone, continental shelf, continental slope, and deep sea	Alluvial fan area, fluvial deposit area, lacustrine deposit area (including lake shore, meare and deep lake), with relatively narrow deposition range
Transverse continuity of sedimentary formation	Long transverse extension and good continuity	Relatively short transverse extension and poor continuity
Sequence thickness and change	Very thick sequence, ranging from tens to hundreds of meters, relatively stable	Relatively thin sequence, ranging from tens to hundreds of meters, with considerable variation in thickness
Change of depositional facies	Continuous and stable depositional facies, and gradually transiting facies change	Rapid change of depositional facies, and abrupt change of common facies
Characteristics of systems tract	Great transgression and regression, obvious characteristics of systems tract, and easy-to-determine condensed section	Frequent small transgression and regression of lake water, less obvious characteristics of systems tract, and difficult-to-determine condensed section
Tectonic influence	Relatively weak influence within a large range	Relatively strong frequent influence
Prediction of difficulty degree	Relatively easy	Relatively more complex and difficult

This shows that the research on continental sequence stratigraphy can be developed into a sub discipline of sequence stratigraphy. On the other hand, it indicates the differences (Table 4.3) between continental sequence stratigraphy and marine sequence stratigraphy, thus a set of complete theory systems and methods related to continental sequence stratigraphy must be built and improved continuously to overcome research limitations (Gu and Fan 2001), especially in terms of practical exploration and oil and gas field development. Continental sequence stratigraphy research is limited by two main reasons: the first is many problems may be caused by the direct application of concepts of marine sequence stratigraphy owing to small continental lake sedimentation areas, considerable stratigraphic fluctuation, rapid facies change, and complex sequence strata distribution; the second is making petroleum workers learn and directly apply sequence stratigraphy division schemes in production, because a majority of oilfield organizations apply lithologic strata division schemes for oil and gas production in direct combination with hydrocarbon exploration and development in practical use.

4.3.1 Characteristics of Lake Level Change

Lake level changes generally refer to the changes relative to the lake level. Lake level fluctuation refers to changes in height between any point in the lacustrine basin and the base level of the basement. In general, the duration and change

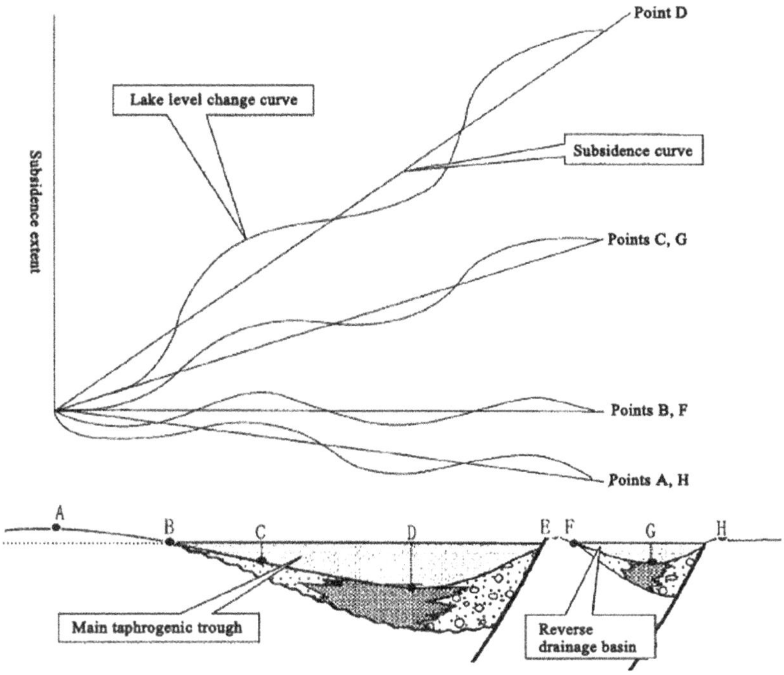

Fig. 4.17 There is a certain relation between base subsidence rate and base level change (by Zhou et al. 2006)

range of the lake level can be determined by coastal onlap, and a lake level fluctuation curve for a fluctuation period can be prepared. The base subsidence rate is a leading factor determining the accommodation of sequence III. The generated sequence unit is fleeting if lake level change is not accompanied by basement subsidence. Sequence accommodation is negative when the basement rises, resulting in denudation without deposition (Fig, 4.17). Relative lake level change decides the periodicity of sequence cycles. In other words, a lowstand systems tract is generated owing to a decrease in lake level at the quarter cycle, transgressive systems tract is formed owing to the rise of lake level at the half cycle, and highstand systems tract (Fig. 4.18) is formed by stable or slow decline of lake level at the quarter cycle.

4.3.1.1 Sensitivity of Lake Level Change

Owing to the much smaller volume of and small quantity of water in a lacustrine basin compared to a sea basin, many factors affect changes in lake level. Dramatic changes in lake level may result from small changes in lake structure, climate, sediment supply, and river injection. On the one hand, this sensitivity may result in very frequent changes in lake level, which are difficult to represent precisely using a lake level fluctuation curve prepared by the general technique. On the other hand, the sensitivity of sea level change may obscure the cycle regularity of sea level changes. Continental basins of different types may have a leading factor affecting lake level changes, and the cyclicity (such as tectonic or climate cyclicity) of lake level change corresponds to that of the leading factor. Moreover, the leading factor affecting sea level changes in different periods in the same basin may be different, which warrants complete analysis of the leading factors when studying lake level changes.

4.3.1.2 Isolation of Lake Level Change

Lake level changes are more complex than sea level changes. The latter can be compared on a global scale because they are not caused by local tectonic events. Because most modern and ancient lakes are not connected, there is no interactivity among the lake levels of different lacustrine basins, and there is no unified lake level. Thus, it is difficult to compare the lake

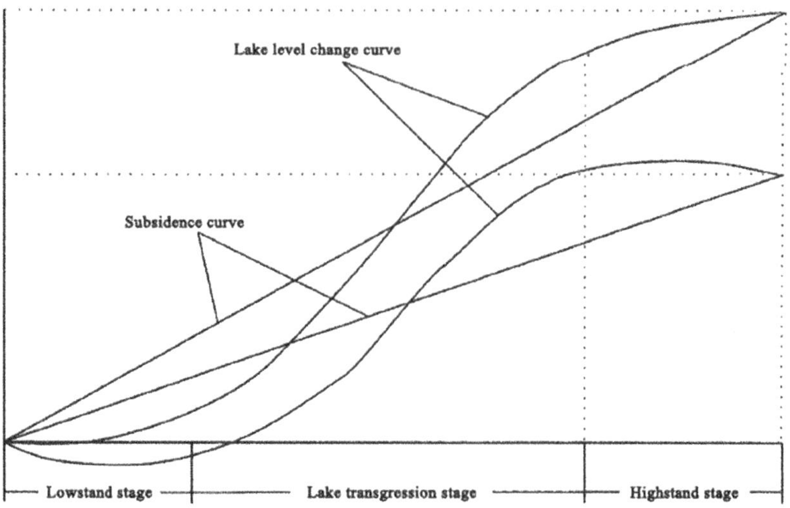

Fig. 4.18 Relative lacustrine level change of level III sequence and phase division within sequence cycle (by Zhou Haimin et al. 2006)

level fluctuation curves of various lacustrine basins and validate them against each other. Moreover, the reliability of the sea level change curve must be increased based on additional evidence to reflect lake level changes such as paleospecies, ichnofossils, and microelements.

4.3.1.3 Differences Between Lake Level Changes of Open and Closed Lacustrine Basins

In hydrogeology, lacustrine basins are divided into open and closed types (Fig. 4.19). The water yield of a closed lacustrine basin is less than the sum of evaporation and underground seepage, and the location of sea level is often lower than the elevation of the lowest overflow port of the basin. The water yield of an open lacustrine basin is less than the sum of evaporation and underground seepage, the location of the sea level is maintained at an elevation that is the same as that of the lowest overflow port of the basin, and surplus water flows out the lacustrine basin via a drainage gallery. Sea level change has different control effects on the two types of lacustrine basins. An open basin rises relative to the lake level when tectonic subsidence increases, whereas there is no change in the lake level of a closed basin (Fig. 4.18) with an increase in the tectonic subsidence distance. The hydrogeologic characteristics of lacustrine basins should be considered when studying the correlation between the sea level change curve and various influence factors. Because the ocean has no overflow port, the influences of tectonic subsidence and sediment supply on sea level change are comparable with those on the lake level of a closed lacustrine basin, however a comparison with open lacustrine basin cannot be simply made. In geological history, open and closed lacustrine basins are transformed mutually due to climate change, resulting in a complex lake level change curve. Thus, it should be noted that hydrogeologic differences and changes are likely to occur over the entire evolution of a lacustrine basin.

4.3.1.4 Basis for Determining Lake Level Fluctuation Curve and Lake Transgression Influences

Based on marine sequence stratigraphy, the most reliable stratigraphic markers for reflecting relative lake level changes are the coastal onlap and toplap phenomena in lacustrine sequence stratigraphy. Landward migration of coastal onlap indicates that the lake level rises relatively, coastal sediment toplap indicates that the lake level is relatively still, and migration of the coastal downlap toward the center of lacustrine basin indicates that the lake level declines relatively. When the lake level fluctuation curve is

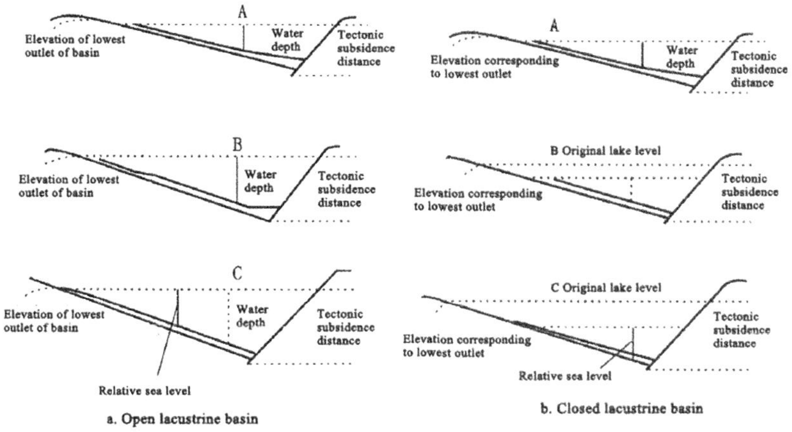

Fig. 4.19 Influences on lake level changes of open lake basin and closed lake basin from tectonic subsidence (Ji 1998). An open lacustrine basin rises relative to the lake level as tectonic subsidence increases (A-B-C), while the relative lake level of a closed lacustrine basin is not affected by an increase in the tectonic subsidence distance

prepared using coastal onlap, changes in stratum thickness resulting from differential subsidence, later diagenesis, and changes in the gradient of the original sedimentary boundary on the gentle slope of the lacustrine basin can be corrected theoretically by referring to the method used to prepare the sea level fluctuation change curve. Unlike sea basins, continental lacustrine basins are strongly influenced by local tectonization. Differential tectonic subsidence often causes erosion of the marginal facies belt, which makes it difficult to determine the original onlap location. Hence, the evolution of structure and control over sedimentary evolution process must be studied fully.

When the elevation of the lowest spill point of a lacustrine basin close to the ocean is lower than the sea level, the sea level may directly influence the lake level. Once the ocean communicates with a lake, the lake is turned into a bay. Because the lake level is equal to the sea level, the factors influencing lake level changes are the same as those influencing sea level changes, and the lake level change curve is identical to the sea level change curve. Furthermore, the lake level change curve is very complicated terms of geological history on the grounds that the lake is sometimes communicated with, or separated from, the ocean, and various data needs to be collected to prove lake level change.

4.3.2 Main Controlling Factors of Continental Sequence Stratigraphy

According to sequence stratigraphy, the ratio of sedimentation rate to the rate of change of accommodation decides the structures of sequence frames and sequence boundaries. The subsidence rate of the sedimentary base level decides the rate of change of accommodation. The underwater sedimentary base level in a continental lacustrine basin is related to lake level, balance level of the river, and its extension in the lacustrine basin. Currently, we believe that lake level changes resulting from tectono-eustatism, climate change, and sediment supply are the main determinants of the scope of sequence development in continental lacustrine basins, sedimentary thickness, evolution stage, and the formation of sequence boundaries. In geologic history, the same lacustrine basin can be controlled by different main factors in different

Table 4.4 Classification of different sequence levels (Vail et al. 1991)

Level	I	II	II	IV	V	VI
Time interval (Ma)	>50	3–50	0.5–3	0.08–0.5	0.03–0.08	0.01–0.03

periods or even in the same period. On the one hand, at different stages of evolution of a lacustrine basin, different factors play major parts in the determination role, and are related to the transformation of the basin type. In other words, open and closed lacustrine basins may be transformed mutually in different periods of basin development. During a period in which an open lacustrine basin is formed, structural factors determine the formation of sequence boundary; during another period, when a closed lacustrine basin is formed, climate plays an important role. On the other hand, different factors jointly affect sequence development in the same period.

As we know, the sequence strata of continental lacustrine basins are sensitive to changes in various factors, and the effects of the main controlling factors can be obscured easily by the effects of other factors. This increases the difficulty of researching the main factors controlling sequence development. Despite all this, many scholars believe that the main factor controlling continental basin sequence stratigraphy is tectonism, which directly determines the change in accommodation and leads the generation, development, and evolution of a continental basin subject to tectonic movements of different orders. Moreover, erosion rate, sediment type, sediment supply rate, and even local climate conditions are affected indirectly by structural factors through changing the geomorphic features.

4.3.3 Classification of Continental Sequence Stratigraphic Level

To accurately determine the level of sequence boundary, the classification of continental sequence stratum requires one to compare the fine sequences of a well section with the macroscopic sequence of a seismic cross section by considering various identification marks, depositional cycles, and tectonic evolution features of the basin.

Vail et al. (1991) divided sedimentary sequences into 6 levels, in which sequence levels 1–3 pertain to structural genesis or global sea level change, while the sequence levels 4–6 are distributed (Table 4.4) within a small scope caused by local effects. Continental sequence strata may be divided into different levels (Table 4.5) based on the properties and distribution of sequence boundaries. The characteristics of continental sequence units at different levels, thickness scope of sequences at different levels, lateral distribution range, and formation time range are different. Regarding technical precision, core and outcrop between the sequence and laminae can be identified for every sequence level. However, logging identification precision can only reach the "stratum" level, and seismic exploration can identify sequences at the parasequence set level at most. Research on fine sequence stratigraphy should be combined with the study of macroscopic seismic stratigraphy to determine any mutual supplementation effects and for contrast verification.

Therefore, the general sequence has three levels, systems tract and parasequence set have four levels, and parasequences have five levels. Continental lacustrine basin systems tracts can be divided trichotomously into lowstand systems tract (LST), transgressive (or extensive) systems tract (TST or EST), and highstand systems tract (HST). Continental lacustrine basin systems tracts can be divided into four types, namely, LST, TST (or EST), HST, and falling-stage systems tract (FFST). However, on the basis of base level cycle, high-resolution sequence stratigraphy can be divided into six levels, namely, megacycle, ultra-long term, long term, medium term, short term, and ultra-short term (Fig. 4.20; Table 4.6).

Table 4.5 Classification of continental sequence strata (Gu and Fan 2001)

Sequence level	Super sequence	First-order sequence	Second-order sequence	Third-order Sequence	Parasequence Set	Parasequence
Sequence boundary	Unconformity area exceeding basin area	Unconformity area exceeding basin area or the vast majority of basin area	Unconformity distributed in basin Vast majority of unconformity distributed in basin	Large portion of unconformity distributed in basin	Between relatively large lake flooding surfaces	Between relatively small lake flooding surfaces
Genesis	Global tectonic movement	Intraplate tectonic movement	Regional tectonic movement	Lake level changes arising from structure and climate	Lake level changes arising from structure, climate, and provenance	Lake level changes arising from climate and provenance
Corresponding sedimentary cycle	Global sedimentation macrocycle	Intraplate sedimentary cycle	Regional sedimentary cycle	Intrabasinal sedimentary cycle (level II)	Lithology group cycle (level IV)	Lithology cycle (level V)
Compared with stratigraphic boundary	Equivalent to stratum boundary or smaller	Equivalent to stratum system or smaller	Equivalent to stratum series or smaller	Equivalent to stratum group or smaller	Equivalent to a group of stratigraphic superimposition modes	One sedimentary cycle
Time (Ma)	200–300	60–120	30–40	2-5	0.1–0.4	0.02–0.04

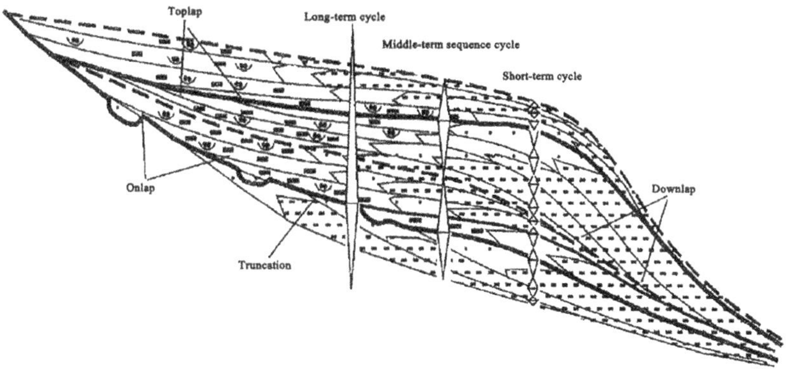

Fig. 4.20 There is a certain correspondence on division between base level cycle and sequence interface cycle (by Deng et al. 2002)

4.3.4 Types of Continental Sequence Stratigraphic Models

On the account of characteristics, such as sensitivity of various sequence control factors, isolation of continental lacustrine basin distribution, and complexity of continental basin tectonic settings, continental sequence strata feature relatively poor sag-crossing comparability, small scope for application of the sequence stratigraphic model, and great difficulty in promotion. However, a corresponding sequence stratigraphic model (Fig. 4.21) can be built to restore the base level change curve or lake level change fluctuation curve for mutual comparison by researching the sequence stratigraphy of continental basins (such as fault depressions and depression basins, offshore and inland basins, basins filled with clastic rocks, evaporite and coal measure strata, and basins in wet and arid climatic zones with different tectonic settings and different rock types). A corresponding sequence model (Fig. 4.22) of a deep

Table 4.6 Grade division of base level cycle and basic characteristics (Zheng et al. 2001)

Base level cycle grade	Boundary type	Time limit (Ma)	Sequence definition	Main controlling factors		Comparison with sequence stratigraphic unit equivalent to Vail's	
Megacycle	Type I	30–100 or greater (depending on basin delay)	Including complete sedimentary filling sequence of prototype basins at all stages of basin evolution	Structural factors	Regional tectonic movement	Unavailable for complete contrast	Equivalent to sequence II
Ultra-long term	Type II	10–50	Tectonic sequence filling (or structural sequence or frame sequence) that takes all stages of basin evolution as units		Stress field transformation at tectonic evolution stage		Equivalent to sequence set III
Long term	Type III	1.6–5.25	Set of regional transgressive-regressive sedimentary sequences formed by genetically related strata with greater water depth variation		Tectonic episode type strength change		Equivalent to sequence III
Medium term	Type IV	0.2–1	Set of transgressive-regressive sedimentary sequences formed by genetically related strata with no substantial water depth variation	Astronomical factors	Long eccentricity	Available for basic comparison	Sequence IV (parasequence set or systems tract)
Short term	Type V	0.04–0.16	Set of transgressive-regressive sedimentary sequences formed by genetically related strata with small water depth variation, or by superimposition of strata with similar lithology and lithofacies		Short eccentricity		Sequence V (parasequence)
Ultra-short term	Type VI	0.02–0.04	Set of superimposition styles that represents single lithology or relative lithology of the smallest genetic stratigraphic unit		Precessional period		Sequence VI (cyclothem)

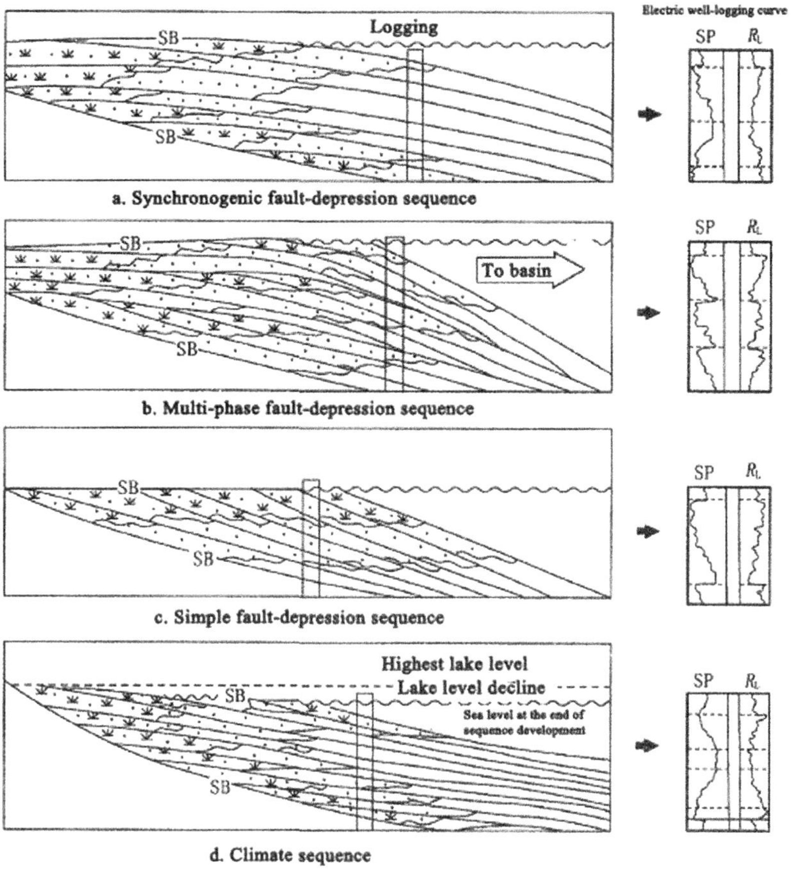

Fig. 4.21 Different types of lake basins have different sequence stratum modes (Ji 1998). SP spontaneous potential, RL lateral resistivity

water fault-depression lacustrine basin can be built according to the type of its slope. Different researchers have built different corresponding lacustrine basin sequence stratigraphic models according to the different geologic features of the areas studied. These models can be classified based on terrain rupture features, such as syndepositional faulted-slope break sequence (Fig. 4.23), syndepositional bending sequence (Fig. 4.24), and lacustrine basin sequence combination model (Fig. 4.25), and they can be integrated. The accuracy and reliability of sequence stratigraphic model interpretation and effective stratigraphic trap prediction (Fig. 4.26) can be improved by establishing sedimentary sequence frameworks and stratigraphic trap distribution models of different types of basins.

4.4 Theoretical System of High-Resolution Sequence Stratigraphy

4.4.1 Concept Introduction

The theory and practice of sequence stratigraphy have been accepted by the majority of geologists. Theoretically, sequence stratigraphy focuses on the important influence of the sea level fluctuation period on the formation of stratigraphic sequences. In practice, on the basis of established chrono-stratigraphic frameworks, sequence stratigraphy is used to interpret stratigraphic distribution models and division of systems tracts in contemporary genetic strata. This provides a

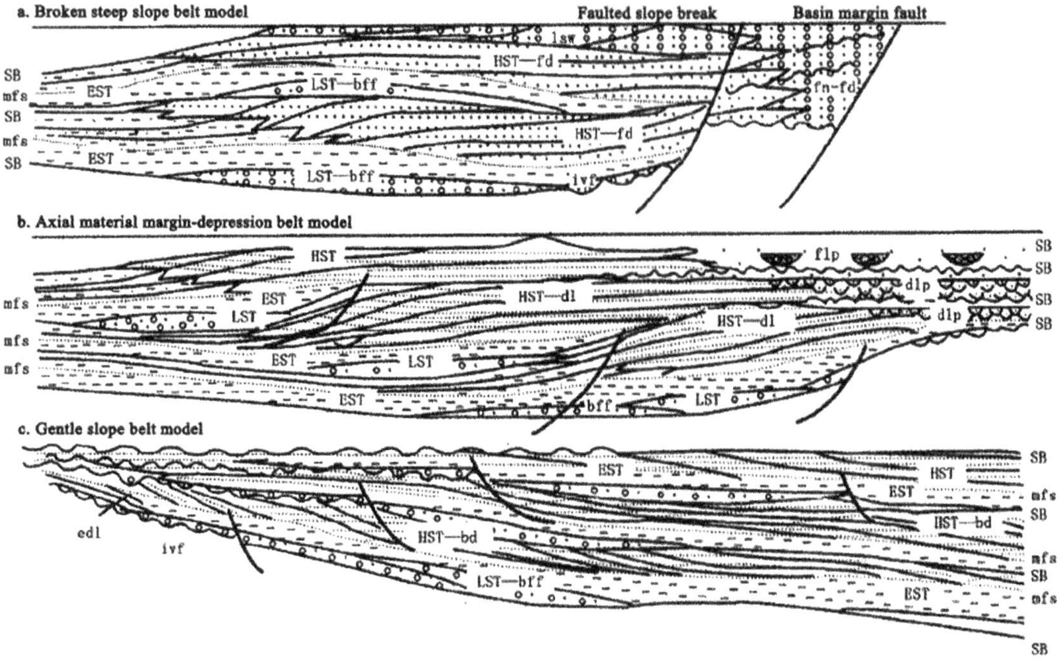

Fig. 4.22 Formation model of deepwater Rift-subsidence basin sequence (Li et al. 2002). SB sequence boundary, mfs maximum flooding surface, HST highstand systems tract, dl delta; flp alluvial plain, bff basin floor fan and other lowstand fans, EST extensive systems tract, fd alluvial fan and fan delta, dlp delta plain, ivf incised valley and incised waterway, LST lowstand systems tract, bd braid delta, lsw lowstand wedge, and edl margin delta during lake extension

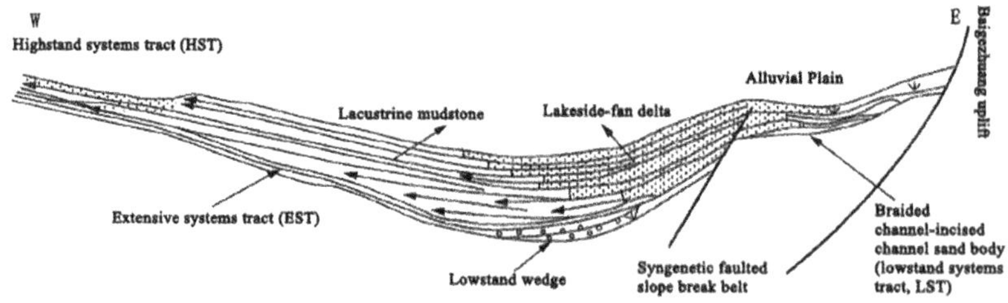

Fig. 4.23 Formation model of Nanpu sag and sedimentary breaking type sequence (Zhou et al. 2006)

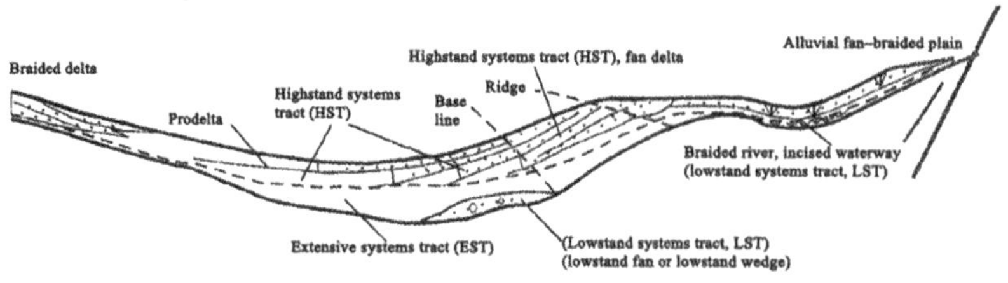

Fig. 4.24 Formation model of Nanpu sag sedimentary structure bending type sequence (Zhou et al. 2006)

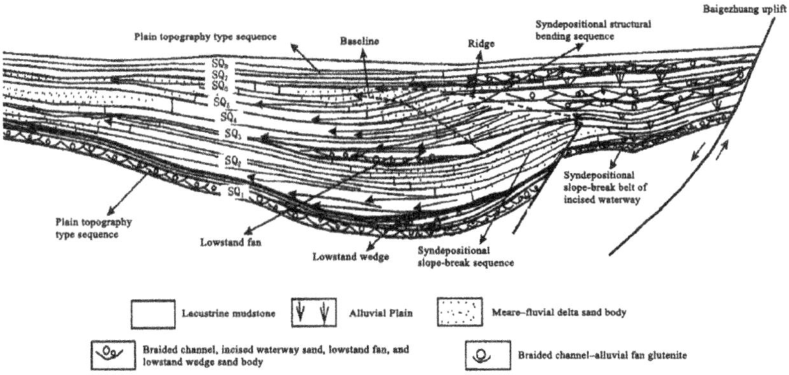

Fig. 4.25 Composition pattern of continental faulted basin sequence model (Zhou et al. 2006)

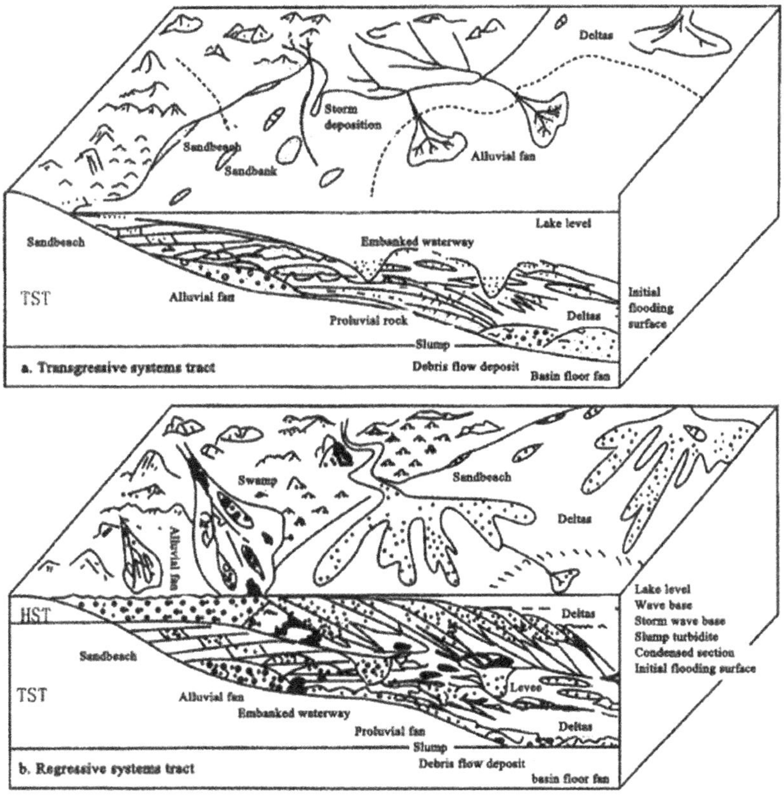

Fig. 4.26 Features of Depression Lake System Tract in Songliao Basin (Wei and Sun 1996)

solid theoretical foundation for reservoir prediction based on stratigraphic analysis of petroleum basins and basin scale, as well as an effective means for hydrocarbon exploration. In addition, it effectively promotes the development of geology, especially the development of petroleum geology. However, theoretical study and practical application prove that the formation and analysis of sequences should have a more comprehensive and logical basis. Meanwhile, as basin petroleum exploration and development advances, petroleum geologists require more precise technology to improve the resolution of sequence stratigraphy analysis and accuracy of reservoir prediction. Accordingly, the high-resolution sequence stratigraphic analysis theory and technique emerged at the right moment. The high-resolution sequence stratigraphy school of the genesis stratum research group in Colorado School of Mines led by T. A. Cross is a representative therein. High-resolution sequence stratigraphy has significant and important effects on petroleum exploration and the development of petroleum companies in America and other countries, which remarkably reflect the concept, method, and progress of high-resolution sequence stratigraphy. For example, the theory of process-response sedimentology is applied for studying high-resolution sequence division and comparison technology, numerical simulation of forward and inversion numerical values, and the control of stratum over flow velocity of fluid in reservoirs and fluid units.

High-resolution sequence stratigraphy is a stratigraphic-sequence division and correlation technique that uses outcrop, drilling, logging, and high-resolution 3D seismic data for building highly precise time-stratigraphic correlation frameworks with reference to a multi-order base level cycle.

4.4.2 Theoretical Basis

Unlike the sequence analysis of a basin or region scale, in high-resolution sequence stratigraphy analysis, a genetic stratigraphic correlation framework of a region, oilfield, or petroleum reservoir is established to evaluate and predict the distribution of reservoirs, barrier beds, and source beds, based on core, outcrop, logging and high-resolution seismic reflection section data, by applying a fine sequence division and correlation technique to change one-dimensional drilling information into a 3D stratigraphic relationship. Owing to the increased time resolution, the accuracy of sequence prediction is greatly improved, and a reliable rock physical model that best simulates fluid flow in a stratum can be obtained. Therefore, sequence stratigraphy has been applied and popularized on a large scale, especially in continental basin research. The theoretical core consists of four parts.

4.4.2.1 Theory of Stratigraphic Base Level

Sequence determination is the basis of sequence stratigraphic analysis. The Colorado School of Mines' genetic stratigraphic team headed by T. A. Cross holds that the stratigraphic base level, which is restricted by the integrated effect of various factors such as sea level, tectonic subsidence, sedimentary load compensation, sediment supply, and depositional topography, is the main frame for understanding the genesis of stratigraphic sequences and sequence division (Deng et al. 2002). Cross et al. quoted and developed the concept of base level proposed by Wheeler and analyzed the process-response principle formed by the base level cycle and genetic sequence formation. However, the base level always tends to move unidirectionally toward the maximum or minimum value of its amplitude to complete a cycle of rise and fall. One cycle of rise and fall of the base level is called a base level cycle. The base level can swing completely above or below the Earth's surface or transit across the upper part of the Earth's surface and then swing to the lower part thereof for return. The latter part is called the base level transit cycle. Since a base level cycle is isochronic, rocks preserved during the change of a base level cycle (also known as time domain) form a genetic stratigraphic unit, that is, a genetic sequence bounded by time surface. Thus, it is also a time-stratigraphic unit.

The base level corresponds to the wavy fluctuation of the Earth's surface, accompanied by changes in sediment accumulation (accommodation) (Fig. 4.27). When the base bevel is above the Earth's surface, sediment accommodation is provided. In the case of sedimentation, any erosion is local or temporary. When the base bevel is below the Earth's surface, accommodation disappears, and any sedimentation is local or temporary. When the base level is consistent (overlapped) with the Earth's surface, only sediment bypass occurs (Fig. 4.16), but neither sedimentation nor corrosion occur. Therefore, in the time domain (note: time is continuous) of

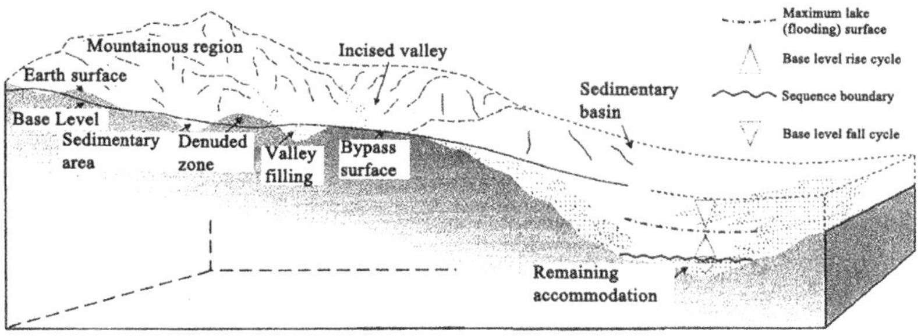

Fig. 4.27 There is a correspondence on division between base level cycle and sequence stratigraphic classification (Deng et al. 2002)

base level change, four geologic processes are presented on different geologic positions of the Earth's surface: ① sedimentation, ② erosion, ③ no sedimentation but only sediment bypass and ④ hungry sedimentation and non-sedimentation owing to non-compensation of sediments (ratio of accommodation to sediment supply $A/S \rightarrow \infty$). Time-space events that represent changes in the base level cycle in stratigraphic records are presented as rock + boundary (discontinuities). Therefore, one genetic sequence may consist of rising semi-cycle and falling semi-cycle of the base level, or it may consist of rocks and interfaces. Its meaning cannot be reflected accurately by "parasequence" in ordinary sequence stratigraphy.

In continuous movement, when the base level is located above the Earth's surface and rises further corresponding to the Earth's surface, accommodation may increase, and the potential accumulation velocity of sediments in the accommodation increases. However, the actual accumulation speed of sediments is further controlled by the geological process of matter transportation. In other words, accommodation controls the maximum value of sediment accumulation in some geographical location within a certain time. When the supply speed of deposited materials remains unchanged, the ratio (A/S value) of accommodation to sediment supply decides the accumulation speed, preservation degree, and internal architectural characteristics of the sediments (effective accommodation) in the accommodation. When the base level is located below the Earth's surface and falls further, the potential velocity of erosion will increase, but the actual erosion rate will be limited by the geological process that removes sediments from the Earth's surface. As a result, the base level describes the process of change between the establishment or disappearance and sedimentation of an accommodation. The base level may serve as a potential energy surface that reflects the imbalance between the Earth's surface and the process of striving for balance. To achieve balance, the Earth's surface must move toward the direction close to the base level by changing its form through sedimentation or erosion.

4.4.2.2 Volumetric Partitioning Principle

The dynamic system of the base level cycle and the accompanying accommodation changes determine stratigraphic and sedimentary characteristics. To further understand this process-response relationship, T. A. Cross proposed the concept of volumetric partitioning of sediments, which is the process where sediments are divided into different phase regions in genetic strata. It is the product of four-dimensional (space + time) dynamic changes in accommodation in different sedimentary environments during base level changes. Sediment volumetric partitioning is an important concept because volumetric partitioning is directly accompanied by many sedimentology and stratigraphy responses such as degree of preservation of the original landform, sediment thickness, and internal architecture. Its connotation can be better reflected by the

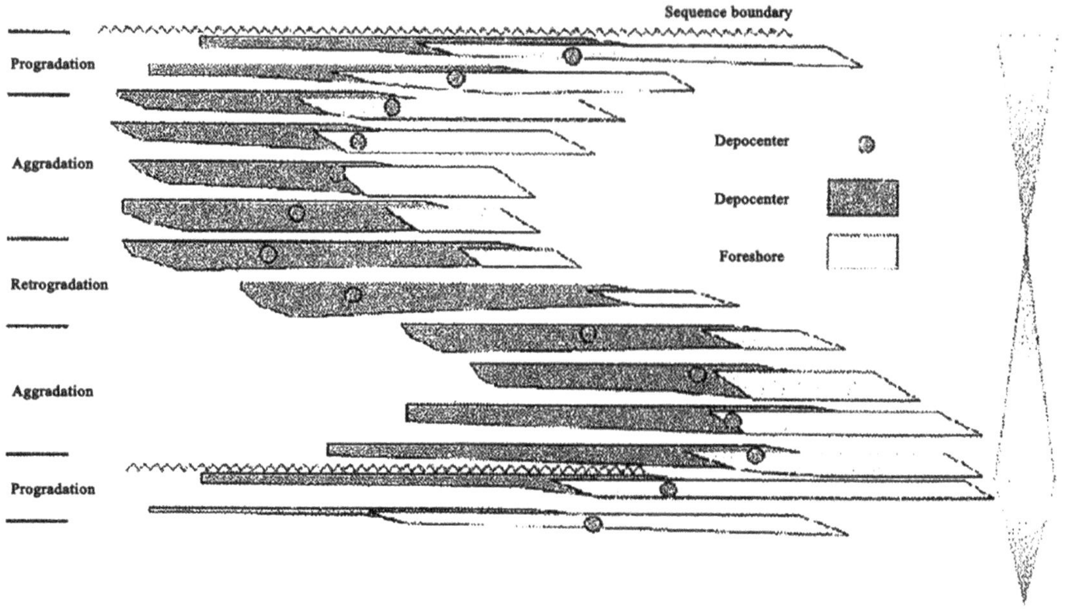

Fig. 4.28 During base level cycle, sediment stacking pattern and volume distribution also have cycling changes (by Cross 1991)

volumetric partitioning of coastal plain-coastal sandstone systems tract and changes in sediment stacking patterns (Fig. 4.28). As the base level declines, the effective position of accommodation is shifted seaward along with an increase in space seaward and decrease landward, leading to a gradual increase in coastal sandstone sedimentation volume and decrease in coastal plain sedimentation volume. In contrast, as the base level rises, the effective position of accommodation is shifted landward along with an increase in space landward, resulting in a gradual increase in sedimentation volume, piled up in a terrestrial phase region on the coastal plain. In the genetic sequence formed by a relatively longer base level transit cycle, the stratigraphic stacking pattern and migration of geographical location are relevant to its position in the base level transit cycle (Fig. 4.28). The seaward-stepping stacking pattern is formed during in the decline part of a longer base level cycle, and the resulting vertical-stepping stratum is formed in the initial stage of the rise part of a base level cycle. The landward-stepping stacking pattern is presented as the base level rises, while the resulting accretion formation occurs in the early stage of decline and the later stage of rise of the base level.

4.4.2.3 Facies Differentiation Principle

Along with the changes in accommodation and volumetric partitioning of sediments (Fig. 4.29), there are significant differences in terms of the sequence, association, type, and diversity of facies preserved in the same sedimentary environment, which are known collectively as facies differentiation. Facies differentiation directly influences the physical characteristics of a reservoir, such as continuity in 3D space, geometry, lithology, and lithofacies type, even rock physical properties (heterogeneity). In addition, remarkable differences exist between channel sand bodies formed by high accommodation and those formed by low accommodation in terms of geometry (width-thickness ratio), sand body connectivity, lateral continuity, mutual clipping level, bedform category, degree of preservation, and thickness and type of bottom sediment (Figs. 4.30 and 4.31). For example, compared with the shoreface sandstone in a

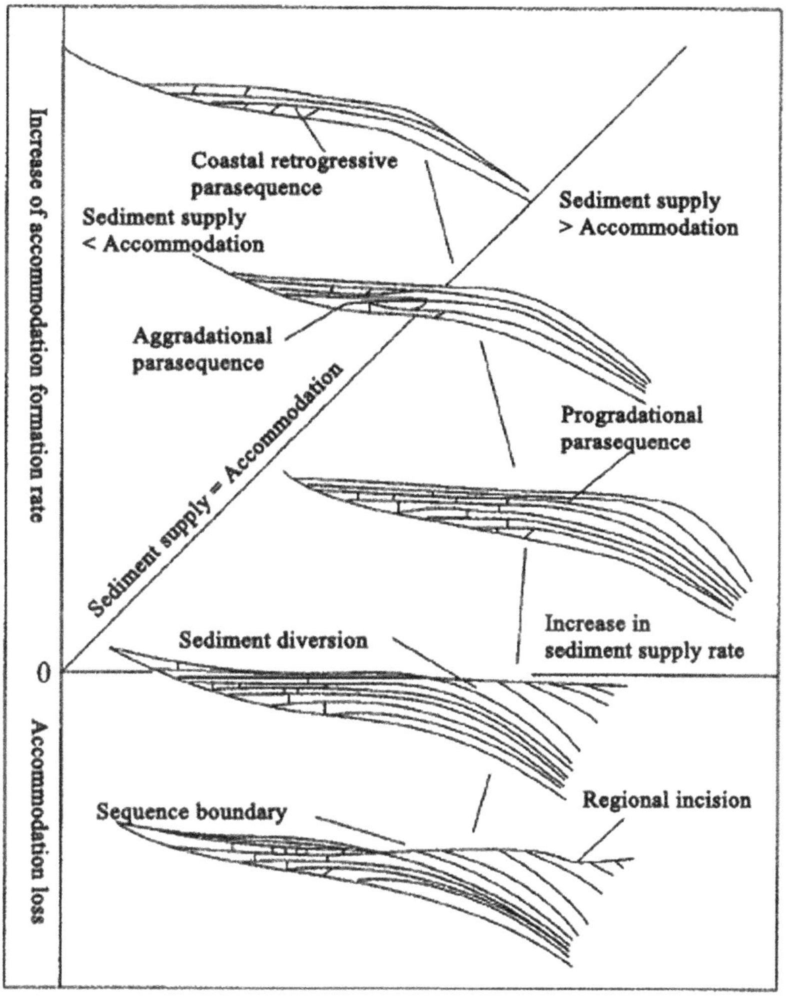

Fig. 4.29 Accompanied by the changes from the accommodation space and sediment supply, phase sequence, combination, type and diversity stored in the same environment have the significant differences (by Shanley 1994)

progradational genetic sequence (low accommodation), the shoreface sandstone of a retrogradational genetic sequence (high accommodation) has lower volume, lower lateral continuity, simple facies type and sequence trend, decreased mud content in sandstone, better sorting, and higher sandstone homogeneity. These facies differentiation features directly influence reservoir properties, or even the flow path and pumping-exhaust system of oil, gas, and water.

From the dynamics viewpoint of base level and accommodation, the volumetric partitioning (volume ratio of sediments in various sedimentary environments) of the same sedimentary systems tract and facies region (stratigraphic record of depositional system), degree of preservation of sediments, strata stacking pattern, facies sequence characteristics, and facies type are not fixed, but are functions of their locations in the base level cycle and accommodation, that is, the functions of space and time. Therefore, it is fundamentally different from the traditional static facies model analogy because it applies sedimentary dynamics to analyze the changes in accommodation due to base level changes during sediment accumulation and to interpret stratigraphic structures and sedimentary characteristics.

4.4.2.4 Cycle Isochronous Correlation Method

High-resolution sequence stratigraphic division and correlation are based on the recorded

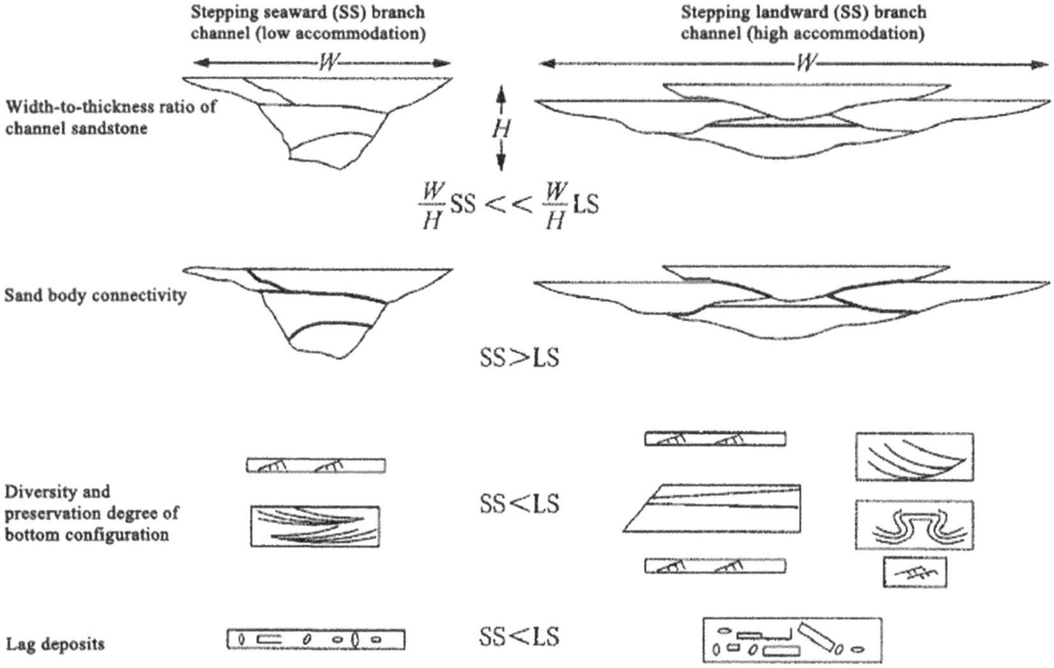

Fig. 4.30 Facies differentiation is significantly reflected in the branch channel in different accommodation spaces (by Cross 1991)

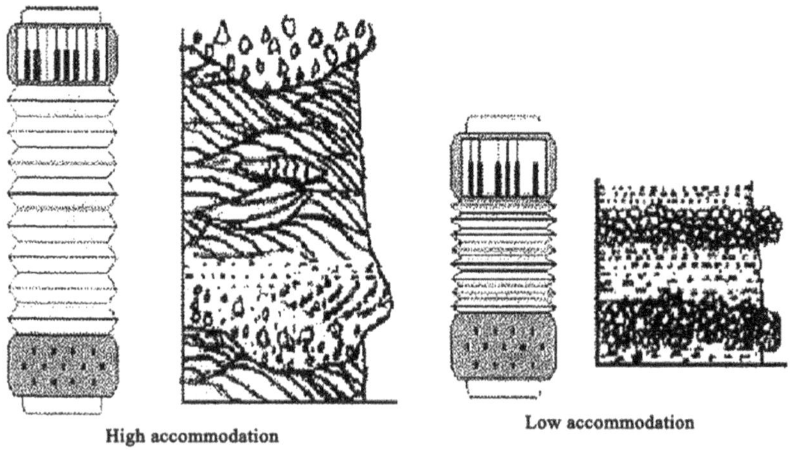

Fig. 4.31 When the ratio of accommodation space to sediment supply changes, the facies differentiation is reflected in the braided river (by Cross 1991)

stratigraphic and sedimentary characteristics of rock resulting from changes in the base level cycle and accommodation. Therefore, genetic sequence correlation is realized by recognizing the location and boundary of a sequence (cycle is on the vertical section) through changes in facies sequence and analyzing the arrangement or sedimentary superimposition style in a continuous sequence space to divide base level cycles into different levels.

The cyclicity of stratum is the bed response of space migration owing to the starvation caused by deposition, erosion, bypass non-sedimentation, and sedimentation non-compensation resulting from changes in the base level relative to the Earth's surface or even the space migration of

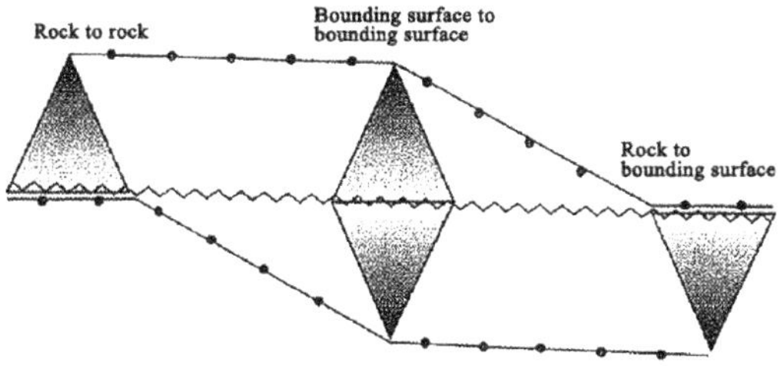

Fig. 4.32 The isochronal principle shall be emphasized by the comparison on base level cycle, involving rock to rock, interface to interface and rock to interface (by Cross 1991)

non-sedimentation over time. A complete base level transit cycle and the increase and decrease of its associated accommodation constitute a complete stratum cycle representing a dichotomous time unit (each part represents an increase or decrease in base level) in stratigraphic records. In some cases, it comprises only an asymmetrical semi-cycle and the interface representing the erosion effect and non-deposition (Fig. 4.32). The cycle is identified on the basis of facies sequence analysis because the vertical facies differentiation of the facies sequence is closely related to the change in accommodation in the base level cycle. Through facies sequence analysis, stratigraphic interface, and facies differentiation, which can indicate the degree of preservation and rate of accumulation of sediments, the symmetry of the stratigraphic cycle, cycle thickening or thinning pattern, amplitude and orientation of facies mismatch at the position where the genetic stratigraphic boundary is crossed, can be recognized, so as to conclude the unidirectional increase and decrease tendency for accommodation that is synchronous with the base level cycle.

Considering the coastal plain-shallow sea sedimentary environment as an example (Fig. 4.33), the declining semi-cycle of the base level is represented by the facies sequence reflecting a sedimentary environment that shallows gradually, and the rising semi-cycle of base level is represented by the facies sequence reflecting a sedimentary environment that deepens gradually. The stratum representing the declining semi-cycle of the base level in the long-term base level cycle is composed of multiple progradation-style short-cycles, and the rising semi-cycle of the base level has a retrogradational structure. On the drilling, outcrop, and electric well-logging curves, stratigraphic stacking patterns can be recognized by comparing the facies components that constitute the facies sequence (short-term stratigraphic cycle).

High-resolution stratigraphic correlation is the correlation of a contemporary stratum and a bounding surface, instead of the connection of cycle amplitude and rock type. T. A. Cross holds that the turnaround point of the base level cycle or the transition position of the base level from decline to rise or vice versa can be taken as the preferable location of the time-stratigraphic correlation. The turnaround point is the extreme position for a one-way change of accommodation to the maximum or minimum value, that is, demarcation of a dichotomous time unit of the base level cycle. Therefore, the turnaround point is presented as a stratigraphic discontinuity surface in some locations in the stratigraphic record and as a continuous rock sequence in other locations. The present position, proportion of rock, and bounding surface are functions of accommodation and sediment supply. As a result, when to conduct the correlation between rock and rock, rock and bounding surface, or bounding surface and bounding surface can be determined by analyzing formation processes. Plotting a time-space diagram is the most effective method for inverting the time-space of a stratigraphic section, which helps with understanding the stratigraphic response (rock + bounding surface) of a geological process (time + bounding surface) and determining the time for correlation between rock and rock, rock and bounding surface, or bounding surface and bounding surface in order to test the reliability of sequence correlation.

Since the stratigraphic record of base level change in an area is presented with multi-level frequency (multi-level cycle) and can cross

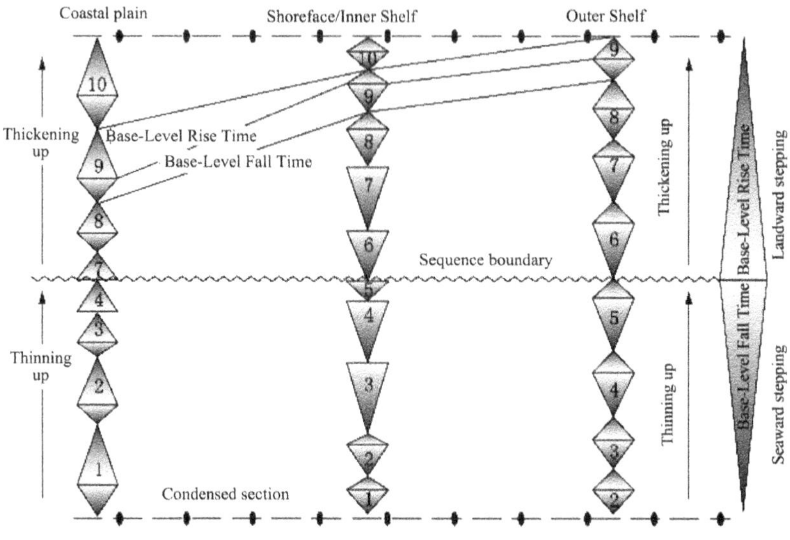

Fig. 4.33 In accordance with comparison on stratigraphic deposition pattern and cycle symmetry, the genetic strata dynamically change in the coastal plain—shallow sea deposit environment (by Cross 1991)

various sedimentary environments, stratigraphic correlation based on stratigraphic base level recognition does not rely on the sedimentary environment, and there is no need to understand the position and direction of movement of the sea level. This is more convenient, hence is used widely in continental stratigraphic correlation.

4.5 Continental Clastic Sequence Stratigraphic Framework Research Method

4.5.1 Basic Data for Continental Clastic Sequence Stratigraphic Research

Continental clastic sequence stratigraphy is a new sub-discipline of sequence stratigraphy, the study of continental sequence stratigraphy is based on seismic, outcrop, well logging and core data, the purpose is to predict lithologic reservoir.

(1) Regional structural background and tectonic evolution of the research basin, along with basic structure and evolution of sedimentation.
(2) Location and particularity of the research area under the action of basin structure and sedimentation.
(3) Fundamental system profile of basin and significant seismic cross section in the research area, preferably with 3D seismic data.
(4) Data of key exploratory wells (including lithology, all types of electric well-logging curves, analysis of depositional cycle and depositional facies of single wells, and unconformities of different scales).
(5) Analysis data relevant to sedimentary environment and paleogeography (including the distribution of depositional facies and vertical development evolution in the entire basin and local area, changes in lake level, palaeoclimate, palaeobathymetry, microelement, and carbon and oxygen isotope analysis).
(6) Data relevant to stratigraphic distribution for outcrop research, main prehistoric life; structure, unconformity, weathering crust, fracture; depositional sequence, depositional facies distribution, facies transition; hydrocarbon generation conditions; source rock distribution, thickness and quality; seal bed and trap, and reservoir formation.

4.5.2 Research Principle

4.5.2.1 Steno's Law

In a stratified rock sequence, any new stratum always lies above an older one. Therefore, rock sequences present an ordinal relation from the bottom to the top or from old to young. This law is applied together with the "lateral accumulation principle" by Weimer because many sedimentary bodies accumulate laterally instead of horizontally.

4.5.2.2 Walther's Law

The corresponding facies association (stratum) can always be found in the vertical direction for horizontally (on a plane) continuous depositional facies (stratum). That is to say, if two depositional facies that cannot develop adjacently horizontally appear and superpose vertically due to rapid migration of the environment, the bounding surface between the two depositional facies is the sedimentary discontinuity. Rapid environment migration can be caused by fast basin subsidence or rapid rise in lake level. Sedimentary discontinuity as defined by this law can be applied to stratigraphic correlation of a parasequence.

4.5.2.3 Fossil Law

A continuous rock unit definitely comprises different fossil assemblages, which can be different or relevant. All other laws relevant to seismic stratigraphy, logging sedimentology, chronostratigraphy, and sedimentology can be applied to sequence stratigraphy research.

4.5.3 General Research Process

Research on continental sequence stratigraphy starts from seismic, outcrop, and subsurface information (including logging and core data) and follows certain steps to ensure that the three sets of information complement each other and converge in some range by establishing a sequence stratigraphic framework in a research area, analyzing the local hydrocarbon distribution pattern, and predicting lithologic stratigraphic traps, which is the goal (Fig. 4.34).

4.5.4 Identification of Continental Clastic Sequence Boundary

In the process of sequence research, the first step is to identify the sequence boundaries of different levels. Based on certain principles and steps, sequence boundaries can be identified by using various boundary identification marks.

4.5.4.1 Response and Identification Marks on Seismic Cross Section

Planes of unconformity are boundaries separating old and new strata. Evidence shows that there is significant sub-aerial erosion truncation or onshore exposure of interruption in deposition along the boundary. Stratigraphical unconformity is presented as a seismic unconformity on the seismic cross section, which can serve as an identifying mark. On the seismic cross section, unconformities are mainly identified in accordance with the reflection-termination of lineups. The typical continental seismic unconformity reflection-termination includes truncation, onlap, and toplap.

1. **Truncation**

Lateral termination of strata due to erosion appears on the top boundary of a sequence. It is both direct evidence of tectonic movement development and the most reliable sequence division index. In addition, it can be reflection-termination between the top of the underlying tilted stratum and the overlying level layer, as well as that of the underlying stratum due to river bottom erosion (Table 4.7).

2. **Onlap**

In the case of increasingly expanding lacustrine basin waters, the bottom of the sequence gradually overlaps upward against the sedimentary slope on the early sequence boundary (Table 4.7). As a reliable mark of the sequence bottom boundary, lakeshore onlap is generally located on the edge of a lacustrine basin and reflects relative rise in lake level.

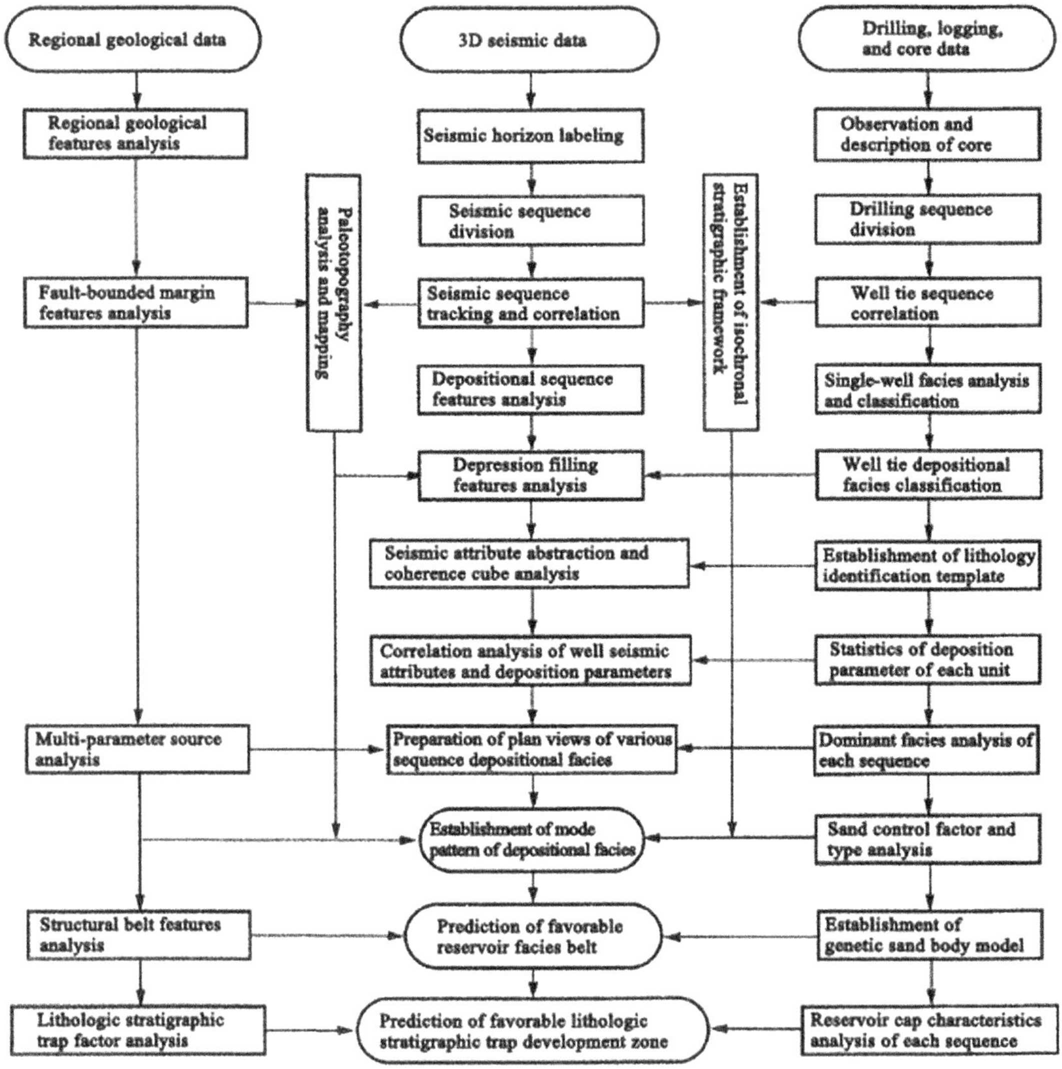

Fig. 4.34 General flow of research on continental sequence stratigraphy and lithologic trap

3. **Toplap**

This means that a top without deposition along the tilted stratum is overlapped by new sedimentation (Table 4.7). In geology, this is a sediment passage phenomenon caused by a very low base level of deposition, and it represents the hiatus of no sedimentation or current-scouring action. It can be seen at the top boundary of the sequence.

In addition, there is less downlap in the lacustrine basin, because continental lacustrine basins have a small area, near the source, and abundant terrigenous detrital supply. Generally, in a lacustrine basin, there is sedimentation of different thickness and fineness. The downlap surface is formed because the deposition rate of far-source mudstone is lower than that of near-source fragments, resulting in the deposition being thick at the edge but thin at the center, and the gradual convergence of reflection lineups moving from the edge to the center. Therefore, the downlap surface in a lacustrine basin is actually a conformity.

Table 4.7 Seismic reflection characteristics of the Jurassic three-level sequence boundary, and maximum lake flooding surface of the Shinan area of the Junggar Basin

Boundary type	Boundary characteristics and classification	Seismic response of boundary	Boundary geological interpretation
Maximum lake flooding surface	Type A apparent truncation	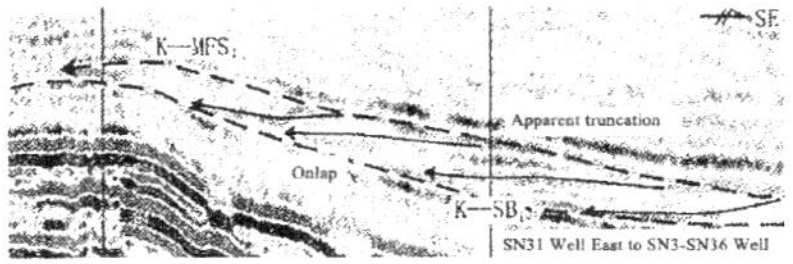	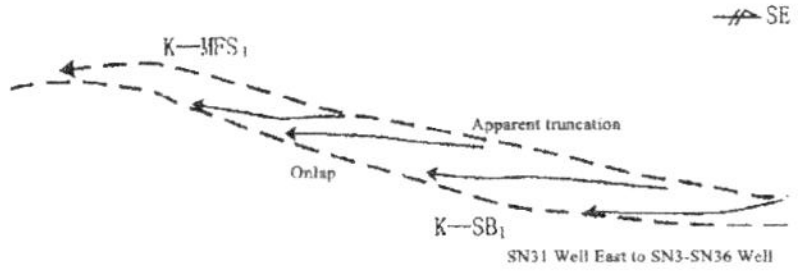
	Type B parallel up and down	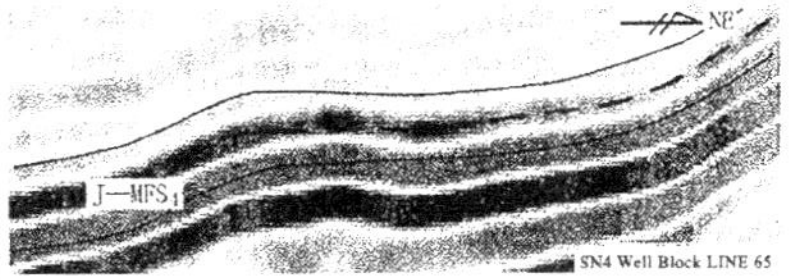	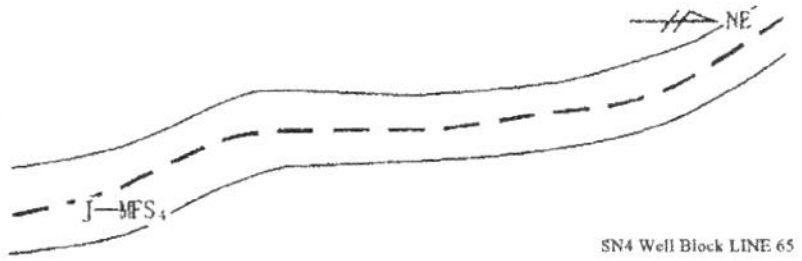
Sequence boundary	Type C parallel/onlap-truncation	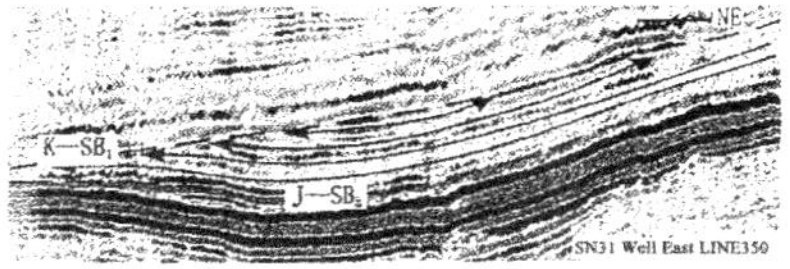	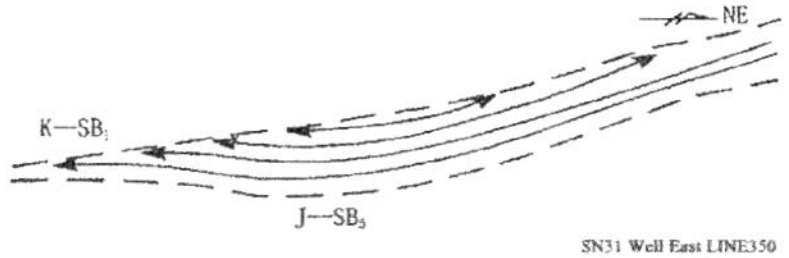
	Type D valley cut	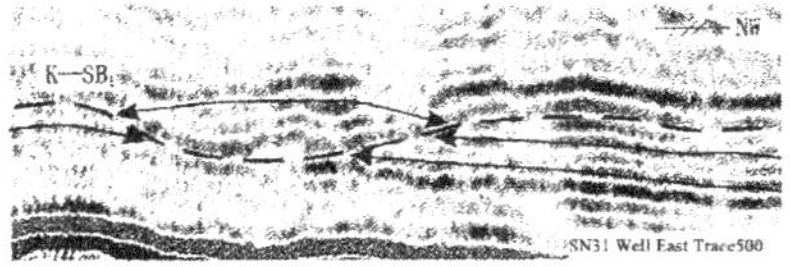	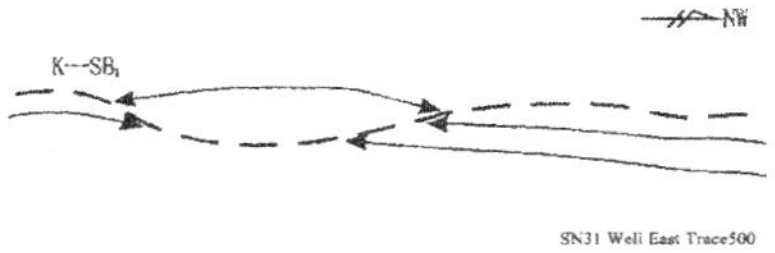

(continued)

Table 4.7 (continued)

Boundary type	Boundary characteristics and classification	Seismic response of boundary	Boundary geological interpretation
	Type E onlap-parallel	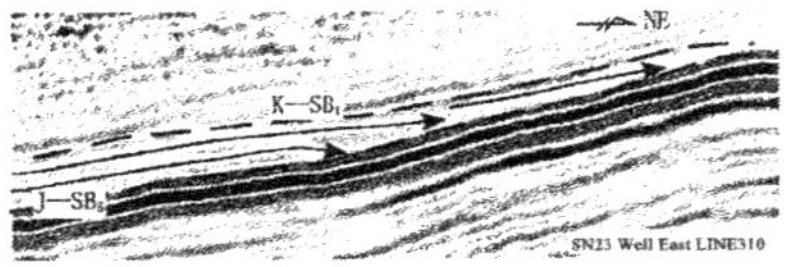	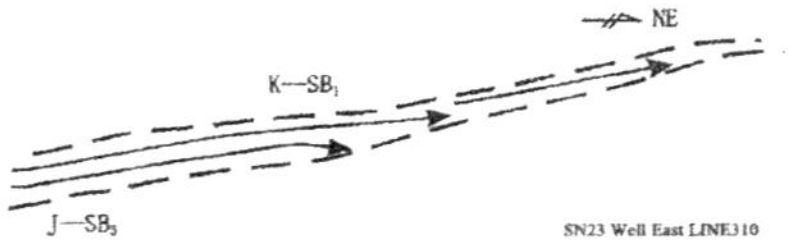
	Type F toplap-parallel	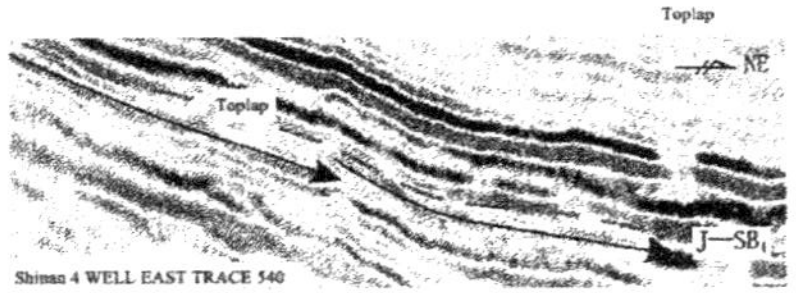	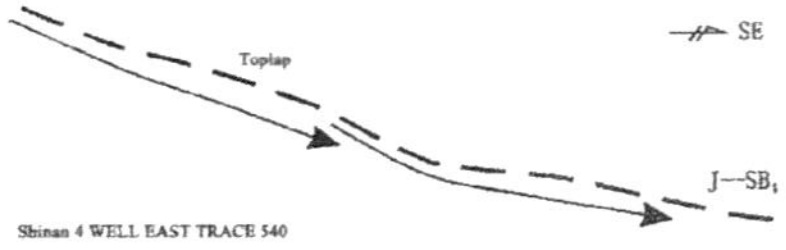

4.5.4.2 Response and Identification Marks of Electric Well-Logging Curve

The key to sequence stratigraphic analysis lies in the identification of sequences at different levels or depositional boundaries. Planes of unconformity, flooding surfaces, and erosion surfaces with isochronal stratigraphic framework characteristics can be identified using logging information. The type of sequence boundary is very complex, and there is a big difference in terms of sequence boundary characteristics between marine and continental basins. In marine basins, the sequence boundary is mainly a large exposed shelf area, entrenched (incised) valley, erosional surface, or facies fault plane. The condensed section in a marine basin is formed in the maximum transgression period and it comprises thin layers of mudstone and limestone with a large distribution area and stable position. In contrast, the condensed section in a continental basin is generally composed of dark pure mudstone of thick lacustrine facies, and coal-bearing deposition is usually developed above and below it. This is to say, different types of sequence boundaries correspond to different logging responses.

1. **Sequence boundary**

On the seismic cross section, the sequence boundary shows truncation, downward migration of the coastal onlap point, and migration of lithofacies toward the basin direction. The identification marks of sequence boundary in a single well mainly include erosion surface, soil layer (root soil layer), entrenched valley, feeder channel and deepwater turbidity accretion, and other submarine erosions. The entrenched (incised) valley is a characteristic symbol of sequence boundary in the shelf area, which has an obvious response on the electric well-logging curve, which is shown mainly as an abrupt erosive basement with a lumpish self-potential curve shape. The entrenched valley is easy and can be identified accurately on the well tie section. Regional unconformity is the fundamental nature of a sequence boundary. A high-resolution dipmeter can directly identify unconformities. Multi-well dipmeter data help determine the presence of widely distributed planes of unconformity. Imaging logging is a more accurate solution for distinguishing unconformities. The sequence boundary is generally an abrupt lithological bounding surface, and the lithologic differences on both sides of the boundary respond significantly on most electric well-logging curves, whose shapes often correspond to a stable abrupt bounding surface (Table 4.8).

2. **Marine flooding surface and parasequence**

As the boundary of a parasequence, a marine flooding surface reflects a sudden increase in water depth. The following factors should be considered in marine flooding surface identification: lithological mutation; sudden increase or decrease in layer thickness; possible scour and erosion; abundance of authigenic minerals, such as glauconite, apatite, and iron pyrite near the bedding plane; and sudden downward increase or decrease in bioturbation. Changes in both lithology and thickness in the parasequence are shown in conventional logging, and they can be determined from the lithologic log and curve shape analysis. The existence of glauconite and bioturbation can be identified by geochemical logging and image logging. On an electric well-logging curve, the responses of different parasequence sets are different. The foreset parasequence set is an upward-coarsening logging combination, the progradational parasequence set is a box-shaped electric well-logging curve combination, and the retrogradational subsequence set is a fining-upward logging combination (see Fig. 4.7).

3. **Condensed Section (maximum marine flooding surface)**

The condensed section is the mark of the top boundary of a transgressive systems tract. It is formed when the deposition rate is considerably smaller than the relative rise in sea level. A very low supply rate of terrigenous sediments leads to the formation of some special rock types, for

Table 4.8 Geological and logging features of a stratigraphic boundary

Boundary	Geological features	Logging mark
Sequence boundary	Identification marks mainly include erosional surface, soil layer (root soil layer), entrenched valley, feeder channel and deepwater turbidity accretion, and other submarine erosions	Abrupt erosion basement and lumpish self-potential curve shape correspond to a stable abrupt bounding surface on the electric well-logging curve
Parasequence boundary (marine flooding surface)	Lithological mutation; sudden increase or decrease in layer thickness; possible scour and erosion; abundance of authigenic minerals such as glauconite, apatite, and iron pyrite near the bedding plane; and sudden downward increase or decrease of bioturbation	Progradational parasequence set is an upward-coarsening electrofacies combination, aggradational parasequence set is a box-shaped electrofacies combination, and retrogradational subsequence set is an electrofacies combination that fines upward
Condensed section (maximum flooding surface)	It has the following special rock types: thin marlstone bed set, thin limestone bed set, glauconite bed set, phosphate rock, thin silt mudstone, and organic-rich mudstone	Thin calcareous mud shale or limestone is presented as a low-natural-potential, high-resistance, high-density, and high-acoustic-velocity layer, mainly as leptokurtic, while relatively pure marine mudstone and lacustrine facies mudstone are presented as low-natural-potential and low-resistance layers

instance, thin marlstone bed set, thin limestone bed set, glauconite bed set, phosphate rock, thin silt mudstone, and organic-rich mudstone. In most cases, the condensed section is tens of centimeters thick and is generally identified by high-resolution geochemical logging. With regard to conventional logging, the condensed section has a significant response, showing high natural gamma, low natural potential, and high uranium content. Logging responses vary with different lithologies. Thin calcareous mud shale or limestone presents a low-natural-potential, high-resistance, high-density, and high-acoustic-velocity layer, which is mainly shown as leptokurtic. In contrast, relatively pure marine mudstone and lacustrine facies mudstone have low natural potential with a low-resistance layer. In addition, the morphological analysis of the corresponding electric well-logging curve indicates that the condensed section is located at the turning point between the upward-fining and upward-coarsening logging responses of a sequence.

4.5.4.3 Sedimentary Geologic Characteristics and Identification Mark

1. **Palaeoweathering exposed surface**

This has a wide distribution and comprises mainly paleosols and plant root soil layers. The weathering crust on an ancient exposed surface is a good mark of a plane of unconformity. The most common palaeoweathering crust is calcareous weathering crust, followed by iron, aluminum, and siliceous weathering crusts, all of which are within the scope of paleosols.

2. **Riverbed lag deposits**

Riverbed lag deposits are the coarse fragments, such as gravel of discontinuous lens, accumulated at the bottom of a riverbed. These coarse fragments are formed upon their movement from upstream or toward the side of an eroded bank nearby. In the process, fine-grained material is moved away selectively. As a result, the bottom of a riverbed lag deposit often shows an apparent erosion boundary, which is the mark of a sequence boundary.

3. **Depositional cyclicity**

This includes fining-upwards cycle sedimentation, coarsening-upwards cycle sedimentation, and complex cycle sedimentation. Different cycle levels are divided based on the rock-electricity feature to determine the sequence boundary position.

4. **Tempestite**

When a lake is located in a vast basin, tempestite sedimentation is produced due to a wide lake surface, shallow water body, and strong storm wave. Sandy tempestite is often developed on the sequence boundary because of large-scale water transgression.

5. **Lithology and lithofacies marker**

The absence and mutation of lithology and lithofacies in a vertical series and the presence of basal conglomerates may be markers of a sequence boundary.

6. **Condensed section**

A condensed section is usually associated with the maximum depth of water in a depositional sequence. It is often formed in the period when the lake level is at its highest and the lakeshore onlap point is the most distant landward, that is, the period when the maximum lake flooding surface is formed. The maximum flooding surface is also the main boundary in the sequence. There is a lacustrine transgressive systems tract above the boundary and a highstand systems tract below the boundary. This is shown as a retrogradational parasequence set below the boundary and a progradational parasequence set above the boundary, respectively, in both well logging and mud well logging data.

7. **Coal bed**

Some scholars hold the opinion that a coal bed with wide distribution can be deemed as a type of sequence boundary because it is analogous to the condensation section in marine basin filling given that peat accumulation conserves coal only when there is a lack of significant clastic deposits, specific construction, and palaeoclimate. At present, the use of the coal bed as an identifying marker of sequence boundary is disputed because different researchers ascribe different levels to the coal bed as a sequence boundary.

4.5.4.4 Paleontologic Markers

1. **Bioclastic (shell) layer**

After the death of creatures with shells that live in shallow water, the shells are moved to the nearby shoreline due to lake waves, and a shell detritus layer is formed under the continuous erosion and crushing effects of lake water. As a result, it is difficult to identify the species involved, and this layer shows disorderly accumulation because the shells have undergone severe crushing. It can possibly reflect a littoral environment. When the stratum above it is a depositional facies reflecting either gradual or abrupt deepening of the water body, these detritus layers can approximately represent the top of a parasequence or a parasequence set, and probably represent the top boundary of a sequence.

2. **Plant root fossil**

Root fossils are ichnofossils, and they are the most easy to identify in cores. Ichnofossils include a great variety of species and complex ecological features. They cannot be deemed as an absolute exposure marker because they are found in mostly products on land or in extremely shallow waters. In the process of sequence boundary identification, the position of a sequence boundary can be deduced based on the vertical intra-stratum changes in ichnofossil contents.

3. **Ichnofossil**

All ichnofossils (except the coprolite) are in situ conservation. They reflect the biological behaviors and the bottom characteristics required for their survival, both of which are directly related to permanent environmental factors and closely related to the depositional environment. By using the sensitivity of biologic traces to environmental conditions, sedimentary discontinuity shown on a lithologic section can be reflected.

4. **Change of biological quantity**

A sequence is a set of strata formed under a certain controlling factor. The quantity of biological materials in a sequence changes gradually from the bottom up, and the sedimentary environment determines whether the quantity increases or decreases. However, there are large differences and mutations in biological quantity between the strata above and below the sequence boundary owing to large differences in lake water depth and depositional environments. Therefore, the mutation of biological quantities in two adjacent layers in a stratum can be used to consider whether a sequence boundary exists.

5. **Change of biological species**

The fossils in upper and lower strata differ greatly in terms of age, or there is mutation of palaeontological oryctocoenosis, which causes discontinuity in organic evolution or mutation of biological species. All these indicate there is a depositional break between strata or long-term erosion and weathering, which evidence the existence of an unconformity (sequence boundary).

4.5.4.5 Geochemical Markers

Due to weathering and exposure in the rock stratum under an unconformity, some elements are often particularly rich or poor, isotopic composition varies, and some salts can be formed. All these can be taken as markers of a sequence boundary.

4.5.4.6 Diagenesis Markers

When clastic rocks are exposed on the Earth's surface, their diagenesis is weak and can be transformed easily by epigenetic diagenesis, thus these markers generally lie in carbonate rock strata. Even so, some special products (for instance, kaolinite and limonite) under the supracrustal diagenetic environment are considered markers of a sequence boundary.

Owing to external interference, some geological and geophysical information shows features similar to that of a sequence boundary with no causality with the sequence boundary. To avoid this situation, mutual corroboration of various data is needed when identifying the sequence boundary to ensure accurate identification.

4.5.5 Establishment of Field Outcrop Sequence Stratigraphic Framework

In a sedimentary outcrop area, sequences and their types are determined and divided according to stratigraphic contact relationships, vertical change in color and lithology, paleosols, stream rejuvenation or land erosion and truncation, and vertical changes in lithofacies. The age of a sequence unit must be determined from the perspective of biostratigraphy, and facies analysis should be conducted on sequence units of different ages by using the concept of accommodation to establish stratum deposit styles of different levels of sequence units and analyze the types and combination features of systems tract types. This would allow one to establish a comprehensive analysis diagram in terms of sequence stratigraphy for certain points in an outcrop area.

4.5.5.1 Identification of Various Levels of Sequence Boundaries of Outcrop

It is necessary to find the markers relevant to an unconformity on an outcrop, and determine the level of the unconformity according to its type and scale in order to identify different types of sequence boundaries and divide the different levels of sequence. For instance, the level of unconformity can be determined based on to whether an unconformity exists because of lithofacies or structure, its impact scope, time of development, and scale of erosion in order to define the sequence boundary meaning represented by the plane of unconformity.

In addition to the identification of planes of unconformity, different sequence boundary types can be identified based on the various characteristics of sequence boundaries.

4.5.5.2 Identification and Division of Sequence Cycles of Different Levels

On the basis of a clear macroscopic sequence framework and depositional system structure, systems tracts, parasequence sets, and parasequences can be divided according to vertical sequence cycle characteristics and the evolution of horizontal depositional facies. The methods used to identify and mark systems tracts, parasequence sets, and parasequences differs based on the depositional setting. Different standard field geological models and depositional facies models for different systems tracts, parasequence sets, and parasequences should be established for regional comparison. Moreover, for field outcrop, visual inspection can be conducted to identify different levels of vertical cyclicity in a stratum in order to give an initial shape to the corresponding macroscopic sequence framework.

4.5.5.3 Outcrop Plane Sequence Stratigraphy and Relevant Geological Research

Stratum exposure permitting, sequence stratigraphy research on multiple outcrop sections should preferably be conducted before a comparative study between regional sequences of different levels in order to establish a regional sequence stratigraphic prediction mode. If multiple outcrop sections are absent, comparison should be performed between an outcrop section and the adjacent well profile to determine the regional regularity of sequence stratigraphy. To this end, one must prepare a plane view of depositional facies, subfacies, and microfacies of sequence-systems tract-parasequence along with a comparison diagram of vertical formations based on the actual requirements in order to analyze the trend of regional sequence stratigraphic changes.

By using the established regional sequence stratigraphic framework and clear distribution and composition of depositional facies in various systems tracts, in combination with the research results of high-resolution sequence stratigraphy, qualitative or quantitative prediction of the distribution characteristics of generation, storage, and capping in the lateral and the vertical directions can be made.

4.5.6 Establishment of Seismic Sequence Framework

The research on seismic sequence frameworks is based on seismic reflection data. Some researchers such as Vail (1977) believe that the seismic reflection boundary is an isochronous surface or an isochronous interface parallel to the internal part of a stratum, and the stratigraphic base level cycle and boundary denote a genetic stratigraphic unit and time surface. Therefore, the seismic reflection boundary is parallel or equivalent to the boundary of the base level cycle. Vail (1987) proposed seven steps for interpretation: ① seismic sequence analysis, ② logging sequence analysis, ③ using synthetic records for well-to-seismic integration and combination with biostratigraphy, ④ seismic facies analysis, ⑤ interpretation of lithofacies and depositional environment, ⑥ forward modeling and inversion of seismic model, and ⑦ final interpretation and industrial drafting.

4.5.6.1 Basis of Establishment of Seismic Sequence Framework

First, it is a requirement to select a standard (backbone) seismic cross section based on regional palaeogeography and tectonic movement. In the new exploration area, detailed seismic sequence analysis of the existing seismic lines should be conducted using comprehensive gravitational, magnetic, and seismic data. Doing so would help select a backbone section net for tracking and identifying seismic sequences from different locations and selecting a preferable sag section in one of those locations as the standard section of the seismic sequence framework. The procedure for selecting the standard section is as follows:

(1) Select sections with complete development, high thickness, less tectonic termination, and ability to extend onto the basin slope as the basis for dividing sequences.
(2) Select sections that have a clear progradational structure and are parallel to the main direction of water flow at the rupture of a lacustrine basin to the greatest extent in order to better identify the sequence and systems tract.
(3) Select one or two measuring lines in each depocenter for analysis in the case of multiple depocenters in order to verify the isomerism of sag sedimentation history.
(4) Select the backbone section that goes through the existing well as much as possible to perform logging-seismic integration and comparative analysis.
(5) Combine the section through the axis of the depocenter and the section transecting the central axis to extend to the depocenter as much as possible.

4.5.6.2 Identification of Seismic Sequence Boundary and Seismic Sequence

The geological properties of a seismic sequence boundary can be determined by drilling (logging) data, while higher levels of a logging sequence boundary can be controlled by the seismic unconformity boundary. In drilling (logging) analysis, although there are many indicators of an unconformity, such as lithology, grain size, sedimentary structure, and electric logging features, they are not as clear and definite as those obtained by seismic analysis. Moreover, the wellhole on the section where an unconformity transforms into a conformity will not show clear evidence of unconformity on logging. Therefore, higher levels of a logging sequence boundary can be demarcated by using the unconformity boundary identified on a seismic cross section.

On the basis of a sequence boundary, the maximum lake flooding surface and the initial lake flooding surface can be determined before the division of a seismic sequence. The maximum lake flooding surface can be determined based on the downlap surface. In the absence of the downlap surface, the onlap apoapsis of a seismic cross section or strong reflection characteristics may be used, while the initial lake flooding surface is the first obvious onlap point at the rupture. After the determination of the sequence boundary as well as the maximum and the initial lake flooding surfaces, the systems tract can be divided. As for a lacustrine basin without rupture, the transgressive systems tract exists below the maximum lake flooding surface and the regressive systems tract above the maximum lake flooding surface. However, for a lacustrine basin with rupture, the lowstand systems tract is below the initial lake flooding surface, transgressive systems tract lies between the initial lake flooding surface and the maximum lake flooding surface, and the regressive systems tract lies above the maximum lake flooding surface. After determination of the systems tract, the parasequence or five- and six-level sequence cycles can be classified and correlated using the principle of high-resolution sequence stratigraphy. However, doing so is difficult because of the resolution ratio of seismic data.

Seismic sequence analysis is performed mainly to identify the seismic sequence, systems tract, and seismic facies according to the reflection configuration of the seismic cross section. The main principles of seismic sequence identification are as follows:

(1) Considering the special reflected wave termination types of a seismic cross section, namely, onlap, toplap, and truncation and erosion, as the main basis for sequence division, and combining the overall internal reflection characteristics. These special reflected wave termination types are key markers of the termination types of stratigraphic geometry.
(2) Identifying various unconformity relationships and their limiting sequences in detail as far as possible on a seismic cross section and combining them to obtain higher levels of sequence sets.
(3) Determining the classification scheme by taking a quality backbone seismic cross section with complete stratigraphic development as the standard, and popularizing it into a working area with poor seismic data.
(4) Utilizing the seismic wave groups that have pronounced features and can be used for tracking contrast in a large scope as the control boundary in order to improve the reliability of vertical seismic sequence division and horizontal comparison.
(5) Taking full advantage of the synthetic seismogram and the converted natural potential curve, apparent resistivity curve, and sonic curve that uses two-way travel time (TWT) as the coordinate for establishing the relation between seismic cross section and drilling section. This would help with analyzing and determining the geological properties of various seismic sequences.

4.5.7 Establishment of High-resolution Drilling (Logging) Sequence Stratigraphic Framework

4.5.7.1 Function of Drilling (Logging) Data in Sequence Stratigraphy

(1) Logging information, especially high-resolution stratigraphic dip logging information, has excellent solution in the vertical direction. With this information, stratum attitude, macroscopic sedimentary characteristics, tectonic features, and lithology and depositional facies features in reservoirs can be researched, and the unconformity characteristics or sequence boundary can be analyzed. In addition, the depositional properties and depositional cycle features of sedimentary strata can be analyzed more clearly through image logging.
(2) Logging information can quantify the vertical division and lateral correlation of a stratum. In the vertical direction, highly accurate and continuous quantitative analysis of an entire well section can be performed using logging information in the vertical direction to reflect the superimposed patterns and the spatial distribution features of a vertical stratum. Horizontally, an electric well-logging curve quantitatively reflects the lithology and combination features of a

stratum. This helps to determine the stratigraphic lateral correlation and analyze the process and form of stratigraphic lateral accretion, thus explaining the deposition process.

(3) The deposition process of a depositional system can be analyzed by establishing the 2D and 3D geometrical morphology of a geologic body through "core calibration logging" and "logging calibration earthquake" in combination with seismic data.

(4) Electric well-logging curves with different investigation depths and based on different fundamental principles can be selected to establish corresponding "electrofacies" and analyze the depositional sequence of a depositional environment.

4.5.7.2 Drilling (Logging) Sequence Analysis Principle

Drilling (logging) sequence analysis interprets a section mainly by using logging information and lithologic strata. Owing to the lack of systematic coring data at work, the focus should be on the following issues:

1. **Selection of key wells**

Given the different positions of wells in a basin, different sequence development characteristics are reflected by wells at different positions in the same basin. Accordingly, when conducting research, it is better to select a well deep in the sag and with complete development to establish a vertical sequence stratigraphic framework for each sag and conduct logging-seismic comparison connection using the depth-time conversion synthetic record of the electric well-logging curve.

2. **Constraint on level of logging sequence by seismic sequence**

Vertically, it is not easy to divide sequences into different levels. Moreover, the sequence number may vary from person to person owing to disturbance from the coexistence of allogenic and authigenic cycles. A reflecting layer is constituted of a formation with wave impedance difference and the mutual interference of different reflection factors. In addition, uncertainty and multiplicity exist in case of comparative interpretation using electric well-logging curves. Interpretation accuracy can be improved by generating synthetic seismic records based on the sonic curve and repeated comparison with actual seismic data from nearby areas. This is a significant step in limiting high-level drilling sequences by using seismic sequences to mark the position of the main plane of unconformity in a logging sequence. Through the analysis of seismic sequences and sedimentary environments, the basin paleogeographic background of wells of different ages can be determined to macroscopically understand the development characteristics of the systems tract of each period. Doing so would help reduce errors when determining the properties of systems tracts. Owing to the different resolutions of seismic and well logging data, the seismic sequence boundary is calibrated during logging, and it corresponds to a depth section on the electric well-logging curve. As for the specific boundary location, further analysis of this depth section on the electric well-logging curve is required. Thus, the reliability of logging sequence analysis can be improved.

3. **Dislocation of base level**

This refers to a condensed section featuring paleontologic development and enrichment of organic carbon. The type of paleontology, degree of abundance, and the amount of organic carbon can help determine the position and primary and secondary sequences of the condensed section.

4. **Comprehensive application of multiple datasets**

Comprehensive judgment is preferable for sequence division by using lithologic sections, rock-mineral analysis, and paleontological data to analyze the formation and evolution history of sequences.

4.5.7.3 Sequence Stratigraphic Analysis Process With Logging Information

Logging sequence stratigraphy is a method for predicting lithofacies and dividing sequence by using logging information. It is characterized by high vertical resolution and can divide most stratigraphic units in sequence stratigraphy (sequence, parasequence set, parasequence, bed set, and bed). Studies have shown that electric well-logging analysis is mainly used for identifying parasequences and systems tracts by employing the isochronal stratigraphic framework established in seismic stratigraphy, biostratigraphy, and sequence stratigraphy studies. Through the analysis of single-well sequences, boundary feature extraction, and sequence boundary recognition, parasequences and systems tracts can be identified better on a through-well seismic profile using the logging information demarcated from the core.

1. **Establishment of isochronal stratigraphic framework**

The objective of sequence stratigraphic analysis is to establish an isochronal stratigraphic framework. A sequence boundary of isochronic surface significance can be identified through seismic stratigraphic analysis. Based on the migration of truncation and the coastal onlap point toward the basin, the seismic sequence boundary can be identified accurately. High-resolution well logging data can be used to divide parasequence and parasequence sets (progradation, aggradation, and retrogradation), and to identify systems tracts. Based on paleontologic data and other dating methods, the age of a sequence stratigraphic framework can be determined to convert lithostratigraphic units into time-stratigraphic units.

2. **Identification of sequence boundary**

The sequence boundary generally has a significant response on the electric well-logging curve, and is commonly shown as an abrupt boundary of lithology. The marker of such an abrupt lithology interface can be found in single-well lithofacies analysis. The regional stable boundary features shown in multi-well comparison on the key section of a depositional strike are necessary for sequence boundary research, and a low-frequency sequence boundary should be available for comparison in the whole basin or on a large scale. High-solution diplogs, under-mine TVs, and micro-resistivity imaging methods can used to effectively identify unconformities. Erosion surfaces have good response on most electric well-logging curves.

3. **Logging in research area and establishment of geological lithofacies knowledge database**

Regional differences exist in rock types in research areas, which mean logging responses differ in terms of lithology and fluid properties. First, a complete logging-geological lithofacies database should be established to realize communication and mutual corroboration of geological data and logging information. The relationships between rock type and electric well-logging curve features are the basis of correct sequence stratigraphy analysis. A logging-geological lithofacies database provides the basic principle for identifying rock type on the electric well-logging curve.

4. **Lithofacies identification of key well and electrofacies analysis**

Lithology can be identified effectively through a comparative analysis of electric well-logging curves and standard lithology libraries. Vertical sedimentary succession from the bottom up can be established by single-well analysis. The transformation of an electric well-logging curve from fine to coarse can be deemed as a mark of parasequence identification and also the foundation of electrofacies analysis. The key well is generally the cored well. By demarcating the lithofacies for electric well-logging interpretation by using core information, and accordingly, establishing an electrofacies template, analysis and research on depositional facies can be carried out.

5. **Establishment of multiple-well key section**

The strike section is established along the center of the main depositional system (e.g., the delta) and the region therein. As for the selection of the depositional strike section, the alluvial fan, fluvial facies, continental slope, and basin plain facies should be avoided to the extent possible; instead, a delta front facies and a continental shelf facies with appropriate sand-mud ratio are preferable.

6. **Prediction of oil and gas distribution**

The horizontal distribution of different units in a sequence stratum is very important for basin analysis. The vertical combination of oil and gas can be predicted according to the distribution characteristics of systems tract for a sequence stratum. In lacustrine basins, it is especially important for determining regional continuity, and in combination with the maximum flooding surface, it is an oil generation indicator for lacustrine facies mudstone. Lacustrine facies mudstone is the most significant regional cap and source bed, and oil-gas accumulation zones can be predicted according to the distribution of systems tract in the sequence. At last, it is essential to predict the oil-gas trap type and development position based on to the interpretation of depositional system and tectonic characteristics.

4.5.7.4 Determination of Short-Term Base Level Cycle of Electric Well-logging Curve

1. **Short-term base level cycle of the core**

To identify the short-term base level cycle of the core, it is crucial to identify the short-term base level cycle by using the core section, and on this basis, to determine the nature of the short-term base level cycle surface. The short-term base level cycle can be identified based on structural changes in the core section, strata termination model, and contact relationships between upper and lower strata (Fig. 4.35).

(1) Channel erosion surface: This surface is formed under the action of erosion when it declines to below the Earth's surface. On the surface, channel block sandstone and lag deposits at the river bottom are often developed, where the difference between the former and the latter lies in the fact that the fragments in the basin are often seen on the water-transgression scour surface with a small range.
(2) As the sequence boundary, onshore onlap migrates toward the basin relative to the underlying strata, shown as coarse sediments of shallow water facies directly covering the fine-grained sediments in deep water.
(3) The turnaround position on the vertical section of rock type or facies association, for instance, the turnaround position from the facies sequence or facies association with water body turning shallow upwards to the facies sequence or facies association with water body turning deep upwards.
(4) The thickness of sand and mudstone presents cyclicity change, for example, under the sequence boundary, the sandstone grain size turns coarser upwards, and its thickness increases upwards. In contrast, mudstone

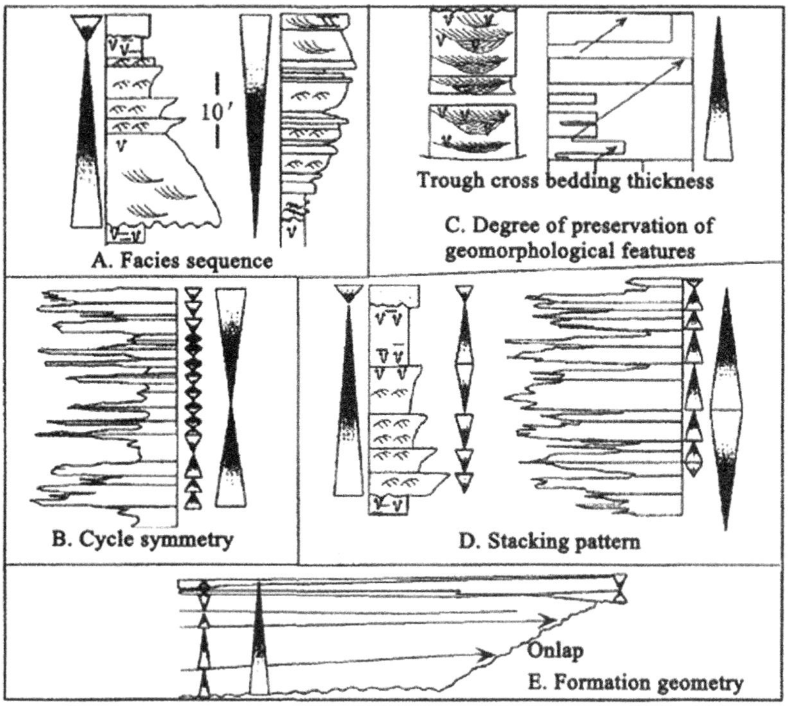

Fig. 4.35 There are mainly five base level cycle identification marks (by Deng et al. 2002)

turns purer and thicker downward, and this situation is opposite above the sequence boundary. The change features of this cycle are often presented in the superimposition mode.

(5) The primary color of mudstone, which reflects water depth, aqueous medium properties, and sedimentary environment, can also be used as the basis to identify the short cycle.

2. **Features of short-term base level cycle of core**

Cored wells located in different depositional environments and phase belts can be identified with different subfacies/microfacies types, for example, subaqueous distributary channels, tidal flat-peat flat, braided rivers, channel bars, point bars, natural levees, and crevasse splays (Fig. 4.36). With specific facies sequences, these subfacies/microfacies constitute a short-term stratigraphic cycle comparable to the short-term base level. The thickness and symmetry of the short-term stratigraphic cycle are the response of different A/S ratio, which presents the specific location of the mid-term base level cycle with a specific A/S ratio. It should be noted that, under the control of the same kind of accommodation (high or low), the lithology, granularity, thickness, and bedding are all different for a short-term cycle comprising different facies sequences.

(1) Short-term cycle with high accommodation: This occurs near the turnaround position from the mid-term base level rise semi-cycle to the base level decline semi-cycle, and the A/S ratio here is high.
(2) Short-term sequence cycle with low accommodation: This is generally developed in two stratigraphic positions of the mid-term base level cycle with a low A/S ratio and the lowest accommodation, that is, near the bottom of the mid-term sequence cycle or near the turnaround point, where the base level transforms from declining to rising.
(3) Short-term cycle with medium accommodation: This is developed mainly under

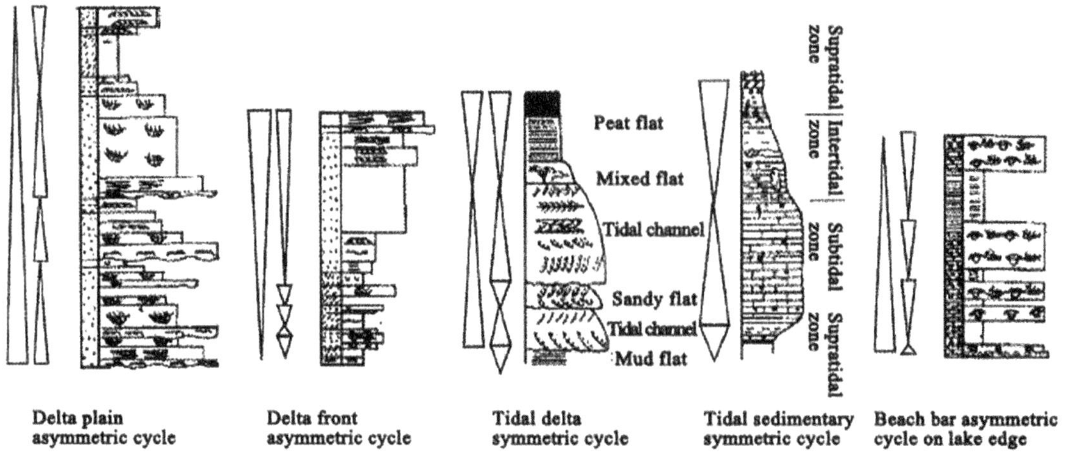

Fig. 4.36 In such special sedimentary environments of delta, tidal deposits and lake, different short-time cycle features can be identified (by Deng et al. 2002)

medium A/S ratios. A short-term cycle with medium accommodation is usually asymmetric or has poor symmetry. A short-term cycle developed by the base level decline semi-cycle is often presented in the mid-term base level rise semi-cycle.

Between the facies sequence or facies association of short-cycle with medium accommodation and facies sequence or facies association of high and low accommodation cycles, a transitional facies association or facies sequence feature is presented often in the form of a change in the thickness of some facies sequence or facies association. For instance, in braided river sandstone deposits, for the short-term sequence cycle with medium accommodation developed by the base level decline semi-cycle of the mid-term cycle, sandstone thickness is greater than that in the short-term sequence cycle formed in high accommodation, but smaller than that of braided river sandstone in the short-term sequence cycle with low accommodation.

3. **Short-term base level cycle of electric well-logging curve**

Sedimentary rock, which is the product of a sediment source and its transport mechanism, as well as depositional environment, can reflect the changes in the physical and chemical conditions of sediments in the process of transportation, deposition, and burial. Its granularity, sorting, and shale content have good corresponding relationships with rock resistivity, natural radioactivity, and natural potential. Due to the correlation between petrophysical parameters and sedimentary characteristics, the electric well-logging curve, which records changes in the petrophysical parameter with depth, can reflect the lithology, thus reflecting the changes in depositional environment factors such as water depth and water energy. This is valuable information from the viewpoint of sedimentary environment identification.

Division of the short-term base level cycle of the electric well-logging curve, particularly determination of the sequence boundary, is conducted by calibrating cored intervals. That is to say, at first, the short-cycle established by a cored interval and the logging response model of the boundary are used to divide the regional electric well-logging curve cycle and determine its base level (Table 4.9).

Based on the results of electric well-logging curve morphological analysis, the base level cycle mainly involves amplitude, shape, smoothness, and contact relationships of the curve.

Table 4.9 Characteristics of base level cycle boundary and superimposition style on electric well-logging curve

Type	Identification features	Electric well-logging curve and base level feature
Identification of short-term cycle boundary	Low gamma and high resistivity	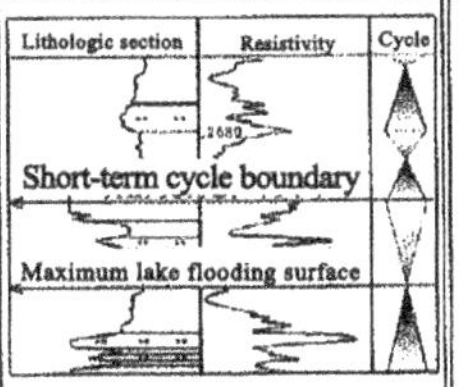
	Base level transformation of channel sedimentation	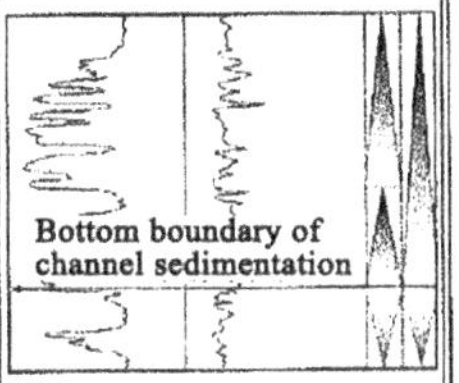
	Contact of shallow-water sedimentation and deep-water sedimentation	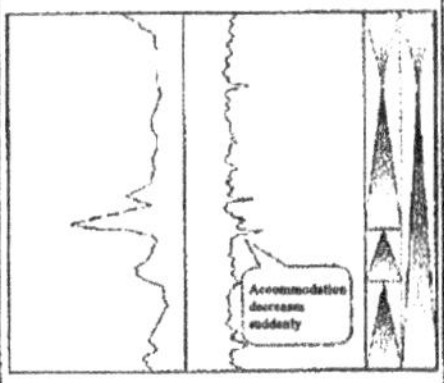
	Transformation of lithofacies type	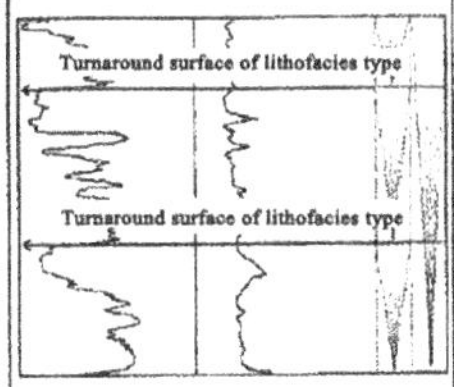
Short-term superimposition-style cycle	Retrogradational superimposition-style cycle composed of retrogradation	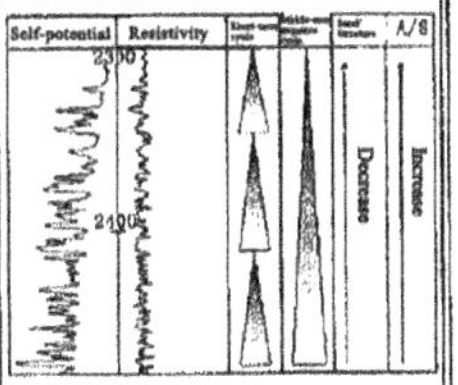
	Retrogradational superimposition-style cycle composed of retrogradation	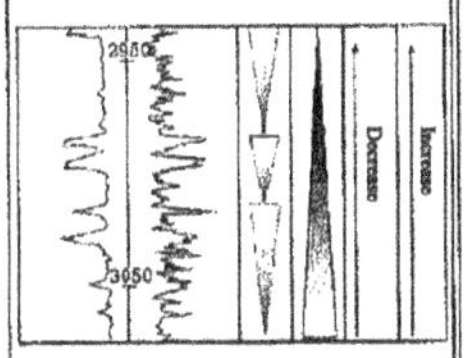
	Retrogradational superimposition-style cycle composed of retrogradation	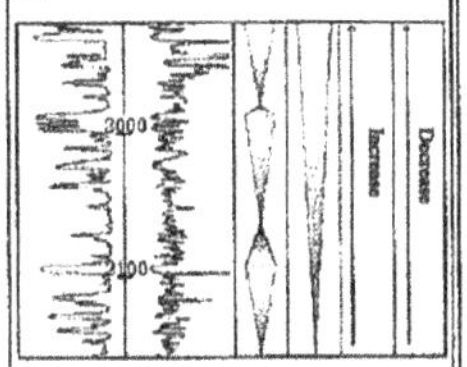
	Retrogradational superimposition-style cycle composed of retrogradation	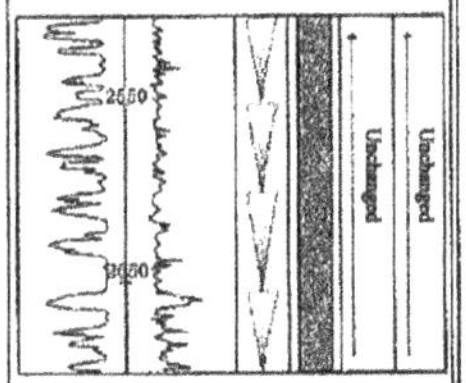
Gradual change in superimposition style	Progradation-retrogradation type	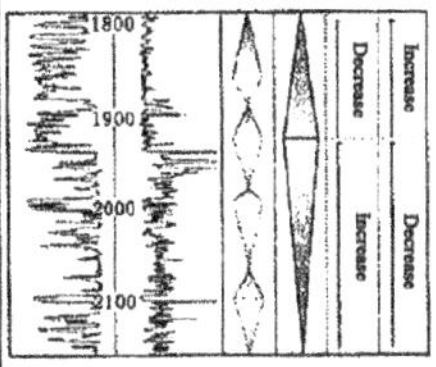
	Progradation-aggradation type	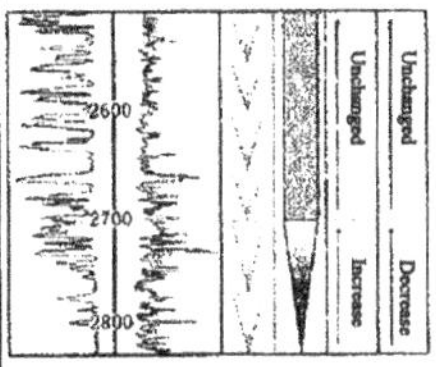
	Retrogradation-progradation type	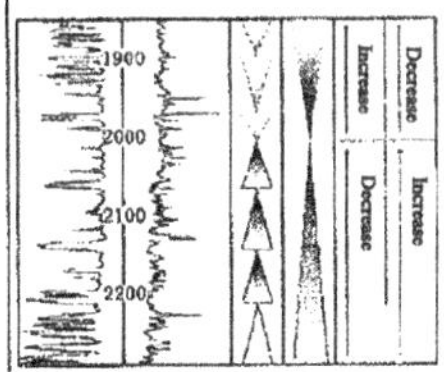
	Aggradation-progradation type	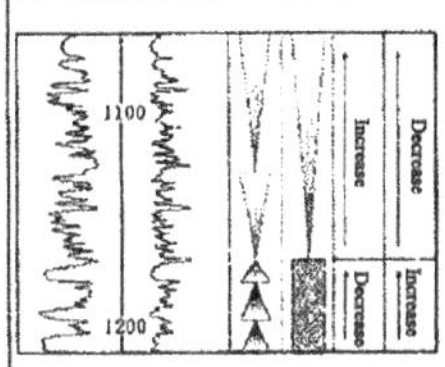
	Aggradation-retrogradation type	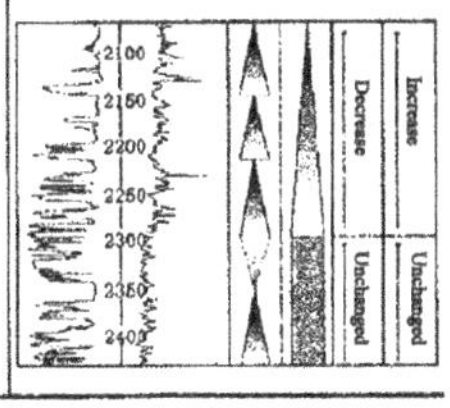

4.5.7.5 Determination and Features of Mid-Term Base Level Cycle on Electric Well-Logging Curve

1. **Mid-term base level cycle on electric well-logging curve**

A set of short-term base level cycles with a specific superimposition style constitute a mid-term base level cycle, which is a set of genetically related rock associations formed on roughly similar geological backgrounds, and these superimposition styles often have distinct logging responses (Fig. 4.37). The superimposition style (progradation) driving seaward (bas-inward) is formed as the mid-term base level declines, A/S < 1, which means the sediment supply rate is greater than the rate of increment of accommodation. As for the adjacent short-cycle, its overlying short-cycle has smaller accommodation compared with the underlying cycle in terms of the sedimentology and petrology properties; the superimposition style driving landward is formed as the mid-term base level rises, A/S > 1, which means the rate of increment of accommodation is greater than the sediment supply rate, and the overlying short-cycle has larger accommodation compared with the underlying cycle in terms of the sedimentology and petrology properties. The short-term base level cycle shows the aggradational superimposition style in the period in which the mid-term base level transforms from rising to declining, A/S = 1, which means the change in accommodation is small as the adjacent short-term cycle is formed. The mid-term sequence cycle can be determined by using the three abovementioned superimposition styles of short-term cycles.

2. Features of logging mid-term sequence cycle

According to the stratigraphic rock features in the research interval, the natural gamma electric well-logging curve-centered, resistivity and natural potential curve-assisted logging system is selected to divide the stratigraphic base level. Determination of the short-term base level cycle from the electric well-logging curve, especially, determination of the sequence boundary properties, is conducted on the basis of calibration of rock-sonic wave properties of the cored interval. That is to say, at first, the cored well is used to establish lithofacies association of the short-cycle and the corresponding logging response model; then, it can be applied to non-cored wells.

Determination of the mid-term base level cycle is based on the analysis of superimposition styles of the short-term base level cycle, which generally corresponds to a four-level sequence. The superimposition styles of short-term base level cycles of the mid-term base level cycle differ in different intervals in different areas. Several typical types and features are described below.

(1) Retrogradational-progradational symmetry: On the electric well-logging curve, this type is shown as a gradual change to the progradational superimposition style from the retrogradational superimposition style. The corresponding lithologic section is shown as gradually thinning single-layer sandstone under the cycle, and mudstone increase, higher A/S, finer granularity, and mudstone (intercalation thickness increased); the part above the cycle features single-layer sandstone with gradually increasing thickness, coarser granularity, lower shale content, thinning mudstone interlayer, and lower A/S. The cycle symmetry axis or the base level cycle is at the turnaround position from rising to declining and at the middle with the thickest mudstone, that is, the position of the condensed section. Based on the analysis of the electric well-logging curve and core information, it can be known that the rising and declining cycles of this combination are composed of river and lacustrine facies sedimentation, whose characteristics are slightly different.
(2) Retrogradational-progradational asymmetry: This type of electric well-logging curve mainly features the retrogradational

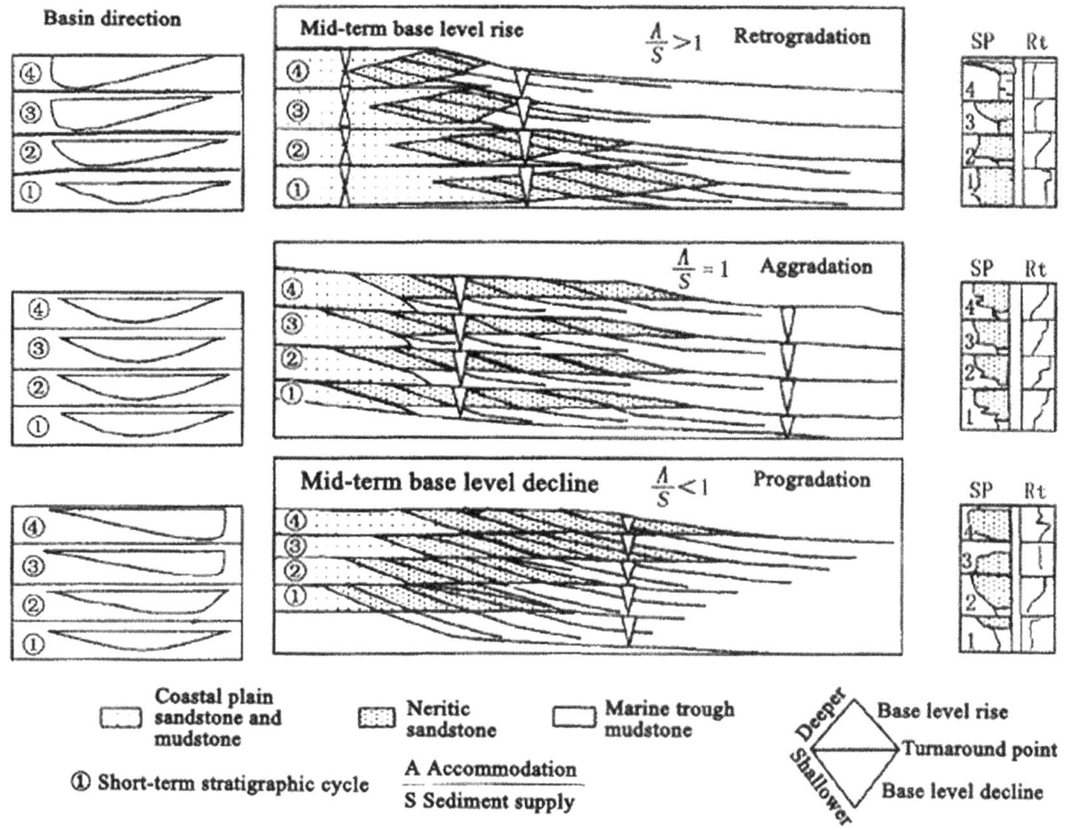

Fig. 4.37 A group of short-term base level cycle combinations with the specific superposed patterns constitutes the medium base level cycle. These superposed patterns often have bright logging responses (by Gardner 1964), SP spontaneous potential, RL lateral resistivity

superimposition style gradually turning into the progradational superimposition style. The curve is divided into the following two types according to the retrogradation: progradation ratio in the mid-term base level cycle.

① Retrogradational-progradational asymmetry dominated by retrogradation. In this type, below the mid-term sequence cycle, sandstone decreases upward, mudstone increases, bottom sandstone increases gradually, mudstone decreases gradually, and the asymmetric axis is located at the location where the mudstone is the thickest. The symmetric axis is located at the bottom, and its short-cycle number only varies above and below the asymmetric axis, with the bottom outnumbering the top.

② Retrogradational-progradational asymmetry dominated by progradation. As for the top and the bottom of the symmetric axis of this type of mid-term base level cycle, progradation is dominant at the top of the symmetric axis. The number of short-term cycles on top of the cycle is considerable larger than that of the bottom of the cycle, top mudstone of the cycle decreases upward, and sandstone turns thicker; all of the above signs are reversed at the bottom.

Additional types include progradational-retrogradational symmetry, progradational asymmetry, and retrogradational asymmetry. Therefore, the types of mid-term sequence cycles and the degrees of development of the various types, which are determined by comprehensive analysis according to specific paleogeographic positions and differences in depositional environments, can be generalized.

4.5.7.6 Determination of Long-term Base Level Cycle of Electric well-logging Curve

The long-term base level cycle reflecting changes in the base level cycle, which is longer than the mid-term base level cycle, is formed by the mutual superimposition of several mid-term base level cycles with the same geological background and sedimentary characteristics. The long-term base level cycle generally corresponds to a three-level sequence, and it can be compared with a seismic sequence.

Different levels of sequence cycles have different combination features. When the single-well sequence cycle is divided, it fully integrates earthquakes, cores, and single-well sedimentation microfacies types, as well as the vertical distribution law of sand bodies. Thus, each level of the sequence cycle can be demarcated comprehensively (Fig. 4.38).

4.5.7.7 Correlation between Seismic Sequence Stratigraphic Framework and Drilling (Logging) Sequence Stratigraphic Framework

Seismic sequence and drilling (logging) sequence are two different approaches for analyzing the sequence stratigraphic framework within a basin. Synthetic seismogram or logging time-depth conversion data should, therefore, be used to demarcate the electric well-logging curve in the well-tie seismic section to test the results of seismic and logging sequence division. This would help to complete the basin seismic-drilling (logging) comparison scheme and establish a comprehensive sequence system within the basin. This is of great significance from the viewpoint of continental sequence stratigraphy research. The reason for this is that many sediments in continental strata, particularly in fluvial strata, are shown as nearly parallel horizontal reflections. Therefore, the plane of unconformity used commonly for continental margin or sea (lake) clinothem deposits or the seismic reflection termination used to reflect stratum discordance can hardly be adopted for identifying a stratigraphic sequence and its boundary. Seismic strata can be determined by converting natural potential and resistivity to two-way transmission time coordinates by using synthetic records in order to demarcate the well-tie seismic section, as well as by combining seismic marks at the stratigraphic cycle boundary and regional variations in seismic facies (Figs. 4.39 and 4.40).

Within a basin, although relatively great variations are noted on the surface for facies association within LSC and MSC, tracing and comparison within the basin can be conducted in the basin association because variations in thickness and cycle symmetry are relatively small. Meanwhile, because stratigraphic units that can be distinguished by high-resolution 2D and 3D seismic sections can also correspond to long-term and middle-term sequence cycle facies that can be identified by logging and drilling, the properties of LSC and MSO can also be used to explain and predict seismic facies and geometrical morphology, and LSC and MSC can be taken as the basis for establishing and comparing high-resolution stratigraphic frameworks. On the seismic cross section, closed tracing of the determined long- and middle-term sequence cycles is performed, and the tracings are projected to the plan view to obtain plots of time-stratigraphic units.

4.6 Application of Clastic Sequence Stratigraphy in Hydrocarbon Exploration

With the development of seismic stratigraphy and sequence stratigraphy, especially high-resolution sequence stratigraphy, hydrocarbon exploration

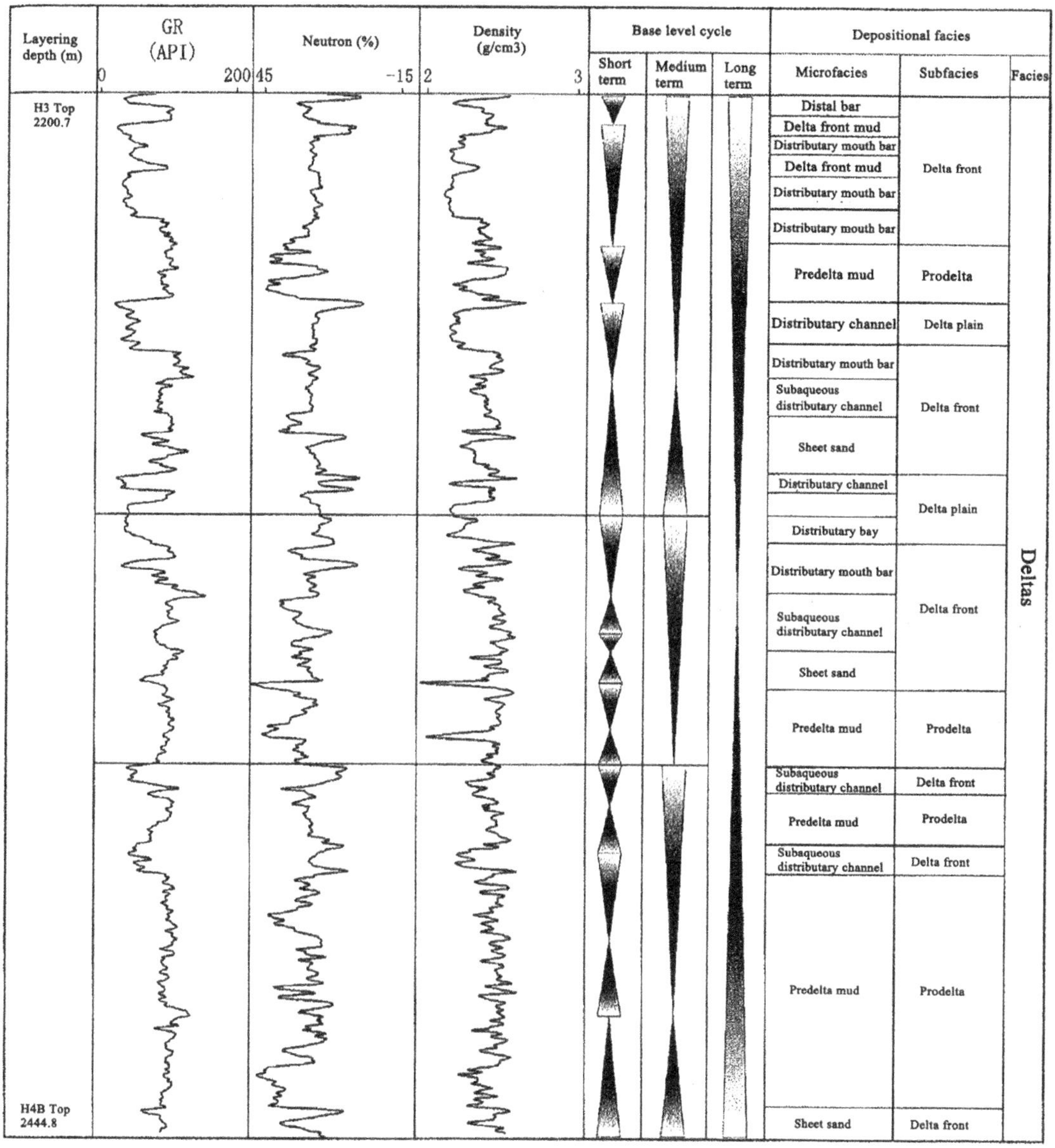

Fig. 4.38 This is complex chart of Miocene II Oil Group LSC2 cycle division and sedimentary microfacies in certain Well of Xijiang 30-2 oilfield

has entered a completely new stage which is characterized by exact prediction and accurate description. At the moment, sequence stratigraphy has gradually developed into an effective tool for studying the evolution of petroliferous basins and sediment distribution patterns, and accurate prediction and description of favorable reservoir bodies, and good application effects have been achieved. Presently, sequence stratigraphy theory and research methods have been applied successfully in many domestic and overseas subtle hydrocarbon reservoir explorations. By extensive research on sequence stratigraphy during analysis of petroliferous basins and hydrocarbon

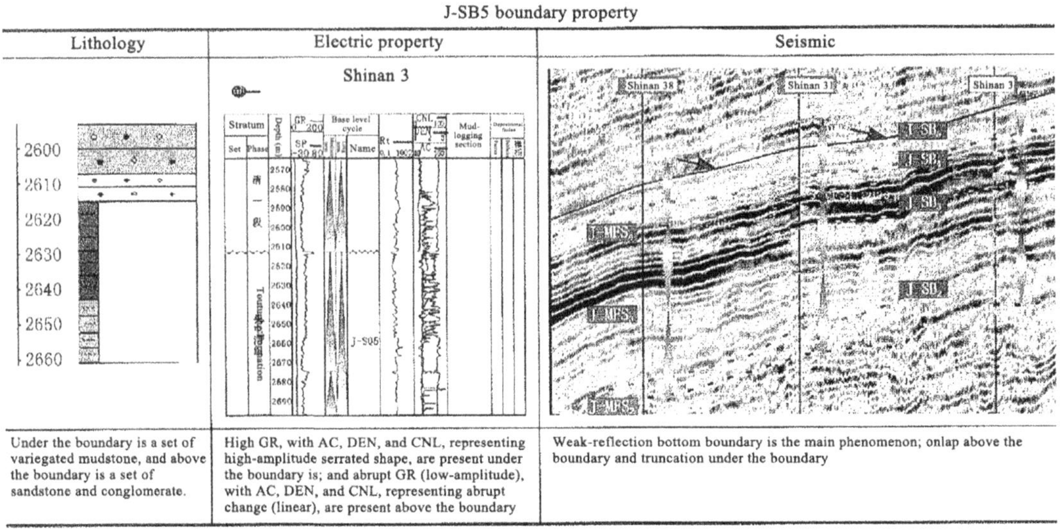

Fig. 4.39 The chart mainly describes well and seismic sequence stratigraphic frameworks in Shinan area of Junggar Basin

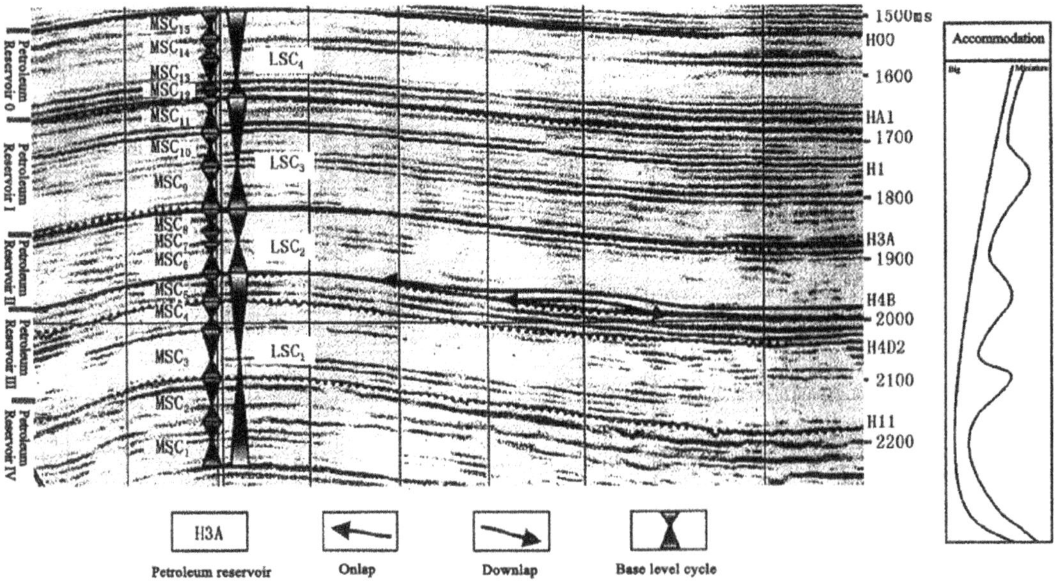

Fig. 4.40 The seismic section mainly describes Miocene seismic sequence identification and division of the certain well in Xijiang 30-2 Oilfield. LSC long-term sequence cycle, MSC middle-term sequence cycle

exploration, China's petroleum geologists have gained a deeper understanding of the filling features of basin depositions and the stratum distribution law.

4.6.1 High-Resolution Sequence Stratum and Isochronal Stratigraphic Framework

4.6.1.1 Isochronal Stratigraphic Framework

Identification of the sequence boundary, division of stratigraphic sequence, and establishment of isochronal stratigraphic framework are the basis of all geological research. Sequence stratigraphy puts emphasis on the comprehensive application of outcrop, drilling/logging, and seismic data to establish chrono-stratigraphic frameworks at the basin or regional scale, and it can be used to analyze formation conditions, development characteristics, and distribution laws in stratigraphic frameworks.

The establishment of a sequence stratigraphic framework requires comprehensive research, which includes analysis of drilling sequence stratum, seismic sequence stratum, and outcrop sequence stratum, and a combination of the three. First, identification and comparison of planes of unconformity must be performed to determine isochronal frameworks of third- or higher-order sequences. Later, with the shortening of the stratum time interval, prehistoric life, depositional cycle, and lithofacies changes can be used to determine the isochronal frameworks of fourth- or higher-order sequences.

4.6.1.2 Depositional System and the Spacing Configuration Relation

In a sequence stratigraphic framework, it is necessary to identify the structure and configuration relationships of depositional systems tracts and recognize the properties and distribution characteristics of each depositional system within each sequence, establish depositional systems, determine the relationships among the systems, and predict the distribution of reservoir rocks and the location of possible stratigraphic lithologic traps.

4.6.1.3 Development Conditions and Combinations of Source-Reservoir-Cap Rock

Sequence development is based on certain rules and it possesses certain characteristics. Different formations and parts of a sequence have diverse depositional conditions, as well as lithofacies and lithologic characteristics, which play different roles in hydrocarbon generation and accumulation, resulting in the formation of predictable distribution rules and configuration relationships of source-reservoir-cap rock.

4.6.2 High-Resolution Sequence Stratum and Hydrocarbon Exploration

All system tracts within a sequence can potentially support reservoir development, and hydrocarbon reservoirs are formed if favorable reservoir-forming assemblage conditions are available.

The highstand systems tract is characterized mainly by sediment progradation with a developed delta, fan delta, and channel. Delta front sandstone and branch channel sandstone are the most common reservoir sand bodies, with their flooding surfaces acting as the seal bed. Although seepage can occur in the lateral direction, hydrocarbon reservoirs can take shape with some faults and structures, and a few sedimentary anticlines and channel sands can also form traps (Table 4.10).

Under wave action and repeated scouring, shoreline sand bodies of the lacustrine/marine transgressive systems tract are characterized by sound sorting and rounding property, and they are parallel to the shoreline. Retrogradation during the water transgression period is beneficial for the formation of sand-mudstone in interbedded combination, and the top and lateral sides transform gradually into the condensed section, which can act as a good seal bed. However,

Table 4.10 Hydrocarbon exploration characteristics of highstand systems tract (Sangree et al. 1990)

Reservoir	Source bed	Seal bed	Migration	Trap
Mainly features discontinuous fluvial facies, delta facies, followed by shoreface facies	Generally acts as deep source bed; shale within the systems tact is poor in oil and rich in raw gas in most cases	Leakage into transgressive systems tact occurs in updip direction with lateral seepage, and the top seal bed serves as the maximum flooding surface	Gas and small quantity of oil generally originate from homochronous source bed, and vertical fault passage is often required for migration of good oil sources	Mainly involves structural trap, and early formation time is the key

Table 4.11 Hydrocarbon exploration characteristics of lacustrine/marine transgressive systems tract (Sangree et al. 1990)

Reservoir	Source bed	Seal bed	Migration	Trap
Lakeshore/coast-shoreface facies sand body is characterized by sound porosity and permeability, and lagoon facies have variability, which can be used for reservoir prediction with linear extension	Good source bed on the top and along lateral direction	Good seal bed on the top and varying seal bed on side and bottom	Downward and lateral migration	Stratigraphic trap for isolated sand body; structural trap is required for continuous transgressive trap at the bottom

closed conditions at the bottom and the nearshore lateral direction change, and an isolated sand beach can form a stratigraphic trap, while a continuous sand body can form a structural trap. During the water transgression period, sediments overlap toward the shore, and they may form an up-dip pinch-out hydrocarbon reservoir (Table 4.11).

The lowstand systems tract is of special significance in hydrocarbon exploration. On the one hand, basin floor fans, slope fans, progradational complexes, underwater canyons and incised valleys (rivers), and other sedimentary bodies (Table 4.12) possess their own hydrocarbon exploration characteristics during the lowstand period.

Distribution of reservoirs on the surface within the lowstand systems tract is limited, and lateral sand body extension is short and generally spliced with mudstone in deep-water environments. Lithologic traps are likely to form under such circumstances. On the other hand, the lowstand systems tract is on the lower basin slope, where the lacustrine/marine transgressive systems tract and the condensed section are developed as a good regional seal bed and oil-source rock, respectively. The presence of the lowstand systems tract within an area is important from the viewpoint of evaluating the hydrocarbon potential of the area.

In recent years, some scholars have held that the structural slope break belt is a critical position favorable for hydrocarbon trap formation. In particular, sublacustrine fans and delta sand bodies of the lowstand systems tract within the faulted slope-break zone are significant sandstone reservoirs. It should be indicated that the main systems tracts rich in oil and gas (particularly lithologic hydrocarbon reservoirs) within different lacustrine basin sequence frameworks may be different. According to some frameworks, it may mainly be the lowstand systems tract; according to others, it may mainly be the lacustrine transgressive systems tract or the highstand systems tract. The Erlian Basin is a typical example within which the highstand

Table 4.12 Hydrocarbon exploration characteristics of lowstand systems tract (Sangree et al. 1990)

Sedimentary bodies	Reservoir	Source bed	Seal bed	Migration	Trap
Basin floor fan	This sedimentary body possesses sound porosity and permeability, and variable continuity; channel sand body in the upper part is a common concern	Shale that may originate from the top and lateral sedimentary section	Good deep-sea shale with slow deposition; no seal bed in case of slope fan cover	Vertical migration from deep source rock, or downward and lateral migration of sedimentary section	Generally stratigraphic trap
Slope fan	Thickness of channel sand is 540 m, that of the floodplain facies sand is (1–3 cm), and the channel sand is discontinuous, while the floodplain may be widely distributed but hard to identify and evaluate	Uncertain, and may come from the deep	Internal shale is sealed, the top can be sealed by shale from the sedimentary section, and floodplain sand is under the restriction of natural levee and skirt-shaped pinch-out	Uncertain, and may be vertical migration through fault passage or vertical migration in lowstand systems tract	Generally stratigraphic trap, but sometimes structural-stratigraphic trap
Progradational complex	Variable overlapped river, delta, and shoreface facies with variable continuity	Source bed in transgressive systems tract in the deep or at the top	Well-sealed top of transgressive systems tract, possibly poor lateral sealing	Deep oil source; migration through fault passage or downward migration from transgressive systems tract	Generally structural trap, possibly with compaction trap
Underwater canyon	Changeable, underwater channel sand and turbidite, poor continuity	Uncertain and homochronous oil source bed may be rich in gas	Partial shale sealing	Uncertain; vertical migration through the fault may be optimal	Stratigraphic pinch-out
Incised valley	Generally braided river sand, with good or perfect continuity	Top oil source may come from transgressive systems tract, possibly with deep oil source	Shale of transgressive systems tract with poor lateral sealing	Downward migration of transgressive systems tract or possible vertical migration through the fault	Closed nose structural trap is generally required

systems tract is more beneficial for hydrocarbon accumulation because the lowstand and lacustrine transgressive systems tracts have poor physical properties from the viewpoint of reservoir formation and lack of effective source rock, while fan deltas and sublacustrine fan deposits are generally formed in the highstand systems tract, with the lower transgressive systems tract acting as the effective source rock.

4.6.3 Sequence-Controlled Reservoir Modeling in Oilfield Development

In oil and gas field development, sequence stratigraphy is mainly applied for research on fine reservoir comparison and reservoir geological modeling, as well as for studying the factors controlling reservoir properties and reservoir heterogeneity.

Reservoir lithology, geometrical morphology, continuity, and physical characteristics are formed during sediment accumulation. Better understanding of these characteristics can be gained in four-dimensional space through accurate stratigraphic correlation, which is an effective method for identifying reservoir heterogeneity. Full consideration should be given to sequence reservoir modeling owing to the differences in the sand body superimposition styles of different accommodations (Fig. 4.41) within a sequence stratigraphic framework. Meanwhile, a stratigraphic boundary with significance in time is generally identical with the petrophysical surface of a fluid flow unit, and accurate stratigraphic correlation should be used to divide such flow units. With the improvement of temporal resolution, predictions of stratigraphic geometry and scale, facies location, and petrophysical characteristics will be more accurate.

Rock characteristics control fluid flow. It was generally believed that depositional environment is the only factor controlling reservoir characteristics, while dynamic processes play a greater role in controlling stratigraphic characteristics. Variations of petrophysical characteristics in a genetic sequence with stratigraphic time can be predicted from stratigraphic geometry. Effective research on spatial and temporal distribution of sand bodies and mudstone barrier beds can be realized by combining high-precision sequence stratigraphy and reservoir structure analysis.

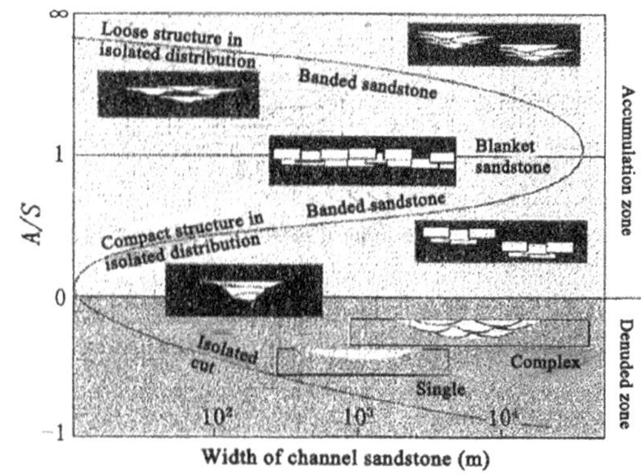

Fig. 4.41 Different accommodation spaces have diverse sand body superposition ways in the sequence stratigraphic framework (by T. A. Cross, 1991)

Yang C.G. (2001) analyzed the reservoir structure characteristics of the Gangdong Development Zone using high-resolution sequence stratigraphy. His results indicate that superimposition of the base level cycle of different grades constrains the distribution of favorable reservoir intervals, and different positions of the short-term base level cycle in the middle-term sequence cycle largely determine the stratigraphy and sedimentology characteristics of sediments within the short-term cycle, including sediment thickness, stratum preservation, stratum stacking pattern, lithofacies distribution and facies type, and petrophysical characteristics. During the early rising stage of the base level cycle, thick and coarse clastic channel sand body sedimentation occurs under low accommodation. The sand body is characterized by lateral superimposition, wide distribution, and sound connectedness, with separation in terms of variations in the heterogeneity of sand bodies and in the action of thin mudstone

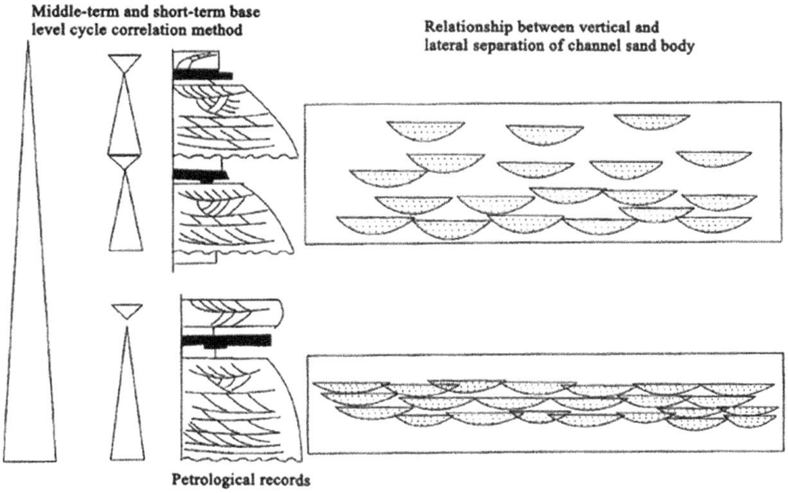

Fig. 4.42 There are different degrees of development of channel sand bodies and connected conditions in different accommodation spaces (by Yang C.G. 2001)

intercalation. During the middle stage of the base level cycle, the channel sand bodies acting as reservoirs are isolated from each other and each single sand body is a reservoir unit. The distribution range is small, and separation is shown in terms of the "sand-in-mud" characteristic. Based on this, a model of lateral distribution and separation characteristics of channel sand body reservoirs in a sequence stratigraphic framework can be established (Fig. 4.42). The characteristics of such a model are as follows: during the early rising stage of the base level, most channel sand bodies are cut by and superimposed with each other, and the connectedness among sand bodies is good, while local residual muddy intercalation separates the petroleum reservoir, which is distributed over a large range. During the middle rising stage of the base level, the formed channel sandstones are isolated from each other, and each single sand body acts as a reservoir unit. Reservoirs are, therefore, confined to individual sand bodies.

Identification of the boundaries at all levels is the key to using high-resolution sequencing to divide a base level cycle into short-term, ultra-short-term, or lower-level units. Therefore, detailed core boundary identification should be conducted in a coring well, and the corresponding logging responses should be used to divide the base level base cycle of the electric well-logging curve. Minimum-level sequence and genetic sand body (depositional microfacies) are simple and practical to use. Thick sand bodies can be divided by combining the requirements of reservoir numerical simulation and the simulation results. Under the condition of small well spacing and dense well patterns, the sand-to-sand and mud-to-mud phenomena are common but not absolute. Large well spacing and variations in structural positions may result from sand-mud within the same period, as has been proven by using the classic stratigraphic sequence model. However, some uncertainty remains in this regard. Changes in depositional microfacies and the superimposition relationships of sand bodies should be considered for correlating large sand bodies. Thus, full consideration should be given to block size, structural position, depositional microfacies, and size of well spacing for sand body correlation during high-resolution sequence division (Li et al. 2004) (Fig. 4.43).

In the figure, GR refers to electric well-logging curve, scale range is 0–200, and S_0 refers to oil saturation. A-correlation mode for a

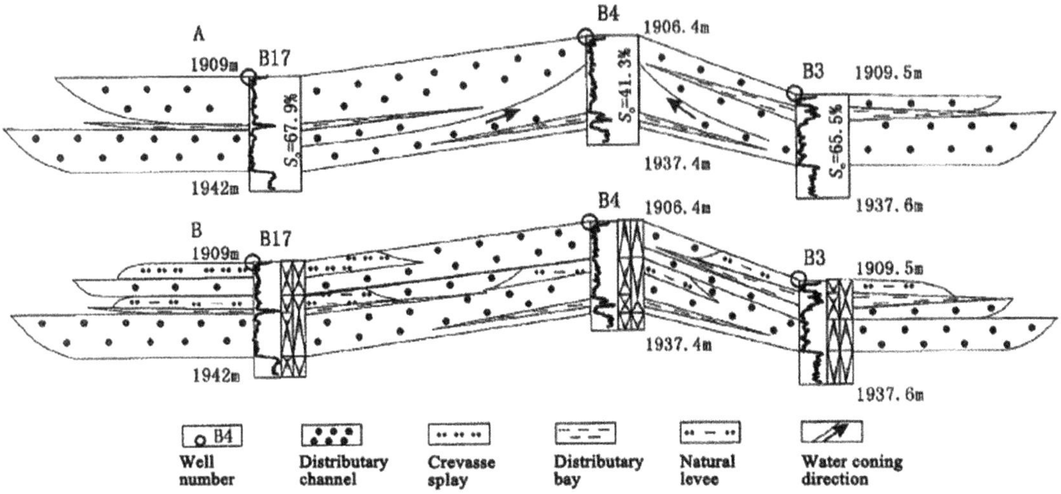

Fig. 4.43 There are different degrees of development of channel sand bodies and connected conditions in different accommodation spaces (by Li et al. 2004)

sand body with conventional lithological characteristics; and B-correlation mode for a sand body with base level cycle and facies control theory.

References

Armentrout, J.M., S.J., Malecek, L.B. Fearn, et al. 1993. Log-motif analysis of Paleogene oppositional systems tracts, Central and Northern North Sea: Defined by sequence stratigraphic analysis[M]// Petroleum Geology of Northwest Europe: Proceedings of the 4th Conference: 45–57.

Cross, T.A. 1991. High-resolution stratigraphic correlation from the perspectives of base-level cycles and sediment accommodation. In Unconformity Related Hydrocarbon Exploration and Accumulation in Clastic and Carbonate Settings (J. Dolson, Ed.), Rocky Mountain Association of Geologists, Short Course Notes. pp. 28–41.

Deng, H.W., H.L., Wang, and Zhu, H.J. 2002. *High resolution sequence stratigraphy—Theory and application*. Beijing: Geological Publishing House.

Gardner, M.H. 1964. *Sequence stratigraphy and facies architecture of the Upper Cretaceous Ferron Sandstone Member of the Mancos Shale, East Central Utah*. Golden, Colorado, School of Mines, unpublished Ph.D. dissertation, 528pp.

Gu, J.Y. 1995. Framework concepts and models of sequence stratigraphy in nonmarine petroliferous basin. *Petroleum Exploration and Development* 22 (4): 6–10.

Gu, J.Y., and Fan, Z.S. 2001. Look back and forward of sequence stratigraphy (in Chinese), Marine Petroleum Geology, 6 (4): 15–25.

Ji, Y.L. 1998. The principle of sequence stratigraphy and the mechanism of sequence genesis. Geological Publishing House.

Jervey, M.T. 1988. Quantitative geological modeling of siliciclastic rock sequences and their seismic expression. In Sea Level Changes—An Integrated Approach C.K. Wilgus, B.S. Hastings, C.G.St.C. Kendall, H.W. Posamentier, C.A. Ross and J.C. Van Wagoner (Eds.), SEPM Special Publication 42: 47–69.

Li, S.T., Y.L., Pan, Y.C. Lu, et al. 2002. Key Technology of Prospecting and Exploration of Subtle Traps in Lacustrine Fault Basins: Sequence Stratigraphic Researches on the Basis of High Resolution Seismic Survey.*Earth Science—Journal of China University of Geosciences* 27 (5): 592–598.

Li, S.L., X.H., Yu, Z. Zhang, et al. 2004. Sedimentary Microfacies and Sequence Stratigraphy of the Miocene of Neogene in Xijiang 30-2 Oilfield, Pearl River Mouth Basin. *Journal of Palaeogeography* 6 (1): 30–40.

Loutit, T.S., J., Hardenbol, P.R., Vail, and G.R. Baum, 1988. Condensed sections: the key to age-dating and correlation of continental margin sequences. In Sea Level Changes–An Integrated Approach C.K. Wilgus, B.S. Hastings, C.G.St.C. Kendall, H.W. Posamentier, C.A. Ross and J.C. Van Wagoner (Eds.), SEPM Special Publication 42: 183–213.

Posamentier, H.W., and P.R. Vail. 1988. Eustatic controls on clastic deposition II–sequence and systems tract models. In Sea Level Changes—An Integrated Approach C.K. Wilgus, B.S. Hastings, C.G.St.C. Kendall, H.W. Posamentier, C.A. Ross and J.C. Van Wagoner (Eds.), SEPM Special Publication 42: 125–154.

Ross, C.A., and J.R.P. Ross.1986. Sea-level changes: An integrated approach [J]. Soc Econ Paleont Miner, 14 (6): 535–535.

Sangree, J.B, P.R. Vail, and R.M. Mitchum. 1990. A Summary of Exploration Applications of Sequence Stratigraphy[M]// Sequence Stratigraphy as an Exploration Tool: Concepts and Practices in the Gulf Coast, 11th Annual Meeting: 321-327.

Siclen, D.W.C.V. 1958. Depositional topography; examples [Louisiana-Texas] and theory[J]. *Aapg Bulletin* 42 (8):1897–1913.

Shanley, K.W., and P.J. Mccabe. 1994. Perspectives on the Sequence Stratigraphy of Continental Strata. *Aapg Bulletin* 78 (4):544–568.

Sloss, L.L, 1963. Sequences in the cratonic interior of North America. Geological Society of America Bulletin, 74:93–114

Vail, P.R. 1987. Seismic stratigraphy interpretation procedure. Atlas of Seismic Stratigraphy Part, 27.

Vail, P.R., F., Audemard, S.A. Bowman, et al. 1991. The stratigraphic signatures of tectonics, eustasy and sedimentology–an overview. In Cycles and Events in Stratigraphy. (G. Einsele, W. Ricken and A. Seilacher, Eds.), Berlin, Springer-Verlag, pp. 617–659.

Vail, P.R., R.M. Mitchum Jr., R.G. Todd, et al. 1977. Seismic stratigraphy and global changes of sea level. In: Seismic stratigraphy application to hydrocarbon exploration, ed. C.E. Payton, Am Assoc Petrol Geol Mem 26: 49–212.

Van, Wagoner J.C. 1990. Siliciclastic sequence stratigraphy in well logs, cores, and outcrops: concepts for high-resolution correlation of time and facies. Annuaire statistique de la Societe des nations. Societe des nations, Service d'etudes economiques, 43–62.

Van Wagoner, J.C., H.W., Posamentier, R.M. Jr., Mitchum, P.R., Vail, J.F., Sarg, T.S., Loutit, and J. Hardenbol. 1988. An overview of sequence stratigraphy and key definitions. In Sea Level Changes–An Integrated Approach C.K. Wilgus, B.S. Hastings, C.G. St.C. Kendall, H.W. Posamentier, C.A. Ross and J.C. Van Wagoner (Eds.), SEPM Special Publication 42: 39–45.

Wei K, and X. Sun. 1996. Sequence Stratigraphic Model of Nonmarine Sediments in Cretaceous, Songliao Basin[J]. *Acta Sedimentologica Sinica* 4: 50–60.

Wheeler, H. E. 1959. Unconformity bounded units in stratigraphy. *American Association of Petroleum Geologists Bulletin* 43: 1975–1977.

Wheeler, H.E. 1964. Baselevel, lithosphere surface, and timestratigraphy. *Geological Society of America Bulletin* 75: 599–610.

Xu, H.D. 1991. Problems in analysis of faulted basins of china based on sequence stratigraphic theory [J]. *Oil & Gas Geology* 12 (1): 52–57.

Xu, H.D., K.S., Wei, W.D. Hong, et al. 1993. Principle of Sequence Stratigraphy (in Chinese), Petroleum Industry Press.

Yang C.G. 2001. A Basic Theory of Clastic Reservoir Stochastic Modeling and Its Practice. Journal of Jianghan Petroleum Institute.

Zheng, R.C., J., Feng, and Wu, C.R. 2001. Grade division of base-level cycles of terrigenous basin and its implications. *Acta Sedimentologica Sinica* 19 (2): 249–255.

Zhou, H.M., Y.X., Dong, B.F. Liao, et al. 2006. Fine exploration practice and potential in the Nanpu Sag. The 2nd China Petroleum Annual Meeting: 38-45.

5 Research Methods of Sedimentary Facies and Sedimentation

The identification of sedimentary environments and sedimentary facies is based mainly on various facies markers that are determined and obtained from geology, logging, and seismic data. However, data analysis and research cannot be separated from discussions on the formation mechanism of these markers or sedimentation. Thus, we can say that sedimentary markers are the foundation, assisted by well logging data and seismic markers.

5.1 Concept of Fluid Dynamics and Analysis on Hydrodynamic Conditions

5.1.1 Concept of Fluid Dynamics

Fluid dynamics, which is a branch of fluid mechanics, deals with the states and rules of fluids (liquids or gases) in movement. It is based on external causes as the condition and internal causes as the core, as is the case with sedimentology. Research on the sediment formation based on the concept of fluid dynamics and the transport mode of sediments is conducive for interpreting sedimentary structures, remodeling paleo-hydrological geomorphological conditions, and describing reservoirs and their heterogeneity.

5.1.1.1 Newtonian Fluid, Non-newtonian Fluid, and Transport Modes

1. **Newtonian fluid and non-Newtonian fluid**

According to hydrodynamic properties, every fluid that obeys Newton's inner friction law is called a Newton fluid; otherwise, it is called a non-Newton fluid. Newton's inner friction law states that the hydrodynamic force viscosity coefficient is always constant within a certain time as the velocity gradient changes.

(1) Traction current

This is any fluid that causes detrital materials to perform traction movement according to Newton's inner friction law, such as a water current (river current, ocean current, lake current, wave flow, tidal current, and contour current) containing a small amount of debris and atmospheric flow. Traction current is a Newtonian fluid. The term debris flow is mainly used in

X. Yu et al., *Clastic Hydrocarbon Reservoir Sedimentology*, Advances in Oil and Gas Exploration & Production, https://doi.org/10.1007/978-3-319-70335-0_5

sedimentology or mostly in geography and geomorphology.

(2) Gravity current or gravity flow

This is a high-density fluid that causes debris to mix with it under gravity but in a manner that does not conform to Newton's inner friction law. It is also called sediment gravity flow. It can be further divided into ① debris flow, ② grain flow, ③ liquefied (sediment) flow, and ④ turbidity current. Sediment gravity flow is a non-Newtonian fluid.

(3) Turbidity current

This is a subaqueous gravity flow formed by the mixing of large amounts of mud, sand materials, and water. It is supported by turbulent flow.

2. **Basic transport modes of fluid**

From the physics viewpoint, there are two basic material transport modes, namely, suspension loading (suspended load) and sandy bedform loading (traction load). Correspondingly, there are two transport modes, namely, suspension transport and traction transport.

(1) Suspension transport

Fine-grained sediments (such as silt, clay particles, and sand-sized particles in different proportions) are diffused and transported as a suspension in fluid by either air or water currents. The most fundamental driving power is turbulence, which may lift particles and suspend them in fluid for transportation.

(2) Traction transport

In sedimentology, the term traction refers to all action processes that transport sediments via sandy bedform loading (traction loading). Traction force results from particle inertia. A low-density, low-viscosity traction current transports sediments via sandy bedform loading (traction load); correspondingly, a density current (gravity flow, block flow) with relatively high density and viscosity that flows under gravity in an entire block transports sediments via suspension loading.

To sum up, based on the significant differences between formation mechanisms, debris can be transported in different ways by both traction and gravity currents. The effectiveness of a traction current is governed by two factors. The first is thrust (i.e., traction force) of fluid on debris particles, where thrust is measured by the numerical value capable of moving the sediment particles, and is determined by the velocity of the fluid. The greater the thrust, the larger is the size of the debris particles that running water can carry. This is because deposited particles can be carried when the thrust is greater than the gravity acting on the particles and the friction between the particles and the bed bottom. The second is loading force (or carrying capacity), wherein the carrying force, which means the numerical value of the total load capacity of sediments carried by the running water, is decided by the flow offluid. The greater the flow, the higher is the carrying force; thus, running water can carry a higher amount of sediments. The thrust of a traction current carries particles mainly as a solution load, suspension load, traction load, or sandy bedform load.

Gravity flow is a high-density mixture (relative density = 1.5–2.0) comprising fragmented materials of different sizes and a fluid. It mainly transports sediments in suspension form. Gravity flow is driven mainly by gravitational acceleration when gravitational acceleration is greater than shearing force in the steep slope zone. As a result, sudden loading occurs, resulting in the formation of various gravity flow sediments when the gradient becomes gentle and the flow velocity decreases.

5.1.1.2 Laminar Flow, Turbulence, and Reynolds Number

The characteristics of fluids mainly include viscosity, density, resistance to shear, and capacity to form turbulence under suitable conditions. For fluid movement, density, which is a measure of fluid inertia, is an important mechanical property. In addition, fluid flow speed and viscosity are important factors. To better distinguish the physical markers of sedimentary environments, we should first have some understanding of the basic concepts of hydrodynamics.

1. **Laminar flow (sheet flow or stratified flow)**

A fluid's resistance to shear varies as the shearing speed changes. When the shear rate is low, a fluid flows in the laminar state. Laminar flow said to occur when "all parts of a fluid slide along a surface in conformity with the fluid boundary."

2. **Turbulence (turbulent flow)**

When the shear rate is high, complex flow routes are developed, typically curved, heliciform, and circinate. "A fluid characterized by eddies" is called turbulence (or turbulent flow). Its mode of flow is characterized by increasing jet velocity due to turbulence and the conversion of laminar flow to turbulence.

3. **Reynolds number**

Reynolds number can reflect flow type (such as laminar flow or turbulence) of a fluid, and it can be affected by fluid properties.

It is defined as "a dimensionless number representing the ratio of inertia force to viscous force." Inertia force is measured as the product of mass and acceleration, $\propto \upsilon^2 d^2 \rho$;

Viscous force is determined as the product of viscosity shear stress and action area, $\propto \upsilon d \mu$.

$$\text{Reynolds Number Re} \approx \frac{\text{Inertia force}}{\text{Viscous force}} \propto \frac{\upsilon^2 d^2 \rho}{\upsilon d \mu} = \frac{\upsilon d \rho}{\mu}$$

where

ρ density (kg/m^3);

d diameter (m);

υ speed (m/s);

μ viscosity (Pa s).

Thus, it can be seen that the density, viscosity, and speed of a fluid can affect its flow properties, and Reynolds number is simply a combination of these factors. The significance of Reynolds number is that fluid flow modes are remarkably different when the Reynolds number representing fluid flow is greater or less than some critical Reynolds number. By this time, the flow is laminar. When Re = 1–40, a dorsal stream wake is formed behind particles in the flow, and it becomes increasingly irregular when Re increases. A "Karman vortex street" is formed when Re is greater than 40, in which case flow is turbulent. As the Reynolds number increases (Fig. 5.1), the dorsal stream wake develops gradually along the dorsal stream direction of spherical particles; and meanwhile, the running way is gradually converted into turbulence from laminar flow.

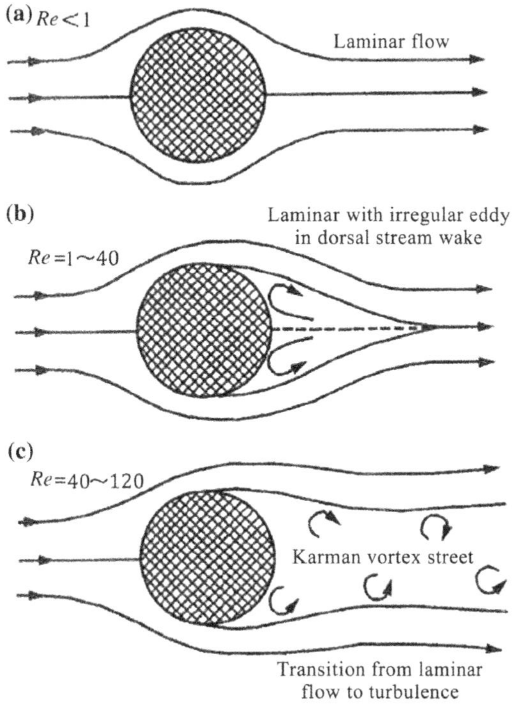

Fig. 5.1 When the fluid is greater than or smaller than a Reynolds number, the flowing pattern is obviously different (Blatt et al. 1972)

Because the energy for moving particles is derived from the upward flow of fluid in a turbulence eddy, particles are kept inside the fluid in a suspended state when the upward speed of the eddy exceeds the subsidence rate of particles.

5.1.1.3 Tranquil Flow, Torrent, and Froude Number

According to flow intensity, fluids can be classified into three flow regimes which are distinguished by the Froude number (Fr): tranquil flow, torrent, and critical flow.

The Froude number (Fr), which is similar to the Reynolds number, is a ratio of two forces or "a non-dimensional number of the ratio of inertia force to gravity."

For the quality of unit fluid (water) whose movement speed is υ, the inertia force is equal to the force necessary for a slowing unit water to a stop within distance L, and the force is equal to υ/t (because the quality of unit water is 1). However, the time needed is $t = L/\upsilon$, so the inertia force is proportional to υ/t, that is, in proportion to υ^2/L. The gravity acting on unit quality water is g. Accordingly,

$$\mathrm{Fr} = \frac{\text{Inertia force}}{\text{Gravity}} = \frac{\frac{\upsilon^2}{L}}{g} = \frac{\upsilon^2}{Lg}$$

Most scholars define the Froude Number (Fr) as the square root of a number, as follows:

$$Fr = \frac{\upsilon}{\sqrt{gL}}.$$

If water flows in an open channel and water depth is D, the Froude Number (Fr) can be defined as

$$Fr = \frac{\upsilon}{\sqrt{gD}},$$

where

υ average flow velocity of fluid (water).

This means that the speed of a gravity wave is equal to $\sqrt{gD}$ and Fr = 1 when the wavelength of the gravity wave is comparable to water depth, or the wavelength of the gravity wave is equal to water depth (D). This brings up the significance of the Froude number (Fr). When the Froude number (Fr) is greater than 1, there would be no upstream propagating waves on a ground where the downstream flow velocity is greater than the upstream propagation velocity. Thus, when Fr is greater than 1, running water acts as torrent or super-critical flow (critical flow). A dynamic state of torrent in a shallow area is also called the upper flow regime. When Fr is less than 1, tranquil flow or critical flow represents a dynamic state of slow flow in deep water, which is also called the lower flow regime. Consequently, Fr is generally used to explain the carrying and deposition of detrital materials as sandy bedform, especially in the analysis of hydrodynamic conditions formed by a sedimentary structure.

5.1.2 Main Types of Current and Sediment Characteristics

According to their effects on bed materials, currents can be divided into four types (according to Friedman G. M. et al. 1979): ① neutral, ② eroding/reworking, ③ depositing/reworking, and ④ depositing–reworking/reforming. Herein, we briefly introduce two types of flow and their transformational effects on bed material.

5.1.2.1 Effect of Eroding/Reworking Current

An eroding/reworking current can erode or transform the current of bed materials. It has four meanings hereunder:

1. **Hjulström's effect**

To erode adherent deposits that are thinner than fine sand, a current should be fast enough to carry pebbles. Resistance to the erosion of fine grain deposits is called the Hjulström effect. The relationships among erosion, transport, and deposition shown in Hjulström's diagram (Fig. 5.2) indicate a close relationship between

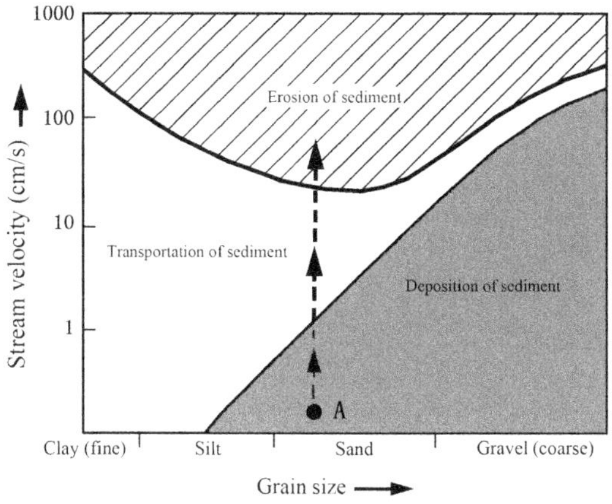

Fig. 5.2 The erosion, transportation and sedimentation manifested by variations of sediment, granularity and flow rate can be represented with Hjulstrom's diagram (Hjulstrom 1935)

particle size and current speed. The mutual relationship between the two determines the type of fluid action on deposit and bedform, such as erosion, transport, or deposition.

2. **Current marks on viscous substrate**

Various marks formed by eroding/reworking currents on a viscous sediment substrate are collectively referred to as current marks, which include scour marks and tool marks. Currents often carry sand level sediments to cover those marks, and the marks generated on the viscous substrate are called sole marks.

3. **Sandy bedform and flow regime**

3.1. Sandy bedform series

The geometric configuration formed between non-viscous sediments and fluid is called the bedform or bed configuration when sediments are carried and deposited. Its size, shape, and type depend on depth of flow, dynamic strength, average flow velocity, flow viscosity, liquid density, size of deposited particles, density of sediments, and gravitational acceleration. Among them, depth of flow, dynamic strength, average flow velocity, and size of deposited particles have greater importance.

For an eroding/reworking current flowing through a bottom bed composed of non-viscous particles, as flow velocity increases, linear stripes parallel to the water current are generated in coarser deposits until a regular bedform series is formed. There are mainly three types of such bedform series:

(1) Ripple: The ripple bedform is small, with ripples having wavelengths of 10–30 cm and wave heights of 0.6–30 cm when water flowing at a speed of about 20 cm/s applies shear force to the top surface of fine sand. All ripples migrate downstream slowly. This bedform has never been found in deposits with median grain diameter greater than 0.6 mm ($d_{50} = 1.25\Phi$), and it is formed mainly in deposits with particle diameter equivalent to that of medium-fine sand or smaller.
(2) Sand waves: These are formed by two mechanisms. The first is transformation from ripper sediments when the flow velocity is 50 cm/s and the second is gradual conversion of plane bed deposits into sand waves when the granularity of deposit is greater than 0.6 mm.

There are four types (Table 5.1) of wavy forms, in which sand waves A, B, and C are characterized by the bedding shear approach to the angle of repose (30°). Sand waves A and B can be formed in shallow-water flows (depth < 0.5 m). Sand wave C is formed in larger flows (15 m or greater in length, 12 m in height), but it is not formed in currents shallower than 4.7 m. Currents are separated after passing over the top points of sand waves A, B, and C. There is a separated vortex band between a sand wave ridge point and the next sand wave ridge point in the downstream direction. The direction of the current is opposite to that of the main current at the bottom of the separated vortex, and its speed about 1/3–1/2 that of the main current. If fine

Table 5.1 Types of large bedform (according to Friedman, modified in 1979)

Type	Description	Remarks
A	Straight ridge sand wave with wave height less than 1 m	"Sand wave" proposed by Southard (1975)
B	Bedform with steep slip slope and concave surface pointing toward the downstream direction (forming a "scour hole"); trough cross bedding formed during migration	"Sand dune" proposed by Southard (1975)
C	Straight sand wave with wave height greater than 1 m; water depth more than 5 m required, large (plate-like) cross bedding can be formed, several grades, and no determined upper limit	A part of "sand wave" proposed by Southard (1975)
D	Transitional form between lower flow regime and upper flow regime, small slip slope dip angle, low dip angle cross bedding formed at about 10° or less than 10°	
E	In-phase sand waves, anti-dunes, and standing wave	

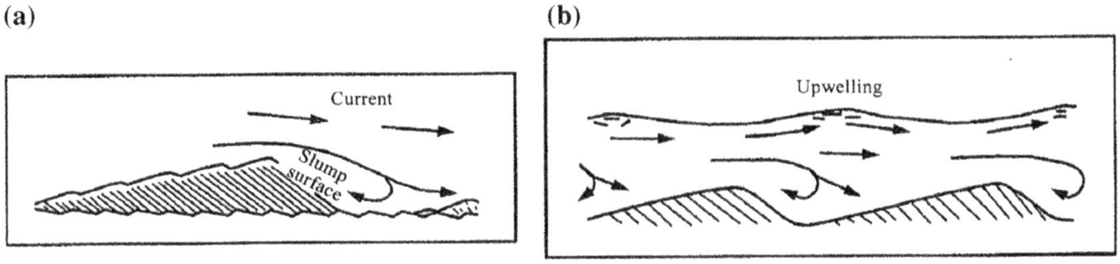

Fig. 5.3 Schematic of cross section of current for formation of sand wave on fine sand in the direction parallel to current. **a** Owing to sand wave migration from ripples of the stream surface to the ridge top, the ripples are destroyed when the sand slumps along the steep dorsal stream wave. The ripples in the downstream trough of the slump surface migrate toward the direction against the current. Reflux ripples buried by sand particles falling from the sand wave slump surface migrating downstream are preserved. The combination of small cross bedding of reflux ripples and large cross bedding of sand waves is a special state in which running water carries fine sand. It is formed by the current separation effect caused by sand waves (according to Boersma 1967). **b** Schematic profile of water and deposits in the vertical plane parallel to the current direction when sand waves are formed by current on coarse sands. When the water level above the sand wave ridge descends, the water level rises frantically above the troughs between sand wave ridges. The ripples shown in this figure **a** are not formed in coarse sands (according to Simons et al. 1965)

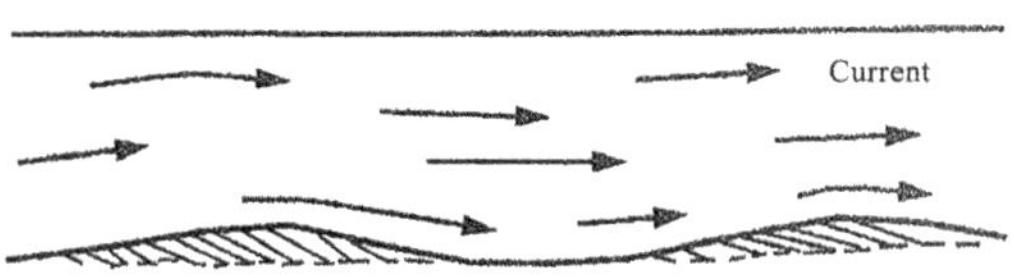

Fig. 5.4 When the shear velocity is greater, D-type sand wave occurs, and all the profiles of the downward shunted water and D-type sand wave keep continuous contact (Simons et al. 1965)

sand is deposited and the speed of eddy reflux is relatively high, the reflux tends to form cross laminations and small ripples opposite to the main current. Meanwhile, different facies (Fig. 5.3) of the first group of sand waves of the water "waves" are caused by mega-dunes. When the shear speed is relatively high, sand wave D is presented. It is not symbiotic to ripples. The water flowing downstream maintains continuous contact with all parts on the section of sand wave D, but the current does not have any separation (Fig. 5.4).

(3) Plane bed or flatbed: If flow velocity increases, the washed-out sand wave D disappears, the boundary between sediments and the water becomes flat, and

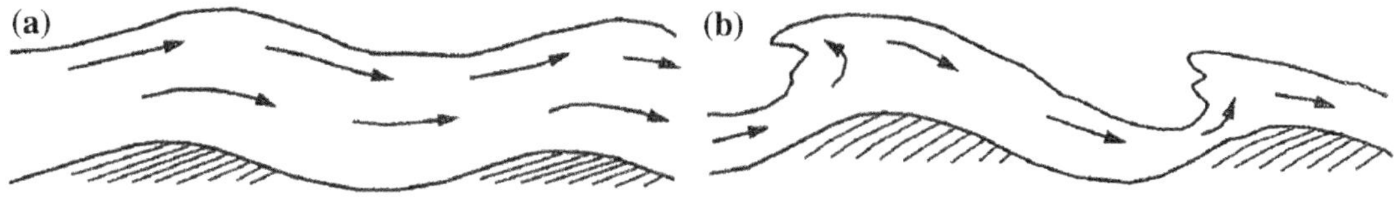

Fig. 5.5 Schematic of cross section of the current forming anti-dunes and sediments in the direction parallel to the current. **a** If the depth of current across a trough is always greater than the height of an anti-dune ridge, a sanding-wave type anti-dune is formed. On the anti-dune stoss side, there is no breaker on the upwelling water surface. **b** If the water depth in the trough is less than the height of the anti-dune, water waves break and collide reversely in the upstream direction and generate severe turbulence while passing over the anti-dune breaking zone (according to Simons et al. 1965; Harms and Fahnestock 1965)

sediments continue flowing downstream rapidly to present a series of flat sheet flows.

(4) In-phase sediment wave: The characteristic of this wave are mainly subject to the Froude number (Fr). When Fr is greater than 0.84 (close to 1), an in-phase smooth sediment wave (Fig. 5.5) of a water wave is presented. In these in-phase waves, sediments migrate with the current, but wave forms can remain unchanged or migrate downstream or upstream. G. K. Gilbert termed sediment waves that migrates upstream as "anti-dunes."

3.2 Flow regime

Flow regime has been put forward when a certain rule is followed between flowing water observed in a flume experiment and non-viscous particle traction load. Flow regime refers to the sum of the phase relationships among current, boundary shape between water and sediments, sediment transport mode, energy dispersion process in current, water surface shape, and water-sediment boundary. Generally, there is a transition flow regime between the lower flow regime and the upper flow regime.

(1) Lower flow regime: This is characterized by the development of small ripples, ripples and sand waves, sand waves and dunes, or only sand waves at the boundary between water and sediments. The transport or capacity of sediments is relatively small, and transport is continuous (Figs. 5.6 and 5.7). The transport is formed by drawing blanket layers to move upwards on the stoss sides of ripples or sand waves and then form gravity slumps on the slip slopes of these steep bedform slopes. In the troughs of sand waves, some reversal transport can be result from reflux ripples, the migration direction of which is opposite to the main current direction. To a certain extent, current energy can be consumed by the roughness and inertial resistance of sedimentary particles, but it is mainly consumed by the morphological resistance of ripples and sand waves, and the reflux vortex at the current-separation part. Thus, sediments with certain sorting along the flow direction and the sediments left on the bedforms are slightly coarser than those flushed downstream.

(2) Transition Phase: The bedform is sand wave D or sand wave (Table 5.1). Sediments trend toward continuous motion, but there is accumulation in the low-dip-angle crossing layer of sand wave D; moreover, sediments do not slump along the edge of the slip slope of sand wave D. Flow energy is dissipated by inertial resistance and roughness of the moving particles. When water is no longer separated, water level tends to flatten and is irrelevant to low and long bedforms.

(3) Upper flow regime: Since the boundary between water and sediment is either a plane or a surface with various ups and downs, the majority of sediments are carried constantly, and gravels are carried downstream at a speed that is about half the average current speed. Flow energy is dissipated owing to roughness of the particles, inertial resistance, breaking of anti-dune waves, and generation

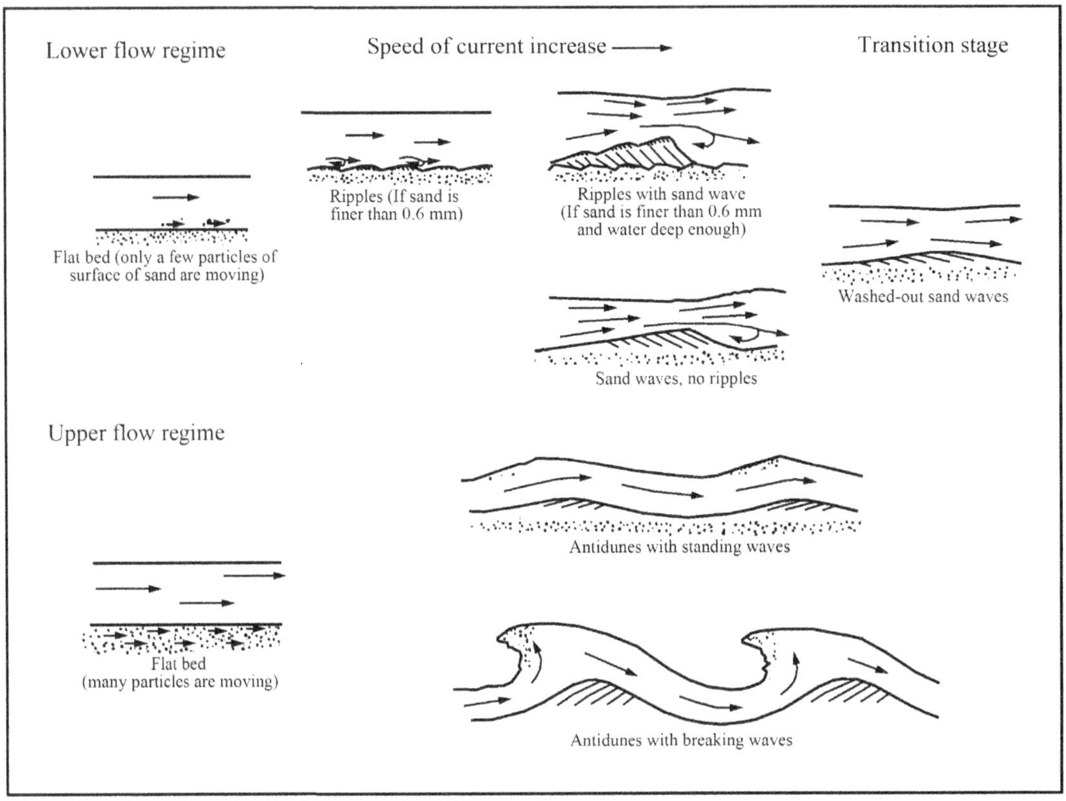

Fig. 5.6 In different flow states, the bed form has different characteristics (Simons et al. 1965)

and disappearance of sediment waveforms. Water level fluctuation has the same phase as that of the uplifted sediment bedform. Owing to the lack of downstream sorting, sediments flowing downstream and sediments forming the bedform have the same granularity distribution (Figs. 5.6 and 5.7).

4. **Selective sediment transport**

Eroding/reworking currents selectively carry sediments on the bedform. The sorting mechanism and degree of sorting change as the flow velocity changes. Therefore, sediments are sorted regularly in the downstream direction. The final products are determined by main sedimentation, grain size, and abundance of particles in bedform materials.

5.1.2.2 Effect of Depositing/Reworking Current

Depositing/reworking currents are slow enough to start depositing sediments at the critical point but fast enough to generate the traction effect in the deposited particles. The main differences between eroding/reworking currents and depositing/reworking currents are as follows: the latter not only deposits sand-level sediments carried in a suspended state onto the bed to form a bedform upwards but also moves the deposited particles in the same state (suspended). Because "reworking" does not mean "reforming" here, the current is of the depositing/reforming type only if there is any transformation. There are no major differences between depositing/reworking and depositing/reforming currents. Actually, at times, freshly deposited sediments may be non-viscous materials or viscous when shear stress is applied by the current. If non-viscous sediments are

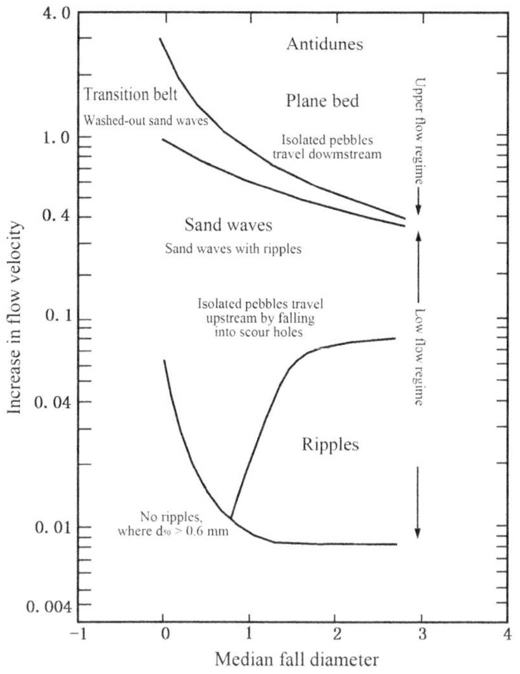

Fig. 5.7 Under different flow energies and particle sizes, there is a certain relationship between the bed form and the flow state (Simons et al. 1965)

dragged, the current is of the depositing/reworking type; if viscous materials are transformed, the current is of the depositing/reforming type. Therefore, the same current can be of the depositing/reworking or the depositing/reforming type depending on the properties of the bottom sediments.

Plane laminae and superimposed ripples are feature products generated by depositing/reworking currents. H. C. Sorby called the latter "ripple drift deposited from above," which is generally called "ripple drift" in short.

The internal laminae of ripple drift sediments are formed by the interaction of current, ripples, and upward construction speed of the bedform. Sediments settled on the internal laminae are separated when small ripples migrate. Meanwhile, very fine sand can spread across the entire ripple surface, silty sand and clay can be deposited on either troughs or slip slopes, and very fine sand can gather on stoss sides and wave ridges. An imbricate arrangement, which is typical for clastic particles, can be formed, especially in gravel deposits, by impelling action in the upstream direction at flat particle-deposition positions.

5.1.3 Flume Experiment-Bedding Formation Mechanism

As mentioned previously, different bedforms may form different types of bedding because different bedforms reflect different flow strengths. Flume experiments in conjunction with the relationships between flow and internal configuration of bedforms or bed configuration can be used for bedform identification. In continuation of Gilbert's (1912) method of applying sand of various grain sizes and different strengths to solve the problem of hydraulic mining of gold, Simons and Richardson, in 1961, carried out a large-scale flume experiment on sand with particle size less than 0.6 mm ($M = 1.25\Phi$). Using the trough width of 2.44 m and trough length of 45.72 m, which were employed in their experiment, one can produce water flow on the flat sand bed of the trough. When the flow velocity (20 cm/s) and gradient increase gradually to the critical value, sediments start moving and forming small concave-convex bedforms (ripples) inclined slightly upstream but steeply downstream owing to flow resistance, where the wave height is 0.5–3 cm and wavelength is generally less than 30 cm, rarely exceeding 60 mm. This type of bedform is called ripple (Fig. 5.8). When the flow velocity (50 cm/s) increases further, large irregular dunes are formed, where the wave length is 10–20 cm and the wave length may reach up to several meters. As mentioned above, ripples and dunes belong to the low flow regime. In general, Fr is less than 1, in which case, water fluctuation is different from bedform fluctuation, resulting in out-of-phase waves.

Either flow velocity or gradient increases continuously until dunes disappear to form a flat bedform, which strictly shows the absence of any bedform type. When the Fr is greater than 0.84, sand waves around the sediments and water surface waves stay in the same phases, and a roughly sinusoidal anti-dune is formed. It refers

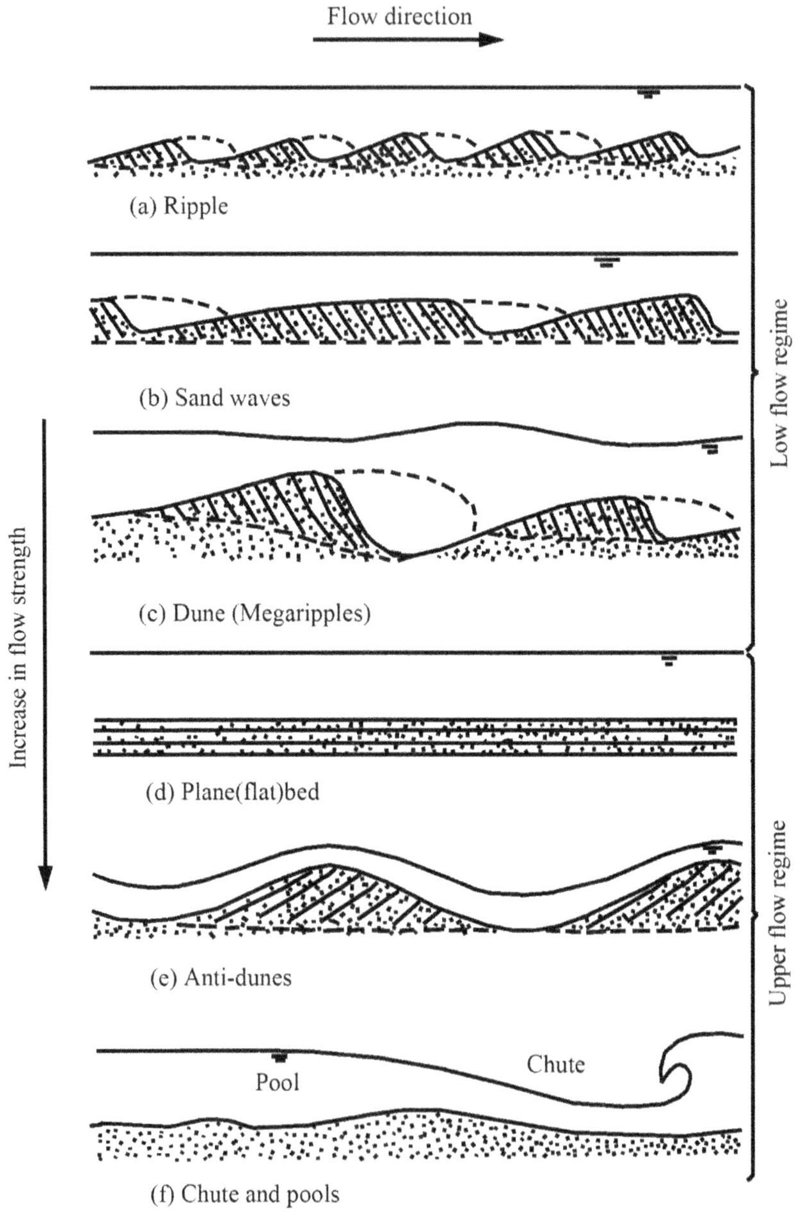

Fig. 5.8 With increase of flow intensity, the bed form gradually changes (D. K. Simons et al. 1961)

to wavy bedform moving upward, which is aggraded on the upstream side and eroded on the downstream side (Fig. 5.9). Chutes and scour pits (Fig. 5.8) are formed with a further increase in flow velocity. Flat beds, anti-dunes, chutes, and scour pits all belong to the upper flow regime, which indicates that the upper flow regime is formed when Fr is greater than 1. Theoretically, Fr is at least 0.84 when anti-dunes are formed.

From the flume experiment carried out on the sand of particle size 0.6 mm, as flow velocity increases, the bedform is developed in the following sequence: flat bed without sediment movement (lower part) → ripple → dune → dune and sand wave → (upper part) flat

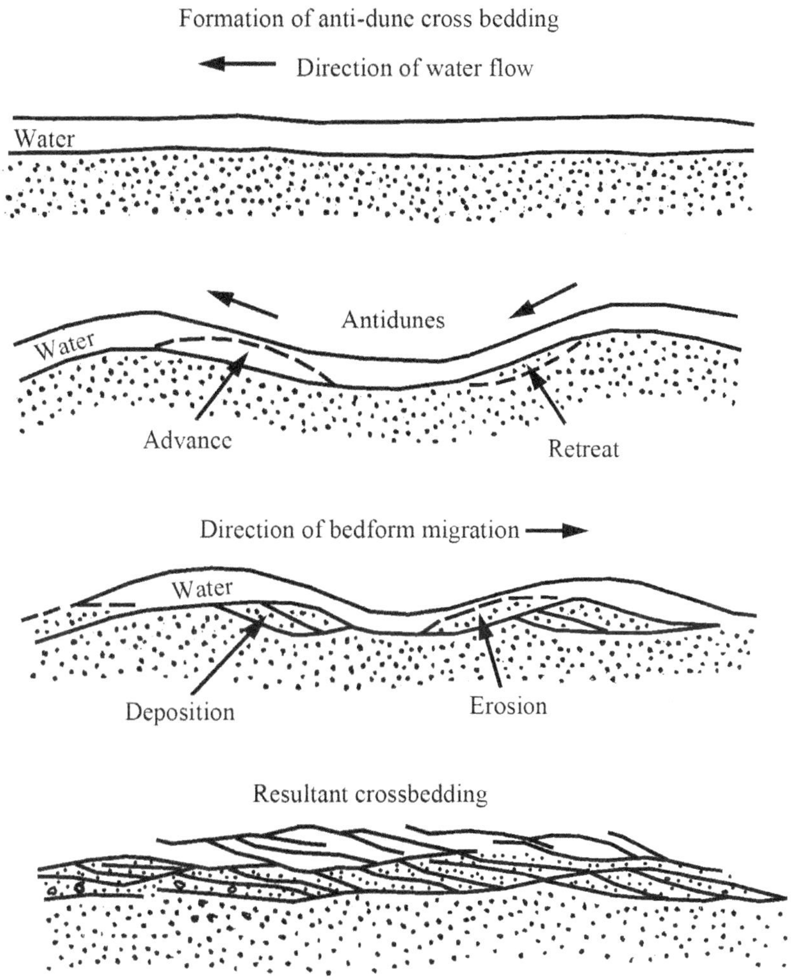

Fig. 5.9 Process of forming antidune (D. K. Simons et al. 1961)

bed → anti-dune → chute and scour pit. The corresponding flow trend is that in the lower flow regime ($F_r < 1$), the bedform and the surface wave present the out-of-phase form, and in the bottom flow regime, the bedform and the surface gravity wave present the in-phase form. In this case, the corresponding bedding sequence is as follows: horizontal bedding → ripple cross bedding → small cross bedding → megascopic cross bedding (groove, plate, and wedge) → parallel bedding → anti-dune bedding. However, the above is only an ideal complete sequence. The types of bedform in a bedding sequence may change, however their precedence order will not change.

Generally, the emergence of planar cross bedding reflects the migration of bars (such as longitudinal bar, diagonal sand top, bar point, longshore bar, and river mouth bar), especially incised planar cross bedding, while truncated planar cross bedding results from accretion and overlapping of bars (such as transverse bars). The emergence of trough cross bedding represents the emergence of channels, incisions, and filling. Parallel bedding indicates the formation background and condition of rapid flow in a multi-order shallow area; ripple cross bedding is mostly a product of slow flow in shallow water, which represents the sedimentation of a natural levee and overbank or sheet flow accretion.

In conclusion, the most important factors that affect the bedform are flow strength, average flow velocity, particle size, and flow depth. J. R. Allen (1964) and Southard (1975) specified the

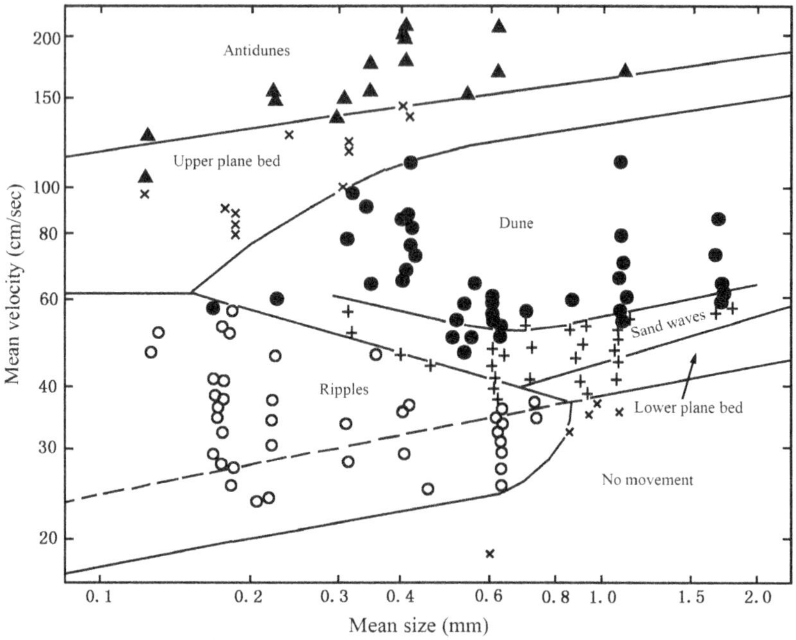

Fig. 5.10 Bed forms manifested by the mean flow rate and granularity in different positions (Southard 1975)

genesis of different bedforms according to various combinations of hydrodynamic indexes. Particle diameters of sediments are plotted for analysis (Fig. 5.10) by applying two variable flow energy values or the average flow velocity. Kinetic energy of a flow is regarded to be equal to the power of the river acting on a unit area of sandy bed. As can be seen from Fig. 5.10, as either the kinetic energy or flow velocity of the river decreases, the bedform presents the following sequence: anti-dune → upper plane bed dune → sand waves → ripples → lower flat bed. This diagram can only separate the locations of bedforms in different forms, which shows that a greater number of variants is needed. Hence, Southard (1975) prepared depth-flow velocity diagrams (Figs. 5.11, 5.12 and 5.13) of fine, medium, and coarse particles, represented by particle sizes of 0.1, 0.45–0.54 and 1.14 mm based on flow depth, flow velocity, and grain size, respectively.

As the average speed increases, the emergence sequence of bedforms formed by sediment particles measuring about 0.1 mm is as follows: no movement → ripple → upper flat bed (Fig. 5.11). There is no dune or ripple between ripple and flat bed, which is the reason for the absence of large cross bedding in the fine sediments. As flow velocity increases, bedforms larger than the ripples are formed by sand particles measuring greater than 0.1–0.6 mm, and the bedform emergence sequence is as follows: no movement → ripple → dune (small sand wave) → upper flat bed (Fig. 5.12). Unlike fine sand, coarse sand measuring 0.6 mm form a

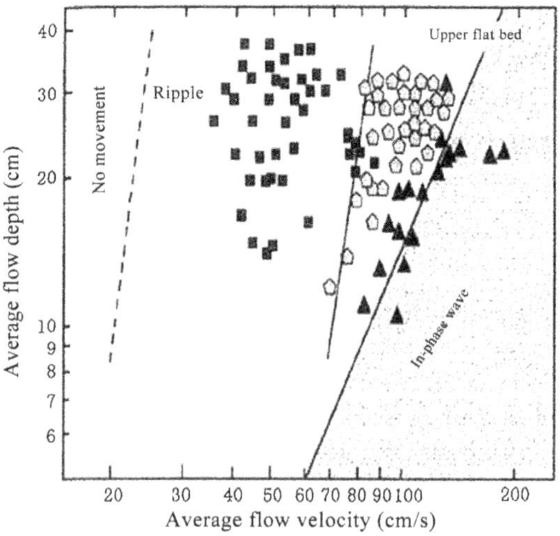

Fig. 5.11 Average depth-flow rate of bed form with sand granularity of 0.1 mm (modified by Southard 1975)

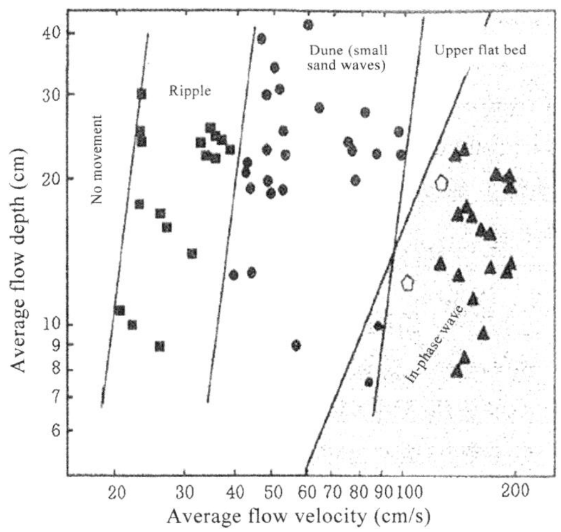

Fig. 5.12 Average depth-flow rate of bed form with sand granularity of 0.45–0.54 mm (modified by Southard 1975)

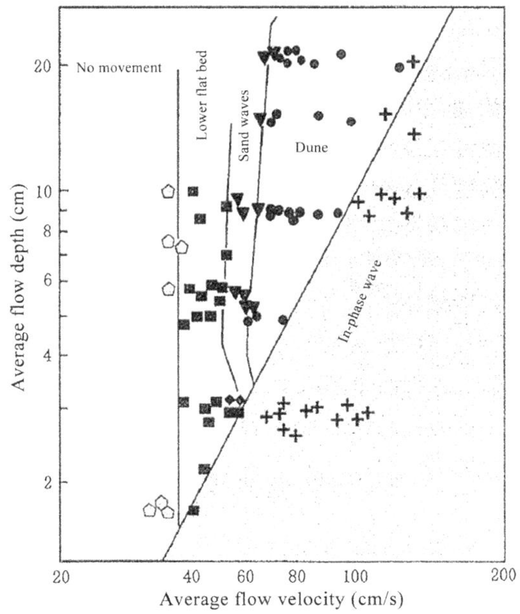

Fig. 5.13 Average depth-flow rate of bed form with sand granularity of 1.14 mm (modified by Southard 1975)

bedform with a flat lower part at relatively low speed. An unstable appearance can be formed by ripples at any speed, which results in the lack of small ripple cross bedding in medium and coarse sand stones (M > 0.6 mm). Within such particle size range, the bedform emergence sequence is as follows as the flow velocity increases: no movement → lower flat bed → sand waves → dunes. In addition, the upper flat bedform is not developed, namely, there is no parallel bedding in coarse sandstone.

5.2 Grain Size Distribution Characteristics and Environmental Significance

The grain size of sediments is called granularity. The method for studying clastic deposition and grain size of clastic rocks, as well as the distribution characteristics of various grain sizes, is called analysis of grain size. Grain size distribution characteristics may directly reflect hydrodynamic conditions and energy when sediments are formed, which is an important physical sign in the analysis of sedimentary environments and hydrodynamic conditions, and is of significance to the evaluation of clastic petroleum reservoirs. Therefore, research on the grain size distribution of clastic reservoirs may serve four functions: ① defining the properties of the transport medium such as wind, water, glacier, debris flow, and turbidity current; ② judging the energy conditions of the carrying medium, such as flow velocity, intensity, and starting capacity; ③ defining transport modes such as saltation, rolling, and suspension; and ④ defining the form of sedimentation such as traction current and gravity flow.

5.2.1 Main Grain Size Analysis Methods

Based on different sample sizes, particle thicknesses, and rock densities, usually three methods are employed.

5.2.1.1 Direct Measurement Method

This is generally used for conglomerate or gravel of hand specimen, and it involves employing a measuring tool to directly measure the diameter or apparent diameter of gravel. In general, no less than 100 gravels (including particles with grain size greater than 2 mm) in a certain area are measured to analyze conglomerates in rivers, coasts, glaciers, and proluvials.

5.2.1.2 Sieving Analysis

This is used for analyzing gravel-containing sandstones-siltstones that are not solidified or poorly cemented, especially modern sedimentary samples. The method comprises the steps of obtaining the weight percentage of all size fractions by using a screen with different apertures of 1/3Φ–1/4Φ spacing to gradually screen clastic particles from coarse to fine. The method is applied mostly to sand-level clastics, where the applicable lower limit of grain size is 4Φ. Sieve analysis is more convenient and precise, and in this analysis, the focus should be on sampling coarse, medium, and fine sands in a complete sequence.

5.2.1.3 Thin Section Grain Size Method

When clastic rocks are cemented tightly and are difficult to loosen, thin section grain size method is the only applicable method. The core in petroleum drilling and rock samples from cutting logging and sidewall core are too small for sieve analysis, and only slice granularity determination can be made in this case. An imaging device can be connected with a computer to calculate the percentages of various size fractions in different fields of vision. Unlike sieve analysis, this method can be used for measuring area percentage rather than weight percentage.

Various methods have their respective characteristics. Sieve analysis has the main advantages of fast analysis speed, accurate result, and small personal error, which objectively reflects the analysis characteristics of granularity. However, the main limitation of this method is its limited applicability, that is, it is applicable only to modern depositions and loose rocks, not to ancient hard rocks and non-coring boreholes. In addition, loose rocks are often loosened incompletely, resulting in crushed and eroded clastic particles. In contrast, the slice method is applicable mainly to ancient hard rocks, where the number of samples is small and both detritus and sidewall coring are used for analysis. Moreover, diagenetic changes in rocks, such as secondary enlargement and tectonic disruption, may be observed, albeit with poor representation and various large errors.

5.2.2 Grain Size Distribution Curve and Grain Size Parameter

In a size division scheme, the value ($\Phi = -\lg 2D$, where D is particle diameter) is mainly applied. The mutual relationship between particle diameter (mm) and the value Φ is summarized in Table 3.1. According to grain size analysis results, various grain size curves can be prepared, and various grain size parameters can be calculated.

5.2.2.1 Grain Size Curve

A grain size curve is a reference mark for analyzing sedimentary environments. Usually, a grain size curve comprises a histogram, frequency curve, frequency accumulation curve, and cumulative probability curve.

1. **Histogram**

This is the most commonly used grain size analysis map, in which the horizontal axis represents the particle size range and the vertical axis shows the percentage of size fraction. Thus, a series of mutually connected rugged histograms (Fig. 5.14, left) capable of intuitively and concisely reflecting grain size distribution characteristics are prepared.

2. **Frequency curve**

A polygonal frequency curve (Fig. 5.14, right) is formed by sequentially connecting the midpoints of the longitudinal and the transverse edges of each column in the histogram, but the area of the defined polygon is still equal to the area sum of the histogram. A frequency curve can clearly indicate the grain size distribution characteristics, sorting degree, symmetry (skewness), and sharpness (kurtosis) of grain size distribution.

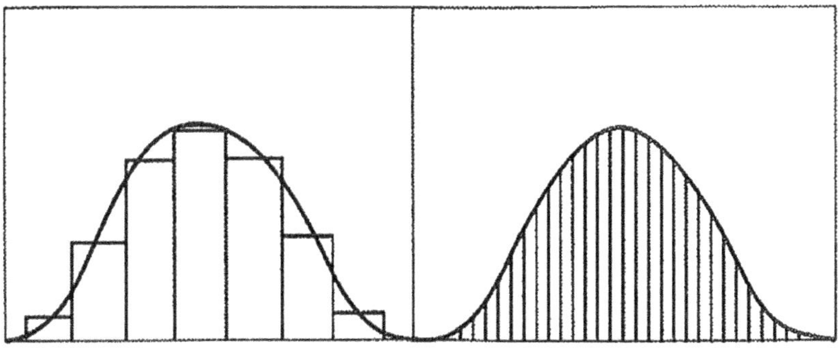

Fig. 5.14 Frequency histogram and frequency curve (according Cromerian 1938)

3. **Frequency accumulation curve**

Beginning with the coarse particle end, the accumulation percentage of each size fraction is marked on a histogram with accumulation percentage as the ordinate and particle diameter as an abscissa. An accumulation curve (Fig. 5.15) is formed by connecting all points with smooth curves. Because accumulation curves generally have an S shape, the sorting degrees of various size fractions can be seen clearly. The form of an accumulation curve can be used to distinguish different sedimentary environments. However, this curve cannot apparently distinguish sub-environments with small differences in frequency because it is insensitive in terms of reflecting the coarse and fine tail parts with lower contents. Consequently, this curve is rarely used for distinguishing environments directly. However, another purpose of the frequency accumulation curve is that it can be used to read the corresponding particle diameter values of some accumulation percentages thereon in order to obtain a few characteristics required for statistical calculations in determining granularity. For example, median (M_d) can be determined as follows. Consider a horizontal line introduced from 50% of the ordinate and intersecting with the accumulation curve. A vertical line is drawn downward from this intersection point and made to intersect with the abscissa. The numerical value (value Φ) at the latter point of the intersection is the median (Fig. 5.15). Moreover, C, Q1, and Q3 are, respectively, the particle diameters corresponding to 1, 25, and 75% accumulation; $\Phi_5, \Phi_{16}, \Phi_{50}, \Phi_{84}$, and Φ_{95} are required for calculating grain size. The above Φ values are particle diameters corresponding to 5, 16, 50, 84, and 95% accumulation, respectively, and are obtained using a frequency accumulation curve. As a result, the frequency accumulation curve is a basic map for grain size analysis.

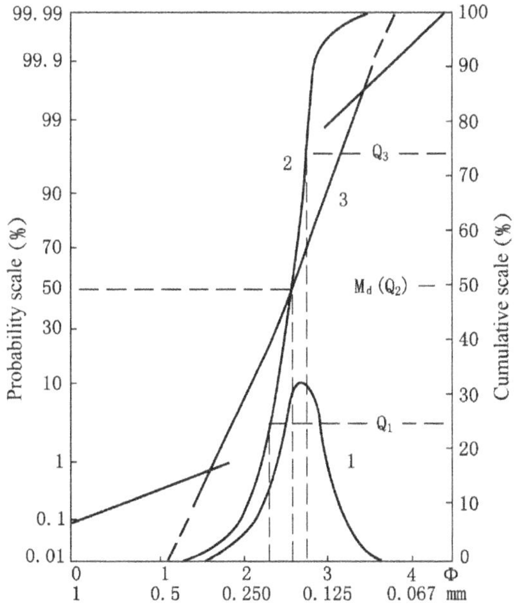

Fig. 5.15 Three common granularity curves (Reinecke et al. 1973). 1—Frequency curve; 2—frequency accumulation curve; and 3—cumulative probability curve

4. **Cumulative probability curve**

This is a typical curve and is used extensively, mostly in sedimentology to analyze hydrodynamic conditions for the formation of sediments in order to help geologists distinguish sedimentary environments by their characteristics; it is a grain size accumulation curve as well. The difference is that it is drawn on a normal probability plot in which the abscissa represents the diameter and the ordinate represents the accumulation percentage of the grain size, scaled by normal probability. The probability coordinate is not equidistant, but it gradually increases the upper and lower ends by taking the 50% point as the symmetric center. This can amplify and clearly present coarse and fine tails, and coarse and fine tails are mostly sensitive to environment reflection. The grain size of clastic sediments is not distributed lognormally on the probability curve but is composed of several secondary populations with lognormal distributions. Generally, three secondary populations show three linear sections on the probability graph, representing three different basic transport modes, namely, suspension transport, saltation transport, and rolling transport (Fig. 5.16).

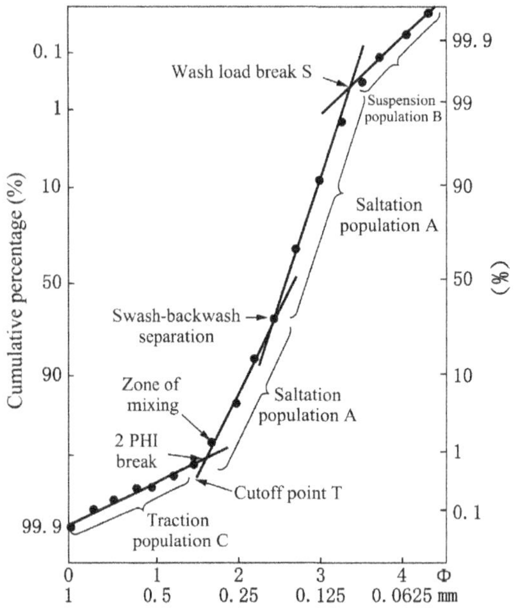

Fig. 5.16 The probability cumulative curve can obviously reflect and distinguish the characteristics of each population, and also provide favorable mathematical treatment methods for environmental analysis (Visher 1969)

5.2.2.2 Probability Graph and Environmental Significance

A probability graph can be used for determining environment type by studying the size distribution in a sample and pinpointing the sediment transport mode. Consequently, the relationship between size distribution and transport mode forms the basis of probability graphs. The three secondary populations on a cumulative probability curve are suspension population, saltation population, and rolling population (traction). In addition, there are other parameters, namely, cutoff point, mixedness, and secondary population percentage.

1. **Key points of probability graph**

(1) Cutoff point: This refers to an intersection between two secondary population lines, represented by the abscissa. Fine cutoff point (cutoff point s) refers to the intersection between suspension population and saltation population, which indicates the thickest particles capable of being suspended; coarse cutoff point (cutoff point T) is the intersection between saltation population and rolling population, which represents the thickest particles capable of saltation.
(2) Mixedness: Some cutoff points are not located on the line through the point of intersection of two secondary populations, but they transit loosely. Hence, they constitute a transition belt, which reflects sedimentary differentiation.
(3) Secondary population percentage: This refers to the percentage of each population in a sample (Fig. 5.15).
(4) Sorting: This refers to the slopes of various population lines, and represented by the angle of inclination of a straight line. Parameters such as development of various secondary populations, particle size range, and sorting are regularly controlled by sedimentary conditions and hydrodynamic conditions.

(5) Aspect ratio: It is important to note that the aspect ratio of a standard Wiechert plate is given, namely, the ratio of the line length of the ordinate with the cumulative probability of 0.01–99.99% to an abscissa with a grain size of 0–4.5Φ is 1.38–1.42. This is because the aspect ratios of this diagram are different when the slopes of some populations (some section) of the sample are analyzed. Although specific data do not change if the aspect ratio of this diagram is different, large errors occur when it is subjected to contrastive analysis, and the standard plate and hydrodynamic conditions or sedimentary environment formed by some other sample is illustrated, resulting in loss of geologic meaning of the probability graph.

(6) Plumbing requirements: The standard probability curve proposed by Visher (1969) is formed by applying the 1/4Φ data point to calculate its percentage content; then, the accumulated percentage contents at various points are calculated and plumbing is carried out on the probability curve. This is mainly because the applied sieve analysis equipment is a screen with an aperture of 1/4Φ. However, most subsequent slice or image methods are not unified because millimeter values are mostly applied in calculation and then converted into Φ values, in which case the absolute equivalent interval of Φ values cannot be ensured. Thus, it should be noted that data should be processed simply to ensure that the relative equivalent interval or difference of Φ is not very large when the probability graph is prepared using the data obtained by the slice method or the image method; 3–4 points should be within the range of unit Φ value, namely, 1/3–1/4 is the interval plumbing. This ensures better accuracy and availability of the probability graph to prevent misinformation.

Hence, the probability curve can apparently reflect and distinguish the characteristics of all populations and provide a beneficial mathematical treatment method for environment analysis. In addition, the probability curve can also reduce personal errors caused by connection, insertion, and extension among size fractions, and it is easy to read, which has led to its extensive use.

2. **Probability graph method**

As mentioned previously, a probability graph can be divided into three sections, each of which represent different transport modes.

(1) Suspension transport: This is composed of superfine particles in the current and is formed by disturbance. As for real suspensions, grain sizes less than 0.1 mm have no changes in vertical direction. However, this numerical value changes greatly in different environments, depending mainly on the disturbance. Thus, cutoff points between suspension and bed load transport may vary greatly (Fig. 5.16), and the maximum grain size of a suspension is a mark of its turbulence intensity. If suspended materials are very fine, they are suspended evenly when the average flow rate is 12 times greater than the particle sedimentation rate, which is also known as complete suspension. If the amount of coarse materials increases, suspension concentration and coarse material proportion increase as the depth increases when the rate of flow is 12 times lower than the particle subsidence rate, which is also called an incomplete or graded suspension (Visher 1969).

A certain level of alteration occurs near the sedimentary boundary between the suspension load and bed load. Mixing of the two aforementioned populations in the same sediments is common. Most sedimentary laminae containing components with a size fraction of 0.1 mm or smaller are deposited directly by suspension transport. Suspensoid, as an independent secondary population, is presented at the fine end of the probability curve, and it is the easiest to identify.

(2) Saltation transport: This refers to the phenomenon in which particles bounce near the

bed while being carried forward, and it belongs to one of the generalized tractions. Particles with grain sizes of 0.15–1 mm can saltate 2 ft away from the bottom surface, and the maximum grain size that can undergo saltation depends on flow velocity, water depth, and bottom layer properties.

The main reason for particle saltation is that particles are uplifted by the vertical component of turbulence and their forward motion is controlled by their horizontal component. Particle saltation occurs when the uplift force is not less than the weight of the particles. An obvious saltation layer is formed when there is certain uplift force and a certain saltation height. Particles that escape the bottom part for saltation are easily sorted in current, and the secondary population with the best sorting is formed in the sample. Sedimentation in rivers and beaches is mainly constituted of saltation populations.

Because it is close to the thick end of suspension populations on the probability curve, under specific conditions, the independent population formed by coarser grains may develop into two saltation populations with slightly different sorting on a beach with active scouring and rip currents.

(3) Traction: This involves coarse particles that roll, slide, drag, and rotate along the bottom surface, excluding traction of the aforementioned saltation in a narrow sense. The factors that affect traction include grain size, provenance distance, driving force of current, gradient and roughness of the bottom layer, current density, and burial rate.

Traction is considerably important because the upstream of rivers and coastal breaker belts are close to the provenance and they feature large gradients in addition to strong breaking and driving forces of reflux. In the presence of a certain gradient and low current density, rolling particles may be transported farther along the bottom surface in some beaches. There are few traction populations in ordinary river sedimentation (midstream and downstream) because of the lack of coarser components. All particles can be carried by the current via saltation. Alternatively, they cannot roll because they are quickly buried by fine-grained sediments once the coarsest particles are deposited with high deposition velocity. Turbidity current is a hyperpycnal current with high current density that can be buried quickly, in which traction is lacked and saltation populations fail to develop. However, rolling transport is presented in a sliding channel in which turbidity current flows down the slope.

3. **Characteristics of probability graphs in different environments**

The explanation of sedimentary environments using a probability graph is the most common graphical presentation method. As can be seen from the probability curves of the various environments mentioned above, various curve graphs can be formed based on various changes in the three lines that respectively represent the three transport modes of suspension, saltation, and traction (rolling). Various changes result from changes in the hydrodynamic conditions of various environments and different sediment transport modes. Moreover, the grain size probability distributions of various sedimentary environments are different (Table 5.2), but the changes conform to certain laws. In total, there are two extreme conditions, namely, flood density current-basin current deposition and wind dune deposition. The former is single total suspension sedimentation with poor sorting, while the latter mostly has total saltation sedimentation with excellent sorting. All other environments are situated between the two. Their change rules include flood density currents (turbidity current), braided rivers, meandering rivers, deltas, barrier bars, shoals, and coastal wind dunes. Furthermore, their suspension population contents decrease gradually (almost from 100% of sample to zero), saltation population increases gradually (almost from zero to 100% of sample), and all fine cutoff points change from coarseness to fineness. Owing to complex traction changes, coarse-grained tractions (less than 1Φ) are presented in all braided rivers, and fine-grained

Table 5.2 Features of grain size probability distribution of sandy sediments in different sedimentary environments (simplified; Visher 1969)

Environment	Characteristics												Main characteristics
	Saltation population (A)				Suspension population (B)				Rolling population (C)				
	Percentage (%)	Sorting	Cutoff point T (Φ)	Cutoff point S (Φ)	Percentage (%)	Sorting	AB mixture	Cutoff point S (Φ)	Percentage (%)	Sorting	Cutoff point T	AB mixture	
Wind dune	97–99	Better	1.2–2.0	3.0–4.0	1–3	Medium	Medium	4.0–4.5	0–2	Poor	1.0–0	Low	Saltation population has extremely high content and good sorting
Beach	50–99	Very good total point	0.5–2.0	3.0–4.25	0–10	Moderately good	Low	3.0–4.5	0–50	Medium	−1.0 to infinity	Medium	Saltation population with high content, presented by two straight lines
River (channel)	65–98	Medium	−1.5–1.0	2.75–3.5	2–35	Poor	Low	>4.5	Change	Poor	To infinity	Low	Great changes, saltation population-dominated, often with suspension population
Flood plain	0–30	Medium	2.1–1.0	2.0–3.5	60–100	Poor	High	>4.5	0–5	Medium	–	n/a	Single suspension population
Turbidity current	0–70	Moderately poor	1.0–2.5	0–3.5	30–100	Poor	High	>4.5	0–40	Poor	To infinity	High	Usually one suspension population with intrastratal grading

tractions (less than 2Φ) are presented in deltas and shoals, although there is generally no traction in downstream rivers (Fig. 5.17).

A probability curve can sensitively reflect changes in hydrodynamic conditions. There are curves of various natures, including several basic types such as density current-type, in which the overall suspension deposition has a single suspension curve; saltation-based river-type, composed of saltation and suspension, product of quick downstream deposition in the river; shoal-type dominated by saltation population with good sorting, composed of small amounts of traction and suspension populations, which are sediments processed by waves; and wind dune-type, composed of single, highly sorted saltation population, which is a deposition processed by wind. These curves are of significance for distinguishing environments, and they may be used as a correction standard for environmental discrimination.

5.2.2.3 Grain Size Parameter and Calculation Method

The various aforesaid distribution curves provide direct impressions of grain size distribution regularity, but it is difficult to express the curves clearly via verbal description. As a consequence, in mathematical statistics, figures (statistical characteristics) capable of reflecting the statistical characteristics of these curves must calculated to depict the characteristics of curve distribution precisely. Above all, such figures can reveal additional statistical properties.

Common grain size parameters include average grain size (M_z), standard deviation (σ_i), skewness (S_k), and kurtosis (K_g). Characteristic parameters for showing the central tendency of grain size distribution include median and mode, in addition to the average grain size.

1. **Mathematical statistics**

With the probability- and statistics-based foundation of mathematics, all size fraction percentages obtained by grain size analysis are directly used for calculation. The common

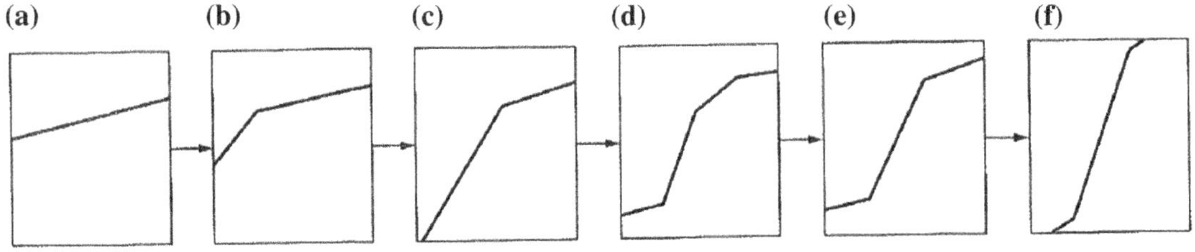

Fig. 5.17 Typical probability curves of several different sedimentary environments (Zheng 1982). A—Flood density current; B—braided river; C—meandering river; D—barrier bar; E—shoal; and F—wind dune

calculation method is moment calculation, which is rather complex and rarely used.

According to the abovementioned parameter calculation method, the calculation formula, i.e., moment calculation, is deduced statistically. The moment calculation formula of grain size parameter is as follows:

$$\text{Average value} = \bar{x} = \frac{\Sigma fm}{100m}$$

$$\text{Standard deviation} = \sigma_i = \sqrt{\frac{\Sigma f(m-\bar{x})^2}{100}}$$

$$\text{Skewness} = S_k = \frac{\Sigma f(m-\bar{x})^3}{100\sigma_i^3}$$

$$\text{Kurtosis} = K_g = \frac{\Sigma (m-\bar{x})^4}{100\sigma_i^4}$$

In each of the formulas, f refers to the weight percentage (frequency) in each size fraction, and m refers to the median (Φ unit) of each size fraction.

This method has the advantages of considering the entire grain size distribution in a statistical calculation, thus leading to increased precision. Limitations of this method include the difficulty in segmenting the fine tail part using the current grain size analysis methods. However, the moment calculation method covers the entire distribution (0–100%) and generalizes the unanalyzed fine tail, thereby improving accuracy.

2. **Graphic method**

This involves reading particle diameters based on some accumulative percentages on a cumulative curve and then calculating various grain size parameters by using simple arithmetic formulas. Grain size parameters and features are introduced below along with the graphic method.

(1) Mean value, median, and mode

Many formulas are available for calculating the average grain size using the graphic method, the simplest of which is P_{50} proposed by Trask (1932). P_{50} is a particle diameter that divides the total frequency into a coarse phase and a fine phase. It is called median and represented by Md in mathematical statistics. However, this method offers relatively low precision. In order to overcome this shortcoming, Folk and Ward (1957) proposed a formula for average grain size: $M_z = (\Phi_{16} + \Phi_{50} + \Phi_{84})/3$, which can accurately reflect the average grain size, thus reflecting the mean kinetic energy of the transport medium.

In addition, mode can be used for showing the average particle size. Also, the particle diameter with the maximum frequency is on the frequency curve, and it is denoted by M_o. If a frequency curve only has one peak, the curve is a unimodal curve, in which the abscissa at the peak is the mode of the curve. If a curve has two or more peaks, it is a double-peak or multi-peak curve, in which the main peak is the first mode, and the secondary peak is the second mode. The approximate solution of the mode is to obtain the peak position of the frequency curve or the inflection point of the accumulation curve.

If sediment grain sizes are distributed normally, the mean, median, and mode are unified. If the curve distribution is asymmetric, the three are different (Fig. 5.18). Mean, median, and

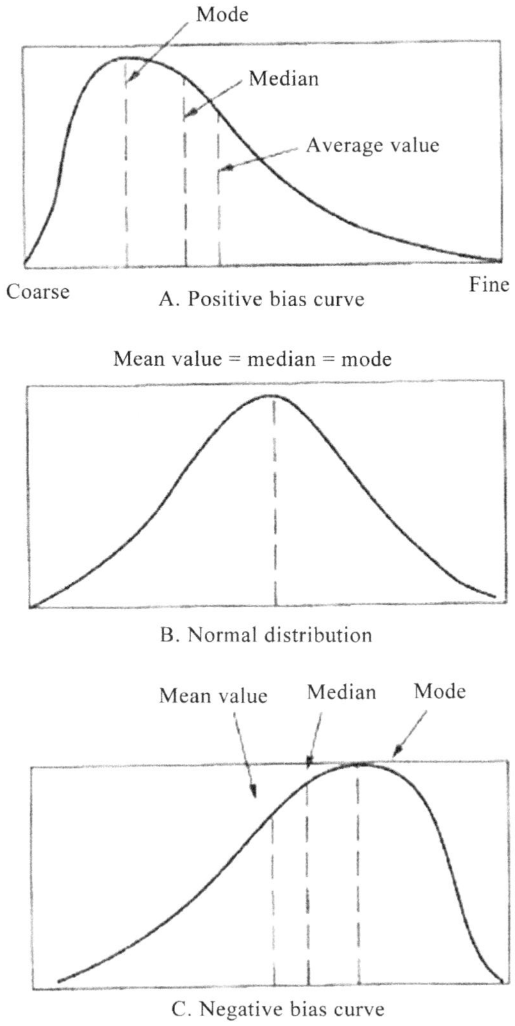

Fig. 5.18 Average values, medians and relationships between modes in three different conditions (Zheng 1982)

mode are grain size parameters used for specifying the central tendency of sediment grain size distribution.

(2) Standard deviation or sorting coefficient

Standard deviation (σ_i) is a characteristic number showing the degree of dispersion of a frequency curve. Grain size analysis mainly reflects the dispersion and concentration of particles, and shows the degree of dispersion of a sample relative to the average particle diameter. Hence, it is also known as the sorting coefficient. As with the mean value, many graphic methods can be used to calculate the sorting coefficient. The simplest among these methods is the Trask (1932) sorting coefficient (S_o), which can be determined as $S_o = \sqrt{Q1/Q_3}$, where Q_1 is the first quartile and Q_3 is the third quartile. Trask stipulates that the sorting is good when S_o is 1–2.5, medium when S_o is 2.5–4.5, and poor when S_o is greater than 4.5. The Trask sorting coefficient has been applied extensively, but its precision is relatively poor. As can be seen from the formula, because sorting changes are calculated only for accumulation frequencies of 25 and 75%, the sorting of sediments in different environments cannot be distinguished well. For example, all S_o of dunes, beaches and river sands are less than 2.5.

In order to make up for the deficiencies of the above formula, four intercepts, namely, $\Phi_{16}, \Phi_{84}, \Phi_5$, and Φ_{95} are applied to the graphic formula of Folk and Ward (1957) to comprehensively reflect changes in sample sorting, as expressed below:

$$\sigma_i = \frac{\Phi_{84} - \Phi_{16}}{4} + \frac{\Phi_{95} - \Phi_5}{6.6}$$

Based on the calculation of the majority of values of samples collected in different environments, the degree of sorting may be divided into seven levels (Table 5.3). For example, a low σ_i which indicates good sorting is typical for aeolian deposits; a high σ_i which suggests poor sorting is for fluvio-glacial deposits; and a moderate σ_i between them which shows moderate sorting is more typical for transitional deposits like delta and coastal systems.

(3) Skewness

Skewness (S_k) is a parameter for expressing the symmetry of a frequency curve, which can be divided into three types (Fig. 5.19, left) according to its symmetric configuration: ① single-peak symmetric curve in a normal distribution with the peak as the symmetry axis, with M_z (mean grain) = M_d (mid-value) = M_o (mode); ② asymmetric positive bias curve with

Table 5.3 Sorting levels based on standard deviation (quoted From Zheng 1982)

Classification	Range of σ_i		Sorting
	Folk and Ward (1957)	Friedman (1962)	
I	<0.35	<0.35	Very good
II	0.35–0.50	0.35–0.50	Good
III	0.50–0.70	0.50–0.80	Better
IV	0.70–1	0.80–1.4	Medium
V	1–2	1.4–2.0	Relatively poor
VI	2–4	2.0–2.6	Very poor
VII	>4	>2.6	Extremely poor

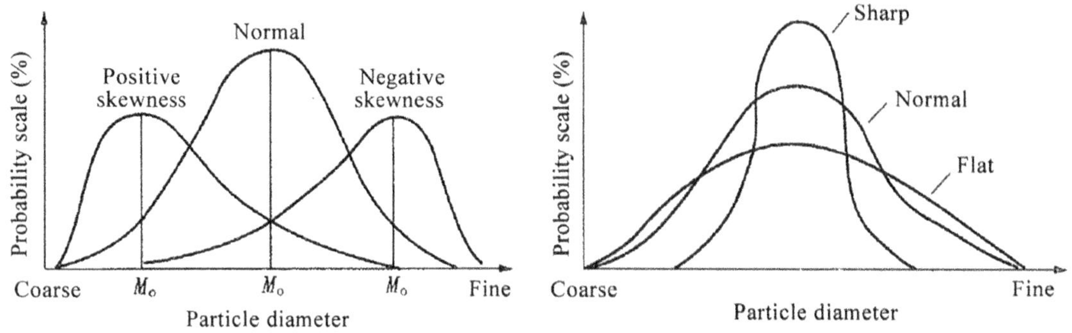

Fig. 5.19 Compared with the normal curve, the left graph is the contrast between positive bias and negative bias, and the right graph is the contrast between sharp peak and flat peak (Selly 1985). The figure on the left shows a comparison between positive skewness and negative skewness, and the figure on the right shows a comparison between sharp kurtosis and flat kurtosis

the main peak toward the coarser side, that is, sediments are based on coarse components; and ③ asymmetric negative bias curve with the main peak toward the coarser side, that is, sediments are based on fine components. Currently, Folk's calculation formula is mostly applied:

$$S_k = \frac{\Phi_{16} + \Phi_{84} - 2\Phi_{50}}{2(\Phi_{84} - \Phi_{16})} + \frac{\Phi_5 - \Phi_{95} - 2\Phi_{50}}{2(\Phi_{95} - \Phi_5)}$$

(4) Kurtosis

Kurtosis or sharpness (K_g) is a ratio of the spread angle of the tail part to that of the middle part of the frequency curve. Compared with the normal frequency curve, it indicates the sharpness or roundness (Fig. 5.19, right) of a curve. The calculation formula for kurtosis or sharpness (K_g) is below:

$$K_g = \frac{\Phi_{95} + \Phi_5}{2.44(\Phi_{75} - \Phi_{25})}$$

Due to different sedimentation control conditions in different sedimentary environments, the grain size distribution features are different (Table 5.4).

5.2.3 Grain Size Parameter Scatter Diagram

5.2.3.1 Conventional Grain Size Scatter Diagram

Although the above parameters have certain genesis significance, various conditions of

Table 5.4 Features of grain size parameters of sandy sediments in various environments (according to Reinecke in 1973; edited by Liu in 1980)

	Grain size parameter characteristics				
Sedimentary environment	Frequency curve pattern	Skewness (S_k)	Kurtosis	Standard deviation (standard deviation σ_i)	Grain size
Glacier	Saddle peak or multi-peak	S_k = 0, low positive skewness to low negative skewness	Broad peak	σ_I = 1.4 to greater than 2.6	Great change
River	Double-peak, common asymmetric curve	S_k = 0, great change, based on positive or negative skewness	Abnormal, generally low	σ_i = 0.52–1.40	Coarse lower part and fine upper part
Beach	Single-peak, symmetric, normal curve-dominated	S_k = 0–0.3, positive skewness, occasionally with negative skewness	Moderate, slightly narrow	σ_i = less than 0.35–0.5	Fine and medium grain-dominated
Wind dune	Single-peak curve, and obvious double-peak	0.13–0.3	Medium	σ_i = 0.21–0.26	Fine sand-dominated
Neritic sea	Single-peak	S_k = <1	Narrow	σ_i > 1–0	Fine sand-dominated

sedimentary environments are very complex on the ground, and there is no simple correlation between any parameter and the sedimentary environment. Thus, some data may not generalize various sedimentary environments. To improve the discrimination of environments by grain size, some scholars distinguish the differences between different environments by applying a combination of pairwise parameters to generate a scatter diagram. Friedman (1961, 1967 and 1979) figured out the grain size parameters of various environments, such as modern beaches, coastal dunes, inland dunes, and rivers, by applying M_z-σ_i, M_z-S_k, σ_I-S_k (Fig. 5.20), σ_I-K_g, and S_k-K_g, and then respectively preparing a scatter point diagram among various parameters to discuss the characteristics between two grain size parameters under different environments, which include 19 grain size scatter diagrams. Therefore, this diagram is used for comprehensively showing grain size parameter characteristics, and it is more meaningful than a single parameter. Sandy sediments with different genesis can be distinguished by preparing scatter diagrams of different parameters. As can be seen from the scatter points of

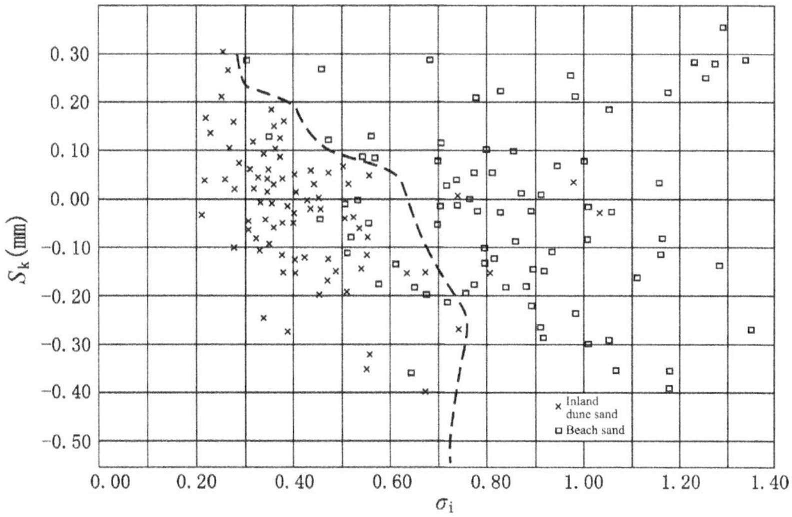

Fig. 5.20 Discrete graph between standard deviation and bias (Friedman 1979)

grain size parameters, although sand in different environments does not have an obvious and unified boundary, we can determine the total tendencies or partitions of different environments. Consequently, in earlier studies on depositional facies in one area, the grain size parameter characteristics of sediments or sedimentary rocks are normally analyzed by this method to provide additional geological basis for determining sedimentary environments.

5.2.3.2 C-M Diagram

A C-M (The C value is a particle diameter with a content of 1% on the accumulation curve, and the M value is a particle diameter value with a content of 50% on the accumulation curve) diagram is a comprehensive genesis diagram (Fig. 5.21) proposed by Passega (1957). It is a sample collection diagram that shows the relationships between sediment structure and sedimentation, and it is a grain size parameter scatter diagram.

1. **Sampling requirements**

Similar to the grain size scatter diagram, a C-M diagram cannot only use one sample, and it requires data about a set of genetic rock strata and a series of samples from coarse to fine must be collected sequentially. More exactly, systematic sampling should be carried out in a complete depositional cycle. In the case of any fault or unconformity, the diagram cannot be drawn by mixing the upper and lower samples. It is best to collect more than 30 samples continuously at certain equal intervals. The collected samples should comprise the cycle or the structure types of all grain sizes in the genetic rock stratum. Thus, respective sampling instead of mixed sampling for different rock strata or lithology should be carried out. Each sample represents a rock stratum with a thickness ranging from several centimeters to 1–2 m. Each C-M diagram, sometimes, can represent an isogenetic section or a stratum with thickness between several meters

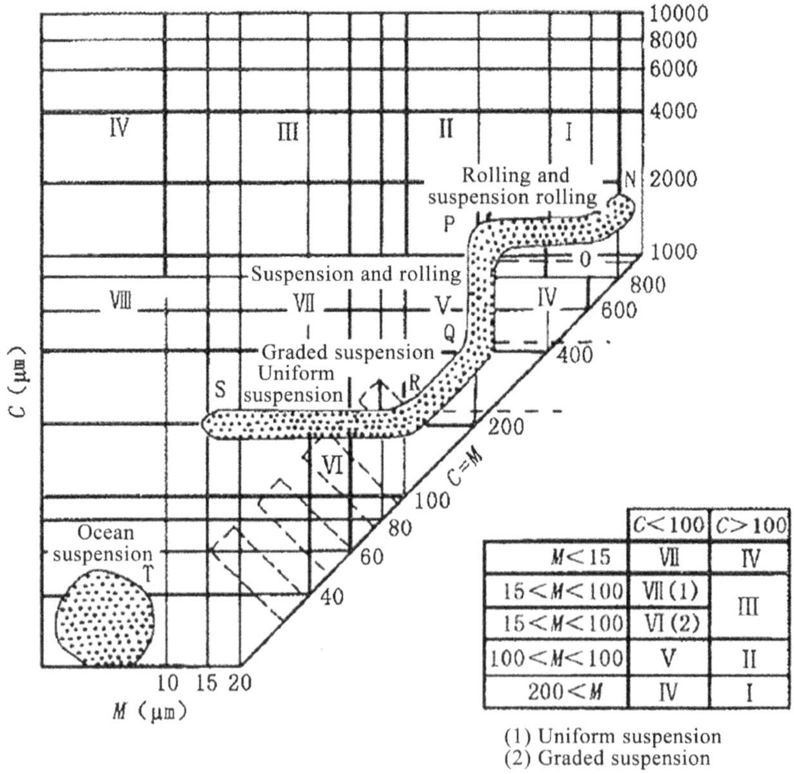

Fig. 5.21 C-M diagram, the comprehensive genetic diagram put forward by Passega (1957), is the sample set graph representing sedimentation relationships, and also is the granularity parameter scatter diagram. It is C-M diagram of the deposits from turbidity current and traction current (Passega 1964)

and dozens of meters. It thus appears that the C-M diagram is rarely made for drilling with little coring.

2. **Key points of C-M diagram**

Passega (1957) believed that the two grain size parameters, namely, C value and M value, can reflect medium transport and sedimentation. Thus, the two parameters are taken as the ordinate and the abscissa of a bi-logarithmic coordinate to form the C-M diagram. The C value is a particle diameter with a content of 1% on the accumulation curve, and the M value is a particle diameter value with a content of 50% on the accumulation curve. The typical C-M diagram can be divided into various sections such as NO, OP, PQ, QR, RS, as well as a region T. The different sections represent products with different sedimentations.

(1) NO section: This represents coarse-grained materials transported by rolling, where C is greater than 1 mm.
(2) OP section: This refers to the mixture of the rolling and suspension components, dominated by rolling transport, where C is generally more than 800 μm, but M varies obviously.
(3) PQ section: This contains a small amount of rolling components, dominated by suspension transport, and C varies but M remains unchanged.
(4) QR section: This represents the graded suspension section, where graded suspension transport means that the grain size of the suspended materials gradually fines from the bottom up and their density decreases gradually in the fluid; C is proportional to M, which causes a section of the graph to be parallel to the C = M base line.
(5) RS section: This refers to a uniform suspension section composed mainly of fine silt deposits, where C shows little change, but M varies considerably.
(6) Region T: This includes ocean-suspended matter, and M is less than 10 μm.

3. **Two fundamental C-M diagrams and environmental significance**

Passega (1957) drew the two most fundamental C-M diagrams, namely, density current (gravity flow) and traction current, by researching C-M diagrams of modern and ancient deposits in various known environments and the relationships between such C-M diagrams and the corresponding sedimentation. Since they respectively represent density current deposition and traction current deposition, which are clearly distinct, we can successfully distinguish the deposits formed by traction and density currents by using a C-M diagram.

3.1 Density current (or sediment gravity flow)

Density difference, which is a cause of density current movement, may be generated by salinity and temperature, but the cause is sediment suspension (Zheng 1982). A density current formed by high-density suspended particles is a mixture of water and suspended matter. It moves downward along the slope because of gravity, instead of fluid flow, and acts on high-density solid matters. Thus, it is also called gravity flow. Based on the particle support mechanism, gravity flow can be divided into ① debris flow, which is a mixture of water and mud-and-sand (matrix) for supporting the sand and gravels therein, with almost no sorting inside; ② grain flow for supporting grains by collision among particles; ③ fluidized flow, supporting particles by upward movement of liquid through pores; and ④ turbidity current, which is a mixture of water and deposits, supporting forward movement of particles by turbulent flow. Regardless of the type of support mechanism in play, gravity flow moves in an entirely suspended way, and particles are carried mainly in a suspended state. When various forces of gravity flow decrease, density flow is deposited gradually in a graded suspension sate. Within a genetic unit, the lithology always changes from coarse to fine, and each sample

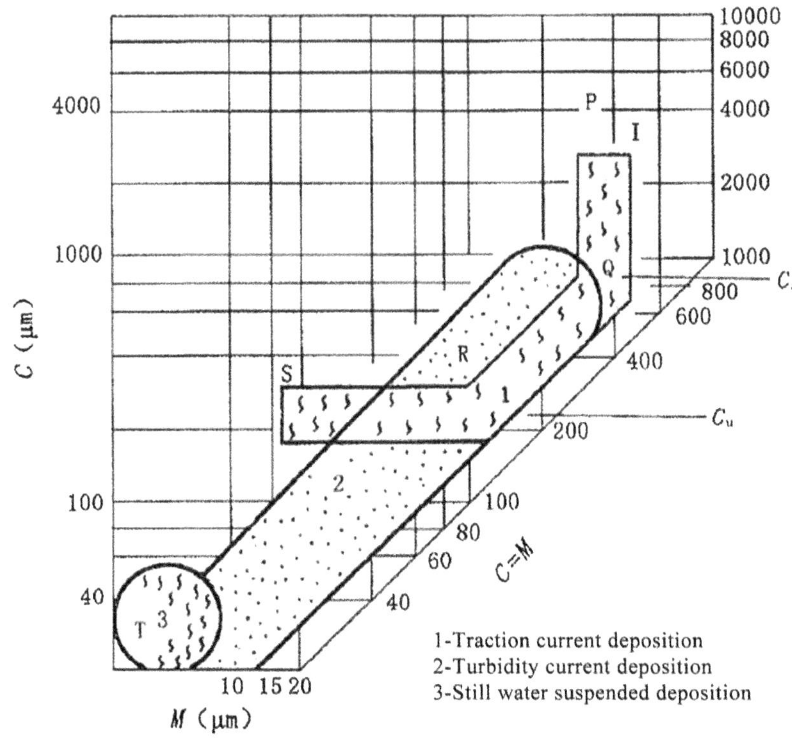

Fig. 5.22 Passega summarizes two most basic C-M diagram types, i.e. density current type and traction current type (modified by Passega 1964). T Pelagic suspension, S homogeneous suspension, R gradational suspension, Q suspension and evolution, P suspension and suspending evolution

varies in terms of the coarse:fine component ratio. In consequence, C and M values at the difference parts of density flow deposition are approximately equal owing to the roughly identical ratios of the maximum particle diameter to the average particle diameter. Because all sample point distributions are almost parallel to the C = M base line, a long strip graph (Fig. 5.22) that is approximately parallel to the C = M base line is formed by density current (gravity flow).

Many scholars at home and abroad have researched (A. Rizzin and R. Passega 1964; W. B. Bull 1972; Zheng 1982) C-M diagrams of various gravity flows (density currents) and analyzed the characteristics of the C-M diagrams (Fig. 5.23) of debris flow, slump gravity flow, deep water turbidity current, and gravity flow channel deposition. In all of these diagrams, the graph is parallel to the C = M base line; the differences lie in graph location, dispersion degree, and C/M ratio, namely, the degrees of closeness with the C = M base line are different: ① debris flow is coarse in size fraction and poor in sorting, C/M ratio is greater than 40, and graph is far away from the C = M base line and very dispersed; ② shallow turbidity current is moderate in sorting, C/M ratio is about 4, and the C-M diagram is relatively concentrated between debris flow and typical turbidity current; and ③ deep water turbidity current and gravity flow channels have better sorting, C/M ratio is 2–3, and the diagram is narrow. The C-M diagrams of various density currents vary regularly from debris flow to typical turbidity current, namely, grain size compositions of the diagrams vary from coarse to fine, dispersion to concentration, broad to narrow, and from far away from the C = M base line to near the C = M base line.

3.2 Traction current

As mentioned above, traction current has three transport modes for clastic particles, namely, rolling, saltation, and suspension. Traction current-type C-M diagrams are formed at the P-Q-R-S sections because of the three different transport modes of traction current, in which section PQ is parallel to the Cv, section QR is parallel to the C = M base line, and section RS is parallel to the M axis (Fig. 5.22).

(1) Section PQ: This represents the coarsest deposits in the bed load, which is composed of graded suspended matters represented by point Q and a small amount of rolling particles. Therefore, it is also known as the rolling section owing to the presence of rolling particles. From upstream to downstream, the C value varies but the M value remains unchanged. The greatly varying C value indicates that the grain sizes of the particles rolling downstream are decreased owing to a decrease in hydrodynamic force, but the M value remains unchanged because the quantity of rolling particles is not in a dominant position.
(2) Section QR: This is also called the graded suspension section, and it represents graded suspension deposition, where the graph is parallel to the C = M base line because C is proportional to M. In graded suspension, the grain sizes of deposits are fine gradually from the bottom to the top, whereas particle densities decrease gradually in the fluid. Deposits are generally located in the lower part of the current, but they may all be transported in graded suspension form when the flow velocity is faster and the lift force caused by turbulence is large. As a result, such graded suspensions are formed mainly at various particle saltation heights arising from various lift forces. The maximum C value in section QR is the value C at point Q, denoted by C_s. This value is called the maximum turbulence index, and it represents the maximum disturbance capacity at the bottom of a traction current.
(3) RS section: This refers to a uniform suspension section with complete suspension transport, in which particle diameter and density do not vary approximately as the depth changes. Because uniform suspension is a transport mode in the upper water flow that is not sorted by bottom current, general grain size in the river does not change from upstream to downstream, but only the coarse grain content decreases gradually. Consequently, the value of C changes little in the RS section, whereas that of M diminishes to the end S, and the graph is approximately parallel to the M axis. The C value at point R is the maximum value V at this section, denoted by C_u, which is the maximum particle diameter of uniform suspension, generally 200–250 μm.

On a C-M diagram, there are obvious different changes (Fig. 5.23d–g) from alluvial fan to river

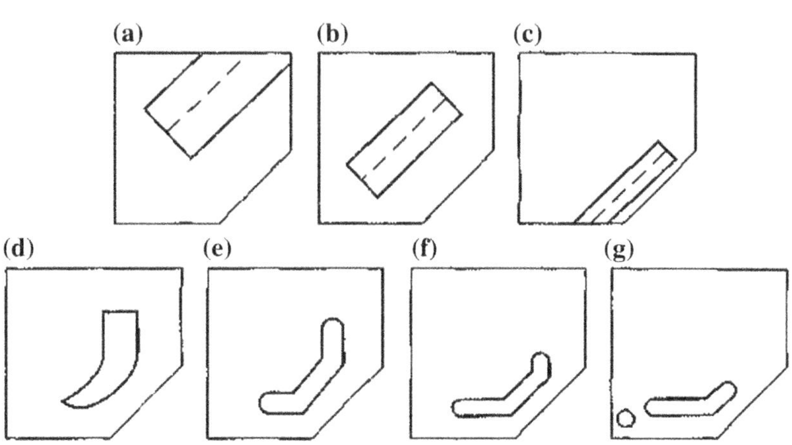

Fig. 5.23 C-M diagram variation of the deposits from different gravity currents and land traction currents (modified by Zheng 1982). A—Debris flow; B—shallow turbidity current; C—typical turbidity current; D—alluvial fan; E—braided river; F—meandering river; and G—delta

Table 5.5 Development of all feature segments in the C-M diagram of continental traction current deposition

Main sedimentary environment	Development of all sections in C-M diagram		
	Section PQ	Section QR	Section RS
Alluvial fan	Dominant structure, wider graph	Well developed	Underdeveloped
Braided river	More developed	Developed	Developed
Meandering river	Based on two sections		Generally absent
Delta or anastomosing river	Underdeveloped	Well developed	Dominant structure

to delta, even from upstream to downstream in a continental environment. Different sedimentary environments have different development statuses (Table 5.5) in all sections of a C-M diagram. The basic law of this change refers to changes in grain size from coarse to thin, where section PQ gradually plays a secondary role, and finally disappears; section RS gradually plays a major role; grain size range of section QR decreases, which reflects that water flows slowly; and transport is based on saltation and suspension instead of rolling and saltation. The understanding of this change tendency is conducive to the termination and acquaintance of sedimentary environments, which is an overall law. Specific graphs for different areas, basins, and horizons have different changes.

5.3 Characteristics and Identification of Sedimentary Structures

Because the sedimentary structure is one of main characteristics that is widely found and easiest to observe directly in both sediments and sedimentary rocks, accurate identification of sedimentary rocks lays a solid foundation for determining sedimentary environments precisely. Both study on sediments or sedimentary rocks and explanation of sedimentary environments necessarily involve sedimentary structures. As a result, correct identification of sedimentary structures is the key to accurately judging sedimentary environments.

5.3.1 Basic Concept

5.3.1.1 Definition

A sedimentary structure represents the overall characteristics presented by the modes of spatial distribution and arrangement of all components in sedimentary rocks, or the sum total of mutual arrangement of particles that form a rock.

5.3.1.2 Effect

Research on sedimentary structures can be performed to determine the sedimentary environment, top and bottom of stratum and sequence, analyze and recover the current system, and indicate current state. Such research is often beneficial for ① determining the top and bottom of a stratum to determine its stratigraphic sequence; ② determining the transport and sedimentation modes of sediments, property of sedimentary medium, and hydrodynamic conditions of fluid; ③ recovering paleocurrent and paleosedimentary environments in the sedimentary basin; and ④ speculating the physical and chemical changes upon sedimentation.

5.3.1.3 Classification

Sedimentary structures of clastic rocks can be classified using many different schemes. According to the classification of development positions and forms, sedimentary structures are divided into intrastratal structure (such as bedding structure) and bedding plane structure; based on the formation of sedimentary rocks, they are divided into sedimentary stage, diagenetic stage, and epigenetic stage; on the basis of

Table 5.6 Classification of sedimentary structure of clastic rocks

Basis	Classification result	
Mechanical genesis	Bedding structure	Horizontal bedding
		Parallel bedding
		Cross bedding
		Compound bedding
		Graded bedding
		Rhythmic bedding
		Homogenous bedding
	Bedding plane structure	Ripple
		Flute cast and other molds
		Mud crack
		Raindrop imprint and hailstone imprint
	Deformation structures	Load structure
		Ball and pillow structure
		Slump structure
		Convolution structure
		Sandstone dike
Chemical characteristics	Salt and crystal imprint	–
	Nodule	–
Biogenesis	Trace fossil	–
	Bioturbation structure	–
	Plant mark and root trace	–

genesis classification, they are divided into mechanical genesis, biogenesis, and chemical genesis. Genetic texture classification has been adopted widely. In this book, genesis classification is employed to divide sedimentary structures into mechanical genesis, biogenesis, and chemical genesis (Table 5.6).

5.3.1.4 Basic Units

1. **Laminar**

This refers to the most basic and minimum unit for forming bedding, in which no layer can be seen visually. In particular, it is a "footprint" of the sandy bedform left during migration, which is usually a few millimeters to several centimeters in thickness. Thicker laminae have only been found in conglomerates.

2. **Set**

This is formed by multiple identical laminae that are similar in terms of ingredient, structure, thickness, and occurrence. They are products with relatively stable hydrodynamic conditions over a period of time and are formed under identical sedimentary conditions. As a result of the migration of the entire sandy bedform, set thickness (thickness between two set boundaries) generally represents the height of the sandy bedform.

Moreover, laminae can be divided into incised laminae and truncated laminae (Fig. 5.24) according to the angle of intersection (such as convergent or linear intersection) between a lamina and the set boundary. The former is characterized by a convergent chamfer-type intersection between laminae and the set

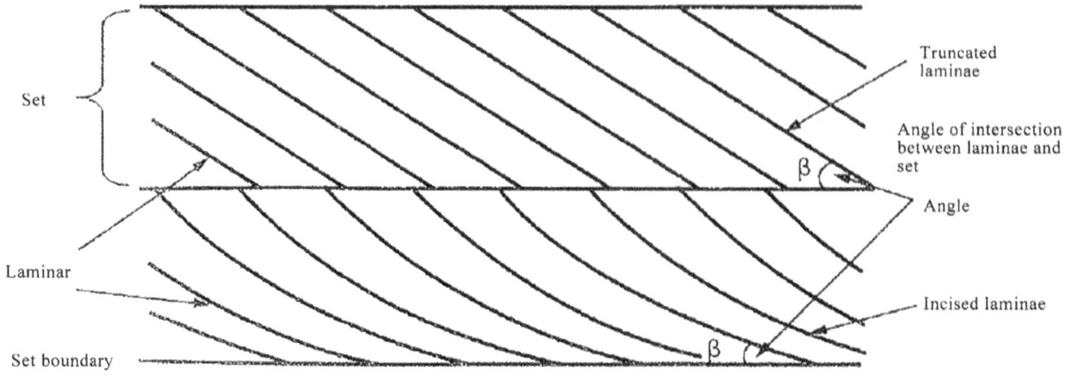

Fig. 5.24 Basic elements of the bedding

boundary, while the latter is characterized by a linear or truncated intersection. As for the hydrodynamic conditions, the former is represented by tranquil flow, while the latter by torrent. Alternatively, the former results from the migration of sand bodies, while the latter results mainly from sand body accretion. Incised laminae are slightly bent upward but convergent downward, and the dip angle varies considerably owing to the forming environment and background. Therefore, they can be used for determining the top and bottom of a rock stratum, namely occurrence. Truncated laminae generally have the characteristics of straightness, large dip angle, and strong current.

This also shows that laminae identification can greatly help with the determination of sedimentary microfacies. Thus, the properties (unidirectional or bidirectional) and direction of a current can be judged by observing the form, dip angle, and modes of laminae. If the angle of intersection between laminae and the set is greater than 15°, it is considered a high angle; otherwise, it is considered a low angle.

3. **Coset**

Laminae can be divided into single groups and multigroups depending on whether the coset is composed of two or more cosets with similar lithology or genetic relationships. Genetically, the coset is a sedimentary unit with no obvious sedimentary depositional break. It is formed under similar hydrodynamic conditions at different times in the same environment.

4. **Bedding**

Bedding represents the general performance of laminae and set in the vertical direction. It is one of the most typical and important sedimentary characteristics of clastic rocks, and a direct reflection of the sediments deposited under hydrodynamic conditions. It is also one of the important marks of sedimentary environments. Bedding is a laminar structure in which rock property changes in the vertical direction. It can show (Fig. 5.24) abrupt or gradual changes in ingredients, structures, and mineral colors.

5.3.2 Sedimentary Structure of Mechanical Genesis

5.3.2.1 Bedding Structure

Because bedding is the most important sedimentary characteristic of sedimentary rocks, the study of bedding is important from theoretical and practical viewpoints. Moreover, it is beneficial for correcting the division and correlation of strata and recovering the original order of a stratum. Strata can be divided and correlated reasonably based on the characteristic that sedimentary rock has a planar bedding structure. Cross bedding is the most important directional structure determining the paleocurrent system.

Table 5.7 Correspondence among different flow regimes, sandy bedforms, and formed beddings

Flow regime	Sandy bedform	Wave ridge form	Bedding type	Scale of cross bedding
Upper flow regime ($F_r > 1$)	Regressive sand wave	Linear	Anti-dune cross bedding	Megascopic-dominated
	Upper flat bed		Parallel Bedding	
Transition type (F_r–1)	Transition type		Transitional bedding	
Low flow regime ($F_r < 1$)	Sand dune (dune)	Egg shaped	Trough cross bedding	Megascopic
		Crescent shaped		
		Wavy		
	Sand wave	Linear	Planar/Tabular cross bedding	
	Ripple	Egg shaped	Ripple cross bedding	Miniature
		Wavy		
		Linear		
	Lower flat bed		Horizontal bedding	

The sedimentary environment can be concluded based on the bedding type. As mentioned previously, different bedforms can be formed in different flow regimes, while different beddings can be formed by different sandy bedforms (Table 5.7). Detailed description and study of various sedimentary structures are important for hydrodynamic force analysis and sedimentary environment identification.

1. **Horizontal bedding**

Laminae are parallel to each other linearly and parallel to the bedding plane. The fine-grained layer is 1–2 mm in thickness. It is also called horizontal laminae owing to the lack of an obvious set boundary. Generally, this bedding is recognized to be formed by the deposition of substances in suspension or solution under relatively stable hydrodynamic conditions. Bedding is caused by various changes in substances entering the deposits, such as grain size change, opaque mineral distribution, and bedding arrangement of sheet mica and carbonaceous fragments.

Horizontal bedding is distributed extensively in fine-grained silt and mud (Atlas I-A). Also, seasonal bedding or varve, which generally shows a low-energy, relatively quiet environment formed by seasonal climate changes, belongs to the low flow regime ($F_r \ll 1$). Hence, it is commonly found in the deep water belts of seas/lakes, closed bays, lagoons, swamps, and oxbow lakes.

2. **Parallel bedding**

Unlike horizontal bedding, parallel bedding is composed of laminar sandy rocks that are parallel but almost horizontal. Laminae measuring 1–2 mm in thickness are formed by flowing water in stronger hydrodynamic conditions, representing static water sedimentation. We can see that the parting lineation (separation line) for representing a product in the case of rapid flow in a shallow area belongs to the upper flow regime ($F_r > 1$).

Generally, parallel bedding comprises overlaying laminae with particles of different sizes, overlaying laminae containing different heavy minerals, or both (Table 5.8, Atlas I-B). It is well developed in fine and medium sand, and occasionally in fine silt. Strata are often divided by erosion surfaces with extremely low angles, together with large cross bedding. Parallel bedding is generally developed in sedimentary areas in shore-shallow sandy environments [lake shores, beaches, turbidite, and rivers (sandy flat in rivers)].

3. **Cross bedding**

This is formed by a series of laminae that intersect obliquely on the set boundary. Laminae and set may be combined in mutually overlaying, staggered, and cutting ways. Cross bedding results from the flow of a sedimentary medium. When the flow velocity is fixed, a series of sand waves may be generated on the sand bed, and an oblique set consisting of a series of laminae is formed on the steep slope side owing to the downstream migration of sand slopes. The incline direction represents the flow direction of the medium. Cross beddings of different stages are formed by mutually parallel or cut sets. Cross bedding may show different characteristics along different incline directions in a section. For example, in the section corresponding to the downstream direction, laminae are inclined obliquely, but they may be staggered in a horizontal, wavy, or even gently inclined manner in the vertical current direction. Therefore, a greater number of sections should be considered for field observation.

According to the shapes and properties of the sets and the upper and lower boundaries, cross bedding can be divided into 8 types:

(1) Tabular or planar cross bedding

Set boundaries are planar and mutually parallel, and laminae and sets intersect (Table 5.8, Atlas II) in the same direction in different ways. Cross bedding can be divided further (Table 5.9) according to the type of intersection between laminae and sets, as well as the number of sets. Megascopic planar cross bedding is most typical in river deposition. Scour surfaces can be seen in the set boundary. Within the laminae, particles change from thicker to thinner moving from the lower to the upper parts, and some laminae converge downward.

(2) Wedge cross bedding

Set boundaries are flat surfaces, but the upper and lower boundaries of mutually parallel laminae are non-parallel planes. In other words, the inner parts of cosets intersect at low angles, set thickness changes in an obvious wedge shape (Atlas I-C), sets are often mutually cut, and the incline direction and dip angle change quickly in seas, shallow lake belts, and deltas. Generally, wedge cross bedding is presented as tabular or planar cross bedding vertical to the current direction.

(3) Trough cross bedding

The set boundary is a grooved scouring surface, and it cuts the tops of laminae. On the cross section, the set boundary and the laminae are trough shaped (Table 5.8); on the longitudinal section, the set boundary is arc shaped, and it intersects obliquely with the laminae; according to the section characteristics of trough cross bedding, it may be divided into concentric groove and exocentric groove (Table 5.10, Atlas I-D, E and F) because the top surface is petaloid; obvious scouring surfaces of large trough cross bedding boundaries have been found commonly in rivers and various channel depositions, with common boulder clays and retained conglomerates in the bottom part.

(4) Wavy cross bedding

The boundary of this cross bedding unit is a wavy curved surface. The upper and lower boundaries may be parallel to each other or they may intersect. Progradation laminae in the wavy cross bedding intersect obliquely with the boundary in sections parallel to the flow direction, while they are approximately parallel to or intersect obliquely (Table 5.8, Atlas III-A) with the wavy boundary in sections vertical to the

Table 5.8 Main sedimentary structure classification of sandstone

Description	Main characteristics	Diagram (Harms 1960)
Parallel bedding	Laminae are parallel to each other, coarse sand fraction is fine-grained, there is usually a parting lineation, and it is formed by unidirectional current in the upper flow regime	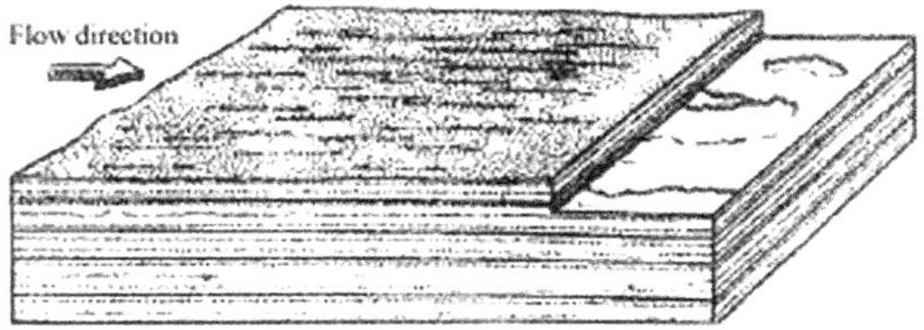
Planar/Tabular cross bedding	Sets are parallel to each other; laminae and sets are staggered; this fraction is fine-grained level or above; and it is a sandy wave formed by a unidirectional current in the low flow regime, mainly owing to the migration of various sand bars (sandy bedforms)	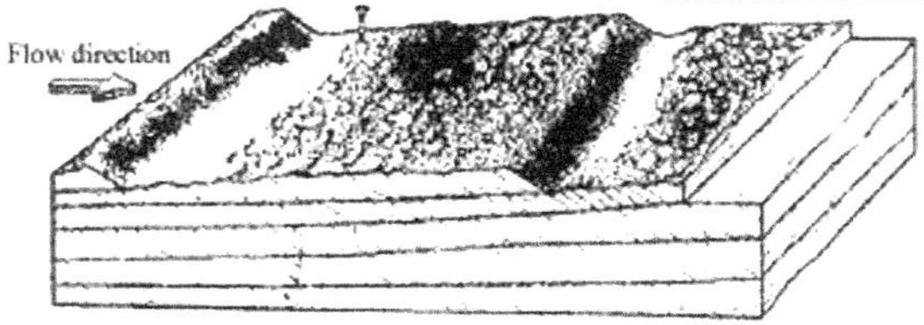
Trough cross bedding	Sets are parallel to each other; laminae and sets intersect, but are concaved simultaneously; this fraction is fine-grained level or above; and it is a sand ridge formed by incising or filling action of unidirectional current and channel in the low flow regime	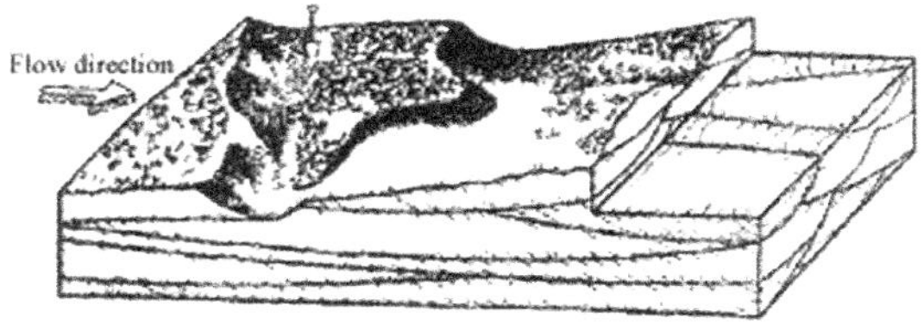

(continued)

Table 5.8 (continued)

Description	Main characteristics	Diagram (Harms 1960)
Wavy cross bedding	Set boundaries fluctuate; laminae and set, and mutually parallel laminae intersect obliquely; this fraction is medium-grained level or below; and it is a low-flow-regime product of unidirectional current with wave influence	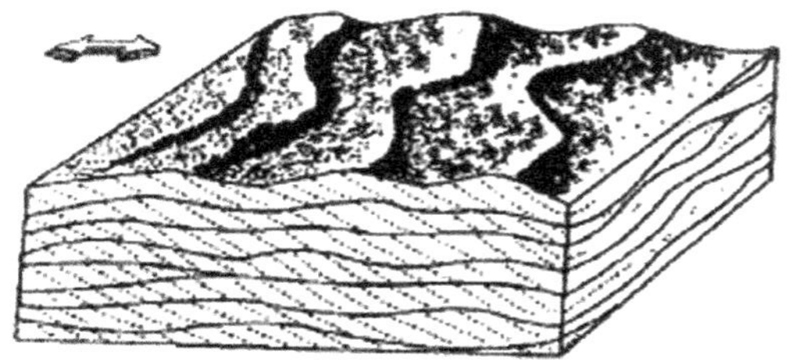
Herringbone cross bedding	Sets intersect obliquely, and thin mud layer is used for spacing oblique sets with opposing tendencies; laminae are flat or bent slightly upward, and incline directions of adjacent oblique sets are opposite. Laminae intersect in a feathered pattern at an acute angle while extending toward the set boundary; and low flow regime is formed in the presence of reversed current	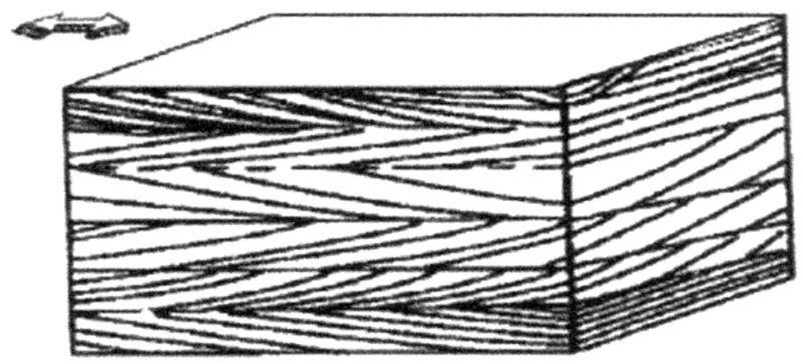
Swash cross bedding	Fine-grained (laminae) layers are in contact with the set boundary at low angles, and adjacent fine-grained layers have different incline directions and dip angles, mainly inclined offward; owing to good grain size sorting, reverse grading may be formed, resulting in an abundance of heavy minerals in the fine-grained layer; most erosional contacts (reactivation surface) are formed between the sets, but non-erosional contacts are formed as well; and owing to small dip angle, fine-grained layer extends farther laterally and has relatively stable thickness	

(continued)

Table 5.8 (continued)

Description	Main characteristics	Diagram (Harms 1960)
Hummocky cross bedding	This is formed by some wide-gentle wavy sets, with a hummocky top; laminae incline around, upper part of the set is eroded, laminae are almost parallel to the boundary of the set, the middle part thereof is of a dispersed-convergent shape, and the laminae have a greatly varying small dip angle (generally smaller than 15°); and it is the main product formed by storm wave oscillation below nearshore zone and normal wave base	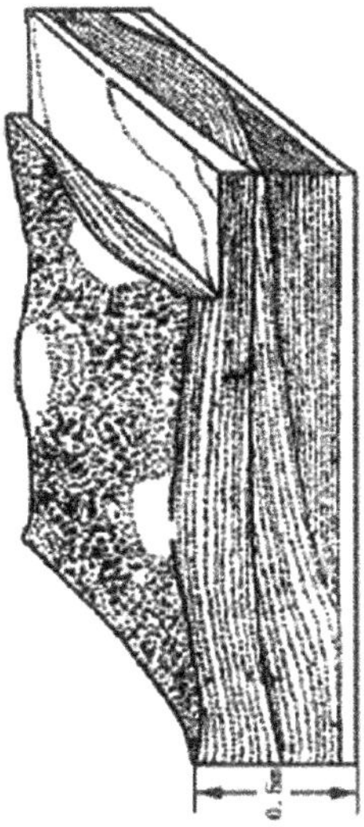

flow direction. The cross bedding formed by wave ripples is typically wave cross bedding.

(5) Herringbone cross bedding

This is a specific type of cross bedding, and it features laminae that are flat or slightly bent upward. The laminae of the adjacent oblique sets are inclined in the opposite directions, and they intersect in a herringbone shape (Table 5.8, Atlas III-B) at an acute angle while extending toward the set boundary. Herringbone cross bedding is formed commonly in tidal deposits, where a flood flows into a tidal channel in the presence of ebb currents, and sometimes, two inclined sets in two opposite directions are spaced by a thin mud layer (mud drapes). Thus, herringbone bedding is a typical record of a tidal environment.

(6) Swash cross bedding

On the seaward slope surface of an oblong and flat beach or longshore bar, internal friction of water particles and friction resistance with the seabed result from wave movement as waves are propagated toward a shallow beach or sand bar, thus leading to their deformation. The movement trace of water particles is a round-elliptic-flat-shoreward-offshore line, which causes the washing action of rush current and reflux. As a result, swash cross bedding is formed owing to the washing action of waves on either a beach or a sand bar. Such cross bedding has the following characteristics: ① fine-grained layer (lamina) is in contact with the set boundary at low angles of 2°–10°; ② adjacent fine-grained layers have different directions and dip angles, but they are mainly inclined offward; ③ owing to good grain size sorting, reverse grading may be formed to result in the presence of lots of heavy minerals in the fine-grained layers; ④ most erosional contacts (reactivation surface) are formed between sets, but non-erosional contacts are formed as well; and ⑤ owing to a small dip angle, the fine-grained layer extends farther laterally and has relatively stable thickness, mostly having large wedge or planar cross bedding (Table 5.8, Atlas III-C) morphology. Such cross beddings

Table 5.9 Classification of planar cross bedding

Characteristics	Classification basis			Genetic explanation	
	No. of sets	Intersection angle between laminae and set	Contact relation between laminae and set	Effect	Product
Planar/Tabular cross bedding	Single-group	High angle (>15°)	Incised	Downstream accretion	Longitudinal bar
			Truncated	Vertical accretion	Transverse bar
		Low angle (<15°)	Incised	Lateral accretion	Diagonal bar or point bar
			Truncated	Vertical accretion	Transverse bar
	Multi-group (≥2)	High angle (>15°)	Incised	Lateral accretion or progradation	Diagonal bar or stream mouth bar
			Truncated	Vertical accretion	Transverse bar
		Low angle (<15°)	Incised	Lateral accretion	Point bar
			Truncated	Vertical accretion	Transverse bar

are mainly distributed in the sedimentary environments of beaches and longshore bars, even the lower parts of intertidal zones. Low-angle bidirectional cross bedding is formed by fine-grained layers consisting mainly of particles and fossil fragments in carbonate rock.

(7) Hummocky cross bedding

This is formed by some wide-gentle wavy sets, with a hummocky top; the laminae incline around, upper part of the set is eroded, laminae are almost parallel to the boundary of the set and the middle part thereof has a dispersed-convergent shape, and laminae have a widely varying dip angle (generally smaller than 15°). Instead, sunken laminae have similar characteristics. The difference is that the upper part of the former is characterized by a mound set, and the lower part of the latter is characterized by a sunken wavy set (Table 5.8, Atlas III-D). Earlier, the occurrence conditions of these beddings were considered as the main marks of storm deposit formed owing to storm wave vibration under the nearshore zone and the normal wave base level. Recent research achievements show that slightly small-scale storm cross bedding can be formed in lakes, such as Lijin Sag of the Bohai Bay Basin, East China (Wang et al. 2015).

(8) Anti-dune cross bedding

This consists of inclined laminae generated by the migration of anti-dunes. The set is of a lens shape, in which the laminae are inclined at low angles (less than 10°). The incline direction depends on whether the anti-dunes migrate upstream or downstream. Fuzzy veins are often symbiotic to parallel bedding. As a mark of high-speed current in the upper flow regime, it has been found only in environments where water flows rapidly in shallow area (beach and river). There are three different cases in terms of the migration direction of anti-dunes and the corresponding inner parts: ① anti-dunes migrate downstream, in which unapparent low-angle laminae are deposited on the slip slope (Fig. 5.25a); ② staying in situ, laminae cover the anti-dunes (Fig. 5.25b); and ③ current moves downstream along sand particles eroded on the

Table 5.10 Classification and characteristics of trough cross bedding

Description		Characteristics	Section diagram	Outcrop pictures
Exocentric groove		① Obliquely intersecting laminae and set; ② Inconsistently concaved set and laminae; ③ Channel incised and migrated; and ④ Developed mostly in meandering rivers		
Concentric groove	Concentric, with same thickness	① Parallel laminae and set; ② Consistent laminae and set; ③ Channel incised and filled evenly; and ④ Developed mostly in anastomosing rivers		
	Concentric, with different thicknesses	① Simultaneously concave laminae and set; ② Inconsistently concaved laminae and set; ③ Channel incised and filled quickly; and ④ Developed mostly in braided rivers	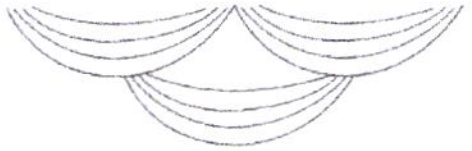	

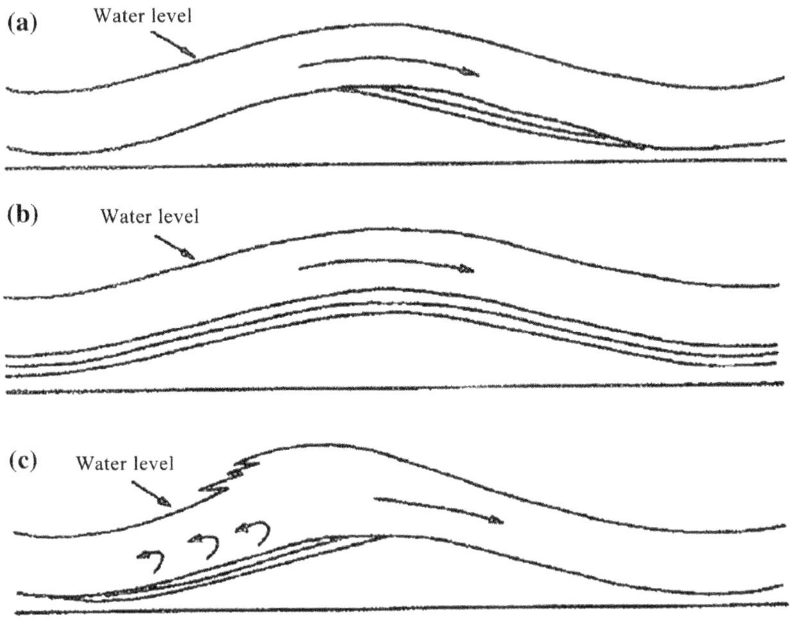

Fig. 5.25 Schematic diagrams of three different deposition patterns in antidune (from Chen Jingshan 1981): **a** the antidune moves downstream, **b** the antidune keeps still, **c** the antidune moves upstream

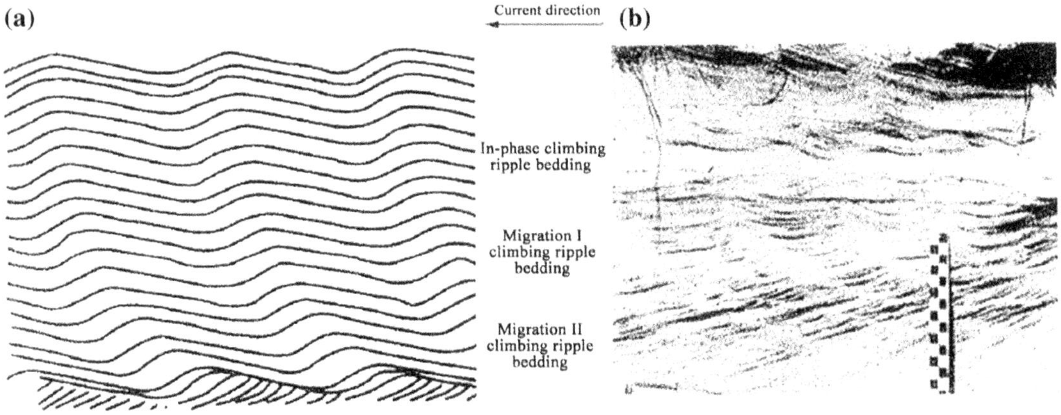

Fig. 5.26 Various types of climbing ripple beddings (according to the editing group of *Sedimentary Structure and Environmental Interpretation*, 1984). **a** Various climbing ripple beddings; and **b** climbing ripple bedding in modern natural levee deposition in Lihe, Hebei

slip slope of the dune to accumulate in the low flow regime of the next dune, which causes the dune to move upstream. The preservation of anti-dune cross bedding is possibly developed by supercritical flows with *Fr* more than 1.

(9) Ripple crossing bedding

Based on genesis, bedding may be divided into current ripples, wave ripples, and wind ripples, which clearly have different characteristics (Fig. 5.26) on the section.

(i) Current ripple cross bedding (Atlas III-E), which is also called climbing ripple lamination, refers to a series of mutually overlaid ripple layers formed by the forward migration and upward growth of ripples. In sediment-rich sedimentary environments, especially suspended sediments, owing to the continuous action of current, ripples not only move forward but also grow upward, forming climbing-ripple lamination. According to the migration of ripple laminae, current-ripple cross bedding may be divided into in-phase and migrated ripple vein beddings (according to Fig. 5.26, Atlas III-F). Ripples have mostly clear asymmetry, but water flows from gentle to steep backlimb. This type of cross bedding is distributed mainly in rivers, deltas, and turbidity currents.

① In-phase current-ripple cross bedding is characterized by ripple laminae directly covering other ripple laminae, and its wave ridge is located on the same vertical line (in-phase) without clear or slight migration in the current direction. Ripples are essentially parallel to each other, with approximately equal laminae thicknesses on the stoss side and the slip slope. The laminae have asymmetric shapes. Ripple crossing bedding is formed when flow velocity, current direction, water depth, and sediment supply remain essentially unchanged.

② Migrating phase current-ripple cross bedding: Ripple laminae have the prominent feature that the wave ridge grows upward and migrates downstream in sections vertically and they even incline in some upstream directions, the dip angle of which decreases as the flow velocity increases.

Migrating phase current-ripple cross bedding may be divided into types I and II (Fig. 5.26) depending on the preservation of laminae on the ripple stoss sides. Type I stoss side and slip slope laminae are preserved well; type II stoss side laminae are eroded, whereas type II slip slope laminae are preserved. The abovementioned in-phase climbing ripple lamination and migrating phase current-ripple cross bedding types I and II are typical examples in a series of continuous changes, and there are various transition types in between the two types. Occasionally, gradual evolution of the two types may be seen in one section. This evolution is related to changes in the ratio of suspended load to bed load in the sedimentary environment. As this ratio decreases, in-phase current-ripple cross beddings migrated from type I to II are formed sequentially until current-ripple cross bedding without climbing is formed.

(ii) Wave ripple cross bedding, which is also called wave ripple bedding, is generated by wave ripple migration. This cross bedding appears similar to trough cross bedding in the section parallel to the wave propagation direction. In sections vertical to the wave propagation direction, the upper and lower boundaries of most cross bedding units are mutually parallel to each other; sometimes, they converge to intersect at one end, and the laminae are approximately parallel to the boundary (Fig. 5.27, Atlas IV-A).

It is more difficult to distinguish wave ripple cross bedding from current ripple bedding. The following features are identifiers of wave ripple bedding: First, the lower boundary of the cross bedding unit formed by wave ripple bedding often has a bending wavy shape because of irregular wave ripple migration. This can be seen

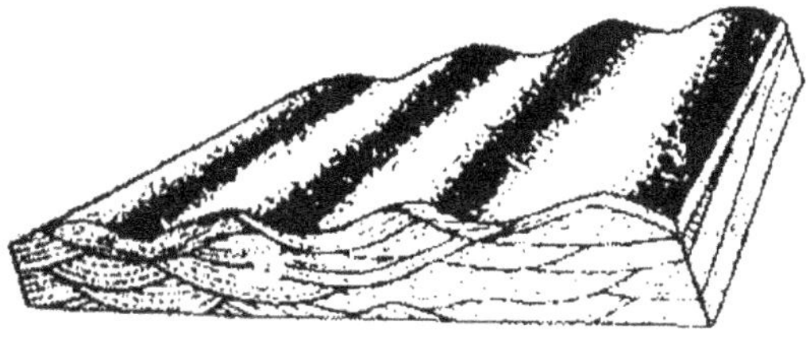

Fig. 5.27 Cross bedding of wave sand ripples (according to Reineck and Singh 1973)

obviously in sections parallel to the wave propagation direction. Second, foreset laminae show distinctive herringbone (in wave ridge) and inverted herringbone (in trough) configurations. Third, the foreset laminae in one cross bedding unit are overlaid in bunches in a branching manner, and they extend to the wings of the adjacent ripples across the trough.

The environmental distribution of wave ripple cross bedding is similar to that of wave ripples, which is seen mainly in shallow environments such as lakeside, tidal flats, and beaches. The development of the bedding is usually ① close to the wave base, ② characterized by intense and frequent wave action, and ③ wave height usually indicates a topographical gradient.

(iii) Wind dune cross bedding: This refers to the cross bedding unit composed of progradational laminae produced due to the migration of wind dunes. It mostly has the clintheriform form, and is rarely seen as wedges or grooves. Wind dune cross bedding is distributed mainly in environments such as deserts, semiarid areas, coastal zones, and flood plains.

The major characteristic of this type of cross bedding is large scale and cross bedding unit thickness varying from tens of centimeters to several meters because of the large scale of wind dunes. The progradational laminae are relatively thick, generally 2–5 cm, and often appear as rectilinear figures, with dip angles greater than 35°. In the case that a large amount of sand is transported in a suspended state, the progradational laminae of sand dunes gradually turn flat downwind, appearing probably as tangent lines (Fig. 5.28, Atlas IV-B). If the lower part of the leeward slope of a sand dune is under erosion and destruction by side wind, the progradational laminae composing the cross bedding unit may become too steep and turn convex shaped.

4. **Compound bedding**

Compound bedding is considered the classical product of tidal action, mainly because its bedding features include vertical uneven superimposition between sand and mud due to frequent advance and retreat of the water body. Because there is no migration of sandy bedforms, laminae and set do not usually intersect. Over nearly 20 years, the bedding is found to develop mainly in lakes and delta fronts, which is mainly because the water body can frequently advance and retreat in these areas. Compound bedding can be of three types based on the size of the sand-mud.

(1) Flaser bedding

This is formed when more sand and less mud are accumulated under strong hydrodynamic forces and there is a better supply, deposition, and reservation of sand. In this type of bedding, argillaceous sediments are distributed mainly in the troughs of sand ripples, but they are very thin or insufficient on wave ridges. Hence, it can be said that argillaceous sediments are distributed in arenaceous sediments in the vein form. Good ripple progradational laminae are usually developed in arenaceous sedimentary formations (Fig. 5.29a, Atlas IV-C).

(2) Lenticular bedding

This is formed when more mud and less sand are accumulated, and there is poor hydrodynamic forces and better supply, deposition, and reservation of mud. In this type of bedding, arenaceous sediments appear as lenticels and are wrapped in argillaceous sediments. These lenticels are shown as interrupted distributions in space, and generally, well-developed ripple progradational laminae are present therein, which are actually the products of isolated ripples (Fig. 5.29b).

(3) Wavy bedding

This is formed when equal amounts of sand and mud form symmetrical or asymmetrical waves, but the overall direction is parallel to the bedding surface. This bedding is formed mainly by wave oscillation of the depositional media and secondly by the progressive motion of unidirectional flow. The former forms symmetric wavy

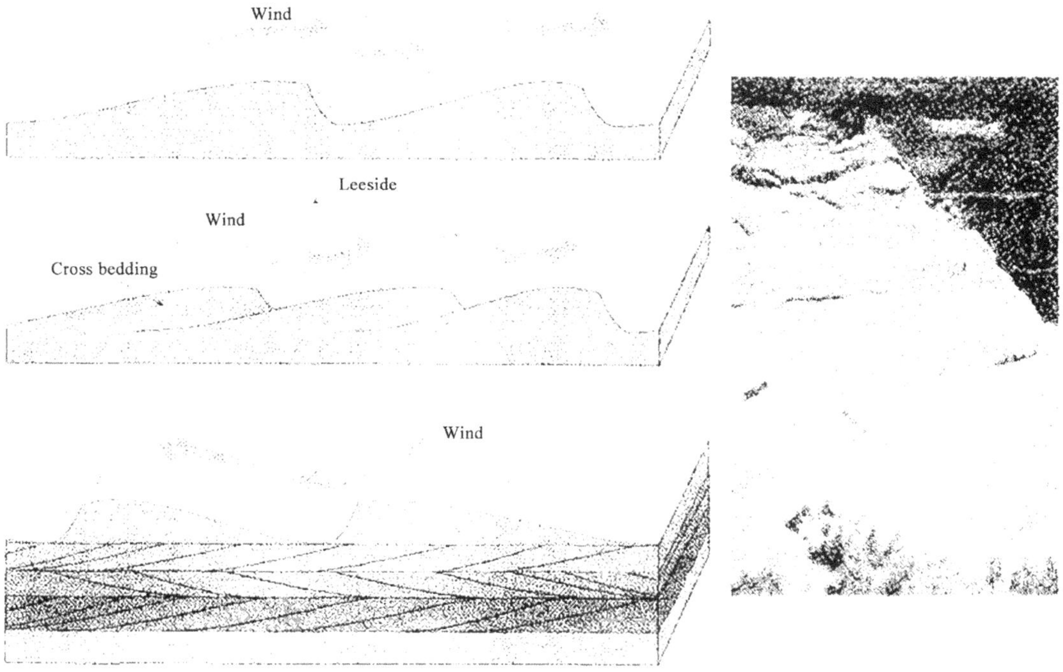

Fig. 5.28 Characteristics and forming process of aeolian dune cross bedding (from E. J. Tarbuck 1997)

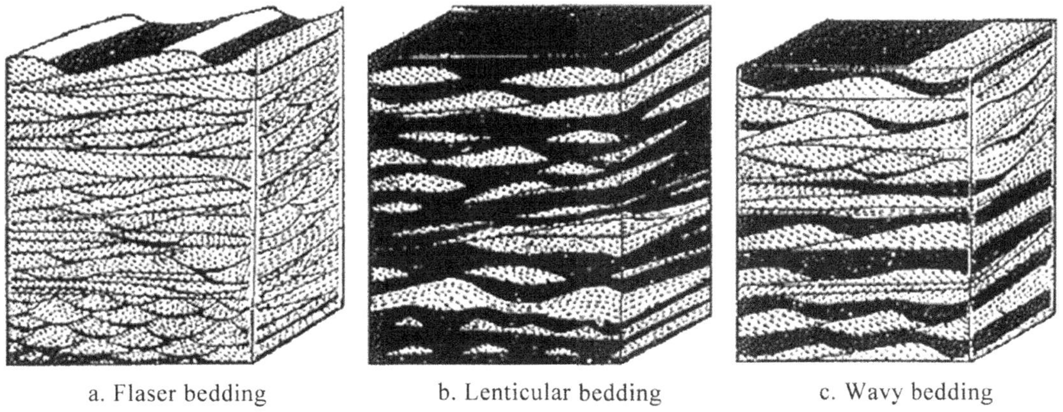

Fig. 5.29 Classification of compound bedding (according to Reineck and Singh 1973)

bedding, and the latter forms asymmetric wavy bedding (Fig. 5.29c).

5. **Graded bedding**

This is a sedimentation unit, and its features include graded change of component particulates. Its bedding surfaces are basically parallel to each other, and there is no intersection or crossing. In general, no interior laminae exist; there is only particle size change. Graded bedding is caused by periodic flow along the slope of a density current comprising mud, sand and water, carrying of sandstone into deep waters, and gradual exhaustion of water flow in the process of sedimentation.

Fig. 5.30 Two types of graded bedding (from E. J. Tarbuck 1997)

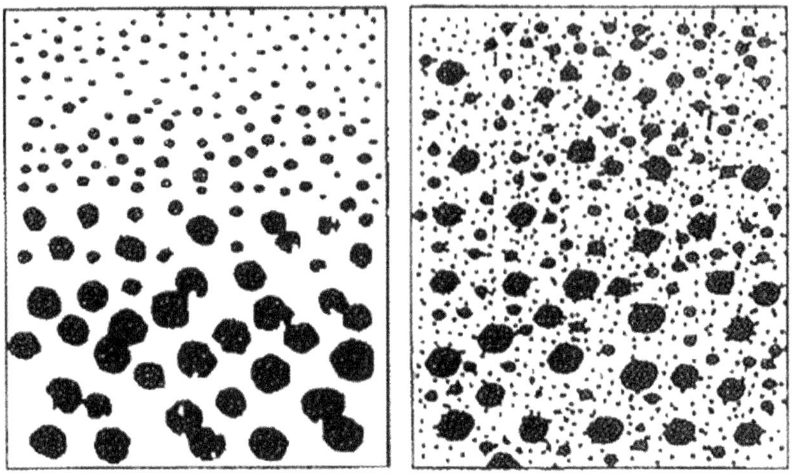

a. Normal graded bedding in forward direction b. Coarse-tail graded bedding

According to the internal structure features or graded bedding types, two fundamental types exist. In first type, particles gradually turn finer upwards, but there are no fine particles at the bottom, which may be the result of sedimentation due to a gradual decrease in water velocity or flow intensity (Fig. 5.30a, Atlas IV-D). In the second type, fine grains are distributed in all layers. Coarse material fines upward, which may be because various particles of different sizes contained in the suspended solids are accumulated due to gravitational differentiation under a low flow rate. This type of graded bedding is also called thick tail graded bedding (Fig. 5.30b). The former is often considered a factor in the mechanical differentiation of traction current, and the latter a gravity flow (turbidity current) factor or storm factor. Most graded beddings are of the second type. Most of scholars hold that coarse-tail grading is the main product of grain flow.

Graded material may be silt, sand, or even gravel. Generally, the thickness of a single graded bed is not high, and the coarser the material, the thicker is the graded bed. The thicknesses of most graded beds of sandstone vary from centimeters to 1 m. Commonly, they measure 20–50 cm in thickness. Graded bedding often has a very clear bottom boundary, which gradually turns finer upward and eventually changes, often, into argillaceous sediment.

6. **Rhythmic bedding**

This is formed by the regularly repeated presence of different laminae from the perspectives of composition, structure, and color. In a narrow sense, the thicknesses of rhythmic bedding laminae are usually lower than 4 mm and are different from that of general rhythmic bedding. The rhythmic repetition is caused by regularly alternating material transportation and generation modes. This variation can be short-term, for instance, tidal rhythms formed owing to changes in tidal environment; or long-term, for instance, seasonal rhythms formed owing to seasonal changes in the climate.

Rhythmic bedding formed in tidal environments in nature comprises alternate sand and mud laminae. The sand layers are deposited by the water activity of flood and ebb tides, while the mud layers are deposited during the retention period of high and low tides. The two alternate to compose a rhythm. This type of rhythm is quite commonly found in intertidal beaches and estuaries, but rarely in an open continental shelf.

The rhythmic bedding produced by seasonal variations in nature is formed by the alternation of dark and light layers. The material composition of each monolayer composing the rhythm is very fine silt or mud particles, and the monolayers are usually divided according to their colors.

Glacier varve is an important type of seasonal rhythmic bedding. Varve is the deposition of ice-melt water in a glacier lake, and it is composed of light layers with coarse particles and dark layers with fine particles in each cyclothem set.

7. **Homogenous bedding**

This is usually called massive bedding, and has a roughly homogeneous appearance without any laminated structure. The interior material is relatively homogeneous and has no differentiation in terms of structure and composition. Therefore, this type of bedding is not observed. Massive bedding, as the product of rapid deposition, is shown in both fine and coarse sediments.

5.3.2.2 Bedding Surface Structure

The various uneven traces of deposition shown on rock surfaces are collectively called bedding surface structures. There are many types of bedding surface structures and they are formed according to various mechanisms. The types mainly include ripples, mud cracks, raindrop and hail imprints, scour surfaces, and gutter and flute casts.

1. **Ripple**

This is a wavelike bedding surface structure formed on the surface of sediments due to the movement of a medium such as wind, water, or wave. Based on continuity, ripples can be classified into three basic types, namely, water, wave and wind, and transitional isolated ripples (Table 5.11). On rock surfaces, they usually form a series of wave crests or wave troughs that are parallel to or divergent with each other. The extensions of wave troughs are perpendicular to the medium movement direction. Ripples are sedimentary structures formed by non-viscous sand grains, often seen in powder fine sandstone.

The ripples formed by unidirectional flow on the surface of a non-viscous material are termed current ripples. These ripples extend perpendicular to the water flow direction, appearing flat on the stoss side and steep on the lee side. The general direction of water flow can be determined by using these features. The ridges of water flow ripples are of different shapes. In general, the wave ridge shape changes from simple to complex and from continuous to discontinuous as the water depth and flow velocity increase. The change sequence is as follows: linear → wave curve → chain shaped → egg and crescent shaped → rhombic (Figs. 5.31 and 5.32). These changes can be continuous or discontinuous.

(1) Wave ripple

This is often seen in the shallow water zones of seas and lakes, and it is characterized by sharp wave crests, smooth wave troughs, and symmetry (Atlas IV-E).

(2) Current ripple

This is formed by directional water flow and is often seen in rivers and coastal zones of seas and lakes with bottom current. The wave crests and troughs are even, smooth, and asymmetric. The abrupt slope direction indicates water flow, and the wave crest direction is roughly parallel to the bank extension on the shoreline of a sea or lake (Atlas IV-F).

(3) Wind ripple

This is formed by directional wind and is often seen in dune deposits on the shoreline of a sea or lake and in the desert. Wind ripples are highly asymmetric (Atlas V-A).

2. **Mud crack**

This is also called sun crack, and it is a shrinkage crack caused by the exposure of sediment to the sun when exposed on the water surface. It is usually formed in clay rock and carbonate rock but not in non-viscous sand. However, some sand bottom covering the surface of mud cracks can have boulder-clay molds, because the sand fills the mud cracks and becomes part of the overlying sandstone during sand deposition.

Table 5.11 Genetic classification of ripples (H. E. Reineck 1973)

Classification	Description	Wave ridge form	Ripple size parameter	Ripple index	Symmetry	Internal structure
Current ripple	Small current ripple	Straight, wavy, ligulate, and rhombic	L = 4–60 cm H can reach 6 cm	Larger than 5, mostly 8–15	Asymmetric	Shape is consistent, inconsistent, or superimposed
	Large current ripple	Straight, wavy, ligulate, bicornuous, and rhombic	L = 0.3–30 m H = 0.06–1.5 m	Mostly greater than 5	Asymmetric	Shape is consistent or inconsistent
	Gigascopic current ripple	Straight, wavy, and forky	L = 30–100 m Rarely 20–30 m H = 1.5–15 m	Mostly greater than 30, can reach 100	Asymmetric or symmetric	Only shape is known to be inconsistent
	Regressive sand wave	Straight	L = 0.01–6 m H = 0.01–0.45 m		Approximately symmetric	Shape is consistent or inconsistent
Wave ripple	Symmetrical wave ripple	Straight and partly forky	L = 0.9–200 cm H = 0.3–22.5 cm	Mostly 6–7	Symmetric	Shape is consistent, inconsistent, or superimposed
	Asymmetric wave ripple	Straight and partly forky	L = 1.5–100 cm H = 0.3–19.5 cm	5–16, mostly 6–8	Asymmetric	Shape is consistent, inconsistent, or superimposed
Isolated ripple	Small isolated current ripple	Small similar current ripples	Small similar current ripples, low H		Asymmetric	Both consistent and inconsistent shapes
	Large isolated current ripple	Straight, curved, and SB	Large similar current ripple, low H		Asymmetric	Both consistent and inconsistent shapes
	Gigascopic isolated current ripple	Similar current ripples	Similar gigascopic current ripples		Asymmetric or symmetric	Inconsistent shape
	Isolated wave ripple	Straight and curved	Similar wave ripples, low H		Symmetric or asymmetric	Shape is consistent or inconsistent
Water-wave synthesis ripple	Longitudinal water flow-wave joint ripple	Straight, ridge without branch parallel to water direction, also seen in ooze	L = 2.6–5 cm		Symmetric and asymmetric	Inconsistent shape
	Horizontal water flow-wave small ripple	Mostly round crest, perpendicular to water flow direction			Asymmetric	Consistent or inconsistent shape

(continued)

Table 5.11 (continued)

Classification	Description	Wave ridge form	Ripple size parameter	Ripple index	Symmetry	Internal structure
Wind ripple	Aeolian sandy ripple	Straight and partly forky	L = 2.5–25 cm H = 0.5–1 cm	Above 10–70	Asymmetric	Laminar sand, little foreset laminae, coarse sand centralized at crest
	Aeolian gravel ripple	Straight and crescent shaped	L = 2.5–20 cm H = 2.5–60 cm	12–20	Asymmetric	Progradational laminae shown in opposite direction, particle enrichment on the ridge

Note L refers to wavelength, and H refers to wave height

On the plane, latticed craquelure is formed due to typical mud crack development (Fig. 5.33, Atlas V-B), and rock is incised into polygons. Most fracture surfaces of mud cracks are V-shaped, and sometimes U-shaped. The scale of mud cracks differs. In very dry conditions, argillaceous layers are shattered into slices with edge upwarping. These mud slices are carried and abraded into flat-cake-shaped boulder clay.

Mud cracks are common on sedimentary surfaces in the shorelines of seas and lakes, intermittent channels, abandoned channels, flooding plains, and intertidal zones. In these areas, mud cracks are often accompanied by raindrop imprints and hail imprints, and these structures are the best markers of sedimentary discontinuity exposed on the Earth's surface, which is of great significance for indicating facies.

3. **Raindrop and hail imprints**

Raindrop and hail imprints refer to the small, striking alveolization formed when raindrops and hail fall on a sedimentary surface. The alveolization is round if raindrops fall vertically or oval if they fall with a slant.

4. **Flute cast**

Flute casts are half-conical protuberant structures distributed at the bottom, and they are formed by the filling of grooves with chiltern due to erosion by directional water flow on an unconsolidated ooze surface. Their morphological characteristics include symmetry, elongated shape and obvious fluctuation, and smooth bulbous shape at the upstream end. The appearance of flute casts indicates strong bottom current and scouring action under the water flow environment at that time, and the paleocurrent direction can be determined based on flute cast shape.

5.3.2.3 Deformation Structures

Deformation structures are also called penecontemporaneous deformation structures, and they refer to various sedimentary structures formed as a result of deformation during sedimentation or

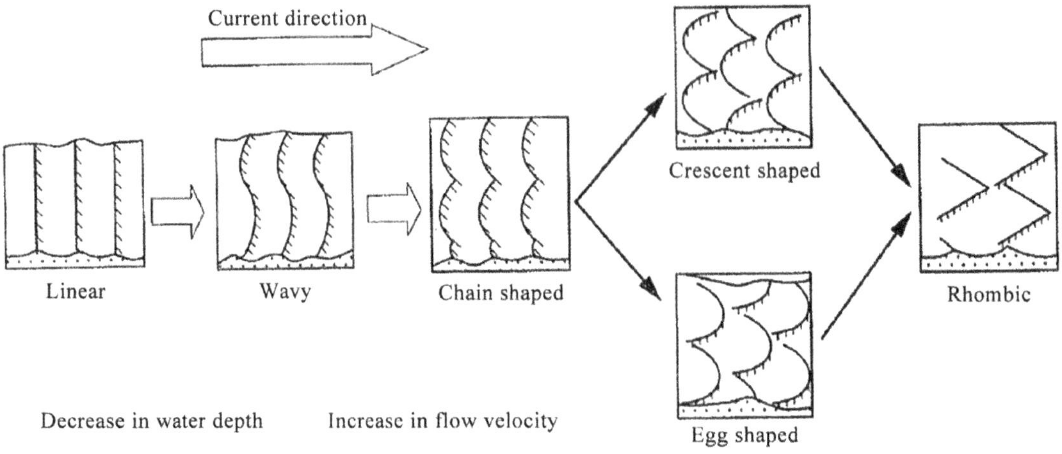

Fig. 5.31 Wave crest shape of current ripple mark and relation with water depth and flow rate (the editing group of *Sedimentary Structure and Environmental Interpretation*, 1984)

Fig. 5.32 Development of ripple mark in loose deposits. **a** Para-ripple formed by wave action; and **b** metaripple formed by fluviation or eolation, with ripples being steeper against the flow direction

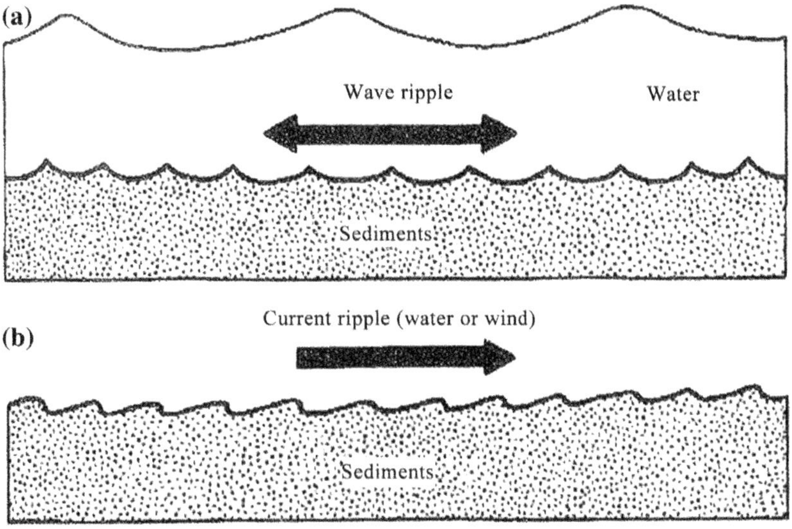

in the plastic state period before diagenetic sediment consolidation. They include load cast, convolute bedding, slump structures, sand volcanos, ball and pillow structures (pseudonodule), and water escape structures (for instance, dish structures) (Fig. 5.34).

(1) Load structure: This refers to the strumae on the bottom of the sand layer covering mudstone. It is formed owing to the overlying sandstone falling into the mudstone when the underlying hydrous plastic mudstone bears uneven pressure loads. The difference between load structure and flute cast is that the former has less regular shapes and it indicates current downstream and upstream directions.

(2) Ball and pillow structure: This refers to the segmentation of a sandstone formation into many close or sparse ellipsoid- and pillow-shaped blocks. This structure is mainly developed at the bottom of sandstone, and it transfers into undisturbed normal sandstone, while the underlying mudstone

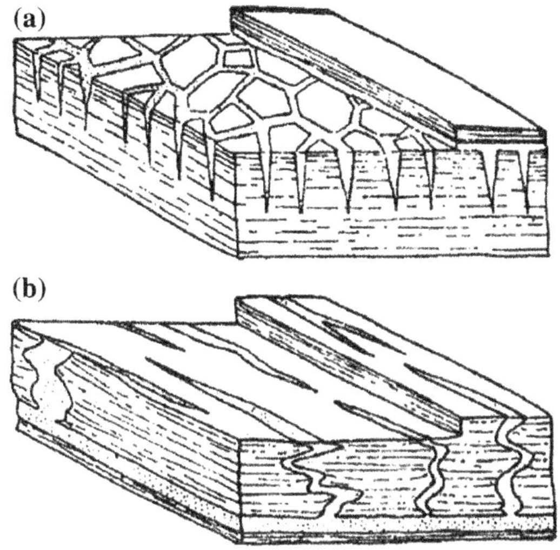

Fig. 5.33 Schematic diagram of desiccation crack and sand vein. **a** Desiccation crack formed by mud shrinkage, with plane arranged as a polygon (craquelure), fracture turning smaller toward F, showing V-shape, and filled with sand; and **b** sand vein showing sausage-shape due to compression, connected with underlying sand layer

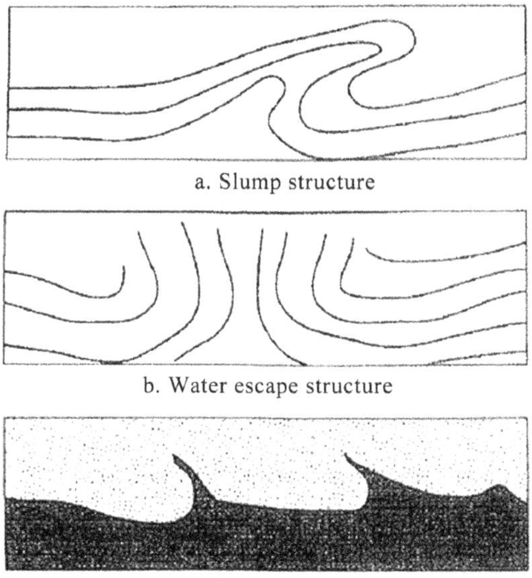

Fig. 5.34 Schematic diagram of main deformation structure

layer is often deformed strongly or even squeezed into the overlying sandstone.

(3) Convolute bedding: This is also called curly bedding, and it refers to the laminae curling back and twisting in a rock stratum (Atlas V-C). It is seen mainly in flagstone. There are many interpretations for the causes of convolute bedding, but it is mainly thought of as a bedding curve caused by the devolatilization of sediments or sediment rolling due to topographic changes.
(4) Slump structure: This refers to the transformation due to the sliding of unconsolidated sediments under the action of gravity on a slope. The sliding of sediments leads to the deformation and scrunching of sedimentary formation, often accompanied by rupture. Generally accompanied with rapid deposition, slump structures are generated and distributed mainly on intertidal beaches, water channels, channel point bars (Atlas V-D), and delta fronts (Atlas V-E).

Generally, the following five mechanisms cause sediment deformation:

(1) Load cast can be formed when high-density sediment covers low-density sediment and they move under gravity.
(2) Liquidation or fluidization of sediment refers to waterlogged fine sand, silt, and mud, which lead to a sharp increase in pore water pressure, decrease in the friction force between particles, suspension or rolling of silt and fine sand along the slope, for instance, to form convolute bedding and sand dykes, under the impact of external factors (for instance, load pressure, seismic waves, and other movements).
(3) On sedimentary slopes, slump structures are formed owing to the sliding of sediments under gravity on a sedimentary slope.

(4) Reverse cross bedding is formed under the shear stress of drag caused by the flow of newly formed sediments.
(5) In sediments, a water escape structure is formed due to liquidation or gravity, which causes pore water to escape. When pore water moves upwards, a dish structure is usually formed. The factors controlling the formation of water escape structures are granularity, accumulation, permeability, sediment strength, and instability of hydrodynamic force.

5.3.3 Sedimentary Structure with Chemogenic Characteristics

This type of structure is formed owing to chemical action in the process of diagenesis or after diagenesis. This sedimentary structure of secondary origin usually results from the action of precipitation and dissolution. Structures of this type include crystal imprints and nodules.

(1) Crystal imprint: Under appropriate conditions, crystalloids of materials such as salt and ice can be formed on soft sediment surfaces. These crystalloids vanish later due to melting and dissolution, leaving behind a special crystal imprint on the surface. Later, these crystalloids are replaced with other materials or the crystal prints are filled with other sediments to form crystal illusions, namely, pseudomorphic crystals. The crystal imprints are well preserved in argillaceous sediments.

Halite crystal imprint is the most common imprint, and its presence indicates increased salinity. It can be seen in both marine and non-marine facies environments, and the latter mainly refers to inland salt lakes undergoing periodic desiccation.

Ice crystal imprints usually appear as needle-like formations, mainly on lakeshores, overbanks, and intertidal beaches in places with cold weather.

(2) Nodule: This refers to the aggregation of authigenic minerals in rocks. In terms of composition, structure, and color, this type of mineral aggregate is represented as irregular blocks that are remarkably different from the surrounding rock. A nodule mainly acts as the dispersed substance in solution state in unconsolidated rock, which is formed after redistribution, concentration, and gradual increase.

Nodules are usually globular, axiolitic, caky, or irregularly shaped, even tubulose at times. Tubulose nodules are plant pseudomorphisms formed by the metasomatism of tree trunks and roots. Nodule sizes vary from less than 1–10 cm.

Nodules are usually composed of carbonate (siderite, ankerite, dolomite, and calcite), iron sulfide (pyrite and marcasite), sulfate (gypsum, barite, etc.), and siliceous materials (opal, chalcedony, etc.), phosphate, and manganese. Nodule composition is often connected to lithology and formation conditions. For example, carbonate nodules are usually found in terrigenous rocks, while caliche nodules are usually seen in carbonate rock. Iron pyrite is usually seen in coal series and dark mudstone strata, while siderite nodules are usually seen in red beds and yellow green sand-mudstone formations.

5.3.4 Sedimentary Structures with Biogenic Characteristics

In addition to the fossils formed by the remains from the death, burial, and preservation of biology, when carrying out activities in or on the surface of sediment, the biology is often destroyed and deforms the original sedimentary structure, leaving active traces. These trace are called biogenic structures, and they include lebensspur structures, bioturbation structures, and plant root traces.

5.3.4.1 Trace Fossils

This refers to a variety of traces with certain shapes formed in or on the surface of sediment owing to biological activity. It includes traces in the process of sports, living, foraging, and feeding during the existence of biological life. Therefore, lebensspur structures are also called trace fossils or ichnofossils. In a sense, trace fossils are the material expressions of the acclimatization of biological behavior. Because they can reflect the living environment at that time, their distribution scope is small, and they are found especially in a stratum with rare hard-bodied animal fossils. Trace fossils are widely distributed and well preserved, which is beneficial from the viewpoints of paleoecology research and lithofacies analysis. Since the 1950s, this new subject has been developed rapidly.

The shapes of trace fossils depend mainly on animal behavior. A primary classification was proposed by Seilacher (1964), who divided trace fossils into five main combinations according to animal behavior: ① resting trace, the form or part of lateral form when necton lives and rests in shallow water environment, with concave traces showing obvious tilt directionality, and showing the direction against current; ② crawling trace, where continuous linear features or intermittent point features are formed when benthos move on the surface of sediment while living in shallow-water environments; ③ browsing traces are channels dug in sediment by benthos for foraging while living in neritic environments, and they have radial arrangements; ④ feeding structures are the regular traces of benthos living in relatively deep water environments eating mud for foraging, and they are shown to have curly and snake-like bent shapes; and ⑤ burrows structures are permanent channel burrows where the benthos living in flooding plains and littoral areas coexist with plankton to avoid the impact of current and waves, and they appear as straight tube shaped, fork shaped, and U shaped (Atlas V-F).

In different depositional environments, benthonic animals have different ecological features for acclimatization owing to differences in environmental factors. In coastal regions, where there are large changes in environment-influencing factors (temperature and salinity, and tidal and wave actions change considerably), most benthonic animals either dig very deep near-vertical burrows or drill holes on rock bottom or rocky coast. In contrast, in shallow sea areas, where the changes in environment-influencing factors are small, the burrows of benthonic animal are shallow, and they are usually tilted or horizontal. In bathyal regions, due to environmental stability, benthonic animals do not need to burrow or drill for protection, but they need to find food systematically. As a result, various curved, reticular, arborization, or spiral foraging and feeding traces or complex burrow systems are left. Seilacher and Hekel et al. (1984). successively divided the following various ichnofacies according to the traces found in different environments (Table 5.12; Fig. 5.35).

5.3.4.2 Bioturbation Structure

Benthos activity destroys sediments and generates new biological structures, among which are trace fossils. However, there is bioturbation without determined forms in nature, which can be identified by the damage to well-developed beddings, usually called bioturbation structures. They can be divided into seven levels based on the effects of bioturbation (Table 5.13).

5.4 Core Description and Lithofacies Division

The core provides the most reliable data for researching lithology, physical properties, electric properties, and oiliness. In addition, it is the most reliable direct evidence for the analysis of underground reservoir sedimentary facies. Therefore, it is significant to observe and describe cores in order to understand geological structures, stratum lithology, sedimentary characteristics, and oil-gas-water content and distribution.

Table 5.12 Classification and characteristics of ichnofacies

Classification	Characteristics and distribution environment	Graphical representation (according to Ekdal et al. 1984)
Scoyenia	① Non-marine sandshale (red bed); ② above and below subnormal extreme-shallow water and fresh water (lake and river coastal zones); and ③ feeding burrow and living, crawling, footprints, and trails	
Teredolites	① Wooden bottom for marine environment, and drill traces combination; and ② peat swamps	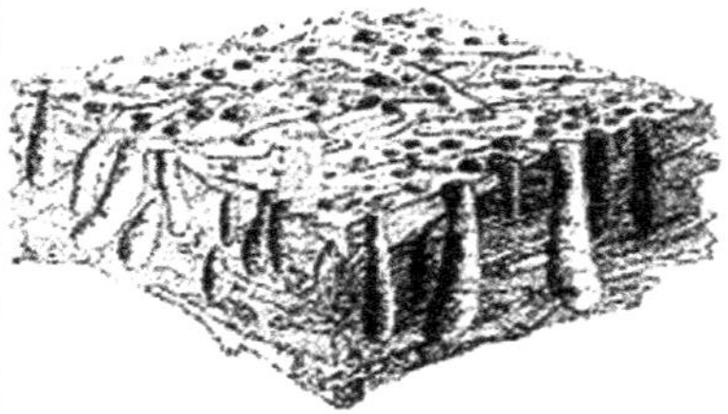
Trypanites	① Drill traces are highly abundant and have low disparity; and ② hard bottom, dwelling traces in rocks, discontinuity surfaces or planes of unconformity	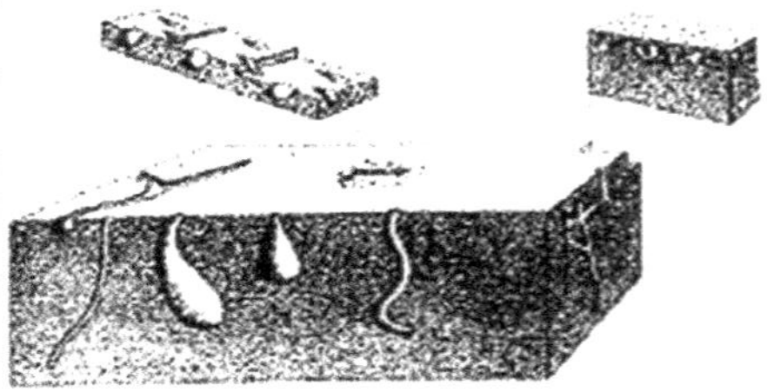

(continued)

Table 5.12 (continued)

Classification	Characteristics and distribution environment	Graphical representation (according to Ekdal et al. 1984)
Glossifungites	① Dwelling trace-dominated, vertical column shape, U-shaped, and arborization burrows; and ② concrete but not fossilized coast and subtidal zone, medium-to-high energy	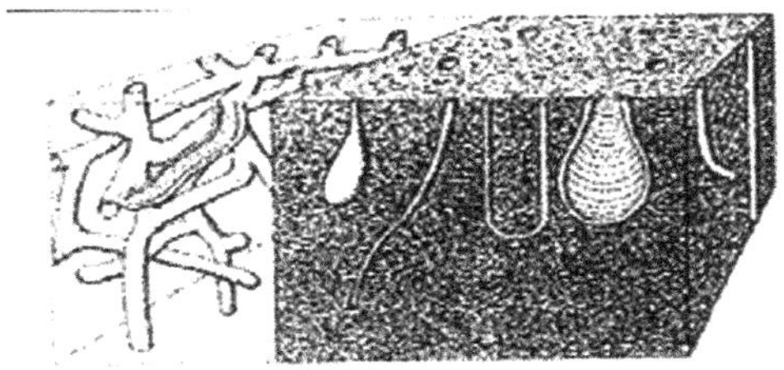
Skolithos	① Dwelling burrow-dominated, long vertical or high-angle columnar burrow, U-shaped and arborization burrows, spherulite lining, and web-shaped structures; ② sand bottom, medium-to-high energy, strong physical modification effect; and ③ seabeaches, sandbanks, etc.	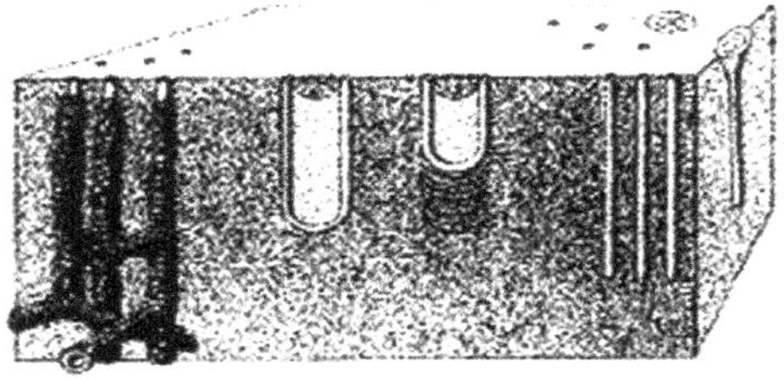
Cruziana	① Relatively wide community, high abundance and disparity, all ecological forms shown as superficial and horizontal; and ② shelves, estuaries, and lagoons	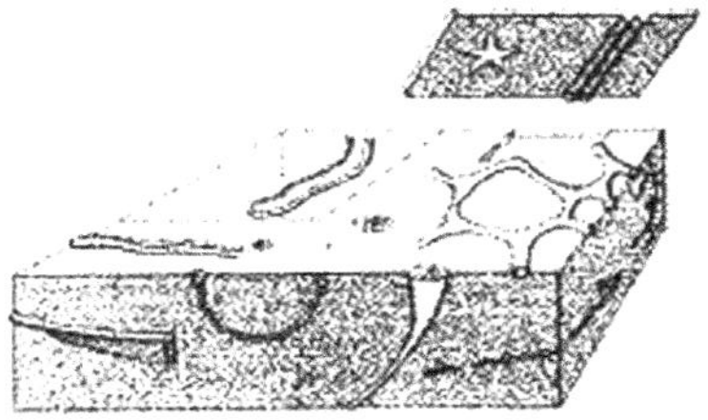

(continued)

Table 5.12 (continued)

Classification	Characteristics and distribution environment	Graphical representation (according to Ekdal et al. 1984)
Zoophycos	① Complex feeding structures, fine sheet, ribbon-like, and wedge-shaped structures from plane to gentle dip, low disparity, and high abundance; and ② mud, plaster, or argillaceous sand rich in organic matter; oxygen deficit in still water	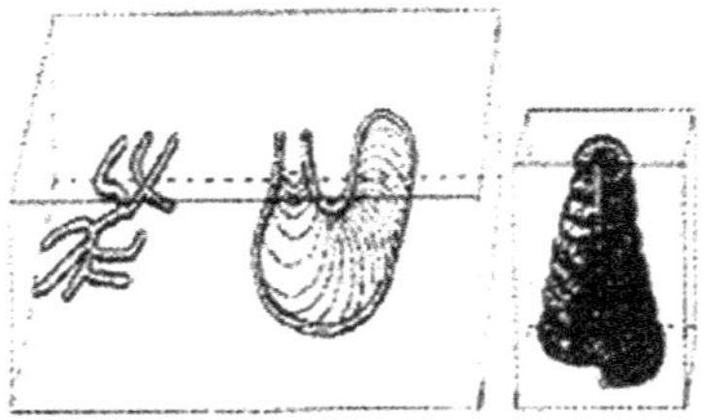
Nereites	① Horizontal and complex foraging traces and patterned agrichnia as features; ②deep water and deep sea turbidity currents	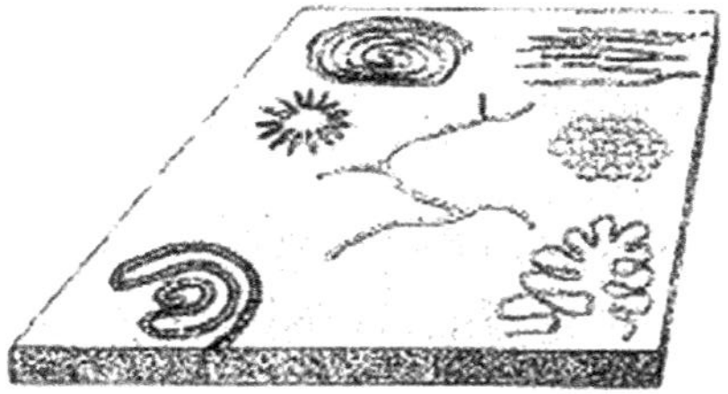

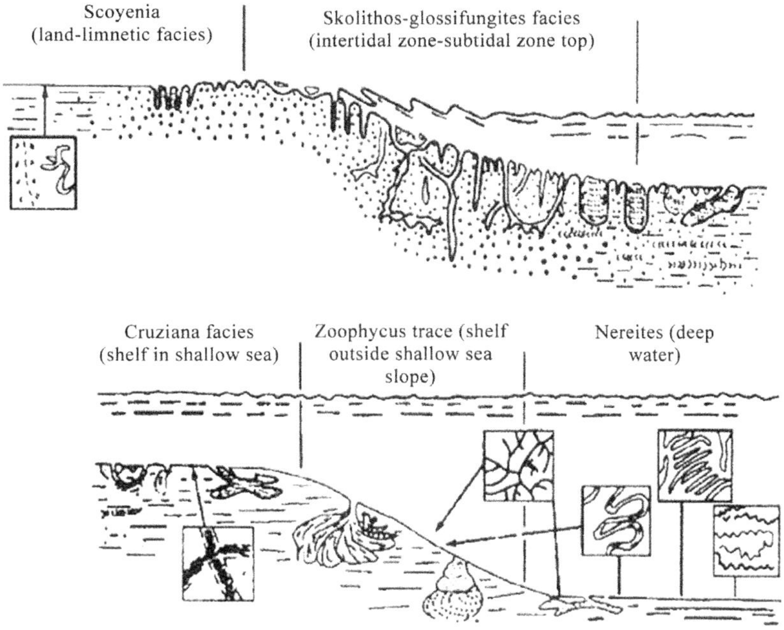

Fig. 5.35 Ichnofacies mode of main sedimentary environment (according to the Japanese *Material and Environment of Epigeosphere*, 1979; also quoted by Liu Baojun 1985)

Table 5.13 Bioturbation levels by reconstruction quantity relative to primary depositional fabric (according to Hu Bin 1997)

Disturbance level	Disturbance quantity (%)	Description
0	0	No bioturbation
1	1–5	Little bioturbation, few clear ichnofossils and escape structures
2	6–30	Low bioturbation, clear bedding, low ichnofossil density, usually seen as escape structures
3	31–60	Medium bioturbation, clear bedding surface, clear ichnofossil outline, rarely seen as superimposition
4	61–90	Bioturbation, unclear bedding surface, high ichnofossil density, seen as superimposition
5	91–99	Strong bioturbation, complete destruction of bedding, low retransformation of sediments, clear traces formed later
6	100	Complete disturbance of sediment and general reconstruction of sediment due to repeated disturbance

5.4.1 Core Description Method and Principle

5.4.1.1 Preparations before Core Description

(1) Check the number of corings, well sections, footages, core lengths, and yield rates.
(2) Check whether core sequence is wrong and whether polished surfaces and broken cores are placed as intended.
(3) Check core number, length mark, correctness of core card, and whether there is omission, repetition, and nonconformity.
(4) If possible, large-scale mud-logging data and logging information should be obtained and

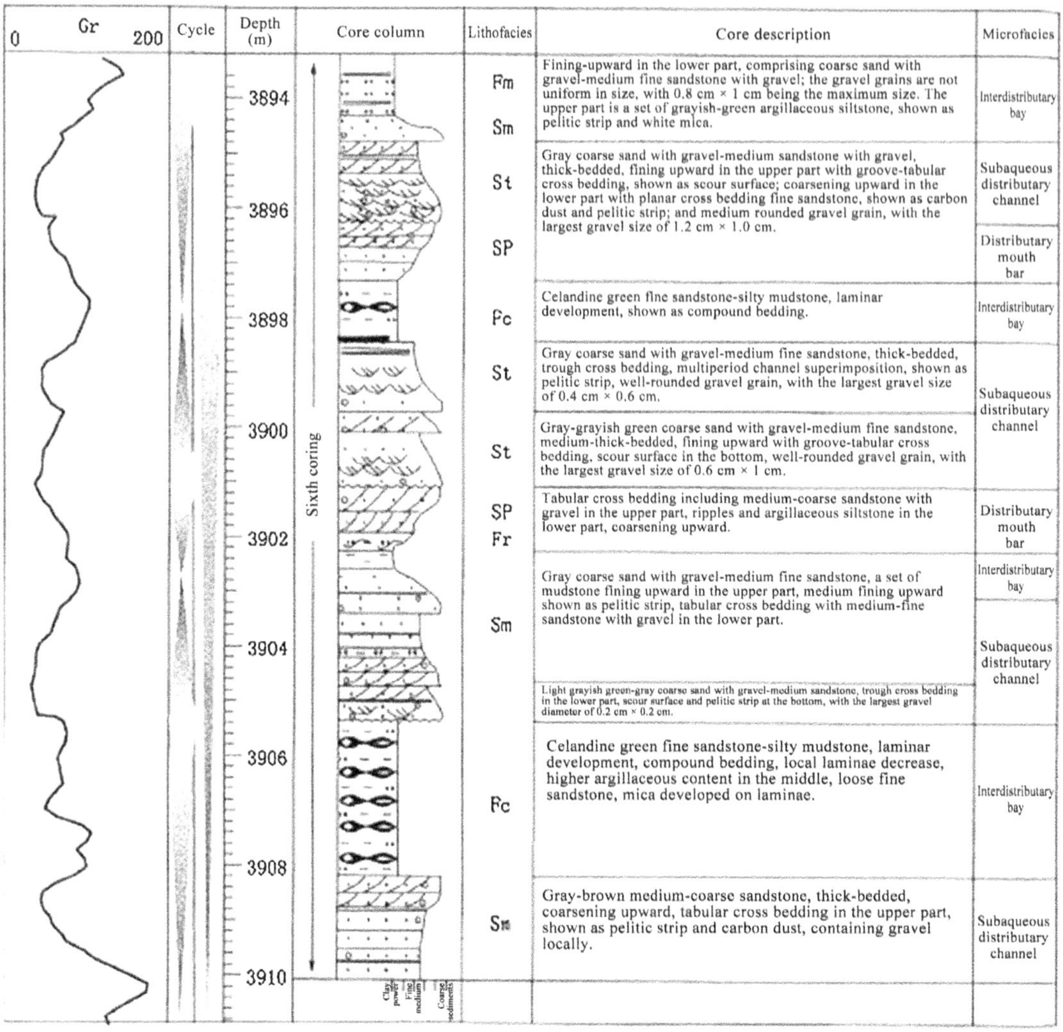

Fig. 5.36 Single-well histogram for typical core description of main production layer in some gas field of the Nanhai Sea

prepared as soon as possible for the sake of comparison during core observation.

As for core description, the most important task is to systematically describe the general and special phenomena represented by a core sample (Fig. 5.36). That is to say, in the process of observation, a core column or sedimentary sequence should be drawn truthfully to make the observations real and objective. If only depth or lithology is recorded onsite, the corresponding drawing will, in a sense, not be real or at least not realistic, resulting in unconscious fraud. Therefore, the key of core description is to draw a core column in proportion onsite on the basis of identification and description of various geological phenomena. The general specification for drawing core columns is to indicate different

Table 5.14 Standard graphic representation of clastic sediments with different granularity

Petrophysics and grain size	Drawing width (cm)	Petrophysics and grain size	Drawing width (cm)
Mud	1	Medium-coarse sand	1.7
Muddy silt and silty mud	1.1	Coarse sand	1.8
Silt	1.2	Coarse sand with gravel (very coarse)	1.9
Silty fine sand (very fine)	1.3	Granule	2.0
Fine sand	1.4	Pebble	2.2
Medium-fine sand	1.5	Coarse gravel	2.4
Medium sand	1.6	Boulder	2.6

granularity with different widths (Table 5.14), thus further reflecting the changes in vertical rhythms or cycles.

5.4.1.2 Core Description Method and Principle

Observing and describing a core sample involves ① determination of lithology, ② identification of sedimentary structure, and ③ establishment of vertical sedimentary sequence. The purpose is to determine the formation environment and facies. Based on years of experience, the author has summarized this process in the following eight steps.

1. **Comprehensive naming**

The essential characteristics of rock should be generalized in the rock name. For this, the structure-composition classification method is usually adopted. Rock name = color + layer thickness (scale) + granularity + composition (structure-composition classification method) + lithology. For example: light gray middle-level medium and coarse detritus quartz sandstone.

Color is the most eye-catching feature of rocks, mainly reflecting the composition of minerals in the rock and the general depositional setting. Therefore, rock color is the first important basis for authenticating rocks, judging depositional environments, and comparing and dividing strata. Monolayer thickness mainly reflects the sedimentation scale, that is, size of accommodation, deposition rate, and source supply, but there should be a relative standard for description. Because there are considerable differences in the sedimentary rates in different strata and different research areas, the standard is not unified. However, it should be the same in a given research area. The generally adopted standard is as follows: mass (huge-thickness bedding) is greater than 50 cm, thick bedding is 30–50 cm, medium bedding is 10–30 cm, thin stratified structure is 1–10 cm, and lamellar structures are less than 1 cm.

Granularity is the main basis for analyzing the strength of hydrodynamic forces, transport distance, and sedimentation mode. Texture maturity should be observed and described after naming in order to analyze hydrodynamic energy and transport distance. The composition reflects not only the nature of lithology in the provenance but also the intensity of weathering and the sediment transport distance, along with the geotectonic environment in a depositional district. However, it is difficult to guarantee the accuracy of composition identification and evaluation, thus the experience of the appraisers is critical. It would be better if slice authentication data, which is the key to analyzing compositional maturity, depositional environment, and reservoir properties, are available.

2. **Boundary division**

Boundary division is the basis for analyzing cycles, and the boundary is represented through the contact relationship of formation lithology. It refers to the different lithologic contact surfaces and their sedimentary change characteristics. Lithologic boundaries are generally divided into three types: ① lithologic gradient, mutation surface or plane of unconformity (Atlas I-A); ② scour surface (morphology, characteristics and boulder clay) (Atlas I-B, C, D and E); and ③ exposed surface (seat earth and plant root system) (Atlas I-F). When describing different lithologic contact relationships, the key is to describe the lithology, color, composition, structure, architecture of different types of contact surfaces, and other sedimentary characteristics of the contact surface. To be precise, boundary division should be made at different levels. In terms of sedimentary reservoirs, Miall (1985a, b; 1988) divided rivers into six levels (see Chap. 9 for details). Furthermore, division should be made in combination with cycle analysis.

3. **Cycle analysis**

In many cases, depositional cycles and rhythms do not belong to the same concept. The former mainly refers to the vertical variation law of all sedimentary characteristics (grain size, thickness, color, and lithology). In contrast, rhythm mainly refers to the vertical variation characteristics of grain size, such as fining upward and coarsening upward. Therefore, the key is to observe and describe rhythm features during core description. Many people consider depositional cycle and rhythm as synonyms because the core of cycle analysis is closely related to sedimentary rhythms.

The depositional cycle is defined as "a combination of thickness, rhythm, and color of sedimentary formation when they change or repeat upward according to certain rules or order," including mainly three basic types, namely, fining-upwards cycle, coarsening-upwards cycle, and complex cycle. For instance, the fining-upwards cycle is the rhythm in which lithology is shown in the order of conglomerate, sandstone, siltstone and pelitic siltstone, and mudstone from the bottom to the top, that is, from coarse to fine. Under normal conditions, the sedimentary thickness of sandstone thins gradually upwards, which means the sedimentary scale decreases and the color darkens moving upward, while a complex cycle is the superimposition of two fining-upwards cycles or two coarsening-upwards cycles. However, in most cases, it is a combination of fining-upwards cycle before coarsening-upwards cycle or the opposite.

The contents of cycle description include vertical changes in grain size, color, and thickness (from the bottom to the top). In combination with the features of composition, color, structure, and sedimentary structure of rock, changes in water body (water depth) can be analyzed further, and the key is to describe in detail the changes in grain size. In addition, rhythm thickness, contact relationships, and variation trends should be described further. When a section of a core is described, the depositional cycle should be summarized, and the vertical variation law and frequency of the compositional characteristics of the cycle should be analyzed. This is useful for determining the number of water body changes during sedimentation or the change features of sedimentation in order to accurately determine the depositional environment. With respect to sequence stratigraphy or base level cycle analysis, the key is to analyze changes in water body and water level depth to further analyze changes in accommodation and control of sediment supply to the depositional system. Doing so can help discuss the changes in sedimentary environments.

4. **Structure identification**

Sedimentary structures, which are formed by the migration of sandy bedform during sedimentation, reflect sediment deposition environments. This is because there are different hydrodynamic conditions under different depositional environments and conditions, and sandy bedforms are different for different hydrodynamic conditions. Hence, the formed structures are different. They mainly include stratified structures (also called bedding) and unstratified structures.

(1) Common stratified structures include horizontal bedding (Atlas VI-G, H), small (water/wave) ripple cross bedding (Atlas VI-I, J), parallel bedding (Atlas VII-A), tabular cross bedding (Atlas VII-B, C), herringbone cross bedding (Atlas VII-D), trough cross bedding (Atlas VII-E, F), compound bedding (lenticular bedding, wavy bedding, and flaser bedding) (Atlas VII-G, H), and the graded bedding formed under the action of gravity (Atlas VII-I).
(2) Common unstratified structures include various deformed beddings and biogenic structures. In terms of sedimentology, what matters is syngenetic deformed bedding, for instance, slump deformed bedding (Atlas VII-J), sliding deformation (convolute bedding) (Atlas VIII-A), crumpled deformation, and liquefaction deformation (Atlas VIII-B, C).

As mentioned before, biogenic structures include bioturbation structures (Atlas VIII-D) and lebensspur structures. The latter mainly refers to ichnofossils, which are also called burrow structures. Attention should be paid to the occurrence (vertical, diagonal, or horizontal) (Atlas VIII-E, F), inner wall characteristics (smooth or not, with backfill or not), size and shape (thickness, length, bending and embranchment of the individual), and intensity. Generally, the bigger and more intensive an individual burrow, the closer is its depositional environment to the shoreline or the stronger are the hydrodynamic conditions. In contrast, this indicates that the water body is relatively deep, and the organic matter is relatively rich.

Sedimentary structure description includes scale (set thickness, usually representing the height of one sandy bedform migration event) and distribution, bedding type and its deformation characteristics, surface features, inclination (mainly referring to the corner dimension and form of laminar and set), and particle arrangement mode.

Since the core is rounded (in China, except for the cores of Chinese-foreign cooperative blocks of offshore oil, which are split into half and polished, most cores of onshore and offshore self-run blocks are not split), it is difficult to identify the laminae of sedimentary structure because the laminae that are straight but not perpendicular to core are shown as arc-shaped on the core. This gives a false impression that the laminae have groove or hummocky cross bedding. The key method for distinguishing them is to understand that cross beddings are in different directions, not only in one direction. In addition, the specific type of megascopic cross bedding is hard to identify on a core, such as tabular or groove cross bedding, because multiple formation bounding surfaces are difficult to identify. For this reason, researchers generally call it "oblique bedding" in the early stages. The author suggests that this concept be abandoned for the following three reasons: ① all laminae are oblique, except horizontal or parallel bedding when a stratum is absolutely parallel; ② it cannot indicate the geological implications and the background at the time of formation; and ③ there is no corresponding terminology in English. The core of a descriptive discipline is to inform the reader about the genesis and implications of the object and content for description in order to facilitate them to analyze; otherwise, its true meaning is lost. In addition, when horizontal or parallel bedding of the core is identified, attention should be paid to distinguish it from scratches made during coring. This can be achieved by washing the surface with waterproof abrasive paper.

5. **Fossil identification**

Fossils can be divided into two types: zoolite (biological medium debris and individual) (Atlas VIII-G, H) and phytolite (ligneous-cane and herbal-fragment) (Atlas VIII-I, J). The descriptions cover name, occurrence, color, composition, size, shape, number, distribution, and preservation.

(1) Occurrence: This refers to fossil distribution modes such as parallel to the bedding surface, perpendicular to the bedding surface, tilted or chaotic distribution, arrangement

mode, direction, and lithology for preservation.

(2) Color: This is specified according to the method for describing rock colors.
(3) Composition: The fillings in zoolite are replaced by gray, siliceous, and dolomitic matters.
(4) Size: Height, width, length, diameter, etc.
(5) Shape: This refers to fossil shape characteristics, for example, ornamentation, degree of clarity, and shape.
(6) Number: Words describing fossil numbers include "occasional," "few," "many," and "rich." In terms of the adjectives for describing the number, it would be better to use a concept corresponding to percentage or density to make the expression more accurate.
(7) Preservation: Words describing preservation include "complete," "relatively complete," "not complete," "broken," or "between the two."

The type, size, and occurrence of fossils are very important for the identification of depositional environments. For instance, an abundance of phytoclasts indicates that the terrain was relatively gentle at the time of formation and water was intermittent, which helps a great deal in determining the environment and plane modality of sedimentary sand bodies.

6. **Special mineral description**

The description of special minerals mainly involves pyrite, glauconite, siderite, nodule (calcium nodule and iron nodule), mud pebbles, mica, charcoal, and coal bed.

As for special minerals, the main description should include density, distribution characteristics, individual size, and crystalline form; the other phenomena to be described include name, color, occurrence, quantity (density), size, shape, arrangement, and distribution characteristics. There is a high demand for people with experience in and knowledge of the recognition and description of special phenomena. It is important to raise questions during observation, for instance, questioning why a phenomenon or mineral is seen or not, which helps in the analysis of the formation environment at hand.

7. **Background thinking (impression)**

This refers to preliminary acquaintance with the following aspects based on all the above observations and descriptions: water body depth, water body energy (hydrodynamic conditions), turbulence and quiet, topography, provenance nature and source distance, transport mode, salinity level, and sedimentary environment. Therefore, it is a process of analysis and discussion, not just simple description, and the key is to take the above descriptions as the basis to analyze problems and draw conclusions.

8. **Environmental determination**

Sedimentary environment should be predicted in combination with features such as lithology, rock texture, sedimentary structure, fossil, and color. When the sedimentary environment is determined at last, the sedimentary facies type should be determined in combination with electric well-logging curve properties and seismic cross section structures. This means "we shall think from the macroscopic to the microscopic, and analyze from the microscopic to the macroscopic." In other words, we must think about the features of electric well-logging curves and cores with respect to the formation background and evolution of the entire depositional system and analyze the spatial distribution law of the macroscopic depositional system and the logging or seismic response characteristics with respect to core features.

It should be noted that the core Atlases (VII–VIII) in this book are typical examples selected from practically thousands of meters of cores, thus the area and depth are not indicated on purpose. However, strictly speaking, during actual observation and description of cores, corresponding Atlas shall be prepared for the parts that can indicate the formation conditions and background in each meter of core (e.g., Atlas IX). As for the atlases, attention should be paid to the following aspects: ① There must be a measuring scale at hand when taking photos of the core, and the top and bottom of the core should be indicated clearly, for instance, the badge on the measuring scale is an upward-pointing mark; ② When a core atlas is made, attention should be paid to the coring round-trip, depth, and position; ③ Special phenomena should be described and explained especially because they are the most direct and reliable evidence and data for underground geological research, especially in sedimentary facies analysis; ④ Do not indicate your identified sedimentary facies. For a rock, you simply need to describe its features, identify the lithology, and analyze its formation conditions. This means that fixing the facies of a rock is unscientific because first impressions are the strongest and they may lead to prejudice against other's analyses. The identification and analysis of depositional facies should be part of integral analysis, and the depositional facies of a certain well position, especially the microfacies, should be determined when preparing a comprehensive column for a single-well core (Fig. 5.36).

5.4.1.3 Lithofacies Classification and Its Significance for Indicating Facies

Most sedimentary structures are occured in more than one depositional environment. For instance, current ripples can be developed in rivers, coasts, or even turbidity currents. Therefore, the existence of a single depositional environment should not be taken as the only mark for environmental interpretation. However, through extensive research, it is known that the vertical sequence of sedimentary structures preserved in a sedimentary sequence is the most typical for special environments, which is very critical for determining the depositional environment correctly.

Miall (1988) divided fluvial sediments into 22 lithofacies types, but later revised the list (1988) to 17 types (Table 5.15). Lithofacies names can usually be expressed as codes, which are made up of two parts. The first part expresses lithology and granularity with an uppercase letter, for example, G-gravel, S-sandstone, F-siltstone and M-mudstone, and the second part reflects a certain sedimentary structure or lithofacies color, which is denoted by one or two lowercase letters, for example, t-trough cross bedding, p-planar cross bedding, and m-massive bedding. For example, in massive gravel lithofacies Gm, G represents gravel and m represents massive bedding. Therefore, the main markers for dividing and identifying lithofacies are lithology, grain size, sedimentary structure, and color. Some scholars call them energy units because the coarseness or fineness of lithology and bedding type can reflect the hydrodynamic conditions and transport mode (Fig. 5.37).

5.5 Identification and Modeling of Electrofacies

In the entire petroleum exploration and development process, the analysis and research of logging information are necessary. Because electrofacies analysis serves to identify depositional facies based on different logging information, it is one of the basic means of underground reservoir sedimentary identification and the most basic and direct basis for detailed correlation of reservoir beds (petroleum reservoirs).

5.5.1 Overview

Microfacies are the most basic component units in a depositional system, reflecting the sedimentary rocks formed when the depositional condition is basically the same. As for the sedimentary characteristics of different microfacies, the reflection and performance vary in logging information, which is the basis for identifying sedimentary microfacies using logging information.

The book *Geologic Well Log Analysis* by Pirson (1977) is the earliest book to

Table 5.15 Lithofacies division of a fluvial system (according to Miall 1988)

Lithofacies code	Lithofacies	Sedimentary environment	Interpretation
Gms	Matrix-supported massive gravel	Graded bedding	Debris flow deposition
Gm	Massive or thick-bedded gravel	Horizontal bedding	Longitudinal bar
		Imbricate structure	Lag deposits and sieve deposits
Gt	Stratified gravel	Trough cross bedding	Creek road filling
Gp	Stratified gravel	Planar/Tabular cross bedding	Longitudinal bar delta
St	Medium-very coarse sand with pebbles	Single or crowds of trough cross bedding	Sand dune (low flow regime)
Sp	Medium-very coarse sand with pebbles	Single or crowds of tabular trough cross bedding	Ligulate and transverse bar sand wave (low flow regime)
Sr	Very fine-very coarse cobble	Ripple	Ripple
Sh	Very fine-very coarse sand with pebbles	Horizontal laminae or parting-plane lineation	Faceted laminar flow (upper flow regime)
SI	Very fine-very coarse sand with pebbles	Low angle ($<10^{\circ}$) cross bedding	Scour-filling, erosion of sand dunes, and anti-dunes (sand waves)
Se	Erosion and scour with intraclast	Protogenetic cross bedding	Scour-filling
Ss	Fine-very coarse sand with pebbles	Wide and shallow scour	Scour-filling
FI	Sand, silt, and mud	Laminite and very fine ripples	Flood plain or concave slope flood deposition
Fsc	Silt and mud	Lamellar-massive	Back swamp deposit
Fcf	Mud	Massive mollusk with fresh water	Back swamp deposit
Fm	Mud and silt	Massiveness and mud crack	Flood plain or drape deposition
C	Coal and calcareous mud	Plant and thin mud layer	Swamp deposit
P	Carbonate rock	Pedogenesis	Paleosols

systematically use logging information in geology and focus on oildom sedimentology research by using logging information and depositional facies research based mainly on electric well-logging curve shape. The French geologist O. Serra et al. (1975) proposed the concept of electrofacies to make this method more systematic and quantitative by summarizing predecessors' works. The fundamental steps in electrofacies analysis include extracting the features of an electric well-logging curve from a set of log responses to reflect stratigraphic characteristics, including amplitude, shape, contact relationship, and combination feature, and divide the stratigraphic section into limited electrofacies in combination with other electric well-logging interpretation conclusions, and verify it with core information to establish a mode for describing sedimentary facies by using logging information. The analysis of core and lithofacies is the geological foundation for depositional facies or microfacies recognition by well logging.

At present, the common logging methods include self-potential (SP), natural gamma ray (GR), caliper log (CAL), acoustic transit time reflecting porosity change (AC), density (DEN), compensated neutron log (CNL), resistivity log (Rt) reflecting oil-bearing property change, and

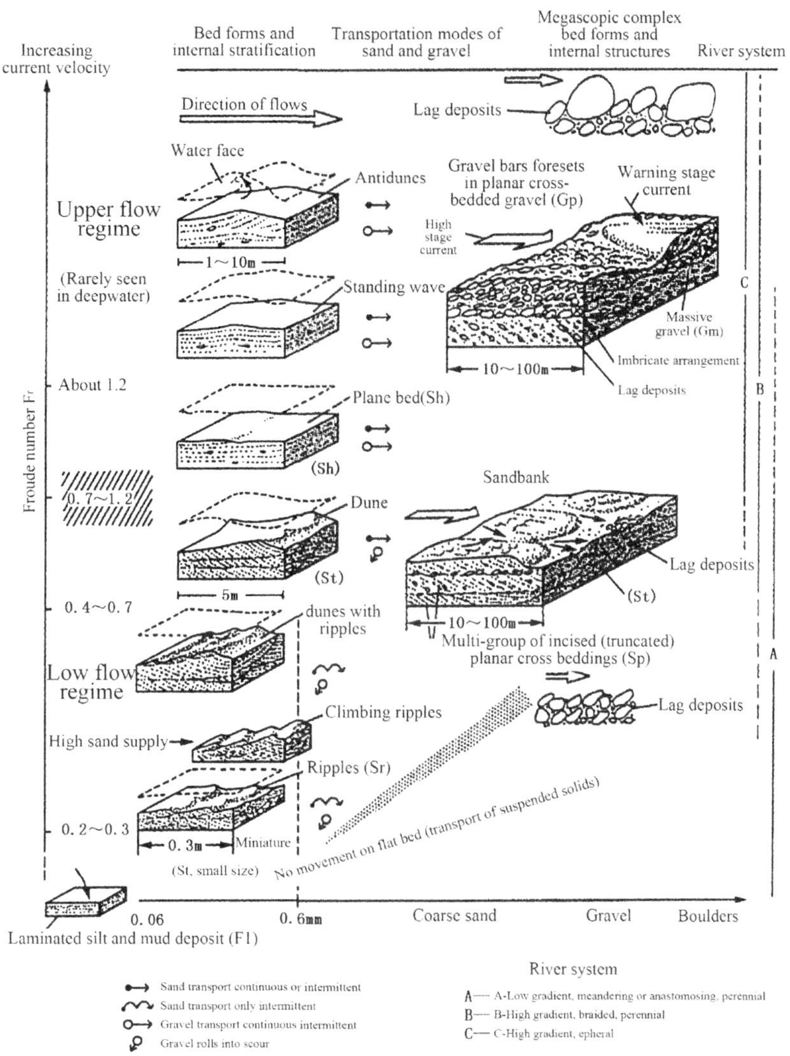

Fig. 5.37 Due to different lithology thickness and bedding types, the lithofacies can reflect the intensity of hydrodynamic conditions and the difference of transportation mode (Harms and Fahnestock 1960; Walker and Cant 1984; Collinson 1986a)

spectral logging and geochemical logging changes in rock composition of a stratum. In addition, SHDT and micro-resistivity imaging logging are widely used in electrofacies recognition. Because various electric well-logging curves reflect different geologic features, their roles in sedimentary microfacies identification vary as well (Table 5.16). For instance, self-potential, natural gamma, and resistivity can reflect vertical graded changes and sediment rhythms, sedimentary structure characteristics, and hydrodynamic energy changes. Geochemical logging and spectral logging can reflect the maturity of rock constituents, which helps with further analysis of the nature of source rock, paleogeographic background, and distance of source region. Diplog and stratum micro-resistivity imaging logging can reflect sedimentary structures, which are important for judging sedimentary environments. In addition, the vertical combination rule of electric well-logging curves is an effective way of determining the sedimentary microfacies combination rule.

Table 5.16 Effect of electric well-logging curve on petroleum reservoir research

Well logging series	Direct action	Indirect supporting role
Self-potential	Calculation of formation water resistivity and permeability instruction	Estimation of shale content, facies instruction, and stratigraphic correlation
Natural gamma	Quantitative calculation of shale content and stratigraphic correlation	Facies and lithology instruction, mineral identification, and unconformity recognition
Sound waves	Quantitative calculation of porosity, seismic interval velocity, and acoustic impedance	Stratigraphic correlation, lithologic identification, source bed evaluation, rock structure analysis, and fracture identification. Degree of compaction and abnormal pressure analysis
Density	Calculation of porosity, and indirect calculation of hydrocarbon density and wave impedance	Stratigraphic identification, mineral identification, source bed evaluation, and stratum abnormal pressure belts
Neutron	Calculation of rock stratum porosity	Identification of atmosphere, lithology, and volcanic rocks
Resistivity/Interaction	Calculation of fluid saturation and lithology identification	Rock structure analysis, stratigraphic correlation, facies analysis and recognition, compaction analysis and overpressure belts recognition, and source bed recognition

5.5.2 Geological Significance of Conventional Electric Well-Logging Curve Features

Well logging refers to continuous whole well measurement in terms of rock geophysical characteristics near the borehole. Well logging data are easier to acquire than coring data. At present, the most common electric well-logging curves for sedimentary facies analysis are self-potential and natural gamma, sometimes in combination with electrical resistivity (generally, a micro-electrode curve is used). This helps with basic elements such as the distinction of lithology and its vertical changes to analyze the shape, amplitude, contact relationship of top and bottom, degree of smoothness, and tooth middle line. Owing to the multiplicity of solutions in analysis, when electrofacies research is conducted, various logging responses should be mastered, and in-depth analysis and research on the corresponding relationship between logging response and depositional facies, or the depositional facies template fit for research area, should be established by applying the electrofacies analysis method in combination with coring data from the research area.

5.5.2.1 Amplitude

The amplitude of spontaneous potential logging mainly reflects the distribution of electric field intensity under natural conditions in a well. In a permeable interval, the amplitude of the SP curve deviates from the shale line depending on the proportion of salt in stratum water to that in the well fluid. In freshwater drilling fluids, the SP curve at the position of a stratum with salt water deviates toward the left, which means a shift in the negative direction. Under the situation that other conditions are the same, the negative excursion amplitude of clean sandstone is the biggest. When there is mud in sandstone, the SP amplitude decreases until the mud rock seam (when argillaceous content is 100%) and the SP curve correspond completely to the base line. When the drilling fluid is saltwater, the SP curve of the stratum with saltwater is under little or no excursion, and there is even a reversal, depending mainly on the saltiness of the formation water.

Natural gamma logging responses mainly reflect the natural radioactive energy of a stratum.

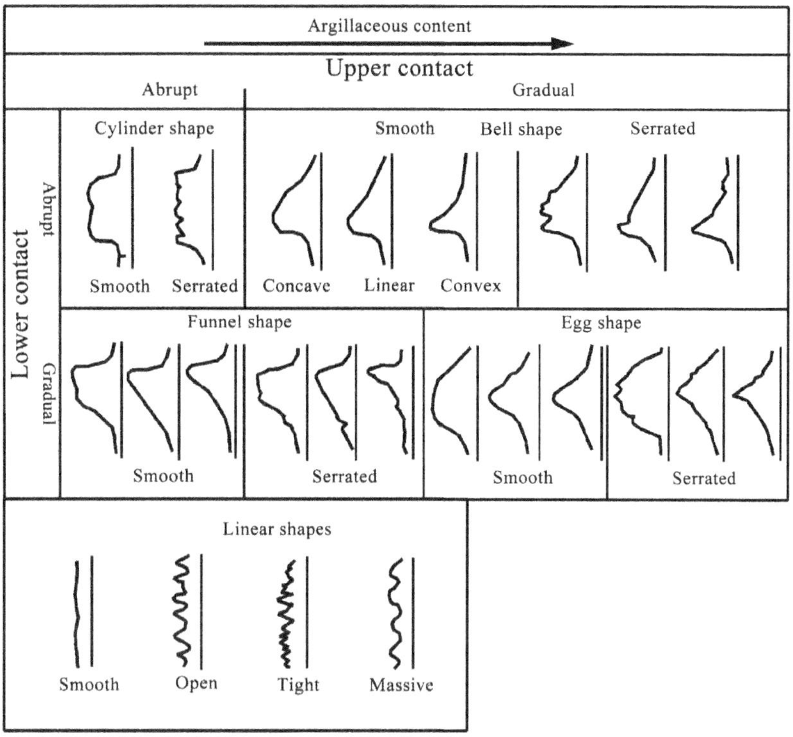

Fig. 5.38 The morphological classification of well log is the basis of analysis of well logging facies (Serra and Sulpice 1975)

Since clay minerals contain potassium radioisotopes (^{40}K), the mudstone layer has strong natural radioactivity, while the radioactivity of sandstone is much lower. Therefore, the natural gamma log curve can well reflect the relative amount of sandstone and shale in a depositional sequence vertically. As the amount of chiltern increases, radioactivity decreases, sandstone grain size becomes coarser, and shale content decreases. Although natural potential and natural gamma rays have similar features, meaning sensitivity to sand and mudstone, the two methods differ to some extent in some special conditions, for example, lamina and dense sand layer. Moreover, natural gamma can make up for the shortcomings of self-potential.

In sand-mudstone sections, the shale content in sandstone is closely related to the depositional environment. In high-energy environments, owing to the jigging of a water body, a fraction of relatively coarse pure sandstone is formed, and its excursion of self-potential or gamma curve is large. In low-energy environments characterized by the detention of flow, fine-grained mud is deposited to form pure mudstone, and its self-potential or gamma curve is in accordance with the baseline. Therefore, the shale content in sandstone and the energy intensity of the depositional environment can be estimated using the SP or gamma curve.

5.5.2.2 Morphological Characteristics

Provenance, water body energy, and water body size vary in different depositional environments, which cause differences in the composition form and sequence characteristics, as shown by different curve shapes on an electric well-logging curve.

In 1975, Serra and Sulpice classified natural gamma and SP curves using the following two classification standards: the first is shape-based, according to which the curves are divided into bell shaped, funnel shaped, cylindrical, egg shaped, and liner shaped; and the second is smoothness-based, according to which they are divided into serrated shape and smooth shape (Fig. 5.38). The classification and description are the basis for electrofacies analysis. However, there may be differences across different well logging series, thus electrofacies analysis can reflect the overall depositional cycle features and rhythms. Concretization of different oilfields is performed according to these rules and characteristics to reflect the features of the research fields and the differences in well logging series.

1. **Morphological characteristics**

(1) Bell shaped

The bottom of the curve covers the largest area, however the curve decreases in size moving upward, which means gradual weakening of flow energy or provenance. Its feature is mutation at the bottom and gradual change at the top, which is the fining upward rhythm, reflecting the normal graded bed sequence of lateral channel migration characterized by the point bar of a meandering river or the filling sedimentation of a channel.

(2) Funnel shape

In contrast with the bell-shaped curve, the funnel-shaped curve presents the coarsened reverse graded bed sequence vertically, reflecting a gradual increase in flow energy or sufficient source supply. Its feature is abrupt contact at the top and gradual change at the bottom, reflecting the reverse graded bed sequence of progradation or downstream accretion characterized by the delta front distributary mouth bar facies.

(3) Box shape

This is also called a cylindrical curve. It reflects the rapid accumulation of sediments or environmental stability under the condition of sufficient provenance and stable hydrodynamic force in the process of sedimentation. Both top and bottom boundaries are subject to abrupt contact, reflecting sufficient source and relatively strong hydrodynamic force characteristic of a tidal sand body or braided channel sand body.

(4) Tongue shape

This is also called an egg-shaped curve. Both the top and the bottom feature gradual deformation, and the thickness is usually not large, thus reflecting the influence of various factors. Moreover, the overall background is the relative weakness or quick change of hydrodynamic force, representing mainly crevasse splays, finger bars, and distal bar sedimentation.

(5) Linear shape

There are no large amplitude variations vertically, thus reflecting mainly a relatively quiet depositional environment, for instance, shallow-deep lacustrine (sea) facies sedimentation with severely lack of provenance supply, dominated by fine-grained sediments.

2. **Smoothness**

(1) Serrated shape

This reflects quick changes in energy or instability of the hydrodynamic environment in the process of sedimentation. It can have both the normal serrated shape and the reverse serrated shape or symmetric serrated shape, which are shown in alluvial fans and turbidite fans.

(2) Smooth shape

This reflects a stable depositional environment and relatively quiet hydrodynamic conditions, as well as stable changes in lithology and the lack of a sand shale interlayer.

5.5.2.3 Contact Relation

Electric well-logging curves can reflect the contact relationships between surfaces. Generally, the contact relationships can be divided into abrupt changes and gradual changes. Abrupt changes can be divided further into top abrupt change and bottom abrupt change, and gradual changes can be divided into top gradual change and bottom gradual change, wherein there are accelerated changes, linear changes, and gradual decline. Bottom abrupt changes usually reflect the scouring action of various rivers or abrupt changes in the environment and the unconformity boundary.

Serra's division laid a solid foundation for electrofacies research, but the combination feature should also be taken into consideration in the case of specific oilfield analysis. Ma (1982a, b), based on the practical situation of the faulted basin oil deposit in continental facies in eastern China, prepared a classification scheme (Fig. 5.39) based on electric well-logging curve amplitude, shape, contact relationship, smoothness, and their combination relationship. This classification has both practical value and realistic significance from the viewpoint of exploring China's continental deposits.

5.5.2.4 Combination Types

The combined form of the electric well-logging curve includes amplitude change combination and form combination (Table 5.17). The amplitude change combination includes accelerating amplitude change, even amplitude change, and decelerating amplitude change, and the form combination includes cylinder shape–bell shape combination, funnel shape–cylinder shape combination, finger shape–funnel shape combination, cylinder shape–bell shape–funnel shape combination, and serrated shape–cylinder shape–bell shape–funnel shape combination (Fig. 5.39), and different combinations can better reflect the sedimentary environment of a stratum.

According to the manifestation of cyclicity, grain size, and lithology thickness on the electric well-logging curve in sedimentology research, depositional facies can be determined (Fig. 5.40) to summarize the log-shape feature of a sedimentary facies belt (Fig. 5.41). Given that the above curve shapes are ideal, core analysis data should be fully utilized in specific areas to summarize the response of the sedimentary facies features on the electric well-logging curve in order to establish an electrofacies model of the research area.

5.5.3 Depositional Facies Research by Diplog and Imaging Logging

5.5.3.1 Sedimentary Environment Research by Diplog

In recent years, the development and gradual improvement of high-solution diplog has provided an effective method for depositional environment research, and this method can provide information about sedimentary structures and paleocurrents.

Bedding is an important sedimentary structure phenomenon and also the ocular proof reflecting paleocurrent dynamic conditions. Dipmeter data can be processed by applying an interactive processing technology (relative contrast or pattern recognition), and the processed results can be analyzed to determine the sedimentary structure and paleocurrent direction.

1. **Vector diagram and reflected bedding features**

On a vector diagram, according to the law that the stratigraphic dip and trend vary with depth, four modes can be formed: green pattern, red pattern, blue pattern, and chaotic model (Fig. 5.42; Table 5.18). According to stratigraphic dip information, various depositional bedding types can be interpreted (Fig. 5.43) and their depositional environments and facies can be interpreted as well.

2. **Paleocurrent direction research**

Diplog data is the most important for paleocurrent direction analysis. There are two ways to determine paleocurrent direction: the first is to statistically analyze the tendency of all laminae in an objective interval and take the main

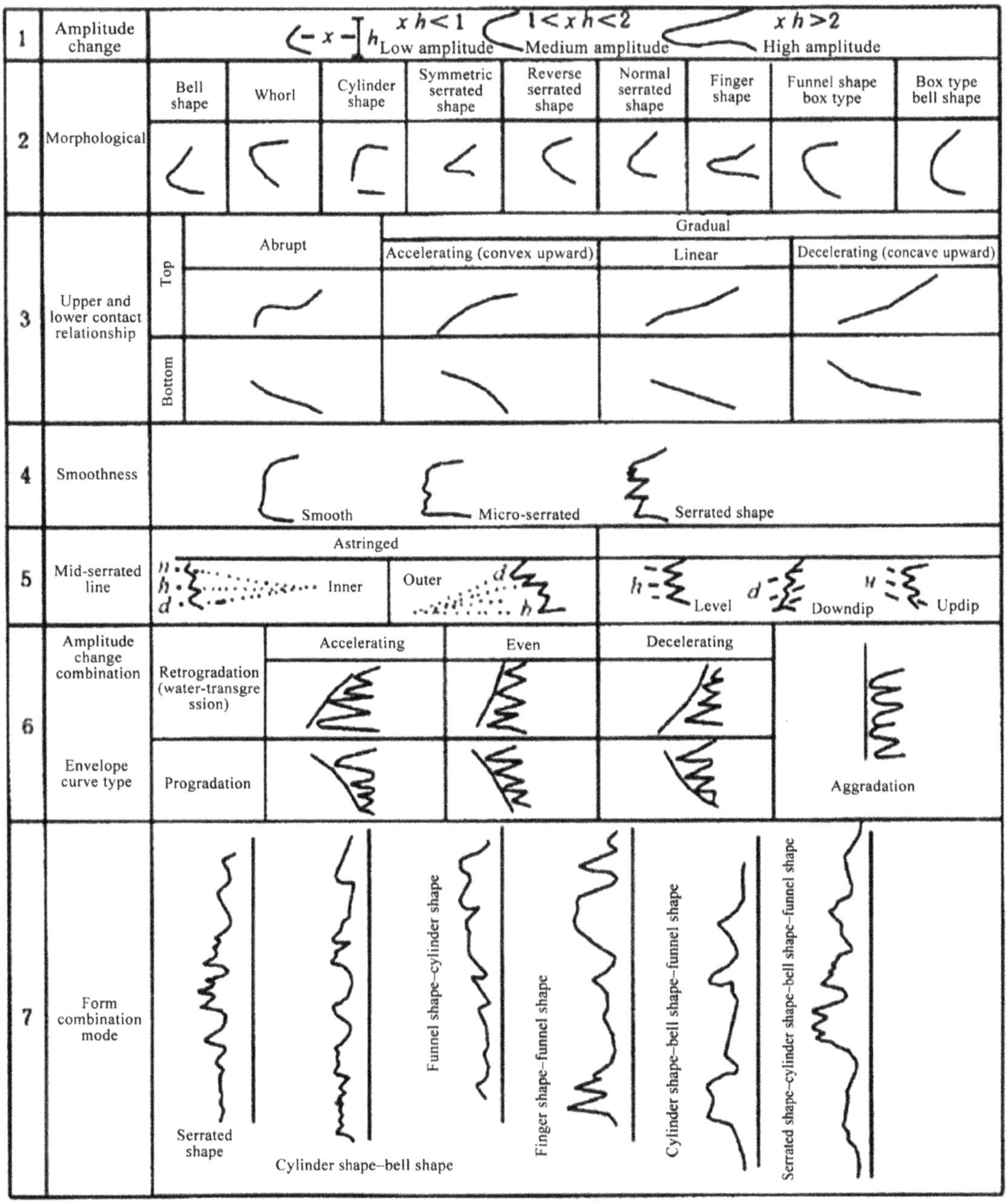

Fig. 5.39 Element chart of natural potential curve (Ma 1982a, b)

Table 5.17 Characteristics of self-potential log for various depositional facies (according to Ma, modified in 1982a, b)

Facies, subfacies, and microfacies			Curve elements						
			Amplitude	Morphology	Contact relationship	Smoothness	Mid-serrated line	Amplitude change combination	Form combination
Alluvial fan	Debris flow		Low amplitude	Serrated shape (reverse)			Updip and parallel	Progradation	Funnel shape
	Inner fan	Main channel	Medium amplitude	Serrated shape (symmetric-normal)			Parallel (level-downdip)		
	Braided channel in fan		Medium-high amplitude	Serrated shape group			Level and parallel	Aggradation and retrogradation	Cylinder shape-bell shape
	Sheet sand of outer fan		Medium-low amplitude	Serrated shape (reverse)			Updip and parallel		
River	Braided river channel sandbank		Medium amplitude	Cylindrical shape	Abrupt change-gradual change	Serrated shape-micro-serrated	Inward convergence	Aggradation	
	Meandering river point bar		Medium amplitude	Bell shape		Serrated shape-micro-serrated	Inward convergence	Lateral accretion	
Deltas	Plain	Branch channel	Medium amplitude	Cylinder shape and bell shape	Top: Accelerating gradual change Bottom: abrupt-gradual change	Micro-tooth-smooth	Inward convergence		
		Crevasse Splay	Medium-low amplitude	Serrated shape			Level		
	Front	Distributary mouth bar	Medium-high amplitude	Funnel shape-cylinder shape		Serrated shape-micro-serrated	Outward convergence-level	Progradation and aggradation	Cylinder shape
		Distal Bar	Low-medium amplitude	Funnel shape group		Micro-serrated	Outward convergence	Progradation	Funnel shape
		Flank facies	High amplitude	Funnel shape and finger shape		Micro-tooth-smooth	Outward convergence and level	Aggradation	

(continued)

Table 5.17 (continued)

Facies, subfacies, and microfacies			Curve elements						
			Amplitude	Morphology	Contact relationship	Smoothness	Mid-serrated line	Amplitude change combination	Form combination
Coastal	Beach		Medium-high amplitude	Serrated-finger shape	Gradual-abrupt change	Smooth	Level-parallel	Aggradation and progradation	Finger shape
	Bar	Center	Medium-high amplitude	Funnel shape	Bottom: decelerating gradual change	Smooth	Outward convergence-level		
		Center (bottom current)	Medium amplitude	Bell-funnel shape	Abrupt at top and bottom	Serrated shape-micro-serrate	Inward convergence-outward convergence	Aggradation	Cylinder Shape
		Inside of barrier	low amplitude	Serrated shape and funnel shape		Micro-serrated	Outward convergence		
		Outside of barrier	High amplitude	Funnel shape and finger shape		Smooth	Outward convergence		
Subaqueous Fan	Main channel of inner fan		Low amplitude	Serrated shape group			Updip parallel	Aggradation	Cylindrical shape
	Mid-fan	Braided channel	Medium-high amplitude	Cylinder-bell shape		Serrated shape-micro-serrated	Parallel (downdip and level)		
		Front	Medium amplitude	Funnel shape		Micro-tooth-smooth	Parallel (level, updip)	Progradation	
	Sheet sand of fan edge		Low amplitude	Funnel-reverse serrated shape		Micro-tooth-smooth	Parallel (level)		
Gravity flow in deepwater	Channel	Bar center	Medium-high amplitude	Cylinder-bell shape	Bottom, abrupt-gradual change	Micro-serrated	Parallel (downdip)	Retrogradation	
		Bar front	Medium amplitude	Funnel shape		Smooth	Parallel (level-updip)	Retrogradation and progradation	
		Bar flank	Medium-low amplitude	Symmetric serrated shape group		Serrated shape	Level and parallel	Retrogradation	Bell shape
	Alluvial fan	Root phase	f	Serrated shape			Parallel downdip	Aggradation	Cylindrical shape-bell shape
		Central facies	Medium amplitude	Cylinder-bell shape		Serrated shape-micro-serrated	Parallel and level		
		Marginal facies	Medium-low amplitude	Funnel-serrated shape		Micro-serrated	Parallel and updip	Progradation	

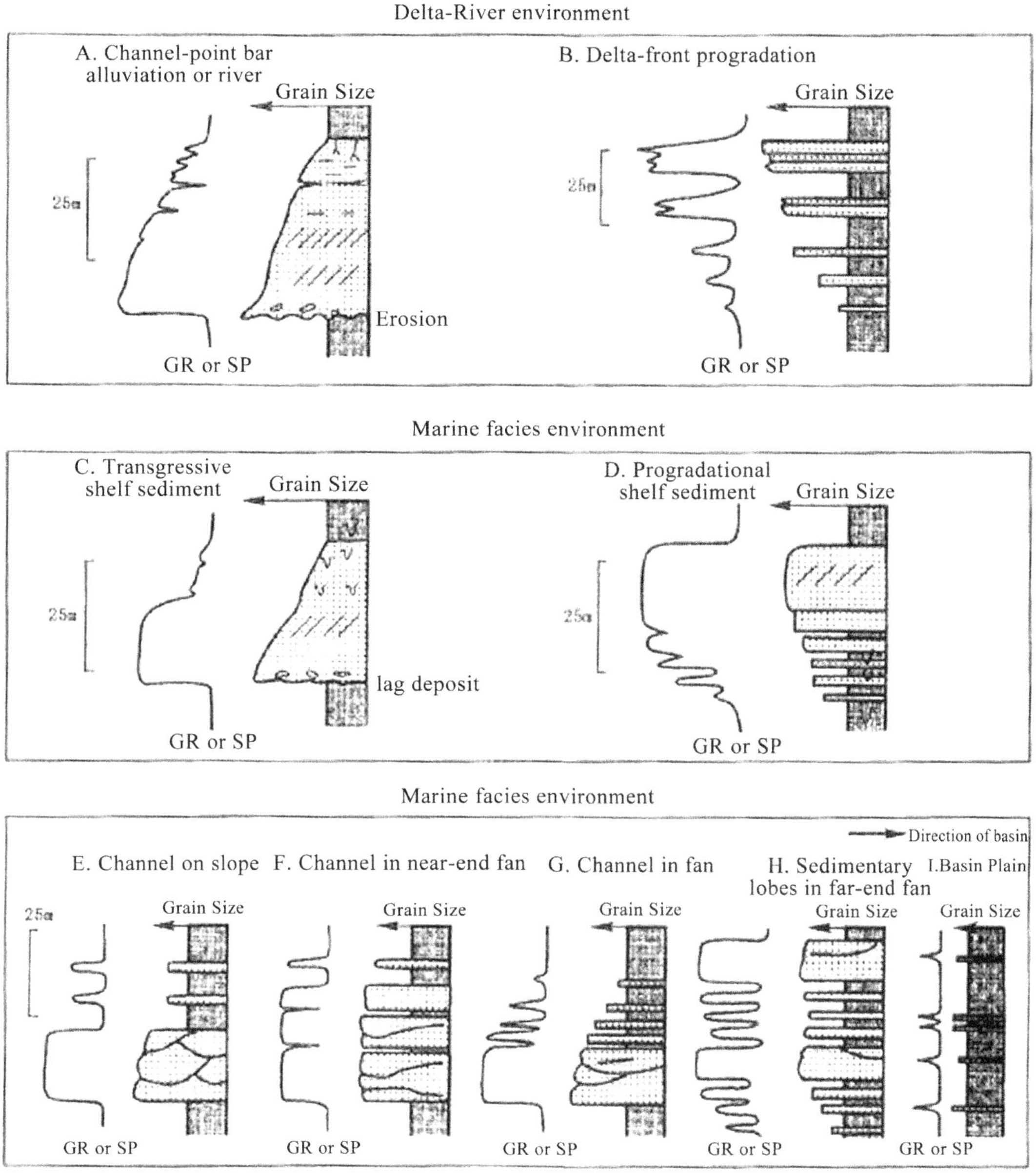

Fig. 5.40 Phase identification of natural gamma (or natural potential) well log (modified based on information from Galloway and Hobday 1983)

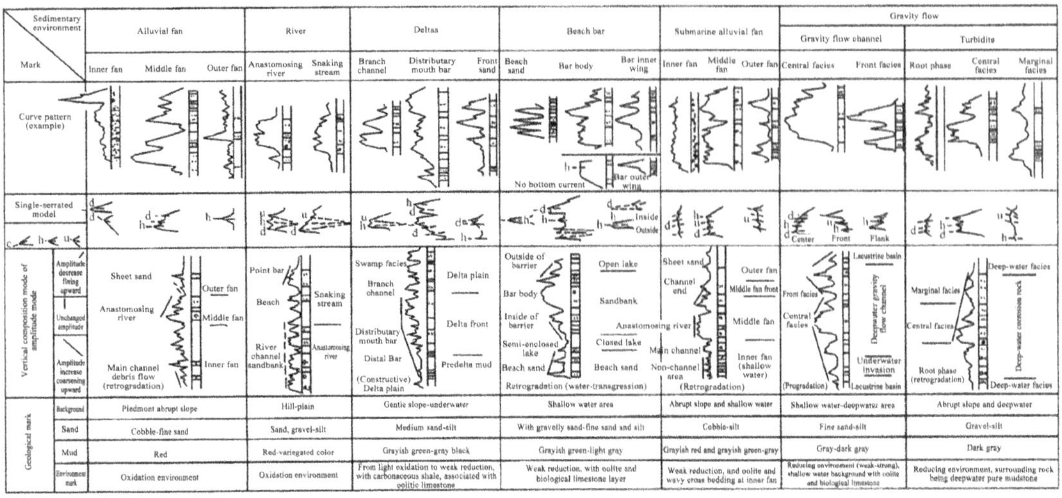

Fig. 5.41 According to the manifestation of the cyclicity, granularity and lithology thickness of deposition sequence on the well log during sedimentological research, the sedimentary facies can be judged, consequently to sum up the curve shape characteristics of different sedimentary facies belts (according to Ma 1982a, b)

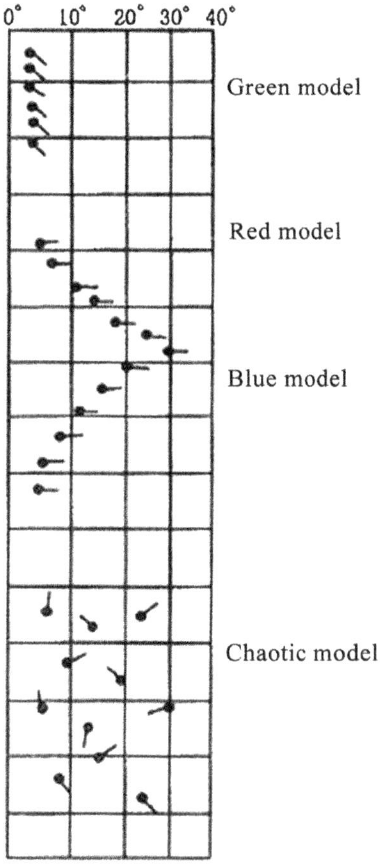

Fig. 5.42 Stratigraphical dip pattern (quoted from Wu et al., modified in 1996)

direction to represent the paleocurrent. The second is to statistically analyze the direction of the blue pattern vector, and take the main direction to represent the paleocurrent. A Schmidt diagram should then be prepared for analyzing the paleocurrent direction of large sandstone sediments, and the advance Schmidt diagram can differentiate the structure, sedimentary tendency, and dip for analysis (Fig. 5.44).

3. **Characteristic Responses of Several Typical Depositional Facies in Diplog**

The characteristic responses of sediments in different depositional environments vary depending on the diplog. This can be illustrated by taking several typical environments as examples, such as meandering river point bars, delta front distributary mouth bars, and turbidity channels (Table 5.19).

5.5.3.2 Depositional Environment Analysis by Imaging Logging and Features

Imaging logging mainly includes electrical imaging and acoustic imaging. Electrical imaging includes FMS, improved FMI, AIT, and ARI. Acoustic imaging includes DSI and CBIL. FMS

Table 5.18 Stratigraphic dip vector mode and reflected geologic features

Vector diagram model	Vector diagram feature	Geological characteristic reflected
Green model	Dip angle and tendency remain unchanged with increasing depth	Horizontal bedding and dip of sedimentary structure laminae
Red model	Dip angle increases with increasing depth, and tendency remain unchanged	Thickening of stratum in the downdip direction, and interpretation of sedimentary characteristics such as sandbank or channel (in combination with SP or natural gamma curve), and maybe folded fault or unconformity
Blue model	Dip angle decreases with increasing depth, and tendency remain unchanged	Thinning of stratum in the downdip direction, which can be interpreted as foreset bedding and may be connected with fold or fault
Chaotic model	Big change in dip and its tendency, no specific law is followed	Fault zone, weathered crust, gravel layer, trough cross bedding, etc.

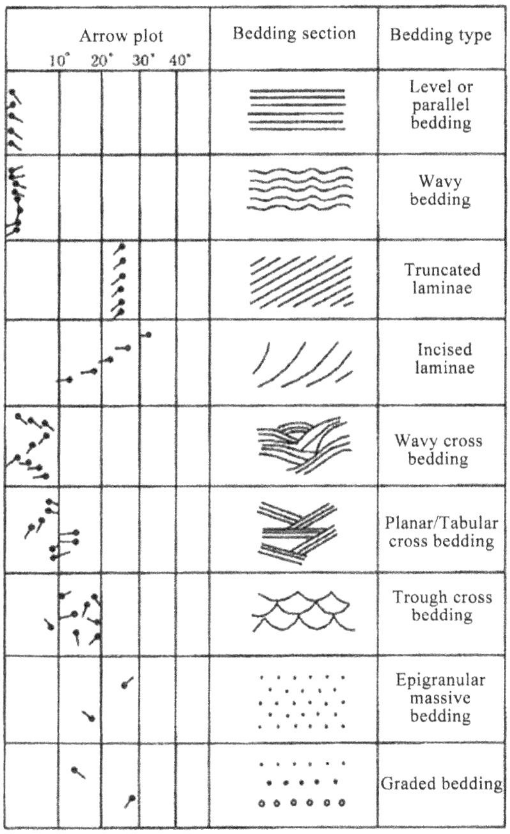

Fig. 5.43 Ideal dip patterns of various beddings (quoted from Wu et al., modified in 1996)

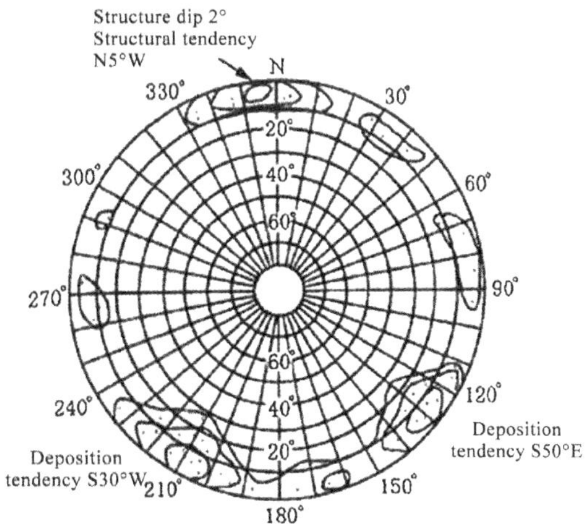

Fig. 5.44 The structure and deposition tendency, dip can be distinguished and analyzed through an improved Schmidt diagram (quoted from Xue et al. 2002)

and FMI data of electric imaging have been used successfully for depositional facies analysis. Generally, imaging logging can reflect clearly vertical sedimentary structures with some scale, for instance, scour surface and megascopic cross bedding (Table 5.20).

5.5.4 Steps in Electrofacies Research

The division of sedimentary microfacies conforms to the method from point to line to plane, which means from core description to single-well microfacies division to microfacies division of section and plane. In this process, accurate observation and description of the core are

Table 5.19 Response characteristics of several typical depositional facies in diplogs

Depositional facies	Response characteristic of diplog
Meandering river point bar	The blue model is inside and the red model outside. Its characteristic tendencies are connected with the well position. There is a difference of 90° between the red and the blue azimuth angles upstream of the point bar, 90° in the middle, and less than 90° at the tail end (Fig. 5.45)
Delta distributary mouth bar	Blue model generally, vector dispersed at top. It is composed of multiple blue models inside the sand layer. If the dip is higher than 10°, accumulation dominates and hydrodynamic energy is weak; if the dip is lower than 10°, hydrodynamic energy is strong. Red model vector on the front sand background (Fig. 5.46)
Turbidity channel	As for the turbidity current supply channel, its bottom is characterized by the red model, and the top by the blue model. The blue model tends to point toward the water (flow) direction, and there is a difference of 90° between the red and the blue positions. The exit of the middle fan channel is characterized by the blue model, and outer fan sheet sand shows accretion features, presented as a low-angle green model vector (Fig. 5.47)

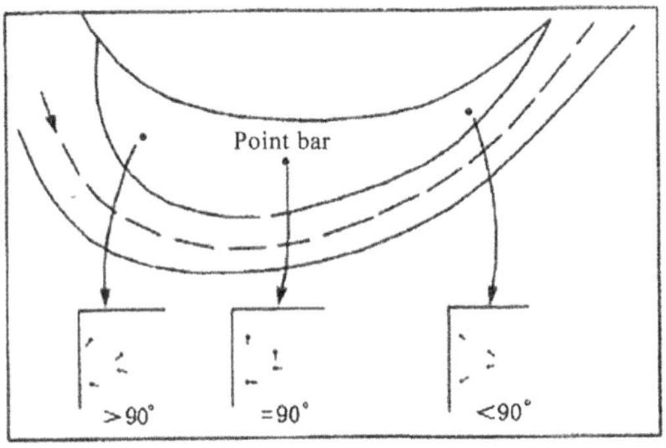

Fig. 5.45 Point bar vector diagram (according to Schlumberger, quoted from Ma 1994)

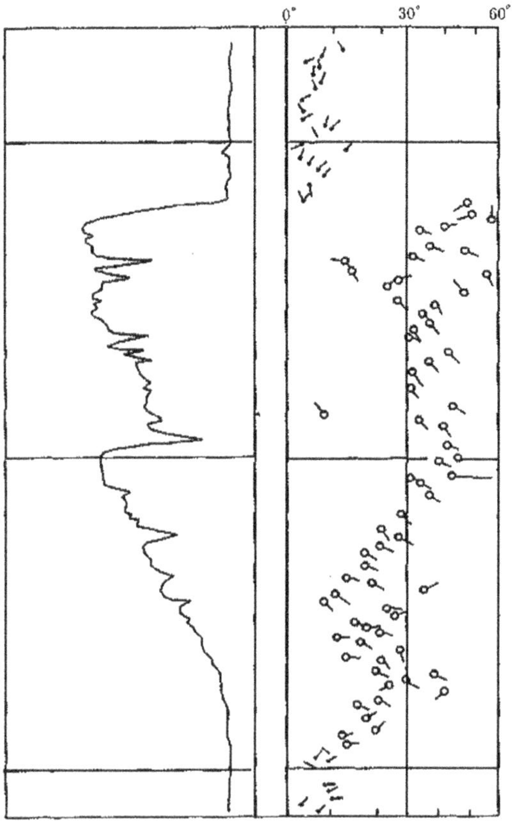

Fig. 5.46 Estuary dam vector diagram (according to Schlumberger, quoted from Ma 1994)

essential, and because the core microfacies and electric well-logging curve correspond, the logging microfacies template can be summarized to obtain the template for the logging microfacies division in the research area and to perform the division and analysis of single-well sedimentary microfacies. Sedimentary microfacies division of the plane and section is performed to compare microfacies and reproduce the plane distribution characteristics.

5.5.4.1 Selection of Key Well

The key well should meet the following conditions: ① favorable location in structure; ② available coring data and complete analysis test data; ③ good well logging condition (good borehole condition and normal drilling fluid), ④ complete open hole logging information; and ⑤ rich formation logging and production well logging data, and abundant production performance data.

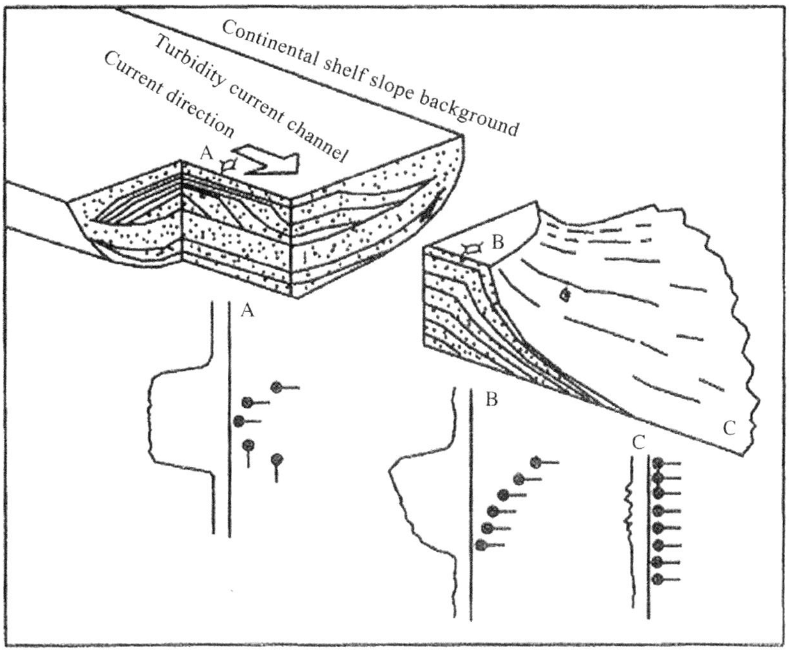

Fig. 5.47 Turbidity channel vector mode (according to Schlumberger, quoted from Ma 1994)

5.5.4.2 Establishment of Electrofacies Atlas

For key wells, single-well facies analysis should be conducted on the basis of core observation and description in combination with analysis and well logging data to divide it into subfacies and microfacies, and prepare a comprehensive column for single-well facies analysis (Fig. 5.36).

The electric well-logging curve characteristics of each microfacies can be compared based on this to determine the relationships therebetween. The electric well-logging curve pattern of the same microfacies may be different in different parts of the microfacies. Microfacies—well logging curve atlas can be prepared as the basis for microfacies division. Taking the electrofacies atlas of a certain oilfield in the Pearl River Mouth Basin, for example, the electrofacies template establishes the logging response of various sand bodies in a braided delta of the electrofacies study (Fig. 5.48).

5.5.4.3 Preparation of Sedimentary Microfacies Map

A sedimentary microfacies map is needed to establish single-well electrofacies and conduct stratigraphic correlation studies, especially correlation analysis of depositional microfacies of main sections by using an electrofacies template in combination with other geological methods to build correlation sections such as section maps of depositional microfacies of Oil Group II in an oilfield in the Pearl River Mouth Basin (Fig. 5.49). Based on this and geological characteristics, as well as parameters of the researched position (such as sand content rate and

Table 5.20 Imaging logging responses of different sedimentary structures (according to Ouyang et al., arranged and modified in 1999)

S. No.	Sedimentary structure and feature	FMI logging response
1	Sequence boundary/Scour surface	There is an uneven surface on the image. The top is dark boulder-clay presented in tabular form and with a slight orientation, and the bottom presents abnormal lithology reflection with high resistance due to gypsiferous mudstone
2	Oblique crossing laminar	It usually corresponds to a set of sine wave curves with light and dark stripes
3	Trough cross bedding	A set boundary is shown by sine curves with different angles on the image, and the cut-off laminae of the upper arc between the two set boundaries are composed of light and dark stripes, and their thickness varies with the core scale
4	Planar/Tabular cross bedding	Several straight set boundaries are usually identified on the image, where each lamina in the set shows that the bottom astringes the light and dark stripes cut-off from the top
5	Ripple cross bedding	On imaging logging around the well, several small short lamina cut-off phenomena are found to be developed locally, and barely any part extends outside of the scope around the well. In terms of comparative interpretation, only ripple cross bedding is shown on parts of typical sections
6	Nodule	Calcareous mottling and sticks are shown as irregular bright blocks and stripes with high resistance characteristics
7	Biological boring structure	Biological boring structure is shown as an irregular bright linear stripe or mottling on the image
8	Herringbone cross bedding	This is shown as a sine curve with reverse tendency of the upper and lower laminae
9	Lenticular bedding	This is shown as a dark stripe with lentoid bright mottling
10	Graded bedding	Coarse lithology (for instance, conglomerate) is shown as a bright color on the image, and fine lithology (for instance, mudstone) is shown as a color gradient from bright to dark
11	Sand-mud interstratified layer	Cyclothem interstratified by sand and mud is presented by parallel light and dark stripes on the image

thickness of sandstone), extrapolation and interpolation of electrofacies can be performed. A sedimentary microfacies map should be prepared based on the required plotting units, such as the plan distribution of the H1 (first Member of H Formation) petroleum reservoir microfacies of an oilfield in the Pearl River Mouth Basin (Fig. 5.50). The vertical sequence and depositional model of the depositional system within the research area can be established on this basis.

The methods and steps described above are based on stratigraphic correlation, under which research should be focused on the vertical and plane distribution laws of depositional microfacies. Therefore, isochronism, reliability,

Electric Well-logging Curve	Morphological	Amplitude	Contact Relationship	Characteristics of Base Level Cycle	Microfacies Interpretation
	The upper part has a small funnel shape; the lower part is mostly cylindrical, sometimes having a bell shape	The upper part has low amplitude, and the lower part has high amplitude	The top shows gradual change, and the bottom shows abrupt change		The top is a natural levee, and the bottom is a distributary channel
	The upper part has a relatively large funnel shape; the lower part is mostly cylindrical, sometimes having a bell shape	Amplitude difference between the upper and the lower parts is small, but some of the lower part is serrated	Abrupt at the top and the bottom		The upper part is crevasse splay, and the lower part is distributary channel
	The upper has a funnel shape, and the lower part has a bell shape	Most of the upper part shows high amplitude, and amplitude of the lower part increases gradually from top to bottom	Abrupt at the top and the bottom		The upper part is distributary mouth bar, and the lower is subaqueous distributary channel
	The upper part has a funnel shape; the lower part is mostly cylindrical, sometimes having a bell shape	Amplitude of the upper and the lower parts is high, while the middle part has a relatively low amplitude	Abrupt at the top and the bottom		Shows successively from top to bottom are crevasse splay, natural levee, and distributary channel
	Funnel Shape	Amplitude of the upper part is higher than that of the lower part	The top shows abrupt change, and the bottom shows gradual change		The upper part is distributary mouth bar, and the lower part is distal bar
	Funnel Shape	The upper part has high amplitude, and the lower part has low-amplitude with serrated shape	The top shows abrupt change, and the bottom shows gradual change		The upper part is distributary mouth bar, and the lower part is sheet sand
	The upper part has a bell shape, and the lower part has a funnel shape	Amplitude of the upper part increases from the top to the bottom, and amplitude of the lower part decreases from the top to the bottom	The top shows gradual change, and the bottom shows gradual change		The upper part is subaqueous distributary channel, and the lower part is distributary mouth bar
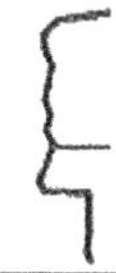	The upper part is cylindrical, and the lower part has a bell shape	Most amplitudes are high	Abrupt at the top and the bottom		The upper part is distributary mouth bar, and the lower part is subaqueous distributary channel
	The upper part is cylindrical, and the lower part has a funnel shape	Amplitude of the upper part is high, and amplitude of the lower decreases from the top to the bottom	The top shows abrupt change, and the bottom shows gradual change		The upper part is distributary channel, and the lower part is distributary mouth bar

Fig. 5.48 Log facies template

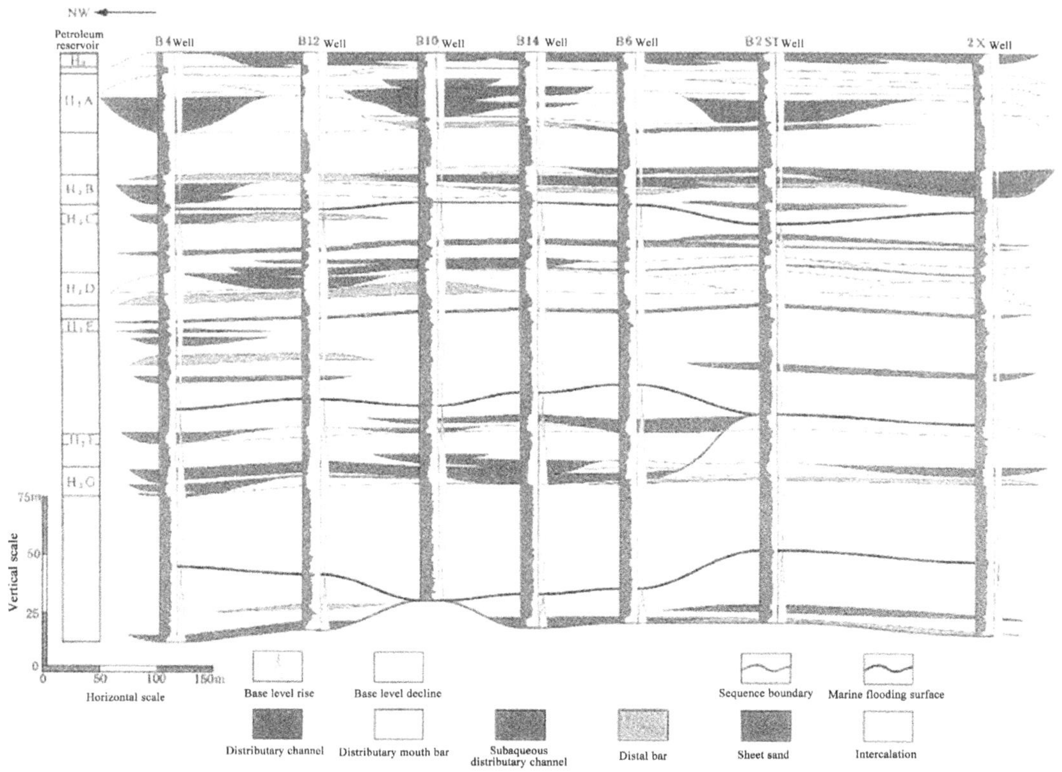

Fig. 5.49 Connecting-well section of II oil formation sedimentary microfacies of some oilfield in Pearl River Mouth Basin

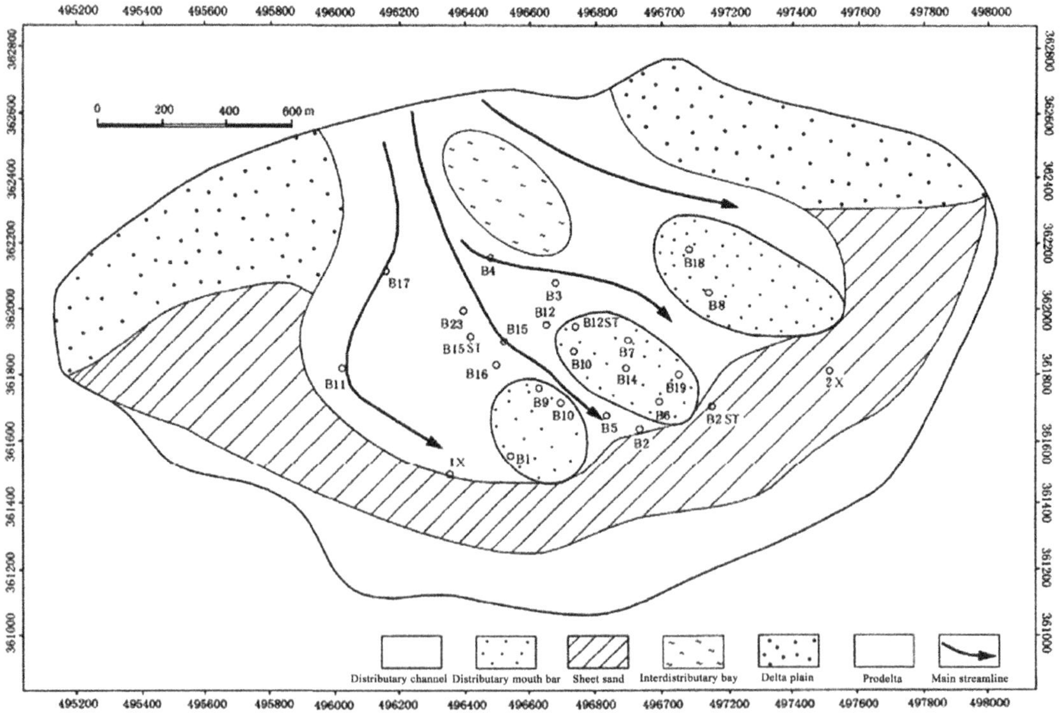

Fig. 5.50 Distribution of sedimentary microfacies of H_1 oil-bearing formation H_1 of some oilfield in Pearl River Mouth Basin

scientificity, and veracity of stratigraphic correlation are of great significance. In recent years, sequence stratigraphy has been developing rapidly, and it has fundamentally changed the ideas and principles of stratigraphic correlation.

5.6 Seismic Facies Analysis

Traditionally, the depositional environment is determined by studying the core and outcrop, while for a wide area with no core or outcrop, reflection characteristics of the seismic section are used to identify depositional facies and predict favorable facies belt.

5.6.1 Basic Concept

5.6.1.1 Definition

Seismic facies, which originate from depositional facies, can be interpreted as the sum of depositional facies on a seismic cross section. Sheriff (1980) defined seismic facies as seismic characteristics resulting from depositional environment (such as marine facies or continental facies) (1982). Seismic facies analysis means "explaining the environmental background and lithofacies based on seismic data" (Vail 1977).

The purpose of seismic facies analysis is identifying the characteristics and form combinations of unique reflection wave groups within each sequence and assigning some geological meaning for depositional facies interpretation. Therefore, studying seismic facies within favorable sequences is beneficial for determining the depositional facies and lateral distribution range of sandstone reservoirs, thus laying the foundation for the comprehensive prediction of sandstone reservoirs.

5.6.1.2 Identification Marks

The main basis for distinguishing the forms of such seismic wave groups is seismic reflection parameters or factors, which, from the perspective of sedimentology, could be called identification marks of seismic facies. These identification marks are the basis for seismic facies analysis, which must be conducted within certain seismic stratigraphic units. Sequence or chronostratigraphic unit is the most important seismic stratigraphic unit. From the perspective of sequence stratigraphy, a systems tract can be divided into a three-order sequence, while remarkable differences in sedimentary systems tracts exist among different parasequence sets. A parasequence set is, therefore, generally taken as the most fundamental seismic stratigraphic unit for seismic facies analysis.

A seismic facies marker means "seismic reflection characteristics within a parasequence set that have significant influence on the features of a seismic section and possess important depositional facies significance." There are many identification marks of seismic facies, and they can be summarized as follows: ① basic attributes and textures of seismic reflection; ② internal reflection configurations; ③ external forms; ④ boundary relationships (including reflection termination and lateral variation); and ⑤ interval velocity. The first three marks are the most common ones, and they comprise the core of this three-level study.

5.6.1.3 Description Principles

It should be noted that a seismic internal reflection configuration means a form combination of several lineups within a seismic stratigraphic unit, and external form refers to the appearance characteristics of seismic stratigraphic units, reflecting the geometry formed by the upper and the lower lineups. The former is an internal property of seismic facies, while the latter pertains to the appearance of seismic facies. Therefore, different description languages should be used. However, the use of "form" for describing seismic facies is confusing, thus the author suggests that "form" should be used for external description, while "shape" should be used for internal description based on Chinese customs. "Form" should be used for the external form of a seismic facies unit, while "shape" should be used for the internal arrangement and combination.

5.6.2 Seismic Facies Markers and Main Characteristics

5.6.2.1 Basic Attributes and Textures of Seismic Reflection

A seismic reflection attribute refers to the physical and seismological characteristics of each component (i.e., lineup) of a seismic cross section, of which the basic attributes include three elements, namely, amplitude, frequency, and continuity (Table 5.21).

1. **Basic attributes**

(1) Amplitude

Amplitude refers to the displacement of a particle away from the equilibrium position. Apparent amplitude is the reflection factor of the corresponding seismic boundary. For the same incident wave, a larger boundary reflection factor will generate a stronger reflection amplitude. The reflection factor is determined by the difference in the wave impedance of the stratum above and below the boundary. The larger this difference, the larger is the reflection factor. Wave impedance is closely related to lithology. In general, the wave impedance of mudstone is relatively low, that of the sandstone is intermediate, and that of carbonate rock is high. Thus, apparent amplitude can be summarized as the lithologic difference of a stratum above and below the boundary.

(2) Frequency

Frequency reflects the space between adjacent reflection boundaries. The larger the space, the longer is the time interval of reflected waves at the upper and the lower boundaries, namely, frequency, and vice versa. When the space between boundaries is smaller than 1/4 of the dominant wavelength of an incident seismic wave, reflected waves on the two boundaries overlap into a composite wave, making it impossible to separate the two boundaries. This is called vertical resolution of a seismic wave (it can be used to determine the minimum space between reflecting surfaces required for two separate boundaries rather than one boundary, with 1/4 of the dominant wavelength).

(3) Continuity

Continuity indicates the extension of amplitude in the horizontal direction and the frequency of lineups, and it can reflect the difference in lithology above and below the boundary, and the lateral stability of the boundary spacing.

2. Reflection Texture

Seismic reflection texture is obviously different from seismic reflection configuration, which is described later. The reflection texture refers mainly to the amplitude, frequency, and continuity of several lineups in a seismic unit or the difference between the intensity and the property of amplitude in a section, which represents the difference in lithology and intensity, and stability of hydrodynamic conditions during sedimentation. Reflection configuration means the arrangement and combination mode of several lineups within seismic units in the section, which is the response of sedimentation and the process.

In seismic facies marks, reflection texture can generally be described and named using two methods: ① When both the upper and the lower stratigraphic units have three homogeneous characteristics, as mentioned above, reflection texture can be described directly and named as "amplitude + frequency + continuity," for example, "strong-amplitude, high-frequency, and high-continuity seismic reflection texture." ② When the characteristics of the upper and the lower stratigraphic units are heterogeneous, reflection texture can be described and named according to characteristics of vertical variation,

Table 5.21 Basic attribute classification and characteristics of seismic reflection

Basic attribute classification of seismic reflection	Geological significance	Classification	Standard	Classification model	Example
Amplitude	Indicates the reflection factor	Strong	Amplitude exceeds one seismic channel	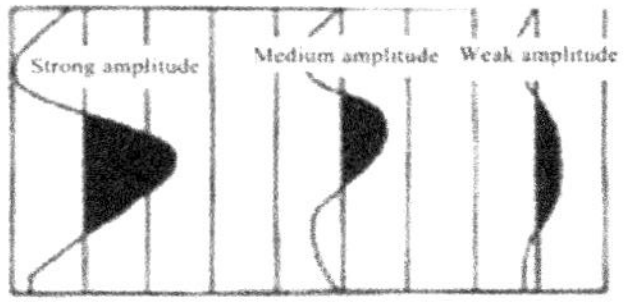	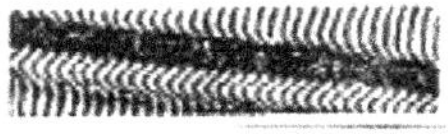
		Medium	Amplitude falls between two seismic channels		
		Weak	Amplitude is smaller than 1/3 of the space between seismic channels		
Frequency	Reflects the space between adjacent reflection boundaries	High	Adjacent lineups are arranged densely	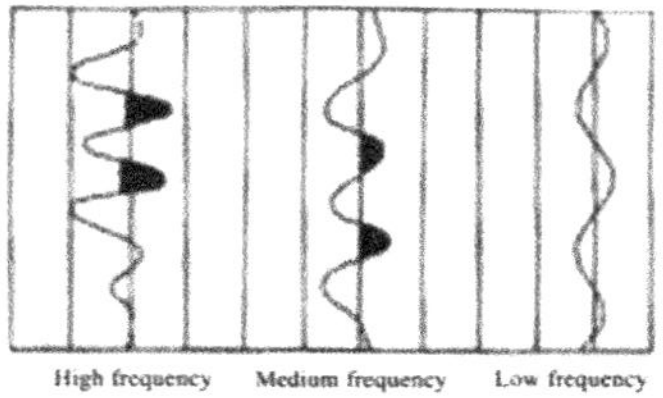	
		Medium	Adjacent lineups are arranged moderately		
		Low	Adjacent lineups are arranged sparsely		

(continued)

Table 5.21 (continued)

Basic attribute classification of seismic reflection	Geological significance	Classification	Standard	Classification model	Example
Continuity	Reflects the horizontal stability of the space between boundaries	Good	Continuous length is longer than superimposition	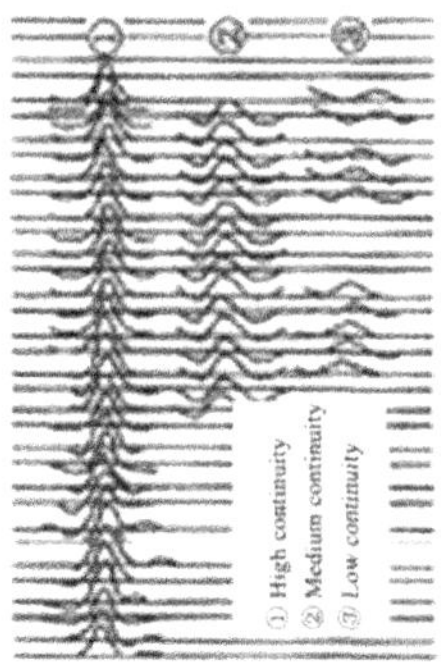	
		Medium	Continuous length is near to 1/2 of superimposition		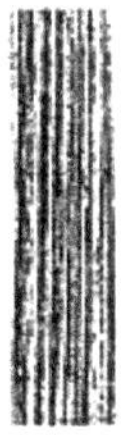
		Poor	Continuous length is shorter than 1/3 of superimposition		

for example, "upward strengthening reflection texture." Common and typical reflection textures are listed in Table 5.22.

(1) Chaotic reflection texture (high-amplitude and low-continuity)

Chaotic reflection texture is essentially characterized by strong but discontinuous amplitude and a disorderly waveform (Table 5.22). The strong amplitude indicates great difference in lithology above and below the boundary. The discontinuity indicates rapid lateral variation of lithology or stratum thickness, and further indicates large lateral variation in the reflection factor. This reflection texture represents the product formed in an environment with high energy and volatile hydrodynamic conditions or the result of later deformation of an original continuous formation. Such textures generally develop in alluvial fans, escarpment turbidite fans, or submarine fans, as well as in a stratum under strong deformation due to gravity sliding or tectonic movement.

(2) Blank reflection texture (texture with very low amplitude)

Blank reflection is essentially characterized by very low amplitude, and the lineups can hardly be noticed. Homogeneous lithology makes it hard for the reflection boundary to take shape, which is the root cause behind the absence of a reflection texture. A depositional environment with relatively stable energy may contain sandstone, mudstone, or limestone with huge thickness or nearly uniform sedimentary formation after bioturbation. From the perspective of sedimentology, the environment may contain thin fine-grained sediments or thick coarse sediments. This reflection texture may develop in deep lacustrine facies mudstone, littoral facies sandstone, continental shelf facies limestone and braided river sandstone extremely poor in muddy sediments. The difference in lithology is great, but the interior is relatively uniform. Blank reflection is also related to the wave impedance of the top unit. When the reflection factor of the top boundary is very large and transmission

Table 5.22 Classification of seismic reflection textures

Reflection texture	Amplitude	Frequency	Continuity	Model diagram	Example
Chaotic	High	–	Discontinuous		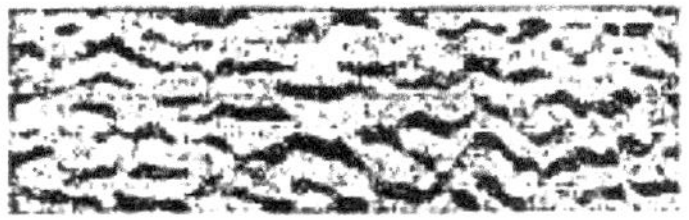
Blank	Very low	May be high or low	Medium continuity		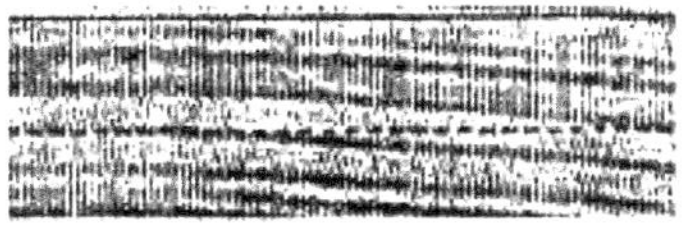
Three highs	High	High	High		
Strengthening upward	Strengthening upward	May be high or low	May be good or bad		

energy is relatively low, the seismic reflection of the underlying strata is shielded, thus greatly weakening the reflection amplitude, even turning into a blank reflection.

(3) Three-high reflection texture (texture with high amplitude, high frequency, and good continuity)

The typical three-high seismic reflection texture is characterized by "high amplitude, high frequency, and good continuity." High amplitude indicates that the differences in lithology above and below the boundary are great. High frequency indicates that stratum depth is small and alternative variation of lithology is frequent. Good or high continuity indicates that the lithology and depth of the stratum are very stable in the horizontal direction. The three-high combination is a typical seismic reflection characteristic of deep-water facies with developed turbidity sand or lakeside swamp facies with stably developed thin coal bed.

(4) Reflection texture strengthening or weakening upward

Its basic characteristic is that the amplitude of the lower part is relatively weak, but it obviously strengthens upward. This indicates that the lithology of the lower part is relatively uniform, while the difference in lithology increases upward. This texture generally develops in a combination of depositional facies in the declining semi-cycle (namely, highstand systems tract), including delta, regressive progradational coasts, and continental shelf deposits. A reflection texture that is the opposite of the abovementioned texture may appear, namely, upward-weakening reflection texture, which generally develops in a combination of depositional facies in the rising semi-cycle (namely, transgressive systems tract), including deltas, regressive retrogradational coasts, and continental shelf deposits.

It could hereby be concluded that the intensity of amplitude corresponds to the difference in lithology above and below the boundary, size of frequency corresponds to stratum thickness, and continuity performance corresponds to the lateral stability of a stratum. Combining the above with the analysis of lithologic difference characteristics, lateral variation rule, and cyclicity of different depositional facies, further analysis and interpretation of depositional facies can be made.

5.6.2.2 Seismic Internal Reflection Configuration

Internal reflection configuration is also called internal reflection texture by some scholars. Within depositional facies marks, sedimentary structure refers to the spatial arrangement mode of each component of sedimentary rock. Similarly, seismic reflection configuration refers to the spatial arrangement mode of each component (lineup) of a seismic cross section, the form of which is similar to the bedding structure (Wang Yingmin 1991).

Seismic internal reflection configuration refers to the spatial arrangement and combination mode of each component (lineup) in a seismic cross section. It directly reflects stratigraphic superimposition style, nature of sedimentation, and sedimentary compensation. As discussed above, seismic reflection configuration is useful for discussing the geometry of and correlation between internal lineups of a seismic stratigraphic unit, and it falls under morphology and geometric seismology.

R. M. Mitchum et al. (1977) divided seismic reflection configurations into 7 categories based on the shape of the internal reflection configuration, namely, sub-parallel, wavy, divergent, progradation clinoforms or foreset, hummocky clinoform, chaotic, and no-reflection. In the author's opinion, the last two categories are not reflection configurations because those arrangements and combinations without lineups cannot explain sedimentation. Hence, the author considers them reflection configurations. All characteristics of reflection configurations possess clear meanings in terms of sedimentation (Table 5.23).

1. **Parallel or subparallel**

This configuration is characterized by parallel or slightly fluctuating lineups. It is the product of

Table 5.23 Classification and characteristics of seismic internal reflection configurations

Internal reflection configuration	Features	Sedimentation	Model diagram	Example
Parallel or subparallel	Parallel or subparallel lineup	Uniform vertical accretion		
Wavy	Fluctuating lineup	Nonuniform vertical accretion		
Divergent	Space between lineups decreases gradually toward one side	Loading subsidence		
Progradation clinoforms or foreset	Lineup is relatively tilted and aggrades toward one side	Lateral accretion	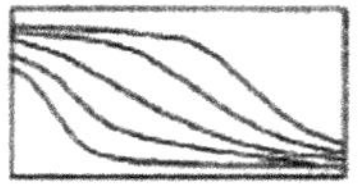	
Hummocky clinoforms	Irregular shape, very poor continuity	Weak flow		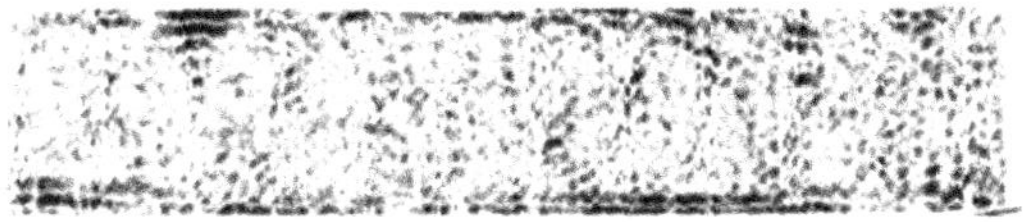

(continued)

Table 5.23 (continued)

Internal reflection configuration	Features	Sedimentation	Model diagram	Example
Bidirectional downlap	Flat-bottom convex-top	Downstream Accretion		
Eyeball	Bidirectional convex	Uneven deposition		

uniform vertical accretion with a roughly equal lateral deposition rate, and it can develop in shelves, deep-sea basins, deep or shallow lakes, swamps, and many other facies. This reflection configuration is characterized by good continuity, while amplitude and frequency may vary according to specific conditions (Table 5.23).

2. **Wavy**

This configuration is characterized by parallel lineups in the overall trend and lineups with a degree of fluctuation in the fine structure. It is the product of non-uniform vertical accretion, which means the configuration is manifested as vertical accretion based on parasequence or genetic sequence; lineups are, therefore, parallel on the whole. However, the lateral deposition rate may be different, and secondary lateral accretion may exist in finer stratigraphic unit levels. It is likely to develop in alluvial plains, shallow seas to semi-deep seas (lakes), and fan bodies with a relatively slow total deposition rate, as well as in contour currents.

3. **Divergent**

The space between lineups decreases gradually toward one end, and some lineups disappear gradually. Hence, the number of lineups decreases toward one end, and the thickness of the corresponding stratigraphic unit decreases accordingly, thus forming a wedge shape. A decrease in stratum thickness cannot be ascribed to truncation or onlap on the top and bottom boundaries, but to a gradual decrease in the space between lineups. It is formed under the differential subsidence, in which case, a decrease in transverse deposition rate will lead to a decrease in stratum thickness toward one end.

4. **Progradation clinoforms or foresets**

Lineups between the top and bottom boundaries of a parasequence set are relatively tilted and aggraded toward one side if the boundaries are taken as references. The typical progradation

Table 5.24 Progradational reflection configuration correlation

Progradational reflection configuration	Foreset	Topset	Bottom set	Water level change	Gradient	Formation environment
SB	√	√	√	Relatively rising	–	Continental slope and mud-rich delta
Toplap	√	×	√	Relatively still	–	Mud-rich delta
Downlap	√	√	×	Relatively rising	–	Alluvial fan, escarpment turbidite fan, and fan delta
Parallel	√	×	×	Relatively still	Large	Delta or fan delta
Shingled type	√	×	×	Relatively still	Small	Wave-dominated delta and depression lake delta

clinoform is composed of a topset, foreset, and bottom set. According to variations in the internal reflection configuration, the morphological characteristics of the foreset, and the degree of development of the topset and bottom set, progradation clinoforms can be subdivided into sigmoid, tangential, complex sigmoid-oblique, parallel, and shingled (Table 5.24). Albeit all progradation clinoforms are products of downstream accretion of sediments, have foresets, and reflect paleocurrent direction, they vary in many aspects. Progradation clinoform is a typical indicator of deltas, fan deltas, various fans, and slope-break transformation.

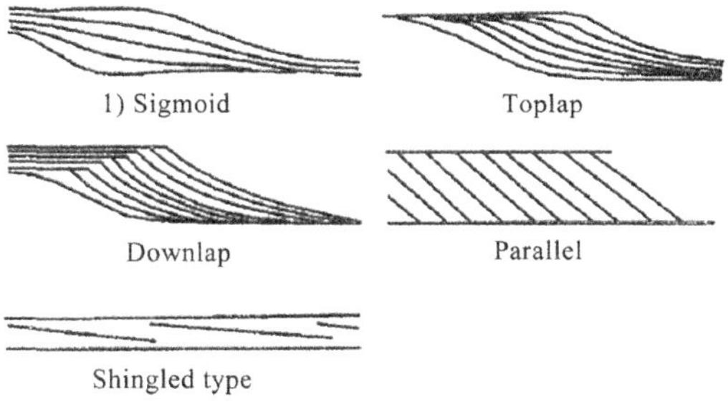

Fig. 5.51 Seismic interpretation mode of foreset structure (according to Berg, modified in 1982)

(1) Sigmoid

This is a typical progradation clinoform consisting of a topset, foreset, and bottom set. A group of superimposed reversed sigmoid reflection lineups are developed inside. On top of the reversed sigmoid reflection is a nearly horizontal topset, in the middle is an inclined foreset, and lineups gradually become gentle downward to form the bottom set (Fig. 5.51). The development of the topset indicates that the water level in the area is in the state of relative rise with an increase in accommodation, leading to vertical accretion of terrigenous matters. The development of the bottom set indicates that a large amount of materials is deposited in front of the sedimentary bodies. According to the principle of sedimentary differentiation, coarse fragments are unloaded in the foreset and the topset, while the area corresponding to the bottom set is composed mainly of fine-grained sediments. Therefore, the bottom set can be deemed as the representation of fine-grained terrigenous matter and abundant muddy sedimentation. In general, this reflection configuration is likely to develop in continental slopes and mud-rich deltas (Fig. 5.52).

(2) Tangential

Some scholars call this tangential oblique, and it is characterized by the absence of the topset and termination of the foreset top on the top boundary of a stratigraphic unit in the manner of toplap (Fig. 5.51). The existence of a toplap indicates that the absence of the topset is not because of later tectonic erosion, but because

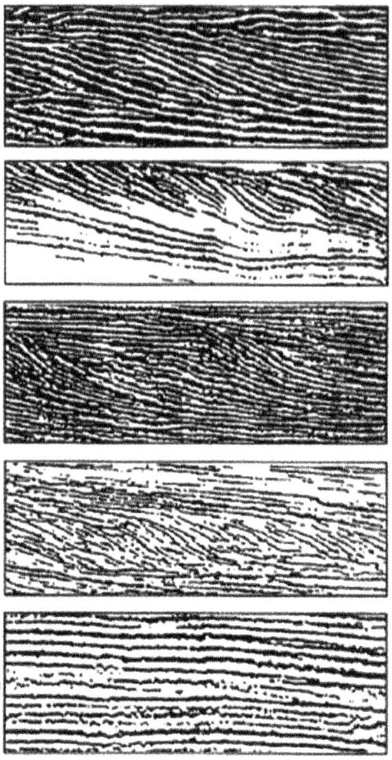

Fig. 5.52 Typical reflection mode of delta (according to Berg 1982)

during the relatively still stage of water level, unchanged accommodation makes vertical accretion over the water level impossible, thus limiting the ability of passing sediments to aggrade at frontal sedimentations. The geological significance of bottom set development is the same as that of sigmoid progradation clinoform. In general, this reflection configuration is likely to develop in mud-rich deltas and during the relatively still stage of water level (Fig. 5.52).

(3) Complex Sigmoid-oblique

This is characterized by the absence of the bottom set, and termination of the foreset bottom on the bottom boundary of a stratigraphic unit in the manner of downlap (Fig. 5.51). The development of the topset indicates that it is formed during the relatively rising water level stage. The absence of the bottom set indicates that the grain size of terrigenous fragments is large and grain-sided sediments are lacking. In general, this configuration is likely to develop in alluvial fans, escarpment turbidite fans, and fan deltas (Fig. 5.52).

(4) Parallel

This is composed of a group of relatively steep reflection lineups and characterized by the absence of the topset and the bottom set. Toplap is noted in the updip direction, and downlap is noted in the downdip direction (Fig. 5.51). It is formed due to the lateral accretion of coarse fragments during the relatively still water level stage. The progradational direction is broadly consistent with the long-axis direction of faulted basins, and this configuration is generally developed in deltas or fan deltas (Fig. 5.52).

(5) Shingled

The characteristics of this type are similar to those of the "parallel," but its foreset dip angle is quite gentle, and the thickness of the corresponding stratum is relatively small, only equal to 1–2x the space between lineups (Fig. 5.51). It is formed due to lateral accretion of sediments when the water is relatively still with a small water depth and a gentle slope. It generally develops in flat slope fluvial-dominated deltas and depression lake deltas. A shingled progradation clinoform is difficult to identify on a seismic cross section owing to its small scale, but it is the most common configuration in lake bases, thus possessing great significance from the viewpoint of research on continental petroliferous basins in China.

5. **Hummocky clinoform**

This is mainly composed of irregular reflection segments with poor continuity, and non-systematic reflection termination and bifurcated lineups are often noted with small amplitude fluctuation, which is near to the limit of seismic resolution. It reflects weak flow sedimentation, which is usually seen in delta and fan delta sedimentations.

6. **Bidirectional downlap**

Its characteristics include lineups that are convex in the middle, while the two ends are under the bottom boundary of a stratigraphic unit in the form of downlap, which is manifested as bidirectional lateral accretion. This is totally different from the aforementioned unidirectional lateral accretion. In fact, it is the cross section of sedimentary bodies without progradation clinoform of the bottom set, and they share the same geological significance.

7. **Eyeball**

Its features include small scale, and it generally develops in a parasequence set. It is characterized by lineups with "convex top and concave bottom," which are eyeball shaped with a thickness equal to that of several lineups. This reflection configuration is likely to develop in a channel sand body with small scale, longshore bar, and superimposed fan lobe.

In conclusion, all reflection configurations possess obvious characteristics that are easy to identify and share close correspondence with depositional facies. Seismic facies analysis can be used to identify and determine sedimentary bodies by combining tectonic settings and regional depositional characteristics.

5.6.2.3 External Form of Seismic Facies

The external form of a seismic facies unit, called unit form in short, refers to the outer shape or body characteristics of a seismic facies unit with the same reflection configuration or texture in 3D space. Similar to seismic reflection configurations, external forms of seismic facies come under the purview of geometric seismology. In practical operations, it refers to external forms constituted of certain seismic reflection textures or configurations on a seismic cross section. It can provide information about geometrical morphology, hydrodynamic conditions, provenance, and paleogeographic background of sedimentary bodies. Most external forms of seismic facies are direct and good reflections of the shapes of sedimentary bodies. For example, "Mound" is the reflection of the cross section of a sandbank or delta. It is, obviously, of great significance to depositional facies interpretation. Several common external forms of seismic facies are shown below on a seismic cross section (Fig. 5.53; Table 5.25).

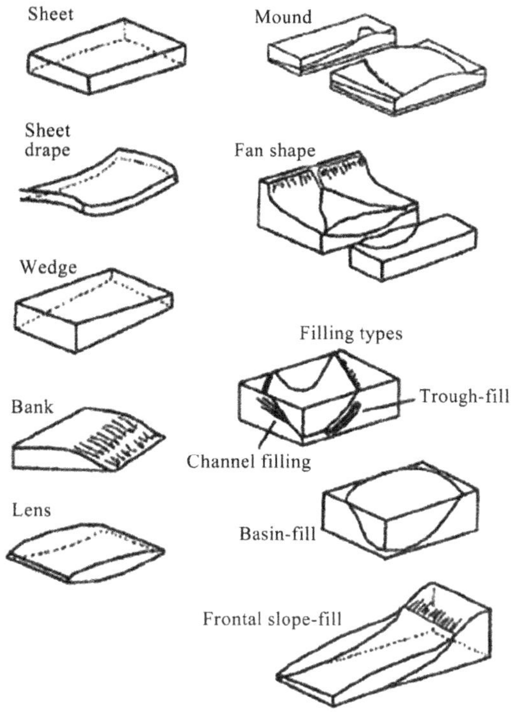

Fig. 5.53 External form of some seismic facies unit (according to Mitchum, Vail and Sangree 1977)

1. **Sheet**

This is the most common external form. The thickness of the seismic facies unit is relatively stable, and the top and bottom boundaries are parallel or sub-parallel to the lineup, with a transverse scope much larger than stratum thickness. It generally corresponds to the parallel (sub-parallel) configuration or the wavy configuration on a section. Sheets are the main product of vertical accretion. Parallel sheets generally represent stable sedimentary environments such as deep seas (lakes) and semi-deep seas (lakes), while the sub-parallel sheets generally represents unstable environments such as shoreline seas (lakes), alluvial plains, and delta plains.

Table 5.25 Division and representation of seismic facies

Contact relationship of sequence top and bottom				Seismic facies identification marks						
				Geometric parameter				Physical parameter		
Top contact	TC	Bottom contact	BC	External form	EF	Internal reflection configuration	IRC	Reflection attribute		RA
Parallel	P	Parallel	P	Sheet	S	Parallel (sub-parallel)	P	Amplitude	Strong	H
Toplap	T	Onlap	O	Drape	D	Divergent	D		Medium	M
									Weak	L
Truncation	Tr	Downlap	D	Wedge	W	Wavy	W	Continuity	Good	G
Representation methods, e.g., seismic facies unit coded as P-D/WF-MMM				Bank	B	Foreset	F		Medium	M
									Poor	B
				Mound	M	Hummocky clinoform	He	Frequency	High	H
				Lens	L	Bidirectional downlap	Bd		Medium	M
				Fill	F	Eyeball	E		Low	L

Seismic facies characteristics representation: *P* parallel contact of top boundary, *D* downlap of bottom boundary, *WF* wedge progradational reflection configuration, *MMM* medium amplitude, medium continuity, and medium frequency

2. **Drape**

This possesses appearance features similar to those of "sheet," but it is covered crookedly on the underlying unconformity terrain. Its form is identical to that of the unconformity terrain, without onlap relations between them. Drape is ascribed to even vertical accretion of suspended sediments in deep-water environments; in other conditions, onlap occurs. Therefore, it is deemed as a distinctive sign of deep-water sedimentation, especially pelagic sedimentation.

3. **Wedge**

Its characteristics include an increase in seismic facies unit thickness along the incline direction in divergent form, thus reflecting differential subsidence or lateral variation of deposition rate during sedimentation. Regarding thickness, it has small changes along the strike direction, representing a parallel (sub-parallel) or wavy external form. It shares the same geological significance as the "divergent" form, and it is developed in basins or slope zones with depression edges.

4. **Bank**

This is characterized by a decrease in seismic facies unit thickness along the incline direction in progradation clinoform, chaotic external form, and no reflection. The thickness of the middle part is large, while that of the two ends is small along the strike direction. It shows bidirectional progradation or the mound configuration, and is manifested as a fan shape on the plane. It is a typical sign of sedimentary bodies such as fans and deltas.

5. **Mound**

"Mound" is characterized by the form of "flat-bottom convex-top," continuous and gentle lineups at the bottom, convex lineups at the top, shaped like a "mound." It is generally interpreted as the product of high-energy deposition and represents quick unloading during sediment transport. Bidirectional downlap reflection is common inside the large two-dimensional mound, which is a characteristic of delta transverse sections in general. Small-scale mounds can be interpreted as proximal subaqueous fans and alluvial fans in combination with structural positions. Medium and small three-dimensional mounds inside the lake basin, especially those with draping reflection on the top, indicate turbidite fans. Moreover, a large mound reflection angle can generally be ascribed to organic reefs or various diapiric tectonic activities that usually develop in deep water.

6. **Lens**

This is characterized by a "bidirectional convex" eyeball shape on a section, which is generally shown as a lens with a thick center and thin edges. Sedimentary bodies represented by this external form can develop in various depositional environments. One form of lens is the result of large subsidence rate and deposition rate in the middle, which decrease toward the two ends; this could be called primary genesis. The other form of lens results from differential compaction in the process of sandstone development in the middle and mudstone development at the two ends; this could be called secondary genesis. In general, the two forms exist simultaneously. This structure possesses a significant facies indication function. A large lens reflection generally indicates delta progradation or a succession main channel, while sedimentary bodies characterized by small lens reflection may be presented in almost all depositional environments.

7. **Channel-fill/Trough-fill**

The section is characterized by a concave "flat-top concave-bottom," sudden increase in the thickness of partial stratum, downward erosion and fill into the underlying stratum, thus forming a mirror symmetry relationship with "mound." This reflection configuration is likely to develop in the cross section of a basin depression axis and taken as a typical mark of local underwater channel erosion. It generally indicates submarine canyon and turbidity current scour, which are formed during the relative decline stage of sea level. For this reflection configuration, there are six internal filling models (Fig. 5.54).

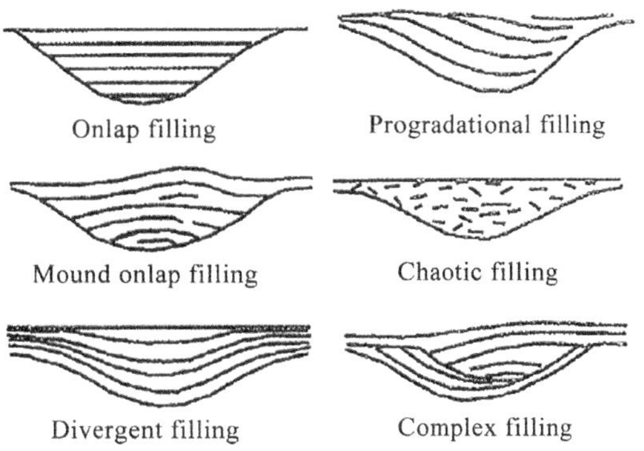

Fig. 5.54 Internal filling mode of filled (concave) type reflection structure (according to Payton 1977)

5.6.3 Seismic Facies Analysis

Seismic facies division refers to the division of different seismic facies units based on seismic facies markers, namely, interpretation and deduction of seismic facies on the basis of seismic facies markers.

Horizontal distribution range and the corresponding level of depositional facies units of different seismic facies markers vary considerably. Therefore, seismic facies are divided into three levels, rather than one level, based on hierarchical relationships. First, it is necessary to divide primary seismic facies units based on external form, then divide secondary seismic facies units based on internal reflection configuration, and at last, divide three-level seismic facies units based on seismic reflection parameters. The divided seismic facies units can be named according to the sequence "external form of seismic facies unit + seismic reflection configuration + seismic reflection attribute (amplitude, frequency and continuity)" (Table 5.25), thus realizing the identification and division of seismic facies on a seismic cross section and obtaining the plan view of seismic facies by demarcating codes on the plane (Fig. 5.55).

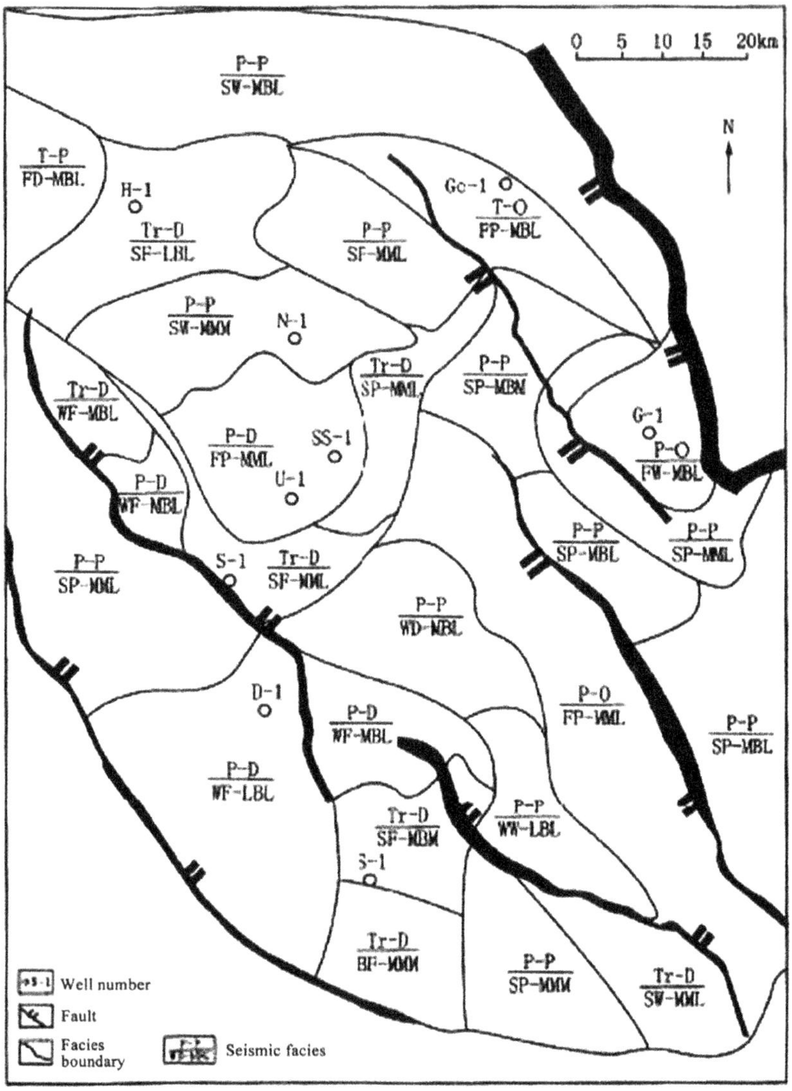

Fig. 5.55 Seismic facies plane for some depression and some sequence

Seismic facies analysis refers to the interpretation and deduction of depositional facies based on various seismic facies. To achieve this, the characteristics of seismic facies must be determined to establish a correct research method.

5.6.3.1 Characteristics of Seismic Facies Analysis

Seismic facies is a comprehensive reflection of a sedimentary body's appearance and of the spatial combination of stratum superimposition modes and lithologic differences, which respectively correspond to seismic external forms and seismic internal reflection configurations.

Multiplicity is a general concern during geological research, and it is extremely obvious in seismic facies analysis. On the one hand, the same seismic facies characteristics may result from totally different depositional facies units. For example, the seismic characteristics of alluvial fans are similar to those of basin turbidite fans, and both appear in bank form, while their internal reflections are presented as foreset, wavy, and hummocky clinoform. As another example, the appearance of a deep-sea basin with developed turbidity sand and an inland sedimentary lake with a coal-containing swamp is "Sheet," and the internal reflection is "Parallel." This is because seismic facies is just a physical response to a sedimentary body's appearance, strata superimposition mode, and lithologic differences, which may be similar for different depositional facies units. In this case, depositional facies units can be divided only on the basis of the characteristics of lithofacies, bio-facies, and electrofacies. On the other hand, different seismic facies characteristics may result from identical depositional facies units. The root cause for this is that seismic facies characteristics are not only related to the depositional facies background but are also affected by the collection and processing of seismic data. Thus, identical seismic data should be provided.

5.6.3.2 Ideas for Seismic Facies Analysis

It should be concluded from the above analysis that seismic facies analysis is started from the identification of sedimentary bodies (matrix facies) for establishing a depositional model of a basin in combination with lithology seismic technology, with drilling as the control point, in order to figure out the spatial distribution law of depositional systems and sedimentary systems tracts of the basin.

Sedimentary body identification is the core and essence of seismic facies analysis. First, from the perspective of sedimentology, sedimentary bodies directly reflect current systems and sedimentary sources, which constitute skeleton facies, which is the most significant component of a sedimentary systems tract. Analysis can be conducted for other depositional facies units filled therebetween based on the properties and distribution laws of skeleton facies. The depositional mode of a basin is a comprehensive and deep summary of tectonic setting, climatic background of depositional basin, and distribution of depositional system, as well as spatial and temporal development and evolution laws. Drilling is taken as the control point to determine which depositional facies a seismic facies shall be classified into. Interpretation of the same seismic facies in other areas should be carried out on the basis of correlation between the area and skeleton facies, correlation between the area and the control well-point, and the depositional model of the basin. By combining facies identification with lithology seismic technology, we can study the lithologic distribution characteristics, find and identify various sedimentary bodies, and determine the significance of the depositional facies of seismic facies units.

5.6.3.3 Seismic Facies Map and Geofacies Interpretation

Depositional facies analysis is based on seismic facies division, which requires a comprehensive analysis of hydrodynamic conditions, differences in sedimentary environments, and specific sedimentation for determining depositional facies by studying regional geological features, sea level change features and analysis results of seismic facies, seismic velocity-lithology of each sequence, and relations between the sequences.

When transforming seismic facies into depositional facies, comprehensive interpretation of various seismic and lithological data should be conducted to the extent possible and priority should be assigned to the analysis of key structural sections to determine the depositional facies represented by seismic facies. This is important for studying the combination relations among seismic facies in order to determine the outline and distribution of various sedimentary bodies, depositional facies, and depositional systems by focusing on seismic facies with special reflection configurations and external forms based on the existing basin tectonic nature and petrology data. Seismic reflection configuration boundary; seismic facies unit with special external form; changes in amplitude, frequency, and continuity of reflection; sand factor contour map; other maps; and extrapolation and interpolation of drilling data are common methods for dividing boundaries of depositional facies and reservoirs in order to determine the horizontal distribution of depositional facies (Fig. 5.56).

Fig. 5.56 Sedimentary facies plane for some depression and some sequence

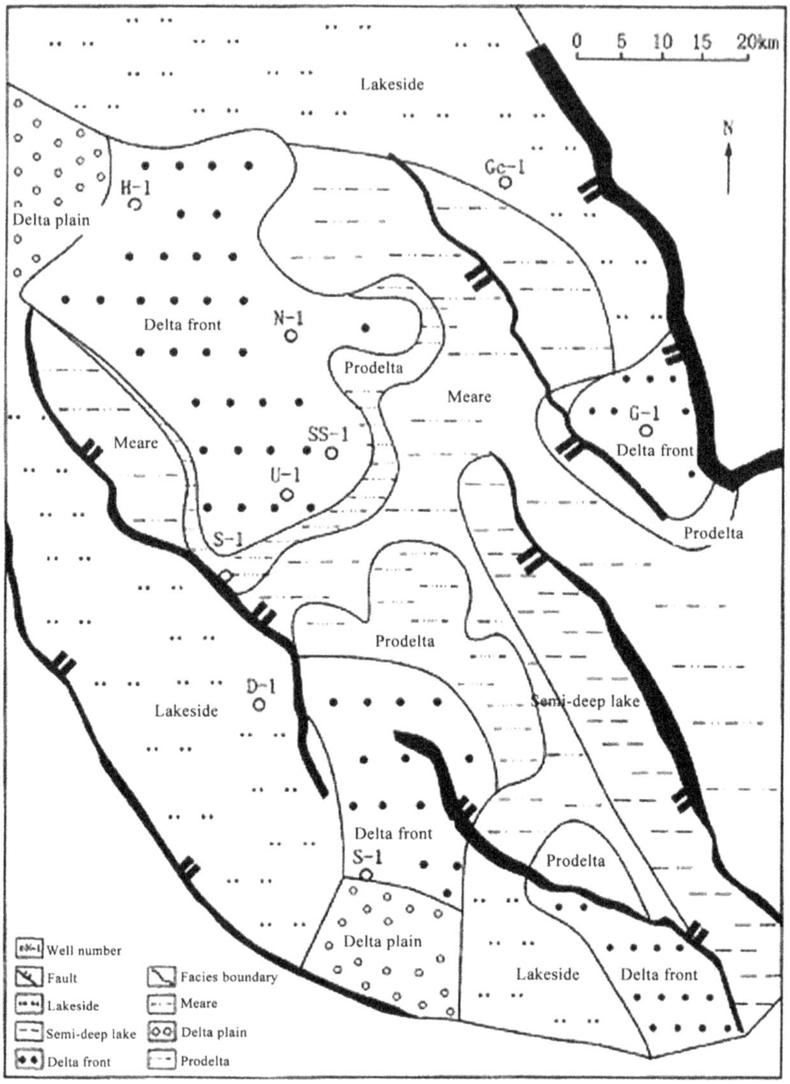

5.7 Sedimentary Process of Clastic Rocks

Petroleum reservoir research is of great significance because petroleum exploration is becoming increasingly difficult, and digging potential old oildoms and increasing petroleum recovery ratio are become difficult. With the improvement of water-flooding development technologies and the practice of various tertiary oil recovery techniques, petroleum geologists have been increasingly aware of and concerned with geological factors that may influence injectant thickness, sweep efficiency, and distribution of remaining oil, which constitutes the contents of intrastratal heterogeneity of a reservoir. Sedimentation of sand bodies shares a sound correspondence with reservoir heterogeneity (Katz 1985; Qiu Yinan 1990; Yu Xinghe 1994 and 1997; Miall 1988). Based on previous studies and years of research on clastic reservoirs, correspondence between eight sedimentations of clastic rock series and sedimentation characteristics, heterogeneity, reservoir characteristics, configuration, and geophysical response is summarized for determining the heterogeneity of clastic reservoirs, configuration characteristics, and factors affecting remaining oil distribution from potential sedimentations of different clastic reservoirs.

Sedimentation refers to main genetic mechanism of various depositional environments. Here sedimentation means depositional modes of a single genetic unit (sand body), which is significant basis and content for reservoir heterogeneity study, because different sedimentations have diverse heterogeneity response relationships. Based on previous studies, sedimentations of clastic rock could be categorized into eight terms, namely, vertical accretion, progradation, lateral accretion, overbank or sheet flow accretion, sieve accretion, winnowing or swashing accretion, aggradation and channel filling, and turbidity accretion.

5.7.1 Vertical Accretion

In a broad sense, vertical accretion means that topographic features on the sedimentation surface directly extend upward without any lateral movement during the entire sedimentation process (Fig. 5.57; Table 5.26). Therefore, it includes bed load during mechanical transport and suspended load transport sedimentation. Sandy sediments are transported as bed load and are deposited and accumulated in the vertical direction when they are too heavy to be carried by running water. Braided river sand body-channel bar sedimentation is the major product of this accretion. The vertical accumulation of sediments is called vertical accretion.

During vertical accretion, the grain size of the sediments carried by the current is different due to the difference in current energy during different periods. This is manifested as grain size difference in the vertical direction, which follows no rule. Coarse-grained sediments are the main sedimentations, unapparent fining upward or abrupt fining upward structures are the most common case in the vertical direction, and troughs and high-angle truncated planar cross beddings are the main sedimentary structure types (Table 5.9, Atlas II). It is not easy for muddy intercalation to be deposited and preserved inside, while muddy intercalation, that is, sludge, can be filled by suspended sediments when the channel is abandoned or during a poor water period. In general, the lateral distribution does not exceed the width of one river valley, and continuity is very poor. The spatial superimposition of a sand body is typical lens or multi-layer wide-stripe type, which indicates the channel is not likely to migrate and bifurcate. However, it is characterized by a high deposition rate and substantial changes.

5.7.2 Progradation

In a broad sense, "progradation" refers to the forward sedimentation of clastics under certain circumstances. Therefore, "progradation" is also called "downstream accretion." In the literature and textbooks, progradation mainly refers to the unloading and forward movement, and accumulation of sediments carried by a flow when the terrain suddenly becomes wide and the gradient becomes steeper (Fig. 5.57). It is often found in deltas and acts as the main sedimentation mechanism for sand bodies of various delta depositional systems. It can develop in other environments. For example, progradation or downstream accretion may appear on a front braided river channel bar when a river flows from the mountains into a wide plain.

In 1985, A. D. Miall proposed the concept of "Configuration" while studying the spatial features of fluvial sand bodies, and eight architectural elements of rivers, one of which is FM-Foreset Macroforms. However, he published a paper on the reservoir heterogeneity of fluvial sandstones at AAPG in 1988 and changed FM-Foreset Macroforms into DA-Downstream Accreting Macroforms, while no alteration was made to the diagram. It could be hereby concluded that "progradation" is different from

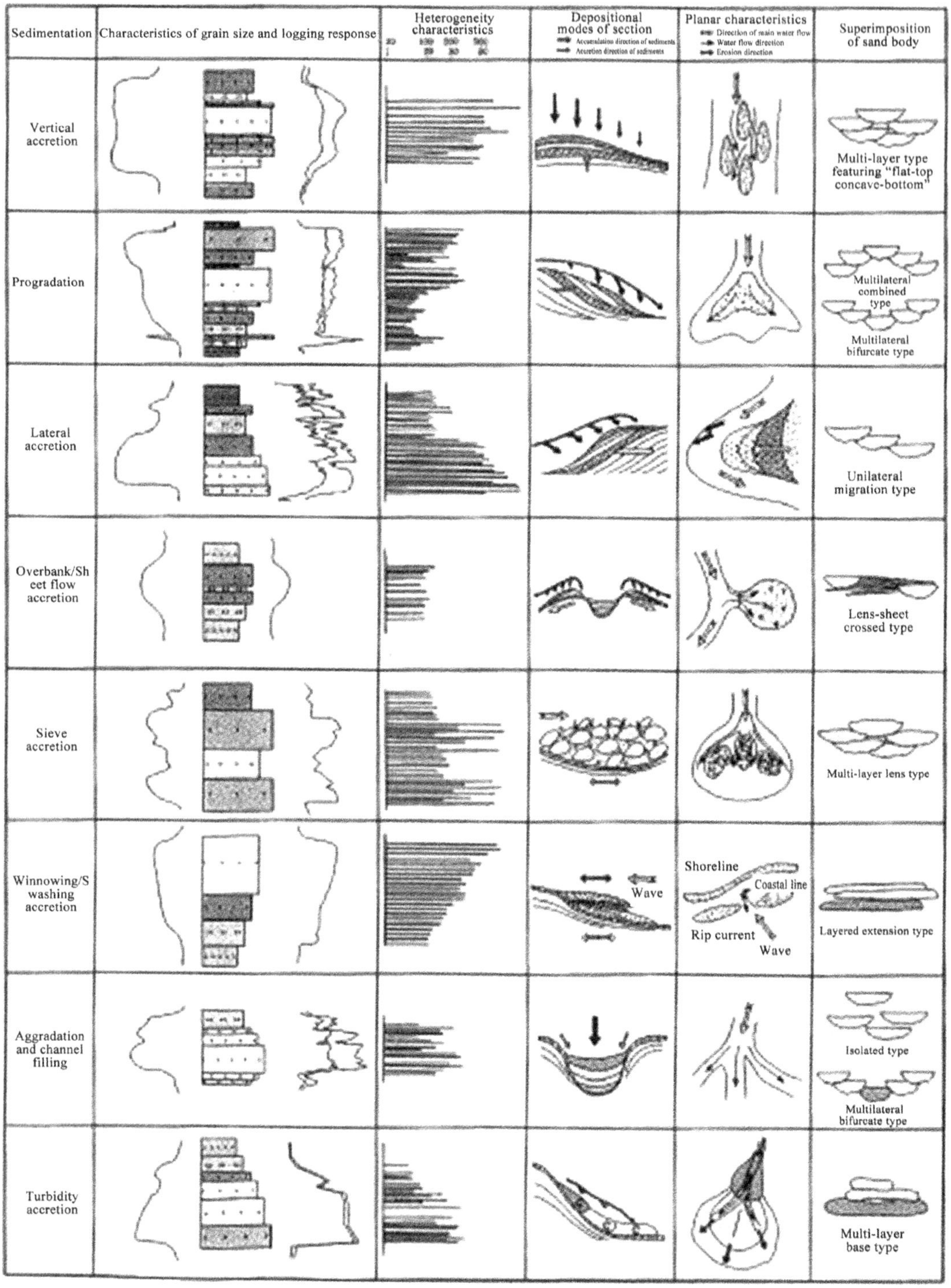

Fig. 5.57 3D configuration and aeolotropic characteristics response to eight sedimentary reservoirs of clastic rocks (according to Yu Xinghe 2004)

"downstream accretion." The former generally refers to sedimentation in deltas, while the latter refers to sedimentation over a wider range.

The depositional characteristics are shown as follows: foreset characterized by coarsening upward; incised planar cross bedding-dominated below the sedimentary structure, current or wave ripple, and trough cross bedding. The distributary mouth bar sequence of a delta front is the representative sequence. The grain size of the upper part is large and that of the lower part is small. Permeability of the upper part is large, while that of the lower part is small, thus showing upward changes with some rules. The correspondence between porosity and permeability is coarsening upward, which provides a good sandstone reservoir for the water-flooding development of an oil-gas field (Table 5.26). Intercalation is generally present in the middle and lower parts of the sand body, with good continuity in the lateral direction, and upward frequency decreases rapidly. Muddy intercalation is not usually developed without an accompanying bottom set. A sand body formed by such sedimentation has a funnel shape on an electric well-logging curve. Spatial superimposition of the sand body indicates that most lateral superimpositions are multilateral and combined, and most vertical superimpositions are of the ladder extension type.

5.7.3 Lateral Accretion

"Lateral Accretion" refers to the accumulation of sediments on a slope in broad terms, and accretion does not change the slope but leads to lateral movement or accumulation of sediments downward the slope. Here, it mainly refers to the redistribution of sediments caused by lateral movement of current due to channel bending, which is the main genetic mechanism for point bar (also called marginal bank) formation. Mixed Load is the main sediment transport method. The sediment grain size can be large or small, and the key is the bending position, which presents concave erosion channel convex bank accretion. The sediments are characterized by typical architectural features of fining upward with gradual change. In the sequence, coarse-grained segments account for 50–70% of the total, with the multi-group low-angle incised planar cross bedding sedimentary structure. It is manifested as a bell shape or a serrated bell shape on an electric well-logging curve. Point bar sedimentation of a sandy meandering river is the representative sequence, with perfect correspondence between porosity and permeability, while intrastratal heterogeneity is strong fining upward. Such reservoir is a flow unit that is very likely to cause water channeling during oil and gas field development. Unidirectional or bidirectional ladder superimposition is the typical spatial superimposition mode of sand bodies, namely, unilateral and multilateral superimpositions, which reflect lateral migration, divergence, and combination of channel, are relatively frequent.

The aforementioned three types of sedimentations are the most basic forms of basin filling and have been adopted by most geologists and petroleum workers. These sedimentations could be deemed as the three most fundamental mechanisms during the formation of clastic rocks as well as the main unloading method for three types of sediments due to traction current sedimentation.

5.7.4 Overbank/Sheet Flow Accretion

Overbank or sheet flow accretion generally refers to the sheetflood sedimentation of an alluvial environment, that is, sheet flood is deposited in an alluvial fan. This book extends such accretion into flooding plain sedimentation owing to the unloading of a large amount of suspended solids. Sedimentation is formed because river water or floods surge over the riverbank and slowly flow away from the channel. This definition is equal to "Sheetflow Deposition" in foreign literature, which is formed by the aforementioned flow carrying suspended solids from a flooding plain, in which case the sediments can accumulate in the vertical direction with suspended load as the transport method. Such accretion, therefore, may lead to the formation of sheet flood deposits in

Table 5.26 Reservoir characterization of eight sedimentations of clastic rock series and problems during exploration and development (according to Yu Xinghe 2004)

Sedimentation	Grain size characteristics	Sedimentary structure	Rhythm and scale	Development of sand and mudstone	Electric well-logging curve pattern	Characteristics of seismic reflection	Correspondence among porosity, permeability, and grain size	Intrastratal heterogeneity	Issues in exploration and development		Injection and production issues
Vertical accretion	Coarse sediment-dominated, generally conglomerate or coarse sandstone gravel	Large planar and trough cross bedding-dominated, with underdeveloped small current ripple	Unapparent fining upward and abrupt superimposition fining ward (complex rhythm); generally in large scale	85% or over 90% are coarse sediments, mud-in-sand	Serrated cylinder with high amplitude	Downward concave with middle or strong amplitude, incised valley	Moderately good, no obvious correspondence rule, with maximum permeability generally at the bottom but there is no obvious rule	Irregular heterogeneity with superimposition fining ward, moderate variation coefficient, and large level difference	A. Desirable seal bed is generally lacked, mudstone and intercalation are undeveloped; B. downstream holds potential resources for exploration	A. Sand flow is be prevented and velocity sensitivity is severe; sand flows out with oil due to overquick exploration; B. Water flow, i.e., water coning or water channeling, should be prevented	It is beneficial for injecting polymers and reducing permeability of the middle and lower parts. Exploration shall be conducted at a slow speed while maintaining stability
Progradation	Grain size may be large or small, while medium-fine sandstone is common; a large amount of mica can be seen	Low-angle planar cross bedding is the main sedimentary tectonics, and current or wave ripples and trough cross bedding can be seen	Architectural feature of coarsening upward; scale depends on the gradient and position of terrain, as well as sediment supply	Good mudstone can develop in the upper and lower parts, mudstone in the lower part is characterized by dark color and pure quality, while that in the upper part is characterized by light color and impurity	Typical funnel shape	Reversed sigmoid or an echelon arrangement with strong amplitude	Good and large, with maximum permeability at the top	Obvious heterogeneity of coarsening upward with medium-to-large variation factors	Favorable exploration area with sand factor of 25–55%	Determining strike and range of sand body	Favorable water-flooding development stratum
Lateral accretion	Grain size can be large or small, dominated by medium-coarse sandstone	Full range of sedimentary structure, characterized by multi-group low-angle incised planar cross bedding	The structure is fine ward characterized by gradual change, and scale is larger than medium level	Sandstone accounts for 50–70% of sediments interstratified between sandstone and mudstone	Typical bell shape or serrated bell shape		Very good, medium-to-large	Strong heterogeneity of fining upward with large variation factors and level differences	Exploration potential is located at two ends depending on the intensity of friction and inertia factors	Water coning and channeling should be prevented; attention should be paid to perforation position, and water injection is not allowed	Favorable stratum for polymer injection

(continued)

Table 5.26 (continued)

Sedimentation	Grain size characteristics	Sedimentary structure	Rhythm and scale	Development of sand and mudstone	Electric well-logging curve pattern	Characteristics of seismic reflection	Correspondence among porosity, permeability, and grain size	Intrastratal heterogeneity	Issues in exploration and development		Injection and production issues
Overbank/Sheet flow accretion	Dominated by siltstone and fine sandstone	It is characterized by small current ripple, and small trough cross bedding can be seen	The rhythm is not obvious, and the majority is small coarsening upward, which is characterized by small thickness and wide scope	A large set of mudstone, intercalated sandstone or thin interbedding sandstone and mudstone, and sand-in-mud	Finger shape or small ligulate		Change in correspondence between porosity and permeability is small and the relation with grain size is not apparent	Similar to homosphere, with small variation coefficient and range	Because thickness of the layer is small, it is generally ignored	Sub-layer can be deemed homosphere during development	
Sieve accretion	Change in grain size is substantial, double or several peaks of grain size distribution, multi-level particle-supported	Massiveness and gravel may be arranged in imbricate shape	No obvious rhythm characteristics	Over 90% of sediments are coarse segments, with insufficient mud; the scope is limited to fans	Serrated projection or wedge with high amplitude	Chaotic reflection and mound superimposition	No obvious correspondence between porosity and permeability; position with maximum permeability is not fixed	Severe heterogeneity of "thief zones," with enormous variation coefficient and level difference	Similar to vertical accretion, lack of mudstone; sand flow and very fast water ejection should be prevented	Attention should be paid to water sensitivity	
Winnowing/Swashing accretion	Grain size is smaller than medium, good sorting	Swash bedding, wave ripples and cross bedding in different directions	Coarsening upward is the main rhythm, fining upward and complex rhythms may be seen	90% of sediments are sandstone, mudstone is lacking	Funnel shape or cylindrical with small amplitude	Strong reflection sheet	Correspondence between porosity and permeability is generally great	Heterogeneity of coarsening upward or complex rhythm	It is favorable for water-flooding development, high energy and high yield	Water coning should be prevented in fining upward	Favorable water-flooding development stratum
Aggradation and channel filling	Grain size is generally small, and exceptionally, grain size can be equal to that of medium-coarse sand	Trough cross bedding is the main sedimentary tectonics, planar cross bedding is not developed	Small fining upward with gradual change	Sandstone is less than 40% of sediments, sand-in-mud	Serrated bell shape with medium amplitude		Good correspondence between porosity and permeability	Weak heterogeneity of fining upward	The thickness is small, reserves are not huge	Difficult to compare with sub-layer	Difficult to explore
Turbidity accretion	Grain size varies greatly, sediments are stacked chaotically	Bouma sequence or coarse-tail grading	Various kinds of rhythms	Numerous sequences, large scale,	Several serrated bell shapes with	Mound downlap,	Changes in porosity and permeability are	No obvious rules, mostly heterogeneity of	Find the distribution and spatial	Focus should be on the effect of	Exploration velocity should be maintained

(continued)

Table 5.26 (continued)

Sedimentation	Grain size characteristics	Sedimentary structure	Rhythm and scale	Development of sand and mudstone	Electric well-logging curve pattern	Characteristics of seismic reflection	Correspondence among porosity, permeability, and grain size	Intrastratal heterogeneity	Issues in exploration and development		Injection and production issues
				more sandstone than mudstone	medium amplitude	relatively chaotic	generally great; it is hard to predict; and the relation with grain size is not good	modifying complex rhythm	superimposition of lowstand fans	intercalation distribution on petroleum, and sand flow should be prevented	stable

Notes In the above table, sand factor refers to the ratio of sandstone thickness to stratum thickness, where sandstone does not contain silt because silt cannot reflect sedimentary frameworks but can be considered as a reservoir. Therefore, density of sandstone = thickness of sandstone (containing silt)/thickness of stratum

various fans and on the two sides of the channel of a (natural) levee, as well as sedimentary bodies of crevasse splays.

This type of sedimentation is characterized by strong intermittent and thin fine-grained sediments are usually formed, dominated by silty fine sand. Sediments may accumulate in the vertical direction with no obvious rhythm characteristics. Small current ripple cross bedding is the main sedimentary structure. When superimposed and connected with other sand bodies, it manifests as a fine-grain thin layer with relatively low permeability (fine sand and silty fine sand); when taken as a reservoir with an independent sand body as the intrastratal homogeneous body, it features little change in porosity and permeability, and weak heterogeneity due to the thin layer. Because this accretion is sedimentation under the condition of a slowly declining base level, the spatial superimposition of most sand bodies is "Lens-Sheet" intersection, which means it is manifested as a strip on the section and as a sheet on the plane.

5.7.5 Sieve Accretion

Sieve accretion occurs mainly in the intermediate plain of an alluvial fan. It is a selective sedimentation process during which the majority of gravels are accumulated and fine-grained solids permeate downward in the transport and unloading process owing to the high permeability of previously accumulated sieve-shape gravels.

It is also called "sieve deposition" based on this definition, and it begins to develop in the early stage of a flood. It forms barriers in a channel or a river valley at the intersection, and bifurcation or selective sedimentation occurs when sheetflood flows through such barriers due to changes in flow energy.

Therefore, sieve sediment is composed of subangular cobble and boulder with good sorting, while gravel is seldom arranged in an imbricate shape. Sand-level sediments are the main fillings among gravels, and obvious double or multiple modes of grain size distribution are presented among the final sediments, which can be categorized as multi-level particle-supported sedimentation. Therefore, sieve accretion manifests as a thin layer with extra-high permeability ("thief zones") in the sandy gravel rock of an alluvial fan. Although it is characterized by small thickness and narrow distribution, this accretion severely complicates the intrastratal heterogeneity of alluvial fan sandy gravel.

5.7.6 Winnowing or Swashing Accretion

Winnowing or swashing accretion refers to sedimentation during which wave action of the catchment basin pushes sandy particles over the wave base to and fro to form a beach bar. Desirable swash bedding can develop owing to this accretion under the combined action of longshore currents, reflux, and waves.

Beach sand, barriers, and longshore bars in coastal environments are products of this depositional mode. Generally, the rhythm features of the sediments are not apparent due to special sedimentation, especially in the case of beach bar sedimentation, which is dependent on the transgression and regression of water level or variation in the base level. While the architectural feature of coastal sandbank is coarsening upward, such action pushes fine-grained matters to the on the top of the sandbank to the edges, thus leaving coarse particles on the top. The grain size may be small or large, with conglomerates (bar gravel) comprising the coarse segment, while most sediments are medium-fine sandstone with perfect sorting and roundness. A large quantity of bioturbation structures may be noted in siltstone and mudstone. The correspondence between porosity and permeability is good, range and variation coefficient of permeability are small, and intrastratal heterogeneity is weak, that is, relatively uniform. When a beach sand channel develops due to reflux, strong reservoir heterogeneity is manifested. High permeability and high porosity are the physical properties of the reservoir, which are generally favorable for high-yield water-flooding exploration. It manifests as a serrated cylinder shape or funnel shape with high amplitude on an electric well-logging curve, and the scale depends on the catchment basin. Because variation in the base level is small, spatial superimposition of the sand body is generally of the layered extension type.

5.7.7 Aggradation and Channel Filling

Aggradation and channel filling mainly refers to filling sedimentation within a channel, during which process, sediments unload and fills the channel when they are too heavy to be carried by the flow. This accretion generally occurs in straight rivers, anastomosing rivers, and distributary channels of a delta, and it is caused by mechanical differentiation. The resulting sedimentary structure is mainly through cross bedding, which reflects incision and filling of channels. Mechanical differentiation leads to fining upward in the vertical direction, which manifests as a serrated bell shape with medium amplitude on an electric well-logging curve. The grain size ranges from coarse sand to fine sand, with moderately poor sorting; porosity and permeability are characterized by fining upward in the vertical direction. The permeability range and moderately small variation coefficients manifest as relatively weak fining upward heterogeneity. Discontinuous muddy intercalation may exist among layers. The sediments are generally manifested as strips or nets, and isolated lens or multilateral bifurcation of the sand body is generally presented on the superimposed transverse section.

5.7.8 Turbidity Accretion

Turbidity accretion refers to the process in which the upward force component of fluid turbulent fluctuation supports particles in a mixture of sediments and water, leading to suspension of the sediments and obvious density difference between the overlaying water the and sediments. The sediments flow along the (underwater) slope to aggrade upward under the action of gravity owing to the aforementioned density difference.

The typical Bouma sequence is characteristic of this sequence, and the hydrodynamic conditions represent the upper flow regime to the lower flow regime from the bottom to the top (Froude number gradually decreases from Fr 1 to Fr $\leq$ 1). Generally, turbidity accretion is the superimposition of several medium-small fining upward structures in the vertical direction, and heterogeneity may be small or large. In most cases, heterogeneity is weaker than the medium level. Correspondence between porosity and permeability is general, without obvious maximum permeability. Modifying complex rhythm heterogeneity is representative of turbidity accretion. Several superimpositions of serrated bell shape or cylinder shape with high amplitude can be seen on an electric well-logging curve. Multi-set and multi-rhythm superimposition is the most important feature of this type of reservoir, and spatial superimposition of sand body is of the multi-layer base type. Moreover, a sandstone and mudstone interbedding sedimentary body of high thickness can be formed if the sedimentation is formed over several stages.

References

Allen, J.R.L. 1964. Studies in fluviatile sedimentation. Six systems from the lower old red sandstone, Anglo-Welsh Basin. *Sedimentology* 3: 163–198.

Berg, O.R. 1982. Seismic detection and evaluation of delta and turbidite sequences: Their application to exploration for the subtle trap. *American Association of Petroleum Geologists Bulletin* 66: 1271–1288.

Blatt, H., G.V. Middleton, and R. Murray. 1972. *Origin of sedimentary rocks*, 2nd ed. New Jersey: Prentice-Hall, Englewood Cliffs.

Boersma, J.R. 1967. Remarkable types of mega cross-stratification in the fluviatile sequence of subrecent distributary of the Rhine, Amerongen, The Netherlands. Report of a preliminary investigation: Geology Mijnbouw, Vol. 46.

CHANGE Harms, J.C., and R.K. Fahnestock. 1960. Stratification, bed forms and flow phenomena (with an example from the Rio Grande).

Ekdal, A.A., R.G. Bromley, and S.G. Pemberton. 1984. *The use of trace fossils in sedimentology and stratigraphy*. S. E. P. M. Tulsa Oklahoma.

Folk, R.L., and W.C. Ward. (1957). Brazos river bar [texas]; a study in the significance of grain size parameters.*Journal of Sedimentary Research* 27 (1): 3–26.

Friedman, G.M. 1961. Distinction between dune, beach, and river sands from their textural characteristics. *Journal of Sedimentary Research* 31 (4): 514–529.

Friedman, G.M. 1967. Dynamic processes and statistical parameters compared for size frequency distribution of beach and river sands. *Journal of Sedimentary Petrology* 37 (2): 327–354.

Friedman, G.M. 1979. Difference in size distributions of populations of particles among sandvS of various origins. *Sedimentology* 26: 3–32.

Galloway, W.E., and D.K. Hobday. 1983. *Terrigenous clastic depositional systems*. New York: Springer.

Gilbert, G.K. 1912. *An introduction of physical geography*. New York: D Appleton & Co.

Harms, J.C., and R.K. Fahnestock. 1965. Stratification, bed forms and flow phenomena (with an example from the Rio Grande). In *Primary sedimentary structures and their hydrodynamic interpretation*, ed. Middleton, 84–115. Spec. Publ. Soc. Econ. Paleont. Miner. 12, Tulsa. 10.1.

Hu, B., G.Z. Wang, and Y.G. Qi. 1997. *Theory and application of trace theory*. China Mining University Press.

Katz, A.J., and Thompson, A.H. 1985. Fractal sandstone pores: implications for conductivity and pore formation. Physical Review Letters, 54 (12), 1325-1328.

Liu, Baojun as chief editor. 1980. *Petrology of the sedimentary rocks*. Beijing: Geological Publishing House.

Liu, B.J., and Y.F. Zeng. 1985. *Geological foundation and working methods of rock facies*. Beijing: Geological Publishing Press.

Ma, Z. 1982a. Decipher the depositional environment by analyzing the curve of natural potential. *Oil & Gas Geology* 13(1).

Ma, Z. 1982b. Interpretation of sedimentary environment by natural potential logging curve. *Petroleum and gas geology*, 13 (1).

Ma, Zheng. 1994. *Oil and gas well logging geology*. Wuhan: China University of Geosciences Press.

Miall, A.D. 1985a. Architectural element analysis: A new method of facies analysis applied to fluvial deposits. *Earth Science Reviews* 22: 261–308.

Miall, A.D. 1985b. Architectural-element analysis: A new method of facies analysis applied to fluvial deposits. In *Recognition of fluvial depositional systems and their resource potential*, ed. Flores. R. S. P. M. short course No. 9. Copyright, SEPM.

Miall, A.D. 1988. Reservoir heterogeneities in fluvial sandstones: Lessons from outcrop studies. *Bulletin American Association of Petroleum Geologists* 72: 682–697.

Mitchum, R.M., P.R. Vail, and Sangree, J.B. 1977. Seismic Stratigraphy and Global Changes of Sea Level, Part 6: Stratigraphic Interpretation of Seismic Reflection Patterns in Depositional Sequences. Seismic Stratigraphy—applications to hydrocarbon exploration.

Ou, Y.J., G.W. Wang, et al. 1999. *Geological analysis and quantitative evaluation of oil and gas reservoirs*. Beijing: Petroleum Industry Press.

Ouyang, Jian, Guiwen Wang, et al. 1999. *Log geological analysis and petroleum reservoir quantitative evaluation*. Beijing: Petroleum Industry Press.

Passega, R. 1957. Texture as characteristic of clastic deposition. *Bull AAPG* 41.

Passega, R. 1964. Grain size representation by C. M. patterns as a geological tool. *Journal of Sedimentary Petrology* 34: 830–847.

Payton, C.E. 1977. Seismic stratigraphy-application to hydrocarbon exploration. *American Association of Petroleum Geologists Memoir* 26.

Pirson, S.J. 1977. *Geologic well log analysis*, 2nd ed, 475. Houston TX: Gulf Publishing Company, 475

Qiu, Y.N. 1990. Process of reservoir sedimentology. Petroleum exploration and development, 17 (1).

Reineck, H.E., and I.B. Singh. 1973. *Depositional sedimentary environments with reference to terrigenous elastics*, 439. Berlin: Springer.

Scott, A.J. 1981. Delta deposition and oil and gas exploration. Petroleum industry press.

Selly, R.C. 1985. *Ancient sedimentary environments*, 3rd ed. London: Chapman & Hall.

Serra, O., and L. Sulpice. 1975. Sedimentological analysis of shale-sand series from well logs. In Trans. Soc. Prof. Well Log Analysts, W1-W23. *16th Annual Logging Symposium*, New Orleans.

Sheriff, R.E. 1980. Seismic stratigraphy: Boston, MA, International Human Resources Development Corporation

Simons, D.B., and E.V. Richardson. 1961. Forms of bed roughness in alluvial channels. *Transactions of the American Society of Civil Engineers* 128: 284–302.

Simons, D.B., E.V. Richardson, and C.F. Nordin. 1965. Sedimentary structures generated by flow in alluvial channels. In *Primary sedimentary structures and their hydrodynamic interpretation*, ed. G.V. Middleton. Soc Econ Paleontologists Mineralogists Spec Publ.

Southard, J.B. 1975. Bed configuration. *Society of Economic Paleontologists and Mineralogists Short Course* 2: 5–44.

The Bool Group. 1984. Sedimentary structure and environmental interpretation. Science Press. SEILACHER A.,1964. Biogenic Sedimentary Structures. In: IMBRIE and N. NEWELL (Editors), Approaches to Paleoecology. Wiley, New York, N.Y., pp. 296–316.

Trask, P.D. 1932. *Origin and environment of source sediments of petroleum*, 323. Houston, Texas: Gulf Pub. co.

Vail, P.R., R.M. MitchumJr, R.G. Todd, et al. 1977. Seismic stratigraphy and globe changes of sea lever. In Payton CE (ed) Seismic stratigraphy application to hydrocarbon exploration. Am Assoc Petrol Geol Mem 26: 49–212.

Visher, G.S. 1969. Grain size distributions and depositional processes. *Journal of Sedimentary Petrology* 39: 1074–1106.

Walker, R.G., and D.J. Cant. 1984. Sandy fluvial systems. In *Facies models*, 2nd ed, ed. R.G. Walker, 171–188. Canada: Geosci.

Wang, Y.M. 1991. Seismic facies analysis lecture (2). *Sedimentary and Tethys geology* (3), 54–60.

Wang, J., Z. Jiang, and Y. Zhang. 2015. Subsurface lacustrine storm-seiche depositional model in the Eocene Lijin Sag of the Bohai Bay Basin, East China. *Sedimentary Geology* 328: 55–72.

Wu, Y.Y, Chen B.Y. 1996. *Oil geology*. Beijing: Petroleum Industry Press.

Wu, Y.Y., Long Xu, et al. 1996. *Oil and gas reservoir geology*. Beijing: Petroleum Industry Press.

Xue, S.H., et al. 2002. *Lake basin sedimentary geology and oil and gas exploration*. Beijing: Petroleum Industry Press.

Yu, X.H., and Y.J. Chen. 2004. The eight sedimentation of clastic rock series and its oil and gas reservoir characterization methods. Experiment of oil geology, 26 (6): 517–524.

Yu, X.H., and D.F. Wang. 1994. Characteristics of braided river delta sand body and sand body exhibition model, modern deltaic sedimentation in Inner Mongolia Daihai lake. Journal of petroleum.

Yu, X.H., and D.F. Wang. 1997. Sedimentation of sand bodies in the basin of the rift basin and the heterogeneity of reservoirs. Earth science, 22 (1).

Zheng, J.M. 1982. *Granularity sign of terrigenous clastic sedimentary environment*. Beijing: Beijing Graduate School of Wuhan Geological College.

6 Reservoir Diagenesis

Diagenetic evolution in petroliferous basins is usually one of the main research focus areas in oil and gas exploration. However, clastic reservoir diagenesis is influenced by numerous factors, including mainly basin tectonic evolution, depositional system distribution, burial history, thermal evolution history, and underground water solution activity. As a result, petroleum geologists should consider the structure, deposition, and diagenesis of a basin and take them as an organic whole for the analysis and research of porosity evolution in a certain period or at a certain position. In addition, multi-disciplinary theories and methods should be applied to discuss the law that the hole varies with the evolution of time, space, and position. Similar to the factors influencing reservoir heterogeneity, as mentioned in Chap. 6, the macro factors determining the reservoir pore development law are ① tectonic evolution stage, ② sedimentary pattern complexity, and ③ diagenesis diversity.

Diagenesis of clastic rock refers to the physical, chemical, and biological changes that occur in the long geological history after the deposition of clastic sediments until metamorphism. These diagenetic changes play a key role in the pore formation, preservation, and destruction of clastic reservoirs, and they decisively influence reservoir properties. Therefore, the diagenetic events causing changes in physical properties (including the destructive effect on pores and formative action generating secondary porosity) are the key to reservoir diagenesis research. The evolution law of pores in reservoirs should be discussed from the viewpoint of predicting favorable petroleum distribution zones.

6.1 Research Methods and Contents

Research on the diagenesis of clastic reservoirs is referred to in numerous fields of geoscience, but it focuses mostly on certain aspects or the analysis of a single element. In terms of petroleum geology or the development of science in the 21st century, single-item analysis and independent research on a single discipline are completely inadequate for interpreting and understanding current geological phenomena. This warrants multi-disciplinary research through comprehensive analysis from various perspectives, from macroscopic thinking to microscopic analysis, from phenomenon analysis to genetic explanation, and from conclusion and understanding to prediction of porosity and permeability changes caused by reservoir diagenesis.

In terms of diagenesis, the research should start from the product-rock, and systematic and detailed observations of the reservoir core (from the macroscopic and slice aspects) should be made. Special attention should be paid to the temporal and spatial changes in reservoir pores to obtain accurate information about lithology, various diagenetic phenomena, and change features of porosity, as well as to predict possible

X. Yu et al., *Clastic Hydrocarbon Reservoir Sedimentology*, Advances in Oil and Gas Exploration & Production, https://doi.org/10.1007/978-3-319-70335-0_6

diagenesis and the associated process. According to diagenetic parameters such as pore-fluid pressure and temperature, the mechanism and conditions of diagenetic reactions can be discussed from physicochemical and thermochemical viewpoints. Moreover, in combination with the basin structure evolution law and the distribution characteristics of depositional facies, the diagenetic evolution sequence of reservoirs can be established to reveal the pore evolution law and pore distribution characteristics, as well as to further evaluate and predict secondary porosity development zones.

Various comprehensive analysis methods are required for reservoir diagenesis research. In addition to the regular research techniques discussed in detail in petrology, numerous advanced testing techniques are involved. The methods of diagenesis update continuously with advancements in technology. At present, common research methods and means can be divided into the petromineralogy method and the experimental test and analysis method (non-petrology method).

6.1.1 Petromineralogy Method

Various methods are used in the laboratory for studying rocks. Visual study of samples, including observation under binocular microscopes and simple chemical testing, is necessary for mineral research. Observation under polarizing and reflection microscopes, including measurement of the basic optical parameters of minerals, cathodoluminescence, and impregnation with blue or fluorescent dye are widely used for mineral identification.

The research techniques used for crystalline minerals include crystal measurement using a reflecting goniometer and observation of crystal surface micromorphology using interference and scanning electron microscopy.

The methods for detecting mineral chemical composition include spectral analysis, conventional chemical analysis, atomic absorption spectroscopy, laser spectroscopy, X-ray fluorescence spectroscopy and polarographic analysis, electron probe analysis, and neutron activation analysis.

In mineral facies analysis and mineral crystal structure research, the most common methods are X-ray analysis of powder crystals and single crystals for phase identification to measure lattice parameters, space groups, and crystal structure. In addition, infrared spectroscopy is used as an auxiliary method in structural analysis to measure atomic groups; the valence state and coordination of iron is measured using Mössbauer spectroscopy; mineral color and internal electronic configuration are quantitatively researched using ultraviolet–visible spectroscopy; molecular structure is measured using nuclear magnetic resonance; lattice imperfections are measured using paramagnetic resonance (e.g., color centers); and the dehydration, decomposition, and phase change of minerals are researched using the thermal analysis method. The high resolution of transmission electron microscopy can be used for direct observation of ultrastructures and lattice imperfections, which is increasingly attracting attention in mineralogical research. To solve special problems in certain respects, some special research methods such as the wrapped rock research method and isotope research method, are used. Also, some aspects of physicochemical property experiments should be performed with minerals as the material.

On the basis of detailed observation and descriptions of outcrops or cores, as well as various analysis methods, microscopic observations and identification are made through sampling. The main purpose is to obtain lithology parameters and the temperature parameter, and observe various diagenetic phenomena that have occurred before. Owing to the different methods employed and different research purposes, different slices and samples are usually prepared.

6.1.2 Regular Rock Slice Research

This method is the most basic, common, and essential laboratory method. In this method, research on diagenesis and porosity involves observing samples under a polarizing microscope; describing the composition, structure, and construction of rock; determining diagenetic changes and porosity characteristics; and rock naming. Dyed casting slices can be prepared. It is more effective to produce multi-purpose slices such that the same slice can be used in multiple analysis methods. Multi-purpose slices are generally categorized into two series: fluorescence series (for observing oil distribution and quality) and cathode luminescence series. In general, the following data are needed for regular rock thin slice analysis for the determination of the diagenesis and pore evolution sequence.

(1) Mineral composition: particle, interstitial material, metasomatic material composition, and their respective contents.
(2) Architectural feature: granular size, shape, sphericity, and sorting.
(3) Compaction feature: contact relationship and compaction rate calculation.
(4) Cementation feature: cementation type, cement components, occurrence, distribution, sequence, and cementation rate calculation.
(5) Dissolution characteristics: type of dissolved minerals, occurrence, degree, stage, secondary pore type, size, content, and distribution.

6.1.2.1 Casting Thin-Section

Pore casting is a direct method for researching the pore structure of reservoir rock, in which the geometry, distribution, and connectivity of pore spaces in clastic rock are observed using a scanning electron microscope. Two methods are usually employed: direct observation of a copy of a pore structure-pore cast under a scanning electron microscope and cutting the rock sample of a casting body into slices for observation under a polarizing microscope. Such slices can not only act as regular rock thin slices, but also be used for analysis and research of pore structure (Atlas X), including pore and throat type, size, shape, distribution, surface porosity, and pore-throat coordination number.

6.1.2.2 Fluorescent Thin-Section

Rock thin-sections can be exposed under fluorescence to identify the presence of oil or oily weight. The research flow of a fluorescent series slice is as follows: polarization analysis → fluorometric analysis → X-ray diffraction analysis → energy spectrum analysis → electron probe microanalysis → the result level plus fluorescence.

6.1.2.3 Cathode Luminescence Thin-Section

This is the main tool for researching debris and cement composition, cement generation, and rock structure and construction, and it is very effective for determining the cementation phenomena of calcite, Fe-oxides, and quartz, which cannot be handled with an ordinary microscope, and a few recrystallization phenomena, as well as pore type identification, especially the identification of ferrodolomite with a girdle structure (Atlas XI).

Cathode luminescence is the visible light generated when a sample is bombarded with an electron beam. Different minerals generate different cathode luminescence owing to the presence of different activator elements. Cathode luminescence devices are designed and built based on this principle, and a cathodoluminescence microscope can be constructed by installing a cathode luminescence device on a microscope. This is an important supplement for polarizing microscopes and an important means for diagenesis research.

1. **Factors influencing of mineral luminescence**

(1) Activator and quencher element contents are related to the existence of luminescence. Quencher elements lower mineral luminescence, for example, iron, cobalt, and nickel. Minerals containing a certain amount of quencher elements will not emit light (extinction), such as some dolomites. Activator elements cause mineral luminescence, for example, manganese, titanium, and other rare earth elements.

(2) Color of light is related to activator type or to different valences of the same activator. For example, feldspar with Ti^{4+} gives out blue light, while feldspar with Fe^{3+} gives out red light, and different trace elements are associated with different emission colors (Chen 1993).
(3) Light intensity is related to the relative amounts of activator and quencher elements, which means the higher the activator element proportion, the higher is the light intensity and vice versa.

2. **Mineral luminescence characteristics**

Minerals can be identified according to the luminescence characteristics, and this method is particularly effective for identifying carbonate minerals and different genetic quartz and feldspars. Most calcites emit orange-yellow or jacinth light, while a few calcites emit blue light. Ferrocalcite is relatively darker compared to calcite and is even non-luminescent. Dolomites emit orange-red or pale violet (pink) red light, while ferrodolomite does not emit light (Table 6.1).

(1) Carbonate minerals

The color emitted by carbonate minerals varies from yellow to kermesinus because they contain activator elements, i.e., trace elements such as manganese and strontium. Ferrous (Fe^{2+}) is a quencher (element preventing mineral luminescence), which is non-luminous. In calcites and dolomites, the iron content has a direct effect on luminescence. Hence, ferruginous carbonate minerals and nonferrous carbonate minerals can be distinguished according to the light intensity of these materials.

(2) Feldspar

Feldspar has many cathode luminescence colors, with blue, red, and green being common. By conjoint analysis using a cathode luminescence microscope and an electronic probe, it can be shown that the emission color of feldspar depends on its activator content (Chen 1993).

For K-feldspar, the most common cathode luminescence color is blue. Through determination, all blue-light-emitting feldspars contain small amounts of Ti^{4+}, whereas all feldspars emitting light of other colors do not contain Ti^{4+}. Hence, the emission of blue light from feldspar is connected with its Ti^{4+} content. Barely any feldspar emits red light, which is related to Fe^{3+}, Cr^{3+}, and Mn^{4+} (Chen 1993). In feldspars that emit red light, the $Fe^{2+}0_3$ content is higher than that in the blue-light-emitting feldspar, usually at around 0.4%. In addition, there is a substantial amount of Cr_20_3 and little MnO_2 in red-light-emitting feldspars (Chen 1993).

In general, orthoclase emits blue light, microcline blue light, or no light. Perthite emits light blue-yellow green light. Plagioclase emits multiple illuminant colors depending on its composition: if it contains Ti^{4+}, it emits blue light (wavelength = 460 nm); if it contains Fe^{2+}, it emits green light (wavelength = 550 nm); if it contains Mn^{2+}, it emits yellow-green light (wavelength = 570 nm); if it contains Fe_2O_3, it emits red light (wavelength = 700 nm); and authigenic feldspar usually emits no light.

(3) Quartz

In 1978, U. Zinkernagel systematically researched the cathode luminescence feature of quartz, including igneous rock, contact metamorphic rock, regional metamorphite, and sedimentary authigenic quartz. The entire luminescence spectrum of quartz reflects two emission maximums: ① wavelength is 350–450 nm, which is within the blue range; and ② wavelength is 600–650 nm, which is within the red range. In different spectral compositions of quartz, the identities of the two abovementioned emission maximums are different (Fig. 6.1). In sedimentary detritus quartz, there are three different luminescence types (Fig. 6.2), thus we can deduce that quartz can emit three colors:

Table 6.1 Compositions and cathode luminescence of carbonate minerals (according to Chen 1993)

Mineral	Cathode luminescence color	Thin-section staining	X-ray diffraction	Composition analysis (electron probe)						
				$FeCO_3$	$MgCO_3$	$CaCO_3$	Ca/Mg	Fe/Mn	Mn/Fe	Fe/(Fe + Mn)
Calcite	Orange yellow-brown	Red	3.035	0.05–0.1	0.4–0.5	99	165–195	0.6–1.4	0.7–1.5	–
Ferroan calcite	Orange-brown	Purplish red	3.02	2.5–2.9	1.2	94	66	0.8–1.4	0.7–1.2	–
Dolomite	Orange red-yellow	Non-staining	2.332–2.338	0.05–1.2	3.96–54	51–28	0.9–1.26	0.13–6.5	0	0.004–0.04
Ferruginous dolomite	Brown-deep brown	Light blue	2.895	3–11.8	34.6–43	51–56	1–1.5	2.4–10.7	0.89–1.41	0.05–0.37
Ferrodolomite	Non-luminescence	Blue	2.907–2.194	14–26	21–30.5	54–57	1.5–2.2	13–93	0.01–0.074	0.48–0.62
High-calcium ferrodolomite	Non-luminescence	Blue	–	14–20	19–25.5	57.5–61.6	2–2.6	13–126	0.008–0.074	0.48–0.64

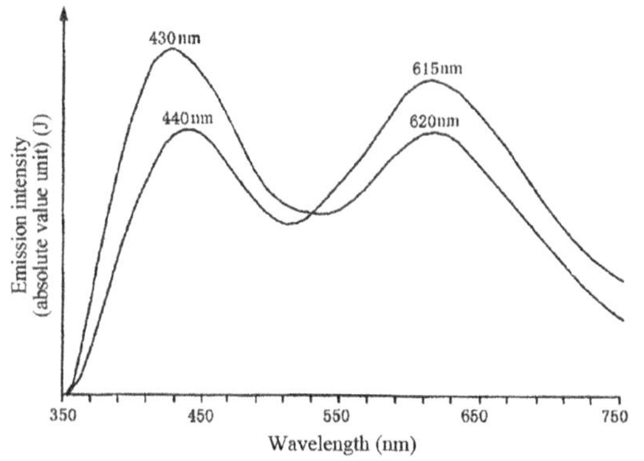

Fig. 6.1 The cathodoluminescence spectrum of quartz presents two maximum emission values (according to U. Zinkemagel 1978; Foster 2004)

I. High-temperature quartz with fast cooling speed, which originates in volcanic rock, plutonite, and contact metamorphic rock, is characterized by the emission of bluish-violet light; II. Mainly high-temperature quartz with slow cooling speed (generally refers to metamorphic quartz in high-temperature areas) and low-temperature quartz (generally refers to the quartz in low-grade metamorphic rock) are characterized by the emission of brown light; and III. In general, authigenic quartz is characterized by the lack of emission of light (Table 6.2).

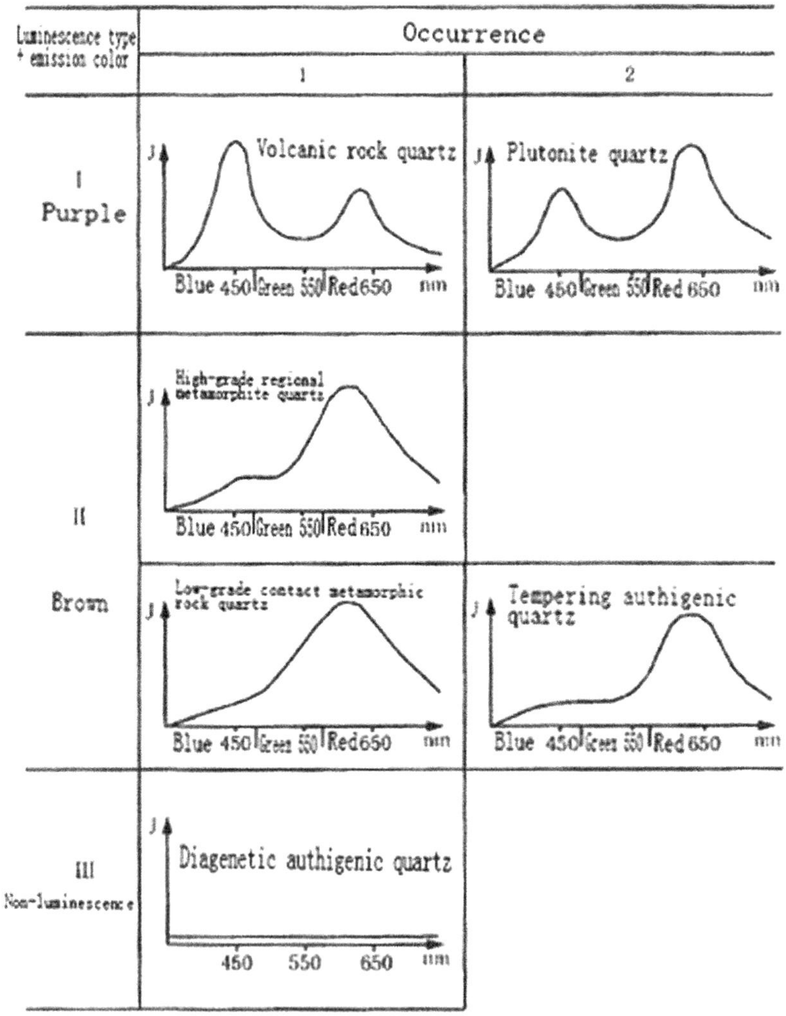

Fig. 6.2 Cathodoluminescence spectrum of quartz with different genesis (according to U. Zinkemagel 1978)

Table 6.2 Luminescence characteristics of quartz (according to Chen Lihua 1993)

Emission color	Temperature (°C)	Crystallization condition	Occurrence
Bluish-violet quartz	>573	Quick cooling	Volcanic rock, plutonite, and contact metamorphic rock
Brown quartz	>573	Slow cooling	High-grade regional metamorphite
	<573		Low-grade metamorphic rock
Non-luminescent quartz	<300		Authigenic quartz

3. **Effect in Diagenesis Study**

(1) Cement generation analysis

Quartz increment, carbonate mineral cementation, and other cements may be presented by generations present the generation of mineral growth, which is hard to distinguish under a polarizing microscope but is obvious under cathode luminescence. In the case of crystal growth, ion differences exist in pore fluid, and the generation accumulated during cement formation results in the formation of different girdles that emit different colors.

The crystal growth history and changes in the chemical characteristics of pore fluid in the diagenetic process can be understood by analyzing cement generation to deduce the diagenetic environment.

(2) Recovery of Primary Structure

After the influence of diagenesis, the primary structure of rock may not be recognizable. For instance, in the case of the quartz sandstone with strong siliceous cementation or after undergoing strong metasomatism, recrystallization, and rupture filling, it can be hard to discern the original structure of the rock under a polarizing microscope. However, a cathodoluminescence microscope can help determine the primary structure of the rock and provide a reliable basis for determining the formation condition during deposition. The structure distortions or pseudomorphs observed under a regular microscope include the following: ① Granularity larger than that of the original; ② roundness lower than that of the original; ③ sorting better than that of the original; and ④ contact is closer than that of the original.

In addition, under a cathode luminescence microscope, the crushing and healing effects of quartz particles can be understood. Often, fractures are seen on a few quartz granules, and a few of the granules are even crushed into pieces; these are later subject to rehealing, which makes it difficult to find the fracture under transmission light. Occasionally, the sample can be mistaken as multicrystal quartz. However, these crushing and healing phenomena can be distinguished clearly under cathode luminescence. Crushing takes place when the lithostatic pressure surpasses the compressive strength of the particles, while healing occurs when the crushed parts undergo no relative displacement. After healing, the particles' optical properties stay the same as that of single-crystal quartz. If the crushed parts have displacements and twists, their properties will be inconsistent with that of single-crystal quartz, which implies that the particles turn into pleomorphic quartz after healing.

(3) Deduction of diagenetic paragenesis

The cement formation sequence is closely related to diagenetic evolution. When there is secondary enlargement of quartz, the diagenetic sequence can be deduced by cathode luminescence.

(i) Lack of enlargement at the contact point of quartz and debris with obvious enlargement at the contact point with cement indicates that the rock first underwent compaction and pressure solution, and later in the presence of a silica source, authigenic enlargement occurred in the un-contacted pores, and re-cemented later by other chemical cementation. This shows that the authigenic enlargement of quartz occurs before late cementation but after mechanical compaction.

(ii) Where the authigenic enlargement of quartz develop around, i.e., there is enlargement between detrital quartz and other particles and cements, it indicates that the authigenic enlargement of quartz occurs before or along with mechanical compaction.

(4) Identification of secondary porosity

The secondary porosity formed due to the dissolution of grains and cements can be identified by using a cathode luminescence microscope, because the dissolved carbonate cement can be shown under the cathode luminescence as long as there is a small amount of residue. The distribution of residual carbonate minerals in pores generally occurs according to the following two patterns: a small amount of residual at the edge part of particles and small residual point-like carbonate minerals distributed in pores. These features are usually difficult to identify under a polarizing microscope.

More commonly developed is redistributional secondary porosity whereby the products of dissolution reactions are locally precipitated (Giles and de Boer 1990). This involves the dissolution of reactive detrital grains or matrix (or even earlier formed cements) and the local reprecipitation of reaction products. This can lead to new porosity being created within the dissolving mineral but the occlusion of intergranular porosity by the newly formed minerals. An example of this is the creation of porosity within dissolving detrital K-feldspar grains and the local precipitation of illite in primary pores and quartz on the free surfaces of detrital quartz grains. Redistributional secondary porosity is probably common in sandstones during burial, although the importance of additional secondary porosity remains unclear.

6.1.2.4 Scanning Electron Microscopy Analysis

Owing to its high magnification and resolution, the phenomena that cannot be seen under a regular microscope can possibly be observed using scanning electron microscopy, for example, clay minerals and micropores. As a result, it is widely used in diagenesis research. Basically, mineral identification using a scanning electron microscope is based on morphology and crystalline form (Atlas XII), but the effect is poor for minerals with similar crystalline forms. Currently, the common practice is to equip a scanning electron microscope with an energy-dispersive spectrometer or WDS (wavelength-dispersive spectrometry). In this way, morphology analysis and component analysis can be combined. During diagenesis and pore evolution research, the scanning electron microscope is mainly used for the analysis of the following aspects:

(1) **Cement type**: identifies the authigenic cement type in the pore and throat, especially tiny clay minerals and different types of zeolite (Table 6.3).
(2) **Cement occurrence**: identifies the occurrence of authigenic clay cement (pore filling, pore lining, and pore bridge-plug) and the level of secondary enlargement in quartz and feldspar.
(3) **Dissolution and metasomatism**: feldspar alteration, mineral dissolution, metasomatism, regrowth, authigenic mineral association, and formation order.
(4) **Pore type (especially, micropore)**: morphology and quantity.
(5) **Throat type**: throat size and morphology.
(6) Connection of the pore and throat.

6.1.2.5 X-Ray Diffraction Analysis

Clay minerals and a few authigenic minerals are hard to identify under a polarizing microscope, and so are minerals with a similar appearance, the relative amounts of which cannot be determined. However, X-ray analysis can reveal a

Table 6.3 Common clay mineral properties (according to Chen Lihua, modified in 1993)

Clay mineral	Chemical formula	X-ray diffraction pattern feature (d) (001) Å[a]	Monomer form under a scanning electron microscope	Aggregation form under a scanning electron microscope	Main distribution mode in pore (occurrence)
Kaolinite	$Al_4[Si_4O_{20}](OH)_8$	7.1–7.2	False hexagonal plate shape	Page-shaped, vermiform and accordion-shaped	Scattered particle (pore filling)
Smectite	$(1/2Ca, Na)_{0.7}(Al, Mn, Fe)_4[(Si, Al)_8O_{20}]nH_2O$	Na-12.99 Ca-11.50	Cotton-shaped, schistose, and cellular	Petal shaped, cellular, and cotton-shaped	Pore lining Pore bridge-plug Pore filling
Illite	$K_{1-1.5}Al_4[Si_{7-6.5}Al_{1-1.5}O_{20}(OH)_4]$	10	Schistose, cellular, and thread-shaped	Flaky, plate-shaped, and feathery	Pore lining Pore bridge-plug Pore filling
Chlorite	$[Mn, Al, Fe]_{12}[(Si, Al)_8O_{20}](OH)_{16}$	14, 7.14, 3.5, 4.72	Needle-shaped, rose-shaped, and fluff sphere-like	Slice and flaky	Pore lining Pore filling

[a] $1\ Å = 10^{-10}$

mineral's crystal structure and it plays a special role in the identification of clay minerals and a few authigenic minerals; it is especially effective for research on clay minerals. Quantitatively, mineral category is determined according to the peak value on the X-ray diffraction pattern, and the diffracted intensity is determined according to peak height and peak area. Some changes in the mineral itself can be reflected according to the peak function, which means a peak-shape change for the same type of mineral. Therefore, X-ray diffraction analysis can be used for the following: ① quantitative analysis of clay minerals; ② mixed-layer clay mineral identification and calculation of mixed-layer ratio; and ③ analysis and identification of authigenic detrital minerals.

6.1.2.6 Scanning Electronic Probe and Energy Spectrum Analysis

Scanning electron microscopy analysis can identify authigenic minerals based on mineral shape, however it cannot identify minerals having different crystalline morphologies or heterogeneous phase minerals. For the mineral luminous element to be found under cathode luminescence microscope and many other minerals which are hard to be identified under the slice, identification can only be made by element composition analysis by using an electronic probe and an energy-dispersive spectrometer. This is because an electron beam irradiating a sample produces different information. The wavelength and energy of the X-rays generated by different elements are different. An electronic probe is used to measure the wavelength of X-rays, and an energy-dispersive spectrometer is used to measure the energy of X-rays. The chemical composition of minerals, especially those composed of a few tiny complicated minerals, such as zeolites and clay minerals, can be determined with high sensitivity by measuring X-ray wavelength or energy. This is a key microscopic component analysis method. The ability and precision of mineral identification can be improved further by using cooperatively scanning electron microscopy, cathode luminescence, an electronic probe, and energy dispersive spectroscopy.

(1) The chemical components of minerals (clay or zeolite mineral) can be measured using scanning electron microscopy, and the mineral type can be determined accurately. Thus, the reliability of mineral identification can be improved.
(2) The composition of cement can be researched to understand the relationships between mineral emitting colors and microelements (Table 6.1) using cathode luminescence microscopy in order to determine the variation of pore water properties in the diagenetic process and interpret the diagenetic environment.
(3) It is essential to measure the K_2O content of clay mineral; distinguish smectite, I/S mixed-layer, and illite; and determine the type of I/S mixed-layer according to its K_2O content (Table 6.4).

6.1.2.7 Fluid Inclusion Analysis

Mineral inclusion is the diagenetic medium solution captured during mineral growth. It records the conditions and fluid properties at the time of mineral formation. Authigenic minerals with inclusion may include calcite, dolomite, quartz, zeolite, gypsum, and halite.

1. **Classification**

According to the physical characteristics, mineral inclusion can be divided into ① pure liquid inclusion (determination of salinity, composition, and periods of primary, secondary and pseudo-secondary); ② gas liquid inclusion (determination of uniform temperature and salinity); ③ multiphase inclusion (refers to the inclusion of more than three phases such as gas, liquid, and solid); ④ organic inclusion (organic liquid, e.g., oil; gas, e.g., methane and ethane; and solid, e.g., pitch); ⑤ successive inclusion (e.g., inclusion in detrital mineral), which can be used to determine the nature of the source rock and the source direction.

2. **Assumed condition**

There are three hypotheses in inclusion research: ① inclusion is captured in a homogeneous system and fills the space at the time of capture; ② the space does not change obviously after inclusion trapping; and ③ there is no transport of foreign substances in and out of the space after inclusion capture.

3. **Research methods**

(1) Homogenization method

It is necessary to place inclusion slices on a heating/cooling stage for heating. When gas and liquid are homogenized in one phase, the temperature at this time is the one when the inclusion is restored until formation, which is also the homogenization temperature that reflects the diagenetic temperature and minimum temperature of formation without pressure correction of the mineral.

(2) Cryoscopic method

It is essential to measure the temperature (ice point) at the point when the last ice crystal in the

Table 6.4 K_2O content in I/S mixed layer (according to Wu Shenghe et al. 1998)

Mixed layer type	Smectite interlayer (%)	K_2O content (%)
Smectite	–	<1
Chaotic mixed layer	>70	<2.2
	50–70	2.2–3.7
Partially ordered mixed layer	35–50	3.7–5.0
Ordered mixed layer	15–35	5.0–7.0
Kalkberg type	<15	7–8.5

inclusion in the saline solution vanishes (or the first ice crystal emerges), determine the concentration of the saline solution based on the ice point, i.e., cryogenic temperature, and obtain the density of the fluid inclusion based on its salinity, eutectic temperature, and homogenization temperature.

(3) Microscope crushing method

When an inclusion is found under the microscope, microanalysis should be performed by extracting the inclusion with a special precision sampling device to determine its composition and property.

(4) Fluorescence microscope method

This is mainly used for the determination of the hydrocarbon composition of organic inclusions.

4. **Research contents and functions**

(1) Research on diagenetic temperature and diagenetic history

The diagenetic temperature can be known by measuring the inclusion temperature in authigenic minerals. The diagenetic temperature in different diagenetic stages can be determined if the homogenization temperature of the inclusion in cements (or outgrowth rims) of different generations is measured. Then, the diagenetic history can be restored in combination with the present ground surface temperature data.

Temperature measurement of fluid inclusions can be useful in analyzing the influence of hydrocarbon emplacement on diagenesis and for predicting hydrocarbon emplacement time. For instance, the albitization of potassium feldspar is restrained obviously by oil and gas invasion because this process requires not only K^+ removal but also a Na^+ supply. In addition, ion exchange in the reservoir and the surrounding rock is obviously restrained by hydrocarbon emplacement. Therefore, the highest formation temperature of albitization can be used as the stratum temperature during hydrocarbon emplacement, and this time can be deduced in combination with the burial history curve or illite K–Ar isotope dating.

(2) Research on diagenetic fluid properties

Conditions and palaeogeothermal and paleosalinity parameters can be determined by measuring the salinity and density of the fluid in mineral inclusion in order to restore the diagenetic environment at that time. In addition, stable isotope analysis of C, H, and O can be performed by extracting H, H_2O, and CO_2 in the inclusion to determine the causes and sources of the diagenetic fluid.

(3) Application to petroleum exploration

The general direction and relative time of petroleum migration can be determined, and the degree and phase of petroleum hydrocarbon evolution can be determined through research on the type, abundance, and composition of organic materials in an organic inclusion in each fracture period and authigenic mineral.

The development tendency of petromineralogy is as follows: ① research area is enlarged from the Earth's crust minerals to mantle minerals to cosmic minerals on other celestial bodies, and from natural minerals to synthetic minerals. ② The research contents develop both in scope and in depth from macro to micro, from key component to microelement, from average crystal structures of atomic arrangement to local and specific crystal structures and the involved interelectronic and nuclear fine structures in atoms. ③ Mineralogy develops rapidly in the application area. Besides the further application of research results in geological studies and exploration, the object of mineral research is not only to use raw mineral materials to extract certain useful compositions but also to obtain mineral materials with various special performance characteristics. The scope for research in this regard is wide.

6.1.3 Experimental Testing Method

Analysis and testing in a laboratory are the most direct and reliable methods for obtaining underground first-hand data. The main aim of diagenesis studies is to obtain fluid and paleogeothermal information, as well as petrophysics data. For different types of rocks and analysis items, it is very important to make reasonable selections of the amount of rock sample. Strict equal interval sampling should be performed when selecting samples for measuring rock porosity, permeability, and saturation. If it is believed that the sampling density is insufficient to represent the average value of a certain type of parameter, the sampling density can be increased without randomly moving the position of the sampling points in order to avoid the loss of average oil reservoir representativeness. In general, this requirement can be met by taking three samples from each meter along the depth of the core (Liangtian et al. 1992). The common methods are as follows.

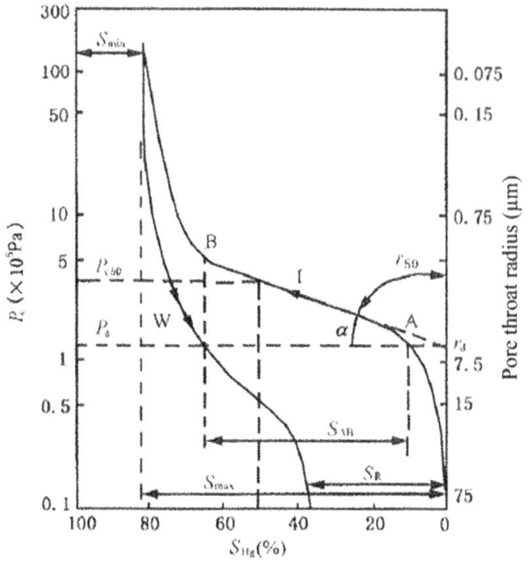

Fig. 6.3 Capillary pressure curve (quoted from He 1994) I-Injection curve and W-Ejection curve P_d: Drainage pressure; P_c50: Saturation median pressure; S_{min}: Minimum wetting saturation

6.1.3.1 Capillary Pressure Method Analysis

There are many methods to measure the rock capillary pressure curve. The following three methods are commonly used in oilfields: ① the semi-permeable baffle plate method; ② the mercury intrusion method; and ③ the centrifuge method.

1. **Quantitative characteristics of capillary pressure**

The quantitative characteristics of capillary pressure are represented mainly by a series of relevant parameters.

(1) Drainage pressure (P_d)

This refers to the highest throat pressure as the non-wetting phase begins to enter into the rock sample, which means that the pressure as the non-wetting phase begins to enter into the rock sample corresponds to the capillary pressure of the largest throat radius in the rock sample. On the capillary pressure curve, it is shown as the corresponding pressure value when the middle flat curve extends to the crosspoint with vertical coordinates when the non-wetting phase saturation is zero (Fig. 6.3).

This parameter is one of the important indicators used for evaluating rock reservoir performance, especially rock permeability, which means that rock with good permeability has a low drainage pressure, and vice versa.

(2) Saturation median pressure (P_{c50})

This refers to the corresponding pressure value when saturation is 50% on the capillary pressure curve, and the corresponding throat radius is the saturation median throat radius, r_{50}. Apparently, the lower the P_{c50} value, the higher is r_{50}, the better is the reservoir physical properties, and the lower is the oil-producing capacity, and vice versa.

(3) Minimum wetting saturation S_{min}

This refers to the volume percentage of the pores that are not intruded into under the highest mercury pressure. It is an indicator of rock pore structure and permeability; the better the physical properties of the rock, the lower is its S_{min} value.

2. **Application of capillary pressure curves**

Capillary pressure curve analysis is one of the main methods of studying microscopic reservoir characteristics, and it is widely used in petroleum exploration and development.

(1) Research on rock pore structure

The capillary pressure curve shows the relationship between capillary pressure and saturation. A certain capillary pressure corresponds to a certain pore throat radius. Therefore, the capillary pressure curve actually includes the distribution law of the rock pore throat. The shape of the capillary pressure curve is influenced by pore throat sorting, skewness of pore throat distribution, and average throat radius, and these parameters can be acquired from the cumulative frequency histogram of pore throat distribution and the cumulative frequency distribution curve of the pore throat.

(2) Evaluation of rock reservoir quality

As mentioned before, the shape of the capillary pressure curve is subject mainly to the sorting and size of the pore throat. Sorting refers to the degree of dispersion (concentration) of the throat size. The more concentrated the throat size, the better is the sorting, longer is the middle flat section on the capillary pressure curve, and closer is it to being parallel to the abscissa. Pore throat size and its degree of concentration mainly influence the curve's skewness, which is a measure of the capillary pressure curve shape inclining toward coarse or fine throat. The bigger the throat, the more number of bigger throats there are, and the closer is the curve to the left lower part, which is called coarse skewness. Otherwise, the curve approaches the upper right part, which is called fine skewness. According to the large number of curve shape features, capillary pressure curves can be divided into six types based on sorting and skewness (Fig. 6.4).

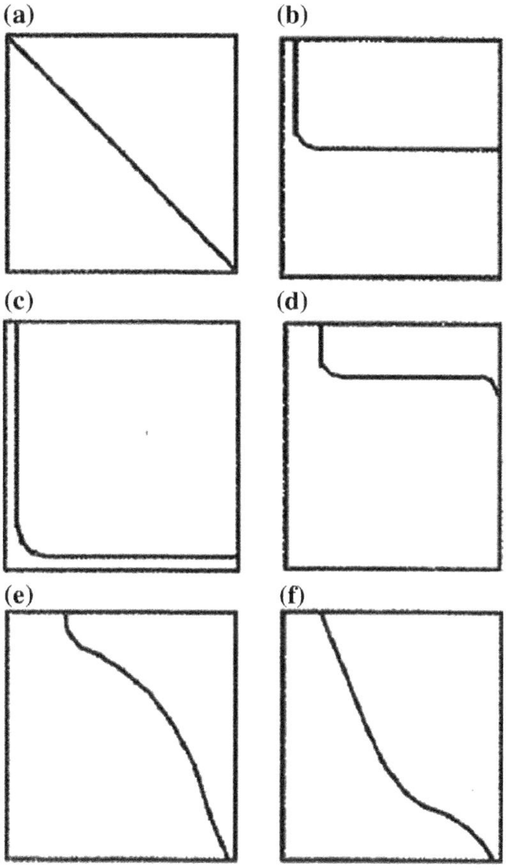

Fig. 6.4 Typical capillary tubes under different degrees of sorting and flexure (according to Chilingar et al. 1972) **a** no sorting; **b** good sorting; **c** good sorting and coarse skewness; **d** good sorting and fine skewness; **e** poor sorting and slightly fine skewness; and **f** poor sorting and slightly coarse skewness

(3) Determination of the average capillary pressure J (S_w) function of oil reservoir

Tiny rock samples are used to obtain capillary pressure curves experimentally, for example, core, sidewall coring, detritus, or a few typical cores. Although it is said to be typical, a tiny rock sample actually represents only one point in a reservoir. In practical work, the issues of averaging and comprehensive research exist.

The J (S_w) function is a computing method proposed to solve these issues. The J function curve synthesizes the surface tension of oil reservoir fluids, wettability, and the influence of rock permeability and porosity to reflect the capillary pressure curve characteristics of the oil reservoir. Therefore, the J function can lead to better evaluation and comparison of an oil reservoir. It is computed as follows:

$$J(S_w) = \frac{P_c}{\sigma \cos\theta}\sqrt{\frac{K}{\phi}}$$

where

J (S_w)	J function non-dimensional;
P_c	Capillary pressure;
σ	Surface tension;
θ	Wetting contact angle;
K	Air permeability; and
ϕ	Porosity.

Since the actual value is hard to measure accurately and when the reservoir wetting core and fluid are used for the measurement, COS can be taken as a constant that does not affect the curve form of the function. Hence, COS is negligible, and the above formula can be simplified as follows:

$$J(S_w) = \frac{P_c}{\sigma}\sqrt{\frac{K}{\phi}}$$

Screening of the J function has two uses: ① acquisition of the average value of the capillary pressure curve for identical types of rock samples (dolomite or finely crystalline limestone); and ② determination of rock petrophysics of different types of rock samples (coarse-grained limestone).

The J (S_w) function curve is different for different reservoirs, thus a uniform J (S_w) function curve cannot be acquired for capillary pressure data with a large permeability difference within the same reservoir. Therefore, the method of systemizing capillary pressure for the J (S_w) function is generally used in the case of a relatively even reservoir. If the reservoir structure is complex and non-homogeneity is considerable, application of the J (S_w) function results in a large error.

6.1.3.2 Analysis of Organic Matter Maturity

In diagenesis research, the thermal maturity analysis of organic matter is performed mainly for division of the diagenetic stage, and three indicators are usually applied to that end.

1. **Vitrinite reflectance (R_o)**

This is also called vitrain reflectance. Vitrinite is a group of oxygen-rich macerals composed of humus relevant to peat which has the characteristics of vitrain. The maturity of organic matter and the maximum rock temperature can be determined by measuring the vitrinite reflectance of the organic matter in the rock sample. Hence, vitrinite reflectance is a good indicator of maturity.

Vitrinite reflectance is closely connected with diagenesis, which means the deeper the thermometamorphism, the higher is the vitrinite reflectance. This is closely related to diagenesis stage (Table 6.5).

2. **Pollen-color and thermal alteration index (TAI)**

The color change of spores and fossil pollens has a great deal to do with thermometamorphism. The extine of sporopollen is complex organic matter composed of elements, such as carbon, hydrogen, and oxygen, and its temperature is mainly temperature controlled. With increasing burial depth, the color of sporopollen fossils turns from pale yellow to orange to light brown to brownish black. According to the Kerogen thermal degradation oil generation theory, the degree of color change of sporopollen fossils in the source rock during thermal evolution directly

Table 6.5 There is great difference between Ro and Tmax in different horizons, thus, particular attention shall be paid to the difference in horizons during research on the maturity of organic matter (according to Demasion 1989; by Hunt, modified in 1977)

Vitrinite reflectance R_o (%)	Paleogeotemperature (°C)	Coal rank	Maturity	Petroleum generation	Diagenetic stage
<0.35	<50	Bituminous coal	Immature	Small amount of liquid hydrocarbon	Early diagenetic stage
0.7	60–80	Highly volatile bituminous coal	Mature	Liquid hydrocarbon	Middle diagenetic stage
0.9	90–100	Highly volatile bituminous coal			
1.10	100–120	Medium-volatile bituminous coals			
1.5–1.9	140–160	Low-volatile bituminous coals			
>2	160–200	Semi-bituminous coal-anthracite	Overmature	Methane	Late diagenetic stage

Table 6.6 Variation diagram for clay mineral with the depth in Panyu gas field reservoir (according to Wu Shenghe et al. 1998)

Maturity indicator		Immature	Unboiled oil	Condensate oil	Wet gas	Dry gas
Vitrinite reflectance (%)		<0.5	0.5–1.3	1.0–1.5	1.3–2	>2
Thermal alteration index		<2.5	2.5–4.5	4.5–5	4.5	>5
T_{max} (°C)	Type I	<437	437–460	450–465	460–490	>490
	Type II	<435	435–455	447—460	455–490	>490
	Type III	<432	432–460	445–470	460–505	>505

reflects the organic matter maturity of the source rock and the petroleum generation phase.

3. **Pyrolysis peak temperature (T_{max})**

T_{max} increases with increasing buried depth of the source rock and with stratigraphic aging. As one of the important parameters of organic matter maturity, T_{max} usually has good corresponding relationships with R_o and TAI (Table 6.6). In addition, the author discussed the change in the relationship between R_o and T_{max} with temperature when researching the Panyu gas field (Fig. 6.5). There is a big difference between different layers of R_o and T_{max}. Therefore, special attention should be paid to the difference between these layers when researching organic matter maturity.

6.1.3.3 Organic Acid Analysis

The organic acid formed in the process of thermal evolution of the source rock is an important underground water solution or fluid from the viewpoint of secondary porosity formation. The types and contents of organic acids in the oilfield water and kerogen can be measured by using ion chromatography or liquid chromatography. In addition, techniques such as nuclear magnetic resonance, infrared spectroscopy, and elemental analysis can be applied to study the structural

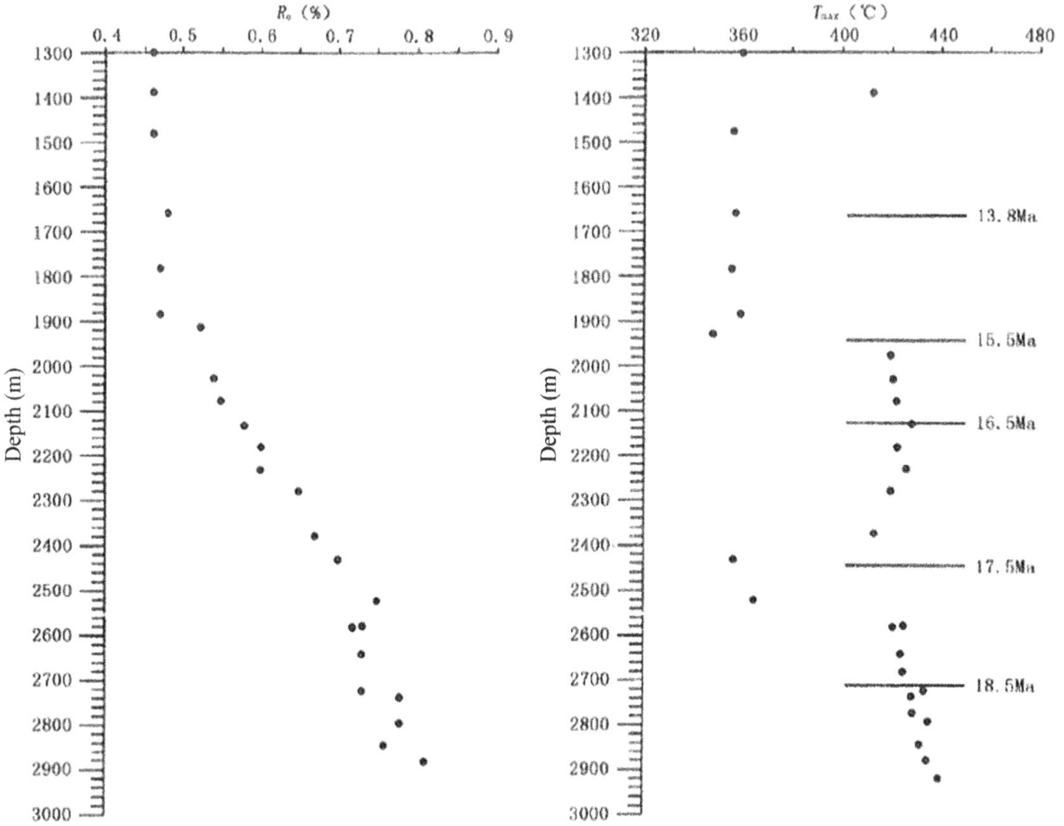

Fig. 6.5 There is great difference between R_o and T_{max} in different horizons, thus, particular attention shall be paid to the difference in horizons during research on the maturity of organic matter (according to Yu Xinghe et al. 2002)

changes and acid-producing abilities of different type of kerogens in the process of thermal maturation.

6.1.3.4 Stable Isotope Analysis

For various reasons, research results pertaining to isotopes in clastic rock (sandstone and mudstone) are scarcer than those pertaining to isotopes in carbonate rock. They mainly include the following: ① the degree of lithological heterogeneity of clastic rock is higher than that of carbonate rock, thus analysis of the whole rock will not mean much; ② the analytical method (BrF_5 and F_2) is both time consuming and error prone, which limits its development; ③ it is still in the exploratory stage as far as the isotope fractionation mechanism of clastic rock is concerned. Even so, a substantial amount of data about research on isotopes with respect to diagenesis is available (Junmao and Ming 1989). Isotope analysis is performed mainly to obtain data such as paleogeotemperature and the formation sequence of authigenic minerals.

1. **Isotope change of water**

Diagenesis is inseparable from the activities of underground fluid or aqueous solution, and pore water or stratum water is the medium by which minerals undergo pressure solution, dissolution, metasomatism, and recrystallization, and secondary porosity is formed. It can transmit isotopic information with some features to solid minerals via deposition and reaction.

Both $\delta^{18}O$ and δD of normal sea water (average water standard) are 0, but this value may vary in some areas owing to fresh water dilution or strong evaporation. The isotope composition of freshwater varies largely from place to place, and $\delta^{18}O$ varies between −50‰ and 0.

2. **Isotopic characteristics of clastic rocks (oxygen)**

The total rock $\delta^{18}O$ of sedimentary rocks is generally high. The mother rock composition of clastic rock controls its isotope composition. In a sense, there is no isotopic equilibrium, thus the isotope composition of clastic rock is heterogeneous. However, in these processes, for instance (low-grade metamorphic) diagenesis and exchange reaction of the stratum fluid, isotopic equilibrium can be realized. As for detrital quartz and feldspar, it is hard to reach isotopic equilibrium with seawater, because their isotope composition is successive.

6.2 Diagenesis and Porosity Evolution

Clastic reservoir rock is one of the important reservoir rocks in the world and, currently, one of the most important types of its kind in China. The diagenesis of clastic rock refers to the physical, chemical, and biological changes that occur after the deposition of clastic sediments until metamorphism, rather than petrification and consolidation of sediments in the narrow sense. These diagenetic changes play a key role in pore formation and preservation and destruction of clastic reservoirs. In short, clastic reservoirs and their properties are subject to both sedimentary facies and diagenesis. The former determines the planar division of reservoir properties, while vertical zonation depends on the latter. Therefore, diagenesis research implies comprehensive analysis of the formation process, mechanism, and evolution law of pore spaces in clastic reservoirs.

6.2.1 Basic Elements of Diagenesis

A variety of possible physical and chemical reactions occur between fragmentary materials and pore fluids prior to metamorphism but after deposition. The mode, process, and degrees of such reactions vary with changes in rock composition, fluid properties, temperature, and pressure. In other words, rock type, fluid, temperature, and pressure are the basic elements of diagenesis (basic diagenetic parameters and conditions). This diagenetic system is generally believed to be open, semi-open, or closed. Thus, one of the purposes of diagenesis study is to determine accurately the reactants, products, and reaction conditions involved in the formation of a rock; then, the laws of pore formation and evolution can be deduced or predicted based on the information obtained in the diagenesis study.

6.2.1.1 Lithology

Lithology of clastic reservoirs mainly involves the composition and structure of clastic particles and interstitial materials (cement and matrix), which depend usually on the sedimentary environment for reservoir formation and various diagenetic changes after deposition. If the presence of authigenic minerals in rock is the result of supersaturation of some ions in pore fluid under certain temperature and pressure, then its composition and structure can directly reflect the environment of diagenesis. Therefore, past diagenetic changes can be analyzed and determined by systematically and accurately observing and analyzing rock composition, especially the composition and structure of authigenic minerals. That is to say, the key to study the causes of diagenetic changes and analysis of the underlying mechanism is deducing the effects of diagenesis on the composition and structure of primary minerals with the aid of a diagenetic product, which due to the dissolution of underground water solution to detrital mineral particles, fresh water leaching on particles, and metasomatism between minerals, which result from difference compositions and structures, enjoy obvious selectivity.

6.2.1.2 Temperature

Diagenesis type and speed, as well as diagenesis direction, can be affected by temperature, which is one of the basic conditions of diagenesis. The key lies in the calculation and recovery of paleogeotemperature during diagenesis. In general, the influence of paleogeotemperature on diagenesis is reflected in the following aspects:

(1) Mineral solubility: The solubility of most minerals increases with an increase in temperature.
(2) Mineral transformation: Actual data show that mineral transformation is different if the geothermal gradient is not the same.
(3) Reaction direction of pore fluid and rock: Because the equilibrium constant of a chemical reaction is determined by temperature, a change in temperature is bound to cause a change in the reaction direction. Secondary pores can be formed by certain diagenetic reactions under one temperature, but under another temperature, authigenic minerals may be produced to block such pores.
(4) Diagenetic evolution sequence of organic matter under the control of paleogeotemperature.

The dissolution of organic acids into mineral particles is one of the key methods for the formation of secondary pores. Organic acids of different chemical compositions, which are clearly different in terms of the dissolved minerals, are derived from the variation of organic matter with temperature.

In summary, paleogeotemperature is one of the main indicators of diagenesis stage division. At present, despite the emergence of various methods, the determination of paleogeotemperature remains a major issue in the study of diagenesis. The commonly used methods include the following: ① fluid inclusion microthemometry, ② vitrinite reflectance, ③ clay mineral assemblage and transformation, and ④ authigenic mineral distribution and evolution. The first two methods have been mentioned above, hence they will not be described in the following.

1. **Clay mineral assemblage and transformation**

Different scholars have a different understanding of the impact of clay mineral assemblage in the same area. One thing almost all scholars agree upon is that temperature is the main factor, thus clay mineral is deemed a geothermometer. In studies of the Panyu Gas Field, the relationships between clay mineral and buried depth are discussed (Fig. 6.6).

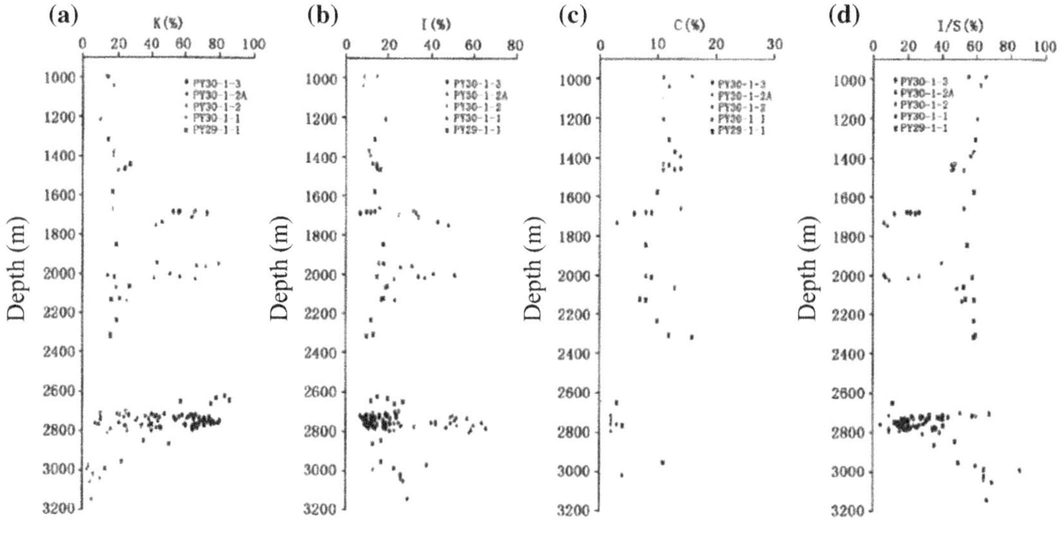

Fig. 6.6 Variation diagram for clay mineral with the depth in Panyu gas field reservoir (Yu Xinghe et al. 2002) K-Kaolinite; I-Illite; C-Chlorite; I/S-Illite/Smectite mixed-layer

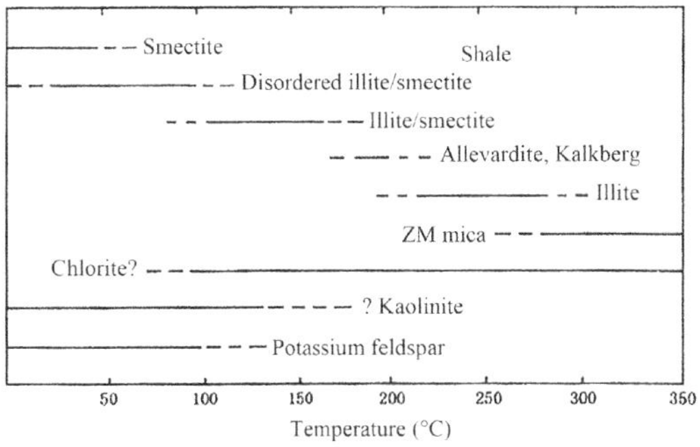

Fig. 6.7 Diagenetic evolutionary sequence of clay minerals (according to Hoffman and Hower 1979)

Hoffman and Hower (1979) proposed a clay mineral diagenesis evolution sequence and estimated the reaction temperature based on a study on Montana mudstone in the United States (Fig. 6.7). These temperature boundaries are roughly similar to the boundaries of diagenetic transformation of clay minerals reported by Japanese scholars. Regarding the continental faulted basin in East China, Ying Fengxiang et al. (1990) conducted detailed research on the transformation of smectite and illite, and divided the illite/smectite (I/S) mixed-layer into four zones in terms of mineral transformation (Table 6.7).

2. **Authigenic mineral distribution and evolution**

In addition to the above clay minerals, the formation and evolution of a few authigenic minerals is related closely to temperature. Hence, authigenic minerals are also called geothermometers, for example, zeolite, authigenic quartz, feldspar, and carbonate mineral, among which zeolite mineral is the most studied mineral.

In a more complete profile, common zeolite belt classification is as follows from top to bottom: volcanic glass belt → clinoptilolite belt → analcime or heulandite belt → laumontite or albite belt. However, different scholars hold different opinions on the formation temperature of each zeolite belt (Table 6.8).

6.2.1.3 Pressure

Pressure controls diagenetic reactions to a certain extent, which is mainly reflected in diagenetic speed and direction. Common pressure parameters include hydrostatic pressure (P_h), pore fluid pressure (P_P), effective stress (P_f), residual fluid pressure (P_e), and lithostatic pressure (P_t, total pressure) (Fig. 6.8). Pore fluid pressure equal to the weight of the overlying water column is called hydrostatic pressure, while that higher than the weight of the overlying water column is called residual pressure. When the pore fluid pressure is greater than the hydrostatic pressure (ultrahigh pressure), the effective stress decreases until it becomes zero. Pressure directly controls mechanical compaction in and pressure solution of the reservoir. In other words, pressure is a key factor for direct property control. The solubility of most minerals increases with an increase in pressure, thus pressure serves as an indirect factor for property control in the middle and late stages of diagenesis.

6.2.1.4 Fluid

The precipitation and dissolution of authigenic minerals found in reservoirs are caused by a large number of dissolved substances in the sedimentary basin. In the process of diagenesis, pore fluid or groundwater solution containing different components exists in the reservoir. This fluid is a dynamic medium for mineral redistribution. Hence, its chemical composition and activity level play an important role in the control over diagenesis. Specifically, pore fluid generally includes pore water and oil and gas, in which the influence of pore water is the most prominent.

1. **Chemical composition of pore water**

Pore water samples are mainly obtained by petroleum drilling. The determination of inorganic ions in pore water is a routine and established analysis technique in oilfield chemistry (six major ions: Cl^-, SO_4^{2-}, HCO, Na^+, Mg^{2+}, and Ca^{2+}). In contrast, the determination of organic ions in pore water generally requires

Table 6.7 Relationships between type and transformation zone of mixed-layer illite/smectite and diagenetic stage and organic matter maturity (according to Ying Fengxiang, revised in 1990; Yang Fan, revised according to standard 2003)

Mixed-layer type	Degree of order	Mixed-layer transition zone	Ratio of smectite layer in I/S mixed layer (%)	Maturity of organic matter (see detail explanation in Sect. 3)	Diagenetic stage	
Disordered mixed layer of smectite	R = 0	Smectite zone	>70	Immature	Early diagenesis	Phase A
	R = 0	Disordered mixed layer	70–50	Semi-mature		Phase B
Partially ordered mixed layer	R = 0/R = 1	Partially ordered mixed layer	50–35	Low mature-mature	Middle diagenesis	Phase A
Ordered mixed layer	R = 1	Ordered mixed layer	35–15			
Kalkberg-type ordered mixed layer	$R \geq 3$	Superlattice ordered mixed layer zone	<15	High-mature		Phase B
Mixed-layer disappears	–	Illite-chlorite zone	–	Overmature	Late diagenesis	

Note R refers to the degree of order of mixed-layer minerals. R = 0, disordered; R = 1, partially ordered; R = 2, ordered; and R = 3, super-ordered

Table 6.8 Formation temperature of zeolite belts published by different scholars (according to Wu Yuanyan 1996)

Researchers and year of publication	Formation temperature of Zeolite (°C)		
	Clinoptilolite	Analcite	Albite
Fujioka, Senkawa	60	88	
VTADA (1971)	36–45	57–72	65–73
IIJIMA, VTADA (1977)	41–49	84–91	120–124
Sato, Hasegawa	40	78	
Kazama, Aoyagi	40	90	120
Aoyagi, Kazama	47	93	110
Shimoyama, Iijima		84–91	120–124
Fujita, Hirai		90–100	
Iijima	30–60	~91	120–124
Aoyagi, Kazama	56	116	138
Sayu Kifuji	57–69	86–117	119–158

instrument analysis, for which commonly used instruments are gas chromatographs, liquid chromatographs, and ion chromatographs. Changes in diagenesis can be reflected by changes in Na^+ and K^+ contents with depth. For example, in the process of diagenesis (usually late diagenetic stage), the dissolution of feldspar can release Na^+ or K^+.

When the buried depth is greater than a certain depth, reduction of Na^+ and K^+ is the result of absorption of K^+ due to rapid transformation from mixed-layer illite/smectite mineral to illite.

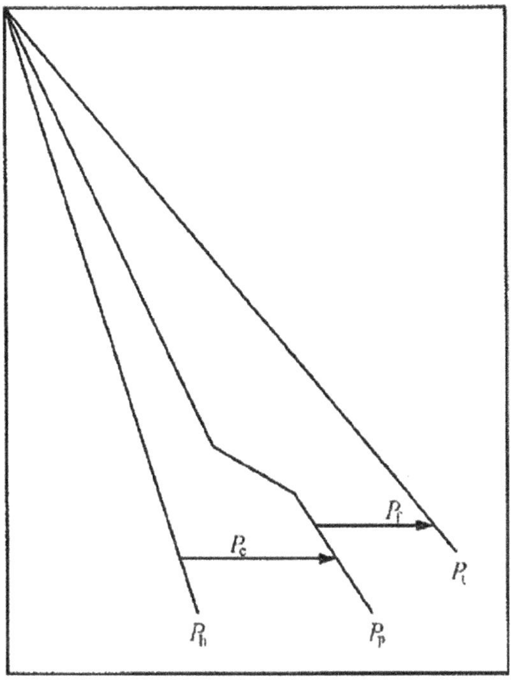

Fig. 6.8 The common pressure parameters include hydrostatic pressure, pore fluid pressure, effective pressure, residual fluid pressure and lithostatic pressure, and the relations are as shown in the diagram

2. **Flow pattern and dynamics of pore water**

Coustau (1977), based on the hydrodynamic characteristics of basins, has divided basins into three stages, namely, "young," "middle-aged" and "elderly," which correspond to compaction-driven flow, gravity-driven flow, and stagnation (no flow), respectively (Fig. 6.9). Compaction-driven flow refers to the compaction of fluid under the load of overlying sediments, which drives the flow from the center to the edge or from deep to shallow. Gravity-driven flow refers to the action of gravity due to terrain elevation pushing fluid flow from a high potential area to a low potential area. Stagnation refers to a situation without any flow.

The hydrodynamic characteristics of sedimentary basins cannot be reflected simply by the flow pattern of a basin. It is essential to divide the dynamic hydrocarbon accumulation fluid system into gravity-driven, compaction-driven, sealed-box, and stagnation types (Yongshang et al. 1999), according to the characteristics of different layers and the source, direction, and evolution of system power. Each fluid system type has a different diagenetic environment (Table 6.9) and diagenetic reactions.

6.2.2 Main Diagenesis

Diagenesis type is determined by four basic elements. Fluid properties and mineral composition determine whether the mineral is dissolved to form secondary pores or precipitation occurs to result in pore plugging. The diagenetic process is also an alternate process of pore formation and disappearance. Based on the study of diagenesis in some areas, a diagenetic and pore evolution model for quartz sandstone has been established (Fig. 6.10). According to this model, main diagenesis can be divided into two categories based on the influence of diagenesis on pore evolution in sand bodies.

The first role is to reduce the diagenesis of sand body porosity and permeability, which is mainly reflected as mechanical compaction and cementation, followed by pressure solution and recrystallization.

The second role is to increase the diagenesis of sand body porosity and permeability, which is mainly reflected as dissolution and leaching. Metasomatism has little effect on pores, but it can provide an increased supply of soluble substances in the later dissolution, which is beneficial for overall dissolution.

6.2.2.1 Mechanical Compaction

Mechanical compaction refers to diagenesis, where under the effect of overlying gravity and the hydrostatic pressure of sediments, water discharge and dense arrangement of clastic particles lead to reduction in pore volume, porosity, and permeability. The compaction of clastic rock is

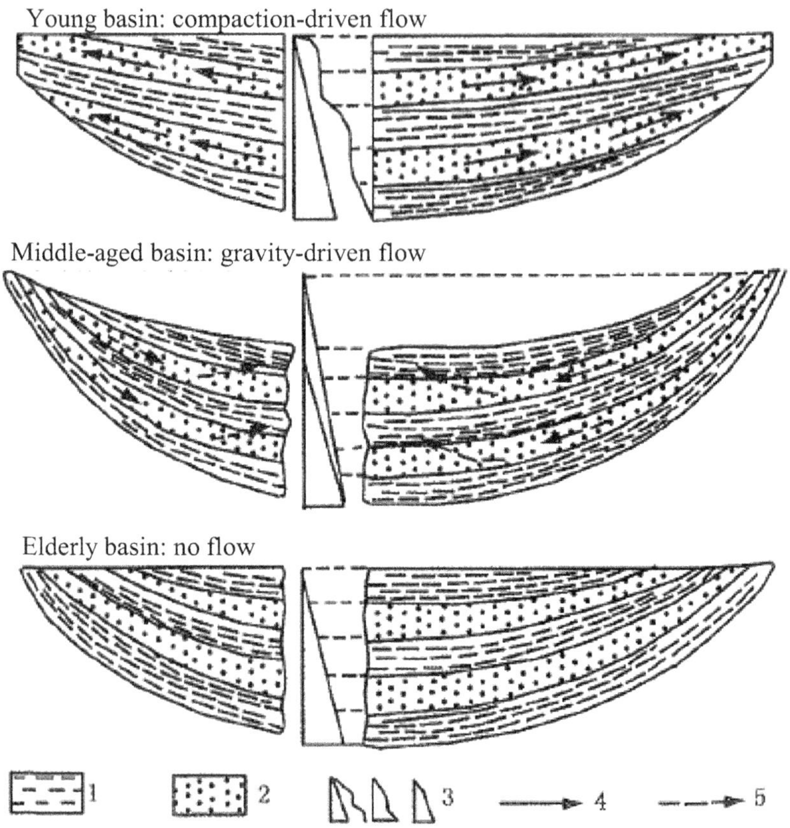

Fig. 6.9 According to different basin evolution stages, the hydrodynamic features can be divided into compaction driven flow, gravity driven flow and no flow. (Coustau 1977) 1—clay; 2—sandstone; 3—vertical pore pressure profile; 4—compaction-driven flow; and 5—gravity-driven flow

Table 6.9 Influence of characteristics, evolution mode, and diagenetic environment of different fluid dynamic systems on diagenesis

Type	Compaction-driven type	Gravity-driven type	Sealed-box type	Stagnation type
Major hydrodynamic characteristics	Fluid flows from basin subsidence center to the edge or from deep to shallow under the action of compaction	Fluid flows from basin center to the edge or from one side to the other side under the action of gravity, including regional and local types	Only periodic fluid exchange exists between the system and the outside, with episodic exchange due to abnormally high pressure	No power source and fluid flow, under normal pressure balance
Evolution mode	Material flow-dominated, and accompanied by energy flow	Material flow-dominated, and accompanied by energy flow	Energy flow-dominated, and accompanied by material flow	Energy flow-dominated, and accompanied by material flow
Diagenetic environment and characteristics	Closed water cycle; organic acid-dominated; and selective reaction	Open water cycle; meteoric water-dominated; and selective reaction	Semi-closed water cycle; organic acid-dominated; and active diagenetic reaction	No water cycle; inorganic acid-dominated; and slow diagenetic response
Factors affecting diagenetic reaction	Pressure, temperature, and abundance of organic matter	pH, Eh, and ion concentration	Overpressured zone and abundance of organic matter	Temperature, pH, Eh, and ion concentration

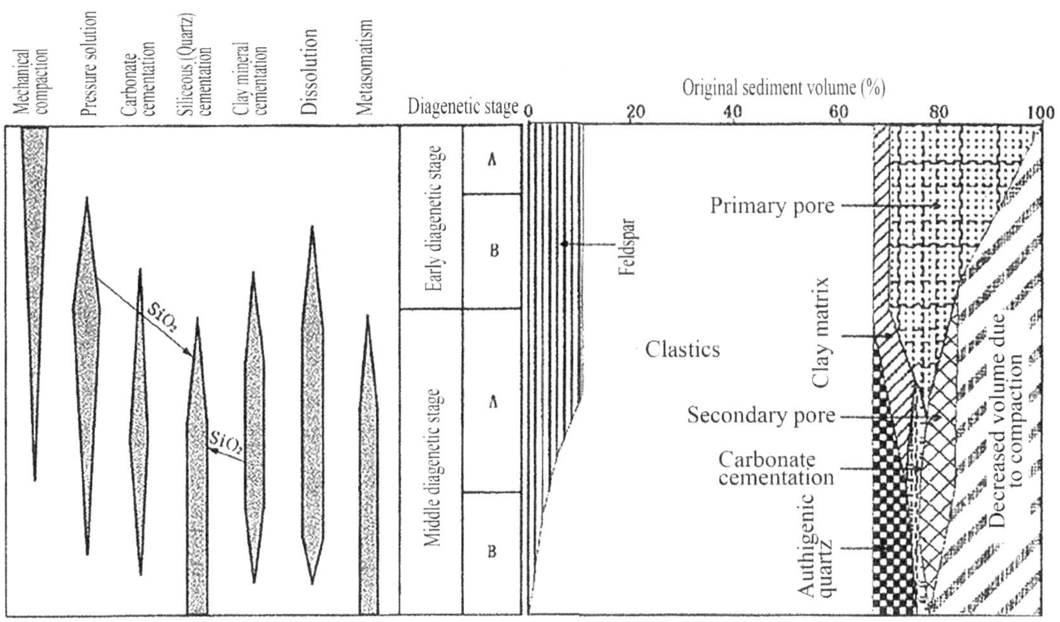

Fig. 6.10 Diagenetic process of quartz sandstone in Panyu gas field and pore evolution pattern (according to Yu Xinghe et al. 2002)

mainly impacted by particle composition, grain-size sorting, roundness, buried depth, and formation pressure. The final result is reduced intergranular volume and primary porosity.

After mechanical compaction, sediments may undergo many changes, mainly including ① rearrangement of clastic particles from free state to near—or the closest packing state; ② plastic detritus extrusion deformation; ③ bending of soft mineral particles to cause compositional variation; and ④ crushing or fracture of rigid detrital minerals.

The degree of consolidation of the sediments is called compaction strength. The method for qualitatively characterizing the compaction strength is describing the contact relations of clastic particles under a microscope. With an increase in the compaction strength, contact among clastic particles is represented successively as point contact → line contact → concavo-convex contact → suture contact. Multiple methods are available for quantitative characterization, but the following two are most commonly used methods

1. **Particle tightness**

Particle tightness, which is usually called packing density thus far, can be calculated by measuring the total length of a particle intercept in any direction under a microscope micrometer or imager according to the following formula:

Particle tightness = (total length of particle intercept/measuring length) × 100%

Apparently, a larger ratio means a tighter particle arrangement, in which case, compaction intensity will be higher. Post-compaction porosity loss can be calculated by the maximum primary porosity in a study area based on the compactness, and pore compaction gradient can be calculated according to a certain well interval to reflect the compaction strength. According to this method, compaction strength can be classified as follows (Table 6.10).

Table 6.10 Compaction strength grading (according to Qiu Yi'nan 1993)

Compaction strength	Weak compaction	Rapid compaction	Strong compaction
Particle tightness (%)	<70	70–90	>90
Porosity loss after compaction (%)	<10	11–27	27
Pore compaction gradient (%/100 m)	<1	>1	0.5

2. **Compaction rate**

Compaction rate can be calculated by comparing the initial pore volume of sand body to the intergranular volume after compaction. It reflects the percentage reduction in the initial pore volume of a sand body after compaction.

Compaction rate = (Initial pore volume − Intergranular volume after compaction)/ Initial pore volume × 100%

The initial pore volume can be estimated by rock grain size and sorting with a Schneider Atlas, and intergranular volume after compaction is usually estimated by using thin slices. The intergranular volume includes pore volume, cement volume, and matrix volume. In calculating the compaction rate, it is better to calculate a different lithologies for each depth segment and establish a depth-compaction rate profile of the different lithologies.

6.2.2.2 Pressure Solution

When the overlying formation pressure or tectonic stress exceeds the tolerable hydrostatic pressure of pore water, it may induce lattice deformation and dissolution at particle contact points. This partial dissolution is called pressure solution. Typically, the pressure solution of coarse sandstone is faster than that of fine sandstone (Junmao and Ming 1989), and the formed depth is larger, mostly greater than 3000 m. Pressure solution, with quartz pressure as the most common pressure, can cause interpenetration between quartz particles to form a line contact, concavo-convex contact, and suture contact, thus reducing rock porosity. In addition, in quartz pressure solutions, soluble SiO_2 is dissolved in pore water, which leads to an increase in Si^{4+} concentration in pore water. SiO_2 precipitation occurs when pore water is oversaturated, which not only provides material sources for the formation or siliceous cementation of quartz secondary enlargement but also reduces rock porosity.

6.2.2.3 Cementation

Cementation, which is a key factor influencing the decrease of reservoir porosity, refers to the minerals that precipitate in clastic sediment pores to form authigenic minerals and consolidate sediments into rock. The diagenesis of cementation is to block pores and lead to smaller intergranular volume, which is different from the diagenesis of compaction.

Essentially, cementation study is the study of the formation of authigenic minerals. In clastic rock reservoirs, the most common minerals include ① various carbonate minerals such as calcite, dolomite, and siderite; ② siliceous rock and aluminum silicate minerals such as quartz, feldspar, and clay minerals; and ③ zeolite and sulfate minerals such as gypsum, anhydrite, and barite. Once clastic sediments are deposited, pore water and particles will react. Whether mineral particles dissolve or new authigenic minerals are formed due to precipitation depends on two factors. The first is mineral saturation, involving the composition of pore fluid and rock particles; the second is the reaction rate between minerals and pore water, which depends on temperature and pressure.

1. **Cementation mode**

Mineral cementation includes pore filling, pore lining, pore bridge-plugs, and particle enlarged edges (Fig. 6.11).

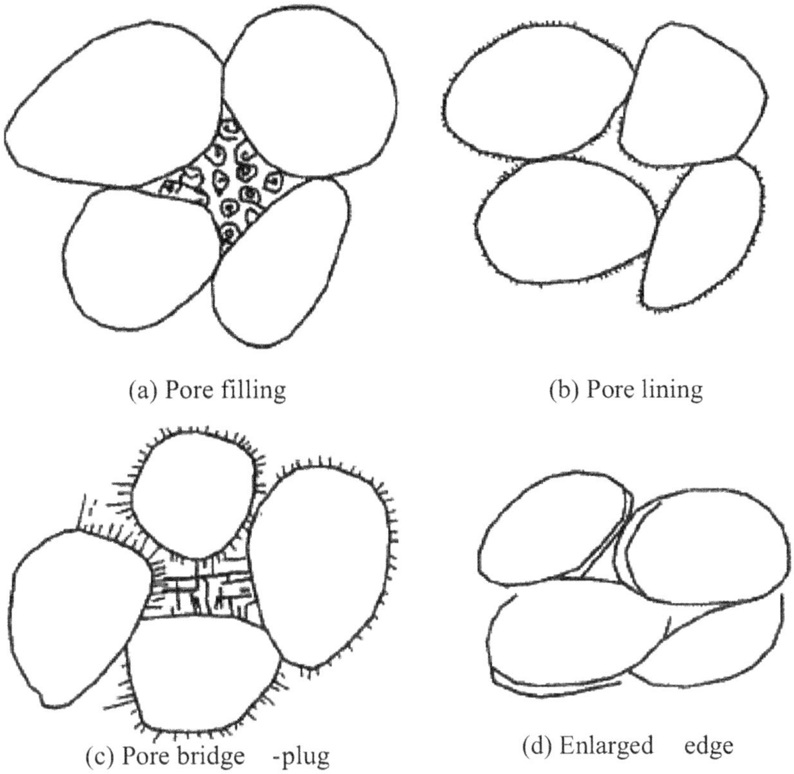

Fig. 6.11 The cementation styles of minerals mainly include pore filling, pore lining, pore bridge plug and side enlarging of particles (according to Wu Shenghe et al. 1998)

(1) Pore filling

This cementation refers to cements distributed in pores in between the particles and it is the most common method. It is found mostly in authigenic clay minerals (especially kaolinite), and in carbonate, sulfate, and zeolite cements.

According to the size of authigenic mineral crystals, the pore-filling methods can be divided into microcrystal filling, inlaid crystal filling, and even crystal filling.

(2) Pore lining

Cements are distributed outside a particle, wrapping the particle entirely. In this mode, cements are grown vertically on particle surfaces or parallel to the direction of the particle distribution, and are attached to particle surfaces, such as illite, coniferous flaky chlorite, and siderite.

(3) Pore bridge-plug

This is also known as the bridge or bypass form, most of which involves authigenic clay minerals. Clay minerals grow from pore wall to pore space and finally reach the other side of the pore space to form a clay bridge. The most common type is flaky and fibrous authigenic illite, which can form a network in pores, split large pores into micro ones, and make the fluid flow channel twist and turn. In addition, smectite and mixed-layer clay minerals can also form a clay bridge in pore throats.

Table 6.11 Characteristics of carbonate minerals in different diagenetic stages (according to Zhou Zili, Zhu 1992)

Main characteristics	Early carbonate minerals	Late carbonate minerals
Period of formation	Before main compaction period	After main compaction period
Type	Calcite, dolomite	Ferrocalcite, ferrodolomite
Structural features	Microcrystalline or ring-shaped edge	Medium crystal, fine crystal, crystal inlay, and crystal stock
Distribution	Less, often in lenticular shape	Layered, featuring wide distribution
Diagenesis nature	Sand pore water precipitation	Burial diagenesis

(4) Enlarged edge

This mainly involves the secondary enlargement of quartz and feldspar.

2. **Cementation**

(1) Carbonate cementation

Carbonate cements are of many types, and calcite is the most common mineral in carbonate cements, followed by dolomite, iron dolomite, and siderite. Pore water containing a certain amount of $CaCO_3$ is a prerequisite for the formation of carbonate cements, and appropriate physicochemical conditions (pH value of the solution, in particular) are the key to the precipitation of carbonate cements (Junmao and Ming 1989). In the process of diagenesis, carbonate cements can be formed in different diagenetic stages and have different characteristics (Table 6.11).

Carbonate solubility is very sensitive to the pH value of the solution. As the pH value increases, solubility occurs to produce carbonate precipitation. In general, an increase in pH value is related to the CO_2 formed in rock by aerobic or anaerobic bacterial decomposition when organic matter is buried. Groundwater down-cycle through strongly altered mafic volcano rock formation is one of the reasons for the increase in pH value. A reduction in pH value can be caused by the downward movement of water through an acidic soil layer. Moreover, temperature of the solution and partial pressure of CO_2 also have a great influence on carbonate precipitation. The partial pressure of CO_2 decreases with an increase in temperature, which is beneficial for carbonate precipitation. For surface water containing carbonates deep underground, the pH value increases with an increase in temperature, and a decrease in CO_2 pressure will result in carbonate precipitation.

Because of the dissolution of anorthite and the formation of clay minerals, Ca^{2+} activity is increased, which triggers the precipitation of calcite, such as the change in anorthite kaolinization, as shown below:

$$CaAl_2Si_2O_8 + 2H^+ + H_2O \rightarrow Ca^{2+} + Al_2Si_2O_5(OH)_4$$

This reaction also increases the pH value but reduces the solubility of calcites. In pore water containing NaCl, ion exchange during the transformation from anorthite to albite is also a reaction that enhances Ca^{2+} activity:

$$CaAl_2Si_2O_8 + 2Na^+ + 2Cl^- + 4SiO_2 \rightarrow 2NaAlSi_3O_8 + Ca^{2+} + 2Cl^-$$

This reaction affects silica activity, hence it can lead to siliceous dissolution, but it does not affect the pH value.

In continental lake sand bodies in China, carbonate cementation enjoys the following features: carbonate content in thin sand layers is higher than that in thick sand layers; carbonate content is higher than that in the middle part (Wu Shenghe et al. 1998); delta-front distributary channel sandstone contains gravel at the bottom or gravel sandstone is often cemented by carbonate minerals; sandstone in open lacustrine shale contains many iron and ankerite cements; and sandstone in oil shale contains many ferrocalcite-containing cements. Here we see that the distribution, content, and mineral species of carbonate cements in sandstone are directly or indirectly under the control of sedimentary facies and are related to changes in the geotemperature.

(2) Siliceous cementation

Silica cements can be presented in both crystalline and amorphous forms. Amorphous forms include opal-A and opal-CT, while crystalline forms include chalcedony and quartz. Opal cements are relatively rare and are often associated with biological siliceous dissolution and precipitation. The most common form of siliceous cementation is optical continuous proliferation of quartz particles, namely, quartz secondary enlargement, which often leads to the formation of a quartz euhedral crystal surface or an interlaced mosaic structure. Usually, the secondary enlargement of quartz is divided into four levels.

Level I: Few quartz crystals with narrow enlarged edges or euhedral crystal faces can be seen under a thin slice.
Level II: Mostly quartz and some feldspars experience secondary enlargement, developed from the euhedral surfaces, and small quartz crystals are visible in some cases.
Level III: Almost all of the quartz and feldspar have secondary enlargement, and the enlarged edge is wider, which presents a mosaic structure.
Level IV: Particles are in the form of suture contact, and the euhedral surface basically disappears.

The impact of quartz secondary enlargement on sandstone porosity varies considerably, depending on enlargement strength. When the secondary enlargement of quartz is very strong, sandstone can be turned into a dense layer or an extremely low permeability layer, which leads to the loss of reservoir properties. However, for arkose in mesozoic and cenozoic continental faulted basins in eastern China, quartz secondary enlargement is usually weak, lowering the porosity by only 3–5%. Moreover, the reductions in pore throat and permeability are not dramatic.

SiO_2 dissolved in porous water can come from different sources: ① dissolution of siliceous biological skeletons; ② volcanic glass alteration and soil water; ③ transformation of clay minerals, during which SiO_2 is released in the transformation to mixed layer minerals with increasing buried depth and rise in temperature; ④ dissolution of silicate minerals and feldspar weathering to kaolinite; and ⑤ pressure solution.

Many factors affect secondary quartz enlargement, including lithologic factors (rock mineral composition and structure), diagenetic environment, and dynamic factors (temperature, pressure, and fluid) (Table 6.12).

(3) Clay mineral cementation

Clay mineral is an important cement originating from sandstone because a certain amount of clay filling is present almost in all sandstones. Common clay mineral cements of sandstone are kaolinite, illite, smectite, and chlorite, all of which involve authigenic and allogenic types. Allogenic clay minerals are derived from the parent rock, while authigenic clay minerals come from in situ formation or regeneration. There are obvious differences in the composition, structure, and distribution of the two (Table 6.13).

In general, clay minerals occur as ① a pore lining (also known as clay or particle cladding), ② pore filling, ③ metasomatic pseudomorph, and ④ fracture and geode filling (Fig. 6.12).

The formation conditions of clay minerals are varied, depending on the composition of minerals in sandstone, properties of pore fluid, temperature, and hydrogen ion concentration.

Table 6.12 Factors affecting strength of quartz secondary enlargement (according to Zhou Zili and Zhu 1992)

Influencing factors		Enhancement factor	Inhibiting factor
Lithologic factors	Support mode	Particle supported	Matrix supported
	Quartz clastic content	Rich	Few
	Primary porosity	High	Low
	Primary pore size	Big	Miniature
	Clastic film cement	Few	Rich
Diagenetic environment and dynamic factors	Geotemperature (°C)	>60	<60
	Ground pressure	Normal	Abnormal
	Fluid properties	Aqueous layer and oil-water zone	Petroleum reservoir

Table 6.13 Comparison of characteristics of authigenic and allogenic clay minerals

Characteristics	Authigenic clay minerals	Allogenic clay minerals
Composition	High purity, single mineral with good transparency	Mixture of multiple ingredients
Morphology	Intact crystalline form	Poor crystalline form
Structure	Coarse particles	Fine particles
Distribution	Cladding irregular distribution on particle surface or around the particles	At particle contact point, filled in pores with directional arrangement

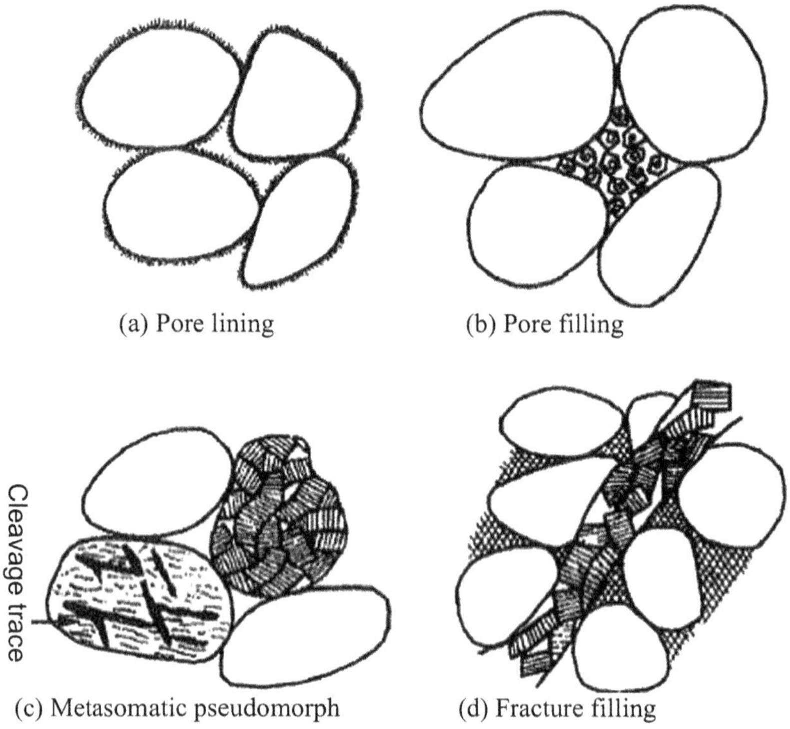

Fig. 6.12 Schematic diagram for occurrence types of authigenic clay minerals in sandstone (M. Wilson et al., 1977)

The growth of authigenic minerals in sandstone reflects the interaction between pore water under seepage and clastic particles. The main controlling factors lie in pore water composition and properties, chemical stability of sand in pore water, and sandstone porosity and permeability. Acidic pore water is favorable for the formation of kaolinite minerals, while alkaline pore water is beneficial for the formation and preservation of other clay minerals. The clasts that usually react most easily with pore water under seepage include volcanic glass, detritus, feldspar, sideromelane, and carbonate particles, which easily form illite and kaolinite in arkose, mainly with illite in lithic sandstone and greywacke. Smectite is formed mainly in volcano clastic rock.

(4) Zeolite cementation

Zeolite cements are rare, but they are the main cement in individual layers of sandstone in some basins. Most cements of this type are laumontite, analcime, and heulandite, which can be formed in each stage of diagenesis. The favorable medium conditions for zeolite formation are high pH value, abundant SiO_2 and calcium, high salinity pore water containing sodium ions, and appropriate CO_2 partial pressure. Zeolite is the most abundant in volcano clastic rocks, and zeolite minerals in sandstone originate mainly from pyroclastic alteration. In addition, sandstone rich in albite easily forms laumontite in a diagenetic environment with high pH value. Laumontite can be formed in high or low temperature. It is related not only to the temperature but also to the nature of rock composition and pore fluid.

Guohua (1985a, b) studied in depth the formation of laumontite in the Yanchang formation in the Shanxi-Gansu-Ningxia basin. This formation has a buried depth of less than 2500 m, with a vitrinite reflectance of 0.5–0.8%. The formation temperature of laumontite is estimated to be 50–80 °C, so obviously, it is formed at low temperatures. Zhu Guohua believes that interaction between the pressure solution of anorthose and pore water is the main mechanism for laumontite formation in the sandstone of the Yanchang formation, as given by the following equation:

$$\underset{\text{(Anorthite)}}{2CaAl_2Si_2O_8} + 2Na^+ + 4H_2O + 6SiO_2 \rightarrow \underset{\text{(Laumontite)}}{CaAl_2Si_4O_{12} \bullet 4H_2O} + \underset{\text{(Albite)}}{2NaAlSi_3O_8} + Ca^{2+}$$

In this area, laumontite cements not only block pores, but also play a supporting role, which can safeguard skeleton particles from strong compaction in order to lay a material base for the later development of secondary pores. Therefore, the laumontite secondary pore sand body is the main reservoir sand body in the northern Shaanxi Yanchang formation.

(5) Cementation of other authigenic minerals

Other authigenic minerals are also formed in the diagenetic process, such as feldspar, gypsum, anhydrite, and iron oxide. They are not significant in quantity but are of great significance to the research on diagenetic history and speculation on the formation and origin of various authigenic minerals.

3. **Influence of cements on reservoir performance**

On the whole, authigenic cements narrow pores and throats and complicate pore shape in a reservoir, thus reducing the reservoir properties or in severe cases, resulting in the loss of rock reservoir capacity. For example, some late calcite cements in rock completely block the pore space to turn it into a tight rock.

It is obvious that the higher the cement content, the higher is the impact on reservoir performance. In practical work, the cementation rate is usually used to quantitatively express the impact of cementation on sand body porosity and cementation strength:

$$\text{Cementation rate} = (\text{Cement content}/\text{Original pore volume}) \times 100\%$$

Table 6.14 Formation conditions of clay mineral (according to He Jingyu and Meng xianghua 1987)

Clay minerals	Primitive matter	Component increase (+) or decrease (−)	pH value
Kaolinite	Pore water	$+Al_2O_3SiO_2{\bullet}H_2O$	<7
Kaolinite	Feldspar	$-(K + SiO_2)$, $+H_2O$	5–7
Smectite	Pore water	Na, Ca	6–8
Smectite	Volcanic glass	$+(H_2O)$, $-(Na^+, K^+, Ca^{2+})$	6–8
Illite	Kaolinite	$+(K^+, SiO_2)$, $-(Al_2O_3{\bullet}H_2O)$	7–8
Illite	Smectite	$+(K^+, Al^{3+})$, $-(SiO_2, H_2O, Na^+Ca^{2+}, Mg^{2+})$	7–8
Chlorite	Smectite	$+(Fe^{2+}, Mg^{2+})$, $-(SiO_2, H_2O, Na^+, Ca^{2+})$	7–9
Chlorite	Illite	$+(Fe^{3+}, Fe^{2+})$, $-(K^+, Al_2O_3)$	7–9

It is noteworthy that the influence of cementation on rock reservoir performance is reflected mainly in the reduction in permeability because pore throat size is directly reduced under the action of cementation. The pore throat is generally small, and throat cementation greatly reduces the permeability of stone (since permeability is inversely proportional to square of the throat radius) (Table 6.14).

4. **Relationships between compaction and cementation**

In the process of diagenesis, compaction and cementation are the main reasons for reduced porosity, but the two are manifested in different ways and are subject to mutual constraints. Compaction reduces the intergranular volume of rock irreversibly. If it moves fast, porosity and permeability decrease rapidly. Then, interlayer water flow is restricted and cementation does not develop. Unlike this, cementation does not cause any reduction in intergranular volume, although it does block pores. Even after the pore fluid–rock balance is destroyed, these authigenic cements can be dissolved to turn into a pore once again. However, compaction will be blocked if cementation moves rapidly. To evaluate the relative importance of these two forces, Houseknecht (1987) prepared an evaluation chart (Fig. 6.13). In this chart, the upper abscissa represents intergranular volume (assuming the primary porosity of the sand layer is 40%), while the lower abscissa represents the percentage of primary porosity eliminated by mechanical compaction and chemical pressolution, which can be determined by the following formula:

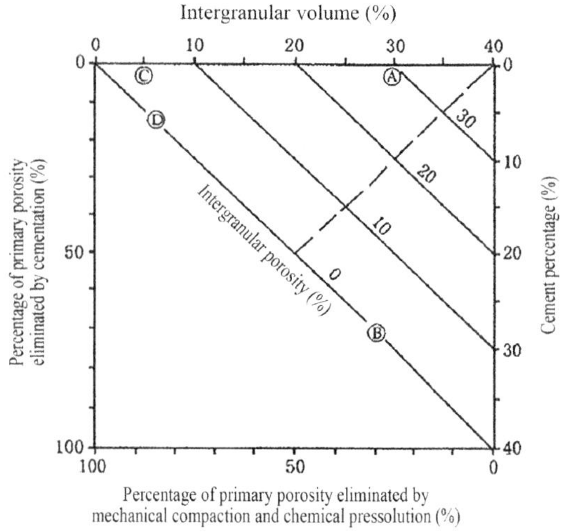

Fig. 6.13 Evaluation for relative importance of compaction and cementation (according to Houseknecht 1987)

$$\text{Percentage of primary porosity decrease due to compaction} = [(40 - \text{intergranular volume})/40] \times 100\%$$

The right ordinate in the figure represents cement, while the left ordinate represents the percentage of primary porosity eliminated by cementation, which can be calculated as follows:

$$\text{Percentage of primary porosity eliminated by cementation} = (\text{Cement}/40) \times 100\%$$

Intergranular porosity is shown in the diagonal form in the figure. Intergranular volume can be determined directly from the graph or calculated according to the following formula:

$$\text{Intergranular porosity} = \text{Intergranular volume} - \text{Cement}$$

The diagonal line therein divides the compaction-dominated zone (upper left) and the cementation-dominated zone (lower right), thus the relative importance of compaction and cementation can be evaluated using this chart. For example, samples A, C, and D in the figure are affected mainly by compaction, which leads to reduced intergranular volume. However, sample B is mainly affected by cementation, which leads to a decrease in porosity.

6.2.2.4 Dissolution and Metasomatism

Dissolution, in a broad way, refers to the dissolution of groundwater solution in rock components, which usually can be divided into two categories: solid phase homogeneous dissolution, which always leaves the fresh surface component of the undissolved solid phase unchanged; and selective dissolution. Owing to the inconsistency of rock component dissolution, the chemical compositions of the new minerals are similar to those of the dissolved minerals, such as feldspar kaolinization. The former is mostly dissolution, and the latter is corrosion. Metasomatism means minerals are dissolved and displaced by the precipitated minerals, wherein the new and the dissolved minerals share no identical chemical compositions, for example, the metasomatism of quartz from calcite.

The dissolution and precipitation of minerals in sandstone is related mainly to the concentration of organic acids, carbonic acid, and CO_2 in pore water, as well as geotemperature and other physicochemical factors.

In a diagenetic environment, inorganic diagenesis in clastic reservoirs is related closely to the organic diagenesis of source rock. Overall, the two stimulate each other and affect the development of many diagenesis processes and the evolution of diagenetic stages. Research shows that organic matter in the sediments can produce large quantities of organic acid during burial diagenesis. When the temperature is lower than 80 °C, the organic acid content is low; when temperature is higher than 80 °C, the organic acid content in oilfield water increases exponentially; and when temperature is higher than 120 °C, organic acid undergoes partial or complete decarboxylation. The decarboxylation of organic acid produces CO_2, which determines the pH value of the aqueous solution to facilitate dissolution. If carboxylic acid formation peaks before the generation of liquid hydrocarbons, the flow of carboxylic acid through the sand layer adjacent to the source rock can cause corrosion of carbonate and aluminosilicate, thus promoting the development of secondary pores and improving the porosity and permeability of the reservoir.

6.2.3 Formation Mechanism of Secondary Pores

In terms of genesis, secondary pores in clastic rocks can be categorized into three types: secondary pores formed by dissolution (corrosion), contraction cracks due to diagenetic contraction, and tectonic fissures as a result of tectonic stress (Fig. 6.14).

6.2.3.1 Genetic Types of Secondary Pores

Schmidt et al. (1979) proposed a detailed classification scheme for the genesis of secondary pores (Table 6.15). The secondary pores formed by dissolution can be divided into three categories from the perspective of genesis.

1. **Pores from sediment dissolution**

This is a very common and important pore type, which is formed by the selective dissolution

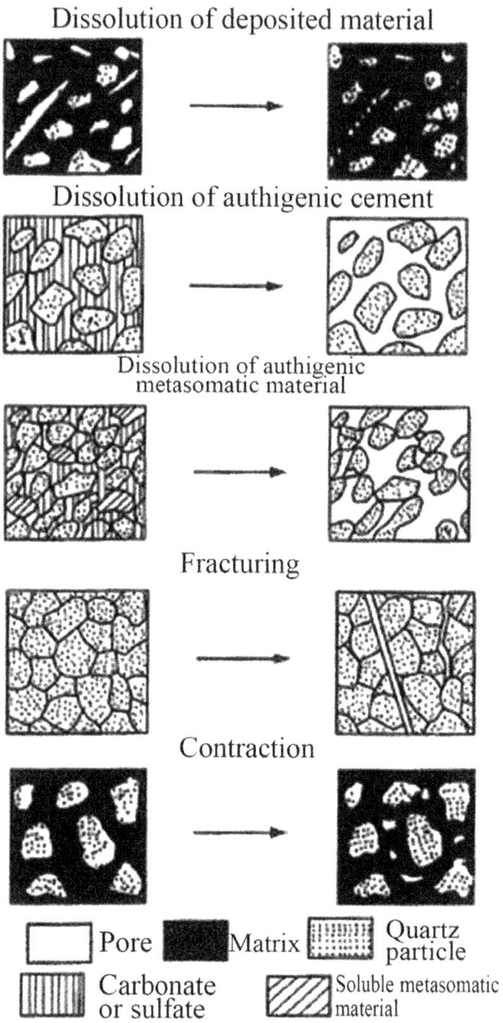

Fig. 6.14 Genetic type of secondary pores in sandstones (Schmidt 1979)

of soluble particles and a soluble matrix (Table 6.16). The dissolution of soluble substances can produce a large number of pores.

2. **Pores from authigenic cement dissolution**

This is one of the most common types of secondary pores. Most dissolved cements are carbonate minerals such as calcite, dolomite, and siderite, as well as chlorite, smectite, zeolite, and anhydrite (Table 6.16). These cements may exist in any primary and secondary pores before dissolution, while the dissolution of cements can re-open and connect the pores.

3. **Pores from authigenic metasomatic mineral dissolution**

These pores result from the selective dissolution of soluble minerals from metasomatism, mainly calcite, dolomite, and siderite (Table 6.16). They occupy a certain proportion in the secondary pores of sandstone.

Some secondary pores have complicated genesis, for instance, they are composed of secondary pores of several different genesis. The combinations of secondary pores of different genesis and that of primary and secondary pores were termed "mixed pore" (Fig. 6.15) by Schmidt (1979). In this sense, the pores in clastic rocks are basically mixed pores.

6.2.3.2 Formation Mechanism of Secondary Pores

Dissolution is the main diagenetic process for the formation of secondary pores. The dissolution of carbonate and silicate is the result of the interaction between acidic aqueous solution or alkaline solution (pore fluid) with rock under a certain temperature and pressure. Under organic acidic conditions, dissolution of the silicate component in clastic rock is more intense than that of the carbonate component (Zhu Guohua 1992). Therefore, analysis of the dissolution of different fluids in various minerals requires one to study the formation mechanism of secondary pores.

As is well known, the main secondary pores in a clastic reservoir are the products of the dissolution of acid-soluble components, which is believed to be caused by two reasons: carbonate-containing solution and organic acids, mainly short-chain carboxylic acid-fatty acid.

1. **Dissolved fluid in pores**

In general, fluids that can dissolve underground rock minerals mainly include meteoric water, organic acids, phenols, and carbonic acid. The chemical properties of fluid and water vary

Table 6.15 Genetic types of secondary pores

Diagenesis		Secondary porosity content
Rock fracturing		Low, but high in some individual strata
Particle fracturing		Very low
Contraction		Low
Dissolution	Calcite	High
	Dolomite	Rich
	Siderite and ferrodolomite	High
	Evaporite	Low, but high in some individual strata
	Feldspar	High
	Other aluminum silicate	High
	Silica oxide minerals	Rich

Table 6.16 Soluble components in sandstone (according to Hayes, Revised in 1979)

Clastics	Authigenic material	
Carbonate	Cement for pore filling	Feldspar (mainly as anorthose)
Biotite		Chlorite
Hornblende		Smectite
Pyroxene		Zeolite
Other heavy minerals, flint		Anhydrite
Carbonate clasts	Metasomatic minerals	
Calcareous and siliceous fossil debris		Carbonate
Volcanic glass and fine volcanic debris		Chlorite
Mixed-layer clay		
Zeolite		

Note Detrital biotite is altered to a mixture of clay and carbonate through diagenesis, which are dissolved later

largely in different reservoirs and diagenetic environments. The effect of meteoric water on rocks and minerals is mainly manifested on the earth's surface and near the surface, and it can extend deep underground through faults and fractures. On the contrary, organic acids, phenols, and carbonic acid act under the ground, and their roles are reflected mainly in underground rock formations.

(1) Meteoric water

When the tectonic movement of a basin uplifts the associated formation for exposure on the surface or shallow bury, the formation is subjected to the leaching of meteoric water or fresh water, and dissolution occurs along a hiatus and unconformity to result in open circulation of an inorganic acid aqueous solution, with mainly impacts shallow strata. Only when the faults and fractures, as well as deep strata, are connected and open, can the effect be extended to the deep strata. Meteoric water is acidic, has a low pH value, and contains CO_2; hence, it can dissolve carbonate, feldspar, biotite, and other minerals.

Krynine (1949) proposed the emergence of secondary pores and the unconformity relationship. He stressed that in the period of crustal uplift and weathering, carbonate cements in sandstone are dissolved by fresh water leaching. For the Denglouku Formation in the Shiwu fault depression in the Songnan area of China, the

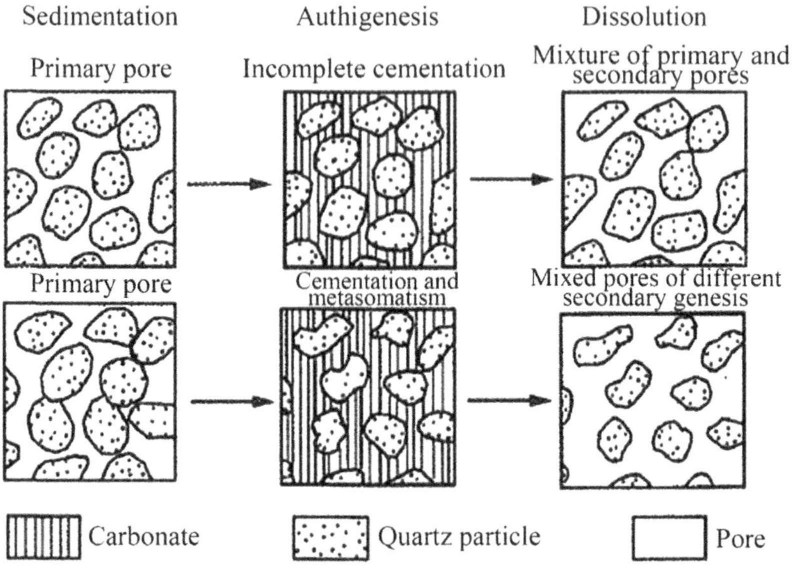

Fig. 6.15 Schematic diagram for development of mixed pore (Schmidt 1979)

basin transforms from fault depression to depression before deposition at the first member of the Quantou Formation, along with tectonic uplift. At the beginning of mechanical compaction in the early diagenetic stage, the third member of the Denglouku Formation caused leaching of meteoric water, which dissolved the feldspar part to form some dissolved pores (Yu et al. 1997). The pore distribution of carboniferous and Permian sandstones in the north China area is also significantly affected by unconformity. In areas that have been sinking or experiencing rock formation away from the denudation plane, sandstone porosity is generally around 5% due to compaction and cementation, and the secondary porosity of sandstone can reach 20–30% in the vicinity of the palaeohigh and paleocrust of weathering (Zheng et al. 1989).

(2) Organic acids and phenols

Organic acids and phenols are key causes of rock component dissolution. Short-chain soluble monocarboxylic acid and dicarboxylic acid, i.e., organic acids, which are also called fatty acid, are usually present in oilfield water. Surdam et al. (1984), Kharaka et al. (1985), and Donald et al. (1988) conducted researched and calculated (Table 6.17) the concentrations of various carboxylic acids in oilfield water. Carboxylic acid in oilfield water mainly comprises acetic acid, propionic acid, oxalic acid, and malonic acid.

Organic acids and phenols originate mainly from the evolution of organic matter. ^{13}C of kerogen: nuclear magnetic resonance and infrared spectroscopy data show that organic matter release massive amounts of organic acids before massive hydrocarbon generation. Kerogen can produce organic acid by thermal degradation to remove oxygen-containing functional groups or form organic acid by the oxidation of mineral oxidants (Fe^{3+} polysulfide in clay minerals and oxidant on particle surface) in rocks.

Studies show that various types of kerogen (Types I, II, and III) can produce organic acids, but kerogen III is the best raw material for the production of carboxylic acid, and kerogen III has a higher conversion rate than those of kerogen II and I. It is also a very important substance for the production of organic acids (and phenols), where the quantity of diacid is greater than that of monoacid (Crossey 1984), however the dissolving capacity of diacid is stronger than that of monoacid. The type of organic matter in coal rocks is usually kerogen, thus the dissolution of

Table 6.17 Highest concentrations of weak acid anions in oilfield water (according to MacGowan et al. 1988)

Carboxylic acid anion		Concentration (mg/L)
Monocarboxylic acid anion	Formic acid (methanoic acid) HCOOH	62.6
	Acetic acid (ethanoic acid) CH_3COOH	10,000
	Propionic acid (propanoic acid) CH_3CH_2COOH	4400
	Butyric acid (butyric acid) $CH_3(CH_2)_2COOH$	44.0
	Pentanoic acid (valeric acid) $CH_3(CH_2)_3COOH$	32.01
Dicarboxylic acid anion	Oxalic acid (oxalic acid) HOOCCOOH	494
	Malonic acid (maleic acid) $HOOCCH_2COOH$	2540
	Succinic acid (butanedioic acid) $HOOC(CH_2)_2COOH$	63
	Glutaric acid	36
	Oxalic acid	0.5
	Pimelic acid	0.6
	Octanedioic acid	5.0
	Azelaic acid	6.0
	Sebacic acid	1.3
	Z-butene diacid (maleic acid) HOOCCH = CHCOOH	26

sandstones on coal measurement strata is rather common.

In the thermal evolution profile of organic matter, the concentration of organic acids peaks at 80–120 °C. Below 80 °C, carboxylic acid anions can be produced, but bacteria consume such short-chain organic groups to reduce their concentration. The concentration of carboxylic acid anions increases exponentially at or near 80 °C. The reason for this is that on the one hand, a variety of organic acids are produced (kerogen thermal degradation) at 80 °C, and on the other hand, bacterial motility decreases (bacteria die out when temperature increases and toxic phenols in the organic matter are released). In the 120–200 °C temperature range, carboxylic acid ions are damaged by thermal decarboxylation, thus the organic acid concentration decreases (Fig. 6.16).

In addition to carboxylic acid, kerogen produces phenol organic matter at temperatures higher than 80 °C. The production of phenols has two meanings: firstly, phenols serve as bactericides, which restrain bacterial activity to be conducive to the storage and increase of organic acids; secondly, phenols serve as solvents available for metal complexation to break the stability of aluminosilicate and dissolve rock minerals to form secondary porosity.

(3) Carbonic acid

The formation of carbonic acid has both organic and inorganic origins.

In 1979, Schmidt et al. from Canada proposed the organic origin. They believe that carbonic acid can be formed by organic matter during the thermal evolution of diagenetic processes in pelite. CO_2 formed by organic acid decarboxylation is dissolved in water to form carbonic acid. On the organic evolution profile, CO_2 content increases as the buried depth (temperature) increases. An aqueous solution containing carbonic acid enters into clastic rocks along with mud-stone compaction drainage, which results in secondary pores owing to the dissolution of acid-soluble fractions in clastic rock.

$$CH_3COOH\text{(acetic acid)} \longrightarrow \text{Temperature } CH_4\ \text{(methane)} + CO_2$$

$$CO_2 + H_2O \longrightarrow \text{Temperature } H_2CO_3\ \text{(carbonic acid)}$$

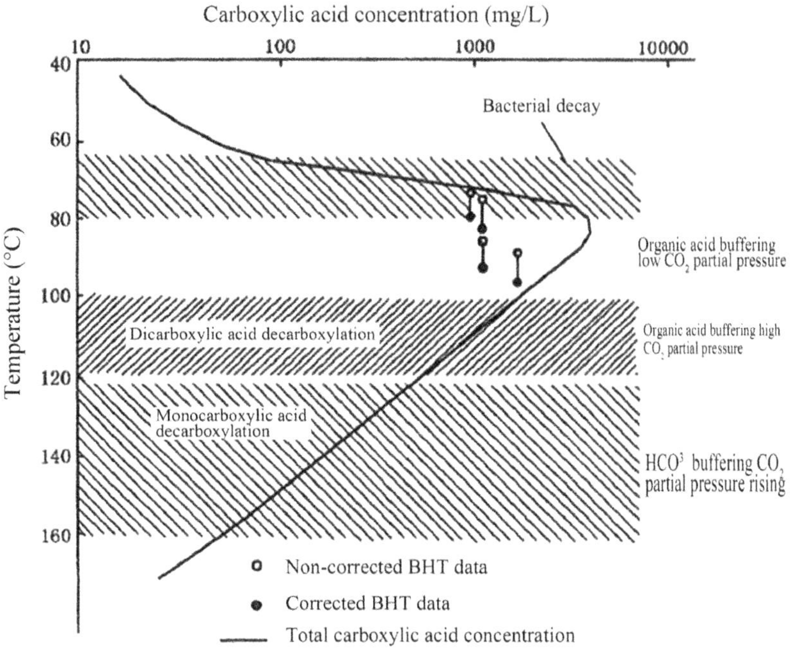

Fig. 6.16 Relations between carboxylate anion concentration and temperature in oil field water (Surdam, 1989)

Hutcheon (1990) proposed a scheme for inorganic CO_2 genesis. He holds that clay minerals react with carbonate to form the majority of inorganic CO_2.

$$5CaMg(CO_3)_2(\text{dolomite}) + Al_2Si_2O_3(OH)_4(\text{kaolinite}) + SiO_2(\text{quartz}) + 2H_2O \rightarrow Mg_5Al_2Si_2O_{10}(OH)_8(\text{chlorite}) + 5CaCO_3(\text{calcite}) + 5CO_2$$

By his calculations, if sandstone contains 10% kaolinite, dolomite, and quartz according to the proportion in the above formula, 1 m^3 of sandstone can generate 2500 L of CO_2 at 25 °C and 1 atm.

Organic acids and carbonic acid are important solvents for the formation of underground sandstone secondary pores, however organic acids, with stronger dissolving capacity, play the main role in generating secondary porosity in sandstone (Surdam 1975, 1984 and 1989) The chemical activity of acid may be measured from the following three aspects: equation of chemical reaction of acid, mainly including the capacity for contributing H^+ (reflecting chemical reactivity); chemical interaction free energy (reflecting thermal driving force); and solubility of salts reacted with an acid in water. Studies show that the capacity of organic acids to contribute H^+ is 6–350× (6, 7, 20, and 350×, respectively, the capacities of propionic acid, acetic acid, formic acid, and oxalic acid to contribute H^+), thus it has very strong reaction capacity. The free energy of organic acid reacted with carbonate and feldspar is lower than that of carbonic acid, which indicates a greater thermal driving force. The solubility of organic acid calcium salts in water is greater than that of calcium carbonate and calcium bicarbonate by three orders of magnitude.

2. Dissolution of carbonic acid to minerals

(1) Dissolution to calcite

$$CaCO_3 + H_2CO_3(\text{water}) \rightarrow 2HCO_3^-(\text{water}) + Ca^{2+}(\text{water})$$

Under normal pressure, calcite is difficultly soluble when the carbonic acid depth is greater than 3000 m, hence the solubility of CO_2 decreases as the temperature increases; however CO_2 is likely to be dissolved in the case of abnormal pore fluid and pore water with higher salinity.

(2) Dissolution to feldspar

$$3KAlSi_3O_8 + 2H_2CO_3(\text{water}) + 12H_2O \rightarrow KAl_3Si_3O_2(OH)_2 + 2K + (\text{water}) + 6H_4SiO_4(\text{water}) + 2HCO_3^-$$

$$KAl_3Si_3O_2(OH)_2 + 2H_2CO_3(\text{water}) + H_2O \rightarrow 3Al_2Si_2O_5(OH)_4 + 2K + (\text{water}) + 2HCO_3^-$$

The dissolution of feldspar increases when the concentration of CO_2 contained in the fluid increases.

(3) Dissolution to silica dioxide

$$SiO_2 + 2H_2O \rightarrow H_4SiO_4(\text{water}) \text{ or } Si(OH)_4$$

$$H_4SiO_4(\text{water}) + HCO_3^- \rightarrow H_3SiO_4 + CO_2 + H_2O$$

Siliceous dissolution follows three main mechanisms: ① in feldspar dissolution, the pH value of the solution rises due to H^+ consumption; ② silica and water are replaced to facilitate quartz dissolution due to the reaction between silica acid and HCO− 3ions in the reaction; and ③ solubility increases as the temperature increases.

6.3 Dissolution of Carbonic Acid to Minerals

(1) Dissolution to carbonate

In the presence of acetic acid, dissolution of calcite is as follows:

$$CaCO_3 + CH_3COOH(\text{acetic acid}) \rightarrow Ca^{2+} + HCO_3^- + CH_3COO^-$$

(2) Dissolution to silicate

We suppose the diagenetic reaction between feldspar and oxalic acid is as follows:

$$KAlSi_3O_8 + H_2C_2O_4 + 2H_2O + 2H^+ \rightarrow 3SiO_2 + (AlC_2O_4 \bullet 4H_2O) + K^+$$

Thus, an adequate amount of organic acid and sufficient water flow are very necessary.

Many scholars (Surdam 1984; Junmao and Ming 1989; Zhu Guohua 1992 and Yu et al. 1997) have found that silicate minerals are soluble in organic acids, but carbonate minerals are difficultly soluble or insoluble. A comparison between the Gibbs free energy values (Table 6.18) of organic acids and carbonic acid for dissolving feldspar and reacting with carbonate also proves that organic acids have better thermal power driving force than carbonic acid, that is, the dissolving capacity of organic acids is stronger. Moreover, compared with carbonates, the thermal drive of carbonates reacting with organic acids is better and is easier to dissolve and convert. It can be concluded that silicate are soluble in silicate but carbonate are less soluble under certain temperatures and pressures (Zhu 1992). While researching the diagenetic features of sandstones in the Liaohe Oilfield, Yu et al. (1999) suggested that organic decarboxylation results in massive dissolution of silicate minerals, for which the insoluble or difficultly soluble ground temperature range of carbonate minerals is 100–130 °C, and the associated pressure coefficient is 1.2–1.3, which is a dimensionless number that describes the relative pressures throughout a flow field in fluid dynamics. This can be proven in many oilfields in eastern China.

Based on the theory of chemical reaction balance, it is not difficult to understand the following reactions:

$$XSiO_3 + 2H^+ + R^{2-} \xrightarrow[\text{and pressure}]{\text{ccertain temperature}}$$

$$XCO_3 + 2H^+ + R^{2-} \rightarrow XR + H_2O + CO_2 \uparrow$$

Table 6.18 Comparison among free energies of carbonic acid, formic acid, and acetic acid (Quoted from Zhu 1992)

Dissolution, conversion	Converting agent (solvent)	ΔGr (kcal/mol)
Microcline → Illite	H_2CO_3	+102.6
	HCOOH	+95.6
	CH_3COOH	+23.6
Microcline → Kaolinite	H_2CO_3	+23.0
	HCOOH	+15.8
	CH_3COOH	−4.28
Anorthite → Kaolinite	H_2CO_3	+15.0
	HCOOH	+16.5
	CH_3COOH	−36.9
Dissolution of calcite	H_2CO_3	+28.37
	CH_3COOH	+11.20

A comparison of the phase states of the two equation products shows that no gas is generated in the first reaction, while in the latter one, CO_2 is present and pressure increases. Based on the theory of chemical reaction balance involving gaseous substances, when the concentration of CO_2 reaches or exceeds a critical value, that is, system pressure reaches and exceeds a critical value, the chemical reaction will tend to stop and proceed in the reverse direction, and carbonate minerals (calcite and dolomite) will be stored in the sandstone.

6.3.1 Factors Influencing Secondary Pore Formation

According to the above, the factors controlling secondary pore formation are as follows:

(1) Sufficient water energy and good permeability are favorable for secondary pore formation. For the dissolution of carbonic acid or organic acid to minerals, sufficient water is needed; the shortcomings of primary water and the solubility of water in acid can be remedied by continuous water flow.
(2) Only when source rock rich in organic matter is as close as possible to being a potential reservoir, does the acid solution generated in mudstone enter into sandstone successfully and transit losses are reduced.
(3) Because the sand-mud ratio is an important index for ensuring adequate acid sources, acid output is insufficient if the proportion of mudstone is too low. Conversely, if the acid output is excessive, a low-energy environment with poor sandstone permeability is formed.
(4) The thermal evolution history of kerogen determines the depth of acid formation, which is a key point for predicting the vertical distribution law of secondary pores.

6.3.1.1 Main Diagenetic Environment for Formation of Secondary Pores

(1) Opening water cycle: based on meteoric water, the formation of secondary pores mainly involves dissolution to carbonate minerals by inorganic acid aqueous solution. Determinants include pH value, Eh value, and ionic concentration.
(2) Closed water cycle: based on organic acid, the formation of secondary pores mainly involves the dissolution to carbonate minerals by organic acid aqueous solution. Determinants include temperature, pressure, and organic acid concentration.

6.4 Division of Diagenetic Stage and Evolutionary Model

The evolution of diagenesis shows differences due to geological conditions of sedimentary basins and historical changes. Owing to the staged influence of tectonic evolution, it is very difficult to divide diagenetic stages. Because of the different objectives and focuses of various researchers, various classification schemes have been developed at home and abroad. In this book, diagenetic stages are divided based on The Division of Diagenetic Stages in Clastic Rocks issued by China National Petroleum Corporation (CNPC) in 2003.

6.4.1 Terms and Definitions

(1) Diagenetic stage: refers to the evolutionary phase of different geological histories from the modification of deposited clastic sediments by various diagenesis processes until metamorphism. It can be divided into syndiagenetic stage, early diagenetic stage, middle diagenetic stage, late diagenetic stage, and epidiagenetic stage.
(2) Syndiagenetic stage: refers to the changes and effects that occur when the deposited sediments are separated incompletely from the overlying water.
(3) Epidiagenetic stage: refers to the stage in which clastic rocks are solidified weakly and those solidified in some diagenetic stage are close to or exposed to the ground surface due to tectonic uplift and then corroded by meteoric water to make changes and take effects.

6.4.2 Basis for Diagenetic Stage Division

6.4.2.1 Features of Authigenic Minerals

This is the main mark for dividing the diagenetic stage involving the distribution formation sequence of authigenic minerals, as well as the homogeneous temperature of inclusions in authigenic minerals. This is because the emergence and distribution of authigenic minerals in the diagenetic process is subject to certain physical and chemical conditions, as well as the specific geological history environment. Its formation and distribution can indicate rock formation and development in combination with rock structures and tectonic transformation. With differences among stratum temperature, pressure, and chemical properties of pore water, different types of authigenic minerals are produced by the interaction between water of different properties and rocks, as well as the interaction between organic and inorganic matter. Therefore, authigenic minerals can provide evolution data with respect to the properties of the water medium in the diagenetic process and have certain significance as geologic thermometers.

Quartz secondary enlargement is generally distributed in reservoirs. With increased development, especially that of the temperature of inclusions in the secondary enlargement part, the formation sequence and stage of quartz secondary enlargement can be judged. For example, the inclusion of weak quartz enlargement was measured to be at 65 ± 5 °C; as the diagenetic temperature rose, the measured temperatures of the enlargement inclusion were 87, 90, 126, and 155 °C. Based on the above result, diagenetic temperature is an important basis for the division of diagenetic stages.

6.4.2.2 Combination of Clay Minerals, and Transformation of Illite/Smectite (I/S) Mixed-Layer Clay Minerals

This is an important basis for the division of diagenetic stages. In Chinese continental clastic rocks, smectite has two evolution modes: the first is that smectite is converted into I/S mixed-layer clay minerals in the presence of a potassium-rich water medium, evolving eventually into illite; and the second is that smectite is converted into a chlorite/smectite (C/S) mixed layer in the presence of magnesium-rich water medium, evolving eventually into chlorite. Of the two evolutions, the former is more common in continental lacustrine basins. The emergence of the C/S mixed layer has certain implications for the properties of the water medium. In general, the degree of mineralization is higher in arid climates or in formation water, and the C/S mixed layer is distributed in the reservoirs with an alkaline water medium.

6.4.2.3 Rock Structure, Tectonic Characteristics, and Pore Types

The tectonic characteristics of rocks, especially cementing mode, generation phenomenon, and cementation type, are used mainly for evaluation. In addition, the evolutionary characteristics of pores on the section can better reflect the lithogenetic evolution stage because pore evolution is a result of diagenetic evolution. In the early diagenesis stage A, based on primary pores, there are essentially no secondary pores; in the early diagenesis stage B, secondary pores begin to appear but are still based on the primary pores, so they belong to a mixed pore development belt; in the middle diagenesis stage A, a great majority of secondary pores develop to form the secondary pore belt; in the middle diagenesis stage B, there are a few secondary pores and cracks; and in the late diagenesis stage, the pores basically disappear, and reservoir space is based on cracks.

6.4.2.4 Maturity of Organic Matter

Because it is a function of time and temperature, the maturity of organic matter is a main geochemical index for dividing diagenetic stages. These stages are usually divided by applying vitrinite reflectance, pollen color, thermal alteration index, and maximum pyrolysis peak temperature. The maturity of organic matter can be divided into immature, semi-mature, low-maturity, mature, highly mature, and excessively mature, which correspond respectively to the six stages for the evolution of smectite into illite via the I/S mixed layer.

6.4.2.5 Paleotemperature

This refers to the homogeneous temperature of fluid inclusions, formation temperature of authigenic minerals, and the evolution of I/S mixed-layer clay minerals.

6.4.3 Marks of Various Diagenetic Stages

The generalized diagenesis can be divided into syndiagenetic stage, diagenetic stage, and epidiagenetic stage. The porosity evolution of a reservoir mainly involves the diagenetic stage, which can be divided further into the early, middle, and late diagenetic stages.

6.4.3.1 Main Marks of the Syndiagenetic Stage

(1) Loose rocks (sediments), and developed primary pores;
(2) Glauconites formed mainly in this stage;
(3) Chamosite formation;
(4) Formation of syngenetic concretion;
(5) Microcrystals, mottling micrite, and siderite distributed along the bedding;
(6) Micritic carbonates distributed among particles and on particle surfaces, shown sometimes by fibriform and microparticle calcites.
(7) Crescent-shaped and gravity cementation occasionally; and
(8) Authigenic minerals separated from alkaline water medium (salt lake basin) have powdery and strawberry pyrite, xenomorphic granular analcite, substrate or mottling cemented gypsum, and glauberite, in which the erosion phenomenon of quartz and other silicate minerals can be found.

6.4.3.2 Main Marks of Diagenetic Stages in Various Diagenetic Environments

According to the different properties of sedimentary water media, the diagenetic environment comprises a fresh water-brackish water medium, acidic water medium (coal-containing stratum), and alkaline water medium (salt lake). As a result, the formed clastic rocks have common rules in terms of diagenetic features and diagenetic stage division, but have their respective particularity. Now, the partition is made hereunder.

1. **Main marks of diagenetic stages in freshwater-brackish water medium environment**

Main marks include petromineralogy, paleogeotemperature, and maturity of organic matter, as in Table 6.19.

1.1 Early diagenetic stage

It can be divided into stages A and B.

(1) Early diagenetic stage A:

① Normal paleotemperature is about 65 °C;
② Organic matter is not mature, of which the vitrinite reflectance R_o is greater than 0.35%, maximum pyrolysis peak temperature, T_{max}, is less than 430 °C, pollen-color is light yellow, and the maximum thermal alteration index, TAI, is less than 2.0;
③ Rock is weakly solidified to semi-solidified, with primary intergranular pores;
④ Mudstone is rich in smectite and more than 70% of the I/S mixed-layer clay minerals (degree of order R = 0) is smectite; the belt is also called smectite belt; and
⑤ Quartz enlargement and low feldspar dissolution usually cannot be found in sandstones, but early carbonate cementation (cellular, drusy and particular) and chlorite rims can be found. Furthermore, smectite, disordered mixed-layer clay minerals, and a small amount of authigenic kaolinite can be found in the clay minerals.

(2) Early diagenetic stage B:

① Paleotemperature range is 65–85 °C;
② Organic matter is not matured, of which the vitrinite reflectance R_o is 0.35–0.5%, maximum pyrolysis peak temperature, T_{max}, is 430–435 °C; pollen-color is deep yellow, and maximum thermal alteration index, TAI, is 2.0–2.5;
③ Owing to compaction and cementation of carbonates, the rock is semi-solidified to solidified, pores are mostly of the primary type, with a few secondary pores;
④ Smectite in mudstone is apparently transformed into I/S mixed-layer clay minerals, in which the smectite layer occupies 50–70%, belonging to disordered mixed-layer (degree of order R = 0) and called disordered mixed-layer belt;
⑤ Level I quartz secondary enlargement, narrow enlargement side or automorphic crystal face can be found in sandstones; small quartz crystallites, fragmentary, or connected incomplete crystal faces, and book-shaped authigenic kaolinites can be found using scanning electron microscopy, but smectite can still be found even if some sandstones are affected by volcanic clastic particles.
⑥ Cloudy flint exists in some sandstone matrices; and
⑦ Some mineral metasomatism and transformation phenomena can be found.

Table 6.19 Division marks of clastic diagenetic stage in fresh water-brackish water medium (Division of diagenetic stages in clastic rocks 2003)

Diagenetic stage: Phase	Stage	Paleotemperature (°C)	Organic Matters: R_o (%)	T_{max} (°C)	Pollen-color	Mature stage	Hydrocarbon evolution	Mudstone: S (%) in I/S	I/S mixed-layer zonation	Sandstone cementation degree	Authigenic Minerals in Sandstone / Dissolution	Particle Contact Type	Pore Types
Syndiagenetic stage		Paleotemperature	① Formation of glauconite and chamosite; ② formation of syngenetic nodule; ③ siderite and mottling microcrystals distributed on parallel layers; ④ micritic carbonates distributed among particles and on particle surfaces; and ⑤ immature hydrocarbons										
Early diagenetic stage	A	Paleotemperature 65	< 0.35	< 430	Light yellow<2.0	Immature	Biogas	>70	Smectite zone	weak-solidification to semi-solidification	(see chart)	Punctiform	Based on primary pores
	B	65–80	0.35–0.5	430–435	Deep yellow 2.0–2.5,	Semi-mature		50–70	Disordered mixed layer	Semi-solidification to solidification			Primary pores and a small amount of secondary pores
Middle diagenetic stage	A	85–140	0.5–1.3	435–460	Orange yellow | 2.5–3.7	Low mature | mature Mature	Based on crude oil	15–50	Ordered mixed layer	Solidification		Punctiform | linear	Primary pores can be preserved, and secondary pores develop
	B	140–175	1.3–2.0	460–490	Brown black 3.7 4.0	High-mature	Condensate oil|wet gas	<15	Superlattice ordered mixed layer zone			Linear | suture	Pores reduce and fractures emerge
Late diagenetic stage		175–200	2.0–4.0	>490	Black>4.0	Over mature	Dry gas	Disappear	Illite belt				Fracture Developed
Syndiagenetic stage		Paleotemperature or Normal temperature	① Minerals (pyrite, siderite, ankerite, ferrocalcite, mica, chlorite, glauconite, etc.) containing low-valence irons; ② limonite impregnation phenomenon; ③ oxidation film containing high-valence irons on clastic particle surfaces; ④ crescent-shaped carbonate cements; ⑤ seepage fillings; ⑥ epidiagenetic calcareoucalcareous concretion; ⑦ anhydrite gypsification; ⑧ epidiagenetic kaolinite; ⑨ dissolved pores and holes; and ⑩ hydrocarbon oxidation and degradation										

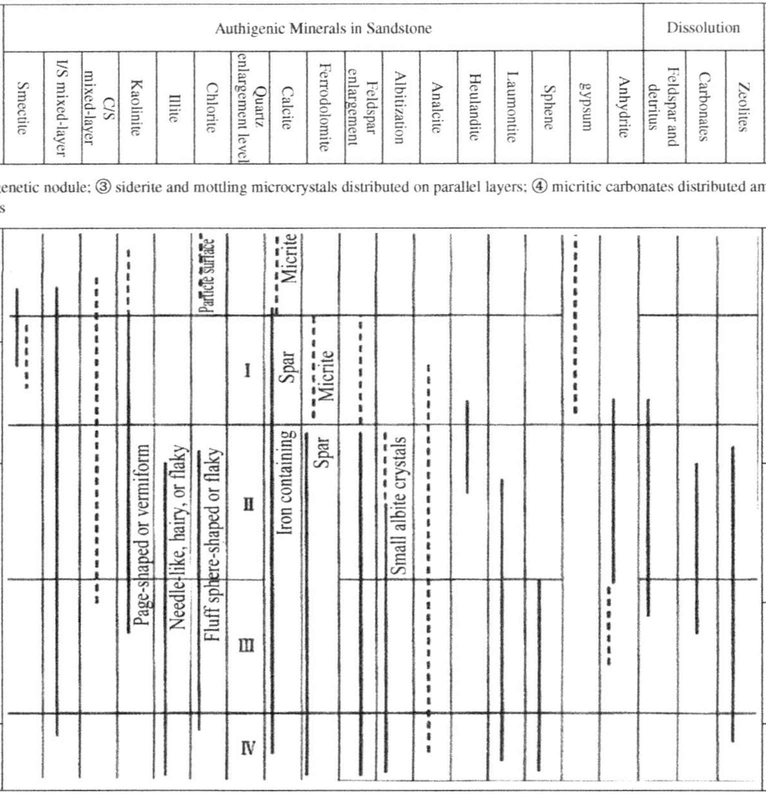

Notes: ① Owing to crust tectonic movement, the epidiagenetic stage is likely to be present or absent anywhere in the early diagenetic stage, middle diagenetic stage, and late diagenetic stage, depending on the specific conditions of all areas; and

②"——" shows a small amount of possibly appeared diagenetic marks.

1.2 Middle diagenetic stage

The middle diagenetic stage can be divided into Stages A and B.

(1) Middle diagenetic stage A:

① Paleotemperature range is 85–140 °C;

② Organic matter is in a low maturity-maturity state, of which the vitrinite reflectance, R_o, is 0.5–1.3%, maximum pyrolysis peak temperature, T_{max}, is 435–460 °C, pollen-color is orange-brown, and maximum thermal alteration index, TAI, is 2.5–3.7;

③ Smectite accounts for 15–50% of the I/S mixed-layer clay minerals in mudstone, 35–50% of which belongs to the partially ordered mixed-layer (R = O/R = I) and 15–35% belongs to the regular mixed-layer (R = 1); in stratum invaded into some igneous rocks or some rocks rich in volcanic clastic materials, the diagenetic stage shall be divided in combination with other indexes, owing to the abnormal transformation and distribution of smectite and the I/S mixed-layer clay minerals in some cases;

④ Late iron-bearing carbonate cements can be found in sandstone. Especially, with further decrease in expandability and commencement of ordering, the iron- and magnesium-rich layers are converted to iUite, and iron and magnesium are released and incorporated in late diagenetic chlorite and carbonates (e.g., ankefite and ferrous dolomite) (Boles and Francks 1979). Ferrodolomite is usually presented as powder crystal-fine grain in the form of metasomatism, enlargement, or cementation; alternatively, other authigenic minerals such as albite, laumontite, heulandites, and analcite can be found;

⑤ Secondary quartz enlargement belongs to level II. Most quartz particles and parts of feldspar particles have secondary enlargement and authigenic crystal face development, in some of which small quartz crystals can be found. Under a scanning electron microscope, most quartz particle surfaces are rather completely covered by authigenic crystal faces, and some authigenic quartz crystals grow up toward porous spaces and are alternately connected to block pores;

⑥ In clay minerals in sandstones, kaolinite, illite/smectite (I/S) mixed-layer clay minerals, silky authigenic illite, flaky or fluff sphere-like chlorite, and chlorite/smectite (C/S) mixed-layer clay minerals can be found, but smectite basically disappears; and

⑦ Feldspar, detritus, and other clastic particles, as well as carbonate cements, are often dissolved. Among the pores, in addition to the preserved primary ones, secondary pores are dominant.

According to the evolution features of I/S mixed-layer clay minerals and organic matter, the middle diagenetic stage A can be subdivided into two sub-stages, A_1 and A_2, bounded by smectite accounting for 35%, vitrinite reflectance, R_o, is 0.7%, and the maximum pyrolysis peak temperature, T_{max}, is 440 °C. In sub-stage A_1, organic matter is in the low-maturity phase, in which the yield of organic acid is high, thus it is a secondary pore producing belt. In sub-stage A_2, organic matter mature to enter the oil generation peak, the concentration of organic acid reduces, and the physical properties are slightly worse than those in sub-stage A_1 owing to the presence of cementation.

(2) Middle Diagenetic Stage B:

① Paleotemperature range is 140–175 °C;
② Organic matter is in the high-maturity phase, of which the vitrinite reflectance, R_o, is 1.3–2.0%, maximum pyrolysis peak temperature, T_{max}, is 460–490 °C, pollen-color is in brownish black, and maximum thermal alteration index, TAI, is 3.7–4.0;
③ There are I/S mixed-layer clay minerals in mudstone, in which the smectite layer accounts for less than 15%, which belongs to the superlattice or the Kalkberg-type ordered mixed layer (degree of order R 3), and is called the superlattice ordered mixed layer belt;
④ Quartz secondary overgrowth is level III; especially, almost all quartz and feldspars in the quartz-rich rocks have enlargement and wide side in the embedded form, in which kaolinites are obviously reduced or are lacking, and some iron-bearing carbonate minerals, namely laumontite, and albitization, can be found; and
⑤ Under an SEM, intergranular quartz idiomorphic crystals can be seen to be connected with each other, and fractures develop among compact rocks.

1.3 Late diagenetic stage

(1) Paleotemperature range is 175–200 °C;
(2) Organic matter is in the excessive-maturity phase, of which the vitrinite reflectance, R_o, is 2.0–4.0%, maximum pyrolysis peak temperature, T_{max}, is greater than 490 °C, pollen-color is black, and maximum thermal alteration index, TAI, is greater than 4.0;
(3) Owing to the presence of extremely compact rocks, particles are in suture contact with suture lines, and there are few pores and developed fractures;
(4) Late carbonate minerals and authigenic minerals, such as albite and titanite, can be found in sandstone, in which the quartz enlargement is of level F, particles are in suture contact, and automorphic crystal faces disappear; and
(5) Sandstone and mudstone include representative clay minerals such as illite and chlorite, in addition to sericite and biotite, where the mixed-layer has basically disappeared; hence, they are called the illite or illite-chlorite belt. According to the crystallinity of illite, the Kübler index (K.I) is 0.25 ($\Delta°2\theta$) < K.I < 0.42 ($\Delta°2\theta$), thus it belongs to the late diagenetic stage.

2. **Main marks of diagenetic stages in acidic water medium (coal-containing lithologies) environment**

These marks include petromineralogy, paleogeotemperature, and maturity of organic matter, as summarized in Table 6.20.

2.1 Early diagenetic stage
The coal-containing stratum rich in aquatic and terrestrial plants easily produces humic acid after burial. The early diagenetic stage has acidic water medium conditions.

Table 6.20 Division marks for clastic diagenetic stage in acidic water medium (Division of diagenetic stages in clastic rocks 2003)

Diagenetic stage: Phase	Diagenetic stage: Stage	Paleotemperature (%)	Organic Matters: R_o (%)	Organic Matters: T_{max}(°C)	Organic Matters: Pollen-color	Organic Matters: Mature stage	Organic Matters: Hydrocarbon evolution	Mudstone: S (%) in I/S	Mudstone: I/S mixed-layer zonation	Sandstone cementation degree	Particle Contact Type	Pore Types
Syndiagenetic stage		Paleotemperature	① Formation of glauconite and chamosite; ② formation of syngenetic concretion; ③ siderite and mottling microcrystals distributed on parallel layers; ④ micritic carbonates distributed among particles and on particle surfaces; and ⑤ immature hydrocarbons									③ Primary pores develop, some of which are compacted;
Early diagenetic stage	A	Paleotemperature 65	<0.35	<430	Light yellow <2.0	Immature	Biogas	>70	Smectite zone	Little solidification to semi-solidification	Punctiform	
	B	65–80	0.35–0.5	430–435	Deep yellow 2.0–2.15	Semi-mature		50–70	Disordered mixed layer	Semi-solidification to solidification		Primary pores and a small amount of secondary pores
Middle diagenetic stage	A	85–140	0.51.3	435–460	Orange yellow-brown 2.5–3.7	Low mature-mature	Based on crude oil	15–50	Ordered mixed layer	Solidification	Punctiform | linear	Intragranular dissolved pore and mold pore development
	B	140–175	1.3–2.0	460–490	Brown black 3.7–4.0	High-mature	Condensate oil|wet gas	<15	Superlattice ordered mixed layer zone		Linear | suture	Pores and a small amount of dissolved holes
Late diagenetic stage		175–200	2.0–4.0	>490	Black>4.0	Over mature	Dry gas	Disappear	Illite belt			Fracture development
Syndiagenetic stage		Paleotemperature or normal temperature	① Minerals (pyrite, siderite, ankerite, ferrocalcite, mica, chlorite, glauconite, etc.) containing low-valence irons; ② limonite impregnation phenomenon; ③ oxidation film containing high-valence iron on clastic particle surfaces; ④ crescent-shaped carbonate cements and gravity cementation; ⑤ seepage fillings; ⑥ epidiagenetic calcareoucalcareous concretion; ⑦ anhydrite gypsification; ⑧ epidiagenetic kaolinite; ⑨ dissolved pores and holes; and ⑩ hydrocarbon oxidation and degradation									

Authigenic Mineral in Sandstone: Smectite, I/S mixed-layer, Illite, Chlorite, Kaolinite, Quartz enlargement, Calcite, Siderite, Ferrodolomite, Feldspar enlargement, Albitization, Barite, Laumontite, Sphene, Anhydrite; Dissolution: Feldspar and detritus, Carbonates

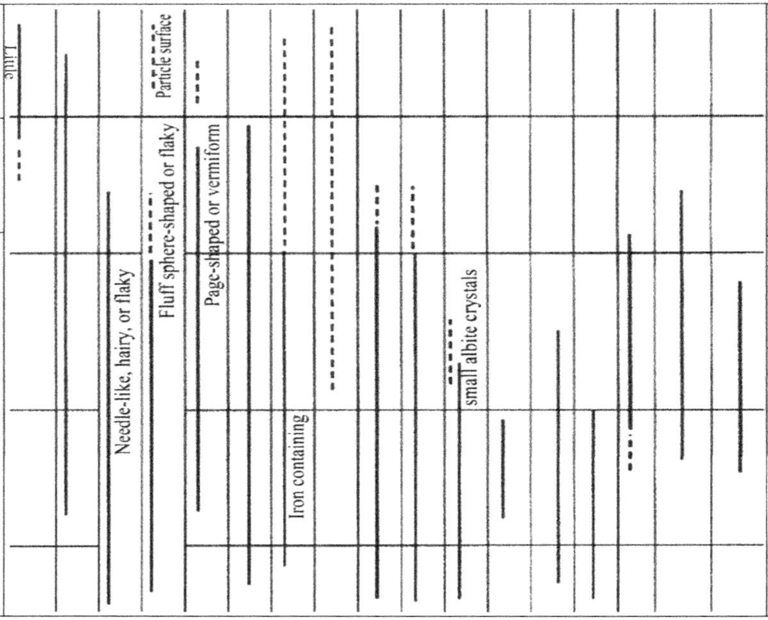

Notes: ① Owing to crust tectonic movement, the epidiagenetic stage is likely to be present or absent anywhere in the early diagenetic stage, middle diagenetic stage, and late diagenetic stage, depending on the specific conditions of all areas;
②"—·—" shows a small amount of possibly appeared diagenetic marks.

(1) Early diagenetic stage A:

① Paleogeotemperature, maturity of organic matter, and I/S mixed-layer evolution indexes are the same as those in the early diagenetic stage A (Table 6.19) in fresh water-brackish water medium environment;
② Rock is weakly solidified to semi-solidified, and primary intergranular pores are developed; and
③ In sandstone, authigenic minerals are not developed, a small amount of calcite or siderite can be found partially, and a few chlorite films can be found around particles.

(2) Early diagenetic stage B:

① Indexes such as paleogeotemperature, maturity of organic matter, and I/S mixed-layer evolution are the same as those in the early diagenetic stage B (Table 6.19) in fresh water-brackish water medium environment;
② Owing to the lack of early carbonate cements, particles are in point-linear contact with strong compaction, and primary pores are apparently reduced due to compaction;
③ A few cements can be found in sandstone, with a small amount of partial early calcite; clay minerals are based on the I/S disordered mixed-layer, with a small amount of chlorite and illite; and smectite can be found in rocks rich in volcanic debris;
④ At the end of the early diagenetic stage B, early quartz enlargement appears, with some crystals having an obvious enlarged side. Hence, particles can be observed to be in linear contact under plane-polarized light, autogenetic kaolinite is relatively developed, and a small number of intragranular dissolved pores and mold pores can be found; and
⑤ Some mineral metasomatism and transformation phenomena can be found.

2.2 Middle diagenetic stage

(1) Middle diagenetic stage A:

① Indexes such as paleogeotemperature, maturity of organic matter, and I/S mixed-layer evolution are the same as those in the early diagenetic stage A (Table 6.19) in fresh water-brackish water medium environment;
② In sandstones rich in quartz and feldspar, the authigenic mineral combination is characterized by quartz enlargement and authigenic kaolinite development, but their degrees of development are related to the contents of quartz, feldspar particles, and interstitial materials. Quartz secondary enlargement does not develop in volcanic debris–rich sandstone containing a few quartz particles. In addition, phenomena such as feldspar enlargement, authigenic albite, calcite, siderite, laumontite, anhydrite, illite, chlorite, I/S mixed-layer clay minerals, and quartz particle crack self-healing and transformation from kaolinite to chlorite can be found;
③ At the end of the middle diagenetic stage A, the water medium is transformed from acidic into alkaline due to PH value change by occurrence of alkaline minerals, which results in the reduction of porosity by the cementation and

metasomatism of late carbonates such as ferrocalcite and ferrodolomite;

④ Particles are mainly in linear contact, and slightly in concave-convex contact; and

⑤ In addition to the dissolution of carbonate parts, on the basis of the dissolution of feldspar and volcanic detritus particles, secondary pores such as intragranular dissolved pores and mold pores are formed, and rocks have the features of large apertures and narrow throats. Moreover, fractures can be found as well.

By reference to the middle diagenetic stage A of fresh water-brackish water medium clastic rocks, the middle diagenetic stage A of acidic water medium clastic rocks can be divided into substages A_1 and A_2.

(2) Middle Diagenetic Stage B:

① Indexes such as paleogeotemperature, maturity of organic matter, and I/S mixed-layer evolution are the same as those in the early diagenetic stage B (Table 6.19) in fresh water-brackish water medium environment;

② Authigenic minerals in sandstone are characterized by the development of ferrocalcite and ferrodolomite; based on metasomatism, quartz enlargement can reach up to level III, and feldspar enlargement, sphene, anhydrite, and barite can be found in authigenic minerals;

③ The main clay minerals are formed by reduction of the contents of kaolinite and I/S mixed-layer clay minerals in sandstone and an increase in the illite, chlorite, and dickite contents;

④ Pores mainly comprise fractures and a small amount of dissolved pores; and

⑤ Particles are either in linear-concave-convex contact or suture contact.

2.3 Late diagenetic stage

(1) Indexes such as paleogeotemperature, maturity of organic matter, and I/S mixed-layer evolution are the same as those in the late diagenetic stage (Table 6.19) in fresh water-brackish water medium environment;
(2) Authigenic minerals in sandstone comprise ankerite, quartz enlargement (reach level IV), and a small amount of sphene; some clay minerals are subjected to siderite metasomatism or illitization, in which chlorite, illite, and biotite are extruded and deformed;
(3) Pores mainly comprise fractures and a small amount of feldspathic lithic dissolved pores; and
(4) Particles are in linear-suture contact, in which the fracture and healing of some quartz particles can be found.

3. **Main marks of diagenetic stages in alkaline water medium (salt lake basin) environment**

The main marks include petromineralogy, paleogeotemperature, and maturity of organic matter, as summarized in Table 6.21.

Three hydrochemical types can be found in the salt lake basins, namely, carbonate, sulfate, and chloride, in which carbonate and sulfate are in the majority. Owing to the differences in water

Table 6.21 Division marks for clastic diagenetic stage of alkaline water medium (Division of diagenetic stages in clastic rocks 2003)

Diagenetic stage			Organic Matters					Authigenic Mineral in Sandstone														Dissolution						
Phase	Stage	Paleotemperature (°C)	(%)	Pollen-color TAI	Mature stage	Hydrocarbon evolution	Sandstone cementation degree	gypsum	Anhydrite	Glauberite	Calcite	Ferroan calcite	Ferruginous dolomite	Dolomite	Ferrodolomite	Siderite	Analcite	Feldspar enlargement	Albitization	Quartz enlargement level	Illite	Quartz	Feldspar and detritus	Carbonate	Analcite	Salts	Particle Contact Type	Pore Types
Syndiagenetic stage		Paleotemperature	① Formation of glauconite and chamosite; ② formation of syngenetic concretion; ③ micritic carbonates (calcite, dolomite, siderite) and mottling microcrystals distributed on parallel layers; ④ micritic carbonates distributed among particles and on particle surfaces; ⑤ erosion of quartz and other silicates; ⑥ formation of matrix-cemented gypsum and glauberite; and ⑦ immature hydrocarbons																									
Early diagenetic stage	A	Paleotemperature 65	< 0.35	Light yellow<2.0	Immature	Biogas	weak-solidification to semi-solidification				Micrite	Micrite	Micrite	Micrite		Micrite											Punctiform	Primary pores and a small amount of secondary pores develop
	B	65–85	0.35–0.5	Yellow 2.0–2.5	Immature		Semi-solidification to solidification				Spar				Micrite					Not developed							Punctiform Main	Secondary pores develop, and primary and secondary pores coexist
Middle diagenetic stage	A	85–140	0.5–1.3	Orange brown | 2.5–3.7	Low mature | mature	Based on crude oil	Solidification						Spar	Spar						I, II							Linear main	Secondary pores develop and fractures appear
	B	140–175	1.3–2.0	Brownish black 3.7–4.0	High-mature	Oil-wet gas condensation														III							Based on concave-convex | suture line	Secondary pores reduce and fractures develop
Late diagenetic stage		175–200	2.0–4.0	Black >4.0	Over mature	Dry gas														IV								Fracture development
Syndiagenetic stage		Paleotemperature or normal temperature	① Minerals (pyrite, siderite, ankerite, ferrocalcite, mica, chlorite, glauconite, etc.) containing low-valence irons; ② limonite impregnation phenomenon; ③ oxidation film containing high-valence iron on clastic particle surfaces; ④ crescent-shaped carbonate cements and gravity cementation; ⑤ seepage fillings; ⑥ epidiagenetic calcareoucalcareous concretion; ⑦ anhydrite gypsification, gypsum separated from glauberite; ⑧ epidiagenetic kaolinite; ⑨ erosion phenomenon; and ⑩ hydrocarbon oxidation and degradation																									

Notes: ① Owing to crust tectonic movement, the epidiagenetic stage is likely to be present or absent anywhere in the early diagenetic stage, middle diagenetic stage, and late diagenetic stage, depending on the specific conditions of all areas; and

②"- - -" shows a small amount or possibly appeared diagenetic marks.

hydrochemistry, the separated salt minerals are also different in different salt lake basins. The precipitation sequence is calcite, dolomite-gypsum belt, halite belt, sodium sulfate and magnesium belt, and sylvite belt.

3.1 Early diagenetic stage

(1) Early diagenetic stage A:

① Indexes including paleogeotemperature and maturity of organic matter are the same as those in the early diagenetic stage A (Table 6.19) in fresh water-brackish water medium environment;
② Particles are mainly subject to point contact;
③ Primary pores are developed;
④ Authigenic minerals include granular analcite and micritic carbonate, without secondary quartz enlargement; when the paleogeotemperature is lower than 42 °C, gypsum and glauberite are separated and presented as a matrix of cemented clastic particles. When the paleogeotemperature is higher than 42 °C, gypsum is transformed to anhydrite, and anhydrite and glauberite are separated; ferrocalcite and ferruginous dolomite contained in the micrite are separated;
⑤ Clay minerals in mudstone are based on the illite-chlorite (I-C) and illite-chlorite-illite/smectite (I-C-I/S) combinations; the I/S mixed-layer mainly comprises an ordered mixed layer and a disordered mixed-layer, in which smectite is rarely found however kaolinite can be found in sandstone; and
⑥ Erosion phenomena for quartz and feldspar can be found.

(2) Early diagenetic stage B:

① Indexes including paleogeotemperature and maturity of organic matter are the same as those in the early diagenetic stage B (Table 6.19) in fresh water-brackish water medium environment;
② Particles are mainly subject to point contact, but partially to linear contact;
③ Secondary pores develop to form a situation wherein primary and secondary pores coexist;
④ Authigenic minerals include sparry calcite, dolomite, ferroan calcite, ferruginous dolomite and micritic ferruginous dolomite, porously cemented anhydrite, and glauberite; quartz secondary enlargement is level I, and the enlargement side is narrow and noncontinuous; automorphic crystal surfaces and parts of feldspar secondary enlargement can be found occasionally;
⑤ Clay minerals in mudstone are based on the illite-chlorite (I-C) and illite-chlorite-illite/smectite (I-C-I/S) combinations, in which kaolinite or smectite is rarely found; the I/S mixed-layer comprises mainly an ordered mixed layer (smectite layer can take up to 20–25%); and
⑥ Feldspar, carbonate, and analcite are corroded.

3.2 Middle diagenetic stage

(1) Middle diagenetic stage A:

① Indexes including paleogeotemperature and maturity of organic matter are the same as those in the middle diagenetic stage A (Table 6.19) in fresh water-brackish water medium environment;

② Particles are mainly subject to point-line contact, but partially to concave-convex contact;

③ A great number of iron carbonate–containing cements are presented in the form of automorphic powder crystal-fine crystals as porous cementation, metasomatism, or enlargement; anhydrite and glauberite are presented as porous cementation or metasomatism, quartz secondary enlargement is of level n, quartz and feldspar show secondary enlargement phenomena, and automorphic crystal surfaces develop. Under a scanning electron microscope, the particle surfaces are completely covered by scanning electron microscopes or automorphic crystals are presented, parts of feldspar have albitization, and analcites are gradually reduced or removed completely;

④ Clay minerals in mudstone are based on the illite-chlorite (I-C) and illite-chlorite-illite/smectite (I-C-I/S) combinations, in which kaolinite can be found occasionally; the I/S mixed-layer mainly comprises an ordered mixed layer (smectite layer is less than 20%); and

⑤ Feldspar and other clastic particles and carbonate are often dissolved, secondary pores develop, and dissolution fractures begin to appear at the end of the period.

With reference to the middle diagenetic stage A of fresh water-brackish water medium clastic rocks, this stage can be divided into substages A_1 and A_2.

(2) Middle Diagenetic Stage B:

① Indexes including paleogeotemperature and maturity of organic matter are the same as those in the middle diagenetic stage B (Table 6.19) in fresh water-brackish water medium environment;

② Due to compact rocks and advanced fracture development, particles are mainly subjected to concave-convex contact and suture contact, while some are subjected to line contact;

③ A great number of iron carbonates and anhydrites are presented as metasomatism or as pore fillings among particles; quartz secondary enlargement is of level I, most quartz and feldspar have secondary enlargement, enlargement side is narrow and continuous, and the automorphic crystal surface of quartz is developed. Under a scanning electron microscope, automorphic crystals are connected mutually, and most of the feldspar has albitization; and

④ Clay minerals in mudstone are of the illite-chlorite combination.

3.3 Late diagenetic stage

(1) Indexes including paleogeotemperature and maturity of organic matter are the same as those in the late diagenetic stage (Table 6.19) in fresh water-brackish water medium environment;

(2) Due to extremely compact rocks, fractures develop and suture lines appear;

(3) Authigenic minerals such as albite and titanite can be found in sandstone, in which quartz secondary enlargement is of level IV, particles are in suture contact,

automorphic crystal faces disappear, and albitization can be found generally; and

(4) Sandstone and mudstone include representative clay minerals such as illite and chlorite, and further include sericite and biotite.

6.4.3.3 Main Marks of Syndiagenetic Stage

(1) Minerals (pyrite and siderite) containing low-valence irons are ferritized or impregnated with limonite;
(2) Oxidation film is present on the surfaces of clastic particles;
(3) Crescent-shaped carbonate cementation and gravity cementation;
(4) Seepage filling;
(5) Supergene caliche nodule;
(6) Anhydrite gypsification;
(7) Supergene kaolinite;
(8) Due to corrosion, dissolved pores and karst caves result in the development of secondary pores under planes of unconformity to improve physical properties; and
(9) Development of faults and fractures provides channels for downward seepage of surface water and convection between deep formation water and surface water; meanwhile, secondary pores are formed.

6.4.4 Study of Diagenetic Sequence

Based on the division of diagenetic stages, there are certain differences among different diagenetic division marks under different water medium conditions. According to the marks, such as paleogeotemperature, R_o, T_{max}, pollen color, TAI, maturity stage, hydrocarbon evolution, S% in I/S, I/S mixed-layer zonation, and sandstone solidification degree under different water medium conditions, the division standards of all diagenetic stages are consistent. According to particle contact types and pore types under different aqueous medium conditions, the division standards of the diagenetic stage are different.

While studying the diagenetic characteristics of sandstones in the 21–13.8 Ma layers of (Pearl River Mouth Basin) the Panyu Gas Field, the author made detailed partitions (Fig. 6.17) to the diagenetic stage in this area by analyzing core data. The author found three secondary pore

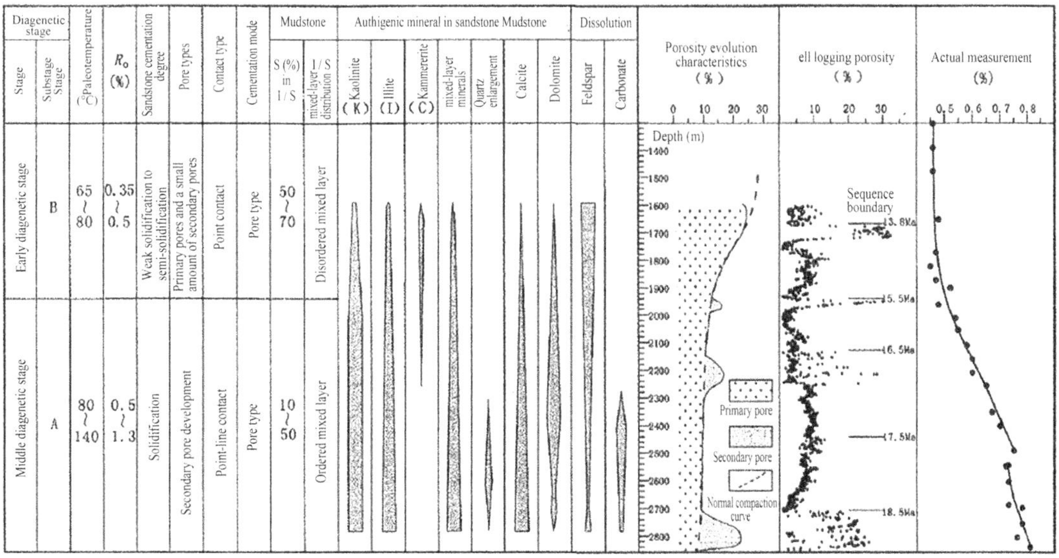

Fig. 6.17 Diagenesis sequence of Panyu gas field (according to Yu Xinghe et al. 2002)

development belts, in which the formation of secondary pores in the 16.5 and 16.5 Ma layers is based on an open weakly alkaline water cycle, and the formation of secondary pores in the 18.5 Ma layer is based on a closed acidic water cycle.

6.4.5 Establishment of Diagenetic Evolution Model in Lacustrine Basin

According to the Dividing Standard for Diagenetic Stage of Clastic Rocks (1990) issued by the former CNPC, Zhao Chenglin et al., diagenetic stages can be divided into the syndiagenetic stage, early diagenetic stage, and late diagenetic stage in combination with the specific conditions of the Dongpu Depression. The early diagenetic stage can also be divided into substages A and B, and the late diagenetic stage into substages A, B, and C. Furthermore, a diagenetic evolution model (Fig. 6.18) is established for the study area by considering the influences of tectonic evolution and sedimentary characteristics of all the stages of diagenesis.

Taking the Houwujiahu Gas Field in the Lishu Depression in the southern Songliao Basin, for example, the author started by analyzing tectonic evolution in the basin and the depositional system, and established an integrated mode of structure, sedimentation, and diagenesis. Moreover, detailed analysis (Fig. 6.19) on reservoir development, genesis, and distribution in this area is given, and some discussions based on the five stages of tectonic evolution are presented.

(1) The third sedimentary period of the Denglouku Formation is when the Shiwu fault depression was located in a basin tension–faulted period, a fan delta with coarse debris was deposited, the north and west edges of the basin were obviously thicker than the east edge, and the southeast was based on shore deposition.
(2) Then, the basin started the transformation from fault to depression, finished tension, started tectonic uplift, and was subjected to partial structural inversion, which led to the top of the Denglouku Formation being denuded. Short discontinuity and filling

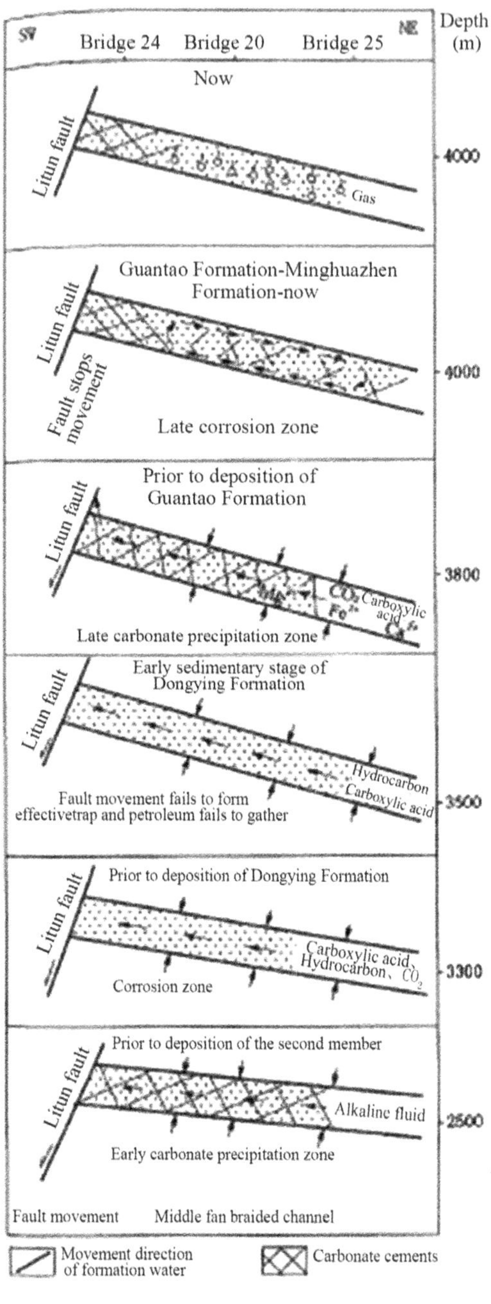

Fig. 6.18 Diagenetic evolution mode of the fourth submember of Shasan member in Qiaokou Region (according to Zhao Chenglin 1992, Yu 1997)

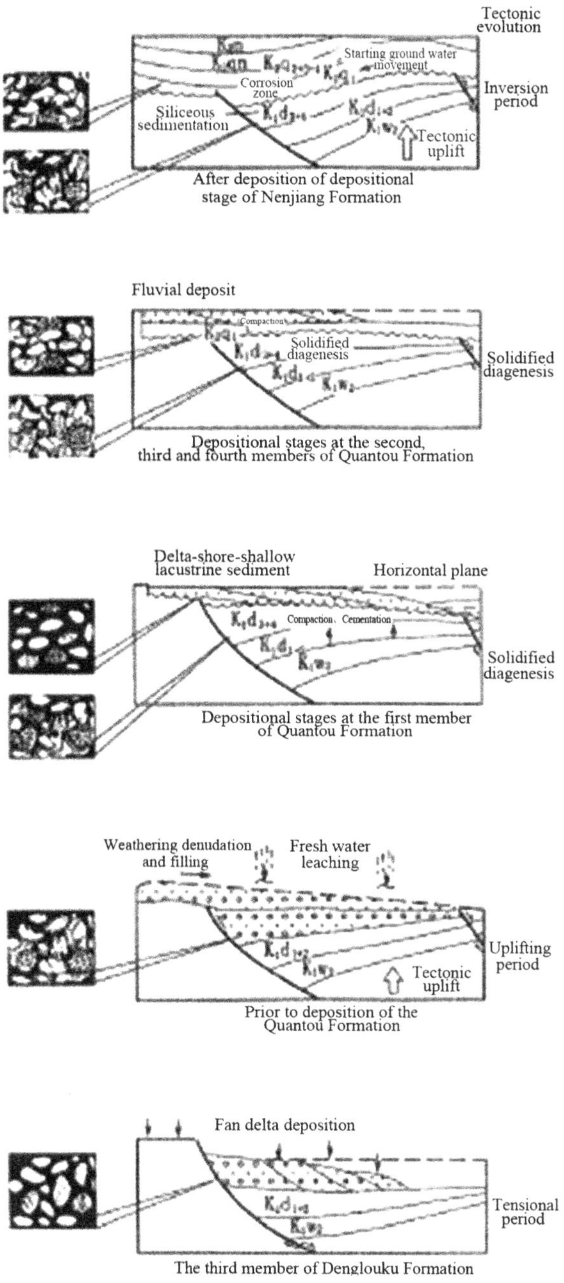

Fig. 6.19 Integrated mode of structure, sedimentation and diagenesis of Houwujiahu gas field

occurred on the basin strata, and the third member of the Denglouku Formation was accompanied by fresh water leaching to result in partial corrosion of feldspar and the production of a few intergranular dissolved holes when mechanical compaction of the early diagenetic stage was initialized.

(3) When the basin was depressed and uplifted, the first section of the Quantou Formation started receiving deposition to form a delta-shore-shallow lake deposit. The upper stratum of the Denglouku Formation was located in a shallow buried stage, and compaction and cementation were further

deepened, water medium was transformed from acidic to neutral-alkaline, of which the pH value was approximately equal to 7. When a large number of pores were reduced, the transformation of clay minerals occurred on the Denglouku Formation, and calcite cementation was formed on a large scale.

(4) Upon the deposition of the first section of the Quantou Formation, the basin was still in the depression stage, and the second, third, and fourth sections of the Quantou Formation were mainly made of fluvial deposition. By this time, the Denglouku Formation had entered the late diagenetic stage, siliceous cementation had started, pressure was raised further as the overlying strata thickened and buried depth increased continuously, and part of the calcite was dissolved. However, when the first section of the Quantou Formation was located in the mechanical compaction and cementation periods of the early diagenetic stage, a large number of pores were reduced, and calcite cementation and clay mineral transformation were presented at the same time.

(5) When the Nenjiang Formation was deposited, the basin started structural inversion to lead massive denudation of the overlying strata and a greater change in formation pressure. Hence, the compaction borne by the Quantou formation was far weaker than the borne by the Denglouku Formation. The change in fluid potential energy in the study section resulted from this tectonization, and an abnormal low pressure area was formed. Greater pressure change was produced between the planes of unconformity of the Quantou Formation/Denglouku Formation, which caused the flow of open acidic water in the first section of the Quantou Formation, hence the main development belts of the secondary pores were formed near the planes of unconformity. Material exchange above the plane of unconformity was far greater than that below the plane of unconformity, thus dissolution of the first section of the Quantou Formation was stronger.

References

Boles, J.R., and S.G. Franks. 1979. Clay diagenesis in Wilcox Sandstones of SW Texas: Implications of smectite diagenesis on sandstone cementation. *Journal of Sedimentary Petrology* 49 (1): 55–70.

Chilingarian, G.V., R.W. Mannon and H. Rieke. 1972. *Oil and gas production from carbonate rocks*. Amsterdam: Elsevier Publishing Company, 408 p.

Chen, Lihua, and Jiang Zaixing. 1993. *Reservoir experimental testing techniques*, 220–232. Beijing: Petroleum Industry Press.

Coustau, H. 1977. Formation waters and hydrodynamics. *Journal of Geochemical Exploration* 7: 213–241.

Crossey L.J., Frost B.R. and Surdam R.C. 1984. Secondary porosity in laumontite bearing sandstones, clastic diagenesis. *AAPG Mem2 oir* 37: 225–238.

Demaison, G., and B.J. Huizinga, 1989. Genetic classification of petroleum basins: *AAPG Bulletin* (United States), 73: 3.

Foster, J.E. 2004. Observation of quartz cathodeluminescence in a low pressure plasma discharge.

Gengsheng, He. 1994. *Oil reservoir physics*. Beijing: Petroleum Industry Press.

Giles, M.R., and R.B.D. Boer, 1990. Origin and significance of redistributional secondary porosity: Marine & Petroleum Geology, 7: 378–397.

Guohua, Zhu. 1985a. Formation of lomonitic sand bodies with secondary porosity and their relationship with hydrocarbons. *Acta Sedimentologica Sinica* 6 (1): 1–8.

Guohua, Zhu. 1985b. Formation of low permeability sandbodies and secondary pore sandbodies in the upper permian Yanchang series of South-western Shan-Gan-Ning Basin. *Acta Sedimentologica Sinica* 3(2).

Guohua, Zhu. 1992. Origin and evolution and prediction of porosity in clastic reservoir rocks. *Acta Sedimentologica Sinica* 10 (3): 114–132.

Hayes J B. 1979. Sandstone diagenesis; the hole truth (in aspects of diagenesis). Special Publication—Society of Economic Paleontologists and Mineralogists (26): 127–139.

He, J.Y. and Meng, X.H. 1987. Sediments and Sedimentary Facies Model and Construction. Beijing: Geological Publishing House.

Hoffman, J., and J. Hower. 1979. Clay mineral assemblages as low grade metamorphic indicators: Application to the trust-faulted disturbed belt of Montana U. S. A. SEPM. Special Publication No 26, 55–80.

Houseknecht, D.W. 1987. Assessing the relative importance of compaction processes and cementation to reduction of porosity in sandstones. *AAPG Bulletin* 71: 633–642.

Hunt, J.M. 1977. Distribution of carbons as hydrocarbons and asphalic compounds in sedimentary rocks. *AAPG* 61(1).

Hutcheon I., Abercrombie H. 1990. Carbon dioxide in clastic rocks and silicate hydrolysis. Geology 18: 541–544.

Junmao, Zheng, and Pang Ming. 1989. *Research on diagenesis of clastic reservoirs*. Wuhan: China University of Geosciences Press.

Kharaka, Y.K., D.J. Specht, and W.W. Carothers, 1985. Low to intermediate subsurface temperatures calculated by chemical geothermometers: *AAPG Bulletin* (United States), 69: 2.

Krynine P.D. 1949. Mineralogical research on oil reservoirs. *Mineral Industries (University Park)* 18(4): 1–3.

Liangtian, Sun, et al. 1992. *Reservoir physics experiment*. Beijing: Petroleum Industry Press.

MacGowan D., Surdam R.C. 1988. Difuetional carboxylic acid anions in oil-field water. *Organic Geochemistry* 12: 245–259.

Schmidt V., McDonald D.A. 1979. Texture and recognition of secondary porosty in sandstones. SEPM.SPEC. Pub.No26, 209–225.

Schmidt V., McDonald D.A. 1979. The role of secondary porosity in the course of sandstone diagenesis (in Aspects of diagenesis). *Special Publication - Society of Economic Paleontologists and Mineralogists* (March 1979), 26: 175–207

Surdam, R.C., S.W. Boese, and L.J. Crossey. 1984. The chemistry of secondary porosity (in Clastic diagenesis). *AAPG Memoir* 37: 127–149.

Surdam R.C., Crossey L.J., et al. 1989. Organic-inorganic and sandstone diagenesis. *AAPG Bulletin* 73: 1–23.

Surdam R.C., Wolfbauer C.A. 1975. Green River Formation, Wyoming; a playa-lake complex, *Bulletin of the Geological Society of America* 86(3): 335–345.

Wilson, M.D., and E.D. Pittman. 1977. Authigenic clays in sandstones: Recognition and influence on reservoir properties and paleo environmental analysis: *Journal of Sedimentary Research*, 47, 3–31.

Wu, S.H. and Xiong, Q.H. 1998. Oil and gas reservoir geology. Beijing: Petroleum Industry Press.

Wu, Y.Y. et al. 1996. Hydrocarbon Reservoir Geology. Beijing: Petroleum Industry Press.

Ying, F.X. and Wang, Y.Q. 1990. Elemental composition of minerals and cathodoluminescence. *Journal of Chinese Electron Microscopy Society*.(3): 244–244.

Yongshang, Kang, and Wang Jie. 1999. Fluid dynamic systems and the formation of oil and gas pools. *Acta Petrolei Sinica* 20 (1): 30–33.

Yu, X.H. 2002. Clastic Hydrocarbon Reservoir Sedimentology. Beijing: Petroleum Industry Press.

Yu Xinghe, Junmao Zheng et al. 1997. The establishment of integrated model on structure, deposition and diagenesis. *Acta Sedimentologica Sinica* 15(3).

Yu Xinghe, Junmao Zheng, Liheng Song et al. 1997. Faulted basin delta sandbody sedimentation and reservoir intrastratal heterogeneity. *Earth Sciences* 22 (1).

Yu, Xinghe, Daojian Zhang et al. 1999. Depositional facies model of deep shahejie formation in eastern and western depressions in Liaohe Oilfield. *Journal of Palaeogeography* 1(3).

Zhao, C.L. et al. 1992. Sedimentary System and Diagenesis of Paleogene Clastic Rocks in Dongpu Depression. Beijing: Petroleum Industry Press.

Zheng, J.M. and Pang, M. 1989. Study on diagenesis of clastic reservoir rocks. Wuhan: China University of Geosciences Press.

Zimkernagel, M.U. 1978. *Cathodoluminescence of quartz and its application to sandstone*. Petrology, Stuttgart.

7 Reservoir Heterogeneity

Description and characterization of reservoirs, a direct target stratum for petroleum exploration and development, are at the center of petroleum reservoir research, whose core lies in the research on reservoir heterogeneity. This is because heterogeneity in reservoir characteristics is closely related to oil and gas reserves, stored energy, yield, and productivity. The first technical problem to be solved is accurately recognizing the various characteristics of petroleum reservoirs during their development. Only through scientific, systematic, and quantitative studies of the heterogeneous characteristics of reservoirs, the benefits of petroleum exploration and development can be improved, the position of development wells selected optimally, and a scheme for improving the hydrocarbon recovery ratio be designed reasonably. In other words, the study of the heterogeneous characteristics of reservoirs constitutes the basis for oilfield exploitation and development schemes while providing an important geological basis for the evaluation of oil reservoirs, discovering productivity potential, and predicting the final recovery ratio.

7.1 Concept and Influencing Factors

Reservoir heterogeneity is absolute, unconditional, and infinite; however, homogeneity is relative, conditional, and finite. Only under certain conditions can a reservoir be approximately regarded as homogeneous within a limited range. The degree of heterogeneity of a marine reservoir is lower than that of a continental reservoir. In China, 90% of its current oil and gas reserves originate from continental sedimentary reservoirs, most of which are subject to waterflooding development. Hence, the study of the level of reservoir heterogeneity will directly affect our understanding of oil, gas, and water distribution laws in reservoirs and development effects.

7.1.1 Concept

In the long history of geology, owing to the combined influence of sedimentation, diagenesis, and late tectonic movement on petroleum reservoirs, the spatial distribution and internal attributes of petroleum reservoirs vary and are distributed unevenly. Regardless of whether we discuss clastic, carbonate, conventional, or special reservoirs, the lithology, physical properties, oil-bearing properties, and electric properties in a three-dimensional (3D) space are often variable, which refers to heterogeneity. Heterogeneity greatly influences the exploitation and development of oil and gas fields, particularly in terms of the movement of underground oil, gas, and water, as well as EOR.

In general, reservoir heterogeneity refers to the uniformity of spatial distribution (anisotropies) and various internal properties (physical properties) of petroleum reservoirs.

X. Yu et al., *Clastic Hydrocarbon Reservoir Sedimentology*, Advances in Oil and Gas Exploration & Production, https://doi.org/10.1007/978-3-319-70335-0_7

The former determines the total petroleum reserves, distribution laws, and well spacing locations for exploration and development, whereas the latter controls the amount of recoverable petroleum reserves, injection-production patterns (such as sweep efficiency), productivity, and distribution of the remaining oil.

In terms of reservoir modeling, the results from the research of the former help build skeleton models, whereas that of the latter help establish parametric models.

In a narrow sense, reservoir heterogeneity involves the nonuniformity of various attributes (lithology, physical properties, oil-bearing properties, and electric properties) of a petroleum reservoir in 3D space.

7.1.2 Main Influencing Factors

The various complex factors affecting reservoir heterogeneity are summarized below (Fig. 7.1).

7.1.2.1 Structural Factors

The effect of structural factors on reservoir heterogeneity depends mainly on tectonism, owing to which faults and fractures are transformed and superimposed on the primary reservoir framework, thus resulting in a fluid block or channel formation.

Fractures usually change the seepage direction and capacity of a reservoir, which leads to considerable differences in permeability in longitudinal, transverse, and vertical 3D spaces. Tectonic movements at different times have different characteristics and properties, which determine the formation and distribution of reservoir fractures, and thus ultimately affect reservoir heterogeneity.

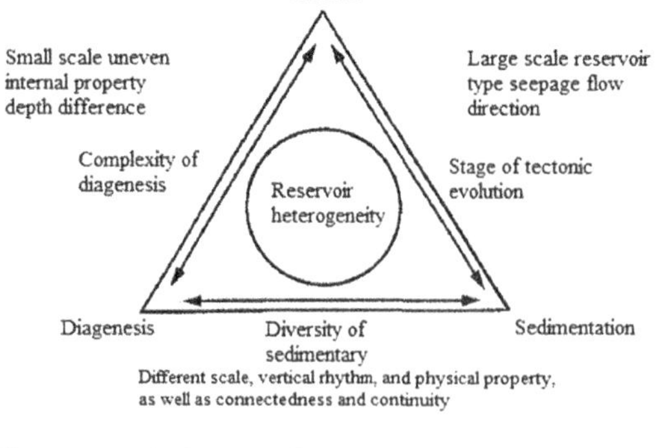

Fig. 7.1 Main factors affecting heterogeneity of oil and gas reservoirs

7.1.2.2 Sedimentation Factors

Sedimentation factors, which are mainly determined by sedimentation or processing, impact the architecture or configuration (spatial form and internal composition of original skeleton and sand body) of a reservoir.

Diverse sedimentary conditions (such as flow intensity and direction, abrupt and gentle paleotopography in sedimentary area, water depth and potential in basin, and supply of debris) result in various types of sediment particles, arrangement directions, bedding configurations, and spatial geometrical morphologies in the sand body. In other words, different distributions of sand bodies in diverse sedimentary facies cause the internal physical properties of sedimentary sand bodies to be different, which results in different levels of reservoir heterogeneity.

7.1.2.3 Diagenetic Factors

Diagenetic factors, which are determined by the characteristics of rock minerals and underground fluids in a reservoir, can lead to the transformation of clay minerals, cementation, corrosion, and leaching, as well as improvement or degradation of basic reservoir properties.

After sediments or sand bodies are deposited, the porosity and permeability of the primary sand body change on account of a series of diagenesis, such as compaction, pressure solution, dissolution, cementation, and recrystallization. Strata at different levels in the basin have different ground temperatures, fluids, pressures, and lithologies, hence their diagenesis is diversified. Moreover, the degree of reservoir heterogeneity is increased owing to the extremely uneven formation and spatial distribution of secondary pores.

The above three points can be generalized as the influences and constraints on heterogeneity due to the tectonic evolution stage, sedimentary pattern diversity, and diagenesis complexity. As for reservoir sedimentology, the latter two are the main factors that affect reservoir heterogeneity.

7.2 Classification of Reservoir Heterogeneity

The structure of a reservoir is incredibly complex, in which the scale of heterogeneity can range from as large as thousands of meters to as small as several meters or as large as several centimeters to as small as several millimeters. According to the intended purpose of research, scholars place extra emphasis on studying the scale, levels, and contents of reservoir heterogeneity. However, holistically, reservoir heterogeneity is mainly classified on the basis of research scale or scope, reservoir genesis or sedimentary boundary, and influence on fluid. Moreover, reservoir heterogeneity aims to convert qualitative descriptions of various reservoir attributes into quantitative indexes for oilfield development in order to better serve the exploration and development of oil and gas fields, particularly reservoir modeling. Reservoir modeling involves the classification, description, and analysis of reservoir heterogeneity.

7.2.1 Pettijohn Classification

In 1973, according to sedimentary genesis and boundary, as well as influence on fluid, Pettijohn Poter and Siever proposed level and classification concepts for the study of reservoir heterogeneity and they built a series spectrum or hierarchical sequence (Fig. 7.2) of heterogeneities with a decreasing trend. This classification is relatively practical because it is carried out based on sedimentary genesis to combine different sedimentary units for genesis research. The correspondence of this classification is as follows:

Level I equivalent to the scale of oil (oil reservoir) formation and the scale of the oil reservoir (1 km × 100 m to 10 km × 100 m);

Level II equivalent to the interlayer scale and the layer scales (100 m × 100 m);

Level III equivalent to the interlayer scale and the scale of the sand body (1–10 m^2);

Level IV equivalent to the core scale and the pore scale (1–10 m^2);

Level V equivalent to the slice scale and laminae scale (10–100 μm^2).

7.2.2 Weber Classification

In 1986, in his design of a qualitative evaluation and development scheme for oil fields, K. J. Weber put forth a more comprehensive classification system based on Pettijohn's classification idea. In the new system, he mainly increased the influences (Fig. 7.3) of structural characteristics, interlayer distribution, and crude oil properties on reservoir heterogeneity. According to the sequence of this classification system, reservoir heterogeneity can be understood quantitatively and studied during oilfield evaluation and development. Different heterogeneity scales have different impacts on oil field evaluation. Large-scale structural systems play a preferential role compared to sedimentary characteristics. The Weber classification can be divided into eight types by scale and genesis.

1. **Sealing and non-sealing faults**

This is a large-scale reservoir heterogeneity property. The sealing degree of a fracture greatly influences fluid seepage in oildom on a large scale. If faults are sealed, fluid seepage between two faults will be cut off; if faults are unsealed, a large seepage channel is formed. Such heterogeneity is mainly in allusion to the sealing of fault-block-type petroleum reservoirs.

2. **Genetic unit boundary**

A genetic unit boundary is essentially a boundary for lithological changes, and is generally a boundary between permeable and impermeable beds or at least a boundary for permeability differences. Thus, it controls fluid seepage on a

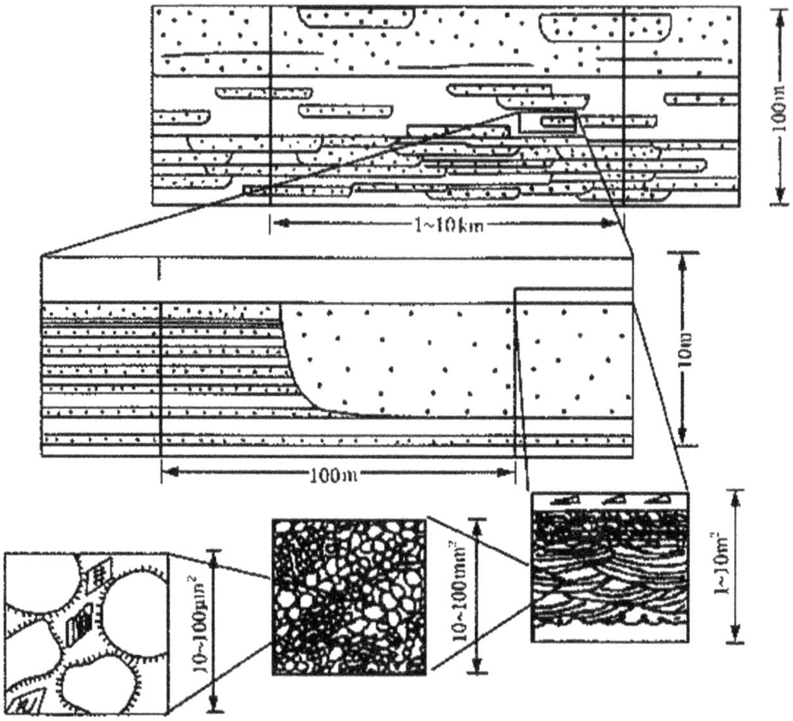

Fig. 7.2 Classification of reservoir heterogeneity of Pettijohn

larger scale. Usually, it refers to the boundary of oil sets or oil reservoirs, which depend on the scale of the genetic unit.

3. **Permeable bed within a genetic unit**

Rock formations with diverse permeability are distributed vertically in a netted manner within a genetic unit, which results in vertical reservoir heterogeneity and directly affects the injection-production process of oilfield development.

4. **Interbed within a genetic unit**

In a genetic unit, interbeds of different scales greatly influence fluid seepage, which not only affects the vertical seepage of fluid but also horizontal seepage, thereby restricting the injection–production position or perforation interval for oilfield development.

5. **Laminae and sets of bedding**

The major difference between sets and laminae within a bedding structure considerably impacts fluid seepage thus affecting the distribution of the remaining oil upon waterflooding development.

6. **Microscopic heterogeneity**

This refers to minimal reservoir heterogeneity, namely, the pore scale reservoir heterogeneity resulting from the differences between the rock structure and mineral characteristics.

7. **Sealing and opening of fractures**

In the case of fractures in a reservoir, the sealing and opening of fractures can lead to reservoir heterogeneity.

8. **Viscosity variation of crude petroleum and tar mat**

This is a specific type. Neither type 7 nor 8 refers to the common heterogeneity in a clastic reservoir. The formation of types 2–4 is affected by the ratio of accommodation to sediment supply (A/S), especially the genetic unit boundary.

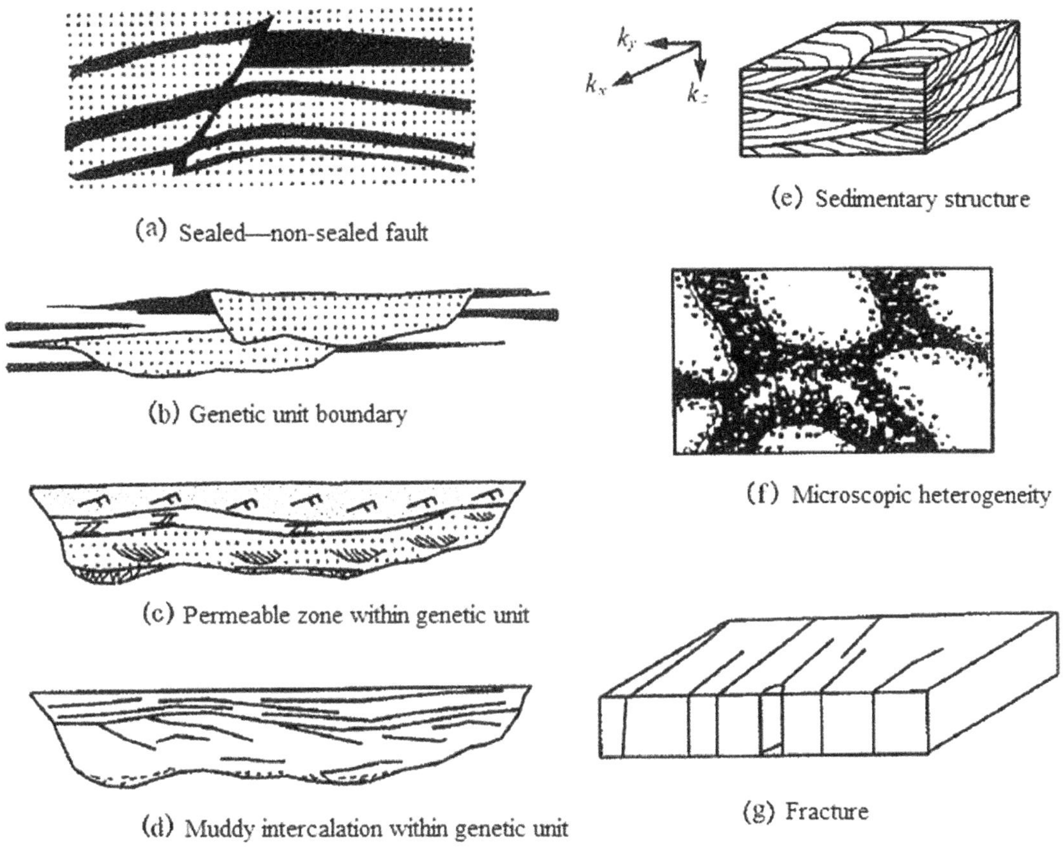

Fig. 7.3 Classification of reservoir heterogeneity of Weber

This classification, which is put forward on the basis of different oil reservoirs, is more comprehensive than the Pettijohn classification. It is stronger in terms of operability and is more convenient to study and use.

7.2.3 Haldorsen Classification

According to the need for geological modeling and the pore characteristics of reservoirs, H. H. Haldorsen (1983) divided reservoir heterogeneity into four types (Fig. 7.4) by level and scale based on volume distribution with respect to pore mean.

(1) Microscopic heterogeneity, i.e., pore and particle size;
(2) Macroscopic heterogeneity, i.e., core scale;
(3) Megascopic heterogeneity, i.e., large net block in a simulation model; and
(4) Gigascopic heterogeneity, i.e., the entire rock formation or regional scale.

In 1987, F. F. Krause et al. extended the characterization and application (Fig. 7.5) of Haldorsen's classification and diagrammatized different levels for various details. In the 2nd International Reservoir Characterization Conference held in 1989, due to the fact that the classification hierarchy was closer to the practical oil reservoir division, four technical topic sessions were named for the classification titles, which were then adopted extensively worldwide.

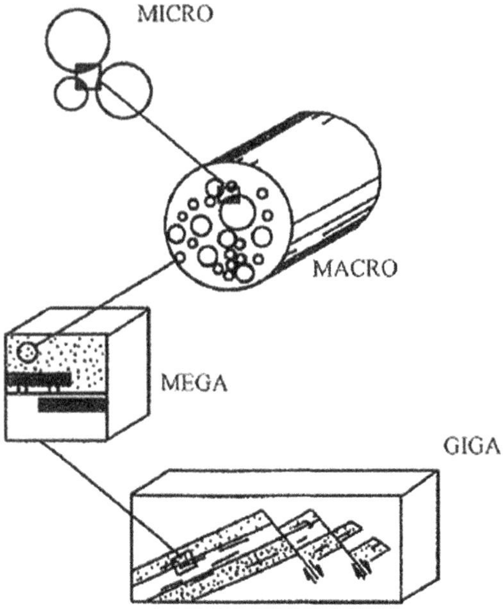

Fig. 7.4 Classification of reservoir heterogeneity (Haldorsen 1983)

7.2.4 Classification by Qiu Yi'nan et al

Based on years of experience and using Pettijohn's idea, Qiu Yi'nan (1987, 1992) divided the reservoir heterogeneity of clastic rocks into four categories in combination with the characteristics of continental reservoirs in China, accounting for both the scale of heterogeneity and the actual conditions of development and production. This classification is suitable for China and convenient to operate. Therefore, it is used extensively in China.

1. **Interlayer heterogeneity**
   This includes cyclicity of strata series, heterogeneity degree of permeability among sand layers, barrier bed distribution, and specific bed distribution.
2. **Plane heterogeneity**
   This involves the degree of connectivity of genetic units, porosity of plane, change in permeability, degree of heterogeneity, and directivity of permeability.
3. **Intrastratal heterogeneity**
   This covers particle size rhythmicity, bedding structure sequence, permeability difference level, location of high-permeability intervals, distribution frequency and size of intrastratal discontinuous thin muddy intercalations, other impermeable barrier beds, level of whole-reservoir scale, and the ratio of vertical permeability.
4. **Pore heterogeneity**
   This refers mainly to the heterogeneity of microscopic pore structures, including the size and degree of uniformity of sand pores and throats, configuration relationship between pore and throat, and degree of connectivity.

In addition to the above classification, there are macroscopic, mesoscopic, and microscopic heterogeneity classifications. Moreover, a few scholars adopt heterogeneity classification schemes from the large, medium, and small perspectives.

7.3 Research and Quantitative Description of Heterogeneity

The objective of the study of reservoir heterogeneity is not only to represent the law of variation and distribution characteristics of various reservoir attributes at different levels, but also more importantly, to build reservoir heterogeneity models. To this end, scientific quantification and indexation are required for various descriptive characteristics. In terms of reservoir sedimentology, reservoir heterogeneity can be divided into macroscopic and microscopic with reference to the various classification schemes for petroleum reservoir heterogeneity at home and abroad, in which the macroscopic type comprises intrastratal, interlayer, and plane heterogeneity.

7.3.1 Microscopic Heterogeneity

7.3.1.1 Intrastratal Heterogeneity

This means reservoir characteristics change in the vertical direction within a single sand layer and it refers to the difference level of intrastratal permeability in the vertical direction, location of the highest permeability section, intrastratal granularity rhythm, permeability rhythm, heterogeneity degree of permeability, and distribution of intrastratal discontinuous thin muddy intercalations. Intrastratal heterogeneity is a key geological factor that directly controls and affects the swept volume of injectant in a single sand layer. Consequently, the core of intrastratal heterogeneity study is the response relationship between sedimentation and heterogeneity. Quantitative indexes include ① difference level of permeability—affects the degree of impact of fluid and water channeling; ② location with high permeability—determines injection-production mode and perforation part; ③ ratio (K_v/K_h) of vertical permeability to horizontal permeability—controls washing effect; and ④ distribution frequency, density, and scope of intrastratal discontinuous thin muddy intercalations—affect injection-production mode and the distribution of oil, gas, and water interfaces.

1. **Granularity rhythm**

The change in the size of clastic particles in the vertical direction within a single sand layer is called granularity rhythm or graded sequence, which is under the control of the sedimentary environment and sedimentation. In general, it is classified into four categories, namely, finning upward, coarsening upward, complex rhythm, and homogeneous rhythm.

1. Finning upward: refers to the change in particle size from coarseness to fineness in a decreasing trend from bottom to top, which leads to poor physical properties from bottom to top (Fig. 7.5). For instance, the granularity rhythm of a channel sand body is often finning upward.
2. Coarsening upward: refers to the change in particle size from thinness to coarseness in a decreasing trend, wherein petrophysics improves from bottom to top. For example, the granularity rhythm of a delta-front distributary mouth is typically coarsening upward.
3. Complex rhythm: refers to the combination of finning upward and coarsening upward. The superimposition of finning upward is called complex finning upward, whereas that of coarsening upward is called complex coarsening upward; particle size with thin upper and lower parts and a coarse middle part is called complex coarsening–finning upward, whereas that with coarse upper and lower parts and a thin middle part is called complex finning–coarsening upward.
4. Homogeneous rhythm or irregular rhythm: unchanged particle size in the vertical direction is called homogeneous rhythm, while

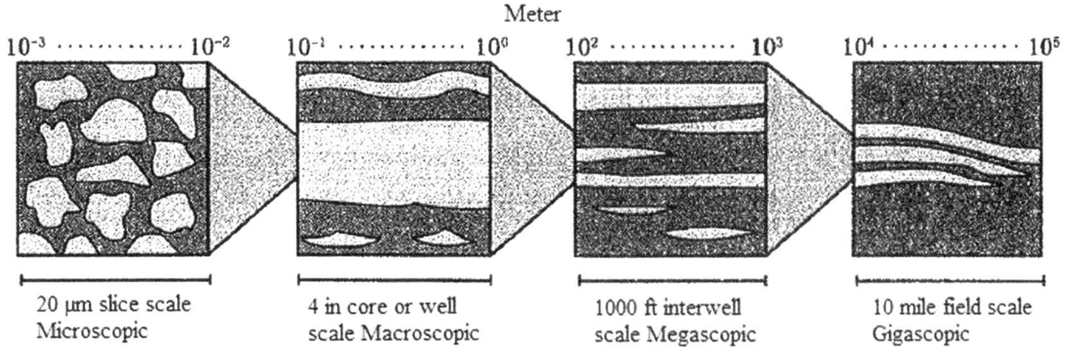

Fig. 7.5 Hierarchical classification of reservoir heterogeneity

Table 7.1 Waterflooding simulation results for sandstones of different beddings

Bedding type	Horizontal permeability (10^{-3} μm^2)	Final recovery ratio (%)
Parallel bedding	816.2	31.8
Planar/tabular cross bedding	723 (parallel lamination direction)	21.3
Trough cross bedding	221.3	42.7

Source Daqing oilfield data

irregular particle size is called irregular rhythm.

2. **Sedimentary structure**

Most clastic rock reservoirs have different primary sedimentary structures, among which beddings are dominant, including mainly parallel bedding, planar cross bedding, trough cross bedding, ripple cross bedding, graded bedding, washing bedding, massive bedding, and horizontal bedding. Bedding types are subject to sedimentary environments and current conditions. Beddings show the characteristics of different sedimentary structures through rock color, particle size, components, and particle arrangement and combination, which result in the anisotropy of permeability (Table 7.1). As a result, the directivity of permeability can be analyzed by studying the laminar occurrence, composition relationship, and distribution laws of various beddings. Reservoir heterogeneity at this level is mainly studied by core analysis and dipmeter logging.

3. **Permeability rhythm**

Rhythmicity comprised of the change in permeability in the vertical direction is called permeability rhythm. Similar to granularity rhythm, permeability rhythm can be divided into finning upward, coarsening upward, and complex rhythm (including complex finning upward, complex coarsening upward, complex finning–coarsening upward, and homogeneous rhythm) (Fig. 7.6). Under normal conditions, changes in the physical properties (porosity and permeability) of a reservoir do not always have a relatively good correspondence with particle size, especially porosity. The vertical change laws of porosity and permeability are affected not only by particle size distribution but also by restriction and modification of rock fabrics, diagenesis, and tectonic activity, especially permeability. This creates the problem that various changes are present in the position with the highest permeability. Generally, the position with the highest permeability is regular in reservoirs with normal granularity rhythm, thus it is easy to determine, but varies in reservoirs with complex graded rhythm.

4. **Ratio of vertical permeability to horizontal permeability (K_v/K_h)**

This ratio has a greater influence on the washing effect in the waterflooding development of an oil reservoir. A low K_v/K_h value indicates that the vertical permeability of fluid is relatively low, and thickness affected by intrastratal washing may be relatively smaller, and vice versa.

5. **Degree of permeability heterogeneity**

Quantitative parameters for representing the degree of permeability heterogeneity include variation coefficient (V_k), permeability max-mean coefficient (T_k), permeability ratio (J_k), and permeability homogeneity coefficient (K_p).

(1) Variation coefficient (V_k)

The variation coefficient belongs to mathematical statistics, and it is used to measure the degree of dispersion of several numerical values for measurements and statistics relative to an average value. Its computational formula is as follows:

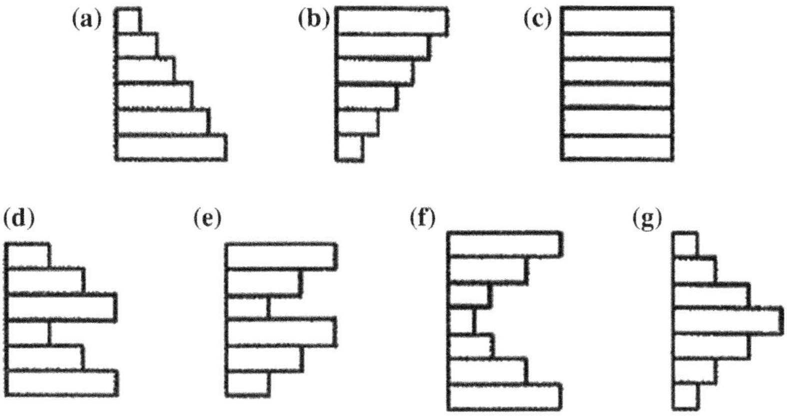

Fig. 7.6 Vertical rhythmic pattern. **a** Finning upward. **b** Coarsening upward. **c** Homogeneous rhythm. **d** Combined finning upward. **e** Complex coarsening upward. **f** Complex finning–coarsening upward. **g** Complex coarsening–finning upward

$$V_k = \frac{\sqrt{\sum_{i=1}^{n} (K_i - \overline{K})^2 / n}}{\overline{K}}$$

where

K_i permeability value of some intrastratal sample, $i = 1, 2, 3, \ldots, n$;

$\overline{K}$ average value of all intrastratal sample's permeability; and

n number of intrastratal samples.

In general, when $V_k < 0.5$, permeability has a homogeneous pattern, which shows that the degree of heterogeneity is weak; and when $0.5 \leq V_k \leq 0.7$, permeability has a relatively homogeneous pattern, which shows that the degree of heterogeneity is medium. When $V_k > 0.7$, permeability has a heterogeneous pattern, which shows that the degree of heterogeneity is strong. Owing to the greater difference in the permeability values of continental clastic rock reservoirs in China, the classification standards are different (Table 7.2) in order to better reflect their heterogeneity characteristics.

(2) Permeability max–mean coefficient (T_k)

This is the ratio of the maximum permeability to average permeability of a sand layer.

$$T_k = \frac{K_{max}}{\overline{K}}$$

where K_{max}-interlayer maximum permeability, which represents the highest permeability in a sand layer relative to the homosphere.

When $T_k < 2$, the permeability has a homogeneous pattern; when T_k is 2–3, it has a relatively homogeneous pattern; and when $T_k > 3$, it has a heterogeneous pattern.

(3) Permeability ratio (J_k)

This is the ratio of the maximum permeability to the minimum permeability.

$$J_k = \frac{K_{max}}{K_{\min}}$$

where K_{min}-minimum permeability value, which usually represents the lowest permeability relative to the permeability in the homogenous section.

The higher the permeability ratio, the stronger the heterogeneity of permeability and vice versa.

(4) Permeability homogeneity coefficient (K_p)

This refers to the ratio of the average permeability to the maximum permeability in a sand layer.

Table 7.2 Classification standards for heterogeneous degree of continental clastic rock reservoirs in China

Reservoir level	Permeability ($\times 10^{-3}\ \mu m^2$)	Homogeneity degree	Permeability variation coefficient	Model code
Extra-high permeability	>1000	Homogenous	<0.25	I1
		Relatively homogenous	0.25–0.7	I2
		Seriously heterogeneous	>0.7	I3
Medium-high permeability	1000–300	Homogenous	<0.25	II1
		Relatively homogenous	0.25–0.7	II2
		Seriously heterogeneous	>0.7	II3
Medium-low permeability	300–100	Homogenous	<0.25	III1
		Relatively homogenous	0.25–0.7	III2
		Seriously heterogeneous	>0.7	III3
Low permeability	100–10	Homogenous	<0.25	IV1
		Relatively homogenous	0.25–0.7	IV2
		Seriously heterogeneous	>0.7	IV3
Ultra-low permeability	<10	Homogenous	<0.25	V1
		Relatively homogenous	0.25–0.7	V2
		Seriously heterogeneous	>0.7	V3

$$K_p = \frac{\overline{K}}{K_{\max}}$$

Apparently, the value of K_p varies between 0 and 1, and homogeneity is better when K_p is closer to 1.

(5) Reservoir quality index (RQI)

To reflect the comprehensive quality characteristics of a reservoir, the concept of reservoir quality index (namely, an index for reflecting the comprehensive characteristics of reservoir porosity and permeability to evaluate the reservoir quality) can be applied in combination with the research characteristics of petroleum reservoir engineering (Amaefule 1993). This is represented by the following formula:

$$RQI = \sqrt{\frac{K}{\phi}}$$

Because this index is a dimensionless relative number, its size difference can reflect reservoir quality and represent the heterogeneity difference of a reservoir in the same area or oilfield.

6. **Distribution frequency (P_k) and distribution density (D_k) of muddy intercalation**

Intercalation refers either to an impermeable bed or a low-permeability bed within a single sand layer with thickness ranging from several

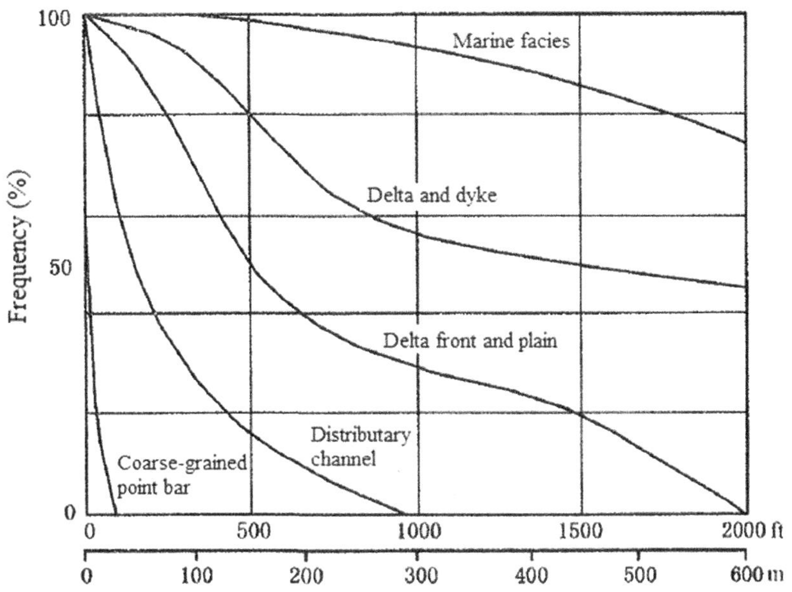

Fig. 7.7 The continuity of shale (silt) interbed is the function of the sedimentary environment (K. J. Weber 1986)

centimeters to dozens of centimeters. It generally consists of mudstone, silty mudstone, or calcareous sandstone. Intercalation is formed by transient and local current status changes, which reflect the phase change of microfacies or sand bodies. Hence, its form and distribution are unstable. Unstable muddy intercalation represents non-permeability or extremely low permeability in fluid flow, which affects the changes in permeability along both the vertical and the horizontal directions. Its distribution and lateral continuity are mainly subject to the sedimentary environment (Fig. 7.7), which is characterized by randomness and difficulty to trace. However, prediction is available through the analysis of the sedimentary environment. Under normal conditions, the distribution characteristics of muddy intercalation can be described quantitatively by the following two parameters:

(1) Distribution frequency (P_k) of intercalation
This is the number of impermeable muddy intercalations in a reservoir of unit thickness.

$$P_k = \frac{N}{H}$$

where

N number of interlayer impermeable intercalations; and
H intercalation thickness, m.

(2) Distribution density (D_k) of intercalation

This refers to the thickness of impermeable muddy intercalations in a reservoir of unit thickness, namely, the percentage ratio of the sum of all intercalation thicknesses to total reservoir thickness.

$$D_k = \frac{H_{sh}}{H} \times 100\%$$

where

H_{sh} total thickness of impermeable muddy intercalations, m; and
H reservoir thickness, m.

The rule of distribution of intercalations on a plane can be reflected by preparing a plane contour map of the two abovementioned parameters. Intercalation mainly acts as a barrier in oilfield development: ① the presence of intercalation clarifies the anisotropy of interlayer permeability; ② intercalation distribution affects oil and water movement rules; and ③ stability of intercalation distribution affects the pressure distribution in a thick oil reservoir.

7.3.2 Interlayer Heterogeneity

Interlayer heterogeneity refers to the difference between reservoirs or sand bodies. It involves the general study of an oil reservoir or oil-bearing series between sand-mudstones, which belongs to the description of a reservoir on a set scale. It comprises either the regularity or cyclicity of sand bodies occurring alternately on sections in various sedimentary environments, and the development and distribution rules of muddy stones as barrier beds, i.e., interlayer differences in a sand body such as the permeability heterogeneity degree between sand bodies.

This is an inherent cause for interlayer interferences, water drive differences, and monolayer breakthrough during waterflooding development. Hence, interlayer homogeneity is the technical basis for the selection of a series of developments and slice mining. In continental sedimentary reservoirs, interlayer heterogeneity is very prominent because of the presence of multiple continental reservoir layers, small thicknesses, fast lateral variations, and poor connections.

1. **Lamination coefficient (A_n)**

This refers to the number of sand layers in either one set or one oil petroleum reservoir. The number of sand layers in the same set on the plane varies due to phase changes, thus its characteristics are represented by the number of sand layers encountered in an average single-well drilling.

$$A_n = \sum N_{bi}/n$$

where

N_{bi} number of sand layers in a well; and
n statistical number of wells.

A greater lamination coefficient results in more serious interlayer heterogeneity and a worse petroleum reservoir development effect.

2. **Sandstone density (S_n)**

This refers to the ratio of the total thickness (silt contained) of sandstone to the total stratum thickness, namely, sand/stratum ratio, which is also known as the net gross ratio (NGR) in the case of an effective reservoir. Because this coefficient is mainly used to reflect the degree of connectivity of a sand body, and siltstone with a certain porosity and permeability may be used as a reservoir, siltstone shall be included in the statistics. As for the concept of the sandstone/stratum ratio, scholars use different names when applying different data for research and there are significant differences between specific meanings. To avoid confusion, the contrast is shown below (Table 7.3).

3. **Heterogeneity degree of permeability among sand layers**

This refers to the interlayer difference in the variation coefficient (V_k), permeability max–mean ratio (T_k), permeability ratio (J_k), and permeability homogeneity coefficient (K_p).

4. **Effective thickness coefficient**

A plane contour map of the ratio of oil-bearing formation thickness to the total thickness of sandstones that can clearly reflect the distribution law of petroleum reservoirs.

5. **Interlayer barrier bed**

Barrier bed refers to the stable and relatively impermeable mudstone, siltstone, or gypsolyte developed among sand layers, with thicknesses ranging from dozens of centimeters to dozens of meters. Its genesis is diverse, for example, the barrier bed is formed mainly by prodelta mud, distributary interchannel, or subaqueous distributary interchannel in a delta development area. The upper and lower sand layers are mutually independent owing to the relatively stable barrier bed distribution, hence they do not belong to the same flow unit. Different development conditions of barrier beds in various

Table 7.3 Concept and function of sandstone/stratum ratio applied in different disciplines in petroleum reservoir characterization

Description	Definition	Discipline	Effect
Sandstone density	Sandstone thickness (inclusive of silt)/thickness	Field development geology	Studies the connectedness of sand body
Sand factor	Sandstone thickness (exclusive of silt)/stratum thickness	Reservoir sedimentology and petroleum exploration	Studies the sedimentary pattern
Sand/stratum ratio	Sandstone thickness (with or without silt is unclear)/stratum thickness	Geophysics	Studies the sand body distribution with seismic attributes
Net/gross ratio	Effective reservoir/oil-bearing formation thickness of total reservoir/total sandstone thickness	Petroleum reservoir engineering	Studies the recovery ratio of petroleum reservoir and layout of development well pattern

Note Some scholars also term sandstone density and sand factor as sandstone percentage. In this case, silt is included in sandstone thickness because its sediments mainly comprise silty-fine sand. It is used to study sedimentary patterns

wells result in different heterogeneities. The types, positions, and plane distribution laws of barrier beds are mainly depicted and analyzed in studies.

7.3.3 Plane Heterogeneity

Plane heterogeneity refers to the geometry, scale, and continuity of a reservoir sand body, porosity within sand bodies, and heterogeneity arising from the level change in permeability. It is directly related to injectant sweep efficiency.

1. **Geometry of a sand body**

Chapter 2 shows that the geometry of a sand body is the relative reflection of its size in different directions. Because it is mainly controlled by the distribution of depositional facies, the geometry of sand bodies in different depositional systems has its own characteristics and rules.

2. **Sand body scale and continuity**

Sand body scale and continuity directly affect reservoir size and well spacing of development well patterns. In general, this study focuses mainly on the lateral continuity of sand bodies. However, the width-to-thickness ratio, drilling ratio, and quantitative geological database are useful and effective methods for reservoir characterization and prediction. Sand bodies can be divided into five levels based on their extended length.

Level I	Sand body extends greater than 2000 m with extremely good continuity.
Level II	Sand body extends 1600–2000 m with good continuity.
Level III	Sand body extends 600–1600 m with medium continuity.
Level IV	Sand body extends 300–600 m with poor continuity.
Level V	Sand body extends less than 300 m with extremely poor continuity.
Drilling ratio	shows the control degree of a sand body under a certain well pattern.
Drilling ratio	(number of drilling sand layers/total number of wells) $\times$ 100%.

3. **Connectedness of a sand body**

The connectedness of a sand body relates not only to the density of development well pattern and the waterflooding development mode but also affects the final petroleum exploitation efficiency. The presence of different sand body

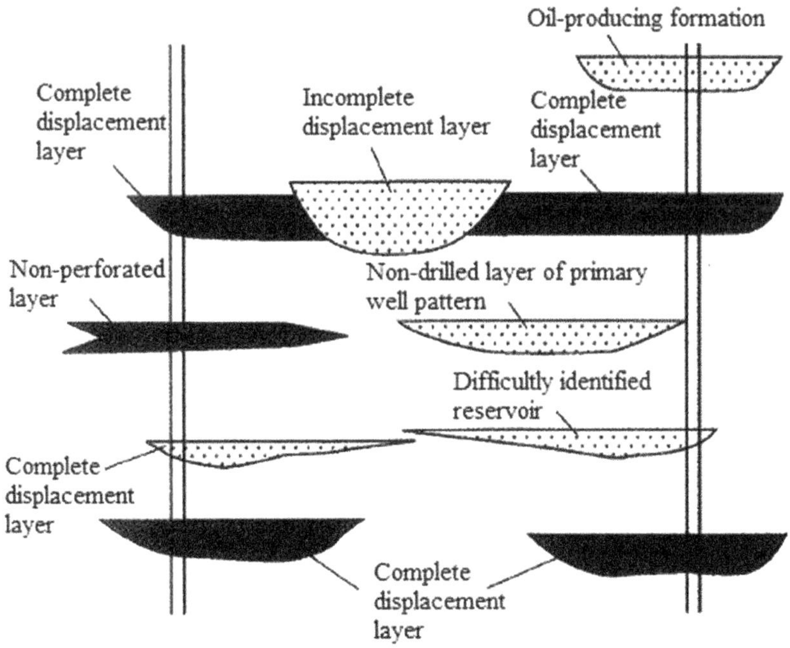

Fig. 7.8 The different connecting modes of sand bodies have greater impact on distribution of the remaining oil

connections has a greater influence on the distribution of remaining oil (Fig. 7.8). The connections of a subsurface sand body can be divided into configuration and deposition by genesis. The former mainly refers to faults or fractures; and the latter refers to mutual contact and communication of a sand body in the vertical and the horizontal directions, which can be expressed by the sand body coordination number, connectivity degree, connectivity coefficient, size of connected components, and permeability of the sand body contact.

1. Sand body coordination number: refers to the number of sand bodies all connected with a certain sand body, which controls oil, gas, and water interfaces, as well as injection-production modes.
2. Connectivity degree: refers to the percentage of area of connected sand bodies to the total area of a sand body.
3. Connectivity coefficient: percentage of the number of connected sand body layers to the total number of sand body layers. The connectivity coefficient can be calculated by thickness, hence it is called thickness connectivity coefficient.
4. Size of connected component: refers to composite sand bodies formed by mutual contact and connections of various genetic unit sand bodies vertically and horizontally. During the evaluation of reservoir development, the number of genetic unit sand bodies contained in a connected component, and the length, width, total area, and thickness of the connected component are studied.
5. Permeability of sand body contact: seepage fluid channel is not always formed on some sand body contacts. For instance, impermeable or low-permeability interfaces may be formed on scouring contact surfaces between sand bodies because of the existence of an argillaceous cladding layer or a calcite cementation layer. We will study the permeability of sand body contact in depth.

The connectivity of a sand body is mainly impacted by sedimentation and structural deformation. Taking a river as an example, its connectivity is usually of the multilateral type

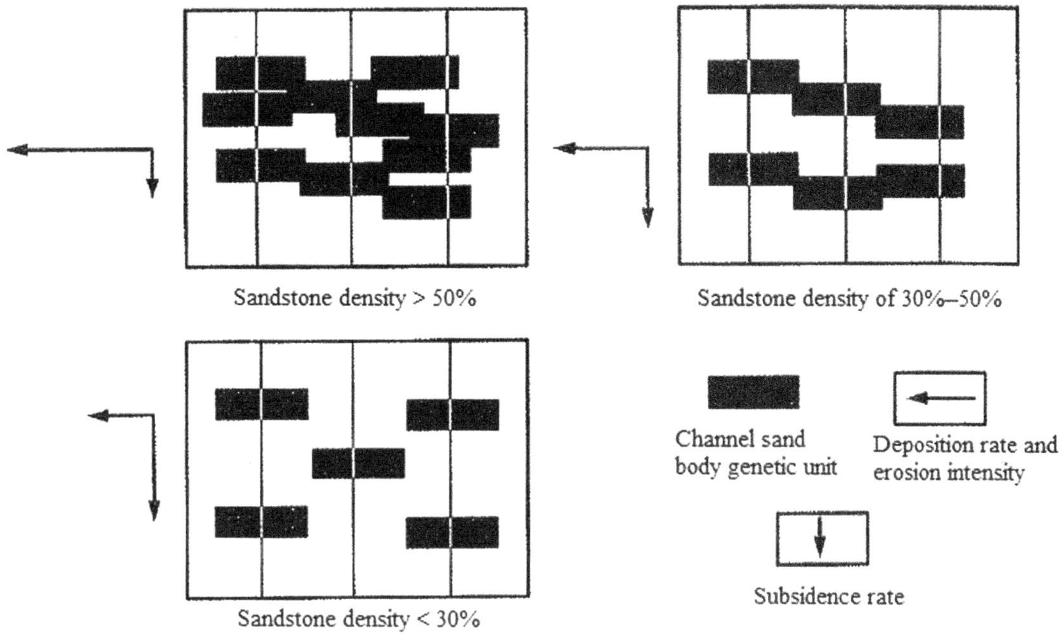

Fig. 7.9 Connection of sand bodies can be evaluated by sandstone density

formed by channel bifurcation and migration; unilateral type resulting from unilateral channel migration; multi-layer type (also known as superimposed type, mostly interconnected in vertical direction) arising out of channel swing and merge; and isolated type (not connected with any other sand body) owing to river overflow and channel bifurcation. In addition, the connectivity of sand bodies can also be evaluated by sandstone density (Fig. 7.9). Therefore, the methods for studying connectivity usually involve ① sandstone density; ② spatial overlay; ③ pressure tests (Fig. 7.10); ④ dynamic detection for production; and ⑤ tracer tracing.

However, owing to fluvial facies reservoir sand bodies of different genesis types, especially the complexity of meandering channel sand (lateral accretion sand) deposition, the simple connection model and the quantitative characterization method are inadequate for depicting reservoir heterogeneity and looking for remaining oil. Lu et al. (1997) classified the connectivity of fluvial facies reservoirs into three types based on long-term research in the Daqing Oilfield: Class I connectivity, which refers to the connectivity of the same channel sand body; Class II connectivity, which refers to the connectivity between different channel sand bodies; and Class III connectivity, which refers to the connectivity between channel sand bodies and interchannel sand bodies (Fig. 7.11).

4. **Porosity within sand body, level change, and directivity of permeability**

The level change law can be represented by preparing a plane contour map of porosity, permeability, and permeability heterogeneity degree. Research focus is on the directivity of permeability, which directly affects the plane sweep efficiency and controls the movement directions of oil, gas, and water. The directivity of permeability can be classified into two categories: ① directivity of macroscopic permeability, which refers to the permeability directivity resulting from lithological changes in a sand body; and ② directivity of microscopic permeability, which refers to the permeability directivity arising from the sedimentary structure and structural factors of a sand body. The influential

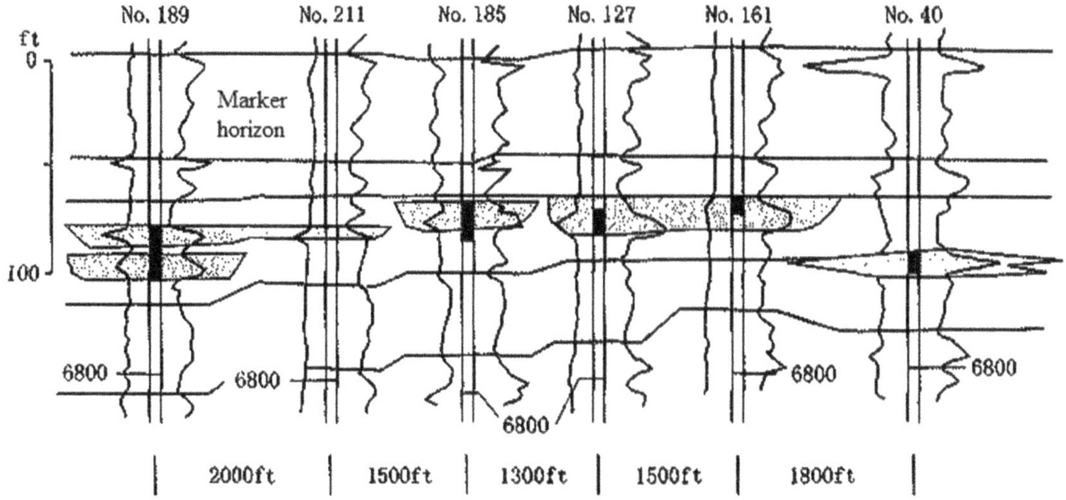

Fig. 7.10 Sand body connection relation determined according to the pressure test information

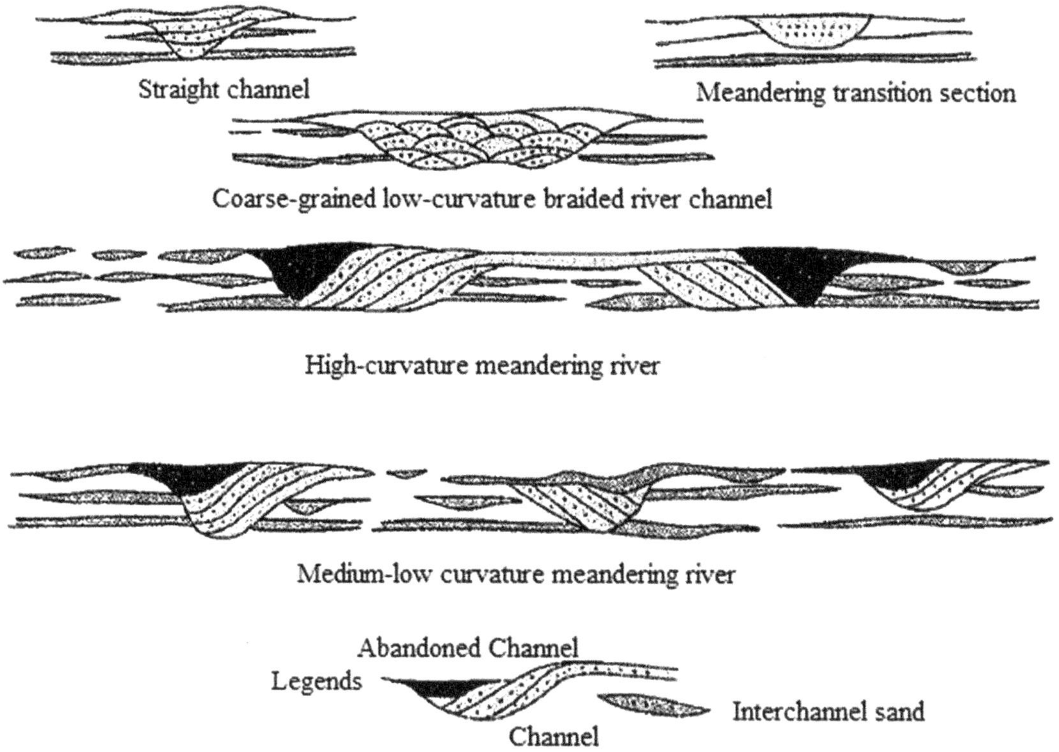

Fig. 7.11 Connection relation between channel sand bodies

Table 7.4 Factors and genesis affecting plane heterogeneity of permeability

Influencing factors	Genesis and influencing results	
Sedimentation	Differences resulting from different facies or microfacies on the plane	
	Primary and secondary flow zones formed by the differences between different parts of the same facies	
	Differences resulting from geometry	
	Differences resulting from palaeocurrent direction	
Configuration (seepage barrier formed by sealing; and seepage channel formed by opening)	Fracture	Microfracture: increase in permeability has little influence on macroscopic directivity
		Local fracture: the extended length is less than well spacing, which Xiaoguangposes certain influences on macroscopic directivity
		Regional fracture: the extended length exceeds well spacing, which can cause serious permeability directivity issues
	Fault	Clay smearing on the fracture surface-formation period
		Cataclasis-later formation period
		Diagenetic seal-after the diagenesis stage
Diagenesis	Differences arisen from the different strengths of rock diagenesis	
	Diagenesis differences caused by fluid property	
	Diagenesis differences due to depth, pressure and temperature	

causes for plane heterogeneity of permeability are rather complicated and are manifested mainly in three aspects (Table 7.4).

5. **Interwell permeability heterogeneity degree**

(1) Interwell permeability variation coefficient
The interwell permeability variation coefficient reflects the overall degree of heterogeneity of sand body permeability on a plane.
(2) Area distribution frequency of permeability at different levels
According to the defined permeability levels on a permeability contour map, the difference level of permeability on the plane can be determined by calculating the percentage of the distribution area of permeability at different levels and preparing a distribution frequency diagram.
(3) Difference level of injection-production interwell permeability
When the injection-production well pattern is determined, it is necessary to describe the difference level of permeability between the injection well and various producing wells. This difference level is an internal cause for horizontal contradiction in waterflooding development.

7.3.4 Microscopic Heterogeneity

Microscopic heterogeneity of a reservoir refers to the geological factors that affect fluid flow in microscopic throats. These geological factors include pore and throat size, connectivity degree, configuration relationship, sorting degree, and heterogeneity of particle and interstitial material distribution. The heterogeneity of this scale directly affects the microscopic displacement efficiency of the injectant. Microscopic heterogeneity comprises three aspects, namely, pore, particle, and interstitial material, among which the latter two are the main reasons for pore heterogeneity.

7.3.4.1 Pore Heterogeneity

Generally speaking, bigger spaces defined by rock particles are known as pores, while narrow parts between two connected particles are called throats. A pore is a basic reservoir space for storing fluid in a rock, while a throat is the main factor controlling the seepage characteristics of fluid in a rock.

1. **Pore and throat size**

Type, size, distribution state, and degree of sorting of pore and throat can be depicted quantitatively by applying pore structure parameters, including the maximum pore radius, maximum pore radius median, maximum connected throat radius, throat radius median, average radius of main flowing throat, peak throat radius, and minimum flowing throat radius.

Notably, the wetting phase fluid forms a layer of liquid film at the edges of particles to diminish the flowable pore channel when pores are filled with liquid. Consequently, the effective pore throat radius is the actual pore throat radius minus liquid film thickness in the presence of a wetting phase fluid.

2. **Throat heterogeneity**

Each throat can connect two pores, and each pore is at least connected by more than three throats, even 6–8 throats, which directly affect the exploitation of an oilfield. The pore throat coordination number quantitatively characterizes pore system connectivity. In a hexagonal grid, the coordination number is 3, while in a triple hexagonal grid, the coordination number is 6 (Fig. 7.12). In the same reservoir, throat type varies owing to the particle contact relationship of a rock, as well as various particle sizes, shapes, and cementation types. There are four common throat types, as listed below:

(1) Pore-narrowed throat

This throat type, which represents the narrow part (Fig. 7.13a) of the pore, often develops in sandstone-dominated intergranular pores, thus it is difficult to distinguish it from pores. The rock is mainly particle supported, with floating particle contact and without cementation. This pore has a large-pore coarse-throat structure. When the pore/throat diameter ratio is close to 1, almost all rock pores are effective.

(2) Necking throat

The throat is the contracted part (Fig. 7.13b) of an intergranular variable section. When sandstone particles are compacted for a closer arrangement, the preserved pores are larger, but throats among the particles narrow down greatly. By this time, sandstone may have higher porosity but lower permeability, hence it is of the large-hole fine-throat type, and some pores are ineffective.

(3) Flaky or curved lamellar throat

Flaky or curved lamellar throat is a striped channel (Fig. 7.13c, d) between particles. When the sandstone compaction degree is stronger or there is crystal regrowth, the pore surrounded by the regrown crystals becomes thinner, and its throat is actually a gap between crystals. Its opening width is generally less than 1 μm, with a few throats spanning dozens of micrometers. When dissolution among particles occurs, a wide flaky or curved lamellar throat can be formed. Therefore, this type of throat varies greatly. It can be a small-pore extremely fine–throat type or large-pore coarse-throat type after dissolution. The pore throat diameter ratio is medium-to-large.

(4) Tube-shaped throat

When the contents of the matrix and various cements are high, primary intergranular pores may be blocked completely at times. Micropores (less than 0.5 pm) in the matrix and various cements are distributed crosswise, akin to microscopic capillaries, to form a tube-shaped throat (Fig. 7.13e). Their porosities are medium or relatively low. Meanwhile, their permeability is extremely low, mostly less than 0.1×10^{-3} μm^2. The pore throat diameter ratio is 1 on the ground, where the pore is the throat itself.

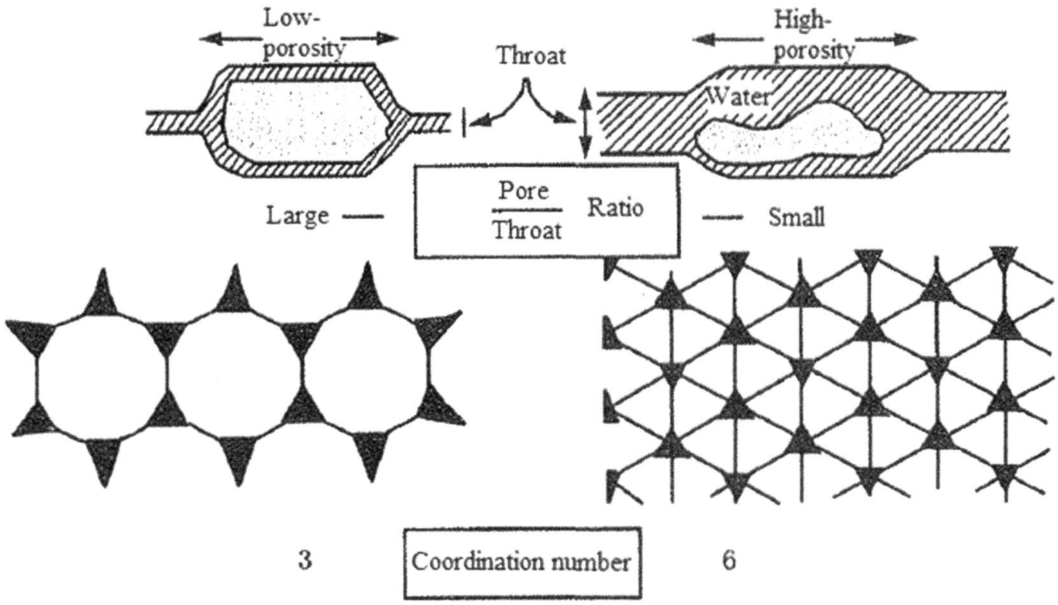

Fig. 7.12 Ratio of pore to throat size and impact of coordination number on the reservoir non-wetting phase recovery efficiency (Wardlaw 1978)

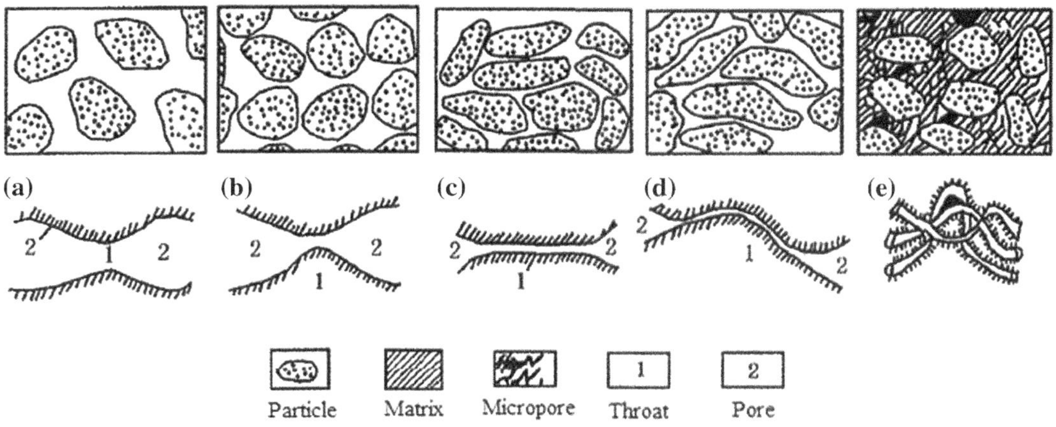

Fig. 7.13 Type of pore throat. **a** Throat as narrowed part of pore. **b** Throat at contracted part of variable section. **c** Flaky throat. **d** Curved lamellar throat. **e** Tube-shaped throat

Briefly, different throat shapes and sizes can result in different capillary forces that affect reservoir properties and pore permeability. All reservoirs are composed of pores of different apertures and pore throats of different sizes, hence their permeability varies considerably. The degree of heterogeneity of pore throat size distribution can be represented using parameters such as sorting coefficient, relative sorting coefficient, heterogeneity coefficient, pore throat skewness, and pore throat kurtosis.

3. **Pore connectivity**

Pores are connected by throats, but different pores may be connected differently. The connectedness can be characterized by pore throat coordination number, pore throat diameter ratio, or pore throat volume ratio. Obviously, better pore connectivity is more conducive to oil and gas exploitation.

7.3.4.2 Particle Heterogeneity

This refers to particle size, shape, sorting, arrangement, and contact relationship. It affects not only pore heterogeneity but also results in anisotropy of permeability while affecting dynamic changes in a reservoir during waterflooding development.

The arrangement directivity of particles, which is an important factor that leads to permeability anisotropy of a reservoir, is mainly a function of the depositional palaeocurrent direction. The long axis of a particle tends to be consistent with palaeocurrent direction, and permeability along this direction is greater than that in the other directions owing to a higher palaeocurrent speed and smoothness of the pores. In contrast, pores on two sides form a tranquil flow zone or stagnant zone, in which more fine-grained particles or clay materials may be formed. Thus, differently unblocked pore channels in different directions are formed, which result in permeability anisotropy.

7.3.4.3 Interstitial Material Heterogeneity

As is well known, interstitial material includes the clay matrix (autogenetic or allogenic) and cements. Its type, content, and occurrence vary greatly in various reservoirs, which lead to disparities in porosity, permeability, and heterogeneity across reservoirs. Data can be acquired to analyze heterogeneity characteristics via various research methods such as microscopic identification, statistics, and experimentation. The characteristics of interstitial materials are important parameters in terms of their effects on porosity heterogeneity and intrinsic causes, and on the material basis of reservoir sensitivity.

7.4 Reservoir Heterogeneity and Hydrocarbon Recovery Ratio

In oilfield development, the final recovery ratio is affected by three main factors: ① reservoir heterogeneity, ② fluid property, and ③ injection-production scheme and production system. Herein, reservoir heterogeneity is the most fundamental and major geologic factor.

7.4.1 Influence of Macroscopic Heterogeneity on Waterflooding Development

During the waterflooding development of a multi-reservoir oilfield, macroscopic heterogeneity directly affects the waterflooding development effect.

7.4.1.1 "Monolayer Breakthrough" Resulting from Interlayer Heterogeneity

Interlayer heterogeneity is an inherent cause of interlayer interference and monolayer breakthrough (known collectively as interlayer contradiction) in waterflooding development. Under multi-layer commingled production, interlayer contradictions are more prominent. In the presence of a greater number of layers, greater interlayer contradictions, and higher single well liquid production, water content is usually higher. Under normal conditions, water drive can be realized easily owing to low water drive start-up pressure in a high permeability reservoir. The water injection well of good oil absorbs higher amounts of water with a fast front advance, which leads to high output from a high-permeability reservoir. In contrast, a low-permeability reservoir has a higher starting pressure, low water absorption, and slow front advance, even without flow. Owing to interlayer

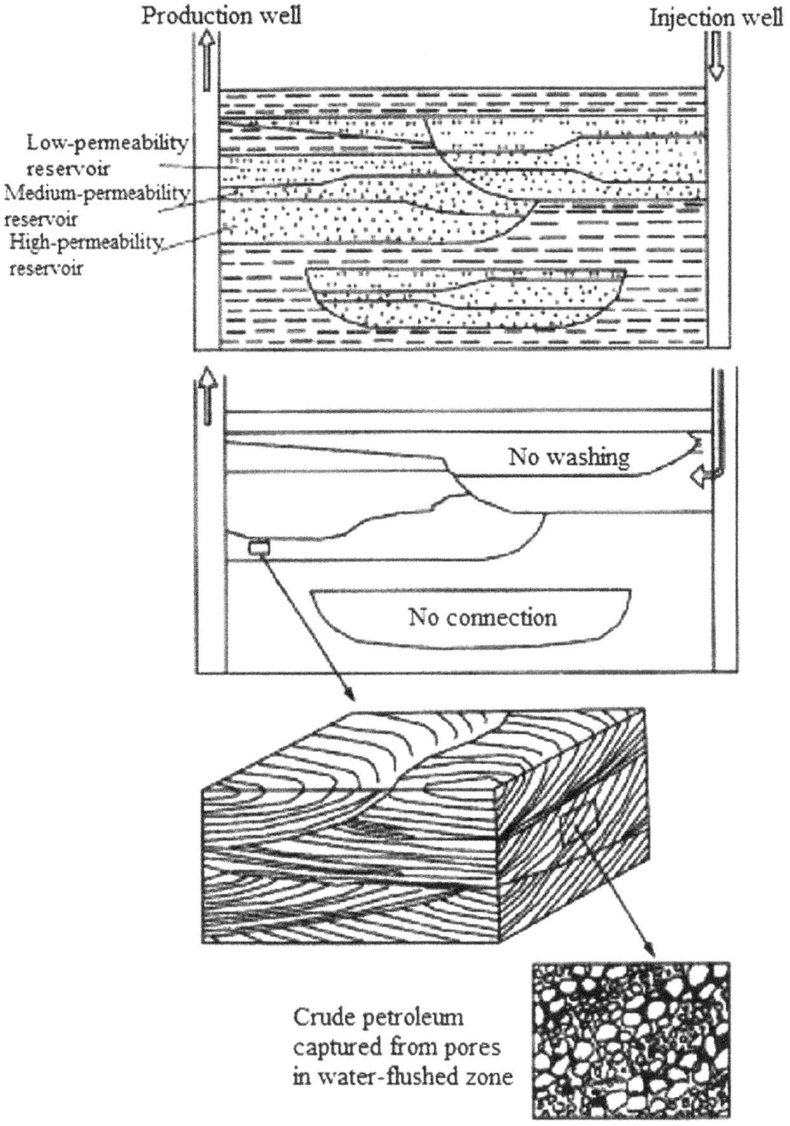

Fig. 7.14 Due to interbed contradiction between high permeability and low permeability, "monolayer breakthrough" occurs and the remaining oil of the low permeability reservoir is obvious

contractions between high- and low-permeability reservoirs, significant interlayer interference is shown in both the producing well and water injection well. As a result, "monolayer breakthrough" of high-permeability reservoirs and outstanding remaining oil (Fig. 7.14) of low-permeability reservoirs occur. The permeability ratio is proportional to the inactive sand body thickness. That is, the higher the ratio, the more inactive the oil reservoir will be (Fig. 7.15). Interlayer interference can be discerned clearly from the water injection profile and the fluid production profile. In particular, significant differences exist in the water-absorption capacity of unit thickness at all levels in the case of commingled production.

7.4.1.2 "Plane Fingering" Arising from Plane Heterogeneity

Plane heterogeneity can decrease the flooded area coefficient by causing a discontinuous distribution across individual reservoirs on the plane. Moreover, "regions of bypassed oil" that are omitted during drilling are formed at the corners

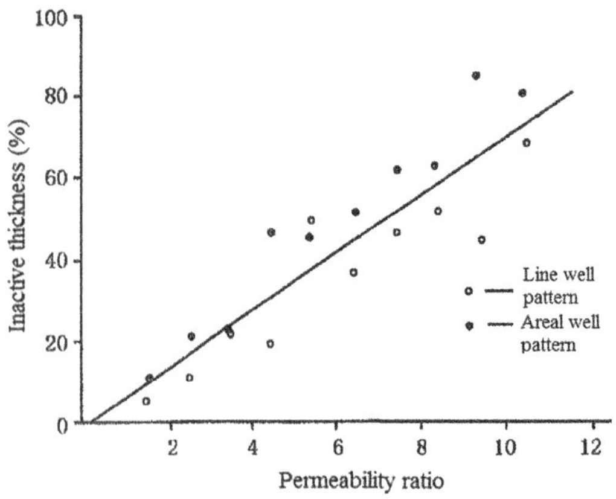

Fig. 7.15 According to Li Bohu's research, the poor permeability is directly proportional to the thickness of the inactive sand body

of an oil reservoir during waterflooding development. In addition, the injected water tongues quickly because of permeability differences along the high-permeability zone on the plane. However, water drive in the middle and low permeability zones is reduced relatively, which leads to a decrease in the flooding area coefficient.

As mentioned above, the continuity of a sand body depends mainly on the spreading of depositional facies, while its connectedness hinges mainly on the spatial superimposition form of the sand body. The former is the geological basis for determining well pattern density, while the latter is the main factor affecting the selection of injection-production well modes. A reasonable injection-production mode and well pattern directly affect the development effect of an oilfield.

Moreover, permeability directivity directly affects the advance direction and speed of various oil displacement methods, and often lead to better oil displacement efficiency in a high-permeability zone than its surrounding areas. Low-permeability zones constitute the major remaining oil distribution area after a period of development. Hence, the main directions of oil displacement include ① trend of high-permeability zone; ② palaeocurrent direction; and ③ fracture development zone.

7.4.1.3 "Region of Bypassed Oil" or "Water Channeling" Resulting from Intrastratal Heterogeneity

Intrastratal heterogeneity reduces the waterflooding thickness coefficient. Heterogeneity among all monolayers is presented mainly as the permeability difference, where the difference may be several, dozens of, or even hundreds of times. In the case of heterogeneity under multi-layer commingled water injection and oil extraction, the injected water firstly tongues quickly along reservoirs with good connectedness and high permeability, hence it quickly enters into the producing well to improve the water cut in the oil well or even brings about flooding production suspension. However, owing to the low degree of production of low-permeability reservoirs, most crude petroleum remaining underground forms "bypassed oil," thus reducing the waterflooding thickness coefficient.

1. **Influence of prosodic features on oil displacement efficiency**

In general, the different prosodic features of permeability manifest as different waterflooding forms (Table 7.5). Moreover, they are the primary cause for the relative concentration of remaining oil and different recovery effects in the low-permeability intrastratal part.

2. **Intercalation influence**

Compared to stable intercalation development, it is beneficial for oilfield development, particularly for thick oil pay. Stable intercalation can divide thick oil pay into several sections, thus restraining vertical fluid channeling in thick oil pay, improving the degree of oil and gas production, and increasing washing thickness. In consequence, the higher the intercalation frequency and density, the better is the oil displacement effect. A rather complex seepage barrier, which affects the oil displacement effect

Table 7.5 Relationship between prosodic features of reservoir permeability, and waterflooding type and oil displacement efficiency

Prosodic features		Waterflooding type	Oil displacement rule	Reservoir development effect
Finning upward		Bottom waterflooding type	High oil displacement efficiency at the bottom, and fast water cut increasing rate	High permeability ratio, small waterflooding thickness, and easily occurring water channeling
Coarsening upward		Upper waterflooding type	Serious upper waterflooding	High-medium permeability ratio, high liquid production capacity, and beneficial for injection-production
		Homogeneous waterflooding type	Basically consistent whole-reservoir oil displacement efficiency	Low permeability ratio, and beneficial for injection-production
		Lower waterflooding type	Large waterflooding thickness coefficient, and strong water washing	Low permeability ratio, and water-wet reservoir
Complex rhythm	Complex finning upward	Segmental waterflooding type	Small washing thickness	Better than finning upward
	Complex coarsening upward	Segmental waterflooding type	Rather uniform washing	Similar to coarsening upward
	Complex coarsening-fining upward	Middle waterflooding type	Medium oil displacement effect	Large and fast liquid production capacity
	Complex fining-coarsening upward	Upper and lower waterflooding type	Relatively poor oil displacement effect	Complicated condition, and usually small in waterflooding thickness
Homogeneous rhythm		—	Oil displacement effect depends on thickness	Extremely small permeability ratio and high recovery ratio

and results in the complex distribution of remaining oil, can be formed in an oil reservoir owing to unstable intercalation.

7.4.2 Relationship Between Microscopic Heterogeneity and Hydrocarbon Recovery Ratio

Microscopic heterogeneity directly affects the microscopic displacement mode and efficiency of an injectant, and the microscopic displacement mode directly affects the distribution and quantity of the remaining oil at the microscopic scale. However, the microscopic displacement efficiency is related to microscopic pore structure, wettability and fluid property, among which pore structure is the most important factor affecting microscopic displacement efficiency.

7.4.2.1 Microscopic Displacement Mechanism in the Pore System

In a pore medium, hydrocarbon detention agents are mainly of the following three types: ① capillary force, which acts as a retention force, and is mainly represented in oil-wet rocks; ② viscous force; and ③ gravity.

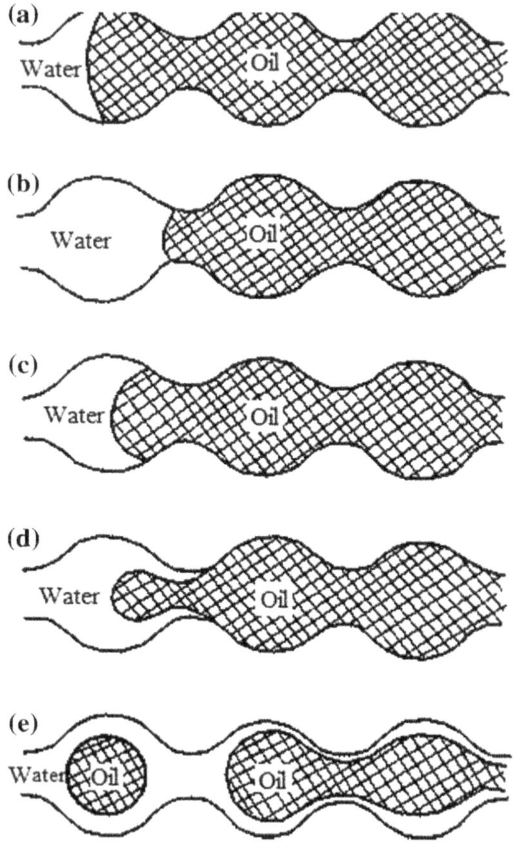

Fig. 7.16 Condition of the oil-water interface passing through converging-diverging capillary tube (Dawe et al. 1978)

During oil-displacement production of an injectant, the power for displacing crude petroleum in pores is mainly an externally applied force, i.e., displacement force. Capillary force serves as water-driving resistance in oleophilic (oil-wet) reservoirs, while it serves as the driving force for sucking water into small pores automatically in hydrophilic (water-wet) reservoirs, which is also known as the spontaneous imbibition phenomenon.

In a single pore, the course of displacing crude petroleum by using an injectant is the process in which the driving force overcomes resistance. However, reservoir pore systems are very complicated, and heterogeneity among various pores in the displacement process can cause interference between holes. Moreover, wettability differences and clay minerals in pores complicate the microscopic displacement process.

Pore channel types in underground rocks are rather complex. Taking a serial pore channel for example (Fig. 7.16), capillary force and driving force jointly push fluids forward under water-wet conditions. However, there may be a blocking action, that is, water automatically wets the pore throat surface. As the water film changes, the throat axis neck is extruded into a filiform shape, and finally, the oil thread may break to form a water bridge in the throat part. The water bridge blocks oil passage, which leads to the accumulation of residual oil right behind it.

Under the oil-wet condition, if the applied pressure is sufficient to overcome the capillary force, liquid will flow. Once the applied pressure is insufficient to promote the passage of the interface through the capillary, seepage stops. Briefly, continuous oil threads may be cut off at the throat pass to form isolated oil drops while passing through a porous medium, depending on the balance between the driving force and the capillary force.

7.4.2.2 Influence of Pore Heterogeneity on Oil Displacement Efficiency

As is well known, the formation of residual oil has a lot to do with reservoir pore structures. In other words, oil displacement efficiency during waterflooding development is closely related to the reservoir pore structure (size and distribution of pore and throat). For petroleum reservoirs with residual oil, the efficiency of displacing residual oil in the tertiary oil recovery process, i.e., the efficiency of tertiary oil recovery, is also related to the pore structure. This is because the re-movement of residual oil hinges on the capillary and viscous forces in the pores. In general, the stronger the pore heterogeneity, the lower is the oil displacement efficiency.

References

Amaefule, J., M. Altunbay, D. Tiab, D. Kersey, and D. Keelan. 1993. *Enhanced reservoir description: Using core and log data to identify hydraulic flow units and predict permeability in uncored intervals/wells*. SPE paper 26436, 205–220.

Dawe, R.A., E.O. Egbogah, and R.J. Wright. 1978. Models of two phase flow in porous media. in *proceedings of 2nd European Symposium on Enhanced Oil Recovery, Held at the Institute of Offshore Engineering*, Heriot-Watt University, Edinburgh, July 5–7, 289–302.

Haldorsen, H.H. 1983. *Reservoir characterization procedures for numerical simulation*. Doctoral thesis, University of Texas at Austin, Austin, TX, United States, 576 pp.

Krause, F.F., H.N. Collins, D.A. Nelson, S.D. Machemer, and P.R. French. 1987. Multiscale anatomy of a reservoir: geological characterization of Pembina-Cardium pool, west-central Alberta, Canada. *Aapg Bulletin*, 71(10): 1233–1260.

Lu, Xiaoguang, Hanqing Zhao, Zhiguo Fu, and Shun Lei. 1997. Elaboration on the plane continuity of fluvial facies reservoir. *Acta Petrolei Sinica* (2): 69–74.

Lv Xiaoguang, Fushi Ma, and Yongzhen Zhang. 1994. Preliminary understanding on reservoir microstructure in Daqingoil field. *Petroleum Geology & Oilfield Development in Daqing* (2): 18–21, 75–76.

Pettijohn, F.J., P.E. Potter, and R. Siever. 1973. *Sand and sandstone*, 618. Springer.

Qiu, Y.N. 1992. Developments in Reservoir Sedinientology of Continental Clastic Roeks in China. *Acta Sedimentologica Sinica*, 10(3).

Qiu, Yinan. 1990. Study process of reservoir sedimentology. *Petroleum Exploration and Development* 17 (1).

Qiu, Y.N., Z.S. Zhang, M.F. Tang, S.J. Yang, and G.X. Feng. 1987. The detailed correlation of fluvial sandbody reservoirs. *Petroleum Exploration and Development*, (2): 46–52.

Weber, K.J. 1986. How heterogeneity affects oil recovery. In *Reservoir characterization*, ed. L.W. Lake and H.B. Carroll, 487–543. Orlando: Academic Press.

8 Alluvial Fan Depositional System

Alluvial fans are developed at valley exits. They are formed mainly by the current scour of temporary flood and are characterized by a limited scope, shape similar to a conical piedmont, and coarse fragmented debris. They scatter radially from the valley entry to the basin, and the plane form of an alluvial fan appears similar to a cone or a fan (Fig. 8.1). It is developed in areas with a large relief amplitude and rich sediment supply. In general, alluvial fan groups in ribbon or skirt patterns distributed along a piedmont are formed because of the interconnection and overlapping of multiple alluvial fans (Fig. 8.1).

In Chinese geomorphology and quaternary geology, alluvial fans are called proluvial fans, which are different from the alluvial cone accumulated at gully entries. The latter comprises small coarse sediments formed by the schistose current resulting from rainwater flowing down the hillside or water from the melting of ice and snow, which scours the slope as it washes down rapidly to the gully entries. Alluvial cones can be fan-shaped (or taper-shaped), but they have small scales and scattered distributions, which can mainly be attributed to the lifting of the denuded zone. Owing to crust rise and enlargement of the denuded zone, very few alluvial cones have been preserved in geological time. Alluvial fans are current-laid in a channelized form, but from the perspective of their specific geological locations of development, typical fan shape, and internal architecture, they are clearly different from the typical fluvial facies or alluvial plains. In addition, an alluvial fan is different from a fan delta because it is developed entirely on the Earth's surface and is a type of continental sedimentary body, whereas a fan delta is a depositional system of continental and underwater transitional type, which results from the lacustrine or marine transformation of an alluvial fan after direct deposition into a relatively stable and independent water body (lake or sea). Fan delta depositional facies are well developed in the meso-cenozoic continental faulted basin in eastern China, and they often form important petroleum reservoirs. Please see Chap. 11 for detailed characteristics.

8.1 Basic Characteristics and Classification of Alluvial Fans

8.1.1 Basic Features

The formation of an alluvial fan requires an adequate terrigenous detrital supply, transition from a mountainous area to a basin, and discontinuous terrain with a large altitude difference. A mountain river limited by a canyon reaches a valley entry with plenty of detrital material eroded from provenance area. Therein, the river's transportation capacity decreases suddenly owing to the open terrain, gentle slope, wide and shallow channels, and decreasing flow rate. As a result, many loads of detrital material

X. Yu et al., *Clastic Hydrocarbon Reservoir Sedimentology*, Advances in Oil and Gas Exploration & Production, https://doi.org/10.1007/978-3-319-70335-0_8

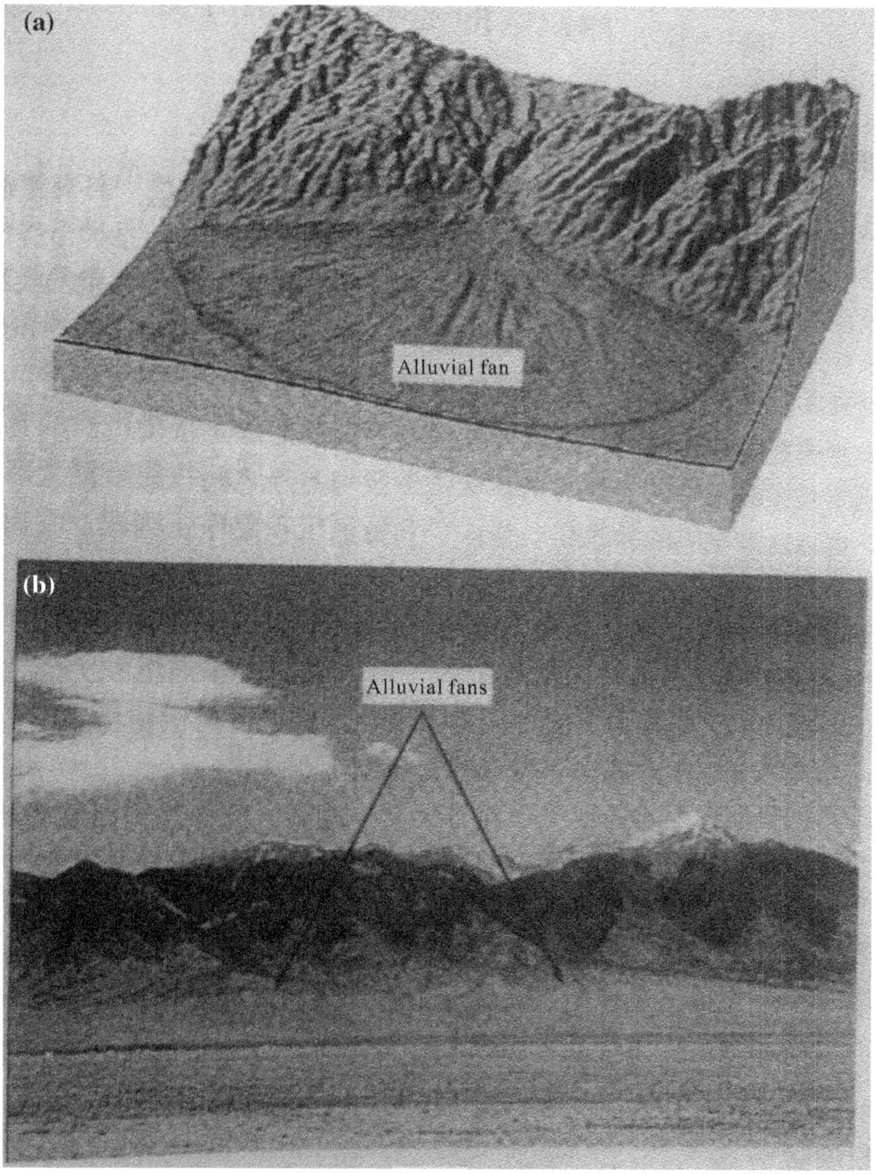

Fig. 8.1 Geomorphic features of alluvial fan and alluvial fan group (according to Tarbuck and Lutgenswrited 1993)

accumulate rapidly to form an outward-radiating fan-shaped sedimentary body with the valley entry as the peak, and this body is called an alluvial fan. In arid-semiarid climatic regions, vegetation is not well developed, physical weathering is strong, rainfall capacity is low even though there are mostly rainstorms, and floods are short but violent. These conditions provide plenty of proximal detrital materials. The open and gentle terrain outside a mountain pass is favorable for receiving the aforementioned sediments, which is essential for the formation of an alluvial fan. Syngenetic faults tend to develop between the uplift zone or mountainous area and the basin. When such a fault is in persistent activity, very thick alluvial fans can be developed to form unique depositional sequences. As a result, alluvial fans are mostly developed along

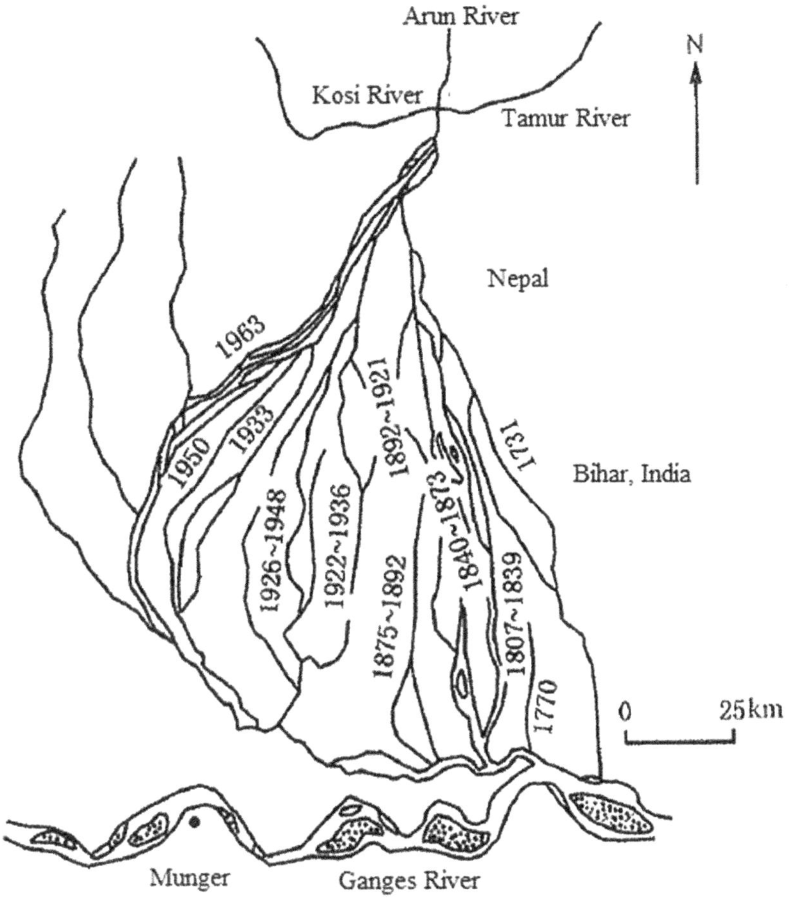

Fig. 8.2 Alluvial fan of the Kexi river in the southern foothills of the Himalayas (according to Gole and Chitale 1966)

the terrain transition zone from mountainous to inland basins or plains.

Alluvial fans vary considerably in size, with radii ranging from smaller than 100 m to larger than 150 km. However, their average radius is smaller than 10 km. Sediment thickness in alluvial fans varies from several meters to approximately 8 km. Alluvial fan deposit, which is the coarsest and worst-sorting proximal deposit in the continental depositional system, generally merges into the fluvial system with fine grain and low gradient from the downdip direction. However, some alluvial fans can enter the water body of the lacustrine or oceanic basin directly to form a subaqueous fan or a fan delta sedimentary body.

The present-day alluvial fans are distributed widely in arid and semi-arid regions around the world. For instance, a series of alluvial fans have developed along the northern foot of the Qilian Mountains-Altun Mountains-Kunlun Mountains in northwest China. They overlap and connect with each other for thousands of meters, and are very spectacular. Owing to the lack of vegetation in arid climate zones, the rock stratum is vulnerable to weathering, which means arid and semi-arid climatic regions are favorable for the development of alluvial fans. Furthermore, large alluvial fans can also be formed in humid or semi-humid climate zones, where there is abundant rainfall and vegetation development, provided the geological structure and terrain conditions are suitable, and there is sufficient material supply. For example, the Kosi River lies within a tropical humid climate zone at the southern foot of the Himalaya mountains, and the famous Kosi River alluvial fan was formed over the past two centuries as the river moved from east to west by 170 km due to sufficient water volume, large gradient, abrupt current, and rapid lateral swing (Fig. 8.2).

8.1.2 Classification of Alluvial Fans

Alluvial fans can be divided into two categories based on climatic conditions: arid alluvial fans, which develop in arid and semi-arid climatic regions, and wet or humid alluvial fans, which develop in humid and semi-humid climatic regions. Often, they are abbreviated as arid fans and humid fans.

The common characteristic of these two categories is that both their plane modalities appear fan-shaped, scattered from the mountain pass to the inland basin or alluvial plain. Surface gradient, deposit thickness, and deposit granularity vary gradually from gentle, thin, to fine when moving from the mountain pass toward the edge (Fig. 8.3). The mountain pass area with the highest terrain is called the inner fan (also, fanhead or upper fan), the transitional edge zone of the inland basin or alluvial plain is called the outer fan (or lower fan), and the part in between is called the middle fan. According to the primary statistics from literature reports, more than 80% of the alluvial fans worldwide are arid.

This large humid alluvia fan was formed due to the riverbed movement from east to west 230 years ago.

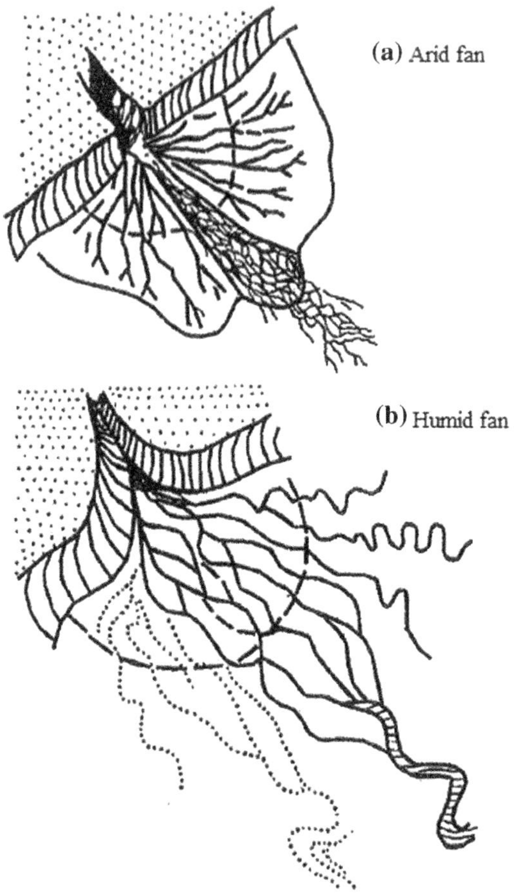

Fig. 8.3 Plane features of the dry fan and wet fan (according to Galloway, 1983)

8.1.2.1 Arid Fan

The major characteristics of an arid fan include the development of a main channel (braided river) and a clear fan boundary (Fig. 8.3; Table 8.1). Coarse clastic sediments become finer rapidly toward the end of the fan, and fan thickness decreases rapidly. The granulometric class can vary from gravel to mud. Mixed conglomerate and imbricate conglomerate layer deposition are mostly present at the source end of the fan (i.e., inner fan), and is characterized by current alluviation and sedimentation of debris flow. The alluviation sediment of a sandy gravel river is developed in the middle of the fan, while silt and argillaceous rock sediments are mainly seen at the end of the fan, which is dominated by sheet flood or unconcentrated flow. It is common to see inverse-cycle depositional sequences with a brownish-red coarse clastic profile and thickness of hundreds to thousands of meters.

8.1.2.2 Wet Fan (Braided Plain)

Wet fans develop in humid areas, where there are perennial currents, and they have unclear sediment fan shapes (Fig. 8.3). The braided plain is mainly composed of a gravel braided river with gentle terrain, which is commonly seen as a superimposed gravel braided river characterized by multiple channels, shallow cutting, and non-fixation. The sedimentary bodies extend long into the basin, which is different from an arid fan in terms of a lack of debris flow. A sequence combination that becomes finer upward is formed

Table 8.1 Geologic feature comparison between arid and humid fans

Characteristics	Type	
	Arid fan	Humid fan
Weather conditions	Arid	Humid
Current feature	Intermittent flow or flood	Perennial flow
Form	Clear fan shape	Unclear fan shape
Channel	Main or single channel	Superimposed channel, braided plain
Sediments	Dominated by paraconglomerate, poor sorting, and mixed sedimentation; fast vertical particle size change, common red bed, and gypsum-salt sediment; no coal bed	Orthoconglomerate development, no paraconglomerate, good sorting; gradual change of vertical particle size, no red bed or gypsum-salt sediment; with coal bed
Sedimentary structure	Few types, less development	Complete and developed
Gravity flow	Debris flow development	Lack of debris flow, available mud flow development
Facies belt	Clear facies belt distribution	Unclear facies belt distribution

in the middle and at the end regions, namely, the sedimentary section of conglomerate-sandstone-mudstone mingled with a carbon layer and the coal bed formed by autochthonous vegetation. As pointed out by Steel (1976), the size and shape of the catchment basin strongly influence the distribution of runoff. Therefore, alluvial fan systems supplied by a large watershed and developed in an area with relatively low rainfall show the features of wet fans. In contrast, alluvial fans formed in areas with abundant rainfall but a small watershed may show all the specific characteristics of arid fans.

8.1.3 Elements Controlling the Geometric Form of the Fan Shape

Alluvial fans have a stable plane and transverse modality, which means that they appear as lobe- or fan-shaped in plane and as wedges, lens with flat bottoms, or bulge top or mound-shaped transversally. Herein, we mainly refer to the vertical profile shape, which is subject to the tectonic settings of the basin edge, and the thickest and coarsest sediment is often next to the fault system at the basin edge. However, it is not rare for thickening to occur in the middle fan or even in the outer fan.

8.1.3.1 Control of Fracture at Basin Edge

Bull (1972) and Heward (1978) researched this topic from different perspectives, mainly involving the morphological characteristics of the longitudinal section of a single fan body. Heward's research focused on the control of peripheral fault features over the alluvial fan and the resulting vertical shape. He identified three different fault structures at the basin edge that control the geometry and scope of an alluvial fan system in the filling basin.

(1) Basin margin fault with persistent activity: This is mainly connected with a strike slip belt, which results in very thick aggradation along a relatively narrow zone (Fig. 8.4a). It is characterized by a high subsidence rate, high deposition rate, and rapid phase change in the side direction.
(2) Limited back-off fracture: In a graben or semi-graben, this type of fracture often forms an alluvial fan system of medium thickness

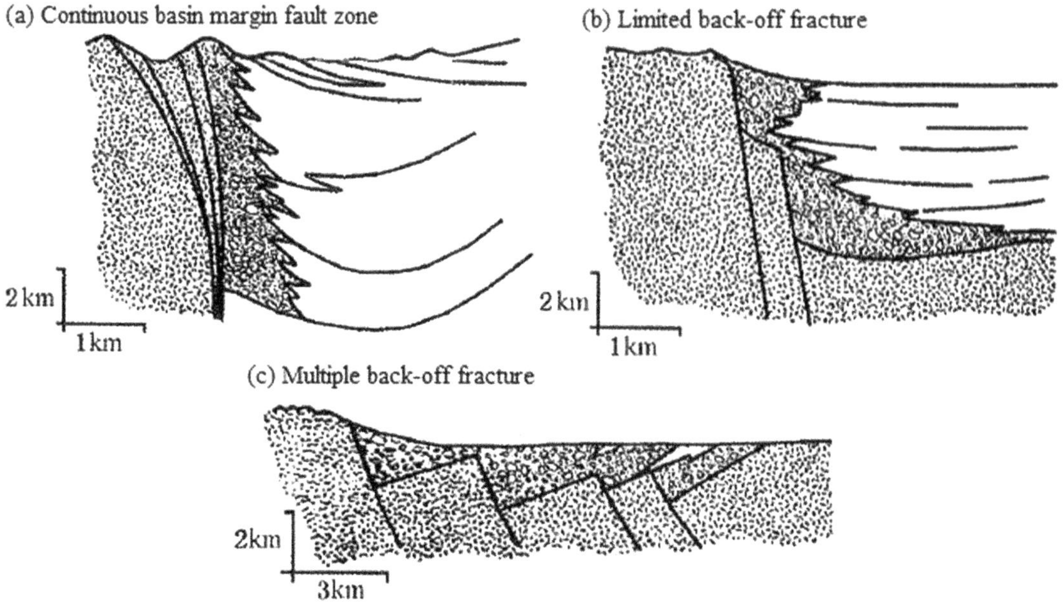

Fig. 8.4 Geometric difference in cross section of alluvial fan caused by different types of faults on the margin of the basin (after Heward 1978)

(Fig. 8.4b). Continuous uplift and incision may lead to the formation of an "overturned bedding" of the clastic type.

(3) Multiple back-off fracture: An alluvial fan conglomerate (Fig. 8.4c) with a wide ladder-like distribution and irregular thickness can be formed, whose prominent feature is a wide extension.

Fans with this cross section have developed widely in the Meso-Cenozoic inland basin in China. This basin is mainly characterized by the Yanshan period fault structure, and it leads to the formation of a faulted basin, dustpan-like basin, or titled basin. Alluvial fan depositional systems of different scales are developed at almost all sides close to the fault zone. The stronger the activity of the fault zone, the greater is the difference between the uplift and subsidence of the bilateral block, and the longer is the geological occurrence. In addition, the wider the basin range, the larger is the formed alluvial fan scale, the more diverse is the geometry, and the more complex is the internal structure and sequence structure.

1. **Christmas tree shape**

This shape is formed mainly due to continuous fault activity in one place. Huge, thick accumulations formed by the vertical superposition of multi-period alluvial fans have a superposed wedge-shaped Christmas tree–like appearance (Fig. 8.4a) owing to the descending basin, continuous superimposition of alluvial fans, and gradually decreasing source supply.

2. **Lens shape**

In terms of morphological characteristics, the thickness of the fan decreases in the middle and outer portion, and the bottom of the fan body is concave owing to the limited back-off fracture of the graben or semi-graben. This reflects the continuous uplift of the mountains during the formation of the alluvial fan. A large amount of sediment is received in the front of the mountain, which results in high thickness. In addition, the area close to the mountain front is eroded and this thins the sediment in the inner fan in order to form a concave-down lens sedimentary body.

3. **Imbricate shape**

Owing to the continuous uplift of mountains at the inner fan and the incision of the inner root, a stepped terrain is formed on the back of multiple back-off fractures, which results in the formation of a widely distributed imbricate alluvial fan body with irregular thickness. With the progress of back-off fractures, the facies decrease in age toward the outer fan.

8.1.3.2 Influence of Drainage Area and Provenance

The morphology and scale of alluvial fans are not only controlled by structural factors but also by the following factors: drainage area, sediment amount carried by supplemental water, granulometric class, composition of mother rock, terrain, and climatic conditions in the provenance.

(1) Alluvial fans with large drainage areas usually have a gentler slope than fans formed by the same material in a relatively smaller provenance.
(2) The gradient of an alluvial fan is steep when the granulometric class of the clastics and the sediment concentration increase.
(3) In areas with high precipitation, the gradient of an alluvial fan is gentle. However, fans in arid areas have steep gradients because debris flow sedimentation in an arid area is better developed than that in fluvial deposit.
(4) If fine-grained material, such as shale or argillaceous rock, flows through an alluvial fan, its area and slope will be larger and steeper than that of an alluvial fan with coarse sedimentary rocks such as sandstone, crystalline rock, and volcanic rock.

It is obvious that the development of arid and humid fans is controlled by geologic structures. Alluvial fans with large thicknesses tend to be widely distributed at the footwall of an active fault zone with large differential uplift and subsidence or at the edge of a faulted basin (Fig. 8.4).

8.1.4 Relationship between Area of an Alluvial Fan and Drainage Basin

Determining the area of an alluvial fan is very important for the prediction of the mineral distribution range, and it is of much concern to sedimentologists. The research on present-day alluvial fans shows that the fan area is mainly controlled by the scale of the catchment basin. As for fans in semi-arid areas, the relationship between fan area (A_f) and drainage basin area (A_d) can be expressed as follows:

$$A_f = C \times A_d^n$$

where C and n are empirical values, n is usually 0.8–1.0 while C is 0.15–2.10 and both values are connected to lithology, tectonic activity, and climate of the provenance area. For example, if the mother rock is argillaceous, its C value is twice that of sandstone, which means that the area of an alluvial fan formed by the former is about twice that of a fan formed by the latter. In other words, the better the stability of mother rock lithology, the smaller is the C value; and the poorer the stability, the larger is the C value.

There is usually a large altitude difference from the inner fan to the outer fan edge similar to the different scale surfaces of alluvial fans. The altitude difference within a large, active alluvial fan can reach several hundreds of meters, which means that there is very thick sediment in the proximal region. The altitude difference increases if the boundary (peripheral) fault continues to sink. However, most alluvial fans have a middle concave-down outline on the radiation profile. The gradient of a large alluvial fan is usually gentle, and the gradient ratio of an alluvial fan formed by a drainage basin dominated by mudstone (35–75%) is much larger than that of an alluvial fan dominated by sandstone. This is one of the features that distinguishes mud flow from alluvial fans.

As a special sedimentary body, mud flow is different from debris flow. Usually, mud flow can

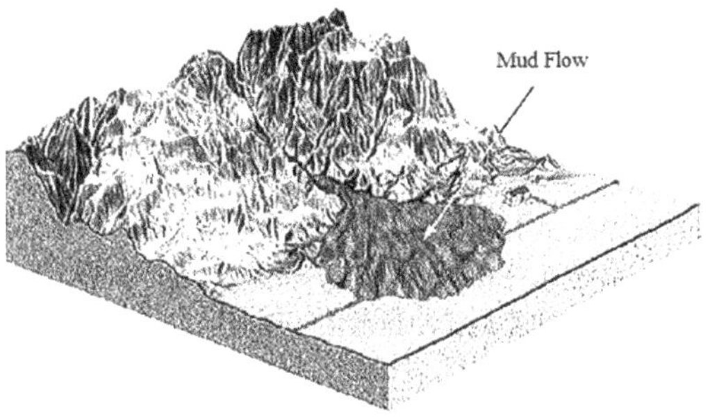

Fig. 8.5 Sketch Map of sedimentary model of the mudflow (according to the *Illustrated Dictionary of Earth Science*, 1999)

form a fan body at the downthrown side of a major fault at the basin edge, in other words, it is a variant of debris flow. Its deposit morphology can vary from thin and wide sheet to thick lobe body or fan body with obvious edges (Fig. 8.5), and it is composed mainly of fine grain and mud, usually with high viscosity. Its main formation conditions are as follows: ① damp climate or formation under water; ② large argillaceous supply and lack of sand and gravel supply; ③ steep slope (usually larger than that of a normal alluvial fan); and ④ provenance area dominated by plastic rock body.

In summary, the geologic features of alluvial fans depend mainly on ① tectonic activity; ② weather conditions; and ③ provenance supply and lithology. Tectonic activity mainly controls the shape of the fan body, deposition scale, deposition rate, and vertical sequence features. Weather conditions mainly control the type of alluvial fan and structure and distribution of sediments, and provenance nature mainly determines the internal composition and the distribution range of a fan body.

8.2 Sedimentation of Alluvial Fans, Sediment Types, and Features

The sedimentation of an alluvial fan mainly includes two types: ① debris flow sediment formed due to gravity and flood effects and ② water-laid sediment resulting from transient or intermittent flow action, which mainly falls into three types: sheet flood deposit, channel deposit, and sieve sediment. The former is mass flow with high viscosity, while the latter is fluid flow with low viscosity.

8.2.1 Debris Flow Deposit

Debris flow, which is a mixture of sediments and water, is a type of fluid with high density and high viscosity. Given its high material density, debris flow moves along the shear plane in the material aggregation. The particles are supported by an intergranular mud and water mixture and move under gravity. Usually, a flow with a sediment content higher than 40% (or even as high as 80%) is called viscous debris flow, whereas that with a sediment content of 10–40% is called diluted debris flow. Because viscous debris flow contains a considerable amount of mud base and its fluid intensity is very strong, it can lift and move giant boulders. By comparison, diluted debris flow is turbulent in nature. When the flow rate of debris flow becomes gentle, loads of different sizes are accumulated rapidly to form sediments comprised of gravel, sand, and mud with poor sorting. As a result, debris flow depositional facies barely have a massive internal structural layer. This means that particle size varies and granularity differs greatly (Table 8.2). Sometimes, inverse grading that becomes coarse upward can be seen (Fig. 8.6a).

8.2.2 Sheet Flood Deposit

As one of the main types of fan sedimentation, this is mainly comprised of schistous sands, silt, and gravel deposited by a wandering river. It can be considered as the product of sedimentation in wide shallow water, and is formed due to overflowing of the riverbed from the end of an alluvial fan river. Sheet flood deposit is composed of sediments such as plate-like sand, silt, and gravel. More specifically, it is distributed at the end of a downstream channel with water flow as shallow slope runoff. Therefore, the features of sheet flood are shallow water and rapid flow, as temporary water flow of the upper flow regime. It

Table 8.2 Sediment types and features of alluvial fan

Types and features		Debris flow	Channel deposit	Sieve sediment	Sheet flood deposit
Formation condition and hydrodynamic force		① Steep slope and rare vegetation ② Plenty of mud and clastics ③ Outburst flood	No vegetation developed and the terrain higher than the base level	Corniform or sub-corniform gravel mainly in the parent rock material supply with few fine grain	Flood deposit with low viscosity, short flow duration but high flow velocity
Development positions		Inner fan and middle fan	Both are distributed, but mainly in the middle fan	Inner fan and middle fan	Outer fan, often accompanied by coarse grain and incising the filling sedimentation of riverbed
Main geologic characteristics	Petrophysics	Mixed size and very poor sorting	Composed of gravel and sand, with medium to poor sorting	Composed of subangular coarse grain, good sorting, and multi-level particle supported	Composed of gravel, sand or few silt with clay, medium sorting
	Sedimentary structure	No bedding developed, mainly shown as bulk, possible with graded bedding	Poor bedding developed, great variation of single layer thickness, probable planar cross bedding, horizontal bedding and imbricate	Massive structure	Massive, cross bedding or parallel bedding can be seen
	Shape and occurrence	Leaf-shaped lingula mandibulae, mixed in the sheet flood deposits	Shown as incised-filling lens, rough and uneven bottom or concave upward, transiting with sheet flood deposit	Lens	Individual sand bodies shown as lens, and joint form plate-like shape

is mainly seen in the downstream zone of the channel down the middle fan. The sheet flood turns into a braided stream and sand bar rapidly after the flood peak. The sand layer has parallel bedding and anti-dune bedding, as well as trough or planar cross bedding (Fig. 8.6b), and a declining torrent can produce a fining-upward sedimentary sequence.

8.2.3 Channel Deposit

Since the channel on the alluvial fan is mostly distributed on its upper half, channel deposit mainly refers to the channel-filling sediments that are temporarily pitched in the alluvial fan. The typical inner fan channel is straight and deep, but it turns shallow gradually, mostly into a braided river. Down the crossing point (between the channel longitudinal section line and the fan surface), the river water tends to overflow the channel to form sheet flood. When there is sufficient groundwater recharge in the channel, the channel is developed down the crossing point to the outer fan. The channels on arid and semi-arid fans are mostly wide, but there are shallow intermittent rivers, of which the main sedimentation process happens in the short flood period

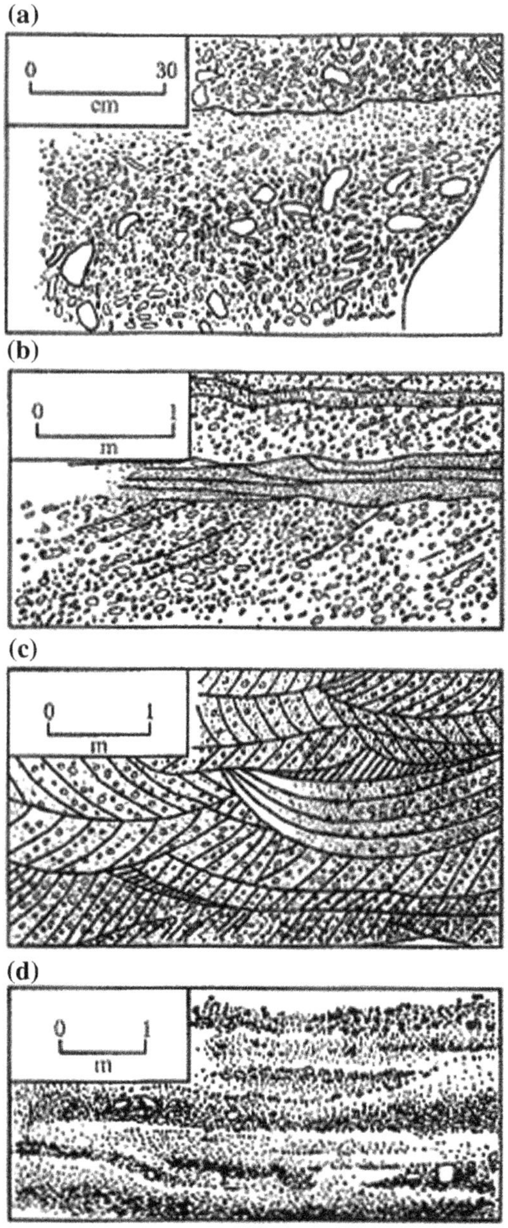

Fig. 8.6 Four kinds of gravel facies of alluvial fan sediments in old red sandstone, Scotland. **a** Paraconglomerate deposited by debris flow. **b** Conglomerate deposited by sheet flood. **c** Trough cross bedding sandy conglomerate due to channel deposit; and **d** Channel sedimentary conglomerate

in the rainy season. The sand layer has a transition flow regime and parallel bedding in the upper flow regime, as well as coarse planar and trough cross bedding (Fig. 8.6c, d), and the gravel is usually arranged in the imbricate form (Table 8.2). The channel on the alluvial fan is very unstable and is often under diversion. As a result, the drainage distribution in each flood period differs considerably, and old channel-filling sediments are often covered by later sheet flood sediments. Therefore, the channel depositional facies transit upward to the sheet flood depositional facies to form a fining-upward cycle.

8.2.4 Sieve Sediment

Sedimentary bodies constituted of gravel are formed when the sediments carried by flood lacking fine-grained materials (silt and mud). Owing to the extremely high porosity and permeability of gravel layers, tongue-shaped gravel layers close to the part under the crossing point act as sieves, which cause the water flow to permeate underground from the gravel layer and lodge the carried fine-grained detritus in the pores between large gravel particles to form sand gravel with a bimodal particle size distribution, which is also called sieve sediment. In general, sieve sediment is a type of local accumulation in the fan body, which causes a further decrease in fan body slope. It can be seen as a tongue-shaped gravel layer sediment developed on the fan body surface.

The abovementioned four types of sediments (or depositional facies) are distributed randomly in the alluvial fan, and their distribution varies with flow change in each flood period and with changes in the sectoral drainage distribution. In most cases, the entire alluvial fan is not submerged in each flooding. Sections of various sizes are always exposed on the water surface. Therefore, channel filling and sheet flood deposit are the most common and widely distributed depositional facies in the depositional district. On an alluvial fan with sufficient fine-grained sources, debris flow sedimentation also occupies most of the region upside of the alluvial fan, and sieve accretion is developed only locally. Strong evaporation leads to the formation of suncracks

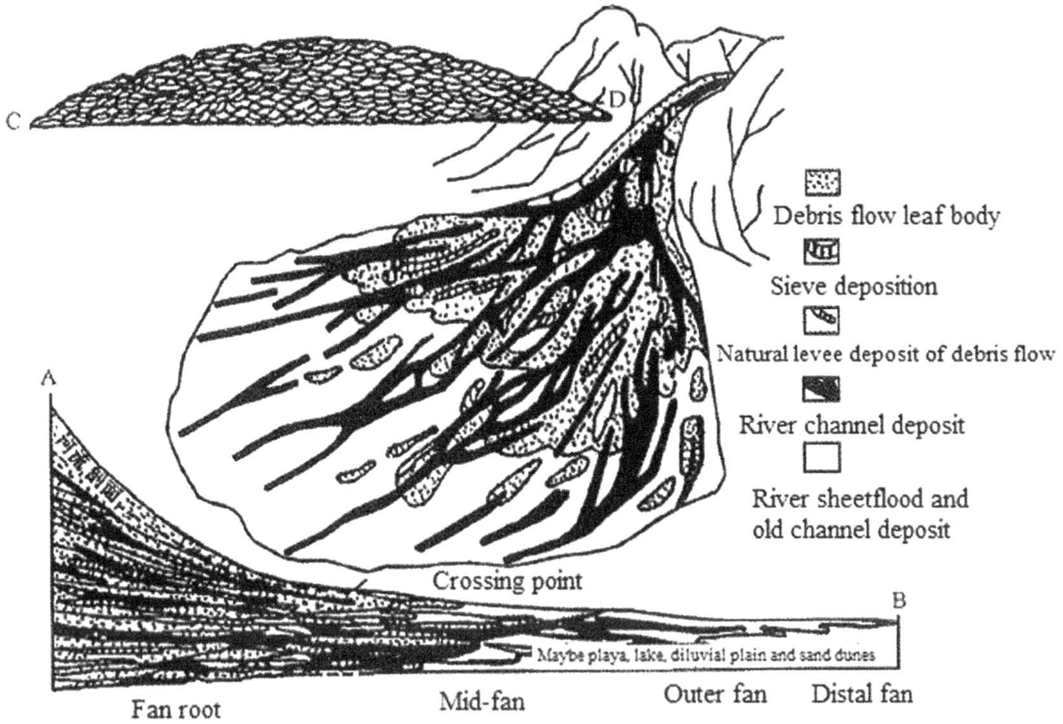

Fig. 8.7 Model and section of sedimentary faces in the ideal alluvial fan. AB longitudinal section, CD transverse section

or calccrust on the surface of fine-grained sediment, whereas strong oxidation results in the formation of dark-colored minerals with iron and magnesium in clay and hematite, which make the sediments red.

8.3 Sedimentary Environment and Sedimentary Sequence of Alluvial Fans

8.3.1 Sedimentary Environment

According to the landform of alluvial fans and distribution characteristics of sediments, alluvial fans can be further divided into three subenvironments (subfacies), i.e., inner fan, middle fan, and outer fan, with no clear boundaries among these subenvironments (Fig. 8.7). However, each overall sedimentary sequence has its own features.

8.3.1.1 Inner Fan

The inner fan is distributed at the bluff adjacent to the top zone of an alluvial fan. It features the largest sedimentary slope angle and 2–3 straight, deep single main channels. Its sediments are composed of conglomerate and gravel with an imbricate structure or of conglomerate with poor sorting, no structure, and mixed size. Therefore, debris flow sedimentation and braided channel sedimentation mainly occur in this area.

8.3.1.2 Middle Fan

As a key component of an alluvial fan, the middle fan is located in the middle part and is characterized by medium to relatively low sedimentary slope and developed braided channels. Therefore, its sediment is composed mainly of sandstone, conglomeratic sandstone, and conglomerate. Compared to the depositional facies of the inner fan, the textural maturity of sand and gravel is better. Trough cross bedding is formed

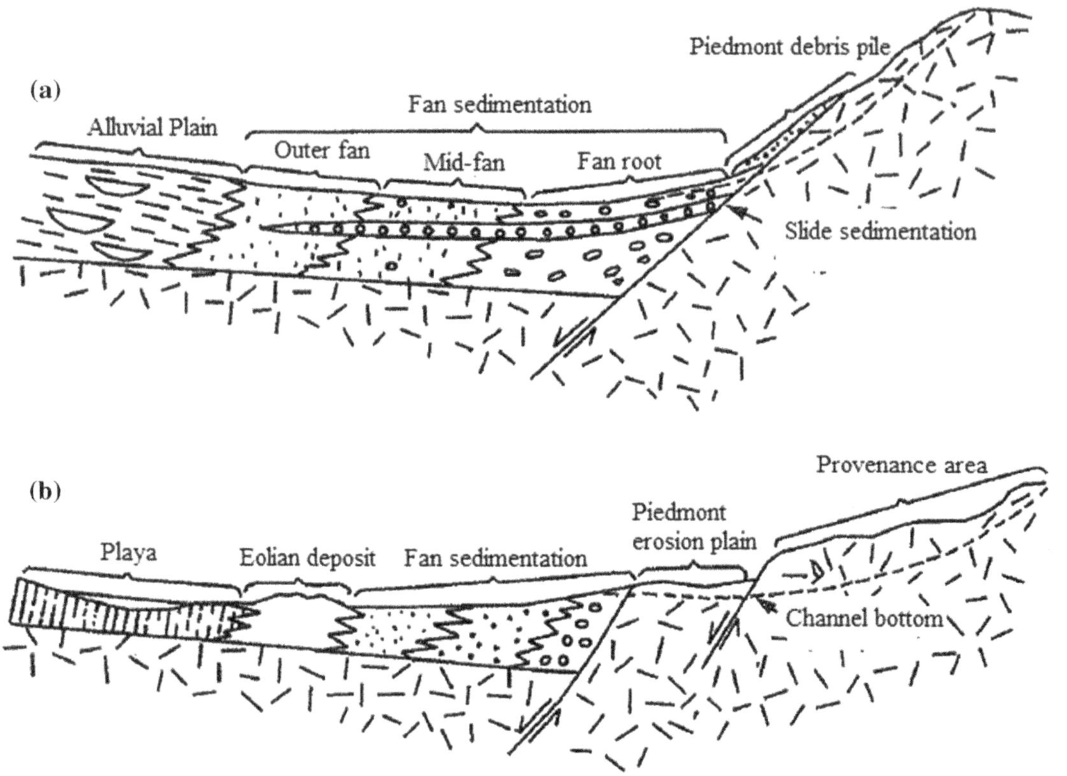

Fig. 8.8 Sketch map of cross section of alluvial fan and adjacent sedimentary environment (after Nilsen and Zuffa 1982). **a** Shown as finger-like alternate variation with alluvial plain; and **b** Shown as finger-like alternate variation with playa and eolian deposit

due to the frequent swing of the braided river and incisions, and the development of antidune cross bedding in the scour structure can be seen. Sieve sediments and braided channels are developed mainly in the middle fan, where the channel bar is clearly lower than that of the inner fan and channel width decreases.

8.3.1.3 Outer Fan

This appears at the toe of an alluvial fan, and its geomorphic features include low sedimentary slope or gentle terrain. Its sediments are usually composed of sandstone and pebbly sandstone, mixed with siltstone and clay rock. In addition, gypsum-salt layers can be seen locally. The finer the sandstone particle size, the better the sorting. Moreover, there may be deformation structures and exposure structures (for instance, mud cracks and raindrop imprints). Thus, sheet flood deposit and braided channel sedimentation are mainly seen in this area.

8.3.2 Vertical Sedimentary Sequence

During the formation and development of an alluvial fan, particle size and thickness change from the inner to outer fan, which is always characterized by coarse-to-fine and thick-to-thin, respectively. The debris flow facies is mainly distributed in the inner fan, while sieve sediments are present in the inner and middle fans, but mainly in the middle fan. Although channel and sheet flood deposits are developed in all fans, the middle fan–outer fan is composed mainly of two facies. In the direction of the basin, the alluvial fan transits into the inland basin (playa and eolian deposit) and flood plain (Fig. 8.8).

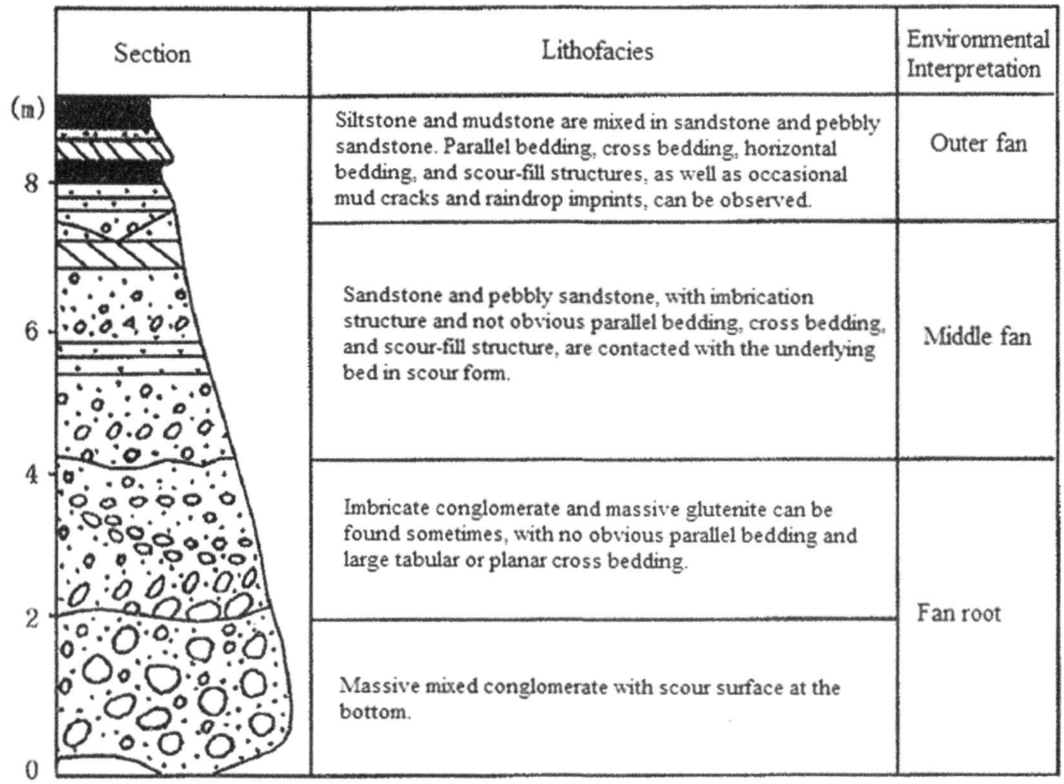

Fig. 8.9 When the sediment accumulation rate is smaller than the basin sedimentation rate, then, the sedimentary sequence characteristics of fining-upwards sequence are formed (according to Yongchuan and Huisheng 1985)

8.3.2.1 Formation Process and Characteristics of Alluvial Fan Sequence of Fining-upwards Cycle

Because the accumulation speed of sediments is different from the subsidence rate of the basin, the sand body of an alluvial fan can undergo progradation and retrogradation or lateral transfer. This process is obviously reflected in each sedimentary sequence of the alluvial fan. When the sediment accumulation speed is higher than the subsidence rate of the basin, the sand body of an alluvial fan will move forward gradually toward the basin, which causes the inner fan sediment to overlap with the middle fan sediment and the middle fan sediment to overlap with the outer fan sediment. This results in the formation of an inverse-cycle sedimentary sequence featuring fine downward and coarse upward. Retrogradation is opposite in the provenance area direction. When the sediment accumulation speed is lower than the subsidence rate of the basin, the sand body of an alluvial fan will be retrograded to the provenance or transferred laterally to result in the formation of a coarse downward and fine upward sedimentary sequence (Fig. 8.9).

8.3.2.2 Distribution Law

The sedimentary sequence differs in different parts of an alluvial fan (Figs. 8.10 and 8.11). The sedimentary sequence of the inner fan is mainly fining-upward sedimentary association composed of massive mixed conglomerates and gravels arranged in the imbricate form. The sedimentary sequence of the middle fan is composed of gravels arranged in an imbricate arrangement with no obvious parallel bedding, cross bedding conglomeratic sandstone, and sandstone from bottom to top. The sectional structure of the outer fan is usually pebbly sandstone with scour-fill structures, cross bedding and parallel bedding

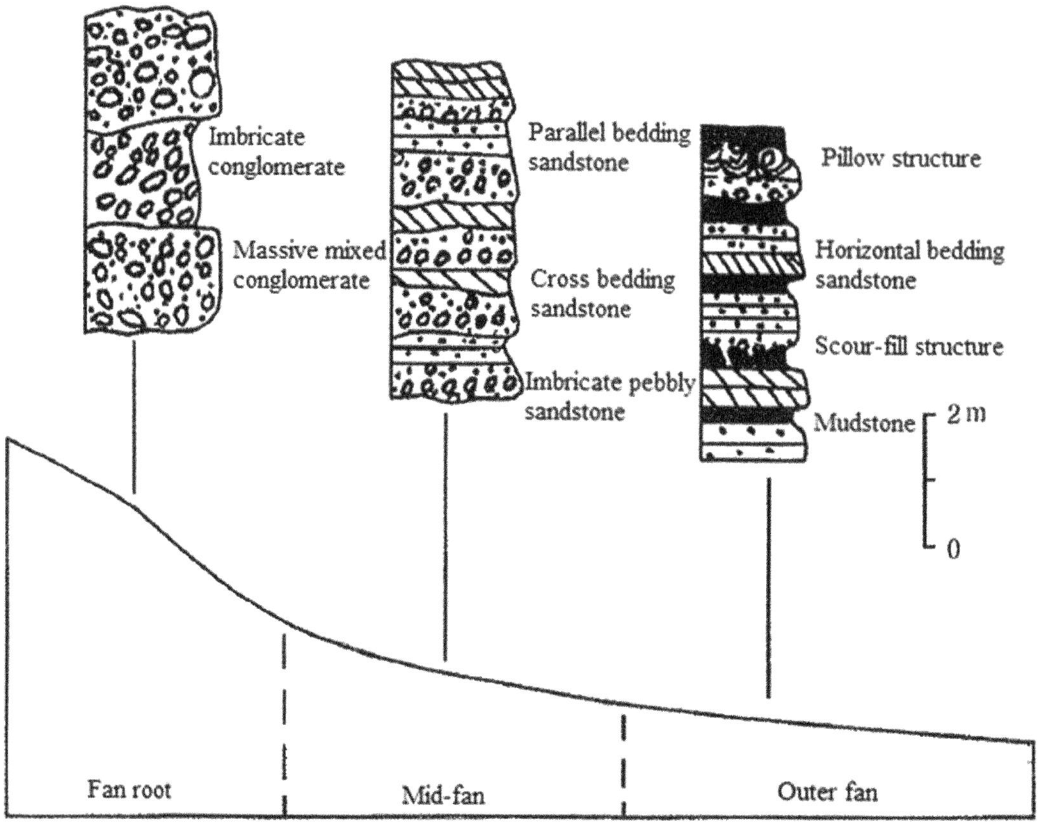

Fig. 8.10 Sedimentary sequence characteristics in different sub-environment of alluvial fan (according to Yongchuan and Huisheng 1985)

sandstone, and horizontal textured siltstone or massive mudstone. However, sometimes, a deformation structure develop, for instance, the pillow structure.

8.3.2.3 Main Influencing Factors

Modern and ancient large alluvial fans are usually developed at the footwall of a peripheral fault. In addition to fluctuations in climate, activity of the underlying geologic structure plays an important control role in the development of alluvial fans and the structure of their internal sequences. In tandem with the activity of the peripheral fault, an alluvial fan migrates, shrinks, or advances continuously. Different periods of adjacent alluvial fan bodies incise or overlay each other to form a sequence and cycle with large thicknesses and a complex structure. They can be either sequence coarsening and thickening upward or sequence fining and thinning upward, or they can be proximal facies overlapping on distal facies or inverse sedimentary sequences. Most commonly, one can find complex megascopic sedimentary sequences constituted of multiple-cycle fining or thinning upward (Fig. 8.12).

The cycles can be divided into two categories based on causes and characteristics: auto cycle and allo cycle. Auto cycles are formed during fan body migration, abandonment, and progradation caused by flooding. They are usually small-scale sequences with thicknesses ranging from one meter to several meters and can be both fining upward and coarsening upward. Auto cycles usually have no direct connection with tectonic activity, or they are formed due to sedimentation when the base level does not change. In contrast, allo cycles are large sequences caused by

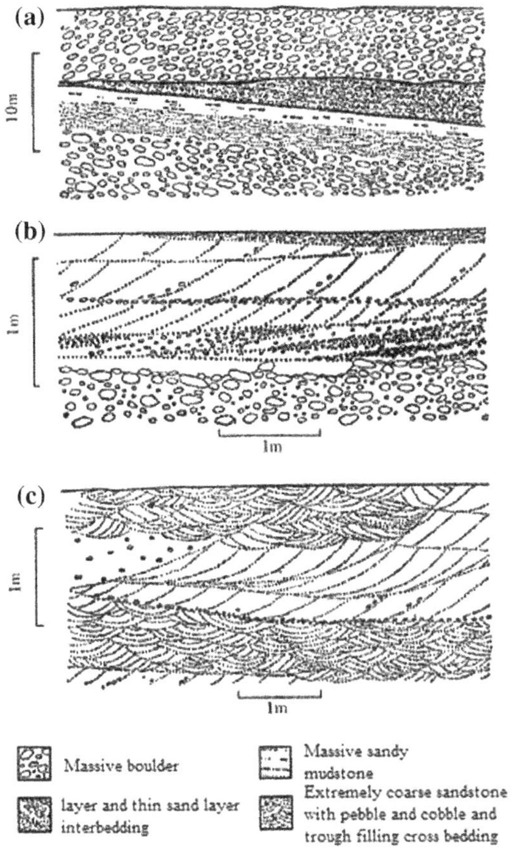

Fig. 8.11 Wetland fan deposits in Van Horne, Texas (after McGowen and Groat 1971). **a** Inner fan: boulder parameter can reach 1 m, and gravel is the key component; **b** Middle fan: conglomerate and cross bedding with conglomeratic sandstone interbedding; and **c** Outer fan: sandstone with planar and trough cross bedding

periodical changes in climate and tectonic activity, with thicknesses ranging from tens of meters to hundreds of meters. Major sequence changes caused by tectonic activity are complex and generally connected with depocenter transfer, change in basin-fill structure and erosion, and overall migration of the fan body. A. P. Heward proposed three cases in 1978:

(1) Megasequence coarsening and thickening upward is formed because periodic uplift of the provenance area or basin subsidence causes river rejuvenation, which leads to migration of the alluvial fan prograding toward the basin to the proximal coarse-grained sedimentary facies on the fan edge and covering the distal fine-grained sedimentary facies, which implies megasequence cycle coarsening and thickening upward due to provenance area uplift or basin sedimentation.
(2) If there is slowing or a short pause in the process of provenance uplift or basin sedimentation, sequence fining and thinning upward are formed on sequence coarsening and thickening upward as provenance turns gentle or razed, and during this process, symmetric or nearly symmetric depositional cycles can be formed.
(3) Simple sequence fining and thinning upward are usually caused by escarpment retreat due to faulting and gradual deplanation in the denuded zone.

8.4 Identification of Alluvial Fan Sedimentation

8.4.1 Sedimentary Marks for Identifying Alluvial Fans

(1) Because fan sediments are formed under oxidizing conditions, which lack organic matter and sediments with reducibility, the rocks in an arid fan generally have a red hue.
(2) Because the fan sediments tend to evaporate, salt sediments are often present, for example, gypsum and calcite, which are distributed in the outer fan and shown as nodular or laminated occurrences.
(3) Fan sediments are usually composed of river sediments of a wandering river (braided river) or debris flow sediments, where debris flow sediments and sieve sediments are good identification marks of an alluvial fan.
(4) Conglomerates account for a significant proportion of the entire sequence, and particle size distribution is more than one mode on the histogram, which is also called multiple kurtosis, which means the mixed

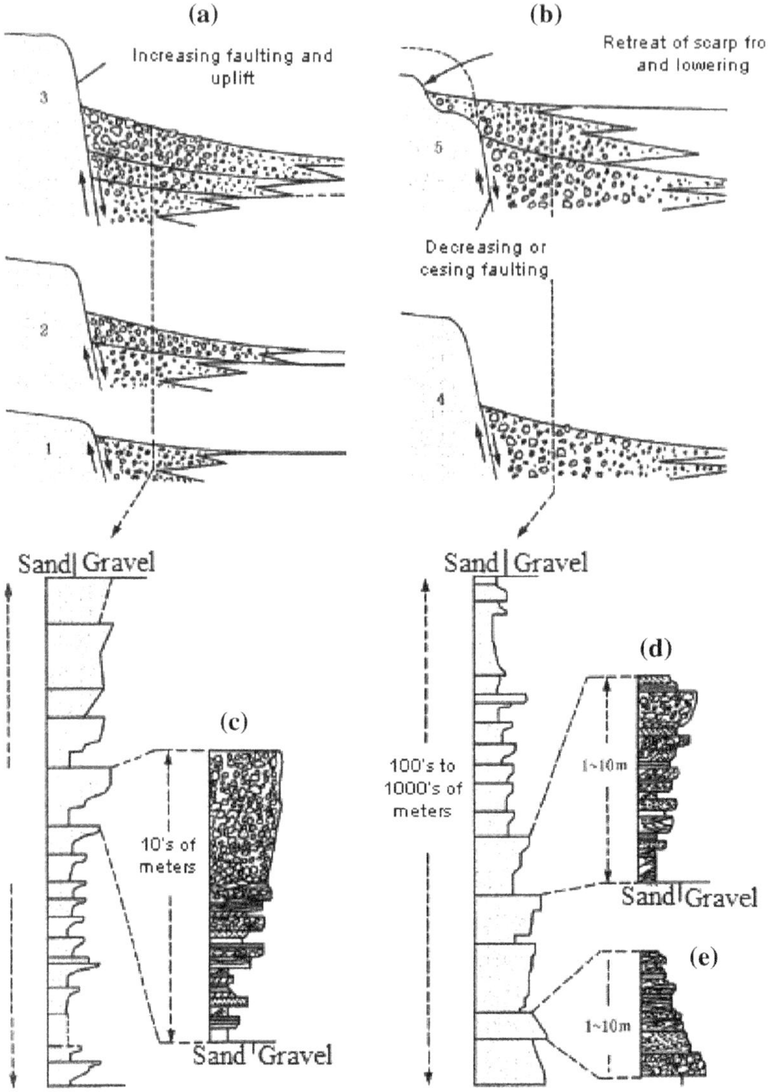

Fig. 8.12 Ideal alluvial fan sedimentary sequence formed in different tectonic evolution stages (after Ethridge and Wescott 1984). **a** Large-scale coarsening-upward sequences are caused by persistent fault activity and progradation of alluvial fan (stages 1–3); **b** Large-scale fining-upward sequences are caused by retreat of provenance escarpment and decline of ground surface (stages 4–5) or lateral movement or migration of fan body; **c** and **d** Small-scale coarsening upward cycles are caused by progradation of individual fan lobes; and **e** Small-scale fining-upward cycles are caused by filling of braided channel and migration of sand bar

accumulated sand and conglomerates have low texture maturity and poor sorting, thus reflecting the near transport distance of sediment.

(5) It is developed under the geographic background of a basin edge with frequent tectonic activity and large relief amplitude, and characterized by near transport, fast sedimentation, and coarse granularity. Therefore, the bedding structure is barely developed, and although there is often a massive structure, it develops not obvious cross bedding, parallel bedding, and scour-fill structure. In addition, no obvious graded bedding can be

found and imbricate gravel arrangement is very common.

(6) On the C-M diagram, two types are often shown: one is the curve of tractive current–type, which represents transient flow deposit, for instance, the T curve; and the other is the linear graph of turbidity current–type, which is approximately parallel to the C = M base line, for instance, the M-curve, which represents debris flow sedimentation. The C value on the shaft of the debris flow is about 45 times the M value.

(7) The water flow mode of an alluvial fan is shown as radiation, and data on historical water direction can reflect the features of radiation flow.

(8) Apart from scattered vertebrate skeleton and phytodetritus, usually, biological fossils are absent, and plant debris is relatively low.

8.4.2 Features of Alluvial Fans on Seismic Sections

Because an alluvial fan is a continental depositional system developed at a basin edge, its overall feature is "chaotic reflection and mound downlap," and the specific seismic recognition marks are as follows:

(1) Developed under the large fault of a basin edge with vertical fault strike development (Fig. 8.13).

(2) The sedimentary bodies form mounds on the cross section and a wedge shape on the longitudinal section, thinning inside the basin and show an obvious taper shape as a whole. However, the horizontal arrangement of multiple alluvial fans along the fault boundary leads to the formation of an alluvial fan skirt.

(3) Chaotic progradation structures are the most common on the longitudinal section, followed by the complex sigmoid-oblique progradation structure and oblique progradation structure, whereas chaotic reflection and the progradation reflection structure with bidirectional downlap are common on the cross section. The common point of the progradation structure is the usual lack of a bottom set, and the foreset and the substratum are shown as a downlap contact.

(4) Its reflection is mainly of the chaotic reflection configuration or no reflection configuration. In general, amplitude is strengthened from the inner to the outer fan and continuity improves.

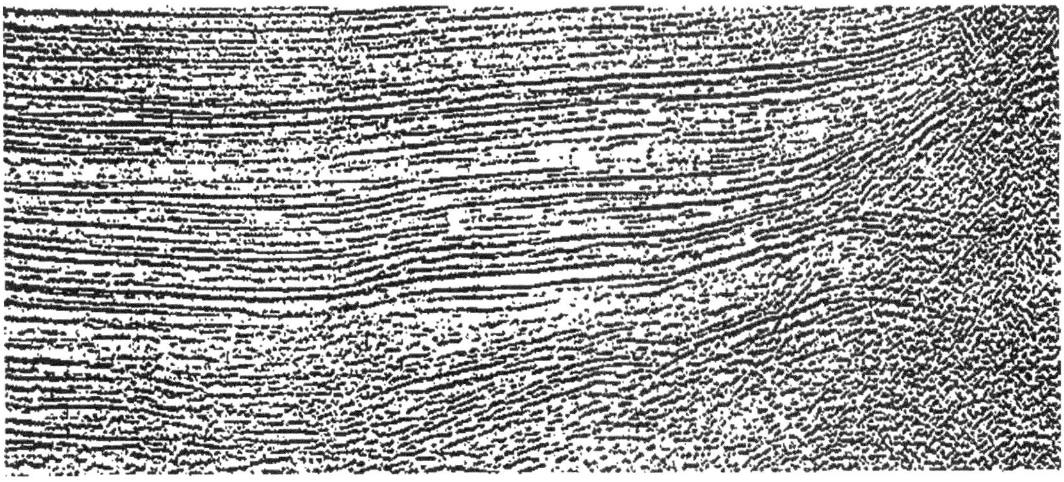

Fig. 8.13 Seismic characteristics of alluvial fan (*Typical Seismic Section Atlas of China*, 1987)

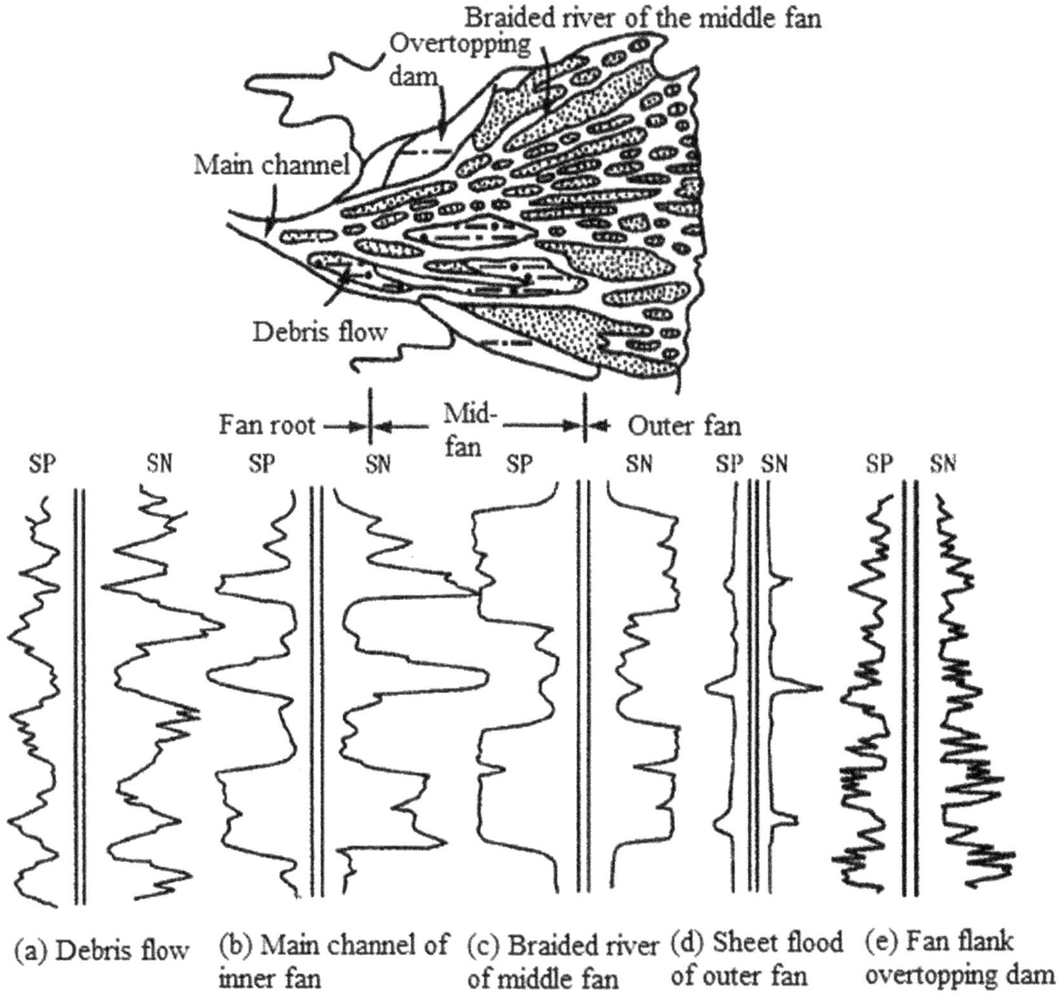

Fig. 8.14 Characteristics of logging curves in different parts of alluvial fan

8.4.3 Electric Well-logging Curve Features for Each Part of an Alluvial Fan

8.4.3.1 Inner Fan

1. **Debris flow**

This can be seen as uneven serrations on the resistivity curve. The peak value is usually high with gradual change at the top and the bottom. However, on the SP curve it can be seen as medium-amplitude serrations, with gradual change at the top and the bottom interfaces (Fig. 8.14). Its feature is a high-amplitude serrated finger shape.

SP—natural potential curve; and SN—potential polar system resistivity curve of short polar distance

2. **Main channel deposit**

On the electrical resistivity electric well-logging curve, the main channel of the inner fan is shown as a large-amplitude curve with a serrated edge, and its abnormal amplitude is usually the largest in the entire alluvial fan. The abnormal amplitude of SP is also large, but not the largest. The interface curve shape is mostly abrupt cylindrical at the top and the bottom. Occasionally, it changes gradually from top to bottom or has a bell shape.

8.4.3.2 Middle Fan

The middle fan is dominated by braided river sedimentation. On the resistivity curve, the braided river sedimentation of the middle fan is shown as a medium-amplitude cylindrical or bell-shaped curve with a serrated edge.

Although the shape of the self-potential log is presented as cylindrical or bell-shaped with a serrated edge, its amplitude is usually not the largest.

In general, the interface forms of electric well-logging curves feature a gradual change at the top and bottom, or an abrupt change at the bottom and gradual change at the top.

8.4.3.3 Outer Fan

On the electric well-logging curve, outer fan sedimentation has very a small amplitude. Sometimes, a low flat curve with a small serrated peak of thin sandstone, or small serrated finger-like and beak shapes can be seen. The amplitude curve of outer fan sedimentation is the smallest across the entire alluvial fan.

8.4.3.4 Flank Overbank Deposit

This is shown as an uneven serrated curve on the electric well-logging curve, with many serrated peaks of small amplitude, which tend to decrease moving upward.

8.5 Geometry and Reservoir Features of Alluvial Fans

8.5.1 Geometry of the Fan Body

The geometry of the vertical profile shape of an alluvial fan system depends mainly on the tectonic setting at the basin edge, and the thickest and coarsest sediment is often next to the fault system at the basin edge. However, thick sediment is not rare in the middle fan or even the outer fan. Both Bull (1972) and Heward (1978) researched this from different perspectives. Bull's research focused mainly on the morphological characteristics of the longitudinal section of a single fan body, while Heward's research focused on control of the peripheral fault features over the alluvial fan and the resulting vertical shape.

8.5.2 Complex Microscopic Pore Structure

Alluvial fan sedimentation is dominated by conglomerates, and it falls into the most proximal depositional system in clastic rocks. In the process of sedimentation, intermittency is the highest and sorting is at its poorest. The composition of rock particles can vary from boulder to mud, and the granularity frequency distribution of rock has multiple peaks. These features fundamentally determine the particularity of alluvial fan gravelly reservoirs.

8.5.2.1 Pore Type

Pore structure features of oil reservoirs, including pore type and combination form, can be studied vividly and quantitatively by analyzing scanned images of rock samples and casting sheets. The following pore types can be seen in conglomerate reservoirs: intergranular pore, intergranular dissolved pore, intragranular dissolved pore, micro dissolved pore, and micro-chink in matrix or cement (Table 8.3).

8.5.2.2 Throat Characteristics and Pore and Throat Combination Types

Different from a sandstone reservoir throat, a conglomerate reservoir throat has a small mean value and non-uniform size. Most capillary pressure curves of the reservoir are steep, and the throat frequency distribution features multiple and low peaks. On the capillary pressure curve, we can see four types of shapes, namely, a multi-peak coarse shape (Fig. 8.15 curve a), multi-peak fine shape (Fig. 8.15 curve b), single-peak fine shape (Fig. 8.15 curves c and d), and single-peak coarse shape (Fig. 8.15 curve e).

Table 8.3 Pore types of glutenite reservoir of alluvial fans

Pore type	Genesis	Pore diameter (μm)	Characteristics
Intergranular pore	Primary pore	59–750	Generally larger than Φ 20%, K > 100 × 10^{-3} μm^2, mostly polygon, with ligancy of 3–4
Intergranular dissolved pore	It is mainly formed due to the corrosion of part of cements or particles	350	Pore size is not limited by particle boundary, featuring embayed edge with irregular shape
Intragranular dissolved pore	Some easily soluble material (feldspar) is corroded into wrack	Generally less than 100	It retains original mineral crystalline form, and is developed in the reservoir with strong diagenesis
Micro dissolved pore in the interstitial material	It mainly includes calcite intergranular, hydromica, black mica, kaolinite and zeolite small dissolved pores	Generally less than 10	In the case of development in groups, it is mainly in the cellular and interconnected form and constitutes a fine mesh channel
Micro-chink	It includes gravel-edge fracture, cleavage fracture, and tectonic fracture	2–35 (gravel-edge fracture) <0.1 (cleavage fracture)	Gravel-edge fractures refer to linear fractures along the edge of coarse gravel; cleavage fracture is mainly developed in mineral; and tectonic fracture refers to micro fractures in a tight reservoir

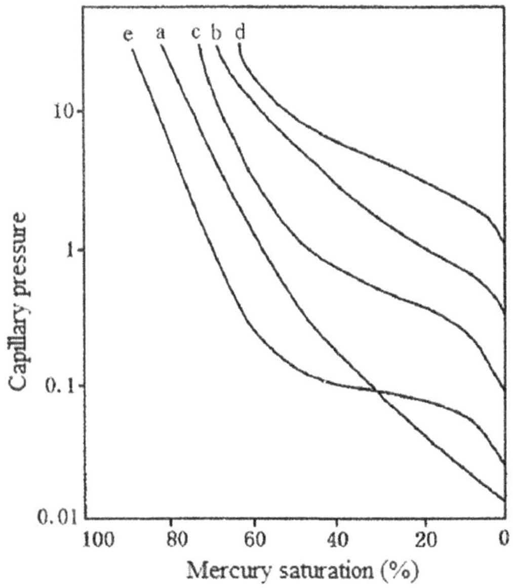

Fig. 8.15 Characteristics of capillary pressure curve (according to Zhang 1991)

In general, four types of pore-throat combination forms are often seen in reservoirs: ① Intergranular pore + intergranular dissolved pore + intragranular dissolved pore; ② Intergranular dissolved pore + intergranular pore + matrix pore; ③ Matrix pore + intergranular (intragranular) dissolved pore + gravel-edge fracture; and ④ Gravel-edge fracture + intergranular (intragranular) dissolved pore + cement dissolution pore (Table 8.4).

8.5.3 Lithology and Physical Characteristics

The rock composition of conglomeratic reservoirs has strong succession in terms of the nature of the source rock in the provenance area, and there are many unstable materials. Specifically, the detritus content is high, while the quartz content is low.

8.5.3.1 Clastic Composition

It is mainly quartz, feldspar, and debris. The quartz content can reflect the compositional maturity of rocks, and the quartz content in a conglomeratic reservoir is generally as low as 6–41%, which relates to the nature of the source rock and the sedimentary environment in the

Table 8.4 Pore and throat combination types and characteristics

Combination type	Main characteristics	Porous development	Pore communication	Pore diameter (pm)	Pore throat coordination number	Physical property		Evaluation
						Porosity (%)	Permeability ($\times 10^{-3}$ μm^2)	
①	Mainly intergranular pores and intergranular dissolved pores	Good-medium	Network shape	50–468 (136)	1–3	21.1	>220	Primary coarse pore group
②	Intergranular dissolved pores account for more than 50%, followed by intergranular pores, matrix pores, and microfractures	Medium-poor	Approximate network shape	18–475	0–3	17.7	179	Secondary coarse pore group
③	Mainly intragranular pores and matrix pores, accounting for 65%, followed by intergranular dissolved pores, small amounts of gravel-edge fractures and microfractures	Relatively poor	Medium	13–375 (65)	0–1	15.5	37.2	Micro-pore group
④	Mainly gravel-edge fractures and microfractures	Poor-extremely poor	Few	10–125 (29)	0–1	12.6	0.5	Fractured pore group

Table 8.5 Types and characteristics of clay minerals in alluvial fan reservoirs

Clay minerals	Content (%)	Shape	Occurrence
Kaolinite	30–50, even up to 65	Book-like, partly wormlike or mainly irregular shape	Mainly filling type
Illite-montmorillonite mixed layer	About 30	Mostly irregular and crumby structure	Mainly packaged type
Illite	10–30	Schistose	Bridge plug type, packaged type and filling type
Chlorite	10–20	Single crystal being schistose	Varied; mainly packaged type
		The aggregation mostly being irregular shape	Filling type

provenance area. A high detritus content is an important feature of alluvial fan conglomerate reservoirs. The detritus content in the Karamay conglomerate can be higher than 70%, and its debris is composed mainly of volcanic rock, metamorphic rock, and sedimentary rock cuttings.

8.5.3.2 Distribution of Clay Minerals

Clay minerals have a direct impact on the waterflooding development effect because of their specific physico-chemical properties and sensitivity to the injection fluid. The clay membrane around clastic particles can increase the particle radius by 1–6% in order to greatly decrease the pore throat (if the particle size is increased by 4%, the throat diameter decreases by 26%) and reduce permeability.

Authigenic clay is the major component in conglomeratic reservoirs. According to X-ray diffraction analysis data, there is mainly kaolinite or an illite-montmorillonite mixed layer or illite, and chlorite increases in the deep part. Expanded clay materials, such as halloysite and montmorillonite, are contained in some reservoirs (Table 8.5).

8.5.3.3 Physical Characteristics

Most alluvial fan conglomerate reservoirs are of medium and low porosity, while their permeability varies greatly. For instance, the mean porosity of the Permian and Triassic Karamay Oilfield conglomerate reservoir is only 15%, while conglomerate porosity in the Caoqiao Oilfield is 11–14%. Permeability is very sensitive to rock structure. The sand/gravel rate and shale content change greatly, and the existence of few intergranular pores with large diameters can cause large variations in permeability. For instance, in the alluvial fan conglomerate reservoir of the Guantao Formation in the Caoqiao oilfield, the permeability is 60–6000 $\times 10^{-3}$ μm^2. The permeability of the Karamay Oilfield is 10–400 $\times 10^{-3}$ μm^2. Lithology has a strong impact on the physical property. Permeability increases with an increase in the median grain diameter. After reaching a certain level, permeability remains the same or declines (Table 8.6).

8.5.4 Reservoir Heterogeneous Characteristics

8.5.4.1 Plane Heterogeneity

The features of alluvial fan reservoirs differ depending on the microfacies belts on the plane. The inner fan subfacies conglomeratic body is shown as a chunk with a large planar extension and relatively stable distribution facies. It narrows down or is even missing in some areas due to its ancient high terrain. The transverse section has a biconvex or up-convex shape, whereas the longitudinal morphology is wedge shaped. The middle fan subfacies conglomeratic body is mostly layered on the transverse section, and the interlayer barrier bed is relatively developed. The braided stream belt microfacies conglomeratic

Table 8.6 Relationship between reservoir lithology and physical property in Karamay Oilfield (according to Qiu Yinan et al. 1997)

Median grain diameter (mm)	Petrophysics	Sorting coefficient	Porosity (%)	Permeability
0.125–0.25	Fine sandstone	2.78	17.68	77
0.25–0.50	Fine-coarse sand	2.80	17.92	174
0.50–2.0	Very coarse sand-conglomeratic sandstone	2.87	15.70	416
2.0–5.0	Fine-grained small conglomerate	3.50	14.51	3062
5.0–10	Coarse-grained small conglomerate	4.43	13.95	238
>10	Big conglomerate	4.72	13.26	85

body on the plane is presented as a curving stripe along the direction of the palaeocurrent, and the braided stream sand island (channel bar) microfacies sandbody is in an islet form inlaid between braided stream belt microfacies sandbodies or braided stream belt sandbodies and sheetflood belt sediments (Fig. 8.16). The outer fan subfacies are mostly sheet flood deposits, mudflats deposition, and alluvial plains, and the inter-channel braided sand body is distributed as a few lens with poor continuity. However, it is basically composed of sandy conglomerate and is a connected component of the reservoir. Therefore, in general, alluvial fan reservoirs have good planar continuity.

In addition, as mentioned previously, the fan body size can vary greatly and be influenced by multiple factors such as drainage area, slope, climate, lithology in the provenance, and paleotopography in the depositional district. As long as the fan body radius is of the order of a thousand meters, reservoir continuity will not be the principal contradiction in oilfield development. For instance, the single oil-bearing conglomeratic body in the Karamay Oilfield is 1–3.3 km^2, and the drilling rate of the development well pattern (400 m) is higher than 65–80%.

However, the alluvial fan reservoir petrophysics, especially permeability, is very sensitive to the rock structure, and plane heterogeneity is still very serious, which is manifested mainly in the difference between the facies and belt. In general, the physical property turns deteriorate from the inner fan to the outer fan, hence the physical property of microfacies in the main channel of each subfacies belt is the best. There can be obvious changes from the proximal to the distal direction in the same microfacies, and flood events with different energies can make the same kind of microfacies belt reservoir property with great differences (Table 8.7).

8.5.4.2 Intrastratal Heterogeneity

In general, the rock granularity of alluvial fan conglomeratic is mostly fining upward vertically. Because of frequent periodic changes in hydrodynamic force during the development of an alluvial fan and gravel braided river reservoir, the formed bedding structure is rather complex, and the development scale differs greatly. Flood bedding and megascopic cross bedding are developed in the alluvial fan, and ripple cross bedding and horizontal bedding can be seen in the last stage of alluvial fan development, however the flood bedding indicates that facies are developed especially in inner fan subfacies. The development and distribution of thin muddy intercalation greatly affects the vertical permeability of conglomerate reservoirs, however it is less developed and is only 1/2–1/100 of the

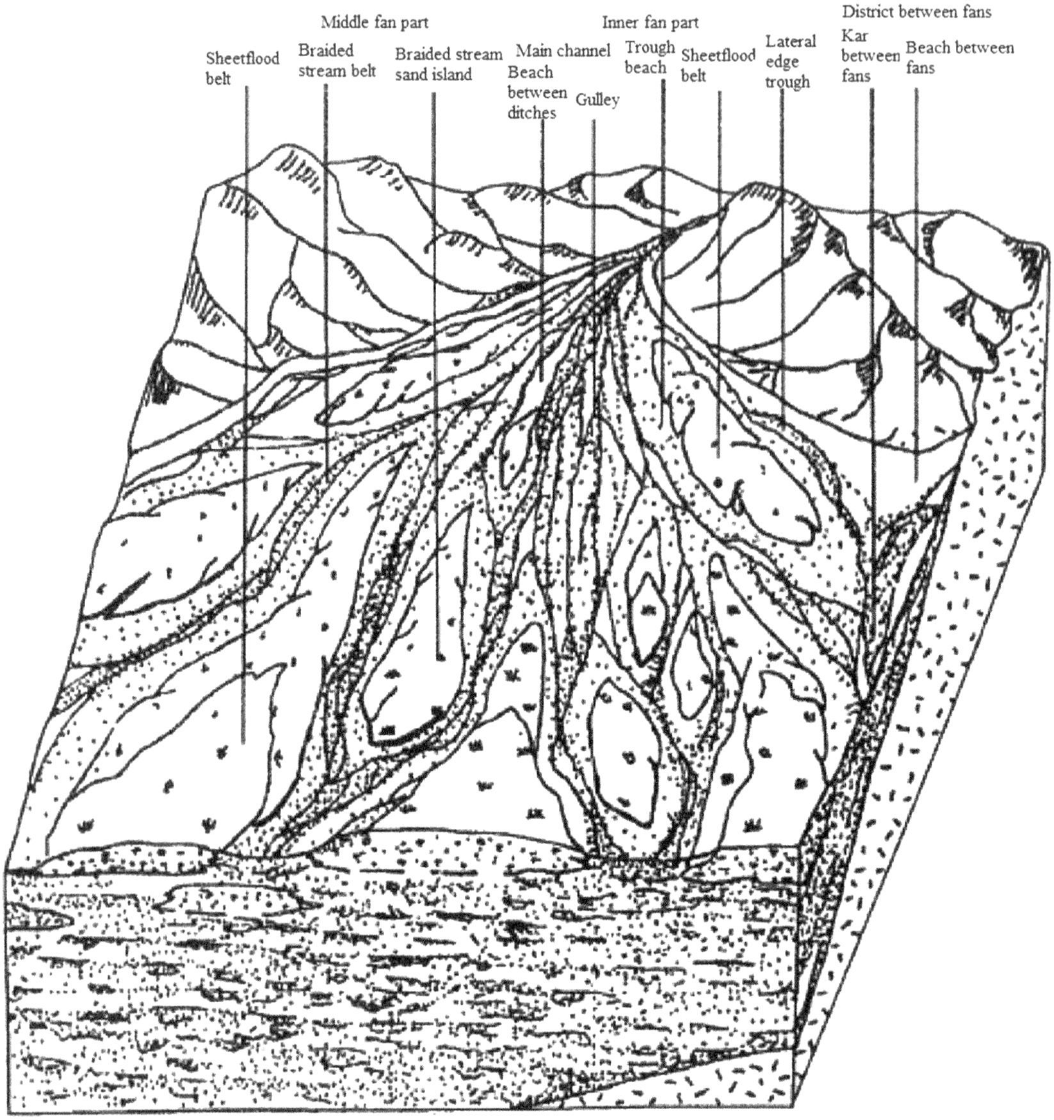

Fig. 8.16 Partition diagram of alluvial fan faces and micro-faces in the karamai oil field (according to Jiyi 1980)

horizontal permeability. The intrastratal bed of the conglomerate reservoir in Karamay is mostly unstable and can be divided into two types: low or impermeable lithology of fine grain, for instance, siltstone, argillaceous siltstone, silty mudstone, and mudstone; and highly dense or hard argillaceous conglomerate with low permeability, the thickness of which is generally 0.2–0.4 m, and there is no far extension to hinder well-to-well correlation. Most interbeds fall into the first type, and the second is mainly developed in the inner fan subfacies of a few fan bodies.

Given the abovementioned complex features, the permeability heterogeneity of alluvial fan

Table 8.7 Microfacies belt reservoir properties of Kexia formation in no. 3 District in Karamay Oilfield (according to Jiyi 1980)

Sedimentary microfacies belt	Main channel	Lateral edge trough	Trough beach	Sheetflood belt	Braided stream belt	Braided stream sand island	Sheetflood belt
Conglomeratic thickness/deposition thickness	0.975	1.0	0.822	0.225	0.845	0.705	0.33
Conglomerate thickness/conglomeratic thickness	0.965	0.6	0.898	0.369	0.94	0.618	0
Median grain diameter (mm)	3.69	1.95	2.75	2.0	1.8	1.88	0.11
Shale content (%)	9.95	8.55	–	35.73	14.56	13.9	39.3
Average permeability ($\times 10^{-3}$ μm^2)	521	323	–	<1	556	283	13.2

reservoirs is very strong. The heterogeneous parameters of type I and II reservoirs are relatively close (Table 8.8), however the various sedimentary microfacies differ considerably. The level difference between the average permeability ratios of the main channel facies is 214×, while the permeability ratio of the braided stream belt and the braided stream sand island is generally dozens of times. Owing to their low average permeability, the permeability ratio of type III reservoirs is low as well, and is only 10–30×. The permeability ratio of gravel braided stream facies is equal to that of middle fan subfacies. This shows that the degree of heterogeneity of the sedimentary sandstone body under strong hydrodynamic conditions is much higher than that under a steady environment (Table 8.9).

8.5.4.3 Interlayer Heterogeneity

The delamination of alluvial fan sedimentation is considerably lower than that of clastic reservoirs in other environments owing to the relatively underdeveloped stable mudstone interlayer. For instance, based on the statistics on petroleum reservoirs in the Karamay Oilfield, the average layering coefficient of alluvial fan conglomeratic reservoirs is 4.0–7.7, and the inner fan subfacies is smaller than the middle fan subfacies. In arid environments, the intermittence of sedimentation is strong or debris flow sedimentation is well developed. As for reservoirs in the middle fan, the argillaceous interlayer is relatively developed; for instance, delamination of the Kongdian Formation in the Zaoyuan Oilfield is relatively good. Although the layering coefficient of alluvial fan conglomeratic reservoirs is small, the interlayer permeability heterogeneity of these reservoirs can be considerable. In conclusion, one outstanding feature of conglomeratic reservoir heterogeneity is the large difference in interlayer permeability. In particular, the interlayer permeability ratio of the conglomerate oil reservoir in Karamay is generally 10–50×, in which the permeability ratio of type I oil reservoirs is low at 9.8–35×; that of type II reservoirs is 7.8–47.7×; and that of type III is larger, at 51.6× on average. Because the reservoir and the interlayer are thin, and some structures with high permeability exist, and the interlayer permeability contrast of conglomerate reservoirs is more obvious.

Table 8.8 Reservoir classification features (according to Hu Futang, Modified in 1997)

Reservoir classification	Macroscopic feature					Microscopic feature			
	Ratio in total reserves (%)	Unit reserve ($\times 10^4$ t/km^2)	Air permeability ($\times 10^{-3}$ μm^2)	Porosity (%)	Connectivity rate (%)	Mean value of throat diameter (μm)	Mean value of pore diameter (μm)	Variation coefficient	Mercury withdrawal efficiency (%)
I	17	428	459	16.7	82	Mainly first type of pore structure			
						2.4–4.4	250–330	1.1–3.6	40–60
II	44	242	100	17.2	65	Mainly second type of pore structure			
						0.5–1.1	100–200	5–10	33–55
III	10	45	59	18.4	31	Mainly third type of pore structure			
						0.34	75	14	25–45
IV	29	2114	0.9	9.0		Mainly fourth type of pore structure			
						0.18	45	15	20–40

Table 8.9 Intrastratal heterogeneity features (according to Hu, 1997)

Facies belt	Type I reservoir		Type II reservoir		Type III reservoir	
	Ratio (times)	Dart coefficient	Ratio (times)	Dart coefficient	Ratio (times)	Dart coefficient
Main channel	214.3	5.0	214.5	5.0		
Trough beach	117.8	4.5	134.3	4.6		
Braided stream belt	140.7	4.7	71.2	4.1	27.8	3.4
Braided stream sand island	69.6	4.1	95.6	4.4	15.9	3.0
Braided channel	165.6					
Braided stream bar	66.6					

References

Bull, W.B. 1972. Recognition of alluvial fan deposits in the stratigraphic record. In *Society of economic paleontologists and mineralogists*, vol. 16, ed. W.K Hambin, and J.K. Rigby, 63–83. Special Publication.

Ethridge, F.G, and W.A. Wescott. 1984. Tectonic Setting, Recognition and Hydrocarbon Reservoir Potential of Fan-Delta Deposits[J]. Cspg Special Publications.

Gole, C.V., and S.V. Chitale. 1966. Inland delta building activity of Kosi River. *Proceedings American Society of Civil Engineers, Journal of the Hydraulics Division* 92: 111–126.

Galloway, WilliamE. 1983. Terrigenous clastic depositional systems: applications to petroleum, coal, and uranium exploration[M]. Springer.

Heward, A.P. 1978. Alluvial fan sequence and mega sequence models: With examples from Westphalian D-Stephanian B coalfields, Northern Spain. In: *Fluvial sedimentology* ed. A.D. Miall, 669–702. *Memoirs Canadian Society of Petroleum 5*, Calgary.

Jiyi, Zhang. 1980. *The diluvial fan coarse clastic reservoir of Karamay*. Xinjiang Petroleum Geology.

McGowen, H., and C.G. Groat. 1971. Ban Horn Sandstone, west Texas: An alluvial fan model for mineral exploration. In *Representative Bureau of Economic Geology*, vol. 72. Austin: University of Texas. (Invest).

Nilsen, T.H., and G.G. Zuffa. 1982. The Chugach Terrane, a cretaceous trench-fill deposit, southern Alas- ka. In *Trencb—forearc geology*, ed. J.K. Legget, 213–227. Special Publication of Geologist Society London.

Steel, R.J. 1976. Devonian basins of western norway - sedimentary response to tectonism and to varying tectonic context[J]. *Tectonophysics*, 36(1–3): 207–224.

Tarbuck, E., and F. Lutgenswrited. 1993. *The Earth: An Introduction to Physical Geology*. 4th ed[M]. Macmillan Publishing Company.

Tinghui, Zhang. 1991. The pore structure characteristics and sort evaluation of conglomeratic reservoir in Karamay Oilfield. *Xinjiang Petroleum Geology*, 37–45.

Yinan, Qiu, Xue Shuhao, and Ying Fengxiang. 1997. *Continental oil and gas reservoirs in China*. Beijing: Petroleum Industry Press.

Yongchuan, Sun, and Li Huisheng. 1985. *Clastic sedimentary facies and sedimentary environment*. Beijing: Geological Publishing House.

9 Fluvial Depositional System

Rivers are not only the main geological agent for corroding and transforming continental topography and carrying weathered materials to lakes or seas but are also important sedimentary agents in continental areas. Given an appropriate structural environment and sedimentary background, sometimes a fluvial sedimentary stratum with thickness of thousands of meters can develop.

Over the years, fluvial processes and fluvial sediments have always been one of the key research entities in petroleum exploration and development in China and elsewhere around the world. Studies on the hydrodynamics, flow patterns, erosion, transport, sedimentation, sediment structure, and sedimentary structure of modern rivers have greatly promoted the understanding and identification of ancient fluvial facies and actively pushed the development of geological prospecting. With the development of production and exploration practices, petroleum reservoirs and various important minerals, such as gold, copper, diamond, quartz, platinum, and aluminum, have been found in succession in fluvial sandstones. Therefore, research on various sedimentation, sediment characteristics, and their distribution laws in rivers is of great significance when looking for sedimentary minerals.

In general, a fluvial system can be divided into three parts, namely, the upstream reach, middle and downstream reaches, and river estuary (Fig. 9.1). Upstream reaches (equivalent to the youth stage of a river) are distributed mainly in mountainous areas. The water source can be mountainous river systems, thawed glaciers, or abundant rainfall in humid areas. Moreover, the area where a network comprising a large number of small tributaries converges is very favorable for alluvial fan development.

Meandering rivers are formed in the middle and downstream reaches (equal to the mature and aged stages of a river), where river networks gather in the main channel. Moreover, floodplain and point bar sedimentation are developed.

River estuaries, which are characterized by the formation of a branch flow network, are formed in the coastal areas of seas and lakes, resulting from repeated channel bifurcation. Eventually, river estuaries become part of the sea or a lake in the form of a delta.

9.1 River Classification

The major geomorphic feature of an alluvial plain is the river valley formed by fluvial processes. River valleys can be subdivided into secondary geomorphic units such as channels (bed), natural levees, floodplains, and abandoned channels, where channels are both the passages for carrying sediments and the main locations for river erosion and sedimentation.

The so-called channel pattern refers to the geometrical shape of channels on the plain. In general, the classification of channel patterns,

X. Yu et al., *Clastic Hydrocarbon Reservoir Sedimentology*, Advances in Oil and Gas Exploration & Production, https://doi.org/10.1007/978-3-319-70335-0_9

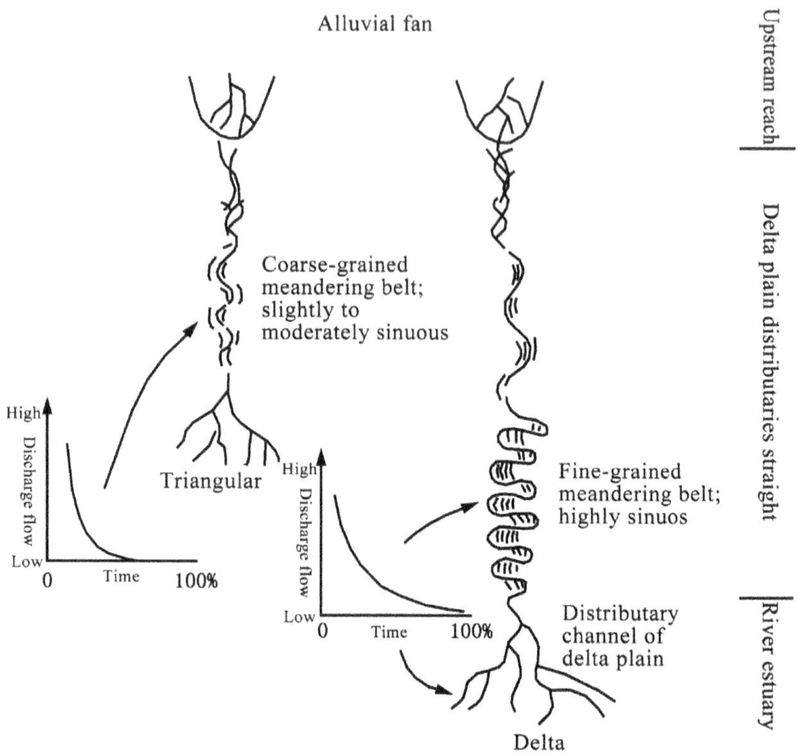

Fig. 9.1 One fluvial system can be divided into upstream, mid-downstream and river mouth, which have different load characteristics

i.e., river classification, is also known as pattern classification. Different channel patterns play an important role in controlling the sedimentary environment combinations of rivers and their sedimentary characteristics. Hence, pattern classification is a precondition and the base for establishing sedimentary facies models. Thus far, geologists and geomorphologists have put forth various river classification schemes using different classification rules.

9.1.1 Early Classification Schemes

9.1.1.1 By Development Stage

At the end of the 19th century, Daws divided rivers into the youth stage, maturity stage, and old stage by river development period.

9.1.1.2 By Structure Determinants

According to genetic and spatial distribution relationships between rivers and regional structures, classified rivers into five patterns, namely, consequent, obsequent, antecedent, subsequent, and superimposed.

9.1.1.3 By Sediment Load Form

By linking geomorphologic forms to flow and sediments, Schumm (1963) classified rivers into bed load, suspended load, and mixed load based on the main load forms of rivers or the main sediment transport mode.

9.1.1.4 By Geomorphic Features

The geomorphologic classification of rivers, with different meanings in diverse articles, has been applied gradually in the 19th century. It was then unified gradually through development from the initial confusion. In 1957, by integrating field work and flume experiment data, L. B. Leopold and M. G. Wolman clearly categorized river patterns as braided, meandering, and straight.

Usually, three parameters are employed for river pattern description: ① sinuosity index (S) = length of channel (l)/length of river valley

Table 9.1 Channel classification (according to Rust 1978)

sinuosity	Type	
	Single channel (channel bifurcation parameter < 1)	Multi-channel (channel bifurcation parameter > 1)
Low sinuosity (<1.5)	Straight channel	Braided river
High sinuosity (>1.5)	Snaking stream	Anastomosing river

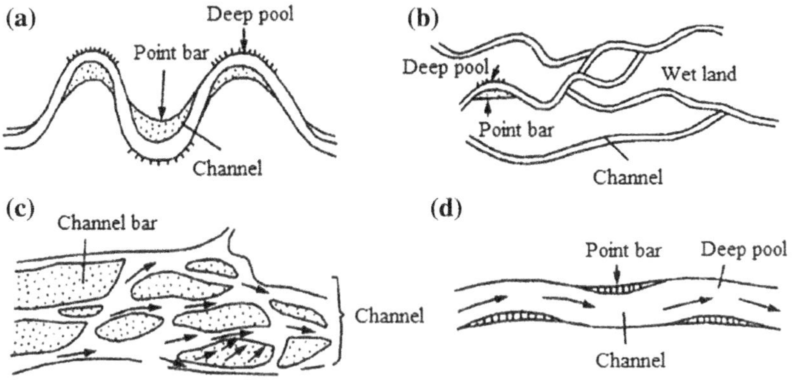

Fig. 9.2 Rust used river bifurcation parameters and camber, finally, purposed the four types of partitioning schemes, i.e. braided river, meandering river, net-like river and straight river (after Miall 1977). **a** Meandering river; **b** Braided river; **c** Anastomosing river; and **d** Straight river

(L), which refers to the ratio of channel length to valley length or channel length per unit river valley length. A higher sinuosity index indicates higher tortuosity. ② Braid index (B) = 2 × total length of all river islands/length of channel (l), snaking stream S > 2.0, B = 0; straight river 2.0 > S > 1.3; braided river S < 1.3, B > 0. ③ Channel bifurcation parameter, which refers to the number of channel bars in each average snaking wavelength (Table 9.1).

Rust (1978) applied the channel bifurcation parameter and sinuosity to supplement and improve the abovementioned three-pattern river classification scheme, eventually forming four types (Fig. 9.2), namely, braided, meandering, anastomosing, and straight. This classification scheme has been recognized widely and continues to be used extensively.

One thing we should point out is that the non-uniform translated names for "braided river" cause some confusion in practice and application because some scholars translate it into braided river, anastomosing river, braid river, and the like, while others consider braided and anastomosing rivers as the same concept. A few scholars confine anastomosing river to the anastomosing distribution of distributary channels on a delta plain.

An anastomosing river, which is a concept introduced in 19th century, has an unclear meaning. For example, T. C. Chamberlin and R. D. Salisbury (1909) called the Platte River, which was defined as a braided river by Davis, an anastomosing river. Schumm (1968) gave a meaning distinct from anastomosing river as follows: "In a stable multi-channel system with low bed load, channels with low slope and high sinuosity are separated by vegetational batture islands." Until the late 1970s and 1980s, Smith (1983) and Rust (1978, 1983) clearly defined an

Table 9.2 Diverse river patterns and associated microtopography

Braided River	Meandering River	Anastomosing river
Unstable, shallow braided river channel	Single channel with high sinuosity	Stable multiple channels with low sinuosity
Mid-channel bar	Meandering point bar	Apparent fixed island and natural levee
Broad braided belt	Oxbow lake	Broad wet land
To sum up, the previous two schemes	related to the classification applied by early (19th century) geologists	are basically no longer in use

anastomosing river as the fourth pattern based on a few modern river deposits in Canada and Australia. J. D. Smith defined it as "a channel of quickly aggraded, mutually connected, laterally limited, stable sandy or gravel river bed with low slope and sinuosity," which is clearly different from braided and meandering rivers (Table 9.2). The topographic form of an anastomosing river is similar to that of a braided river, both comprising multiple channels separated by channel bars. However, the important difference is that the channel bar and channel of an anastomosing river are stable, so the main sand body is constituted of deposition within a restrictive channel. The channel bar and channel of a braided river are unstable, and the main sand body is formed by deposition of the mid-channel bar. In 1979, A. D. Miall, a Canadian editor-in-chief, affirmed this classification in the summary of the report of the 1st International Conference on Fluvial Deposits. This independent anastomosing river concept and river pattern quartering have been gradually accepted by the majority of researchers. This river pattern quartering scheme was formally applied in *Sandstone Depositional Environments*, which was edited mainly by the Association of American Petroleum Geologists (AAPG) in 1982.

The third and fourth classification schemes, with many commonalities, can be contrasted with each other, but these two schemes are still associated in a few documents. For instance, a bed load river is equivalent to a braided river, a suspended load river is equivalent to an anastomosing river, and a hybrid river is equivalent to a meandering river with high sinuosity (Galloway 1983).

As A. D. Miall et al. analyzed and pointed out, all rivers have beds and suspended loads, and they cannot serve as the main basis for river pattern division due to differences between bed load particle diameters and proportions resulting from different sizes and changes in water flow. In addition, A. D. Miall also holds that the river pattern quartering scheme proposed by Rust (1978) is particularly unsuitable for establishing fluvial deposition facies modes. The two terms, namely, braided and snaking, involve different parameters and can be used for defining the same river pattern. However, anastomosing rivers show a great change in landform, especially sinuosity. A few researchers listed a series of river patterns to reflect the great changes in channel sinuosity, bifurcation, channel stability, and occurrence frequency of sandbars and channel bars. These geomorphological parameters are isolated to a certain degree, but they can be connected to constitute a few channel types between the four river patterns suggested by Rust (1978).

Valley gradients, sedimentary loads, river bank materials, climate, and tectonic mechanism affect downstream changes in channel landforms, and with time, the same control factors can cause changes in the form of some specific reach. Under the influence of topographical gradient, provenance supply, and climate, any river can present diverse river patterns at different reaches. In particular, the sequence of braided reach, meandering reach, and straight reach can be completely exchanged from upstream to downstream, i.e., a braided river can change into a meandering river

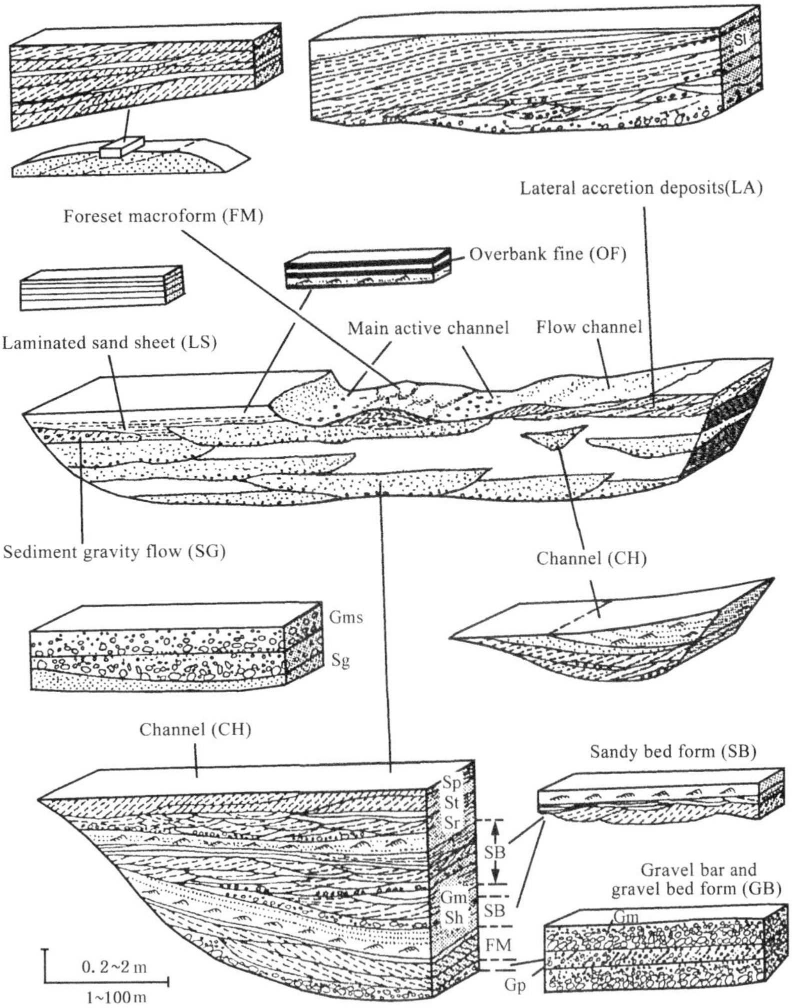

Fig. 9.3 In the second international river conference (1985), A. D. Miall purposed configuration elements, and divided the river sediments into 8 basic configuration elements (after Miall 1985a, b)

and vice versa. In the case of multiple rivers, some areas can have a braided river on one side and meandering river on the other side due to different landforms and provenances.

9.1.2 Classification by A. D. Miall

In 1985, A. D. Miall divided rivers into 12 categories based on the results of years of research and proposed a research method called "Architectural Element Analysis." In addition, he pointed out that every river has its specificity, whether in modern or ancient times. Hence, traditional river classification schemes and facies models have more limitations. Modeling is only the simplification of components, and only basic architectural elements can reflect the essential features of rivers. Therefore, he proposed eight basic architectural elements of rivers (Fig. 9.3). This analysis method has been applied extensively for reservoir description in various sedimentary environments.

Bounding surface hierarchy, lithofacies type, and architectural elements constitute the basic framework and research content of this analysis method. Architectural elements can be interpreted as

$$\text{Architectural element} = \text{lithofacies association} + \text{sandbody geometry}$$

9.1.2.1 Preconditions for This Method

① This method emphasizes on the genetic theory of sedimentation; ② there is a certain probability between sedimentary environment and depositional facies; and ③ research on reservoir sedimentology is necessary for establishing geological models.

9.1.2.2 Related Concepts and Definitions

The analysis of architectural elements involves considering the primary depositional facies association and its geometric shape.

(1) Research content: Refers to study of the lithofacies composition, appearance (geometric shape), and internal structure (fabric-lithofacies composition) of a depositional sand body.
(2) Research method: Refers to the analysis of sedimentation scales, bounding surface hierarchy, element comparison, and petrophysics of different sedimentary units by researching surface outcrops or oilfields (oildoms) with a high degree of exploration, thereby realizing and satisfying the above three preconditions.
(3) Definition of architecture scale: Sediments are composed of lithofacies of various scales and structures, from ripples to the entire depositional system. Each sedimentary unit scale is produced along with sedimentation in a specific time range, and it can be distinguished by different boundary levels (series).
(4) Architectural element: Refers to the lithologic body represented by geometric shape, facies association and its scale, and ability to represent specific sedimentation in its sedimentary system or a set of sedimentary processes.

9.1.2.3 Bounding Surface Hierarchy

A bounding surface hierarchy system for fluvial deposits was developed on the basis of Allen's (1983) research on Devonian brown sandstones in Wales, who also proposed the concept of bounding surface hierarchy for dividing fluvial deposit into three boundaries. When Miall

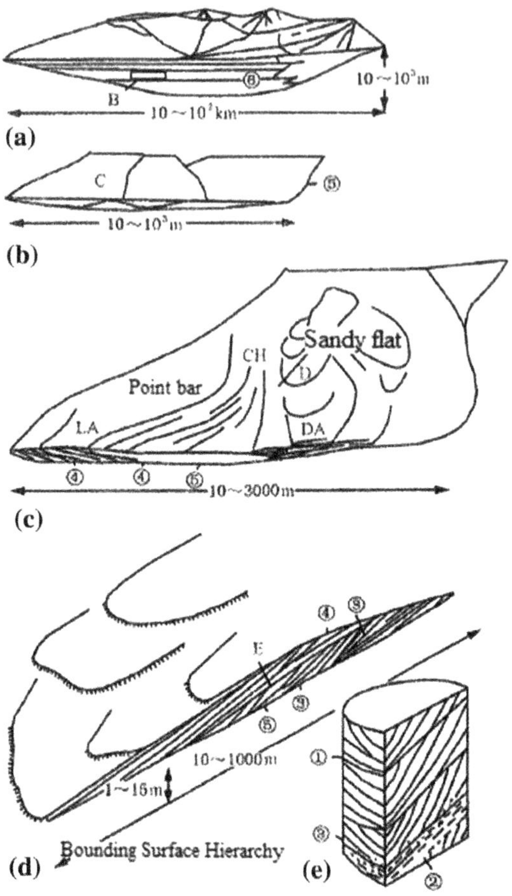

Fig. 9.4 In 1988, A. D. Miall purposed configuration factor analysis, and divided river interface into six grades. For ①–⑥, refer to the bounding surface (according to Miall 1988)

(1985a, b) proposed the architectural element analysis, he divided river boundaries into four levels but modified and expanded the scheme to include six levels (Fig. 9.4; Table 9.3) in 1988.

1. **Identification method**
   Level 1 boundary-set boundary of cross bedding;
   Level 2 boundary-coset boundary;
   Level 3 boundary-equivalent to small boulder clay scour surface developed between sand layers;
   Level 4 boundary-equivalent to top and bottom surfaces of large sedimentary bed forms, usually including two types: ① bottom scour

Table 9.3 Scale of sedimentary units in fluvial facies sand bodies (according to Miall 1988)

Boundary level	Lateral extension of sedimentary unit	Thickness of sedimentary unit (m)	Area of sedimentary unit (km^2)	Genesis	Underground mapping method
7	200 km × 200 km	0–30	4×10^4	Section or subsection, subtle structure control	Regional electric well-logging curve correlation
6	1 km × 10 km	10–20	10	Sheet sand body of channel genesis	Electric well-logging curve correlation in oilfield; three-dimensional seismology
5	0.25 km × 10 km	10–20	2.5	Strip-shaped channel sand body	Difficult to map unless well spacing is small; three-dimensional seismology
4	200 m × 200 m	3–10	0.04	Gigascopic bed form unit (lateral accretion and downstream accretion)	Level 3–4 boundary dip angle can be identified on the core
3	100 m × 100 m	3–10	0.01	Activation of gigascopic bed form	Level 3–4 boundary dip angle can be identified on the core
2	100 m × 100 m	5	0.01	Coset of lithofacies similar to cross bedding	Core lithofacies analysis
1	100 m × 100 m	2	0.01	Single cross set	Core lithofacies analysis

surface of a small channel (water channel); and ② top surface of lateral accretion or foreset;

Level 5 boundary-megascopic sand body boundary of channel-filled complex; and

Level 6 boundary-bottom surface of a group of large water channels (channel) or ancient river valleys, which may be reflective surfaces of seismic units.

Level 2–4 boundaries are included in the second classification proposed by Allen (1983), and Level 3–4 boundaries are equivalent to the "secondary" boundary proposed by Bridge and Diemer (1983).

Level 5–6 boundaries may be the easiest to identify and compare underground because channels have obvious scouring and fining upward structures, with wide lateral extensions. Furthermore, simple, straight, or slightly bent channel geometries are easier for contrast. The identification and contrast of various boundaries are apparently useful for explaining the complexity of a fluvial sedimentary system.

2. **Principles of boundary division**

Boundary divisions should ① reflect the reservoir heterogeneity of various scales: ② facilitate easy geological and well logging identification, with representativeness; and ③ have significance from the viewpoint of sedimentary genesis.

3. **Lateral change in boundary level**

① A boundary can be truncated by another boundary at the same level or at a higher level, but it cannot be truncated by a boundary at a lower level. ② Because a boundary often records erosion events, it can be determined logically based on post-erosion characteristics as opposed to previous characteristics. For instance, the top surface of a megascopic sandy bed form is confined by a Level 4 boundary. Unless cut by a large channel, the bottom surface is usually constituted of Level 5 boundaries. ③ Level 1–3 boundaries can change horizontally or laterally, for instance, a set boundary can transform into a coset boundary.

9.1.2.4 Architectural Elements

A. D. Miall proposed this concept at the 2nd International Rivers Conference in 1985 and divided fluvial deposits into eight basic elements (Fig. 9.3) composed of lithofacies associations (Table 9.4).

1. **Channels (CH)**

Multiple channels separated by a flat or a concave-up erosion surface exist in a fluvial system. Larger channels usually contain complex fillings composed of one or more architectural units.

2. **Gravel bars (GB) and bed forms**

Tabular or cross bedding gravel constitutes a simple longitudinal or transverse bar.

3. **Sandy bed form (SB)**

Lithofacies formed by the bed form in a low flow regime are St, Sp, Sh, Si, Sr, Se, and Ss. They are mutually combined to form a series of architectural units of different geometric shapes. The architectural unit that can best represent a sandy bed form is the tabular or sandy bed form, which is usually located at the bottom of a channel or top of a sandbar or crevasse splay.

4. **Downstream accreting (DA) macroform or foreset macroform**

In 1985, A. D. Mall defined it as foreset macroform (FM). However, in 1988, he changed FM into downstream accreting macroform (DA) without any alteration (Fig. 9.3) in the figure in Fluvial Sandstone Reservoir Heterogeneity issued by AAPG. It could be hereby

Table 9.4 Tectonic units in fluvial deposits (according to Miall 1988)

Architectural Unit	Symbol	Major lithofacies association	Geometry or Interrelation
Channel	CH	Any type	Fingered, lens; concave-up erosion basement, with large changes in scale and form; common secondary internal erosion surface
Gravel bar and bed form	GB	Gm, Gp, and Gt	Lens, blanket type, generally tabular body mixed with sandy bed form
Sandy bed form	SB	St, Sp, Sh, Si, Sr, Se, and Ss	Lens, sheet, blanket, and wedge type in channel filling; crevasse splay, sandbar fan, sandbar top, and small sandbar
Downstream accreting macroform	DA	St, Sp, Sh, Si, Sr, Se, and Ss	Lens located above flat or channel basement, with convex level 3 boundary in the inner and top parts
Lateral accretion deposit	LA	Rare St, Sp, Sh, Si, Sr, Se, Ss, G, and F	Wedge, sheet, and tongue type, with internal lateral accretion features
Sediment gravity flow	SG	Gm and Gms	Tongue and sheet type, generally mixed with SB
Laminated sand sheet	LS	Sh, SI, little St, Sp, and Sr	Sheet and blanket type
Overbank fines	OF	Fm and Fi	Thin-thick blanket type, generally mixed with SB, and possibly filled with abandoned channel deposits

concluded "progradation" is different from "downstream accretion." The former generally refers to delta sedimentation, whereas the latter extends to a wider range.

This architectural unit (DA) has convex inner and top boundaries. All components of large bed forms (DA) are correlated under hydrodynamic conditions, which indicate that the incline direction of the boundary is parallel or subparallel to the palaeocurrent direction. Thus, this architectural unit represents complex sandbar depositions formed by downstream accretion.

5. **Lateral accretion deposit (LA)**

Bed form shows that the architectural unit is developed to form well-known point bars by means of lateral accretion when the angle between the palaeocurrent direction and the incline direction of the internal accretion surface is large. Point bars are mainly developed in meandering rivers, and are characterized by erosion at concave banks and accretion at convex banks.

6. **Sediment gravity flow (SG)**

This is gravel deposition formed by debris flow. Lithofacies Gms is a main lithofacies type with a fast deposition rate and high shale content.

7. **Laminated sand sheet (LS)**

This is mainly composed of the association of Sh and SI, which is a flatbed product in a high flow regime.

8. **Overbank fine (OF)**

This is composed mainly of mudstone, siltstone, and little fine sandstone formed in both floodplains and abandoned channels. It is usually developed in crevasse splays, natural levees, and floodplains.

9.1.2.5 River Classification Characteristics and Model

Architectural element analysis emphasizes the lateral distribution of sedimentary bodies and association of geometric shapes and internal sedimentary structures, and highlights the dynamic mechanism of the formation process and sedimentary events.

Miall combined eight basic architectural elements into 12 fluvial deposit modes (Fig. 9.5).

1. **Model 1**

This river model, which is composed of SG, SB, and GB, represents sedimentary models of proximal gravelly braided rivers developed only by the alluvial fan root of sediment gravity flow deposition.

Gravity flow (debris flow) deposition is usually cut by distributary channels (Fig. 9.5a). Channels are often filled with gravel bar and its bed form. Thin sandy bed forms can be found in abandoned channels.

2. **Model 2**

Model 2 is composed of GB, indicating the sedimentary model of a distal gravelly braided river formed in an alluvial fan or a braided plain.

It can be commonly found in river environments dominated by alluvial fans and other gravel deposits. The difference between Models 2 and 1 is that Model 2 has no gravity flow deposition. Its main deposition includes gravel bar and its bed form and sandy deposition or even alluvial plain deposition can be formed in abandoned channels. Occasionally, a small sandy delta (Fig. 9.5b) can be found at the edge of a gravel bar.

3. **Model 3**

This is composed of GB, SB, and OF, and represents gravelly multi-step (terrace) fluvial

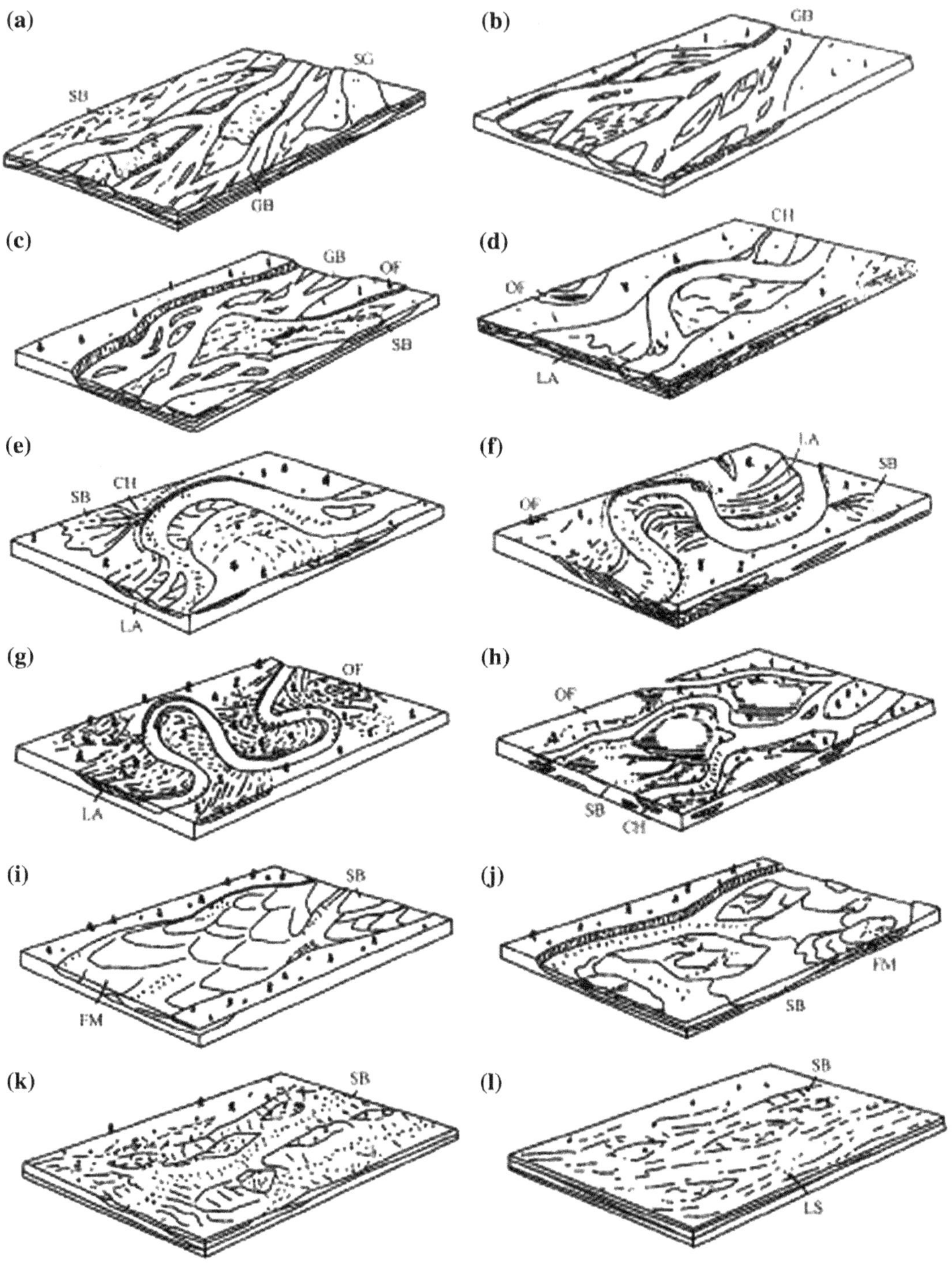

Fig. 9.5 A. D. Miall combined 8 basic configuration factors of river into 12 river sedimentary models (according to Miall 1985a, b)

deposit with low sinuosity downstream of the alluvial fan. It belongs to gravelly rivers with confined channel terraces formed by a braided river rushing out of a piedmont region, namely, shore gravel fluvial deposit with low sinuosity.

Model 3 is similar to the previous two models, but its difference lies in that the river under this mode is wider and deeper and there is a terrain height difference of 3–4 terraces. Gravel bar mainly deposited in the lowest part of the river is the most active main channel deposition. The primary to secondary sandbars are deposited at medium height during flooding, but overbank flooding muddy deposition can be found occasionally in the highest places during a catastrophic flooding period. As a result, the sedimentary sequence of its vertical section shows fining-upward grain size (Fig. 9.5c).

4. **Model 4**

This is the sedimentary model of a gravelly meandering river (high-sinuosity river) composed of CH, LA, and OF.

Both this model and the three models mentioned after belong to meandering river deposition, while its main difference lies in the grain sizes of sediments. Rivers represented by this model may have multiple channels, including a main channel and a few small channels. Moreover, a few chutes result from excavation during the flooding period, and abandoned channels and overbank fines may be formed in the inactive lowstand period in normal times. Gravels are the main sediments in the river, and small gravel bars or bed forms (Fig. 9.5d) can be found on large gravel bars.

5. **Model 5**

Composed CH, LA, and SB, it represents the depositional model of "coarse-grained meandering river" for the sand and gravelly sand layers.

Sediment is comprised of sand and gravels. In general, gravels are mainly deposited in a channel to form gravel bars or sandy dunes, while sand is deposited in the higher part, and it likely develops chutes during flooding. Sand is mainly deposited on crevasse splays on the floodplain, with a small channel present (Fig. 9.5e) at the crevasse splay end.

6. **Model 6**

This is a typical meandering river model with a sandy mixed load composed of SB, OF, and LA.

A meandering river with large sinuosity consists completely of sand, thus its sandbar is rich in various bed form configurations. This model is similar to the last model, with only a few detailed differences (Fig. 9.5f).

7. **Model 7**

This is composed of LA and OF, and represents the depositional model of a fine-grained muddy meandering river.

The corresponding river has larger sinuosity than the abovementioned three types, in which sediments mainly include fine sand, silt, and mud, occasionally with a coarser grain size. Point bars that are relatively wide and steep have a clear broad ridge line and are more developed; the blank spaces on the two sides of Fig. 9.5g show floodplain depositions. Lateral accretion is the main architectural element of this kind of river, which is formed mainly because of gentle terrain.

Such rivers may be developed near an estuary, in which marine ichnofossils can be found. Therefore, when researching this river type, we distinguish it from the marine environment.

8. **Model 8**

This is composed of CH, SB, and OF, and represents the typical meandering river depositional model.

Anastomosing rivers with multiple channels differ from braided rivers, because their channel is fixed and difficult to migrate. River diversion is caused by large-scale crevasses. Reaches with low or high sinuosity can be developed in the channel. Small point bars may be formed at the bending part, but they are not the main sedimentary characteristic of anastomosing rivers. As a result of a stable channel, channel depositions are superimposed vertically, resulting in thick channel deposition, and thick overbank deposition characterized by "sand-in-mud" is formed between channels. Swamps usually developed on the floodplain can be good coal-forming places. The channel deposition is rich in cross beddings and sandy bed forms, but overbank fine-grained deposition (Fig. 9.5h) may be formed in its upper part once the channel is abandoned.

9. **Model 9**

This is composed of SB and DA, and represents the depositional model of bank-free sandy low-sinuosity sheet rivers with longitudinal accretion sandbars.

It belongs to typical sandy braided river deposition, composed mainly of two depositions owing to both wide and shallow rivers: the river bottom is occupied by dunes with trough cross bedding, while the river top is distributed with planar cross bedding constituted by large flat sand waves. Some islands can be formed in this river deposition, but we are not sure whether they can be preserved (Fig. 9.5i).

10. **Model 10**

This is composed of SB and FM, and represents the depositional model of a typical sandy braided river with relatively deep water, which produces the "mixed effect" of vertical and lateral accretion.

Relatively good development of sandy flats and shoals, open terrains, and gentle valleys are the prerequisites for the formation of such rivers. Such rivers are wide and deep, and are characterized by the development of large and complex sandbars. Miall calls this sandbar deposition megascopic foreset, which is interbedded with sandy bed forms (Fig. 9.5j), possibly with various sedimentary structures.

11. **Model 11**

This is formed by SB, and represents the typical depositional model of flood intermittent or seasonal rivers near the piedmont.

Intermittent fluvial deposits occur only in high-energy flood seasons, and no sedimentation occurs for a long time in the other seasons. Therefore, there are obvious flood beddings. They do not have any special channels, and the channel boundary is unclear. Megascopic sheet sand and gravel deposits can be commonly found, where megascopic low-angle cross bedding and imbricate structures can be seen. Floodplain muddy deposition on the vertical section may be very thin or, sometimes, even absent (Fig. 9.5k).

12. **Model 12**

This formed by SB and LS, and represents sheet flood river plain deposition with instantaneous high flow. It is a variant of Model 11, and is dominated by sand and mainly developed downstream. The sedimentation of intermittent fluvial deposit is only produced during the flooding period, and often developed as laminar sheet sand, each layer of which represents one flooding event. In a low flow regime, a sedimentary structure may be formed at the edge of the flood. When the flood attenuates quickly, these low flow regime structures can be developed in vast regions (Fig. 9.5l).

9.1.3 River Structure-Generic Classification

In view of the division principles generally applied in the existing literature reports with respect to rivers at home and abroad, and in international river meetings, rivers can be classified into two categories by grain size, and then subdivided by sinuosity. This principle seems to be applied extensively. According the author's years of research and application practice, a specific division scheme (Table 9.5) is proposed herein in combination with Miall's river classification.

Table 9.5 Generic classification of fluvial system structure

Fluvial system	Gravel river	Low sinuosity	Braided	Proximal gravelly braided river
				Distal gravelly braided river
		High sinuosity	Snaking shape	Proximal gravelly meandering river
				Distal gravelly meandering river
		Intermittent gravel river		
	Sandy river	Low sinuosity	Braided	Proximal sandy braided river
				Distal sandy braided river
			Straight river	
		High sinuosity	Snaking shape	Proximal sandy meandering river
				Distal sandy meandering river
			Anastomosing river	
		Intermittent sandy river		

9.1.4 Lithofacies Association Classification

As mentioned above, lithofacies division is an important base for sedimentary facies identification, especially microfacies discrimination, because lithofacies reflect the hydrodynamic conditions, transport modes, and formation backgrounds of sedimentary facies. However, an important approach to sedimentary facies identification is studying the vertical sedimentary sequences of different sedimentary facies, while the method for accurately showing the vertical sedimentation sequence is analyzing association features or types of various lithofacies in the vertical direction. Lithofacies (code) is marked on sedimentary sequences and lithofacies association features under different sedimentary environments are summarized in a variety of related studies, for example, Miall (1985a, b), Walker (1984a, b, c), and Galloway (1985). The author (1992) divided Permian fluvial facies into 13 facies (Table 9.6) while studying Carboniferous-Permian sedimentary characteristics in North China, systematically studied the important sedimentary sequences of Permian fluvial facies in North China in combination with the 12 models summarized by Miall (1985a, b), and summarized 8 lithofacies association features (Table 9.7) of Permian fluvial facies in North China by observing and describing eight wild outcrop sections and cores in dozens of wells that are approximately 1000 m deep.

9.1.4.1 Type 1: Gms → Gt, Gp → Sh → Fl, and M

This is composed of coarse-grained lithofacies and overlain M facies, and characterized by a small amount of sandstone mixed in large amounts of sandy gravels in the section. Mainly exhibiting planar cross bedding, coarse-fined particles take up 90% of the entire cycle, whereas the lower part of the sequence often has graded bedding. It reflects the features of nearer provenance and larger channel slopes and represents the proximal reach (Fig. 9.6) of a braided river on the alluvial fan. A majority of wide and shallow channels with low sinuosity are bifurcated and merged at different places to form eroded concave banks and sandbars. During the discharge period, sandbars may emerge from the water and be cut by small channels.

9.1.4.2 Type 2: Gm ↔ Sp → Sh → M

This is mainly comprised of coarse-grained stones or may be occasionally developed with fine-grained substances. It is characterized by a thinner sedimentary body and extremely frequent

Table 9.6 Classification of Permian fluvial facies and lithofacies in north China (according to Yu Xinghe et al. modified in 1992)

Code	Description	Code		Description
Gms	Debris flow deposit gravel lithofacies	Sr		Wavy—intermittent wavy cross bedding—fine sand lithofacies
Gm	Lag deposit or batture massive gravel lithofacies	Fr		Ripple crossing bedding silt lithofacies
Gp	Planar cross bedding gravel lithofacies	Fl		Red, variegated mudstone and algal silt lithofacies
Gr	Trough cross bedding gravel lithofacies	C		Carbonaceous mud lithofacies and coal bed
Sp	Planar-wedge cross bedding gravel lithofacies	M	Md	Dark mudstone facies
Sf	Trough cross bedding gravel lithofacies		Mg	Greenish gray or grayish-green mud lithofacies
Sh	Parallel bedding gravel lithofacies		Mr	Brick-red piebald mud lithofacies

Table 9.7 List of lithofacies association types for Permian fluvial facies in north China

SN	Lithofacies association type	River property
Type 1	Gms → Gt, Gp → Sh → Fl, M	Proximal gravelly braided river
Type 2	Gm ↔ Sp → Sh → Sr → M	Intermittent gravel river
Type 3	M → Sh → Sp → Gm	Distal gravelly braided river with downstream accretion sandbar
Type 4	Gp → Sp, St → Sh → M	Distal sandy braided river with multiple terraces
Type 5	Gt,St ↔ Sp → Sh → Mr ↔ Fr and FI	Distal sandy braided river with "mixed effect"
Type 6	Gp, Sp ↔ St → Mr ↔ Sr and FI	Distal gravelly meandering river
Type 7	M → St → Sp → M and C	Anastomosing river
Type 8	St → Sp → Sh → M → Fr	Distal gravelly meandering river

scouring action. Sedimentary structure is dominated by planar cross bedding and horizontal bedding, without trough cross bedding. Each cycle is only 3–4 m thick, or even less (Fig. 9.7). In a distal braided river, especially in the shallow and interlaced channels with indistinct boundaries formed by transient or intermittent runoffs in the drought area, sediments are dominated by sheet, lens, and wedge sand bodies, and overbank fine-grained depositions are relatively rare.

Brian and Turner believed that "a sequence developed by parallel bedding and isolated planar cross bedding is a temporary sandbar in low-sinuosity river due to its relatively stronger hydrodynamic energy." They also hold the opinion that frequent scouring (erosion) is a

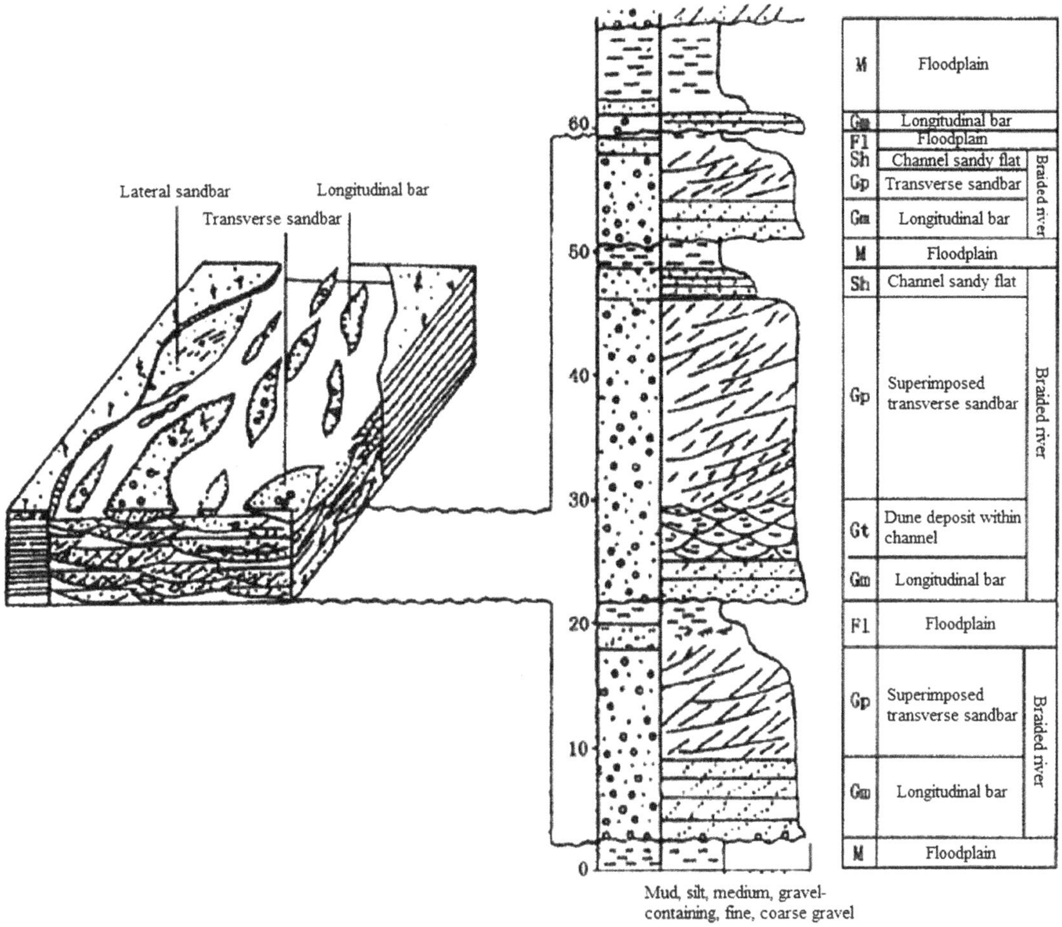

Fig. 9.6 Gravel braided river has the features of near provenance and larger river slope, representing the near source section of the braided river in the alluvial fan (Yu et al. 1992), with the field section P2s35-38 of western hills of Beijing taken as an example

typical feature of temporary rivers, which represents the conditions of water level fluctuation and change.

9.1.4.3 Type 3: Mr → Sh ↔ Gp → Gm

Type 3 shows a coarsening upward trend in the lower part. The sedimentary structure is composed mainly of planar cross bedding and parallel bedding, generally without trough cross bedding. The entire upward coarsening sequence is 5–10 m thick. The sequence reflects that channels in the form of rapid flow in shallow water are not clear, with a gentle topographic slope, dominated by sheet flood. A graded texture is developed, and charcoals are enriched on the textured surface (sand waves cannot exist) to show a process in which fluvial overflow gradually transits to channel deposition in a wide and shallow low-sinuosity river (Fig. 9.8) carrying a large number of sandy bed loads. Channels are filled with large-area transverse bars, and sandbars are gradually aggraded forward, which led to them being called forest sandbars. A sequence exactly opposite to the abovementioned one can often be found in the upper part of the sequence. The two can transit directly or carry a small amount of fine sandstone and mudstone.

There are two upward coarsening sequences: ① grain size coarsens upward in single layer; and ② grain size of the entire sequence coarsens

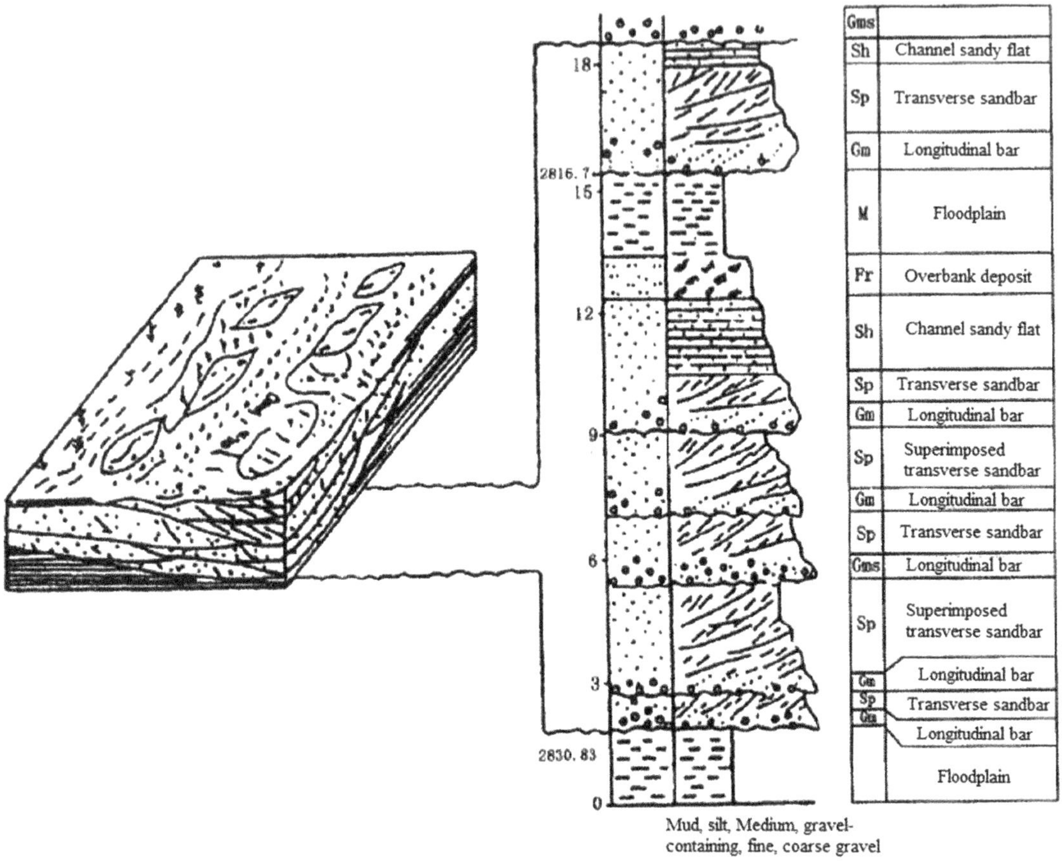

Fig. 9.7 Intermittent gravel river is far source braided river plain, with features of horizontal bedding and isolated plate-like cross bedding (according to Yu et al. 1992) taking P2S of well Ge 2 in the Renqiu oilfield as an example

upward from the bottom to the top. The former sequence is the result of the hydrodynamic sorting during a period of stable current, whereas the latter is formed owing to the unloading of sediments by a current as the water level decreases. Here it belongs to the latter one.

9.1.4.4 Type 4: Gp → Sp, St → Sh → M

Compared with Type 1, the biggest difference is that the planar cross bedding sandstone facies is more developed, whereas the fine sand lithofacies (siltstone and mudstone) is generally difficult to find but has relatively high texture maturity. According to, A. D. Midi (1985), "floodplains cannot become a significant part of this system, completely depending on the width of river valley and the stability of channel," and this represents a gravel bedding river (such as main rivers) with a significant difference in terrain height and low sinuosity. A river valley has three or four different terrain heights, the higher terrain of which is often covered by plants, while the lowest has movable channels; deposition in the higher terrain occurs only in the highstand period (Fig. 9.9).

9.1.4.5 Type 5: Gt, St↔ Sp → Sh → Mr → Fl and Fr

Trough cross bedding and planar cross bedding are developed simultaneously. Compared with fine-grained lithofacies, sandy lithofacies is not well developed, and it lacks the distinctive upward fining structure. The scale of trough cross bedding decreases gradually from the bottom to the top, and bank and flooding depositions are

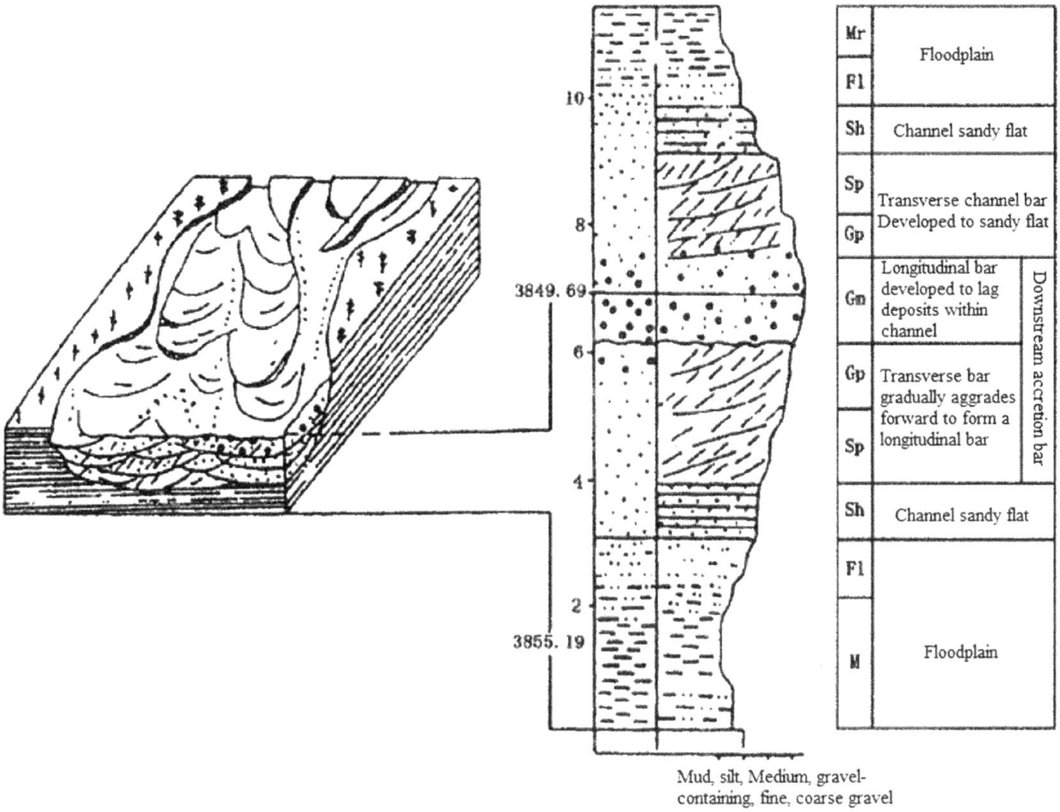

Fig. 9.8 Typical downstream and accretion stream reflect the sedimentary from washland to the river channel gradually, with an upward thickening of the reverse rhythm (according to Yu et al. 1992) taking P2s of well Su 401 in the Renqiu oilfield as an example

often found in its top part, which reflects that the filling effect of the channel is stronger, and its sandy deposition is more developed than that of Type 4. The degrees of development of St, Sp, and Sh depend completely on the sequences of cross impacts of sandy flat, sandbars, and channel depositions, which embody the features of a low-sinuosity river with large sedimentary differences among channels, sandbars, and the tops of sandbars. Thus, it is a typical sequence of a sandy braided river (Fig. 9.10).

A majority of researchers call fine sandstone with parallel bedding developed on the transverse bar sandy flat. The prototype of a new sandy flat is controlled by the location and direction of the transverse sandbar of a channel and the high point position of a ridge. Sandy braided rivers in the south of Saskatchewan, Canada, are typical examples of this type.

9.1.4.6 Type 6: Gp, Sp $\leftrightarrow$ St $\rightarrow$ Mr $\rightarrow$ Sr and Fl

This has the obvious structural characteristic of fining upward. Both coarse-grained deposits and fine-grained suspending deposits are developed, at the bottom of which channel lag deposits can usually be found, with clear lateral accretion. The geometric shape of lateral accretion and lithofacies association can vary greatly depending on the geometric shape and sedimentary load of a channel. This sequence is a typical coarse-grained (gravel-containing sandstone included) meandering river, the main part of which

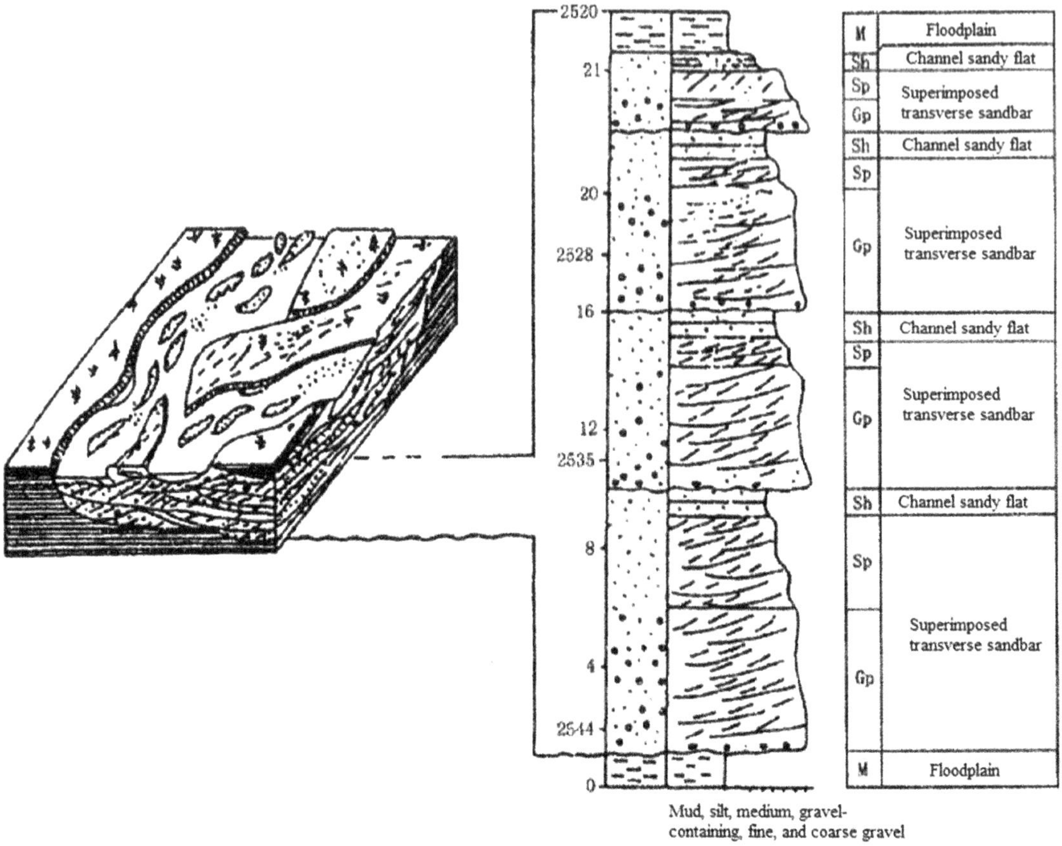

Fig. 9.9 Far source sandy braided river with multiple terraces represent low sinuosity gravel river with obvious terrain height, and valley includes three to four different terrain height (according to Yu et al. 1992) taking P2s of well Fengsen 1 in the Dagang oilfield as an example

constitutes a point bar. Moreover, the accretion surface of the sandbar is cut transversely by a variety of sand waves and dunes (Fig. 9.11). The most striking difference between this section and the previous types is the lack of abrupt grain size change in this type of section.

9.1.4.7 Type 7: M → St ↔ Sp → M, and C

This is characterized by mudstone interbeded with thick conglomerates orgravelly sandsotne, and "sand-in-mud" sequences in its section. The thickness ratio of sandstone to mudstone and argillaceous siltstone is often greater than 1/3. Usually, there is no sand and mud interbedded transition zone on the sequence, which indicates the absence of lateral migration. The sediments aggraded in the channel belong to filled channel deposition. The upper part of the sequence is often developed with facies I and facies K, which shows that flood swamp and flood lakes are more developed. Both multiple stages and grain size sorting reflect the sedimentary features of anastomosing rivers relatively well. It is a low-energy complex (Fig. 9.12) composed of several mutually connected channels with variable sinuosity.

9.1.4.8 Type 8: Sf → Sp → Sh → M → Fr

The biggest difference between Types 6 and 8 is the lack of gravel in the latter. Moreover, the fining upward characteristic of Type 8 is not as clear as that of Type 6. The small scales of point bar and point bar top show that river sinuosity is

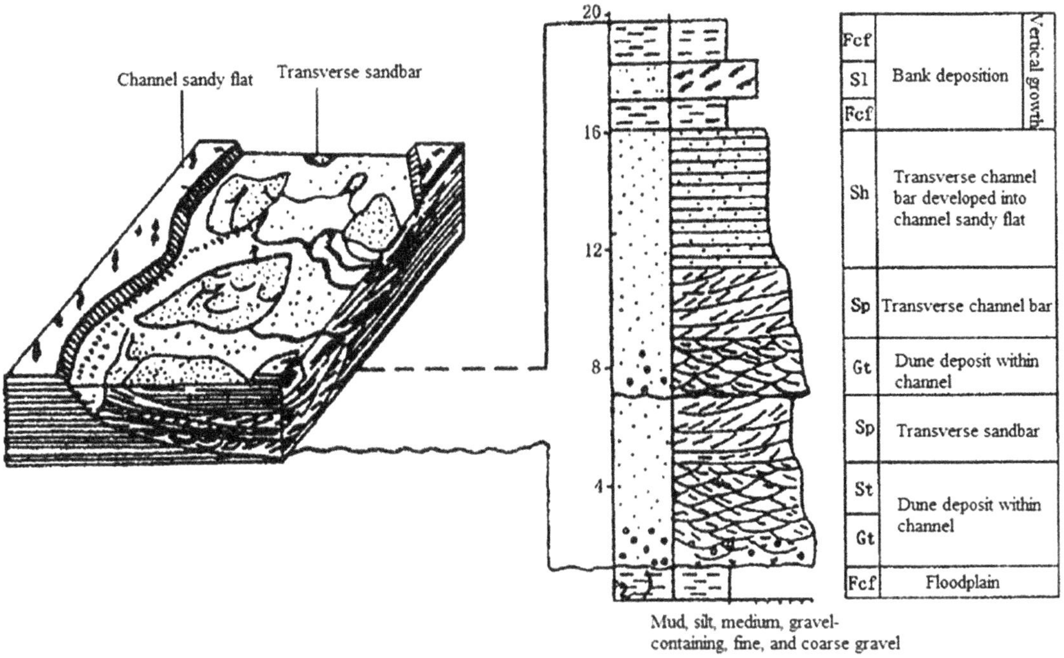

Fig. 9.10 Trough cross bedding and tabular cross bedding occurs in the far source sandy braided river with typical "mixed effects", reflecting the characteristics of low Sinuosity River, i.e. very different sedimentation of river channel, sand bar and the tops of the sand dam (according to Yu et al. 1992) taking field section P1x4-68 formation in Zibo, Shandong, as an example

smaller, so it belongs to sandy meandering river deposition. Mudstone in the upper part of the sequence is K or H, depending completely on position and climate changes. In the south, usually, delta distributary channel deposition can be found. It is difficult to summarize a typical sequence and model owing to the massive changes in the distributary channel. The two are extremely similar in sequence, but differ mainly in terms of upper and lower lithology association. Hence, they are placed in the same class (Fig. 9.13).

9.2 Basic Features of Different River Patterns

The channel pattern classification scheme proposed by Rust (1978) has been applied mostly at home and aboard. Below, we introduce and analyze the basic features of this scheme.

9.2.1 Straight River

A straight river is a single channel with low sinuosity and relatively stable banks. Its formation usually needs a few specific structural settings and geological conditions, for instance, fault troughs or vegetation development to form firm banks. Delta deposition into a lake occupies a leading position due to river energy. The delta distributary stream is usually almost straight or has a shape characterized by low sinuosity, and its sedimentation is dominated by aggradation. It should be noted that the river bottom of a straight river is usually bent to form alternately distributed abysses and shoals. In a straight channel, the flow velocity at the bottom is usually the lowest, but that at the midline is the highest (i.e., main streamline). Two symmetric circulations are formed at the two sides of the main streamline, and surface water flows from the two banks to the middle part. Hence, the main streamline

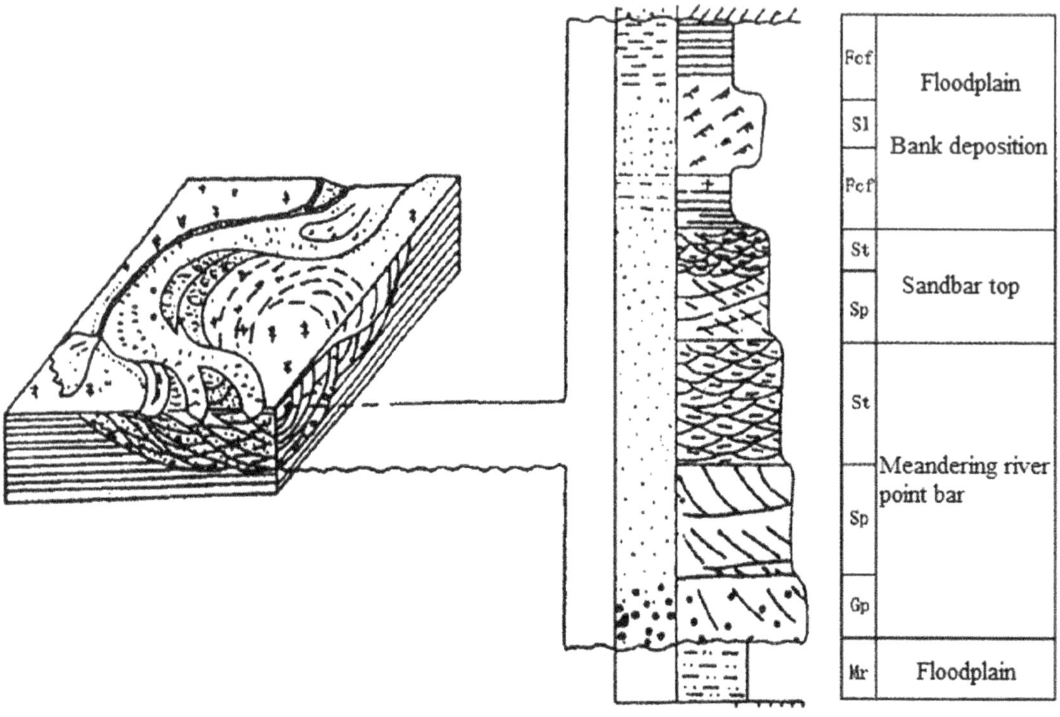

Fig. 9.11 Typical far source gravel meandering river is mainly composed of a meandering point bar, and the accretion surface of sandbar is cut by many ripples and dunes (according to Yu et al. 1992) taking field section P1x61-63 formation in Zibo, Shandong, as an example

moves in a straight line along the channel center (Figs. 9.14 and 9.15).

A straight river has low sinuosity. Hence, straight rivers and meandering rivers are generally bounded by a sinuosity index (channel length/river valley length) of 1.3. A certain curvature between the deep trough line and the maximum flow line in the channel can be found by carefully observing a straight river. Therefore, a few scholars believe that there are only high-sinuosity and low-sinuosity meandering rivers, and no straight rivers. In general, straight channels are rarely found. Even when they are found, they have short lengths. This is because straight rivers are dominated by erosion but sedimentation is not obvious. In contrast, a meandering river is in a balanced state in terms of erosion and sedimentation, while a braided river is dominated by sedimentation accompanied by erosion (Fig. 9.16). Hence, researchers often ignore straight rivers.

9.2.2 Braided River

A braided river is a wide and shallow river, in which channels are cut by many channel bars, and water is cut into multiple channels that constantly bifurcate and converge again (Fig. 9.17) around a variety of channel bars. Because channel bars and channels are unstable and banks are scoured easily, the reach is broadened quickly and shallowed under the action of current, and the majority of irregular channel bars (sometimes, there are point bars) develop to disperse the current at the river bottom. The river's mainstream swings in an unfixed manner, and channel movement varies

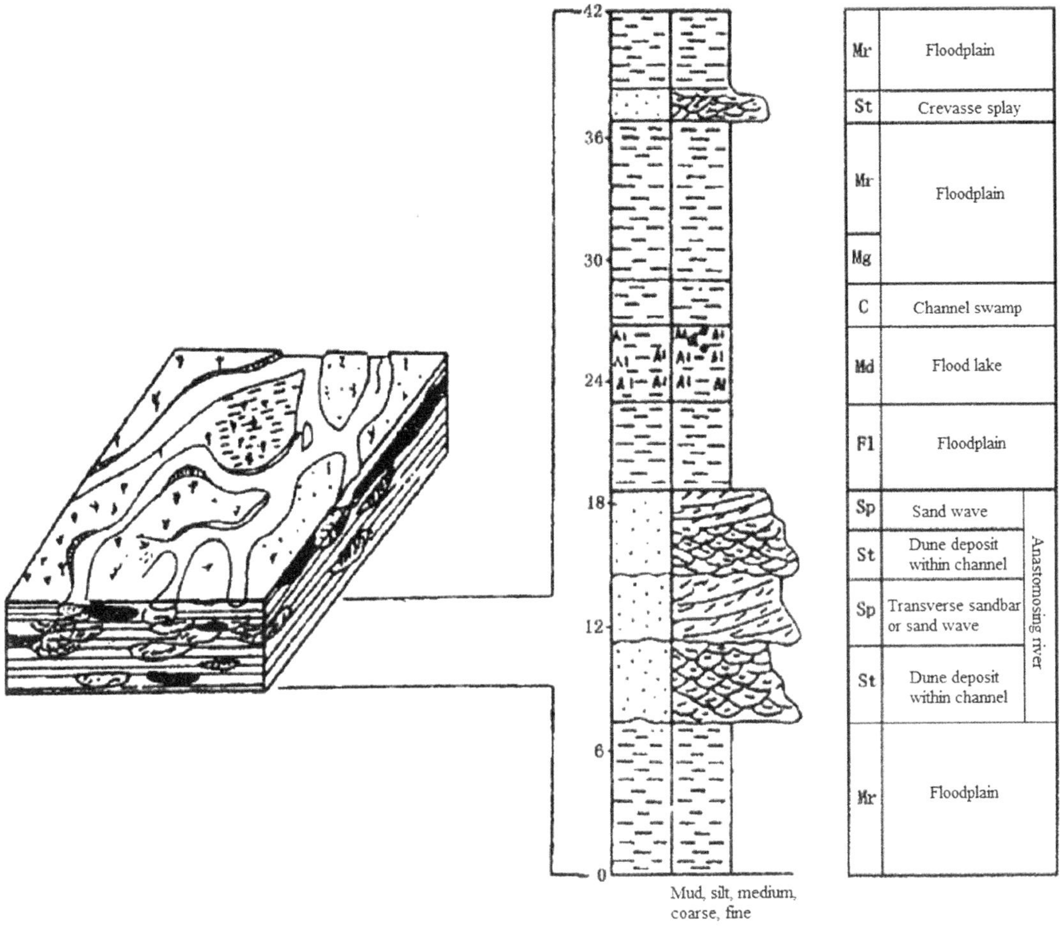

Fig. 9.12 Typical net-like river is a low-energy complex composed of a few camber changing, interconnected river, forming the sequence characteristics of "sand included by mud" in the section (according to Yu et al. 1992) taking pore G511 in Woyang, Anhui, as an example

greatly. Meanwhile, owing to the high flow velocity of the river, high sand carrying intensity at the river bottom, and quick swing and modification of channel bars, the geomorphic features of the riverbed vary rapidly.

Braided rivers are formed in an environment with high gradient, fast flow change, poor bank corrosion resistance, and large river bed load/suspended load ratios. They are developed easily upstream of a river with steep gradient, high flow, and coarse debris transport. Hence, braided rivers are developed mostly between alluvial fans and meandering rivers.

Braided rivers are relatively flat, feature low sinuosity, large gradients, large flooding intermittency, high flow changes, and coarse debris dominated by bed load. The majority of channel bars are formed in the entire river valley, whereas a large number of channels are bifurcated and combined around the channel bars, which are interlaced like a "braid" to move through the river valley. Their channels and channel bars are unstable, even with constant swing and rechanneling (Fig. 9.18) during deposition.

Braided river patterns and anastomosing river patterns are usually used interchangeably in early foreign literature. However, Lane (1957) noticed that the same braided river pattern appears to have abrupt and gentle slopes. In particular,

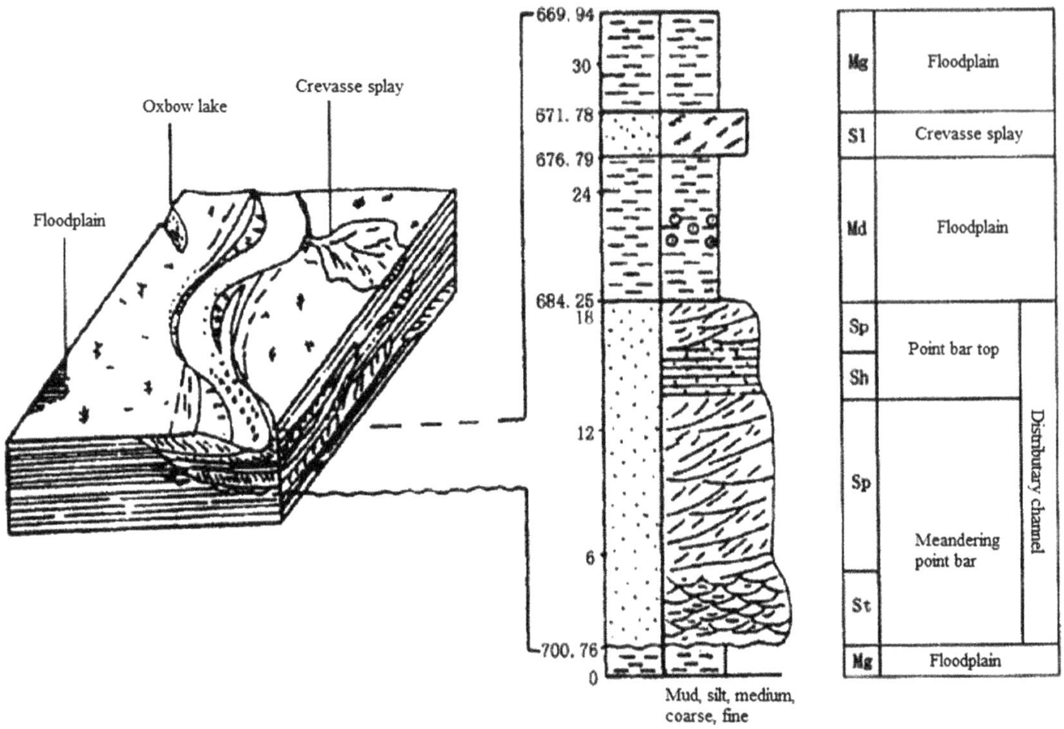

Fig. 9.13 Sequence characteristics of the far source sandy meandering river is very similar to that of the deltaic distributary channel, and the largest different is gravel lacking with the type 6 (according to Yu et al. 1992) taking field section P2s82-92 formation in Zibo, Shandong, as an example

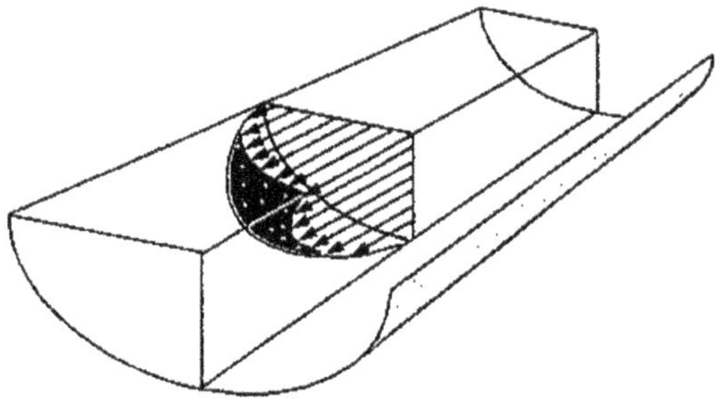

Fig. 9.14 In the straight river channel, the current speed is generally lowest in the base, while highest in the middle. The arrow shows flow direction, and arrow length shows flow velocity (according to Laing 1991). The arrow shows flow direction, and arrow length shows flow velocity

BHce (1982) determined the differences between anastomosing rivers and braided rivers: the former is cut by a central bar, but the latter is cut by a majority of sandbars (channel bars). Relative to river width, the central bar has a larger size. All channels of anastomosing rivers are separated by significant distances, and their positions are relatively fixed. A few channel sections may not have water flow at the normal level, but they are active and clearly distinguishable troughs not blocked by vegetation. Schumm (1963) also noted the dynamic characteristics of rivers and claimed that anastomosing rivers are relatively stable and differ from wandering braided rivers in terms of frequent mainstream swings.

Braided river deposition is dominated by battures (bars). Battures are formed by vertical accretion during the course of the movement of multiple flooding events downstream. A sand body does not have the typical upward fining grading. Instead, it is characterized by the development of megascopic planar cross bedding and parallel bedding in the upper flow regime. Another type of sand body is the sand filled in an abandoned channel. A braided river channel is abandoned at slow speeds and is

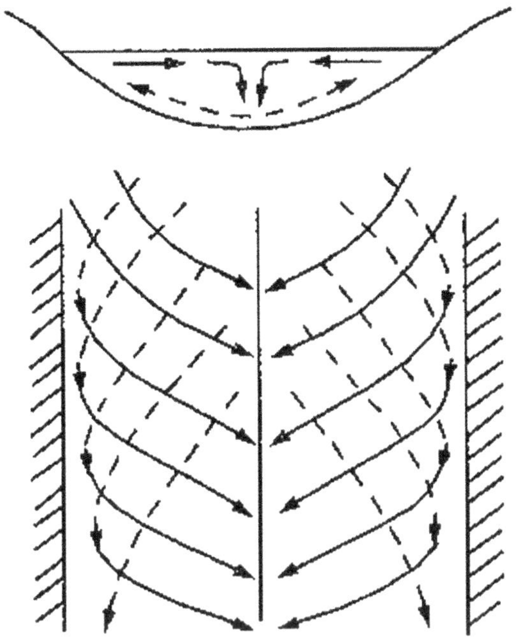

Fig. 9.15 Two symmetrical circulations fomed in two sides of the mainstream line of the straight river, and surface flow concentrate to the middle from the two banks, which cause that the main line moves straight along the center of the river (according to Allen 1964). The arrows show flow direction, the solid lines indicate main current, and dotted lines indicate secondary current

interconnected with movable channels for easy "recovery," thus it is generally filled with coarser debris. Owing to the low amount of suspended loads carried by braided rivers, there are only a few muddy or silty subfacies deposits on the top and few muddy intercalations.

Owing to coarser braided river deposits, gravelly braided rivers and sandy braided rivers are separated in different physico-geographical environments, in which gravel deposits are mostly contained. A braided river is loose in bank substance and its channel swings frequently, resulting in the vertical superposition or lateral connection of a variety of genetic unit sand bodies to form a large-area connected sand body. In addition, a large-scale sandy flat can be developed in a distal sandy braided river.

The three geologic features of braided rivers are summarized below.

1. **River bed form**

This has a variety of channel bars in the wide and shallow river bed and looks like a vast expanse of water with roaring waves in floods. In the drought period, it has densely distributed channels with dispersed currents, and sometimes it is difficult to even find the mainstream.

2. **Current**

Water flows rapidly in shallow channels, hence the current is often unstable, and sometimes sinusoidal water waves and sand waves may be formed on the water surface or river bed close to the critical flow.

3. **Sedimentary characteristics**

Since there are few argillaceous deposits and plant growth is difficult in a braided river, channel bars can migrate easily and change quickly, and their transformation into a central bar is impossible. However, the point bar of a meandering river is deposited quickly, and it can likely be transformed into a floodplain. However, it is unstable in the normal season, hence vegetation cover is relatively poor. The thickness of vertically aggraded fine-grained deposit is low and discontinuous, thus the amount of sand is apparently greater than that of mud in the entire braided river deposition. Most of the mudstone here is too thin to be confined because it is mainly composed of sludge formed during the arid period.

9.2.3 Meandering River

Meandering rivers are characterized by a curved single channel. Compared with braided rivers, they have a lower gradient, width/depth ratio and ratio of bed load to suspended load in the carried debris, and relatively slow flow change, but large depth. Of course, the flow change is substantial nevertheless, and long-time low water and short-period flooding occur. Meandering rivers

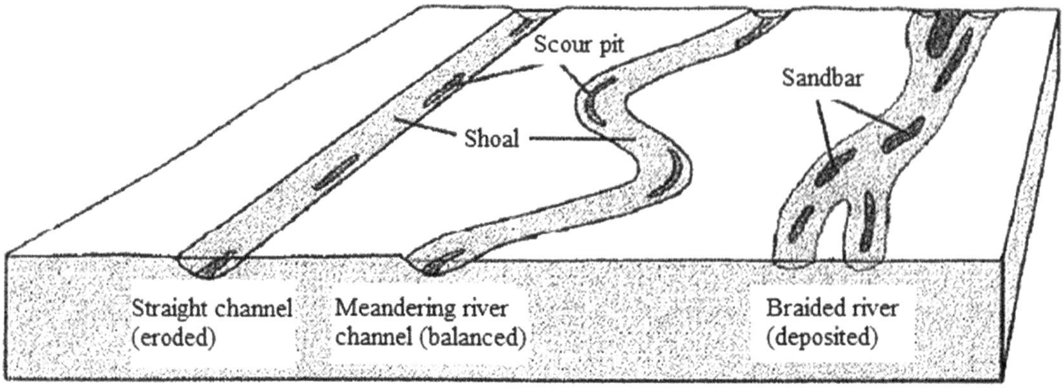

Fig. 9.16 Erosion effect is dominant in the Straight River, while the sedimentary is not obvious; the Meandering River is in the balance state, i.e. with erosion and sedimentary; sedimentary is dominant in the Braided River (according to Laing 1991)

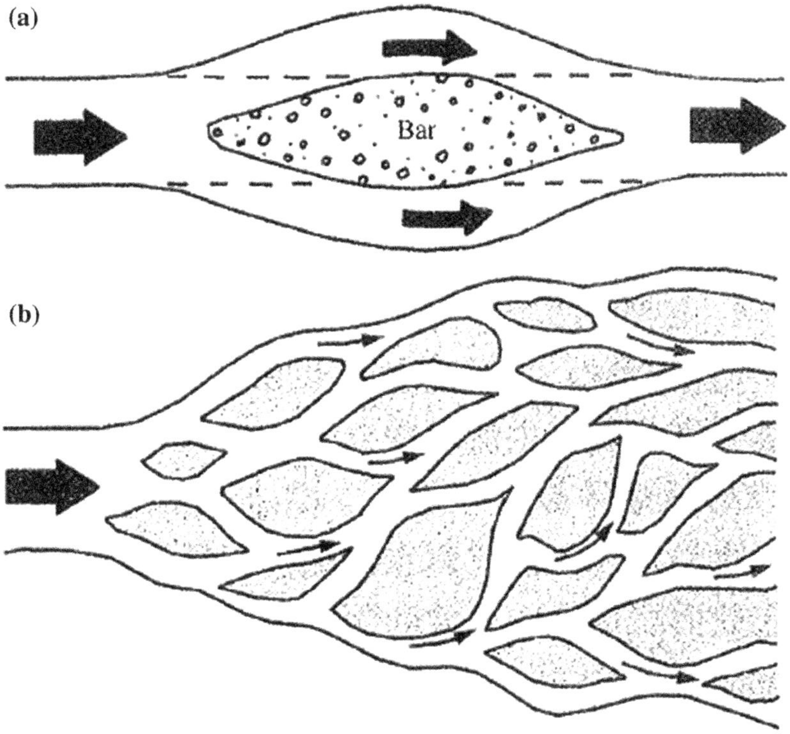

Fig. 9.17 Braided river is a wide and shallow river, and channel is separated by many heart beaches. The stream forms many channel, and keep branching and re-converging along many river island (quoted from C. C. Plummer and D. McGeary 1996). **a** Dividing channel bar current into two streams; and **b** variety of small channels formed due to channel bar development

are usually developed downstream alluvial plains, above delta deposits, but below braided rivers.

A meandering river is a wandering single channel with higher curvature, lower gradient, relatively low flooding intermittency, little flow change, finer debris, and low ratio of bed load to suspended load. The corrosion resistance of the banks is enhanced owing to the presence of natural levees. In the entire sedimentation process, the concave bank (steep bank) is increasingly exploited or eroded, and the convex bank

Fig. 9.18 Modern braided river in Alaska, USA

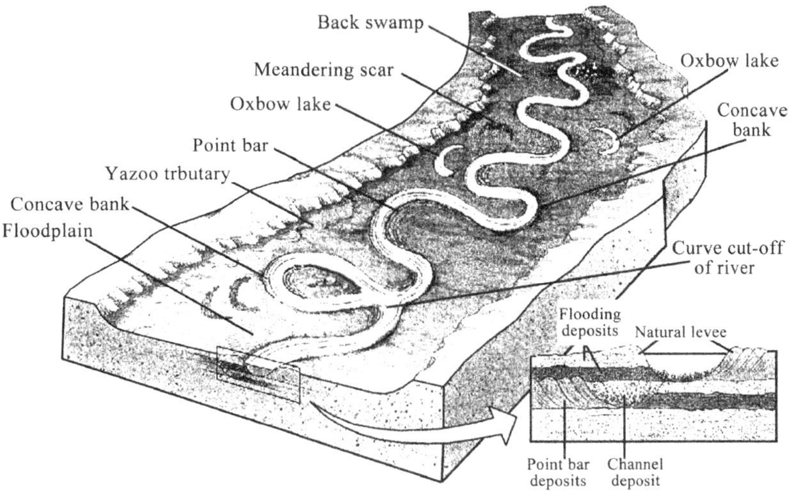

Fig. 9.19 Patterns and microtopography characteristics of meandering river

(gentle bank) is constantly accreted to form a "point bar" (Fig. 9.19) in a sedimentary sand body or in geomorphology.

The most important sedimentary process of a meandering river is related to its lateral migration. The concave bank collapses due to lateral erosion, while deposition occurs at the convex bank causing an increase in channel sinuosity. A point bar is deposited at thc convex bank of each meandering section throughout this process, and it is the main sedimentary sand body of a meandering river. The point bar is formed at the lateral accretion of the convex bank, which has a certain upward fining grading. Moreover, the sedimentary structure evolves from large (planar/trough) cross bedding to current ripple, and this is the common "point bar sequence". Various lateral accretion bodies may be scoured and contacted in the point bar or covered with thin argillaceous formations during flooding. All of these constitute the specific internal structure of the point bar and are important identification marks.

The sinuosity of a meandering river is subject to terrain and provenance supply. In general, the gentler the terrain slope, the lower is the

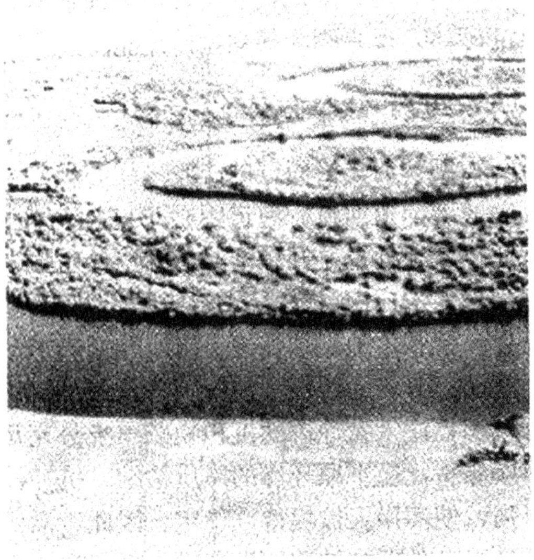

Fig. 9.20 High curvature meandering river in grassland of Hulun Buir League, China

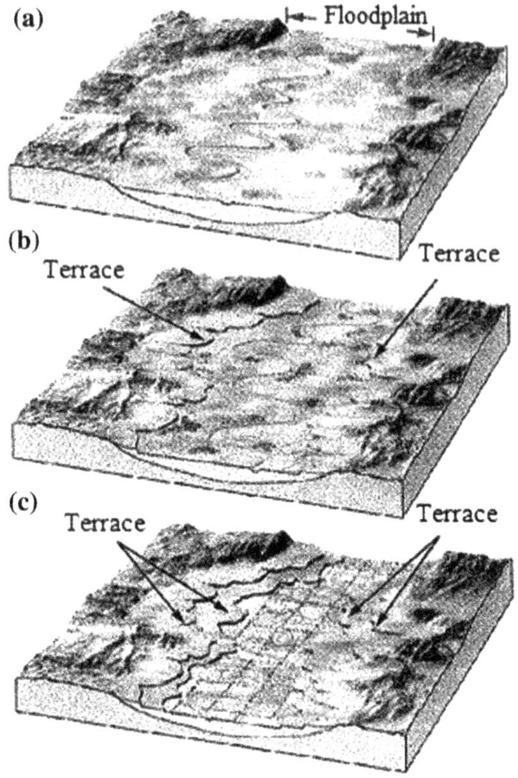

Fig. 9.21 The formation process and model of river terraces in depression basin

provenance supply and the higher is river sinuosity (Fig. 9.20). Especially, sediment supply decreases significantly during the basin depression period. In addition, the terrain is in a very open area. Hence, the sinuosity of a meandering river is high. When the base level is low or regionally raised, multi-terraces (Fig. 9.21) can be formed due to river incision, but the floodplain narrows gradually. Neogene deposition in Mesozoic-Cenozoic faulted basins in eastern China has such characteristics, and the trend of river deposition is usually parallel to the long-axis direction of the basin.

A meandering river is developed further with overbank deposits such as natural levees, crevasse splays, and erosion ditches. The upper and large parts of abandoned channel deposits are usually filled with mud. Although there are large differences in size and development degree among these sand bodies, the point bars of all reaches can be connected to form a meandering belt sand body. This sand body and the argillaceous deposits on the widely developed floodplain constitute a sedimentary set with frequent phase change of sandstone and mudstone both on the interstratified layer and the plane.

From the viewpoint of petroleum reservoir studies, genetic unit sand bodies deposited by a meandering river are meandering belt sand bodies with homochronous meandering deposition. Because their lateral continuity is related to channel width and sinuosity, the old meandering belt is abandoned and a new meandering belt sand body starts deposition when the river is subject to avulsion for diversion. The degree of connectivity of the meandering belt sand bodies for different genetic units (Fig. 9.22) depends on the relative magnitude of the subsidence rate, subsidence speed, and frequency of river avulsion and diversion. When the subsidence rate is relatively high and subsidence speed is relatively low, mutually connected sand bodies are formed

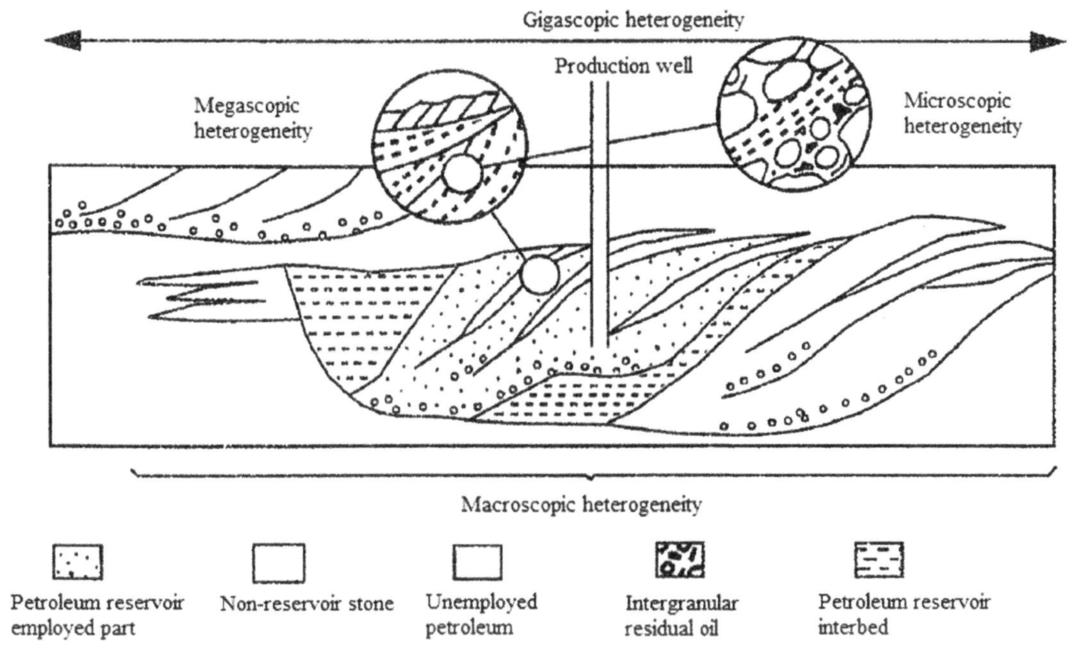

Fig. 9.22 Characteristics of reservoir heterogeneity at different levels reflected in the meandering point bar

easily on a large scale; in addition, isolated sand bodies are formed.

9.2.4 Anastomosing River

An anastomosing river is a multi-channel river flowing along a fixed batture (central bar). Its channels are firm and stable owing to the channel bar and bank, which is the main difference between anastomosing and braided rivers. Because an anastomosing river needs specific sedimentary conditions, the sedimentary basin must descend constantly or control the local base level rise for ensuring quick and continuous channel filling. An adequate amount of sediments must be injected into the basin to retain the sticky bank soil of the floodplain and strong erosion resistance, thereby stabilizing the natural bank levee. Hence, a region with vegetation growth under wet conditions is more suitable for the formation of an anastomosing river, and they are formed generally in gentle downstream areas.

Smith (1979, 1983) found that channels and channel bars are stable under the climatic conditions suitable for growing a variety of plants while studying some modern anastomosing rivers in the west of Canada. Therefore, the associated sand facies in an anastomosing channel has some peat swamps. However, Rust (1978, 1983) found an anastomosing river in arid continental Australia; in an article published in 1983, Smith pointed out that the condition under which a warm-wet climate is suitable for growing a variety of plants shall be reconsidered, assuming that only a stable bank condition can develop an anastomosing river.

An anastomosing river is generally found in the river downstream. Its sediments are mainly transported through suspended loads. Its channel is narrow, deep, and has multiple curves, presented as retiform downward. In general, its channels are separated by semi-permanent alluvial islands and floodplains or wet lands. Alluvial islands and floodplains or wet lands, with relatively stable positions and size, are composed

Fig. 9.23 Modern net-like river in Yunnan, China

mainly of fine-grained substances and peats. Compared with narrow channels, they occupy wide areas (60–90%). Owing to the warm and humid climate in southwest China, such rivers have developed in a few areas in that region (Fig. 9.23).

Channel sand bodies are the main sediment of an anastomosing river. Multiple formation superimposed complex fining upward is formed by constant aggradation in the channel. The sand body has megascopic cross bedding, hence the channel may evolve into a small meandering river to deposit a small point bar before being abandoned eventually. Channel sand bodies are distributed in a narrow-thick strip shape, and other associated sand bodies constitute natural levees and small crevasse splays, which do not have dominating roles here.

The development of the above four channel patterns is controlled by factors, such as topographic slopes, river flow, riverbed sections, load carrying process, and clastic property, and it varies (Fig. 9.24) as these factors change. As a consequence, the channel pattern may vary within the same river. In fact, a channel pattern is a continuous spectrogram. Rivers usually follows the evolution law (Fig. 9.25) that a braided river evolves from a low-sinuosity

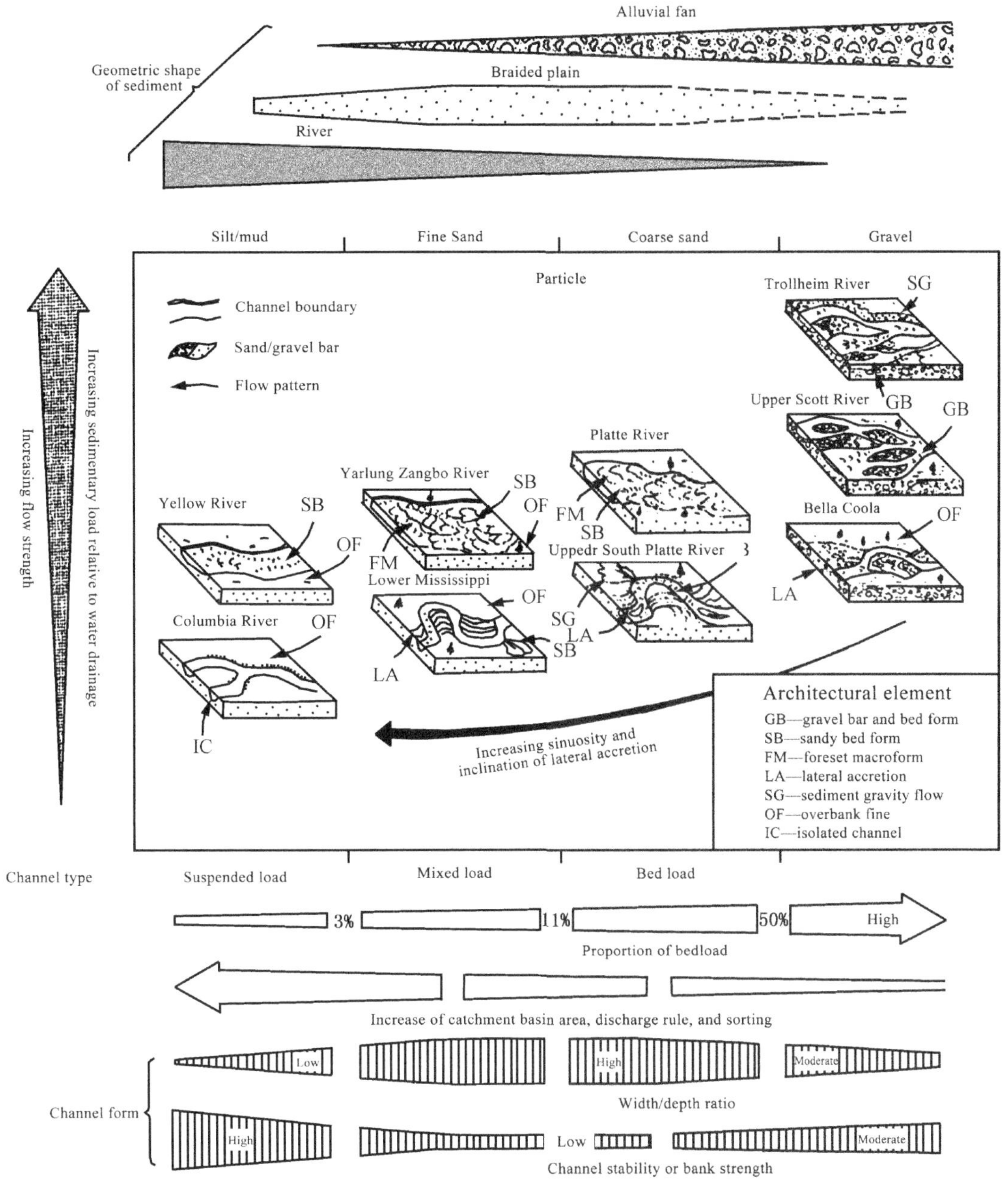

Fig. 9.24 Relationship between different forms of load, slope, load and river pattern (after Orton and Reading 1993)

meandering river into a high-sinuosity meandering river, from upstream to downstream, along with the evolution of the aforementioned factors controlling channel pattern. As flood varies, the former and latter rivers are transformed into each other, for instance, D changes to E or vice versa.

9.3 Sedimentary Environment and Sedimentary Sequence

Meandering and braided rivers are the most common river patterns in nature, thus research on these two river types is extensive. Typical

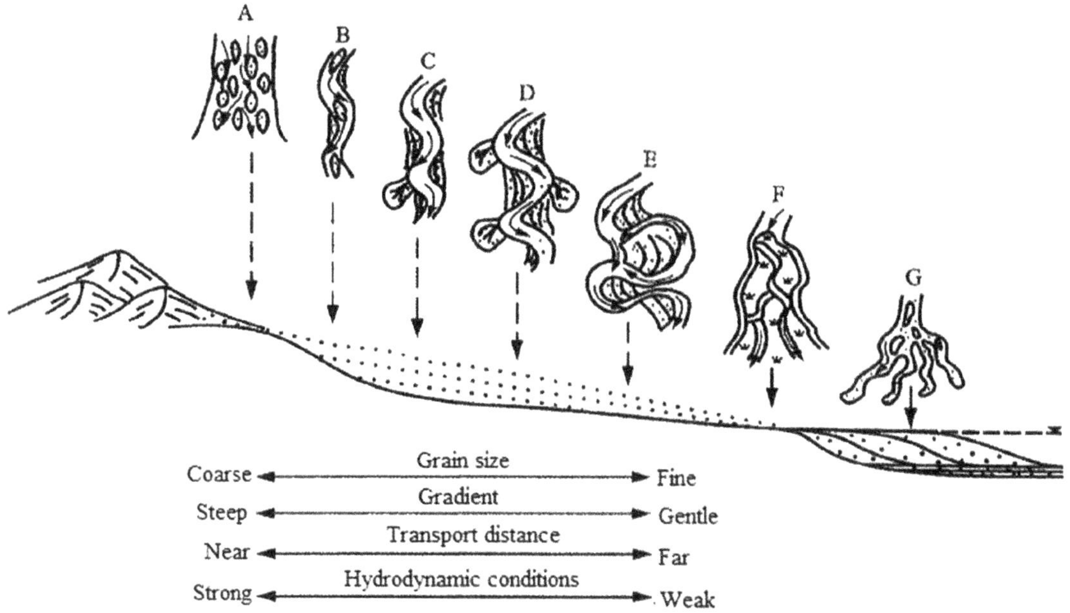

Fig. 9.25 The river type is actually a continuous spectrum, from upstream to downstream, the evolution law of a river is generally from braided river to low sinuosity meandering river, then to high sinuosity meandering river

anastomosing rivers are rarely found; hence, related studies are scarce. Although a few straight reaches are frequently found in some rivers, no detailed studies and reports on straight rivers have been published thus far, and straight river deposits have not been identified, even in ancient sediments. As mentioned above, straight rivers are dominated by erosion and therefore, have little sedimentation, which makes their identification in ancient sediments difficult.

9.3.1 Environment Type and Model of Meandering Rivers

9.3.1.1 Environment Type of Meandering Rivers

The main sub-environment types (namely, microtopographic unit) of meandering rivers include riverbed lag deposits and point bars, Oxbow lakes, natural levees, crevasse splays, and floodplains.

1. **Channels**

(1) Riverbed lag deposit

This is also known as channel lag deposit, and it is composed of coarse gravels and bottom riverbed lag conglomerates. Riverbed lag deposit is a product of short-distance transport when a river's flow remains at its highest level. Under normal flow, fine-grained substances are carried constantly, but gravels lag at the bottom of the riverbed owing to flow scouring and sorting. Such deposits are usually found at the bottom of the sedimentary section of a river, and are characterized clearly by (Table 9.8) transiting thinning-upward particles to point bar deposition.

(2) Point bar

Point bar deposition is the major geomorphic unit of a snaking river and also the necessary product resulting from lateral riverbed migration. River water advances spirally in the channel and scours the outer bank (or concave bank) constantly, and transverse circulation carries particles in the river to the downstream inner bank (or concave bank) for deposition. This leads to the formation of sedimentary characteristics dominated by lateral accretion, but only shoals are

Table 9.8 Comparison of all sub-environment sedimentary characteristics of meandering rivers

Environment Type		Main lithology and grain size	Sedimentary structure and feature	Vertical sequence characteristics	Morphological characteristics
Channel	Lag deposits	Mainly include coarse conglomerate and gravel-containing coarse sandstone, and a few medium and fine-grained sandstones	① Gravels are arranged directional in imbricate form, and the largest flat plane is inclined upstream ② Scour surface is formed in the bottom part; and ③ The lower part is massive, and the middle and upper parts can be developed into a large trough and planar cross bedding, respectively	① Less obvious fining-upward structure is formed; and ② It is transited to filling deposits of point bars or oxbow lake upwards	① It has a lens shape on the section; and ② it has a strip shape on the plane
	Point bar	① The variation range of grain size is large; ② It is mainly composed of gravels, sand, and silts; and ③ It can be divided into coarse- and fine-grained point bars	① The lower part has megascopic trough or (ε type) planar cross bedding, the upper middle part has miniature trough cross bedding or climbing ripple lamination, and horizontal bedding may appear in the top part; and ② Reactivation surface is commonly found	① Typical fining upward structure; and ② The bottom part has riverbed lag deposits, and the top part is transited to natural levee	① It has a tabular sand body on the section; and ② It is of elliptic or arc shape on the plane
Natural Levee		① Mainly including thin siltstone and mudstone; ② The two are of thin interbed shape	① Miniature current-ripple cross bedding or climbing ripple bedding and horizontal bedding are developed; ② Plants and bioturbation structures can be found in upper mudstone	① Sandy and muddy thin interbed; ② Bottom part is in contact with point bar in a transitional way, while top surface is in contact with overbank fines abruptly	① It has wedge shape on the section ② It has strip shape on the plane
Oxbow lake		It is composed mainly of fine-grained siltstone and mudstone	① Developed with horizontal bedding; and ② Mudstone usually has a massive structure	① Generally rhythm less structure; and ② Bottom part is quickly in contact with lag deposits in a transitional manner	① The section is surrounded by point bars; and ② It is of arc and semicircular shapes on the plane
Crevasse Splay		Mainly includes fine sandstone and siltstone, and thin medium and coarse sandstone can be found at the bottom of crevasse channel.	① Mainly includes miniature cross bedding, but developed locally with medium cross bedding; and ② Scour-fill structures can be found.	① Coarsening upward is common; and ② Local scouring is relatively distinctive, and lateral extension is limited.	Tongue, lens shape
Floodplain		Mainly includes siltstone and mudstone	① Developed with horizontal bedding; ② Massive structure is common; and ③ Bioturbation structure	General rhythmless structure	Planar

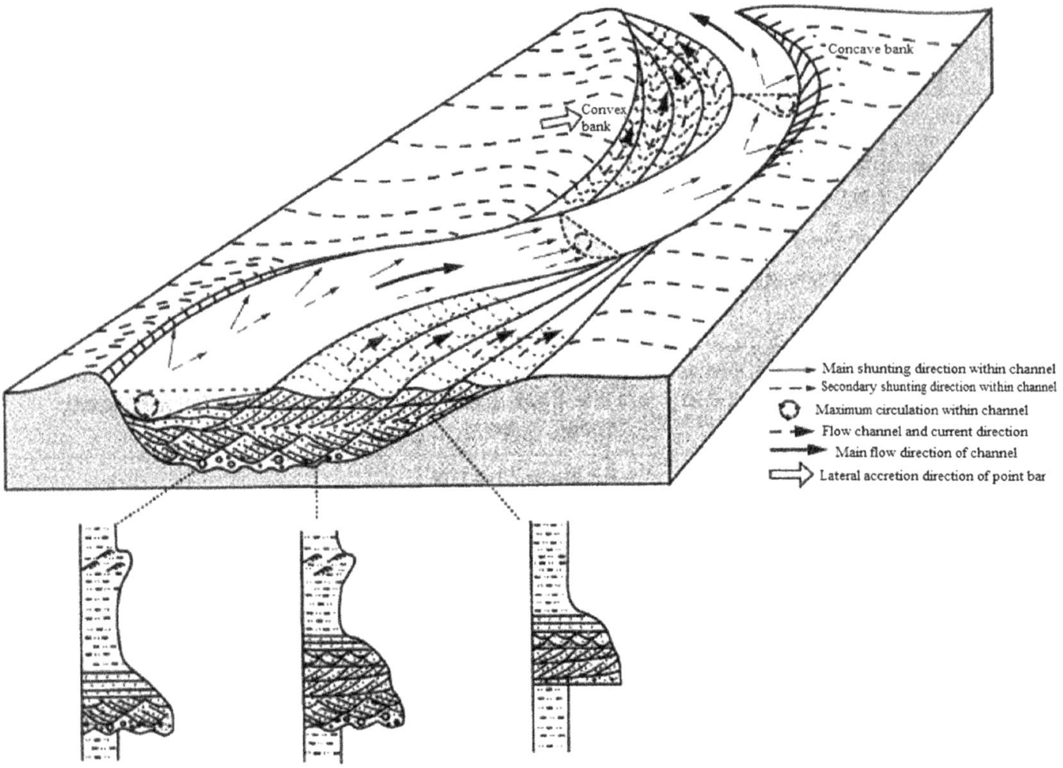

Fig. 9.26 Typical plane and profile structures and the vertical sedimentary sequence in different parts of meandering river

formed at the beginning of sedimentation. As the river constantly migrates laterally, shoals increase constantly, and a wide point bar is formed in the curved convex part of river. In the case of deep water at the concave bank, flow velocity is high; in the case of shallow water at the convex bank, flow velocity is low (Fig. 9.26). The bottom current decreases progressively with lateral movement, causing sedimentation and differentiation of debris. Therefore, the sediments have significant characteristics (Table 9.8). Coarser particles (mainly including sand) are deposited at the lower part of the point bar close to the river center, and finer particles are deposited in the upper part of the point bar far away from the river center.

The size and form of a point bar vary when the size and sinuosity of the associated river change. In small rivers, a simple point bar is located on one side of the convex bank of a river bend, and it is inclined gently toward the channel. In large rivers, the situation is relatively complex because chutes and chute bars are formed when parts of the river flow over point bar surfaces in the flooding period. Each chute bar represents the result of one channel migration in the primary flooding period.

The thickness of a point bar is similar to the depth of a riverbed, whereas its width reflects river size and sinuosity. The larger the channel sinuosity, the broader is the developed point bar and vice versa. In other words, the width of a point bar is controlled mainly by channel sinuosity rather than the channel width, but when channel sinuosity is identical, the greater the channel width, the larger is the point bar.

The deposited grain size of point bars varies greatly owing to their different distances from the provenance and different development positions. Point bars can be divided into coarse-grained (gravelly meandering river) deposition and fine-grained (sandy meandering river) deposition (Figs. 9.27 and 9.28) based on grain size characteristics. The characteristics of coarse and fine point bars, as determined using the cumulative relative frequency curve, show two populations in grain size distribution, including saltation and suspension populations, wherein a rolling population having medium to higher slope and medium to better sorting is not developed. Suspension population is low in slope and content, and poor in sorting. The cutoffs of the two are abrupt, about 3^{Φ}–3.5^{Φ}.

In terms of vertical sedimentary sequences, it has distinctive features, namely, obvious dual structure characteristics (Fig. 9.29) with a coarse lower part and a thin upper part.

2. **Oxbow lakes**

Oxbow lake deposits are abandoned channel deposits, which are mainly fine-grained materials generated in the flood period, such as silt and mud. Some sections of the channel or the entire channel can be abandoned by snaking owing to the cut-off effect or avulsion in the development process, thus leading to the formation of oxbow lake deposits or abandoned channel deposits in the early bending parts.

The formation of an oxbow lake is connected to the stream cut-off effect, while channel abandonment is not only related to the cut-off effect but also to avulsion, in which the cut-off effect can be divided into chute cut-off and neck cut-off (Figs. 9.30 and 9.31).

(1) Chute cut-off

Multiple downstream chutes are formed on the point bar of meandering rivers owing to the change in water flow energy between flood and the non-flood periods. The chute cut-off effect occurs once a chute is changed as the main channel, provided that the channel has high sinuosity. Therefore, chute cut-off is a phenomenon in which the old depression on a point bar is occupied gradually by the channel (shallow slot or chute), further compressing the short-term process. The water flow in the abandoned river is reduced gradually with a decreasing trend, which leads to the deposition of thicker silt and fine sand. Hence, a sequence structure comprising multi-sand and low amounts of mud is produced. Notably, when the water flow speed is high, a chute bar turns into a channel bar and a meandering river turns into a braided river, and water flow action is stronger in the early main channel and the later chute.

(2) Neck cut-off

When the narrowest part of the neck at the meandering ring is suddenly cut off to form a new channel, the outlet and inlet of the truncated meandering ring are choked quickly with sand. In this case, water flow is stopped soon. Later, silt, clay, and other fine-grained suspended sediments are accumulated. Thus, the sedimentary sequences are different for different cut-off effects (Fig. 9.31).

(3) Avulsion

The channel of a meandering river with large sinuosity is truncated repeatedly and abandoned, and it accumulates fine-grained materials that are generally difficult to erode away. To some extent, these materials enclose the meander belt. At this point, the altitude of the entire meander belt is higher than that of the floodplain owing to vertical accretion. A great crevasse in a natural levee can result in river avulsion during the flood period. Such a crevasse is formed most easily at the tangent of the running water and the sinuosity part of the channel, and the crevasse forms crevasse splays on the floodplain nearby.

3. **Natural levees**

Natural levee deposits, which are also one of the main characteristics of meandering rivers, are distributed on both sides of a meandering river. Natural levees are formed because large amounts of suspended materials are deposited on the river bank quickly when water overflows the bank during the flood period and the river flow rate is

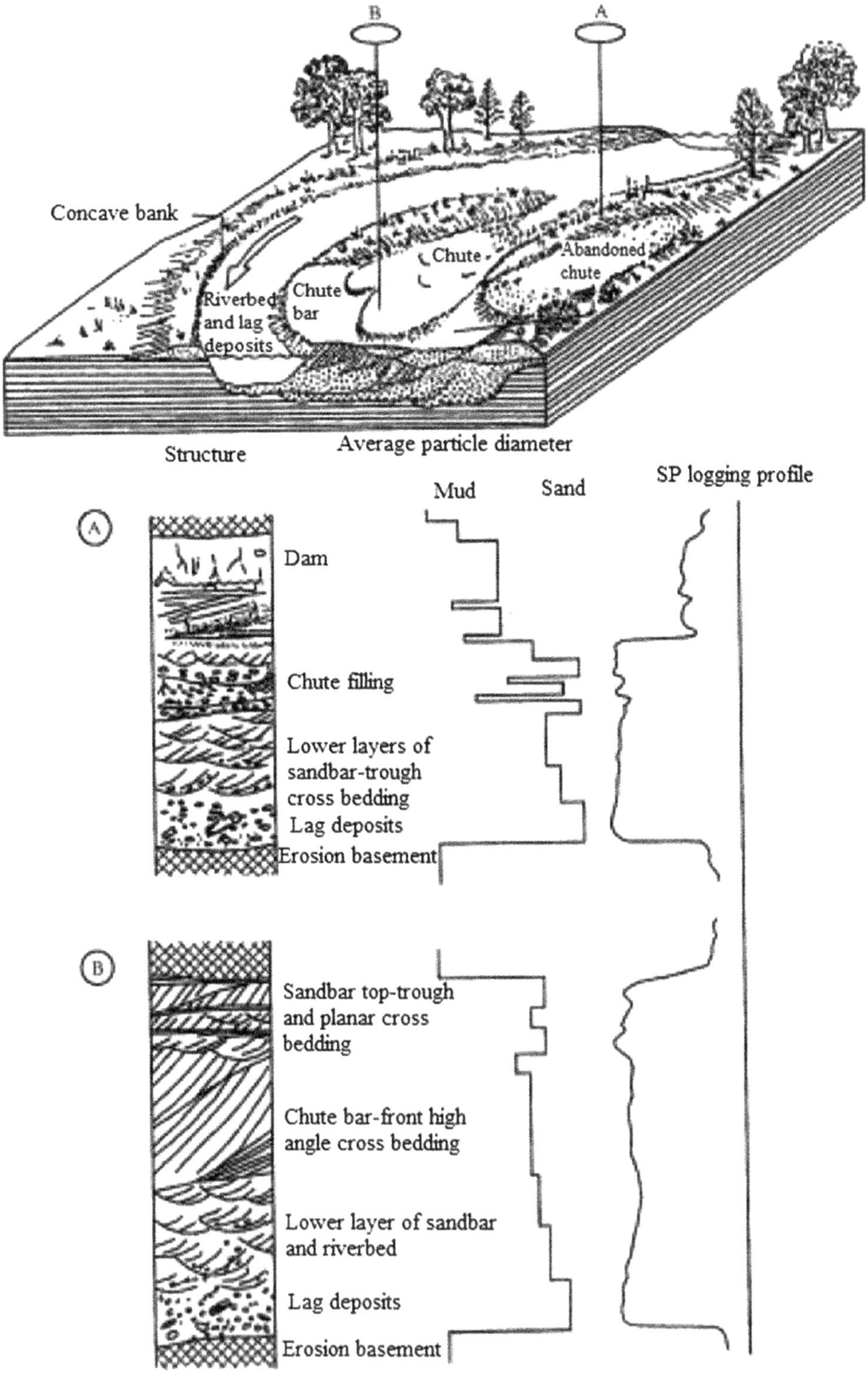

Fig. 9.27 Sedimentary model and sedimentary sequence characteristics high sinuosity with washed river point bar (gravel meandering river) (after Galloway 1983)

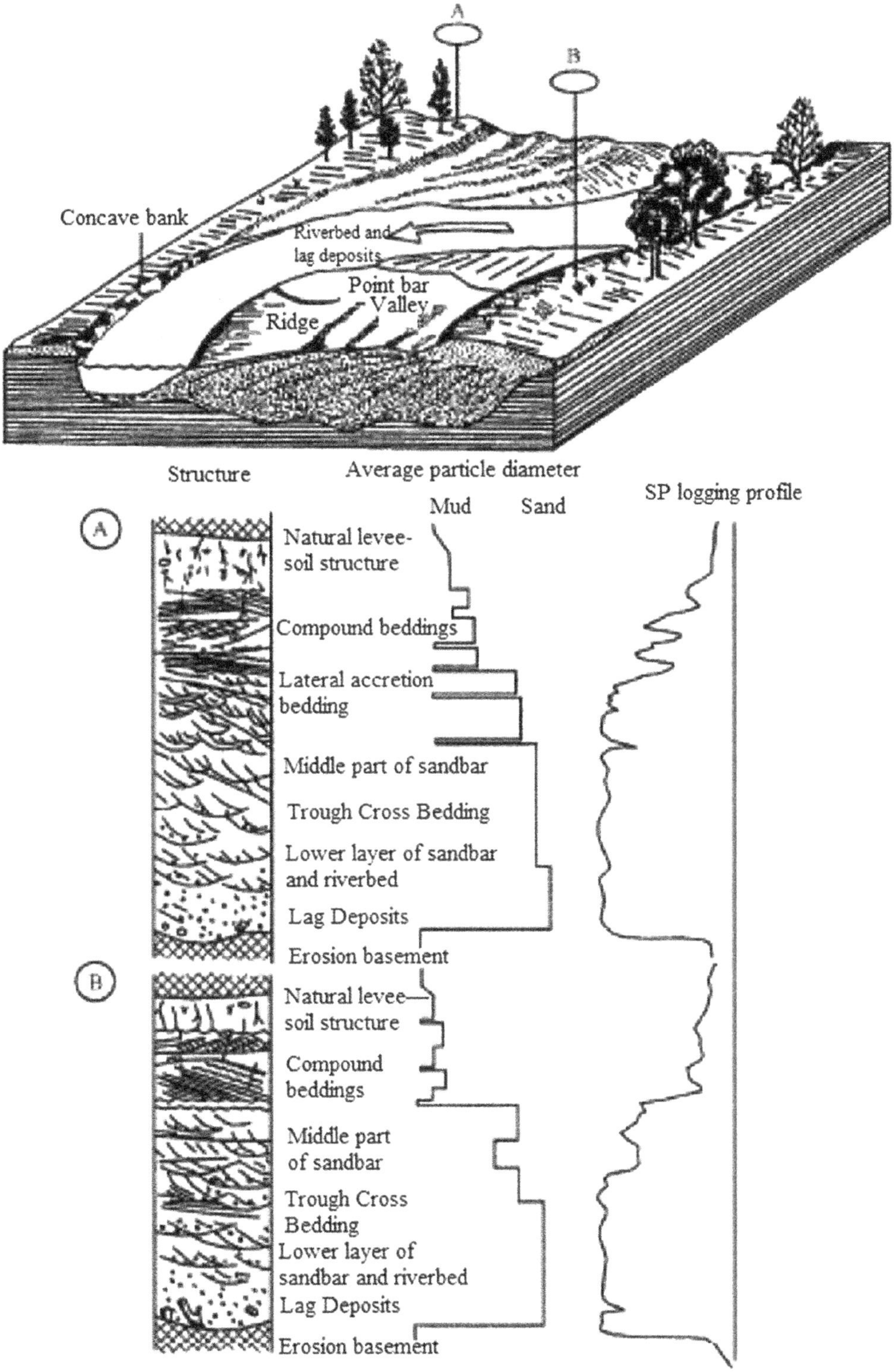

Fig. 9.28 sedimentary model and sedimentary sequence characteristics of the high sinuosity river with complete point bar (sandy meandering river) (after Galloway 1983)

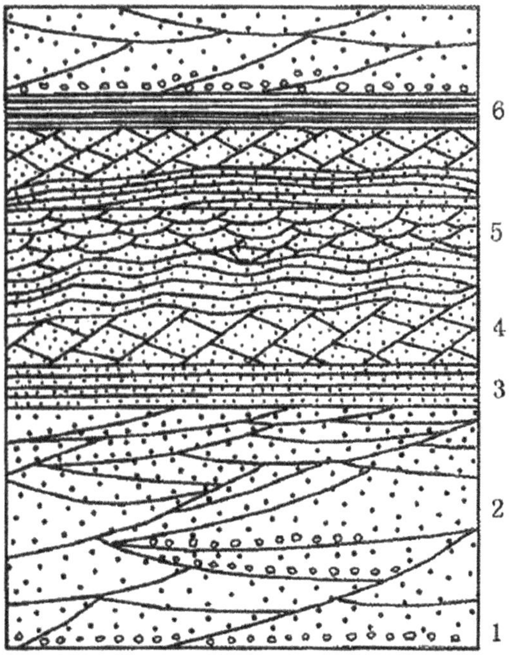

Fig. 9.29 In terms of vertical sequence, the point bar deposit has obvious dualistic structure, i.e. lower coarse and upper fine (according to Reineck et al. 1973). 1—Bottom lag deposits; 2—megascopic cross bedding; 3—parallel bedding; 4–climbing bedding; 5—current ripple; and 6—horizontal bedding

decreased. The deposited materials are finer than those constituting a point bar, and their grain size is decreases with increasing distance from the channel.

A natural levee on the plane is strip-shaped, with a body that bends as the channel bends. It belongs to wedge ridge sediment, which is distributed mainly on the concave side of a meandering river. An asymmetric triangle is shown on the cross section, featuring a steeper slope on the river side, flatter slope on the floodplain, and gradual transition backward. Therefore, the top surface of a natural levee represents the maximum flood water level that can be reached during the flood period, and the grain size of its rock is finer than that of a point bar but rougher than that of a floodplain far away from the channel (Table 9.8). A natural levee is close to point bar deposits in the vertical direction. Furthermore, it has a developed sedimentary structure, which makes possible the visualization of all types of ripple beddings, irregular horizontal laminae, mud cracks, and plant roots.

4. **Floodplain**

An overbank, which is also called a floodplain, is the part with the most widely distributed area in a meandering river. In modern geomorphology, floodplains and alluvial plains are synonyms, but in sedimentology, alluvial plains mostly refer to the plain on an alluvial fan or around a braided river. Thus, a floodplain is the most low-lying part when river water overflows on the both sides of the plain of a meandering river during flooding. Hence, it is also called a flood basin. Sediments here are mainly suspended loads arising from floods, and sedimentary rocks including mudstone, siltstone, and thin argillaceous limestone are the typical sheet flow accretion products. It is possible to see mud crack structures, as well as calcareous and iron nodules, in dry climate areas. Plants in this part flourish under humid conditions. Furthermore, swamps housing considerable amounts of organic materials, or even peat horizons and coal bed deposits, can be developed (Fig. 9.32).

A floodplain is located in the upper part of an alluvial cycle on the section. The higher the river sinuosity, the thicker the fine clastic sediments that represent a floodplain (Fig. 9.33).

In addition, a floodplain can form a flood basin, which is the most low-lying part in a floodplain. Given its flat terrain and poor drainage, it plays the role of a sedimentary basin. During flooding, when the water overflows the bank and passes through the natural levee and crevasse splays, thicker materials are deposited; suspended fine materials are deposited mainly in the flood basin, but the deposition speed is relatively slow, with a deposition thickness of 1–2 cm in one flood period.

5. **Crevasse splays**

In high water levels, the fan deposits of arborization drainage formed on the slope of the

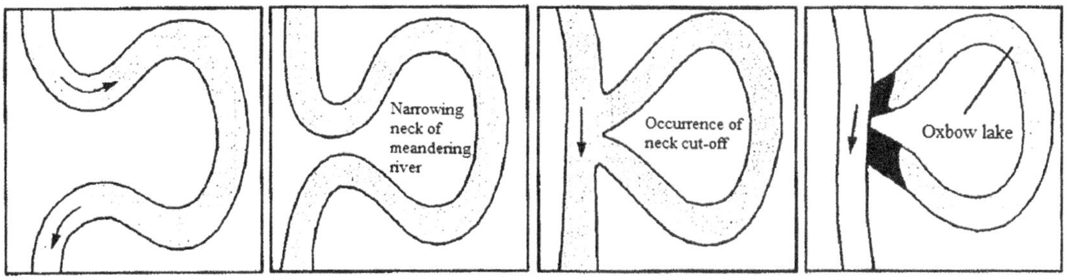

Fig. 9.30 The process of straightening and the formation of Mork lake in meandering river (after Graham R. Thompson et al. 1991)

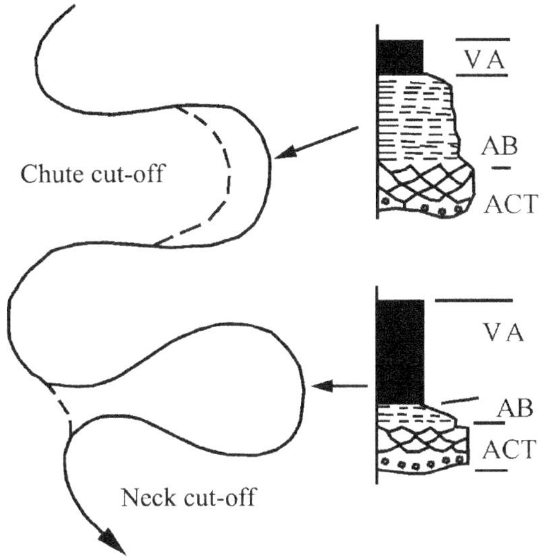

Fig. 9.31 The string straightening and the neck straightening and the sedimentary sequence (after Walker 1976). *VA* vertical accretion; *AB* abandoned channel; *ACT* activity channel

plain direction close to the embankment when excessive flood bursts a natural levee is called a crevasse splay. This is one of the main products of avulsion.

Crevasse splay sedimentation is presented as ligulate sand bodies, and it appears to have a lens shape in the section. The thickness is generally small, ranging from a dozen centimeters to several meters, and the grain size is often rougher than that of connected bank deposits (Table 9.8).

9.3.2 Vertical Sedimentary Sequence of a Meandering River

The appearance sequence (Fig. 9.26) of the microfacies of a meandering river in the section can be divided into four sedimentary units from bottom to top:

① Lag deposits of channel bottom; ② point bar deposits, common ε (multi-group low angle incised type) planar cross bedding; ③ point bar top or ridge top with well-developed climbing ripple bedding, parallel bedding, or small trough cross bedding (product of chute); and ④ vertical accretion or sheet flow accretion, specifically as natural levees and floodplains. It is mainly constituted of intermittent wavy (miniature current ripple) cross bedding siltstone, horizontal textured silty mudstone, and massive mudstone.

9.3.2.1 Depositional Model of a Meandering River

The sedimentary model of meandering rivers proposed by Allen (1964) (Figs. 9.34 and 9.35) is still regarded as the classical depositional model characterized as follows:

① Terrigenous clastic rocks dominate, with conglomerates, sandstone, and mudstone from bottom to top. The lithology changes frequently, and it is especially unstable, occasionally with argillaceous limestone lamina in the upper parts. ② Sedimentary structure decreases in size upward, cross bedding type changes from large

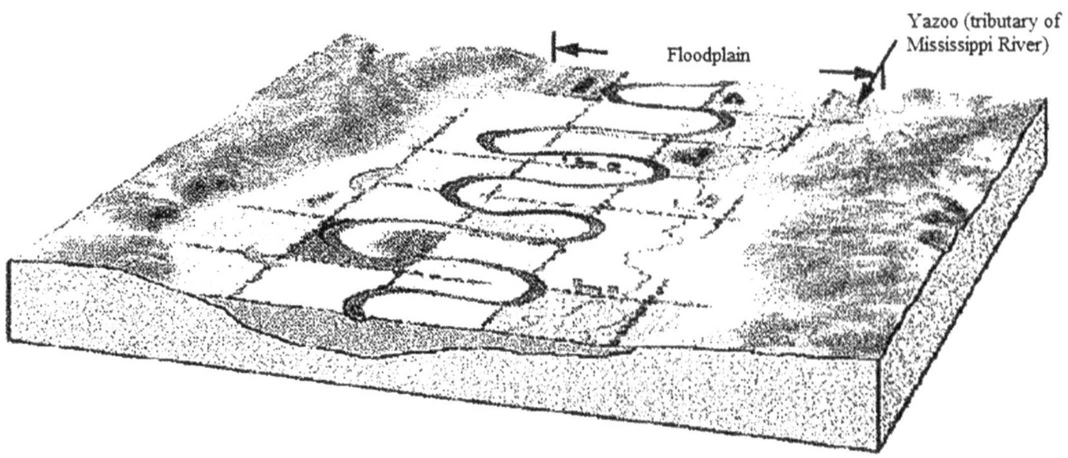

Fig. 9.32 Geomorphic features of flood plains (according to the *Illustrated Dictionary of Earth Science*, 1999)

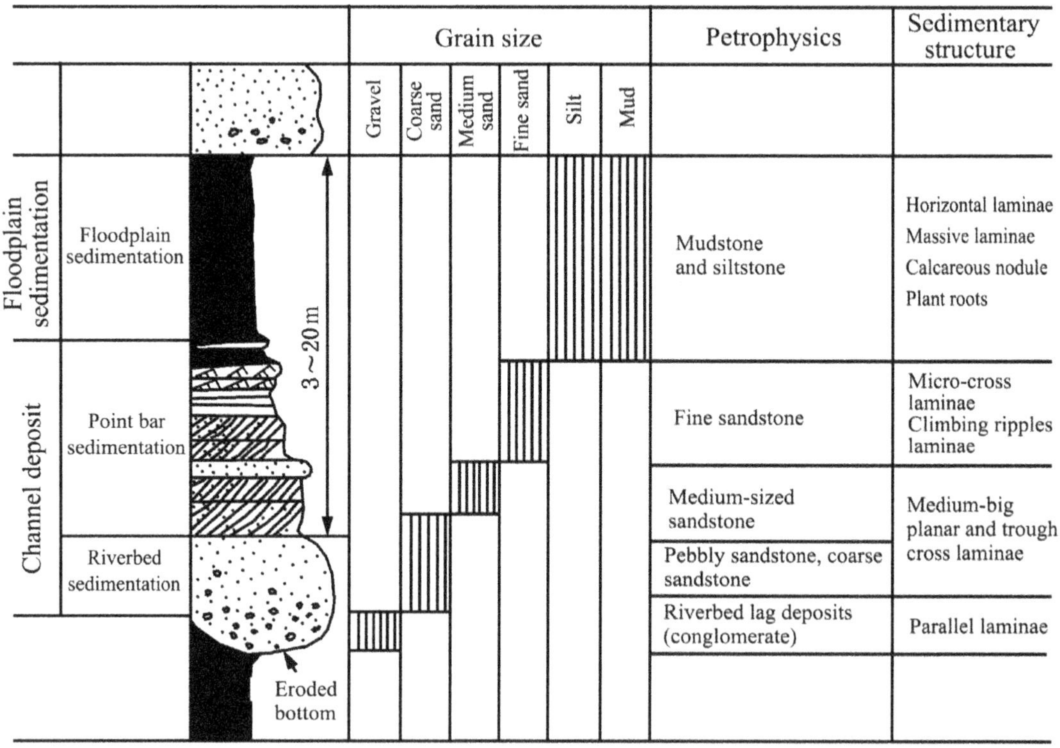

Fig. 9.33 Sedimentary sequence of meandering river (after Klein 1972; Allen 1970)

to small, and it gradually emerges as miniature current ripples and climbing wavy cross beddings upward, sandwiched with horizontal bedding and developed with mud cracks and raindrop imprints. ③ Plant imprints and plant roots are common. Mainly, peat and coal may possibly be seen, occasionally with a small number of freshwater mollusks such as gastropods and bivalves. The above depositional model reflects the macroscopic marker

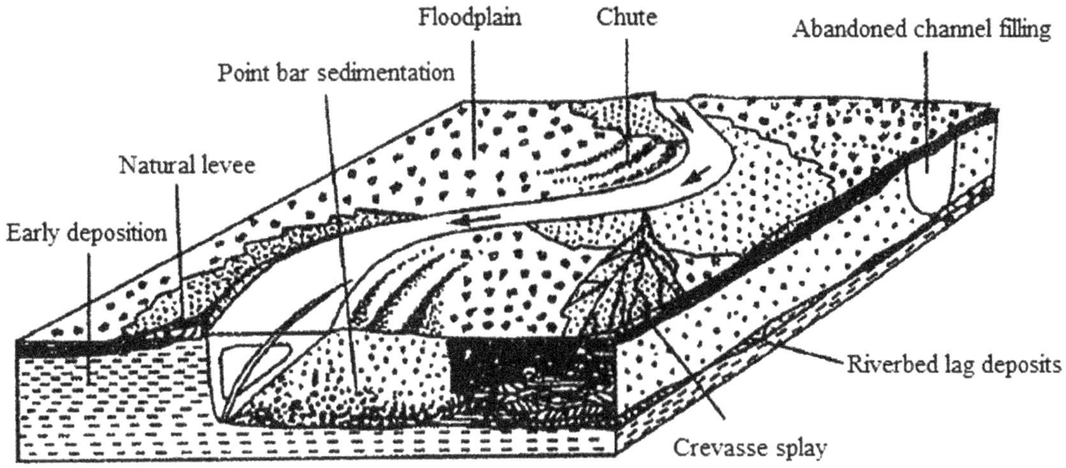

Fig. 9.34 Sedimentary model of meandering river (after Allen 1964)

combination of meandering rivers. At the micro level, these rock strata, especially sandstones, have an obvious granularity distribution, allogenic and authigenic mineral composition, and combination characteristics.

9.3.3 Environment Type and Model of Braided River

Braided rivers are those with sinuosity less than 1.5. They are characterized by the development of a large number of channels separated by sandbars within the entire width range (or valley) of the river. The channel, which is wide and shallow with frequent swing and uncertain wandering, is also known as a wandering channel. As mentioned previously, a braided river usually appears in a place with a large slope. Change in the channel slope is large, flow rate is high, and they often erode the banks quickly, so point bars and natural levees are not developed normally, with the exception of channel bars (channel sand bodies and gravel bars).

In general, braided rivers are often seen in climatic zones with significant humid or damp seasonal changes or floodplains. The net river flow varies with season, and the flow rate is unstable. In spring and autumn, flood often occurs, which can submerge the channel bar of a braided river because of an increase in the flow rate owing to sufficient rainfall or snowmelt supply. However, in drought seasons, the water is limited to flow in the narrow channels between the channel bars owing to decreased flow rate and the fact that the channel is located above the water. Each alternate flood and drought period changes the form and layout between the channel bar and the channel. Therefore, the frequent migration of channel and channel bar is the most important feature of braided rivers.

Some scholars also call braided rivers "over loaded" given their large loads. Braided rivers can be subdivided into two types according to sedimentary composition and grain size of the channel bar: ① gravelly braided river (Fig. 9.36); and ② sandy braided river dominated by sandy deposits (Fig. 9.37). These two types differ substantially in terms of sedimentary characteristics and vertical sedimentary sequence.

9.3.3.1 Environment Type of Braided River

The riverbed of a braided river is wide and shallow. Many channels repeatedly bifurcate and converge, so the channels are easy to abandon and revive, and such activity is caused by frequent changes in geomorphic units. From the macroscopic perspective, the environment types

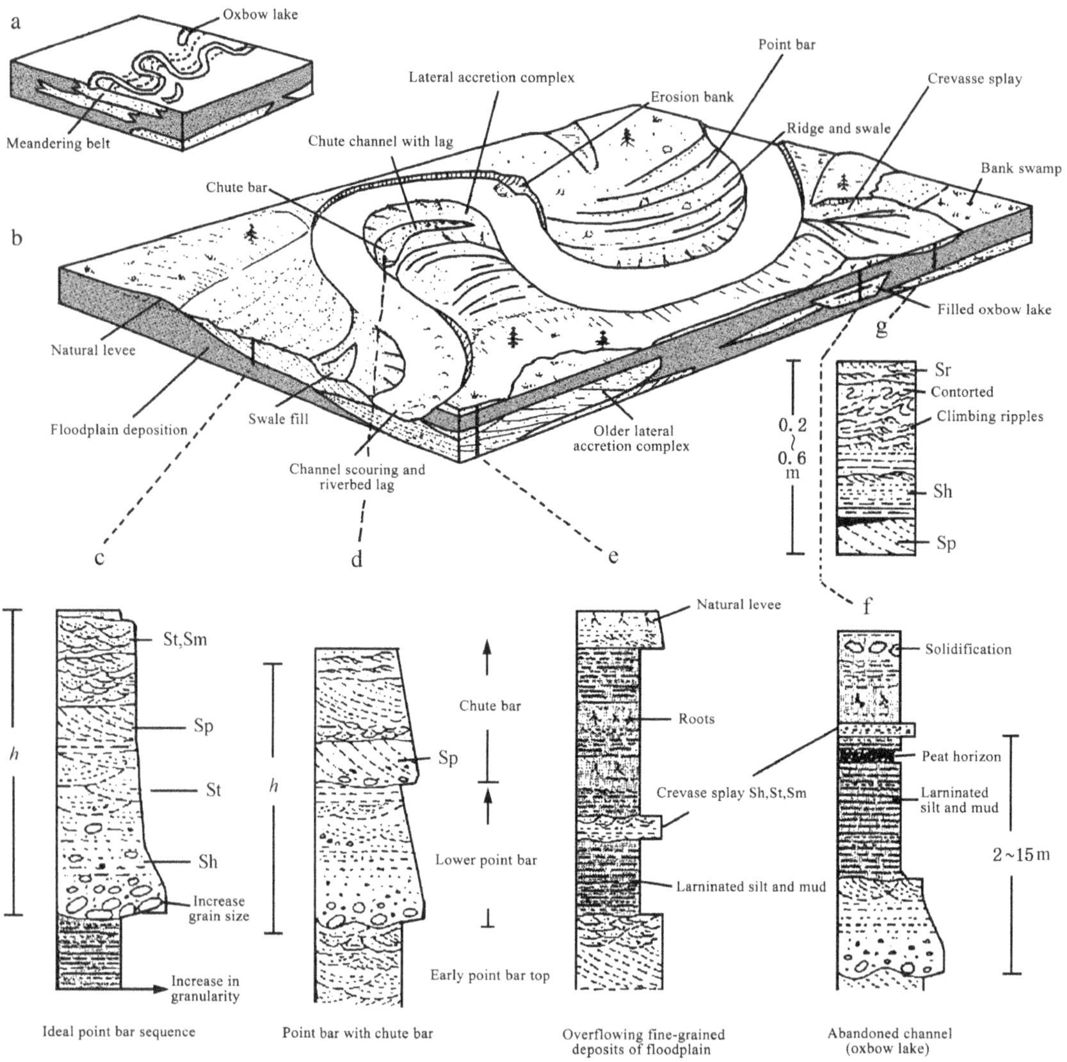

Fig. 9.35 Sedimentary model of the meandering river system and sedimentary sequence of micro geomorphic units (after Cant and Walker 1984; Galloway 1985; Miall 1985a, b)

of braided rivers can be divided roughly into channels and floodplains, and subdivided according to the secondary geomorphic units of the channel and the floodplain.

Channel bars are an iconic geomorphic unit of braided rivers. In a narrow sense, a channel bar is an eyot (beach, island) above water under normal circumstances (poor water season). More broadly, it is generalized as a positive geomorphic unit with a certain scale within the channel of a braided river, including various sandbars in addition to river islands, namely, longitudinal bars, transverse bars, and diagonal bars.

In general, a channel bar is constituted of rough-grained materials. Although, its deposits are fine and can represent the nature of a braided river in the middle and downstream reaches due to the high load content in the river, such as the Brahmaputra River in India and the middle and downstream reaches of the Yellow River in China. At this time, its sedimentary characteristics are not easily distinguishable from those of a

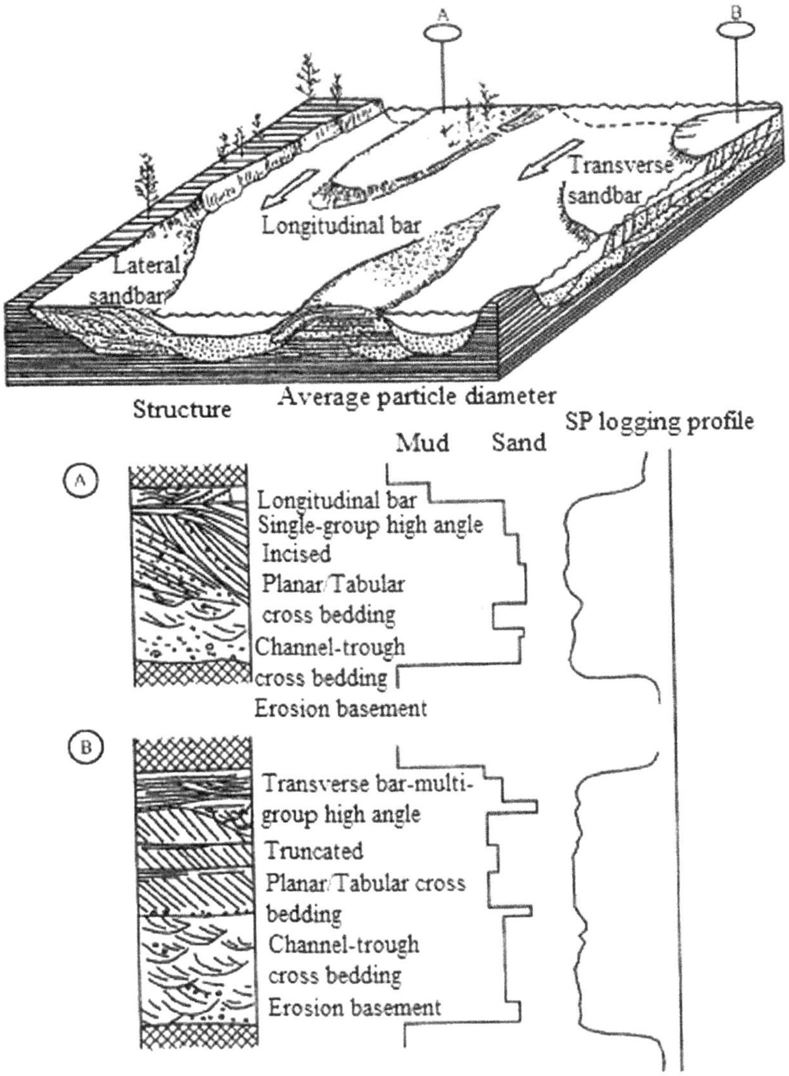

Fig. 9.36 Low sinuosity river with sand enrichment (braided river) sedimentary model and sedimentary sequence characteristics (after Galloway 1983)

meandering bar. Thus, these rivers are defined by some scholars as "mixed" rivers (Walker and Cant, 1984). The South Saskatchewan River, which was formed mainly by the combined effect of vertical accretion and lateral accretion, is a typical example of this.

Embankments and crevasse splays are generally not developed in braided river sediments. It is difficult to maintain embankment deposits and crevasse splay deposits owing to the strong erosion, frequent swinging, or rapid migration of braided rivers. In most cases, the channel of a braided river is diverted directly upon the occurrence of crevasse. Therefore, a crevasse in a braided river means the formation of a new channel and abandonment of an old channel; in such a case, the river does not generally leave crevasse splay deposits. Point bars sometimes are developed in braided rivers, but they are much smaller than those in meandering rivers in terms of scale and degree of development and are often subject to stronger reformation. Herein, we focus especially on the formation and characteristics of all types of sandbars in braided rivers.

There are two types of classification schemes for channel bars: by sedimentation or the

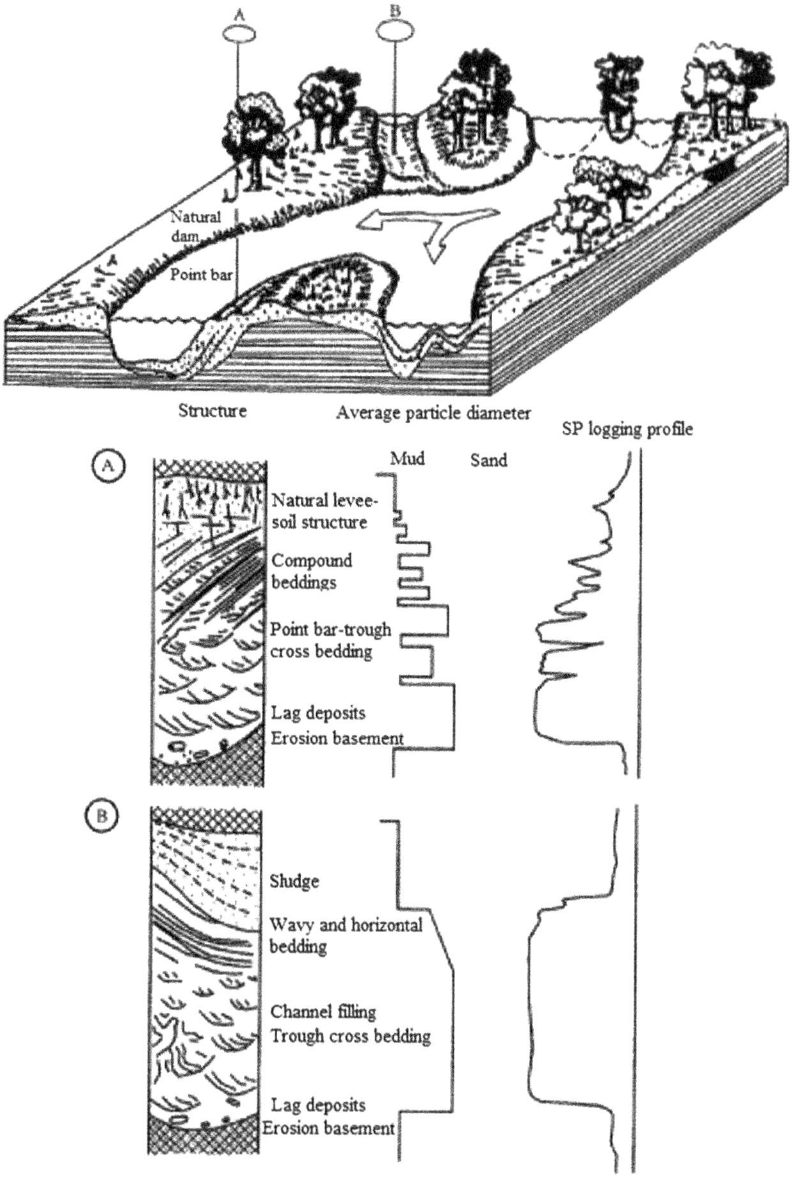

Fig. 9.37 Low sinuosity river with sand enrichment (braided river) sedimentary model and sedimentary sequence characteristics (after (Galloway 1983)

presence or absence of bar migration and based on the relationship between the morphology of the sandbar and river bank.

1. **Active and inactive channel bars**

Channel bars can be formed due to sedimentation, erosion, or the combined action of erosion and sedimentation. Most of them are lens or planar sand bodies (Table 9.9), and they can have a variety of different geomorphic features depending on the changes in current conditions and geometric shapes of channels.

(1) Active channel bars

Sedimentation on an active channel bar is controlled mainly by sandbar formation, migration of river bed form, and accretion effect in flooding and discharge. The upstream reaches

Table 9.9 Sandbar types and sedimentary characteristics of braided rivers

Sandbar type		Sedimentary structure	Morphological characteristics
Active channel bar	Longitudinal bar	Dominated by single or multiple groups of low-high angle incised planar cross beddings with parallel bedding on top	External feature of flat bottom and convex top; long-axis direction parallel to water flow direction, and mostly rhombic or tongue shaped on the plane
	Transverse sandbar	Dominated by single or multiple groups of high angle truncated planar cross bedding, with trough cross bedding in the upper part	External features of flat bottom and convex top; long axis vertical to water flow direction, and wide rhombus appearing on the plane
	Diagonal bar	Megascopic single or multiple groups of low angle planar cross bedding and massive bedding, with trough cross bedding in the upper part	Lens and wedge-shaped sand body with flat top and concave bottom on the section; the long-axis direction skews the water flow direction; long strip shape appearing on the plane as an echelon arrangement
Inactive channel bar	River island	Single or multiple groups of low angle incised planar cross bedding in the lower part; mainly trough cross bedding in the middle-upper part, with locally visible parallel bedding; miniature trough cross bedding and parallel bedding in the top part with phytoturbation and bioturbation structures	External feature of flat bottom and convex top; long axis vertical to water flow direction; generally evolved by early longitudinal bar or diagonal bar

of a sandbar are steeper due to erosion and scouring action, but the downstream reaches are relatively flatter due to sedimentation, which causes the channel bar to migrate gradually toward downstream reaches. The parts close to the upstream reaches are constituted of coarse materials, while those close to the downstream part are constituted of fine-grained materials. This complete process is called vertical accretion in a braided channel.

Lateral migration may occur sometimes in a channel bar with erosion on locally steeper concave banks, but deposition in locally flatter convex banks. Thus, as for the channel bar or batture, there exists lateral and vertical erosion, and the accretion effect simultaneously, and foreset laminae can be generated in the downstream direction and on the convex banks of the sandbar. Accretion of fine-grained suspended materials can occur in the low water period, flood period, and after channel abandonment, but still they are not much less developed than the meandering river.

(2) Inactive channel bar

The positions of inactive channel bars in the channel are relatively fixed. Most inactive channel bars form inter-channel alluvial islands with vegetation growth. The face of such a sandbar is usually developed with fine-grained deposits from floodplains and is covered by the residues left after sandbar erosion, which lead to an increase in accretion, higher terrain, and the growth and development of plants.

2. **Morphological classification of channel bars**

Four types of main sandbars were proposed by Smith (1974) according to their geomorphic features, scales, and relationships with current direction and river banks, namely, longitudinal bars, transverse bars, diagonal lateral bars, and meandering bars (Fig. 9.38), among which the meandering bar is rarely found in braided channels.

Fig. 9.38 The types of gravelly sand bars in Kicking Horse River, British Columbia (after Walker and Cant 1979)

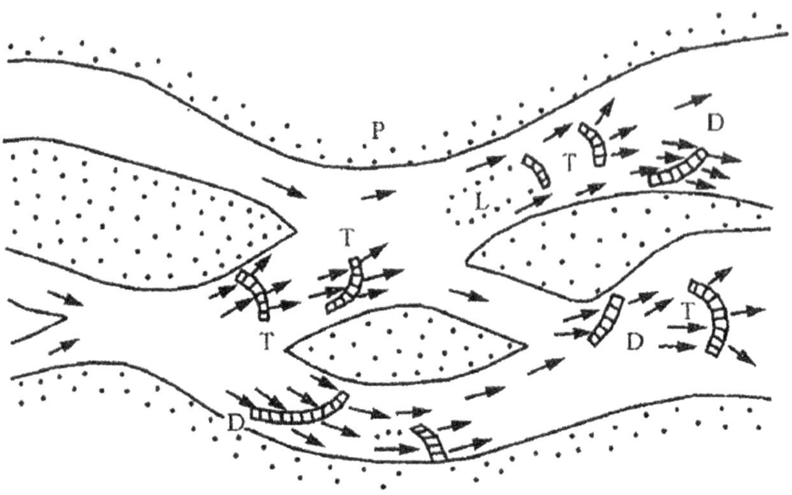

P-meandering bar, D-longitudinal bar, L-diagonal bar and T-transverse bar.

(1) Longitudinal bars

A longitudinal bar is a long sand body parallel to the water flow direction. With a distribution direction almost the same as the channel extension direction, and it is formed by one-way water flow parallel to the sandbar in shallow water areas and is seen commonly in gravelly braided rivers. Its deposits are generally constituted of coarse-grained gravel materials (Fig. 9.39). The upper end of a sandbar is eroded, but the lower end receives deposits. A longitudinal bar is shaped like a diamond or rhombus, and is a thin diamond-shaped gravel sheet body formed over multiple iterations. Interior sandbars are composed mainly of gravel and sandy planar cross bedding. If their sediments are mainly gravel, inconspicuous horizontal bedding and high angle incised planar cross bedding are formed. This results in a mostly massive sandbar with a fining-upward sequence, but the rhythm is not too obvious. As for sedimentation, A. D. Miall regards longitudinal bars as the main products of downstream accretion.

(2) Transverse bars

The long axis of a transverse bar is vertical to the channel axis, and its front edge has several shapes such as tongue-shaped, straight shape, and curved shape.

It is generally believed that the extension direction of its front edge is almost vertical to the direction of water flow, and transverse bars often come into being in streamline divergent regions formed owing to an increase in channel width and a sudden increase in depth. Most of them are formed in an isolated form, sometimes, in an echelon distribution. During the formation of a transverse bar, firstly, it is aggraded to the equilibrium state by sand and gravel sediments. Then, it grows through the downstream extension of the sliding face and generates multi-group high angle truncated planar cross bedding (Fig. 9.29), which is the typical product of vertical accretion.

(3) Diagonal bars

Diagonal bars, whose long axis extension direction skews with that of the main water flow direction, are produced mostly as a result of main river channel sinuosity and asymmetric water flow. The cross section of this bar is roughly triangular and it has a downstream sedimentary edge composed of sliding faces or shallow shoal. The bars formed in different periods frequently have thin sandy seal bed deposits (Fig. 9.40). Diagonal bars are similar to longitudinal bars.

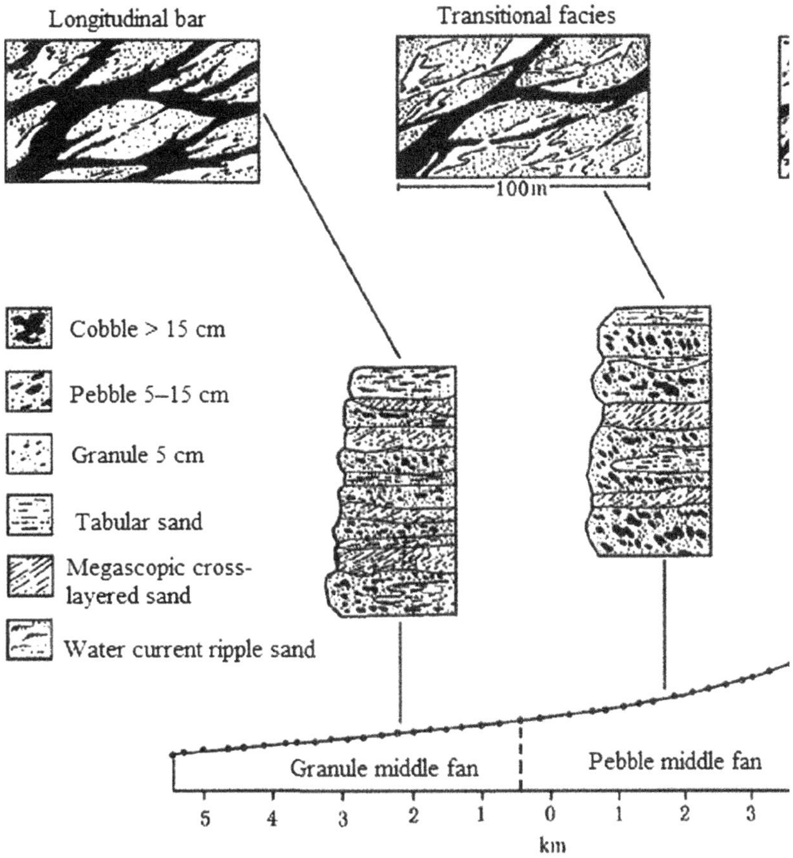

Fig. 9.39 Slope variations, types of gravel dams and sedimentary sequences of braided river in alluvial fan, Alaska (after Boothroyd 1972)

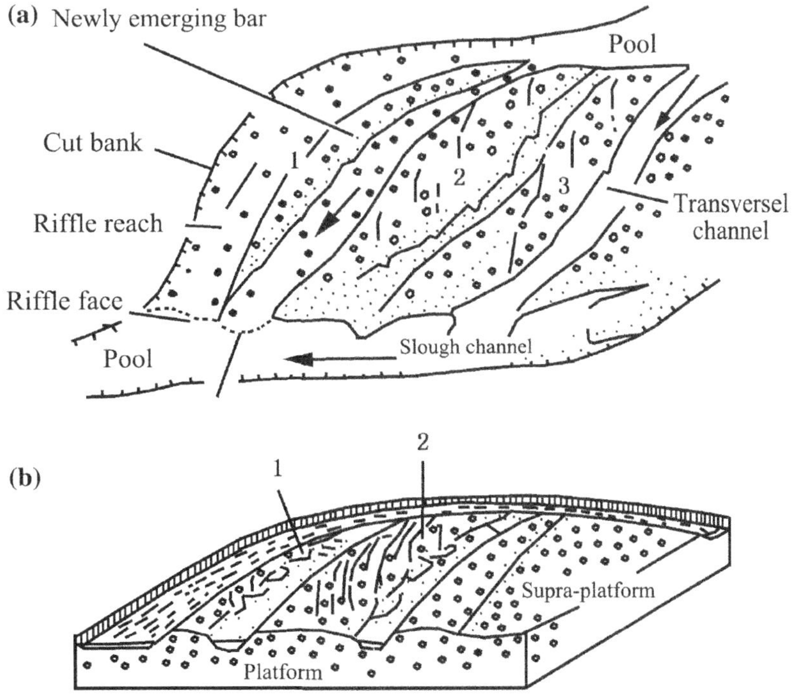

Fig. 9.40 Microtopography and sediment distribution of diagonal gravel dam in the braided river (after Bluck 1974). The numbers 1, 2, 3 in the figure represent the formation sequence of each single ba

Planar cross bedding can be formed upon the occurrence of an avalanche of sliding faces or migration of shallow shoal. Diagonal bars show an imbricated fabric when the sediments are coarse, which can lead to the development of inconspicuous horizontal bedding or multi-group low angle incised planar cross bedding with the characteristics of lateral accretion.

9.3.3.2 Sedimentary Characteristics of a Braided River

(1) A braided river is dominated by gravel and sandy deposits with silt and clay in local parts, and characterized by the macroscopic sedimentary feature "mud-in-sand" on the vertical profile, which is represented mainly by lens and planar batture bars or channel bars, as well as a clear scour surface.
(2) Most of them are reaches close to the water source; thus, their rock components are complex, mineral maturity is lower, variation range of grain size is wide, and sorting is poor. The typical grain size distribution characteristics of braided rivers comprise three populations on a probability graph, among which traction population and suspension population are developed, but in most cases, jump population is lacking and rare. Therefore, grain size shows a bimodal distribution. The two corresponding cutoff points (coarse cutoff point and fine cutoff point), S and T, are abrupt in nature.
(3) They have various types of bedding. The representative bedding is massive trough cross bedding, the bottom interface of which is an obvious scour surface; it can also be massive wedge cross bedding or massive planar cross bedding, sometimes with anti-dune cross bedding.
(4) The most prominent characteristic of batture deposition, which is different from that of a point bar, is the lack of bank deposits on the upper parts, which is determined by the wandering feature of its riverbed. In addition, the channel is unstable, so bank deposits cannot be saved.

9.3.3.3 Vertical Sedimentary Sequence of Braided River

The sedimentary sequence of a braided river is usually rather complex. A typical example is the Devonian braided river in the Gaspe Peninsula, Quebec, Canada (Fig. 9.41). It has the fining upward structure from bottom to top, which

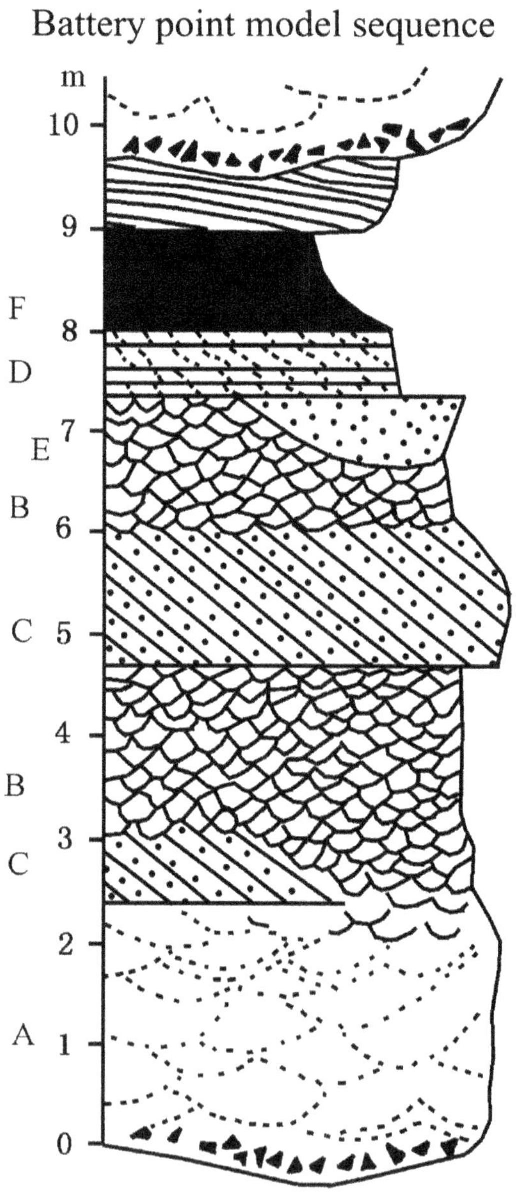

Fig. 9.41 Sedimentary sequence of the devonian braided river in the Gaspe Peninsula, Quebec, Canada (after Cant and Walker 1976)

reflects the gradually weakening deposition of hydrodynamic energy.

① Gravel deposited by channel lag and megascopic or gigascopic trough cross bedding, with cobble–granule conglomerate, conglomeratic—medium-coarse sandstone, and obvious scour surface between the bottom of the sequence and the underlying sediments as lithology. ② Coarse sand in megascopic trough cross bedding or megascopic single-group high angle truncated planar cross bedding. ③ Parallel bedding sand or fine sand in miniature planar and trough cross beddings, as well as climbing ripple lamination. ④ Fine sand and argillaceous sediments of horizontal bedding, along with drought structure in the mud layer.

The sequence characteristics of a braided river can be summarized as follows:

① Thicker grain size and conglomeratic development; ② various bedding types in the lower parts of the sequence formed by the migration surface of channel bar, such as massive or inconspicuous parallel bedding and megascopic high angle (truncated type) planar cross bedding; ③ trough cross bedding development of large scale; and ④ thinner or undeveloped fine-grained sediments in the floodplain.

9.3.3.4 Depositional Model of Braided Rivers

The sedimentary sequence of the Devonian braided river in the Gaspe Peninsula, Quebec, Canada, which was published by D. J. Cant and R. G. Walker (1976), has been widely viewed as the standard vertical sedimentary sequence of all braided rivers since its disclosure. Later, they drew further conclusions based on the vertical sedimentary sequence of sandy braided rivers and proposed three basic sedimentary sequence models (Fig. 9.42), which are dominated by beach bars, the mixed effect, and channels. However, when compared with meandering rivers, the changes in a braided rivers are especially complex. According to Miall (1985a, b), "each braided river is different". It is impossible to summarize the sedimentary characteristics of various braided rivers with only one model, thus he proposed six types of depositional models by using the concepts of lithofacies, lithofacies association, and architectural unit, as well as by considering the geological geography background and climate characteristics of braided rivers (Table 9.10; Figs. 9.43 and 9.44). The sedimentary sequence and model established by Walker and Cant constitute a subset of Miall's braided river models.

1. **Trollheim**

This sediment is dominated by debris flow under the normal conditions of abrupt slope, large sediment supply, and occasional flash floods. These conditions are typical of alluvial fans in arid and semi-arid areas, and the Trollheim fan can be considered a typical example (Fig. 9.43). The features of Trollheim facies association are as follows (Hooke 1967; Rust 1978; Miall et al. 1978):

(1) It comprises mainly superposed debris flow deposit (lithofacies Gms) in the vertical direction, and a single debris flow layer can be 3 m in thickness.
(2) The debris flow layer is usually flat (non-channel), generally as a steep undersurface with a lobate geometric shape.
(3) The interbedded unit of Gm lithofacies is usually fine-grained rock stratum, which can fill the obvious washed-out pit. This reflects the fact that the slope required by river runoff is lower than the debris flow, and sheet flood sediments of river runoff may exist in the primitive fining-upward cycle.
(4) The St, Sp, Fl, and Fm lithofacies may be present as secondary units under the Scott mode. On the adjacent alluvial fan, size variation of debris flow deposit can reflect the difference in sediment supply.

2. **Scott**

This is developed on an alluvial fan lacking debris flow and in other proximal gravel rivers. This sediment mainly comprises Gm lithofacies forming superposed longitudinal bar deposition (Fig. 9.43), and it represents the flooding

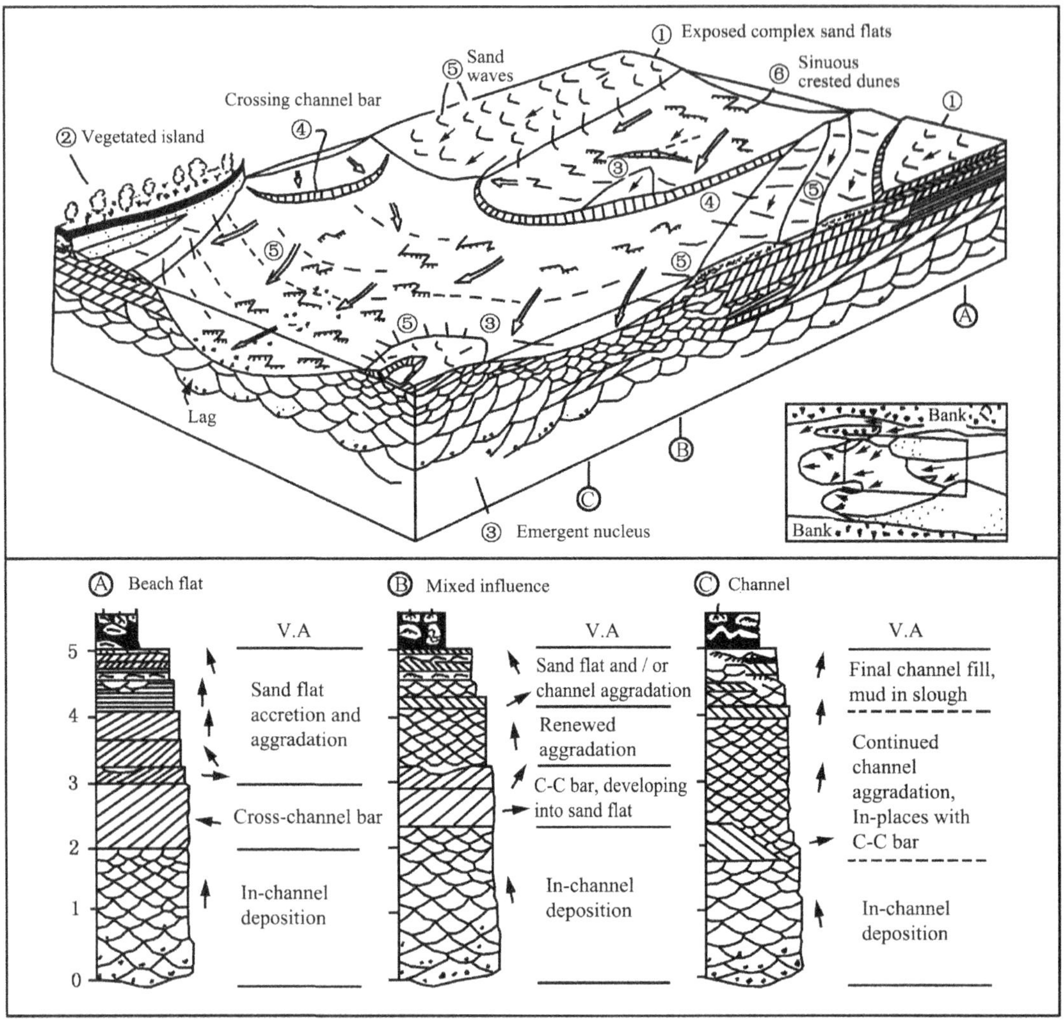

Fig. 9.42 Sketch map of geomorphic units, bed sand bodies and vertical sedimentary sequence in the sandy braided river (after Walker 1984a, b, c). The single-headed arrow indicates the flow direction; and the double-headed arrow represents the bed migration direction

Table 9.10 Six sedimentary lithofaces association models of braided rivers dominated by conglomeratic (1978)

Description	Sedimentary Environment Position	Major lithofacies	Secondary lithofacies
Trollheim	Proximal river with debris flow (mainly developed on alluvial fan)	Gms and Gm	St, Sp, Fl, and Fm
Scott	Proximal river with channel runoff (including alluvial fan)	Gm	Gp, Gt, Sp, St, Sr, Fl, and Fm
Donjek	Distal gravel river (developed with cycle deposit)	Gm, Gt, and Gs	Gp, Sh, Sr, Sp, Fl, and Fm
S. Saskatchewan	Sandy braided river	St	Sp, Se, Sr, Sh, Ss, Sh, Gm, Fl, and FM
Platte	Sandy braided river	St and Sp	Sh, Sr, Ss, Gm, Fl, and Fm
Billot stream	Seasonal stream withstanding violent flood	Sh and Sl	Sp and Sr

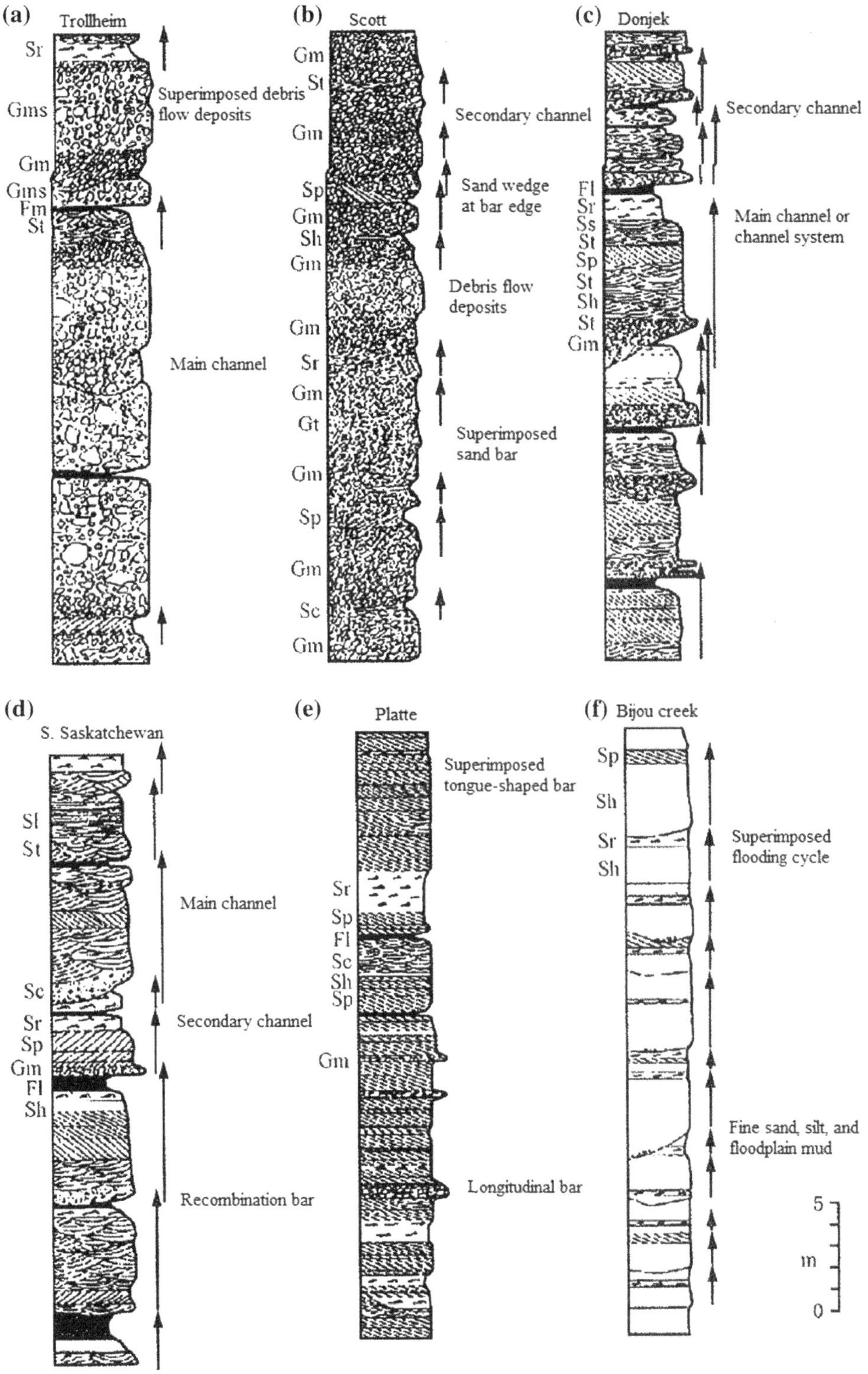

Fig. 9.43 Six vertical sedimentary sequence models in braided river (after Miall et al. 1978)

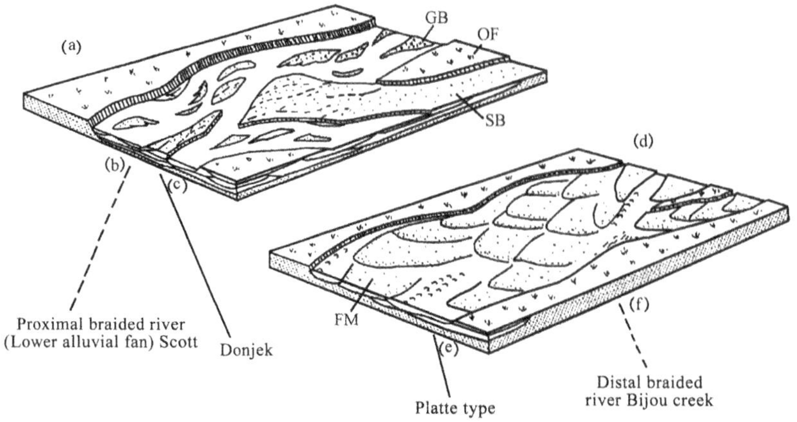

Fig. 9.44 Models and different sedimentary sequence characteristics in the braided river system (after Miall 1985a, b). **a–c** Indicate gravelly proximal-Mesogene reaches; **d** is a sandy braided river; and **e**, **f** indicate distal sandy braided river

process. As a deposit in the retreating flood period (or interflood period), the secondary lithofacies include Gp, Gt, Sp, St, and Sr. They can form miniature-cycle fining upward with thickness of about 1 m. The sand layer is disposed in the abandoned channel or it progrades outward from the gravel bar to form a micro triangle wedge set when the water level drops and the gravel bar is exposed. This lithofacies association is named after the Scott glacier-fed stream in Alaska (Boothroyd and Ashley 1975).

3. **Donjek**

This kind of lithofacies association is developed mainly under the following conditions and background: ① There is an active channel with a clear boundary and parts of the river reaches lack water due to orographic uplift. The terrain elevation difference on both sides of the river system is 3–7 m; and ② the bed load is composed of large amounts of sand and gravel.

This type of river, with the fining-upward cycle, was clearly recorded first by Williams and Rust (1969) in the Donjek River in Canada. Hence, this lithofacies association is named after the river. The coarsest sediment is generated in relatively deep channels, including probable lithofacies such as Gm, Gp, and Gt. The section without running water above the deep channel receives gravel deposits, and the highest part of the river system may be covered by dense vegetation, which can block some superfine sediments from infiltrating floods.

The lithofacies associations of Scott, Donjek, and S. Saskatchewan may exist in the same alluvial system according to the relationship of gradual change from proximal to distal. A few researchers have proposed to distinguish the three types based on gravel content (percentage of gross thickness in the total thickness of section): Scott is more than 90%, Donjek 10–90%, and S. Saskatchewan is less than 10%.

4. **S. Saskatchewan**

This is a typical sandy braided river and the river first proposed to be represented by "mixed effects." In the sandy braided river, owing to the joint development of the channel and sand bars, the fining-upward cycle may be developed (Cant and Walker 1976, 1978), and this lithofacies association is named after the river in the south of Saskatchewan in Canada.

A stranded conglomerate (lithofacies Gm) is developed at the bottom of the channel. Above this layer, sand is transported as a coarse-grained bed load. The bed form in the deep channel (3 m or deeper) is usually a sand dune with a bent ridge to form the lithofacies St. Sand waves (lithofacies Sp) are very common in shallow river reaches, and various types of downstream-accreting sand bars migrate downstream. In the South Saskatchewan River (but not necessarily in all sandy braided rivers), there are many small sand bars embracing aggregation to develop a big sandy flat. Owing to the wide angle of the sand bar with a straight edge (transverse bar) to the channel direction, these big sandy flats will aggrade upstream and may also grow downstream. In the flood period, lithofacies Sr and Sh may be formed at the top of the sand bar, and fine-grained floodplain deposits may exist in the area without water running (Fig. 9.44).

This combination is similar to a few meander bar sequences. Through careful research on paleocurrents, this can be determined from the vertical section because it is markedly different from the meander bar of the river.

5. **Platte**

It is a variant of S. Saskatchewan. In this type of river, a very wide and shallow channel may be present, and there is no obvious difference between sections with and without running water. The sediments are composed mainly of tongue-shaped sand bars and transverse progradation bars or sand waves. Therefore, a superimposed set of the lithofacies Sp is formed. If the lithofacies is rarely seen, there will be few deep channels. Similar to S. Saskatchewan, the bar crest is composed of sand and silt (lithofacies Sr, Sh, Fl and Fm).

6. **Bijou Creek**

Under the upper flow regime, a flat riverbed is formed, and current lineation may develop on the flat bedding surface (lithofacies Sh). This type of extra upper flow regime is seen rarely in a yearlong stream, but it is common in seasonal streams, which are characterized by occasional flash floods. The current can form a channel or a sheetflow deposit. Examples include the Bijou Creek in Colorado, America (Mckee et al. 1967), after which this type is named.

Facies models of seasonal streams have been discussed by Miall (1977, 1979), Rust (1978), and Tunbridge (1981). All of them emphasize the importance of the Sh lithofacies in sedimentary association. The Sp, Sr, Fl, and Fm lithofacies can be formed in the flood falling period, generating a thin fining-upward cycle. Sediments with thickness greater than 1.5 m can be formed in one flood. In addition, an erosion surface (lithofacies Se) and the filling washed-out pit can be seen.

9.3.4 Environment Type and Model of Anastomosing Rivers

9.3.4.1 Environment Type of an Anastomosing River

Modern rivers are very rarely of the typical anastomosing type, and it is very difficult to correctly identify anastomosing river deposition in ancient sediments because it can only be achieved with very dense well pattern control. Therefore, anastomosing rivers and their sediments have not been researched extensively.

An anastomosing river is a low-energy complex composed of several mutually connected channels of varying sinuosity levels. Its sediments are carried mainly as suspended loads. In the anastomosing river system, alluvial islands (river islands) and floodplains or wet lands account for most of its area (60–90%). Therefore, thick silt and clay with substantial amounts of mud are the dominant sediments of anastomosing rivers (Fig. 9.45).

The subenvironment types (or microfacies) of anastomosing rivers can be divided into channels, natural levees, and floodplains.

1. **Channels**

Unlike meandering rivers, the channel of an anastomosing river is shown as an obvious lateral migration and increase in point bar, which is characterized by simple vertical accretion or aggradation. The erosion competency of the channel toward both banks is very weak, thus the

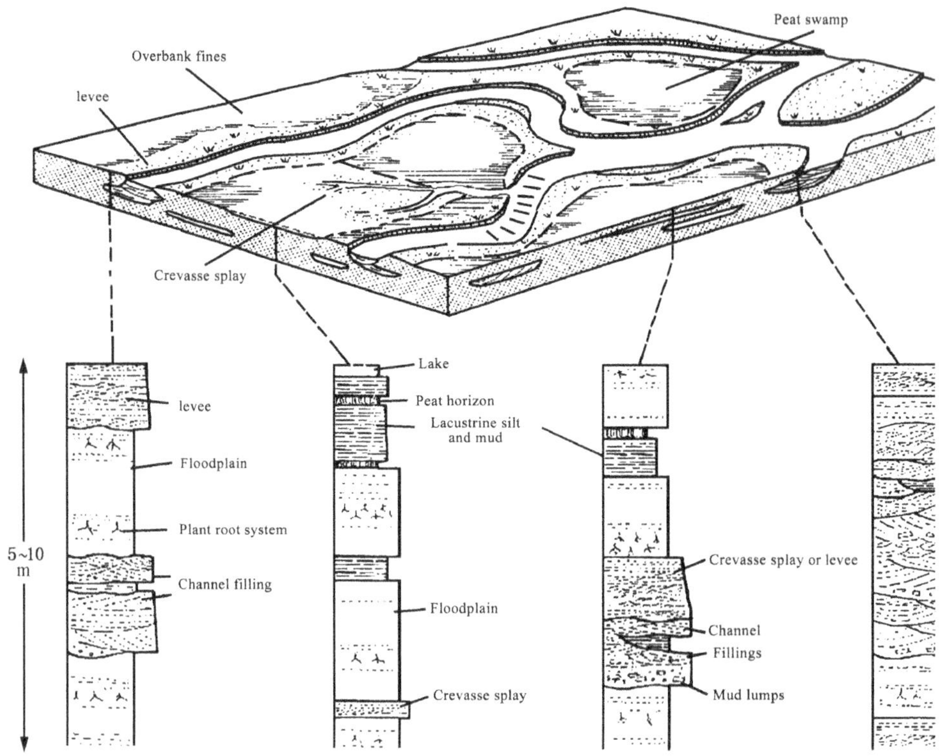

Fig. 9.45 Sedimentary sequence and model of net-like river system with low-high curvature (after Smith 1983; Miall 1985a, b)

channel has very high lateral stability and the channel sand shape of an anastomosing river is different from those of the former two kinds on both the plane and in the section (Table 9.11). On the vertical sequence, it is shown as a fining-upwards sedimentation fining upward, but without obvious zonation.

2. **Natural levees**

Natural levees are usually developed on both sides of an anastomosing river channel. Its sediments mainly include siltstone, and they are in transitional contact with floodplain sediments.

3. **Floodplains**

Floodplain deposit is an important part in the anastomosing river depositional system, and a wide range of swamps, wet lands, floodplains, ponds, and peat swamps are distributed between narrow anastomosing river channels. The horizontal bedding is developed widely with floodplain sediments. Near the channel, the grain size is slightly coarser, the amount of siltstone increases, that of mudstone decreases, and thin-layered fine sandstone is mingled.

9.3.5 Sedimentary Sequence and Model of Anastomosing Rivers

Owing to the stability of anastomosing rivers, both narrow strip-shaped channel sand bodies and widely distributed fine-grained floodplain deposits emerge in thick-bedded form. Hence, it is very difficult to integrate the two into one

Table 9.11 Sedimentary characteristics of anastomosing rivers under various sub-environments

Environment type	Main lithology and grain size characteristics	Sedimentary structure and feature	Morphological characteristics
Channel	Mainly sandstone of various grain sizes; granule and thin-layered conglomerate usually at the bottom	Massive structure in lower part; developed with megascopic trough cross bedding; mainly miniature trough cross beddings in the middle and upper parts; and horizontal bedding at the top	Bent shoelace-shaped on the plane; narrow and thick wall-like sand body on the section; nearly vertical contact surface between the two sides and fine-grained sediment in floodplain
Natural levee	Dominated by siltstone and mudstone; mingled with thin-layered fine sandstone	Developed with horizontal bedding and ripples bedding	Wedge-shaped or triangular on the section; strip shaped on the plane
Floodplain	Dominated by mudstone, argillaceous rock, sandy mudstone, and siltstone; mingled with peat horizon	Developed with horizontal bedding	Bulk surrounded by strip-shaped channel sand body

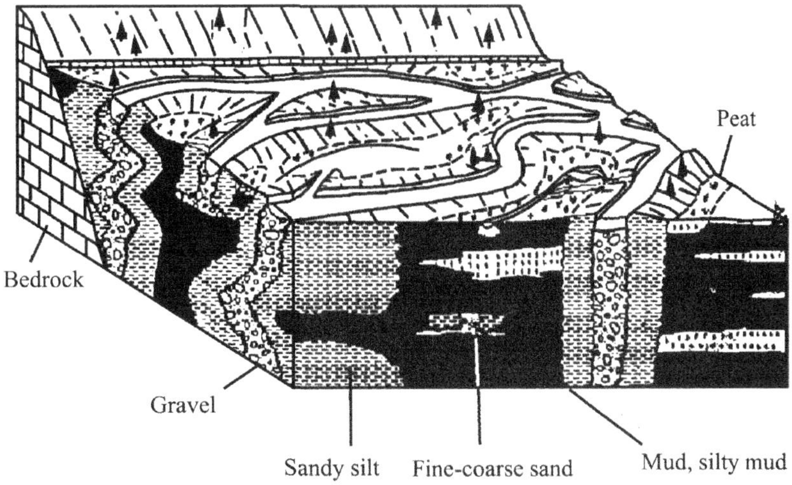

Fig. 9.46 Lithologic distribution and sedimentary model of net-like river (after Smith et al. 1980)

sedimentary sequence. Fining-upwards sedimentation featuring fining upward but without obvious zonation is presented on the vertical sedimentary sequence of an anastomosing river channel. By comparison, the floodplain sedimentary sequence comprises simple thick-bedded fine-grained sediments mixed mostly with a peat horizon (Figs. 9.45 and 9.46).

9.4 Identification of Fluvial Deposit

9.4.1 Based on Geological Features

Based on the above analysis and discussion, the general features of fluvial deposits can be summarized as follows:

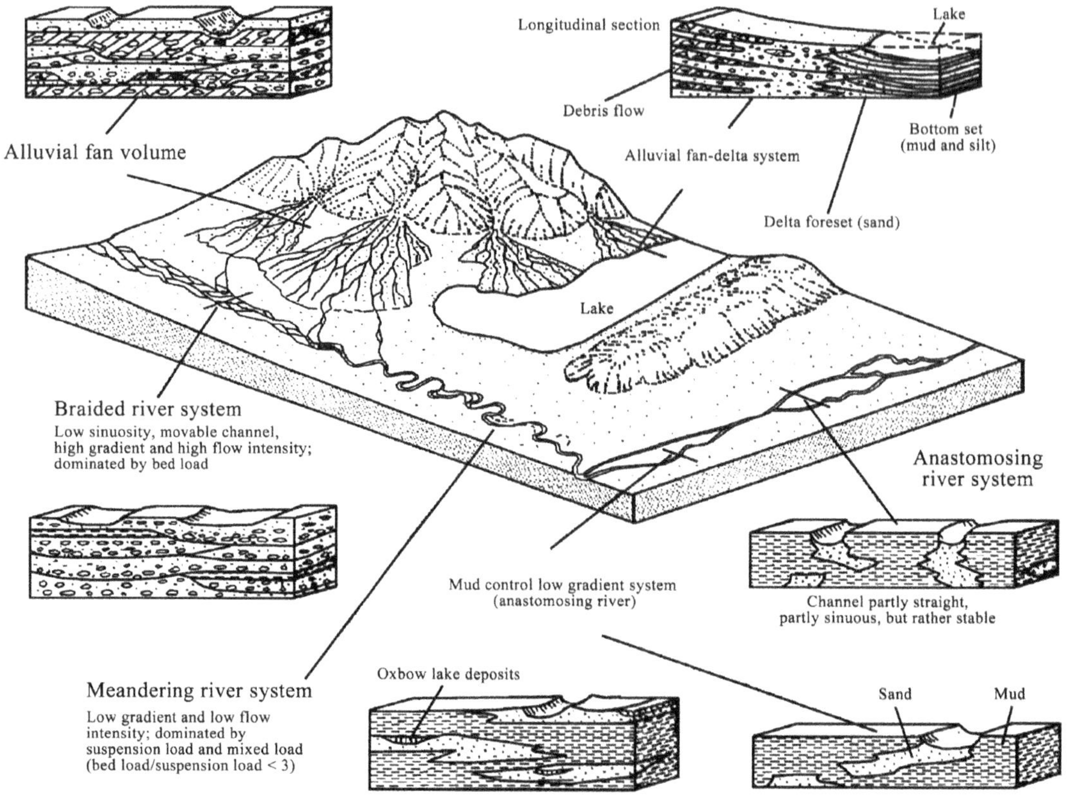

Fig. 9.47 Main types, distribution, and cross sectional features of the river system (after A. D. Miall 1999)

(1) On the vertical sedimentary sequence, it is shown mainly as fining-upward deposition, and a scour surface is commonly seen at the bottom. In terms of the depositional environment, from bottom up, the complete fluvial deposition sequence shows lag deposits at the river bottom and channel deposit and floodplain or overbank deposits. However, in special cases, the coarsening-upward sequence can be seen.

(2) In the sedimentary structure direction, the main sedimentary type is cross bedding caused by water ripples, which reflects the features of unidirectional potamic transport, and the corresponding changes in combination features with changes in grain size. The basic order from bottom to top is as follows: scour surface → megascopic trough cross bedding → megascopic planar cross bedding → parallel bedding → anti-dune cross bedding → climbing ripple lamination → discontinuous wave (ripples) cross bedding and horizontal bedding.

(3) As for size distribution, the C-M figure is of the typical tractional current type, and the probability graph is composed mainly of saltation and suspension populations.

(4) Regarding biological characteristics, plant roots and clastic plant fossils are mainly shown, especially the emergence of plant roots and carbonaceous mudstones in overbank deposits. A few limnobios fossils can be seen, and there are no marine fossils.

(5) The petrologic characteristics are manifested as medium-poor maturity, mainly with kaolinite as the clay mineral, which reflects an acidic environment.
(6) The sand body geometry constitutes a bent strip with a vertical depositional strike, which is shown often as lens and clintheriform-lens on the section. There is a dual structure (especially meandering river) in the vertical direction, with gravelly and sandy sediments at the bottom, and silt and argillaceous sediments at the top.

Different channel patterns have different hydrodynamic conditions, and migration and evolution laws. Hence, not only are the geomorphic features different but obvious differences also exist from many perspectives such as lithology, granularity, sedimentary structure and association, vertical sequence, spatial configuration and distribution of various sediments (Figs. 9.47 and 9.48; Table 9.12).

9.4.2 Based on Logging Features

As mentioned in Sect. 9.4, with respect to electrofacies, Sp, Gr, and Rt, especially Gr and Sp, are generally used to analyze thc vertical prosodic features and sedimentary sequence.

9.4.2.1 Difference Between Different Rivers

① Braided rivers are characterized by higher cylindricity; ② Meandering rivers are characterized by a serrated bell shape with gradual changes; and ③ Anastomosing rivers mostly feature a low-amplitude serrated miniature bell shape. The silt content increases from ① to ③, with the granularity fining and the serration number increases, mainly depending on river sinuosity. In the vertical combination, braided rivers feature "mud-in-sand," meandering rivers are characterized by the "sand-mud interbed," and anastomosing rivers are mostly of the "sand-in-mud" type. In terms of the main

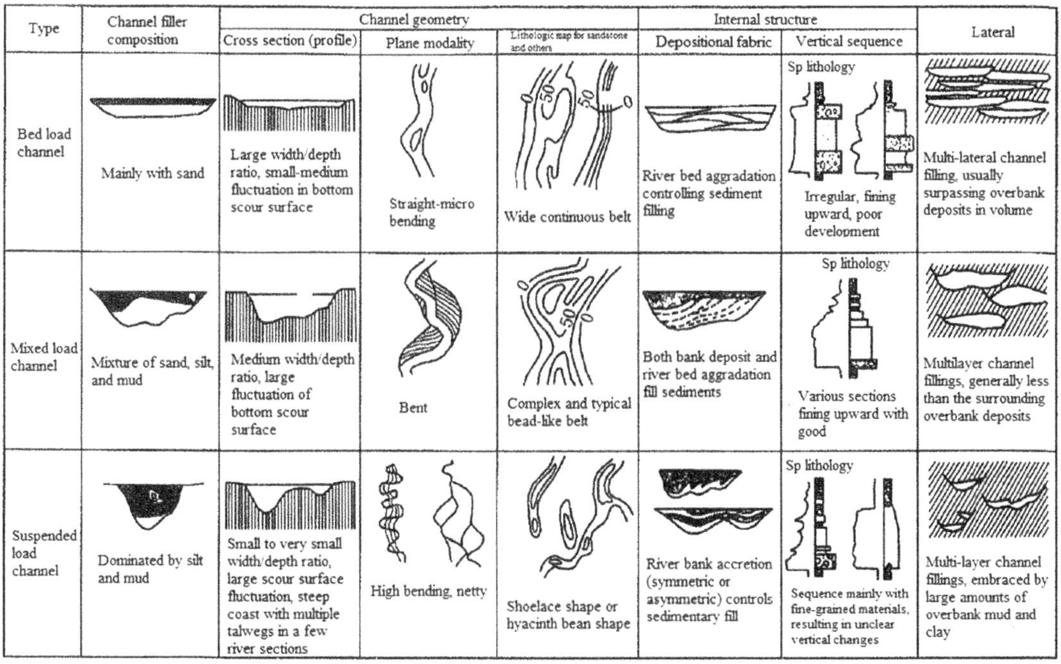

Fig. 9.48 Contrast map of landform and sedimentary characteristics of different river patterns

Table 9.12 Comparison of identification marks of different types of fluvial deposits

Characteristics	Braided river	Meandering river	Anastomosing river
Subfacies	Batture (bar)	Point bar	Anastomosing river
Petrophysics	Dominated by sand and conglomerates, often developed with thick-bedded conglomerate and gravel-containing coarse sandstone	Dominated by sand and mudstone; generally thin conglomerate layer	Dominated by siltstone and mudstone, followed by sand and conglomerate
Profile combination	Mud-in-sand	Sand-mud interbed	Sand-in-mud
Vertical Sequence	Fining-upward structure, thin fine-grained deposit or without deposit	Typical fining-upward structure	No obvious fining-upward structure
Sedimentary structure	Developed with various megascopic trough and planar cross beddings; common massive bedding; general lack of ripple bedding	Great diversity; typically marked by incised planar cross bedding	Dominated by trough horizontal bedding and horizontal bedding
Size distribution (probability graph)	Dominated by three-phase	Dominated by two-phase	Dominated by two-phase
Sand body morphology	On the plane: single sand body is strip shaped with low sinuosity; channel belt sand body is on the tabular or wide strip section, and single sand body and channel belt sand body are of lens form	Single sand body is presented as bent strip; meander belt composite sand body is tubular	On the plane: narrow strip shape; interweaved and twisted together into nets. On the section: upright or inclined narrow and thick wall shape, separated and far from each other
Thickness scale	Medium thick-bedded to thick-bedded Scope: several meters to tens of meters	Medium thick-bedded; Scope: several meters to more than ten meters	Medium-bedded Scope: several meters to more than ten meters
Superimposition of sand body	Multi-layer vertical superimposition	Unilateral or multilateral lateral superimposition	Isolated type

transport modes, braided rivers mainly carry bed loads, meandering rivers mainly mixed loads, and anastomosing rivers mainly suspended loads.

9.4.2.2 Common Characters of Different Rivers

There is a trend with upward amplitude decreasing for the electric well-logging curve of all rivers, which means there are obvious-to-unobvious fining upward structures, and abrupt contact exists between the bottom and the underlying bed. All muddy intercalation takes place in the upper part of the sequence for key development.

9.5 Reservoir Characteristics and Prediction of Fluvial System

9.5.1 General Characteristics

Owing to the complex types of fluvial systems, a great diversity of reservoir systems can be formed, and bed load rivers are rich in reservoirs, but lack good intercalation. On the contrary, suspended load rivers may only contain small reservoirs surrounded by large quantities of mudstone. All fluvial systems have the following features: ① The main reservoirs are channel sand and various sand bars, and crevasse splay is the secondary reservoir body; ② The overall sand body strike is parallel to the channel strike, but there are obvious local changes; ③ Reservoirs parallel to the channel direction have good continuity, while those vertical to the channel direction vary substantially; ④ Reservoir heterogeneity is usually serious; and ⑤ Mudstone facies of the fluvial system may contain a large number of plants or humic-type organic matter.

9.5.1.1 Bed Load (Braided) Rivers

Sand bodies with permeable frameworks form rich reservoirs with good interconnections (Fig. 9.49a). Bed load channel-filling usually forms reservoirs having good physical properties. Such reservoirs can produce uniform fluid flow. The physical properties at the channel edge are usually poor, and floodplain mud is generally discontinuous silt or sand. Therefore, the possibility of the fluvial facies becoming a stratigraphic trap is limited, and hydrocarbon accumulation in traps with unconformable structures or regions is very common.

9.5.1.2 Mixed Load (Meandering) Rivers

Point bars are the main reservoir framework sand body in this case. Crevasse splay sediments form isolated wedges or lobes in the floodplain. In a big fluvial system, their volume is large enough to form a small reservoir of economic value. The sand body of a meandering river has very complex and serious heterogeneity. The decrease in permeability reflects the structural trend of fining-upward granularity. On the plane, the permeability grid element is mostly arc-shaped, and such reservoirs can be separated by the mudstone part (Fig. 9.49b). Therefore, this river sand body can form a lens lithologic trap.

9.5.1.3 Suspended Load (Anastomosing) Rivers

The structural features of "sand-in-mud" lead to the formation of permeability channels of isolated shapes. Thus, suspended load rivers are ideal from the viewpoint of lithologic trap exploration in fluvial systems. An isolated meander belt or interlaced "shoelace-shaped" sand body often shows a large change in the direction relevant to the paleoslope and structural tendency (Fig. 9.49c). In large big oilfields with limited structure, the development of channel-filling facies reservoirs and the associated crevasse splay sand bodies is very difficult because their size is limited and direction is changeable, and there is no law of intercalation. The fining-upward structural trend reflects that the reservoir properties worsen moving upward. Given that the suspended load fluvial system generally appears at the lowest point in a sedimentary basin, stratified alluvial plain swamp coal or lignite may emerge in the surrounding flood basin sediments. Therefore, there is great potential for the formation of natural gas of biogenetic and pyrolytic genesis inside the suspended load fluvial system.

9.5.2 Heterogeneity Features

9.5.2.1 Braided River Reservoir

1. **Geometry and continuity of sand body**

Braided rivers are characterized by wide and shallow channels. Multiple channel bars can emerge in one channel cross-section. The channel waste filling is composed mainly of sand, thus the geometric shape of one time unit sand body reflects the scale of the ancient river.

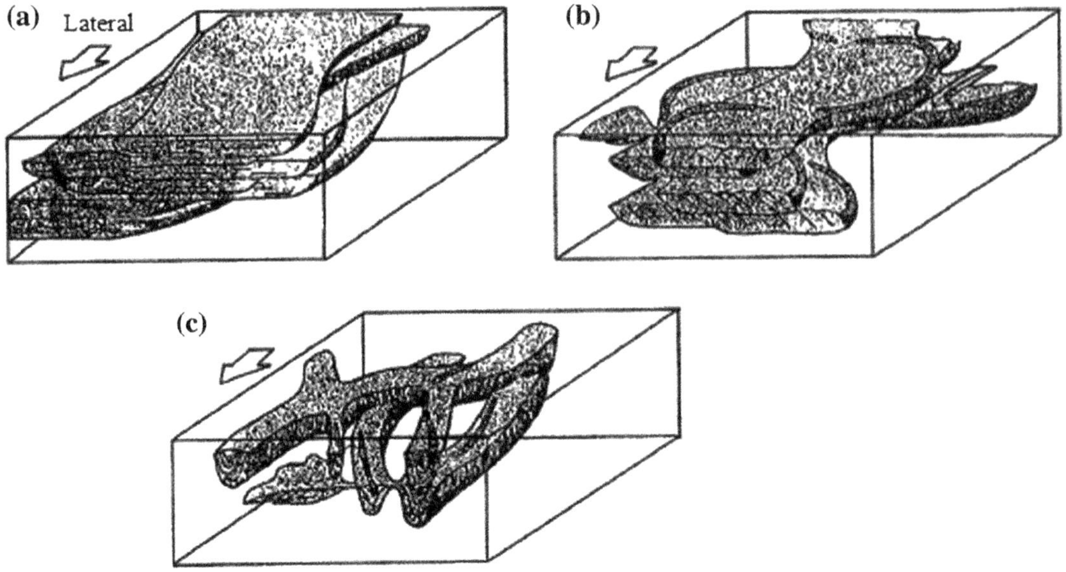

Fig. 9.49 Sketch of geometric shape, lateral relation and internal configuration of different types of rivers reservoirs (according to Galloway et al. 1983)

Table 9.13 Comparison of the physical characteristics of sedimentary reservoirs for different river types

Type	Braided river reservoir	Meandering river reservoir	Anastomosing river reservoir
Reservoir thickness	Medium thick-bedded to thick-bedded Scope: several meters to tens of meters	Medium thick-bedded Scope: several meters to more than ten meters	Thick-bedded Scope: more than ten meters to tens of meters
Sand body superimposition	Multi-layer vertical superimposition	Unilateral or multilateral lateral superimposition	Isolated type
Transverse continuity	Wide, good continuity	Relatively wide, good continuity	Narrow, poor continuity
Connectedness	Good sand body connectedness	Relatively good sand body connectedness	Relatively poor sand body connectedness
Intercalation	Rare and uncontinuous	Common and continuous	Many and common
Vertical physical property	No obvious law, shown mainly as high and low alternating segment characteristics	From bottom to top: low porosity and poor permeability	Change features similar to those of meandering river reservoir
Planar physical property	Zonality and segmentation; good reservoir blocks in belts	Zonation and partitioning; possible to divide into several good reservoir zones	Narrow strip-shaped reservoir interweaving into net

River thickness determines the lateral continuity of the genetic unit sand body. However, braided rivers have loose banks, mainly with sandy deposits, and they generally feature very frequent lateral swings. The braided river sand body is within the scope of certain alluvial plains, and the probability of lateral connection to sand bodies spanning multiple time units is high (Table 9.13).

2. **Features of Microscopic Pore Structure**

The pore structure features of gravelly braided river deposit are similar to those of alluvial fan conglomeratic, and they can also have a bimodal structure. A sandstone reservoir deposited by a sandy braided river has the general sand pore structure; its difference from other river sand bodies is that the shale content in a few braided river sand bodies is low, resulting in the sandstone having a vertical permeability value that is very close to the horizontal permeability value, which enhances the full play of gravity caused by fluid density differences during mining.

3. **Intrastratal heterogeneity**

As mentioned above, the basic deposition mode of braided river channel bars is vertical accretion. Intrastratal vertical graded changes can reflect the energy fluctuations of each flooding event and the thickness of the carried debris, thus "irregular" grading is usually shown. Although there are a few small rhythms, the whole sand layer still has a coarse-fine interbed. This lays the foundation for irregular vertical changes in permeability within the channel bar sand layer of a braided river. In addition, owing to the lack of fine-grained sediments, the permeability is heterogeneous and lower than that of the meandering river sand body, especially in terms of the ratio. Suspended load deposit is rarely seen, resulting in scarcity or even the complete absence of unstable muddy intercalation, which is a main feature of the channel bar's innerlayer heterogeneity, thus leading to high a vertical/horizontal permeability ratio of the entire sand layer.

4. **Plane heterogeneity**

Similar to other channel sand bodies, the braided river sand body shows permeability orientation along the channel body belt. However, a large area of the sand body can be connected, and its planar permeability heterogeneity has no significant influence on the sweep efficiency of the final injection area. The injected water spreads quickly on the plane. In China, sufficient natural water–driven energy often appears in large areas with connected braided river sand body reservoirs.

9.5.3 Meandering River Reservoirs

1. **Geometry and continuity of sand body**

As mentioned before, meandering river deposit is mainly found in point bar sand bodies, which are distributed on the convex bank side of each meander section. The point bar sand in a meander belt can be connected through other associated sand bodies, especially abandoned channel sand bodies, which can be deemed connected components. As a fluvial sand body, it is always distributed with the strip-shaped geometry. Its continuity, mainly lateral continuity, can be studied from the perspective of development. The lateral continuity of a meander belt sand body depends on the river scale (mainly river width) and sinuosity. The bigger the river width and sinuosity, the better is the lateral continuity. However, there is a limit on the width of the meander belt, and it is rarely more than 15–20× the river width. Therefore, the width-to-thickness ratio of a sand body in a meander belt is rarely more than 200. The aeration zone of the Permian Shihezi formation in Da Niu Di, Erdos, China, has this feature, and the Da Niu Di gas field He 2–1 is the transitional product from a braided to a meandering river (Figs. 9.50 and 9.51). The part with the thickest sand is not the center of the channel. As for both braided and meandering rivers, the part with thickest sand tends to be the sand bar, that is, battures and point bars. The best way to predict reservoir sandstone is to find sand according to facies, as opposed to drawing facies according to the sand.

2. **Intrastratal heterogeneity**

Taking a point bar as an example, its physical property typically deteriorates moving vertically upward. The section with the highest permeability is always located in the retention region at

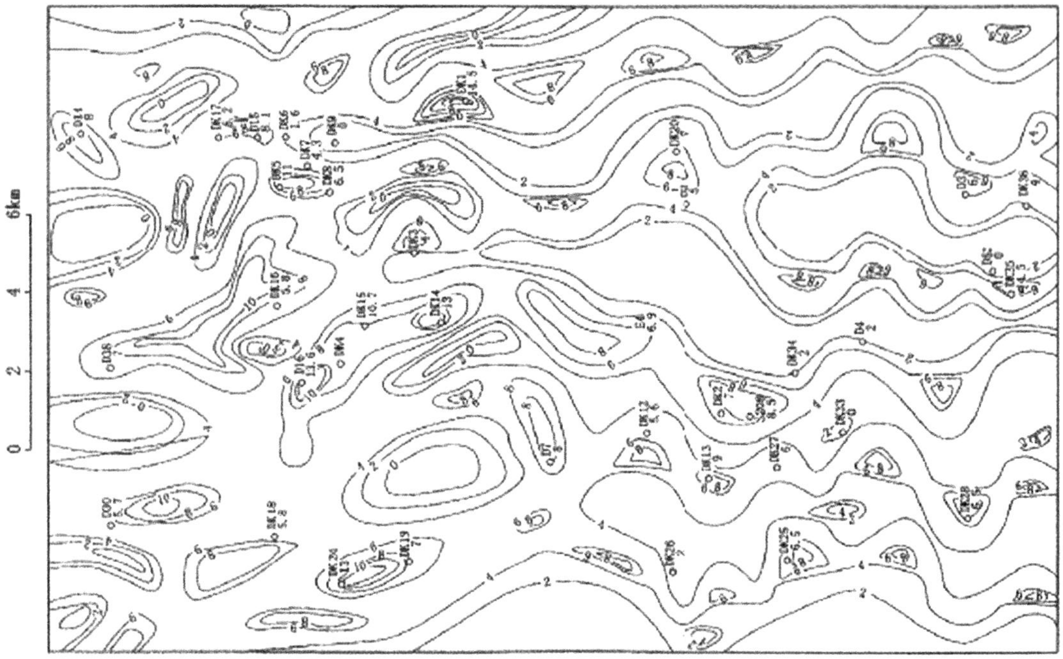

Fig. 9.50 Sandstone thickness map in He 2–1 section of the Daniudi gas field

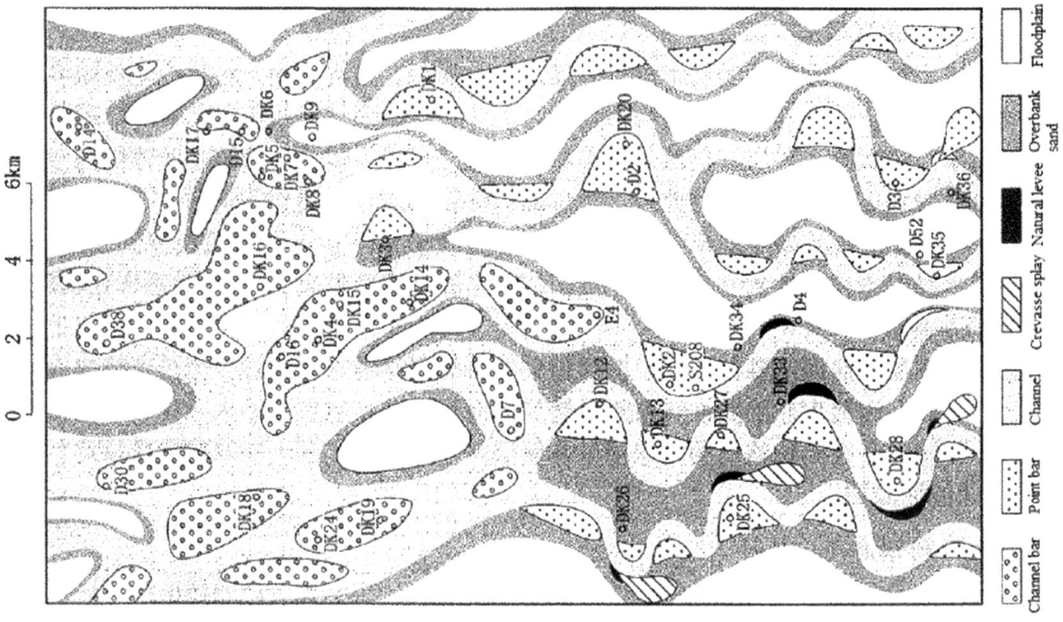

Fig. 9.51 Sedimentary facies diagram in He 2–1 section of the Daniudi gas field

the very bottom, which decreases evenly and gradually upward until it reaches its lowest value at the silt section of the top ripples laminae. Its permeability ratio can be more than 40x, and the variation coefficient is generally from 0.7 to more than 1.0. It usually has the highest degree of heterogeneity among various river sand bodies. Therefore, water channeling is more likely to occur during development.

The other important feature of the innerlayer heterogeneity of point bars is the complex contribution of the unstable thin argillaceous interlayer. The following cases may occur for the unstable thin argillaceous interlayer in the point bar sand body: ① Inundant mud on the upper part of sand body has poor continuity; ② Lateral accretion mudstone between the sand bodies is coated on the point bar; and ③ Dense boulder clay at the bottom turns the scour surface into an impermeable barrier, which we will focus on when a multi-period point bar sand body is superimposed atop it.

3. **Plane heterogeneity**

The heterogeneity of plane permeability for meandering river sand bodies actually reflects the permeability heterogeneity of the point bar part between different sedimentary bodies for two reasons: ① The argillaceous interlayer distribution in the upper part of the sequence is complex which results in complex connecting conditions, and ② connectedness at the bottom of the sequence is very good. Especially, in the early-to-mid waterflooding period, the heterogeneous sweep degree of the injected water level is controlled mainly by the lower part of the sequence. Therefore, its plane heterogeneity is shown specifically as the obvious directionality of permeability along the course of the paleochannel.

9.5.3.1 Anastomosing River Reservoirs

1. **Geometry and continuity**

The anastomosing river sand body comprises the fillings in the channel, but almost no sand is deposited in the wet land part (central bar) of the batture, and crevasse splays and natural levees only have a little sheet sand sedimentation. Therefore, as a reservoir body, its geometric shape is strictly controlled by the geomorphic features of the channel, which is presented as a very narrow strip-shaped sand body with a small width-to-thickness ratio and interweaved with each other in the form of a net. The sand body is filled and superimposed in the channel for a long time, and its width-to-thickness ratio has nothing to do with the breadth-depth ratio of the river. This means that the width-to-thickness ratio cannot be predicted by recovering the breadth-depth ratio of the paleochannel. This is also one of the differences of this body from the other channel sand bodies.

The typical filling sand body of ancient restrictive river valleys has already been found in China in the early Jurassic Fuxian Formation in the Shan-Gan-Ning Basin and in the lower Yan'an Formation. The dense well pattern in the oilfields verifies that the geometry of the entire sand body is a thin belt with a width of 200–400 m, which is characterized by poor lateral continuity.

2. **Intrastratal heterogeneity**

The anastomosing river channel sand body is superimposed by the rhythm layer deposited by multiple flooding events, and sand layers spanning tens of meters in thickness can be formed. The deposition of a flooding event shows a small positive rhythm, with a thickness of approximately 1 m, and the size difference of each small positive rhythm is very small, which reflects the feature of being higher down and lower up in terms of permeability variation. However, the degree of heterogeneity is low. In addition, from the occurrence to abandonment of the channel, the entire sand layer also shows a coarse positive rhythm. The permeability is heterogeneous to some extent, but weaker than that of point bar sands in meandering rivers. Thin intrastratal muddy and silty interlayers, which are the products formed between two flooding events, are common.

3. **Plane heterogeneity**

The planar physical property of heterogeneity has no practical significance for this type of strip-shaped sand body.

References

Allen, J.R.L. 1964. Studies in fluviatile sedimentation. Six systems from the lower old red sandstone, Anglo-Welsh Basin. *Sedimentology* 3: 163–198.

Allen, J.R.L. 1970. Sediments of the modern Niger Delta. In *Deltaic sedimentation: Modern and ancient*, vol. 15, ed. J.P. Morgan, 138–151. Society of Economic Paleontologists and Mineralogists Special Publications.

Allen, J.R.L. 1983. Studies in fluviatile sedimentation: Bars, bar-complexes and sandstone sheets (low sinuosity braided streams) in the Brownstones (L. Dexonian), Welsh Borders. *Sedimentary Geology* 33: 237–293.

Bluck, B.J. 1974. Structure and directional properties of some valley sandur deposits in Southern Island. *Sedimentology* 21: 533–554.

Boothroyd, J.C. 1972. Coarse grained sedimentation on a braided outwashfan, Northeast Gulf of Alaska, 127. Tech, No. 6. C. R. D. Coastal Research Division, University of South Carolina, Columbia.

Boothroyd, J.C., and G.M. Ashley. 1975. Process, bar morphology and sedimentary structures on braided out—wash fans, North-eastern Gulf of Alaska. In *Glaciofluvial and galcilacustrine sedimentation*, vol. 23, 193–222. Tulsa: Society of Economic Paleontologists and Minerologists.

Bridge, J.S, and J.A. Diemer. 1983. Quantitative interpretation of an evolving ancient river system. *Sedimentology* 30 5: 599–623 (Blackwell Science).

Cant D.J, and Walker R.G. 1976. Development of a braided-fluvial facies model for the Devonian Battery [J]. *Canadian Journal of Earth Sciences*, 13(1): 102–119.

Cant D.J., and Walker R.G. 1978. Fluvial processes and facies sequences in the sandy braided South Saskatchewan River, Canada[J]. *Sedimentology*, 25(5): 625–648.

Chamberlin, T.C., and R.D. Salisbury. 1909. *A college text-book of geology*. H. Holt and Company

Galloway, W.E. 1985. Meandering streams-modern and ancient. In *Recognition of fluvial systems and their resource potential*, vol. 19, ed. R.M. Flores, F.G. Ethridge, A.D. Miall, W.E. Galloway, and T.D. Fouch, 145–166 (SEPM Short Course).

Galloway, W.E. 1983. Terrigenous clastic depositional systems: applications to petroleum, coal, and uranium exploration[M]. Springer.

Hooke R L. 1967. Processes on Arid-Region Alluvial Fans[J]. *Journal of Geology*, 75(4): 438–460.

Klein, G.V. 1972. A sedimentary model for determining Plaeotidal range. *Geological Society of America Bulletin* 82: 2585–2592.

Laing, David. 1991. *The earth system: An introduction to earth science*. William C Brown Publishers

Lane, Z.W. 1957. *A study of the shape of channels formed by natural streams flowing in erodible material*, vol. 9. Sediment Series, 106. Omaba, Nebraska: U. S. Armycops of Engineers Missouri River Division.

Mckee, E.D., E.J. Crosby, and H.L. Berryhill Jr. 1967. Flood deposits, Bijoll Greek, Colorado, June 1965. *Journal of Sedimentary Petrology* 37: 829–851.

Miall, A.D. 1977. A review of the braided-river depositional environment. *Earth-Science Reviews* 13: 1–62.

Miall A D. 1978. Facies types and vertical profile models in braided river deposits: a summary [J]. *Fluvial Sedimentology*, 5.

Miall, A.D., and M.R. Gibling. 1978. The Situro-Devonian clastic wedge of somerset island, Arctic Canada, and some regional paleogeographic implication. *Sedimentary Geology* 21.

Miall, A.D. 1985a. Architectural element analysis: A new method of facies analysis applied to fluvial deposits. *Earth Science Reviews* 22: 261–308.

Miall, A.D. 1985b. Architectural-element analysis: A new method of facies analysis applied to fluvial deposits. In: *Recognition of fluvial depositional systems and their resource potential*, ed. Flores. R. S. P. M. short course No. 9. Copyright, SEPM.

Miall, A.D. 1988. Reservoir heterogeneities in fluvial sandstones: Lessons from outcrop studies. *Bulletin*

Association of American Petroleum Geologists 72: 682–697.

Miall A.D. 1999. Perspectives; In defense of facies classifications and models [J]. *Journal of Sedimentary Research*, 69(1): 2–5.

Orton, G.J., and H.G. Reading. 1993. Variability of deltaic processes in terms of sediment supply, with particular emphasis on grain size. *Sedimentology* 40: 475–512.

Plummer, C.C., and D. McGeary. 1996. Physical geology with interactive plate tectonics. Dubuque in United States: William C Brown Publishers

Reineck H.E., Singh I.B. 1973. Importance of Sequence in Environmental Reconstruction [J].

Rust, B.R. 1978. A classification of alluvial channel systems. In *Fluvial sedimentology*, ed. A.D. Miall, 187–198. Canadian Society of Petroleum Geologists.

Rust B.R., Masson G., and Dilles S, et al. 1983. Fluvial style and its influence on coal deposition in the Carboniferous Sydney Basin of Nova Scotia [J]. *Archives of Dermatology*, 95(4): 402–7.

Schumm, S.A. 1963. Sinuosity of alluvial rivers on the great plains. *Geological Society of America Bulletin* 74: 1089–1100.

Schumm S.A. 1968. Speculations Concerning Paleohydrologic Controls of Terrestrial Sedimentation [J]. *Geological Society of America Bulletin*, 79(11): 1573–1588

Smith, N.D. 1974. Sedimentology and bar formation in the upper Kicking Horse River, a braided river. *Journal of Sedimentary Petrology* 42: 624–634.

Smith, D.G., and P.E. Putnam. 1980. Anastomosed river deposits, modern and ancient examples in Alerta, Canada. *Canadian Journal of Earth Sciences*.

Smith, D.G. 1983. Anastomosed fluvial deposits, modern examples from Western Canada. In *Modem and ancient fluvial systems*, vol. 6, ed. J.D. Collinson and J. Lewin, 155–168. International Association of Sedimentologists Publications.

Thompson, G.R., and T. Jonathan. 1991.*Modem physical geology*. Saunder College Publishing, A Division of Holt, Rinehart and Winston, Inc.

Tunbridge I.P. 1981. Sandy high-energy flood sedimentation — Some criteria for recognition, with an example from the devonian of S.W. England[J]. *Sedimentary Geology*, 28(2): 79–95.

Walker, T.R. 1976. Diagenetic origin of continental red beds. In *The continental Permian in Central, West, and South Europe*, ed. H. Falke, 240–282. Dordrecht: Reidel.

Walker, R.G. 1984a. *Facies models*, 2nd ed. Toronto: Geological Association of Canada. Business and Economic Service Limited.

Walker, R.G. 1984b. *Facies models*. 2nd ed. Ainsworth Press Limited.

Walker, R.G. 1984c. Shelf and shallow marine sands. In *Facies models*, vol. 1, ed. R.G. Walker, 141–170. Waterloo, Ontario: Geological Association of Canada.

Walker, R.G, and D.J. Cant. 1979. Facies models 3. Sandy fluvial systems. In *Facies Models*, vol. 1, ed. R. G. Walker, 23–31. Geoscience of Canada (Reprint series).

Walker, R.G., and D.J. Cant. 1984. Sandy fluvial systems. In *Facies models*, 2nd ed, ed. R.G. Walker, 171–188. Geoscience Canada.

Williams, P.F., and B.R. Rust. 1969. The sedimentology of a braided river. *Journal of Sedimentary Research* 39: 649–679.

Yu, Xinghe, Defa Wang, and Junmao Zheng. 1992. Lithofacies association types or sequences and depositional system of permian sandstones in North China. *Acta Sedimentologica Sinica* 10 (1): 27–34.

10 Lacustrine Depositional System

Lakes are low-lying areas that collect running water on land. Currently, the total area of lakes all over the world is approximately 250×10^4 km^2, which accounts for 1.8% of the global land area. In China, only 1% of the land area is covered by lake water. Lakes were scattered like stars in China during the Meso-Cenozoic era. More than 10 large oilfields, such as Daqing, Shengli, Liaohe, and Dagang, were found in these lacustrine strata. They constitute a feature of Chinese petroleum geology and occupy a unique place in the global petroleum industry.

Lakes are important places for accumulating continental deposits, blocking a variety of deposits transported by rivers, and may act as the main locations (particularly in interior drainage basins) for chemical precipitation. Deposits formed by lakes that are not affected by the seawater homogenization effect (in terms of chemical properties) are sensitive to climate variation. Deposits in ancient lakes (closed lakes in particular) have an important place in paleoclimate studies and in comparative studies on global climate variation. They are the most favorable locations for the enrichment, burial, and transformation of organic matter in terrestrial facies into petroleum oil and gas. More than 90% of the petroleum reserves found in China are continental lake deposits from the Meso-Cenozoic era. These lacustrine deposits are also rich in minerals such as salt, iron, coal, and oil shale. Therefore, the study of lacustrine deposits is important from the viewpoints of geological theory and national economy.

10.1 Classification and Basic Characteristics of Lakes

Owing to complicated genesis and numerous influencing factors, lake classification schemes are diversified. The common classifications are as follows:

(1) By genesis: fluviatile lakes, glacial lakes, volcanic lakes, barrier lakes, wind lakes, karstic lakes, and tectonic lakes;
(2) By geographical location and landform: plateau lakes and plain lakes;
(3) By property of deposits and climatic environment: 6 types (Fig. 10.1); and
(4) By tectonic nature, salinity of lake water, and geographic position: 12 types (Table 10.1).

Herein, we focus on three classification schemes: by lake structure, salinity, and geographic position.

10.1.1 By Structure

In geological history, tectogenesis lakes feature long time, large areas, more minerals, and high research value. Tectonic movement is the most

X. Yu et al., *Clastic Hydrocarbon Reservoir Sedimentology*, Advances in Oil and Gas Exploration & Production, https://doi.org/10.1007/978-3-319-70335-0_10

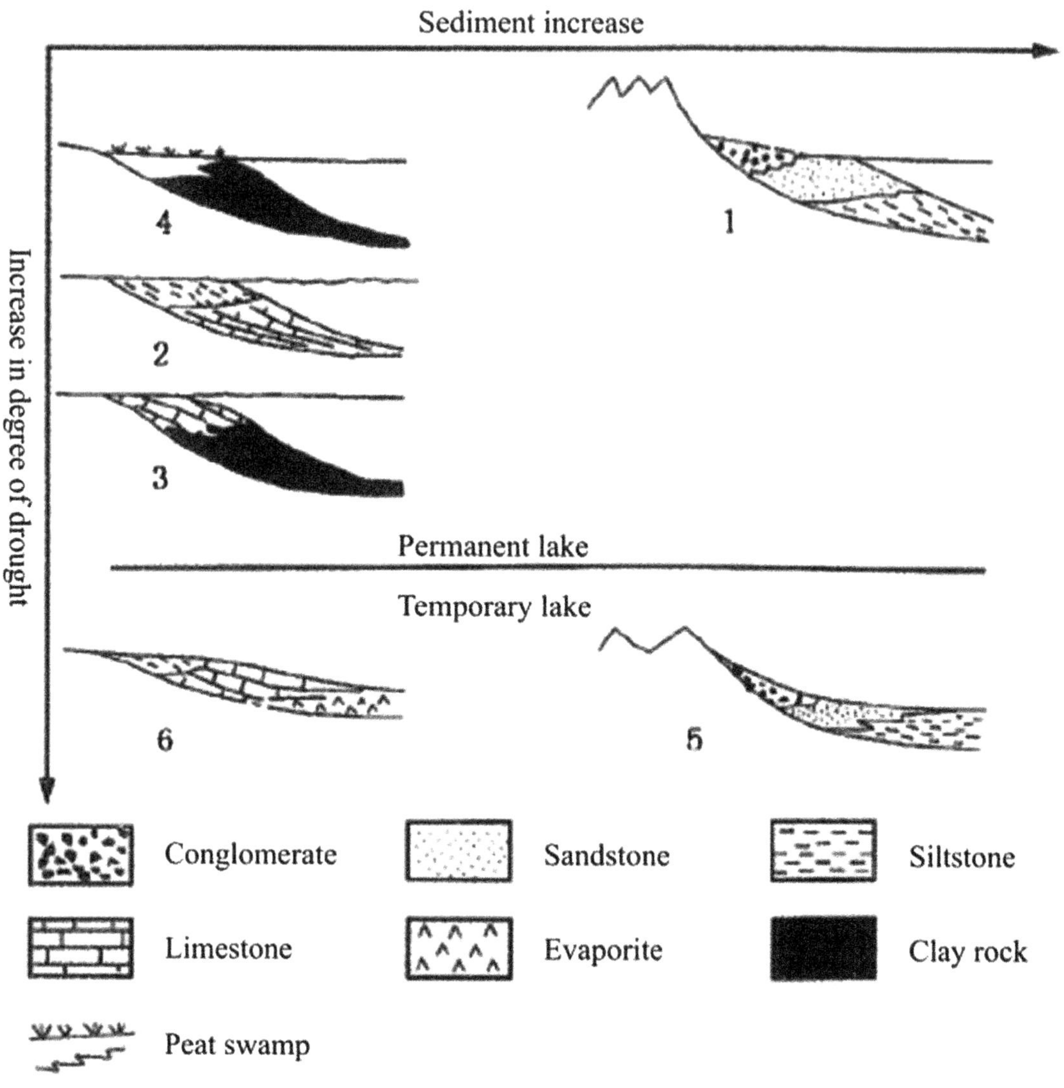

Fig. 10.1 Lakes can be divided into six types according to the sediment properties and climate environment (according to Selley 1976), 1—permanent lakes of terrigenous deposit; 2—permanent lakes of endogenous deposit (deposited mud in deep water area); 3—permanent lakes of endogenous deposit (muck in lake center); 4—permanent lake with swamp edge; 5—alluvial apron, playa; and 6—inland sabkha

fundamental factor controlling the distribution of depositional systems in such ways that the lake structure determines landform, deposition, and reservoir. By extensional basins, lakes can be divided into three types (Table 10.1), namely, faulted lacustrine lakes, depression lakes, and transitional lakes based on the tectonic movement characteristics of the area in which the lake is located.

According to the different structural types, basins can be divided into extensional lacustrine basins, compressional lacustrine basins, and craton lacustrine basins. Different basins have different sedimentary characteristics. Xue Shuhao (2002) summarized the main sedimentary characteristics of typical ancient lacustrine basins in China from the perspective of their structural types (Table 10.2).

Table 10.1 Types of extensional lakes in eastern China during the Meso-Cenozoic era (according to Chongjun 1983)

Salinity of lake water[a]	Faulted lacustrine lake		Depression lake		Transitional Lake	
	Offshore lake	Inland lake	Offshore lake	Inland lake	Offshore lake	Inland lake
Freshwater lake	Offshore faulted freshwater lake	Inland faulted freshwater lake	Offshore down-warped freshwater lake	Inland down-warped freshwater lake	Offshore transitional freshwater lake	Inland transitional freshwater lake
Salt lake	Offshore faulted salt lake	Inland faulted salt lake	Offshore down-warped salt lake	Inland down-warped salt lake	Offshore transitional salt lake	Inland transitional salt lake

[a]Based on the influence of climate on lakes in terms of the salinity of lake water, lakes can be categorized as freshwater lakes (brackish water lakes) and salt lakes (including salt water lakes)

10.1.1.1 Extensional Lacustrine Basins

1. **Faulted lacustrine lakes**

Faulted lacustrine lakes are distributed in the depressions of faulted basins. Tectonic activities in depression areas, which are dominated by fault depression, are mostly inhomogeneous tilted fault-block movements. There are steep normal faults at one or two sides, with fault dips as high as 30°–70° and dozens to thousands of kilometers of falls. In general, faulted lacustrine lakes have contemporaneous fault properties for faulting while depositing. The long axis of a lacustrine basin is consistent with that of a depression, parallel to the main boundary fault. A small amount of dustpan-like depressions show steep cross sections on one side but gentle on the other side and few are steep at both sides. Dustpan-like depressions, with one steep-sloped side as the contemporaneous fault and another gentle-sloped side, account for the majority of petroliferous fault lacustrine basins in China. Lakes in a few Paleogene petroliferous basins (Bohai Bay, Nanxiang, Jianghan, and Northern Jiangsu basins) in eastern China are all faulted lacustrine lakes of this type. Moreover, a few inland faulted lacustrine lakes in inter-mountainous or preorogenic small depressions are mostly distributed along regional major fractures, often at the intersections of a secondary fault and a main fault, such as some inland lakes in central and western China.

Faulted sedimentary basins are usually cut into plenty of depressions, and the deposition of each depression can develop its own system. Hence, the area of such lakes is not large. Lakes in petroliferous faulted lacustrine basins from the Meso-Cenozoic era, which generally cover about hundreds to thousands of square kilometers, have deeper fault depression and better succession. Hence, such lakes are characterized by deep waters, high sedimentary thickness, and high source bed thickness. The provenance adjacent to such lakes varies rapidly in terms of lithology and thickness. When the deep-water area tends to be on the steep-sloped side, the depocenter may not necessarily be consistent with the subsidence center. Most such lakes have an irregular elongated shape, and feature a steep slope along the short-axis direction and clearly differentiated gentle slope. In most cases, depocenters and subsidence centers tend to be on the steep-sloped side. Hence, the distribution in lacustrine subfacies zones is asymmetric. The steep side has deep water and a short slope, subfacies is narrow and varies quickly, even with the phase-jumping phenomenon, and the deepwater area is straight through the precipice edge. On the contrary, the gentle side has shallow water and a long slope, wide subfacies that varies gradually, and large-area coastal shore-shallow lake subfacies.

The sedimentary characteristics are different in the different phases of faulted lacustrine lake development because regional tectonic movement and other sedimentary conditions vary. The

Table 10.2 Sedimentary characteristics of lacustrine basins based on structure (according to Shuhao et al. 2002)

Sedimentary characteristics		Basin structure type				
		(1)	(2)	(3)	(4)	(5)
		Craton down-warped lacustrine basin	Intracontinental rift lacustrine basin	Foreland lacustrine basin	Lacustrine basin in collisional orogenic belt	Continental margin lacustrine basin
1	Basin geometry	Determined by basement structure and deep fracture, presented as open lacustrine basins of different forms such as rectangular and rhombic	Determined by deep fracture, mostly presented as elongated sedimentary depression	Presents elongated shape parallel to folded zone in piedmont subsidence zone	Lacustrine basin parallel to two folded belts, elongated in shape	Elongated sedimentary depression for early continental faulted lacustrine basin
2	Internal architecture of lacustrine basin	Simple internal architecture, with large uplifts and depressions or gentle slope	Complex internal architecture, arranged randomly in multi-ridge, multi-depression, multi-projection, and multi-concave, strewn randomly	Cross section of lacustrine basin sedimentation has dustpan-like shape, including thrust belt, subsidence zone, slope zone, and forebulge	Cross section of lacustrine basin sedimentation has dustpan-like shape, including thrust belt, subsidence zone, and slope zone	Internal architecture of early continental faulted lacustrine basin is similar (2)
3	Surrounding geological evolution and main provenance direction	Control of surrounding long-time uplift determines the main sedimentary system, dominated by longitudinal (axis) provenance	Mainly transverse provenance, accompanied by longitudinal provenance	Transverse provenance plays the leading role, from craton in the early period but folded zone in the later period, or from both	Transverse provenance plays the leading role, and the source supply direction is determined by the developmental phase of fold mountain systems on two sides	From continental direction and uplift area in the basin
4	Main depositional system type and facies belt	Fluvial-delta depositional system with a long history, complete facies differentiation, wide facies belt, single subsidence-sediment centers, consistent position in basin center, and 10–15% deep water area	Proximal short-distance fan delta, braided river delta and subaqueous fan, incomplete facies differentiation, and narrow facies belt. Subsidence center is adjacent to one side of growth fault of abrupt slope, sedimentary center shifts to lake direction, and deep-water	Fan delta is situated on one side of folded zone, and fluvial-delta is located on one side of craton direction; the former has narrow facies belt, while the latter has wide facies belt; subsidence center is located in piedmont subsidence zone,	Subsidence centers of braided river delta and fan delta are situated on one side of active orogenic belt, and depocenters are located below the slope zone	Depocenters and subsidence centers of fluvial delta and fan delta are identical to (2)

(continued)

Table 10.2 (continued)

Sedimentary characteristics		Basin structure type				
		(1)	(2)	(3)	(4)	(5)
		Craton down-warped lacustrine basin	Intracontinental rift lacustrine basin	Foreland lacustrine basin	Lacustrine basin in collisional orogenic belt	Continental margin lacustrine basin
			area in deep-depression period may account for 1/2 of this type.	while depocenter shifts toward the craton direction		
5	Deposition rate (mm/a)	0.02	1.25	0.029–0.148		
6	Sedimentary cross section form					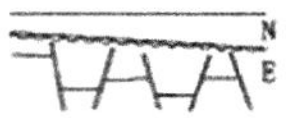
7	Example	Ordos Basin (Mz)	Bohai Bay Basin (E)	Longmenshan Basin in western Sichuan plain (Mz)	Turpan-Hami Basin (J)	Pearl River Mouth Basin (E)

early and middle periods of lake development span in time from an initial rift to the largest deep-depression period. Deep-water deposition develops to form very thick dark mudstone that is rich in organic matter, and characterized by deep color and pure quality, which indicate a good source bed. In the later period of lake development, the lake shrinks and transforms gradually into a depression lake.

2. **Depression lakes**

Depression lakes are distributed in down-warped sedimentary basins. Such basins are dominated by relatively uniform overall uplift and subsidence activity. They have a large area, smooth terrain, and gentle margin slope, without large raised segmentation in the middle part. As a result, a concave and convex limited separation phenomenon occurs after most continental faulted lacustrine lakes enter the depression stage. The area is large, but does not necessarily have deep water because unified open lakes are formed by tectonic movement. The subsidence center of the basin, which is generally consistent with the depocenter, migrates slightly close to the lake center during evolution. The most typical depression lake with the most abundant petroleum reserves in China is the lake in the Songliao Basin (Fig. 10.2), which was formed during the Early Cretaceous epoch. The Daihai Lake, which is the most typical modern fault lacustrine basin, has significant steep and gentle sedimentary differences (Fig. 10.3). Wing deltas, which are steep in the north and gentle in the south, are distributed with the characteristics of small but steeper slopes and large but less steep slopes, including fan-shaped slopes, lobate gentle slopes, thick steep slopes, and thin gentle slopes (according to Yu et al. 1992, 1994).

In a depression lake, coarse grains and facies belts rich in fragments are distributed at the edge of the lake, whereas thin deposits (calstic) are developed at the center (Fig. 10.4) of non-compensatory basins.

The distribution of a down-warped lacustrine depositional system is significantly different from that of a faulted lacustrine lake in the later period. In particular, the meandering river is vertical to the long axis of the basin during the faulted period and then turns parallel to it. Hence, the flooding plain is developed extensively. In terms of different lacustrine basin development periods and sediment distribution characteristics, it is described as "depressions along the basin, with fault depressions in the vertical direction" (according to Yu et al. 2007).

Three types of depression lakes, namely, offshore depression lakes in humid zones, concave depression lakes in humid zones, and inland depression lakes in arid zones, mainly developed in the Meso-Cenozoic era in China.

Offshore depression lakes in humid zones developed mainly in eastern China and southwestern China, for example, the Sichuan Basin developed in the Late Triassic epoch, the Songliao Basin developed in the Upper Cretaceous period, and the Bohai Bay Basin developed in the Neogene period. All of these lakes are in areas with a humid climate, were developed in a river system, are abundant in source, have broad basins, small topographical gradients, and are stable in subsidence. The main depositional system of this type of basin is as follows: an ever-lasting stem river is developed along the long axis of the basin with broad alluvial plains; delta sand bodies are formed in the shallow water area of the lacustrine basin, and turbidity depositions are distributed in a small proportion in various depositional facies.

Inland depression lakes in humid zones developed extensively in the late Triassic epoch-Early Jurassic period, such as the Junggar Basin, the north of the Tarim Basin, and the Ordos Basin.

Inland depression lakes in arid zones, which are represented by the Tarim Basin and the Qaidam Basin in northwest China, have little but concentrated rainwater, rare vegetation, and an intermittently developed water system. The

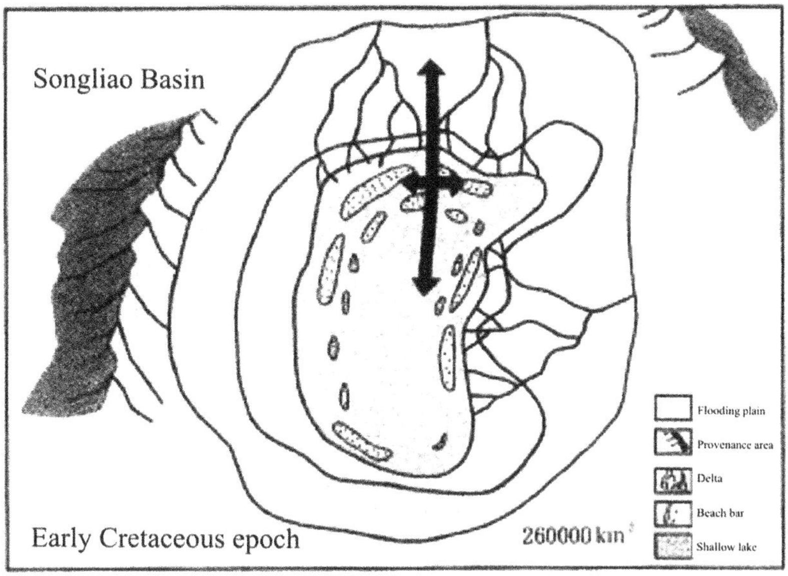

Fig. 10.2 The form of early Cretaceous most typical depressed lake with most abundant oil and gas in the Songliao Basin, China

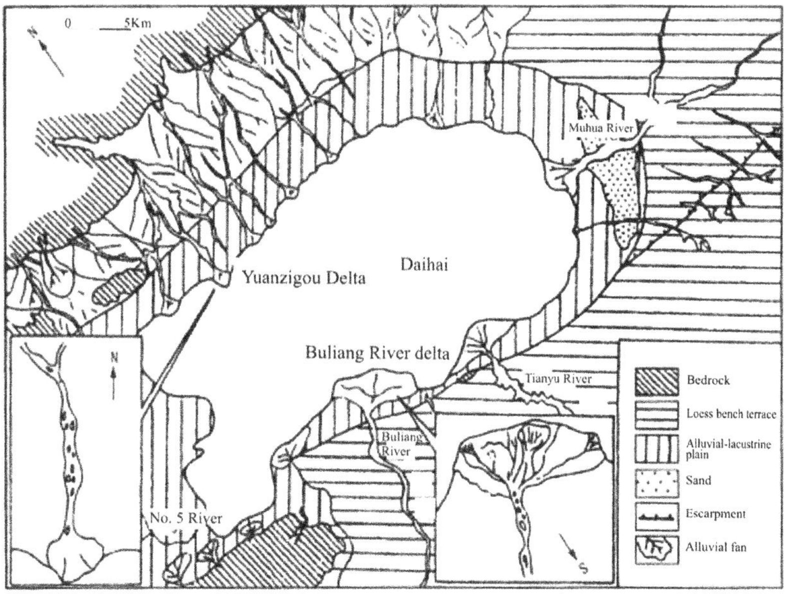

Fig. 10.3 The most typical modern depression lake basin-the Daihai Lake landform has obvious steep sedimentary difference (according to Yu et al. 1994)

stable deep lake area accounts for a small proportion of the basin area, and water and shallow water deposits are distributed extensively.

3. **Transitional lake**

A transitional lake and its sedimentary basin have both fault and depression features. For example, small faulted basins in the north and west margins of the Qaidam Basin in the Jurassic Cretaceous period developed into down-warped basins in the Eogene, with a typical binary structural pattern (Fig. 10.5).

10.1.1.2 Compressional Lacustrine Basin

A compressional lacustrine basin is an inland basin formed by tectonism and compression.

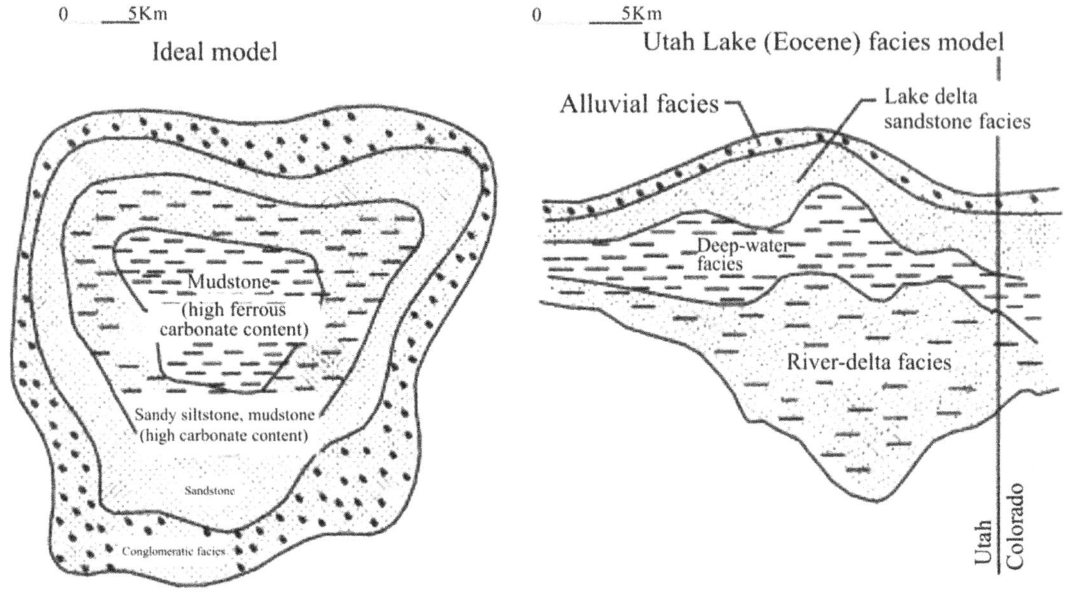

Fig. 10.4 In the depression lake, coarse grained and detrital facies belts are located in the lake margin, and the fine sediments occur in the central areas of the clastic sediments uncompensated basins (according to Picard and High 1972)

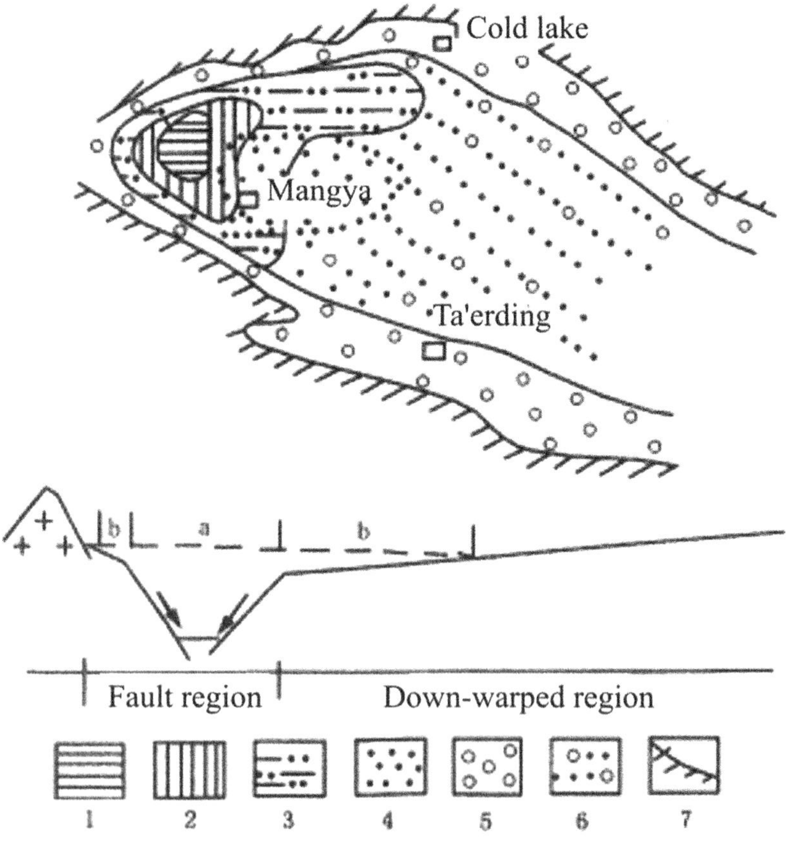

Fig. 10.5 Fault-depression lake and its deposit basin has both fault and depression characteristics with typical structural pattern of two element structure (according to Chongjun 1982), 1—deep lake region; 2—shallow lake region; 3—expansive lake region; 4—delta; 5—proluvial fan; 6—flood plain; 7—basin boundary; a—stable lake region, and b—expansive lake region

Table 10.3 Sedimentary characteristics of foreland basins

Sedimentary section	Asymmetric dustpan shape
Sedimentary characteristics	Sedimentary thickness and deposition rate change greatly
Provenance and paleo-drainage	Bidirectional: Craton and thrust belt direction
Evolution phase	① Early abyssal-bathyal flysch; ② medium marine molasse; ③ late continental facies molasse
Source-reservoir combination	Abyssal-neritic-continental depositional source reservoir combination from top to bottom

A typical example is a foreland basin. Foreland basins are sedimentary basins distributed along the outside of orogenic belts and in transitional zones distributed between an active orogenic belt and a stable craton. The subsidence amplitude decreases gradually toward the craton direction, and the sedimentary bottom surface is sloped if strong subsidence belts are present in front of the mountain. Thrust belts, subsidence belts, slope belts, and front uplift (Table 10.3) can be divided from the near orogenic belt to the craton.

Given that the western areas of China have been under the control of the Eurasian Plate collision for a long time, compressional foreland basins are rather common there. Especially, in the Late Hercynian, the northwestern region was basically separated from extensive transgression, so most depositional sequences of foreland basin are continental deposits. The Yunan Chuxiong Basin is a Mesozoic foreland basin, the sedimentary formation of which ranges from marine facies to marine-continental transitional facies to continental facies. This basin is divided into the west thrusting belt, middle depression belt-slope belt, and east uplift belt from west to east. Depositional facies changes regularly from west to east. The west thrusting belt in the late Triassic epoch ranges from marine facies to marine-continental transitional facies in the west, the east uplift belt comprises offshore continental deposit, and the middle part may be marine-continental transitional facies (Fig. 10.6).

10.1.1.3 Intercratonic Down-Warped Lacustrine Basin

"Craton" refers to continental crust that remains stable for a long period and deforms only slightly. A large intercratonic down-warped basin refers to a large continental down-warped basin developed above the Paleozoic platform (with a crystalline basement in the Precambrian platform). Since Meso-Cenozoic times, eastern China has been in a tension stress field, and western China in a compressional stress field. However, central China is suited for the development of a large continental down-warped basin owing to the weakened action of the two tectonic stresses and relatively stable tectonic activity. Its basin characteristics differ significantly from those of extensional basins and compressional basins (Table 10.4).

10.1.2 By Salinity

Modern lakes can be divided into freshwater lakes, brackish lakes, salt lakes, and playa (Fig. 10.7), which correspond to lake water salinity levels of less than 0.1%, 0.1–1%, 1–3.5%, and greater than 3.5%, respectively. In terms of the formation environment, evaporite can be formed in playas, salt ponds, mud flats (inland sabkha), shoreline supratidal mud flats (marine sabkha), near-shore lagoons, and salt flats. Ancient lakes can be classified based on their sediment components. To be specific, lakes containing mainly clastic deposits, such as silt and small amounts of carbonate, are equivalent to freshwater lakes and brackish lakes. Lakes containing mainly carbonate deposition are equivalent to salt lakes. Lake containing mainly sulfates and chlorides are equivalent to playas.

Salt minerals are evaporative minerals, and their precipitation is controlled strictly by the

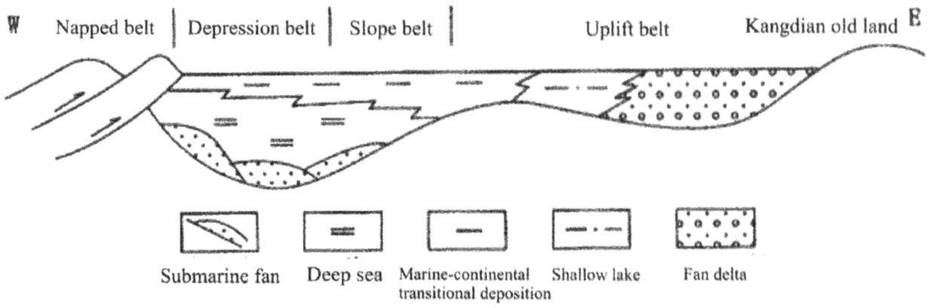

Fig. 10.6 From west to east, the basin can be divided into thrust belt in the west, depression-slope belt in the middle and the dome belt in the east. The sedimentary facies have regular variation, i.e. Marine-land sea facies in the west, offshore continental deposit in the east, land sea facies in the middle (according to Xue Shuhao et al. 2002)

Table 10.4 Main characteristics of intercratonic down-warped lacustrine basins

Basement characteristics		Paleozoic platform (with crystal basement in the Precambrian platform)
Basin controlling factors		Controlled by long-term inherited lifting movement
Tectonic characteristics	Subsidence rate	Based on slow subsidence
	Subsidence and source supply	Underbalanced state
	Subsidence center	In the middle of lacustrine basin
Sedimentary characteristics	General rules	Broad basin, multi-drainage lake inlet
	Thickness	Thinner with little change
	Sediment differentiation	Rather sufficient
	Thickness gradient and deposition rate	Smaller but little change on the plane
	Sedimentation type	Continental clastic rocks and coal series
	Water depth	Shallow water environments can be found frequently in lake region

climate in addition to provenance factors. With regard to oil generation capacity, biological growth is affected by the extremely high salinity of lake water, hence the oil generation index of salt lake deposition is not as good as that of a freshwater lake. In addition, most kerogens are of the humic type, which is not good for oil generation. However, a salt layer is generally a good seal bed, which is favorable for the preservation and transformation of organic matter for oil generation in the underlying deep lake. Therefore, abundant oil and gas resources can be found in basins containing salt deposition.

Among the Mesozoic-Cenozoic petroliferous basins in China, a few Paleogene-faulted basins in the east contain more salt minerals in oil bearing formations, such as the Kongdian Formation and the Shahejie Formation in the Bohai Bay Basin, wherein S3 and S4 members can be found commonly. Most salt deposits are distributed in the center of a lacustrine basin, and are frequently interbedded with thick dark mudstone, mostly in a waxing and waning relationship with sandstone development, such as the S3 member of the Dongpu depression. Salts are formed mostly according to the deep-basin deep-water salt formation model, and only a few salt formations can be attributed to meare and playa depositions.

The parent rocks mainly contain carbonate rock in the provenance area of the sedimentary basin with more salts, and they are characterized by poor sand body development, small particle size, small amount and volume, and more

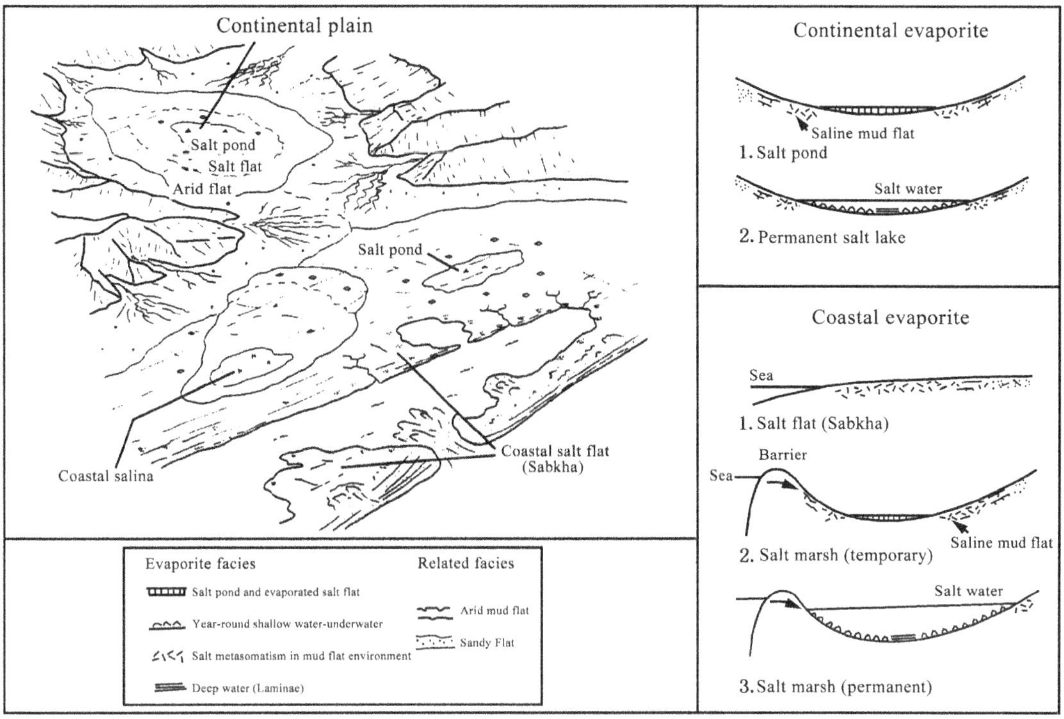

Fig. 10.7 Moden lakes can be divided into freshwater lakes, brackish water lakes, salt water lakes and dry saline Lake, according to salinity of lake water (<0.1%, 0.1–1%, 1–3.5% and >3.5% respectively) (according to Kendali 1992)

chemical cements such as calcium; hence, oil storage property is affected. In this basin, the focus should be on looking for reservoirs of various lithologies and oil reservoirs of various types.

10.1.3 By Geographic Position

Lakes can be divided into offshore and inland by geographic position. Affected by marine climate, offshore lacustrine basins are relatively humid, which is good for biological growth. The abundance of organic matter in a lake can result in the formation very thick oil-bearing series because the subsidence amplitude of areas with organic matter is high.

In China, offshore lakes are distributed mainly in the Bohai Bay basin, Jianghan basin, Northern Jiangsu Basin in eastern China in the Paleogene-Neogene, and in the Songliao Basin in Northeast China in the Cretaceous. In the Tarim Basin in the west during the Upper Cretaceous and the Paleogene, bays stretched into the Paleotethys to affect the lakes in west China. A variety of other middle and west basins are inland sedimentary basins with no relation to the sea, and the lakes therein are of the inland type.

10.2 Classification of Lacustrine Facies Belts and Characteristics

Although lake types are diversified, their subfacies division principles are essentially identical. Subfacies are classified on the whole according to two fundamental conditions, namely, deposit position in lake and lake water depth. Subfacies are defined on the basis of wave base, low water surface (low average water level), and flood surface (high average water level) (Fig. 10.8). i.e., in general, lakes can be divided into several subfacies (areas), including deep lakes, shallow

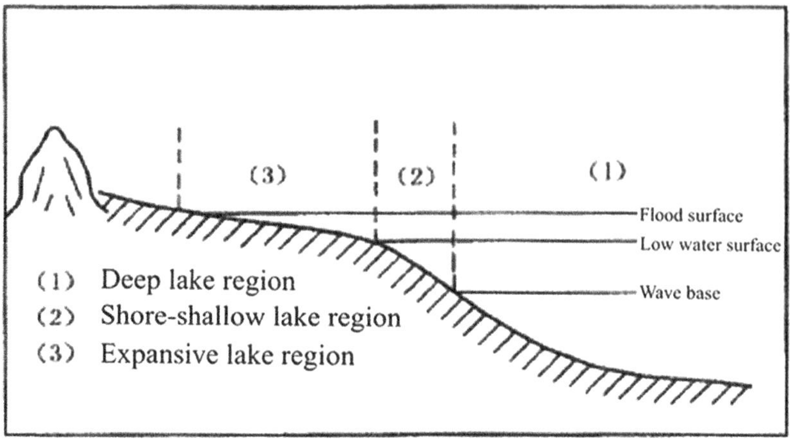

Fig. 10.8 According to wave base surface, dry water surface (average low water surface) and flood surface (average high water surface), the lakes can be divided into four sub-faces, i.e. deep lake, shallow lake, coastal lake and spreading lake

lakes, lake shores, and expansive lakes, which are characterized by annular distribution on the plane (Fig. 10.9). Bay subfacies can be divided as well. These three boundaries reflect not only subfacies distribution positions and lake water depth of a lake but also hydrodynamic conditions. Moreover, the source-reservoir-cap distribution is closely related to the three aforementioned boundaries. For example, good source beds are distributed below the wave base. Also, most reservoir sand bodies are constituted of nearshore shallow water sand bodies or delta sand bodies from the part above the wave base to the flood surface. Turbidity and storm flood sand bodies are located below the wave base.

Based on the characteristics of lacustrine sediments, lakes can be classified as clastic and chemical. In a clastic lake, terrigenous debris carried to the lake is transformed and redistributed by the lake effect, and sediment distribution and structural characteristics of the sedimentary system are determined mainly by the hydrodynamic conditions of the lake. Hence, the lake facies belt is divided mainly based on hydrodynamic factors. According to the types of chemical deposit, mainly various evaporite minerals, such as carbonates, sulfates, borates, and chlorides, lakes can be divided into carbonate lakes and salt lakes. Herein, the emphasis is on the classification and characteristics of clastic lakes and salt lakes.

10.2.1 Clastic Lake

Clastic lakes mainly contain clastic sediments, with little or almost no chemical sediments. Such lakes, though sometimes developed in arid inland intermontane basins, are distributed mainly in low-lying areas with abundant rainfall and developed overland runoff in damp climates. They are high in freshwater injection, low in lake water salinity, abundant in nutrients, and prosperous in biont. The deposits include mainly detrital materials (sand and mud) that are weathered and eroded by bed rocks in the source region carried by rivers. Only a small fraction of deposits are carried in solution form. In addition, a small quantity of detrital materials originate from the erosion of lakeshore bedrocks and volcanic eruption products (volcanic debris and cinerite) by lake waves, and a large number of ice-carried substances can be mixed near the front edge of a glacier. Clastic lacustrine sediments are essentially exogenous, and endogenous sediments include only organic matter produced by the remains or decay of animals or plants growing or living in the lake. Chemical deposits are very rare.

10.2.1.1 Deep Lake Subfacies

This quiet zone is not stirred by lake waves and lake currents below the wave base. As the deepest part of a lacustrine basin, it is distributed

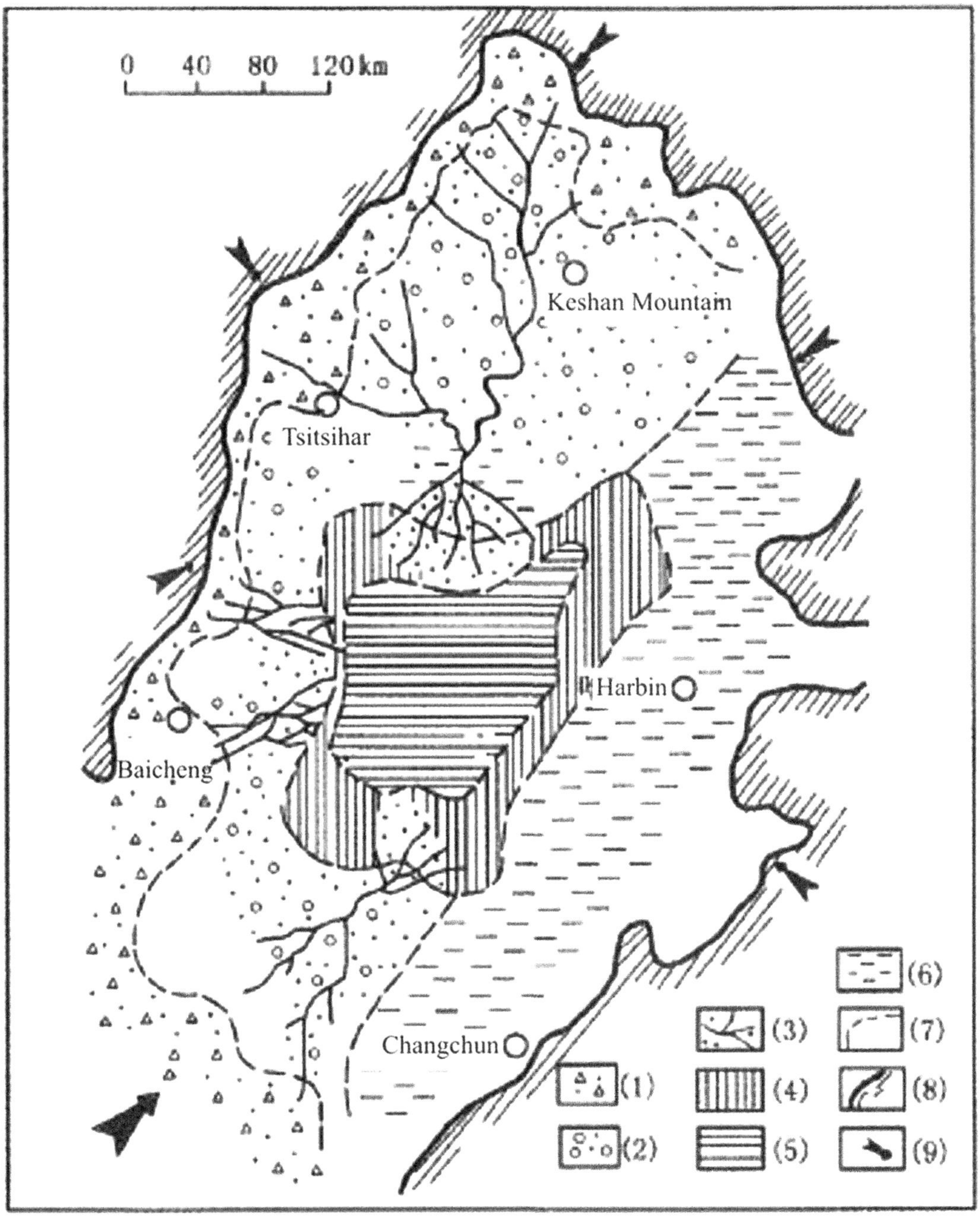

Fig. 10.9 The four sub-faces of lake (deep lake, shallow lake, coastal lake and spreading lake) are broadly distributed in a ring on the plane (according to Tian Zaiyi et al. 1983) (1) Alluvial fan facies; (2) River facies; (3) Delta facies; (4) Shallow-lake subfacies; (5) Deep lake facies; (6) Silty mud sedimentary area; (7) Facies boundary; (8) Basin boundary; and (9) Provenance direction

mostly in depression areas of faulted lacustrine basins close to the boundary fault. Deep lake subfacies are an anoxic reducing environment that lacks benthos but is abundant in plankton and nektons, which are completely preserved. However, they are monotonous in species and have a small size.

General lithological features of this zone include fine particle size, deep color, and high organic matter content. Rocks mainly include mudstone and shale of high purity, and limestone, marlstone, and oil shale may be developed. In addition, bedding is horizontal and fine horizontal. Pyrite, which is one of the common authigenic minerals, is distributed mostly in clay rock in a dispersed manner. Deep lake subfacies are the most favorable oil-generating facies belt because their lithological distribution is stable in the transverse direction and its sedimentary thickness is large.

Turbidity currents are formed in many deep lake facies belts. They have good Bouma turbidity sequences similar to those of turbidity deposits in the sea, which can form turbidite fans, and also include various sedimentary units such as turbidity channels, natural levees, and tongue-shaped bodies.

The natural potential curve of a deep lake subfacies section is a smooth line adjacent to the base line. Seismic facies is sheet shaped, with internal architectures reflecting in parallel as well as top and bottom contact relationships. When sediments are mudstone-mingled with siltstone lamina, the stratification is relatively good, which present strong reflection with high frequency, medium-strong amplitude, and good continuity. Very thick massive mudstones with poor stratification present discontinuous weak reflection or no-reflection (Fig. 10.10) with low frequency and weak amplitude.

10.2.1.2 Semi-deep Lake Subfacies

A weak reducing- environment exists in the part with the deeper water body below the wave base, which is actually a transitional zone between the shallow lake facies belt and the deep lake facies belt. The sediments are affected mainly by the

Fig. 10.10 Seismic facies shape of deep lacustrine subfacies is sheet, the internal structure is parallel reflection with conformable contact between the top and bottom (according to Chongjun 1983)

lake current, thus wave action barely affects the sediment surface. In particular cases, sediments can be affected by storm currents above the storm wave base. Sediments are distributed adjacent to the innermost part of the lake on the plane, but close to the side of the boundary fault in a faulted lacustrine basin.

Rock mainly includes clay rock and siltstone, and thin intercalation or lens of chemical stone can be found frequently as well. In general, clay rock refers to dark mud, shale, or silty mud and shale rich in organic matter. Furthermore, horizontal bedding develops with fine wavy bedding. Here abundant plankton fossils are preserved better, but benthos are not developed. Authigenic minerals including siderite and pyrite can be found.

When the lacustrine basin area is small and sedimentary characteristics are not distinctive, it is difficult to distinguish this facies belt.

10.2.1.3 Shallow Lake Subfacies

Shallow lake subfacies refer to the region between the lowest water level and the wave base surface in the dry season. Strictly speaking, in terms of classification, it refers to the region between the zone affected by waves and lake currents and the wave-breaking zone near the low water level. This facies belt, which is located near the lakeshore on the periphery of a semi-deep lake facies belt, has shallow water, but it is always below the water. It has the characteristics of strong wave and lake current action, good water cycle, sufficient oxygen, good light transmission, and an abundance of various aquatic organisms. Plants, including various algae and water plants, and animals, mainly freshwater gastropods, bivalves, fish, insects, and arthropods, are presented in the strata in good shape. The lithology is constituted of mudstones of different sorts of colors from light gray and grayish-green to green gray and sandstones. Oolitic limestone and bioclastic limestone are common in arid areas. Oolite, which is a good mark of a neritic shallow lake region, is best developed at depths of 2–3 m under water. Carbonated plant debris is also a common component. Sandstone, with higher texture maturity and mainly calcareous cementation, appears to have parallel bedding, wave ripple bedding, medium-small cross bedding, etc. In addition, wave ripples, vertical or inclined burrows, underwater shrink joints, and other sedimentary structures can be seen frequently.

The distribution of shallow lake subfacies depends on the lake area, water depth, and lakeshore terrain. Shallow lake facies of a depression lake with flat terrain is relatively wide. For this type of faulted lacustrine lake, the shallow lake facies belt on the gentle slope side is relatively wide, while that on the steep slope side is narrow or even absent. Some lakes with small depths can be located entirely above the wave base, and almost all such lakes are characterized by shallow lake facies belts, except the shore lake facies belt. This type of lake has no deep lake facies, hence it can be called shallow-water oxygen-rich lake facies. Shallow lake facies, which has a weak-oxidation-to-weak-reducing environment, also has some oil-generation capability, but the quality and abundance of the source rock are very inferior to that of deep lake facies. Many types of sand bodies are developed in the shallow lake facies belt, such as deltas, fan deltas, and beach bars, which are especially favorable for the accumulation of oil and gas. Many mud flat or sulfate particles can be deposited in a shallow lake facies belt, which lack a supply of terrigenous materials.

10.2.1.4 Shore Lake Subfacies

Shore lake subfacies are located between the flood shoreline and the low-flow shoreline, namely, the zone where the lakeshore is washed repeatedly by the flow and back flow of the surf formed by the breaking of waves for reducing water depth. The width of shore lake subfacies depends on the water level difference between flood and low water level, and the slope of the lakeside shore. A shore lake facies belt with a steep shore and low water difference is very narrow, with a width of only a few meters. However, the width of such a belt with a flat shore and high water level difference can reach several kilometers.

A shore lake facies belt is an important area for lacustrine sediments, the composition and distribution of which are affected by shore terrain, water conditions, prevailing wind conditions (speed, wind direction, etc.), and lake currents. Sedimentary types are very complex, with main sediments such as gravel, sand, mud, and peat. Gravel deposits are generally developed on a steep lake shore with bedrock, and such gravel originates from the exposed bedrock. A gravel bank with a good degree of grinding and sorting feature can be formed owing to repeated winnowing by waves and lake currents because the bedrock is peeled and collapsed due to long-term weathering erosion and storms and waves, and is then accumulated on the shore in situ. This gravel bank appears as a lenticular layer in the stratum. The gravel is in imbricate arrangement. The largest flat surface of flat gravel is inclined along the lake, and its long axis is mostly distributed parallel to the shoreline. Sandy sedimentation, which is the most widely developed deposit in shore lake facies belts, is the main product accumulated when it is brought by a river to a lake during flood season and then transported to a shore lake by waves and lake currents. It generally has higher maturity, as well as better sorting features and grinding degree, as a result of long-distance transportation and repeated scouring by lake waves. Its main components include quartz and feldspar interspersed with a few heavy minerals. The sedimentary structure mainly displays all types of current cross beddings and ripples. For shore lake sand, a thicker beach bar is often formed around the lake periphery. The width and particle size of the sand body depend on the strength of the prevailing wind conditions and wind direction. A weather shore has higher wave energy, wider sand body, coarser particle size, and high sorting features, whereas a lee shore is relatively poorly developed. There are fewer fossils in its sandy sediments, but plant debris, fish bones, and shell debris or sometimes bivalve shell beaches can be seen. Burrows are seen frequently in fine sand and silt layers. Muddy and peat sediments, mud rich in organic matter and peat layers—often sandwiching a few thin silt layers—are distributed mainly on the gentle leeward shore and low-lying wetland swamps. Muddy layers have horizontal beddings, while silt layers have miniature ripple cross beddings. Peat swamps are extremely well developed in a few lakes. Especially, in the late stages of lake evolution, the entire lake can be subject to swampiness. Hence, shore lake facies belt is an important coal-accumulating environment.

The sediment types found in shore lakes are the most abundant lacustrine sediments. In terms of particle size, gravel and sand beaches, as well as mud flats, can be seen in shore lakes. Normally, a steep shore is composed mainly of gravel, flat shore mainly of sand, and bay mainly of mud. However, shore lake facies belts are periodically exposed to the environment. Because parts of a lake are exposed above water in the dry season, many mud cracks, raindrop imprints, vertebrate footprints, and other exposed structures are retained. Thus, the appearance of various exposed structures and swamp intercalation are important indicators of shore lake sedimentary facies belt, and these features distinguish it from other types of facies.

This special lithology caused by the special hydrodynamic conditions of a shore lake subfacies is a good mark for determining the lake shoreline. However, shore lake subfacies are very narrow. It is hard for ancient lacustrine sediments to be separated from shallow lake subfacies, thus they are collectively referred to as shore-shallow lake subfacies or classified into shallow lake subfacies.

Shore-shallow lake subfacies are characterized by the frequent interbedding of sand and mud with good differentiation and obvious stratification. They are also characterized by rapid lateral changes in lithology and thickness, poor continuity, and inconsistent development in different places, thus the electric well-logging curves and characteristics of the seismic facies change considerably. In general, seismic facies are wedge shaped, with top truncation and toplap in the nearshore zone, and downlap or onlap at the bottom, which is constituted of divergent lineups with poor-medium continuity and medium-weak amplitude. Toward the near edge of the slope, the lineup has non-systematic lateral termination. However, toward the lake center, the frequency increases and the phases increase. A shore-shallow lake region with an underdeveloped sand body may present sheet reflection with better continuity, medium amplitude, and flatness, but with local waves. The area, particularly the location of sand body development, is characterized by medium-high amplitude, low frequency, continuous and intermittent, obvious reflection, sporadic and no reflection, and other phenomena, depending on the type of sand body and the scale (Fig. 10.11).

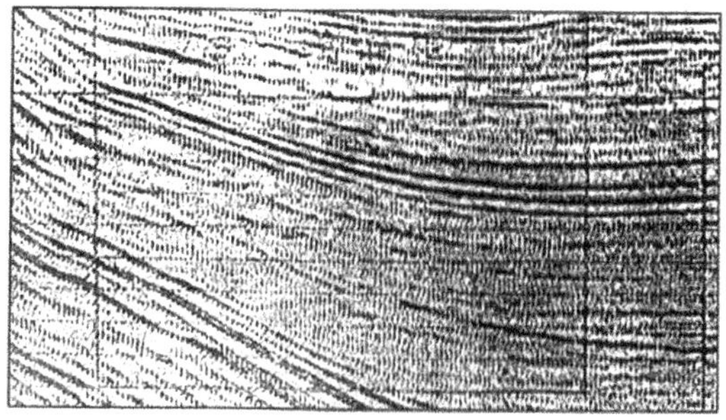

Fig. 10.11 In the coastal-shallow lake zone with sand body undeveloped, sheet reflex occurs with good continuity, moderate amplitude and flat but locally undulating (according to Chongjun 1983)

10.2.1.5 Lake Bay Subfacies

A lake bay is a zone with a semi-closed water body, where the water near the shore cannot be well exchanged with the water in the lake due to a barrier. The water body of a lake bay is calm and lacks oxygen at its bottom. Its sediments are dominated by fine-grained mud shale owing to poor circulation in the water body, weak effect of waves and lake currents, and lack of injection from a big river. In damp climates, plants grow in lake bays to form a peat swamp. The mudstone here is black and dark blue, sandwiched with carbonaceous shale and coal seams. Some small fining-upward sand bodies can be contained in sediments, and graded bedding, parallel bedding, wave-built miniature ripples, and low-angle cross bedding can be developed in a lake bay with intermittent source injection.

10.2.2 Salt Lake and Playa

A salt lake is dominated by sedimentary evaporite minerals, and characterized by sulfate and chloride saline minerals (Fig. 10.12). Under arid weather, when water evaporation is greater than rainfall in the lake region, and surrounding surface runoff and groundwater input are low, lake water concentrates gradually and its salinity increases. As a result, certain salt minerals are precipitated when a threshold salt saturation level is reached (Fig. 10.13). Saline minerals are frequently divided into carbonate, sulfate, and chloride according to their anion, which also roughly represents the extent of dissolution of different salts and the precipitation sequence.

Salt lake sediments can appear in the deep-depression period and the decline phase in the development of a lacustrine basin. Many salt lakes are formed at some stage or a late stage in lacustrine basin development owing to lake water drying up or an increase in salinity. Based on the salt distribution environment, a few salts are deposited deep in the lake, while others are deposited in the shallow lake or are even playa sediments.

10.2.2.1 Deep Salt Lake Facies

Deep salt lake facies are developed in several sedimentary basins in eastern China. Salt accumulation formations belong mostly to the deep-depression period or earlier stage, as well as to the oil generation period or its earlier stages, such as the S3 and S4 members and the Kongdian Formation in the Bohai Bay Basin. Salt sediments, as well as mud and sand deposits, have an obvious zonality on planar distribution, appearing as gypsum-salt sedimentary zones to

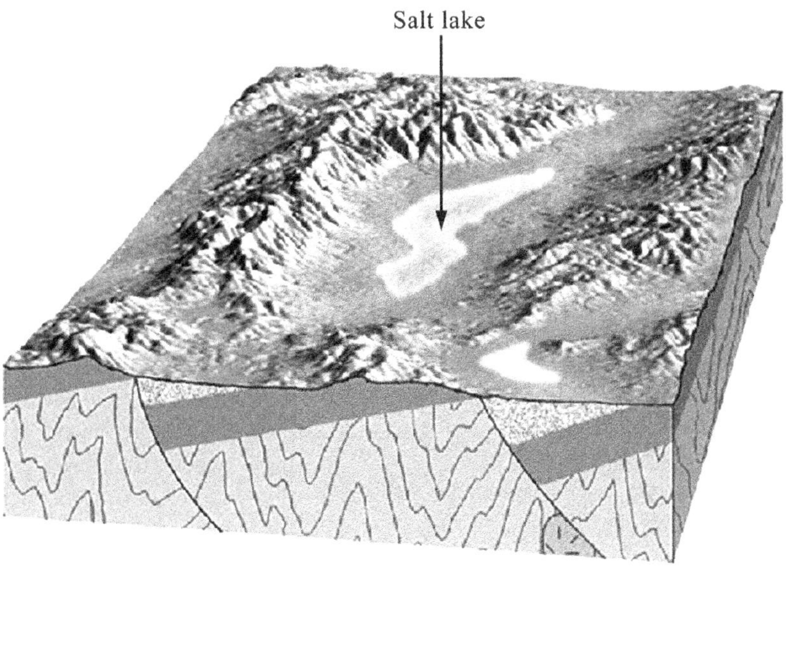

Fig. 10.12 Sketch map of Saline Lake distribution position (A lake dominated by sedimentary evaporite minerals, characterized by sulfates and chlorides) (according to Tasa Graphic Arts 1999)

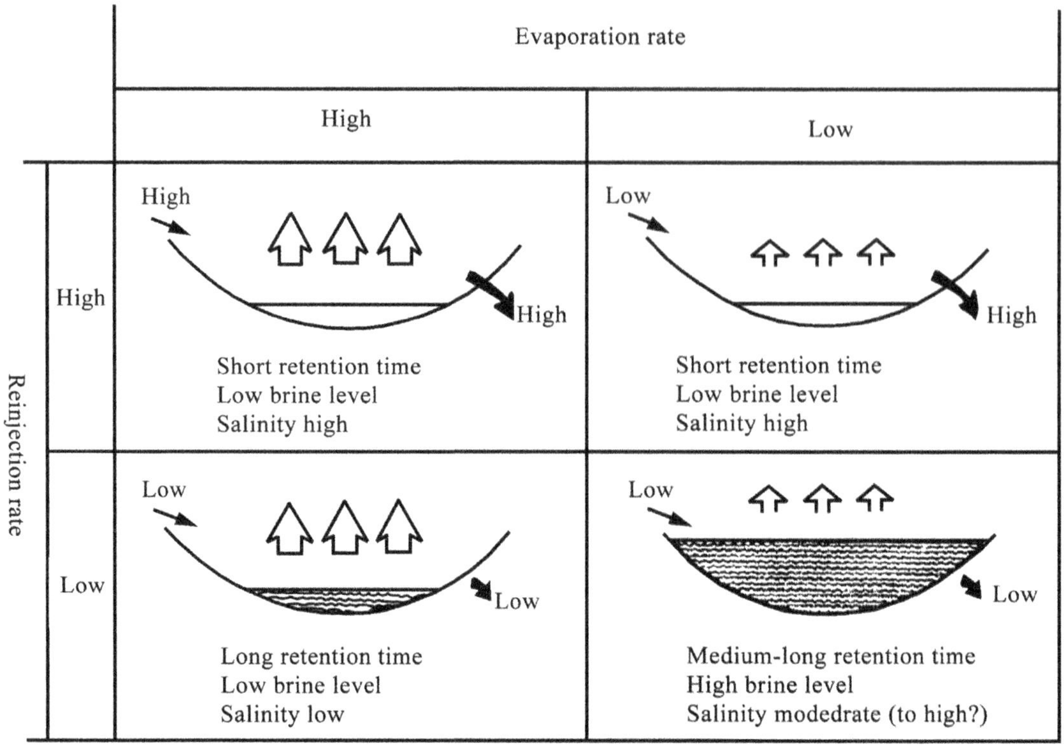

Fig. 10.13 In arid climate, when the evaporation of lake water is greater than that of the lake area, and the amount of surface runoff and groundwater input is small, the lake gradually became concentrated and salinity increased. When a certain salt saturation is reached, then this certain salt mineral is precipitated (according to Kendali 1992)

gypsum-salt and argillaceous sedimentary areas and sand mud sedimentary areas successively from the center of a lake to its shore (Fig. 10.14).

10.2.2.2 Shallow Salt Lake Facies

Shallow salt lake facies are developed mostly in the depression and decline stages during the evolution of some basins. Here, crowds and rainfall are low, and the average water depth is relatively low due to the natural geographical environment. For example, the salt lakes in the Qaidam Basin are all very shallow, with an average depth of tens of centimeters (Fig. 10.15). In these salt lakes, the distribution of salts with different compositions is not in the typical concentric ring shape.

10.2.2.3 Playa Facies

Playas, which are usually distributed at the periphery of salt lakes and formed in the later period of salt lake development (Fig. 10.16), are dry salt flats, exposed on the Earth's surface. They are formed owing to the evaporation of a lake to dryness. Surdam and Wolfbauer (1975) regarded the Gosiute Lake as a playa complex based on the research of the Green River group. Gosiute Lake and its surrounding sedimentary rock layers can be divided into three types of lithofacies: ① marginal silt and sandstone facies; ② carbonate mud flat facies; and ③ lacustrine facies. Each facies has the assemblage characteristics of carbonate minerals. Marginal facies are characterized by calcite nodules and calcareous cements, mud flat facies by calcite and dolomite, and lacustrine facies by trona (sodium carbonate) or oil shale (calcite or dolomite). The plane distribution of facies is in a concentric belt shape: the lacustrine facies in the center are surrounded by mudflat facies, which are encircled by marginal facies. Oil shale is

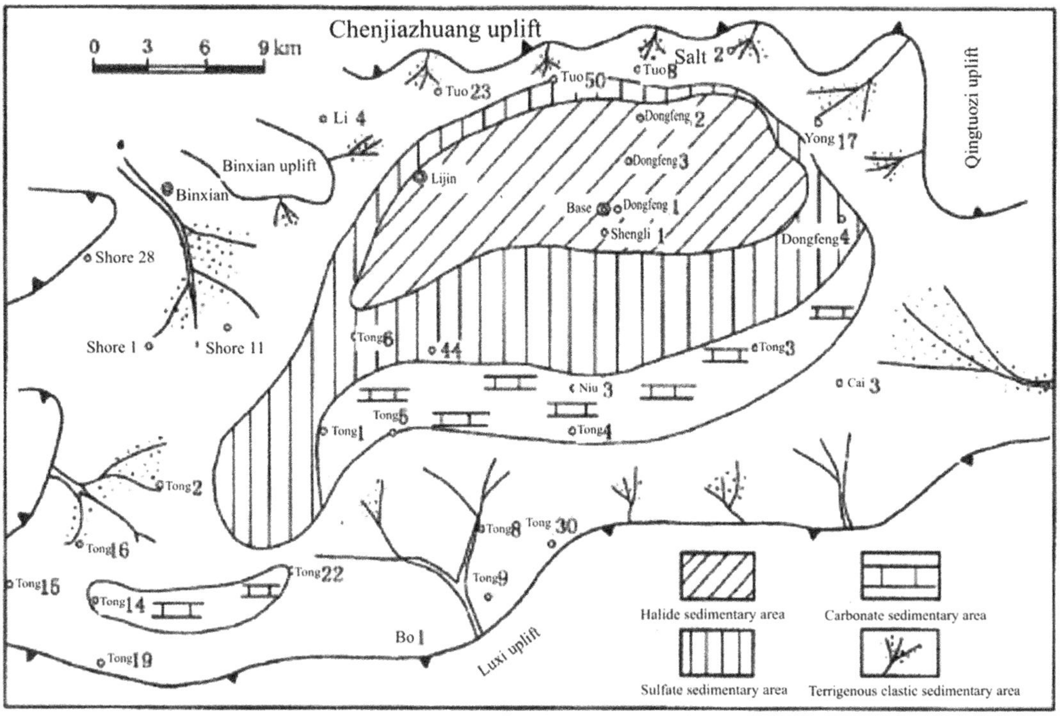

Fig. 10.14 Salt deposits and sand mud deposits have obvious zonation in plane distribution, from center of lake to bank, gypsum salt sedimentary area, gypsum salt and muddy sediment area and sand mud sedimentary area would appear by turn (according to Chongjun and Shuhao 1993)

formed in a lake during high stand, whereas trona is formed during low stand (Fig. 10.17). Liu Qun et al. (1987) also divided playas into three lithofacies zones based on the characteristics of most Mesozoic and Cenozoic salt lakes in China: I-clastic rock zone; II-marginal initial salt formation depression zone (including gypsum mudstone); and III-salt formation zone in concentrated brine center (Fig. 10.18).

10.3 Vertical Sedimentary Sequence of Lakes and Evolution Models

The structure of lacustrine facies changes along with changes in texture, terrain, provenance, and climatic conditions during different phases of evolution. Even lakes with different facies types are in the same evolution phase, the structure features of sedimentary facies and the relations between facies also have obvious differences because of the lacustrine basin size, water depth, lake bottom, shore terrain, etc. Therefore, the vertical sequences and facies association models of the sediments filling a lake vary greatly.

10.3.1 Vertical Sedimentary Sequence

The sediments filling a lake are complex and variable, depending mainly on regional tectonic activity, climate, and provenance. In China, there are single- and multi-cycle basin filling sequences in Meso-Cenozoic petroliferous basins. Lakes with a single cycle only have one period for sag extension. Hence, there just exists one best development phase for the source bed. The Meso-Cenozoic-faulted lacustrine basins in China, such as the Hailaer Basin, Kailu Basin, Tieling-Changtu Basin, and Zhangwu-Heishan

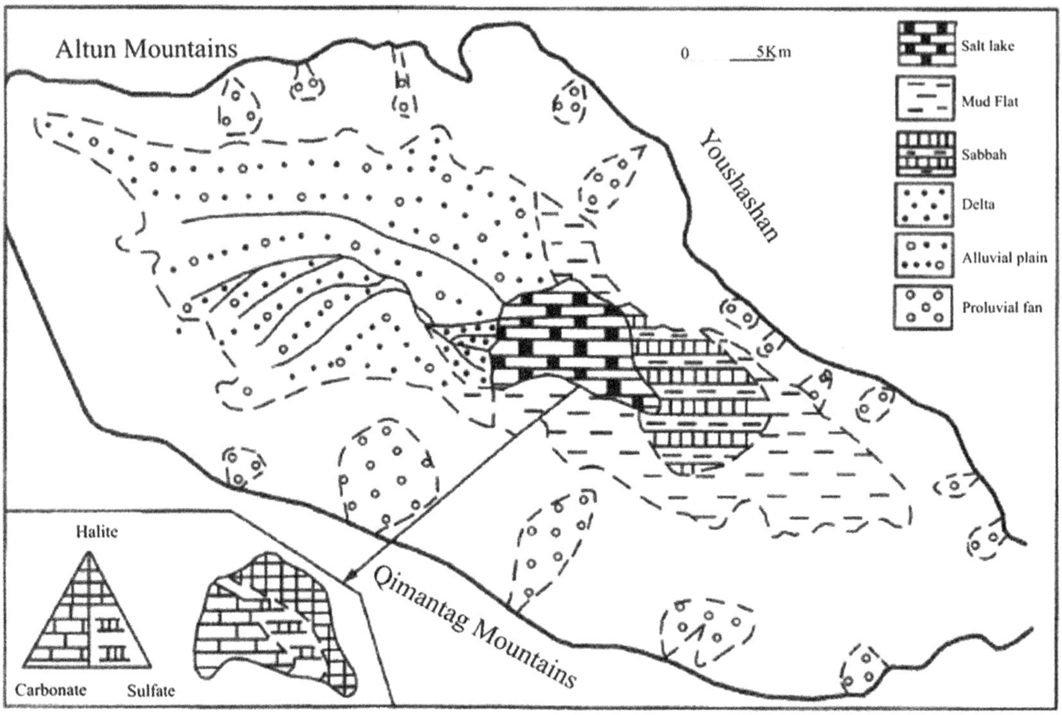

Fig. 10.15 The salt lakes in the Qaidam Basin are very shallow and the water depth is generally tens of centimeters

Basin, have the characteristics of single-cycle filling. The vertical sequence of the Wuerxun depression in the Hailar Basin is red-black-red and coarse-fine-coarse from bottom to top, and the shallow-deep-shallow characteristic of the lake water represents a complete tectonic cycle (Fig. 10.19). The Biyang depression in the Nanxiang Basin represents a single sedimentary cycle as well.

Multi-cycle lake-filling sedimentary sequences are also normally seen in the Meso-Cenozoic petroliferous basins in China, which indicate that the period of sag extension was developed vertically many times and many sets of source beds were formed. For example, the Qing-1 member and the Nen-1 member sedimentary periods in the Songliao Basin are the sag extension periods of the lacustrine basin, forming good source rocks with an organic carbon content of 1.5–2.4% and soluble organic matter content of 0.3–0.5%. The parent material of the source is Type I kerogen, and the atomic rate of H/C is 1.70–1.98. Multi-cycle filling deposition can form many types of sand bodies and multi-layer petroliferous systems (Fig. 10.20).

10.3.2 Evolution Model

As mentioned earlier, from the perspective of petroleum geology, tectonic lakes can be further divided into fault, down-warped, and transitional types. The former two are commonly seen in the Meso-Cenozoic petroliferous basins in eastern China, and they present different sedimentary patterns during different evolution phases, thus forming different lake models.

A fault basin features alternating concave and convex, large topographical gradients and strong separation in its internal terrain. According to the different geological characteristics of each tectonic zone along with the climate and provenance compensation, the sedimentary filling features of each tectonic zone should be analyzed

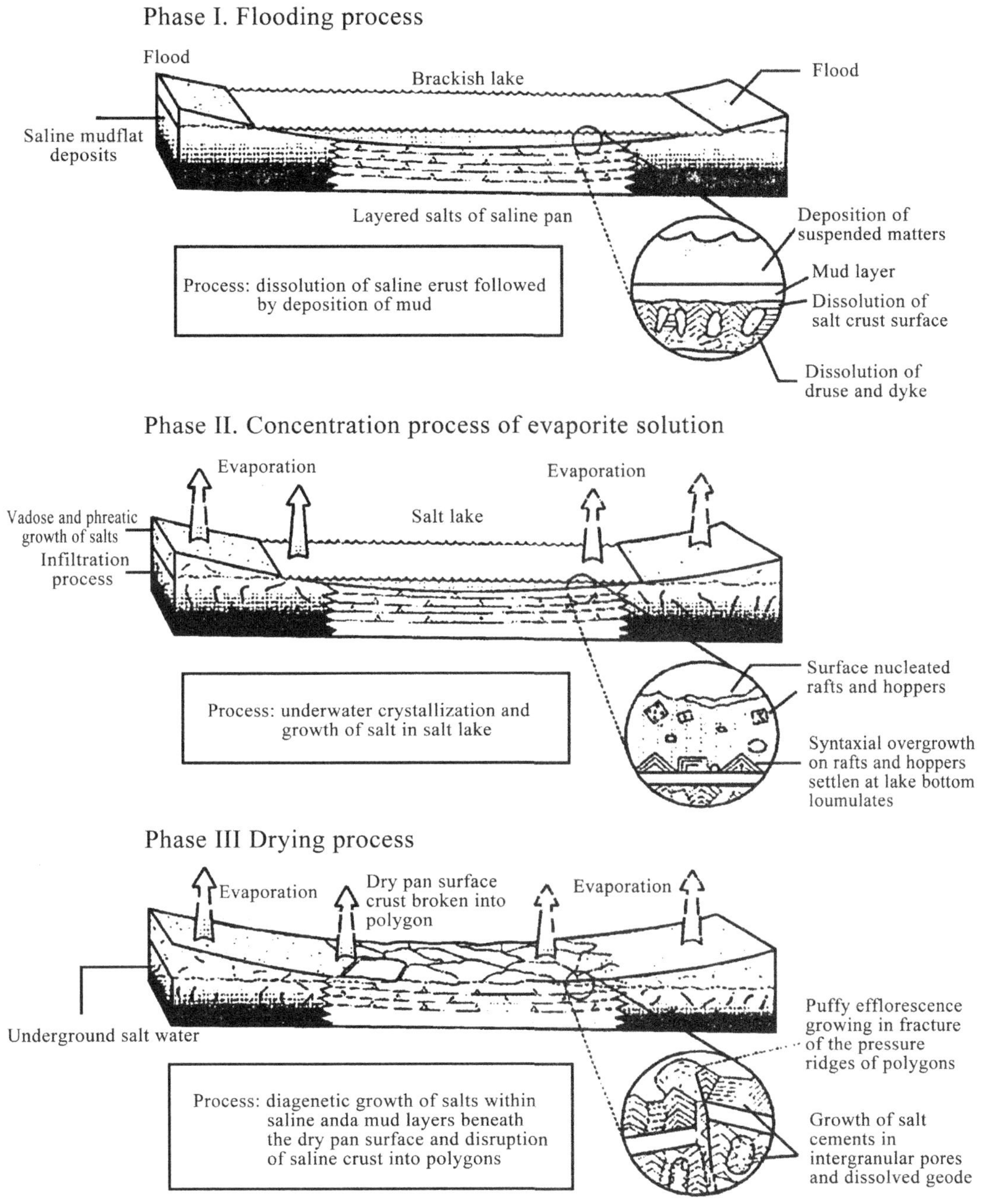

Fig. 10.16 Dry saline lake is generally distributed in the periphery of Saline Lake or the late development of Saline Lake, and the exposed to the surface dry salina when the water in Saline Lake is steamed or steamed (according to Lowenstein and Hardie 1985)

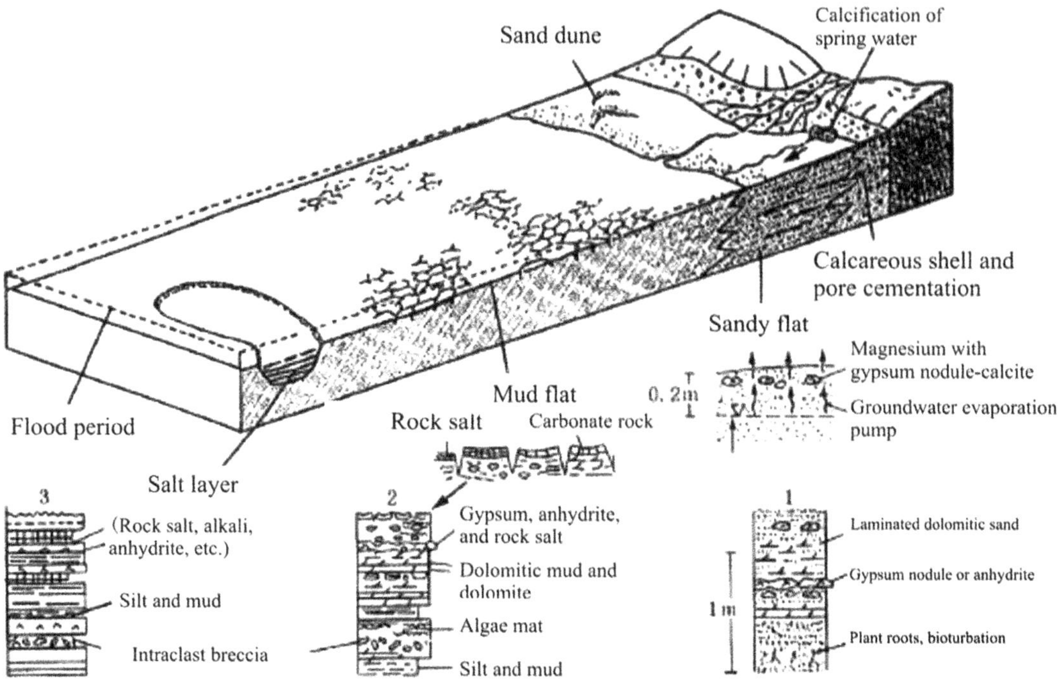

Fig. 10.17 The plane distribution of lithofacies is concentric: the central lake facies are surrounded by mud flats, while the mud flats are surrounded by marginal facies. Oil shale is formed during the high water stage of lake, while natural alkali is formed at low water level of lake (according to Einsele 1992)

systemically to determine its sedimentary characteristics and the control effects of determinants on sedimentary filling (Table 10.5).

Slope area refers the slope part with four concave edges, which can be divided into steep slope and flat slope based on slope degree. Then, slope area can be subdivided into three subclasses based on the presence of terraces or slope breaks: "slope type (single)" without slope break, "single-step type (single folded type)" with a terrace or slope break, and "multi-step type (multi-folded type)" with multiple terraces. The slope area (steep and flat slope area) sedimentary filling is mainly subject to tectonic setting (terrace type and slope degree). In addition, climate conditions and provenance compensation have important effects on the filling model. Hence, faulted lacustrine basin formations follow various sedimentary filling models and the associated filling characteristics; further, various depositional systems are formed under the effects of many factors.

A depression area is generally located in the sedimentary center of a lacustrine basin. The sedimentary filling in the depression area is affected mainly by climate and source supply (dominated by climate), which can be divided into compensation type and under-compensation type in terms of source supply. Compensation type can be subdivided into muddy and sandy compensation type. However, depression areas are mainly classified as arid, semi-arid, and humid based on the characteristics of change in climate. Different sedimentary fillings are developed in a depression area under the common control of climate and source supply. The filling materials are mainly coarse clastic rocks in early times and, occasionally, salt lake sediments, deep lake to semi-deep lake facies and, sometimes, deep turbidity sediments are deposited with an increase in lacustrine basin area, thus deepening of the water body, and increasing the distance of the depression center from the provenance in the late period.

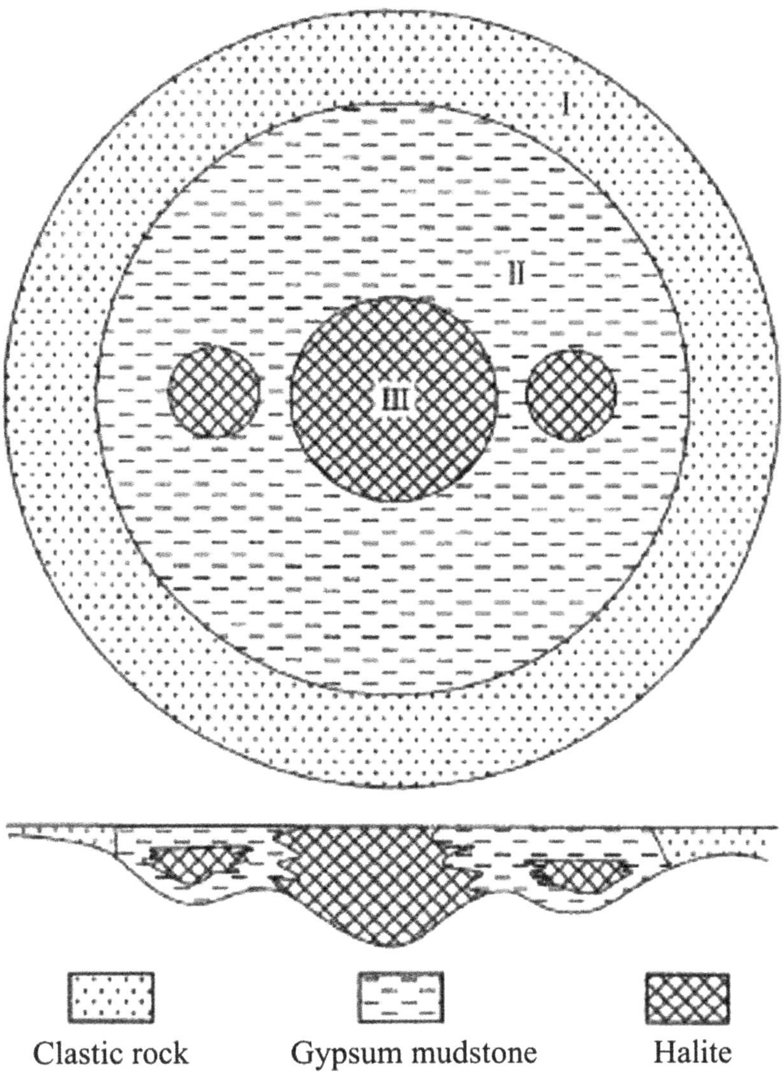

Fig. 10.18 According to the features of the most Mesozoic and Cenozoic salt lakes in China, three lithofacies can be also divied: I—clastic rock zone; II—marginal initial salt forming zone (Gypsum bearing mudstone); III—central brine forming zone of concentrated brine (Halite) (according to Liu and Chen 1987)

Central uplift areas refer to the bulge distributed in the ancient terrain of a lacustrine basin, which usually does not contact with the provenance directly and the sediments above are reached via transport through the filled depressions since the central uplift area is surrounded by depressions. In addition, because its water body is shallower than that of a depression, it is easily winnowed by waves and formed into coarse clastics due to stronger energy. Sedimentation thickness of the deposits at the bottom increases toward both sides from the uplift, and the higher it is, the smaller is the thickness difference.

10.3.2.1 Faulted Lacustrine Lake

Faulted lacustrine lakes generally experience three sub-stages, namely, initial rifting, sag extension, and lifting contraction, in their development process, and they have different characteristics in each development period (Table 10.6).

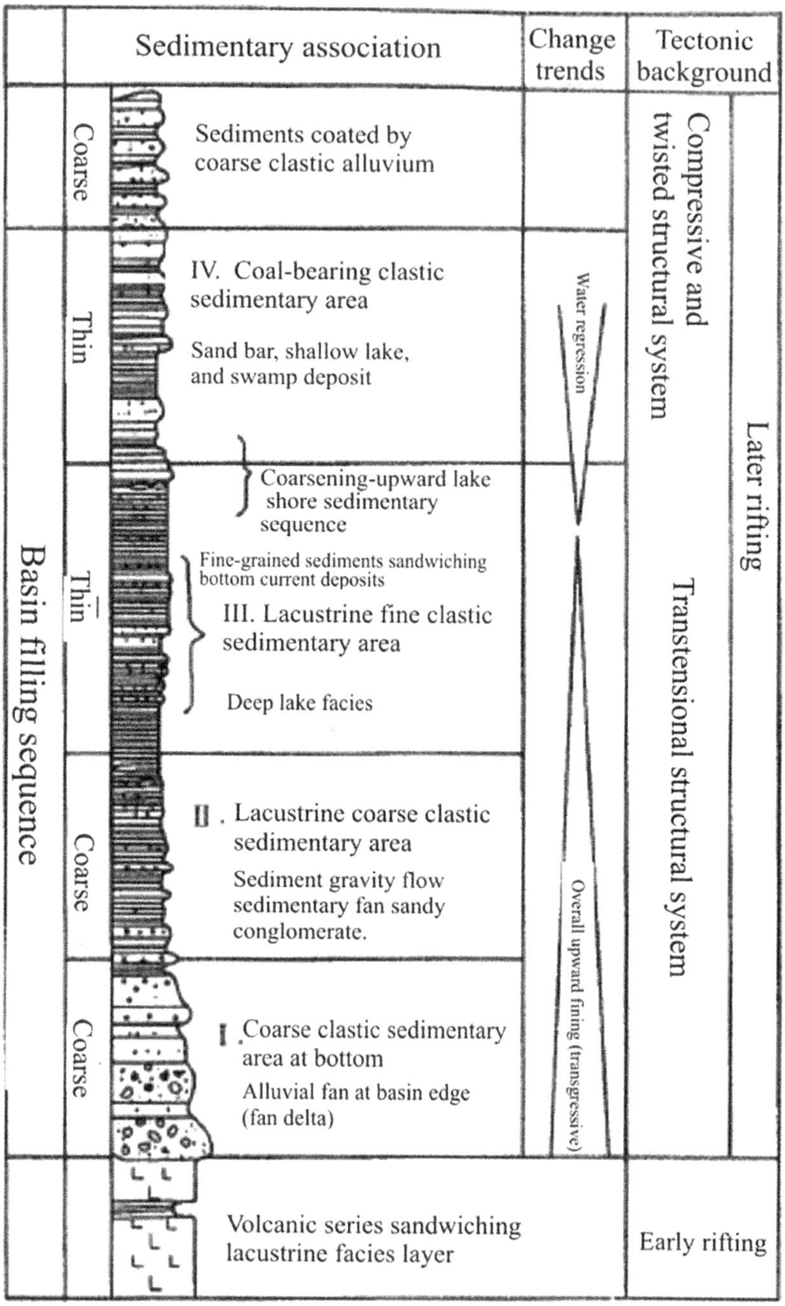

Fig. 10.19 Vertical sequence of Wuerxun depression in the Hailar Basin shows red-black-red, and coarse-fine-coarse from lower to upper. Lake water has the feature of shallow-deep-shallow, representing a complete tectonic cycle

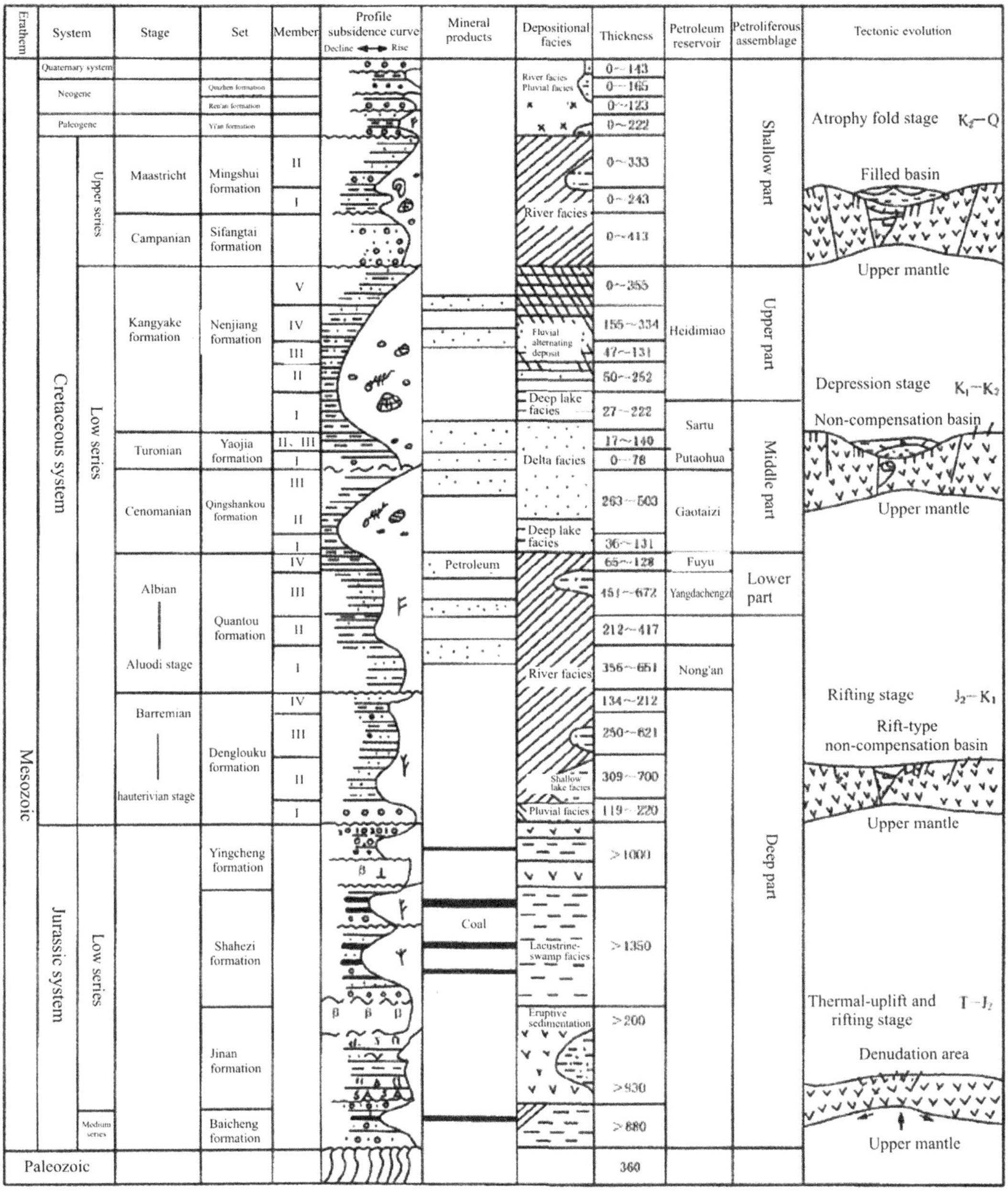

Fig. 10.20 Polycyclic infilling sedimentation can forms a large number of sand body types and multilayer oil-bearing systems (according to Wang et al. 1983)

Table 10.5 Sedimentary filling characteristics and determinants of continental facies faulted basins (according to Yu et al. 2007)

Influencing factors structural zone		Structural factors	Climatic conditions			Source supply		
			Arid	Semi-arid	Humid	Compensation type		Under-compensation type
						Sandy	Muddy	
Slope area	Steep slope belt (>10°)	Slope type	Gypsum salt, rock salt	Fan delta	Subaqueous fan	Alluvial fan, fan delta	Mud flow	Sand, gravel beach
		Single-order type	Alluvial fan	Fan delta, braided delta	Normal delta	Coarse-grained delta, fan delta	Mud flow, subaqueous fan	Sand, gravel beach
		Multi-order type	Gypsum salt, rock salt, alluvial fan	Normal delta, braided delta	Alluvial fan	Normal delta, coarse-grained delta	Subaqueous fan	Sand, gravel beach
	Flat slope belt (<10°)	Slope type	Salt flat	Sandy bench, swamp	Mud swamp	Salt flat	Mud swamp	Carbonate rock
		Single-order type	Slope fan	Normal delta	Mud swamp	Incised valley	Mud swamp	Carbonate rock
		Multi-order type	Salt swamp, alluvial plain	Normal delta	Alluvial fan	Normal delta	Fine-grained delta	Gypsum salt, acid salt rock
Depression area	Depression belt		Salt swamp, salt flat	Beach bar	Alluvial fan	Salt flat, beach bar	Salt swamp, mud, shale	Carbonate rock, swamp
Uplift area	Central uplift belt		Incised valley	Delta	Sublacustrine fan	Coarse-grained delta	Shore-shallow lake	Salt swamp, mud swamp

Table 10.6 Filling characteristics of the flat zone in each evolution phase of continental fault basin (according to Yu et al. 2007)

Zone period		Steep slope belt	Gentle slope belt	Depression belt
Uplift contraction period	Early stage	Nearshore subaqueous fan	Developed slope fan	Semi-deep lake mud (light color, with impurity)
	Medium term	Fan delta	Normal or pedate delta, beach bar	Shore-shallow or semi-deep lake
	Late stage	Dominated by alluvial fan	Some beach bars	Delta, alluvial plain
Sag extension period	Early stage	Nearshore subaqueous fan	Slope fan, distributary mouth bar–type delta	Dominated by semi-deep lake
	Medium term	Subaqueous fan, nearshore turbidite fan	Gilbert-type delta	Many source rocks and a considerable amount of oil shale in deep to semi-deep lake mud (dark color and pure quality)
	Late stage	Gilbert-type delta	Normal fine-grained alluvial-dominated delta, beach bar	Silty mudstone and argillaceous siltstone
Initial rift stage	Early stage	Dominated by alluvial fan	Shoreline and carbonate rock sedimentation	Shore-shallow lake or salt lake
	Medium term	Dominated by nearshore subaqueous fan	Miniature coarse-grained delta, sandy beach bar	Developed alluvial fan
	Late stage	Transgressive delta, far-shore turbidite fan	Fine-grained fluvial-dominated delta	Source rock formed in the center of depression; rock salt layer on the edge

1. **Initial rift stage**

In this period, sedimentary distribution patterns are complex and affected greatly by tectonic activity, climate, and provenance. Lake-filling sediments comprise volcanic eruptive rock and volcaniclastic rock, and sandwiching lacustrine mudstone, sometimes with poor coal seams in a few places (Wang Defa et al. 1991).

Although the terrain is steep, the supply of fragmentary materials is insufficient in the early formation times because the formation of fragmentary materials needs weathering erosion for some time or for a long time. Moreover, deep and large cracks incise into the basement in the early formation of rift basins, with the upwelling of mantle heat flow causing the local climate to be hot-arid for quite some time. In the late stage of the arid period, the climate gradually transforms to a damp one with an obvious deeper water body. In the meanwhile, the source supply is mainly along the short-axis direction. The author subdivided the source supply into the following three steps according to the determinants and sedimentary characteristics of sediments, and established corresponding filling models (Fig. 10.21).

(1) Early-rift period

This is equivalent to the early sedimentary stage of the Kongdian formation in the Jiyang depression. In this period, the basin starts rifting and there is subsidence with dry weather, full sunshine, increasing salinity, and no large injection of debris, especially in the gentle slope belt.

The steep slope belt mainly forms an alluvial fan, whereas the gentle slope belt forms shoreline and carbonate deposits because the water level rises gradually and the water body begins to become

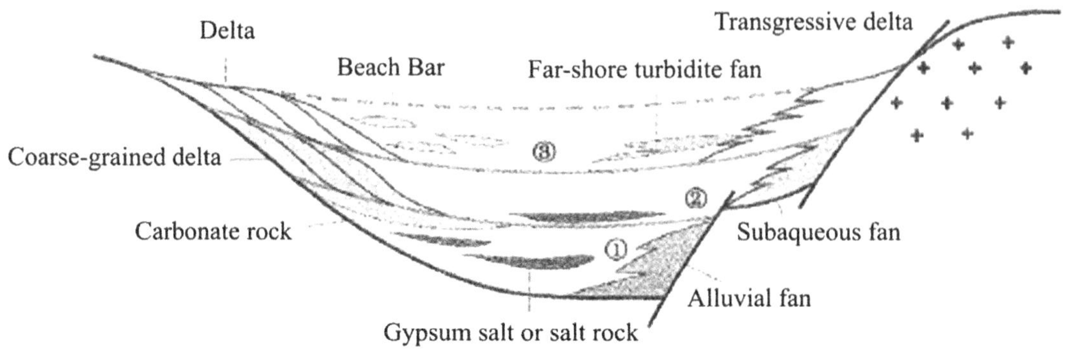

Fig. 10.21 According to control factors of sediment and characteristics of sediments, the sediments can be divided into three stages of initial rifting (①–③)

deeper, which is good for the formation of carbonate rocks. Although sediment supply is inadequate in the depression area, full sunshine on the water body is sufficient. Hence, the environment is mainly of the shore-shallow lake type or even salt flat or salt lake sedimentation sometimes, which leads to the formation of a gypsum salt layer.

(2) Middle-rift period

This is equivalent to the late period of the Kongdian formation deposition. In this period, the tectonic subsidence rate increases significantly, climate changes from arid to semi-arid, and water body deepens continually with an increasingly wider scope. Moreover, sediment accumulation increases and a substantial amount of terrigenous clastics are injected into the basin.

Nearshore subaqueous fans are formed mainly on the steep slope belt. Turbidite fans can be formed in the depression area when the terrain slope increases and further forms a fault terrace, while deep rock salt layers can be formed when the source supply is inadequate and the terrain is gentle. Miniature coarse-grained (fan or braided) deltas can be formed on the gentle slope belt, and numerous sandy beach bar sediments can be formed under the effect of waves and longshore currents.

(3) Late-rift period

This is equivalent to the late period of the S4 member. In this period, structural subsidence is higher, sometimes accompanied by strong eruption of alkali basalt and characterized by a semi-arid climate, which is transformed to a humid one in the later period with a deeper water body, thus expanding the range and decreasing the sediment supply.

Transgressive fan deltas are formed mainly on the steep slope belt of a basin. Although the terrain slope on the delta front is not steep, a few far-shore turbidite fans can be formed owing to sufficient source supply, gravity currents, effects of turbidity channels, and, sometimes, the presence of alkali basalt around the fault zone. Normal fine-grained fluvial-dominated deltas or braided deltas can be developed on the gentle slope belt, and a few beach bars can be formed along the front edge because of gentler topographic slopes, which increase the range of sand bodies and the effects of waves. Good source rock, namely, a thick dark mudstone layer, is formed in the depression area because of insufficient supply of source and rich organic matter, whereas a salt rock layer is formed in the basin front owing to the arid climate.

In short, the association patterns of different facies can be seen in a Paleogene lacustrine basin in the early depression. The tectonism of a few lakes is stronger in their initial rift, which results in obvious terrain elevation differences, thus creating conditions for the formation of coarse-grained sediments. Turbidite fans, subaqueous fans, and fan deltas are distributed along the lake edge, and shallow lake facies sandy

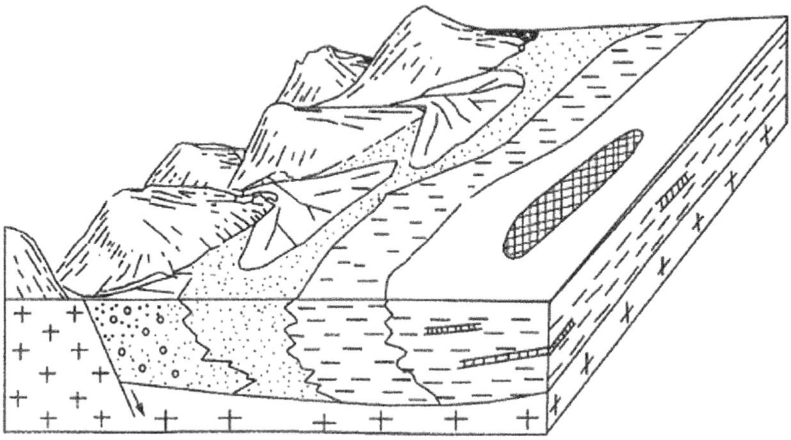

Fig. 10.22 There are turbidite fan, underwater fan and fan dalta in the lake margin, and shallow lacustrine facies, sand, mudstone deposits or gypsum saline Lake deposits would appear in the basin-ward direction

mudstone or gypsum salt lake sedimentation can be found along the basin direction (Fig. 10.22). The tectonic activity of a few faulted lacustrine basins (Dongpu Depression) in their early depression periods is weak, and terrain fluctuations are smaller. Lacustrine basins exist in relatively well-oxygenated shallow lake environments, thus forming shallow sand bodies with large areas.

2. **Sag extension period**

The sag extension period of a faulted lacustrine basin mostly coincides with the early and medium phases of fault basin formation, which are the results of the continuous activity of basin-controlling deep fractures. In this case, the subsidence of basin is the deepest and topographic relief is large. Most of the deepest lake areas are located in the depression at the downthrown side of the boundary fault, where the position is stable, duration is long, displacement is not large, and the sedimentary center is basically consistent with the center of subsidence. The source bed here is very thick, with a thickness of up to over thousands of meters, owing to its non-compensatory (under-compensatory) sedimentary status and the sedimentation of dark mud shale with massive thickness. Kerogen, which is mostly of the transitional and sapropel types, is the main sedimentary period of the source bed in a fault basin. Although the area of a single lake is not large, the percentage of lake area in the depression area and that of the deep lake area in the lake area are relatively high. For instance, in the Paleogene fault basin in eastern China, the percentage lake area is 60–90%. There are many depressions and lakes in one basin, and their cumulative area is quite large; the thickness is high, usually over thousands of meters, so petroleum reserves are abundant, such as the lower-middle part of the S3 member in the Bohai Bay basin.

Owing to strong block faulting and steep topographic slopes, shore-shallow lake subfacies are not very narrow or well developed. According to the different tectonic characteristics of different positions in the basin, sediments can be divided into three systems, namely, transverse steep slope system, transverse gentle slope system, and longitudinal system (Fig. 10.23).

The author established the sedimentary fill mode for the sag extension period of a faulted basin according to the depositional features of the period (Fig. 10.24).

This evolutionary phase is mainly caused by the continuous activities of basin-controlling deep fractures, and it results in the formation of depressions in lacustrine basins. In addition, the climate would have changed from semiarid to humid, thus resulting in a rapid rise in the lake water level and deepening of the water level. The tectonic characteristics of graben-like depressions lead to the formation of the deep lake area, which is located mainly in the deep depression at the downthrown side of the boundary fault with stable

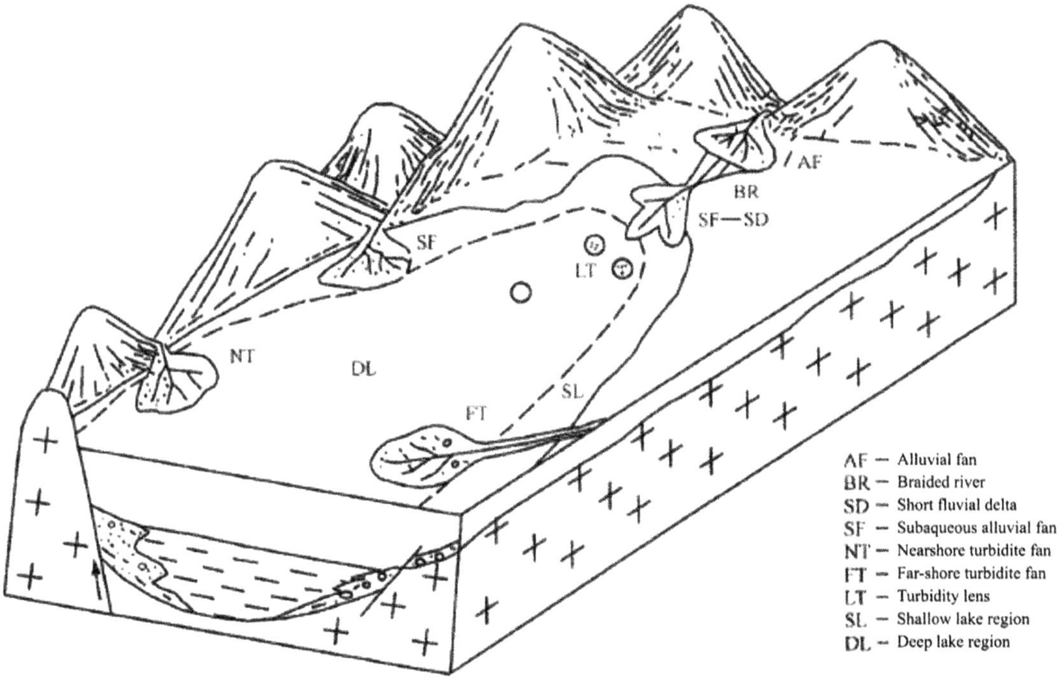

Fig. 10.23 According to the different positions of the basin with different structural characteristics, the sediments can be divided into three systems, i.e. transverse steep slope system, transverse gentle slope system and vertical system (according to Chongjun 1986)

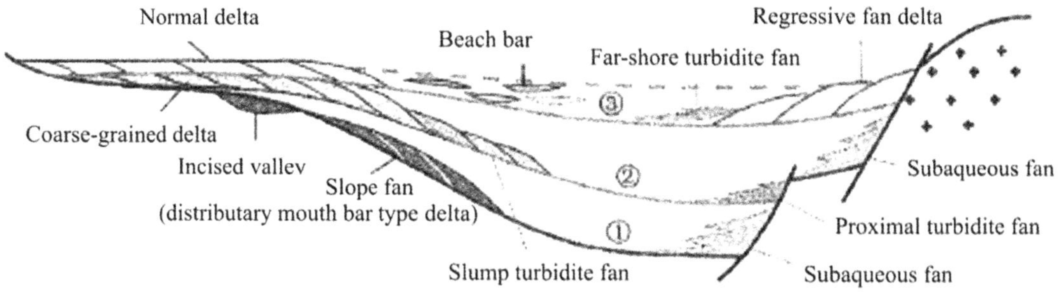

Fig. 10.24 According to the continuous activity, the basin-dominant deep large faults can be divided into three stages of deep depression expansion, i.e. from ① to ③

position and long duration. The area of the deep to semi-deep lake is large, usually accounting for 50–70% of the entire sag. Owing to the strong block faulting and a steep topographic slope, deepwater sedimentation is relatively developed, but the shore-shallow lake subfacies are not well developed. According to the determinants and sedimentary characteristics of sediments, the following three stages can be distinguished.

(1) Early extension

This is equivalent to the early deposition stage of the S3 Member, when the sedimentation activities are strong, climate is humid, and lacustrine basin water is relatively deep and extensive.

Along the edge of the steep slope zone, nearshore subaqueous fans are formed; slope fans or far-shore turbidite fans are formed on the gentle slope belt, where an incised valley can be developed; and semi-deep lake mudstone deposits are developed in low-lying areas. Owing to an insufficient source supply and the presence of humidity, semi-deep lake mudstone is mostly characterized by a dark color and pure substances.

(2) Medium extension

This is equivalent to the middle and late periods of the deposition of the S3 member. This period is the heyday for basin development, and tectonic activity in this period is very strong.

Subaqueous fans are formed mainly in the steep slope zone. When late fault activity bifurcates with that of earlier stages or there is a fault terrace, and offshore turbidite fans are mostly formed at the front end of subaqueous fans or fan deltas. On the gentle slope belt, coarse-grained to Gilbert-type deltas are formed, and they feature a clear three-layer structure. When the topographic slope breaks at the front end of a delta, or the topographic slope turns steep again toward the basin, a few turbidite fans can be formed under slump gravity flow. Because the water body is the most extensive and deepest in this period, the resulting rock is dark in color and constituted of pure substances. Moreover, the basin center is obviously in starvation, thus the depression area is favorable for the formation of source rock and oil shale.

(3) Late extension

This is equivalent to the early deposition stage of the S2 member, when the sag caused by tectonic activities has stopped, accommodation decreases gradually due to the filling of sediments, and the climate turns from humid to warm humid.

With the advance of sediments, a regressive fan delta is developed in the steep slope zone. Similar to the late initial depression, far-shore turbidite fans can sometimes develop at the front end, and normal fine-grained fluvial-dominated deltas are formed mainly on the gentle slope belt. Under the combined action of waves and longshore currents, the distributary mouth bar at the front end can be transformed into multiple beach bars. Owing to warming of the climate, the amount of wind and sand increase, water body shallowing occurs, and silty mudstone and argillaceous siltstone are mainly formed in the depression zone, generally accompanied by thin pure mudstone.

Turbidite sand bodies are among the most featured parts of the sand bodies developed during the sag extension period in a fault basin. Oil and gas can be seen in various turbidite sand bodies, in which coarse-grained turbidite sand bodies are the richest in oil. For example, the far-shore turbidite fan in the Liaoxi sag and the nearshore turbidite fan in the Zhanhua and Miyang sags are distributed on a short-axis gentle slope and a steep slope, respectively. Subaqueous fans or nearshore turbidite fans dominated by debris flow and grain flow are mostly developed under faulted terraces on a steep slope, while far-shore turbidite fans are superposed in layers and distributed in the lobate form under a rupture in the basin on a gentle slope side. When the slope is very steep and close to the delta front, slump turbidite fans dominated by fluidized flow are formed. In addition, subaqueous alluvial fan sand bodies may be formed in shallow places. Substantial changes occur on the long-axis side. Some have no sand body, and others have various types of sand bodies depending on the specific provenance and tectonic activity (Fig. 10.25).

4. **Uplift contraction period**

As tectonic activity stabilizes, the nature of a lacustrine basin turns from fault to depression, and the lakeside environment begins to develop. The commonest shoreline depositions are deltas, fan deltas, and beach bars.

Through sedimentary filling and basin uplift in the sag extension period, the lake terrain changes. Later, the lakebed becomes flat, the gradient of the original barranca belt decreases, the lake becomes shallower, and the deep lake area decreases or even disappears. The

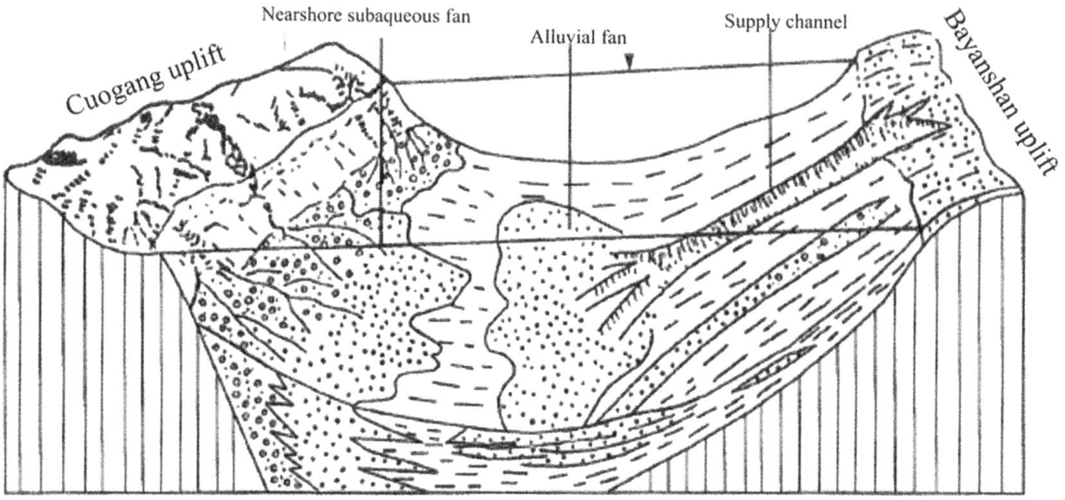

Fig. 10.25 The sand bodies of rift lake are characterized by turbidite during the deep extension, and the underwater fan or near shore turbidite fan with debris flow and particle flow mainly developed under steep slope fault terrace; But under the slope of the gentle slope, it forms a layered superimposition, fan like far shore turbidite fan. When the gradient is very steep and close to the delta front, forming a liquefied flow dominated slump turbidite fan

sedimentary and subsidence centers gradually move away from the barranca belt, and the proportion of shore-shallow lake facies increases. In this period, the sand body is developed, and it is of the shallow-water type (Fig. 10.26). This is the main reservoir development phase for faulted lakes. In addition, the sand body is directly located above the underlying source bed, and this layer has oil sources. Hence, it is suitable for of oil and gas accumulation.

As for the sedimentary-filling mode of faulted lacustrine basins during the uplift contraction period, see Fig. 10.27, this indicates that the basin fault activities in this period have come to an end. Soon after, the basin is uplifted as a whole, and the lacustrine basin clearly changes from a fault to a depression, or is transitional. The basin water body is relatively shallow, climate is warm, terrain elevation difference decreases apparently, and the weathering and denudation rate decreases apparently. As a result, the source supply is relatively scarce. According to the determinants and characteristics of sediments, the following three depositional phases can be formulated.

(1) Early contraction

This is equivalent to the late period of Member S2. In this phase, because sag extension has just been completed, when the sediment supply is insufficient, the basin starts accepting sediments again.

In this phase, the sedimentary characteristics of gentle and steep slope belts are similar to those of the early extension. The difference is that mudstones in the lake mainly feature a light color and are composed of mixed substances, and no evaporite is developed because of the differences in source supply and climate warming.

(2) Medium contraction

This is equivalent to the sand-sedimentation period. In this phase, short-term regional settlement with a small range, but large scope occurs in terms of tectonic movement. The climate is warm, debris injection decreases, and organic matter is abundant.

The joint normal delta–beach bar depositional system is developed on the gentle slope belt of a

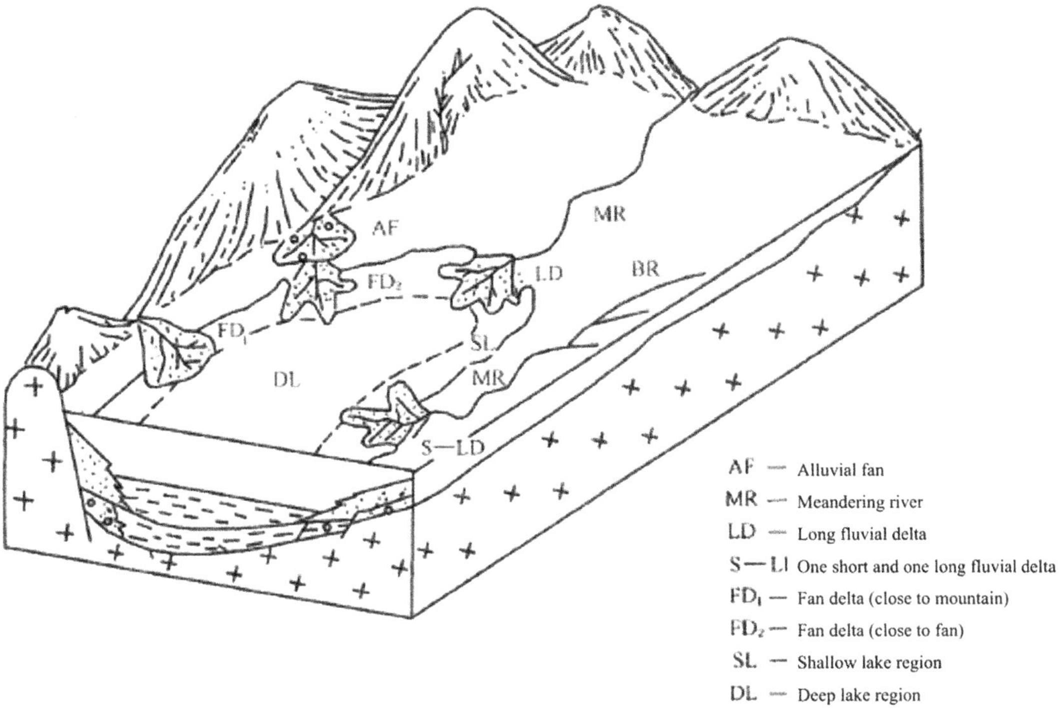

Fig. 10.26 The topography of the lake changed obviously after the filling and uplift of the basin during the deep enlargement period. The bottom of lake changes to flat, and the slope of the steep lake bank decrease, and the lake became shallower, the deep lake zone decrease and even disappear. The center of deposition and subsidence is gradually far away from the steep shore zone, with increase of the proportion of coarsal-shallow lake facies. During this period, sand bodies developed and characterized by shallow sand bodies (according to Chongjun 1983)

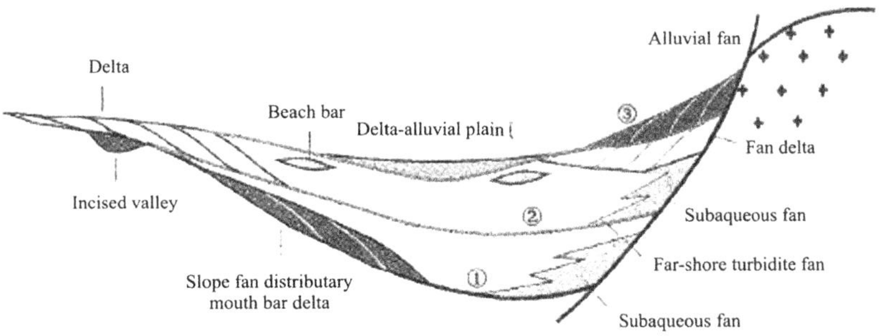

Fig. 10.27 At this time, the basin fault subsidence activity is at the end, then the basin is uplifted, and the lake basin changes from fault to depression, suggesting that the basin is in the transition period. Based on the control factors of sediments and characteristics of sediments, the sediments can be divided into three stages of ascending systole, i.e. from ① to ③

basin, while subaqueous fans transform into fan deltas on the steep slope belt, which reflect that the color of the water body turns from dark to light. In addition, shallow to semi-deep lake deposits dominated by oil shale and dark mudstone are formed in the depression area.

(3) Late contraction

This is equivalent to the sedimentation period of the Dongying formation. Because the structure is uplifted again, the climate is dry, the lacustrine basin is in a large contraction area, and lake water becomes shallow gradually.

On the basin edge of the steep slope belt, because the water body decreases and the water surface is exposed, which is dominated by alluvial fan sedimentation, a few beach bars are developed on the gentle scope belt. The depression center is dominated by delta-alluvial plain facies sedimentation. As the water body continues to decrease, alluvial plain sedimentation occurs parallel to the long axis, especially fluvial facies sedimentation. In this period, the main flow direction of the river changes clearly through basin evolution (Table 10.7).

10.3.2.2 Depression Lake

Regarding Mesozoic large depression lacustrine basins, for instance, the Songliao Basin, which features a large area and flat terrain, changes in depositional facies, lithology, and thickness are gentle; sand type is relatively simple, mainly nearshore shallow water sand body, and the river provides sediments, mainly siltstone and fine sandstone, along the long-axis direction. The lacustrine basin depocenter is located in the central lacustrine basin, which is consistent with the center of subsidence.

1. **Extension period**

In this period, the deep lake area is large, but lake water is not always very deep. Shore-shallow lake facies are relatively narrow and distributed in a ring-shaped form around the deep lake area. Source rock is distributed widely and is superior in quality. The main sand body types are fan deltas and deltas, and bedded turbidite can be distributed in the middle of the lake. In contrast to sag ponds, the slope at the edge of the depression lake is very gentle in the period of sag extension. Hence, turbidite sand bodies are underdeveloped, meaning only small turbidite sand bodies are formed by some small marginal sand bodies, and there are not many shallow water sand bodies either. A subaqueous alluvial fan-fan delta may form on the steep slope along the short-axis direction. Circumstances change considerably on the short-axis gentle slope and along the direction of the long axis. In the presence of abundant detrital origin, the deposit velocity is higher than the rising speed of the lake surface, and various shallow offshore sand bodies or even small deltas may be developed; if not, some mud flat sedimentation activity will occur (Fig. 10.28).

A typical example is Member 1 of the underground Cretaceous Qingshankou formation in the Songliao Basin. This member was deposited during the largest sag extension period of the Songliao Basin. The basin area is approximately 26×10^4 km^2, and the lake area is 8.4×10^4 km^2, of which the deep lake area accounts for 80%, with an overall deposition thickness of 500 m. The lake water depth is usually 30 m, and it peaks at 60 m. The generative kerogen consists of lake creatures, and the kerogen is mainly of the sapropel type with huge petroleum potential.

2. **Uplift contraction period**

In this period, the terrain is flat, lake water is not deep, and nearshore shallow water sand body is well developed. In particular, normal river-delta sand bodies along the long-axis direction are most commonly found, and the delta front is formed by the superimposition of delta subaqueous distributary channels. A beach bar sand body can be developed at the edge of the lake, which is an area lacking a stable source supply.

A typical example is the Daqing Changyuan Delta complex developed in the sedimentary period of Members 2 and 3 of the underground Cretaceous Qingshankou Formation in the Songliao Basin, with an area of 2×10^4 km^2, a thickness of 500 m, and an accumulated cycle number of 38–40. The delta developed at the southern long-axis end, while the Yingtai fan delta developed on the short-axis steep slope in the west. Most of the eastern short-axis gentle slopes are shore-shallow lake mud flats, and there

Table 10.7 Meso-Cenozoic evolutionary characteristics of faulted lacustrine basins in the Bohai Bay

Period		Evolution characteristics	Sedimentary characteristics
Uplift contraction period	Late	Since the structure is uplifted again, the climate is dry, lacustrine basin is in a large contraction area, and lake water turns shallow gradually	Steep slope area is dominated by alluvial plains, and depression center is dominated by delta alluvial plains
	Medium	Short-term regional settlement with small range but large scope occurs. The climate is warm, debris injection is minimal, and organic matter is abundant.	Steep slope transfers from subaqueous fan to delta; the sedimentary association of a normal delta-beach bar is developed on the gentle slope; and shallow lake to semi-deep lake deposition dominated by oil shale and dark mudstone occurs in the depression area
	Early	Sediment supply is insufficient, climate is warm, and basin starts to receive sediments again	Steep and gentle slopes are similar to those in the early extension, and light-colored impure mudstone exists in the depression area
Sag extension period	Late	Sag stops, accommodation decreases, and climate is warm humid	Regressive delta is developed on the steep slope; normal fluvial-dominated deltas and beach bars are developed on the gentle slope; and silty mudstone and argillaceous siltstone are formed in the depression area
	Medium	Tectonic activity is very strong, and basin development continues to peak	Subaqueous fan is formed on the steep slope, and nearshore turbidite fan is developed in the late period; coarse-grained delta to Gilbert-type delta is formed on the gentle slope; and favorable source rock and oil shale are formed in the depression area
	Early	Sedimentation activities are strong, climate is humid, and lacustrine basin water is deep	On the steep slope, nearshore subaqueous fan is formed; on the gentle slope, slope fan or far-shore turbidite fan is formed, where an incised valley is developed, and dark-colored semi-deep lake mudstone composed of and pure substances is developed in the depression area
Initial rift stage	Late	Larger subsidence, accompanied with the eruption of alkali basalt; semi-arid or humid climate, deeper water body, and decreased sediment supply	Transgressive deltas and far-shore turbidite fans are mainly formed on the steep slope; fine-grained fluvial-dominated deltas and braid deltas are formed on the gentle slope; good source rock is formed in the depression area; and salt is formed at the basin edge
	Medium	Large subsidence, arid–semi-arid climate, deeper water body, increased accommodation, and sufficient terrigenous detrital	Nearshore subaqueous fan is formed on the steep bank; a small group of fine-grained deltas or sandy beach bars sedimentation is developed on the gentle bank. In the presence of a steep slope in the depression area, turbidite fans are formed; in the case of a gentle slope, deepwater salt formation takes place
	Early	Basin sinks and settles, climate is dry, salinity increases, and there is no large debris injection	Alluvial fan is formed on the steep slope, shoreline or carbonate rock is deposited on gentle slope, and shallow lake or gypsum rock is deposited in the depression area

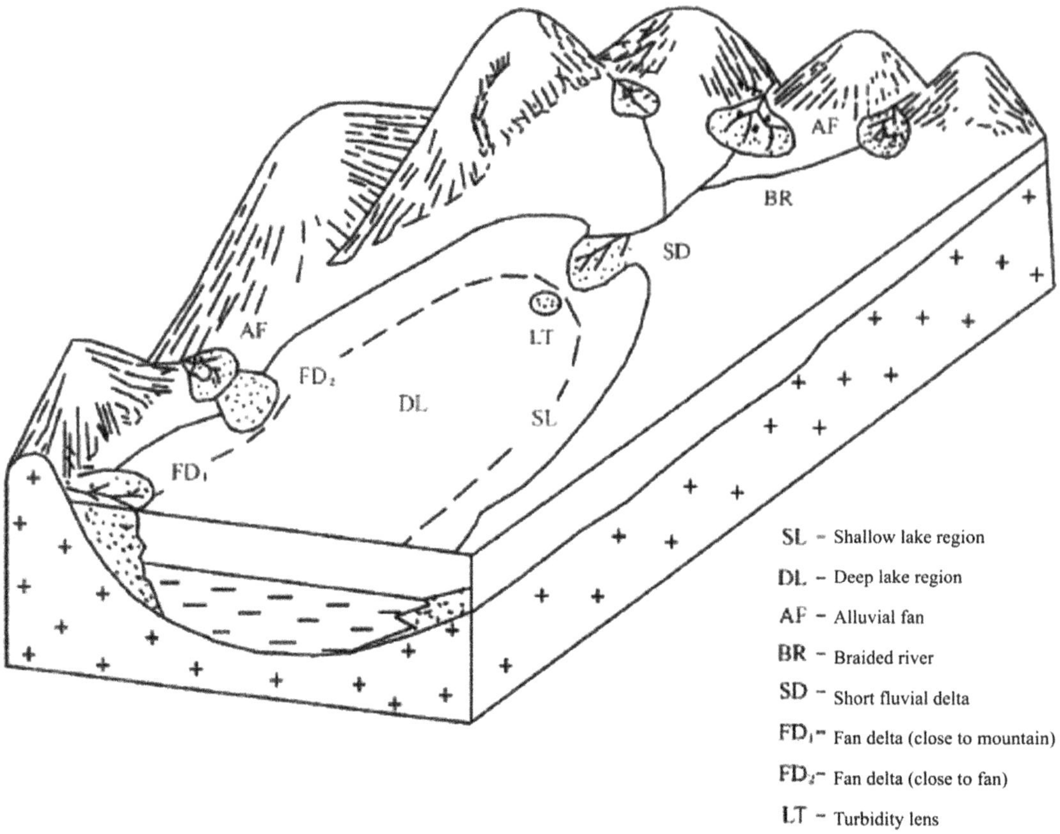

Fig. 10.28 The coastal-shallow lake facies belt is narrower and distributed around the deep lake district as ring. The main sand body is fan dalta and dalta, and there is stratiform turbidite in the centre of lake. Small slump turbidite sand body occurs in the margin slope (according to Chongjun 1983)

is a delta sand body in the north (Fig. 10.29). The oil fields found presently are mainly concentrated in the Daqing Changhuan Delta, followed by the Yingtai sand body on the west coast. The Baokang sand body has little oil because its depocenter is located in the north, which is west of the basin.

Most of the Meso-Cenozoic sedimentary basins in China have undergone the process from fault depression to sub-depression, and the lake types in these basins have evolved thereupon. For example, in the Jurassic period, the Songliao Basin comprised a few small, shallow fault depressions, as the swamp was developing into a lake. There was little oil and coal-formed gas. In the Cretaceous, it sunk by a large margin on the whole, forming a down-warped basin, which developed a large, deep lake with large amounts of oil and gas. For example, the Bohai Bay Basin is a deep faulted basin in the Eogene, in which period, many deep lakes developed and large amounts of oil and gas generated. However, in the Neogene, the basin was transformed into a flat down-warped basin. A river alluvial plain was deposited therein, and the region's oil generation capacity weakened. Therefore, both faulted lacustrine basins and down-warped lacustrine basins can generate large amounts of oil and gas. In specific applications, a problem-specific comprehensive analysis is required.

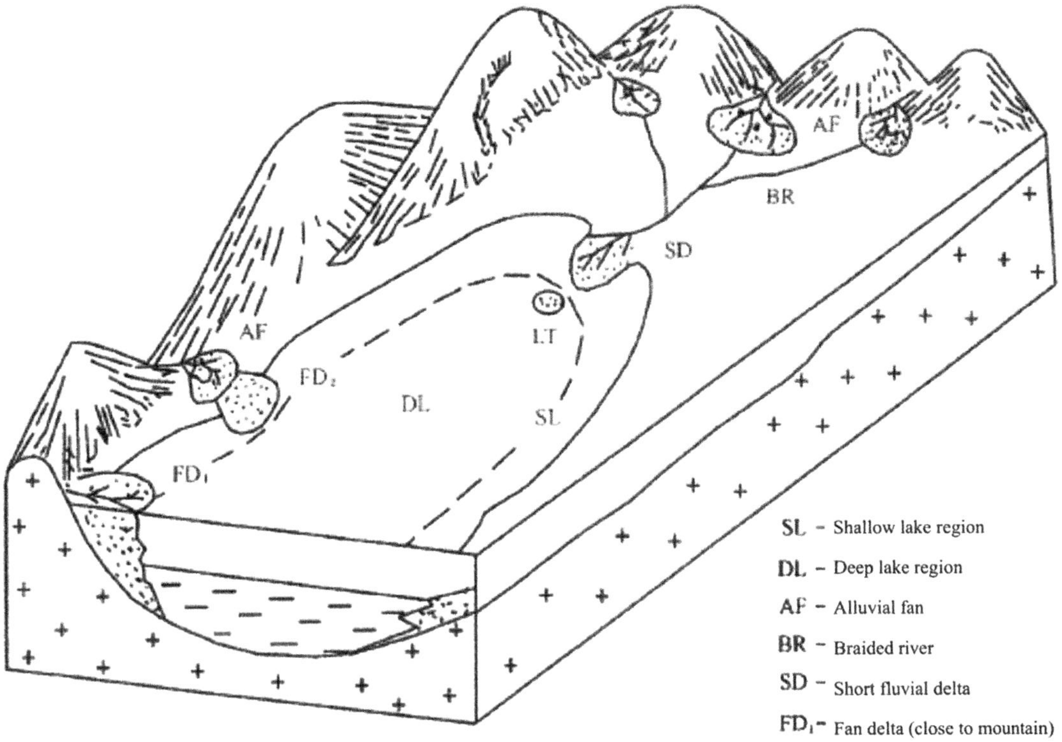

Fig. 10.29 Dalta occurs in the south end of the shaft, and the Yingtai fan dalta occurs in the west short axis steep slope. The eastern part of the minor axis of the gentle slope is coastal-shallow lake mud beach, and dalta sand body occurs near the north (according to Chongjun 1982)

10.4 Identification of Lacustrine Deposit

The terrigenous clastic lake is the most common lake type, and its ideal lacustrine depositional model shows annular distribution on the plane. From edge to center, the granularity of sediments changes from coarse to fine, and the ideal distribution series is as follows: lakeshore gravel outer belt → sandy sedimentary belt → silty and argillo-calcareous inner belt → lake center ooze sedimentary belt. However, each belt is not always continuous, and their distribution is irregular as well. The ideal mode corresponds roughly to the changes in the hydrodynamic conditions of lake water, that is, wave belt → upper belt of wave base → lower belt of wave base. However, lacustrine deposit can be seen mainly as gravel on the steep slope, sand on the gentle bank, and mud in the bay.

The actual deposits in the lake are much more complex than those corresponding to the ideal mode. For instance, owing to the effects of river flow into the lake, a delta sand body may be formed at the lake edge. When the delta front presents obvious subaqueous topographic changes (rupture in the basin), turbidity sediments of coarse-grained materials are developed in the deep lake area. Briefly, various factors can destroy the regularity of annular sediment distribution. If complex factors, such as flow of a river into the lake, are considered, the sediment distribution pattern from the land to lake center is as follows: lacustrine delta deposition (or lake-bay sedimentation) → lakeside deposition → shallow lake deposition → deep lacustrine or turbidity deposition.

The terrigenous clastic lake facies differ from other types of depositional facies, which are often distinguished by the following indicators.

10.4.1 Rock Type

Rock type is dominated by clay rock, sandstone, and siltstone, and conglomerate is rarely seen, and these are distributed only in the lake shore area with escarpment, mainly due to denudation of the surf (Table 10.8). The sandstone is generally more complex than sandstone of marine origin, and all types of sandstone can be seen. Compared with river sedimentary facies, its mineral maturity is high, and the quartz content can reach more than 70%. Arkose, feldspathic quartz sandstone, and lithic arkose sandstone are distributed mostly in the Meso-Cenozoic lacustrine sedimentary sandstone in eastern China. The granularity of this sandstone is finer than that in river facies, and its sorting is relatively good, which presents difficulties in distinguishing it from marine facies because its granularity probability curve is also similar to that of marine-origin sandstone.

Clay rock, which is distributed widely in debris lacustrine deposit, increases in quantity from the lakeshore to the center. Lacustrine clay rock formed in a reducing environment in relatively deep waters is usually rich in organic matter and forms a good source rock system. The source rock system in China's oil and gas fields is mostly clay rock of lacustrine origin.

Various chemical rocks and biochemigenic rocks may be formed in debris lacustrine deposits, for instance, limestone, muddy limestone, diatom earth, and oil shale, for which the sedimentary thickness and distribution range are relatively limited.

10.4.2 Sedimentary Structure

Bedding types are varied, but horizontal bedding is the most developed type (Table 10.8). Because

Table 10.8 Essential characteristics of lacustrine deposit (according to Chongjun 1982)

Facies	Rock	Sedimentary structure	Biologic fossils
Shore lake facies	Dominated by sandstone and siltstone; conglomerate rarely seen; good debris sorting and roundness, with biodetritus or oolitic limestone	Miniature wave ripple cross bedding, wavy bedding, monoclinic cross bedding, variant cross bedding, mud cracks, raindrop imprints, hailstone imprints, worm trails, and scour structures are mainly developed	Bivalves, fish, gastropods, and ostracoda
Shallow lake facies	Dominated by siltstone and mudstone; a few lenticular fine sandstones, with fuchsia microcrystalline limestone or dolomite	Irregular horizontal bedding, miniature cross bedding, wavy bedding, and wave ripples can be seen	Fossils are abundant and well preserved, dominated by thin-shell gastropods, bivalves, ostracoda, conchostraca, fish, and phytolith
Deep lake facies	Dark gray–grayish black argillaceous rock, shale, marlstone, and oil shale; high organic content, sometimes with gravity flow deposit	Minute horizontal bedding and seasonal lamination can be seen at times	Fish, ostracoda, bivalves, gastropods, etc.

the lake scope is limited, wave base depth is small, and a vast area of lake is located under the wave base, the clay rock mostly develops horizontal bedding or, occasionally, massive bedding. Megascopic cross bedding and wave ripple cross bedding can be seen in the inshore region.

In addition, relatively developed ripples may be formed in lacustrine deposits. It was believed that para ripples are an indicator for distinguishing lakes from rivers, but according to Picard et al., ripple symmetry is not unique to lakes. Moreover, lakes also develop asymmetric ripples, in which the wave crest direction is mostly parallel to the shoreline, and the steep slope of the asymmetric ripples is aligned with the bank direction. Also seen commonly are mud cracks, raindrop imprints, and mixed structures.

10.4.3 Biologic Fossils

An abundance of biologic fossils is an important feature of clastic lacustrine deposits. The common biological species include ostracoda, lamellibranch, and gastropods, in addition to algae, which are usually developed in lakes (Table 10.4). Stoneworts are specific to freshwater environments, and blue-green algae, diatoms, and a few green algae are commonly seen as well. The blue-green algae in this case are different from the laminated structure seen in marine facies, often as arborization or separated-nodule massive structures. Red algae are never seen in lacustrine facies. In addition, an abundance of roots, stems, leaves, spores, and pollens of terrestrial plants are also important features of lacustrine facies. Although phytolith occurs in marine facies, lacustrine facies can be identified by the gradient variation of decreasing species and amount far from the shoreline.

Large amounts of biological fossils can be found in Meso-Cenozoic clastic lake deposits in eastern China. For instance, the fossils of ostracoda, gastropods, stoneworts, spores, and pollen are abundant in the Palaeogene lacustrine mudstone and shale of the Jiyang depression, which is an important indicator of stratigraphic division and comparison, as well as an identifier of sedimentary facies.

10.4.4 Physical Process and Chemical Features

Their physical process is usually similar to that of the facies seen in the ocean basin, but lacustrine sediments lack the evidence of tidal currents. Compared with marine littoral sediments, the wave activity of shore lake sediments is considerably lower. As for composition, the maturity of shore-land deposits is low, and there is little quartz sandstone. Lake terraces are very common, which reflect a frequent or even annual rise and fall of the lake surface. In consideration of the long-term frequent fluctuations of lake surfaces, various shore-land facies intersect with each other rapidly, because of which single sedimentary bodies or facies, such as lake shores, are not very thick. In abyssal regions or in a lake with salinity lower than that of normal sea water, gravity flow can generate a nearly continuous density current flowing down toward the delta front and form turbidites characterized by semi-developed Bouma sequences, compared with the marine environment.

As for changes in the chemical components and salinity of lake water over the entire geological time, both speed and quantity are many times larger than those in the sea, and all these changes are recorded in sediments, with varve being a particularly common type of sediment in this regard.

10.4.5 Vertical Sedimentary Characteristics

Clastic lacustrine deposits mostly present the coarsening-upward sequence of a deep lake to shore-shallow lake, which is distinguished from the fluvial facies sedimentation of the intermittent normal fining-upward cycle. The sedimentary structural characteristics are dominated by ripples or wave cross bedding, granularity is dominated by coarse and fine sand, and sedimentary layer thickness is relatively stable.

10.4.6 Distribution Range and Deposition Thickness

The distribution range of lacustrine deposits is larger than that of river facies but smaller than that of marine facies. The facies belt, lithology, and thickness are distributed roughly in an annular pattern. Moreover, transverse changes in lithology and thickness are more stable than those of river facies, but the degree of stability is poorer than that of marine facies.

10.5 Types of Lacustrine Sand Bodies

Petroleum generated in lakes accumulates preferentially in lacustrine sand bodies. The vast majority of reservoirs in the Meso-Cenozoic oil and gas fields in China are sand bodies of different types deposited in the lake. Broadly, there are many types of sand bodies, and all sand bodies deposited in a lake can be called lacustrine sand bodies.

Given that the edge of a lake is close to terrigenous detrital provenance area, sand bodies in lakes are rather well developed and their proportion is much higher than that of marine strata. Sand bodies are distributed from the lakeshore to the lake center. These sandy sediments often constitute very good hydrocarbon reservoir sand bodies. Lacustrine sand bodies are obviously controlled by lacustrine facies belts. However, as for sand bodies at different places, their distribution location, shape and size, lithology, physical properties, relationship with adjacent lithology, and petroleum source-reservoir-cap configuration differ considerably owing to differences in lake bottom slope, water depth, distance from the provenance, hydrodynamic conditions, and formation mechanism. In addition, their petroleum source-reservoir-cap differs in quality, which influences the abundance of oil. Therefore, the focus should entirely be on the basin, classification of sand bodies should be based on their location in lakes, and the geographical conditions of sand bodies should be emphasized. First, shore sedimentation should be distinguished from sedimentation in the lake. In the latter case, the type of lacustrine subfacies, whether in shore-shallow lake or semi-deep to deep lake, or in shallow water sand body or deep water sand body, should be known. Then, it is necessary to subdivide sand body types according to the actual conditions such as the distance from the bank, central lake, distance between the river mouth and provenance area, as well as transgression and regression of lake water. This division can not only reflect both the hydrodynamic conditions and the formation mechanisms of sand bodies but also help identify the sedimentary characteristics of various sand bodies. Moreover, it is easy to predict the distribution of various sand bodies and the advantages and disadvantages of source-reservoir-cap association based on the internal architecture of basins and lacustrine subfacies. Meanwhile, it is also easy to find other types of sand bodies from a given one or from a type of sand body based on the facies association and combination law in order to improve predictability.

In consideration of the above classification principles, when studying sand bodies, both the depositional features of the sand body itself and the features of the surrounding rock (mostly as argillaceous rock, better with special rock type) should be researched to determine the location of sand body deposition or the position of the depositional environment. In fact, such a

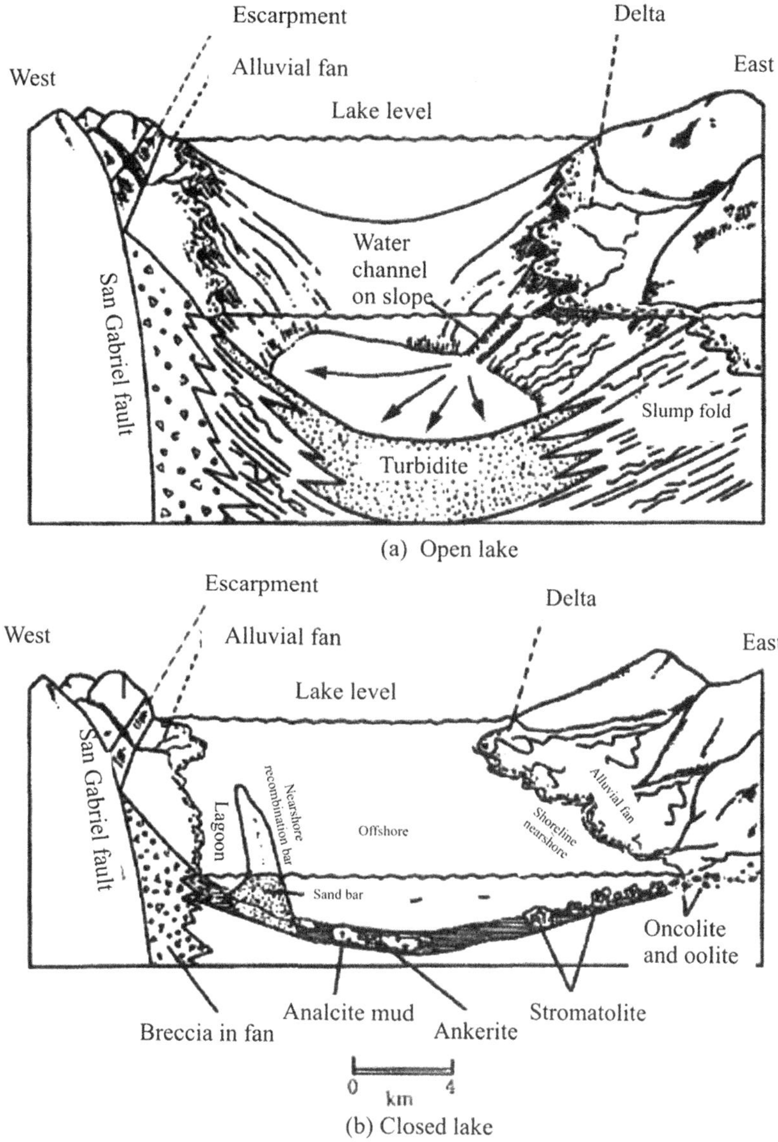

Fig. 10.30 Lacustrine sandbody can be divided into dalta, fan dalta, beach bar, underwater fan and turbidite sand body, and the former three types are nearshore shallow water sand body and the latter two types are deep water sand body (according to Link and Osborne 1978)

comprehensive analysis of depositional facies using the principles of paragenetic association of facies is more efficient than researching sand bodies discretely.

According to the above principle of classification, lacustrine sand bodies are categorized as deltas, fan deltas, beach bars, subaqueous fans, and turbidite sand bodies. The former three are nearshore shallow-water sand bodies, while the latter two are deep-water sand bodies (Fig. 10.30).

Since each facies belt sand body is discussed in the relevant chapters of this book, here, the distribution locations, hydrodynamic conditions,

Table 10.9 Sedimentary characteristics of main sand body type in lacustrine basins

Sandstone type		Turbidity current	Delta	Fan delta	Subaqueous fan	Beach bar
Sedimentary Environment		Deep lake to semi-deep lake	Gentle slope of lacustrine basin, place for river entry into the lake	Steep slope of lacustrine basin, place for entry of alluvial fan into the lake	Steep slope of lacustrine basin, subaqueous deposit	Shore-shallow lake region
Sedimentation		Turbidity current	Dominated by tractional current, accompanied by downstream accretion (progradation)	Mutualism of tractional current and gravity flow	Dominated by gravity flow	Shore current winnowing accretion
Mudstone characteristics		Dark color with pure quality	Light color with impurity	Mixed color with impurity	Dark color with pure quality	Light color with impurity
Sand body features	Lithology and bedding	Normal graded bedding gravel is mixed with dark-colored deep lake mudstone, and the Bouma sequence and water escape structure are seen commonly	Sandstone is mingled with mudstone. Often, the three-layer (belt) structure is seen, dominated by planar cross bedding, wave ripple cross bedding, and mixed bedding	Conglomeratic is mingled with mudstone, with the three-layer (belt) structure, megascopic cross bedding, wave ripple cross bedding, and graded bedding	Conglomeratic is mingled with mudstone, without the three-layer structure; melange accumulation, megascopic planar cross bedding, and graded bedding	Sandstone and siltstone are frequently interbedded with mudstone, presenting wave ripple crossing bedding
	Distribution features	Layered superimposition; sand and mud interbedded in lobate distribution	Layered superimposition, lobate distribution	Massive accumulation, fan-shaped distribution, and mud-in-sand	Massive accumulation, fan-shaped distribution, and sand-mud mix	Layered extension, mud layer mingled with sand
	Type and rhythm	Nearshore or offshore turbidite fan and turbidity channel sand body, Bouma sequence	Long fluvial delta, short fluvial delta; dominated by coarsening upward	Short-axis steep bank of lacustrine basin, regressive delta, transgressive delta, either fining or coarsening upward	Steep lacustrine basin bank, dominated by fining upward	Edge of lacustrine basin, both fining upward and coarsening upward
Lacustrine basin development stage		Largest deep-depression period	Later phase of fracture or depression period	Uplift period after deep-depression of lacustrine basin	Water-transgression phase in fault depression period	Atrophic phase or depression period of fault depression of lacustrine basin
Adjacent sand body		Nearshore shallow water sand body, cliff lakeshore	Shoreward: river flood plain; Lakeward: turbidity lens; Sideward: beach bar	Shoreward: Laoshan and alluvial fan; Lakeward: turbidite sand body	Shoreward: raised Laoshan; Lakeward: turbidite sand body	Sideward: nearshore sand body such as delta; Lakeward: shallow lake Lakeward: swamp and river

sedimentary characteristics and types, oil storage properties, and source-reservoir-cap association and their identification marks are described only briefly (Table 10.9).

References

Chongjun, Wu, and Xue Shuhao. 1993. *Sedimentology of petroliferous basins in China*. Beijing: Petroleum Industry Press.

Chongjun, Wu. 1981. The Division of Lake Sedimentary Facies. *Petroleum Exploration and Development* (2)

Chongjun, Wu. 1982. *Sandstone oil-gas oilfield geomorphology. Developments in petroleum science*, 4. Beijing: Petroleum Industry Press.

Chongjun, Wu. 1983. Structural lake deltas and oil-gas distribution. *Acta Sedimentologica Sinica* (1).

Chongjun, Wu. 1986. Sedimentation types of lake sand body. *Acta Sedimentologica Sinica* (4).

Einsele, G. 1992. *Sedimentary basins: Evolution, facies, and sediment budget*. Berlin Heidelberg: Springer-Verlag.

Kendali, A.C. 1992. Evaporites. In *Facies models: Response to sea level change*, ed. R.G. Walker, and N.P. James, 375–407. Waterloo, Omtario: Geological Association of Canada.

Link, M.H., and R.H. Osborne. 1978. Lacustrine facies in the Pliocene Ridge Basin, California. In *Modern and ancient lake sediments* ed. A. Matter, and M.E. Tucker, 167–187. Spec. Publ. Int. Ass. Sediment.

Liu, Qun, and Yuhua Chen. 1987. Salt sediments of terrigenous clastic-chemogenic type in meso-cenozoic of China. Beijing: Beijing Science and Technology Publishing House.

Lowenstein, T.K., and L.A. Hardie. 1985. Criteria for the recognition of salt pan evaporites. *Seclimentology* 32: 627–644.

Picard, M.D., and L.R. High Jr. 1972. Criteria for recognizing lacustrine rocks. In *Recognition of ancient sedimentary environments*, vol. 16, ed. J.K. Rigby, and W.K. Hamblin, 108–145. Spec. Publ. Soc. Econ. Paleont. Miner. Tulsa.

Selley, R.C. 1976. *An introduction to sedimentology*, 408. London: Academic Press.

Shuhao, Xue, et al. 2002. *Sedimentary geology of lakes and basins and hydrocarbon exploration*. Beijing: Petroleum Industry Press.

Surdam, R.C., and C.A. Wolfbauer. 1975. Green river formation, wyoming; a playa-lake complex. *Bulletin of the Geological Society of America* 86 (3): 335–345.

Wang, Hengjian, et al. 1983. A study of fresh water carbonates in the cretaceous Qiangshankou and Nenjiang formations in Western Songliao Basin. *Petroleum Exploration and Development (Geology)*.

Yu, Xinghe, Defa Wang, and Junmao Zheng. 1992. Lithofacies association types or sequences and depositional system of Permian sandstones in North China. *Acta Sedimentologica Sinica*, 10(1): 27–34.

Yu, Xinghe, Defa Wang, et al. 1994. Braided river delta sandbody characteristics and sandbody distribution model-modern delta sedimentary study in Daihai lake of inner Mongolia. *Acta Sedimentologica Sinica*, 15 (1): 27–36.

Yu, Xinghe, Hui Jiang, et al. 2007. Depositional filling models and controlling factors on mesozoic and cenozoic fault basins of terrestrial facies in Eastern China—A case study of Dongying Sag of Jiyang depression. *Lithologic Reservoirs*, 19(1): 39–45.

11 Deltaic Depositional System

As early as 1885–1890, G. K. Gilbert studied the Pleistocene lacustrine delta deposits in Lake Bonneville, America, and first found that the delta deposit body has a three-tier structure. Later, according to Gilbert's description, J. Barrell (1992 and 1914) studied the characteristics of the sedimentary facies of the Devonian Catskill Delta in the Appalachian Basin, identified the topset, foreset, and bottom set, and respectively depicted the characteristics of the lithology, beddings, and fossils of all formations thus pioneering the study of the sedimentary facies of ancient marine deltas. Although Barrell noted that not all delta deposits have the three-tier architecture of a Gilbert-type delta, the delta depositional models established by Gilbert and Barrell have influenced researchers' awareness about deltas in the 20th century. Previously, researchers used to consider a megascopic foreset as an important sign for identifying an ancient delta. Moreover, because abundant energy source deposits (including coal, petroleum, and natural gas) having large economic value had not yet been found in delta sedimentary formations, studies on deltas were lacking. As a result, we barely understand the complex and changeable sedimentary characteristics of a deltaic depositional system. Nonetheless, these early research achievements have been of great significance.

Since the early 20th century, it has been found that a variety of energy and mineral resources are contained in delta sedimentary formations. Worldwide, there are a number of large oil and gas fields (such as the Bulgan Oilfield in Kuwait; Bolivar Oilfield in the Maracaibo Basin and Venezuela; Mesozoic and Cenozoic oilfields in Mexico; oilfields in the Niger River Delta; West Tuscola Oilfield in Texas, America; asphaltic sand in Athabasca, Canada; and most sandstone oilfields in the Daqing Oilfield and Pearl River Mouth Basin), Zambia-Zaire copper ore belt, Witwatersrand Gold-uranium ore belt in South Africa, Lake Huron and Lake Agnes gold-uranium ore belt in Canada, and most important coal fields in the sedimentary environment of a delta. Since these mineral resources have been exploited and developed, researchers have attached great importance to the study of the sedimentation, environment, and sedimentary facies of deltas, which reveal much information.

Since the 1940s, the sedimentary environment, sedimentation, and depositional systems of the modern deltas of the Mississippi River, Rhone River, Niger River, Yangtze River, and Yellow River have been systematically and comprehensively investigated, and this has laid a theoretical basis for the establishment of a sedimentary facies model and sedimentary system along with the analysis of the hydrodynamics in an estuary region. It is worth mentioning that the Coastal Studies Institute, Louisiana State University, has conducted systematic coastal research since the 1960s. Their studies revealed a variety of geological factors that control the delta development and the correlations between them and the forms of delta sand bodies, summarized

X. Yu et al., *Clastic Hydrocarbon Reservoir Sedimentology*, Advances in Oil and Gas Exploration & Production, https://doi.org/10.1007/978-3-319-70335-0_11

the characteristics of the tri-dimensional depositional framework of different deltas, and established and identified transverse and vertical sequences of various deltas. Through these studies, researchers' ability to identify ancient delta sedimentary formations has improved greatly.

Since the 1970s, many studies have focused on the deltas in China. Owing to the exploitation and development of oil and gas fields in the offshore shelf, China has began systematically studying the characteristics and sedimentary environment of modern deltas in the Pearl River, Yangtze River, and Yellow River, resulting in many findings. Oil and gas fields have been found in many deltas; in particular, most oil and gas fields in continental faulted basins in eastern China are related to the deltas deposited in the Meso-Cenozoic periods. As a consequence, the main clastic oil-producing formations of all large oil and gas fields (regions) that can produce millions of tons of oil and gas currently in China are related to deltaic depositional systems.

11.1 Basic Characteristics, Classification, and Models of Deltas

11.1.1 Basic Characteristics

A delta refers to a protruding triangle-like sand body with a discontinuous coastal line which is formed when a large number of deposits are carried by a river into a relatively static and stable catchment basin or region (such as sea, lacustrine basin, semi-enclosed sea, and lake). The speed of deposit supply is higher than the redistribution speed of the local basin action.

A delta is usually a deltaic depositional system with a protruding geometric shape in which the fixed water supply system (which finally forms a main river) supplies deposits to the coastal line (seacoast or lakeshore) and merges into water deposits in the local area. Furthermore, it constantly advances toward the sea or lake foreset. A delta is a deposit accumulation system formed by the common effect and interaction of fluviation and oceanization, which can extend underwater from the land, and therefore, it is a transitional deposit between a continent and a sea (or lake).

11.1.2 Delta Classification

Many factors affect delta building, including the property of the impounding body, hydrodynamic conditions, gradient, and material distances. As a result, researchers worldwide have developed their own classifications based on different factors when studying deltas. Overall, there are roughly eight classification approaches (Table 11.1), which make careful and in-depth analyses of the sedimentary controlling factors of deltas at different angles.

According to the type of impounding body, two types of deltas can be defined: if the impounding basin is an intra-continent lake, the delta is called a lacustrine delta or continental

Table 11.1 List of delta classification schemes

Scheme	Results
Property of impounding body	Lacustrine and marine deltas
Hydrodynamic conditions	Fluvial-dominated, wave-dominated, and tide-dominated deltas
Morphological characteristics	Pedate, beak, and estuary deltas
Property of source supply body	Fan, braided, and normal deltas
Developed part	Continental slope, continental shelf, and Gilbert-type deltas
Type of river mouth process	Friction-dominated, inertia-dominated, and suspension-dominated deltas
Grain size	Coarse-grained and thin-grained deltas
Structural genesis	Six types of deltas are defined by synthesizing all types of factors

Table 11.2 Types of deltaic depositional systems (according to Galloway 1975)

Features	Fluvial-dominated delta	Wave-dominated delta	Tide-dominated delta
Form	Stretched-lobate	Arched	Estuary-irregular
Type of distributary channel	Straight-curved	Snaking	Expandable-curved
Main sedimentary deposits	Muddy-mixed	Sandy	Variable
Framework facies	Distributary mouth bar, channel filling sand, and marginal sand sheet	Barrier bar and sea ridge sand	Estuary filling and tidal sand ridge
Framework orientation	Parallel to sedimentary slope tendency	Parallel to slope	Parallel to slope tendency

delta; and if the impounding basin is a shallow sea (bay or lagoon), the delta is called a marine delta. In China, lacustrine deltas are greatly related to oil and gas. Compared with a marine delta, a lacustrine delta is unique in the following respects: ① it has a small scale, and its size ranges from less than dozens of square kilometers to thousands of square kilometers; ② it is usually developed in a faulted or depression basin; ③ in a faulted basin, the development of the delta is characterized by small and more steep slopes, and in a depression basin, it is characterized by large and less gentle slopes; ④ in a large depression, a delta has a larger scale, but the distributary mouth bar is not developed enough and is dominated by the superimposing subaqueous distributary channels to form a subaqueous delta plain; and a ⑤ fluvial-dominated delta dominates compared to a rare wave-dominated delta and tide-dominated delta.

The formation, development, and morphological characteristics of a delta are mainly determined by the fluviation and relative strength of the impounding body's energy. A delta is mainly formed by rapidly accumulating a large amount of mud and sand carried by a river. Various actions (waves, tides, and longshore currents) of seawater play a role in transforming, destroying, and redistributing a delta. As a consequence, various deltas can be produced under the interaction of river and sea water. Most scholars divide the genetic types of deltas according to the relative strengths of the river, waves, and tides. In 1975, W. E. Galloway proposed a ternary classification scheme according to the correlation of the above three actions based on a comprehensive analysis of the data of 34 modern deltas in the world. Each end member has its unique sand body framework characteristics (Table 11.2). Other different transitional deltas can be marked on the corresponding positions of the delta diagram according to the three main control actions of the river, waves, and tides. This diagram can systematically represent the classification scheme of a continuously changed delta, thereby not only being applicable to the classification of a modern delta but also to ancient deltaic depositional systems (Fig. 11.1).

Coleman and Wright (1975) emphasized that some of the factors affecting and controlling the form, composition, structure, and sand body distribution characteristics of a delta only play a leading role in some parts of the delta, and all factors are mutually combined with different strengths to jointly control the formation and development of a delta, all of which jointly combine with the overall background environment of the delta system. As a consequence, the final structure of any deltaic depositional system is determined by its overall background environment, rather than by a single factor, which is why previous studies usually applied a structural-genetic classification. Coleman and Wright eventually synthesized six representative

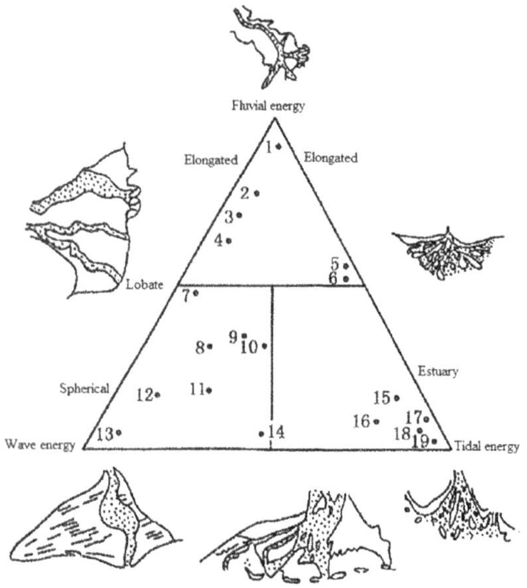

Fig. 11.1 Trigonometric diagrams about the classification of delta (after W. E. Galloway 1975). Various delta names are specified below: 1—modern Mississippi; 2—Sankt Bernhard Mississippi; 3—Po River; 4—Danube; 5—Yukon; 6—Mahakam; 7—Ebro; 8—Nile; 9—Orinoco; 10—Niger; 11—Burdekin; 12—Rhone; 13—Sao Francisco; 14—Coppermine; 15—Yalu; 16—Colorado; 17—Fly River; 18—Ganges River; and 19—Klang-Ranga River

deltas by comprehensively analyzing and statically comparing data of 400 environmental parameters from 55 modern fluvial deltas. Each type has its unique sand body form and distribution characteristics (Fig. 11.2, Table 11.3).

According to the various abovementioned classifications, each classification highlights close correlations between the form characteristics of the deltaic depositional system and various actions. Unlike W. L. Fisher and Galloway, who emphasized the correlations between a single leading factor (river, wave, and tide) and the whole form of the delta, Coleman placed more emphasis on control actions in terms of the form, distribution, and thickness change of a delta sand body by the background environment with a variety of interacting factors. It is very important to place emphasis on the characteristics of the sand body, because it is of significance to the classification of the delta and constitutes the main petroleum reservoirs in the delta. One of the main purposes of petroleum exploitation is to determine the sand body distribution laws; in other words, the establishment of sand body

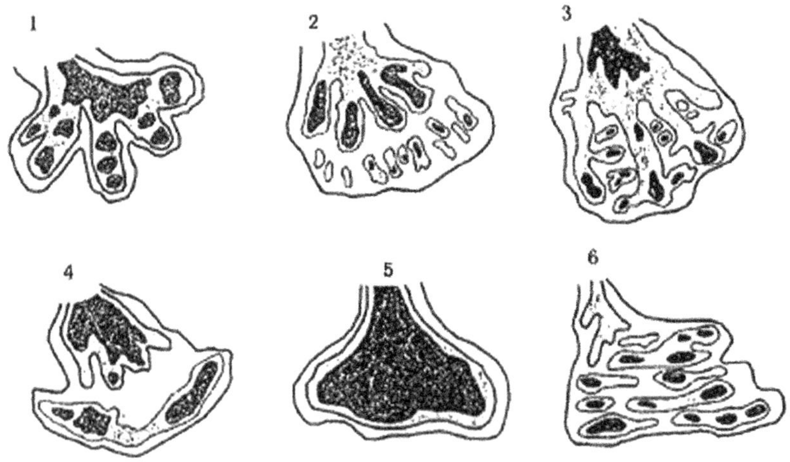

Fig. 11.2 Six delta sand body shape distribution types according to multi parameter analysis (according to Coleman and Wright 1975)

Table 11.3 Geometric conditions, characteristics, and examples of six delta types (according to Wright and Moseley, modified in 1975)

Type	Condition	Features	Example
1	Low wave energy, small tidal range, weak littoral drift, gentle offshore slope, and fine-grained deposit load	Finger-like channel sands distributed vertical to shoreline vertically	Modern Mississippi delta
2	Low wave energy, high tidal range, weak littoral drift, and narrow basin	Finger-like channel sand, transited to striped tidal ridge sand offshore	Oder, Indus, Colorado, and Ganges-Brahmaputra River deltas
3	Medium wave energy, high tidal range, low littoral drift, and stable shallow basin	Channel sand distributed vertical to shoreline, laterally connected with barrier beach sand	Burdekin, Irrawaddy, and Mekong deltas
4	Medium wave energy, small tidal range, gentle offshore slope, and low deposit supply	Channel and estuary sand bar connected by offshore barrier island	Apalachicola and Brazos River deltas
5	Persistent high wave energy, low littoral drift, and steep shoreline steep slope	Sheet, laterally stable barrier beach sand with updip channel sand	San Francisco and Grijalva deltas
6	High wave energy, strong littoral drift, and offshore steep slope	Multiline striped barrier beach sands arranged parallel to shore line, with squared channel sands	Senegal River delta

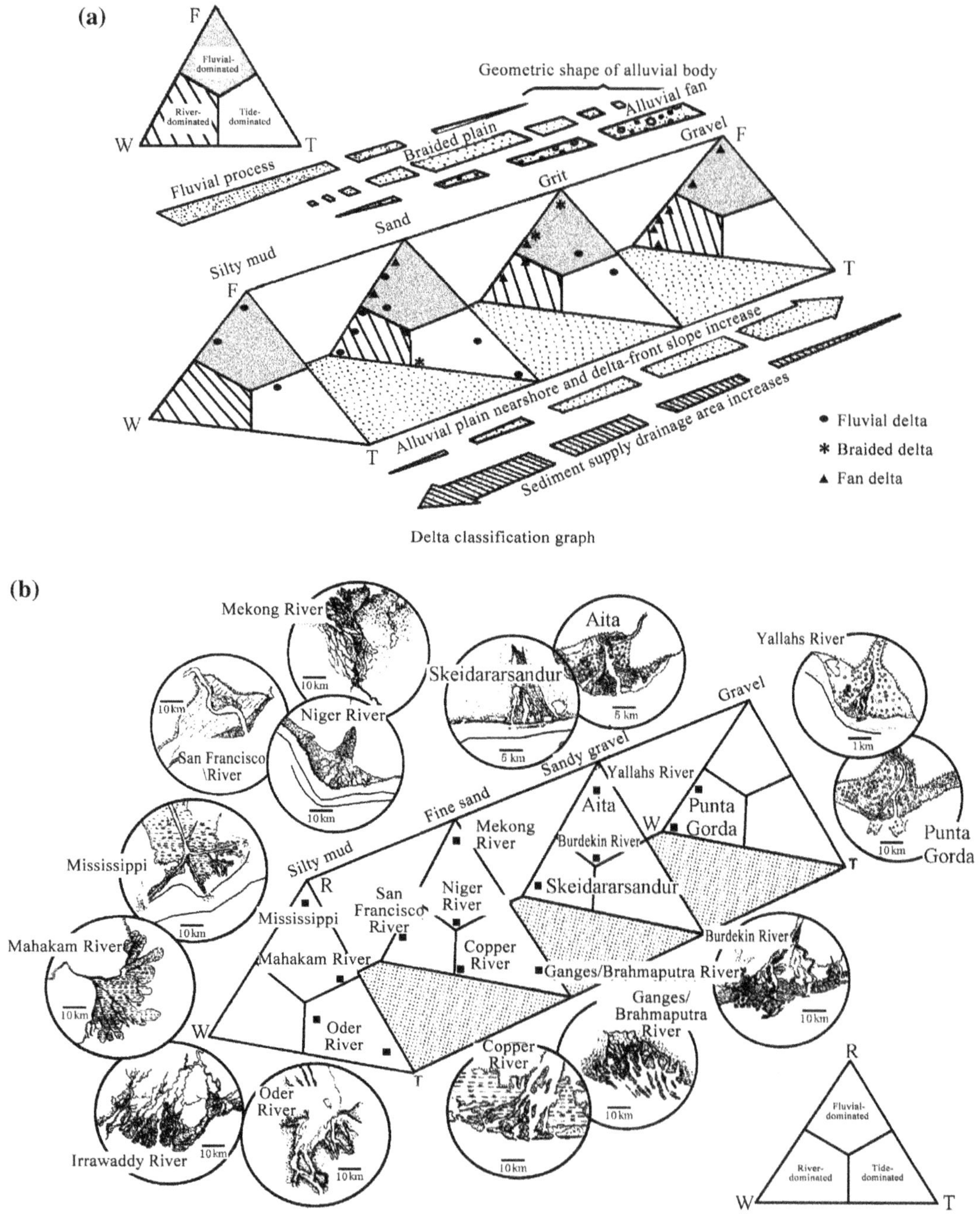

Fig. 11.3 Delta classification diagrams (after Orton 1988, 1993). **a** Structural-genetic classification scheme (1988); and **b** examples from all over the world, considering the grain size of the delta (1993)

distribution models is the basis for delta research and sand body prediction.

As studies on delta research and petroleum exploitation and development have advanced in depth and breadth, researchers are increasingly realizing that there are different types of deltas. Following a study on the properties of 34 modern deltas in the world, Orton (1988, 1993), based on the main control factors (grain size, geometric shape, slope, source supply property, and drainage area of deposit supply) of the delta, proposed a detailed delta classification scheme, i.e., structural genesis classification [the author believes that this classification scheme reflects the characteristics of different deltas better (Fig. 11.3)]. Accordingly, deltas can be divided into two categories and five types (Table 11.4).

In nature, the development background and deposit characteristics of a coarse-grained delta and fine-grained delta vary, but the evolution is regular. Actually, they are end-member components in one continuous spectrum.

It is thus clear that a coarse-grained delta can be divided into two categories: fan deltas and braided deltas (Fig. 11.4). In the past, fan delta deposits have been explained as a delta formed in such a way that the continental alluvial fan directly enters a sea or lake. In *Fan Delta*, chiefly edited by W. Nemec and R. J. Steel in 1988, a more accurate definition for a fan delta was proposed, namely, a fan delta is one formed by an alluvial fan as a supply source. This book defines it as "a proximal pebbly delta formed by an alluvial fan as the source supply transported in the form of a bed load." A braid delta is a "coarse-grained delta formed by a braided river as the source supply transported through a bed load." Furthermore, some scholars have further divided braided deltas into a delta formed by a single braided channel and one formed by a braided plain. As a result, according to the above definitions, a fine-grained delta is a delta dominated by a mixed load formed by a normal river (meandering river or straight river) as a source supply. On the basis of the different distribution positions, a coarse-grained delta can be divided into three categories and 12 types (Table 11.4, Fig. 11.5).

Table 11.4 Structural-genetic classification scheme and facies classification of deltas

Grain size classification	Genetic classification					Subfacies classification	
Coarse-grained delta	Fan delta	Marine facies	Continental shelf	Continental facies	Transgressive	Quartering	Upper delta plain Lower delta plain Delta front Prodelta
			Slope type		Regressive		
			Gilbert-type		Gilbert-type		
	Braided delta	Braided delta					
		Braided plain delta					
Fine-grained delta	Fluvial-dominated delta	Pedate delta				Trichotomy	Delta plain Delta front Prodelta
	Wave-dominated delta	Beak delta					
	Tide-dominated delta	Estuary delta					

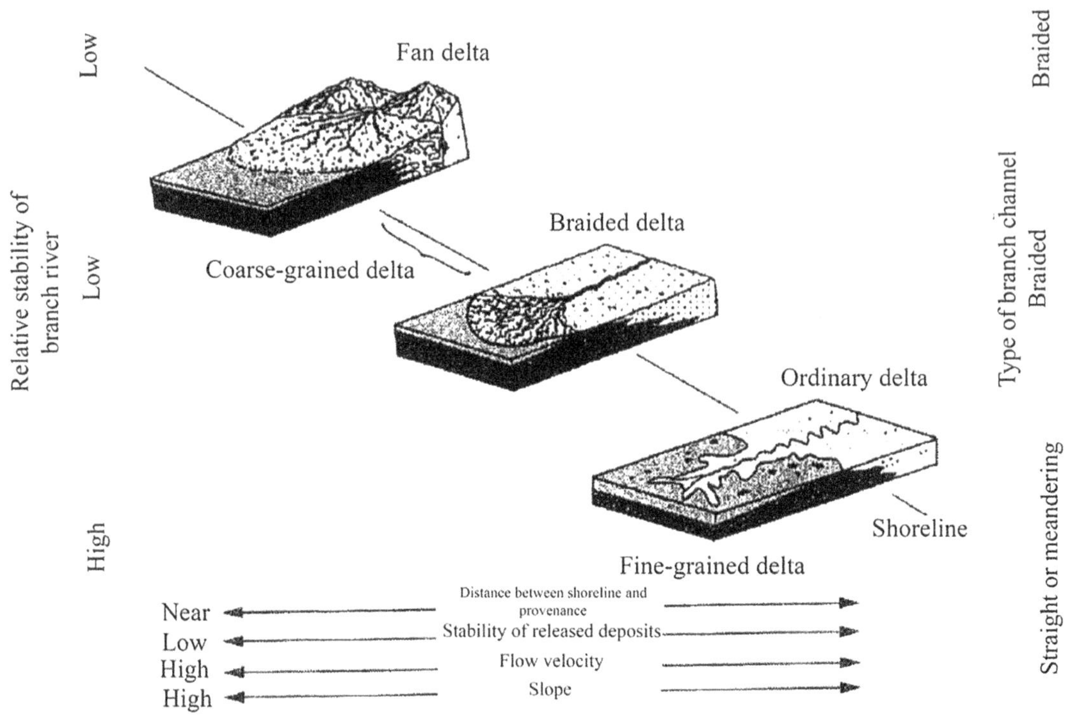

Fig. 11.4 Contrast maps of the coarse-grained delta (fan delta, braid delta) and fine-grained delta (Branch channel shape and stability, sediment load and size, river bed gradient, stream velocity) (after McPherson et al. 1988)

11.1.3 Depositional Model and Characteristics

In light of the diversity of modern deltas and the demands for petroleum exploitation and development, requirements cannot be met with only one delta pattern. Previous researchers proposed several representative delta models (Fisher et al. 1969; Cloeman and Wright 1975; Galloway 1975; Nemec and Steel 1988) that have been widely applied. Based on the quantitative analysis and comparison of modern deltas, Fisher et al. (1969) divided deltas into fluvial-dominated constructive deltas and wave-controlled destructive deltas. Constructive deltas can be of lobate and pedate types, and destructive deltas can be of wave-dominated and tide-dominated types (Fig. 11.6). Each type of delta has its specific forms and sedimentary characteristics, which can be depicted based on its vertical sedimentary sequence, facies area distribution, and geometric shape of it sand body. This method places emphasis on the correlation of sedimentary facies, which can be directly used for studying an ancient delta's facies sequence. It should be noted that the sedimentary facies mode of a single factor cannot be applied to generalize all characteristics of a complicated deltaic depositional system, and therefore, it is very necessary to build a multifactor multifacies sedimentary facies model. In accordance with the background environment of a multifactor interrelation, J. M. Coleman (1975) discussed the deltaic deposit rule, which was undoubtedly a very

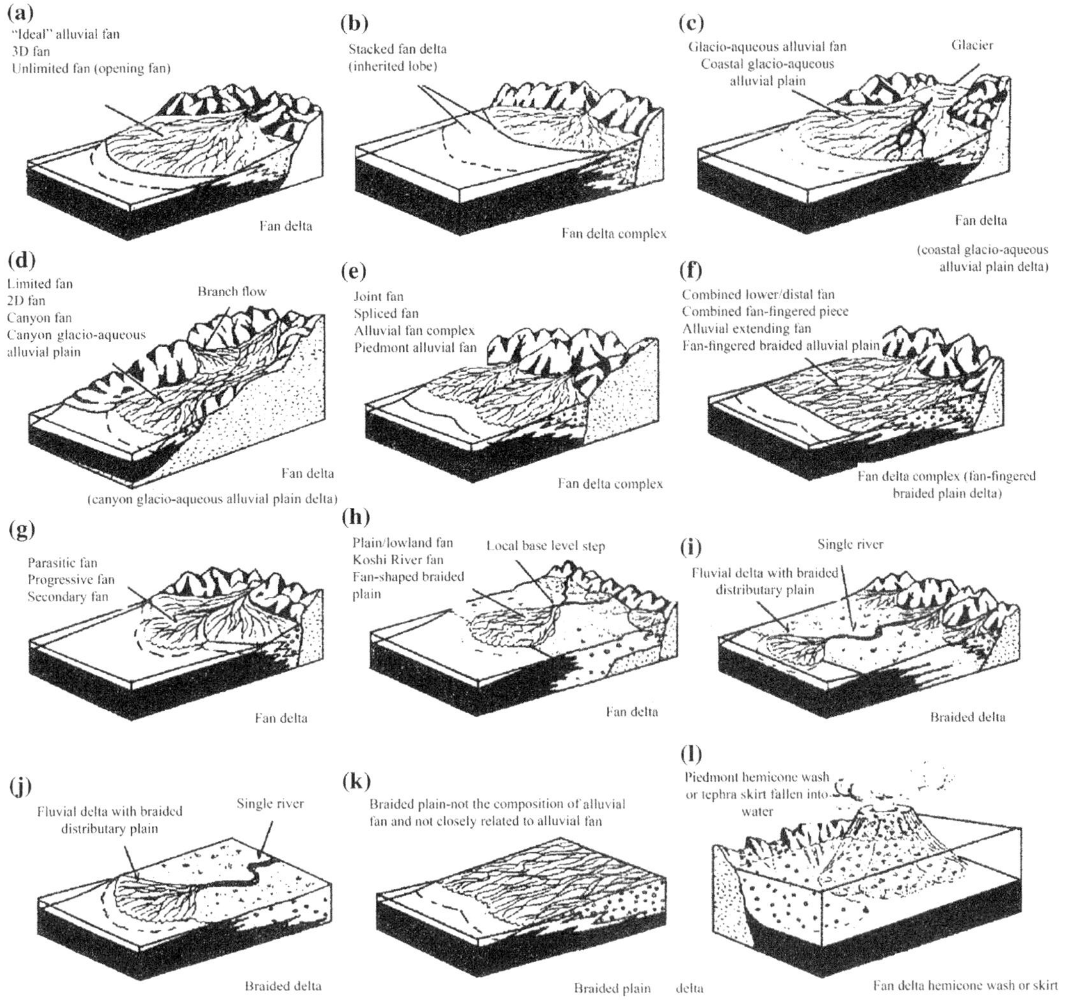

Fig. 11.5 According to the different distribution positions, the coarse-grained delta can be divided into three large classes and 12 patterns (after Nemec and Steel 1988)

meaningful attempt. However, unfortunately, the popular classification still considers three types of deltas divided by a single leading factor: fluvial-dominated, tide-dominated, and wave-dominated. On the other hand, the universally accepted deltaic deposit rule subject to a systematic and comprehensive study is still a fluvial-dominated, highly constructive delta as represented by the modern Mississippi delta. Studies on other types of deltaic deposit models have realized some progress; however, they are not very mature. Therefore, Orton (1988) suggested a triangular diagram of the structural-genetic classification of a delta;

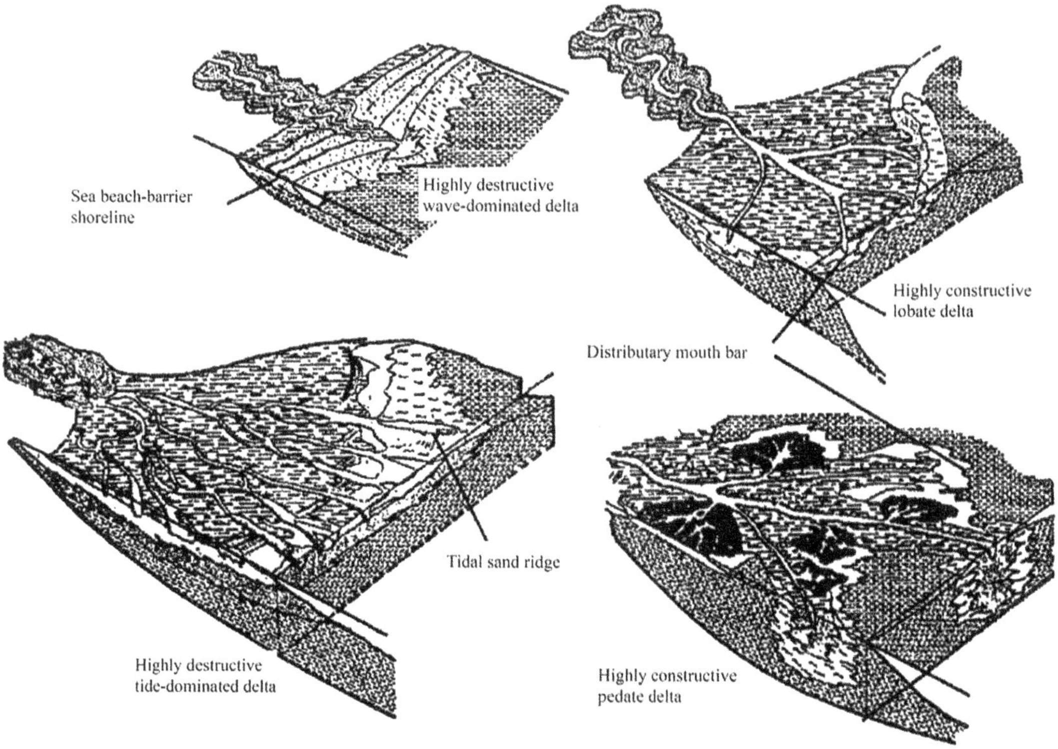

Fig. 11.6 Fine-grained delta can be divided into river-dominant constructive delta and wave-dominat destructive delta (after W. L. Fisher 1969)

furthermore, Orton (1993) provided planar geometric characteristics (Fig. 11.3) on the basis of examples from all over the world, in order to provide a model and basis for the planar prediction of a delta sand body.

11.2 Hydrodynamic Conditions and Sedimentation Characteristics of Delta Building

11.2.1 River Mouth Process

Friedman and Sander (1978) explained and classified a deltaic sedimentary environment as an estuary accompanying a transitional environment, which indicates that river mouth processes play an important role during the building up of a deltaic sedimentary system. An estuary is not just a place where running water mixes with water in a catchment basin; it is a dynamic dispersion place for deposits and also a distribution center of terrigenous clastic sediments.

A river transports deposits to the estuary and then disperses them to surrounding lakes/seas. The distribution status of these deposits and the formation of various sand bodies are determined by the hydrodynamic conditions in a river estuary. However, it is more important to lose river transportation during river mouth processes. Deltaic depositional systems of different types and models are built based on the different energy loss forms and speeds.

Coleman (1976) proposed three basic factors for determining an estuary. The dispersion of the outflow and deposit diffusion mode is determined by their interactions:

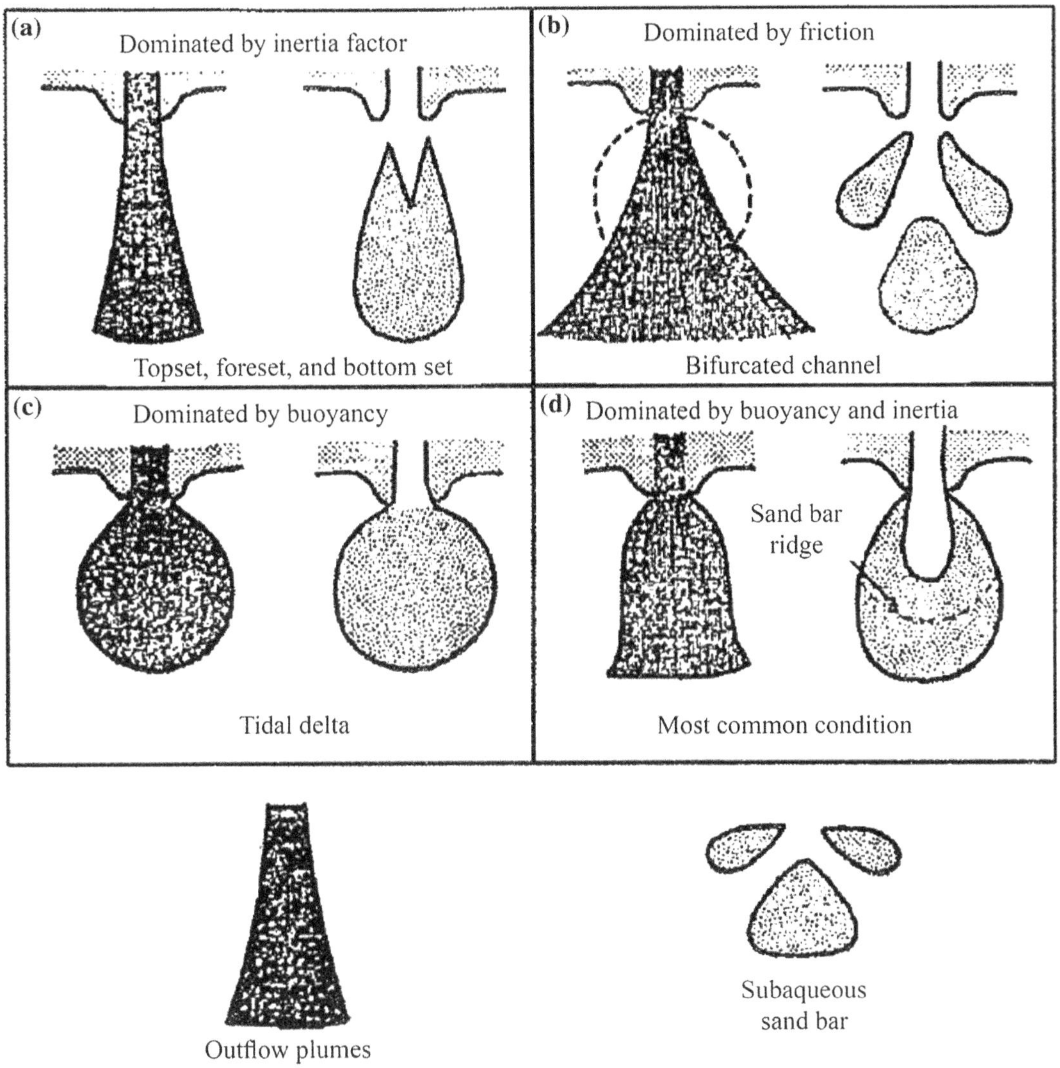

Fig. 11.7 According to three basic factors determining estuarine action, J. M. Coleman purposed four formation mechanism of the delta Estuary (after Coleman 1976)

(1) Inertia-inertia force of river water flowing into a catchment basin and accompanied turbulence diffusion;
(2) Friction-friction force between outflow water and estuary bedform; and
(3) Buoyancy-buoyancy resulting from density difference between outflow water and basin water. Furthermore, he proposed the genetic mechanisms of four estuaries (Fig. 11.7).

In combination with inertia, friction, and buoyancy and in consideration of the depth of the impounding body, gradient of slope, falling speed, and tidal energy, Postma (1990) proposed a detailed and systematic classification and description (Fig. 11.8) based on the type of river mouth processes. It is thus clear that the shallower the water, the more gentle is the landform, and the stronger the friction effect, the more

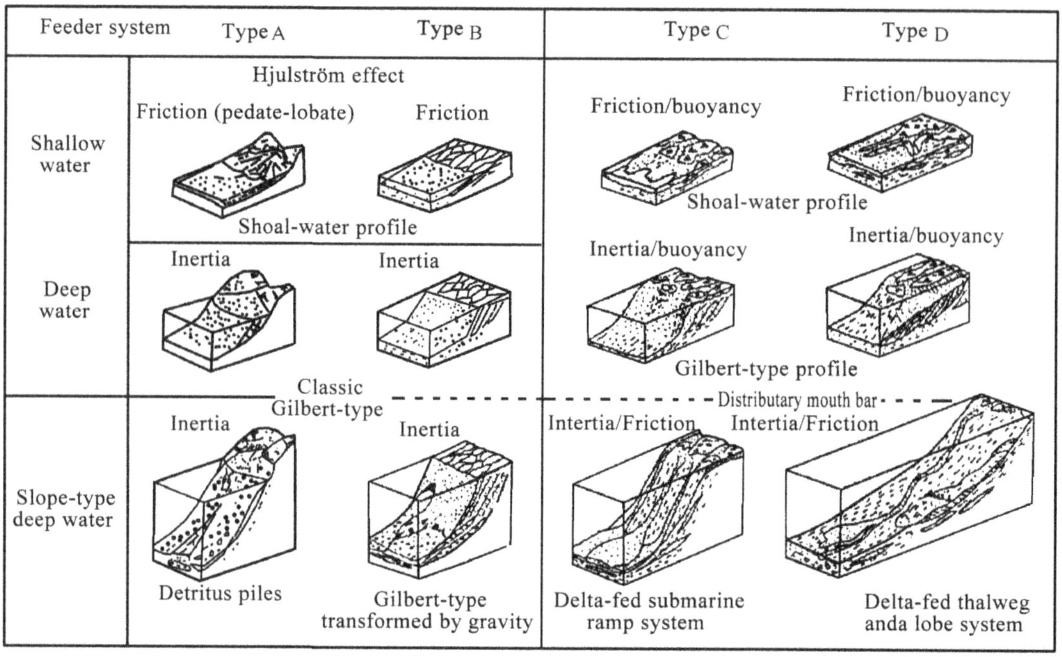

Fig. 11.8 Based on inertia, friction and buoyancy, coupled with the depth of water storage, the steepness of topography, the speed of injection and the magnitude of tidal energy, Postma purposed a detailed classification and description of the types of estuarine action (after Postma 1990)

developed is the distributary channel; conversely, the deeper the water, the steeper is the landform, and the stronger the inertial flow, the more similar the delta plane form is to that of the fan. A Gilbert-type delta is dominated by inertia; a pedate delta, by the friction factor; and a distributary mouth bar, by buoyancy. Regarding the vertical section structure, the Gilbert-type delta has a typical three-tier architecture. The other two types of deltas have a two-tier architecture; the pedate bottom set is mostly not developed, and the distributary mouth bar delta often loses the topset (Yu Xinghe 2007).

11.2.2 Hydrodynamic Conditions

11.2.2.1 Hydrodynamic Conditions of Normal Fine-Grained Deltas

When one river flows into a relatively static catchment basin, its flow type is determined by the density difference between the two water bodies if no greater effects arise from waves and tides. Based on the study by Bates (1953), Fisher et al. (1969) determined the hydrodynamic conditions according to the relative density of the water body and divided it into three types:

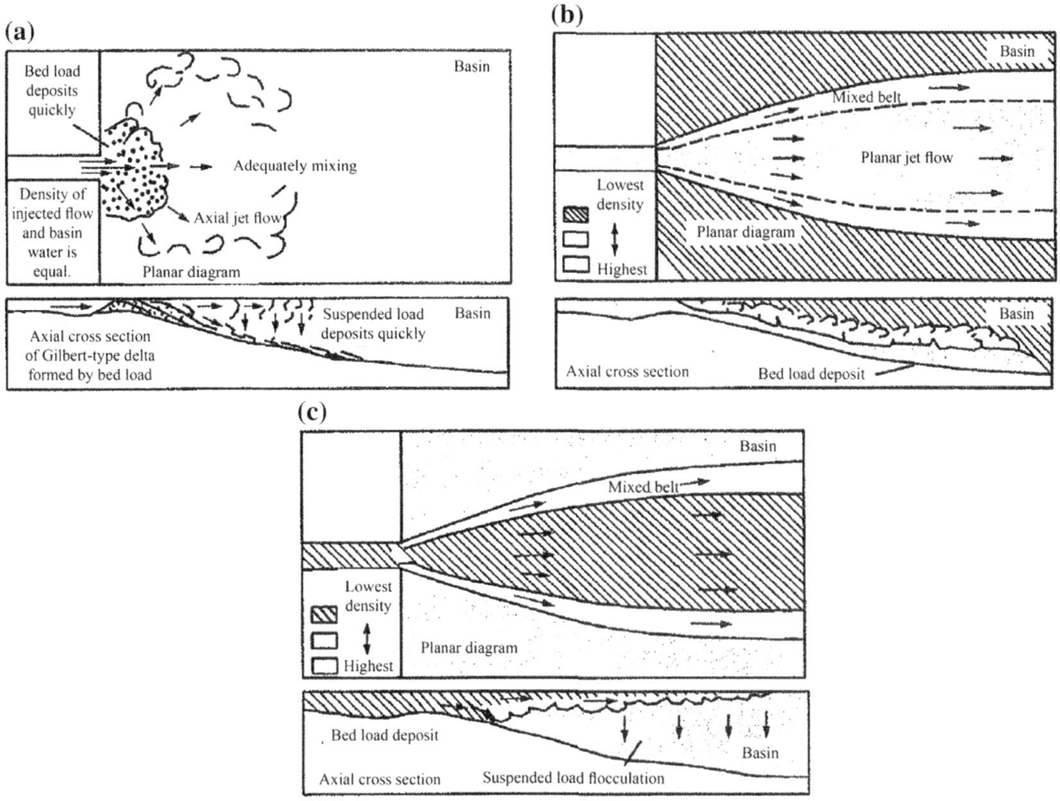

Fig. 11.9 Hydrodynamic state should be identified by relative density of river water and water storage body, and the hydrodynamic state can be divided into three types (Bates 1953)

① homopycnal flow, ② hyperpycnal current, and ③ hypopycnal current (Fig. 11.9).

(1) When the density of the inflow water and the impounding body are equal, it is called a homopycnal flow or equal density current. When the river falls into a freshwater lake, the two types of water are subject to 3D spatial mixing, and the water velocity reduces quickly. When the bed load quickly unloads accumulation near the estuary, the suspended load may be deposited at a farther place to form a lacustrine delta (or Gilbert-type delta).
(2) When the density of inflow water is high, being greater than even that of the basin water, it is called a hyperpycnal flow or super-gravity flow, and the inflow water is ejected along the bottom of the basin to form a planar diffusion. This condition can be usually found on the continental slope, and unconsolidated bottom sediments slump or slide to result in a gravity (turbidity) current due to gravity or other exogenous processes.
(3) When the density of inflow water is rather low, being less than even that of the basin water, it is called a hypopycnal flow or low-gravity flow. Most coastal deltas are filled with this type of fluid, because the density of fresh water is always lower than that of sea water, and deposits generally diffuse in such a manner that sea water transports river water and suspended solids by floatation. The density of freshwater is only 6% that of salt water. This low-density flow outflows on the salt water surface, which has a planar jet flow. The water in a river with large water flow can spread

outward to form a fluvial-dominated coastline delta.

Bates (1953) studied the hydrodynamics formed by a delta. He compared a delta estuary to a nozzle in hydraulics. When river water flows into the impounding body (catchment basin) via an estuary, two free jet flow types can be formed according to the type of mixed river water and impounding body: axial jet flow and planar flow (Table 11.5).

1. **Axial jet flow**

When the density of the injected water and water in the impounding basin is equal, an axial jet flow is formed. As mentioned above, the axial jet flow is a homopycnal flow. When the axial jet flow forms, the mixing effect of the river water and the water in the impounding body is similar to three-dimensional mixing in 3D space, which is characterized by a fast decrease in the flow velocity and rapid and sufficient mixing speed. Deposits transported by the bed load are accumulated in the estuary; however, fine-grained deposits transported by suspension are deposited in the peripheral area near the estuary. In nature, a river flows into a freshwater lake by jet flow, which often forms a lacustrine or Gilbert-type delta (Fig. 11.9a).

2. **Planar jet flow**

This refers to the mixing effect of river water and water in an impounding body only in planar mixing in a 2D space, which is characterized by keeping a higher flow velocity farther away from the basin direction when the flow velocity reduction mixing effect is less.

When there is a density difference between the fallen water and the water in the impounding basin, a planar jet flow is formed. A planar jet flow can be divided into the following two types: ① hyperpycnal current—when the density of fallen water is greater than that of the impounding body, the fallen water flows along the bottom of the basin in a planar shape, and two types of water are mixed in the bottom 2D space with slow mixing speed and low speed reduction rate. In nature, ice water falls into warmer lake water, and turbidity currents in the deep lake and deep sea are hyperpycnal currents (Fig. 11.9b); and ② hypopycnal current—when the density of fallen water is less than that of the impounding water, the fallen water flows on the surface of the impounding body owing to its lower density, and the two types of water mix in a planar shape at a place where they make mutual contact. This mixing process is slow, and the flow velocity of the fallen water reduces slowly; therefore, higher flow velocity and density stratification can be seen over a large area. In nature, a river has a low-density flow (Fig. 11.9c) while entering the sea.

11.2.2.2 Hydrodynamic Conditions of a Coarse-grained Delta in Continental Faulted Lacustrine Basins

In fact, the hydrodynamic conditions of a delta are not static, and the hydrodynamic states at

Table 11.5 Comparison between axial jet flow and planar jet flow

Characteristics	Type		
	Axial jet flow	Planar jet flow	
		Bottom jet flow	Top jet flow
Density ratio of inflow water and impounding body	Equal	>1	<1
Flowing type	Homopycnal flow	Hyperpycnal flow	Hypopycnal flow
Characteristics of jet flow	Adequate mixing in 3D space	Fluid flows along the bottom of impounding body	Fluid flows along the top of impounding body
Result	Easily built Gilbert-type delta	Easily built turbidity current	Easily built coastal delta

Table 11.6 Hydrodynamic conditions within a coarse-grained delta in continental faulted lacustrine basins

Hydrodynamic conditions	Planar jet flow	Sedimentary graded flow	Sedimentary smooth flow	Sedimentary wave flow
Flow direction	High energy, changed from sarciniform to divergent shape	Unidirectional flowing	Low-energy overflow	Multiway flow affected by wind
Flow velocity changes	Flow velocity reduces drastically	Flow velocity decreases slowly	Extremely low velocity	Changes along with the size of wind
Sediment characteristics	Coarse grain size, dominated by sand-gravel	Dominated by silt-fine sand, with better sorting	Dominated by silt, with better sorting	Dominated by silty-fine sand
Microfacies formed	Sheet conglomeratic	Distributary channel	Overbank between distributary channels	Beach sand/distal bar

different parts show great changes. There are four main types (Table 11.6) of continental coarse-grained deltas.

1. **Planar jet flow**

This refers to a high-energy fluid capable of transporting more deposits. It is reflected by the water load changing from a beam flow to a divergent flow dramatically after passing through a jet cross section (also called an injection orifice). Considerable fluid energy is released, flow velocity decreases sharply, and the load carried under a high-energy state is unloaded and deposited. The sediments form a sand-gravel dominated sheet conglomeratic with a coarse grain size, which is indicated by the PQ section in the C-M diagram, and therefore, the upper delta plain is a product of this flow pattern.

2. **Sedimentary graded flow**

Water moves unidirectionally, and therefore, water energy is released slower in a flowing process, with relatively uniform attenuation gradient. When the transportation energy of a carrier decreases to the sedimentary static boundary of the load, the load is deposited, and the loaded deposition sequence is determined by the weight (grain size and specific gravity). As the carrier energy decreases, sedimentary particles are gradually reduced, thus showing the sorting effect of the current carrier in the depositional process. It is characterized by silt and fine grains with better sorting. It develops as a QR section in the C-M diagram, which shows a deltaic distributary channel deposit.

3. **Sedimentary smooth flow**

This is one current process in a flood peak in the convergence area of a moving fluid, with weak water energy and the flow velocity reduced to the minimum limit. It is mainly characterized by silt, even fine grains, with better sorting. It is developed as the RS section in the C-M diagram, which is composed of overbank fine-grained deposits between delta distributary channels.

4. **Sedimentary wave flow**

This is a multidirectional moving hydrodynamic condition, and the main energy is derived from the actuating pressure of the wind force on the main energy. Lake waves generate various flows, such as washed-out flow, backflow, and longshore flow. The load is transported and transformed by several flows and also deposited, and particles can not only be transported back and forth but also move transversely along with the longshore current. Deposits mainly comprise silty-fine sand, with good sorting. Beach sand is a main product of this flow pattern.

Overall, coarse-grained delta deposition is a product of the combined actions of the above four hydrodynamic conditions, and it is dominated by one or two hydrodynamic conditions at different facies belts or parts. In general, the upper delta plain is dominated by a planar jet flow, whereas the lower delta plain is presented in two types—sedimentary graded flow and sedimentary stationary flow—and the delta front is based on a sedimentary wave flow and sedimentary stationary flow (when the topographic slope is steep, the lower delta plain may be dominated by a planar jet flow). An evolution pattern of the planar jet flow-sedimentary graded flow-sedimentary wave flow is formed in the delta system, which is the scientific basis (Table 11.4) for quartering coarse-grained delta subfacies or facies belts by the author.

11.2.2.3 Fluvial Action Force

In an estuary, there are three action forces affecting and controlling the water flow rate change and diffusion mode: inertia force, friction force, and buoyancy. The inertia force is related to the incident velocity; the higher the flow velocity of injected water, the greater is the inertia force. In an estuary dominated by inertia force, sand bodies are mainly deposited within a smaller scope near the estuary with high axial flow velocity of the injected flow. Because a friction force occurs between the water body and the bed deposits, a small basin gradient and shallow depth are conducive to the friction force, and the increased friction force forms a subaqueous distributary channel and an outward radiating sand body farther away. The greater the density difference between the river water and the sea water, the more will it contribute to buoyancy. In an estuary with dominant buoyancy, suspended matter can be transported to peripheral sea areas farther from the estuary owing to the large deposit diffusion scope. In any estuary in nature, the inertia force, friction force, and buoyancy always coexist to influence one another and interact, and a more complicated hydrodynamic background is formed owing to the sea/lake water dynamic conditions (wave, tide, and longshore currents), basin form, and gradient change. Therefore, the sand bodies in the estuary are distributed in a changeable form. Generally, the inertial flow dominates in a steep slope area, with a gentle slope caused by the friction effect. The different inertia forces and friction forces directly affect the development of the delta lobe and its scale. When the inertia force is stronger, there are a small number of delta lobes with great thicknesses, and the deltas are small in the planar scale but large in the vertical scale. When the friction force is stronger, more delta lobes are developed, the deltas laterally swing, and distributary channels are developed. As a consequence, the planar scale is large, however the vertical scale is relatively small. This is why the deltas in the continental faulted basin in China are presented in such a manner that steep slopes are "small and many" and "thick and coarse," whereas gentle slopes are "large and less" and "thin." In addition, the plane forms are also characterized by a "fan-shaped steel slope and lobate gentle slope." The Daihai Lake is a typical example of modern deposition in Inner Mongolia (Fig. 10.3).

11.2.3 Deposition Rate

The deposition rate is one of the quantitative markers of a sedimentary environment. Owing to greater differences in various environments, the delta deposition rate is highest in the shoreline area, and fast accumulation is one of the important conditions that make deltas the main petroleum reservoir type. During delta development, rapid deposition, slow deposition, depositional break, and erosive destruction occur alternately, and the deposition rate balances the main quantitative indexes of these changes and results in these actions.

According to the deposition rate of the Yangtze Estuary of 30–120 mm/a calculated according to nautical charts in 1842 and 1865, the deposition rate of a small copper sand shoal, which is being formed, can reach up to hundreds of millimeters every year. The deposition rate of the Mississippi is 300–450 mm/a, and it even reaches up to 3000 mm/a in the flood season. It

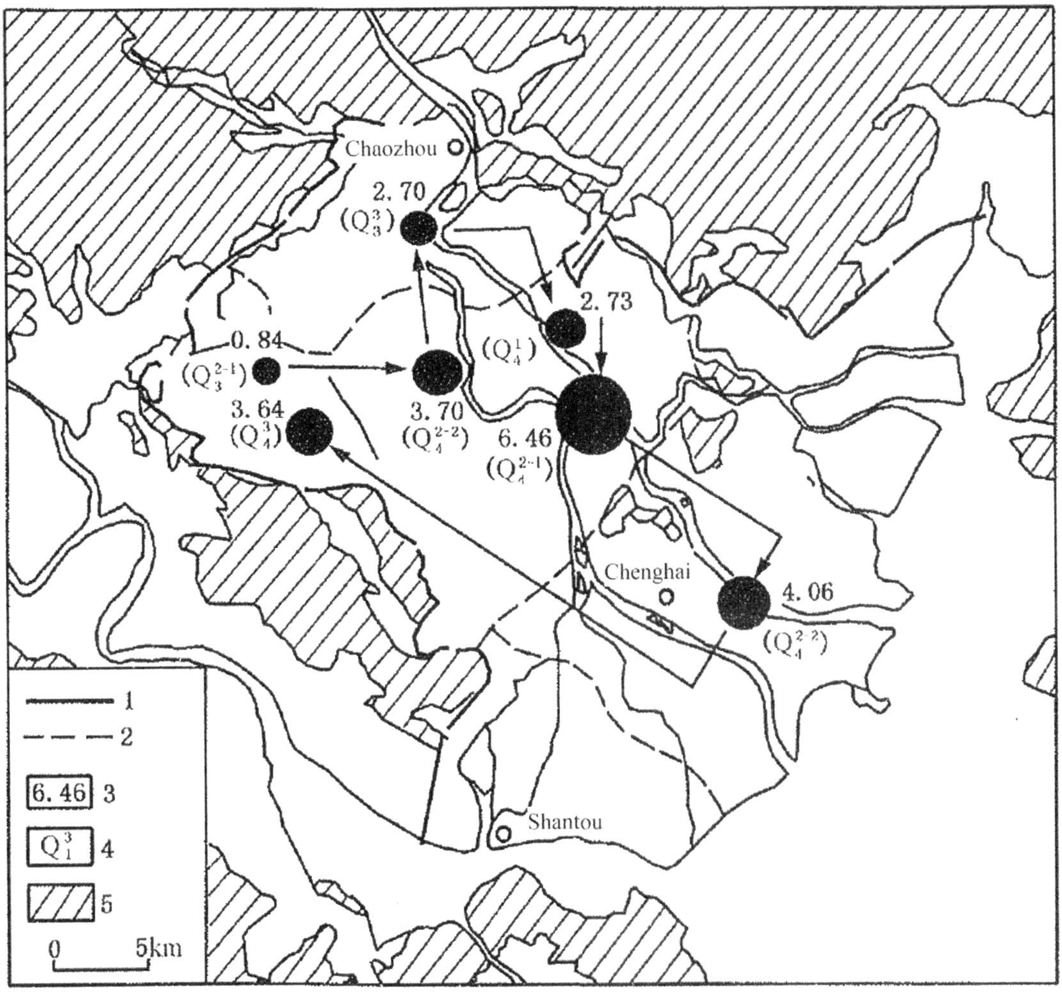

Fig. 11.10 Transfer of the maximum deposition rate in the Hanjiang river delta. Black points in the figure represent the deposition rate, thus the deposition rate is higher when the black point is larger. 1—Delta boundary; 2—Sedimentary area boundary; 3—Deposition rate (mm/a); 4—Sedimentation stage; and 5—Mountain and hill

must be noted that the deposition rate of the delta only represents the local deposition rate of the main distributary estuary area in the early development stage of the delta, rather than the entire delta deposition. Because the deposition rates vary greatly at different parts of the delta, the deposition rate of the main estuary area is high. However, it gradually reduces from the estuary to the open sea. In addition, as the estuary changes continuously, the maximum deposition rate transfers gradually (Fig. 11.10).

11.2.4 Sedimentation of Delta

The sedimentation of the delta, which is dominated by progradation or downstream accretion, is mainly characterized by coarsening upward reverse graded deposition. However, there are different sedimentations at different parts of the delta, such as aggradation of the distributary channel, interdistributary bay overbank accretion, and winnowing accretion of the front sand body. However, regarding the development of the whole delta, the overall retrogradation of the delta is presented during the transgressive period, whereas the overall progradation of the delta is presented during the regressive period.

In the basin stabilization period, the water energy of the deltaic distributary channel is related to the migration of the sedimentary center. As a consequence, the interaction between the deposition rate (R_d) and the subsidence rate (R_s) of the sedimentary basin may result in sedimentations (Fig. 11.11) of different cycles and different superimposed patterns of the delta. Owing to the faster deposition rate of the delta, the distributary plain advances to the delta front-prodelta, whereas the deposition rates in the basin and on the delta plain are rather low. On account of covering high-energy deposits above low-energy deposits, the sequence characteristic of upward coarsening is formed. The inactive portion of the delta can be destroyed or transformed through wave action to form a front sheet sand during transgression.

In combination with the analysis theory of sequence stratigraphy, the sea level rises but the delta retrogrades, and vice versa. A more comprehensive analysis of the ratio of the deposition rate to the subsidence rate (R_d/R_s) reflected by the tectonic movement and deposit supply should be performed.

11.3 Formation, Development, and Abandonment of Deltas

An integrated deltaic deposit body is a comprehensive product evolved by a delta in the historical category. In the geological history, each delta has its complicated history of occurrence, development, and extinction, and this results in the migration and superimposition of the delta.

11.3.1 Main Factors Affecting the Formation and Development of the Delta

The main factors affecting the formation and development of a delta are very complicated, and they generally comprise the following:

1. Flow velocity of river, drainage, quantity of carried mud and sand, and ratio;

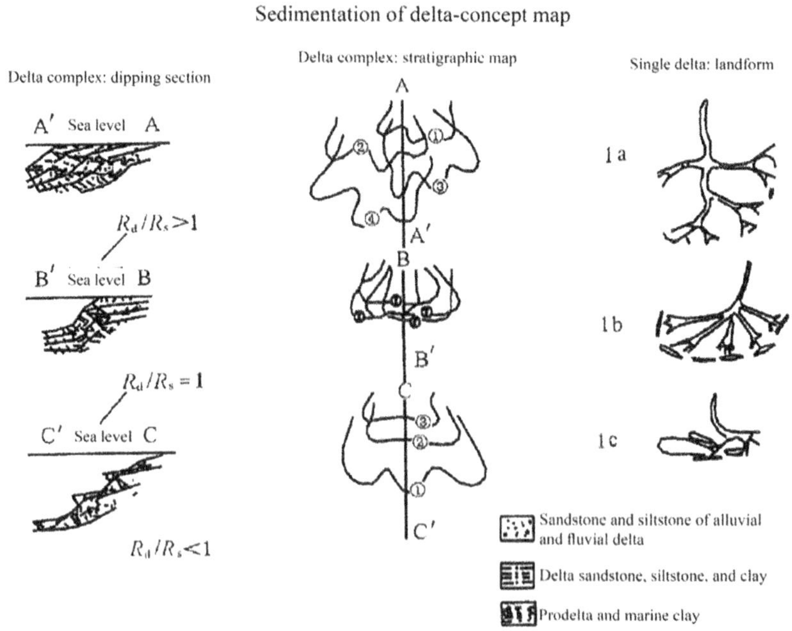

Fig. 11.11 Delta distribution map by the ratio of deposition rate to sedimentation rate

2. Properties of drainage and impounding bodies, especially the size of relative density;
3. Types (wave, tide, and ocean current) and intensity of impounding body agent, especially a correlation with the input of the deposits;
4. Shape and landform of the drainage basin;
5. Tectonic activity and property of the sedimentary basin, including the stability, subsidence speed, and transgression and regression of the sedimentary basin;
6. Topographic gradient; and
7. Climate and wind.

These influencing factors can be classified into four categories: ① river property (including flow velocity, drainage, quantity, and proportion of mud and sand; ② basin property (including tectonization, terrain, and gradient); and ③ climate and wind (Table 11.7). The first three are the main influencing factors.

Table 11.7 Factors affecting deltaic sedimentation (according to Morgan, modified in 1970)

River stage (variant affecting sedimentary load and transport capacity)	Flooding phase	Sedimentation load	Suspended load and bed load (i.e., stream capacity) increase during flooding phase
		Grain size	Suspended load and bed load grain size (i.e., stream starting capacity) increase during flooding phase
	Low-stand river phase	Sedimentation load	Stream capacity reduces
		Grain size	Stream starting capacity reduces, and grain size decreases
Coastal sedimentation	Wave energy		High wave energy results in turbulence and current erosion to retransform and screen delta deposits
	Tidal range		Due to high tidal range distribution, wave energy crosses shore zone and generates tidal stream
	Current intensity		Strong coastal current is generated to transport coastal, offshore, and inshore deposits through waves and tides
Tectonism, basin topography, and shelf slope (corresponding to the change of sea level)	Stable area		Hard bottom bed prevents delta subsidence and forces delta plain to be built upward
	Subsidence area		Subsidence and sedimentary compaction are connected by structural depression, so that delta is established as an overlapped sedimentary lobe similarly to progradation
	Uplifted area		Land uplift (or sea level declination) may result in river distribution incision and transform the deposits
	Basin area		Landform change controls the development of vegetation, weather denudation depth, water distribution density, and longitudinal profile sloping of river. The greater the landform of the local area, the stronger is the erosion of the river
	Steep slope		Quick accumulation, relatively small delta scope, large number, and coarse grain size
	Gentle area		Slow progradation or retrogradation, relatively great delta scope, small number, thin thickness, and fine grain size

(continued)

Table 11.7 (continued)

Climate	Humid area	Hot or warm	High temperature and moisture are beneficial to forming dense vegetation coverage and good for capturing deposits transported by river or tide currents
		Cool or chilly	Seasonal characteristics of plant growth have little effect on the capture of deposits. Plant fragments are seasonally accumulated to form delta plain peat in cool winters
	Arid area	Hot or warm	Rare vegetation coverage has little effect on the capture of deposits and makes eolian sedimentation on the delta plain significant
		Cool or chilly	Rare plant coverage has little effect on the capture of deposits; winter freeze interrupts river sedimentation; and snow and ice melt in spring, and eolian sedimentation affects deposit transportation and deposition

11.3.1.1 River Property

1. Flow

In the river valley of a river with unstable flow and the delta formed by it, there are more distributary channels that often migrate quickly and frequently. However, a river with stable flow change rule tends to develop a snaking channel and shoelace-type sand body. Furthermore, the flow distribution also affects the grain size change and sorting of deposits transported to the delta. Although the flow is small and unstable, rivers concentrated in a short flooding period more easily transport coarse sediments to the delta, whereas rivers with large flow but stable change flow transport clastic substances with significantly better sorting to the delta.

2. Quantity, property, and sand factor of accompanying articles

Many factors control river-carried deposits, and the number of deposits is also a function of the drainage basin area and flow. A greater number of deposits more easily result in large delta plains, for example, Ganges-Brahmaputra, Yellow River, and Yangtze Delta Plains. The property (including gradients, grain size, sorting, roundness, etc.) of deposits carried by the river and sand factor directly affect the size of the delta building scope and the property of delta, which reflect the distance from the provenance and the type of river. In general, rivers with more fine-grained suspended matter are meandering rivers, and a broad fine-grained subaqueous delta is easily formed at a place farther away from the provenance; fine-grained muddy prodelta rich in water is poor in stability, which easily results in various deformation structures (slump, diapir, puncture, and mud volcano); and a river with more coarse-grained substances (high sand factor) that is closer to the provenance is generally a braided river, hence it easily forms a coarse-grained braided delta and fan delta.

11.3.1.2 Property and Agent of Impounding Body

The property of the impounding body mainly refers to the size of the water area, density of the water body, etc., and the size of the water area can reflect the size of the accommodation and affects the scale of the delta. The density of the water body determines the hydrodynamic characteristics when the delta is built, namely, the characteristics (agents depicted in the impounding water mainly refer to waves, tides, and ocean currents) of the estuary jet flow significantly influences the formation and transformation of the delta.

1. **Wave**

Wave action is very important to the formation and development of the delta; in particular, it greatly influences the retransportation of estuary sand and the change in the delta shoreline. Its main action is reflected in modifying the river-carried deposits. In an estuary subject to wave action, the distribution and form of the sand body is mainly determined by the mutual waning and waxing correlation between the capacity of the deposits supplied by the river and transformation and redistribution for deposits through waves. As the wave action is affected by wave energy, the greater the wave action, the stronger is the transformation to deposits. When deposits are continuously transported to the estuary by the river, a fluvial process always tends to distribute sand bodies in a direction that intersects the shore line at a high angle without the interference of waves, whereas the wave effect forces sand bodies to be arranged parallel to the shore line (Fig. 11.12). With the fluvial process decreasing and wave energy increasing, the form of the delta sand body shows a series of regular changes, generally transiting from pedate extending to the sea remotely to lobate, and then becoming pointed (Fig. 11.13). As the wave action can also greatly improve the maturity of deposits, a high-energy wave can form pure quartz sand with higher texture maturity. A low-energy wave generally has a small transformation to sand bodies that usually contain more clays, with lower texture maturity and permeability. Of course, the actual effect and result of the delta are further affected by other factors at the place where the delta is, such as the landform, quantity, and property of river-carried deposits.

2. **Tide**

The tidal range indicates the change in water level, and it also reflects the size of the tidal current. In each tide period, bidirectional flows are formed by the water current in both the estuary with a strong tide and the downstream river section. The deposits transported by the river are usually transformed into a series of subaqueous tidal sand ridges parallel to the flow direction (vertical shoreline) by virtue of the bidirectional tidal movement. In a flood-current-dominated estuary, the tidal current can make its way upstream to directly affect the upstream transportation of marine microfossils and marine authigenic minerals, the tidal sand ridge can expand to the channel, and most channels with strong tides have horn shapes (Fig. 11.14).

Ocean currents affecting the delta include the deep sea current impacting the continental margin, various nearshore currents, longshore currents, and rip currents derived from waves and tides, all of which can transform and redistribute the deposits to varying degrees. Longshore currents with various genesis can result in drifting deposits coastwise on a large scale, which greatly change the trend of the sand body in the estuary, and even force the channel to change the entering direction (Fig. 11.15).

11.3.1.3 Basin Structure, Landform, and Slope

1. **Basin structure**

For a basin with a stable structure, such as an epicontinental sea on a craton, the delta system changes slowly to form a shallow water delta. In a basin with quick structural subsidence, the subsidence and sedimentary compaction are connected by structural depression, and therefore, the progradation of the delta results in overlapping sedimentary lobes and a delta system with large

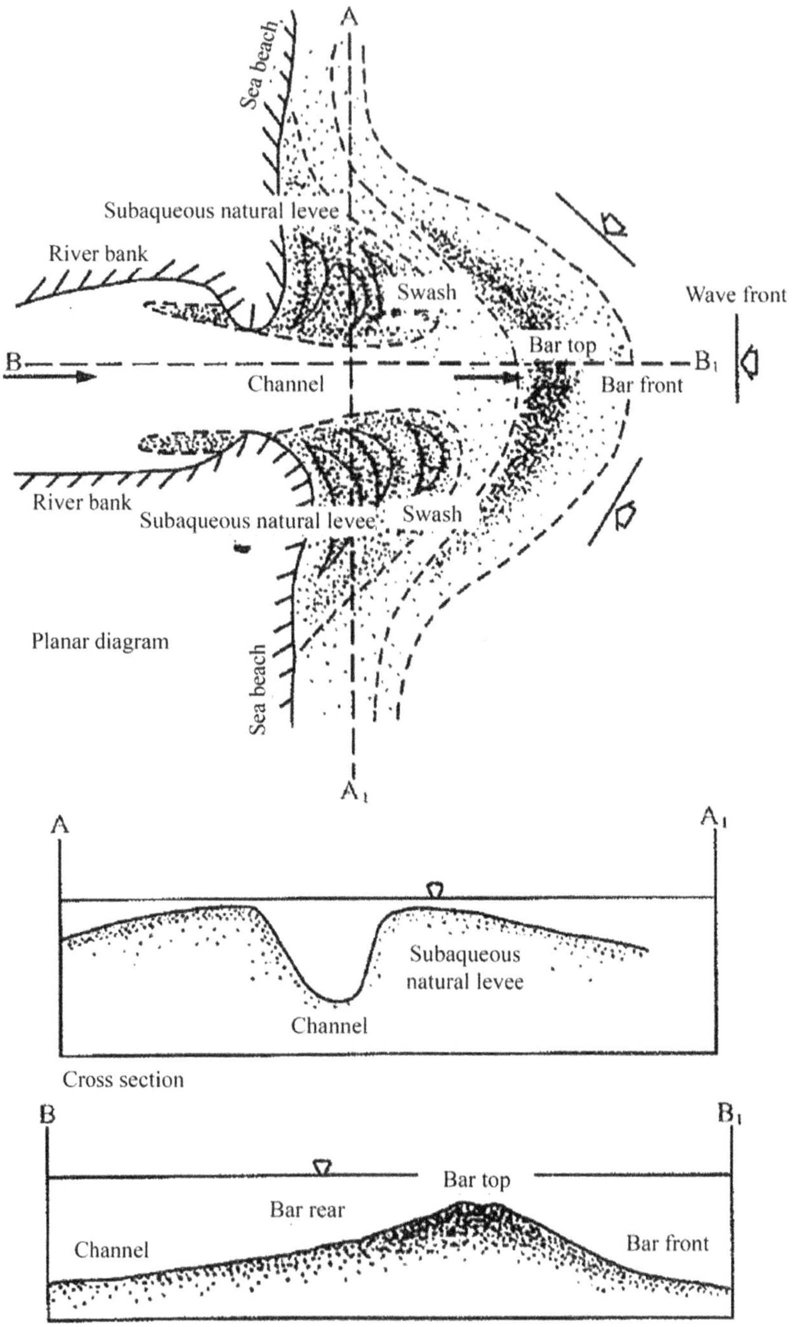

Fig. 11.12 Distribution pattern of estuary dam in wave-dominated delta (after Wright and Walker 1977)

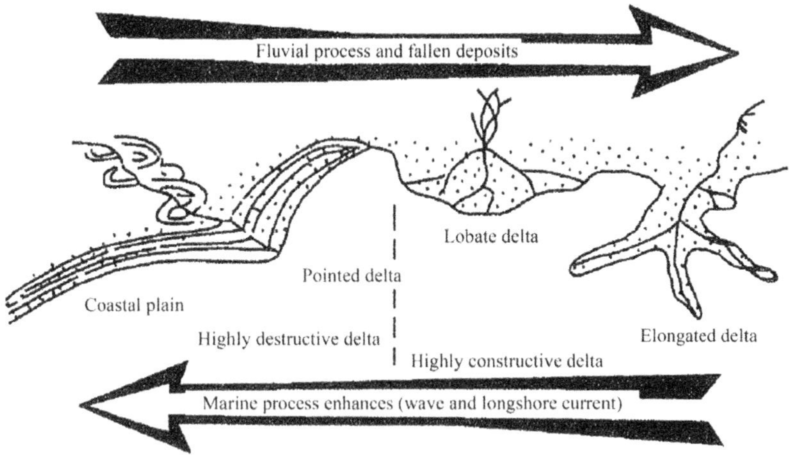

Fig. 11.13 The change of delta form under the interaction of rivers and oceans (waves and coastal currents) (after A. J. Scott 1969)

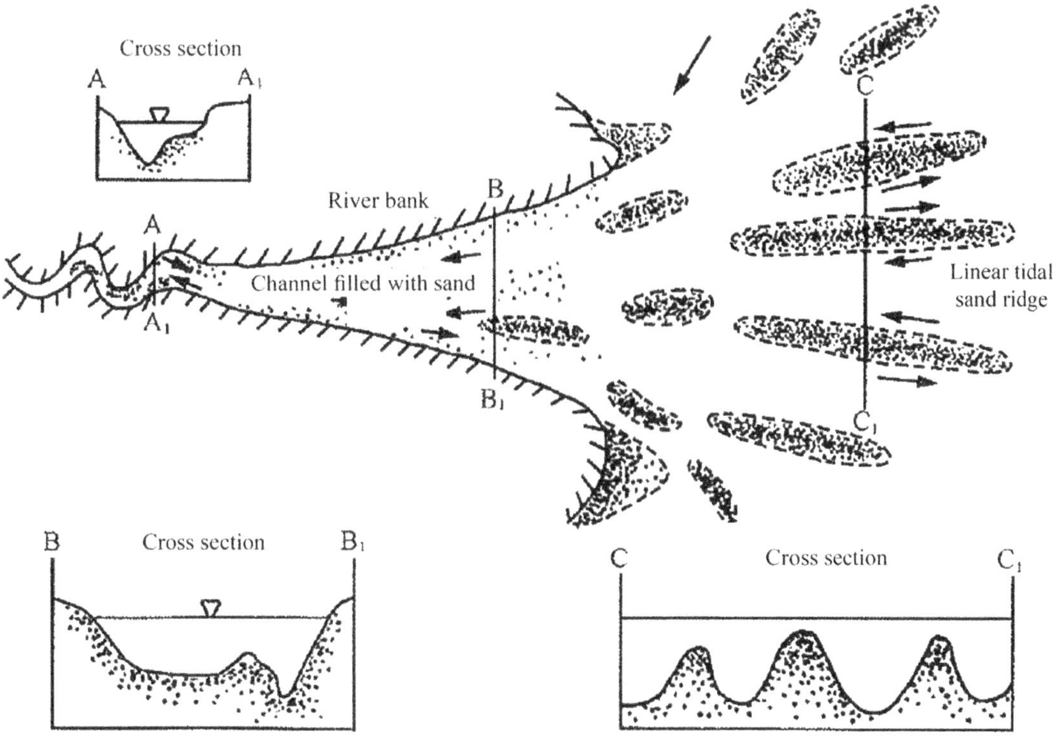

Fig. 11.14 Distribution pattern of deltas and tidal ridges (after Wright and Walker 1977)

thickness is formed. In an area with structural uplifting, sea/lake level declination may result in river distribution incision and transform its deposits. Specifically, sea/lake level declination is the main agent and process forming the subaqueous distributary channel of a delta.

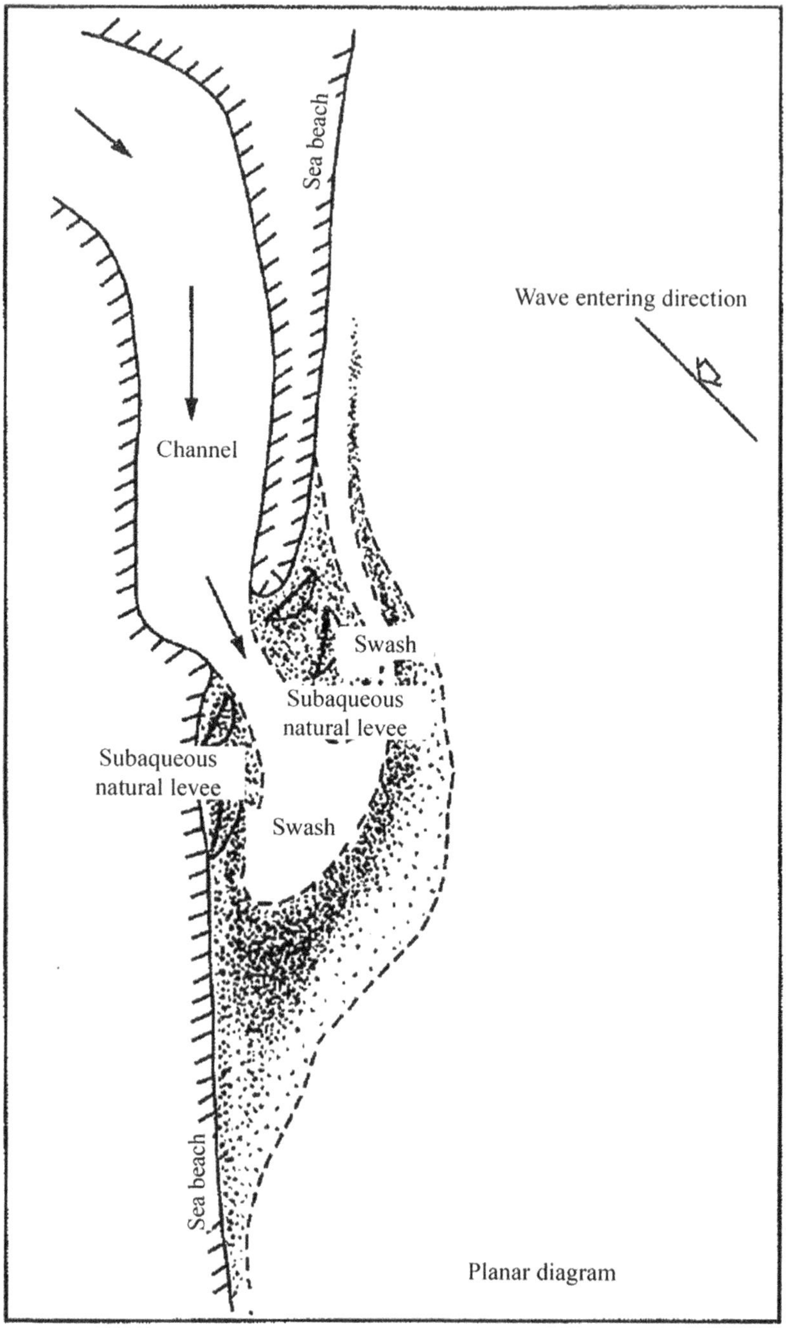

Fig. 11.15 Distribution pattern of delta sand body dominated by alongshore current (after Wright and Walker 1977)

2. **Basin shape and landform**

The basin shape greatly influences the delta forms, distribution characteristics, and change types of deposits. In combination with tectonic characteristics, Coleman and Wright (1975) divided the catchment basin into five morphotypes (Table 11.8), each of which has diverse structural diagrams (Fig. 11.16) for delta development.

Morphotype I refers to an elongated slot with openings at the two ends. The tectonic subsidence axis is parallel to the marine trough in the middle of the basin. The river supplies deposits to the marine

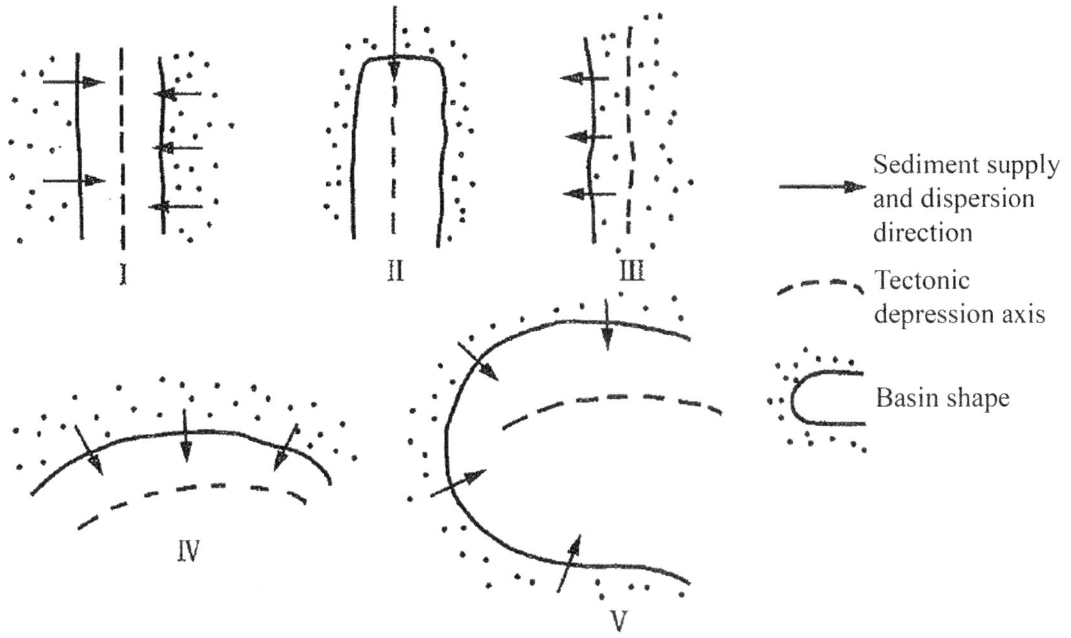

Fig. 11.16 Types of the water storage basin (after Coleman and Wrighr 1975)

Table 11.8 Types and characteristics of impounding basins for building deltas (according to Coleman and Wright 1975)

Type features	I	II	III	IV	V
Structural type	Elongated slot with openings at the two ends	Elongated slot with opening at one end	Inland down-warping area in basin	Tectonic subsidence belt in basin	Closed or semi-closed basin
Provenance direction	Supplying to marine trough from the two sides of the basin	Falling into the sea from the close end of basin		Several vertical river shorelines falling into the basin	
Fan characteristics	Tide-dominated or current-dominated asymmetric coastal bodies are developed	Mostly with large linear offshore tidal ridges	Freshwater deltas and megascopic swamps are developed	Usually controlled by high wave energy, mostly developed wave-dominated deltas	Deltas has quick subsidence characteristics

trough from the two sides of the basin, and it flows along the axis vertical to the basin. As the tidal current in a marine trough is confined by the shape of the marine trough, most strong tidal currents flow in the direction parallel to the axial direction. The delta sand bodies formed in such marine troughs are usually tide-dominated or current-dominated asymmetric coastal bodies.

Type II refers to an elongated slot with an opening at one end. The tectonic subsidence axis is parallel to the basin axis. The river falls into a marine trough from the closed end of the basin, most coastal currents and offshore currents flow parallel to the axial direction, deposits are diffused along the marine trough axis, and the delta has a large linear offshore tidal (sand) ridge.

Type III refers to an inland down-warping area in the basin. The structural axis is parallel to the shoreline, and the depositional site of the active delta is a stable platform. A freshwater delta and megascopic swamp are usually developed in the basin.

Type IV refers to the basin that is a tectonic subsidence belt. The tectonic depression axis is

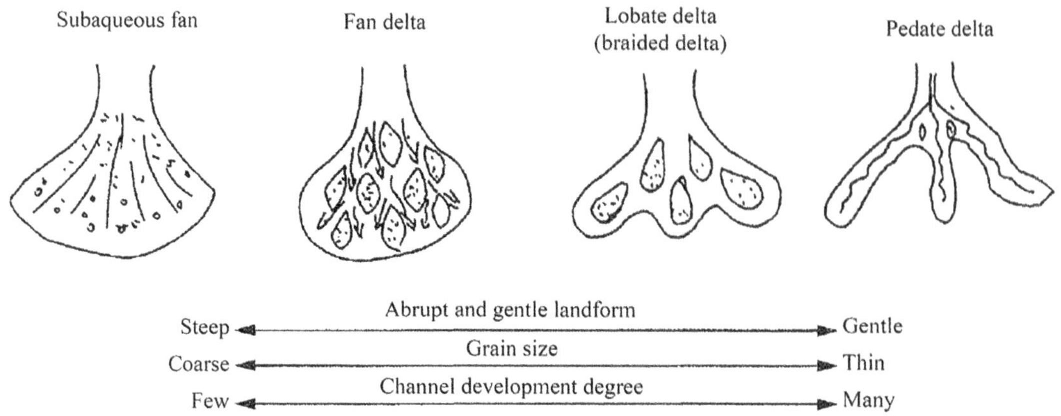

Fig. 11.17 Plane shape of delta is controlled by the paleo topography of sedimentary background. As the topography becomes slower and the distance of transportation increases, the bifurcation of the delta is stronger

located at one side of the shore line toward the sea. Facing a broad ocean basin, several rivers fall into the basin such that they are vertical to the shore line; however, the basin cannot be filled with deposits. Deltas are usually controlled by high wave energy, and wave-dominated deltas mostly develop.

Type V refers to a closed or semi-closed basin. The tectonic subsidence belt stays close to one side of the basin. The delta developed here has the specific characteristics of rapid subsidence, such as the Mississippi Delta at the north of the Gulf of Mexico, whereas the delta at the other side of the basin develops on the stable platform.

The watershed topography has a great effect on defining the number of deposits supplied. The landform change controls the development of vegetation, weather denudation depth, water distribution density, and longitudinal profile sloping of the river. Under the same climate conditions, the greater the topographic relief, the stronger is the erosion effect of the river and the greater are the sorting differences and supply amount of coarse-grained clastic deposits and vice versa, the contents of fine-grained deposits and clays are improved.

3. **Topographic gradient**

With regard to the sea basin, the total amount of deposits carried to the shelf through the river is generally higher than that of deposits transported away through waves and tides. As a result, the delta shelves are sedimentary shelves. The fast wave energy consumption of a low-gradient shelf is good for a delta to advance to the sea; delta deposits in a steep shelf are easily transformed by the sea; a narrow and deep shelf is not good for forming a large delta, and deposits are easily transported to the deep sea along the subaqueous river valley at the front edge of the delta to form a submarine fan. All large deltas are developed in the passive continental margin and wide and shallow shelves in the world. As a consequence, a delta formed in a steep slope, dominated by inertial flow, is characterized as many and small; however, a delta formed in a gentle slope, dominated by friction, is characterized as less and large. Therefore, the distributary channel of the latter is more significantly developed than that of the former.

The plane modality of the delta is significantly dominated by the paleotopography of the depositional background. The steeper the landform, the coarser is the grain size and the more similar to the fan is the planar form; as the landform becomes gentle, the transportation distance increases and the grain size thins gradually, resulting in stronger branching of the delta. When the distributary mouth bar changes from a rhombus to an egg shape (Fig. 11.17), the deeper the branching channel is cut, the more significant is the bank, and a crevasse splay will result during the flooding period.

11.3.1.4 Influence of Climate and Wind

The influences of climate on a delta are mainly presented as follows: various deposits can be provided to the river owing to physical and chemical weathering arising from the development of plants and soil within the drainage basin; the changes in rainfall capacity and evaporation capacity determine the change in the river runoff.

1. **Humid area**

In a hot or warm humid area, especially in a tropical or subtropical humid area, dense vegetation coverage is proven to be conducive to capturing deposits transported by a river or tidal current owing to high temperature and moisture; in the meantime, as the rainfall capacity is greater than the evaporation capacity, plants grow with strong chemical weathering as well as large and stable river runoff. The high content of suspended deposits in the river load forms a large fine-grained delta.

The delta plains in hot-wet areas and polar regions are beneficial to marsh plant growth, a large number of organic matters can be preserved in the deposits, and peat formations can be developed. In temperate zones, swamps can develop; however, organic matter is easily destroyed and is generally less than in the tropical belt, and peat formations are thin and unstable and are usually intersected with fine-grained deposits in a finger shape.

In chilly humid areas, the seasonal characteristics of plant growth have little effect on the capture of deposits. Plant fragments are seasonally accumulated to form a delta plain peat in cool winters.

2. **Arid area**

In rivers flowing through arid zones, aboveground vegetation is usually not developed owing to small rainfall capacity and high evaporation capacity; however, the ratio of the bed load/suspended load is high in river water owing to the strong physical weathering action, and a coarse-grained delta can develop easily. In a deltaic water plain, the climate affects the components and quantity of autochthonous deposits. In a hot or warm arid area, the rare vegetation coverage has little effect on the capture of deposits, and it makes eolian sedimentation on the delta plain significant.

In a cool or chilly arid area, the rare plant coverage has little effect on the capture of deposits; the winter freeze interrupts river deposition; snow and ice melt in the spring; the water activity of the distributary channel on the delta plain strengthens; and deposits are transported increasingly. Owing to the arid climate and inadequate current, the eolian sedimentation affects the deposit transportation and deposition.

In the arid belt, delta plain deposits usually contain evaporated salt deposit intercalations. For a smaller influence of the climate on the sea basin, the redistribution of deposits in the estuary is controlled by the change in meteorological flow (wind, etc.) affecting seawater dynamic conditions.

3. **Influence of wind**

Wind action is significant in broad and gentle delta areas. In a deltaic water plain, strong wind can retransport deposits to form broad wind dunes; a series of wind dunes parallel to the shore line can be formed by strong onshore wind in a deltaic coast affected by high-energy waves. In addition, wind is also the power for wind; in particular, storm waves have greater influence on the sedimentation in an estuary.

11.3.2 Correlations Between Fluvial Process, Marine Process, and Delta Form

A delta can be scoured, transformed, and redistributed by seawater waves or tides. When the fluvial process is greater, more mud and sand are transported; however, the marine process is weak, and the delta develops and advances to the sea (Fig. 11.13). Conversely, when the fluvial process is greater, less mud and sand are transported; however, the marine process is strong, and the delta is destroyed. As a consequence, the delta is a product of the long-term interaction of the fluvial process and the marine process.

11.3.3 Abandonment of Delta Lobe, Bifurcation of Distributary Channel, and Formation of Distributary Mouth Bar

11.3.3.1 Abandonment of Lobe

The formation and development of a delta is essentially a process in which a distributary channel bifurcates continuously and advances to the sea. Near the river estuary entering a sea/lake, a large number of bed loads are accumulated to form a distributary mouth bar owing to the reduced subaqueous slope, dispersed current, and suddenly decreased flow velocity. During the formation and development of a delta, deltas are abandoned owing to river migration, rechanneling, and source supply shortage, and an abandoned delta is eroded and destroyed owing to the strengthened basin agent of the impounding body; therefore, some corresponding sand bar-lagoon facies are formed, such as in the Mississippi Delta and Luanhe Delta.

11.3.3.2 Formation of Distributary Mouth Bar and Bifurcation of Distributary Channel

First, the cross-sectional area of the estuary area is greatly decreased by the accumulation of the sand bar. A river is usually divided into two branch flows by the upstream end of the sand bar. As the cross-sectional area of the estuary is reduced, the current will be blocked and will scour the outsides of the two branching flows to make the distributary channel expand outwards, and subaqueous natural levees will be formed outside them (Fig. 11.18).

Later, a distributary channel develops forward, and two new distributary mouth bars are formed at the two distributary channels to divide the distributary channel again, so as to form

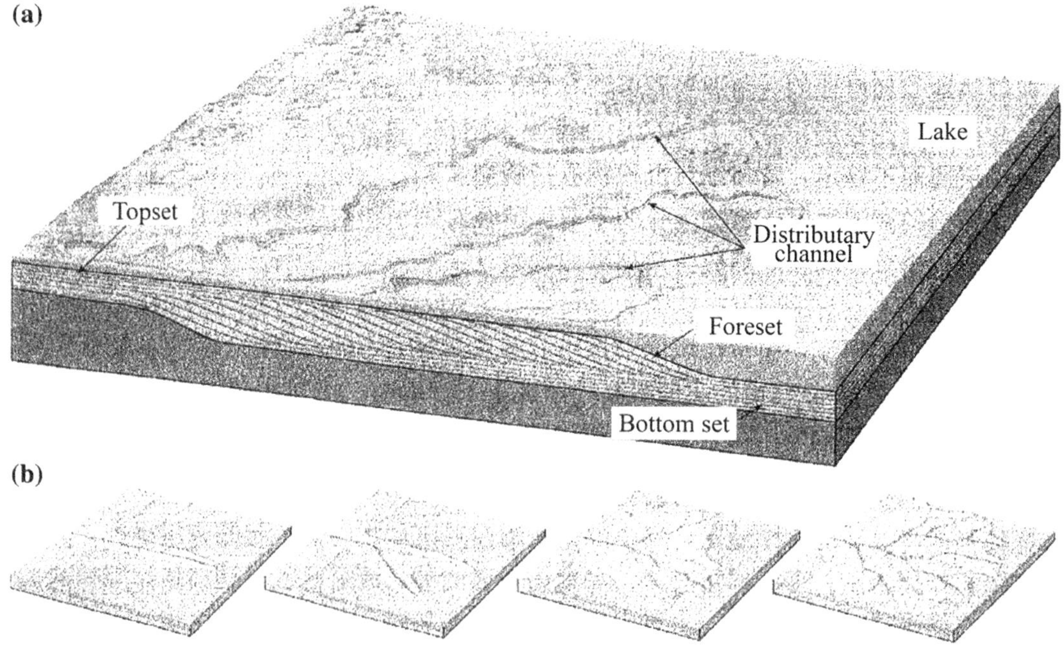

Fig. 11.18 The formation of the delta distributary channel is the distributary course of distributary channel (from J. T. Edward and K. L. Frederick 1997)

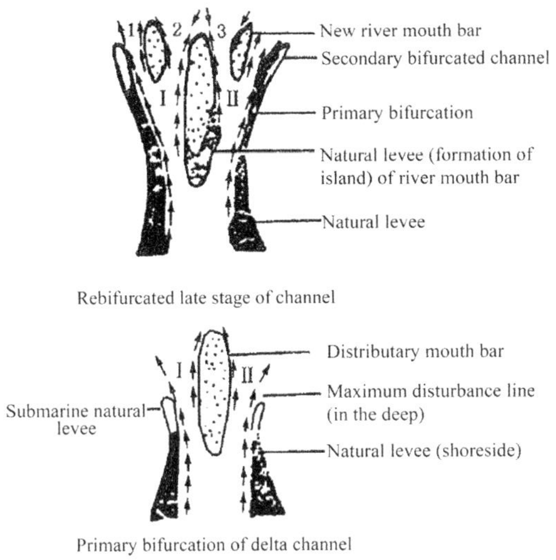

Fig. 11.19 Formation of the bifurcation of distributary channels and estuary dam (after Russell 1907)

secondary distributary channels expanding outward. As this process develops continuously and repeatedly, the delta advances to the sea constantly to form a lower delta plain or delta front developed by the subaqueous distributary channel and distributary mouth bar, which causes the delta to advance longitudinally and expand (Fig. 11.19) transversely (or laterally).

11.3.3.3 Migration of Channel and Formation of Crevasse Splay

The main reason for transversely expanding the delta is that the channel laterally migrates and swings. It is mainly dominated by the topographic slope and friction. In addition, another reason for transversely expanding the delta is the formation of a crevasse splay. The heights, widths, and stabilities of natural levees at the two sides of a distributary channel are gradually decreased downstream, and therefore, the natural levees of the lower deltaic distributary channel are developed poorly and even easily broken by flood water to form a crevasse and make the mud and sand flush out of the crevasse to form a crevasse splay, which leads the continental area of the delta plain to expand constantly in a transverse direction.

11.3.3.4 Abandonment and Evolution of Deltas

The deltaic distributary system advances to the sea/lake direction; however, it does not develop without limitations. Furthermore, the branch flows excessively expand to finally result in a river rechannel, and therefore, the water flows into a lower area. Alternatively, owing to a crevasse, the main river is rechanneled and the primary delta is abandoned.

When the sea/lake water invades, upper deposits are transformed by the sea/lake water action to start a delta destruction period. In the meantime, a new delta also grows near it. Sometimes, one delta is not finished and another one starts. After some period of time, the main channel can also return to the abandoned place of the primary delta to produce a new delta. In conclusion, the above phenomena can repeatedly occur more than once, so that all deltas are staggered and overlaid with each other to form a complicated delta system.

Regarding the abandonment and evolution of deltas, the most typical example is the Mississippi Delta in America. Between the 1940s and 1950s, it was believed that seven delta lobes (Fig. 11.20) were formed in succession in the Mississippi Delta over 6000 years. Delta activities can be known from the recently abandoned Lafourche Delta. Numerous distributary channels are distributed on the delta plain, and they frequently intersected toward the shore area to form a series of spatially connected distributary mouth bars at intersections between distributary channels and shoreline. Any residual beach ridge that cannot be found near the distributary mouth bars may belong to the wave-dominated delta front sheet sand. By the end of the 1960s, Fisher and McGowen (1969) found that a series of large delta complexes occurred in the ancient Mississippi on the modern Mississippi and shallow shelves formed at its east and west sides. Four large delta complexes were identified, amounting to 15 delta lobes (Fig. 11.21), which were abandoned gradually over the past 6000 years. Various delta complexes that can be seen in the present are in diverse abandonment stages, and therefore, various transitional

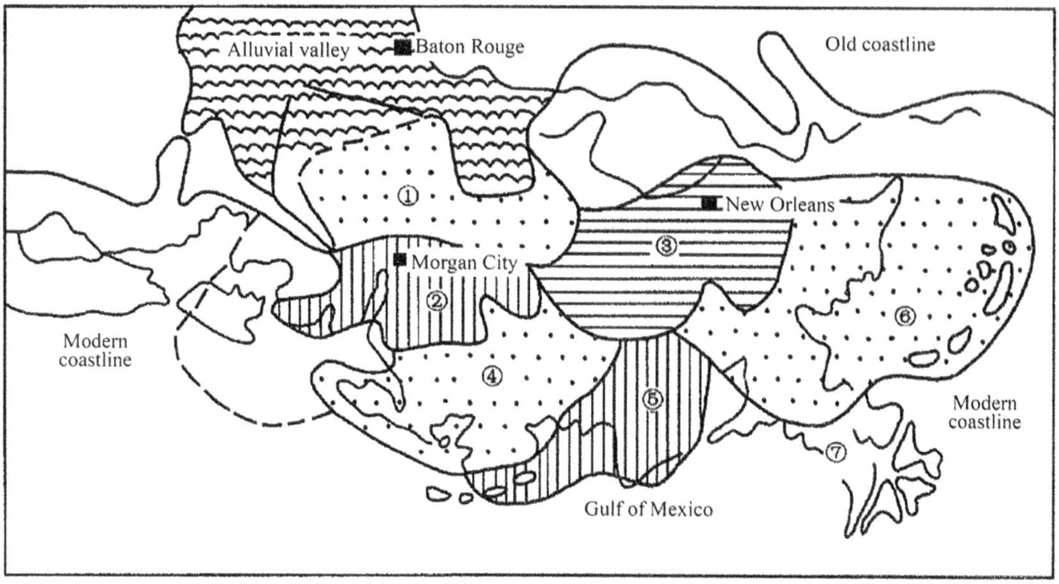

Fig. 11.20 Mississippi delta plain, consisting of seven delta lobes, with the sequence from old to young (from ① to ⑦) (after H. N. Fisk 1944; J. P. Morgan et al. 1951)

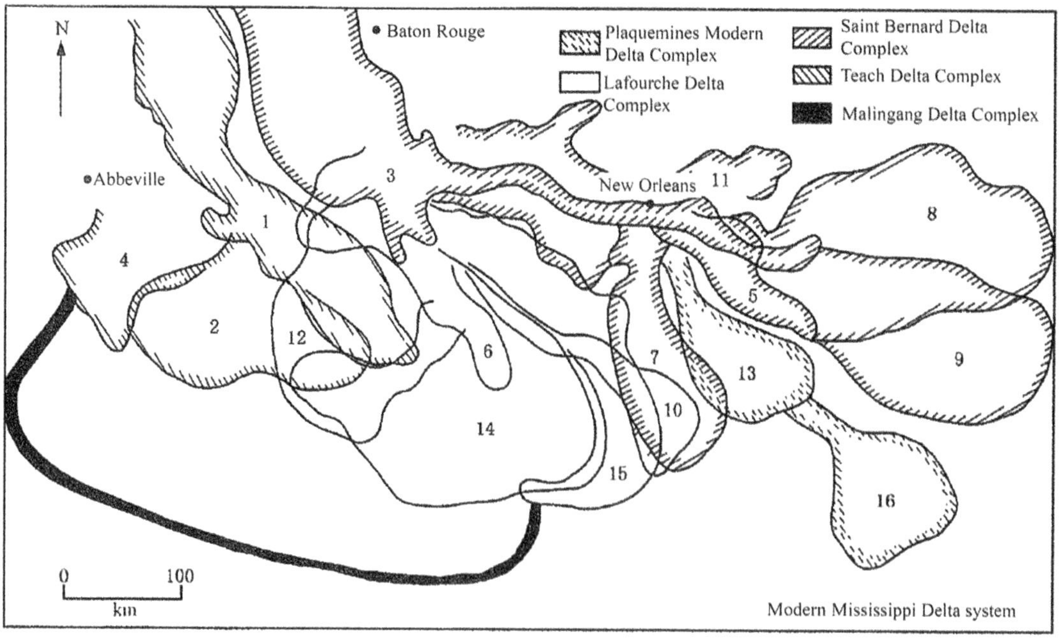

Fig. 11.21 Regional distribution of abandoned delta complexes and lobes in Mississippi river (since 6000 years) (after W. I. Fisher and J. H. McGowen 1969)

characteristics presented in the abandonment process may be determined by studying these delta complexes.

Some scholars believe that the Mississippi Delta had five delta lobe evolutions over the past 7500 years, and the delta evolved in (Fig. 11.22)

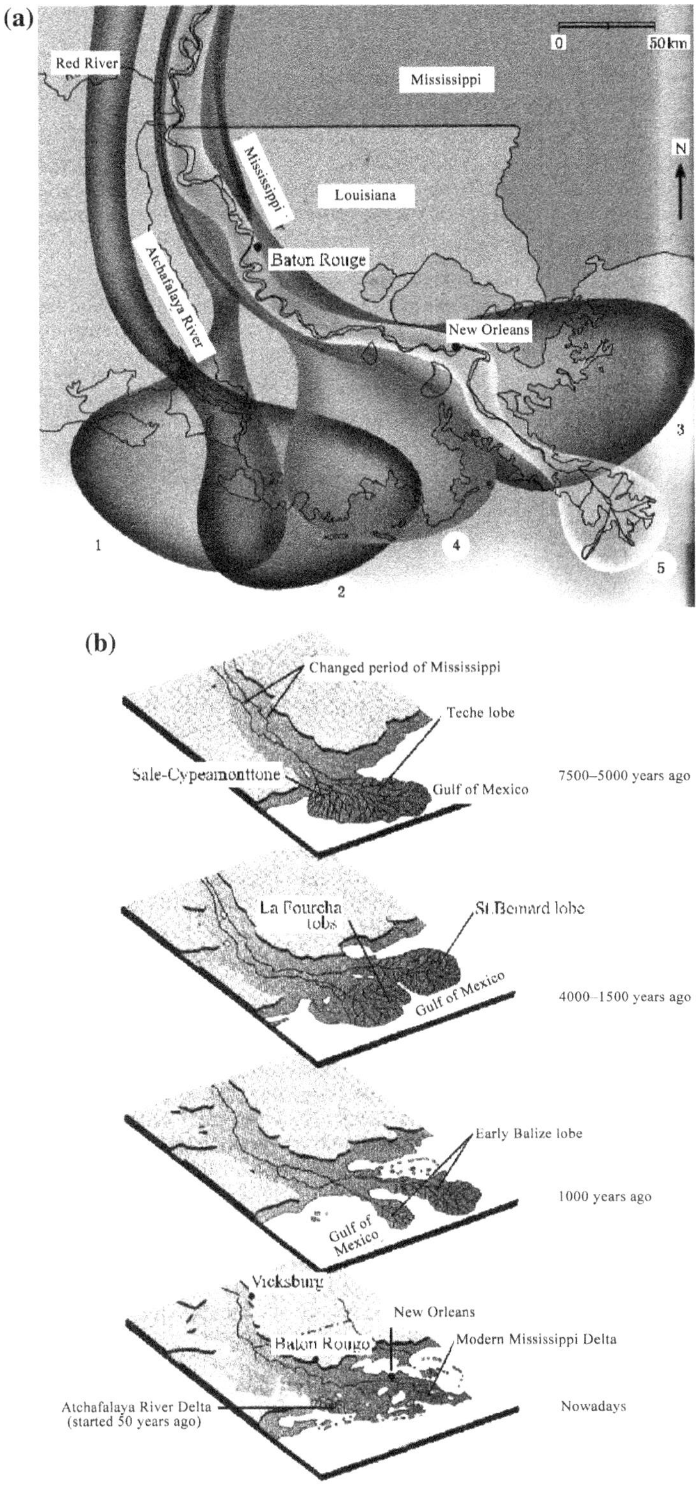

Fig. 11.22 Evolution of delta in Mississippi river since 7500 years (from Kathi Prancan 1998). **a** Evolution of Lobe; and **b** Delta characteristics and forms in all periods

four main periods. Irrespective of which viewpoint is more reliable, it is indicated that the Mississippi Delta was subjected to lobe abandonment and evolution during its geological history.

11.4 Division of Deltaic Sedimentary Environment and Facies

The zonality of the delta sedimentation on the plane is the main basis for depicting the sedimentary environment of the delta, and therefore, scholars divide the three facies belts or facies areas sequentially from the shore to the water below while studying a delta. As oil and gas exploration and development progresses and continental coarse-grained deltas are discovered, quartering is employed when the geometric shape of the sandstone reservoir is discovered. In other words, the delta is divided into four facies areas or subfacies, including the upper delta plain, lower delta plain, delta front, and front delta.

Therefore, when the delta subfacies is divided, quartering is applied for coarse-grained deltas and trichotomy, for fine-grained deltas. Of course, if the boundary between the upper and the lower delta plains is not clear and does not show obvious sedimentary characteristic differences, trichotomy may be applied uniformly. There are many types of delta microfacies (Table 11.9). There are diverse and specific methods for dividing different areas; however, more microfacies can be generally found in the distributary channel, natural levee, crevasse splay, swamp, interdistributary bay, subaqueous distributary channel, distributary mouth bar, distal bar, sheet sand, and prodelta mud. The lithology and sedimentary structures of these microfacies are different from each other.

11.4.1 Delta Plain

The delta plain is the upper sequence of the delta, and it mainly consists of continental deposits on the delta. In general, the delta plain is a wide low-flat area on which there are a number of distributary channels. A large amount of corresponding microfacies are formed between the distributary channels (Fig. 11.23). The coarse-grained delta can be generally divided into upper and lower delta plains.

11.4.1.1 Upper Delta Plain

In practical use, the part between the highest tidal level or highest level and estuary is defined from the estuary to the bifurcation of the main channel. As a consequence, it is generally an onshore part of the delta. The interdistributary bar is also developed in the coarse-grained delta, and it is dominated by braided channel deposition. In any type of delta, its channel has the characteristics of small tortuosity, stable location, relatively coarse grain size, and vertical fining upward. In one fining upward, there are above 60% of

Table 11.9 Summary of delta subfacies and microfacies classification

Delta plain		Delta front	Prodelta
Upper delta plain	Lower delta plain		
Main channel, braided channel, abandoned distributary channel, and flooding plain	Distributary channel, natural levee, crevasse splay, interdistributary bay, external flooding plain, lagoon, and swamp	Subaqueous distributary channel, subaqueous interdistributary bay, distributary mouth bar, distal bar, sheet sand (or storm sheet sand), barrier sand bar/lagoon, tidal sand bar/tidal flat, and tidal channel	Prodelta mud, turbidity sand, slump gravity flow deposits, and turbidity channel

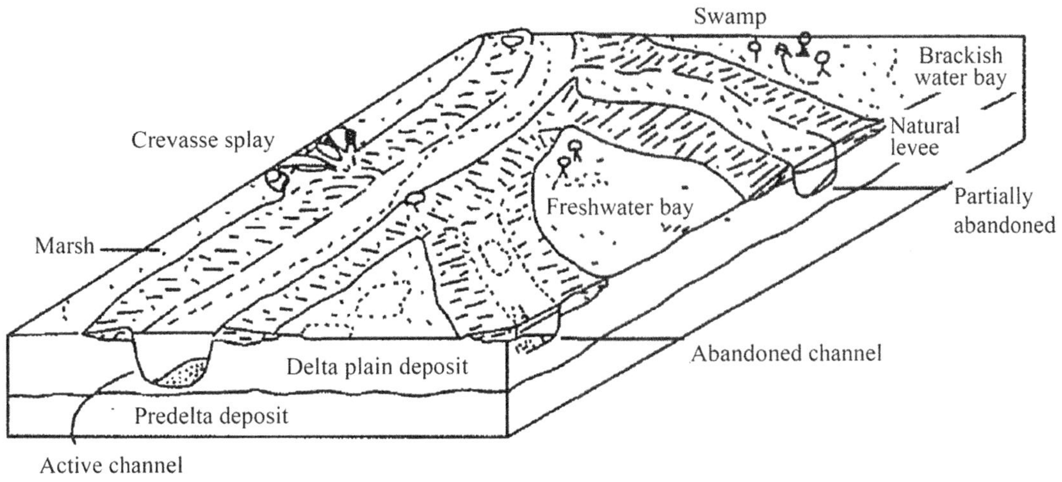

Fig. 11.23 Inner origin unit stereogram in the deltaic plain environment (after I. D. Meckel 1972)

sandstones and a relatively small amount of mudstone with light color.

11.4.1.2 Lower Delta Plain

This is situated between the highest and the lowest tidal levels. In practical use, it is defined between the bifurcation of the main channel and the disappearing of the distributary channel or distributary mouth bar, sometimes below water or above water. The distributary channel (coarse-grained delta is a braided distributary channel) is more developed, also with partially abandoned channels. The cross section of the sand body is embedded into the surrounding interdistributary bay muddy deposits in the form of a symmetric lens. As the interdistributary bar is close to the lakeshore or seacoast, the groundwater level is high for plant growth to form an interdistributary bay swamp. Therefore, there is frequently carbon, or even a coal seam, in addition to silt and muddy deposits. In addition, landforms including the interdistributary bay, natural levee, crevasse splay, and oxbow lake are generally developed on the lower delta plain. Their sedimentary structures and sequences are similar to those of a sandy meandering river, anastomosing river, and straight river, but with a smaller scale and thinner thickness.

The main microfacies characteristics of a delta plain are presented as follows:

1. **Distributary channel or abandoned distributary channel**

The distributary channel is the main microfacies of a delta plain, and a large amount of mud and sand for forming the delta are carried by the distributary channel. The distributary channel, with the characteristics of an ordinary channel, has a unidirectional current with periodic water level changes. Its property may be presented by a meandering river, braided river, anastomosing river, or straight river. The upper delta plain is dominated by a braided river and the lower delta plain, by a meandering river. The deposits dominated by the sand of each distributary channel belong to an upward thinned fining upward graded bed sequence formed by a unidirectional current with periodic water change; however, they are thinner than fluvial deposits at the middle and upper streams and show better sorting. Medium-thin sands are deposited at the bottom part in such a way that they change upward gradually into silts, muddy silts, or silt mud, and the top part includes silt and clay formations containing a large amount of plant roots. There is usually a scour surface at the bottom boundary. There is a trough, planar, or wavy cross bedding in an upward decreasing scale in the sand layer.

During complicated delta development, the distributary channel may play a part in abandonment to produce an abandoned distributary channel resulting from some abrupt factors such as natural substance blockage, slope reduction, or channel flow change. Under normal circumstances, the lower sequence of the abandoned channel includes normal channel deposits, above which fine-grained substances are covered, mostly in the manner of abrupt or quickly or gradually changed contact. As time goes by and the subsidence lasts continuously, the whole channel is eventually filled with fine grains, biodetritus with poor sorting, charcoals, and clays with an extremely high water content. MacMillan (1974) divided the filling of the distributary channel into three types: active channel, partially abandoned channel, and abandoned channel (Fig. 11.24).

2. **Interdistributary bay**

The interdistributary bay, the front end of which is usually connected with a sea or lake, is the deposition between the distributary channels. Its lithology is mainly presented by mudstone, silty mud, and muddy siltstone, or even a fine-siltstone lens-type sand body. When the delta advances to the sea, it is finally filled with deposits carried by crevasse splays or floods. Interdistributary bay mudstone gradually changes into prodelta mudstone downward and into swamp deposits that are rich in organic matters on the sequence. The sedimentary structure is dominated by horizontal, lens, and wavy beddings with strong bioturbation, and marine fossils may be found accidentally.

3. **Natural levee**

The natural levees at the two sides of the deltaic distributary channel are similar to those of a normal river; however, they are closer to the sea or lake. The natural levees situated at the two sides of a distributary channel become steep in the channel direction but gentle in the outside direction. Furthermore, natural levees, the heights of which decrease but widths increase, are developed better on the upper delta plain. Dominated by silt and silty mudstone, their grain sizes are finer than those of the distributary channel and crevasse splay. Horizontal and wavy cross beddings are developed, and ripples, clastic plants, stems, roots and burrows can be found commonly. Its bioturbation structure is very developed, and raindrop imprints, cracks, etc. can be found sometimes. On account of the alteration of the flooding period and water-shortage period, the sedimentary sequence of a natural levee presents the characteristics of interbedding of silt and silty mudstone.

4. **Crevasse splay**

Owing to the influences of increasing channel tortuosity, tectonic activities, and flooding action, the river water in a distributary channel breaks the natural levees and falls into the interdistributary bar, and its deposits are accumulated in a fan shape so as to form a crevasse splay, which is also called a secondary delta. Such crevasse splays have similar sedimentary characteristics to those of a normal river, and they are superimposed in a transverse direction and longitudinal direction to form a delta complex, which typically include the crevasse splay (Fig. 11.25) of the interdistributary bar in the Modern Mississippi Delta.

5. **Swamp and mud swamp**

Swamp deposits are distributed most broadly on the delta plain, and they generally account for above 90% of the delta plain area. Sand bodies are not developed; however, dark mudstones, peats, or brown coal deposits as well as siltstones carried by flood are usually developed. The common sedimentary structure comprises blocky bedding, horizontal bedding, and bioturbation, and sometimes, burrows can be found. Usually, there are clastic plants, charcoals, roots, ostracoda, gastropods, and siderite.

A mud swamp is similar to a swamp. The main difference between the two is that the former usually contains higher plants, including trees and shrubs, whereas large-area mud swamp deposits are usually developed on the upper delta plain behind the swamp.

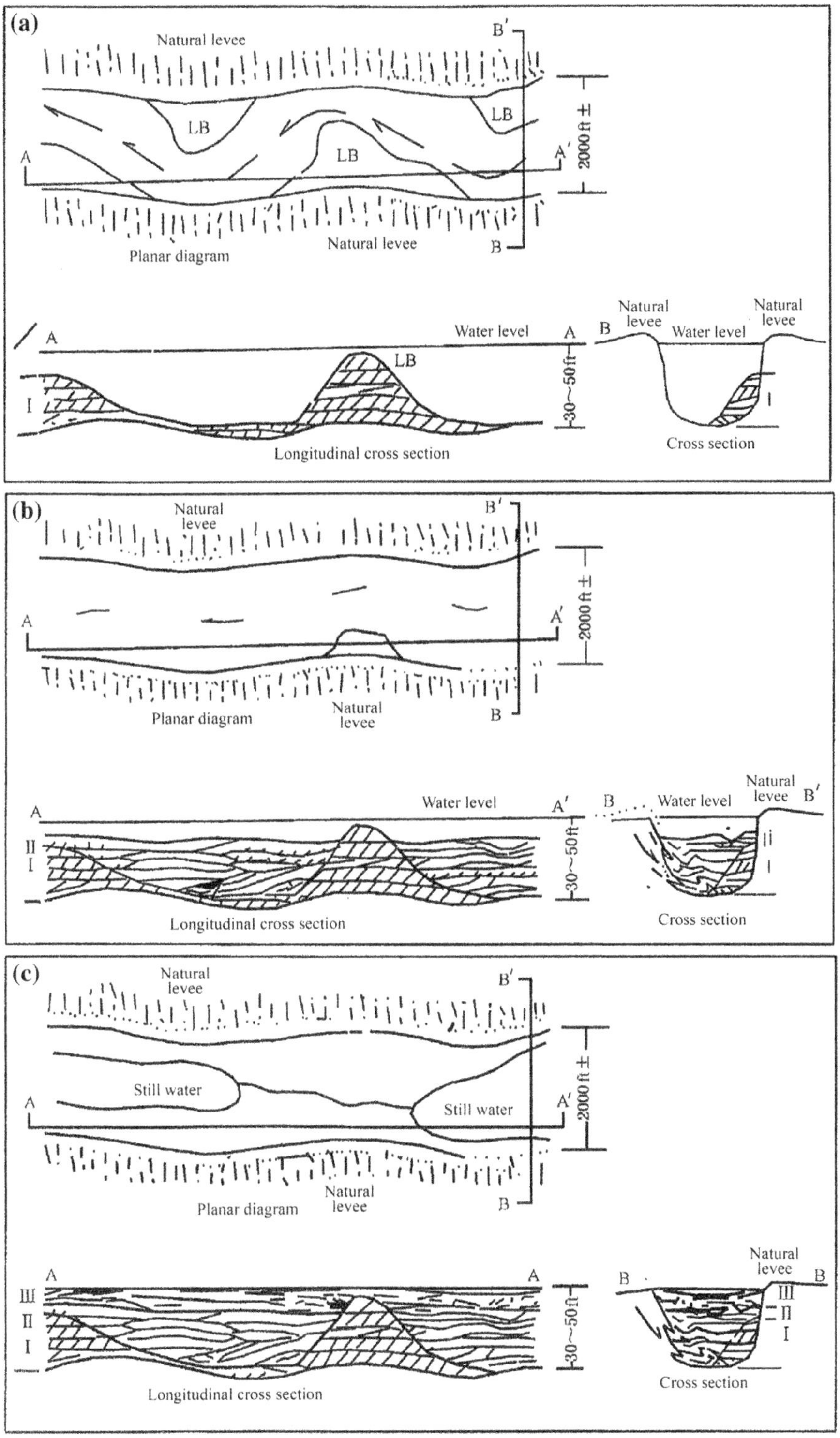

Fig. 11.24 Filling profile map of active, partial abandoned and abandoned river channel (after MacMillan 1974). **a** Active channel filling model; **b** partially abandoned channel filling model, with the lithology of fine-extremely fine grained sandstone and thin-corrugated bedding; and **c** abandoned channel filling model, with the lithology of extremely fine sandstone, siltstone, and clay, and laminar-thin bedding

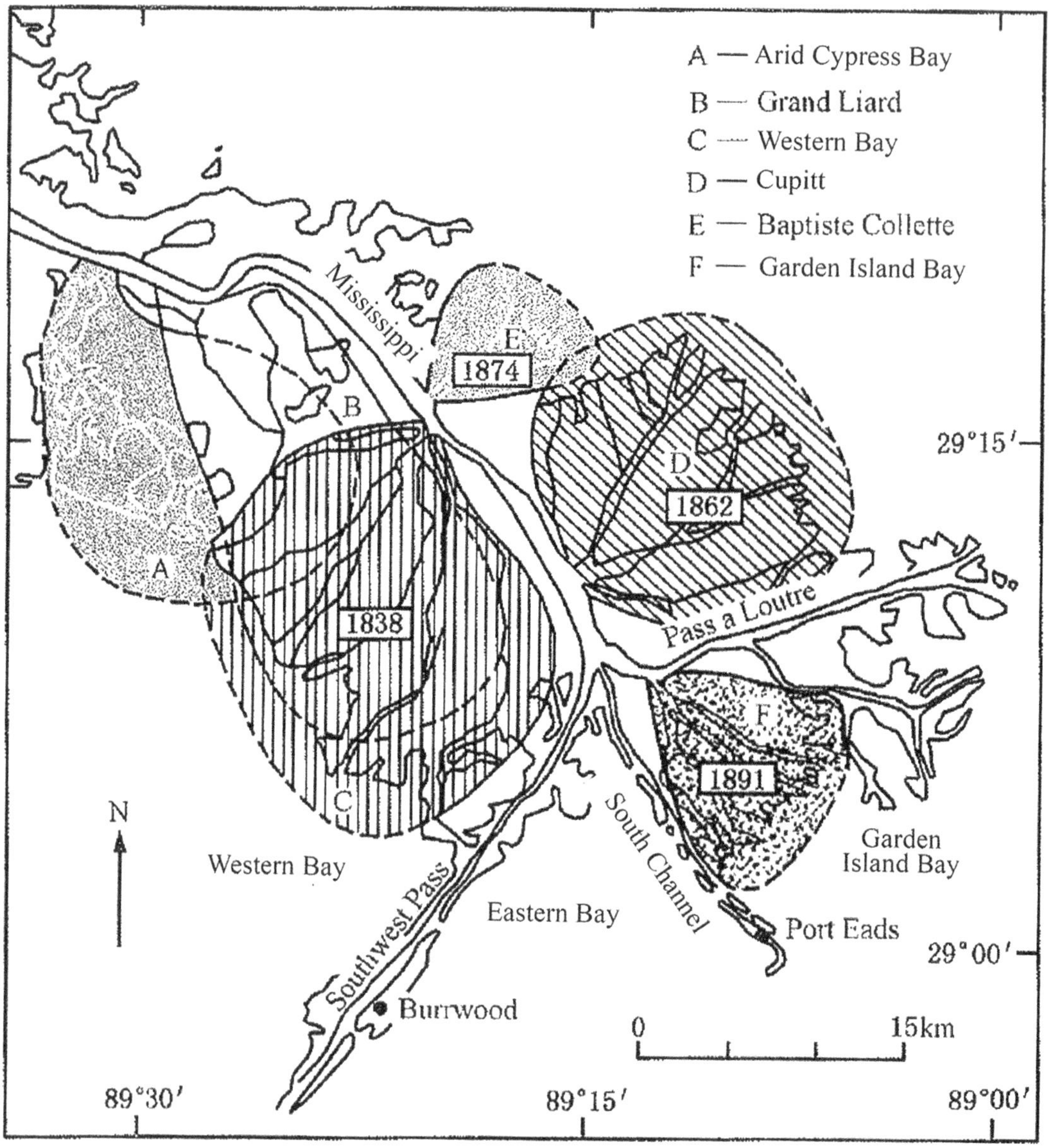

Fig. 11.25 Crevasse splay deposit of distributary Bay in modern Mississippi river delta (after J. M. Coleman et al. 1904)

11.4.2 Delta Front

The delta front is a centered development zone for sandstones in the delta. The slope zone situated between the lowest horizontal plane and the wave base, which is called the lower shoreface-meare gentle slope zone, is the most unique zone formed by the combined action of a river and lake. In the direction from this facies area to the catchment basin, subaqueous distributary channel, subaqueous natural levee, subaqueous interdistributary bay, distributary mouth bar, distal bar, and front sheet sand can be found. The development degrees of all microfacies vary according to the specific circumstances and conditions.

11.4.2.1 Subaqueous Distributary Channel

The subaqueous distributary channel refers to a delta plain distributary channel that continuously extends to the lake/sea. Since it is located underwater, it is called a subaqueous distributary channel. The stronger the fluvial process, the longer is the subaqueous channel, and therefore, it will be distributed in a striped vertical shoreline. The lithological section is a sand body formed by overlaying multiple layers of small fining upward sandstones, surrounded by gray or grayish green-dark shore-shallow lacustrine/marine mudstones containing lacustrine/marine fossils. It differs from the distributary channel in the lower delta plain. When a subaqueous distributary channel is very developed, microfacies including an interdistributary bay and a subaqueous natural levee can be found between the channels. It should be noted that the distributary precondition and reason for the distributary channel result from the continuous formation of the distributary mouth bar, namely, forming the bar and then bifurcating.

11.4.2.2 Distributary Mouth Bar

A distributary mouth bar, which is the most unique sand body in the delta, may be presented at the following places: in a delta without a developed subaqueous distributary channel and the distributary mouth bar situated at the estuary (Fig. 11.26) of the distributary channel of the delta plain entering a lake/sea; conversely, it is located at the terminal of the subaqueous distributary channel, the importance and characteristics of which are greatly inferior to those of the former.

The distributary mouth bar in the estuary of the distributary channel entering a lake has a slightly crescent, round, rhombus, ellipse, or egg shape (long axis is vertical to the lakeshore) on the plane. If there is a stronger lake current or lake wave transformation, sand particles can be laterally migrated so that all distributary mouth bars are connected to form striped sandy embankments parallel to the shoreline. The longitudinal section (vertical lakeshore) is an asymmetric lens, one thick end of which toward the estuary is the main body of the distributary mouth bar with thick substances. For the reverse-graded thick layer, blocky sands are formed owing to the winnowing of lake waves; multidirectional inclined middle tabular cross beddings can be found, and sandstones are usually formed by chemical cementation (such as of calcium, iron, and silica), with higher mineral and texture maturity. There may be a small amount of gravels and ooids at the top of the middle tabular cross bedding. The thickness and grain size of the sand layer, which usually belongs to the thin bedding between silt or silty fine-grained bedding and muddy bedding, are gradually reduced at one end toward the center of the lake. Mudstones transition in color from light gray and grayish-green to gray. When the slope is relatively steep, a miniature cross bedding, wavy bedding, lenticular bedding, flaser bedding, and horizontal bedding may be presented, and the movement energy of the lake water is weakened; a burrow and biological agitation structures are developed, including iderite strips or nodules. This zone is called the tail end (also called the front end or distal end) or tip of the distributary mouth bar, namely, the distal bar.

11.4.2.3 Distal Bar

The distal bar is situated in the area in front of the bar at one seaward side of the distributary mouth bar, with the slope slowly inclined toward the impounding basin. The deposits, which mainly include silts and a small amount of clays on the distal bar, are thinner than those on the distributary mouth bar, and clay silt beddings are usually formed. During the flooding period, fine sand deposits, and sometimes, graded bedding may be found. The sedimentary structures comprise horizontal bedding, ripple cross bedding, wavy cross bedding, and ripples as well as wavy-flaser-lenticular combined bedding. More clastic plants and charcoals are distributed on the bedding surface. As benthos may live in this area, biological fossils and

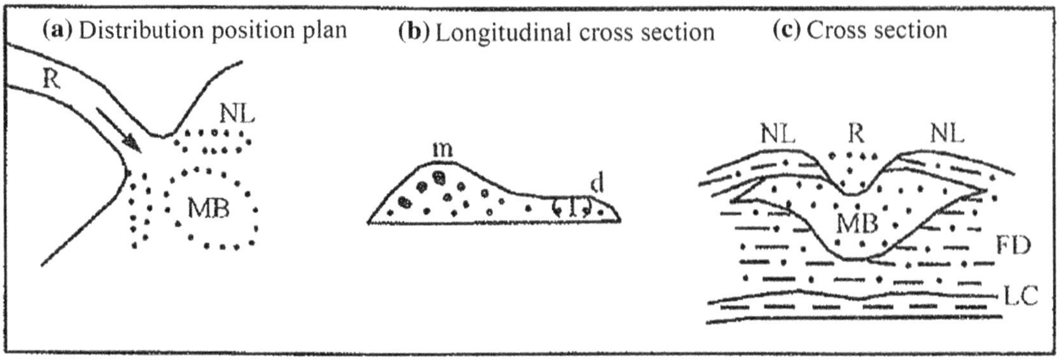

Fig. 11.26 Sketch map of position and shape of estuary dam. R distributary channel; NL subaqueous natural levee; MB distributary mouth bar; m sand bar body; d sand bar tip; FD prodelta mud; and LC lacustrine mud

burrow traces can be found, with developed bioturbation structures and shells distributed sporadically. Owing to the transformation of the wavy action, the long axis on its plane is of an elliptical shape and is usually parallel to the shoreline. On the plane, the deposits on the distal bar are usually of laminar shapes that extend farther and that are vertically distributed below the distributary mouth bar so as to form an upward thickened complete sequence together with the distributary mouth bar.

11.4.2.4 Sheet Sand

Because the distributary mouth bar and distal bar on the delta front are scoured by sea water or are subjected to storms, the front sheet sands are distributed on their side wings to form thin and large sand layers. This sand layer with good sorting and pure quality can form an excellent reservoir, and parallel bedding and wave ripple cross bedding can be found on its sedimentary structure. When the delta is dominated by inertial flow, sand is stored easily but large-area sheet sand is uneasily formed; when the delta is dominated by friction, delta lobes are developed; and when the delta advances to the sea, sand covers the thin mud layer, and sheet sand is formed under the wavy action of sea water. When the action of sea water is stronger, the sheet sand and distributary mouth bar are connected; when the destruction of sea water is very strong, the distributary mouth bar may disappear completely; and when large-area sheet sand is formed on the delta front, it is difficult to distinguish the sheet sand from beach sand. In seasons with more intense storms, various sand bars and beach sands form sheet sand all around the sand bar under the action of wind; this is also called storm sheet sand.

11.4.3 Prodelta

The prodelta is situated in front of the delta front, i.e., the part below the wave base, and it mainly consists of dark mudstones mixed with silty sandstones. Actually, it belongs to a normal shallow lake or shallow marine deposition, and therefore, it is called an offshore deposition; it usually contains slump turbidite lens sand bodies.

Overall, the development of all facies belts in the delta varies according to the factors of specific areas, such as tectonic subsidence, landform, and hydrodynamic conditions, all of which interact with each other. For instance, the development degree of a distributary mouth bar is comprehensively affected by various factors, such as the comparison between the subsidence rate of the distributary mouth bar, accumulation speed of mud and sand carried by the river, comparison between river energy and lake wave intensity, sand-mud ratio, and estuary landform. If the mud content in the substances carried by the river is high, the mud deposits of the prodelta are thick, and the subsidence and sedimentary speeds are equal or slightly greater; the distributary mouth bar edge and sedimentary edge are settled in the muddy belt and kept below water, and a distributary mouth bar is developed and preserved best. When the river energy of the onshore distributary channel is intermediate and the subaqueous gradient in the distributary mouth bar is slightly gentle, the distributary mouth bar will gradually break the surface of the water when it thickens and rises, and therefore, the continuous flow of the distributary channel entering the lake is blocked, and the distributary river is bifurcated to bypass it so as to form new distributaries. If the river water energy is strengthened, the distributary channel directly cuts through the top of the distributary mouth bar (Fig. 11.24c), and therefore, the deep trough at the top of the distributary mouth bar is filled with distributary channel sands to form the superimposition of two genetic sand bodies. When this action continues, narrow and straight finger-like striped sand bodies will be formed, called finger bars, and they may even be distributed vertical to the bank or slightly radially. When the river energy is stronger and the gradient of the estuary is slightly high, the distributary river entering the lake undercuts continuously and advances forward to form a subaqueous distributary channel that extends farther; by this time, the subaqueous distributary channel sand body, rather than the nearshore estuary sand bar, can be found on the shoreward side of the delta front, and some small and thin sand bars and front sheet sands are developed at the place where the channel disappears in front of it.

11.5 Sedimentation Mechanism of Different Deltas

Owing to the effect of the interaction strength of the impounding body and river, the sedimentation mechanisms and reservoir characteristics of different deltas vary. Hydrodynamic conditions and geological agents are the most basic factors for forming diverse deltas, thus the formation mechanisms cannot be discussed without mentioning these conditions and agents.

11.5.1 Sedimentary Mechanism of Fluvial-Dominated Deltas

Such deltas are formed when the quantity of deposits input by the river is much greater than the seawater energy, i.e., these deltas are formed when sufficient deposits are supplied. The sedimentary structure of fluvial-dominated deltas is a series of massive beddings with erosion surfaces, and parallel or wavy beddings of thin-layer lenses or clays; siltstones and clay intercalations are also usually found. These deltas are also rich in small to large symmetric or asymmetric ripples, dunes, and trough cross beddings. In addition, high-angle and unidirectional current distributions are the most common sedimentary characteristic types. They are associated with clay in the prodelta. Moderate to strong bioturbation may be found in the shell bed of a scouring structure rich in ferric nodules. The planar shape of the fluvial-dominated delta is mainly lobate or pedate, with lens sandstones produced by lateral accretion and distributary mouth bar deposits formed by progradation. The lentoid shapes of the two are just opposite, i.e., the former is a concave lens and the latter is a convex lens. Both can evolve into continuous rock strata. According to their shapes and characteristics, they can be further divided into two types: elongated (namely, pedate) and lobate delta.

11.5.1.1 Sedimentary Mechanism of Elongated (or Pedate) Delta

Owing to weak seawater action, the large injection of mud and sand carried by the river, especially the low sand-mud ratio and many suspended loads, natural levees can be beneficially built on the fluvial-dominated delta, and the distributary channel will remain stable; meanwhile, very thick prodelta mud can be deposited. Distributary mouth bars also develop and extend forward to cover the thick prodelta mud, and they can quickly settle in it; however, most are preserved to form a typical "finger sand bar." The maximum turbulent line and turbulence in a distributary channel are located at the deep part of the channel and are close to the two sides, especially in the riverbed. These turbulent lines are associated with the maximum cutting and scouring; therefore, most deposits are thrown to the parts with smaller turbulence. Considerable river load is pushed to the center to form the central shoal of the channel. In an estuary entering the sea, water in the channel continues flowing forward and enters the sea in the form of a jet flow. When water is out of the control of fixed river banks, it scatters to the two sides to a certain degree to make the flowing energy disappear eventually by widening the jet flow and lowering the speed. Near the front ends of limited river banks, jet flows are concentrated and move forward to enter still water; the scattering of the jet flow results in an increase in the distance between turbulent lines with maximum strengths. There is a tendency to cut and scour below each turbulent line, and a majority of the input substances are delivered to the marginal still water at the two sides during the turbulent exchange process. As a consequence, a subaqueous natural levee deposition is formed outside each turbulent line. Meanwhile, deposits are introduced into relatively broad still water in the center of the channel, so as to deposit a shallow shoal. The channel is bifurcated into two distributary channels along the central shallow shoal. A new distributary channel also produces its own marginal maximum turbulent line, and a new bifurcation process is produced at each new distributary channel mouth. Only if the wave and longshore current do not produce converse destructions, this process will continue to increase the channel in the form of a geometric progression, grow forward, and return to the delta continuously (Fig. 11.27). A pedate delta, which is formed by a sedimentary process, not only makes most distributary channels on the delta plain develop into a straight river and deposit elongated distributary channel sands but also characterizes the delta front sand body by the elongation of the vertical shoreline, so that a so-called "finger bar" (Fig. 11.28) is formed, especially a "striped lingual bar" in most cases.

11.5.1.2 Sedimentary Mechanism of Lobate Delta

The platform of this delta has a semicircular or fan shape protruding to the sea. On its front, some parts are convex (distributary port), and the other parts are concave (interdistributary bay) with a slightly zigzag shape (Fig. 11.29). Compared with an elongated or pedate delta, the input of mud and sand in this delta is low, the mud-sand ratio is rather high, and seawater wavy action is strengthened. As a consequence, the distributary mouth bar covers the thinner prodelta and settles slower to scour the sand of the delta front and transform it easily to be redistributed among all distributaries, so as to form sheet sand layers connected with the distributary mouth bars of all distributaries.

The distributary mouth bars of these deltas, which are far less than in the development of the pedate delta, usually have a "rhombus" shape on the plane. The later transformation of waves is significantly strengthened, and the topographic slopes are steeper than that of the former. Because of this, the vertical thickness is relatively greater; however, the planar spreading scope may be relatively smaller. Owing to the effect of the stronger wavy and tidal transformation, the connectedness between sand bodies is better. The main deposits of these deltas are lobes, with the lens unit produced by very important downstream accretion. Owing to the high content of sand in the belt with high sand content, the lobate profile (Fig. 11.30) vertical to the basin margin may be drawn.

11.5.2 Sedimentary Mechanism of Wave-Dominated Delta

This delta has the following characteristics: one or two main rivers enter the sea generally, with neither too many nor too large distributaries, low mud and sand quantity are input into the sea by the river, and the sand/mud ratio is higher and wavy action is greater than those in the fluvial process. As a result, the sand and mud input by the river are rapidly redistributed under the wavy

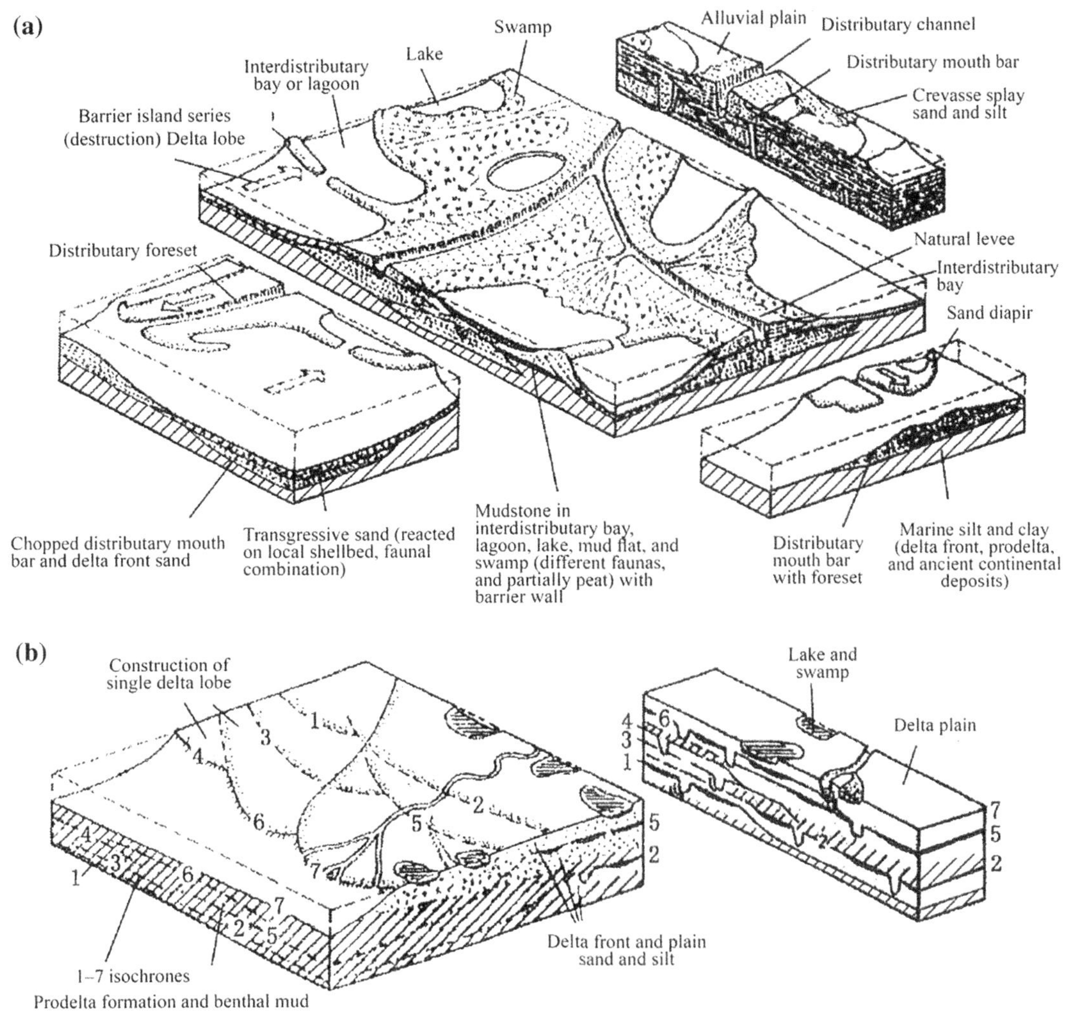

Fig. 11.27 Model maps of the bird-foot delta (after Einsele 2000)

action, hence a series of beach ridge sands or barrier sand bars parallel to the coast are formed at the two sides of the estuary. Furthermore, more sandy deposits are deposited on the estuary to form the estuary in an arc and beak (pointed) shape protruding to the sea.

In case the wavy action is further strengthened, the fluvial process is almost completely overcome; meanwhile, a unidirectional strong longshore current will result in the estuary shifting, even parallel to the coast. A linear barrier island or barrier bar is built in front of the estuary to block the estuary from forming a closed beak delta.

When the ocean has enough wavy and longshore current actions, the clastics carried into the sea by the distributary channel can be scattered coastwise to prevent the growth of a natural levee and bifurcation of the distributary channel. Therefore, the delta keeps a stable and smooth front to form an arc-shaped delta. In some cases, only a pair of natural levees is deposited forward to form a pointed delta. The formation process of an arc-shaped delta is essentially as follows:

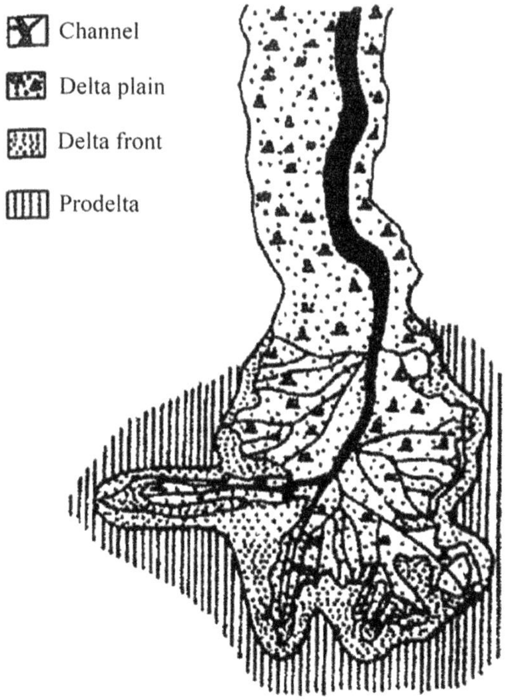

Fig. 11.28 Bird-foot delta in Mississippi

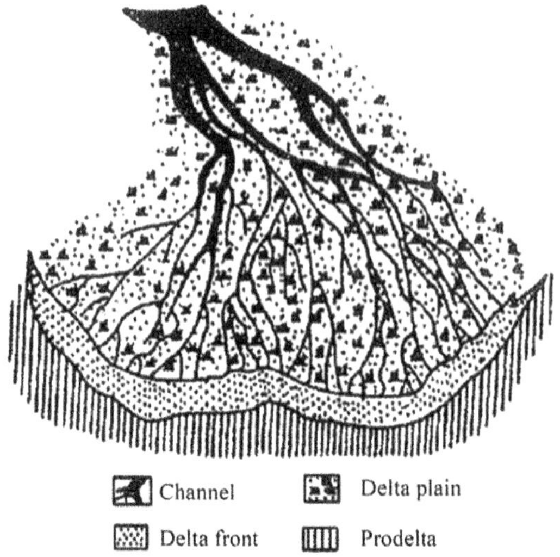

Fig. 11.29 Mississippi river Holocene lobate delta

At the preliminary stage, one smaller distributary mouth bar is formed on the estuary of the distributary channel entering the sea, and the distance between the distributary mouth bar and the primary shore is around four times longer than the width of the channel. The distributary mouth bar is usually cut into several sections through some subaqueous channels radially scattered from the estuary. There are sandy deposits at the bar top, as well as clay, silt, and a small amount of sands behind the bar. Deposits from the bar top to the sea (bar front) are changed into silt and mud from sand.

Later, the distributary mouth bar grows constantly upward and seaward; when it is higher than the sea level, plants grow to further catch fine-grained deposits. Therefore, the distributary mouth bar gradually connects with the land, as a result of which the early distributary mouth bar changes into part of the delta plain, and the distributary channel passes through the old distributary mouth bar cut to form a new distributary mouth bar outside the shore. When the secondary distributary mouth bar aggrades to break out of the marine surface, the distributary channel passes through it and then deposits the third distributary mouth bar to the sea, thus continuing to make the delta aggrade to the sea continuously. The later distributary mouth bar is always deposited above the early predelta deposits. In the new delta plain formed by the distributary mouth bar, some small shallow swamps are usually formed between the top ridges of the two distributary mouth bar deposits.

The delta is mainly characterized by the formation of an elongated and coastal distributary mouth bar sand body belt along the front belt of the whole delta plain. There are significant differences in the distributary mouth bar sand bodies as well as in the distributary plain deposits between the arc-shaped delta and the pedate delta. A swamp environment that is affected by tides as well as large amounts of small meandering rivers can be found on the arc-shaped delta plain. These small meandering rivers are connected to the main distributary channels. Owing to the effect of the greater tidal differences, small tidal channels branch from the main distributary channel and freely move to form a meandering river on the delta plain, so as to deposit a large amount of deposits filled with a small point bar and abandoned channel.

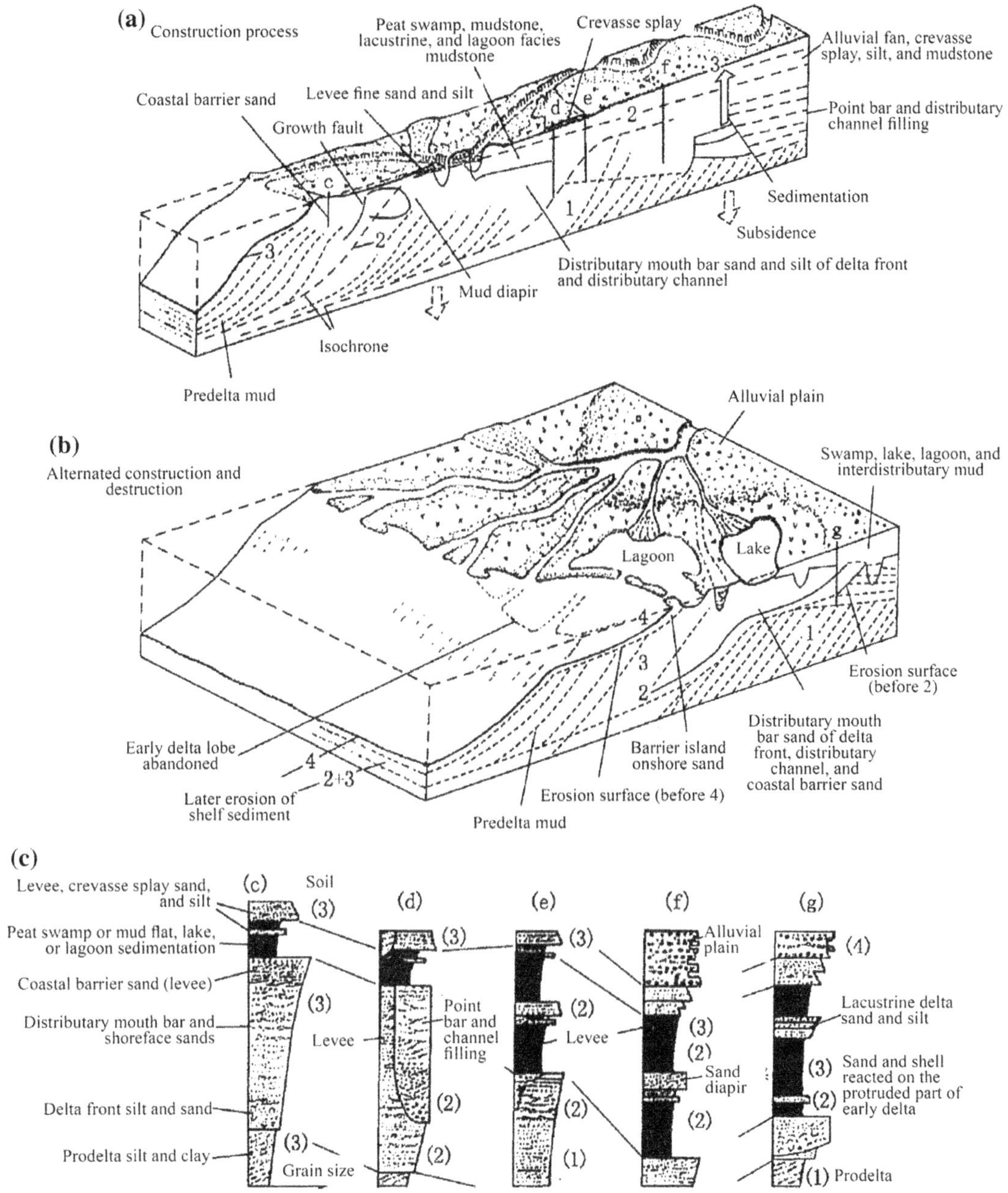

Fig. 11.30 The relationship of the typical lobate delta faces form coastal erosion and fluvial deposits due to constant subsidence (similar to the Niger delta, after Allen 1970, from Einsele 2000). **a** During the construction process, the unchanged scenario is that the prodelta advances constantly, and the forward advancing of the isochronous surface should be focused on; **b** discontinuous progradation is interrupted by destruction (erosion surfaces 1 and 2) of the phased local delta, and the destruction not only affects the shallow covering the interdistributary bar, lagoon, and parts of deposits in the swamp and lake but also moves sand and silt deposits of the distributary mouth bar and subaqueous levee; **c–g** show vertical sections (positions marked in **a** and **b**)

However, the low migration of the pedate delta distributary channel results in a very small point bar sand.

11.5.3 Sedimentary Mechanism of Tide-Dominated Delta

Some rivers fall into areas with stronger tide effects, hence the strength of tidal action plays an important role in controlling the formation of delta sand bodies. Owing to the tide influences and bay control, the area with stronger tidal action is also called an estuary delta. When the tidal action is strengthened, deposits carried by the river are usually transformed into a linear tidal sand bar through the scouring actions of the bidirectional tidal current and river flood. These sand bars parallel to the tidal direction are radially distributed in front of the estuary in the form of fingers and are filled in the tide or the two sides thereof.

On account of the diverse sedimentary characteristics of various delta fronts, the sedimentary thickness of the distributary mouth bar in a fluvial-dominated delta is greater. The paragenesis of a wave-dominated delta distributary mouth bar and beach sand ridge results from wave transformation. A beach sand ridge, with good sorting, has symmetric ripples. Undoubtedly, the characteristic beddings of tidal effects and double mud layers are found in the tide-dominated delta front deposits.

The tide-dominated estuarine deltas (Fig. 11.31) are built on the estuary with great tidal ranges. For instance, the tidal range of the estuary reaches up to 8 m, and the distribution of clastics is obviously affected by the tidal current in the modern Colorado Delta. The reverse current formed in the distributary channel between the rising and the falling tides may become the main means by which energy is diffused from the deposits. In the distributary channel and in the seaward direction, the deposits can be repeatedly transported and deposited to form a series of linear or finger-like sand ridges parallel to the tidal direction.

The speed of the delta foreset is determined by the interrelation of the river flooding period and marine storm period. When the marine process is small (without storms) during the river flooding period, a delta foreset is formed quickly; the strong ocean current process and wave action during the storm period, especially during the non-flooding period, will destroy the delta deposits partially to slow down the foreset speed (Fig. 11.32) of the delta.

11.6 Sequence Characteristics of Deltas

The most common vertical sequence of a delta is a three-tier architecture, including the topset, foreset, and bottom set (Fig. 11.33); the Gilbert-type delta is a typical example. When the delta advances to the impounding basin and the provenances are sufficient, the delta on the section is presented as an obvious inverse-cycle superimposed sequence, namely, in the sequence of prodelta mud belt-delta front belt-delta plain belt from down to up, wherein the coarsening upward consists of the prodelta mud belt and delta front belt, and the fining upward constitutes the plain part. This situation mainly occurs when sand bodies advance to the lake/sea constantly under lake/sea regression. When the lake/sea transgresses, retrogradation results in the delta to form a fining-upwards cycle in the vertical section, namely, delta plain-delta front-prodelta from bottom to up; by this time, the distributary mouth bar is not developed, subaqueous distributary channel deposits are mainly formed, and the delta is easily transformed into sheet sand. If the lake/sea regression is further improved, the delta may stop developing and may be covered by lacustrine/marine mud. Overall, however, most deltas are dominated by lake/sea regression phases, and large oil and gas fields are easily formed in this type of delta. The different types of deltas have diverse sedimentary sequence characteristics. Among these, the sedimentary sequence of the fluvial-dominated delta is most easily preserved, and it has been widely studied in detail with the geometric history. Usually, the sedimentary sequence of a wave-dominated delta cannot easily be distinguished from the sedimentary sequence of coast sand.

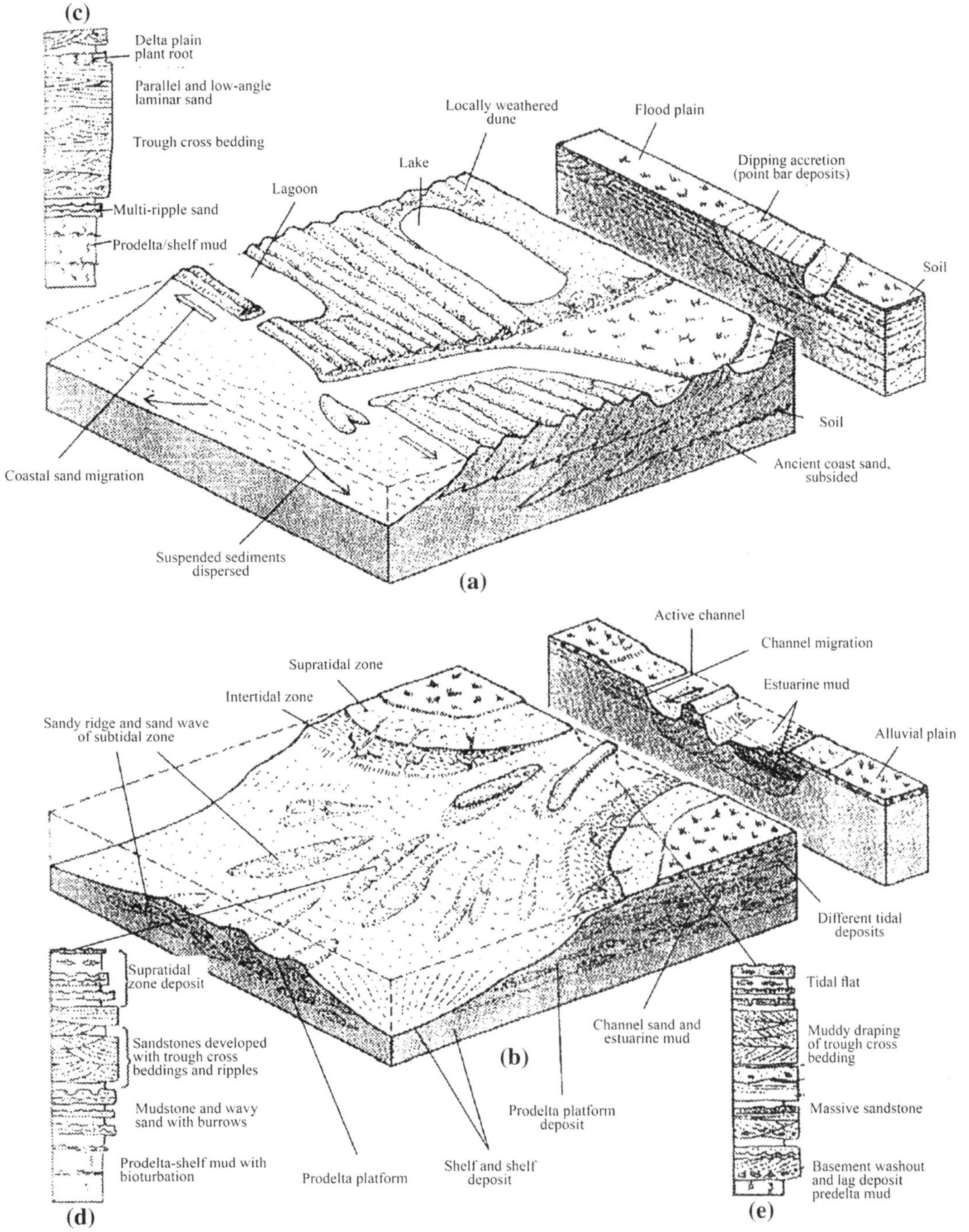

Fig. 11.31 Models of the wave-dominated delta and tide-dominated delta (after Galloway and Hobday 1983). **A** Wave-dominated delta; **b** tide-dominated delta; **c** vertical section of foreshore/coastal sandy ridge wall continuously moving forward; **d** estuarine channel filling; and **e** tidal sandy edge on prodelta platform of tide-dominated delta

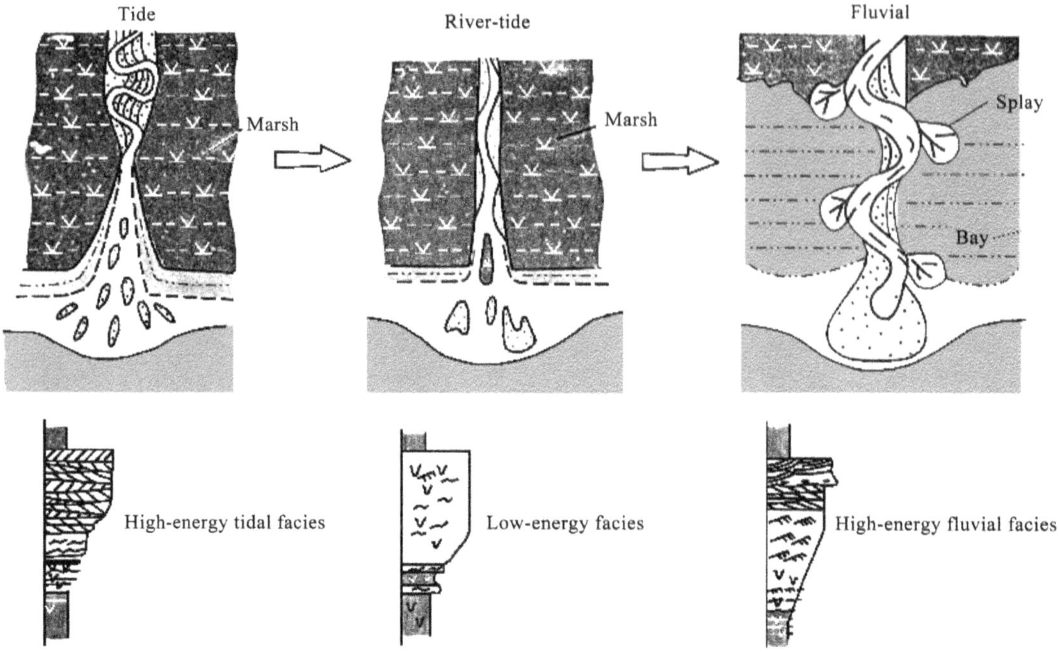

Fig. 11.32 Tide-dominated delta in different tide intensity (after Coleman and Wright 1975)

Fig. 11.33 The three-layer structure of the Delta (after F. S. Pazzolia 1986)

11.6.1 Sequence Characteristics of Fluvial-Dominated Deltas

Because the formation of large cyclic sedimentation is available, a fluvial-dominated delta continually moves forward from the land to the sea/basin, resulting in the formation of a unique progradational vertical sedimentary sequence. Generally, its bottom is predelta mud with delta front sand and silt sedimentation successively upward, and the top is covered with coarse-grained distributary channels and fine-grained swamp sedimentation of the delta plain, which is roughly the coarsening-upward inverse cycle sedimentary sequence, i.e., progradational sequence.

This delta can normally be divided into two large parts, namely, underwater and land parts. The main characteristic is that the grain size at the lower part features coarsening upward until the deposition site of the distributary channel, which begins fining upward owing to the feature of the fluvial cycle and constitutes a complex sequence characteristic that shows coarsening upward before fining upward from the bottom to top (Fig. 11.34). The formation of this typical sequence is the result of the continual movement of the delta front sedimentation toward the sea, arising mainly from the fluvial process.

Overall, this sequence has the following three characteristics:

(1) The grain size shows coarsening upward deposit; however, secondary fining upward may appear locally at the upper part of the sequence. The sedimentary structure changes along with the change in the grain size.
(2) Mixing or alternating occurrence of marine and continental biological fossils as well as marine fossils decrease upward, which mainly distribute in the mudstone of the neritic shelf at the bottom; plant fossils

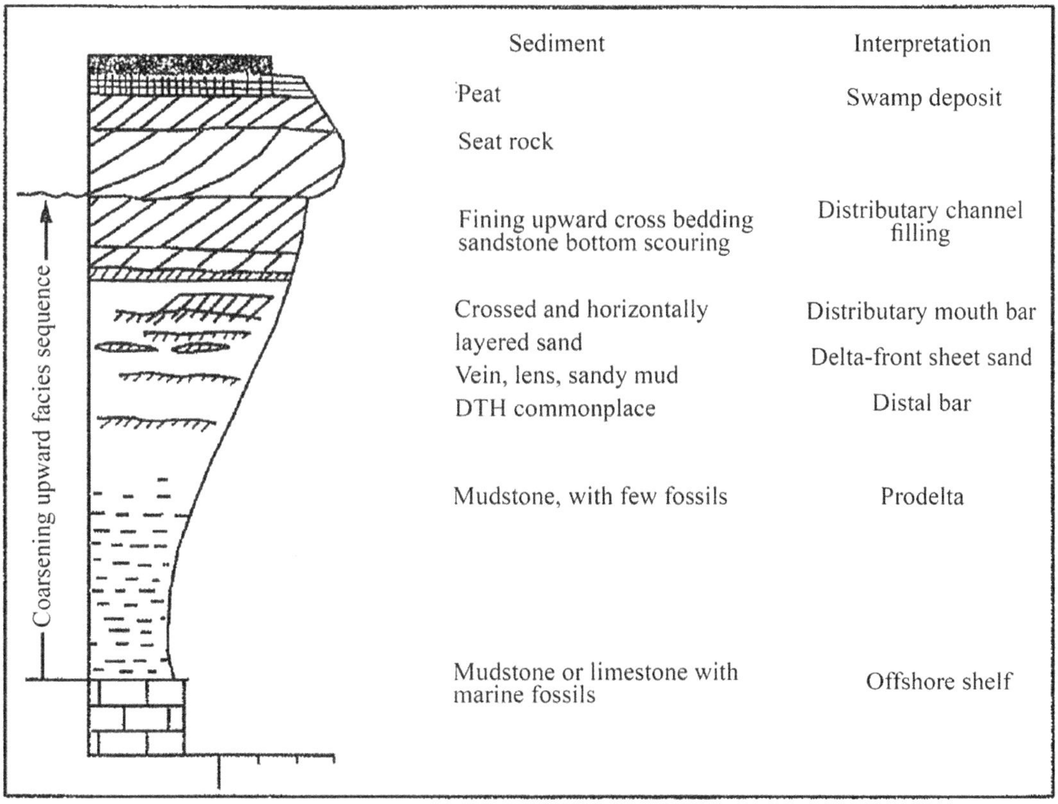

Fig. 11.34 Typical sequence of river-dominated delta (after Tucker 1981)

increase upward, sometimes even with carbonaceous mudstone or a coal seam.
(3) Coexistence of various cross beddings of current and wave genesis.

The sequence characteristics of such deposition are caused by the progradation formed by the gradual movement of the delta toward the sea basin (or lacustrine basin). The distributary channel, which carries a lot of sediments, flows from the delta plain to the sea basin or lacustrine basin. Owing to the sudden decrease in the flow energy, most of the coarse-gained load is deposited, resulting in the formation of a distributary mouth bar. The remaining fine-grained materials, in a suspended state, are carried far away from the river mouth by the water flow, as well as the silt and mud of the prodelta, resulting in the formation of a distal bar. The farther away the materials are from the river mouth, the thinner will be the deposits.

The sequence of the fluvial-dominated delta has a cyclic feature, which can generally be divided into the construction and destruction phases. In the construction phase, the sludge in the prodelta will be recovered by the silt and sand of the delta, which will, in turn, be recovered by the river mouth bar owing to the effect of active progradation (Fig. 11.35). It generally consists of sandstone and is then covered by a lignite bed that is sometimes contained in the swamp deposits of the delta. In contrast, in the destruction phase, the delta lobes may be abandoned if a gap is caused, and then, the top stratum will be eroded by the wave and water and may be completely retransformed and redistributed.

Elliot (1974) summarized the vertical sequence and sedimentary structure type of the distributary channel in the fluvial-dominated delta (Fig. 11.36). There are a series of massive beddings of the erosion surface in the fluvial-dominated delta. The parallel bedding and wavy bedding formed by the interaction of thin lens or clay and siltstone are seen commonly, and the miniature-megascopic and symmetric or asymmetric ripples and trough cross beddings are very rich.

The vertical sequence of the fluvial-dominated delta cannot be simply regarded as coarsening upward. It turns complex along with the increasing or decreasing of two distributary mouths in some cases, such as being affected by two distributary mouths simultaneously. From a typical example, such as the carboniferous delta in the north of England (Fig. 11.37), we can see that the vertical sequence of the delta plain changes along with the change in the condition.

The abovementioned sequence is formed when the deposition rate (R_d) of sediments is larger than the subsidence rate (R_s), namely, $R_d/R_s > 1$. In addition, the delta migrates and swings around instead of fixing at one place during its development process, resulting in the delta often showing a cyclic characteristic (Fig. 11.38).

11.6.2 Sequence Characteristics of Wave/Tide-Dominated Deltas

The sequence of the wave-dominated and tide-dominated deltas is usually coarsening upward; however, they are destructive deltas, hence the delta sequence formed is not as complete as that of the fluvial-dominated delta (Figs. 11.39 and 11.40).

The vertical sequence features of the ancient delta front subfacies also reflect the change in the water body and energy. The sequence of the fluvial-dominated delta front is presented as megascopic and coarsening upward, and it becomes a nearshore deposit upward dominated by sandstone from the beginning of the fine-grained infralittoral deposit, which results from the delta front movement to the sea under the effect of the fluvial process. A fine-grained region is not well developed at the bottom of the sedimentary sequence in the wave-dominated delta front, because the suspended matter near the delta front is scattered to a farther area owing to the wave action. Sandstone with good sorting on the medium or upper part has a symmetric

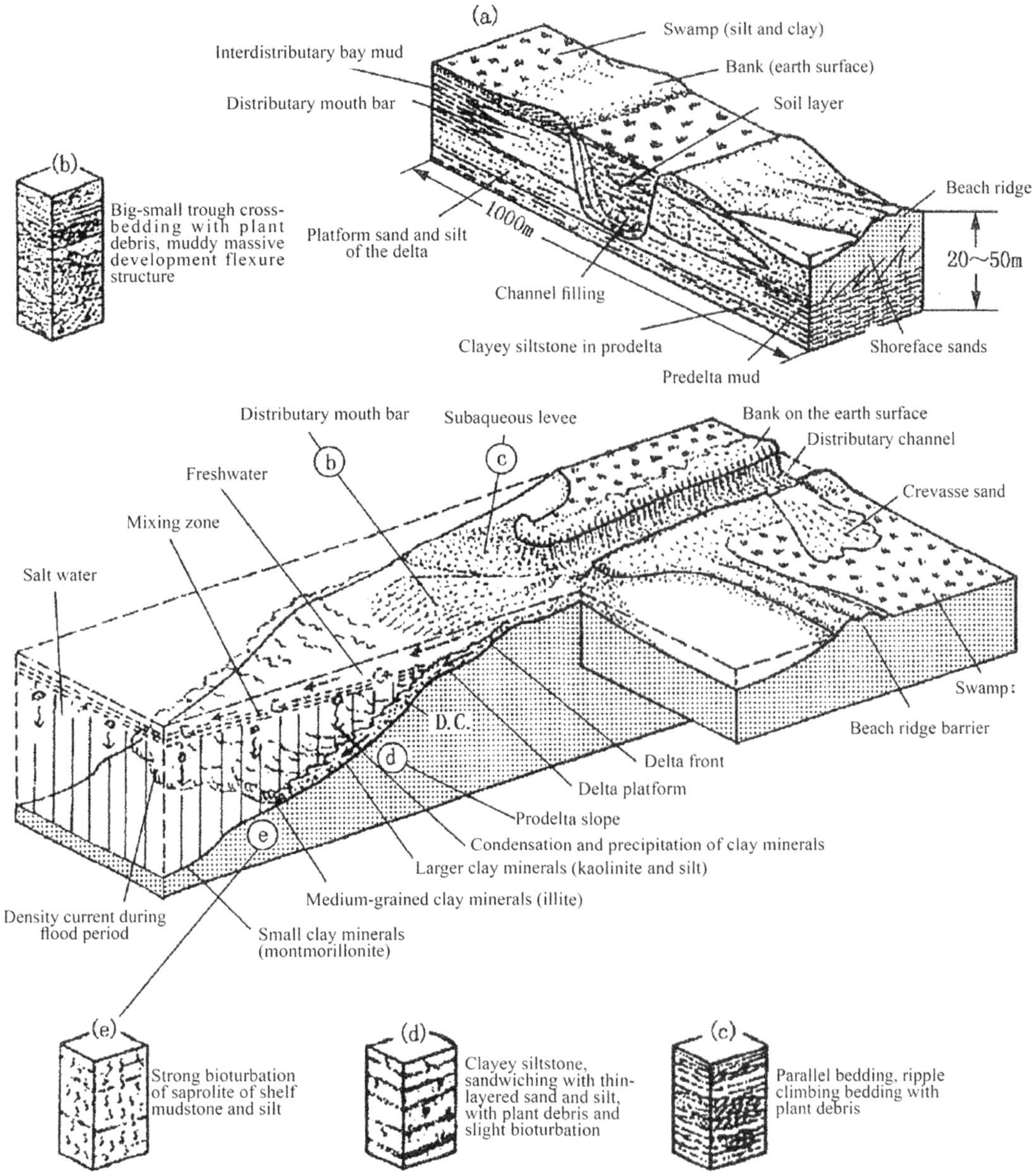

Fig. 11.35 Model of the distributary channel (pedate or lobate delta) and estuarine sedimentation process (after Einsele 2000). **a** Profile of the channel and surrounding sediments; and **b–e** some sedimentary structure characteristics in some sub-environments

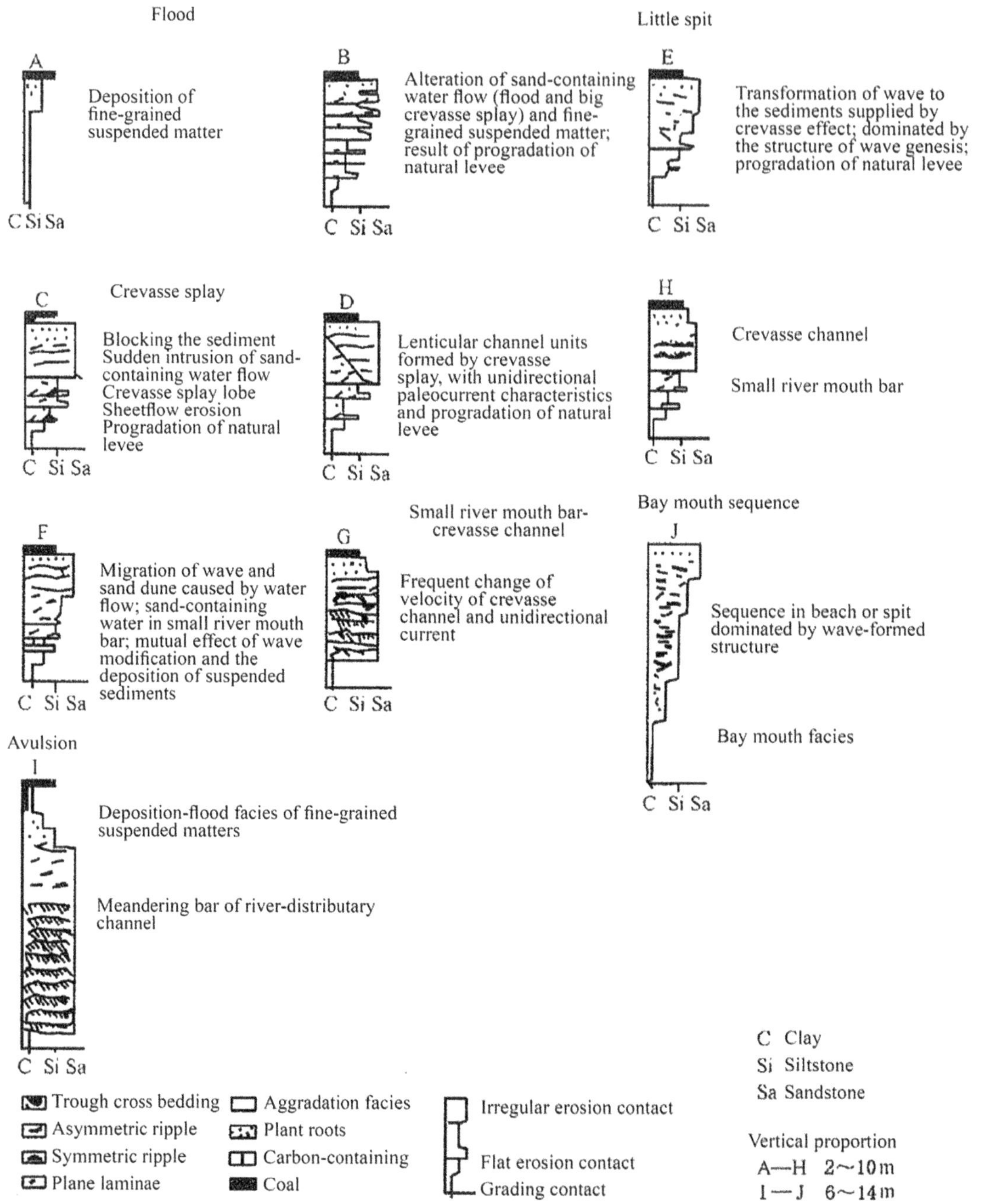

Fig. 11.36 Vertical sequence and sedimentary tectonic of the zones between river-dominated delta distributary channels (after Elliot 1974)

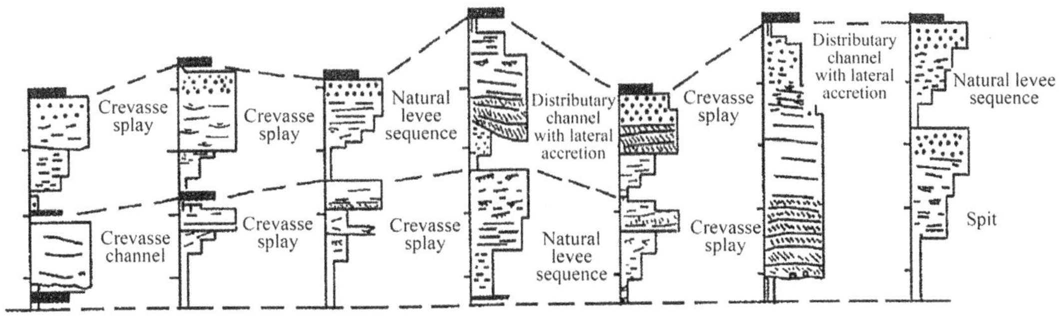

Fig. 11.37 Combination change of plain facies of river-dominated delta—Carboniferous in northern England (after Elliot 1975)

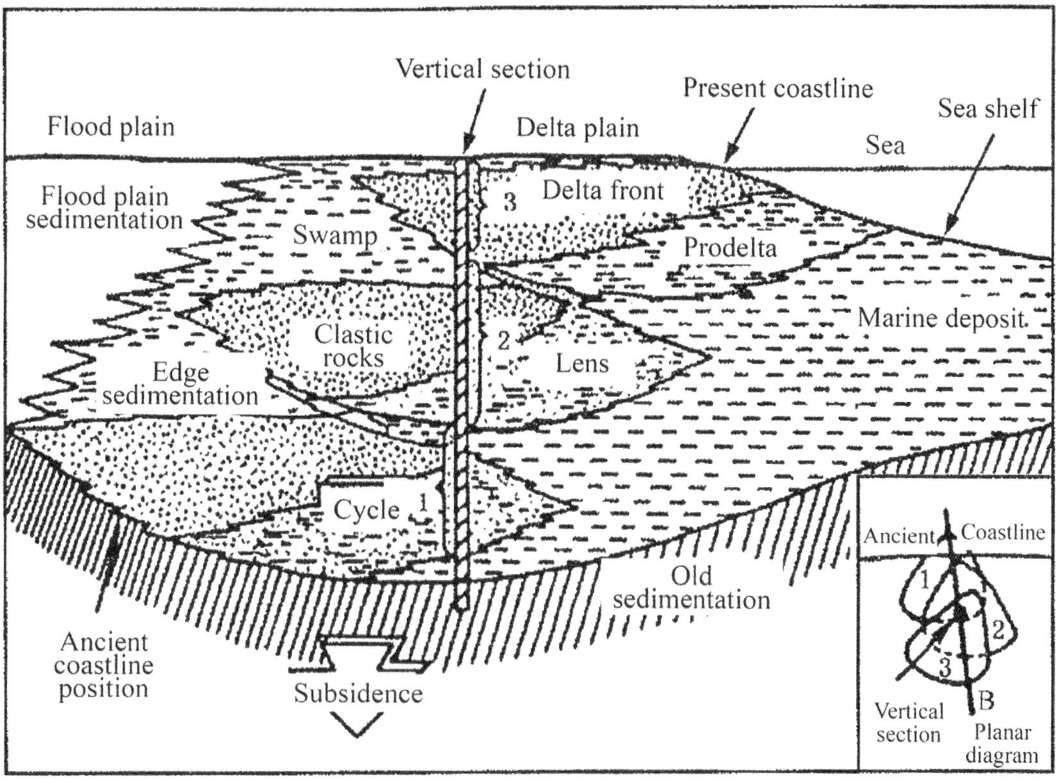

Fig. 11.38 Features of the multicycle sedimentary sequence of delta sedimentary caused by location migration (after J. M. Coleman et al. 1904)

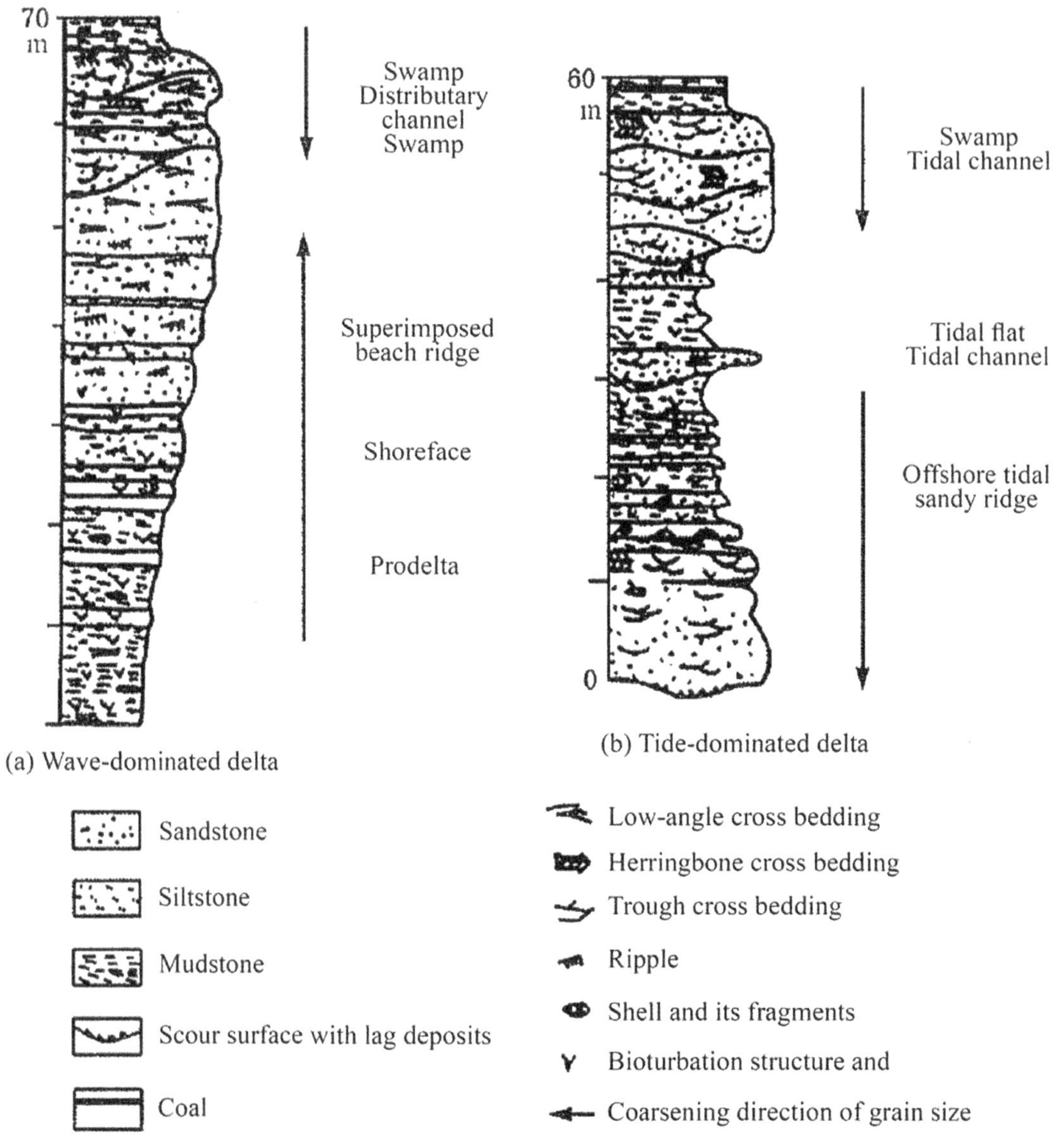

Fig. 11.39 Sedimentary sequence of the wave-dominated delta and tide-dominated delta (after Walker 1978)

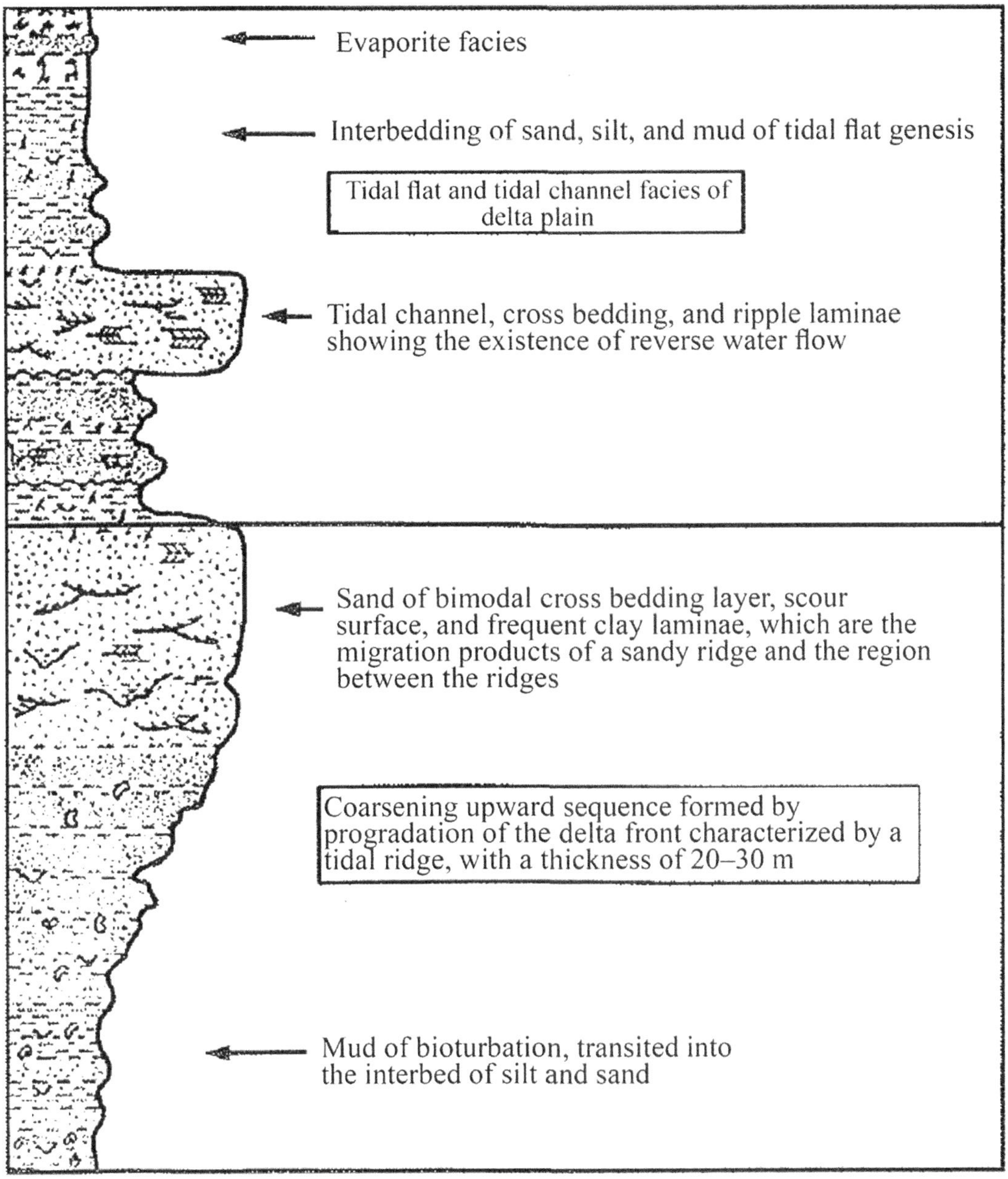

Fig. 11.40 Vertical sequence of tide-dominated delta in Oder river (after Coleman and Wright 1975)

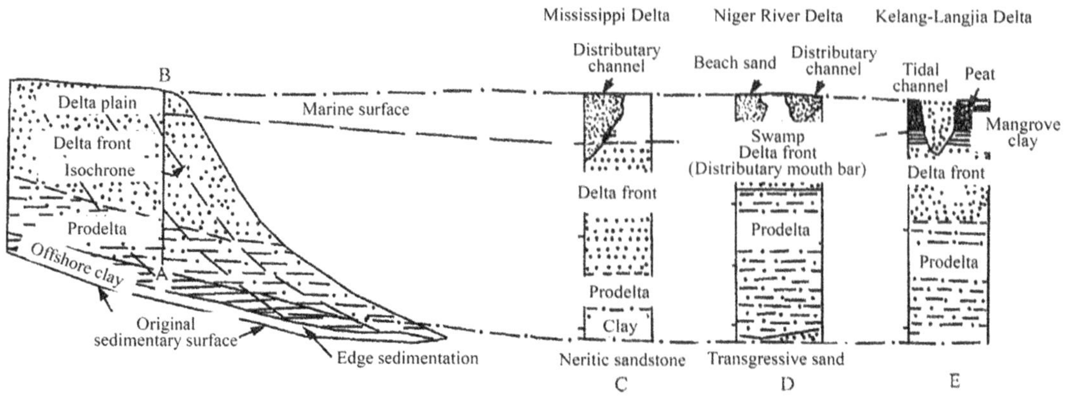

Fig. 11.41 Sequence and structural section of different types of delta (from G. D. Klein 1974). A–B—Position of C, D, and E columns in ideal delta; C—Mississippi Delta; D—Niger River Delta; and E—Kelang-Langjia Delta in Malaysia

ripple and scouring structure, which completely proves the transformation effect of the wave. It is generally covered by the coast sand deposits on the delta plain.

However, the biggest characteristic of the vertical sequence of the ancient delta front of the tide is that bimodal cross-bedding, complex bedding, reactivation surface, and other structures can be seen near the top part, and the alternation of the tide bar and channel sedimentation is seen between bars (Fig. 11.41).

11.6.3 Vertical Sequence Comparison of Deltas

The sequences and profiles of fluvial-, wave-, and tide-dominated deltas are different (Fig. 11.41). The fluvial-dominated delta plain mainly consists of the distributary channel, swamp of interdistributary bay, and gulf, which has thicker delta front sedimentation and is dominated by the distributary mouth bar. The wave-dominated delta plain consists of a channel, beach, sandy ridge, barrier island, and swamp sedimentation, and its front is less thick because the sand carried by the river into the sea is redeposited onto the beach or sandy ridge by wave transformation (Table 11.10). The tide-dominated delta plain is dominated by the sand and mud flat of the tidal channel and peat swamp sedimentation.

11.7 Fan Deltas

The fan delta was first proposed by Holmes (1965). It was originally defined as an alluvial fan that directly moved forward toward a stable water body (lake or sea) from nearby mountains. Initially described by McGowen (1971) and Galloway (1976), a fan delta is considered as an alluvial fan that has arid and humid alluvial fans, having two climate types, i.e., arid fan delta and humid fan delta. Because the latter delta plain generally has a braided water system, Galloway defined a fan delta as a sedimentary system formed by the injection of an alluvial fan and braided river into a stable water body.

Nemec and Steel (1988) viewed a fan delta as a coastal sedimentary system deposited in the border regions of an active fan body and stable water body with an alluvial fan (including arid and humid alluvial fans) as the provenance. This sedimentary system can be partially or completely deposited under water, which represents the products formed by the mutual effect of the alluvial fan with considerable sedimentary loads and a sea or lake. The sedimentary system that is partially deposited under water is called a fan delta by domestic scholars, and one that is completely deposited is called a subaqueous fan. A fan delta can be divided into a lacustrine fan delta and marine fan delta. A fan delta complex

Table 11.10 Vertical sequence comparison of fluvial-dominated, wave-dominated, and tide-dominated deltas

Type	Fluvial-dominated delta	Wave-dominated delta	Tide-dominated delta
Plain development situation	Mainly consists of a distributary channel, swamp of interdistributary bay, and gulf	Mainly consists of a channel, beach, sandy ridge, barrier island, and swamp	Consists of a tidal channel, tidal flat, and swamp
Front development situation	Large thickness; dominated by distributary mouth bar	Small thickness; transformed by wave	Small thickness; sandwiching with tidal flat sedimentation
Complete sequence	Relatively complete construction phase; incomplete destruction phase	Mainly destruction; incomplete	Mainly destruction; incomplete
Grain size variation	Generally coarsening upward; locally fining upward	Coarsening upward; undeveloped fine-grained section; good sandstone sorting in the middle part	Coarsening upward
Sedimentary structure	Water flow ripple and various cross beddings	Symmetric ripple, scouring structure	Bimodal cross bedding, complex bedding, tidal channel

consists of a series of fan deltas that are connected mutually and stacked vertically. Obviously, Nemec and Steel especially emphasized that an alluvial fan is treated as the provenance supply system rather than the geomorphic features. In addition, they also used a simple stereogram (Fig. 11.5) to demonstrate the definition of the fan delta and relative system terms.

As mentioned earlier, the fan delta and braided delta are both a type of coarse-grained delta. Here they are defined as follows: a fan delta is a proximal gravel delta formed with an alluvial fan as the provenance and transported in the bed load, and a braided delta is the distal coarse-grained delta formed with a braided river as the provenance and transported in the bed load/mixed load (according to Yu et al. 1995 and 1997). From the distance from the provenance and the type of delta facies, we can clearly distinguish a normal delta, fan delta, and braided delta (Fig. 11.42).

11.7.1 Conditions for Formation and Development

The important conditions for the formation of a fan delta are that the terrain elevation difference of the lakeshore (coast) is large, the slope at the basin edge is relatively steep, the distance from the provenance is close, and the source supply is sufficient. The distribution of the fan delta is relatively limited. It can only be formed and developed under specific terrain, structural, and climatic conditions, among which the structure is the main controlling factor (Table 11.11).

11.7.1.1 Terrain Conditions

The necessary and sufficient condition for the development of a fan delta is that the basin edge is near the mountain with a large elevation difference and steep slope. Generally, the closer the distance away from the mountain and the steeper the slope, the easier is the development of the fan delta.

11.7.1.2 Structural Conditions

A fan delta is mostly developed in active structure regions owing to its fan shape, and it often accompanies a synsedimentary megascopic fault zone. From the geotectonic background, the fan delta will be most favorably developed along the continental collision coast, island arc collision coast, pull fringing coast, and rift basin in the craton or steep coast edge of another faulted basin, such as the famous Copper River fan delta (continental collision coast type) in Alaska, Yallahs fan delta (island arc collision coast type) in Jamaica, some fan deltas (pull fringing coast type) in the Ridge Basin edge in California, fan deltas (rift valley type) on the west coast of the Dead Sea, and many fan deltas developed in the

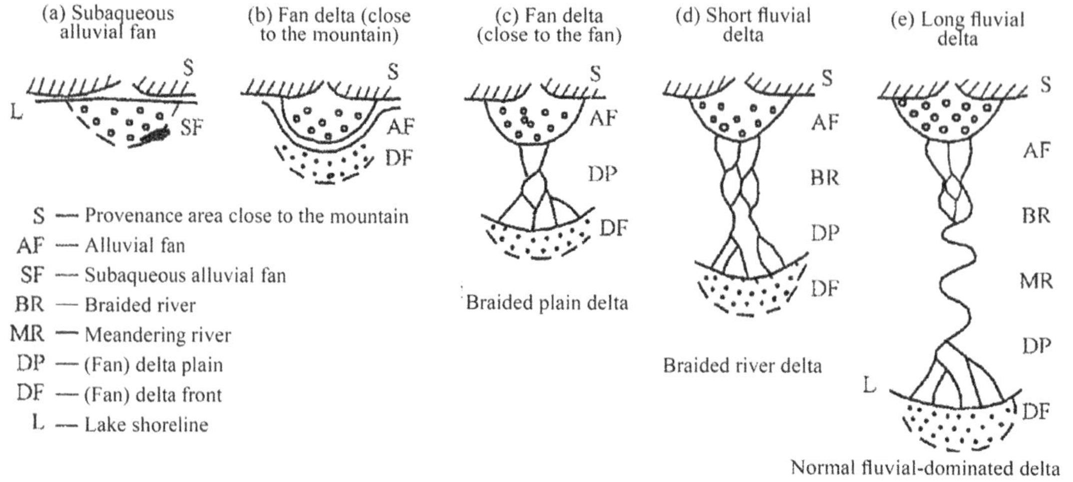

Fig. 11.42 Sketch Map of lakeshore location, sand body type and evolution relationship

Table 11.11 Main factors affecting the formation and development of fan deltas

Terrain conditions	Structural conditions	Climate conditions
Basin edge near the mountain, with large elevation difference and steeper slope	Mostly developed in the regions with more active structures; relative to the activity of rift valley or fault depression; and larger scale of hanging-wall fan delta than that of the footwall fan delta in fault	Arid fan delta is developed in arid area; and humid fan delta is developed in humid area

Meso-Cenozoic faulted basins in China (dustpan-like concave steep coast type). Periodic rift activities have an important influence on the shape of a fan delta. A fan delta is shown as a fan shape in the active tectonic period and as a wedge sheet with wide distribution during tectonic stability.

11.7.1.3 Climate Conditions

A fan delta can develop under various climatic conditions. However, the climate will affect the development of vegetation, weathering type and strength of the provenance area, surface hydrological conditions, and supply of terrigenous materials through the change in temperature and rainfall, which in turn will affect the formation of the fan delta. Therefore, fan deltas of different types are formed in different climate regions. An arid fan delta mostly develops in arid or semi-arid regions (such as the fan delta on the north coast of the Dead Sea), and a humid fan delta is easily formed in humid tropical and warm regions.

11.7.2 Main Characteristics and Sedimentary Facies Classification

11.7.2.1 Main Characteristics of Fan Deltas

Overall, the geological characteristics of a fan delta are as follows:

(1) A fan delta is generally located around the piedmont, and it usually accompanies the

boundary fault of a lacustrine basin. In a faulted lacustrine basin, it is mainly located on the steep side of the basin short axis, and a homochronous braided delta is often distributed at the flat side of the basin short axis. The onshore part of a single fan delta is generally small, and the plane modality is mostly shown as a fan shape.

(2) Fan delta sedimentary bodies are usually bounded by a fault in the direction of the land, and their proximal sediments (fan root or upper fan section) overlap the ancient bedrock with angle nonconformity. The thickness and extension scope of the fan delta sequence are controlled by the fluctuation amplitude of the edge fracture difference. For the sedimentary sequence of a single fan delta, the thickness can reach tens of meters, and the thickness of the fan delta sequence developed on a plate edge over a long geological time can reach thousands of meters or even tens of kilometers.

(3) A fan delta consists of gravel, gravel-containing sand, sand, and other coarse clastic sediments. The low maturity of the components and texture reflect that it is close to the provenance area, the transport distance is short, and sedimentary rate is high. In general, the vertical sequence of a single fan delta is coarsening upward.

(4) The development of the distributary mouth bar at the fan delta front is poor, and bars may even be lacking; this is determined by the hydrodynamic conditions of the estuary. The distributary mouth bar is not easily formed owing to the effect of the planar jet flow, namely, the distributary channel is not developed. In other words, the sediment supply is quick and short, which makes it difficult to form a stable distributary channel. Even though the river mouth bar is formed at the temporary distributary channel, it is easily destructed by later sedimentation.

(5) The geometric shape and change in the grain size of the fan delta sedimentary body are generally shown as a clastic wedge, which will be thinner from the direction of the piedmont to the basin (sea or lake) and gradually turns into basin facies and then disappears.

11.7.2.2 Sedimentary Facies Classification

The facies classification of fan deltas applies the quartering of coarse-grained deltas (Table 11.12), i.e., the upper and lower fan delta plains are dominated by the alluviation of continental water, the fan delta front is transformed by the basin hydrodynamic force, and the front fan delta is dominated by basining. Each part has different sedimentary sequence characteristics and sedimentary facies associations (Fig. 11.43 and Table 11.12).

1. **Fan delta plain**

The fan delta plain is the onshore part of a fan delta, and it covers the coastal plain area from the outer fan to the shoreline. In most cases, its plane modality is shown as a fan shape inclined toward the direction of the basin. The modality of the fan delta often changes as a result of the influence of its surrounding bedrock terrain, fluvial process, waves, and tides (Fig. 11.44). In strong wave regions, a symmetric fan delta can be formed in a bed load river with high gradient, and an elongated fluvial-dominated fan delta can be formed when the wave energy decreases. A bayhead beach ridge plain can be formed in a low-gradient river with low input of sediments under the impact of waves, and a tide-dominated delta can be formed under a protective bay environment in a bay surrounded by bedrock. If the catchment basin is a lake, only a fluvial-dominated delta can generally be formed in the lake because there is no tidal action in the lake environment and the lake wave effect is much less than that of the sea wave (some large lakes can also have strong waves).

The sedimentary facies characteristics of a fan delta are mainly determined by the type of sediment supply system. The fan delta plain is divided into arid and humid fan delta plains owing to the impact of climate; these show obvious differences in sedimentary background,

Table 11.12 Subfacies classification, sedimentation characteristics, and sedimentary structure of fan deltas

Subfacies and microfacies classification	Fan delta plain				Fan delta front	Front fan delta
	Arid fan delta plain		Humid fan delta plain			
	Upper delta plain	Lower delta plain	Upper delta plain	Lower delta plain		
Sedimentary characteristics	With arid fan characteristics; complex gravel component, low maturity, unobvious stratification; frequent cross superimposed channel sedimentation and gravel bed of sheetflow deposition; developed with braided channel, debris flow, sheetflow, and sieve sediments		Developed with gravelly braided river system sedimentation; repeated and quick lateral migration of channel; poor sorting and lake of fossils; developed with braided channel, channel lag facies, and gravel bar facies		Available for fluvial-dominated type, wave transformation type, tidal transformation type, wave and tide interaction type, etc.	Front fan delta deposits in fan delta on wide shelf is mainly shoreface-offshore silty and muddy deposits
Sedimentary structure	Lack of regular and clear megascopic cross bedding; developed with scour-fill structure		Lack of clear bedding; fining grain size towards the direction of the basin; developed with mud crack, cross bedding, parallel bedding, and burrow, or occasionally with stromatolite		Various sedimentary structures transformed by rivers, waves, and tides	Ripple developed in the sand layer; frequently with benthic biological agitation; marine facies frequently with carbonate rock sedimentary structure

sedimentary facies association, and sedimentary stratigraphic structure.

(1) Arid fan delta plain

The fan delta plain in arid or semi-arid areas has the sedimentary characteristics of an arid alluvial fan, and it is generally distributed near the active fault zone with a large slope degree and small fan body. The surface water system belongs to quasi channelized floods with intermittent and sudden features. In addition to the frequent cross-superimposed channel and gravel bed with sheetflow sedimentation, the sedimentary facies is also accompanied by a large number of debris flow and sieve sediments, sometimes with collapse sediment near the root of the cliff. The gravel component is complex with low maturity, inconspicuous stratification, insufficient regular and clear megascopic cross bedding, and scour-fill structure development.

The typical arid fan delta plain, such as the fan delta plain of the Ridge Basin in the Pliocene Epoch in California (Fig. 11.45), has a rock stratum that features insufficient clear bedding, poor sorting, insufficient fossils, and few sedimentary structures. The thickness in the proximal fault zone is very large (1100 m), and its grain size becomes thinner toward the direction of the basin center. Mud cracks, cross bedding, parallel

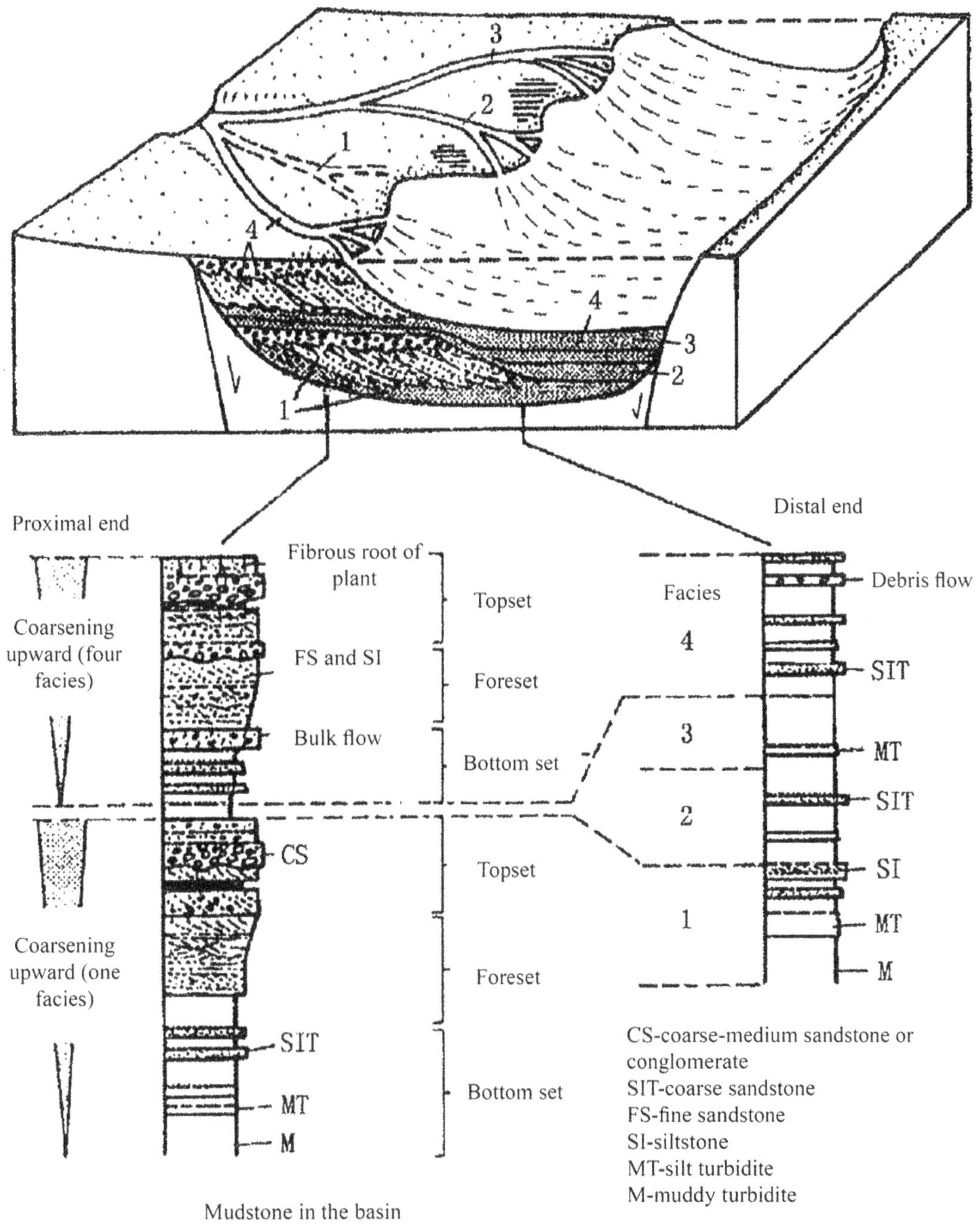

Fig. 11.43 Related facies of fan delta deposits (after Einsele 2000). The topset, foreset, and bottom set in the lacustrine basin are included; they are easily affected by the changeable estuary and subsidence of the lake bottom. Emphasis is laid on several facies at the delta front, coarsening upward sequence, and gravity flow movement near the delta front

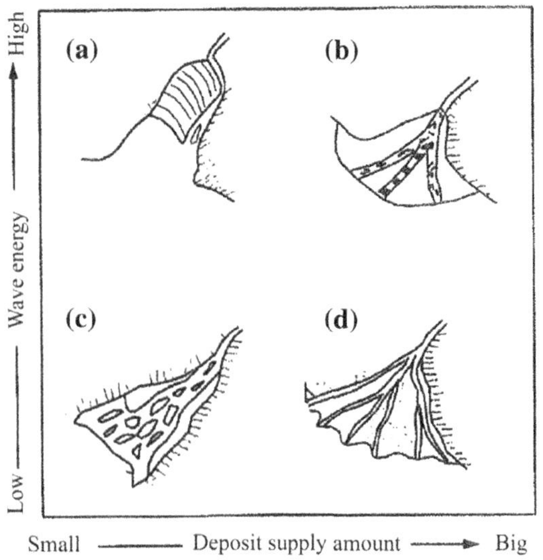

Fig. 11.44 Affection to the fan delta shape by the river action and wave action (after Wescott and Ethridge 1990). **a** Bayhead beach ridge plain; **b** Symmetric lobate fan delta; **c** Fluvial-dominated fan delta with elongated distributary; and **d** Tide-dominated fan delta

laminae, and burrows are developed, and occasionally, stromatolite is seen. Overall, a finger-like cross transition with lacustrine strata is gradually shown eastwards.

(2) Humid fan delta plain

A fan delta plain in a humid region is characterized by the development of pebbly braided river system sedimentation, and the braided river system of a fan delta plain in a humid region has a channelized flow, in which the channel lag facies and gravel bar facies are the main sedimentary facies associations. The formed sequence structure has obvious multicyclic features owing to the repeated and rapid lateral migration of the channel.

The modern Yallahs fan delta in the southeast of Jamaica is a fan delta developed in the northeast trade-wind belt with a warm tropical humid climate. The fan delta plain consists of a braided distributary channel, abandoned channel, flood plain, grass swamp, tree swamp, and some small lakes (Fig. 11.46). The braided channel mainly consists of a longitudinal gravel bar with poor sorting and imbrication structure, and the coarsest gravel is clustered together at the central axis part of the bar. The abandoned channel forms linear lowlands with the plant root system and scattered gravel muddy sand covering over the homogeneous mud. The flood plain is mainly gray silt and extremely fine sand sedimentation with miniature ripple cross bedding, climbing bedding, miniature scour-fill structure, rich burrows, and plant root systems. The remaining small lakes are surrounded by mangroves and grass swamps, and the lithology of the grass swamp is argillaceous sand and organic mud with isolated gravel or a gravel layer (Wescott and Ethridge 1990).

In other words, the different sedimentations and sedimentary facies formed thereby on the delta plain can be compared with those of the alluvial fan or braided river under similar structural backgrounds and climate conditions.

2. Fan delta front

The fan delta front, which is also called the transition zone (Wescott and Ethridge 1990) and is located in the shallow water area from the shoreline to normal wave base, is a zone with mutual effects of continental currents, waves, and tides. Because the strengths of the various influencing factors are different, a fan delta front can be divided into fluvial-dominated type, wave transformation type, and tidal transformation type, among which the strength of wave and tidal energy and their common effects govern the formation of these three types of deltas.

Fluvial-dominated fan delta fronts, which have the characteristics of lacustrine fan deltas, are formed mainly by the continental current effect. In addition, the development of fluvial-dominated fan delta fronts can be seen in some bay edges with a small tidal range and low wave energy. River water continues to flow to form many subaqueous distributary channels after flowing into the lake, extends toward the depth of lake along with the channel, and gradually becomes shallower till it disappears. The sediments of a fluvial-dominated fan delta front are mainly of sand and silt of different sizes, often with gravel deposits. The grain size fines

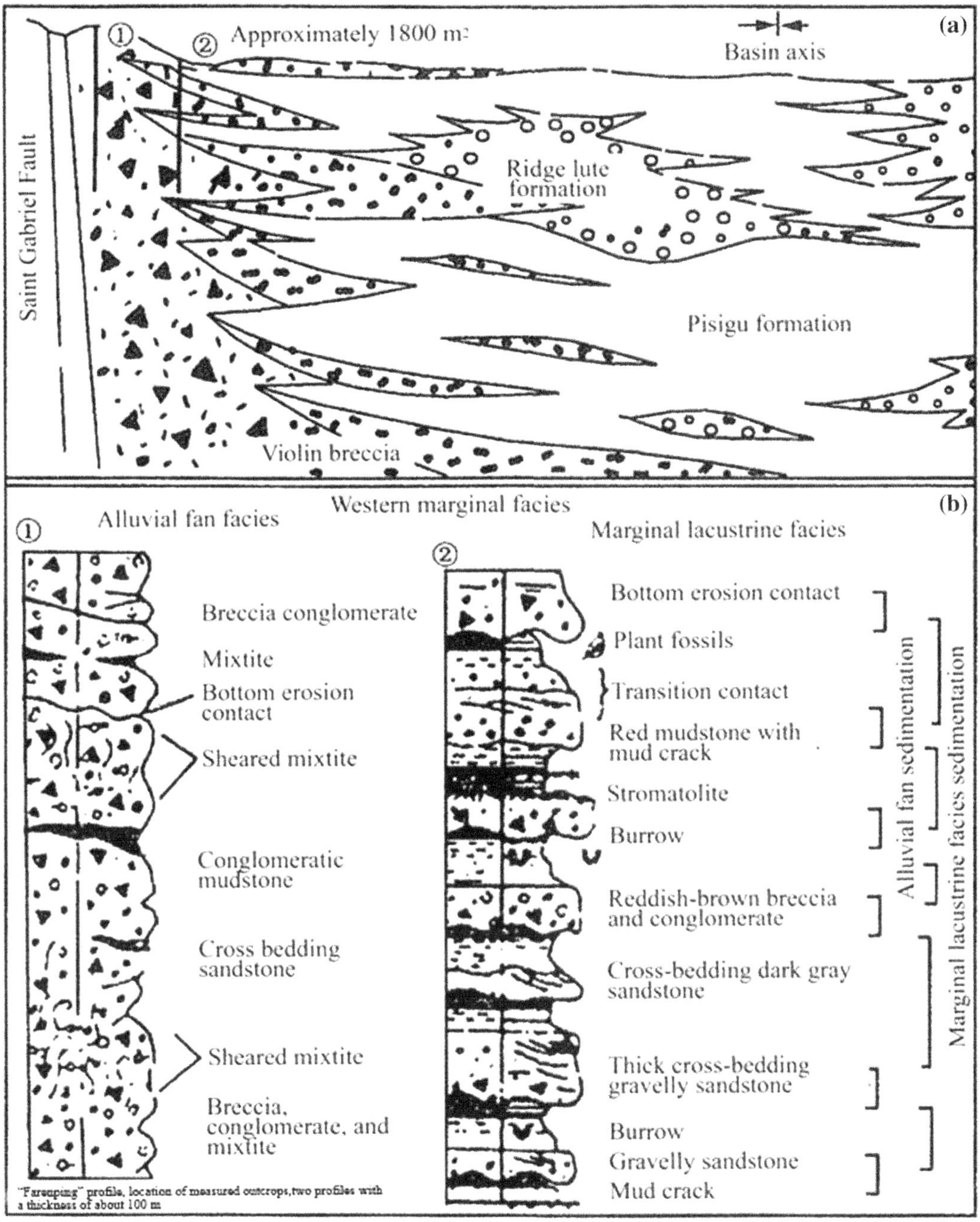

Fig. 11.45 Geological section of plain facies and marginal lake facies of fan delta in the Faneling Breccia Fm. of the west margin of the Ricci Basin (after Link and Osborne 1978). **a** Cross-sectional profile; and **b** stratigraphic columns

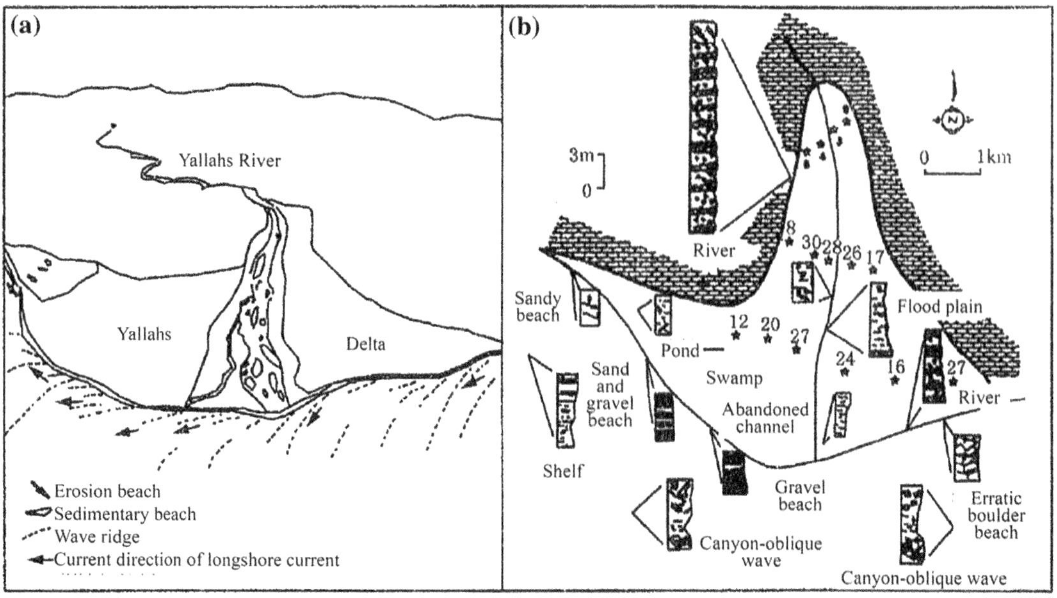

Fig. 11.46 Sedimentary environment and sedimentary facies map of modern Yeras fan delta in Jamaica (after Wescott and Ethridge 1990). **a** Sedimentary environment of the Yallahs Fan Delta; and **b** water plain, transition zone, and underwater sedimentation characteristics of the Yallahs Fan Delta; the number next to the pentagramme shows the depth from the bedrock

toward the direction of the basin. Cross bedding is developed in the sand layer, and it is presented as a finger-like cross transition accompanied by a fan delta front. For example, the fan delta front in the western coast of the rift valley in the Dead Sea is composed of sand and mud interbeds. The sand layer here has current ripple bedding, which may have formed by sudden flood, and the mud layer may have under calm lake water. On the whole, the sand and mud interbed shows a tendency of fining upward with a continuous rise (transgression) in the lake level (Sneh 1979).

In a longshore zone characterized by small tidal range and strong wave energy, a wave-dominated delta front is formed by the river sediments transformed by waves with the development of beach sedimentary association. The fan delta in Yallahs can be taken as a typical example (Fig. 11.46). This fan delta can be divided into erosion beach and sedimentary beach according to its landform and structural features. The sedimentary beach, where beach berms, backshore, and foreshore zones are well developed, is relatively wide and sandy. All of the abovementioned structures have parallel beddings with the foreshore zone inclined toward the sea and the backshore zone toward the land. Because of the influence of storm waves, sandwiched imbricated gravel layers are frequently found in the beach sand layer, which exists commonly in beach berms and backshore zones. The erosion beach with a steep and narrow foreshore zone is constituted of especially coarse sand and boulders. These sediments are lag deposits formed after the river sediments consisting of beach scarp collapse and screen under the impact of waves. The Yallahs fan delta has a steep gradient, hence the shoreface belt is not developed and the foreshore zone transits quickly into canyon-steep slope sedimentation toward the direction of the sea.

3. **Front fan delta**

The front fan delta refers to the lower part of the wave base of a fan delta. It transits downward

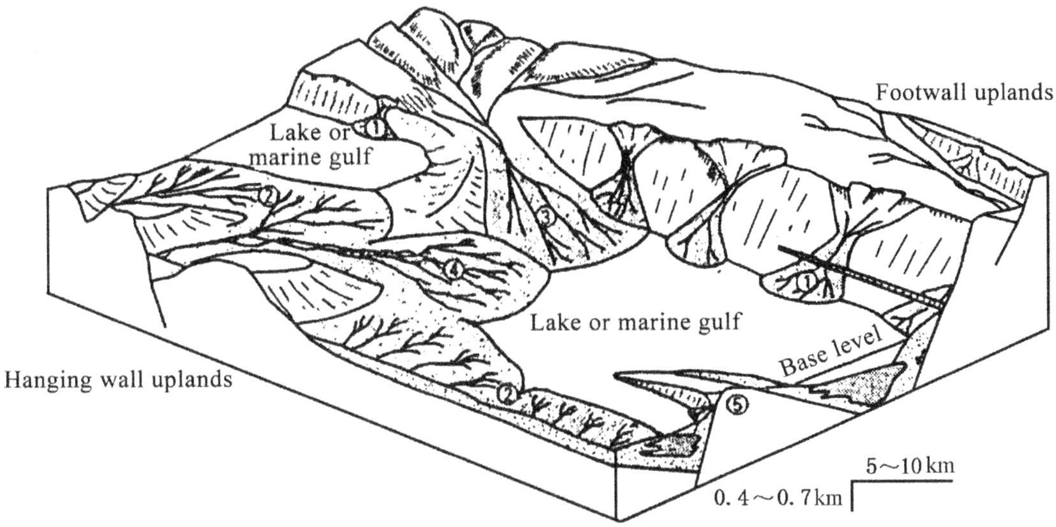

Fig. 11.47 Distribution and development of fan delta in rift basin (after K. I. Gawthorpe, A. Colella 1990)

with shelf mud or deep water basin sedimentation, and there is no obvious lithology boundary. The structural characteristics of the basin edge significantly influence the sedimentary features and sediment distribution in a front fan delta.

On a wide shelf, front fan delta sedimentation is developed as shoreface-offshore silt and mud, presented with interbedded shelf mud. Its sedimentary facies is stable with a wide distribution area, generally with obvious and complete coarsening-upward sequence. For example, some fan deltas in the southeast of Alaska are represented by the Copper River fan delta. In addition, some fan deltas are directly moved to the shallow-water carbonate shelf. The front fan delta often has the characteristics of a shallow-water carbonate sedimentary interbed along with transgression and regression of the fan delta because the development of carbonate sedimentation is hindered by coarse-grained terrigenous clastic sedimentation.

11.7.2.3 Sedimentary Model of a Fan Delta

The sedimentary compositions and structures of all fan deltas are different owing to the influence of geologic structure, climate, interaction between the river and ocean or lake, shelf width and slope gradient, as well as the related sedimentary system. As mentioned previously, there are various fan delta models, thus it is difficult to use a comprehensive model to summarize the characteristics of all fan deltas. Ethridge and Wescott (1984) believed that the fan delta front is the key to distinguishing the fan delta front from the fan delta of different types in four sedimentary belts of fan delta. The geometric shape and sedimentary facies association of the fan delta front reflect the basin edge sedimentary state and structural background. Thus, they developed three types of sedimentary models for fan deltas according to the geological geography background characteristics of fan delta fronts, namely, shelf-type model, steep slope model, and Gilbert-type model (Wescott and Ethridge 1990). In a rift valley, deep-water slopes and Gilbert-type fan deltas are usually developed along a fault footwall, while shelf-type fan delta rich in sand is frequently developed on the uplifted side of the fault (Fig. 11.47).

11.7.2.4 Shelf-Type/Flat Slope Type Model

Shelf-type fan deltas, also called the flat-slope-type model, develop on a continental shelf with a flat gradient but wide area, such as

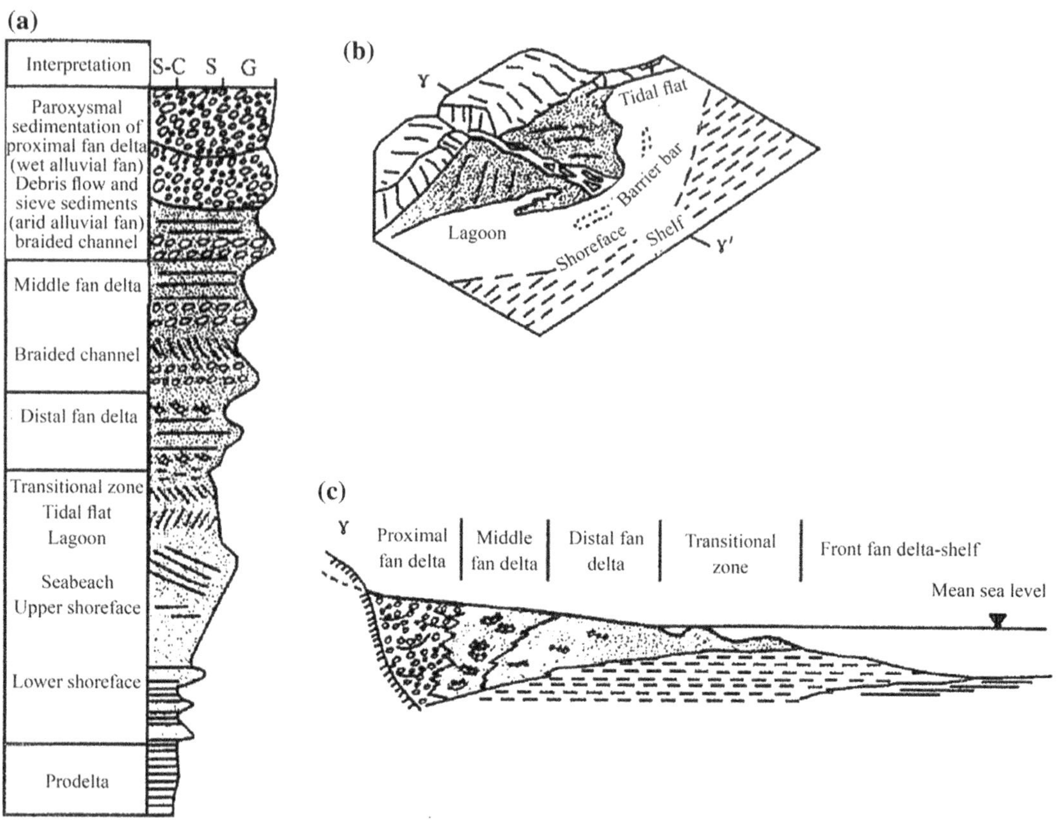

Fig. 11.48 Sedimentary model of shelf type fan delta (after Wescott and Ethridge 1990). **a** Vertical sequence; **b** planar graph; and **c** longitudinal section

the flat slope area of a dustpan-like depression. The shelf-type fan delta model was established based on the sedimentary characteristics of some fan deltas represented by the fan delta in the Copper River on the southeast coast of Alaska (Fig. 11.48). Given its wide shelf, shelf-type fan delta sedimentation is not interrupted by gradient change, hence the resulting sedimentary body can move forward a long distance toward the basin direction. The upper and lower fan delta plains, fan delta front (transitional zone), and pre-fan delta with obvious differentiation form a continuous transitional progradational sequence of gravel-sand-mud, with good development and an obvious coarsening-upward sequence.

The land plain of this fan delta, which is the same as the other types of fan deltas, has the characteristics of alluvial fan and braided river sedimentation. A beach rich in sand, and shoreface belt sedimentation or tidal flat-lagoon-barrier island and other sedimentary systems can be developed in front the transitional zone owing to the frequent influence of the effects of waves and tides. Front fan deltas mainly contain mud that is rich in biological fossils and remains, which is transited with shelf mud.

11.7.2.5 Steep Slope Model

The sedimentary model of a steep slope fan delta is established mainly according to the sedimentary characteristics of the Yallahs fan delta on the southeast coast of Jamaica (Fig. 11.49). This model is applicable the fan deltas advancing toward island slopes, land slopes, or steep slope edges in a faulted basin. The normal sequence of fan delta sedimentation (land plain zone + front zone + front fan delta zone) is often truncated because of the suddenly steepening slope in the

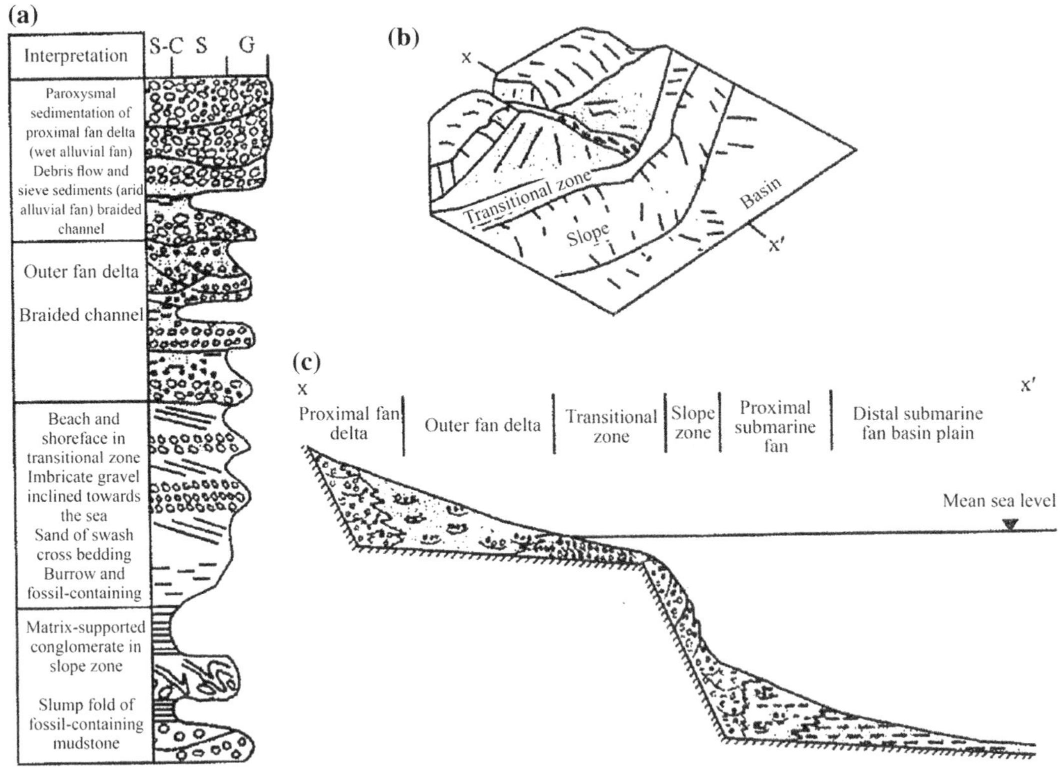

Fig. 11.49 Sedimentary pattern of ideal slope fan delta (after Wescott and Ethridge 1990). **a** Vertical sequence; **b** planar graph; and **c** longitudinal section

slope breaking area of the shelf edge. Coarse-grained clastic sediments can cross over the shelf edge to be deposited on the slope and in the basin directly instead of the head part of the canyon, and the slump and gravity flow effect in the canyon result in a complex coarsening-upward sequence.

The proximal part of the fan delta plain (namely, the plain part of an upper fan delta) is mainly clumpy conglomerate with poor sorting of paroxysmal torrent deposit in the restrictive channel, which is often interbedded with the conglomerate and gravelly sandstone characterized by coarse parallel bedding in shallow longitudinal bars of a braided river. The largest flat surface of gravel is inclined toward the land. The distal part of the land plain is also composed of conglomerate and gravelly sandstone of the braided river, and it has coarse parallel bedding and an imbricated structure with a few trough cross beddings. If it is in an arid or semi-arid region, development of the braided river in the proximal part is not as good as that in a wet alluvial fan plain, nevertheless high proportions of debris flow sedimentation and sieve sediment interbeds can be seen.

Beach and shoreface belts are developed in the transitional zone of a fan delta front, and the main sediments in them are conglomerate and sandstone with good sorting. The sandstone is characterized by parallel bedding and swash cross bedding. The conglomerate has an imbricate structure inclined toward the sea. A small proportion of fine-grained organic sediments may represent isolated coastal lagoons and interdeltaic bay sedimentation containing fossils and burrows.

Slope sedimentation is mainly constituted of marine mudstone and matrix-supported conglomerate, and it often has the slump

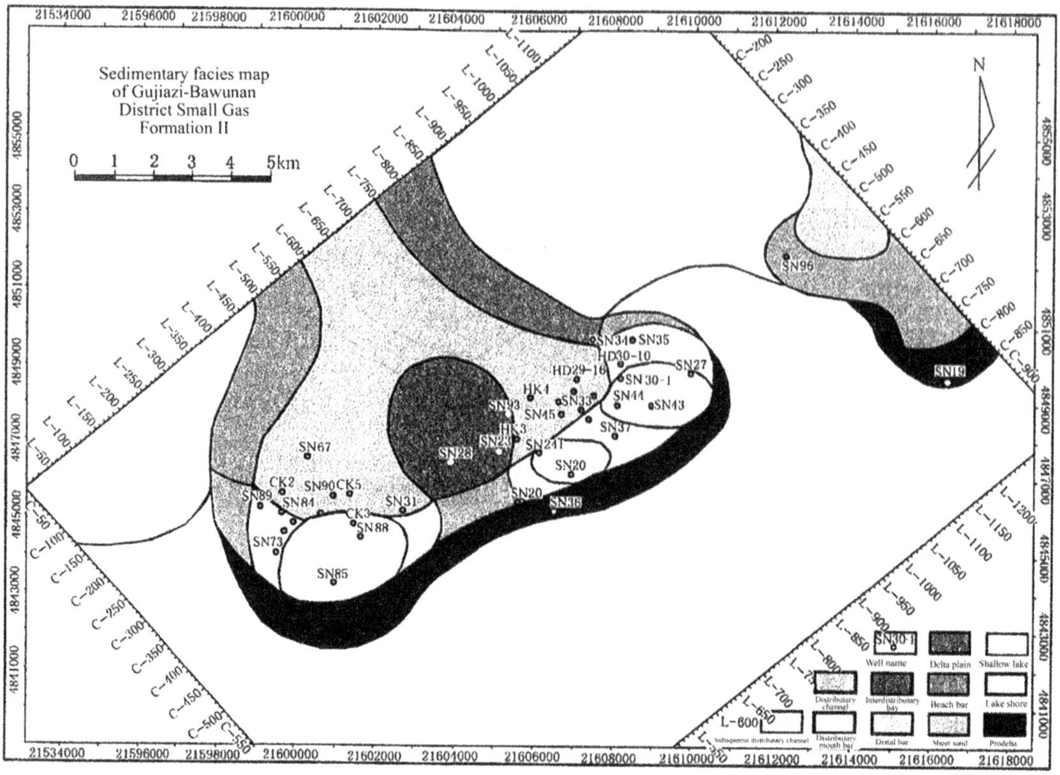

Fig. 11.50 Fan delta pattern of Xiao-2 gas formation in Shiwo fault of the Songliao Basin

deformation structure. In the areas from the slope belt to basin edge, debris flow, mud flow, grain flow, fluidized flow, and the sliding structure of sediments are very active, which can result in the formation of submarine fans. Marine mudstone is rich in fossils, accompanied by the development of bioturbation structures.

The Gujiazi-Bawunan Small Gas Formation II in the Shiwu Fault Depression in the Songliao is of this type (Fig. 11.50). This fan delta developed in the sedimentary period of a faulted lacustrine basin. The gradient of the terrain where the river flows into the lake is relatively steep, and sediment deposition in the provenance occurred at a rapid accumulation rate. Slope sedimentation shows a lobate shape, which transited from the fan delta plain to the delta front in the south-north direction, mainly with the development of distributary channels, interdistributary bays, subaqueous distributary channels, distributary mouth bars, sheet sand and prodelta mud, and other microfacies.

11.7.2.6 Gilbert-Type Model

Lake Bonneville in America, which was first studied by Gilbert, is a typical lacustrine fan delta. This fan delta features a three-layer structure of topset, foreset, and bottom set. Therefore, Ethridge and Wescott termed fan deltas characterized by this feature as Gilbert-type fan deltas. Gilbert-type fan deltas can be developed in both lacustrine and marine environments, especially in a protected fjord background with low tidal range and small wave energy (Fig. 11.51).

The topset of a Gilbert-type fan delta is formed due to the migration of the downstream channel of an alluvial fan. Large bed load is accumulated in the river mouth area, and it

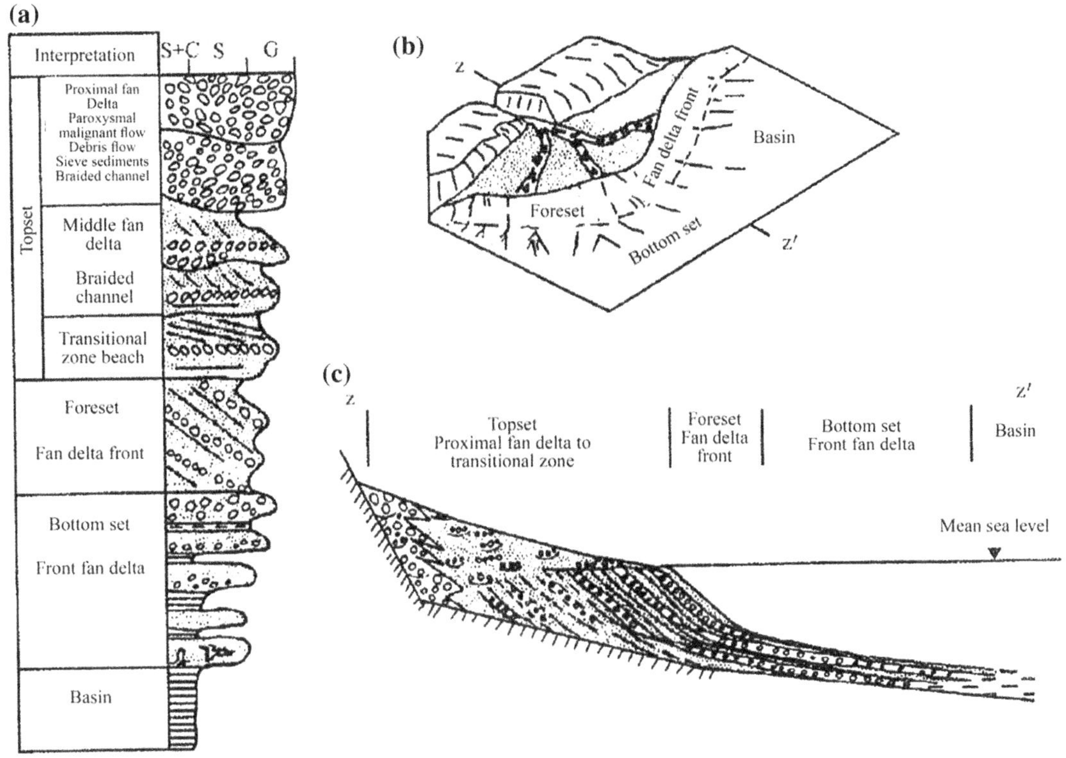

Fig. 11.51 Sedimentary model of gilbert type fan delta (after Wescott and Ethridge 1990). **a** Vertical sequence; **b** planar graph; and **c** longitudinal section. Bottom set is missing in the proximal part of the foreset

avalanches from the bar crest under the action of gravity, forming a steep foreset composed of coarse-grained sediments. Without the effect of an external force, the downward inclination of the foreset can reach about 35°. Fine-grained sediment is transported into the basin by the river as suspended load and is deposited at a farther place as the bottom set. Because the slope is steep and water is deep, the foreset formed by a Gilbert-type fan delta is often transformed into high-density turbidity due to gravitational sliding. Therefore, debris flow and turbidity current sedimentation intercalation often coexist in the bottom set.

The plane modality of a typical Gilbert-type fan delta is usually fan-shaped, and its vertical sequence clearly coarsens upward. The grain size range of a Gilbert-type fan delta sediment prograding into the lacustrine basin ranges from coarse (gravel) to fine (silt and mud) because equidensity axial jets and low-energy stable lake water are favorable for the sedimentation of fine-grained sediment. Mainly coarse clastic sediments prograde from a Gilbert-type fan delta into the sea because most of the suspended fine particle loads are brought to place farther from the fan delta for sedimentation. In conclusion, the common characteristics of Gilbert-type fan deltas include coarse fraction, river control type, high construction, rapid progradation, and dominance of inertial flow. Compared with other types of fan deltas, the most important difference of Gilbert-type fan deltas is the presentation of megascopic gravel foreset with a steep slope. The bottom set is missing in the sequence in the proximal part at the rear end of the foreset (Massari and Colella 1988; Colella 1990), and it can be seen on the basement strata as unconformable downlap of a higher angle (Fig. 11.52).

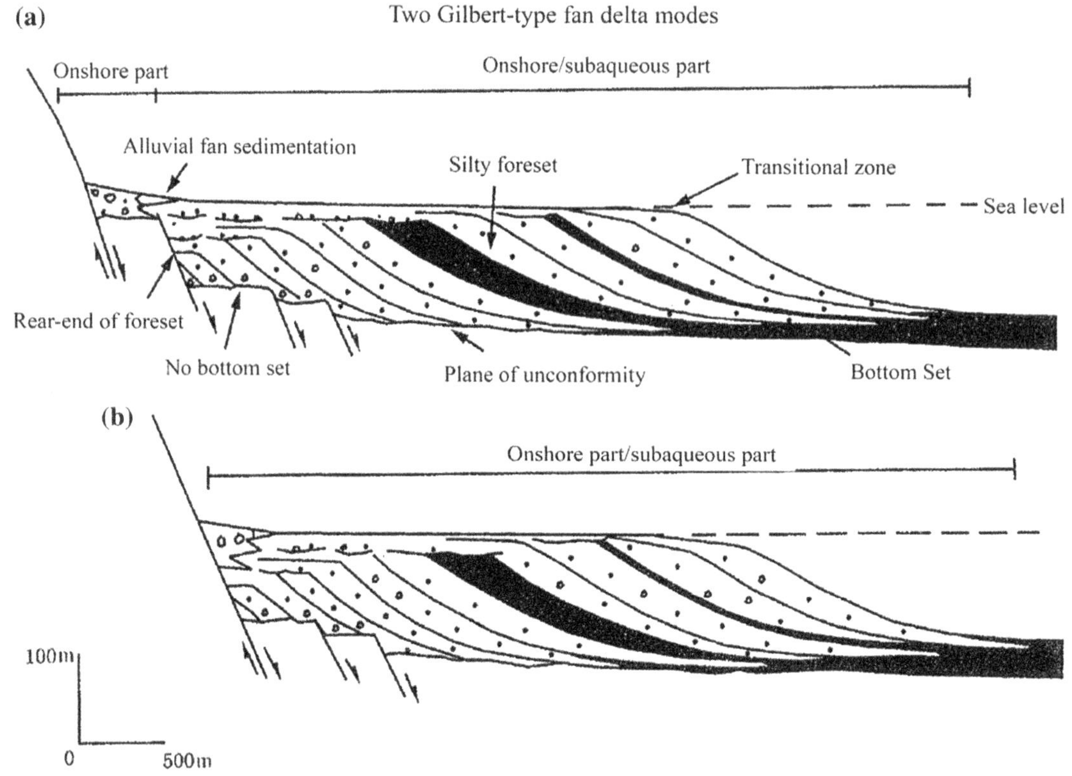

Fig. 11.52 Two Gilbert types sedimentary model of fan delta (fault-dominant, Pleistocene) of Krut Basin, south of Italy. **a** With continental alluvial fan; and **b** without the development of a continental alluvial fan and with the shoreline next to the fault escarpment

11.7.3 Type and Sedimentary Characteristics of Lacustrine Fan Deltas

In Meso-Cenozoic continental faulted lacustrine basins in China, fan deltas are very common, and often, important petroleum reservoirs are formed in them. Lacustrine fan deltas are distributed mainly in nearshore waters, and are very sensitive to changes in structure and climate. These changes mostly induce changes in lake water level and lake shoreline, thus impacting the differences in sedimentation and depositional facies. Through the analysis of many sand bodies of lacustrine fan deltas in China and by referring to foreign fan delta sedimentation models, sand bodies of lacustrine fan delta sedimentation facies in some oilfields in China can be divided into three types, namely, transgressive, regressive, and Gilbert-type. The fan deltas of these three types differ in terms of the classification of subfacies and microfacies, and names (Table 11.13).

11.7.3.1 Transgressive Mode

This type of fan delta is formed because a mountain river enters the shallow water area of a lacustrine basin after flowing out of a mountain pass. Almost all transgressive fan deltas are subaqueous fan-shaped conglomeratic. Such a fan delta is usually developed on the steep slope side of a faulted lacustrine basin that is close to the provenance and lacks an onshore fan delta environment. The color of the surrounding mudstone is grayish-green or light gray, and shallow water biological fossils are contained, which indicate a lake shore-shallow lake environment. This type of fan delta is generally

Table 11.13 Comparison of the classification of lacustrine fan deltas and division of subfacies and microfacies

Classification of lacustrine fan delta	Transgressive fan delta		Water-regression fan delta		Gilbert-type fan delta	
	Subfacies	Microfacies	Subfacies	Microfacies	Subfacies	Microfacies
Subfacies and microfacies classification	Inner fan	Main channel and interchannel	Upper fan delta plain	Braided channel	Topset (fan delta plain)	Braided channel
			Outer fan delta plain	Braided distributary channel		Braided distributary channel and interdistributary bay
	Middle fan	Braided channel, front of middle fan, and middle fan inter-channel	Fan delta front	Subaqueous distributary channel, distributary mouth bar, and sheet sand	Foreset (fan delta front)	Subaqueous distributary channel and front bar
	Outer fan	Shallow lake, semi-deep lacustrine mud, and silty-fine sand	Front (fan) delta	Shallow lake, semi-deep lacustrine mud, and silty-fine sand	Bottom set (front fan delta)	Front fan delta mud or silt

Table 11.14 Subfacies and microfacies characteristics of transgressive lacustrine fan deltas

Subfacies	Inner fan	Middle fan			Outer fan
Microfacies	Channel and interchannel	Channel of middle fan	Front of middle fan	Interchannel of middle fan	Shallow lake, semi-deep lake
Lithology	Conglomeratic, sandstone mixed with grayish-green or gray mudsand	Dominated by conglomeratic and sandstone, mixed with grayish-green and gray mudstone	Dominated by sandstone	Grayish-green-gray mudstone, mixed with flagstone	Dominated by gray mudstone, mixed with flagstone
Grain size	Coarse	Relatively coarse	Relatively fine	Thin	Very fine
Sorting	Poor	Relatively poor	Better	Better	Good
Composition	Complex, relatively high mud content	Unstable, high mineral content		High mud content	High mud content
Sedimentary structure	Massive bedding, normal or inverse graded megascopic cross bedding	Scour structure, parallel bedding, medium-megascopic cross bedding	More cross bedding	Massive and miniature bedding, slump and deformation structure	Parallel bedding, current ripple, transmutation, and massive bedding
Vertical sequence	Normal cycle, fining upward	Normal cycle, fining upward	Normal or inverse cycle	No obvious or normal cycle	
Electric well-logging curve features	Self-potential serrated with medium-low amplitude, high resistance	Self-potential serrated cylinder shape with gradual change at the top	Self-potential serrated funnel shape or bell shape	Self-potential finger shape or serrated bell shape	Finger shape or serrated shape

formed in the lacustrine transgressive stage, and the fining-upward sedimentary sequence is dominant. However, there is no obvious microfacies feature for each facies belt of the delta. In addition, the lithology, morphology, and zoning characteristics of this type of fan delta are similar to those of piedmont alluvial fans, thus transgressive fan deltas can be subdivided into sand body subfacies belts of the inner fan, middle fan, and outer fan (Table 11.14).

1. **Inner fan**

The inner fan is mainly composed of a limited number of channels and sediments between channels. Channel deposits can be characterized by conglomerates, conglomeratic, matrix- or particle-supported sandstone fining upward, coarse clastic grain size, complex composition, high mud content, poor sorting, and mixed granular size. The thickness of a single layer is up to tens of meters, and often, a mix of grayish-green and gray mudstones is found. The bedding is generally unclear. The common bedding types are massive bedding, inverse-normal grading, and unclear megascopic cross bedding. The rock bottom is often a scour surface or is in abrupt contact with the underlying bed. The SP curve mostly shows not obvious or medium-low amplitude serration with high resistance.

2. **Middle fan**

The middle fan is in front of the inner fan. As the channel development zone, it is the main body of fan delta sedimentation, as well as the most developed part of the sand body, with a high content of sandstone and low conglomerate content. The associated sedimentation microfacies can be subdivided into three: middle fan (braided) channel, middle fan front, and middle fan inter-channel, among which the middle fan channel and middle fan front are the main sand microfacies.

(1) Microfacies of a braided channel

Features such as composition and structure of sediment are controlled by the provenance. For instance, in some areas, the gravel content is high and the lithology is finer, but overall, the lithology is coarse, dominated by conglomeratic and sandstone. The sorting of sandstone and conglomerates is poor, and their compositions are complex, with high contents of unstable mineral detritus and matrix. A scour structure is often seen on the undersurface of the sand layer, and massive bedding, parallel bedding, and medium-megascopic trough cross bedding are seen commonly. The thickness of the fining-upward sequence is usually tens of centimeters to meters. Often, gray and grayish-green mudstone layers can be found between layers. Superimposed sandstone measuring several meters in thickness is often found. The SP mostly shows a serrated cylinder shape and cylinder shape with gradual changes at the top and the bottom.

(2) Middle fan front microfacies

The middle fan front is actually an extension of the middle channel toward the basin direction, with the grain size fining; the bottom scour is less developed than that in the channel area. Cross bedding is developed to a greater extent than that in the channel area, but its thickness is generally smaller. The SP curve usually shows a serrated funnel—bell shape and bell shape, which reflect the weakening of channel scouring action and certain wave transformations.

(3) Interchannel microfacies of the middle fan

This is mainly composed of grayish-green to gray mudstone mixed with flagstone. Here sandstone develops massive bedding and ripple cross bedding, and slump and deformation structures are seen commonly. Because a braided river has strong scouring force, frequent diversions can be found. In case of a diversion, sediments are scoured and thinned or even scoured out completely. The SP curve mostly shows a finger shape or serrated bell shape.

The middle fan subfacies are the most important and form relatively good petroleum reservoirs in transgressive fan deltas; thick petroleum reservoirs can be formed by the overlapping of each sand body microfacies of middle fan subfacies.

3. **Outer fan**

This is located at the front of the middle fan and enters shallow to semi-deep lake areas. The lithology is mainly gray mudstone mixed with flagstone, and the sandstone can develop parallel

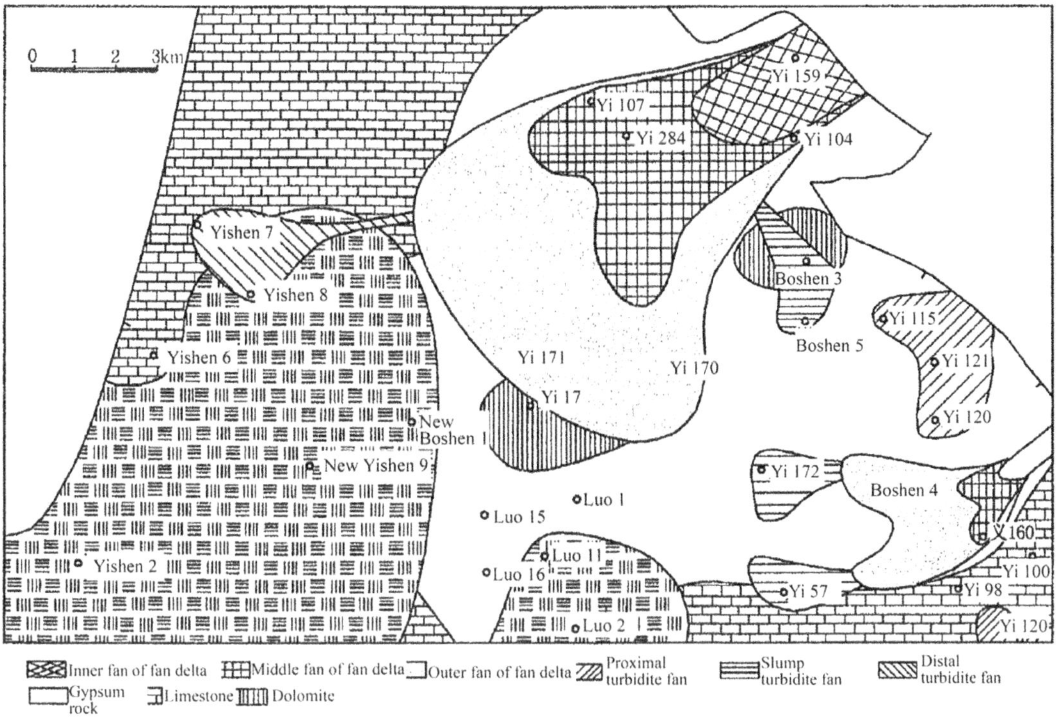

Fig. 11.53 Sedimentary facies model of upper sub-section of Shasi section in Bonan depression

beddings and current-ripple cross bedding. Graded bedding and massive bedding can be seen as well. The SP curve is shows finger shape or serrated shape.

The subsection fan delta on the S3 Member in the Bonan Sag of the Zhanhua Depression (Fig. 11.53) is a typical transgressive fan delta. Transgressive fan deltas can be found in wells Yi 159 and Yi 160 on the upper sub-member of the S3 Member in this area. Offshore turbidite fans and slump turbidite fans may occur in the front part of the fan delta.

11.7.3.2 Regressive Type

A regressive fan delta is formed mainly in the initial stage of the return of a lacustrine basin after depression. This type of fan delta is commonly developed on the short axis and the steep bank side of Meso-Cenozoic faulted lacustrine basins in China, and a typical example is the fan delta (Fig. 11.54) in the S3 Member of the Oubei-Dawan area in the Liaohe Basin, which houses two regressive fan deltas. Regressive fan deltas are formed mostly in the lake regression period. With the gradual shrinkage of a lake, the regressive fan delta is carried forward toward the center of the lake, and the part exposed over the water surface increases in size and transforms into fluvial facies. The facies belt of this kind of fan delta generally shows complete development, and well-developed coarsening-upward sequence can be seen. The regressive fan delta can also be divided into fan delta plain, fan delta front, and front (fan) delta (Table 11.15 and Figs. 11.15, 11.55).

1. **Fan delta plain**

Fan delta plain subfacies are mainly composed of a mixture of conglomerate and conglomeratic, mingled with red, yellow, grayish-green, and variegated mudstone and dominated by braided river sedimentation. The single sequence is fining upward. Relatively megascopic planar cross bedding and parallel bedding can be seen in conglomeratic. Mixed

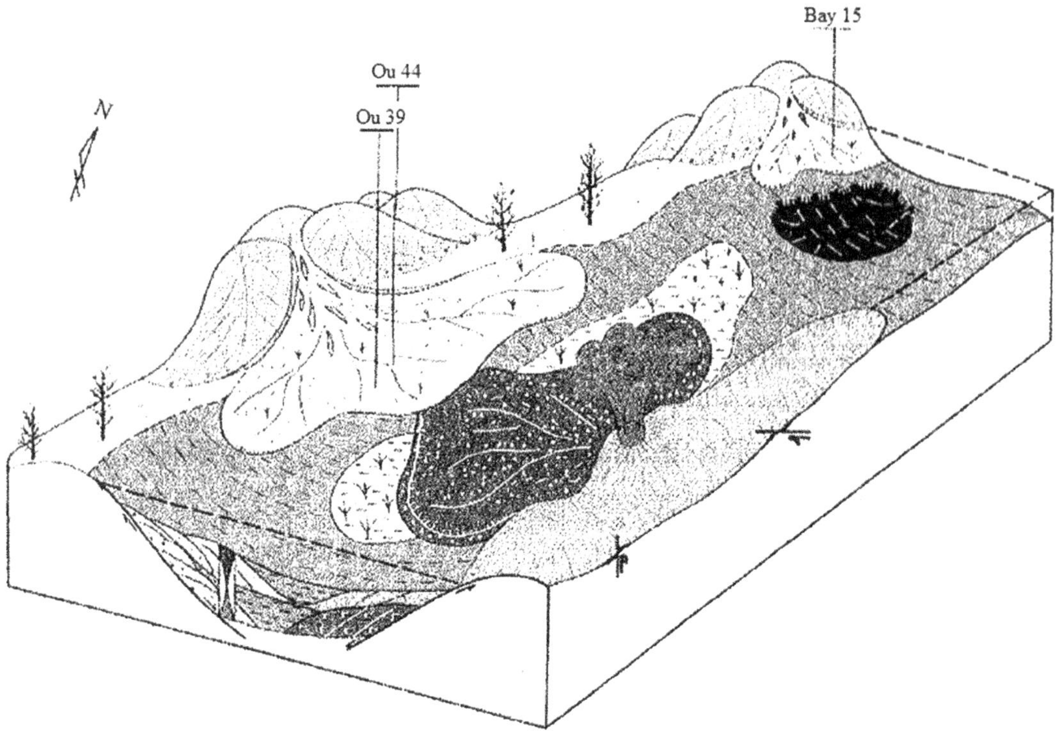

Fig. 11.54 Sedimentary facies model of upper sub-section of Shasan section in Oubei-Dawan, Liaohe Basin

Table 11.15 Subfacies and microfacies characteristics of regressive lacustrine fan deltas

Subfacies	Fan delta plain	Fan delta front	Front fan delta
Lithology	Mixed conglomerate, conglomeratic mixed with red, yellow, celadon, and variegated mudstone	Conglomeratic and sandstone, mixed with grayish-green mud sand and a few inferior oil shales	Light and dark gray mudstone mixed with a little sandstone, siltstone, calcareous shale, and oil shale
Grain size	Dominated by coarse grain	Relatively fine	Thin
Sorting	Relatively poor	Better	Good
Sedimentary structure	Relatively megascopic oblique bedding, cross bedding, parallel bedding, also massive debris flow structure	Megascopic oblique bedding, parallel bedding, and cross bedding	Miniature cross bedding and wavy bedding
Main microfacies	Braided channel	Subaqueous distributary channel, distributary mouth bar, and sheet sand	Predelta mud
Vertical sequence	Fining upward	Coarsening upward as a whole, subaqueous distributary channel fining upward, distributary mouth bar coarsening upward, and sheet sand shown as finger shape	Coarsening upward
Electric well-logging curve features	Self-potential shown as serrated low amplitude cylinder shape	Dominated by subaqueous distributary channel cylinder shape and distributary mouth bar funnel shape, sheet sand finger or serrated shape	Serrated or low-amplitude flat shape

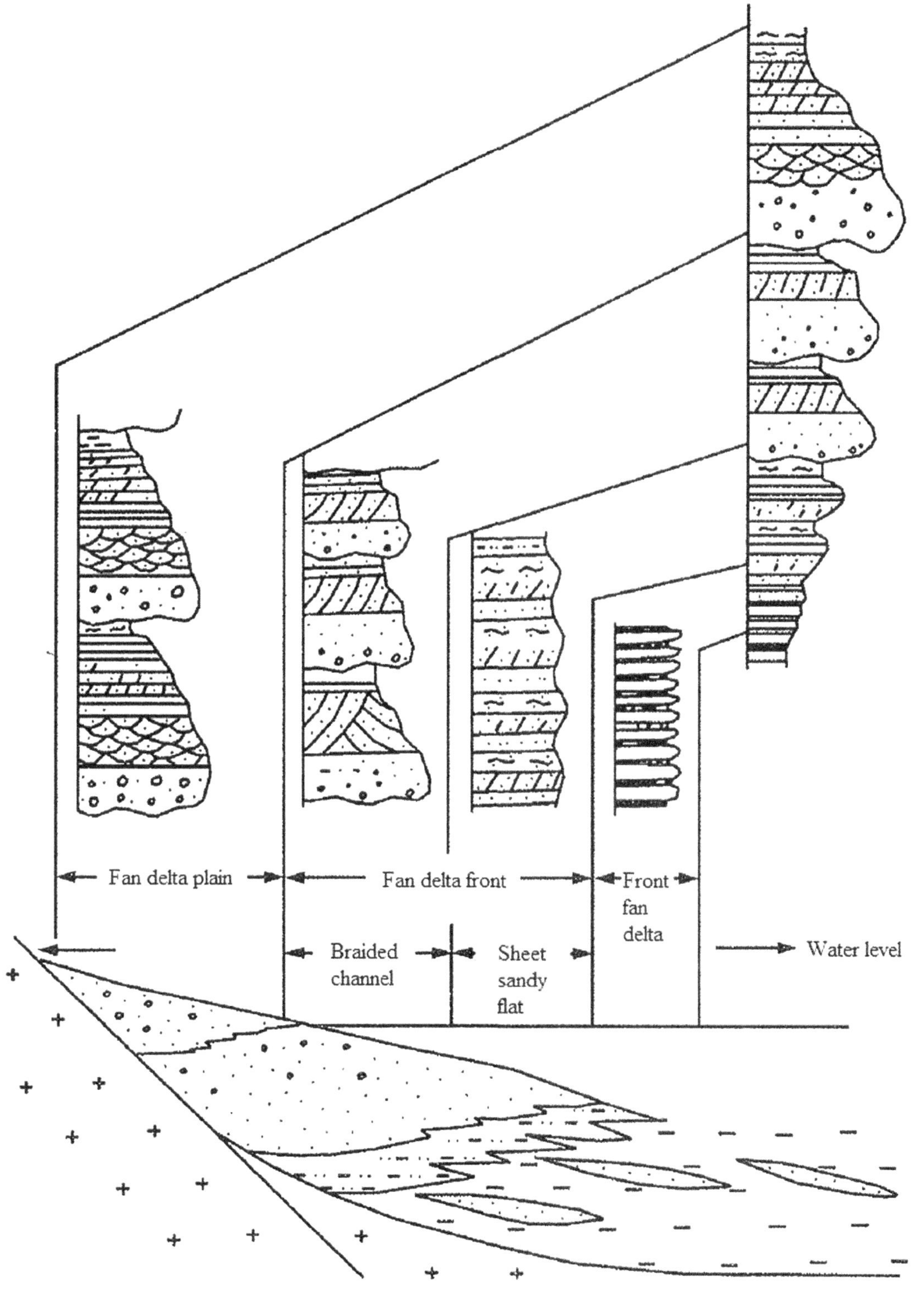

Fig. 11.55 The face zones of fan delta occur generally completely, with well developed upward coarsening sequence characteristics

massive argillaceous conglomerates or conglomerates supported by matrix can also be seen in the coarse clastic profile, belonging to onshore detrital flow deposit. The SP curve of the fan delta plain facies belt mainly shows a serrated low-amplitude cylinder shape.

2. **Fan delta front**

The facies belt of a fan delta front mainly comprises conglomeratic and sandstone mixed with grayish-green mud sand and a little inferior oil shale. The lithology of this facies belt varies considerably, and this facies belt is the best developed part in a fan delta sand body, which can be further divided into a subaqueous distributary channel, subaqueous distributary mouth bar, and sheet sand. The subaqueous distributary channel is a combination of conglomeratic and sandstone, mixed with thin layers of mudstone, and it develops megascopic trough cross bedding, parallel bedding, and cross bedding. The thickness of a single sequence is 0.3–2 m, and positive rhythmic layer fining upward is presented. The superimposed sand body thickness of a multilayer river can reach tens of meters, and its SP curve mostly shows a cylinder shape. The river mouth bar is formed by the interbedding of well sorted pebbly sandstone and sandstone with gray mudstone. Bedding, dominated by low-angle planar cross bedding and parallel bedding, is developed. The SP curve shows a funnel shape-bell shape or foreset finger shape. In certain fan-delta front facies, the river mouth bar is poorly developed or not developed at all. Sheet sand, which is a sand body, can be seen as a thin layer distributed at the outer edge of a river mouth bar, with lithology fining and dominated by sedimentary sandstone. The SP curve shows finger and serrated shapes.

3. **Front (fan) delta**

Front (fan) deltas enter semi-deep lake areas, and their lithology is light and dark gray mudstone mixed with a little sandstone, siltstone, calcareous shale, and oil shale. Many ostracoda and pyrites are contained in the mudstone. The sandstone mainly develops miniature wave ripple cross bedding and ripple crossing bedding. The SP curve shows a serrated or the low-amplitude flat shape. The underlying layer of a regressive fan delta is mainly deep lake subfacies. Front contemporaneous deposit is mostly deep lake subfacies also. Its main sequence is coarsening upward, the sand body plane is shown to be fan shaped, and the section form is lenticular.

11.7.3.3 Gilbert-Type Model

Gilbert-type fan deltas are also common in ancient lakes located at the crack edge of a lacustrine basin and steep slope zone. In the Meso-Cenozoic continental faulted lacustrine basins in China, there are many examples of this type of delta, for instance, the fan delta in the Shuanghe Oilfield He-3 Member in the Biyang Depression and the fan delta of the Lower Member 9 in the Majiapu area in the Luxi Sag (Shouqing 1987). This type of fan delta has an obvious three-tiered structure, comprising topset, foreset, and bottom set (Table 11.16).

1. **Topset (fan delta plain)**

This is mainly the braided river sedimentation of a fan delta plain, which mainly contains coarse-grained sediments including gravelstone, conglomeratic, and sandstone supported by matrix or particles. Channel sand with a thickness of up to tens of meters is deposited by the longitudinal bar and transverse bar in the channel. The interchannel sediments are mostly amaranth and variegated mudstone; nodules may be contained, and mud cracks can be seen. In addition, high-angle planar cross bedding can be seen in the channel.

2. **Foreset (fan delta front)**

The foreset is the most developed part of a fan delta sand body, which is composed mainly of high-angle conglomeratic and foreset sandstone, and it can be divided further into sand microfacies such as subaqueous distributary channels and front bars.

Table 11.16 Subfacies and microfacies characteristics of Gilbert-type lacustrine fan deltas

Subfacies	Topset (Fan delta plain)	Foreset (Fan delta front)		Bottom set (Front fan delta)
Microfacies	Braided channel and interchannel	Subaqueous distributary channel	Front bar	Predelta mud
Lithology	Dominated by conglomerate, conglomeratic and sandstone, mixed with amaranth and variegated mud sand	Conglomeratic and sandstone, and large amounts of boulder clay	Conglomeratic	Gray-to-dark-gray mudstone, shale, and oil shale, with a few thin layer of siltstone possibly mixed in
Grain size	Coarse	Relatively coarse	Relatively coarse	Thin
Sorting	Poor	Ordinary	Better	Good
Sedimentary structure	Nodule and mud crack, high-angle planar cross bedding	Planar cross bedding	High-angle planar and wedge cross bedding	Horizontal bedding
Vertical sequence	Fining upward	Fining upward	Coarsening upward or repeatedly coarse and fine	
Electric well-logging curve feature	Serrated low-amplitude cylinder shape	Dominated by cylinder shape and serrated cylinder shape	Serrated funnel shape-cylinder shape combination and progradational finger shape	Flat straight line

(1) Subaqueous distributary channel

This is mainly composed of conglomeratic and sandstone, and high-angle planar and trough cross beddings are developed. Local conglomerates are shown as directional alignment, and many intrastratal scour surfaces, lag gravels, and boulder clays are present. The interchannel sediment is mainly sand shale interbed, formed because of an overflowing channel or crevasse in the flood period. The SP curve of distributary channel deposit is dominated by cylinder and serrated cylinder shapes.

(2) Front bar

In the depositional sequence of Gilbert-type fan deltas, the high-angle foreset gravel bar is well developed. The thickness of a single layer of conglomeratic is usually more than 1.5 m. It is in abrupt contact between the bottom and the underlying layer, and conglomerate is mostly shown as having imbricate or directional alignment with a foreset angle of 25°–45°, coarsening upward or with alternating intrastratal grain size change. Toward the basin, gravel foreset gradually changes into sandy foreset. The top and bottom of the sand layer mostly show gradual change with the fine-coarse-fine sequence characteristic. Outside the river mouth bar, thin frontal sheet sand that transits into the bottom set can be developed. The SP curve of the front bar shows a serrated funnel shape-cylinder shape combination and a progradational finger shape.

3. **Bottom set (prodelta)**

The bottom set comprises fine-grained sediments of the front (fan) delta, and the lithology is gray to dark gray mudstone, shale, and oil shale, possibly mixed with a few thin siltstone layers. The SP curve mostly shows a flat straight line.

11.8 Braid Delta

A braid delta is a type of coarse-grained delta commonly formed in the short-axis direction of a lacustrine basin. It can also be developed when the slope of a basin is narrow in the long-axis direction and the provenance is near. Nemec and Steel (1988) divided braided rivers into single braided river deltas and braided plain deltas according to the quantity, in which the former takes the single braided river as its source and the latter takes the braided plain as its source. There is a large gradient on the shoreside and the subaqueous slope of this type of delta, the lake-shore is close to the piedmont, and current is short, which means it enters lake water only when developed in the braided river stage. Hence, a braid delta is formed, which shortens the flow path of the river before it enters the lake. Accordingly, it is called a short flow path delta. In sag ponds, the braid delta is commonly developed and is mainly distributed on the gentle slope side of the short axis (or narrow steep slope in the long axis direction). On the short-axis steep side of a lacustrine basin, where the slope is steep and close to the mountains, the alluvial fan directly enters into the lake to form a fan delta, but in the continuous foreset of the fan delta, the slope increases and becomes gentle, resulting in a gradual transformation to a braid delta (Fig. 11.57). Hence, braid deltas fall under the scope of fan deltas. A single braided river delta refers to a coarse-grained delta rich in sand and gravel, and is formed due to the braided river plain by the foreset of a single bed load river entering a stable water body (Yu et al. 1995; Fig. 11.56).

11.8.1 Differences Among Braid Deltas, Normal Fan Deltas, and Fan Deltas

The sedimentary characteristics of a braid delta lie between those of a normal delta and fan delta.

11.8.1.1 Difference from Normal Delta

The sedimentary characteristics of braided river deltas are similar to those of normal deltas, however the biggest differences between them are their source and particles. In general, a braided river delta or short flow path delta is supplied by a braided river, while a normal delta or long flow path delta is mainly supplied by a meandering river. In addition, the granularity of a braid delta is usually coarse; hence, it is called a coarse-grained delta. However, the granularity of a normal delta is finer than that of a braid delta; hence, it is called a fine-grained delta. Braid delta subfacies can also be divided into three sections, namely, delta plain, delta front, and prodelta sedimentation. However, it is commonly divided by quartering, which means dividing the delta plain into upper and lower parts, and each subfacies and microfacies are different from those of a normal delta.

(1) The distributary channel of a braided river plain has the characteristics of a braided river, which means channel deposit is tabular and has high width/thickness ratio; clastic particles are relatively coarse; the contents of sand and gravel are high (normal delta is dominated by sand and silt); channels have no typical "dual structure" feature, which means few topset subfacies or overbank sediments; and channels are not stable and easy to migrate, thus coarse clastic sand bodies are usually distributed in pieces on the plane.
(2) A braided river is developed in a subaqueous distributary channel. Owing to the large flow magnitude of a braided river and the abundance of fragmentary material, the bed load/suspended matter ratio is high. Therefore, after entering into the water body, the channel depositional facies is relatively developed, followed by the distributary mouth bar, which is very different from the pattern observed in a normal delta. When the source supply is adequate, the terrain slope is medium, distributary mouth bar is relatively developed, and its plane modality is mainly shown as a rhombus sand bar.
(3) The size of a braided delta is smaller than that of a normal fine-grained delta, however braided deltas are often distributed in groups, especially in steep terrain.

11.8.1.2 Difference from Fan Delta

Braided deltas and fan deltas are coarse grain deltas. Some scholars hold the opinion that they can be merged into fan deltas, but their sedimentary characteristics are obviously different. The main difference lies in their development states of source supply and gravity flow. The supply source of a braided delta is a braided

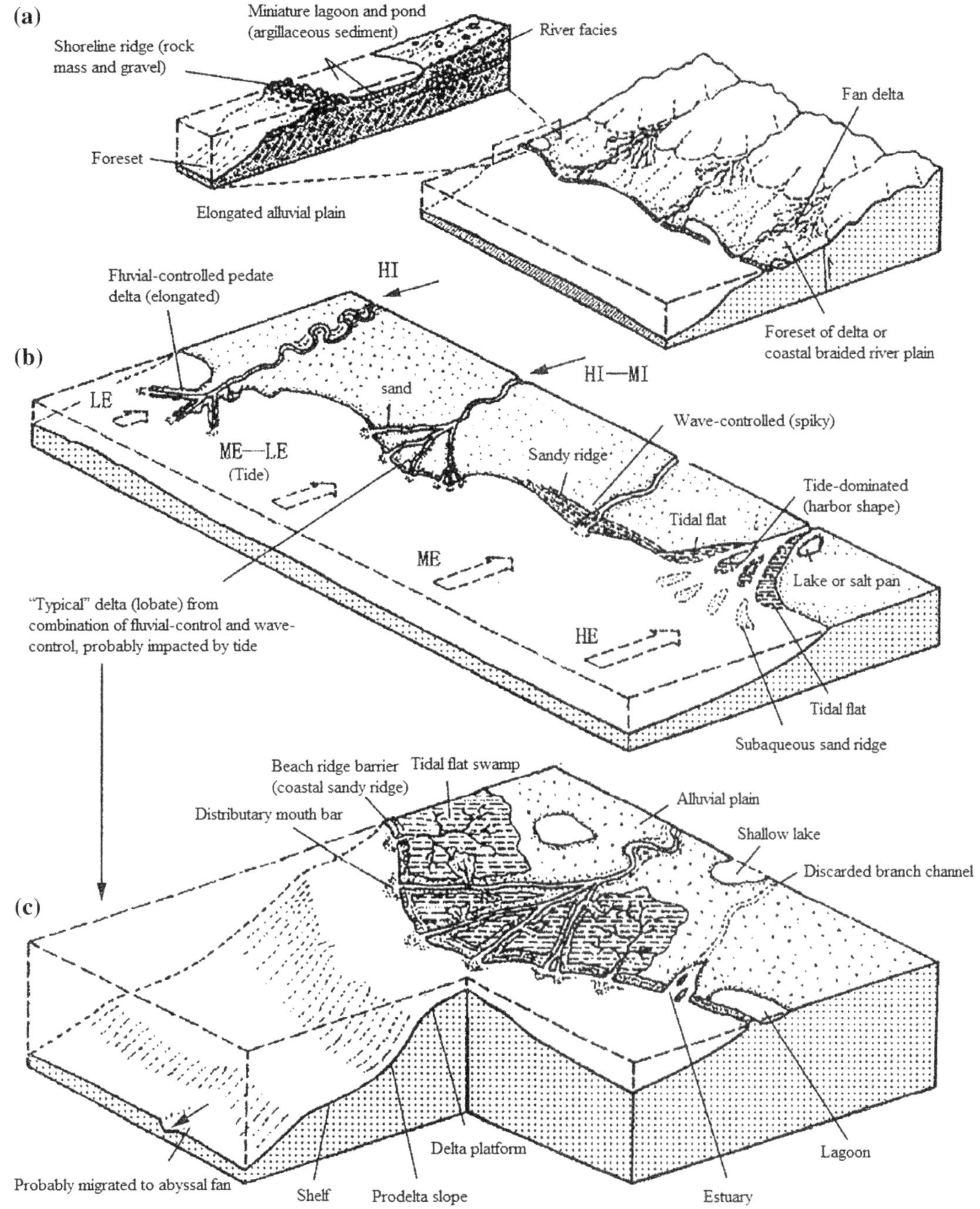

Fig. 11.56 Sedimentary sketch map of all types of delta (after Einsele 2000). **a** Marine fan delta formed due to alluvial fan or braided river plain prograding seaward. Notably, a coarse-grained beach ridge (rock river gravel) can form a lagoon or pond free from waves and air flow, and the silt and mudstone of an ancient beach ridge and lagoon facies may be covered by fluvial deposits. **b** It is not only affected by sedimentary supply (HI high supply; MI medium supply) but also divided into different forms, namely, wave-controlled and tide-dominated (LE low-energy; ME medium-energy; HE high-energy) megascopic marine delta. **c** Different sub-environments of megascopic lobate wave-controlled to tide-dominated delta system (similar to the modern Niger Delta)

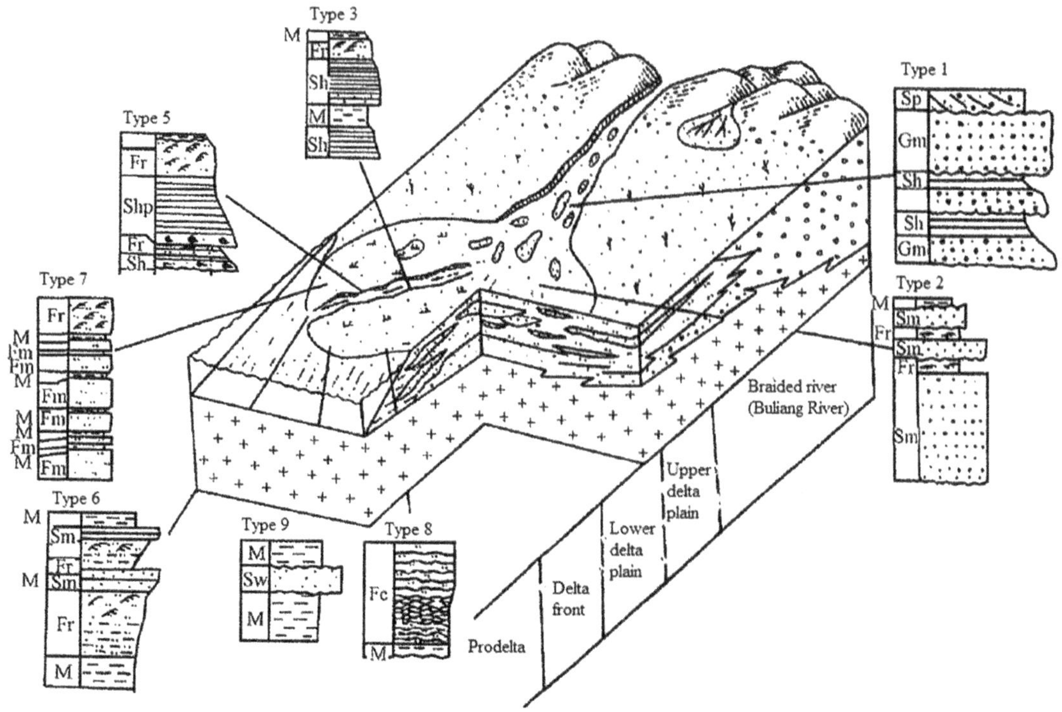

Fig. 11.57 Sedimentary model map of braid delta in Buliang river, Daihai

river, while that of a fan delta is an alluvial fan (including dry fan and wet fan), and no debris flow is developed on a braided delta plain. However, debris flow is commonly seen on a fan delta plain, and dry arid fan delta debris flow is especially developed. Specifically, the following points should be noted (Table 11.17):

① Gravity flow sedimentation of a fan delta is usually better developed than that of a braid delta, and debris flow is particularly common in a fan delta plain. ② The granularity of a fan delta is much coarser than that of a braided (river) delta. ③ Fan deltas mainly take Gm, Gp, Gi, and Gt as their main lithofacies, while lithofacies Sm, Sh, and Fh are less developed compared to those of braid deltas. ④ The vertical sedimentary sequence of a fan delta is dominated by conglomerates with rapid granularity change, while the sequence granularity change of a braid delta is relatively slow; a fan delta shows a relatively wider range of granularity variation than a braid delta. ⑤ The distributary channel of a braid delta is a fine-grained straight river or a meandering river with low sinuosity. ⑥ Both deltas do not develop the coarsening-upward sequence well, however conglomerate facies may occur on certain sequences of a fan delta, while the coarsest particle in a braid delta is medium sand, because a braid delta mostly comprises fine sand-silt.

11.8.2 Sedimentary Facies Belt and Depositional Model of a Braid Delta

In terms of granularity, although a braid delta is a coarse delta, its granularity is finer than that of a fan delta. Its lithology may be from conglomeratic to pure mudstone, and it is characterized by relatively poor to relatively good sorting. Its sedimentary structure includes massive bedding, various cross beddings, horizontal bedding, and complex beddings (Table 11.18).

As mentioned earlier, the coarse-grained delta facies belt is usually divided by quartering, and the upper and lower delta plains are divided

Table 11.17 Comparison of the formation conditions and recognition features of fan deltas, braid deltas, and normal deltas

Characteristics		Type		
		Fan delta	Braided delta	Ordinary delta
Provenance features		Source supply from alluvial fan, near provenance	Source supply from braided river, relatively near provenance	Source supply from meandering river, distal source
Formation background	Terrain conditions	Steep slope	Steep slope-medium	Gentle slope
	Structural conditions	Abrupt slope in fault zone	Mostly away from fault zone	Away from fault zone
External shape	Form	Wedge-shape and lens shape	Sheet	Sheet and lobate distribution
	Area	Around tens of square kilometers	Hundreds of square kilometers	Tens or thousands of square kilometers
Internal architecture	Facies sequence characteristics	Single cycle commonly fining upward; both fining upward and coarsening upward for multi-period superimposition	Commonly fining upward, uncommonly coarsening upward	Commonly coarsening upward as the overall feature
	Sorting and roundness	Poor sorting: corner angle and sub-rounded	Medium sorting: sub-rounded and round shape	Good sorting: sub-rounded and round shape
	Matrix content	High	Low	Medium-low
	Delta plain	Brown and yellowish-brown conglomerate, pebbly sandstone; megascopic cross bedding, convolute bedding, and top horizontal bedding Occasional plant roots or burrows	Yellowish-brown conglomerate, with conglomerate and sandstone; megascopic and medium planar-trough cross bedding, parallel bedding, lateral accretion cross-bedding, and scour surface structure	Bedding plane structures such as sandstone, siltstone and mudstone raindrop imprint, sun crack, and footprint; miniature cross bedding Few biological fossils, dominated by freshwater zoolite and plant residue
Sub-environment division and feature	Delta front	Yellowish-brown and gray pebbly sandstone and middle-fine sandstone Miniature cross bedding, flaser bedding, and horizontal bedding Phreatic water ostracoda and freshwater stonewort, few mica	Gray coarse conglomeratic, medium-sized and coarse sandstone Large and medium-sized cross bedding, lateral accretion cross bedding, scour surface structure, parallel bedding, with mica development	Medium and fine-grained sand and siltstone, rare mud Cross bedding, wavy bedding, and scour-fill structure Burrow and bioturbation structures developed, with large amount of mica

(continued)

Table 11.17 (continued)

Characteristics		Type		
		Fan delta	Braided delta	Ordinary delta
	Prodelta	Dark gray silty mudstone and mudstone, common silty lens mixed with horizontal bedding Relatively rich ostracoda and ichthyolite	Grayish green and dark gray fine-grained mudstone, silty mudstone mixed with silt-fine sandstone debris flow, fluidized flow and turbidity current sediment, and weak inverse graded and scour surface at the bottom	Dominated by dark mudstone, mixed with siltstone developed with horizontal bedding and massive bedding, and common ostracoda
Identification features	Lithology	Dominated by gravel, gravelly sand, and sandstone	Dominated by fine sand-medium sandstone	Dominated by siltstone and mudstone
	Well logging	Serrated, serrated cylinder shape or funnel shape, and finger shape	Serrated cylinder shape	Bell shape
	Earthquake	No obvious progradational reflection configuration	Relatively obvious progradational reflection configuration	Obvious progradational reflection configuration
Liquid type		Prominent gravity flow, limited action of traction current	Dominated by traction current, minor gravity flow	Dominated by traction current
Reservoir distribution features		Poor lateral continuity of reservoir, poor performance, and good vertical continuity	Relatively good lateral continuity of reservoir, good or relatively good performance, and average vertical continuity	Good lateral continuity of reservoir, very good performance, and poor vertical continuity
Occurrence probability in geologic history		Very high	High	Extremely high

Table 11.18 Features of main lithofacies type and genesis of braid delta sediment (according to Li Yanfang 1998)

Lithology facies type	Structure	Main sedimentary structure	Other characteristics	Genetic interpretation
Massive bedding gravel lithofacies	Dominated by fine grain, heterogranular texture	Massive bedding	Subangular-shape, complex composition	Braided river channel
Cross bedding gravel lithofacies	Fine-grained texture	Directional alignment of gravel, with cross bedding shown	Subangular to sub-rounded, complex composition	Braided river channel
Massive bedding gravel lithofacies	Sand mixed with mud, heterogranular texture	Massive bedding	Subangular to sub-rounded, complex composition	Rapid deposition of braided river channel in flood period
Cross bedding conglomeratic lithofacies	Sand mixed with mud, heterogranular texture	Mostly directional alignment of gravel, cross bedding shown	Subangular to sub-rounded, complex composition	Braided river channel sedimentation
Massive bedding coarse sandstone lithofacies	Coarse-grained texture, gravel-containing supported	Massive bedding	Subangular to sub-rounded, complex composition	Deposition of braided river channel and subaqueous braided river channel in flood period
Cross bedding coarse sandstone lithofacies	Coarse-grained texture	Directional alignment of particles or bedding shown by flaky minerals	Subangular to sub-rounded, complex composition	Deposition of braided river channel and subaqueous braided river channel
Massive bedding medium sandstone lithofacies	Dominated by medium-grained sand, and coarse or fine sand present	Massive bedding	Subangular to sub-rounded, complex composition	Deposition of distributary mouth bar and subaqueous braided river channel in flood period
Cross bedding medium sandstone lithofacies	Medium-grained texture	Massive bedding, wedge bedding and trough bedding	Subangular to sub-rounded, complex composition	Deposition of distributary mouth bar and subaqueous braided river channel
Crossing bedding fine sandstone lithofacies	Fine-grained texture	Wavy bedding and wedge bedding Trough bedding and wavy bedding	Subangular to sub-rounded, complex composition	Distributary mouth bar and distal bar sedimentation
Massive bedding siltstone and Argillaceous siltstone lithofacies	Silt shape and mud-bearing silt texture	Massive bedding	Grayish-green and amaranth	Flooding deposits of distributary mouth bar or braided river
Ripple cross bedding, Flaser bedding siltstone, and Argillaceous siltstone lithofacies	Silt shape and mud-bearing silt texture	Ripple crossing and flaser bedding	Grayish-green	Distal bar sedimentation

(continued)

Table 11.18 (continued)

Lithology facies type	Structure	Main sedimentary structure	Other characteristics	Genetic interpretation
Massive silty mudstone and Mudstone facies	Mud or muddy texture with silt	Massive bedding	Grayish-green or amaranth	Distal bar prodelta sedimentation and braided river overbank sedimentation
Lenticular bedding silty mudstone facies	Muddy texture with silt	Lenticular bedding	Grayish-green	Distal bar sedimentation
Massive mudstone facies	Muddy texture	Massive bedding	Grayish-green or amaranth	Prodelta or braided river overbank sedimentation
Horizontal bedding mudstone facies	Muddy texture	Horizontal bedding	Grayish-green or amaranth	Prodelta or braided river overbank sedimentation

according to the location (overwater and underwater) at the time of sedimentation and the features of the distributary channel. Similar to fan deltas, braid deltas can be divided into steep-slope type and gentle-slope type, as exemplified by the braid delta in the Daihai Lake, Inner Mongolia (Figs. 11.5 and 11.58), and the Pearl River Delta (Fig. 11.59).

11.8.2.1 Braid Delta in the Daihai Lake

Daihai Lake is located in Liangcheng County, 80 km southeast of Hohhot, in the Inner Mongolia autonomous region, where there are more than 10 delta sand bodies of various sizes around the lacustrine basin, and a single sand body mainly shows a fan or tongue shape toward the lacustrine basin. The steep coast delta northwest of the short-axis direction of the lacustrine basin features a small size and large amount, whereas the gentle coast in the southeast features a large size and small amount. The Bulianghe Delta is taken as an example of a gentle slope and the Yuanzigou Delta as a steep slope.

1. **Gentle-slope Type-Buliang River Delta**

The facies belt in the Buliang River Delta can be divided as follows:

① Upper delta plain—This is composed of gravelly river facies, and its subaerial deposition is flood intermittent braided river channel sedimentation composed of sand and gravel. ② Lower delta plain facies belt—This refers to the onshore and subaqueous transition part. When the lake level is high, it is located under the water and vice versa. Its basis is dark gray-blue gray lake mud facies sedimentation seen at the section bottom. When rivers enter a flat fan-shaped plain, due to shallow water and rip currents, many parallel bedding sand layers are formed constituting sheet sand, and a sandy distributary channel is formed at the bottom. Few coarsening-upward sequences are developed in facies belts close to the slope zone. Owing to the rapid upload of a large amount of silt carried by the river, in combination with the strong scour of current and flat terrain, the inverse rhythm of a distributary mouth bar is not very obvious. ③

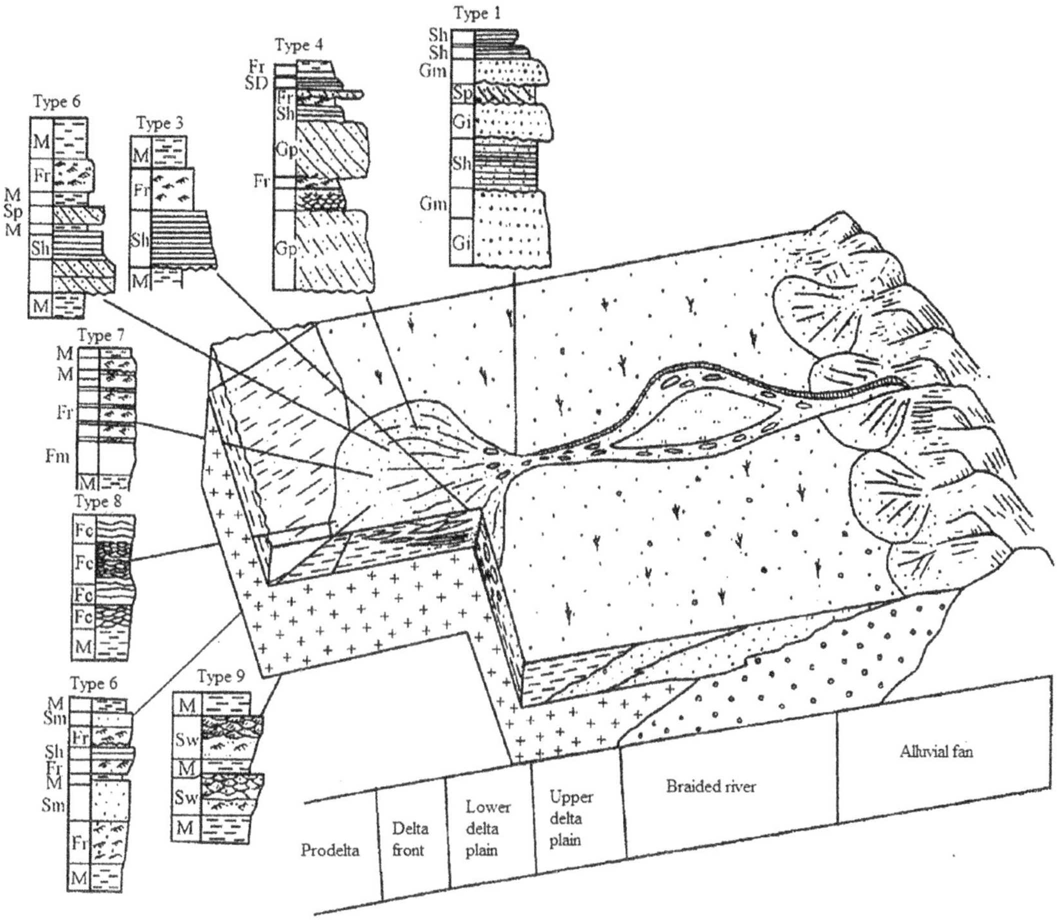

Fig. 11.58 Delta sedimentary model map of braided river of Yuanzigou, Daihai

Delta front slope facies belt—This is mainly dominated by interbedding between thick silt layers and thin mudstone layers. The interbedding is composed of thin silt and blue gray mud on the front slope edge, and the thickness of the mud layer is high. The internal architecture is mostly composed of wave ripples in the front thin layer, and it is distributed along the bank in line. ④ Prodelta mud facies—This is shown as blue gray or dark gray mud.

2. **Steep slope type—Yuanzigou Delta**

The facies belt classification of the Yuanzigou Delta is roughly the same as that of the Buliang River Delta, however the sedimentary characteristics of the former are different, as described below:

① Upper delta plain—This is characterized by gravelly river facies belt. There is no obvious difference between this facies belt and the Buliang River Delta in terms of sedimentary characteristics or sequence types, and only its granularity is coarser than that at the same location in the Buliang River Delta. ② Lower delta plain facies belt—The lithofacies association type of the facies belt is roughly the same as that of the Buliang River Delta, but trough and planar cross bedding sandstone is developed, which is closely connected with the topographic slope of the delta. When the river enters a flat lower delta plain, the sand body is dominated by braided distributary channels and intermittent

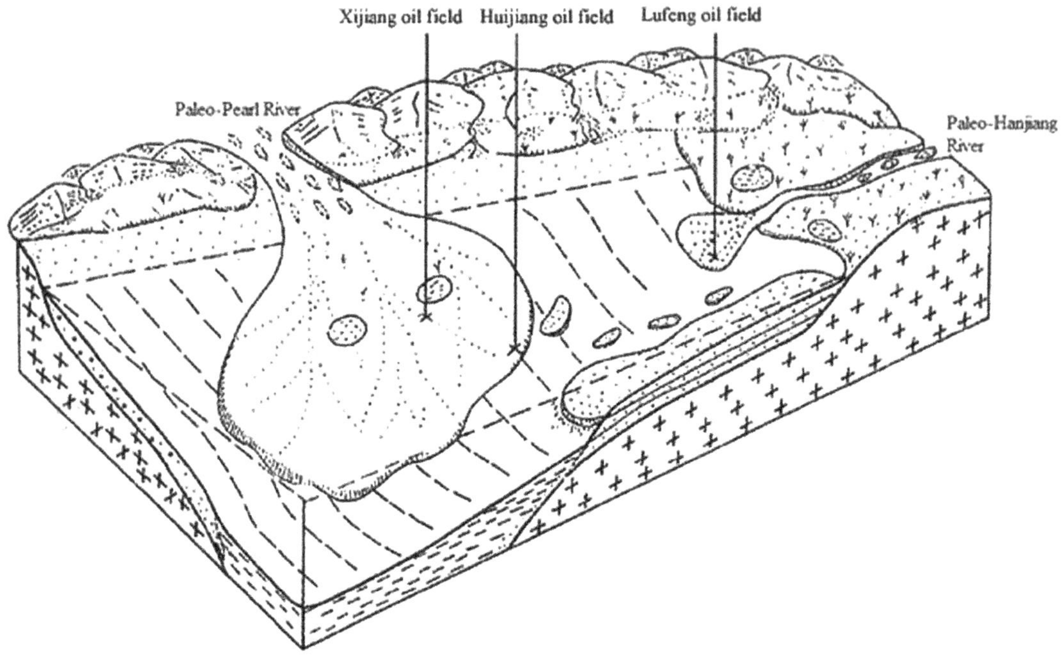

Fig. 11.59 Sedimentary facies model map in the east of Zhujiangkou basin

flood sheet deposit; and a megascopic sheet sand facies belt is present as well. ③ Delta front slope facies belt—This can be divided into two zones. The inner zone is similar to the front slope facies belt of the Buliang River Delta, and it is less affected by waves. However, the outer zone is strongly affected by waves, and complex bedding is well developed. ④ Prodelta mud facies—This is shown as blue gray or dark gray mud.

11.8.2.2 Pearl River Mouth Delta

The Pearl River Mouth Basin is a continental marginal tensional rift basin, in which a series of basement rift-controlled horst-graben are arranged in echelon. In early-medium Miocene, the Earth's crust was under continuous and stable subsidence. This zone is in the sea-land transitional zone, which deposited the Zhujiang Formation-Hanjiang Formation multi-cycle delta compound system (Fig. 11.59). This delta is a typical marine shelf gentle-slope type braid delta, and its several important oildoms are located mostly in the delta front. Subfacies division of this braid delta is also performed by quartering the coarse-grained delta (Table 11.19).

11.9 Delta Identification

Similar to other depositional systems, delta identification is mainly conducted according to geology and geophysical characteristics. However, there are many types of deltas, and multiple types of fans have similar morphologies and structures, making identification difficult sometimes. Its geologic features differ from other fans, in that it is partly over water and partly underwater, and is dominated by traction current (Table 11.20). Regarding the various types of fan bodies, all are submerged underwater or all are onshore and dominated by gravity flow. In terms of vertical structure and section features, the delta takes coarsening upward (inverse cycle) as its major characteristic, and there is a foreset/retrogradation structure, which is shown as a funnel shape in terms of well logging but as

Table 11.19 Sedimentary characteristics and sedimentary facies division of Zhujiang formation delta in (Eastern) Pearl River Mouth Basin

Delta subfacies	Microfacies	Vertical sequence characteristics	Sedimentary characteristics
Upper delta plain (from estuary to the highest water level)	Main channel, braided river channel, and braided distributary channel	Abrupt superimposition fining upward, featuring mud-in-sand dominated by sandstone, mixed with mudstone	Coarse granularity, dominated by conglomerate and pebbly coarse sandstone, relatively complete sedimentary structure
Lower delta plain (from average high water level to average low water level)	Distributary channel, natural levee, crevasse splay, abandoned channel, and interdistributary bay	Medium fining upward, featuring mud-in-sand or sand-mud interbed, roughly equal sandstone and mudstone	Medium granularity, dominated by sand, mixed with silty-fine sand and sand-mud interbed, medium to good sorting, and relatively complete sedimentary structure; relatively light mudstone color and large amounts of plant debris and mica
Delta front (from average low water level to average wave base)	Distributary mouth bar, distal bar, and subaqueous distributary channel	Dominated by medium coarsening-upward sequence, possibly mixed with minor fining-upward sequences, featuring sand-mud interbed and more sandstone than mudstone	Dominated by medium-fine sand, better sorting, and large amounts of mica and carbon dust
Prodelta (below wave base)	Offshore sedimentation (normal shallow sea)	Large amount of mudstone mixed with silt	Mudstone dominance, dark color with pure quality, no burrow developed

various types of progradation reflection seismically.

11.9.1 Geology

(1) Constructive fluvial-dominated delta sedimentation is mainly composed of relatively thick (hundreds of meters, or even thousands of meters) interbedding sediments such as sandstone, siltstone, silty mudstone, and mudstone. Hence, its compositional maturity and texture maturity are higher than those of rivers. Dark organic fine-grained sediments, peat beds, or coal seams are usually mixed in the delta plain. A few to no conglomerate and chemical rocks are present in the delta, which is an important feature of the delta. In the coarse-grained part, there are a few foraminifer and molluscs, but wood fibers are very rich and ostracoda is common as well. However, there are large amounts of sheet mica in fine-grained sediments because excessive sheet mica is transported by the river as float.

(2) Generally, grain size varies over a small range. From land to sea/lake, detrital grain size and sorting in sandstone turns finer and better gradually as a whole. On a C-M diagram, traction current type, namely, the QR Section and RS Section, is shown. Such grain size distribution characteristics mainly reflect the dominance of saltation and suspension transport, accompanied by minor rolling transport. On the probability graph, grain size distribution characteristics are

Table 11.20 Comparison between formation conditions and recognition features of each type of fan and delta

Features		Fan	Delta
Classification		Alluvial fan, slump fan, turbidite fan, basin (sea, lake) floor fan, slope fan, and subaqueous fan	Fan delta Braided delta Meandering river delta
Geological features	Lithology and grain size	Dominated by conglomeratic, relatively coarse granularity	Dominated by middle-fine sandstone, large span of granularity
	Development location	Over water or underwater	Partly over water and partly underwater
	Fluid type	Dominated by gravity flow	Dominated by traction current
	Provenance	Close to provenance	Away from provenance
	Transportation load	Bottom transportation and suspension transport	Mixed transportation, dominated by rolling
Vertical structure and section features	Rhythm	Dominated by fining upward	Either fining or coarsening upward, but mostly coarsening upward
	Tracing pattern	Serrated cylinder shape and bell shape with high amplitude	Dominated by funnel shape and bell shape
	Seismic response	Chaotic and discontinuous reflection, no obvious progradational configuration, many bidirectional downlaps	Relative obvious progradational reflection configuration and good continuity

generally not typical, and the grain size distribution of distal bar sediments is mainly composed of single suspension populations. The probability graph of the distributary mouth bar sandstone is composed of three populations, of which saltation population is dominant. Current action is not very strong, however there is certain wave transformation.

(3) The sedimentary structure type of a delta is complex, which means both sedimentary structures formed by fluvial processes and those formed by wave or tide action can be seen. Current ripple, wave ripple, planar, and trough cross beddings are found in sandstone and siltstone. Horizontal bedding is relatively developed in mudstone. In addition, various undulations; vein-like, lenticular, and convolute beddings; scour-fill structures; deformation structures; and bioturbation structures are developed as well.

(4) With regard to biology, a delta has biota mixture of land and sea, which indicates that its salinity is normal at the time of formation, and both brackish and freshwater biology are developed. Autochthonous organisms, mainly euryhaline organisms, such as lamellibranch, gastropods, and ostracoda, can be accumulated, buried, and preserved as fossils. Mainly, terrestrial organisms and phytoclasts are transported and buried. Moreover, in a vertical depositional sequence, the proportion of marine creatures decreases gradually from the bottom up, while the proportion of freshwater organisms and phytolith increase from the bottom up.

(5) On the depositional sequence, a coarsening-upward structure or feature can be found, and on the delta plain part, at the top of the sequence, a fining-upward structure or feature can be found.

(6) The geometric shape of the sand body is lobate or finger-like on the plane, and it is distributed vertically or obliquely with respect to the coastline. The section is of the lens or diverging broom shape, crossed with prodelta or sea/lake mudstone in a serrated shape.

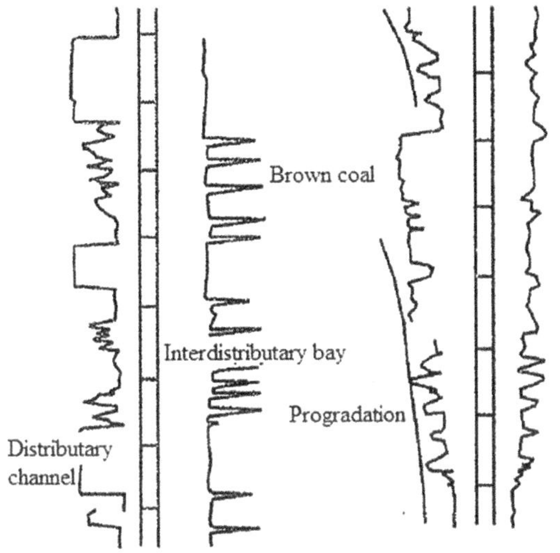

Fig. 11.60 Log facies of typical delta

11.9.2 Logging

From the perspective of lithology and electric well-logging curves, coarsening upward is shown from the bottom to the top with progradational structures or sequences. This is shown as a bell shape or funnel shape on an SP curve. On the electric well-logging curve of delta microfacies, generally, the distributary channel has a bell shape or cylinder shape fining upward, while the distributary mouth bar and distal bar have a funnel shape coarsening upward (Fig. 11.60). Coleman et al. (1982) studied the sedimentary sequence, grain size, lithology, and electric well-logging curve of an abandoned distributary channel distributary mouth bar (Figs. 11.61 and 11.62); according to their results, the bell shape or cylinder shape of the channel and funnel shape feature of the distributary mouth bar are very obvious.

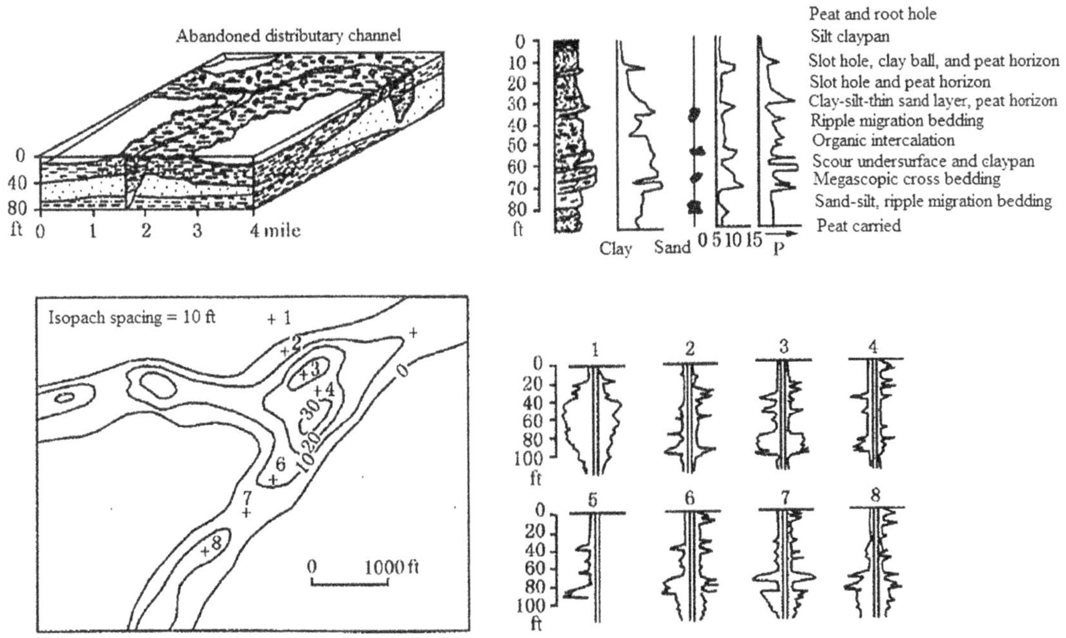

Fig. 11.61 Characteristics of sedimentary filling and electrical logging curves of the abandoned channel

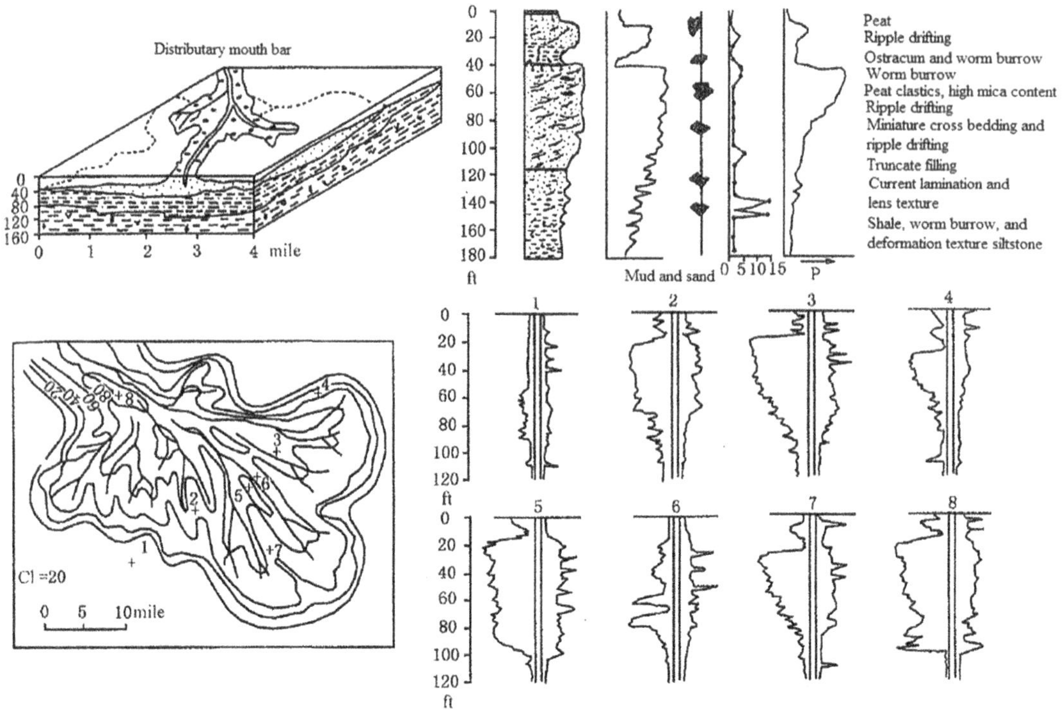

Fig. 11.62 Sedimentary characteristics and logging facies of logs in estuary dam (after Coleman 1982)

Deltas of different types in different areas are characterized by different electric well-logging curve responses. There are many serrated cylinder-shaped curves in a distributary channel or a subaqueous distributary channel, for example, some oilfield braided river deltas in the Pearl River Estuary. However, the electric well-logging curve response of a distributary channel may be dominated by bell shape curves.

11.9.3 Seismology

The sea basin is an open deep-water basin, and its accommodation is nearly unlimited relative to lateral river injection. Hence, the progradation slope of a delta will advance in the long term and will be stable, as opposed to being flattened owing to its front being filled by sediments. In addition, the relative change rate and frequency of the sea level in a sea basin are much slower than those in a continental basin. Hence, the delta can usually advance continuously toward the basin to form a substantial amount of sediments and develop various megascopic foreset structures to make it easy to be identified on the seismic cross section.

11.9.3.1 Sea Basin Fluvial-dominated Delta

Given the large accommodation of a sea basin, the process of sedimentation is manageable and stable, thus the progradation of the delta is obvious, the scale of sedimentary bodies is large, and seismic facies markers are obvious and easy to identify. The fluvial-dominated delta therein is a high-constructive delta, hence the influence of transformation and destruction is small, resulting in the following prominent delta features:

(1) Far away from the basin boundary, and free from the control of the basin edge boundary fault.
(2) The seismic facies unit has a taper or wedge shape. Owing to its large scale, its length and width can be tens to hundreds of thousands

Fig. 11.63 Seismic reflection characteristics of river-dominated delta of continental shelf basin in north of the South China Sea

of meters. Its external physical characteristics may be not obvious on a seismic cross section. It should be noted that its external physical characteristics can be analyzed using an isopach map of sedimentary bodies, especially, a sand factor contour map.

(3) The most important symbol is the development of various progradation structures, among which the sigmoid, oblique, and complex sigmoid progradation structures are the most common (Fig. 11.63). Their common characteristics are well developed bottom set—reflecting fine terrigenous matter—rich argillaceous sediment, and marked difference from alluvial fan and proximal subaqueous fan.

(4) Different parts of a delta are characterized by different seismic reflection textures. In general, the topset is characterized by the medium continuity structure with medium amplitude. When high-precision seismic data is not available, at times, it is difficult to recognize the delta topset. On the foreset part (roughly the same as delta front facies), owing to a decreasing sand-mud ratio and increasing mudstone thickness, the wave impedance difference between sand and mudstone increases, leading to increases in amplitude and continuity. In terms of the bottom set, there are two cases. The first is that the delta downstream accretion rate increases because of which the steep slope of the delta front can easily induce turbidity current to form a turbidite fan in the periphery of the delta front, featuring a three-highs structure. The second is that the delta downstream accretion rate decreases, resulting in no development of turbidity current on the gentle slope of the delta. Stable mudstone with massive thickness is mainly developed in the prodelta, featuring weak amplitude or even no reflection configuration and poor continuity. In general, the first case is more commonly found.

(5) The foreset mostly occurs in the case of seismic facies along the direction of the delta strike (accretion), while the seismic facies unit configuration is mound shaped in the direction transecting its strike; the corresponding internal reflection configuration is bidirectional downlap.

From the perspective of vertical change in a delta sequence, the reducing-upward reflection configuration is generally shown in a delta developed within a front turbidite fan. In contrast, the increasing-upward reflection configuration characterized by the presence of stable mudstone can be observed in the sedimentary bodies of a prodelta.

11.9.3.2 Seismic Facies Characteristics of a Sea Basin Wave-Dominated Delta

When strong wave and longshore current energy transport sediments to both sides of an estuary for deposition, a wave-dominated delta is

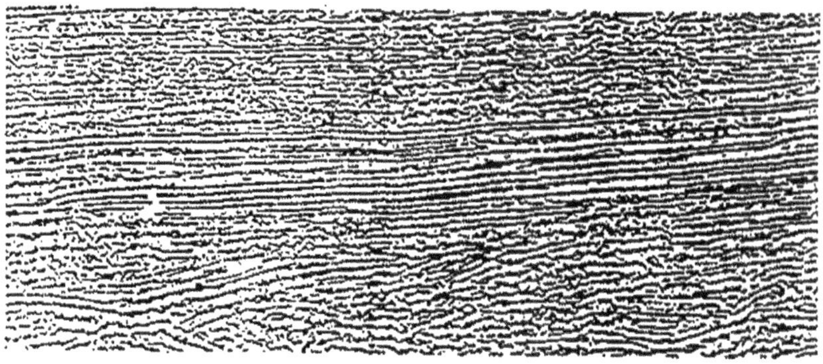

Fig. 11.64 Seismic facies characteristics of marine shelf wave controlled delta

formed. This strong transformational and destructive structure decreases delta length, increases delta width, and greatly weakens delta advance. Therefore, large-scale foreset structures can barely be found on wave-dominated deltas, and an imbricate foreset structure is the basic feature. Moreover, the plane modality of the delta is no longer elongated lobate but lobate with width much greater than length (Fig. 11.64). This also indicates that when the slope is gentle, lateral migration of the distributary channel or lateral spread of the delta increases. Its main sedimentary characteristic is weakened longitudinal downstream accretion.

11.9.3.3 Delta Seismic Facies Feature of a Fault Lacustrine Basin

In a lacustrine basin, the actions of wave, longshore current, and tide are considerably weaker than those in a sea basin. Hence, the resulting delta is mostly of the constructive fluvial-dominated type. However, the shape, depth of water, gradient, and capability of accommodating sediments of a lacustrine basin and sea basin differ greatly. Hence, the resulting deltas are different as well.

A lacustrine basin is basically closed (there is water drainage, however it is insignificant from the perspective of sedimentation), its water depth is also shallow, and its capability to accommodate sediments is limited. Therefore, when coarse clastic sediments carried by a river are unloaded at the river mouth, usually, large amounts of argillaceous sediments are deposited in the middle of the lacustrine basin, and the effect on the deposition rate in the part between the river mouth and the middle part of the lacustrine basin middle part is minimal. In addition, the lateral accretion rate of the delta decreases, which hinders the development of a foreset structure. Relative fluctuation in the horizontal plane of a lacustrine basin is much stronger than that in the case of a sea basin, and frequent advance and retreat of the shoreline lead to frequent changes in delta location. This hinders the development of a foreset structure. Therefore, in a faulted basin, a large-scale foreset structure delta may be formed in the mid-term stage of the fault depression when the basin is in a strong undercompensated condition. The megascopic foreset structure is generally not developed in lacustrine basin deltas. The imbricate foreset structure is common, and the foreset structure may even be absent (Fig. 11.65). The plane modality of this type of delta is mostly pedate or lobate.

11.9.3.4 Seismic Facies Feature of a Fan Delta

A fan delta is formed below a basin marginal large fault, and an alluvial fan provides the source supply and transport through the bed load for the formation of a proximal pebbly delta.

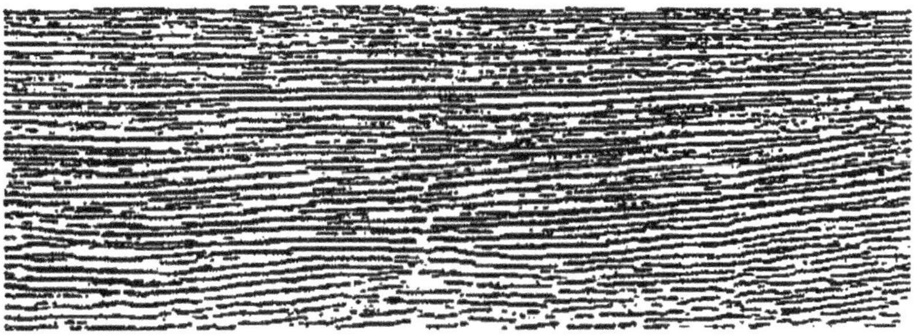

Fig. 11.65 Seismic facies characteristics of rift basin delta

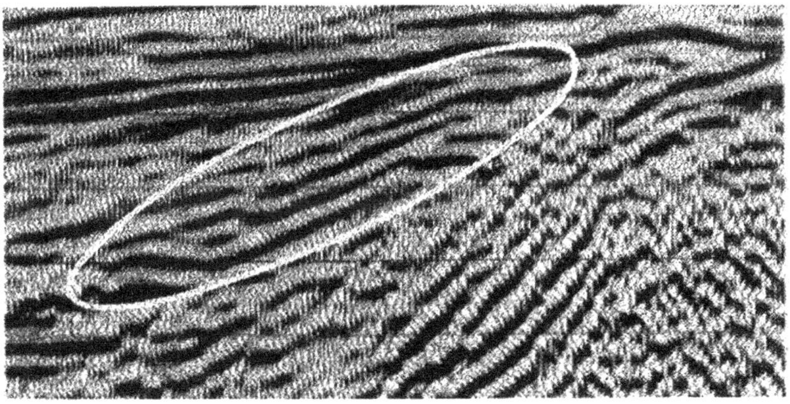

Fig. 11.66 Seismic facies characteristics of fan delta

Therefore, a fan delta has the seismic facies characteristics of both alluvial fans and of deltas (Fig. 11.66).

(1) It is developed under a big fault at the basin marginal boundary.
(2) It has a tapered shape with a scale generally larger than that of an alluvial fan and length ranging from thousands to tens of thousands of meters.
(3) On the longitudinal section, the seismic reflection configurations of the front and rear parts of its sedimentary bodies have obvious differences. The front part (alluvial fan part, generally short) is mainly shown as a chaotic foreset structure or a wave structure, while the rear part (delta front and prodelta part, generally long) can develop various foreset structures. In particular, the complex sigmoid-oblique foreset structure and the oblique foreset structure, shown below, are common. The bidirectional foreset structure or the wave structure may be seen as well.
(4) From root (upper delta plain) to the front, the seismic amplitude and continuity increase gradually.

However, it is very hard to distinguish a fan delta from an alluvial fan based only on their seismic cross section, which means that the two can be distinguished only indirectly based on the associated facies belts. Alluvial fans develop on land and are associated with alluvial plain facies and swamp facies. Associated with lacustrine facies, fan deltas move from onshore to subaqueous levels and lack alluvial plain facies. This is the significant difference between the two. Accordingly, the respective associated facies belts can be used in seismic facies analysis.

The seismic facies characteristics of alluvial plains vary a great deal, however the amplitude is not very strong in general, and continuity is not very good. A wavy reflection structure with medium amplitude and medium continuity is common. In contrast, coal-bearing swamp facies and deep lacustrine facies with relatively developed turbidite sandstone are characterized by strong amplitude and relatively good continuity.

With regard to the depositional model of Meso-Cenozoic faulted basins in eastern China, the alluvial fan is mainly developed in the early stage of fault depression, but the fan delta is mainly developed in the middle and later stages of fault depression (when the lake area is at its largest).

11.10 Characteristics and Examples of Delta Sand Body Reservoirs

11.10.1 Sand Body Reservoir Characteristics of a Delta

11.10.1.1 Sand Body Reservoir Characteristics of a River-dominated Delta

1. **Geometry and continuity of the sand body**

Because marine action has little influence on distributary channels, they are generally straight. The channel width-to-depth ratio is very small, thus the width-to-thickness ratio of the distributary channel sand bodies is small as well. For instance, the breadth-to-depth ratio of the modern Mississippi Delta distributary channel is 3–50, and its plane modality is typical digitate (referring to Figs. 11.27 and 11.28). The width-to-thickness ratio of the straight distributary channel sand body in the ancient delta of the Yaojia Formation in the Daqing Oilfield is 20–40, while the actual width of many sand bodies is only 100–150 m. Poor lateral continuity is the principal driving force of the development of this type of distributary channel sand body (Fig. 11.67). Since the topographic slope is usually very gentle, the channel undergoes frequent migration and swing, resulting in the development of a distributary mouth bar. The formation of the distributary mouth bar is one of the reasons for channel swing.

A distributary mouth bar sand body is wider than a distributary channel sand body. Owing to certain permeability of the bar lateral margin and the presence of fine-grained sediment between bars, multiple distributary mouth bar sand bodies can be connected to form a sand body with good lateral continuity. However, the core of the distributary mouth bar is composed of relatively pure coarse-grained sand, which is seen as a narrow strip, and it has strong directionality and features such as a vertical shoreline.

The distributary channel sand and distributary mouth bar compose finger-like ridges, which extend seaward with a strip shape. When fluvial action is much larger than the action of sea waves and tides, such ridges can extend to dozens of kilometers.

2. **Plane distribution characteristics of the sand body**

A distributary mouth bar (or a diversion tongue-shaped sand body) can compose a physical property distribution unit. It is a relatively high permeability strip in the main body of a distributary mouth bar, and it dips gradually on both sides. Changes in physical properties are even, such as gradual change in granularity. The distributary mouth bar sand body data pertaining to Member 1 in the Yaojia Formation in the Daqing Xingshugang Oilfield indicate (Fig. 11.68) that it is normal to describe the physical properties contour diagram via the interpolation technique for this type of sand body, which reflects the gradual change in the planar physical property of this type of distributary mouth bar sand body. In early water flood recovery, the injected water rips through preferentially along the core of the distributary mouth bar main body, but it is not as obvious as a channel sand body. When the main body has a

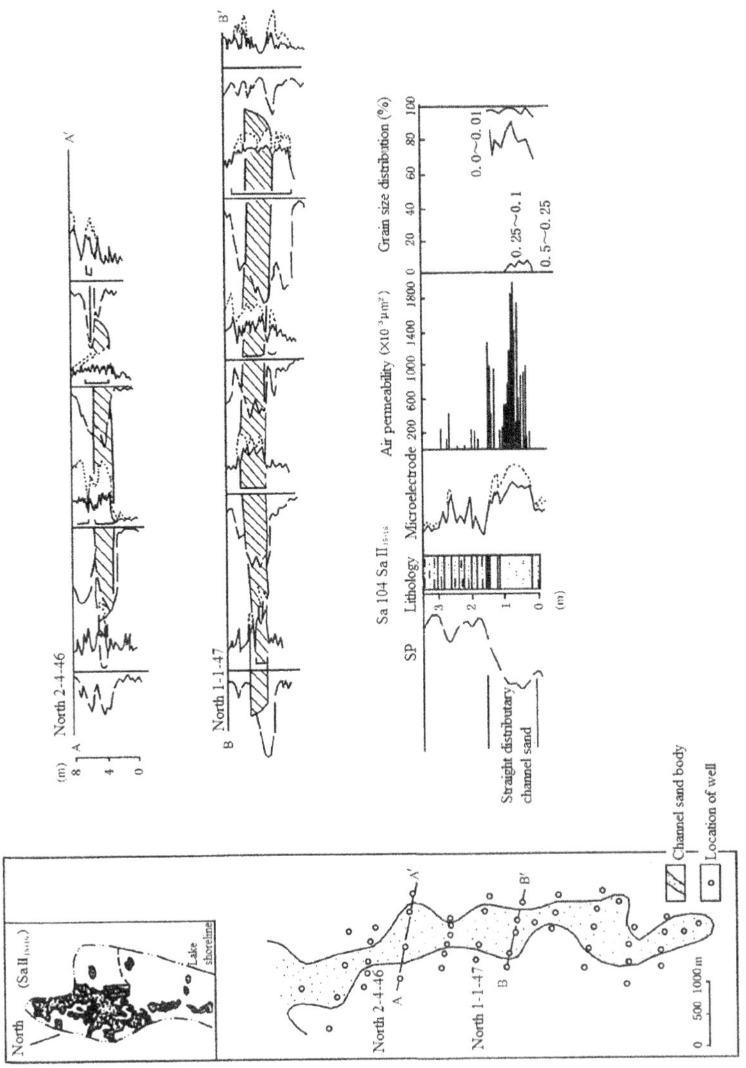

Fig. 11.67 Straight distributary channel of ancient delta of Yaojia formation in Daqing Oilfield, China, has poor lateral continuity, as shown in the sketch map

high water content, it transforms into a water injection well point. Thus, the injection water effect is very strong around lateral oil wells.

Owing to the geometric shape of a distributary channel sand body and the fining upward of intrastratal permeability, the occurrence of water channeling is easier than the formation of a distributary mouth bar. Premature water channeling at the point of superimposition of the distributary channel sand and distributary mouth bar is affected mostly by the water injection response of the upper distributary channel sand body.

3. **Sand inside structure**

The straight distributary channel sand body on the bird-foot delta plain is subordinate to filling sedimentation in the channel, in other words, dominated by aggradation. The activity lifetime is very short after the distributary channel is

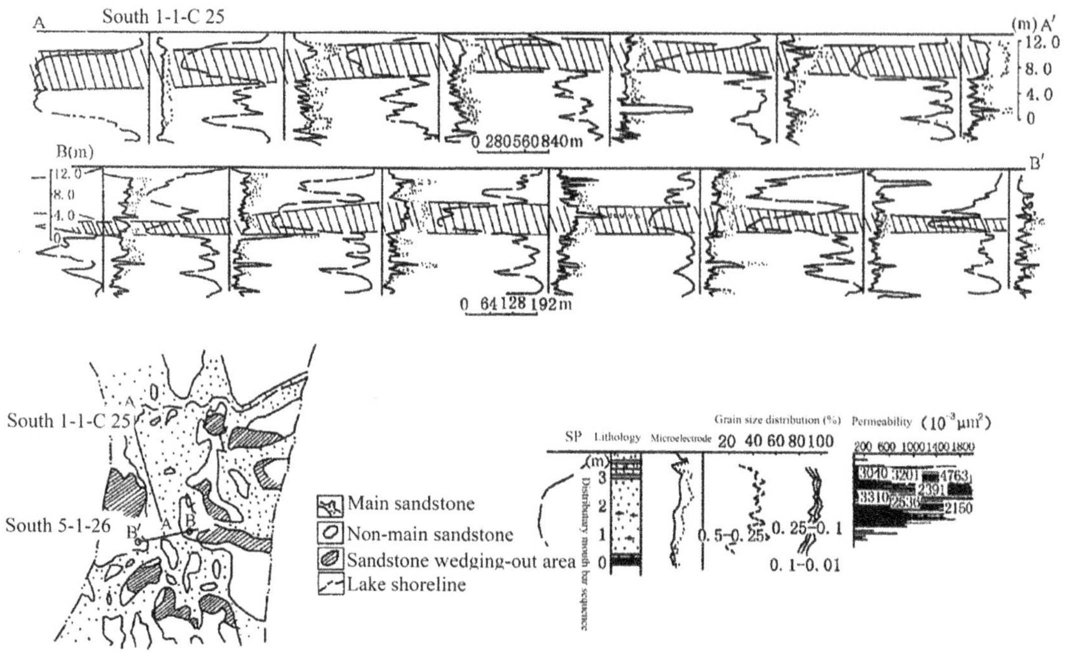

Fig. 11.68 Physical equivalence map drawn by interpolation method reflects the gradual change of plane physical properties of the sand body in estuary dam

formed. Channel formation, sedimentation filling, and cutoff and transfer are completed within a short time. Sedimentation filling of clastics is usually the primary cause of channel cutoff-abandonment-transfer. Regarding this type of fast-filling sedimentation, debris is differentiated roughly only by gravity, with the coarse fraction settling first, followed by the fine fraction. This results in the formation of a roughly fining-upward sand body. Its intrastratal permeability heterogeneity fines upward as well, similar to the common channel sand body on an alluvial plain. However, its degree of heterogeneity is low. Owing to the frequent rechanneling and distributary channel swing, two-period or multi-period distributary channel sand bodies are often formed and superimposed on certain parts. The upper and lower sand bodies can be in "cutting" contact, but most of them are in "plaster" contact. This is because the incision ability of the distributary channel is weak. Therefore, the argillaceous sediments at top of the early sand body are preserved easily. The thick sand layers of thin argillaceous layers are mixed in some layers, and these thin argillaceous layers have good continuity. During water flood recovery, the distributary channel sand is generally bifurcated into independent oil and water flow units.

The granularity in the distributary mouth bar sand body layer coarsens upward, and the highest permeability section is located in the upper pure sand part. Intrastratal thin muddy intercalation is not developed, and it occurs at the top with high possibility. When the distributary channel sand body incises and superimposes on the distributary mouth bar, the incised and superimposed part has a complex rhythm with the highest permeability in the middle. This may be one of reasons that the typical coarsening-upward sequence is not seen often in the distributary mouth bar sand bodies of ancient bird-foot deltas.

11.10.1.2 Sand Body Reservoir Characteristics of a Wave-Dominated/ Tide-Dominated Delta

1. **Geometry and continuity of the sand body**

In terms of the arc-shaped delta distributary channel sand body, both main straight river and small-scale bend distributary channel sedimentation are similar to the distributary channel sand body of a bird-foot delta, since it is distributed in a strip shape with poor continuity. The extension direction of the main distributary channel is along the paleocurrent direction perpendicular to the shoreline, while the small-scale bent partial sandstone bodies have no directionality and form a network distribution with the main distributary channel. The distributary mouth bar sand body is of the prolongation shape, but the perpendicular paleocurrent is parallel to the shoreline, which is opposite to the distributary mouth bar of a bird-foot delta and similar to a littoral area. This feature is one of the main distinctions between them.

2. **Plane characteristics of the sand body**

Changes in the physical properties of the distributary channel sand body and distributary mouth bar sand body show consistent directivity with the sand body geometric shape. Among the parietal crests of each period of the distributary mouth bar sand body, there are depressions in which fine grains filled and deposited. Hence, plane heterogeneity may result from differences in the physical properties of time cells.

3. **Intrastratal features of the sand body**

Similar to the distributary channel sand body of a bird-foot delta, the straight distributary channel sand body formed by filling sedimentation fines upward. The small point bar sand body resulting from miniature bend distributary channel sedimentation is the same as the intrastratally heterogeneous point bar on the flood plain, although its scale is smaller. However, there is an important difference between the miniature bend channel sand body formed by tidal current and point bar on a flood plain. The laterally accreted mudstone in lateral accretion bodies may extend from the top to the bottom of a river. This is because these miniature tide channels may become absolutely dry in a low tidal period, resulting in the preservation of all mud laminae coated on the lateral accretion bodies.

In the process of accretion, the distributary mouth bar is winnowed and transformed repeatedly by wave action. Therefore, the sand layer on the bar crest is characterized by fine granularity, purity, and good sorting. However, fine-grained clastics are transported to the bottom for sedimentation along the bar front. The continuous foreset causes the distributary mouth bar to have the typical coarsening-upward intrastratal granularity sequence, with a high bottom mud content. At the top, it turns into pure sand with very good physical properties. The bottom set fine-grained clastics in the front of the bar usually transform into mud intercalation. The distributary mouth bar is distributed around the bank. The distributary channel sand directly incises and superimposes on a part of distributary mouth bar, and it is much less than that of a bird-foot delta. Therefore, the intrastratal permeability heterogeneity increases from the bottom to the top, with the top having the highest permeability. An intrastratal argillaceous interlayer is developed at the bottom, and it has better continuity, decreasing upward to zero.

11.10.2 Delta Structure Features

When an oil and gas field is in the development stage, various levels or scales of reservoir heterogeneity research are prioritized in reservoir evaluation and prediction according to the type of sand body in question. Accordingly, the research is in the microfacies analysis stage. Sand body types must be divided according to microfacies combinations or genetic units to study their heterogeneity discretely. Architectural element has been defined in the chapter on fluvial systems, hence it will not be described here.

11.10.2.1 Bounding Surface Hierarchy

Through research on continental delta sedimentation, the bounding surface hierarchy of continental deltas can be divided into 6 levels (Fig. 11.69): the boundaries of levels ④, ⑤ and ⑥ should be considered when the genetic unit is divided because the former three levels of bounding surface mainly control second or third developments but have no significant influence on the evaluation, prediction, and primary recovery of reservoirs. All existing reservoir geological models established in China employ levels ④ and ⑤ or the latter three levels for demarcation, especially in interstratified and intrastratal heterogeneity research. The last level of surface is usually the standard for the division of whole oilfields or depositional system surfaces.

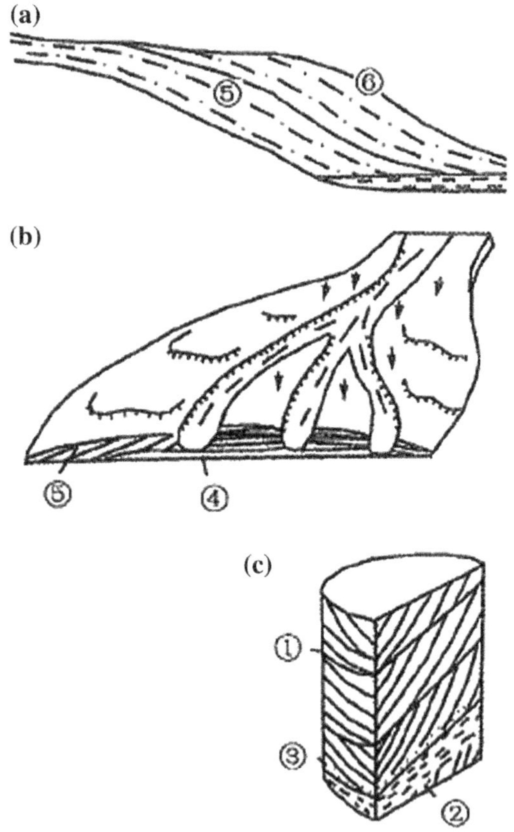

Fig. 11.69 Based on the study of continental deltaic deposits, different deposition scale interface can be divided into six grades. **a** Depositional system scale; **b** sand body scale; **c** core scale; and ①–⑥ represent the six surface levels (see text for explanation)

11.10.2.2 Delta Lithologic Facies Classification

The division of interfaces causes some difficulties in coring, however research on lithofacies association types compensates for the aforementioned difficulties. It not only reflects the differences of genetic sand bodies due to different sedimentations but also clarifies their vertical rhythms, physical properties, and heterogeneity.

Continental lacustrine basin deltas can be divided into 14 types of basic lithofacies: ① massive conglomerate facies (Gm); ② imbricate conglomerate facies (Gi); ③ flood bedding conglomerate facies (Gf); ④ planar cross bedding conglomerate facies (Sp); ⑤ parallel bedding conglomerate facies (Sh); ⑥ trough cross bedding conglomerate facies (St); ⑦ massive conglomerate facies (Sm); ⑧ swash bedding conglomerate facies (Ss); ⑨ wave cross bedding conglomerate facies (Sw); ⑩ wavy-interrupted wavy cross bedding, fine-sand facies (Fr); ⑪ parallel bedding silt lithofacies (Fh); ⑫ massive bedding silt lithofacies (Fm); ⑬ silt and muddy thin interbed complex beddings facies (Fc); and ⑭ mudstone facies (M). Mudstone facies can be subdivided into two types based on genesis and color. The first is dark gray mudstone facies (M_1), and it is usually the product of lacustrine mud; the second is purplish gray, brownish red massive silty mudstone facies (M_2), and it is the product of overbank deposits on a delta plain.

1. **Architectural element**

By combining A.D. Miall's classification of the features of architectural elements of rivers with delta bounding surface hierarchy and lithofacies division, the basic architectural elements of a delta can be determined (Table 11.21). These architectural elements are the basic genetic

Table 11.21 Architectural elements of continental delta sedimentation (according to Yu Xinghe, modified in 1997)

Structural elements	Code	Lithofacies composition	Geometry			Environment interpretation
			Horizontal	Longitudinal	Planar	
Channel Deposit of Estuary Upstream	CH	Gm, Ci, Cf, Sp, St, Sh, and Sm	Flat-top concave-bottom lens	Sheet	Irregular strip	Braided river (coarse-grained delta) Meandering river (fine-grained delta)
Distributary Channel	DC	Sm, Sp, St, Sh, Fr, and Fm	Flat-top concave-bottom lens	Lens or sheet	Irregular strip	Distributary channel
Gravel Bed Form	GB	Gm, Gi, and Gp	Lens shape	Lens sheet	Ellipticity or potato shape	Longitudinal bar or transverse bar
Downstream Accretion	DA	Sp, St, Sh, Sm, Fr, and Fm	Lens shape with concave-top and flat-bottom	Lens shape	Lobate	Distributary mouth bar or distal bar
Sedimentation Gravity Flow	SG	Gm and Sm	Lens shape	Wedge	Irregular	Gravity flow
Overbank Deposit	OF	Fr, Fh, Fm, and M	Linear	Linear	Sheet	Distributary interchannel flooding plain
Laminated Sand Sheet	LS	Sh, Sw, Fr, and Fc	Finger shape	Linear	Sheet	Lake shore shoal or shoreline sand mud flat
Flood Sheet Deposit	FS	Gp, Sh, and Sm	Linear	Linear	Sheet	Gravel sandy flat
Wave Sand Sheet	WS	Ss, Sw, Sh, and Ml	Lens shape and finger shape	Finger shape and linear	Sheet or lobate	Coastal sandbank and coastal sand sheet
Lake Mud	LM	Ml, Fm, and Fr				Lacustrine facies

units of continental deltas, and they are superimposed variously to form the skeleton of reservoir geological models.

The distributary channel (DC) is the unique product of delta deposit, and it is formed in three ways: the first is vertical accretion, formed only on the plain of a coarse-grained delta and features a braided river channel; the second is lateral accretion, formed by sandy meandering river sedimentation on the (upper) delta plain or the front of a delta; and the third is aggradation and channel filling, developed mostly on the delta front by straight distributary channel sedimentation.

Gravel bed form (GB) is mainly distributed in distributary channels or on the delta plain of a coarse-grained delta, usually as a lens or sheet of intermittent flood sedimentation or longitudinal and horizontal gravel bar.

Downstream accretion (DA) mainly refers to distributary mouth bars or distal bars formed by progradation in terms of the delta.

Sediment gravity flow (SG) is mainly developed in fan deltas and subaqueous fan depositional systems as the product of gravity flow or subaqueous gravel fluvial deposit.

Overbank fine (OF) is the overbank deposit of a delta and distributary channel, and it occurs as intercalation in or on top of a distributary channel.

Laminated sand sheets (LS) are mostly flat bed stripe-shaped sedimentary bodies distributed along the shoreline. Owing to the low deposition thickness of this element, it is barely taken into consideration as an efficient genetic unit in reservoir modeling.

As the typical product of plane jet, flood sheet deposit (FS) is mainly flood sheet deposit, and it is developed mostly in the upper delta plain and the estuary of a coarse-grained delta.

Wave sand sheet (WS) is the sedimentation of beach sand (wave sand and swash sand), and its appearance is distributed as a sheet with lakeshore gradient.

Lake mud (LM) is lake mud sediment and is mostly a good source bed and (or) overlying strata and body of separation for the genetic unit.

These architectural elements are the basic genetic units of continental lacustrine basins, and they are superimposed in various ways to form the skeleton body of a reservoir geological model. Therefore, they constitute the primary content for analysis to establish reservoir geological models, that is, the analysis of genetic units. The features of geological models in each phase area can be obtained by combining them organically with the facies classification of continental lacustrine basin deltas.

11.10.3 Model Feature and Main Structure of a Continental Delta Sand Body Reservoir

Establishing a conceptual reservoir geological model involves establishing the main framework of the reservoir sand body distribution in 3D space, which means preparing a skeleton pattern or sand body distribution model. This is the basis on which reservoir geological models or static models are established. Based on a skeleton pattern, reservoir properties (including porosity model and permeability model) are added to petrophysics data in order to quantify models and determine spatial distribution laws pertaining to different reservoir properties, which is the second step in establishing reservoir models.

To detail the spatial features of reservoirs, reservoir structures are divided into three categories based on spatial superimposition and efficient sand layer distribution: ① layer cake structure, ② jigsaw puzzle structure, and ③ labyrinth structure. The three aforementioned structures are found mainly in marine clastic rock reservoirs. The author holds that this classification is not comprehensive enough to be generalized to continental clastic rock reservoirs. To better describe the common lenticular sandstone in continental strata, the author proposes a fourth structure type (Yu et al. 1997), namely, stuffing pie. The abovementioned reservoir structural models simplify reservoir description and constitute a good basis of classification from the viewpoint of reservoir simulation. Based on the combination of research on continental delta sedimentary reservoirs in China, especially through the comparison of ancient and modern delta features, and the features of architectural elements, the characteristics of reservoir geological models of the continental delta sedimentary sand bodies in each facies area are summarized below (Table 11.22).

Upper delta plain: Based on elements DC, GB, FS, and OF, it is characterized by multi-layer jigsaw puzzle and caky sand body (Table 11.22) owing to the superimposition of upstream channel gravel stone bar and plane jet of flood.

Lower delta plain: It is mainly composed of DC, FS, and OF. In a faulting period or during steep topographic slope formation, because the flood plane jet holds the dominant position, in combination with large topographic slope and rapid deposition, it is hard for a channel to migrate laterally. Hence, the structure is dominated by a caky sand body structure. During basin depression, channel migration and swing are the main deposition modes in the facies area. Accordingly, the reservoir structure is mostly characterized by multilateral jigsaw puzzle transiting to labyrinth type.

Delta front: Its main skeleton is composed of DA, LS, and WS. In the basin faulting period, SG may be formed. Owing to the formation of various sandbars (delta progradation bar-distributary mouth bar, distal bar, shoreline sandbar), the reservoir structure is shown to be of isolated pie shape to layer cake shape. When the distributary mouth bar is developed to a lesser extent but beach sand is relatively well developed, a layer cake–shaped sand body constitutes the main skeleton structure of a reservoir.

Prodelta mud: Its main architectural element is LM. LS may be mixed in the faulting period of the basin and at the time of steep slope formation. It is usually a good sand body isolation layer. The sand body takes the layer cake shape as its predominant feature, and a few isolated stuffing pie structures can be seen.

11.10.4 Research Examples of Delta Reservoirs

11.10.4.1 Configuration Features and Intrastratal Heterogeneity of Modern Braid Delta in the Daihai Lake, Inner Mongolia

Through sedimentary study of the modern delta of the Daihai Lake in Inner Mongolia (Yu et al. 1995), the two braided river sand bodies along the short axis of the basin were divided into 12 lithofacies and 9 lithofacies associations, of which 8 are efficient reservoirs and the representative architectural elements are given in Table 11.23.

11.10.4.2 Delta Reservoir Characteristics and Heterogeneity in the Gujiazi-Bawunan District

Through observation and precise description of the cores of 14 wells in the Gujiazi-Bawunan District of the southern Songliao Basin, 14 lithofacies types were defined (Table 11.24).

According to the above lithofacies types, and core descriptions and microfacies analysis of 14 wells, 10 types of lithofacies associations are summarized for Member 3 of the Denglouku Formation and Members 1 and 2 of the Quantou Formation in the Gujiazi-Bawunan district (Table 11.25).

Table 11.22 Features of structure types of reservoir geological models in each facies area of continental deltas (according to Yu Xinghe 1997) (abbreviations see Table 11.21)

Facies area	Main architectural element	Structural type features
Upper delta plain	CH + GB + FS + OF	Multi-layer jigsaw puzzle and cakey sand body
Lower delta plain	DC + FS + OF	Multilateral jigsaw puzzle transiting to labyrinth
Delta front	DA + LS + WS + (SG)	From isolated stuffing pie to layer cake
Prodelta mud	LM + (LS)	Layer cake mix with few stuffing pie shapes

Table 11.23 Sedimentary parameters and properties of various genetic sand bodies of two braided deltas in the Daihai Lake, Inner Mongolia

Genetic type	Median grain diameter (mm)	Standard Deviation	Mud Content (%)	Porosity (%)	Permeability ($\times 10^{-3}$ μm^2)	Geometry
CH	>2	0.3–0.5	0.5–2.5	>35	100–720	Sheet and lobate
GB	0.25–2	0.25–0.5	1–7.5	>35	200–2800	Sheet and lobate
FS	0.125–0.25	0.35–0.45	3.5–4.5	30–40	200–2000	Sheet
DC	>0.125	>0.4	0.5–4.8	35–40	100–800	Sheet, ellipticity
FM	0.05–0.125	0.25–0.35	1–5.8	35–40	450–1500	Linear
OF	0.05–0.125	0.25–0.35	2–3.5	30–40	50–150	Sheet, ellipticity
LS	0.045–0.0625	0.35–0.4	>4	30±	120–150	Finger shape
WS	0.05–0.176	0.2—0.3	<3.5	>35	–	Linear

According to the vertical combination features of lithofacies, eight combination types or genetic units can be synthesized (Table 11.26), and the changes in their degrees of heterogeneity can be divided into five categories based on the permeability variation coefficient, where V_k: $V_k >$ 1.5, very strong; 1.0–1.5, strong; 0.8–1.0, medium; 0.4–0.8, weak; and <0.4, very weak. Accordingly, the response relationship between sedimentation and heterogeneity and the heterogeneity of vertical and lateral accretion vary from strong to very strong. The intersection result of recombination action may weaken heterogeneity, for instance, the medium heterogeneity of types III and VI. Although the maximum permeability value of foreset and winnowing accretion action may be very large, the associated heterogeneity is not strong. The degree of homogeneity of overbank accretion is the best.

11.10.4.3 Lithofacies and Reservoir Feature of the Pearl River Mouth Delta

The Pearl River Formation-Hanjiang Formation in the Pearl River Mouth Delta is mainly a multi-cycle compound system of a braid delta, in which 21 types of lithofacies can be identified (Table 11.27). Its lithology includes conglomerates, conglomeratic, coarse sandstone, medium sandstone, fine sand, siltstone, and mudstone. Their sedimentary structures are diversified but reflect the features of delta facies as a whole. There are relatively complete microfacies types for this delta, including braided distributary channels, distributary channels, abandoned channels, interdistributary bays, natural levees, crevasse splays, distributary mouth bars, distal bars, subaqueous distributary channels, storm sheet sand, and prodelta mud. Sedimentation modes are mainly foreset and vertical accretion. There are many types of sand bodies, and the overall physical properties are good (Table 11.27 and Fig. 11.70).

11.11 Relations Between Delta Deposit and Petroleum

There is a close relationship between delta deposition, and petroleum generation and accumulation. As is well known, a prospective oil and gas field must provide the following basic geological conditions: "source, reservoir, seal, trap, migration, and preservation." The existence, quality, and cooperative relationship of these conditions directly affect the formation and scale of petroleum reservoirs. Delta depositional systems usually have general conditions such as source-reservoir-cap, and it is shown in petroleum exploration that delta deposits lead to petroleum accumulation. According to domestic and overseas research on petroleum exploration data, petroleum is mainly accumulated in the sea-land transitional zone developed on delta deposits, that is, in the area near the coastline. In addition, there are many megascopic and super-huge oil

Table 11.24 Division standards for and features of reservoir lithofacies in the Gujiazi-Bawunan District

S/N	Lithofacies code	Lithofacies name	Major lithology	Sedimentary structure and facies marker	Sedimentary environment and genetic interpretation
1	Gm	Massive conglomerate facies	Granule conglomerate or conglomeratic	Massive, particle supported	Braided channel
2	Gf	Graded bedding gravel lithofacies	Conglomeratic	Flood-graded laminae	Rapid intermittent flow
3	Sm	Massive sandstone facies	Medium-coarse sandstone	No obvious massive sedimentary structure	High deposition rate and strong fluviation
4	St	Trough cross bedding sandstone facies	Fine-coarse sandstone	Trough cross bedding	Migration of sand dune, channel filling
5	Sp	Planar cross bedding sandstone facies	Fine-coarse sandstone	High and low angle planar cross bedding	Migration of meandering point bar
6	Sh	Parallel bedding sandstone facies	Fine sandstone	Parallel bedding	Upper flow regime, rapid flow in shallow area
7	Ss	Swash cross bedding sandstone lithofacies	Fine medium sandstone	Swash cross bedding	Shore-shallow lake beach bar
8	Sg	Coarse-tail graded bedding with gravel sandstone facies	Medium-fine sandstone contained	Coarse-tail graded bedding	Grain flow deposition
9	Sw	Wave crossing bedding sandstone facies	Silt-fine sandstone	Wave cross bedding	Coastal sand sheet
10	Fr	Miniature wavy or discontinuous wavy cross bedding siltstone facies	Fine siltstone or mud siltstone	Miniature wavy or discontinuous wavy cross bedding, miniature wave ripple, and miniature current ripple	Migration of ripples: ① Product of wave action ② Product of fluviation
11	Fh	Horizontal bedding siltstone facies	Siltstone or mud siltstone	Horizontal bedding	Stillwater environment
12	Fm	Massive bedding siltstone lithofacies	Siltstone or mud siltstone	No obvious sedimentary structure, mainly massive	Stillwater or rapid deposition
13	M_1	Dark mudstone facies	Dark mudstone	Dark color and pure property, horizontal bedding	Lake mud or predelta mud
14	M_2	Variegated mudstone facies	Purplish gray or brown massive silty mudstone	Visible plant debris and occasional burrows and bioturbation structures	Overbank deposit, delta plain, and interdistributary bay

Table 11.25 Geologic features and lithofacies combination types in the Gujiazi-Bawunan District (according to Yu Xinghe et al. 1998)

Microfacies	Lithofacies association feature	Main sedimentation mode	Vertical rhythm	Electric well-logging curve features	Geometric shape of sand body	
					Vertical	Planar
Braided channel	Gm, Gf $\leftrightarrow$ Sh $\leftrightarrow$ Sm $\leftrightarrow$ Fr $\leftrightarrow$ M_2	Vertical accretion, aggradation, and channel filling	Superimposed fining upward	Cylinder shape	Flat-top concave-bottom planar shape	Wide belt shape
Braided distributary channel	Gm $\rightarrow$ Sp $\rightarrow$ Sm $\rightarrow$ Fr $\rightarrow$ M_2	Lateral and vertical accretion	Abrupt fining upward	Cylinder shape or serrated bell shape	Flat-top concave-bottom bulk	Belt shape
Distributary channel	Sp $\rightarrow$ St, Sm, Sh $\rightarrow$ Fr, Fm $\rightarrow$ M	Lateral and vertical accretion	Gradual fining upward	Bell shape	Flat-top concave-bottom lens	Strip
Interdistributary bay	M_2 $\rightarrow$ Fh $\rightarrow$ Fr, Fm $\rightarrow$ M_2	Overbank flow accretion	No obvious rhythm	Miniature serrated or finger shape	Laminated	Sheet
Crevasse splay	M_2 $\rightarrow$ Fr, Fm $\rightarrow$ Sm, Sh	Overbank flow accretion	Miniature coarsening upward	Finger shape or beak shape	Wedge	Fan shape
Distributary mouth bar	M_1 $\rightarrow$ Fr $\rightarrow$ Sh, Sp $\leftrightarrow$ Sm, Gm	Foreset	Gradual coarsening upward	Funnel shape	Flat-bottom concave-top planar shape	Potato shape Ellipticity
Distal bar	M_1 $\rightarrow$ Fr $\rightarrow$ Sh, Sp $\rightarrow$ M_1	Foreset	Miniature coarsening upward	Serrated funnel shape	Lens shape with concave top and flat bottom	Finger shape or potato shape
Front sheet sand	M_1 $\rightarrow$ Sh, Sw $\rightarrow$ M_1	Overbank flow accretion and winnowing accretion	No obvious rhythm	Finger shape	Laminated or belt shape	Blanket or sheet shape
Lake shore shoal	M_1 $\rightarrow$ Sh $\leftrightarrow$ Ss $\rightarrow$ M_1	Winnowing accretion	No rhythm or minor coarsening upward	Miniature funnel or tongue shape	Medium-thin stratified structure	Belt shape
Gravity flow sediment	M_1 $\rightarrow$ Gm, Sg $\rightarrow$ Sh $\rightarrow$ Fr— M_1	Turbidity accretion	Not obvious or fining upward	Serrated bell shape or finger shape with high amplitude	Lens shape	Ellipticity

Table 11.26 Heterogeneity parameters and sedimentation features of intrastratal permeability of a gas field in the Songnan Houwujiahu area (see Table 11.21 for the abbreviations)

Sandstone types	Ratio	Variation coefficient V_k	Dart coefficient	Degree of heterogeneity	Highest permeability location	Vertical rhythm feature	Sedimentation	Reservoir type
CH	2044	1.9003	7.0241	Very strong	Irregular	Superimposed gradual fining upward	Vertical accretion	Medium porosity and medium permeability
FS	272.29	1.1975	4.5899	Strong	Irregular	Superimposed abrupt fining upward	Vertical and lateral accretion	Medium porosity and high permeability
DC	12.143	0.8609	2.9565	Medium	Often at the bottom	Gradual fining upward	Aggradation and channel filling, and lateral accretion	Medium porosity and low permeability
FM + DC	160	0.7639	2.9090	Weak	Top or middle part	Complex and coarsening upward	Foreset	High porosity and low permeability
WS	11.67	0.7500	2.2415	Weak	Often in the middle part	Coarsening upward or complex rhythm	Winnowing accretion	High porosity and high permeability
LS	42.666	0.9127	2.7089	Medium	Irregular	Fining upward or irregular	Overbank flow accretion and winnowing accretion	Medium porosity and low permeability
OF	1.1667	0.0509	1.0687	Very weak	Irregular	Similarly homogenous	Overbank flow accretion	Low porosity and low permeability
FM	6.233	0.4677	1.4308	Weak	At the top	Coarsening upward	Foreset	Medium porosity and low permeability

Table 11.27 Features of sandstone lithofacies found in the delta of the Pearl River Mouth Basin

S/N	Lithofacies code	Description	Sedimentary structure feature	Major lithology	Sedimentary structure and facies marker	Sedimentary environment interpretation
1	Gm	Massive conglomerate facies		Conglomerate or glutenite	Massive structure and particle supported	Braided distributary channel and beach sand
2	Gf	Graded bedding gravel lithofacies		Conglomerate or glutenite	Conglomerate mostly in imbricate arrangement	Longitudinal bar of intermittent braided distributary channel
3	Gms	Matrix-supported conglomerate facies		Sandy conglomerate	Massive structure and matrix supported	Gravity flow sediment and gravity flow sediment
4	Sph	High angle planar cross bedding sand lithofacies		Fine-coarse sandstone	Higher angle planar cross bedding	Migration of longitudinal or transverse bar
5	Spl_1	Truncated low-angle planar cross bedding sand lithofacies		Fine-coarse sandstone	Truncated low-angle planar cross bedding or wedge bedding	Strong current effect, distributary channel or shoreline
6	Spl_2	Incised low-angle planar cross bedding sand lithofacies		Fine-coarse sandstone	Incised low-angle planar cross bedding	Migration of meandering river point bar
7	St	Trough cross bedding sandstone facies		Fine-coarse sandstone	Trough cross bedding	Channel-filling sedimentation
8	Sh	Parallel bedding sandstone facies		Fine sandstone	Parallel bedding	Sand flat with upper flow regime, rapid flow in shallow area
9	Sr	Ripple cross bedding sandstone lithofacies		Medium-fine sandstone	Ripple cross bedding	Migration of sand waves, found in various environments
10	Sm	Massive bedding sandstone lithofacies		Fine-coarse sandstone	Massive, no obvious sedimentary structure	High deposition rate and strong fluviation
11	Sf	Herringbone cross bedding sandstone facies		Fine-coarse sandstone	Herringbone cross bedding	Tide channel
12	Sw	Swash cross bedding sandstone lithofacies		Fine-coarse sandstone	Swash bedding	Beach sand
13	Shm	Hummocky cross bedding sandstone facies	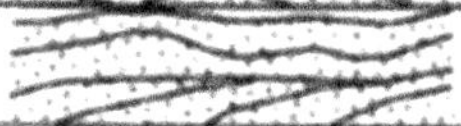	Fine-coarse sandstone	Hummocky cross bedding	Eolian deposit
14	Sbc	Biological debris calcium sandstone facies		Fine-coarse sandstone	Biological debris	Distributary mouth bar, storm and platform margin
15	Fr	Miniature wavy or discontinuous wavy cross bedding siltstone facies		Siltstone or mud siltstone	Wavy or discontinuous ripple cross bedding	Migration of miniature sand waves, occurs in various environments
16	Fh	Horizontal bedding siltstone facies		Siltstone or mud siltstone	Horizontal bedding	Stillwater environment
17	Fm	Massive bedding siltstone lithofacies	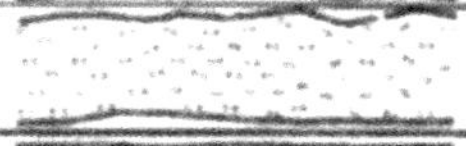	Siltstone or mud siltstone	No obvious sedimentary structure, mainly massive	Stillwater or rapid deposition
18	Fc	Complex bedding siltstone lithofacies	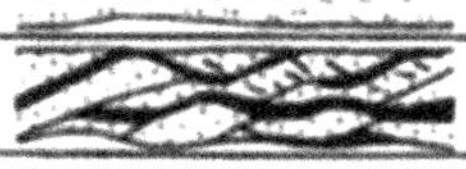	Siltstone or mud siltstone	Compound beddings	Intertidal zone or delta front environment
19	M_1	Dark mudstone facies		Dark mudstone	Relatively pure mud with horizontal bedding or massive	Marine transgression or prodelta
20	M_2	Light mudstone facies		Light mudstone	Siltstone and common plant debris	Delta interdistributary bay or delta front
21	C	Gray lithofacies	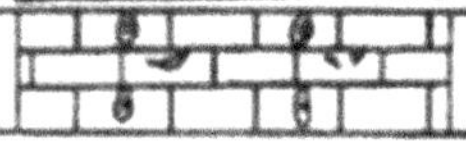	Limestone	Biological debris	Clean, shallow water carbonate platform or reef

and gas fields. Typical examples include the Kuwait Burgan oilfield and the Venezuela Maracaibo Basin Bolivar oilfield. Very thick delta sedimentation can be formed due to the gradual slow subsidence of the basin and repeated advance and retreat of this belt in geological development history. This sedimentation creates favorable conditions for petroleum formation and accumulation, and oil is mainly accumulated in the transitional zone and in the indented stratum of neritic deposit.

11.11.1 Source Bed

Mudstone with a dark color and pure quality in the prodelta is a good unboiled oil facies belt. The river brings large amounts of mud sand to the delta and also organic substances. These organic substances deposited in a prodelta with floating mud, bring rich nutrients for organisms in a lacustrine basin or a sea basin, thus driving their reproduction and growth. Therefore, the mudstone in prodelta contains rich terrestrial sources and organic matters in situ. In addition, a prodelta is composed mainly of argillaceous sediment characterized by very thick, wide distribution, and the prodelta environment is generally a reduction or weak reduction environment, which is favorable for the preservation and transformation of organic matter, as well as rapid deposition and burial of deltas, because waves cannot reach this environment. Therefore, predelta mudstone and silty mudstone can be considered good source rocks. In addition, lagoon depositions associated with deltaic depositional systems are important source rocks. This geological analysis has been proven using organic geochemical analysis indicators.

11.11.2 Reservoir

In delta deposits, good reservoir sand bodies are usually well developed, such as the distributary mouth bar of fluvial-dominated deltas, front sheet sand, distributary channel sand, beach sand and barrier bar of wave-dominated deltas, all of which have good reservoir properties. Distributary channel sandstone is generally not favorable compared to other sand bodies owing to its distance from the oil source area. Therefore, in ancient delta sediments, the main reservoir is constituted of delta front sand and coastal sand, which are closely associated with delta destruction. There, the predelta adjacent to the delta front sand is crossed as finger shape to form a compound reservoir, leading to favorable petroleum accumulation conditions. This is a typical feature of many delta fronts in major oil and gas fields worldwide. However, in the delta system, many large structures are not petroliferous. This is because they are located in the bay argillaceous sediment of poor sandstone between deltas. In addition, petroleum reservoirs are not distributed in the development zone of coastal plain and barrier sand bar.

11.11.3 Seal Bed

In delta deposits, large-scale seal beds, such as swamp deposits, interdistributary bays, continental shelves, and predelta mud, can be considered seal beds. In the process of transgression, mudstone overlaps on the reservoir to form a regional seal bed. Moreover, in the process of advance and retreat during delta formation, the seal bed, source bed, and petroleum reservoir together compose a good source-reservoir-cap.

11.11.4 Trap

In delta deposits, in terms of petroleum reservoir sandstone, most sands are yielded in the shape of lens, except sheet sand and distributary channel sand. This makes it easy to form stratigraphic-lithologic traps, and of course, structural traps can be formed as well. In fluvial-dominated delta deposits, for instance, they are often associated with the contemporaneous fault and traction structure formed therefrom, diapir structure, and salt-dome structure. Hence, many types of traps can be formed. Most traps are formed in the process of deposition in

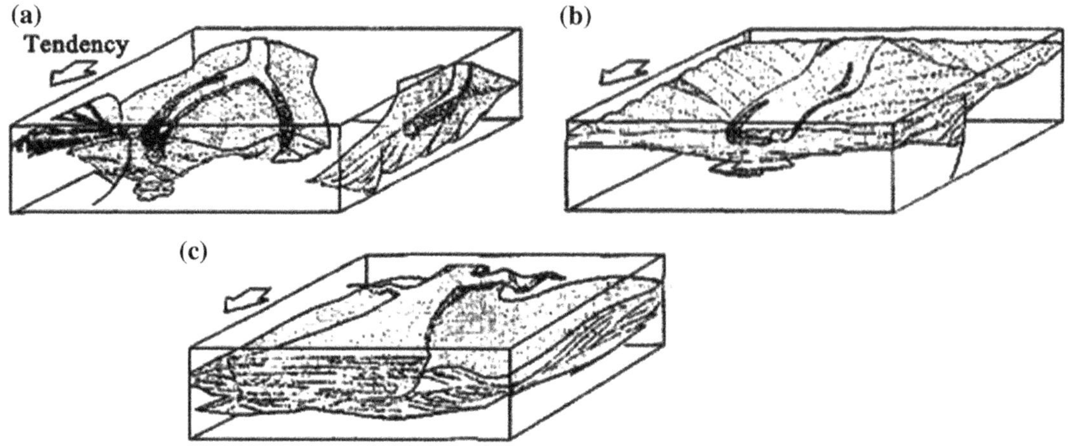

Fig. 11.70 Geometric morphology and lateral overlap relation formed by sand bodies in different delta regime has its own characteristics. **a** Fluvial-dominated, **b** Wave-dominated, and **c** Tide-dominated

earlier times, which is conducive to petroleum accumulation and formation. For example, the Meso-Cenozoic oil and gas fields in Mexico Bay in the United States were formed thusly.

To summarize, delta deposits not only have very thick source beds, but also have high-quality reservoirs with good sorting. In addition, owing to frequent local marine transgression and regression and large amplitude in the delta sedimentation process, many good source-reservoir-cap associations can be formed, leading to the formation of a rich petroleum accumulation belt. Therefore, it is very important to study the sedimentary characteristics of deltas when searching for petroleum.

11.11.5 Comparison Between Source-Reservoir and Delta Trap

The geometric shapes and horizontal superimposition relationships of different delta systems have their own characteristics (Fig. 11.70), which leads to differences in reservoirs. Moreover, the characteristics of deposition determine whether source beds are different from trap types (Table 11.28). This is because the formation diversity of deltas leads to huge differences in the productive capacities of various reservoir types.

11.11.5.1 Fluvial-dominated Delta System

Given the popularity of the Mississippi Delta mode, many descriptions and explanations of fluvial-dominated delta reservoirs have been reported and published. In the pay bed from the Pleistocene to the early Paleozoic, distributary channels, distributary mouth bars, and delta-front sheet sand are recognized reservoirs (Table 11.29).

Reservoir sand bodies in the lower delta plain are usually multilateral, branched, isolated, irregular, and lens-shaped on the cross section (Fig. 11.70a). Interdistributary crevasse splays are secondary in volume, but they may be partially or completely isolated from other sand bodies to form potentially important reservoirs. In large deltas, a separate crevasse fan (lobate) can accommodate up to millions of barrels of oil in isolated sand body traps. In addition, thin locally destructive sand bars can also form small but high-yield isolated sandstone reservoirs. Toward the direction of land, the upper delta plain facies association is generally a channel-filling facies reservoir of the suspended load type.

Carbonaceous sediments of delta plain mud and prodelta rich in organic matter surround potential reservoirs and form finger-like intersections with them. The inherent strong river

Table 11.28 Reservoir sedimentology features of each lithofacies association of (eastern) a sandstone oilfield delta in the Pearl River Mouth Basin

Sedimentary microfacies	Deposition modes	Lithology facies type	Prosodic features	Electric well-logging curve features	Geometry		
					Planar	Longitudinal	Horizontal
Braided Distributary Channel	Vertical accretion	Gi, Gm, Gms, St, Sph, Sm, and Fm	Superimposed abrupt fining upward	Serrated Cylinder shape	Irregular strip	Sheet or massive	Flat-top concave-bottom lens
Distributary Channel	Lateral accretion	Gm, Sm, St, Spl_2, Sh, Sr, Fm, and Fr	Gradual fining upward	Bell shape	Irregular strip	Lens or belt shape	Flat-top concave-bottom lens
Natural Levee	Overbank flow accretion	M_2, Fr, Fc, and Fm	No rhythm	Finger shape	Irregular strip	Sheet	Small wedge
Crevasse Splay	Overbank flow accretion Aggradation and channel filling	M_2, Sm, Sr, and Fm	Minor fining upward or irregular	Beak shape	Irregular strip	Sheet	Lens shape with concave top and flat bottom
Abandoned Channel	Vertical accretion	M2 and Sm	Abrupt fining upward	Serrated bell shape	Long strip shape	Lens shape	Flat-top concave-bottom lens
Interdistributary Bay	Vertical accretion	M_2, Fh, Fc, and Fr	No rhythm	Serrated	–	–	–
Distributary Mouth Bar	Foreset	M_1, Fc, Sm, Spl_1, Sh, Sr, Sbc, and Gm	Gradual coarsening upward	Serrated funnel shape	Ellipticity, potato shape	Lens, sheet	Lens shape with concave top and flat bottom
Distal Bar	Foreset and winnowing accretion	M_1, Fc, Fh, Fm, Sm, and Sr	Minor coarsening upward	Miniature funnel shape	Potato shape	Lens, sheet	Lens shape with concave top and flat bottom
Subaqueous Distributary Channel	Aggradation and channel filling	M, Sm, Shm, Sbc, Sh, Fm, and Fc	Minor gradual fining upward	Little bell shape	Strip	Sheet	Flat-top concave-bottom lens
Storm Sheet Sand	Foreset Vertical accretion	M, Fc, Fm, and Shm	No obvious rhythm	Serrated shape, finger shape	Sheet	Planar	Planar
Offshore/Predelta Mud	Vertical accretion	M_1, Fh, Fm, Fc, and Sm	No rhythm	Serrated shape with small amplitude	–	–	–

(continued)

Table 11.28 (continued)

Porosity (%)			Permeability ($\times 10^{-3} \mu m^2$)		
Minimum	Maximum	Average value	Minimum	Maximum	Average value
Porosity (%)			Permeability ($\times 10^{-3} \mu m^2$)		
Minimum	Maximum	Average value	Minimum	Maximum	Average value
10.3	32.4	22.8	11.0	7650	1706.66
10.0	34.0	23.2	10.0	19580	1405.84
10.0	23.7	16.4	0.13	1714	113.29
15.5	34.1	21.7	5	3020	460.92
2.5	27.5	20.8	0.06	13260	2019.32
1.4	23.8	11.5	0.01	409	12.75
12.5	33.6	23.1	13	5020	1045.06
1.9	32	16.1	0.01	1690	76.29
10.2	28	22.1	8	5540	652.78
15.9	30.2	24.2	11	2210	479.26
1.9	28.4	13.1	0.01	374	10.25

Table 11.29 Comparison among source, reservoir, and trap features of three types of deltas

	Fluvial-dominated delta	Wave-dominated delta	Tide-dominated delta
Source bed	Delta plain and prodelta mud rich in organic matter, gas formed easily in terrestrial sources, dominated by organic matter	Marine organisms mixed with terrestrial creatures, quality source rock, easily formed	Predelta mud and delta plain marsh
Reservoir	Distributary channel, distributary mouth bar, and sheet sand as main reservoirs, followed by isolated crevasse splay and distal bar as secondary reservoirs	Dominated by distributary channel, distributary mouth bar, beach ridge, and longshore bar	Dominated by distributary channel, crevasse splay. and tidal current ridge sand; mostly discontinuous reservoirs
Trap	Abundance of stratum-lithology traps	Abundance of structural traps	Dominated by stratigraphic traps

superimposition of a delta system is conducive to the formation of terrigenous and herbaceous organic matter, which generate gas easily. Local or regional destructive transgressive mudstone, which offers good sealing, can be the best seal bed.

Reservoirs in a delta may be composed of a channel-filling sand body, distributary mouth bar, distal bar, and delta-front sheet sand. Each sand body has distinct characteristics in terms of reservoir, permeability, and geometry. Later burial and diagenesis may enhance these differences. In older delta systems, perhaps only coarse channel-filling sand has the permeability required to produce oil and gas. On the contrary, in a different diagenetic system, a well-sorted distributary mouth bar is probably the only reservoir, while symbiotic channel fillings are dense and oil-free sand bodies.

On vast and flat delta plains, a variety of isolated or partially isolated sand bodies show very different trends, which provide a variety of forms of potential stratigraphic traps in fluvial-dominated deltas.

11.11.5.2 Wave-dominated Delta System

When the transformation effect of waves is enhanced, isolated mouth bars are connected, forming transversely vast, interconnected beach ridges and coastal barrier sand bodies (Fig. 11.70b). Owing to multiple high-permeability, well-sorted, directionally arranged delta fronts and river channel sand bodies, wave-dominated deltas are the best places for the formation of high-quality reservoirs. However, the high degree of connectivity of sand bodies, funnel-shaped distribution of sand, and connection with the river system in the updip direction decrease the possibility of forming stratigraphic traps. Structural traps, including growth faults and diapir structures, are the main trap types in this type of delta.

The dominant effect of oceans leads to a low deposition rate in the prodelta, adjacent shelf, and slope environment. Therefore, most marine organic matter can mix in the region, and the opportunities for bacterial decomposition of herb detritus materials from the river increase. All these factors are conducive to the formation of relatively good source rocks. However, in proportion, the volume of potential oil source rock facies is usually smaller, and owing to slow accumulation, and some organic substances may be oxidized completely. As in the case of fluvial-dominated deltas, it is possible to produce oil-rich sources by providing petroleum source materials from a nearby depositional system.

11.11.5.3 Tide-dominated Delta System

Only a small amount of data has been proven for the development geological characteristics of tide-dominated delta systems, however they can be deduced reasonably as characteristics of petroleum reservoirs.

The potential sources include predelta mud (generating gas and some oil) and organic matter in a delta plain marsh (generating gas). Although the reservoirs are products of transformation by the ocean, they might be complex and discontinuous, except those in the most sand-rich systems. Potential reservoir sandstones include ① large-scale distributary channel filling and the associated delta plain crevasse splay opened toward the basin direction; ② those found in the area between the delta front and the isolated tidal current ridge sand body (Fig. 11.70c). These two types of reservoirs can show the external shape and internal separation of dip orientation. The potential for stratigraphic traps ranges from good to poor, depending on the sand-carrying capacity of the river system and the range and intensity of delta edge transformation. In tidal channel sands and bar sands formed due to tides, a large amount of mud sediments and good preservation may result in complex permeability levels and heterogeneity features.

The petroliferous characteristics of each delta system can be summed up in terms of the following four aspects:

(1) For a long time, deltas have been deemed as the biggest "machines" for generating petroleum. Although there are a few simple ideas about the mechanism of petroleum formation in delta deposits, the fact that deltaic depositional systems have the potential for generation, and reservoir and cap formation holds.
(2) In each delta system, there are two large reservoirs. The distributary channel forms a discontinuous reservoir network with irregular dip orientation, while the delta front sand provides a wide sand collection with fine granularity and good sorting. They are concentrated along the front zone of the delta sedimentary body. Crevasse splays, prodelta slip fault structures, and partially destructive sand bodies constitute quantitatively unimportant but locally important oil and gas producing reservoirs.
(3) Source bed quality generally ranges from low to medium, and it contains a large number of type III kerogens. Therefore, a delta system with an internal oil source is often dominated by gas yield. However, prodelta sediments and continental shelf and slope sediments are crossed in a finger shape, causing the prodelta produce a greater amount of source rock. The abundance of source rock is basically affected by the deposition rate and the abundance control of offsite herbal organic matter or shelf plankton in prodelta environments.
(4) In each delta system, the structure in same sedimentary period is the most common trap type. Their economic importance is determined by the progradation scale of the delta. All Neogene delta basins are characterized by progradational sediments with growth fault or diapir trap measuring hundreds of meters in thickness. Although similar features appear in the delta systems of simple or complex craton inner basins, the scale is generally too small to form petroleum traps of economic value.

References

Allen, J.R.L. 1970. Sediments of the modern Niger Delta. In *Deltaic sedimentation: modern and ancient*, vol. 15, e.d. J.P. Morgan, 138–151. Society of Economic Paleontologists and Mineralogists, Special Publication.

Bates, C.C. 1953. Rational theory of delta formation. *AAPG Bulletin* 37: 2119–2162.

Coleman, J.M., and L.D. Wright. 1975. Modern river deltas: variability of processes and sand bodies. In *Deltas, models for exploration*, ed. M.L. Broussard, 99–149. Houston: Houston Geological Society.

Coleman, J.M. 1976. *Deltas: processes of deposition and models for exploration*, 102p. Champaign: Continuing Education Publ Comp. Inc.

Coleman, J.M. 1982. *Deltas processes models of deposition for exploration*, 2nd edn. Boston: International Human Resources Development Corporation.

Einsele, G. (2000). *Sedimentary basins-evolution, facies, and sediment budget*, Second, Completely Revised and Enlarged Edition, 1–792. Springer, Berlin

Elliott, T. 1974. Interdistributary bay sequences and their genesis. *Sedimentology* 21: 611–621.

Elliott, T. 1975. The sedimentary history of a delta lobe from a Yoredale (Carboniferous) cyclothem. *Proceedings of the Yorkshire Geological Society* 40: 505–536.

Ethridge F.G., and W.A. Wescott. 1984. Tectonic setting, recognition and hydrocarbon reservoir potential of

fan-delta deposits, Sedimentology of Gravels and Conglomerates
Fisk, H.N. 1944. Geological Investigation of the alluvial Valley of the Lower Mississippi River, 78pp. Mississippi River commission, Vicksburg
Fisher, W.L., L.F. Brown Jr, A.J. Scott, and J.H. McGowen. 1969. Delta systems in the exploration for oil and gas. *Texas Bureau of Economic Geology*, 78p.
Fisher, W.L., and J.H. McGowen. 1969. Depositional systems in the Wilcox Group (Eocene) of Texas and their relationship to occurrence of oil and gas. *Bulletin American Association of Petroleum Geologists* 53: 30–54.
Friedman, G.M., and J.E. Sanders. 1978. *Principles of sedimentology*, 729p. New York: Wiley.
Galloway, W.E., and D.K. Hobday. 1983. *Terrigenous clastic depositional systems*. New York: Springer.
Galloway, W.E. 1976. Sediments and stratigraphic framework of the Chopper River fan delta, Alaska. *Journal of Sedimentary Research* 46: 726–737.
Galloway, W.E. 1975. Process framework for describing the morphologic and stratigraphic and stratigraphic evolution of deltaic depositional systems. In *Deltas, models for exploration*, ed. M.L. Broussard, 87–98. Houston, TX: Houston Geological Society.
Gawthorpe, R.L., and A. Colella. 1990. Tectonic control on coarse-grained delta depositional systems in rift basins. In: Colella, A., Prior, D. B. (Eds.), *Recent and Ancient Nonmarine Depositional Environments: Models for Exploration*. Soc. Econ. Paleontol. Mineral. Spec. Publ. 31: 127–155
Holmes, A. 1965. Principles of physical geology. London: Thomas Nelson and Sons Ltd.
Link, M.H., and R.H. Osborne. 1978. Lacustrine facies in the Pliocene Ridge Basin, California. In *Modern and ancient lake sediments*, ed. A. Matter and M.E. Tucker, 167–187. Special Publication of International Association of Sedimentologists.
Massari, F., and G.C. Parea. 1988. Progradational gravel beach sequences in a moderate-to high-energy, microtidal marine environment. *Sedimentology, Blackwell Synergy*.
McGowen, H., and C.G. Groat. 1971. Ban Horn Sandstone, west Texas: An alluvial fan model for mineral exploration. In *Representative Bureau of Economic Geology*, vol. 72. Austin: University of Texas. (Invest).
Mcpherson, J.G., G. Shanmugam, and R.J. Moiola. 1988. Pan deltas and braid deltas: conceptual problems. In *Fan deltas: sedimentology and tectonic settings*, 14–22, ed. W. Nemec and R.J. Steel. London: Blackie and Son.
Morgan J.P. 1951. Report on the mudlumps at the mouths of the Mississippi River, Partal, the occurrence and origin of the mudlumps at the mouths of the Mississippi River: New Orleans District, U. S. Army corps of Engr, Unpub, Rept, 127pp
Morgan, J.P. 1970. Deltas—a resume. *Journal of Geological Education* 18: 107–117.
Nemec, W., and R.J. Steel. 1988. *Fan delta*. Glasgow: Blackie and Son Ltd.
Orton, G.J., and H.G. Reading. 1993. Variability of deltaic processes in terms of sediment supply, with particular emphasis on grain size. *Sedimentology* 40: 475–512.
Orton, G.J. 1988. A spectrum of Middle Ordovician fan deltas and braid-plain deltas North Wales: a consequence of varying fluvial clastic input. In *Fan deltas: sedimentology and tectonic settings*, ed. W. Nemec, and R.J. Steel, 23–49. London: Blackie.
Postma, G. 1990. Depositional architecture and facies of river and fan deltas: a synthesis. In *Coarse—grained Deltas*, ed. A. Colella and D.B. Prior, 13–27. Special Publication of International Association of Sedimentologists.
Scott, A.J., and J.H. McGowen. 1969. Deltas systems in the exploration for oil and gas: A Research colloquium: Univ. Texa (Austin). *Bureau of Economic Geology* 178.
Shouqing, W. 1987. A heterogeneous model describing the heterogeneity of the production reservoir in Shuanghe oil field. *Petroleum Exploration and Development* (1).
Sneh, A. 1979. Late Pleistocene fan—deltas (sic) along the Dead Sea (sic) rift. *Journal of Sedimentary Petrology* 49: 541–552.
Tucker, M.E. 1981. *Sedimentary petrology introduction*. Blackwell Scientific Publication.
Walker, R.G. 1978. Deep-water sandstone facies and ancient submarine fans: models for exploration for stratigraphic traps. *AAPG* 62: 932–966.
Wescott, W.A., Ethridge, F.G. 1990. Fan deltas—alluvial fans in coastal settings. In *Alluvial fans: a field approach*, 195–221, ed. A. H. Rachocki and M. Church.
Wright, J.V., and G.P.L. Walker. 1977. The ignimbrite source problem: significance of a co-ignimbrite lag fall deposit. *Geology* 5: 729–732.
Yu, Xinghe, Junmao, Zheng, Liheng, Song, et al. 1997. Faulted basin delta sandbody sedimentation and reservoir intrastratal heterogeneity. *Earth Sciences* 22 (1).
Yu Xinghe, Jiang Hui et al. 2007. Depositional Filling Models and Controlling Factors on Mesozoic and Cenozoic Fault Basins of Terrestrial Facies in Eastern China-A Case Study of Dongying Sag of Jiyang Depression. *Lithologic Reservoirs*, 19 (1): 39–45.
Yu, Xinghe, Wang Defa, Sun Zhihua. 1995. Lake braided river delta feature and reservoir geological model-modern delta sedimentary study in Daihai Lake of Inner Mongolia. *Acta Sedimentologica Sinica* 13 (1).

12 Sandy Coast (Shore) and Neritic Depositional System

As we know, the marine environment is quite different from the continent. In addition to physical and chemical conditions, there are great disparities between coastal and submarine deposition conditions and processes. Because submarine environment characteristics are related to seawater depth, the sea area can be divided by seawater depth into four zones, shore (coast), shallow sea, semi-deep sea, and deep sea. The continental shelf (also called shelf), continental slope, continental rise (also called continental apron), and ocean floor (also called abyssal plain) (Fig. 12.1) can be divided by the geomorphic features of the continental margin. Shelf break is the turning point (or zone) of the continental shelf and continental slope.

The neritic zone below the normal wave base is also called the shelf zone. Continental shelf waters are 0–500 m in depth with an average greater than 130 m. This is a flat seabed that has a simple topography with an inclination or slope of several minutes to several degrees, generally less than 4°. A continental slope, with a greater slope of generally 4°–7°, even 13° sometimes, is situated outside the continental slope and the water is usually less than 2.5 km deep. The continental rise is located between the part outside the continental slope and the ocean floor.

A coast with shallow water, directly connected to the continent, is situated between the shallow water wave zone and the euphotic zone in the shelf area. Owing to complex hydrodynamic conditions, waves, tides, and longshore currents cause disturbance, transportation, and transformation of deposits. The weather effect is significant, and seawater temperature and salinity vary greatly. The oxygen content of seawater is high, and ecological differentiation varies due to the presence of enriched nutrients and biological prosperity. Therefore, the variability of sedimentary conditions is the most prominent characteristic of a coastal zone that distinguishes it from other sedimentary environments. In geologic history, a coastal zone is also an area that is most sensitive to crustal movement, sea-level fluctuation, transgression, and regression, as well as most active in terms of geologic processes.

In more than one century, geological prospection has proven that coasts contain abundant energy, metals, and non-metallic mineral resources. Hence, analysis of the geologic features of the coastal zone, particularly its sedimentary environment, is important from theoretical and practical viewpoints.

12.1 Basic Characteristics and Classification of a Sandy Coast

A sandy coast (shore) generally refers to a zone spanning from the average wave base to the highest wave surface. More simply, it is the littoral area above the normal wave base (generally within 20 m of water depth).

X. Yu et al., *Clastic Hydrocarbon Reservoir Sedimentology*, Advances in Oil and Gas Exploration & Production, https://doi.org/10.1007/978-3-319-70335-0_12

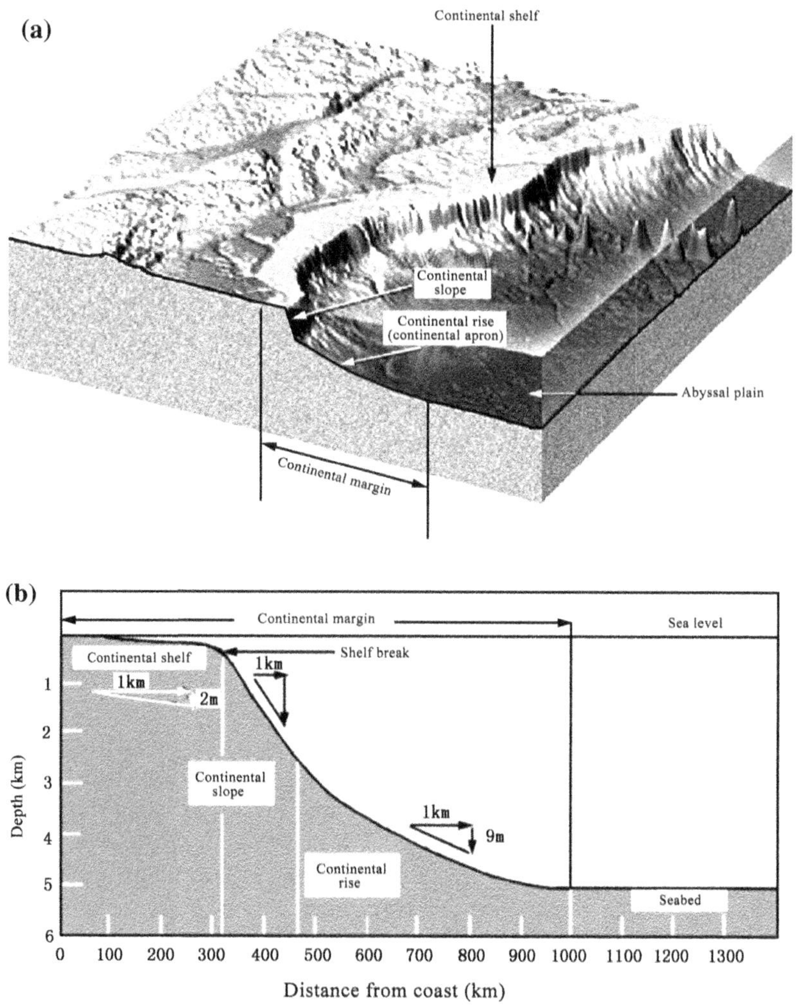

Fig. 12.1 Ideal models of the continental margin. **a** Including the continental shelf, continental slope, continental rise, and the ocean floor, the section map as shown in (**b**) (after Earth Science illustrated dictionary of America, 1997)

It is similar to terms such as "shore zone," "shore line," "shore environment," "longshore zone," and "coastal region," but the terms for defining the specific scope and describing this zone have not been unified. Some scholars confine the shore zone to the zone between the mean low tidal level (or maximum low tidal level) and the mean high tidal level (or maximum high tidal level). Actually, it means an intertidal zone, which is a narrow concept of shore zone. However, most sedimentologists, oceanologists, and geomorphologists employ the generalized shore zone concept. The upper limit is that the shore zone can extend landward to zones that can be affected by the maximum storm surge or to zones with natural changes in geographic features, such as sea cliffs, coastal dunes, and permanent plant growth zones. The lower limit is seaward until the wave base depth in normal weather.

12.1.1 Geological Features

Shore zones, the widths of which vary according to different coast slopes, wave intensities, and tidal range sizes, are distributed continuously around the margins of continents or islands in belts. In zones with gentle slopes, strong tide areas, and powerful shorewardstorms, shore zones can range from thousands to tens of thousands of meters in width (for example, the Dutch Coast in Northwestern Europe and the

Table 12.1 Basic geological features of shore zones

Item	Features	Influencing Factors
Depth range	Greatly changes within 20–30 m; unstable hydrodynamic conditions	① Wave and tide; ② slope
Width	Not fixed—from dozens of meters to thousands of meters	① Gradient; ② geomorphic features
Sedimentary characteristics	Composed mainly of sandy deposits. Gravel deposits might develop, but they are rather small in scope and scale	① Terrain; ② lithologic characteristics
Environment characteristics	Repeated marine transgression and regression, abundant sunlight, various organisms, greatly changed temperature and salinity, rapidly varied biogroups, and strong adaptability	Tide and wave intensity

northern Suzhou shore zone in China). However, in steep rock coasts, shore zones are very narrow, only dozens of meters wide (including some zones in Norway and the southeastern coastal waters of China).

The basic geological characteristics of shore zones are as follows: ① depth range; ② width; ③ sedimentary characteristics; and ④ environmental characteristics (Table 12.1). The depth range is dominated mainly by the wave intensity in normal weather. The broad seaward shore zone has high wave intensity and a deep wave base, while the shore zone confined to the bay has low wave intensity and a shallow wave base, only several meters deep. Since wave intensity usually varies from season to season, and sea level fluctuation is affected by tidal cycles, and the upper and lower limits of the shore zone change frequently. Thicker and wider coastal deposits can be formed when the coastal line migrates constantly. In addition to various terrigenous clastic sediments, more developed littoral carbonate deposits can be formed as well.

12.1.2 Environment Classification

On account of different purposes, scholars from different disciplines have diverse classification schemes for coastal environments. In summary, the coastal environment is classified according to the main control factors (such as hydrodynamic conditions, sediment supply status, changes in climate, topographical features, sea level fluctuations, and geotectonic backgrounds). From the perspective of sedimentology, the coastal environment can be divided into terrigenous clastic coasts (sandy coast) and carbonate (including evaporite) coasts based on sediment type. Sandy coasts can be divided further into non-barrier coasts and barrier coasts based on the presence of a barrier terrain. Based on whether the hydrodynamic conditions are dominated by wave action or tidal action, it can be divided into beach coast and lagoon-tidal flat coast or wave-dominated and tidal-dominated coast, respectively.

No barrier island exists between the non-barrier coast, which is also called the open sea coast, and open sea shelf. Hence, the connectedness between them is very good. Barrier terrains, including sand bars or bioherms in the coastal zone, are not developed. Consequently, because of obvious wave and coastal current actions, seawater can be fully circulated.

Barrier coasts are also called localized coasts. Coastal sea is isolated or partially isolated from the open sea owing to the presence of barrier terrain. Because such seawater has limited circulation, its wave action is weak, and because the influence of tidal action is greater, hydrodynamic energy is generally low and the salinity of the seawater is abnormal (Fig. 12.2). As the sea level rises, a barrier sea can communicate with the open sea. As the sea level declines, a lagoon-tidal flat is formed.

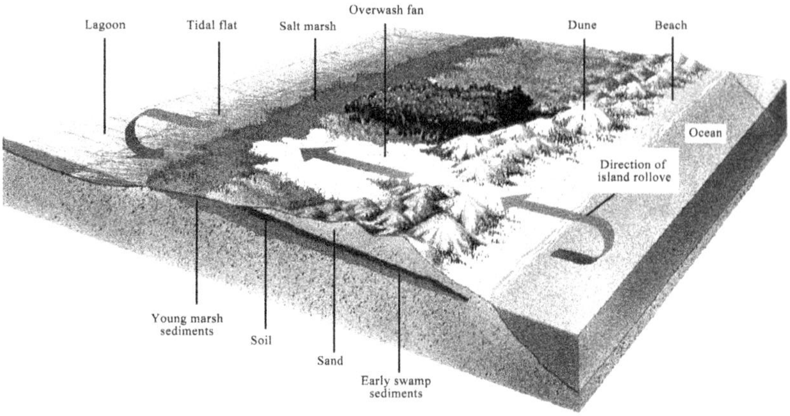

Fig. 12.2 Barrier island coast is also called limited coast. When sea levels rise, barrier sea connects with open sea, and when sea levels drop, lagoon-tidal flat forms (from Falcon 1998)

12.2 Hydrodynamic Conditions of Coasts

Coastal regions can be affected by wind, sea waves, waves, tides, and sea flow actions, as well as fluvial processes. The interaction of these different agents influences coastal landforms and sedimentary characteristics. Sea waves and near-shore currents have dominant influences on the sedimentary characteristics of a non-barrier coast, while tide has a dominant influence on a barrier coast.

12.2.1 Wave Action and Sediment Transportation

Based on the cause of their generation, ocean waves are termed wind waves, storm waves, and tsunamis. Wind waves, which often play a continuous role, constitute the main factor affecting sedimentation and beach processes. Both storm waves and tsunamis are abrupt temporary events, however most tsunamis are huge waves resulting from seismic activities, hence their impact is huge.

In the case of a normal climate, sea (lake) level fluctuation occurs when wind blows over the sea level. Due to episodic wind flow with eddies, wind acts on the water surface with irregular tangential stress and transmits energy to the water surface. Originally, ripples are blown on the water surface, and these ripples then continuously develop into waves. Waves, which are the main means of transmitting energy coastwise from the coastal zone, not only erode the coast and transport and transform sediments but also cause longshore and rip currents. Longshore currents give rise to longshore drift, while rip currents lead sediments to migrate seaward. Wind waves are propagated from their forming region to the littoral area with constantly changing spectra. As seawater gets shallower, wind waves, swells, build-up, breakers, beach combers (combers), and surf and swash (Fig. 12.3) can be presented successively. Consequently, during energy transmission, waves can show varied (Table 12.2) characteristics owing to their diverse forms at different depths and in different coastal zones.

12.2.1.1 Control of Slope Over Wave Form

In the shelf or deep water area, images of long and gentle waves with a small gradient show a sinusoid form. The oscillation diameter of waves decreases (Fig. 12.3a) like an exponential function downward. When the water depth is greater than half-wavelength of a wave, forward propagation of the wave will result in a series of sinusoidal image sections of ridges and troughs. As a result, the turbulence of the water at the bottom of a shelf or deep water area is very small, generally without sediment movement. Hence, water depth equal to half-wavelength is

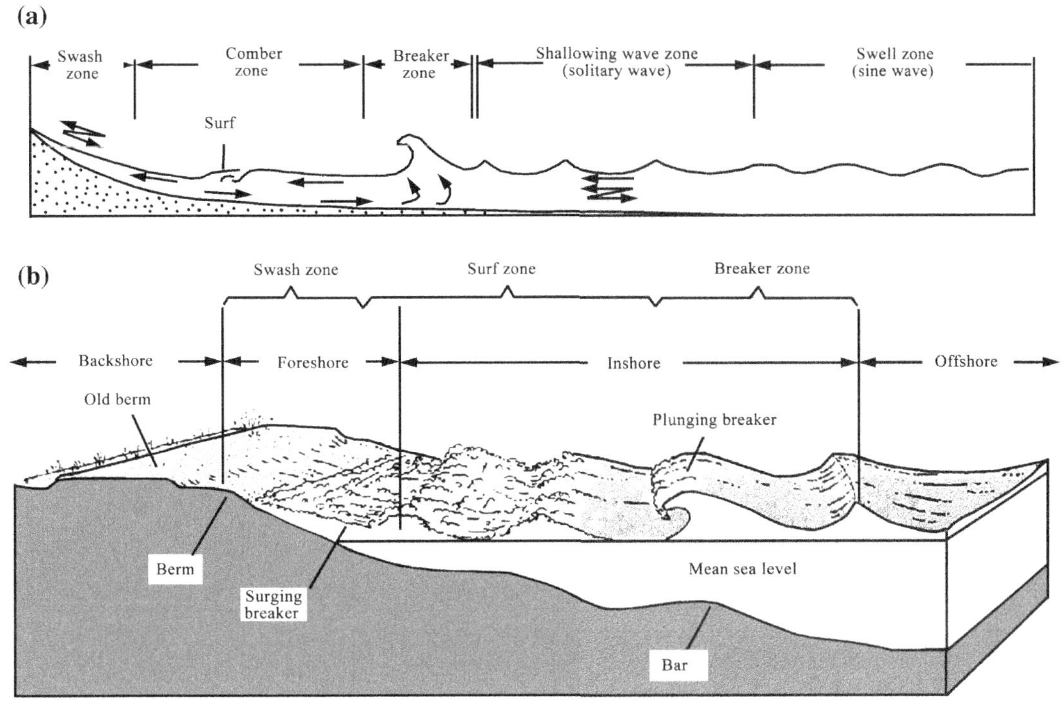

Fig. 12.3 During the process of shore wave propagation, as the sea water becomes shallower, wind waves and swell, surge, waves, surf (surf) or surf and surf appear successively (after Shepard and Inman 1950)

often taken as the wave base. When the water depth is greater than the half-wavelength, disturbance of sediments resulting from waves is weak. This is the reason for selecting this depth as the wave base in geology.

When deep water waves enter a shallow water area, in addition to wave refraction produced by waves, the following changes occur as the wave velocity decreases: the gradient (wave height/length) increases when wavelength decreases, i.e., waveform changes from symmetric into asymmetric, and wave breaking occurs when the depth is close to 4/3 wave height; and the motional orbits of water particles change from circular to elliptic movement, eventually to reciprocating rectilinear movement. The hydrodynamic zone is formed by only one wave characteristic, that is, the wave period remains unchanged.

The width of the hydrodynamic zone varies according to the slope of the beach face and the state of tide. Beaches with steep slopes and wave breaker zones occur near the shoreline, and such breakers are narrow but not developed. Breakers are produced to widen the breaker zone when the gently inclined beach face is rather far away from the coast. The changes in low tide and high tide levels influence the water depth to affect the position change of breaker zones.

12.2.1.2 Formation of Longshore and Rip Currents

There are two situations when waves adjoin the coast: ① waves are parallel to the shoreline; and ② waves intersect with the shoreline at a certain angle. Because components (or component forces) in two directions can be produced by the actions of waves and the coast, the breaker zone intersects obliquely (Fig. 12.4a) with the shoreline when the waves are parallel to the shoreline. In other words, the breaker zone is parallel (Fig. 12.4b) to the shoreline. Consequently, longshore currents are produced by the component parallel to the shoreline, so sediments are

Table 12.2 Genesis, functions and characteristics of various coastal waves

Features	Storm flow	Swash	Comber or surf	Breaker	Build-up and swell
Location	Backshore	Foreshore	Nearshore		Offshore
Genesis	Shallow and high-speed abrupt waves, near the shoreline	Formed by forward flushing of water particles, whose movement speed is greater than that of waveform, at the wave crest	Formed by higher and steeper crest, forward dumping and breaking into spray	Formed by swell height increase and steeper crest in shallow water	Propagation of storm waves in deep water
Water movement	Flushing	Generating component, longshore current and backflow	Spilling	Oscillation	Oscillation
Effect	Climbing up the sea beach to transport sediments shoreward	Most important agent on the coast, dominated by transportation	Strong transportation and transformation, dominated by erosion	Shoreward movement of sediments, dominated by accretion	Has little influence on seafloor sediments
Characteristics	Upper flow regime ($Fr \geq 0.84$), with high speed and strong waves	Breaking wave propagation shoreward in sequence	Water depth roughly equal to wave height	Water particle propagation to waves, without backflow	Large wave length, approximately equal wave height; short period, similar to sine wave

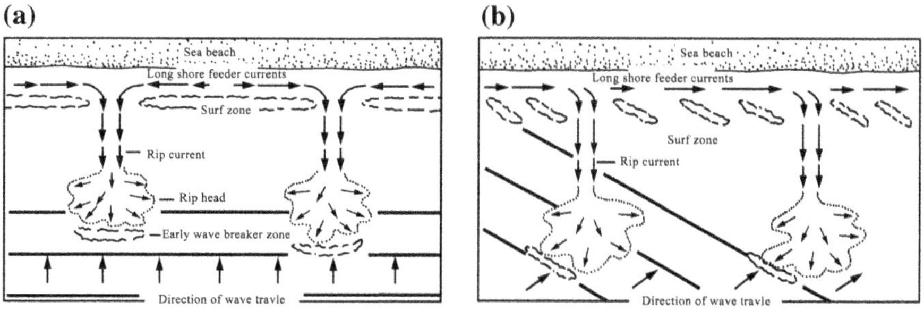

Fig. 12.4 The formation mechanism of coastal current and offshore flow and the characteristics of the surf zone (after Komar 1976)

subject to movement along the coast. The stronger the wave energy and the smaller the angle between sediments and the shoreline, the more obvious is the longshore current. Rip currents (Fig. 12.4), formed by the component vertical to the shoreline, are terminated in the breaker zone and then transport fine-grained sediments to the sea. Baymouth bars and spits (Fig. 12.5) can be formed by longshore current.

12.2.1.3 Influence of Climate and Seasonal Variations

1. **Storm periods and less storm periods**

Sand movement in a sea beach is dominated by wave action, as is also reflected in the changing storm and poor storm periods. In good weather, that is, the storm-free period, most

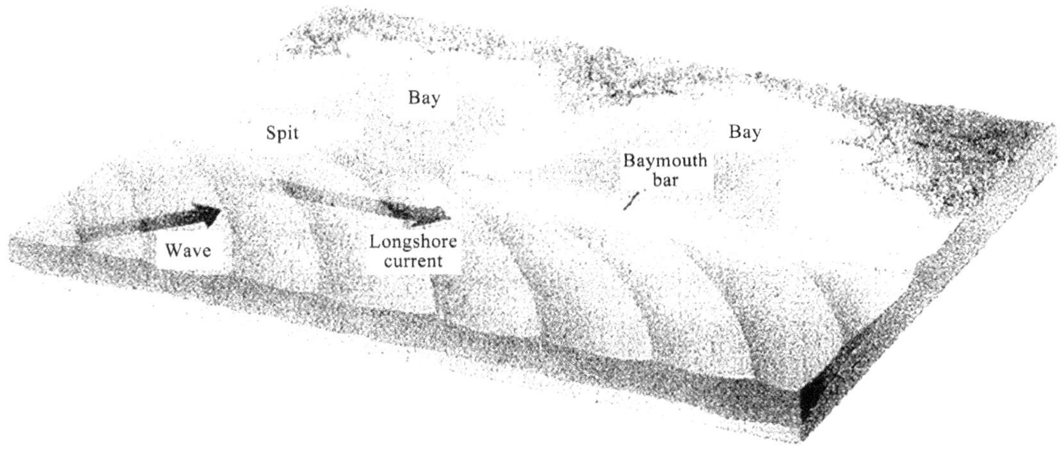

Fig. 12.5 The coastal current usually can form the bay mouth dam and spit

waves have low amplitude and long period, resulting in the direct production of landward bottom pouring and transportation of only a modest amount of sediments in shallower water. If the longshore current is weak, a small amount of sediments taken to the sea by the action of a rip current is accumulated on the beach face for expansion into the berm. However, in stormy weather, a large amount of sediments can be transported by waves of large amplitude, short period, high energy, and low wave base. However, the landward effective bottom swell reduces with increasing wave gradient, thus the amount of sediments supplied to the sea beach can be decreased, but that taken seaward by the rip current can be increased. Hence, distal bars can be formed easily.

It should be noted that the erosion effect is mainly produced on the sea beach by waves during the storm period, and sedimentation occurs after the storm.

2. **Seasonal difference**

In different seasons, the sea beach range varies and wave forms have obvious differences, which are represented as follows: ① in the summer, waves are characterized by long wavelengths, low wave heights, and wide beach faces, and sand is composed mainly of shore deposits; and ② in winter, waves are characterized by short wavelengths, high wave heights, and narrow beach faces, and sand is composed mainly of offshore deposition (Fig. 12.6).

12.2.2 Tidal Action and Sediment Transport

Tides and tidal currents greatly influence coastal zone sedimentation and sediment distribution, particularly in coastal zones with low wave energy and large tidal ranges, horn-shaped estuaries, and bays.

12.2.2.1 Features of Tides

Tide refers to a movement phenomenon in which sea water rises or falls periodically due to the action of the attractive forces of the sun and the moon on the Earth. Oceanic tide, which is a long-period global sea level fluctuation, refers to the phenomenon in which sea level rises or falls periodically owing to tide-generating forces (resultant force of the attractive forces of the sun and the moon acting on the Earth and the centrifugal force produced by the Earth's rotation). A tidal cycle is formed by the rise and fall of sea level, that is, a falling tide is closely followed by a rising tide. The difference between the rising level and the falling level is called the tidal range in a tidal cycle. When the tidal level fluctuates, water flows laterally on the plane due to water

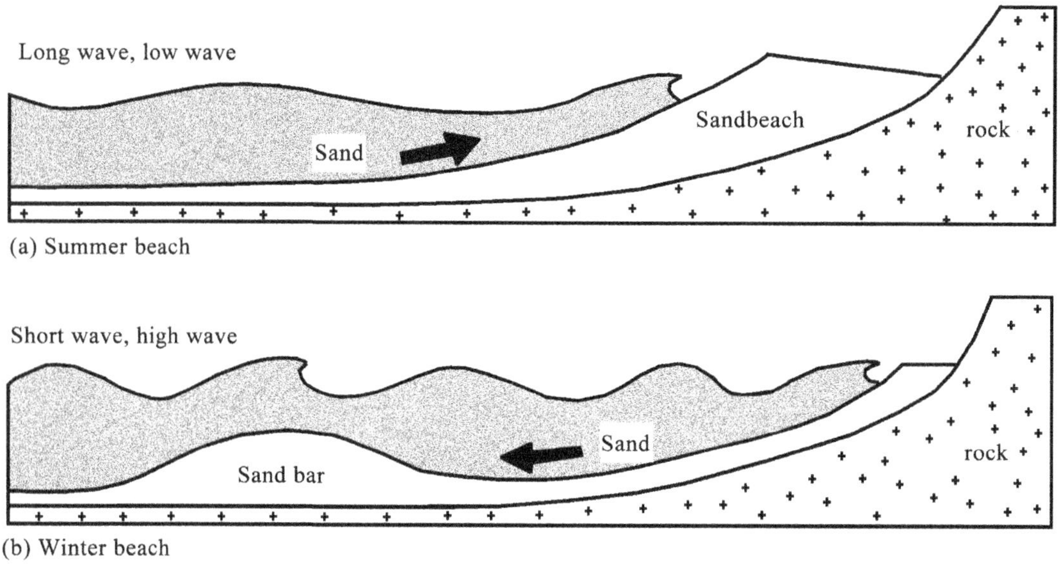

Fig. 12.6 Beaches and waves in different seasons have different characteristics, the summer wave is characterized by long wave and wide high-low beach face of wave, while the winter wave is the opposite

level difference, which is called tidal current. In the open ocean and sea, the speed and direction of tidal currents vary regularly at different times in the tidal cycle. In nearshore zones or bays, the back and forth movement of tidal currents is confined along the direction vertical to the shoreline owing to topographic influence. Tidal activity and the induced hydrodynamic conditions are very complicated, however they follow obvious rules, with the general characteristics as follows: ① due to the globalization of tidal activity, only those sea areas interconnected with the ocean have tidal activity; ② tidal activity possesses obvious periodicity; ③ its flow direction has typical bidirectionality; and ④ the size of energy is asymmetric.

12.2.2.2 Sediment Transportation and Deposition

Tidal action has significant influence on the transportation and deposition of sediments in the shore zone, distribution of depositional facies zone, and structure of the sedimentary sequence. Barwis and Hayes (1978) pointed out that the forms of coast and the distribution of sand bodies are closely related to the tidal range (Fig. 12.7). For instance, a coast zone with an elongated barrier island is developed easily owing to weaker tidal kinetic energy in the area with a small tidal range. Owing to a small amount of tidal inlets, a lagoon is usually developed behind the barrier island when a washover fan is developed. In an area with medium tidal range, the barrier island is cut into smaller islands by a greater number of tidal inlets, while the tidal delta is more developed, and the tidal flat behind the barrier island is developed extensively. In an area with a large tidal range, tidal sand ridges parallel to the tidal direction are usually formed in the estuary owing to high tidal energy, and broad tidal flats can be found in the coastal zone.

Tidal energy decreases gradually from the subtidal zone to the intertidal zone and the supratidal zone. Hence, tidal action is contrary to wave action and tends to transport fine-grained

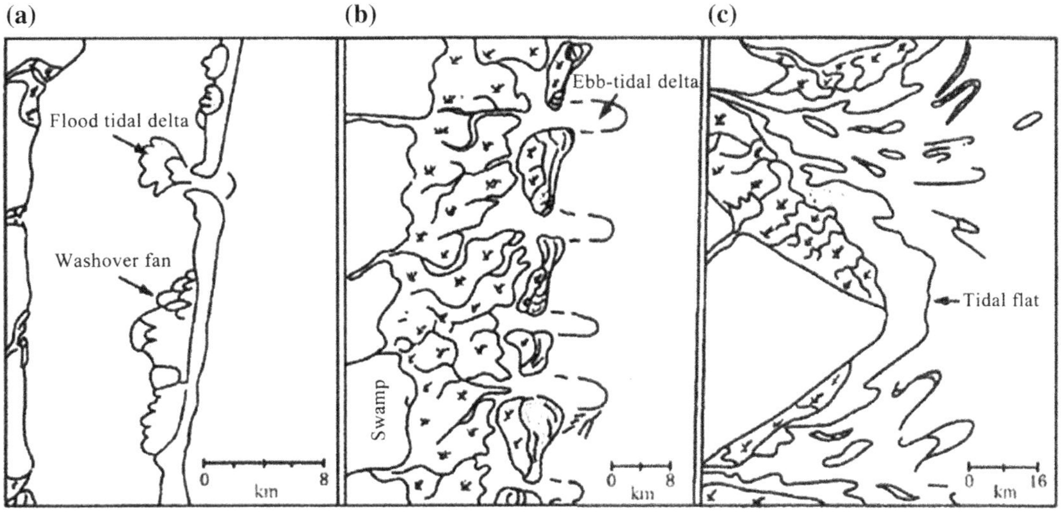

Fig. 12.7 Coastal sand body types formed by different types of tidal range (after Barwis and Hayes 1978). **a** Elongated barrier island formed by small tidal range; **b** Short tidal range formed by medium barrier island; and **c** Tidal sand ridge formed by large tidal range in estuary

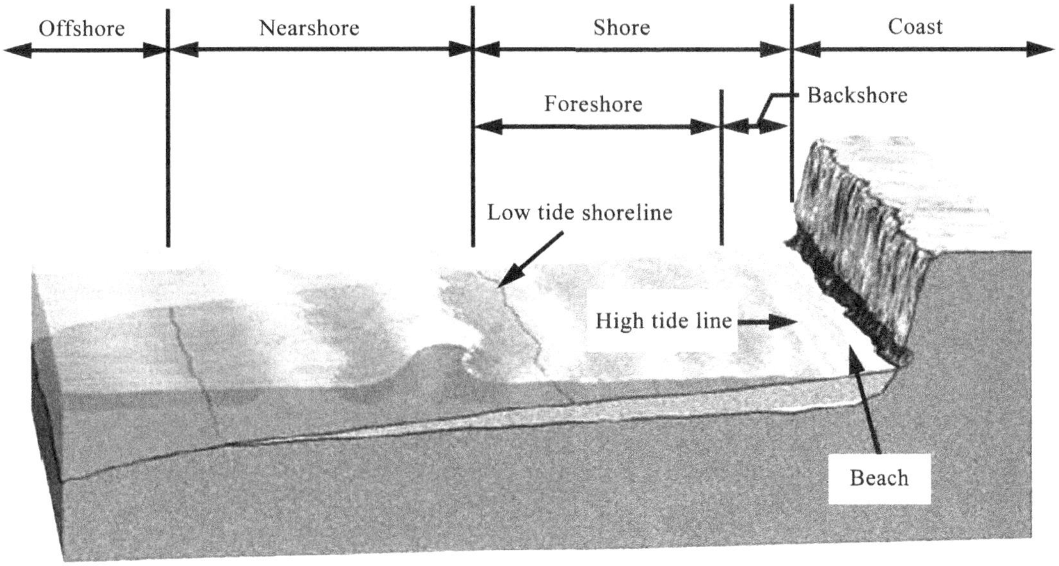

Fig. 12.8 Geomorphic units of non-barrier coastal (after H. E. Reineck, 1973)

sediments to the coast, mainly in the form of suspended fine-grained deposits. Because tidal velocity decreases constantly and the slack tide period of high tide is longer than that of low tide, suspended argillaceous sediments are deposited continuously and sand sediments are confined to the tidal channel or the lower part of the intertidal zone.

12.3 Non-barrier Coastal Environment

12.3.1 Geomorphic Feature

A non-barrier sandy coastal environment is generally characterized by the development of beach topography. Based on coastal landforms, hydrodynamic conditions, and sediment characteristics, a typical non-barrier sandy coastal environment can be divided into several units: coastal dunes, backshore, foreshore, nearshore (or shoreface), and offshore (Fig. 12.8 and Table 12.3), in which foreshore and nearshore are also called shorefaces (zones). In addition, it can be further divided into an upper shoreface and lower shoreface.

The lower bound of a coastal dune is located approximately near the maximum storm rising tide. The supratidal zone, in which geomorphic features such as beach ridges and berms can be developed, lies between the lower bound of the backshore dune and the average high tide line. The area between the average high and low water levels is the foreshore, which is the intertidal zone. The foreshore is not usually developed in

Table 12.3 Sedimentary characteristics of the subfacies of non-barrier sandy coasts

Subfacies	Microfacies	Development location	Sedimentary characteristics	
			Lithology	Sedimentary structure
Coastal dune	Shoreline dune	Above the maximum storm rising tide	Fine-medium particles, high texture maturity, lack of mud; single component, mainly composed of quartz; containing a small amount of heavy minerals and biological fossils	Eolian bedding (trough type), thick set, steep angle (>30°)
	Beach sand ridge	Extreme position of storm wave in backshore zone	Mainly composed of coarse sands, gravels and biological shells	Parallel bedding and cross beddings in different directions, with laminar dip angle of 7°–28°
	Chenier	Coastal swamp area	Mainly composed of fine-grained sands and shells rich in plant debris, locally with marlstone	Strong bioturbation
Backshore	Wind dune	Between the lower bound of sand dune and the average high tide line	Dominated by sand, and shell beds are usually found	Horizontal bedding and cross beddings at different angles
Foreshore	Sea beach (beach sand)	Between the average high and low tide lines	Dominated by sand on sandy beaches, and gravel on gravel banks, with high texture maturity	Swash bedding is developed
Nearshore	Longshore bar	Between the average low tide line and the wave base	Slightly coarse sediments with good sorting	Wedge or planar cross bedding, low dip angle between set and laminae, horizontal laminae, available with running water and wave ripple
Offshore	Offshore sand bar or sand beach	Below the wave base	Fine sediments with tempestite or biological shell formation	Horizontal laminae, hummocky bedding, complex bedding, and cross beddings in different directions

tide-less areas. The backshore can be transited directly to the shoreface or nearshore. The nearshore is located in areas between the average low tide line and the wave base, which lies in the subtidal zone. Longshore bars are usually developed in the foreshore and nearshore areas.

12.3.2 Sedimentary Characteristics of the Main Sub-environments

12.3.2.1 Shoreline Dune Subfacies

The landward side of the supratidal zone is the highest water level that tide water can reach in a terrible storm. It includes sedimentary units such as shore dunes, sea sand ridges, and Chenier.

1. **Shoreline dunes**

Shoreline dunes are low dunes or dune regions formed by the modification of the shoreline by wind action. These small dunes are distributed along the coast as long ridges or crescents, which occupy considerably wide areas. Coastal (shore) dunes have typical wind dune cross bedding, the form of which usually belongs to large planar (trough) cross bedding. The foreset laminar dip angle of the cross bedding is steep (30°–40°), and is similar to the ordinary wind dune cross bedding structure (Fig. 12.9). In addition, convex foreset cross bedding with a set boundary and small-scale deformation structures can usually be found on shore dunes.

2. **Beach sand ridge**

A beach sand ridge is a continuous linear uplift geomorphologic unit (including banks) composed of coarse-grained sediments near the high tide line. It is located at the extreme position that a storm can reach in the backshore zone. These linear sand ridges can be found as individuals or as a group, with a trend parallel to the coastal line. They are several meters high, dozens of meters wide, and hundreds to thousands of meters long. In general, there is a scour surface between sand ridge deposits and the underlying deposits. A sand edge is usually formed by multiple wave transformations of wide and gentle sand bars.

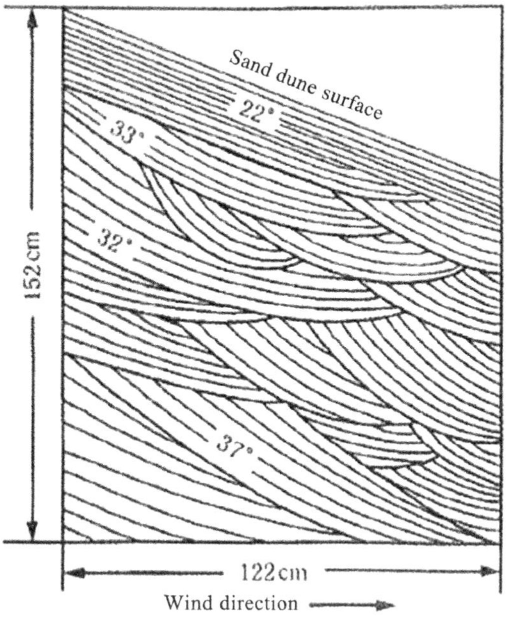

Fig. 12.9 Eolian stratification in coastal dunes

3. **Chenier**

The word (Chenier) sand hill is derived from the French word Le chene (meaning oak, i.e., covered by plants). It is a rather low (3–6 m), long and narrow beach sand ridge with flourishing trees in the seaward direction in the coastal swamp (and mud flat). Its long-axis direction is parallel to the coastal line. Shoreline swamps may be separated from the sea due to the formation of sand hills of this type.

12.3.2.2 Backshore Subfacies

Owing to its location above the average high tide line, backshore subfacies can be subject to wave action in terrible storms and extraordinarily high tides. It is a flat and dry corridor, and its deposits are dominated by sand, biologic debris, and shells. Sometimes, even a shell bed can be formed by a storm, usually with the convex surface of shell facing upward (Table 12.3).

12.3.2.3 Foreshore Deposition

The foreshore, which is located in the zone (intertidal zone) between the average high tide line and the average low tide line, is the main part of beach deposition. It comprises a gentle slope zone with flat terrain inclined gradually from the lower part of the beach to the sea. Foreshore zone development is related to coastal landform (or slope) and tidal action. If the coast is steep and without tidal action, a foreshore is not developed, but a backshore can be transited directly to a nearshore and vice versa.

The main bedding type of the foreshore is beach swash cross bedding, that is, megascopic gentle wedge cross bedding with laminated beddings or laminae that are parallel and sets intersecting at low angles (Fig. 12.10). Laminae can extend for 30 m in the direction parallel to the coast and 10 m in the direction vertical to the coast. All laminae, with specific sorting features, have reverse beddings. With regard to lamina composition or the dip angle of sets, the gradient of the dip angle of laminae is closely related to the size of deposited particles and the magnitude of the beach slope angle. In general, the coarser the particles and the greater the beach slope gradient, the greater is the dip angle of the laminae. Therefore, a foreshore zone is characterized by the swash action of waves.

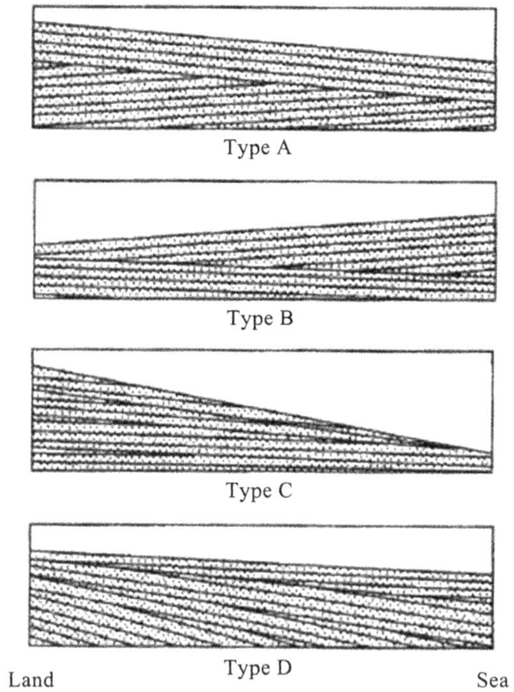

Fig. 12.10 Four types of flushing cross bedding in beach sediments

12.3.2.4 Nearshore Deposit (or Shoreface Deposit)

This refers to the area (subtidal zone) between the average low tide line and the intersection of the wave base with the seabed, which is usually immersed underwater. There are transitional zones with different widths between the nearshore deposit and the offshore zone. One or more asymmetric coastal sand bars distributed parallel to the coastal line can be usually found on its surface. This sand bar is formed on the breaker zone, with slightly coarse deposits derived mainly from the bank and land. Its degree of development is related to the wave energy at the coast. The weaker the coastal wave energy, the lower is the number of coastal sand bars. A groove, as a channel for coastal flow, is also present on the landward side of coastal sand bars.

12.3.3 Sedimentary Sequence and Model of a Non-barrier Coast

In the geologic history of coast development, progradational and retrogradational coastal vertical sedimentary sequences can be formed as transgression and regression occur. In general, the most common is the progradational vertical sequence in ancient stratigraphic sections. When the sea level is relatively stable and adequate sediments are supplied or the deposition rate exceeds the rate of sea level rise, sea beach advances seaward owing to continuous accretion, and nearshore sediments are superimposed successively on offshore sediments to form a gradually coarsened sedimentary sequence from

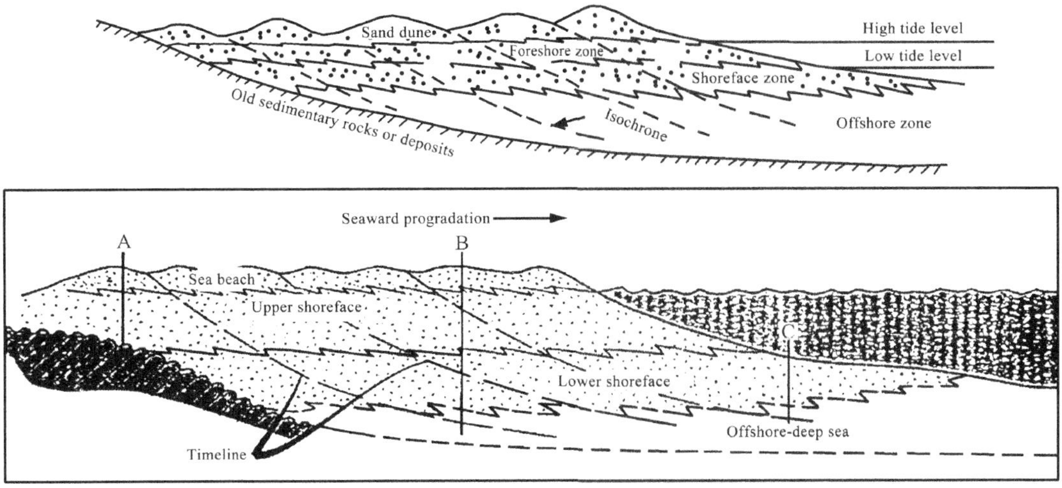

Fig. 12.11 From bottom to upper, the progradation beach appears offshore mud (continental shelf mud), lower beach face (near beach), upper beach face or before beach, and the most top is coastal eolian dunes or beach sands

bottom to top. Furthermore, sedimentary facies vary correspondingly to result in offshore mud (shelf mud), a lower shoreface, upper shoreface, and foreshore from bottom to top. Coastal aeolian dunes or beach sands (Fig. 12.11) are located in the uppermost reaches.

12.3.3.1 Shelf Mud Facies

This is also called an offshore, since the offshore shelf neritic deposit is below the wave base in normal weather. It is composed mainly of silty mud and mud, with horizontal bedding, but is often destroyed due to strong disturbance by bentonic organisms. Bioturbation structures are well developed, containing normal neritic fossils and abundant ichnofossils, mainly zoophycos (crawling trace) and the occasionally transient trace (burrow). As a result of storms, storm deposit intercalation, such as silt, fine sand, or shell beds, can be found, and storm graded bedding and hummocky cross bedding (Fig. 12.12) can be developed.

12.3.3.2 Transitional Facies

In normal weather, the main deposits include silty mud and muddy silt near the wave base, dominated by horizontal bedding and miniature wave ripple bedding. Strong bioturbation is often mixed with silt and fine-sand bedding (storm wave product, also called storm sheet sand), and ichnofossils and normal marine fossils can be usually found. It should be noted that mud flow deposits are usually formed in the transitional region between the shelf and slope. These deposits have an obvious characteristic, that is, mud diapirs usually exist below the mud flow development area. Owing to the influence of these mud diapirs, mud flow is aggravated to produce mudflow gullies. These mudflow gullies not only exist in the shoreface, but also in offshore deeper water environments. This deposit has been found in the Neogene in the Yinggehai Basin. For instance, the sedimentary facies of the Dongfang 1-1 Gas Field is severely affected and transformed (Fig. 12.13) by mud flow deposits, which led to transformation of the lower shoreface sand body in the form by mud flow, with variable spreading direction. However, mudflow gullies lie mostly along the northeast to southwest direction; thus, these sand bars are mostly distributed along the northeast to southwest direction.

12.3.3.3 Shoreface Facies (Shoreface)

This is also called the nearshore, and it can be further divided into three subfacies, namely, upper, middle, and lower shorefaces. In countries other than China, it is mostly called shoreface or

Section	Phase	Representative primary sedimentary structure	Symbiotic sedimentary structure	Ordinary lithology	Environment interpretation
	7	Aeolian trough cross bedding	Plant root marks, deformed structure	Medium-fine sandstone	Coastal dune
	6	Horizontal bedding	Small ripple bedding, low-angle cross bedding, fine current mark, adhesive ripple, and bubble sand structure	Sandstone	Backshore
	5	Swash cross bedding	Current ripple, wave ripple, antidune, interference ripple mark, transformation ripple mark, current and wave ripple bedding, parallel bedding, graded bedding, washout structure, swash mark, fine current mark, adhesive ripple, parting lineation, and burrow	Medium-fine sandstone	Foreshore
	4	Current ripple bedding	Chopped wave ripple, weak bioturbation	Fine sandstone	Nearshore
	3	Horizontal bedding	Symmetric wave ripple, large current ripple, large ripple bedding, and medium-strong bioturbation	Fine sandstone	
	2	Sand and mud interbedded, bedding and bioturbation structure	Strong-complete bioturbation, homogeneous bedding	Interbedded mudstone, siltstone and sandstone	Transitional zone
	1	Horizontal bedding	Medium-strong bioturbation structure, ichnofossil, homogeneous bedding, and graded bedding	Silty mudstone, muddy siltstone, mixed with fine sandstone	Shelf

Fig. 12.12 Vertical sedimentary sequence of the sandy barrier free coast

is divided into two zones (upper and lower shorefaces). Its deposits, which are dominated by silt and medium-fine grained sands, coarsen from the bottom to the top. Beddings include horizontal bedding (dominant in lower shoreface), asymmetric miniature wave ripple cross bedding, and megascopic wave cross bedding (usually found in middle and upper shorefaces). It should be pointed out that nearshore sand bars (Fig. 12.3) distributed coastwise can be formed

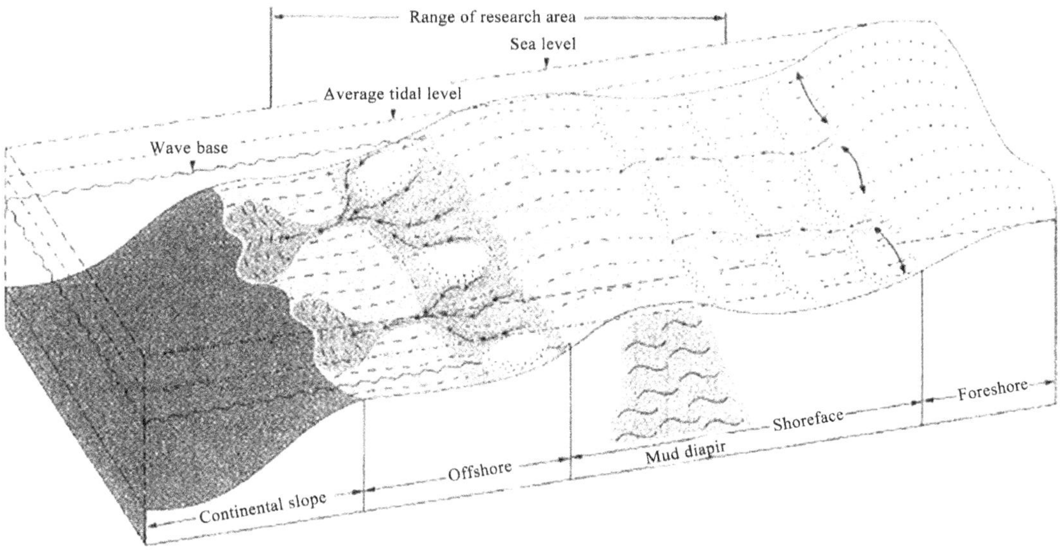

Fig. 12.13 The sedimentary facies model transformation of mudflow in the Neogene Ying'er section of the Yinggehai Basin, China

in this region when wave energy is strong (as in the storm period). Bioturbation structures weaken gradually moving upward. Lebensspuren are dominated by oblique crossings and vertical backfill burrows used for living or as temporary residence, and sometimes zoophycos are found.

12.3.3.4 Foreshore Facies

This is located above the shoreface zone, and is equivalent to the intertidal environment. Deposits mainly include fine-medium sand. In a high wave energy coast, medium-coarse sand deposits can be mainly found with gravel beddings (particularly gravel beaches). On account of the formation of longshore and rip currents (Fig. 12.14), longshore bars and rip channels (also called beach sand channels when developed on the foreshore) can be usually developed. Both in high- and low-energy environments, longshore bars are characterized by coarsening upward, current ripple is the typical model in the channel, and wave ripple is mainly presented in the sand bar. The former is asymmetric, while the latter is symmetric. The typical characteristics of the foreshore zone include the development (Fig. 12.15) of swash bedding and cross beddings along different directions in the sea beach, as well as the development of a large number of exposed structures. Rare biological fossils can be found, occasionally along with broken biological shells.

12.3.3.5 Backshore Facies

Sediments are dominated by fine sand, mixed with silty mud, and swamp formation. Lenticular interbed and beach sand ridge mixed and accumulated with various biodetritus can be formed under the influence of storm waves, with backshore facies undeveloped sometimes.

12.3.3.6 Coastal Dune Facies

This is the best pure quartz sand body with the best maturity, and is characterized by megascopic wind dune cross bedding. It lacks fine-grained mud and biological fossils

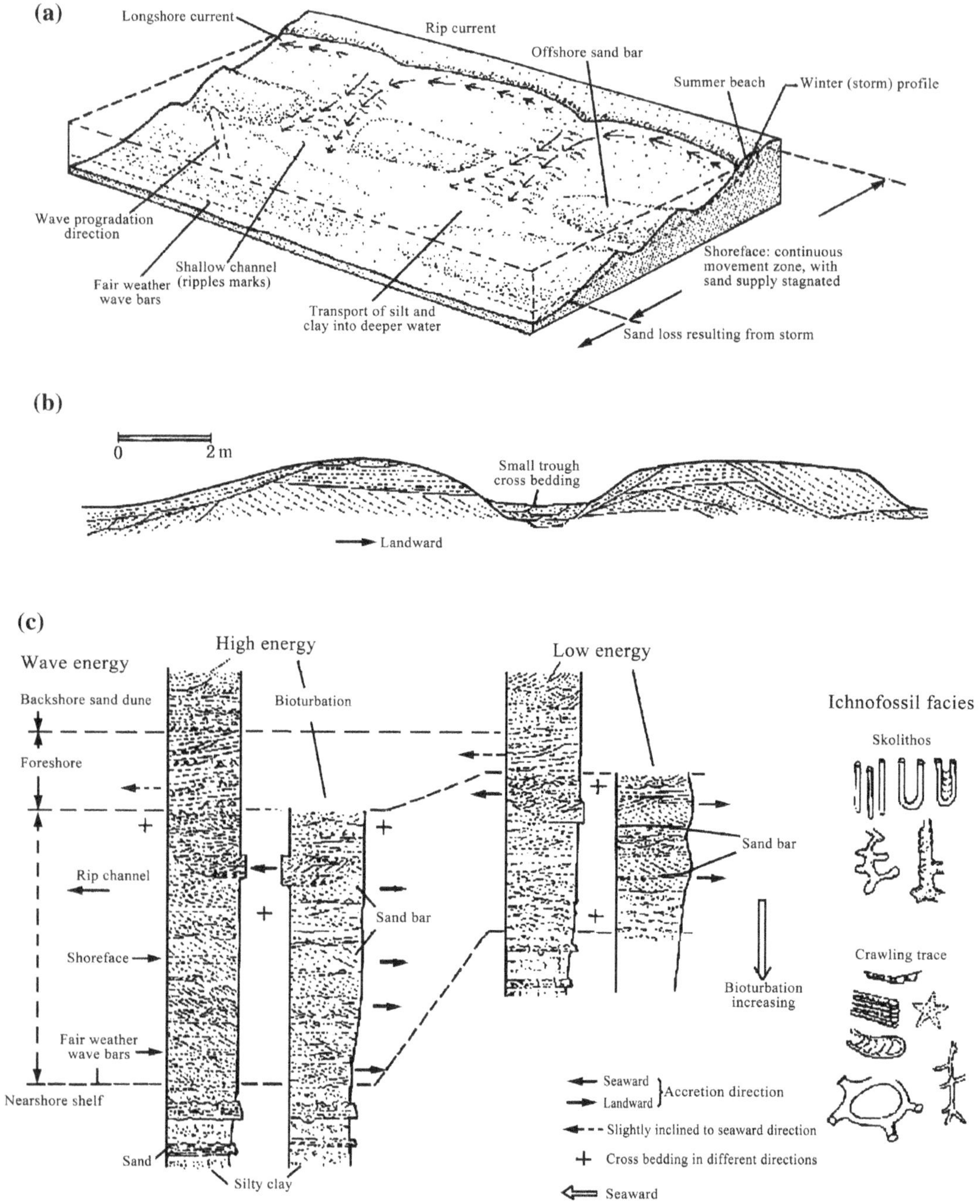

Fig. 12.14 Hydrodynamic distribution and sequence models of various sand dams in the beach profiles

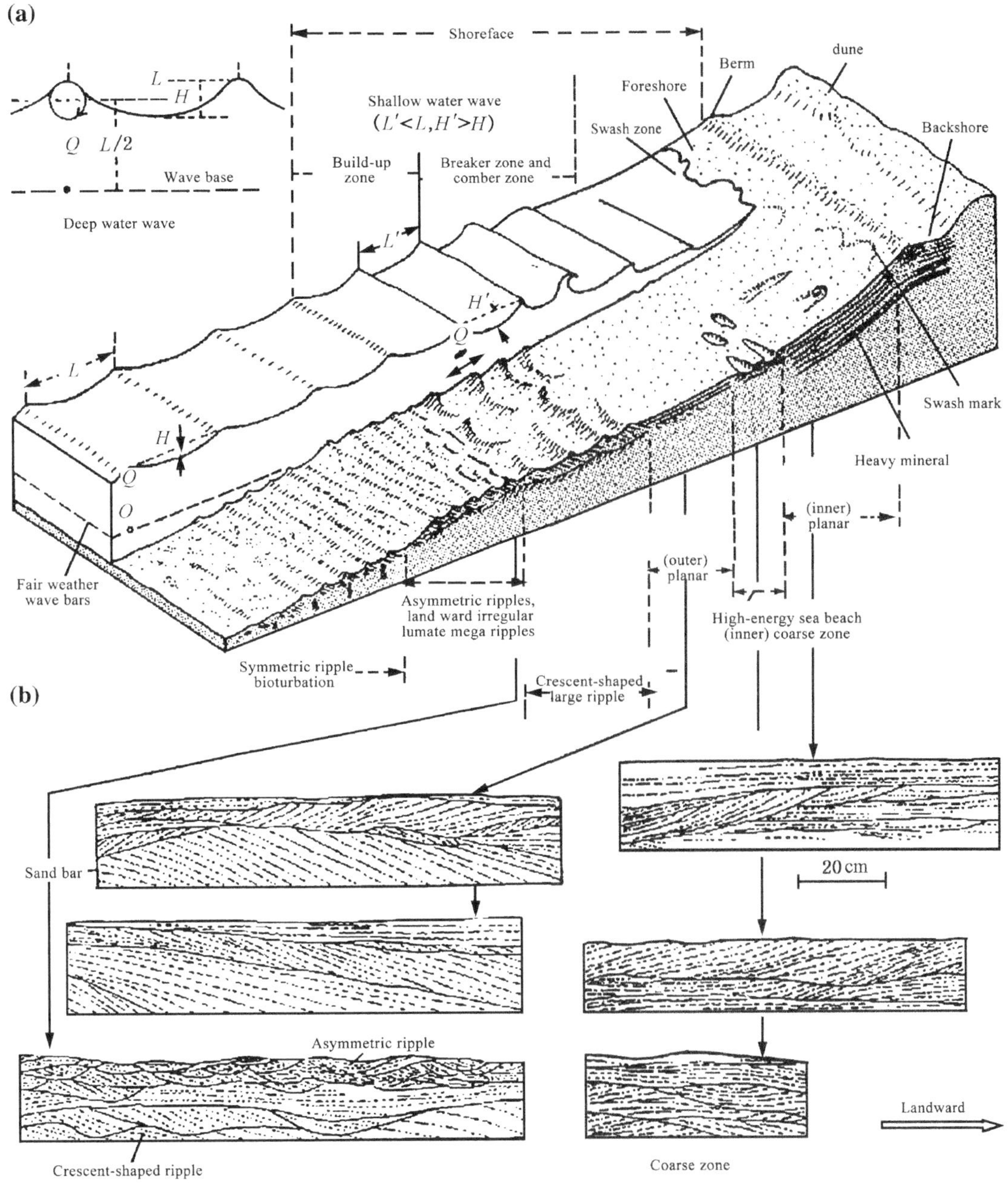

Fig. 12.15 Sedimentary model and sedimentary structure characteristics of shallow water-deep water in the riparian facies zone (after Clifton et al. 1971)

12.4 Barrier Coastal Environment

In coastal regions, in the presence of barrier terrain (generally long and narrow geologic bodies), such as barrier islands, barrier sand bars, and bioherms, to isolate or partially isolate the nearshore sea and open sea to lead sea water in partial circulation, a barrier coast is formed (Fig. 12.16). This zone dominated by tidal action is characterized by special hydrological conditions such as weak wave action, generally low

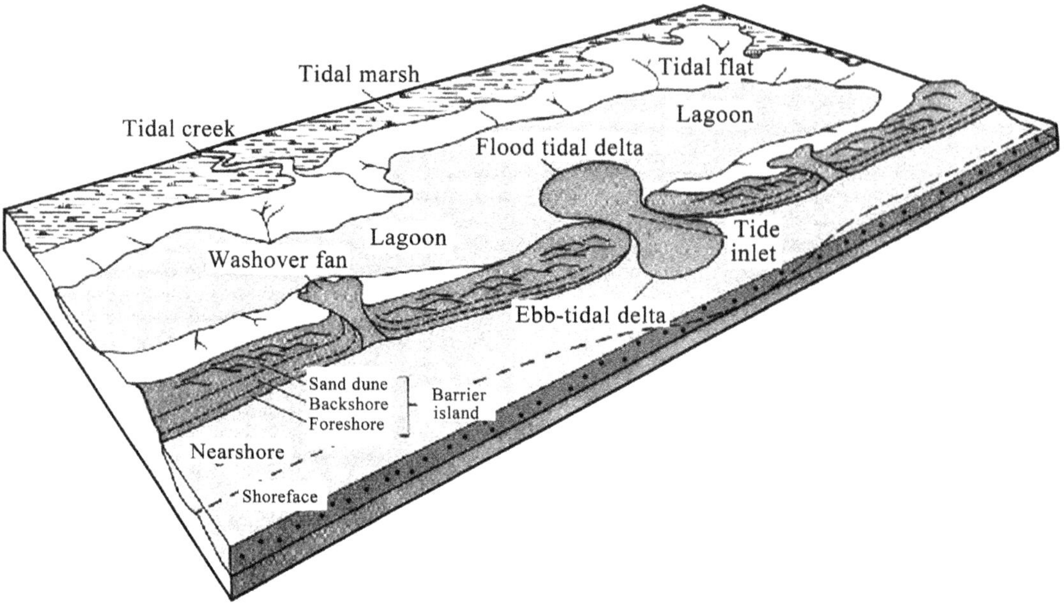

Fig. 12.16 The main force of the barrier coast is tidal action, and the combination of the sedimentary facies are mainly tidal flat, tidal channel, barrier island lagoon, and tidal delta

hydrodynamic energy, and abnormal seawater salinity. The sea water can be salinized or desalinated, which is related to whether the climate is arid or humid, and the injection of continental freshwater or supply of freshwater. As a consequence, it is unique in terms of the properties of its deposits, type of sedimentary structure, and biological characteristics.

The main agent of barrier coast zone formation is the tide effect. The sedimentary association of a barrier coast zone mainly includes tidal flats, tidal channels, lagoons, sabkha salt marsh, barrier islands, tidal deltas, and washover fans. Areas with tidal action can be divided into subtidal zone, intertidal zone, and supratidal zone. When the subtidal zone is situated below the low water level, deposits are always covered with water. Owing to the influence of various tidal currents, subaqueous sediments are affected by both rising and falling tides. The intertidal zone is located in the region between the rising and falling tide levels. The lower part of the intertidal zone is greatly influenced by the rise and fall of tides. The supratidal zone is affected by extraordinary tidewater above the average high tide line.

On the side facing the open sea, the barrier is mainly affected by wave action, similarly to the sea beach or non-barrier coast, and the other landward side is dominated by tidal action behind the barrier. Therefore, whether the sedimentary association is a barrier coast-tidal flat lagoon system or non-barrier coast of an open sea-epeiric sea system can be distinguished by an important indicator, that is, whether the sedimentary process is dominated by waves or tides. According to statistics, 13% of modern coasts are barrier coasts.

12.4.1 Tidal Flat Environment

Tidal flat refers to a coast zone with periodic tidal action and an extremely gentle gradient. The supratidal zone above the high tide line is dominated by fine-grained muddy deposits (muddy flat) in a low-energy environment; the high-energy environment below the low tide line is

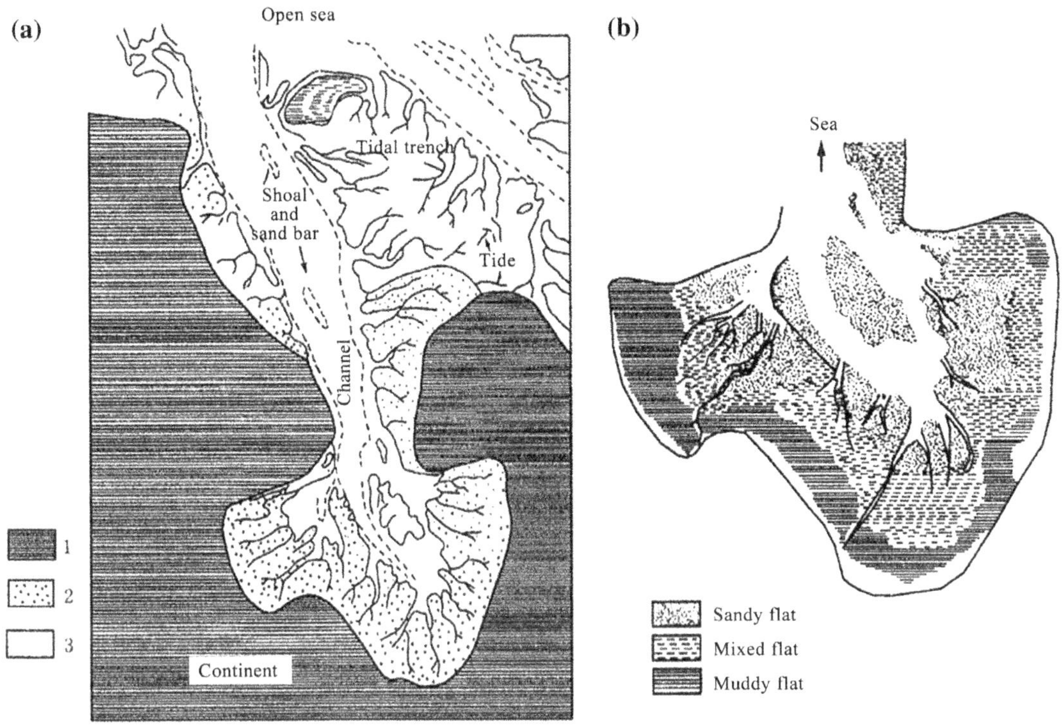

Fig. 12.17 The microtopography of tidal flat and the distribution of sediment grain size on the plane in the Yade bay, Germany. 1—Upratidal zone; 2—Intertidal zone; and 3—Subtidal zone

dominated by sandy deposits (sandy flat) with bed load; the moderate energy zone between the two is dominated by sand and mud mixed deposits (mixed flat) (Fig. 12.17). Of these, tidal channels and tidal trenches on the tidal flat have the highest energy. The hydrodynamic conditions of tidal environments are dominated by tidal action. As mentioned above, four apparent distinct characteristics of tides determine the sedimentary characteristics of tidal flat environments: global distribution in all coast zones, reciprocating current due to bidirectionality, energy difference produced by asymmetry, and obvious zoning characteristics as a result of periodicity.

12.4.1.1 Tidal Flat Development Mechanism

The development of tidal flats is dominated by tides, accompanied by waves. The strength of climate and hydrodynamic conditions, as well as organism and provenance supply significantly influence tidal flat development. Tidal flat deposits are dominated by fine grains, with particle diameter changing from fine to coarse (in contrast to sea beach grading dominated by wave action) from land to sea. This is caused by hydrodynamic energy distribution and scouring and silting mechanisms. When wave action is strong near the low tidal line, action time is long; dynamic energy decreases gradually in the landward direction, hence particle size diminishes gradually. Tidal flat erosion and siltation, which take place when the flow velocity of the rising tide is greater than that of the falling tide, occur repeatedly. Therefore, the mechanical sedimentary differentiation of fining of tidal flat particle diameter shoreward is very significant.

Tidal flat development is affected by the magnitude of hydrodynamic force. In a bay area, a sea beach is developed when the hydrodynamic force is strong, and a tidal flat is developed when it is weak. In Putuo Island in China, a striking contrast is presented between the east and the west coasts.

Climate significantly influences tidal flat deposits. In frigid zones, peat is formed by bryophyte in the form of autochthonous deposition; in tropical zones, flourishing organisms, algae development, and strong photosynthesis are favorable for the formation of carbonate rock, and algae swamps are mainly distributed on the tidal flat. Dry hot climate can promote the formation of evaporite; in humid climate, vegetation is usually developed to form salt swamps, and mangrove forests are mainly distributed in the intertidal zone.

12.4.1.2 Sedimentary Characteristics of Tidal Flats

1. **Sediment distribution**

The sediments in muddy tidal flats are mainly clay, silt, and sand; coarser sediments of gravel size are rarely seen. The sediments may also contain mud gravel and biological shell clastics. Especially in the high tidal channel, mainly coarse clastic sediments are deposited with very small amounts of suspension transport mud. Fine-grained materials brought by the flood tide are deposited mainly in regions usually submerged by sea water in the supratidal zone, mostly during the high-tide slack water period, which results in the formation of argillaceous sediments (mud flat) with the occurrence of silty laminae in the case of high tides. In humid climate zones, swamp peat deposits can be developed in the mud flat, while evaporite sediments such as gypsum and rock salt are developed under arid climate conditions.

The intertidal zone is the main sedimentary zone of the tide. Here the tidal range is generally 2–3 m and sometimes even 10–15 m. Because the terrain is flat, the intertidal zone can be very wide. The main sediments of the intertidal flat are mud, silt, and sand, and occasionally there exist shell clastics and sandy lens sedimentation on the eroded surface. Mud and sand or silt are in an interbedded shape or shown as muddy laminar sand (silt) or sandy (silty) laminar mud, which reflect the changes between normal and storm tides.

The main sediments in the subtidal zone at or below the low tidal line are sand, and these sediments are affected by tide and wave action. Because the sediments are affected frequently by the winnowing effect of wave action, mainly sand is deposited, forming sandy flats.

Washover fans can be developed in tidal flats toward the direction of land in barrier land, which is a type of fan deposit because the wave washes over the barrier during huge storm wave periods. Thus, the particle size in washover fans can be very coarse, even gravel size, and this is determined mainly by barrier island genesis; the sedimentary type is inclined to the sea in the early stage. Swamps can often be developed in tidal flats around lagoons. If the climate is arid, salt marsh, salt flat, or Sabkha salt marsh can be developed due to strong evaporation.

2. **Structural features**

The biggest characteristic of tidal flat sediments is that the zonation of grain size on the plane is obvious, i.e., the closer to land, the finer is the grain size; and the closer to sea, the coarser is the grain size. That is to say, the average grain size on a tidal flat gradually changes from medium-fine sand to fine sand and clay from the subtidal zone to the supratidal zone. The grain size distribution of sandy flat sediments is very characteristic. Its probability cumulative curve is dominated by saltation and suspension population, with little rolling population existing in individual samples. There are mutation relationships between the two, without the occurrence of the mixing phenomenon. The particle size distribution in the tidal creek and of the tidal channel sediments is similar to that of fluvial deposits. Overall, saltation is well sorted, whereas suspension is bad. There mostly exists an abrupt relationship, and gradual change between them is partial. The mixedness is about 1Φ.

3. **Sedimentary structural characteristics**

There are various types of beddings; however, different types of sedimentary structures are developed in different parts of the tidal flat due to different hydrodynamic conditions. Megascopic wave cross bedding is formed in the tidal channel or in the sediments of the shoal in the subtidal zone because of strong currents and wave energy, while herringbone cross bedding and reactivation surface structures are formed owing to the bidirectional current effect. Miniature current ripple cross bedding and a minority of parallel beddings and herringbone cross beddings can be seen in sandy flat sedimentation. Among them, the main characteristic of cross bedding is its bidirectional feature. Therefore, trough cross bedding is usually not developed, which can be seen only occasionally in tidal channels. The characteristic of mixed flat is that it has flaser, wavy, and lenticular bedding; horizontal bedding with sand-mud interbed; and other complex bedding structures.

12.4.1.3 Sedimentary Sequence and Model

In general, the tidal flat sequence can be developed as a regressive prograding sequence, but it can also be developed as a transgressive retrograding sequence. The former is more common in ancient tidal flat sedimentation. It is similar to the rhythm structure of fluvial sedimentation in its vertical direction, but its sedimentary structure and lithological characteristics are clearly different from those of rivers (Fig. 12.18).

Klein (1970) proposed a type of special B-C sequence while studying the tidal deposits of the fine-grained quartz sandstone layer under the Precambrian system Latin series in Islay, Scotland (Fig. 12.19). Section A is the massive sandstone of subtidal sedimentation, and sections B and C are low tidal flat sedimentations.

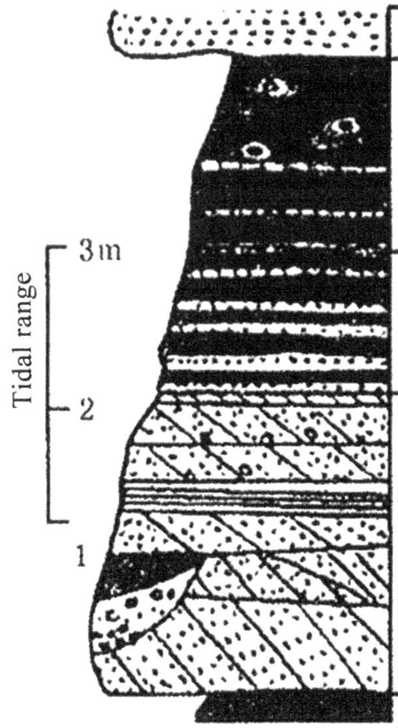

Lithology	Sedimentary structure	Interpretation
Reddish brown mudstone	Nodule	Supratidal flat
Reddish brown and brown mudstone	Horizontal and wavy siltstone laminae	High tide mud flat
Mudstone and quartz sandstone interbed	Arid cracks, cross laminae, flaser, lenticular, and wavy bedding	Middle tidal flat
Quartz sandstone	Parallel bedding, flow cinch mark, ripple and cross bedding, chevron structure, and reactivation surface	Low tidal flat
	Megascopic cross bedding, massive sandstone, tidal channel, chevron structure, and reactivation surface structure	Shallow subtidal zone

Fig. 12.18 Tidal flat deposition can develop into a regressive progradational sequence, and also develop into transgressive retrogradational sequence type, but the former is more common

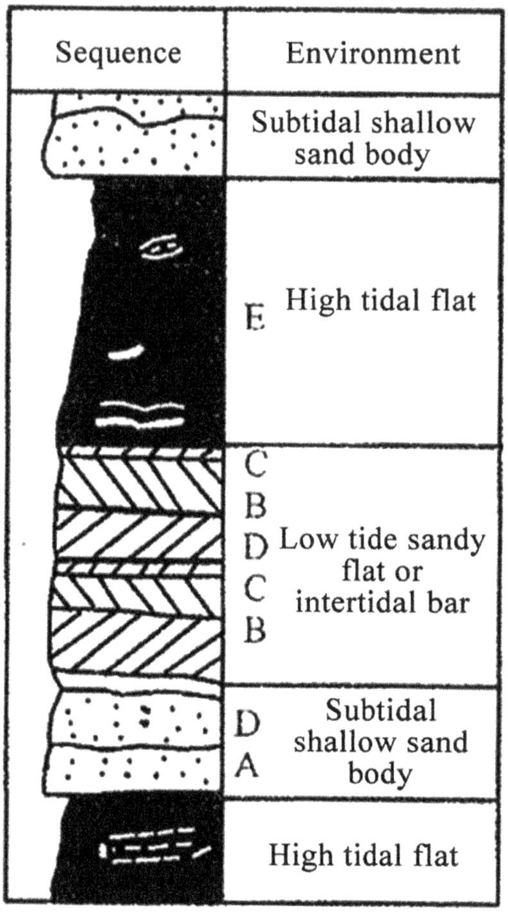

Fig. 12.19 During the study of tidal flat sedimentary sequence of fine quartz sand in lower Latin of Precambrian of Ailai, Scotland, Klein purposed a special B-C sequence (after Klein 1970)

Section B is formed by the migration of big ripples in high water levels with the development of medium-large planar or herringbone cross bedding. Section C is formed when the water level of the ebb tide falls and sediments are above the water. Current ripples can be formed when ravine water in the big ripple flows along with the surface downslope. Cross bedding sandstone can be formed along with the migration of such small ripples. Section C covers Section B, and their laminar dipping is vertical or oblique crossing. This combination relationship is called the B-C sequence.

According to the main sediment types of tidal flats, their sedimentary models can be divided into clastic (siliceous) type, calcareous type (carbonate type), and storm type (Figs. 12.20, 12.21, and 12.22, respectively). They are restricted by climate conditions and energy, that is, different sedimentary sequence characteristics can occur under different climate conditions (Fig. 12.23): ① the energy of a siliceous clastic tidal flat is relatively moderate, and its sedimentary structure is well developed. ② Carbonate tidal flats are usually formed under warm and wet conditions. Many megascopic tidal channels are developed and natural levees are developed in the flank of the channel when the energy is especially high. ③ Storm tidal flats, also called arid evaporative tidal flats, are formed in warm and arid low-energy environments, in which algae and evaporite development are relatively better. Sabkha salt marshes can be developed between the intertidal zone and shoreline, in which a bay with high salinity, considerable amount of gypsum, and hard rock are formed. Only storms can alter its sediments.

12.4.2 Barrier–Lagoon Environment

Because the barrier island–lagoon system (also called barrier–lagoon system) is separated from the open sea by a long and narrow sandy body—barrier island or barrier bar with distribution parallel to the shoreline—only the outlet can interlink with the open sea. This restricts the free flow into and circulation of water with seawater, resulting in isolation or half-isolation (Fig. 12.24). Therefore, lagoon formation and development is mainly controlled by barrier islands (or bars), and they constitute a sedimentary system of symbiotic origin. This system comprises many genetically related sub-environments, and the sedimentary facies characteristics of each sub-environment are different. They can be divided into three sedimentary facies associations: ① Barrier island-beach sedimentary facies association, which is a long and narrow sand body stretching parallel to the shoreline on the seaward side of a

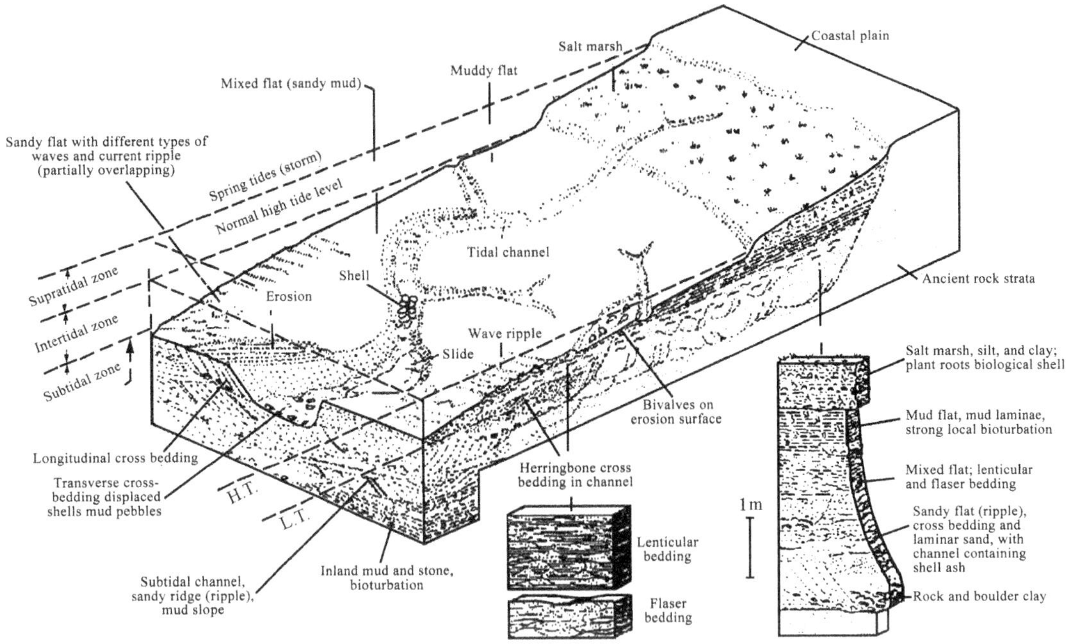

Fig. 12.20 Model map of the Siliciclastic tidal flat (after Klein 1970)

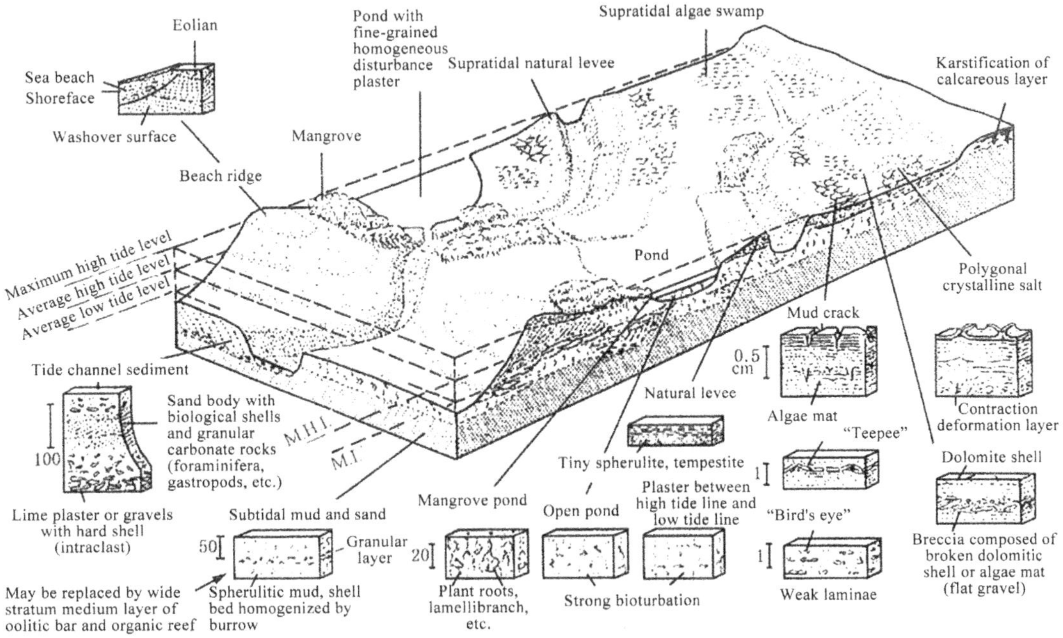

Fig. 12.21 Model map of the Carbonate tidal flat (after Ginsburg and Hardie 1975; Shinn 1983; Sellwood 1986)

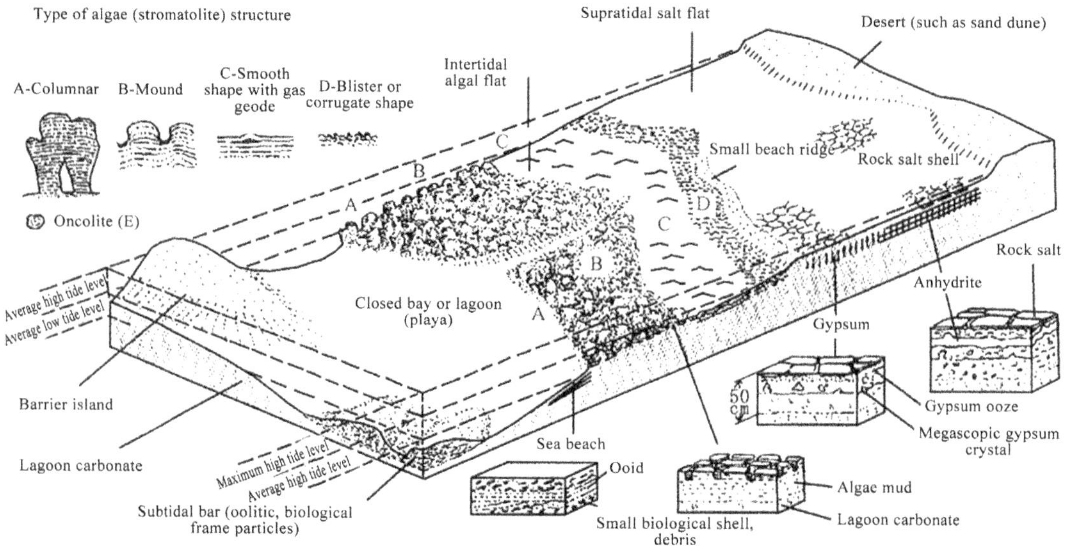

Fig. 12.22 Sedimentary model map of the storm type (or dry, steaming) tidal flat (after Purser 1973; Schwarz et al. 1975; Butler et al. 1982)

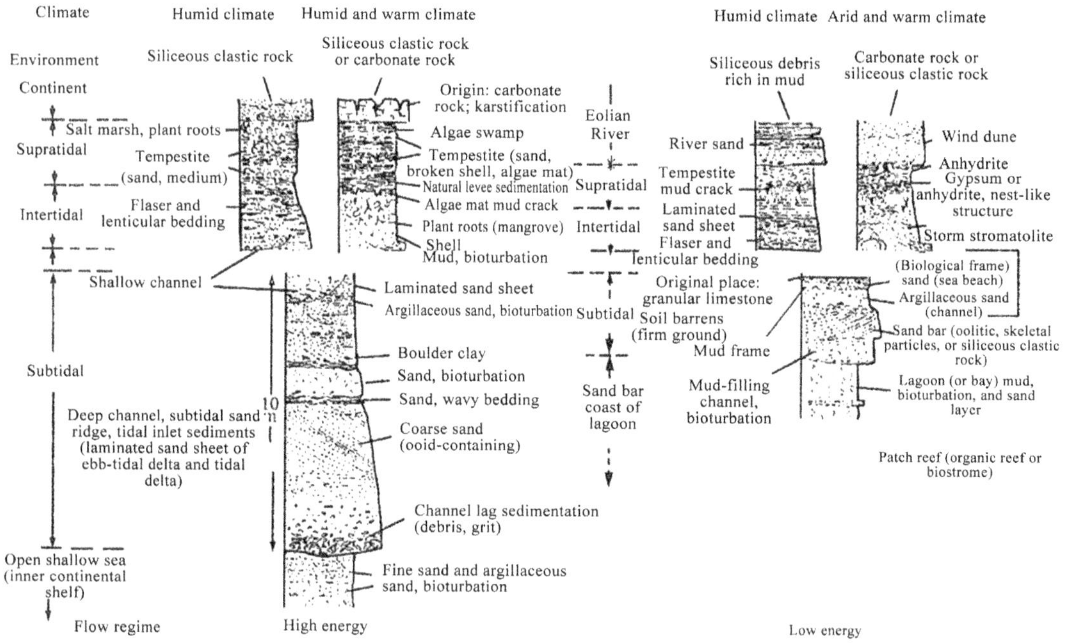

Fig. 12.23 Sedimentary sequence characteristics of tidal flat under different climatic and energy conditions (after Ginsburg 1975; Shinn 1983; Sellwood 1986)

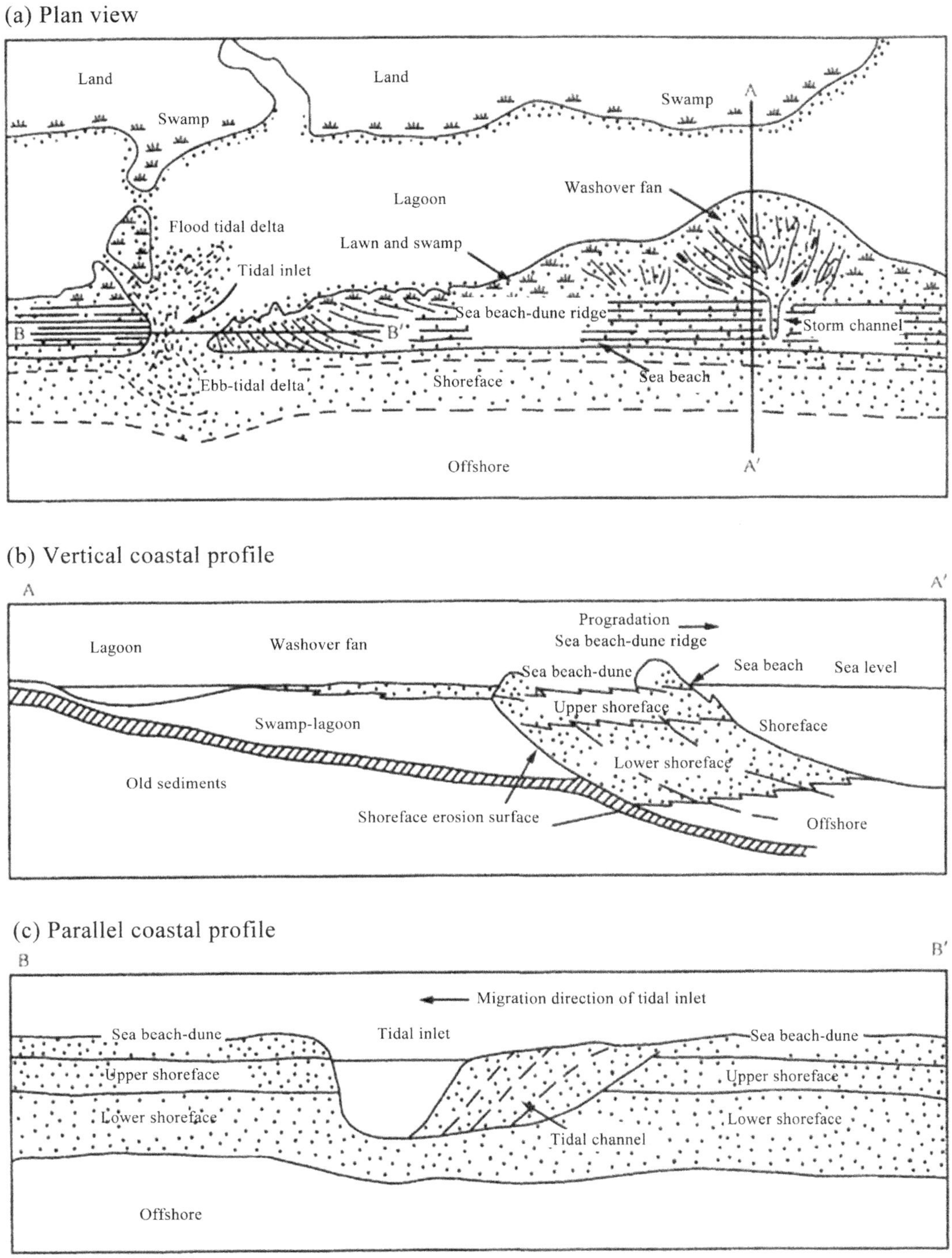

Fig. 12.24 Sedimentary environment of the terrigenous clastic barrier coast is mainly the barrier island-lagoon system

Fig. 12.25 Morphology of coastal barrier islands in Florida, USA

barrier island. Similarly to a non-barrier coast, it is mainly controlled by waves and longshore currents. ② Tidal channel-tidal delta sedimentary facies association, which is a sand body formed when the strike developed on the landward side of the barrier island is vertical to or obliquely intersected with the coast. It gradually stretches into the lagoon toward the land direction, which is mainly dominated by tidal action. ③ Barrier–lagoon sedimentary association, which is surrounded by tidal sedimentation after being distributed in the barrier island, is mutually superimposed with washover fans, tidal deltas, and other sand bodies. Similarly, it is controlled by the tide and dominated by suspended sedimentation. Since the former two associations are the same as those of non-barrier coasts and tidal environments respectively, it is not necessary to repeat them.

12.4.2.1 Characteristics and Formation Mechanism of Barrier Islands

1. **Characteristics of barrier islands**

Barrier islands and barrier bars block lagoons: barrier islands (bars) are distributed parallel to the coastal line in the near-shore areas, which can be straight, slightly bent, or with weak branches. Typically, multiple barrier sand bodies can be connected together to form a row of barrier islands (bars). Two or more rows of barrier islands (bars) roughly parallel to each other can be developed in some areas. In general, the thickness of a barrier sand body is 10–20 m, with its width ranging from hundreds to thousands of meters, and the length from a few meters to more than a dozen kilometers (Fig. 12.25). The height of such a body depends on wave height: the greater the wave height, the higher is the barrier island. Moreover, the width is relative to the time and direction of wave action: the longer the time, the wider is the barrier island.

2. **Genesis of barrier islands**

Barrier island genesis is of four types: ① Change of beach: Originally, as sand dunes on a continental beach, it is separated gradually from the continent with the occurrence of transgression. In the late stages of transgression, in which the regions behind the barrier island are changed into a lagoon or a tidal flat, while the barrier island continues to grow upward along with the rise of sea level. ② On the basis of longshore bar and spit: A longshore bar is generally under the sea level, except in the period of the lowest tide, and it grows upward gradually under the effects of waves and currents along with a continuous supply of sediments. However, longshore bars will be above the sea level and be eroded partially when the Earth's crust rises and the sea level falls. The original longshore bar will be developed into the barrier island little by little and the residual sea water behind it will form a lagoon when the sea level continually falls. ③ By the abandonment of deltas in early times owing to the falling of sea level and transformation of waves. ④ By bioherms: When a bioherm grows in littoral-neritic sea, particularly to a certain scale, the fall of sea level results in the barrier island being above the water, resulting in the barrier island formation. ⑤ As local highland caused by tectonism or volcanism at the sea bottom.

Regardless of the type of genesis, there is a common effect, that is, a barrier islands separate the sea from the continent, forming a lagoon environment that is occasionally linked with the sea, but with poor circulation. It should be pointed out that the lithology and grain size of a barrier island determine the components and grain size of its surrounding sediments. Generally, the grain size of a washover fan is coarser, reaching up to gravel size.

12.4.2.2 Barrier–Lagoon Sedimentary Association

1. **Barrier islands**

Sediments of barrier islands are the products formed by redistribution after long-term scouring and winnowing of seawater. They are dominated by pure sandy sediments with good sorting, accompanied by coarse-grained clastic gravels and biological shells. A barrier island is usually divided into three parts: island beach, barrier flat, and wind dune. ① Island beach is the narrow and long terrain at the seaward side of the barrier, that is, beach sand of the open sea. ② Barrier flats, which are opposite to the island beach, are the wide and flat slope zones on the landward side, and it gradually transits to the lagoon. Its sediments, which are dominated by fine silt and sand, are finer than those of the island beach with relatively poor sorting. Relatively coarse debris can only be seen in the washover fan and the tidal inlet. Wave cross bedding and complex bedding are developed because it is located in the intertidal zone. ③ Wind dunes, located in the center of a barrier island, are formed when beach sand is transformed by the wind. Therefore, their sedimentary characteristics are the same as those of coastal dunes.

Based on the genesis of barrier islands, they can also be divided into transgressive and regressive types (stable type), as well as aggradation type (Fig. 12.26). The transgressive type comprises narrow striped sand bodies with broad storm slopes covered by plants, gently located on the lagoon side. The regressive type generally has wide beaches, and the aggradation type is transformed mainly by storms without substantial transgression and regression changes. Therefore, the three types correspond to island beaches, barrier flats, and wave dunes.

2. **Lagoon**

Lagoons, which are semi-closed water areas formed when water is blocked by barrier islands, are shallow low-energy environments with weak wave action and obvious tidal action. Sediments include fine sand, silt, mudstone, peat layers, etc. Lagoons can be divided into salinization lagoons and desalination lagoons. Salinization lagoons are usually developed in arid climate conditions, where evaporation is greater than the amount of fresh water injection. Biological species are simple, mostly bivalves, gastropods, and shell-shaped organisms with broad salinity. Various salt sedimentations can be formed with very single sedimentary structures, such as gypsum and rock salt. However, desalination lagoons are formed under humid climate conditions with more fresh water injection than evaporation. Organism species are simple and fewer, have morphological abnormalities, small bodies, and thin shells. The grain size of sediments does not differ significantly from that of salinization lagoons. Iron-manganese nodules, siderite

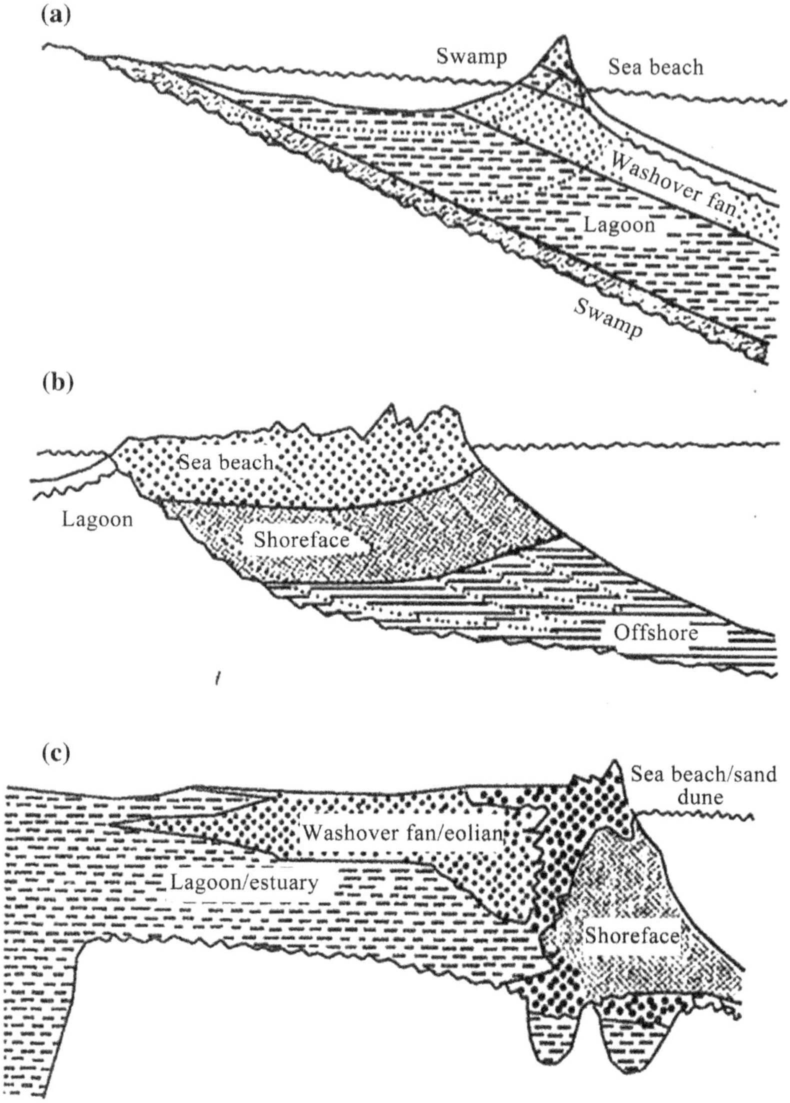

Fig. 12.26 Differences of barrier island with the changes of sea level (after Reinecke and Singh 1979). **a** Transgressive type (according to Reinecke and Singh 1979); **b** Regressive type; and **c** Aggradation type (according to Fisk 1960)

nodules, siliceous minerals, pyrite, and chamosite can be seen, and mainly, horizontal bedding is developed.

3. **Tidal channel**

Tidal channel refers to a tidal inlet cutting through a barrier island. It links the open sea with a lagoon. Tide water floods into the lagoon during the flood tide period, stays for a while during the slack tide period, and is then discharged from the tidal inlet during the ebb tide period. The tidal inlet is not well developed in microtidal range areas. The barrier island can extend dozens to hundreds of kilometers without a tidal inlet. The tidal inlet is mostly developed in mesotidal range areas, where the barrier island is separated into many parts. The width of the tidal inlet usually can reach from hundreds to thousands of meters. Its depth can reach 10–20 m, and its length is limited by the width of the barrier island. The tidal inlet mostly continually migrates downstream under the influence of longshore currents. The sediments in the tidal inlet are saved when the tidal inlet cuts below the sea level. If migration speed is quick and stable,

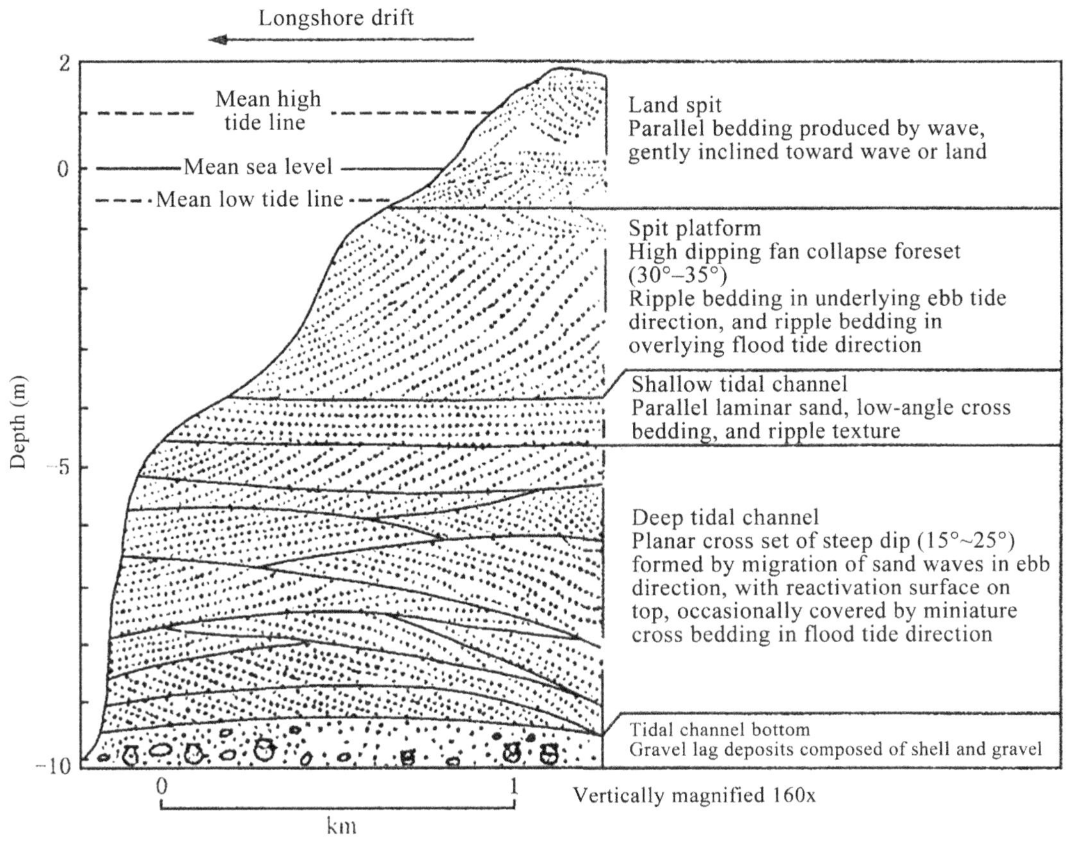

Fig. 12.27 Vertical sedimentary sequence of tidal inlet in Phil Island and Long Island, New York (after Kumar and Sanders 1974)

the entire barrier island can be constituted of tidal inlet deposition.

The sedimentary sequence of filling sedimentation in the tidal inlet is as follows: there is an erosion surface at the bottom, above which lag deposits composed of shell and gravel are formed. Above the lag deposits are deeper tidal channel sedimentations fining upward, having herringbone megascopic cross bedding, and medium trough cross bedding. Above the deeper tidal channel sedimentations, are shallow tidal channel sedimentations with medium and fine sand layers constituted of bidirectional small-medium trough cross bedding and current ripple bedding. The grain size decreases gradually and the bedding becomes thinner in an ascending order, which represents a fining-upward sequence (Fig. 12.27). Deep tidal channel sedimentation is mainly affected by the reciprocating flow of flood and ebb currents. The shallow tidal channel is usually above the wave base and is affected by both tidal current and wave action. Salt and brackish water mixed biocenosis is often contained in tidal inlet deposits.

4. **Tidal delta**

An ebb tidal delta is formed on the seaward side of the tidal inlet due to the impact of ebb current, while a flood tidal delta is formed on the landward side due to the impact of flood current. The ebb tidal delta sand body is covered above the nearshore zone deposits, but it is mostly transformed by longshore currents and waves, forming bed forms with changeable directions.

In the ebb tidal channel and the central lobe of an ebb delta, tidal action is dominated by the development of planar cross bedding, whereas the limbs of the lobe mostly show cross bedding in multiple or different directions owing to the strong transformation of waves. Flood tidal delta deposition coexists with lagoon deposition, the complete sequence of which is maintained because of the low-energy environment and diminished wave transformation. The flood tidal delta is dominated by megascopic planar and trough cross beddings mainly landward, sandwiched by sandy layers with bimodal cross bedding formed in ebb tides. Its sequence features thinning and fining upward.

The sedimentary structure of ebb and flood tidal deltas is particularly similar, but the most important difference is that ebb delta deposition coexists with shoreface zone deposition, and its ripple and cross bedding sets are multidirectional. However, the bidirectional tidal bedding of flood tidal deltas is obvious and it coexists with lagoon facies. In addition, ebb deltas are represented mostly as coarsening upward in the vertical sequence, namely, thicker and coarser in ascending order.

12.4.2.3 Sedimentary Sequence and Model

The profile structure of the barrier island-lagoon system has different changes as a result of the influences of sea level fluctuation, subsidence velocity of sedimentary basin, and sediment supply rate. The entire facies belt moves toward the sea, forming a transgressive or regressive barrier island-lagoon sedimentary sequence (Fig. 12.28a) or, sometimes, a sedimentary sequence in which lagoon facies overlap barrier-beach facies, when the sediment supply is continuous and sufficient, the sea level is stable, and the sedimentary basin has small-medium subsidence velocity. The entire facies belt migrates toward the land to form a transgressive sedimentary sequence that is more complex than the regressive type, as sediment supply decreases and sea level rise increases (Fig. 12.28b).

The profile structure is different in different parts of the barrier island-lagoon sedimentary system. For example, the strata model of a barrier-tidal inlet is usually represented as the fining-upward sedimentary sequence. The cross bedding thickness, also thins upward, and there is an erosion surface at the bottom of the profile. From the bottom up, the successive order is deep tidal channel, shallow tidal channel, and spit sedimentary sequence.

12.5 Identification Marks of Terrigenous Clastic Coasts

12.5.1 Sedimentary Marks of Terrigenous Clastic Coasts

In combination with the sedimentary characteristics of modern and ancient coasts, the main identification marks of terrigenous clastic coasts are as follows.

12.5.1.1 Lithology

In general, coastal deposits contain pure sand, high content of stable components such as quartz, relative enrichment of heavy minerals, and high compositional maturity and texture maturity, but there are great differences between different terrains and slopes.

Various marine biological shells and their debris in different quantities are often contained in coastal deposits, sometimes with laminar shell layers in the shoreline area, which are mostly biological assemblages formed by organisms in different ecological environments. Biological shells are usually broken, worn, and rounded. In addition, the biological trace fossils are very rich, and the energy is different in different subfacies belts. There are large differences in their size, intensity, and occurrence. Under normal circumstances, the burrow body is larger, density is higher, and their occurrence is more upright when the energy is higher.

12.5.1.2 Grain Size Characteristics

The grain size distribution of coastal sand is uniform. The probability graph shows that the saltation population is developed with a large

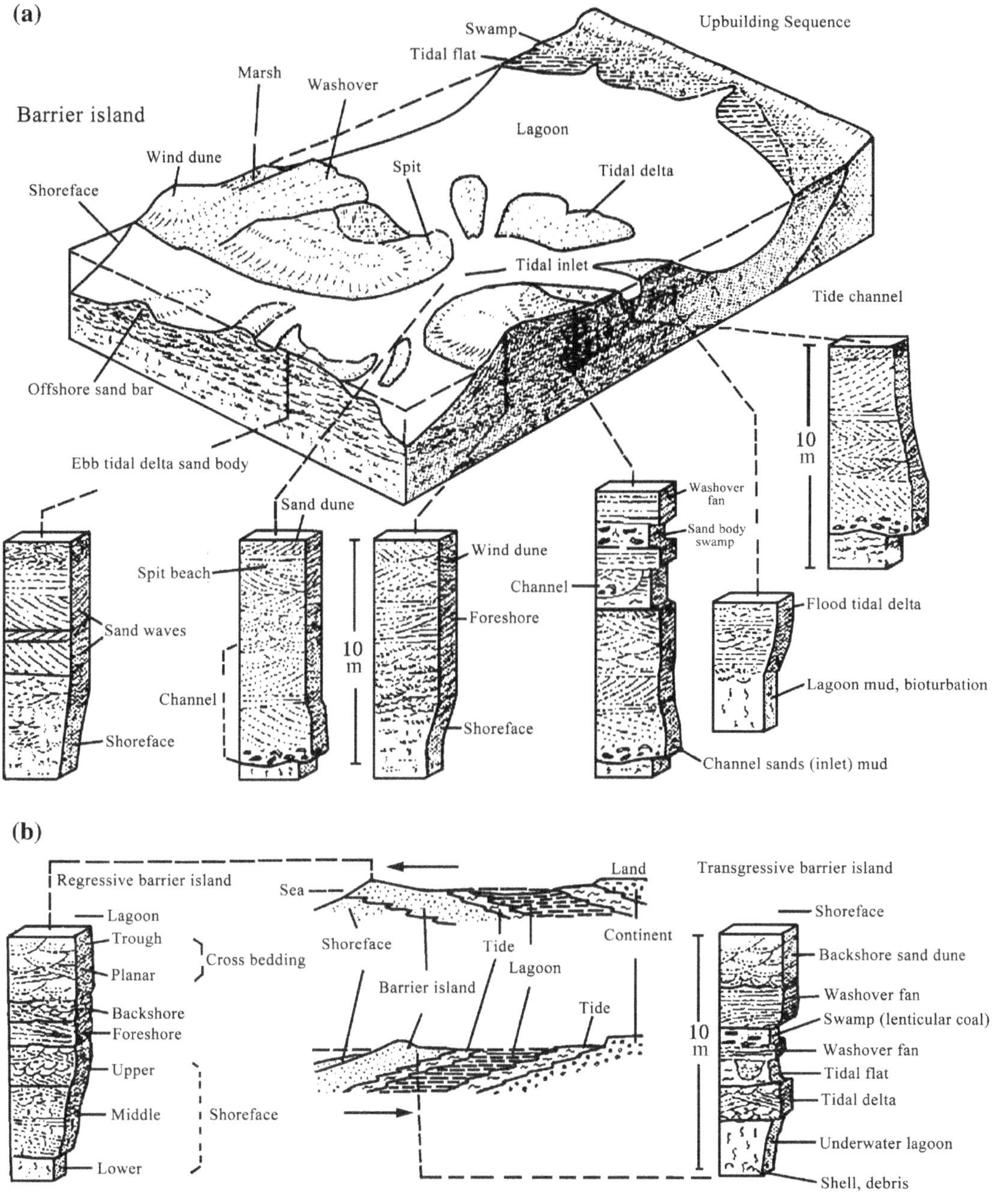

Fig. 12.28 Sedimentary model, vertical sequence of each geomorphic unit and the characteristics of the transgressive and regressive barriers of the barrier-lake system (after Reinson 1984)

slope and good sorting, and two secondary populations exist due to the mutual effect of bidirectional (multidirectional) currents, such as the scour and rip current effect of waves, and the difference between flood and ebb tides. Notably, the gradient and form of coastal terrain directly affect the strength of waves and tidal action. Therefore, the factors mentioned above determine the grain size and compositional maturity of sediments, which are generally characterized

Table 12.4 Sedimentary characteristics and sedimentary facies division of the Zhujiang Formation shoreline in the (Eastern) Pearl River Mouth Basin

Facies	Subfacies	Microfacies	Vertical sequence characteristics	Sedimentary characteristics
Non-barrier coastal facies	Foreshore	Beach sand, beach sand channel	Superimposed gradual fining upward	Coarse grain size, dominated by medium-coarse sandstone, occasionally with gravel; high texture maturity, with swash bedding; lack of trough cross bedding; no mudstone
	Shoreface	Longshore bar, upper shoreface	Coarsening upward dominance, fining upward subordination, mud-in-sand as main characteristic; sandstone $\gg$ mudstone	Larger grain size range, dominated by middle-fine sandstone, medium-good texture maturity
		Lower shoreface	No rhythm, sand < mud	Silty mudstone deposition
	Offshore	Offshore deposit	Large set of mudstone mixed with siltstone	Mudstone dominance, pure quality with dark color, no burrow development
Barrier coastal facies	Barrier bar	Barrier flat	Dominated by coarsening upward, sandstone > mudstone	Medium-fine sand dominance, medium maturity, with biological debris, dense lithology
	Tide channel	Tidal channel	Gradual fining upward, featuring mud-in-sand, sandstone $\gg$ mudstone	Dominated by middle-fine sandstone, medium maturity, often with herringbone cross bedding
	Tidal flat	Mixed flat	No rhythm, sandstone $\approx$ mudstone	Muddy siltstone or silty mudstone dominance, with obvious complex bedding, containing siderite nodule and mica
	Lagoon	Lagoon, tide Delta	No rhythm	Dominated by dark gray laminated silty mudstone, obvious bioturbation, with plant debris, local parts mixed with laminar coarsening upward sandstone

by gravel on steep shore, sand on flat shore, and mud in the bay.

12.5.1.3 Sedimentary Structure Characteristic

Planar cross bedding in the nearshore zone and cross bedding in different directions are rather developed; trough cross bedding is generally not developed, and horizontal bedding and biological burrows can be seen at the bottom of the nearshore. Megascopic beach swash cross bedding is developed in the foreshore zone. Current or parting lineation can be seen along the bedding plane; in addition, various small wavy ripples or waves, rhomboid ripple marks, rill marks, and other bedding plane structures can be developed. Of these structures, megascopic swash cross bedding is the typical indicator of coastal sedimentation.

12.5.1.4 Vertical Sedimentary Sequence

The transgressive sedimentary sequence is mostly developed with the coarsening-upward characteristic and a thinner lower part; offshore-nearshore-foreshore-backshore occur successively from the bottom to the top (Table 12.4).

12.5.1.5 Sand Body Morphology

Coastal sand bodies usually show a linear or striped distribution and appear in rows. Their

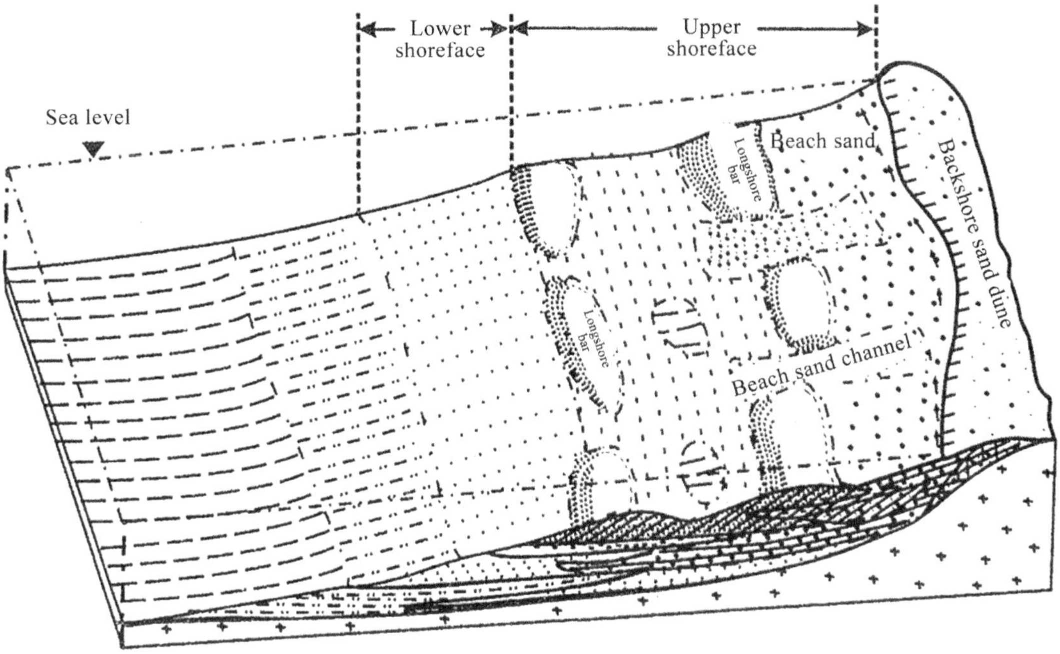

Fig. 12.29 The most typical ancient barrier free marine deposits in China is the Lufeng 22-1 oil field in the Pearl River Mouth Basin

profile is often lens- or sheet-like with a flatter lower part and a convex upper part.

Because most of the oil and gas fields in China are continental deposits, there are only a few studies about coastal deposits in the Chinese literature. Here we take the author's research in the Pearl River Mouth Basin as an example to discuss their sedimentary characteristics (Table 12.4). A typical example of ancient non-barrier marine sedimentation is the Lufeng 22-1 Oilfield in the Pearl River Mouth Basin (Fig. 12.29).

12.5.2 Features of Electric Well-Logging Curves

12.5.2.1 Non-barrier Coast

This electric well-logging curve is given priority to its high-amplitude and smooth gradual thinning upward cylinder shape, sometimes bell or funnel shape; however, its thickness is often large, which is different from other sedimentary environments, whether in amplitude or the thickness of the sand body. It features a large set of sandstone mixed with thin mud, that is, its characteristics are similar to those of braided rivers. "Mud-in-sand" is one of its main characteristics, although it is different. Due to its good sorting, the serrature of the electric well-logging curve is not too obvious, and the mud content in sandstone is low.

12.5.2.2 Barrier Coast

This is clearly different from non-barrier coasts. Although its electric well-logging curve is dominated by a shell shape and can develop a funnel shape, its amplitude is relatively smaller, that is, it is closer to the baseline. Because the change in energy and mud content of sandstone is large, barrier coasts can present obvious serrature. Its sand-mud interbed is different from that of non-barrier coasts in the vertical direction; the entire structure can feature "sand-in-mud." Because ferrous manganese ore and siderite nodules, as well as a thin sand layer, can be developed in such an environment, the electric well-logging curve can be tongue shaped.

A finger-like electric well-logging curve can also be seen because of the characteristics of tidal flat deposition.

12.5.3 Seismic Feature

It is generally difficult to distinguish different shoreline (coast) types in a seismic cross section, which depends on the sedimentary characteristics of the shoreline environment. The change in sand bodies deposited on the shoreline is not as large as those deposited in a delta, in both the vertical and lateral directions. However, attribute extraction with an isochronal stratigraphic frame is conducted using high-resolution 3D seismic data. Hence, facies belt (subfacies) distribution characteristics are often reflected on the plane. Practically, it is mainly concluded based on geological and logging identification, as well as analysis of structure and sedimentary background, but the coastal seismic reflection texture usually shows a higher amplitude and better continuity. There are no obvious characteristics of the reflection texture and shape, which appears mostly as parallel or sub-parallel reflections.

12.6 Reservoir Features of Coastal Environments

The most typical regions with coastal deposits as reservoirs are Lufeng Oildom and Huizhou Oildom in the Pearl River Mouth Basin. As for their oil-bearing reservoirs, the former is mainly non-barrier coastal deposit, while the latter is mainly barrier coastal deposit. Typically, their holes and permeability are better (Table 12.5), particularly for the open sea environment without a barrier. However, this type of oilfield will have water channeling in fining-upward sandstone during its development because its interbed is not well developed, while the development of the sea bed should be considered and studied during the exploration process given the lack of mudstone. Such oilfields are formed frequently in the highstand systems tract of the third-order sequence. This type of reservoir forms oil and gas reservoirs, particularly for the sedimentary system of open sea, which often has high productive capacity and stored energy. The lateral continuity of their reservoirs is also very good.

The sedimentary system at the basin edge usually contains a potential sand body interbedded with shallow sea (lake) mud facies or mud facies affected by seawater. This sand body is generally parallel to the sedimentary strike (shoreline), most of the coastal zone systems are wedged out in the updip direction, and a stratigraphic trap may be formed. The best reservoir rock with high structural maturity is produced by the inherent extensive transformation in the coastal zone. Porosity and permeability may be decreased owing to the biological burrow effect and the generation of calcite cement because of the leaching and re-precipitation of shell materials.

12.6.1 Transgressive Barrier Sand Body

Transgressive barrier sand bodies often occur in a strip that is isolated, narrow, long, and parallel to the strike (shoreline) (Fig. 12.30a), which is the ideal target for oil and gas exploration. A thick, complex sand body is produced by the superimposition of transgressive sedimentary units owing to shore line swing. The sand body is generally accumulated in a paleotopographic gap or a structural slope break.

12.6.2 Sand-Rich Coastal Plain Sand Body

Prograding sand in sand-rich coastal plains forms widely distributed and excellent reservoirs (Fig. 12.30b), which show the trend of oil generation in structural traps. The best reservoirs are generally in sand bodies in the upper shoreface and beach deposits. The prograding sequence may appear in the lower shoreface, whose productive capacity may be affected in fully saturated reservoirs.

Table 12.5 Reservoir sedimentary features of each lithofacies association in the (eastern) sandstone oilfield shoreline system of the Pearl River Mouth Basin (according to Yu et al. 1997)

Sedimentary microfacies	Depositional modes	Prosodic features	Feature of electric well-logging curve	Geometry			Porosity (%)			Permeability ($\times 10^{-3}$ μm^2)		
				Planar	Longitudinal	Horizontal	Minimum	Maximum	Average value	Minimum	Maximum	Average value
Beach sand	Winnowing accretion	Superimposed fining upward or not obvious	Cylinder shape	Sheet	Planar, massive	Planar, massive	12.3	31.8	21.8	348	10,944	3408.06
Beach sand channel	Winnowing accretion Vertical accretion	Gradual fining upward	Semi-trapezoid	Strip	Flat-top concave-bottom lens	Flat-top concave-bottom lens	14.4	27.8	18.6	840	13,310	2686.82
Longshore bar	Winnowing accretion	Abrupt coarsening upward or complex rhythm	Low-amplitude funnel shape	Ellipticity	Wide and flat lens shape	Lens shape of concave-top and flat-bottom	15.1	28.2	21.4	33	15,250	3640.85
Upper shoreface	Winnowing accretion vertical accretion	Superimposed abrupt fining upward	Cylindrical	Sheet	Planar	Planar	14	30	24.2	11	6690	1382.8
Lower shoreface		No rhythm	Negative beak shape	Sheet	Strip	Strip	1.1	28	15.5	0.01	806	
Barrier flat		Coarsening upward	Serrated Funnel shape	Arc-like	Lens shape of concave-top and flat-bottom	Lens shape of concave-top and flat-bottom	11	21		<50		
Tidal channel	Aggradation and channel filling Winnowing accretion	Gradual fining upward	Low amplitude bell shape	Strip tree shape	Flat-top concave-bottom lens	Flat-top concave-bottom lens	10.1	18.9	15.7	7.8	870	118.07
Lagoon mud	Vertical accretion	No rhythm	Low amplitude Serrated shape	–	–	–	2	15.8	6.4	0.01	7.7	0.33
Mixed flat		No rhythm	Low amplitude Serrated shape	Sheet	Strip	Strip	1.3	9	5.8	0.01	0.07	0.04
Platform shoal	Winnowing accretion	No rhythm	Cylinder shape	Sheet	Planar	Planar		15.5	6.9	0.01	2	0.11

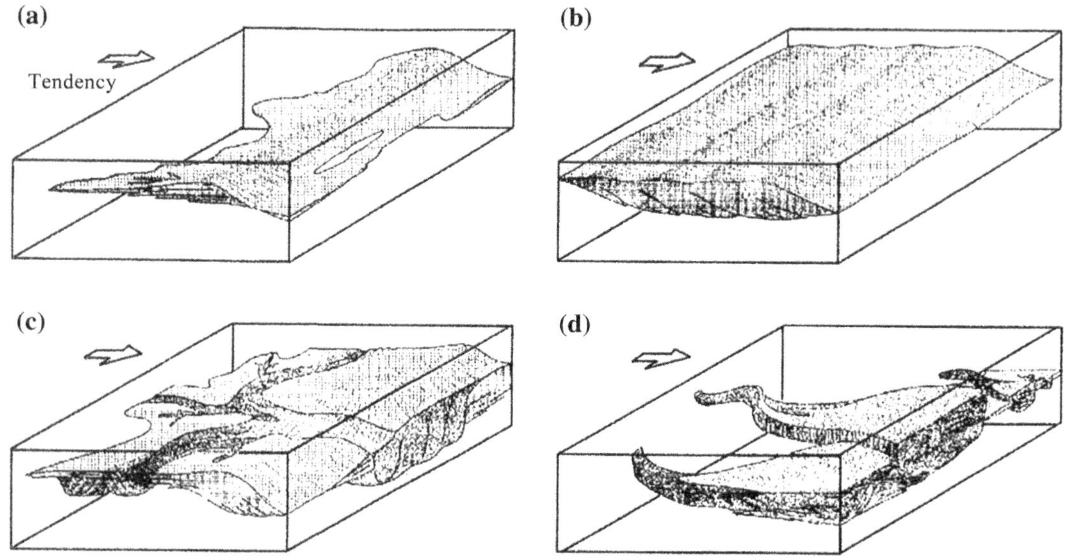

Fig. 12.30 Three dimensional geometric shapes, lateral relations and internal configurations of typical reservoir sand body in littoral zone system. **a** Coastal barrier bar; **b** prograding coastal plain rich in sand; **c** barrier bar and tidal inlet complex of small-medium tide transitional coast; and **d** estuary of strong tide coastal zone and subtidal sand bar complex

12.6.3 Barrier Bar Sand Body

Multiple barrier bars are shown as having a horizontal cross shape, and the resulting sand body is surrounded by muddy deposits, resulting in the formation of a narrow and long terrain. The tidal inlet and washover (fan) sand wedged out irregularly in the lagoon mudstone can usually form a series of potential stratigraphic traps (Fig. 12.31), which constitute the oil and gas channel parallel to the strike. Certain gas-containing intervals of the South China Sea Y13-1 Gas Field and the Taiyuan Formation in the Tabamiao Gas Field in the Ordos Basin (Fig. 12.32) have these typical reservoirs.

12.6.4 Tidal Channel Sand Body

This sand body is crossed by multiple tongue shapes with its tail toward the land, and it stretches along the direction vertical to the shoreline, forming an irregular and discontinuous sand belt. Large funnel-shaped sedimentary bodies are formed by filling materials from the estuary, which flow into the marine mud toward the direction of the sea and landward wedged out into the impermeable swamp, tidal flat, and gully sediments, forming ideal stratigraphic traps in the basin with a simple structure. In tide-dominated deltas, because distributary channels in the estuary are interconnected with tidal channels rich in sand, the sand body is not easily wedged out toward the land direction (Fig. 12.30c). The interior sand body in the tide-dominated coastal zone is particularly complex and highly separated and crosscut by supratidal mud and silt seal bed (Fig. 12.30d).

12.7 Neritic Deposition and Vertical Sequence Characteristics

As mentioned before, a neritic environment generally refers to the flat wide shallow water area from below the normal wave base to depths of 0–500 m. The hydrodynamic effect is complex and diverse, including ocean currents, waves, tidal currents, and density currents. Their comprehensive effects produce great changes in

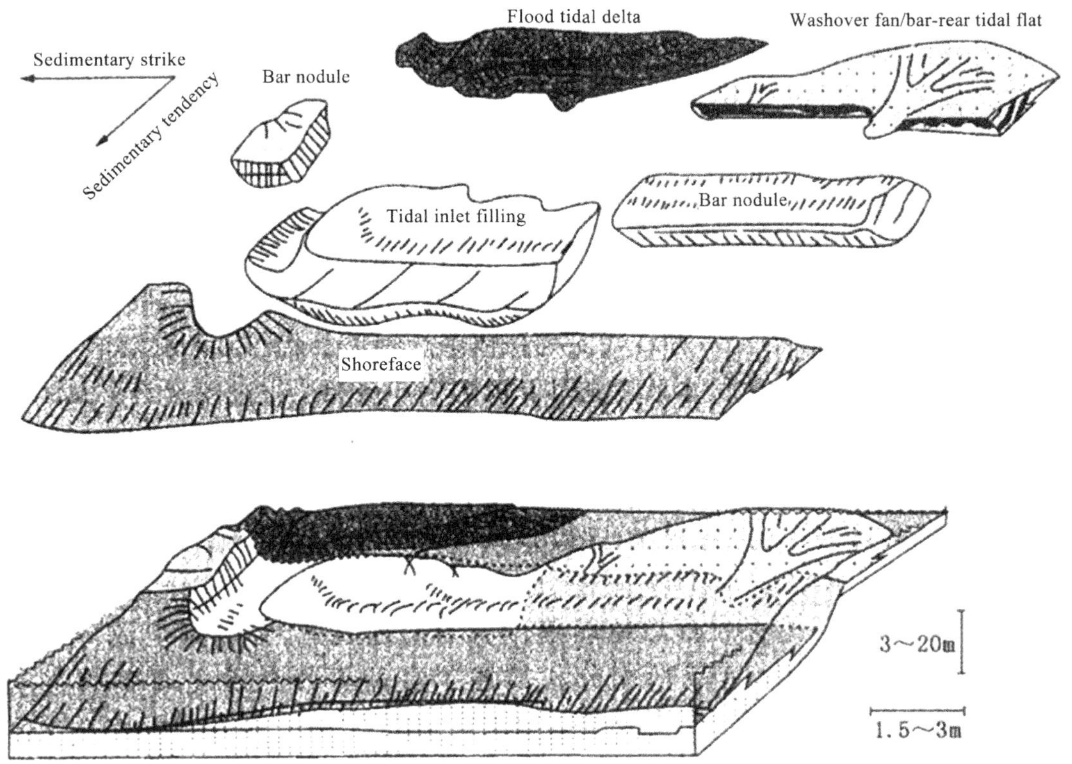

Fig. 12.31 Tidal inlet sand and overflow fan sand with irregular cusp in the lagoon shale can identify a series of potential stratigraphic traps

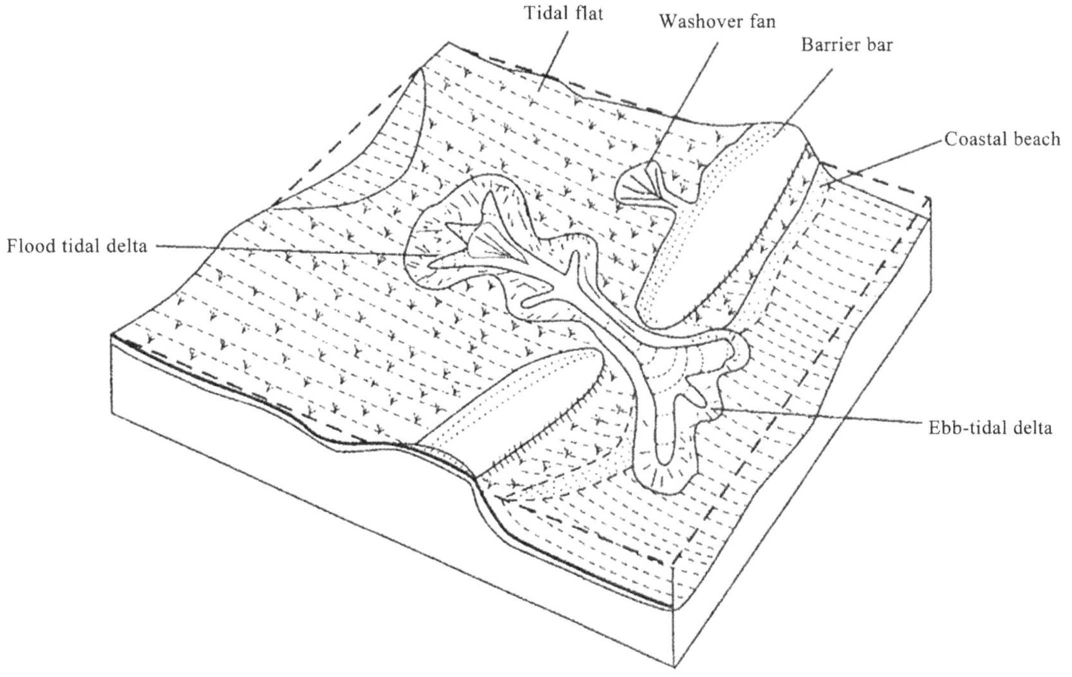

Fig. 12.32 This kind of reservoir, such as sand bodies of barrier bam occurs in the Taiyuan formation of Tabamiao gas field, Ordos Basin

the properties, strength, and current direction of the ocean current system in a neritic environment. However, in normal cases, the flow velocity under a neritic environment is relatively, which will not exert great influence on sediment surfaces. In case of a strong storm, strong waves will affect the bottom of the sea, thus transporting sediments for dozens of kilometers toward the sea in suspension status. Additionally, the flow velocity of tidal currents, density currents, and other meteorological ocean currents in the shelf of a narrow sea or a strait can exceed 150 cm/s. According to the Hjulström effect, this can cause erosion and transportation of sediments.

12.7.1 Transitional Zone Deposition

The transitional zone is the area between the coastal shoreface and the neritic sea. Its average gradient is only several minutes, and the depth of the water body is determined by coastal energy. The lower the coastal energy, the shallower is the transitional zone. Water depth in the transitional zone is 2–20 m, averaging 8–10 m. The transitional zone sediments are usually clayey silt and silt. In addition, a sandy layer can also be deposited during a strong storm period. The individuals and species of organisms in the transitional zone are various, and the bioturbation structure is well developed, which can sometimes seriously destroy the bedding structure, resulting in the formation of homogeneous bedding.

12.7.2 Neritic Shelf Deposition

12.7.2.1 Sediment Types and Sources

Neritic deposits can be divided into clastic materials, biological debris, pyroclast, and authigenic minerals, where priority is given to clastic sediments constituted mainly of silty clay. The muddy sediments of the neritic shelf originate mainly from fine-grained suspended matter transported from rivers. However, the shelf sand is composed mainly of residual genesis. A few sandy layers are formed directly by tidal currents, storm currents, density currents, contour currents, and flood currents. Carbonate shelf sediments of biological and bioclastic genesis can be developed in tropical or subtropical regions, which lack a supply of terrigenous detrital material. Volcanic sediments and eolian sediments are also important sources in local areas, and there are some authigenic minerals, mainly glauconite and apatite.

12.7.2.2 Main Factors Affecting Sediment Characteristics

(1) Type and speed of sediment supply: The sediments on the neritic shelf originate from the adjacent continent through the effect of river mouths, glaciers, and wind, which dominates the river mouth process. Mostly fine-grained suspended matter is transported from the estuary to the shallow sea. The coarse clastics deposited on the neritic shelf are carried mainly from the shoreface belt by tidal currents and storm backflow. Residual sediments occur mainly at the beginning stage of transgression, and they are transformed completely and covered by fine-grained sediments. Moreover, volcanic materials are mixed near the volcano activity area. In local areas, eolian sediments are important sources as well.

(2) Hydrodynamic conditions: The hydrodynamic conditions are complex, but tidal currents and storm currents play a leading role in sediment distribution.

(3) Sea level change: Sea level change affects the hydrodynamic energy at the bottom of the sea through changes in water depth. In contrast, sea level changes also determine the variation of the continental erosion base level, which further determines the quantity and speed of river sediments. Long-term sea level fluctuations determine the vertical sequence characteristics of neritic deposition and the spatial distribution of facies belts.

(4) Climate: Climate mainly acts on deposition in the neritic shelf through its impact on the inland. It determines the type and speed of weathering and erosion, further affecting the

type of sediment transported to the shallow sea. Moreover, it determines the modes of sediment transportation toward the sea, such as water, ice, or wind.

(5) Biological action: The remains of organisms and bioturbation structures can be involved directly in sediment composition. A spate of biological traces can affect sediment characteristics. However, the strength of bioturbation and the type of biological traces are different in different environments.

(6) Chemical action: Neritic chemical action is mainly involved in the formation process of authigenic minerals, such as glauconite, chamosite, phosphorite, and it is of great significance during seabed cementation and the diagenetic process.

12.7.2.3 Sedimentary Characteristics of the Neritic Shelf

1. **Sediment**

Its sediments are composed mainly of silty mud and muddy silt, partially silt. Silt-size sedimentation is usually formed during strong storm periods, so such sedimentation is well known as a "storm sand layer" or "storm sheet sand," which is shown as a massive or a graded bedding structure.

2. **Texture and structure**

They include the re-transformation of coastal sand bodies, high mineral and texture maturity, chemical cementation, good sorting, high rounding degree, and high ratio of particles to matrix. Glauconite, biodetritus, and collophanite are often present. Cross beddings in different directions, and symmetric and asymmetric wave ripples can be developed, and occasional scour and groove filling structures can be seen.

3. **Biological characteristics**

It is rich in organisms that are frequently found in the shallow basin of a continental shelf.

4. **Mineral features**

Glauconite, chamosite, and phosphorite are key authigenic minerals in the shelf sediments. Glauconite, which is a cold-water mineral, is mainly formed in the sea at depths of 10–1800 m. Chamosite, which is a warm-water mineral, is formed mainly in the tropical zone and in the sea at depths of 10–150 m. Conditions for the formation of a substantial amount of phosphorite are the most favorable at water depths of 30–300 m.

12.7.3 Sedimentary Sequence and Model of the Neritic Shelf

Under normal circumstances, sediment transport forces in the neritic shelf are of four types: ① storm current; ② wave current; ③ tidal current; and ④ ocean current. In special cases, contour current and gravity flow may appear (such as mud flow). The neritic sedimentary sequence can be divided into sandy high-energy neritic sequence, gravelly high-energy neritic sequence, and muddy low-energy neritic sedimentary sequence according to the different energy levels and sediment compositions. It can also be divided into high-injection basin and low-injection basin according to the injection (supply) of sediments, as suggested by G. Einsele (1992). On the whole, the vertical sedimentary sequence of a low-injection basin is dominated by fining upward, while the high-injection basin is dominated by coarsening upward. According to the energy level, the neritic sedimentary sequence

Table 12.6 Genetic classification of the neritic sedimentary sequence

Neritic sedimentary sequence of high clastic injection basin				Neritic sedimentary sequence of low clastic injection basin			
Low energy		High energy		Low energy		High energy	
Muddy type	Sand and mud	Storm type	Tide-dominated type	Muddy type	Sand and mud	Storm type	Ocean current type
Coarsening upward	Coarsening upward	Coarsening upward	With slight rhythm	Coarsening or fining upward	Unclear rhythm	Fining upward	Rhythm not apparent

can be subdivided into high-energy neritic sequence and low-energy neritic sequence (Table 12.6). The differences in sediment structure (grain size) should be focused on for the low-energy neritic sequence, while for the high-energy sequence, analysis of the differences of its sedimentary agents should be considered.

12.7.3.1 High-Injection Basin

1. **Low-energy muddy neritic shelf**

In this shelf, mudstone and muddy siltstone with continuous low wave energy dominate. The shoreface muddy sediments are frequently covered by thin beach bars and foreshore bars. A part of the mud comes from the estuary, and the remainder is carried offshore by storms. Here, biological debris is not well developed (Fig. 12.33a).

2. **Low-energy sandy and muddy neritic shelf**

The shoreline sand is thicker than that of the former shelf type. It includes offshore sand or proximal sandy storm sedimentation in the deep water transitional zone. In addition, a biological shell layer is usually seen, and bioturbation is strong in offshore sediments. In the vertical direction, muddy storm sediments with bioturbation and autogenetic silty clay are interbedded (Fig. 12.33a).

3. **High-energy storm neritic shelf**

The shelf is flat and an obvious facies belt is formed by sandy deposition. It is dominated by sand, possibly beach sand channels and storm deposits. In addition, it has hummocky cross bedding, and the storm deposits can be stretched to deep water. The sandy deposition occupies a considerable part of the entire section; partial mud is brought into the deep water by storms and transformed by storms and ocean currents (Fig. 12.33b).

4. **Tide-dominated neritic shelf**

The water depth is shallow here, with tidal action dominating, possibly with sandy ridges and tidal channels. This type of shelf has obvious tidal bedding and can form the barrier-lagoon sedimentary sequence with unclear rhythm in the vertical direction (Fig. 12.33b).

12.7.3.2 Low Injection Basin

1. **Low-energy muddy neritic shelf**

Compared with the low-energy muddy neritic shelf with high injection, the main difference is that it is rich in biogenetic products. Therefore, the biological shell bed is developed, plankton can be seen, and bioturbation is strong. Given that the water body is relatively quiet, carbonate rock can be formed outside the shelf (Fig. 12.34a).

2. **Low-energy sandy and muddy neritic shelf**

Compared with the high-injection status, its biggest characteristics are deposition of bioherms and potential for lagoon sequence development; a biodetritus layer, ocean mud, and carbonate

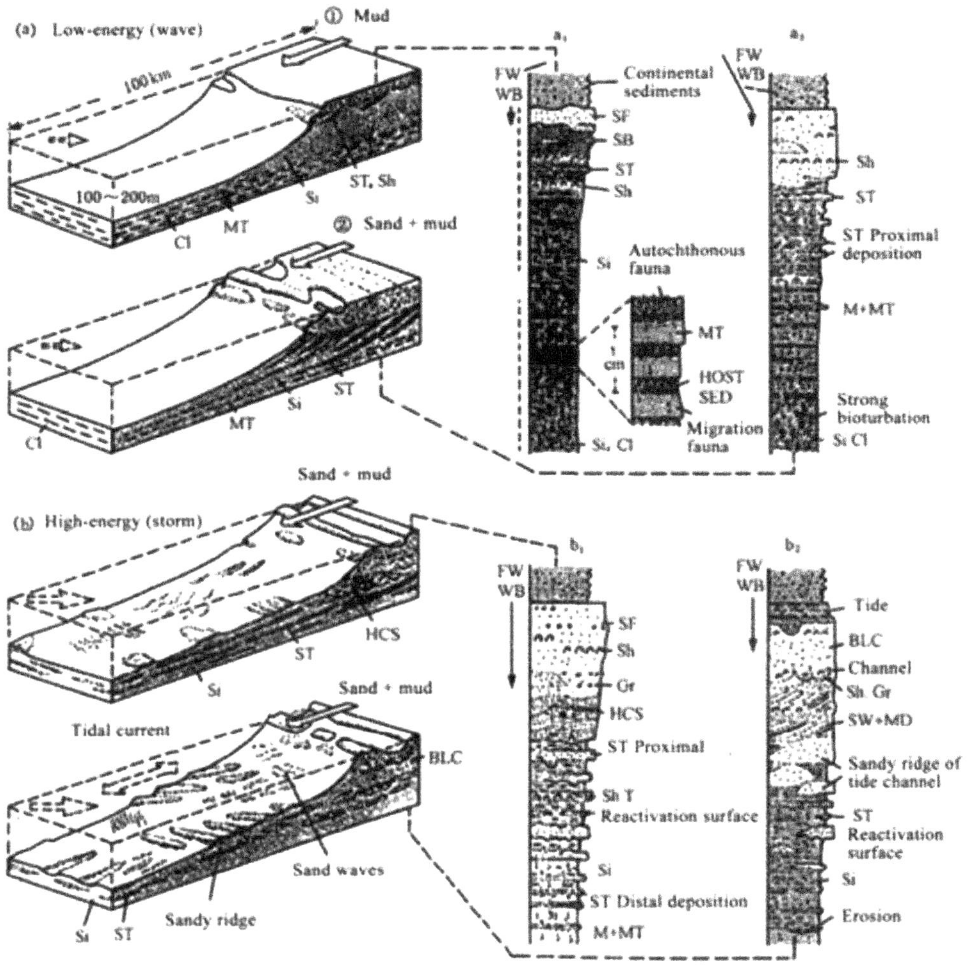

S-main sand body	HCS-hummocky cross bedding
Si-dominated by silt	FS-foreshore sand body
Cl-muddy clay	SF-shoreface sand body
Sh-shell bed	BLC-barrier island-lagoon complex (tidal deposition)
Gr-gravel layer	SB-offshore bar
ST-storm sand	SW-sand wave (cross bedding in different directions)
Sh T-storm shell bed	MD-coating mud
MT-storm argillaceous layer	FW-normal climate
WB-wave base	HOST SED-main sediments

Fig. 12.33 Sedimentary model and sequence characteristics of the shallow shelf in the high clastic injection basin (after Einsele 1992)

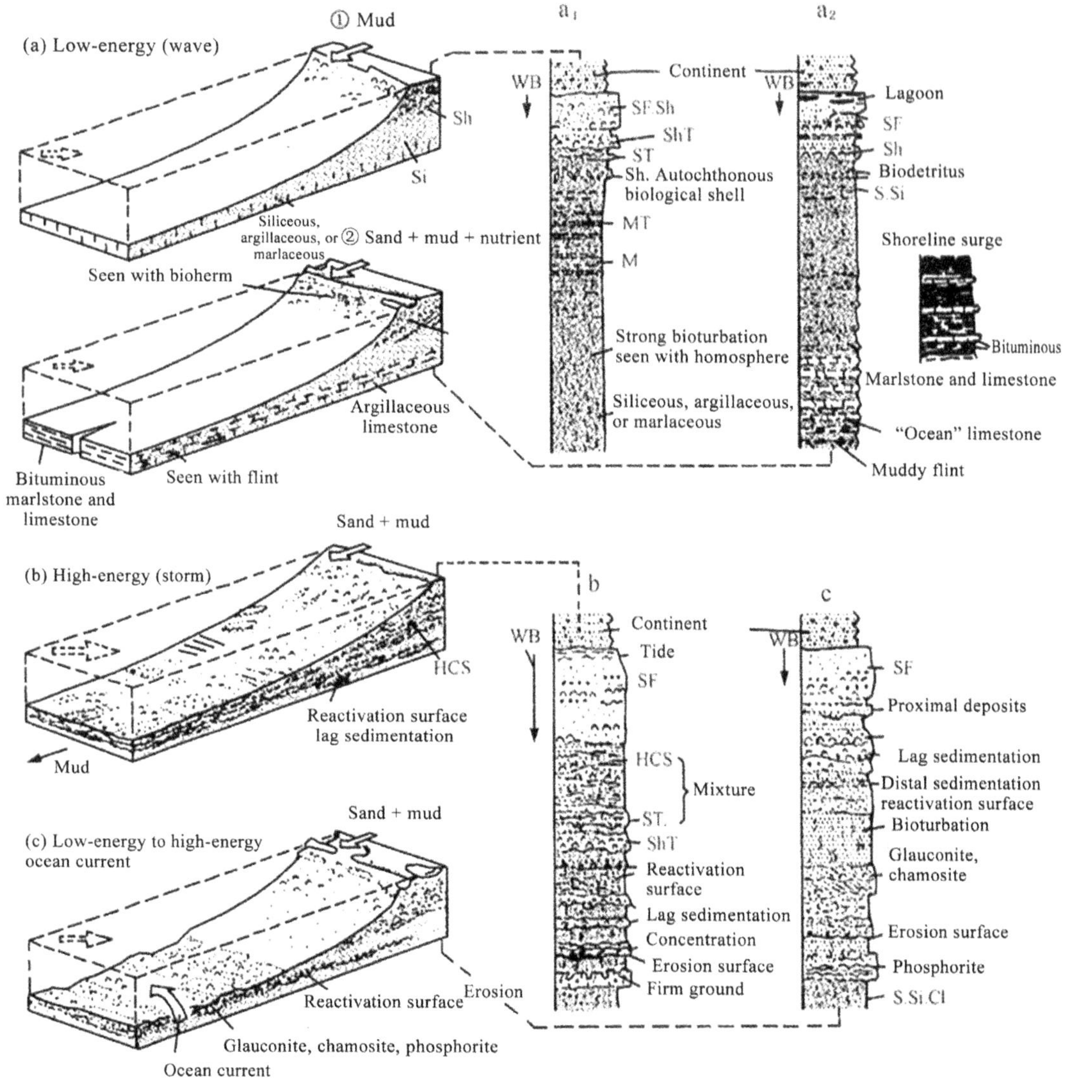

Fig. 12.34 Characteristics of coastal sedimentary model and sequence in the low clastic injection basin (after Einsele 1992)

rock can be seen; and bioturbation is not strong. The progradation is not obvious owing to the low injection volume (Fig. 12.34a).

3. **High-energy storm neritic shelf**

Neritic progradation is often limited because of the low injection volume and gentle shelf slope. The shell bed and fine-grained biogenic deposition are rather developed when compared to the high-injection status, which occupy an important position in the entire section. Obvious scour and lag deposits can be seen at the bottom of the sandstone in the offshore-to-shoreface transitional zone. A dense layer (concentration section) can be found in deep waters. Similarly, hummocky cross bedding can be seen (Fig. 12.34b).

4. **Ocean current neritic shelf**

Sedimentation dominated by fine sand can be formed by continuous ocean current activity. Multiple scour surfaces can be developed in the vertical profile, on which conglomerate lag sedimentation can be formed on the surface. Glauconite and chamosite can be developed at the sediment–water interface, and phosphorite can be formed in such an environment owing to the slow deposition rate (Fig. 12.34b).

12.8 Storm Current and Sedimentary Characteristics

In coast and open neritic shelves, as well as in meare-semi-deep lakes of continental facies lacustrine basins, the storm tides arising out of typhoons and hurricanes can lead to a rise in sea (lake) level, increase in sea (tide) surface flow velocity, and increase in the depth of wave propagation, which will destroy the sediments formed under normal circumstances and accumulate them into storm current sedimentation.

12.8.1 Storm Action and Sedimentary Feature Characteristics

Storm action has obvious seasonality and periodicity, and it can produce two results: ① a common event formed by a typhoon of magnitude greater than 7-strom tide; ② events generated by a hurricane with a huge speed-storm current.

12.8.1.1 Storm Development Site

Storm deposits are developed mainly in the continental shelf (shelf), which lies between the normal wave base and the storm wave base. However, the continental shelf is characterized by a medium-lower slope, the distribution range of which is from the lower shoreface (depth of about 5 m) to the continental slope break area, which is about 200 m below the sea level on average. The change in the shelf depends on the plate tectonic setting (Shepard and Inman 1950). A lower shoreface, interior shelf, and possibly the deep water area of the outer shelf will be subject to strong influence and transformation by rare storm events in the neritic (shallow lake) environment dominated by storm action. Tempestite is often found as widely distributed sheet sand and muddy layers in the lateral direction. Tempestite has been found in many places around the world within the long geologic age.

Storms can usually be deposited in two different zones: ① supratidal and backshore zones (supratidal storm layer) and ② deep water environment outside the shoreface (tempestite). As for lakes, storm deposition is found is mainly in shallow and semi-deep lake areas.

12.8.1.2 Storm Type and Frequency

1. **Storm type**
(1) Large storm

Extra large storms affect not only the deposition process of coastal and shoreface belts but also the deposition in the inner continental shelf and potential deposition in the outer continental shelf. Strong storms are formed in certain climates and geographies. Not only can large storms produce a deep wave base and a large range of waves, but its surface flow will also cause the net water flow to move toward the shoreline (Fig. 12.35a). Owing to low pressure, abundant rainfall can heighten the water surface on a wide neritic shelf and convergent shoreline. Excess water can carry suspended sediments through rip currents toward the sea side.

The bottom return current of the breaker and shoreface zones and the rip current are usually locally channelized to form a rip current channel and can generate a breaker channel under the sea. The non-closed and non-channelized bottom return current formed by the descending stream in the inner continental shelf is deemed to be generated due to the Earth's rotation. It can also be deflected under the action of the Coriolis

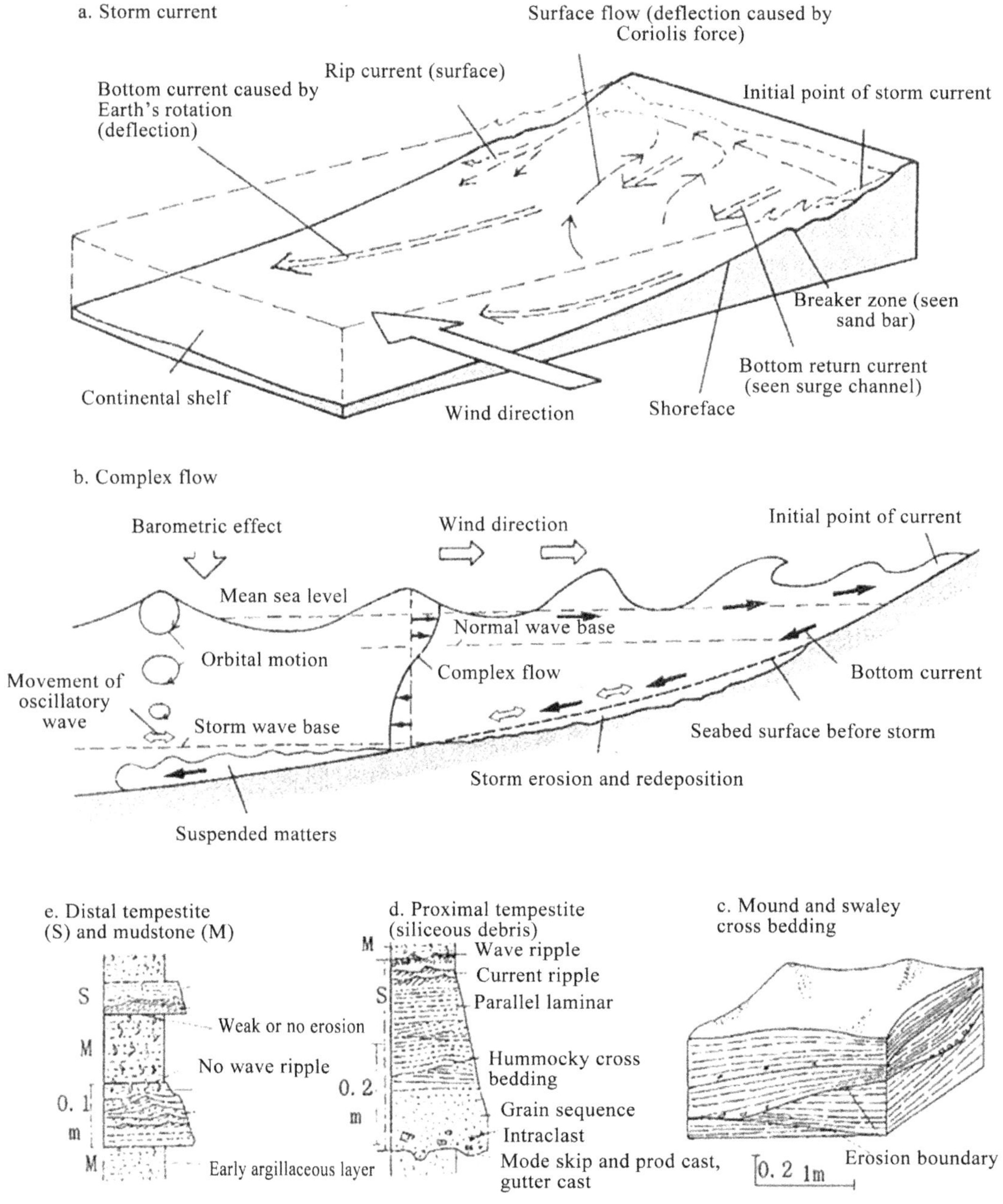

Fig. 12.35 Conditions, classifications, characteristics, and vertical sedimentary sequences of the formation of the storm (after Walker 1984)

force, so the current direction intersects obliquely with the coastline and is away from the coastline. In the high tide belt area, the current produced by the Earth's rotation enhances the highest tidal wave. Storm waves and currents induced by storms and those generated by the Earth's rotation occur simultaneously, resulting in a complex flow (Fig. 12.35b). The reciprocating oscillation of surface waves is superimposed on the relatively stable bottom return current. The oscillating shear stress induced by waves is dominant in the shoreface and a part of the inner continental shelf area. However, stable and inclined longshore current is dominant in the deep water environment.

(2) Tropical storms

These are also called hurricanes, and they usually develop in low-latitude areas of the monsoon zone. They move toward the west and are deflected to the polar region under the action of the Coriolis force (which is described as the deflection of linear motion due to inertial flow relative to the rotation system, and it is the result of the combined action of gravity and magnetic force). Through geological observation records, we know that storms prevail in warm climates in the presence of a small temperature gradient between the polar region and the equator. On land, they have strong low pressure in summer and strong high pressure in winter. Hence, the monsoon is formed due to the difference in the pneumatic pressures of the two aforementioned states. The summer monsoon comprises strong onshore wind, which brings a considerable amount of rainwater landward, and the best example is the monsoon zone along the rims of the Indian Ocean and the Pacific Ocean in Asia.

(3) Extratropical storm

Extratropical storms are common in mid-latitude regions, where general atmospheric circulation moves along the polar front. They move toward the east in a roughly consistent direction, so their intensity can be the same as that of tropical storms. The other mechanism for producing storms toward the direction of land is the monsoon in the summer. Extratropical storms prevail in the relatively cold season because of the larger temperature gradient.

2. **Frequency of storms**

The formation of a widely distributed thick tempestite deposition sequence requires a large amount of sandy source from continuous siliceous detritus or biodetritus or both. They are often provided by a river delta, and the river rapidly erodes sandy coastal cliffs (sea/lake) or the carbonate sediments at the ramp edge. In addition, the basin gathcring tempestite subsides relatively, which is essential to avoid erosion of the old tempestite. Accordingly, the storm-producing tempestite in geologic records is less frequent than in the records of modern rare storms. Nowadays, the once-in-a-century storm seems to be an extra large storm event. The time interval for the occurrence of most tempestite reserved in geologic records is longer than tens of thousands of years. On the inner continental shelf (water depth of 20–30 m), the time interval for the occurrence of tempestite in modern seaward stepping sediments is from decades to hundreds of years (Nelson and Maldonado 1988) In sharp contrast, the ancient tempestite sequence with 2–5 tempestite in every 1 m of section shows that the storm reappearing period ranges from 1 ka to more than 10 ka on the assumption that the speed range of average sedimentation (or subsidence) is 20–100 m/Ma. It has been reported that the reappearance period of storms reaches 100 ka.

The difference between modern observations and the limited tempestite reserved in ancient records indicates that even a significant amount of tempestite can be transformed or obscured by later very rare extra-large storms. The early weak storm action can be judged indirectly by mechanical erosion and coarse materials from different sources in the tempestite formation reserved ultimately.

12.8.1.3 Sedimentary Characteristics of Storms

1. **Storm deposition mechanism**

Storm deposition is the product formed by storm currents or storm tides caused by hurricanes and seasonal typhoons, and the sediment depth that could be affected by strong storm action can reach tens of meters. In the period of an extra-large storm, the greatest depth of wave propagation can reach 200 m and elevate the sea level by 5–6 m. The strong force of storm action scours the coastal and nearshore deposits. When the wind power decreases, the storm forms a density current flowing seaward, which carries a large amount of suspended sediments to the sea (Fig. 12.35). This high-density current scours the bottom of the sea, forming a clear erosion surface and erosion marks. Between the normal wave base and the storm wave base, storm waves can affect the bottom of the sea and sedimentation occurs because of the density current, resulting in the formation of hummocky cross bedding sandstone. If the density current flows under the storm wave base, normal neritic turbidite with the Bouma sequence will be formed.

2. **Storm deposition period**

Storm current is a rare event, and it is composed mainly of high-density gravity flow directed seaward that is produced by backflow and oscillating flow caused by a hurricane or a strong typhoon (over 9-class winter monsoon in mid-latitudes). Unlike sediment gravity flow, the wave action of a storm current is strong. Storm currents occur mainly in shallow seas with water depth less than 200 m, commonly at a water depth of 30 m, that is, mainly between the normal and storm wave bases (Fig. 12.35b). A storm current is characterized by large energy and long duration. The deposition process of a storm current can be divided into three stages (Fig. 12.36).

(1) Peak period: When the water body is under strong disturbance of the storm current, the seabed suffers strong erosion, and the formed hyperpycnal current (storm current) moves downward. The carried material is mainly from the sediments below the normal wave base or sometimes from the backflow of storm tide.
(2) Decay period: When the medium-coarse debris (including biodetritus) carried by the storm current is first accumulated, it fills, and scours the pits, and laminar (aftermath effect) fine-grained material (often developed with hummocky cross bedding) is deposited thereon.
(3) Quiet period: When the fine grained sediments are recovered back to the state in normal weather, biological activity is enhanced again.

3. **Sedimentary characteristics of storms**

(1) Material source: Most storm deposits are from or near in situ sources. During a storm, high-energy violent storm surge events occur. They incise and scour the shelf sediment surface (located below the normal wave base), and some organisms, biological shells, and silt on the surface of or inside the seabed sediments are suspended in the current as a result.

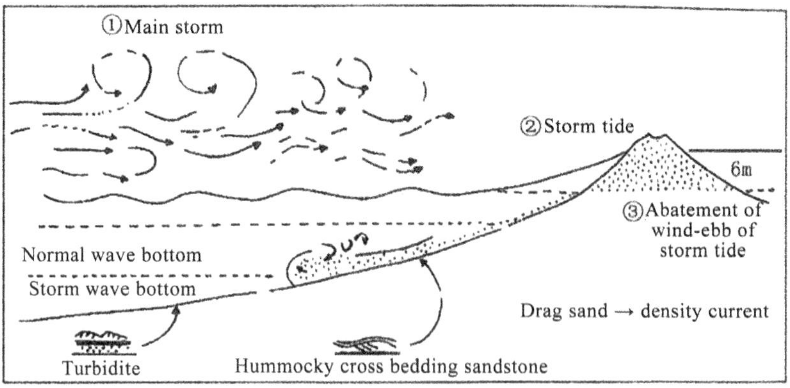

Fig. 12.36 Diagram for the formation of the storm flow

(2) Storm events: These are wave-dominated turbulence events. Wave action is usually fixed and limited in certain area. Lateral transport during the storm is often secondary, and both its transport distance and velocity are less than those of turbidity.
(3) Storm deposition process: This is a transformation process from erosion to redeposition. From the peak to the decay period, storm deposits are shown as changes from high energy to low energy and as the process from erosion to redeposition.
(4) Storm sediments: Such sediments feature a local concentration of coarse debris, enhancement of rhythm, and discontinuous distribution of coarser debris. The coarse and fine-grained materials are in the form of a suspension in the peak of the storm. Once the storm recedes, it immediately has differentiation. Specifically, coarse debris (lag deposits) first accumulates in the erosion pits; then, fine-grained materials are deposited gradually to constitute obvious graded bedding.

12.8.2 Characteristics of Tempestite

12.8.2.1 Sandy Bedform and Vertical Sedimentary Structure Sequence

Storm currents often wash out the seabed and the shell of fauna living at the bottom of the foreshore area and gather them on the normal beach or in a block area to form the supralittoral or backshore storm layer. Owing to little preservation potential, this will not be discussed further.

1. **Bed configuration under different environments**

Offward, the sandy bedform formed owing to storms presents an obvious tendency from the breaker zone to the abyssal region (Figs. 12.35 and 12.37), meaning the formed sandy bedform has different characteristics at different water depths.

(1) Breaker zone and upper shoreface

Controlled by wave action in normal climate, it usually destroys the imprint formed by the early storm and forms megascopic waves on top of offshore bars, that is, flat swash laminae. Trough and planar cross beddings, as well as low-angle swash laminae, are the main structures.

(2) Middle shoreface

A few structures formed in the storm period may be reserved, for instance, the flat nearly horizontal bedding or low-angle swaley cross bedding (Fig. 12.35c). Coarse particles often forms lag deposits with the mud sorted out and deposited in deep waters, while the creatures at the bottom are only those that can survive on suspended biomass.

(3) Lower shoreface

The water depth is 5–20 m and the force of two flows (oscillating flow and bottom current resulting from the Earth's rotation) may stir sand and mud, and they may be redeposited at or near the same position to form graded bedding (proximal tempestite, Fig. 12.35d). In the peak season of storm action, asymmetric wave marks and grooves (caves and troughs) are formed at the bottom of the sea, which are later filled by very coarse siliceous debris and biodetritus. These molds and etch marks sometimes indicate bidirectional or multi-directional flow action. The typical internal sedimentary structure produced by the complex flow is low-angle hummocky cross stratification (HCS). It usually appears on the top of the graded bed with lag deposits at the bottom, followed by parallel pattern laminae and current ripple cross bedding. In the ideal situation, the hummocky cross stratification is covered by wave ripple cross bedding and wave ripples, which indicates the last stage of storm

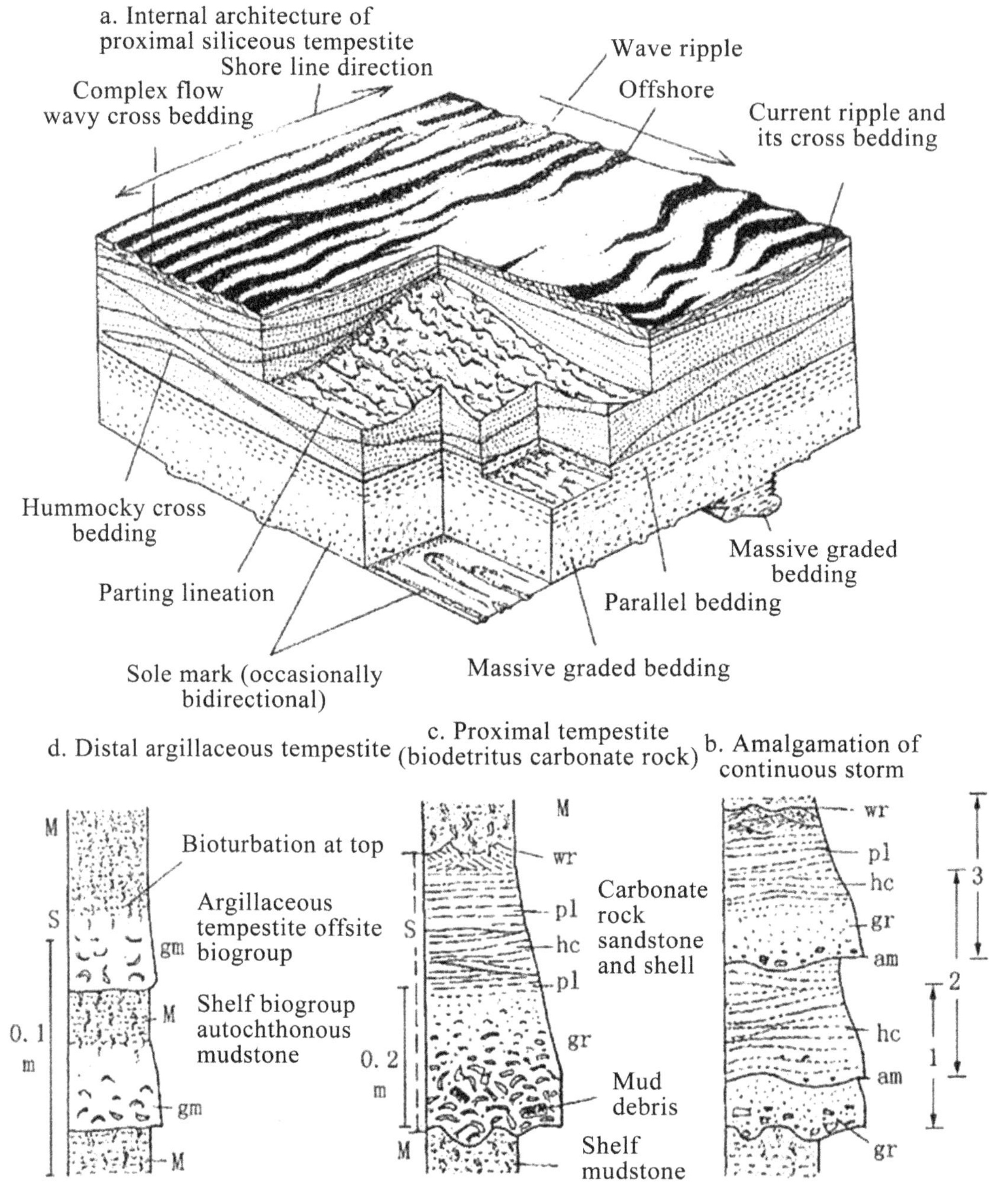

Fig. 12.37 Idealized sandy storm rock model and vertical sedimentary sequence. wr wave ripple; pl parallel laminae; hc hummocky cross bedding; gr graded bedding; M shelf mud; gm massive biological debris; and am abrasion surface

recession. Therefore, the top ripple may also be formed by a subsequent big wave.

(4) Inner continental shelf and partial outer shelf

As for the formation of sandstone or thin graded sandstone and a silty layer (distal tempestite) with hummocky cross bedding, their typical feature is cross bedding and occasional ripples on the top (Fig. 12.35e). The formation of a muddy interbed can not only be ascribed to landward movement of windstorms and erosion of fine-grained materials but also to the slow redeposition of the suspended load of a river. In areas with water depths of 30 m in the North Sea, modern distal tempestite may trace for hundreds of kilometers.

(5) Outer shelf with larger water depth

The current component of a complex storm is dominant, hence current ripple (ripple) fine sand and silt layer are formed (Fig. 12.35b). After a long stationary phase, rare storm events occur in this zone. In high-energy shelf waters, discontinuous distal tempestite occurs only when the water depth is more than 50 m.

2. **Vertical sedimentary structure sequence**

Complete and ideal tempestite at a moderate distance from the coastline has the following sedimentary structures (from top to bottom, Figs. 12.35d and 12.37a, c): ① redeposited shelf mud (mud tail of complex flow); ② wave ripple and its cross bedding; ③ low-angle hummocky cross bedding; ④ parallel laminae and current ripple cross bedding; ⑤ graded bedding with bottom lag deposits; ⑥ erosion surface with sole marks (bidirectional or multi-directional) and animal burrows; and ⑦ normal shelf mud and strong bioturbation.

3. **Difference in water body depth**

In an area with relatively shallow waters, graded bedding usually decreases or vanishes. On the contrary, large-scale hummocky cross bedding, swaley bedding, and coarse-grained scour-fill structures are the main features. In areas with relatively deep waters, hummocky cross bedding is not obvious and is replaced more or less by parallel laminae, wave ripple cross bedding, and lenticular-wavy bedding with wave ripples (Fig. 12.37b). These structures, which are formed by the subsequent storm to some extent, transform the formerly deposited silt-finestone.

In conclusion, many scholars have described these structures in detail, for example, Craft and Bridge (1987) and Krassay (1994), and are not confused by the lenticular and wavy beddings in sandy and muddy tidal flat. The calcareous tempestite with wave ripples found at the top of the marine facies from the Jurassic period in southern Spain indicates that the water depth of this basin cannot be larger than the wave base, hence the tempestite was likely to be formed in the period of lowstand relative to the sea level.

In general, lag deposits are formed by tiny rock debris formed by the debris of gravel, mollusks, and other organisms after they are broken down. In the Proterzoic and Cambrian systems, tempestite is mainly composed of re-transformed argillaceous intraclasts or microbial mats.

12.8.2.2 Other Characteristics of Tempestite

1. **Change in grain size and composition**

There are significant changes in the grain size distribution and composition of tempestite. They vary from coarse grain (sand and gravel) to silt and mud. Tempestite is almost of the pure siliceous type without biological remains in sharp contrast to calcareous bioclastic sandstone, packstone, and wackestone (Fig. 12.35a). The composition of the latter comes from previous littoral benthal or adlittoral biogroups.

2. **Erosion and reactivation**

In proximal areas, amalgamation or transformation are common (Fig. 12.35d). These terms indicate that the previous thick tempestite is truncated or thin tempestite is totally transformed by the later storm, and new tempestite is formed. This process may occur repeatedly until a very large storm produces a rock stratum at last, which will be preserved at the bottom of the rock stratum.

Amalgamation transforms and promotes many times over the physical deterioration and fracture of unstable minerals, in addition to possibly accelerating the dissolution of carbonate and other minerals. Therefore, amalgamation enhances the maturity of lag deposits, including placer deposits. Skeletal remains of autochthonous vertebrates, particularly teeth and coprolite, gather at the bottom of the tempestite (skeletal fragment layer), and the latter undergoes phosphorization before re-transformation.

3. **Ichnofacies features and biogroups**

In the long intermission of a big storm, ichnofacies of the argillaceous sediments of the middle-to-distal tempestite is usually marked by bioturbation. The bottom mark of sandy tempestite possesses neritic environment ichnofossil assemblage.

Burrows are dug out by storm action, the washed out parts of these burrows are subsequently filled by the deposited sediments. The corrosion depth scope of storm action can be determined by the animal burrow form before the storm event. However, even with no storm deposit, sediments that fill the burrow will occur.

Bioturbation in argillaceous sediments before the storm comprises limophagous and herbivorous animals with high disparity (i.e., Skolithos-Cruziana). The ichnofacies in sandy tempestite after a storm is composed mainly of vertical and U-shaped burrows with low disparity, and the burrows are formed only at the top of adequately thick tempestite. The epifauna after the storm is re-combined at the top of tempestite or on an erosion surface not covered by tempestite, for instance, the uplifted proximal area.

The solid and coarse basement attracts oysters, brachiopoda, and stromatolite, accompanied by the burrowing of solid basement. In the distal area, tempestite is very thin and there is barely any re-transformation owing to the subsequent storm and burrowing action of marine benthos after the storm. Hence, the tempestite stratum can show an obvious increase or be obscured completely.

12.8.2.3 Vertical Sedimentary Sequence

There is a close connection between the storm sedimentary sequence and the process of storm action. A sedimentary sequence of tempestite represents the sedimentary process from storm peak to abatement of wind and flow regimes from high-energy to low-energy. In different phases of storm activity, different depositions occur, which lead to different results. A complete storm process can form a vertical sequence with a regular pattern (Fig. 12.38). The features from the bottom to the top of an ideal storm sequence are as follows.

(1) Eroded bottom: At the peak of storm action, the depth of wave propagation increases, and large amounts of sand and mud carried in the water body are transported out of the sea with the storm, and they strongly scour the seabed to form an obvious scour surface. The eroded surface is in abrupt contact with the underlying normal fine-grained shelf sediments.
(2) Coarse-grained lag layer: If there is coarse-grained material in a storm area, for instance, gravel, boulder clay, and shells, they will form coarse-grained lag deposits at the bottom on the scour surface in the storm peak period. They are the products remaining after storm winnowing. The shell beds usually have a preferred orientation, and most of them are convex upward parallel.
(3) Graded bedding: When a shell bed is not developed, massive sandstone with graded bedding occurs on the erosion surface directly; this is very common, particularly in the section below the storm wave base (Fig. 12.39). It is the fining-upward normal graded bedding formed by storm currents.

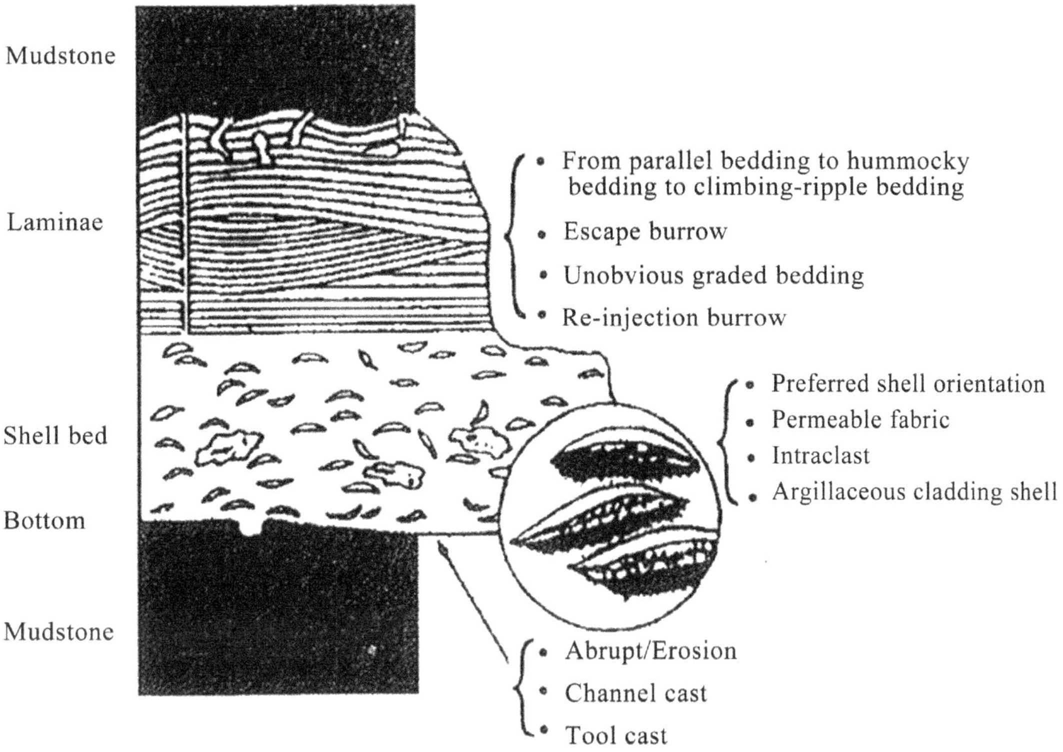

Fig. 12.38 Ideal sedimentary sequence model for storm current deposits

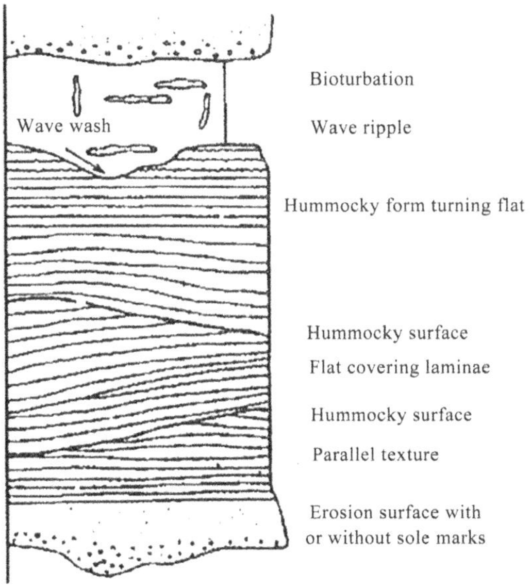

Fig. 12.39 Sequence of storm deposits with a granular layer

(4) Laminar section: After the storm peak, the material carried in the water body is accumulated largely owing to the gradual degradation of sea water energy, which results in fining-upward graded bedding. This is in abrupt contact between the laminar section and coarse lag deposits during the storm peak. The laminar section is composed mainly of fine sand and silt, on which there is often miniature wavy cross bedding and hummocky cross bedding, which are gradually transited to climbing-ripple bedding upward. The most typical beddings associated with storm current deposition are parallel and hummocky cross.

(5) Mudstone section: When is the storm ceases and the shelf recovers to its normal state, the sediments are fine silt and mud deposited by suspended sediments. In this case, the

benthos resettle on the seabed to cause strong disturbance in the bottom substrate.

In general, the storm depositional sequence is fining upward, which is similar to turbidite.

12.8.2.4 Lateral Change of Tempestite

Sediment features differ in different locations. A storm in the nearshore shallow water zone is stronger than that in a deepwater area. Storms can, therefore, be divided into proximal and distal. As for proximal storms, the grain size of sediments is coarse, thickness is high, and the lag layer is relatively developed. Graded bedding is usually dominated by bioclastic limestone or calcarenite, shell bed is composed mainly of peduncles and bivalves, and erosion filling structures and shallow channels are developed at the bottom. The distal tempestite is dominated by fine-grained sediments with thin layers and a clear bottom margin, however no bottom erosion structure is developed.

Dott and Bourgeois (1982) proposed a proximal-to-distal facies model of tempestite under normal conditions (Fig. 12.40a). Muddy intercalation is usually lacking in swaley and large hummocky cross beddings in the nearshore zone. Toward the basin, muddy intercalation, in addition to sediment filling in the washed-out pit, can be formed within the thick-layer belt owing to the amalgamation of storm currents and occur in the passing zone (Fig. 12.40b). In any case, the thickness of the tempestite is usually changed laterally and often wedged out. Toward more distant seas, the number of individual tempestites increase in a specific time period for the first time, however they can decrease as a result of amalgamation. This is ascribable to the suspended fluid formed by relatively weak or moderate storm action moving only over a limited distance. Distal tempestite is characterized by thin bedding and fine grain, and it has an inorganic sedimentary structure similar to that of distal turbidity, however its animal characteristics and vertical sedimentary sequences are different from those of distal turbidite (Einsele 1992). It is hard to identify argillaceous tempestite with thin bedding and strong bioturbation, hence it is often neglected.

The facies model shows that a continuous tempestite sequence from the foreshore to deepwater (Fig. 12.40a) is not common, whereas the advanced model takes into consideration the fact that tempestite deposits occur only in the inner and the outer continental shelves (Fig. 12.40b). In shallow water, as a result of storm action, materials are often either transformed in situ (amalgamation) or transported toward the basin from the middle area without sediments (passing area, passing model). The typical passing area is characterized by isolated caves and gutter cast, as well as by continuous rare tempestite bedding. In some basins, two tempestite models can occur successively. The passing model is likely to be more suitable for environments with increasing subsidence rate from the basin margin to the basin center.

12.8.2.5 Tempestite Feature of Basins in Different Evolution Periods

In a subsiding basin, a large number of continuous storm events can establish storm shale sequences more or less with a rhythmic structure, regardless of the occurrence of differential sediment transportation. These sequences can reflect the following three different tendencies of basin evolution.

(1) Steady state: In the foreshore-shelf environment, the average subsidence rate more or less compensates for subsidence. In this case, a relatively thick tempestite sequence and shelf mudstone are developed alternately, and the paleo-water-depth remains the same in some parts of the basin.
(2) Deep-water basin: Its average deposition rate is less than the subsidence rate, and the vertical sequence is shown as a change from thick but relatively coarse proximal tempestite to thin, fine-grained distal tempestite, and it ends with fuzzy argillaceous tempestite or pure protogenic shelf mudstone (Fig. 12.40c). On a vertical section, the thickness of the transition zone from proximal to distal tempestite is 20–50 m.

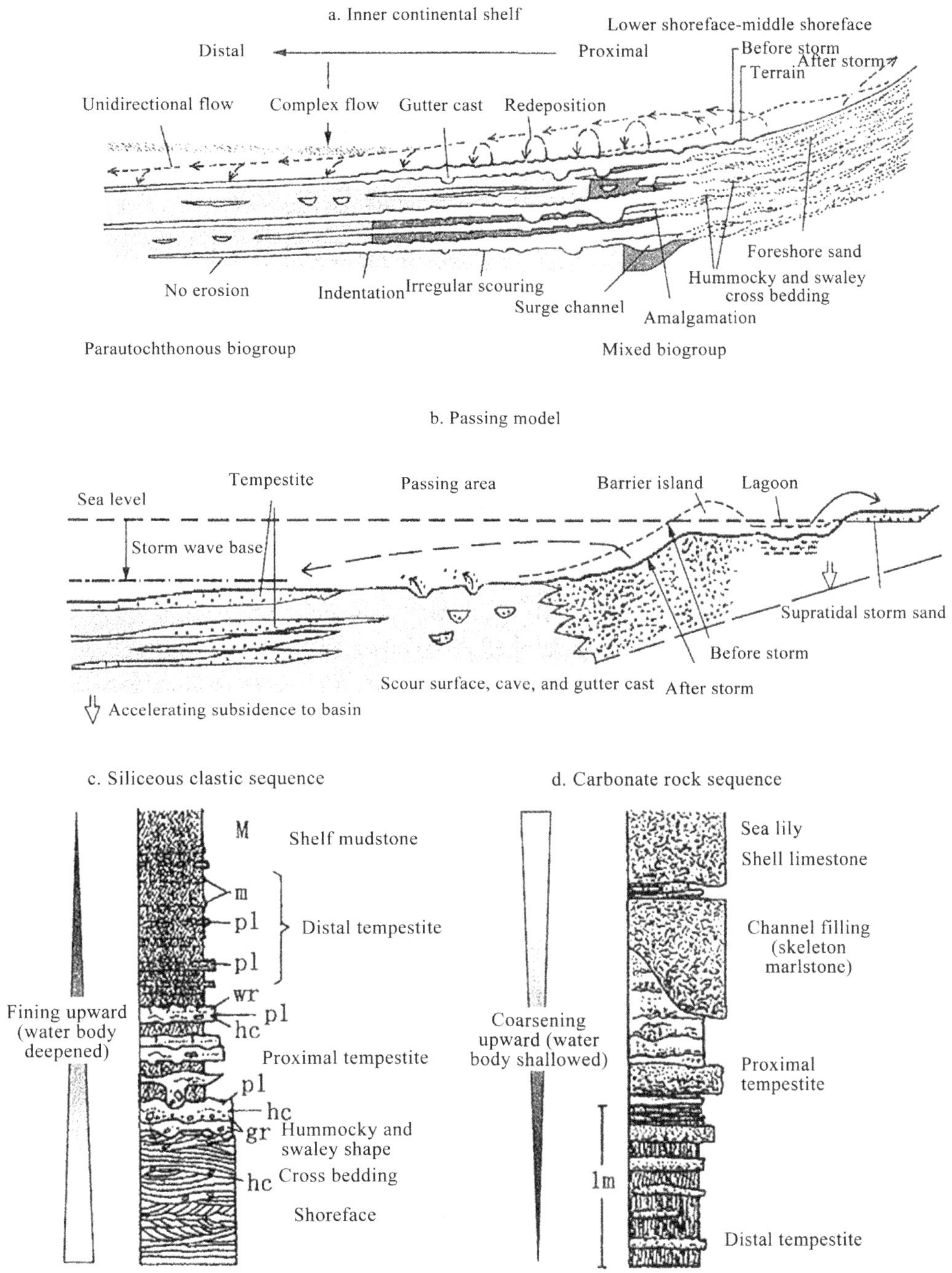

Fig. 12.40 Model map of the storm sedimentary facies. **a** Profile model of storm sedimentary facies from shoreface to inner continental shelf (proximal-distal); **b** Passing model, recording distal lenticular-wavy bedding; **c** Fining-upward vertical sedimentary sequence of siliceous clastic tempestite; and **d** Coarsening-upward calcareous tempestite sequence (number of layers reduced). M shelf mud; m massive sand; pl parallel laminae; wr wave ripple cross bedding; hc hummocky and swaley cross bedding; and gr graded intraclast

(3) Shallow-water basin: Its deposition rate is higher than the subsidence rate, resulting in coarsening-upward granularity in the tempestite sequence and an increase in thickness (Fig. 12.40d). The thickness is not high in the transition zone of the foreshore and the littoral environment from shelf mud to siliceous debris or biodetritus. This regressive action is favorable for the amalgamation of tempestite.

12.8.2.6 Identification Mark of Storm Deposition

(1) Erosion structure: It is varied, including bag shape, asymmetric swaley shape, symmetric ditch shape, wave shape, as well as microwave and flat shape structures, among which the bag-shaped structure is unique to storm current scouring. An erosion surface-shaped structure is filled with the retention material whose component is close to the underlying substance, which is the basic feature of storm deposits.
(2) Wave structure: After storm decay, oscillation of the aftermath can form both hummocky cross bedding and wave ripples. Hummocky cross bedding can be formed in all places with current oscillation, hence it cannot be taken as the only mark for identifying storm deposits, although it is important.
(3) Multi-directional flow: This means that the change in the direction of tool marks is large or it indicates the opposite direction. Biological bone (bidirectional extension) has no preferred direction.
(4) Special rock stratum: Mud composition of shell polycondensation layer, in which creatures show the "mud drag" phenomenon, is consistent with the underlying deposition.

12.8.3 Classification of Tempestites

In recent years, an increasing number of domestic and foreign reports on tempestite have been published, and the deposits resulting from tempestite storm events can be divided into two types based on their composition. The first is clastic storm deposit, and the second is carbonate storm deposit, however the two are also mixed sometimes. In accordance with the water depth and environment, they can be subdivided into seabeach backshore and foreshore storm deposits, shoreface and nearshore storm deposits, offshore continental shelf shallow sea storm deposits, and bay lagoon and inland lake storm deposits. From the perspective of description, they are classified according to the vertical sequence (fining and coarsening upward), grain size (sand or mud), transport distance (proximal and distal), and model (rip current type, agitation type).

12.8.3.1 Rip Current Type

This is a neritic shelf area with certain gradient, which is formed because of storm rip current and the eddy effect. Owing to differences in environmental conditions, such as water depth at different parts, the vertical sedimentary sequence of tempestite is not the same. It can be classified into three types accordingly, which represent the sediments below the normal wave base to near the storm wave base (Fig. 12.41).

1. **Distal tempestite (Type I)**

(1) Development site: It is mostly developed under or near the storm wave base of a continental slope.
(2) Major characteristic: More mud and less sand (thus it is called argillaceous tempestite), dominated by plaster matrix support; weak or no bottom erosion, generally low thickness, and probable development of laminated sand sheet and graded bedding (Table 12.7).
(3) Vertical sequence: from the bottom up: ① It is generally a hummocky erosion surface with a gentle slope at the bottom. ② The lower unit is a biodetritus limestone lag layer, often composed of relatively complete brachiopoda, pyramidal corals, fusulinida with large curvature,

and other fossil fragments, with obvious normal graded bedding sequence or complex normal graded bedding sequence. Its top is the homogeneous biological shell "framework" layer, and the microscler pieces are intertwined vertically or obliquely. A 5-cm-thick plaster matrix support is formed by storm peak deposition. ③ The upper unit is plaster limestone with biodetritus, and it is composed of relatively pure microcrystal calcite and a few fossil fragments. Large and regular zoophycos are developed with thicknesses of 5–10 cm, as deposits after the storm period.

2. **Proximal tempestite (Type II)**

(1) Development site: This is formed in a storm-normal wave base area, that is, offshore (lower shoreface).

(2) Major characteristics: Mainly sand-mud interbed with mixed support; obvious scour surface at the bottom due to erosion, usually large thickness and distant expansion, possibly with hummocky cross bedding and parallel laminae and visible gutter cast (Table 12.7).

(3) Vertical sequence: This is usually composed of three layers, and its vertical sequence features are as follows: ① The bottom is a wavy-trough hummocky erosion surface. ② The lower unit is a biodetritus limestone lag layer, the bottom is thick and contains complete brachiopoda, molluscs, coral bone grain, and fossil fragments, and the content and granularity of biodetritus decreases gradually upward, with the normal graded bed sequence. In addition, its superface is mostly made of undulations, which reflect the fact that it was subjected to the eddy

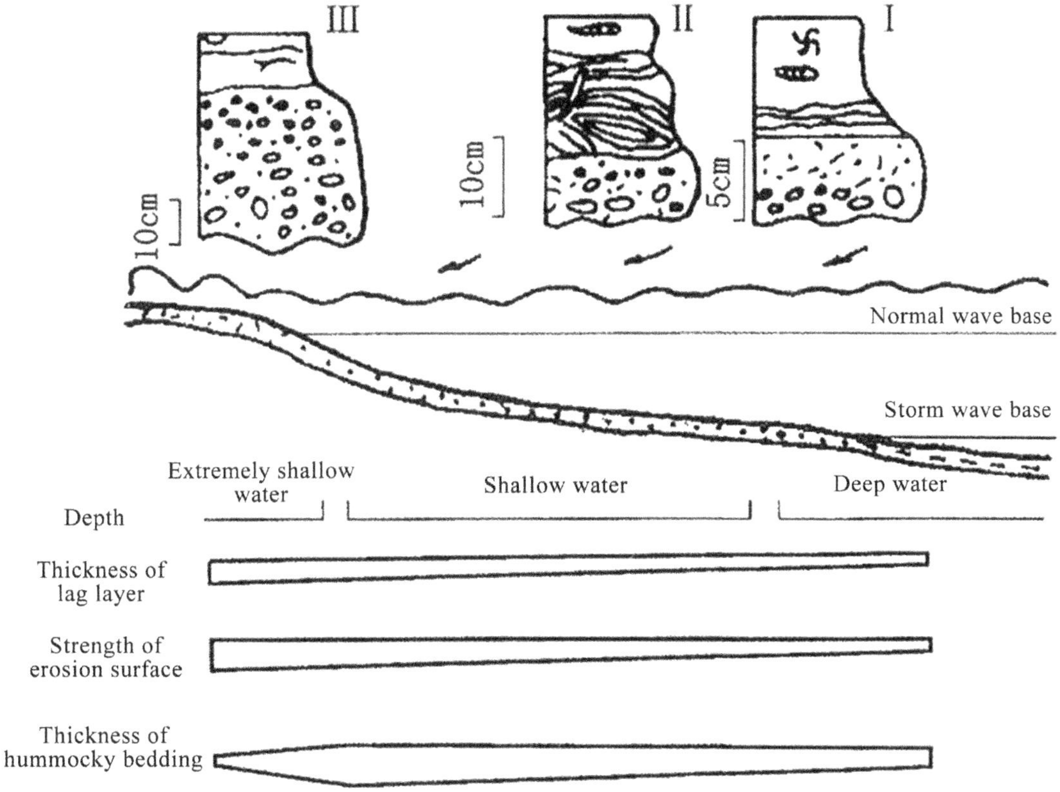

Fig. 12.41 Offshore flow type storm rock is formed by storm offshore flow and eddy currents in shallow shelf area with certain slope

Table 12.7 Characteristic contrast between proximal and distal tempestite

Features	Proximal tempestite	Distal tempestite
Development depth	Relatively shallow water body and near wave base	Relatively shallow water body and near storm wave base
Developed part	Upper shoreface (lower shoreface)	Offshore
Vertical sequence	Normal graded bed sequence	Normal graded bed sequence (or complex normal graded bed sequence)
Lithology	Sandstone	Sand-in-mud
Sedimentary structure	Miniature current ripple, hummocky cross bedding, wave ripple, parallel laminae, graded bedding, visible gutter cast, skip mark, and prod cast	Laminated sand sheet and graded bedding developed
Scale	Usually thick with large extension	Usually thin
Biological characteristic	Autochthonous biogroup and offsite biogroup	Biological shell and intraclast
Bottom	Obvious erosion with scour surface	Weak erosion

effect after formation. It is formed due to material deposition by rip currents in the storm peak, and its thickness is generally 10–15 cm. ③ The middle unit is biodetritus plaster limestone developed with miniature-medium hummocky cross bedding, on which there are biological escape traces, with a general thickness of 10 cm. It is formed because eddies stir the seabed in the storm decay period. ④ The upper unit has a horizontal texture or thin limestone with biodetritus or biodetritus plaster. Sometimes, it has miniature altered zoophycos, parallel burrows, and rhizocorallium. It is formed by deposition after the storm period, and its thickness is 10–15 cm.

According to the development characteristics of the profile, this type can be subdivided into Type II_1 and Type II_2. Type II_1 is formed below the storm-normal wave base area, and Type II_2 is formed in the upper part. The difference between the two is that the scale of the Type II_2 sequence of hummocky cross bedding is relatively large, while for Type II_1, the upper unit, is usually not very developed, with the rhizocorallium ichnofossils, inclined and parallel burrows, and no miniature zoophycos.

3. **Proximal shallow-water tempestite (Type III)**

(1) Development site: It is generally developed in a storm period over the normal wave base-subtidal zone, where it is eroded and scoured intensively to form a relatively deep erosion surface. It is formed owing to the proximal deposition of gravel-sized grains carried by undifferentiated storm rip currents. Hence, this type of storm can form in lacustrine basins as well.
(2) Major characteristics: Mainly sand (thus it is called sandy tempestite) and particle support; obvious bottom scour with a scour surface, usually of large thickness and distant expansion; development of miniature current ripples, wave ripples, and hummocky cross bedding; and visible gutter cast.
(3) Vertical sequence: ① It is a swaley hummocky erosion surface at the bottom. ② The lower unit is a lag layer composed of lenticular gravel-sized grain, in which many erosion surfaces with a general thickness of 30 cm are developed as the deposit formed in the storm period. ③ The upper unit is thin biodetritus plaster

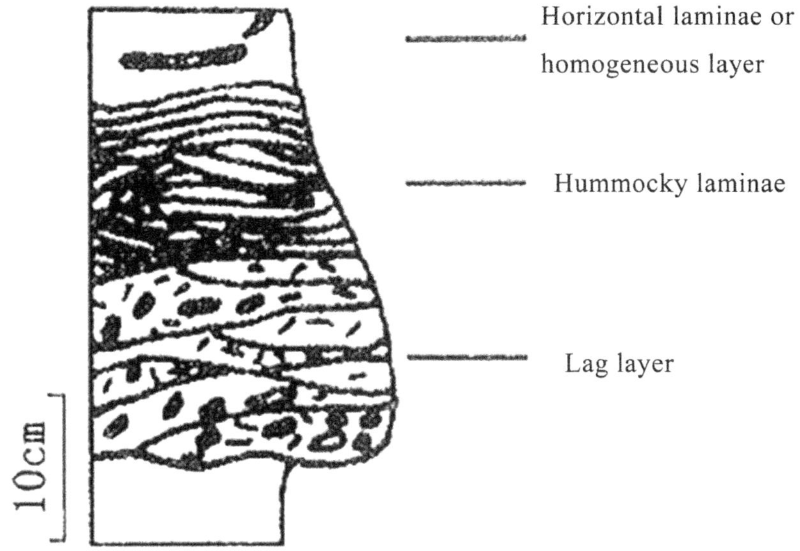

Fig. 12.42 Sedimentary sequence of agitated storm rock

limestone, with inclined and slightly inclined biological burrows and the occasional vertical burrow. Its thickness is about 10 cm, and it is deposited under the subtidal turbulent environment in the quiet period of a storm.

From Type I to Type II_1, the water depth decreases, whereas storm action, fluctuation of erosion surface, and thickness of the lag layer increase.

12.8.3.2 Agitation Type

(1) Formation conditions and features: Submarine topography is flat and wide in the paleogeographic environment of the epicontinental sea. Hence, the formation of a storm rip current along a certain direction is difficult in the strong storm action period, but the reciprocal action of storm eddy is dominant. Storm eddies cause erosion, transformation, suspension, and close transportation of sediments to form a special tempestite called agitation-type tempestite. In the paleogeographic environment of epicontinental sea, both rip current–type tempestite and agitation-type tempestite are very important.
(2) Vertical sequence: ① It is generally a wavy-swaley hummocky erosion surface at the bottom. ② The lower unit is hummocky fluctuant biodetritus limestone, developed with multiple hummocky surfaces and a graded bed sequence, which is formed because of stirring and redeposition of eddies in the storm peak. ③ The middle unit is hummocky bedding section. ④ The top is plaster limestone after the storm period, developed with ichnofossil assemblage, which reflects a quiet environment. Occasionally, hummocky bedding is developed in the lower and upper units, with hummocky fluctuation throughout the layer (Fig. 12.42).

12.8.3.3 Proximal Storm Sandstone

This type is deposited in proximal relatively shallow water areas of the rip current formed by storm-washed shallow shoal, and it is subjected to the eddy effect in the storm decay period. It usually has an erosion bottom margin with large fluctuations, and the sandstone in the lag layer has obvious spot properties with high thickness. Medium-megascopic hummocky cross bedding is developed, which has the features of coarse massive graded bedding.

In addition, fine-grained thin-sheet quartz sandstone formations are often developed in the lagoon facies mudstone after the barrier island, with a wavy erosion bottom margin and a normal

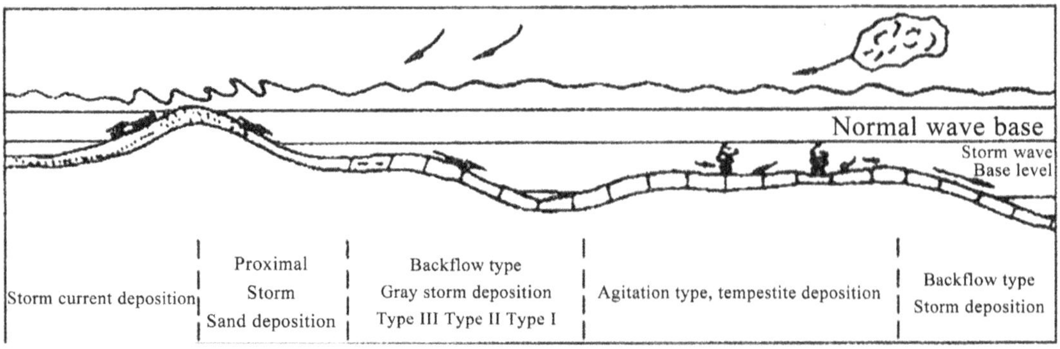

Fig. 12.43 The storm rock is mainly formed between the normal wave datum and storm wave datum, and also formed near or above the normal wave datum in special situation

graded bed sequence. A horizontal feeding structure is often developed as the product deposited by storm sheetflood.

In summary, tempestite is mainly deposited between the normal wave base and storm wave base, and it can be formed near or on the normal wave base under special circumstances (Fig. 12.43).

12.8.4 Similarities and Differences Between Turbidity Current and Flood Flow Deposition

Storm deposit, which is the same as turbidity current and flood flow deposition, also belongs to turbulent flow deposit, a kind of high-energy and high-speed current event characterized by a mostly violent mechanism and abrupt or stimulating events, leading to the features of short time (hours to days), large destructive effect, and strong constructive effect (rapid deposition). The sediments deposited by each high-energy event are usually shown as the changes in the transition process from high-energy to low-energy to the ordinary state. Hence, the features of rapid deposition (deposition in the peak period of an event), to slow deposition (deposition in the decay period of an event), to suspended deposition (deposition in the intermission of an event) are shown respectively from the bottom up, and the sedimentary sequence is from coarse to fine. However, each event deposit differs in its formation mechanism, provenance, and sedimentary environment, resulting in different sedimentary sequences and formations.

12.8.4.1 Common Characteristics of Storm Deposit and Turbidity Current Deposit

(1) Erosion basement surface: Because of the strong washing-out and destructive effects of flow into the basement in high-energy and high-speed events, the bottom surface is an irregular, wavy, or uneven basement, developed with various sole marks or erosion structure marks.
(2) Graded bedding or lag deposits in the lower sequence: An event resulting from turbulent flow can both carry sediments of different grain sizes, and disperse and stir up unconsolidated or semi-consolidated underlying sediments in the event area to form high-density sediments dominated by the action of gravity. Therefore, the lag deposits formed by turbidity currents and floods are usually coarse fragments composed mainly of gravel, while the tempestite is shell debris, often forming a shell lag layer.
(3) Parallel texture with upper flow regime in the middle sequence: Owing to the strong disturbance of normal current in the basin by an event current, both lateral flow (linear flow)

Table 12.8 Comparison of characteristics of tempestite and turbidite deposition

Features	Tempestite	Turbidite
Graded characteristics	Non-uniform thickness, poor lateral extension, dominant normal graded bed sequence, sometimes reverse graded bed sequence, irregular change, without Bouma sequence	Uniform thickness, distant lateral extension, normal graded bed sequence, regular change, with Bouma sequence
Contact relationship	Abrupt change of granularity between graded bedding and parallel bedding	Gradual change of granularity between graded bedding and parallel bedding
Sedimentary structure	With hummocky cross bedding	Without hummocky cross bedding
	Lack of flute cast, with erosion filling structure	Different flute casts
	Coarse lag deposits, without gradual change of granularity	Gradual change of granularity common at the bottom
Formation environment	Neritic shelf	Dominated by deep-water continental slope
Development	Superimposition of individual or several graded beddings	Superimposition of multiple or tens of normal graded bed sequences

and longitudinal flow (oscillating flow), as well as a complex system with the upper flow regime in uncertain directions, are generated, leading to the formation of parallel bedding, horizontal bedding, ripple bedding, deformed bedding, and climbing ripple bedding.

(4) The upper part or top of the sequence is the biological disturbance layer or argillaceous layer with creatures, and it is composed mainly of argillaceous rock in suspended deposition as the turbulence intermission deposition. When the next event occurs, this layer will suffer varying degrees of damage.

12.8.4.2 Differences Between Storm Deposit and Turbidity Current Deposit

(1) Different sedimentary environments: Turbidity current events occur in semi-deep sea to deep-sea environments (basin slope-basin).
(2) The turbidite formed by turbidity current deposition is the rhythm interbed of sand and mudstone, and the sedimentary sequence is Bouma.
(3) Turbidity current sediments are distributed similar to turbidite fans in basinward submarine canyons of the basin slope.
(4) Different ichnofossil community compositions and characteristics: Because the water depth of turbidity current events is different from that of storm events, the associated creatures and ichnocoenosis are different as well.

Therefore, the difference between tempestite and turbidite can be determined by both the specific sedimentary structure (particularly, hummocky cross bedding and wave ripple) and the fauna characteristics of individual rock stratum, as well as rapid variation of lithofacies in vertical sedimentary sequences. Coarsening-upward storm deposits have been reported for each part in different times, and these cycles can be interpreted based on the relative fluctuations of the sea level. Both storm current and turbidity current are density currents, in which the storm current is the ebb current or backflow of storm wave and storm tide, while only the turbidity current has the flow of a density current. There is an obvious difference between the two (see Table 12.8).

References

Barwis, J.H, and M.O. Hayes. 1978. Regional patterns of barrier Island and tidal delta deposition applied to hydrocarbon exploration. In *Stratigraphic concepts in hydrocarbon exploration.*

Butler, G.P., P.M. Harris, and C.G.S.C. Kendall. 1982. Recent evaporites from the Abu Dhabi coastal flats. In

Deposition and diagenetic spectra of evaporites, ed. C.R. Hand-ford, R.G. Loucks, and G.R. Davies, Society of Economic Paleontologists and Mineralogists Core Workshop 3, Calgary, 33–64.

Clifton, H.E., R.E. Hunter, and R.L. Phillips. 1971. Depositional structures and processes in the non-barred, high-energy nearshore. *Journal of Sedimentary Petrology* 41: 651–670.

Einsele, G. 1992. *Sedimentary Basins: evolution, facies, and sediment budget*. Heidelberg: Springer-Verlag.

Falcon, M. 1998. *Essentials of physical geology*. Boston, New York: Houghton Mifflin Company.

Fisk, H.N. 1960. Recent Mississippi river sedimentation and peat accumulation. In *4th Congress Pour Iavancement des' etudes de stratigraphic et de geologic ducarbonifere, Heerlen, Compt Rendu*, ed. Ernest Van Aelst.

Ginsburg, R.N., and L.A. Hardie. 1975. Tidal and storm deposits, northern Andros Island, Bahamas. In *Tidal deposits*, ed. R.N. Ginsburg, 201–208. Berlin: Springer.

Klein, G.V. 1970. Depositional and dispersal dynamics of intertidal sand bars. *Joural of Sedimentary Petrology* 40: 1095–1127.

Komar, P.D. 1976. *Beach processes and sedimentation*. Englewood Cliffs, NJ: Prentice-Hall Inc.

Kumar, N., and J.E. Sanders. 1974. Characteristics of near-shore storm deposit; examples from modern (sic) and ancient sediments (abstract); In American Association of Petroleum Geologists, Annual Meetings Abstracts, San Antonio, TX, V. L. P55.

Nelson, C.H., and A. Maldonado. 1988. Factors controlling depositional patterns of Ebro turbidite systems, Mediterranean Sea. *The American Association of Petroleum Geologists*, 698–716.

Purser, B.H. 1973. *The Persian Gulf: Holocene carbonate sedimentation and diagenesis in a shallow epicontinental sea*, 471p. Berlin: Springer.

Reinecke, H.E., and I.B. Singh. 1979. Depositional sedimentary environments with reference to terrigenous clastics. Changming Chen and Jiliang Li (Trans.). Beijing: Petroleum Industry Press.

Reinson, G.E. 1984. Barrier island and associated strand plain systems. In *Facies models*, 2nd edn, ed. R.G. Walker. Geoscience Canada Reprint Ser 1, Geological Society of Canada, 119–140.

Schwarz, H.U., G. Einsele, and D. Herm. 1975. Quartz-sandy, grazing-contoured stromatolites from coavStal embayments of Mauritania, West Africa. *Sedimentology* 22: 539–561.

Sellwood, B.W. 1986. Shallow marine carbonate environment. In *Sedimentary environment and facies*, 2nd edn, ed. H.G. Reading, 283–342. Oxford: Blackwell.

Shepard, F.P., and D.L. Inman. 1950. Nearshore circulation related to bottom topography and wave refraction Trans. *American Geophysical Union Publication* 31: 163–198.

Shinn, E.A. 1983. Tidal flat environment. In *Carbonate depositional environment*, vol. 33, ed. P.A. scholle AAPG, D.G. Bebout, C.H. Moore, 171–210.

Walker, R.G. 1984. *Facies models*, 2nd ed. Toronto: Geological Ass. of Canada Business and Economic Service Ltd.

Yu, Xinghe, Junmao Zheng, et al. 1997. The establishment of integrated model on structure, deposition and diagenesis. *Acta Sedimentologica Sinica* 15 (3).

13 Deep-Water Depositional System

As global petroleum exploration constantly goes deeper underground, exploration based on architecture has matured, and all architectures that can be found have been explored in petroliferous basins with relatively high degrees of exploration. Thus, in the future, exploration should focus on looking for lithostratigraphic reservoirs. The presence of a bathymetric or lowstand fan is the main condition for the formation of stratigraphic traps. As a result, studying deep-water depositional systems, particularly deep-water gravity flow depositional systems, is very important. One of the main scientific challenges in the 21st century is understanding resource formation and exploration mechanisms in deep water (sea).

Deep-water deposition has been studied constantly for over half a century. Although propelled by sustainable development, marine science, and other factors in coastal regions, deep-water deposition studies are driven mainly by significant scientific and actual problems based on deep-water offshore petroleum exploration and development. P. H. Kuenen et al. (1950) proposed the turbidity current theory. His student, A. H. Bouma, developed the famous "Bouma Sequence" in 1962. Ever since, with the publication of hundreds of articles on ancient and modern abyssal fans (submarine fans and slope fans), the development of deep-water petroleum exploration has been promoted greatly. The reason for this is that these fan models are beneficial for predicting internal structure units, superimposition patterns, and depositional architecture of submarine fans, for instance, the ancient submarine fan proposed by Walker (1966), modern upper fan tongue pattern proposed by W. R. Normark (1970), and ancient fan pattern with channel and tongue-shaped fan proposed by E. Mutti (1977). By combining the modern fan model proposed by W. R. Normark (1970) and the facies concept of ancient submarine fans, Walker (1978) proposed a comprehensive fan model containing a single feeder channel in the upper fan area and a superimposed fan lobe in the middle/lower fan area. Given its strong forecastability, this model has greatly influenced the petroleum industry.

13.1 Deep Water Sedimentation

The term "deep water (sea)" generally refers to the deep water environment situated on the seaward side of the continental slope break, including the deep water environments (water depth >200 m) of continental slopes, continental rises, and abyssal plains, in which the action mechanisms of sediment gravity flow (slide, slump, debris flow, grain flow, turbidity current, and mud flow) and bottom current are the leading sediment dynamic mechanisms. In this book, "deep water" refers to petroleum reservoirs formed in deep-water sedimentary environments rather than the existing petroleum reservoirs or ancient depositional systems in deep water, which mostly refer to exploration fields based on

X. Yu et al., *Clastic Hydrocarbon Reservoir Sedimentology*, Advances in Oil and Gas Exploration & Production, https://doi.org/10.1007/978-3-319-70335-0_13

exploration difficulty, particularly in terms of the water depth for drilling, for example, an exploration field located at a sea water depth greater than 600 m.

The study of deep water sedimentation was started with a focus on the Atlantic earthquake in 1929. Breaking of the trans-Atlantic submarine cable was initially considered to be caused by earthquakes; however, the sequential time delay phenomenon of cable breaking proved that it was caused directly by turbidity current activity. In addition, a study on deep water sedimentation proved that the potential for reservoir formation in deep water environments is huge (Stow and Mayall 2000).

Gaining knowledge about the sedimentation of deep water (sea) and its sedimentary facies is a long process. In the past, researchers have always recognized that deep water (sea) is a very calm, dark, and lifeless environment, where only extremely fine-grained muddy deposits are formed at extremely slow rates. Since the 1950s, starting with "Turbidity Currents as a Cause of Graded Bedding," which is a famous paper by Kuenen and Migliorini (1950), the era of studying deep-water sediment gravity flow was initiated to break the dominance of traditional mechanical differentiation in sedimentology. Thus, the important role of turbidity current mechanisms in deep-water depositional systems was gradually established in geoscience. Turbidity currents, even the classic Bouma sequence, are applied as examples for explaining the general adaptation of coarse-grained deposits in deep water backgrounds. In recent years, with studies on deep water deposition, researchers have gradually recognized that deep water deposition is subject to contour currents, tidal currents, bottom currents, and internal waves, in addition to sediment gravity flow.

13.1.1 Deep Water Action Mechanism and Concept

13.1.1.1 Gravity Flow

Since the deep-water slope area has relatively steep landforms, gravity can easily induce sediment gravity flow activities on submarine superficial sediments, including slide and slump, under the cooperation of other incentives (such as earthquakes, natural gas hydrate melting, overpressure relief, diapir activity, and burst ocean currents).

Sediment gravity flow was divided into four types by Middleton in 1973: fluidized flow, grain flow, debris flow, and turbidity current. While studying subaqueous deposition types in the Gulf of Mexico in America, Coleman (1976) discovered mud flow–type gravity flow sedimentation.

13.1.1.2 Contour Current

Heezen et al. (1959) proposed the concept of contour current after studying continental rise sediments in the North Atlantic Ocean. They defined contour current as a thermohaline circulation formed by the Earth's rotation. This circulation, parallel to the submarine contour line, flows stably at a low speed (5–20 cm/s), mainly in continental rise areas. In practical use, however, some scholars call other types of deep-water or shallow-water bottom currents, even bottom currents in lakes, contour currents. Therefore, Faugères et al. called the thermohaline circulation in relatively deep water environment due to the Earth's rotation a narrowly defined contour current.

13.1.1.3 Bottom Current

George Wust (1956) first proposed the capability of deep-sea bottom current to transport deposits. Hollister and Walter (1963) provided evidence for the existence of deep-sea bottom current and believed that this bottom current flows parallel to the submarine bathymetric contour. Stow et al. (2002) determined several related concepts and assumed that bottom current refers to the ocean currents that act in deep water and are driven for circulation by thermohaline or sea wind in the ocean and its marginal sea. It does not strictly follow the bathymetric contour, however contour current is still widely considered to be similar to bottom current. According to the size of the particles that can be transported by bottom current, contourites, with grain size ranging from mud to sand level, have a series of transitional facies composed of sand, silt, and clay mixtures, presenting a typical varicolored appearance. When the seabed is denuded by extremely strong water, contourites detained by gravels can be formed.

13.1.1.4 Internal Wave

Gao Z. Z (1990) and He Y. B (1997) among others studied internal wave sedimentation. According to their studies, an internal wave is a type of underwater wave on the boundary between two water courses with different densities or within a water course with a density gradient. Internal waves can be found in all oceans, with greatly varying amplitudes, periods, propagation velocities, and existence depths. When the period of internal waves is the same as that of a tide, the waves constitute an internal tide.

While studying the Ordovician in the middle section of the Appalachian in America, Gao et al. (1995) first identified internal-tide deposits in the channel, applied the term "internal-tide deposit" for the first time in the world, and systematically described and genetically interpreted internal-tide sedimentary facies. Based on worldwide findings thus far, internal wave deposits can be summarized into five sedimentation types: ① bidirectional cross bedding sandstone; ② unidirectional cross bedding and cross lamination sandstone; ③ thin rhythmic sand-mudstone interbed; ④ foraminiferal limestone with flaser, wavy, and lenticular bedding; and ⑤ sandy oolitic limestone with flaser, wavy, and lenticular beddings, and lateral accretion cross bedding.

In addition, in special circumstances, deep water may be affected to produce stormy bottom current reflux by storm action, resulting in the formation of distal tempestite (for details, please see Chap. 11). Most storm actions are usually weak in deep water; they can perform sediment transformation but not have the capacity to transport sediments on a large scale.

13.1.1.5 Combined Action of Two Deep Water Action Mechanisms

In most cases, with the joint interaction of deep water action mechanisms in two or more directions, sandy bedform and a sedimentary structure with complex genesis are formed in abyssal environments. For instance, the sediments on the continental slope (Biscayne Bay) on the Aquitaine of the Landes platform have resulted from the dual influences of traction and gravity, as opposed to only the previously considered slope failure. In 2003, Shanmugam et al. systematically summarized various action mechanisms of deep water environments and possible transforming relationships between corresponding sedimentation types (Fig. 13.1).

Overall, tides and waves are the main agents in shallow shelf environments, while deep water environments (canyons, slopes and basins) are characterized by mass flow (sliding, slump, grain flow, debris flow, turbidity current, or mud flow), various bottom currents, and pelagic/hemipelagic deposits. Turbidity current is rare in canyons, but common in basins. The uniqueness of underwater canyons is presented in the process, in which the action in the canyon cannot be controlled by shelf break. More importantly, tide bottom current and mass flow move through canyons, so attention shall be paid to the up-and-down motion of tide bottom current in a canyon. The circular movement of bottom currents and wind-driven bottom currents that follow the bathymetric contour is an important mechanism in deep water environments outside of canyons.

13.1.2 Determinants and Architectural Elements of Deep-Water Depositional Systems

13.1.2.1 Influences of Submarine Geomorphic Features and Tectonic Movement

Tectonic movements of land and sea area, and (structural) landforms formed by such movements

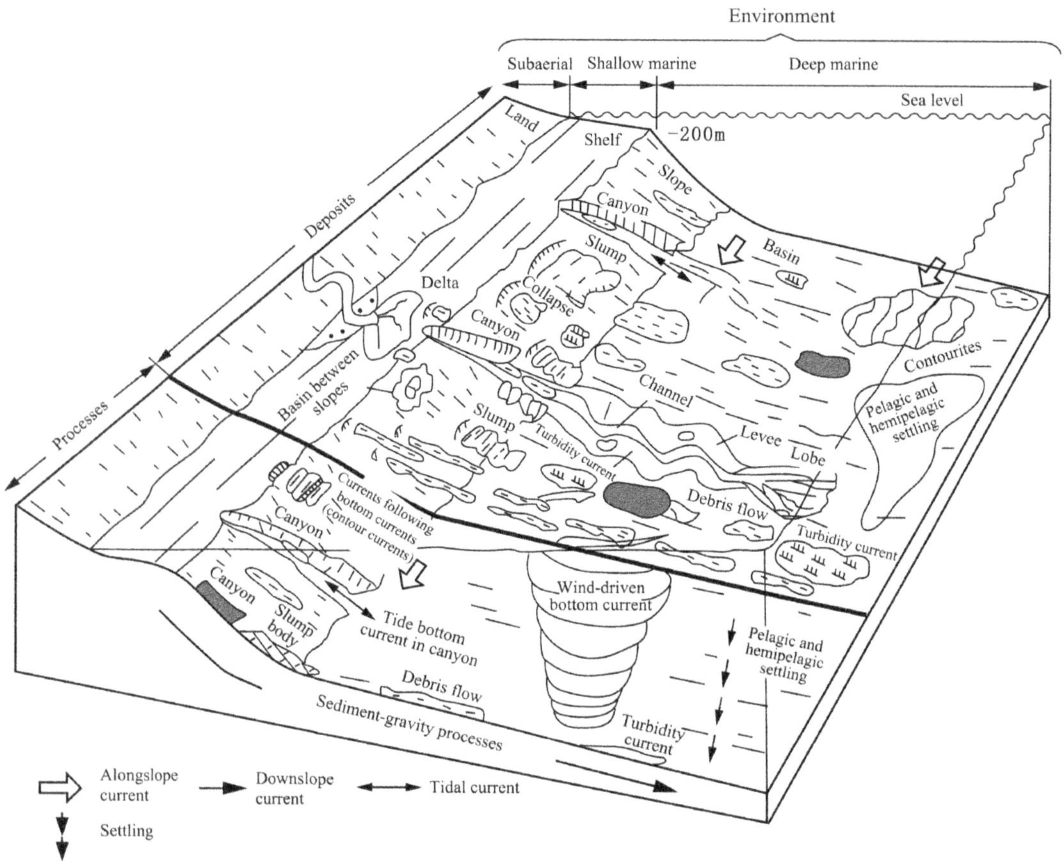

Fig. 13.1 The deep sea environment often has the interaction of two or more different directions of deep water mechanisms, forming bed sand topography and sedimentary structure with complicated genesis (after Shanmugam 2003)

have direct and great influence on deep water sedimentation.

1. **Influences of submarine landforms**

These influences are specifically presented as follows: submarine landform affects water current to influence deep water sedimentation and propagation of sedimentary systems; protruding continental margin landforms can interact with ocean current to produce the water current aggregation effect, resulting in strong erosion on the seabed, so that a flat erosive terrace (similar to the Albacora terrace built by the Santos Basin in the continental margin of southeast Brazil) is developed. Moreover, changes in submarine landform cause oceanic eddy currents when current velocity varies. This is a key factor controlling bottom current on the outside shelf-upper slope boundary. They sweep deposits from the shelf to the slope, acting as a huge rotary brush, which is called "seabed winnowing."

2. **Influence of tectonic movement**

Tectonic movement not only controls (affect) deep-water sedimentation filling but is also the main inducing factor for the formation of deep water gravity flow. Deep canyons built by basin margin slopes or seabeds are more or less related to tectonic movement, resulting mostly from the induction of gravity flow to erode the seabed due to bottom current or turbidity current upon early faulting. With the evolution of tectonic movement, filling deposits vary. Generally speaking, the stronger the tectonic movement, the deeper is the canyon erosional cutting. In addition, structure may result in submarine vulcanian eruption and tephra deposits.

13.1.2.2 Feature Influences of Continental River Basin System and Shelf Shallow Water Area

The states and changes of continental river basins and shelf shallow water areas directly influence the substance supply possible in deep water environments and play dominant roles (Fig. 13.2) in deep water sedimentation and sedimentary systems. However, submarine canyons are the main channels connecting the shelf and slope, and for transporting coarse clastic deposits.

1. **Influence of continental river basin system**

This is presented mainly in terms of sediment supply capability and supply quantity. When a continental river basin is a large river, a large or oversized delta system is often built on the continental slope, which provides sufficient provenance for deep water deposition. On account of diverse climates, the development states of continental river systems differ greatly, followed by changes in sediment supply, with deep water sediments presenting diverse characteristics. Concentrated debris supply forces the shelf to produce foreset and deep water turbidity current deposition on a large scale, and suppresses the homochronous high sea level effect. When provenances are insufficient, the basin is starved. Hence, the basin can be dominated by muddy

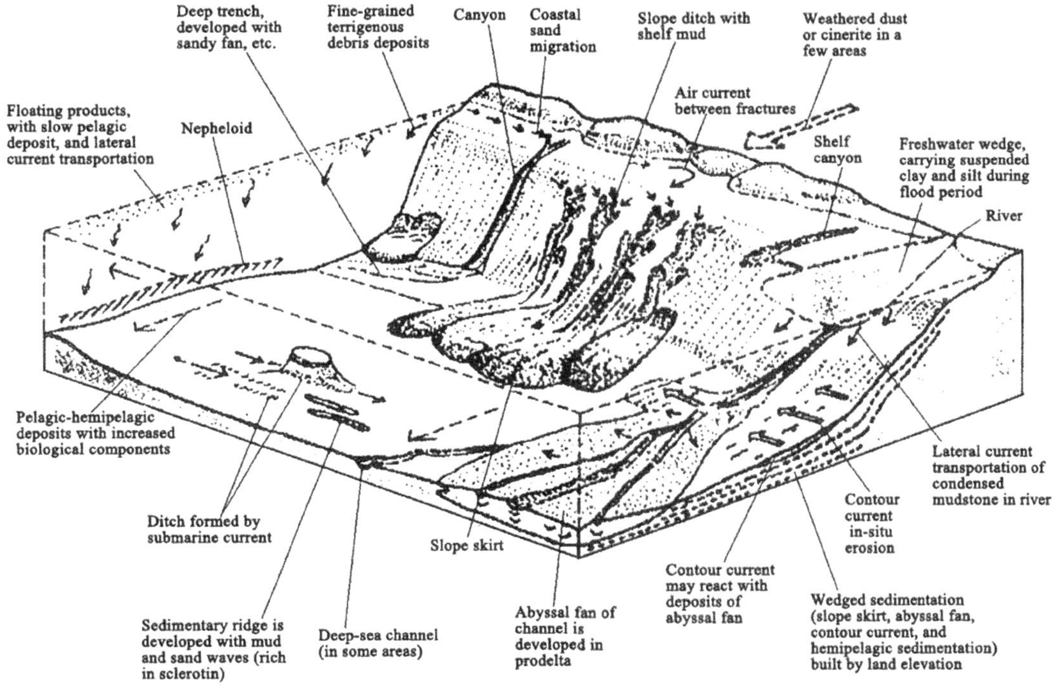

Fig. 13.2 The situation and change of continental basin system and shallow water of continental shelf directly influence the possible material supply of deep water environment, and plays an important role in controlling the sedimentation and sedimentary system of deep water (after Einsele 2000)

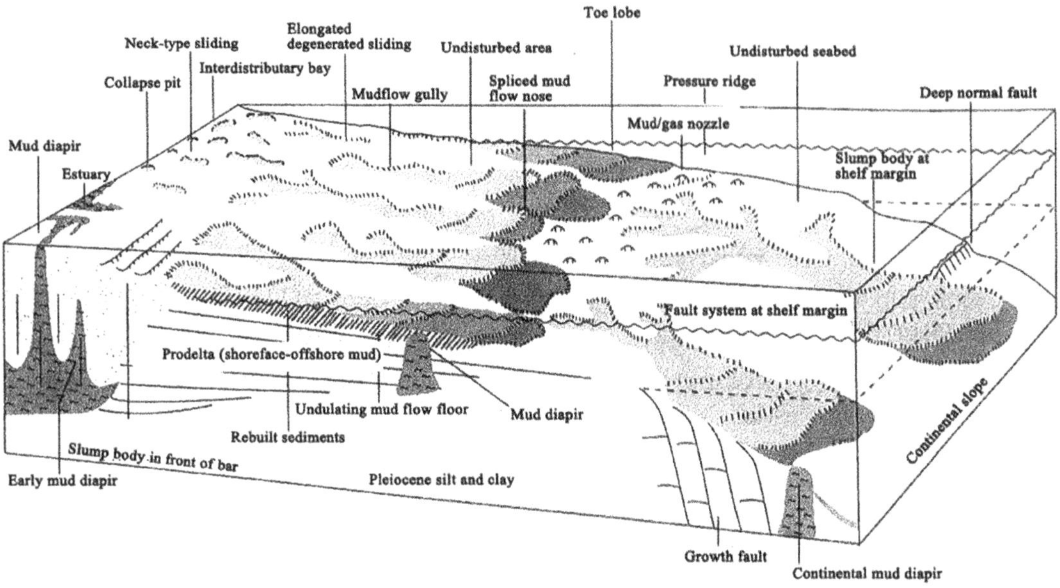

Fig. 13.3 Typical underwater sedimentary types and their distribution in Mexico Bay (after Coleman 1976)

fillings, and muddy deposits are usually formed to produce a "muddy carpet" (Fig. 13.3) under tectonism.

2. **Feature influences of shelf shallow water area**

When a shelf shallow water area is very wide and gentle, sediments must be transported to deep water area over a long distance, however gravity flow deposition is not easily produced in deep waters. Large-scale gravity flow deposition needs a long-term and persistent sediment supply. However, when the shelf shallow area is narrow but steep, sediments are transported easily to deep water areas, so various gravity flows are formed easily, particularly turbidity channels are more developed.

13.1.2.3 Effect of Sea Level Fluctuation

Sea level eustacy has great influence on deep-water action mechanisms. In a continental margin (such as northwest Europe) of high-latitude glaciers, continental glaciers advance seawards during the glacial period, and the deep water environment is dominated by gravity (flow) action. On the contrary, when the sea level rises relatively and gravity (flow) decreases in the interglacial period, the position of the contour current is relatively optimal. Rapid sea level rise is one of the main reasons for contourite formation. In the northwest of Okhotsk, sea level fluctuation plays a critical role in deep-water action mechanisms and sedimentary patterns. In a quaternary, strong bottom current actions along the continental slope cause contour accumulation and sediment waves to form drifts.

13.1.2.4 Architectural Elements of Deep-Water (Sea) Depositional System

Identifiable architectural elements of deep-water depositional systems include the following:

(1) Absence of erosion, and other boundaries;
(2) Erosion sliding and slump scour;
(3) Canyons, oceanic troughs, channels, and gullies;
(4) Channel natural levees and overbank sediments;
(5) Sedimentary lobes (isolated, clustered, and crevasse);

(6) Irregular mound bodies (slide body, slump body, and debris flow mass);
(7) Contour drift (elongated mound, irregular fragment, and contour current fan);
(8) Sheet and drape on slope, basin, fan body, and drift;
(9) Megascopic turbidity accretion and other megascopic strata; and
(10) Tectonic characteristics (growth fault, bottom diapir, and compression fault).

These architectural elements are usually hundreds to thousands of meters wide and several to hundreds of meters thick, usually with irregular overall landforms and equally axial or stripped shapes. Each element may have a series of different scale ranges with similar features. The sedimentary components that form each element include combined vertical sequences, which are all changeable but have typical features (Fig. 13.4) within a certain range.

13.2 Sediment Gravity Flow

In well-known shallow-water environments, most sediments, including gravel, sand, silt, and clay, are transported via fluid, in conformity with Newton's fluid law, downward by virtue of the fluid. Hence, such flow is called fluid gravity flow. Moreover, sediment gravity flow is the general term accounting for the simultaneous flow of gravel, sand, silt, clay, and other sediments with water. Hence, it is a movement in which sediment particles push sedimentary bodies. Sediment gravity flow refers to a high-density fluid diffused with a number of sediments that flows under gravity. In the literature, references to gravity flow usually mean sediment gravity flow.

13.2.1 Characteristics of Sediment Gravity Flow

Sediment gravity flow is the product of paroxysmal, instant, and transient rapid sedimentary deposition events. Because it contains a plethora of suspended matter, the fluid is high in density, peaking at 1.5–2.0 g/cm^3. The suspended matter includes sand, silt, and mud, and sometimes even gravels. In other words, sediment gravity flow is mass flow, mainly including fluidized gravel, silt, and clay, which flow in downward with the aid of gravity. Sediment gravity flow is non-Newtonian fluid flow, thus it does not conform to Newton's inner friction law. The relationship between shear stress thereof and shear deformation rate is as follows:

$$\tau = \tau_B + \eta \frac{du}{dy}$$

where

τ_B yield stress;
η plastic coefficient of viscosity or rigid coefficient of viscosity.

Sediment gravity flow can be divided into subaqueous and onshore. Subaqueous sediment gravity flow refers to the hyperpycnal flow formed by mixing the water and deposits flowing at the bottom of the water body. It is not difficult to define this flow:

(1) This flow must involve the entire massive movement, with poor sorting or without sorting. Spectrum evolution must occur in sedimentary gravity flow. Only the turbulence (eddy) mechanism of turbidity current, as well as the high aquosity thereof, is the transitional type between sediment and fluid gravity flows.
(2) In this flow, clastic particles are necessarily transported by virtue of matrix intensity, collision and jumping among particles, or turbulent suspension and combined support of several support mechanisms. The grain size probability graph of gravity flow sediments must present a uniformly suspended linear segment or an abruptly changed arc line segment in the graded suspension. The C-M graph of gravity flow sediments must be a specific graph parallel to the C = M base line.
(3) This flow must produce erosion scour on the slope or gully through which the fluid flows.

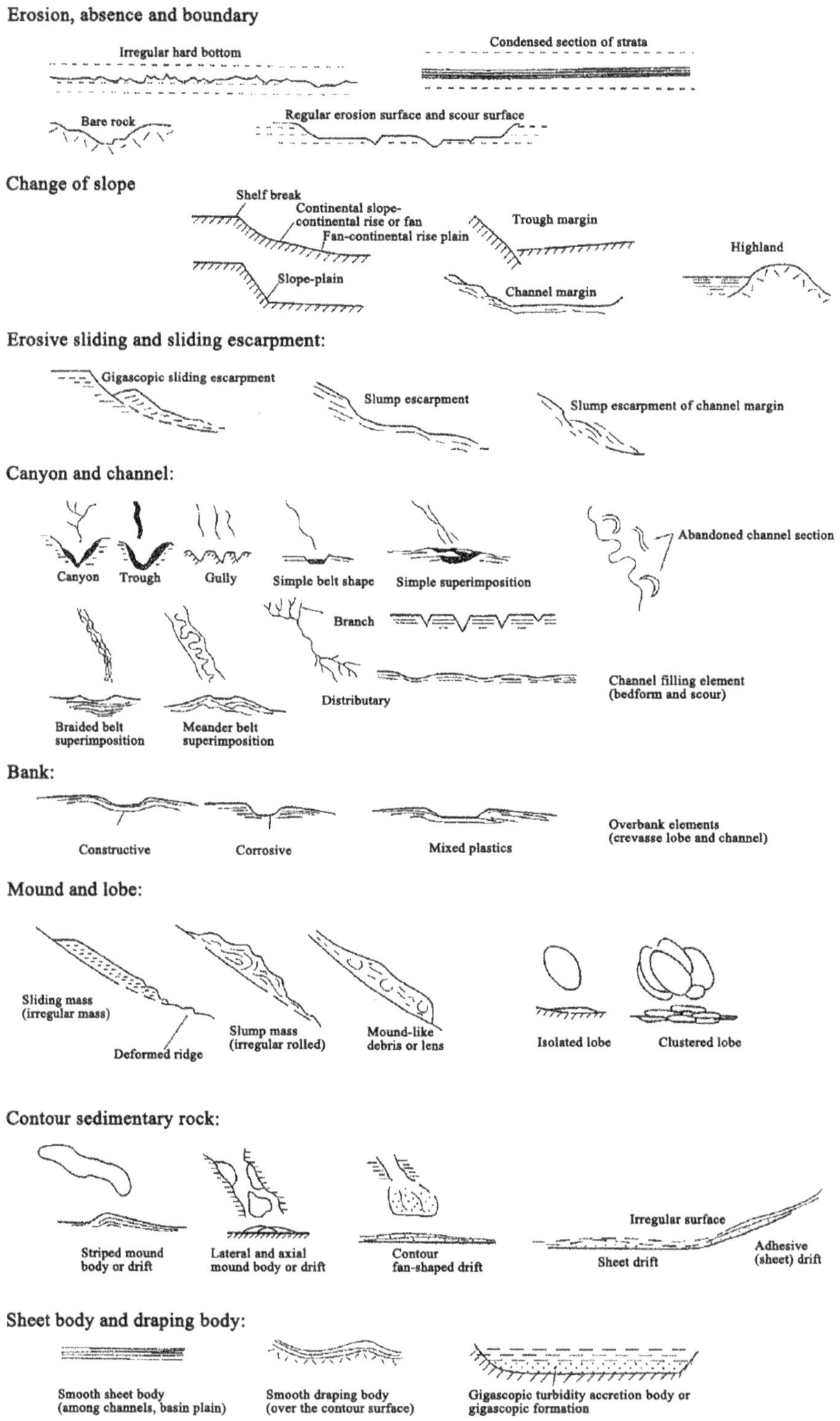

Fig. 13.4 Main configuration elements of deep-water sedimentary system (D. A. V. Stow 1999)

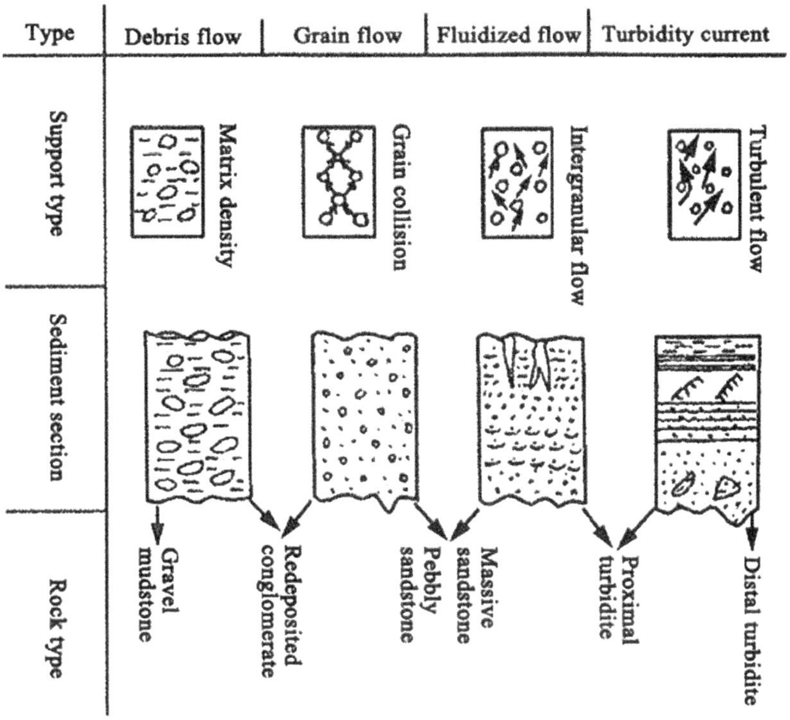

Fig. 13.5 Gravity flow deposits can be divided into debris flow, turbidity flow, particle flow and liquefaction flow (after Middleton Hampton 1976)

A truncated or crumple structure is usually formed by vertical or lateral erosion on a slope or in a gully.

(4) This flow can occur both onshore and underwater. Onshore sediment gravity flow, based on debris flow, is easily destroyed and cannot be well preserved due to the weather denudation action of the Earth's surface. Subaqueous sediment gravity flows, formed in low-lying areas of the Earth's surface, such as lakes and oceans, are preserved for a long time without denudation.

13.2.2 Classification of Gravity Flow

According to the degree of disintegration of migrated sediments, mass-gravity transportation and sedimentary deposit areas are classified into several categories as follows: rock fall, sliding, and sediment gravity flow. On the basis of debris support mechanism, i.e., the mechanism of debris in suspended state, Middleton and Hampton (Hampton 1976) divided gravity flow into four types: debris flow, grain flow, fluidized flow, and turbidity current (Fig. 13.5; Table 13.1). The abovementioned action mechanisms may occur together or transform mutually (Fig. 13.6) at a mass transportation event.

13.2.3 Basic Sedimentary Characteristics of Different Gravity Flow

13.2.3.1 Debris Flow Deposit

This is the most important type of carbonate rock gravity flow and can be usually found in clastic rock gravity flow. The associated debris flow deposition is usually composed of deposits with wide particle diameter ranges (from several millimeters to several meters), often massive in type, without sorting and graded bed sequence. Sometimes a normal graded bed sequence can be shown at the top. Debris flow deposits may be both the filling body of channel and sheet output. According to the matrix content, debris flow can be divided into mud-rich and mud-poor types.

Table 13.1 Internal architecture and sequence characteristics of the four types of gravity flow

Classification of gravity flow		Mechanical property	Sediment transportation and support mechanism	Sediment structure	Vertical sequence	Lithologic section
Mass flow	Debris flow	Plasticity	Shear action is distributed in entire sediment mass, with the matrix's support strength generated mainly from adhesive force, secondly from buoyancy; inviscid deposits supported by dispersive pressure present inertia when flowing at high concentration, but adhesion at low concentration. It is generally developed on steep slopes	Supported by substrate and matrix, with greatly changeable grain size of debris, unequal matrix content, flow structure, and rip up structure	Graded difference, possibly with reverse size grading	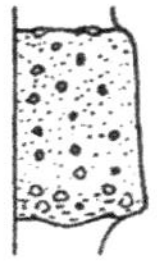
	Grain flow			Basically without continuous deformation of bedding, but with some plastic deformation on outer margin and bottom part	Without grading, but with reverse grading near bottom surface	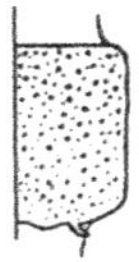
Fluidized flow	Fluidized flow	Viscidity	Loose tectonic framework is changed into tight framework due to destruction, and fluid moves upward to support non-viscous sediments when gradient is greater than 3°	Dish structure, water escape structure, flame-shaped to load cast structure, convolute bedding, etc.	Poor grading (coarse-tail grading)	
	Turbidity current		Supported by turbulence	Matrix supported, with greatly changeable random fabric and grain size of debris, unequal matrix content or even reverse grain size grading, flowing structure, and rip up structure	Grain size grading, Bouma sequence, etc.	

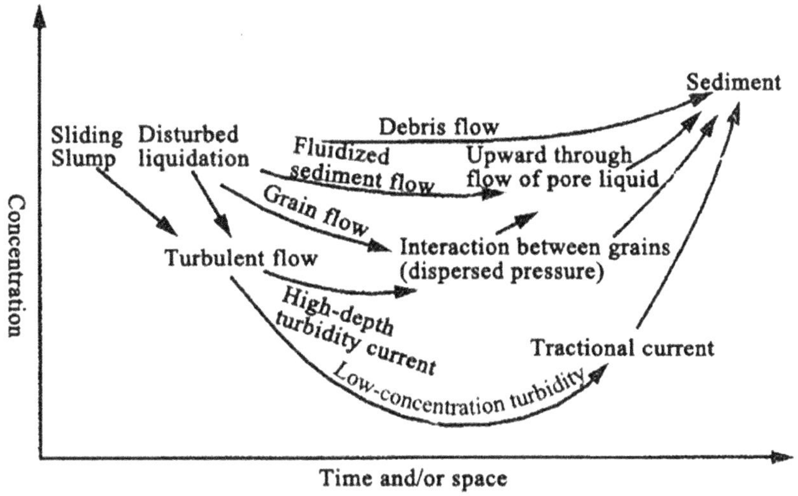

Fig. 13.6 The presumed interaction of different weight events in a primary weight flow transport event (after Middleton and Hampton 1976)

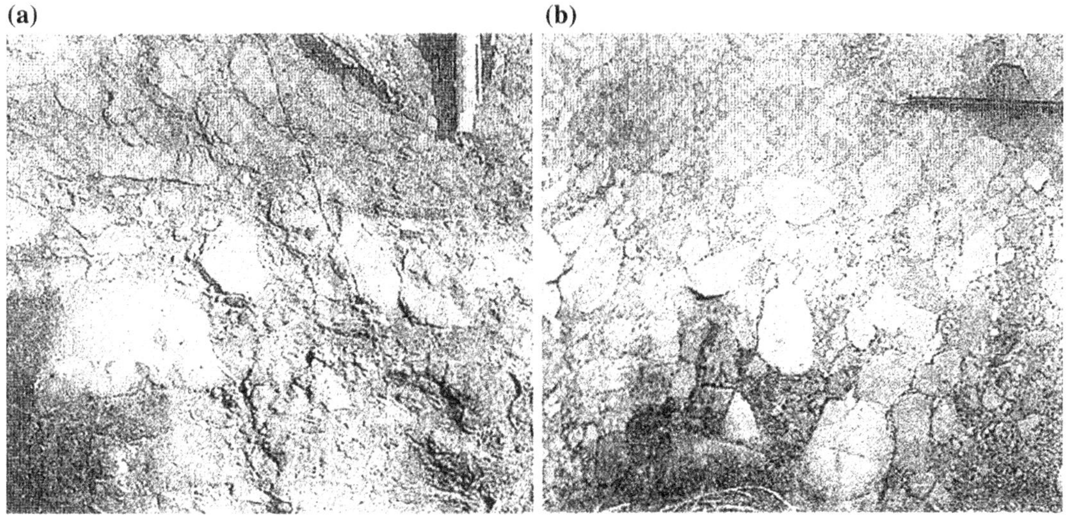

Fig. 13.7 Classification of debris flow

1. **Mud-rich**

Debris flow is rich in clay or plaster matrix, typically supported by a matrix (Fig. 13.7a) in which gravel particles float. This reflects that debris flow is supported completely by matrix intensity and buoyancy during flow, which is typical.

2. **Mud-poor**

The mud content of such a debris flow deposit is rather low, generally with particle support (Fig. 13.7b). This reflects that mutual contact between particles is a support factor during debris flow. Mud and water matrix without a high content can also play a role in lubrication, in addition to the provision of buoyancy and yield strength.

13.2.3.2 Grain Flow Deposition

Because the formation of grain flow requires a rather high gradient, grain flow deposits cannot be usually found in sedimentary basins without a high gradient; even when they are present, it is on a small scale. This is because grain flow is example specific (Table 13.2; Fig. 13.8).

The thickness of grain flow is usually only several centimeters, and the thickness of gravel-containing grain flow deposits is only dozens of centimeters. One of the most significant features of grain flow deposition is the development of reverse grading or thick-tail graded bedding, however this development is generally confined by the middle and lower parts of the sequence. A normal sequence can still be usually found at the top, while most dilacerated

Table 13.2 Classification of mass weathering (according to Barbara, modified in 1995)

Landslide event		Sediment flow		Weathering of subaqueous mass
Slump		Mud flow	Thawed mud flow	Slump
Collapse	Rock fall		Debris flow	Sliding
	Clastic collapse		Mud flow	Flowing
Slide	Mass sliding	Granular flow	Creeping	
	Rock mass sliding		Mud flow	
	Debris sliding		Grain flow	
			Clastic collapse	

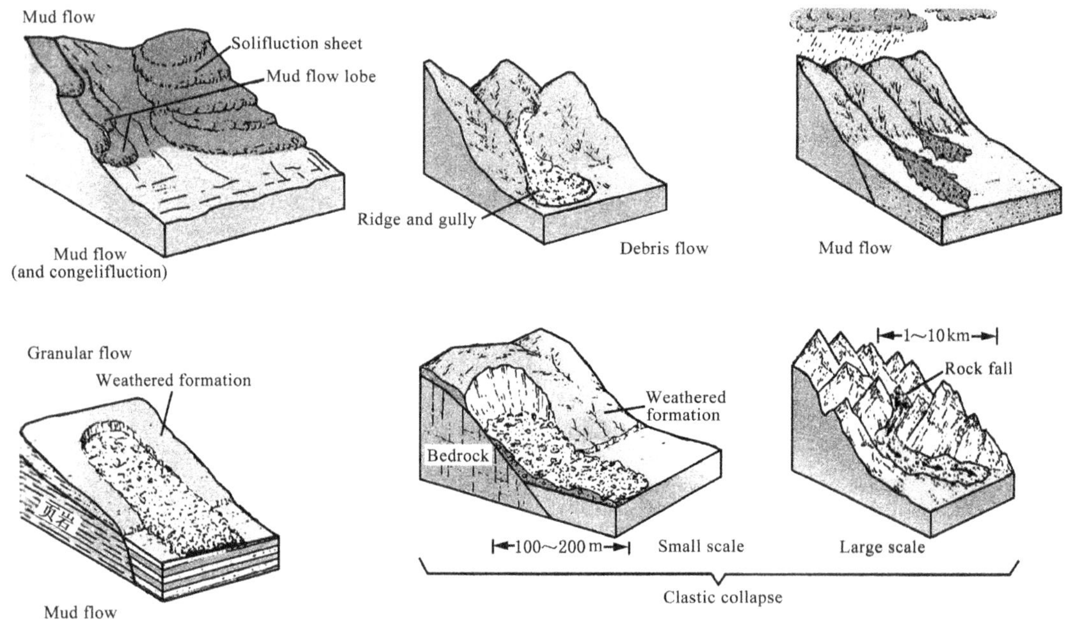

Fig. 13.8 Depositional model of mud flow and granular flow (after Barbara et al. 1995)

gravels are found in the middle coarsest layer. The second feature is that the matrix content is low, and sparite cement can be found generally in carbonate grain flow. Rare deposits are of great significance to ancient landforms.

13.2.3.3 Fluidized Flow Deposition

The critical condition for the formation of fluidized flow deposition is that deposits full of water are rapidly accumulated when the supplied deposits are thin. The small number of fluidized flow deposits found in China are mainly terrigenous debris deposits, often with entire massive formations, normal graded bed sequences in the bottom of unit formations, and dish structure development sections (Fig. 13.9) above the underdeveloped parallel texture upward; the "disk" width decreases and tortuosity increases from the bottom to the top. As a consequence, a deformation structure, particularly the dish structure, is one of the main marks. Upward dish structures disappear gradually to form non-structure sections, however water escape structures can be found when liquidation is strong. The boundary between the top and the bottom parts of the unit layer is in abrupt contact with the upper and lower layers. However, there is no obvious erosion surface, and the bottom part may include gutter cast. Dominated by middle-fine sandstone, it has few components and its texture maturity is low.

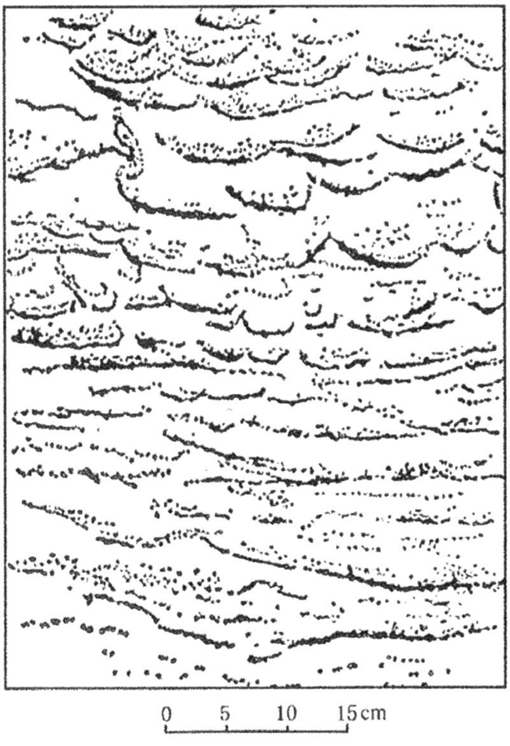

Fig. 13.9 Dish structure (according to Liu et al. 1985)

13.2.3.4 Turbidity Current Deposition

Turbidity current sedimentation or turbidite is the earliest and the most thoroughly studied gravity flow sedimentation type. Typically, turbidity current deposits have been found to have developed in the continental strata in China. The most prominent and apparent feature inside turbidite, the vertical association of which is the Bouma sequence, is graded bedding. The Bouma sequence represents the graded bedding of grain sizes (from strong to weak) formed by turbidity current sedimentation. The classic Bouma sequence consists of five formations, each of which emphasize the varying clastic particles therein confined by the boundary between the top and the bottom parts such that the lower part is thicker than the upper part (Fig. 13.10). Therefore, the law indicates that large particles are first deposited when the energy of turbidity current decreases, and then small particles are deposited sequentially to form the aforementioned graded bedding.

A turbidite body generally has a head, body, and tail. The head, with a thickness greater than those of the other parts of turbidity current, and flow velocity slightly lower than that of the body, is the erosion area composed mainly of coarse debris deposits. The body part behind the head has uniform thickness and flows almost stably, and fluid with higher density therein usually and gradually supplements the head to remedy energy loss due to turbulence. The tail part of turbidity current, or the terminal of turbidity current, thins quickly, or even has rarefied thickness. Several parts of a turbidite body are usually advanced forward beyond the deposits, for example, upon head deposition, the body deposits can move beyond the head deposits, while the tail part is deposited in the uppermost

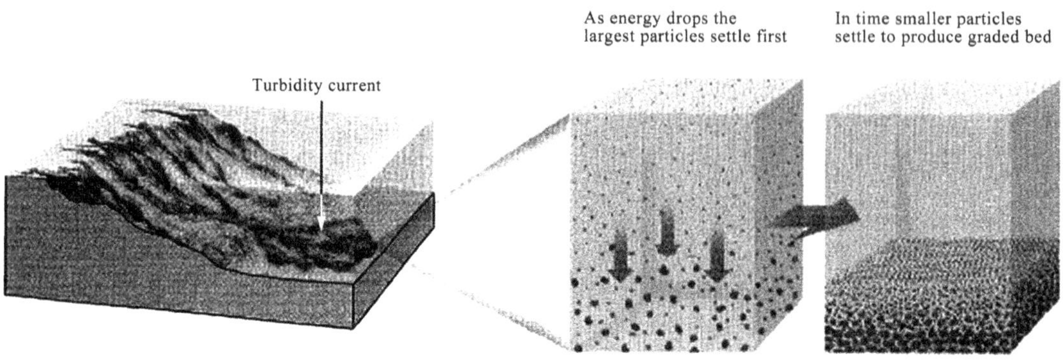

Fig. 13.10 Diatactic structure (from Tarbuck and Edward 1984)

and farthest places. Thus, on the one hand, turbidity sediments can form a vertically upward fining sequence; on the other hand, they fine gradually from proximal to distal turbidite particles. Graded bedding, accompanied with a bottom mould, can be considered as the feature for identifying turbidites.

According to the physical features of turbidity current sediments, turbidity current sedimentation can be divided into two categories.

1. **Low-density turbidity current deposit**

This is generally called turbidite or typical turbidite, and its characteristics can be described by the Bouma sequence. It is the most widely distributed gravity flow deposit, developed both in terrigenous detrital and carbonate rocks. The AE and ABCE combinations are the most common. Terrigenous clastic rock particles mainly include silt to medium and fine particles; and carbonate rock particles mainly include carbonate debris with certain roundness from the platform margin.

2. **High-density turbidity current deposit**

Such deposits are turbidites that cannot be ignored. They have obvious features and good multi-period repetition, although they are distributed in a limited way. A high-density turbidity current deposit generally includes medium and coarse sands, and fine gravels. Coarse tail graded bedding, parallel bedding, and medium-large cross bedding are regularly presented in the vertical direction, resulting in the formation of similar Bouma sequences such as ABCE, ABE, and ABC (Fig. 13.11a).

Sometimes, a sedimentary structure combination (Fig. 13.11b) is formed under the upper flow regime, with coarse tail graded bedding replaced by anti-dune bedding at times, which reflects the characteristics of hyperpycnal flow. Such deposits are usually developed in seabed channels, and they coexist with debris flow sedimentation.

D. R. Lowe (1982) pointed out that low-density turbidity currents including clay, silt, and fine to medium-fine sand particles are transported in a suspended manner through fluid turbulence, regardless of their concentrations; high-density turbidity currents containing many types of particles from clay to fine grit are transported jointly by subsidence resulting from turbulence and high concentration and the suspension force of fine-particle substances, which is effective when the particle concentration is greater than 20–30%.

13.3 Formation Mechanism of Sediment Gravity Flow

13.3.1 Gravity Flow Formation Conditions

According to numerous scholars' studies and practical experiences at home and abroad, there

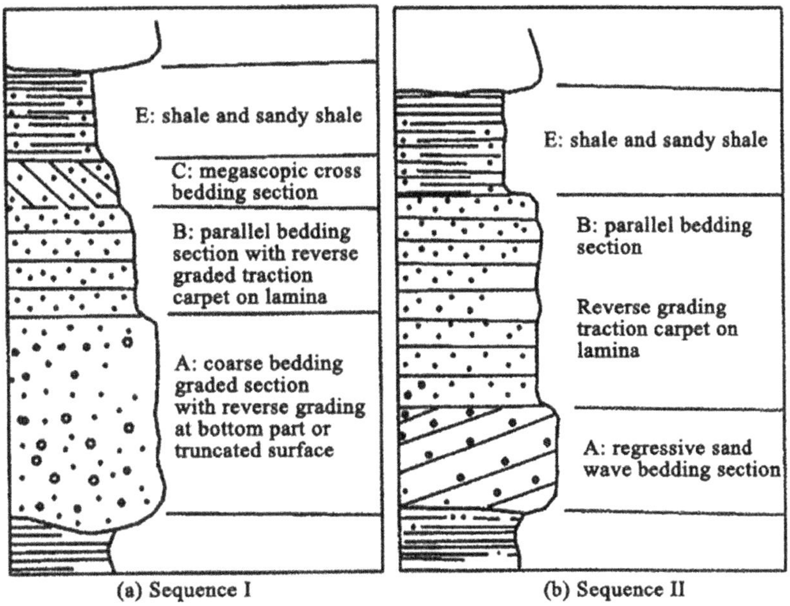

Fig. 13.11 Sedimentary sequence of high density turbidity currents (according to Feng et al. 1994)

are five necessary conditions for gravity flow: sufficient water depth, adequate gradient and density differences, equivalent water regression, abundant provenance, and certain trigger mechanisms.

13.3.1.1 Sufficient Water Depth

Sufficient water depth is a necessary condition to prevent gravity flow deposits from being scoured and destroyed after they are formed. Usually, the water depth of gravity flow deposits is 1500–1800 m, with the minimum and maximum being about 100 and 8000 m, respectively. Galloway (1996) believes that continental slope and slope bottom depositional systems, which are important features of gravity flow deposit, are built mainly in relatively deep water areas below shelf breaks. However, debris flow, as an exception, can be developed in onshore areas, where the water depth is not considerable. As for sufficient water depth in the relative sense, oceans differ from lakes. Regardless of the actual water depth, the forming depth must be below the storm wave base.

13.3.1.2 Adequate Gradient and Density Differences

To maintain constant turbulent flow in a fluid, energy needs to be supplied stably. Adequate gradient is an objectively necessary condition for triggering the mass movement of unstable deposits. Wesere (1978) assumed that the minimum gradient is 3°–5°, while the typical terrigenous clastic slope is 2°–5°. However, according to many practices, the minimum gradient for forming gravity flow is 2°–3°. The submarine slump gradient of the Mississippi Delta is only 0.5°. The sufficient condition for the formation of gravity flow is the adequate density difference between the gravity flow and lake water. In other words, the density of gravity flow obviously compensates for the gradient of gravity flow.

13.3.1.3 Equivalent Water Regression (Unstable Background)

Equivalent water regression includes equivalent marine regression and equivalent tidal regression, with the objective action of enabling deposits to be in the touch-and-go unstable state with greater gradient. Equivalent water regression may be regional structural water regression resulting from crustal ascent, partial structural water regression caused by contemporaneous fault movement, or sedimentary water regression arising out of constant accretion of delta and alluvial fan deposits.

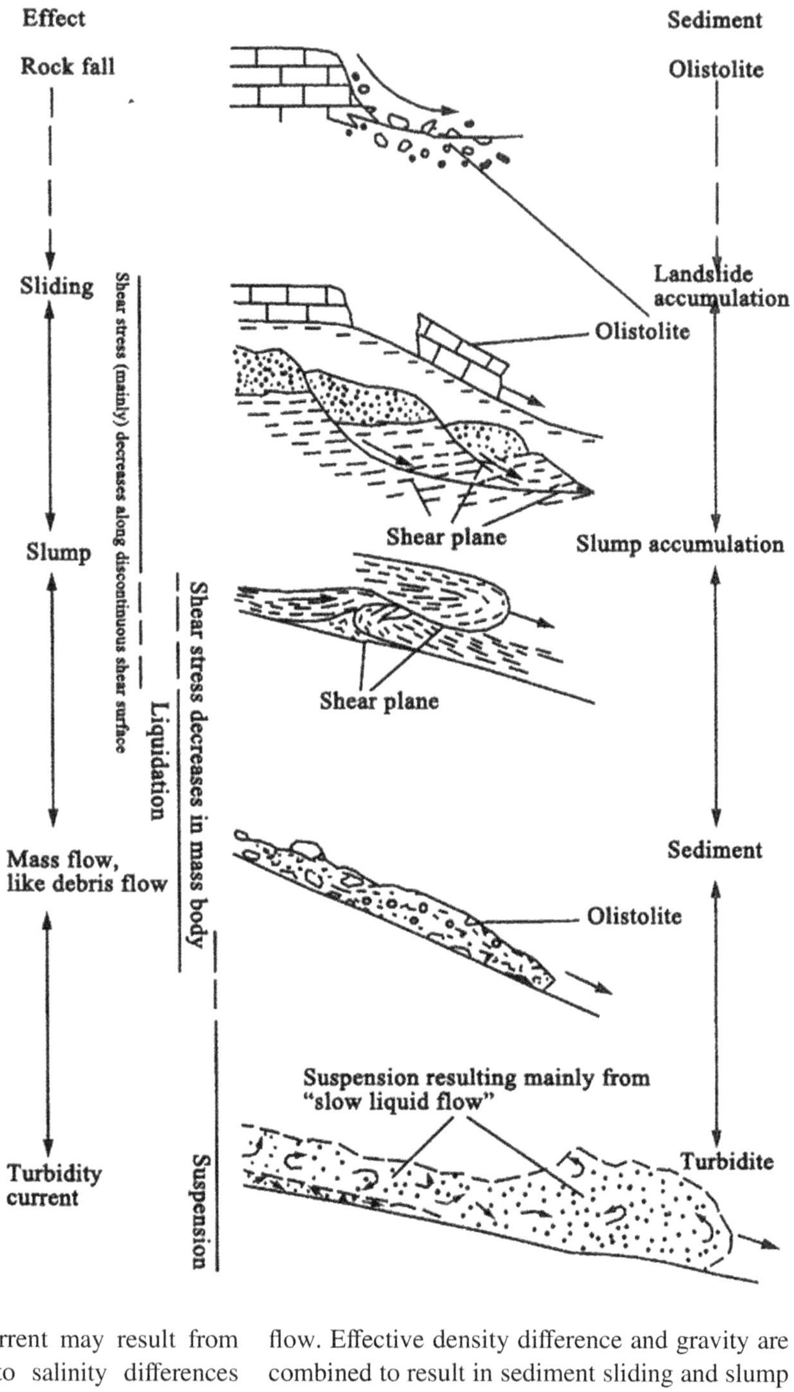

Fig. 13.12 Combination between the effective density difference and gravity face, sliding and slumping of sediments would occur under certain slope, which would cause sediment gravity flow (after Kruit 1955)

In water, a density current may result from density differences due to salinity differences (saltwater wedges in estuaries), temperature differences (cold currents formed by the melting of flowing ice and snow in a lake, as well as cold and warm currents in the ocean), or sediment concentration. Gravity flow diffused with a number of deposits constitutes a type of density flow. Effective density difference and gravity are combined to result in sediment sliding and slump under certain gradients; sediment gravity flow reverses to produce turbulence therein to support sediments suspension in order to prevent them from being deposited by diffuse turbidity current, which leads to turbidity (Fig. 13.12) at the front end of a slump body.

13.3.1.4 Abundant Provenance

Abundant provenance, which provides the material basis for gravity flow, is one of the necessary and sufficient conditions for the formation of gravity flow. Fragmentary materials felled by floods, vulcanian eruption materials, adlittoral fragmentary materials, and carbonate substances are material sources for sediment gravity flow. The provenance component determines the type of gravity flow sediments, which vary regularly as the provenance component changes.

13.3.1.5 Certain Trigger Mechanisms

The formation of gravity flow deposit falls under eventful sedimentation produced according to certain trigger mechanisms. For instance, mass flow and hyperpycnal flow are produced under the direct or indirect induction of paroxysmal factors such as floods, earthquakes, tsunami surges, storm tides, and vulcanian eruption. For example, on November 08, 1929, an earthquake on the Newfoundland coast resulted in the slump of continental materials and formed large-scale turbidity current deposition (Fig. 13.13) on the abyssal plain. The development conditions of turbidity currents are recorded in different periods.

13.3.2 Formation Phase

Meeting the above conditions lays the foundation for sediment gravity flow formation. Different phases of sediment gravity flow development differ greatly in terms of sedimentary model, sediment type, and spatial distribution of sediment bodies. Turbidity current, as an example, is introduced below.

Bouma (1962) believed that the formation and movement of turbidity currents can be divided into four phases.

(1) Delta phase: Because land is an important turbidity current substance, rivers transport most denuded materials to a basin margin to form a delta. Owing to the influences of earthquakes, tsunamis, storm winds, or unstable steep slopes (liquefied due to excess pore pressure) formed by the accumulation of a large number of levee deposits, large quantities of materials move integrally.
(2) Sliding phase: A quantity of materials slides downwards and then accelerates gradually as the moisture content improves, and the number of particles decreases gradually, resulting in the formation of slide deposits.
(3) Slump phase: When the sliding materials fail to mix completely with water, parts of the materials remain in highly polymerized and adhered states, and coarse grains are not concentrated at the bottom front. Deposits may be accumulated after they are stopped under this circumstance, and such deposits are called fluxoturbidite. However, with a certain gradient, moving materials will not stop under gravity, but flow continuously toward the basin center with an increasing

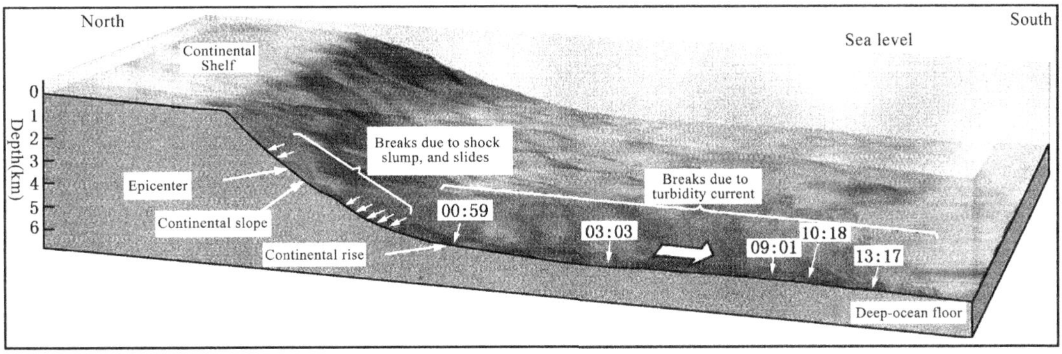

Fig. 13.13 Simulation map of turbidity currents induced by earthquakes (after Heezen and Ewing 1954)

speed, resulting in the formation of slump deposits, creating the conditions necessary for mass flow (debris flow) formation.

(4) Debris flow phase: Abrupt change of terrain from steep to gentle transforms the slump body into terrane flow, that is, an intermediate state between slump body and high-density turbidity current (Fig. 13.14). Kruit (1955) and Shanmugam et al. (1995) found that usually, a debris flow phase is formed between slump and turbidity current. While studying several deep-water sandstone reservoirs in the North Sea, Shanmugam (2000) clearly pointed out that thick-layer deep-water massive sand is debris flow deposit rather than high-density turbidity current. This provides a new viewpoint for explaining deep-water deposition and reservoir prediction because sandy debris flow with good lateral continuity may be a good reservoir.

(5) Turbidity current phase: Flowing substances may form a complete turbidity current in a suitable environment. In a turbidity current, coarse-grained substances are concentrated near the bottom front, and flow velocity may increase continuously. Based on the magnitude of gradient and length of slope, a turbidity current may reach its highest flow velocity. As the gradient becomes gentler and flow velocity decreases, deposits start unloading (Fig. 13.13) to form a turbidity current.

13.3.3 Formation Mechanism and Environment

Generally, sediment gravity flow is characterized by a large scale and high speed. In particular, deep-water gravity flow has very strong corrosion and transportation, playing an important role

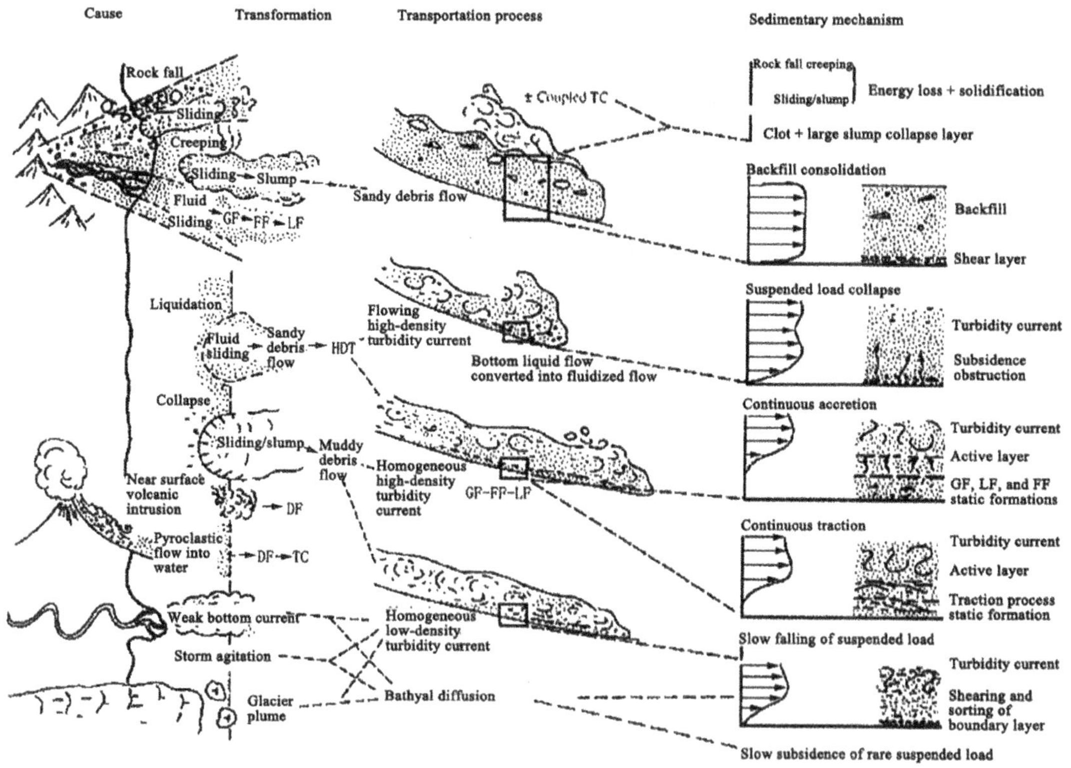

Fig. 13.14 A comprehensive composition model for the formation, transformation and transport of the slope gravity flow deposits (after D. A. V. Stow et al. 1996). GF grain flow, FF fluidized flow, LF liquefied flow, HDT high-intensity turbidity current, TC turbidity current, DF debris flow

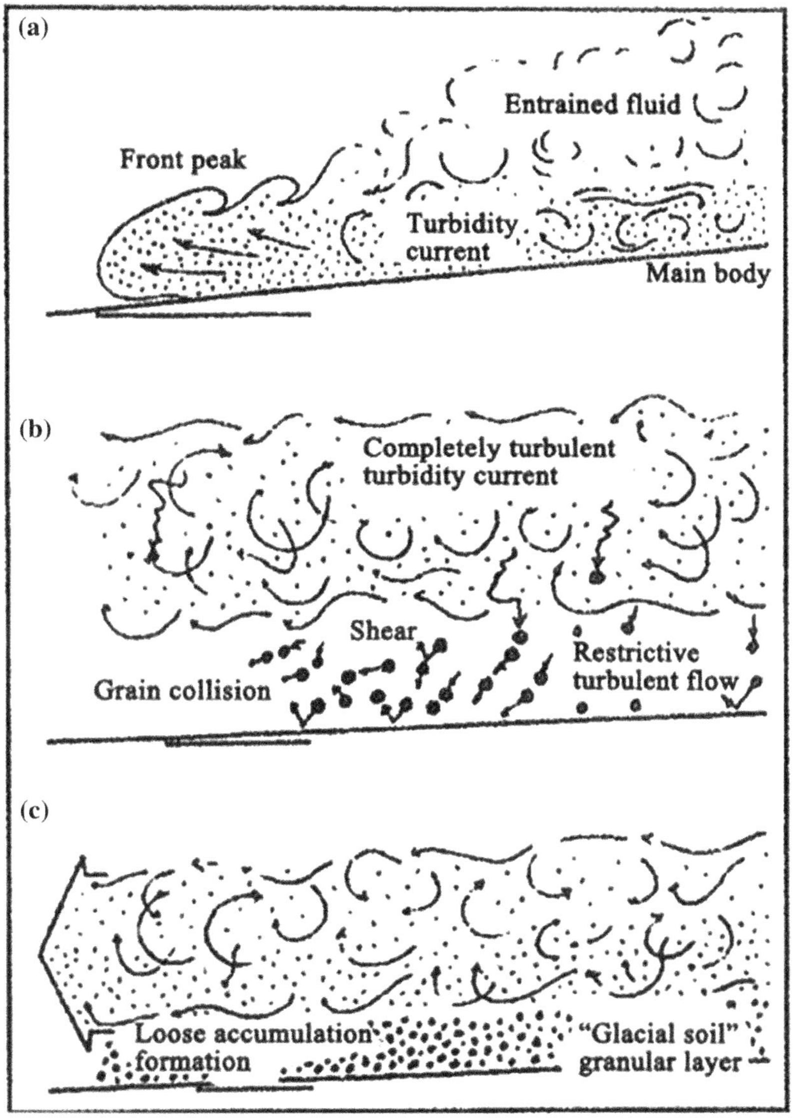

Fig. 13.15 Unloading pattern of turbidity currents (after Walker and Cherry 1992)

in modeling submarine deposits and submarine geomorphic features. If the continental shelf or continental slope is transected and terminated in a submarine canyon on a continental rise, the submarine canyon or feeder channel is said to be a product of sediment gravity flow corrosion, in addition to being a channel (Fig. 13.16) for the flow of gravity flow deposits. According to discrepancies in factors such as width-depth ratio, tortuosity, dyke height, and channel stability, there are different patterns (Fig. 13.17) of channel erosion degree and later filling. Gravity flow moves down a slope to erode a continental margin, resulting in an increase in the scale of a submarine canyon. The density currents of these load deposits are unloaded eventually to form a submarine fan after losing power. A single layer formed by each incident is characterized by fining upward grain size, resulting in the formation of graded bedding (Figs. 13.15 and 13.18).

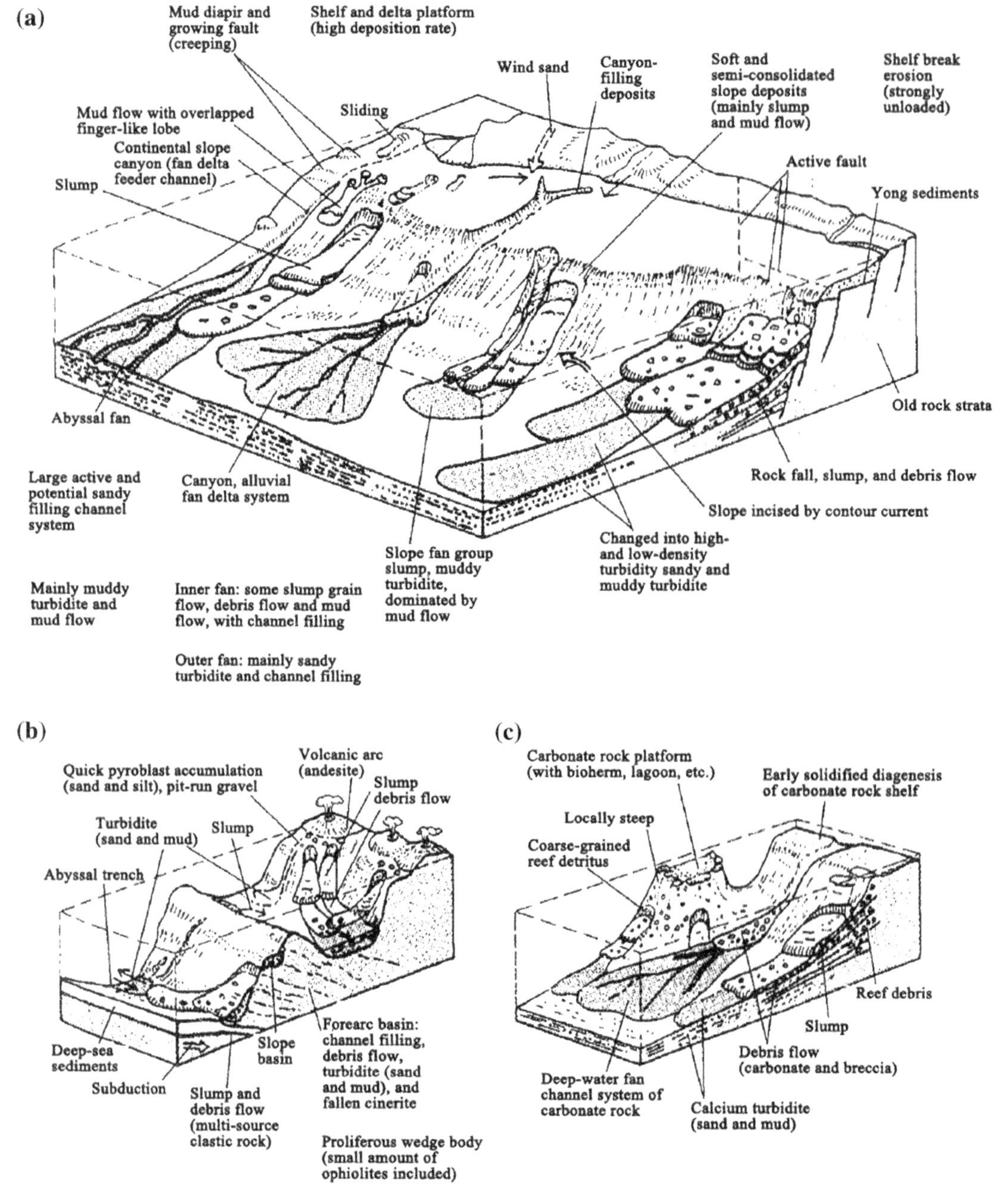

Fig. 13.16 The depositional environment of large gravity block flow and turbidity (after Stow 1986; Einsele 2000). **a** Early phase of rift expansion of passive continental margin basin; **b** convergent margin of forearc basin; and **c** mass flow and abyssal fan with debris supply from carbonate rock shelf or platform

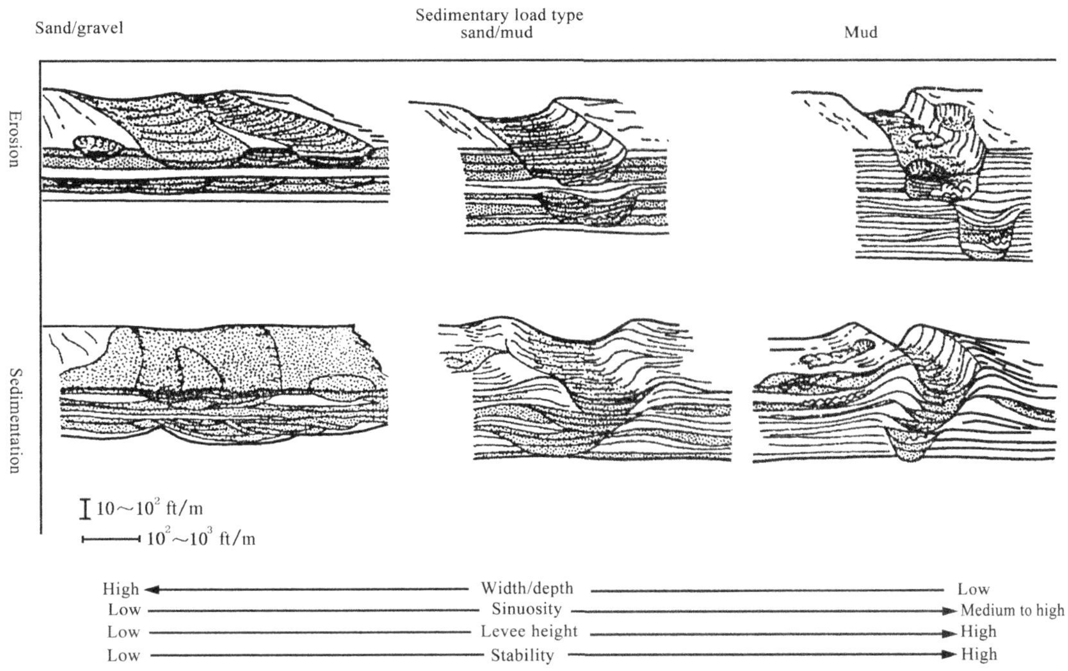

Fig. 13.17 Erosion and filling models of the turbidite channel (after Galloway et al. 1996)

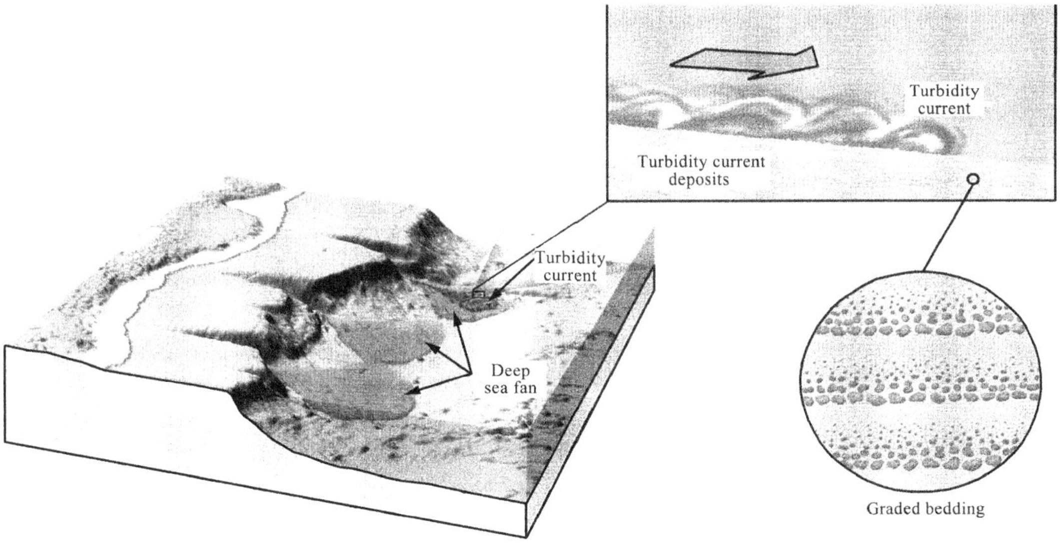

Fig. 13.18 Deep sea bottom fan (turbidity) model and flow state (from E. J. Tarbuck and F. K. Lufgens 1997)

13.3.4 Turbidity Current Depositional System

According to the components in turbidity current deposits, the turbidity current sedimentary system can be subdivided into three categories (Fig. 13.19).

Type I (laminar lobe): Its important feature is that the continental margin is eroded on a large scale, facilitating the formation of large, thin

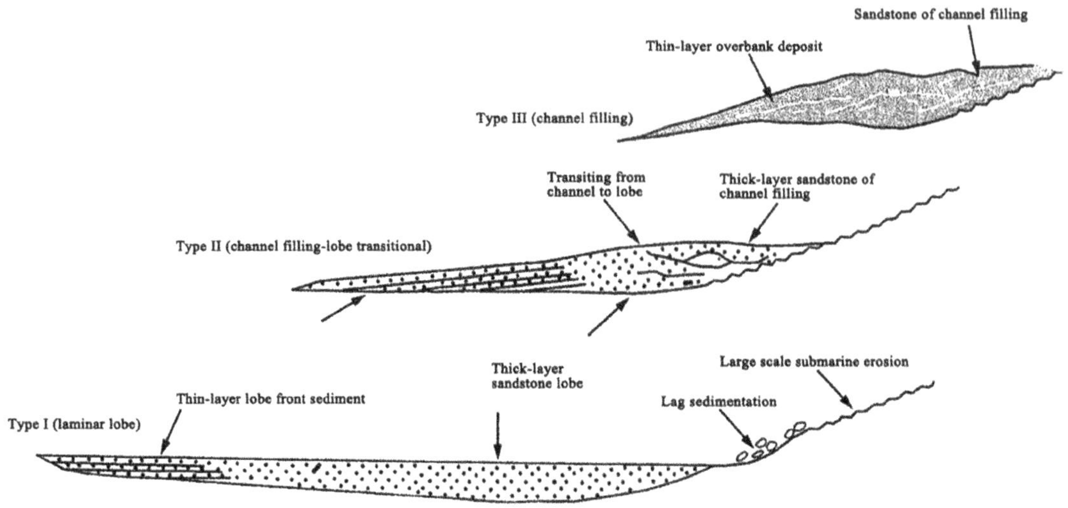

Fig. 13.19 Turbidite sedimentary system types (after Mutti 1985)

laminar lobes and interaction of sand and mud mainly on flat terrain.

Type II (channel filling-lobe transitional type): The basin margin is eroded on a large scale, and thick-layer sandstones are deposited in turbidity current channels near the basin edge; they evolve to the channel filling-lobe transitional type toward the inner part of basin, with gradually decreasing thickness.

Type III (channel filling): The body, including filling deposits and overbank deposits of turbidity current channel, forms small and thick lobes, and is also called "channel-natural levee composite."

However, the above models are only ideal classifications. Most turbidity current depositional systems are represented as composite depositional systems (Fig. 13.20). The above three models can be identified on relatively small scales in a composite system. Each small model internally includes an erosion part and a corresponding sedimentary part that may be composed of different sedimentary rocks.

In addition, there are important relationships between the scale and the spatial geometric form of a depositional system and slope stability (Fig. 13.21). When the slope is steady, it is less likely to be denuded; when the scale of slump gravity flow is smaller, it is less likely to form a lobe, instead producing a channel-natural levee or even being distributed as a sheet. As channel instability improves, the scale of sediments improves gradually to form mutually superimposed lobes, mostly in the form of mounds. When channel instability is very strong, a turbidity current with large scale and strong scour erosion is usually developed on the slope fault zone, which strongly erodes even the continental slope or continental rise. A deposit passing zone is developed in front of the channel, facilitating channel separation from the lobe and superposition of lobes formed in multiple periods. However, unlike the lobe formed by Type II, such lobes are sheet shaped (Fig. 13.22).

13.4 Bouma Sequence and Features

Turbidite, in a broad sense, refers to the sum of various gravity flow deposits formed in deep-water sedimentary environments and the sedimentary rocks formed by the same. Typical turbidite refers to sedimentary rocks presenting the Bouma sequence and are formed by the action of turbidity. A complete Bouma sequence consists of five continuous sedimentary formation sections, namely, A, B, C, D, and E (Fig. 13.23).

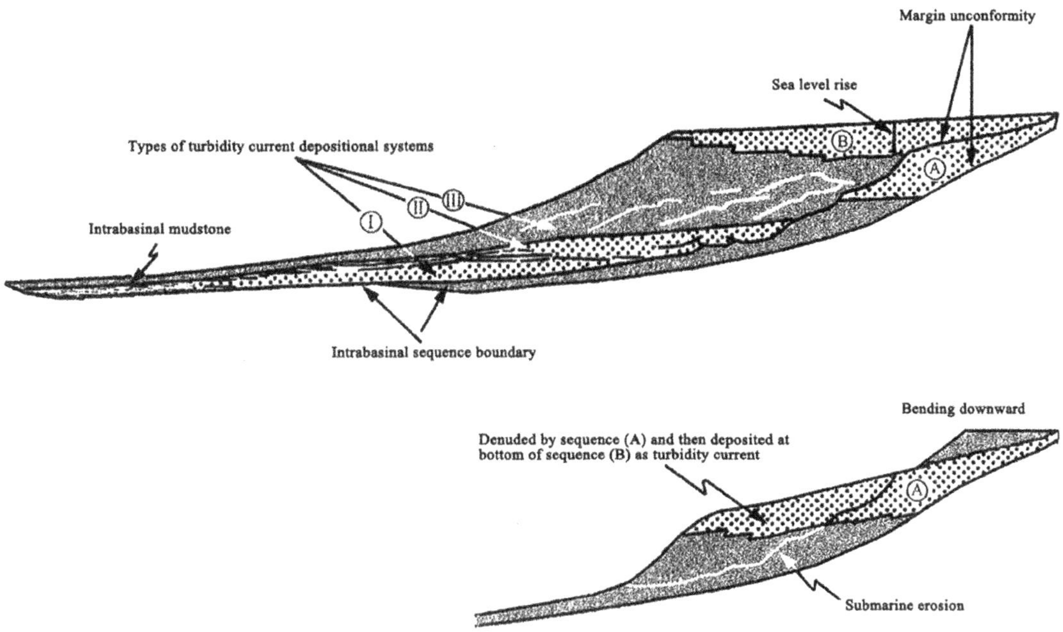

Fig. 13.20 The relation between sea-level rise and turbidity current sedimentary system (after Mutti 1985)

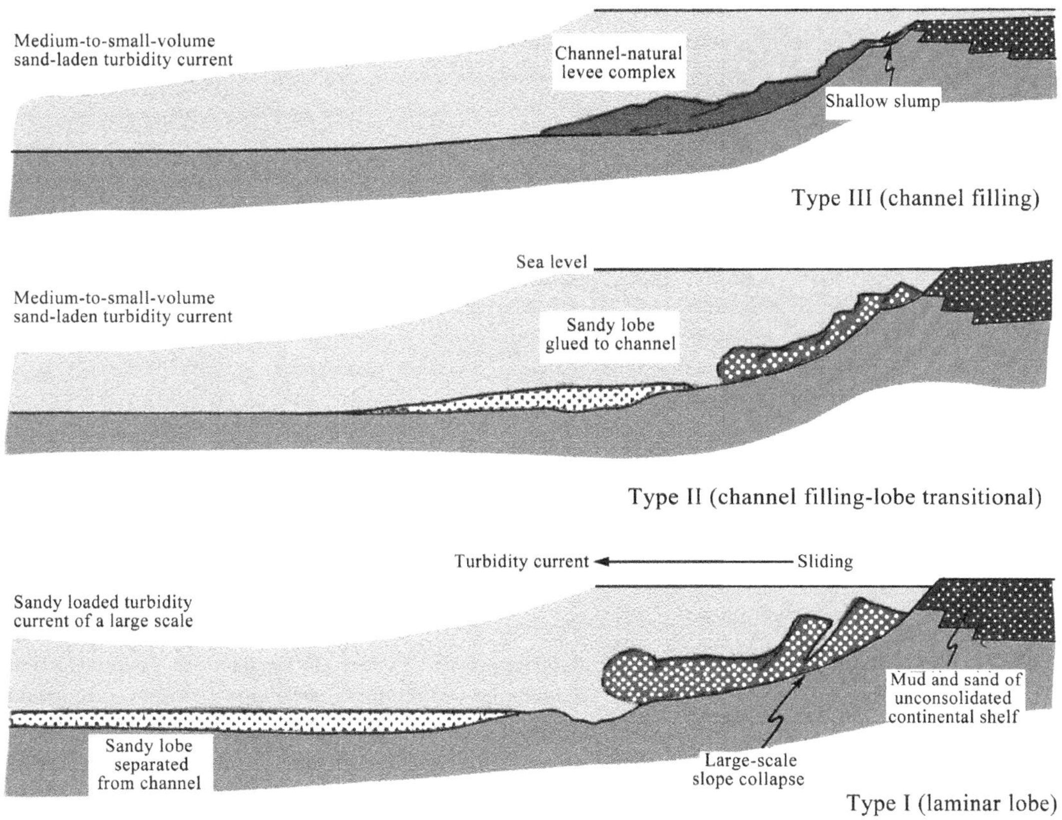

Fig. 13.21 The relation between the scale of turbidity current sedimentary system and slope stability (after Mutti 1985)

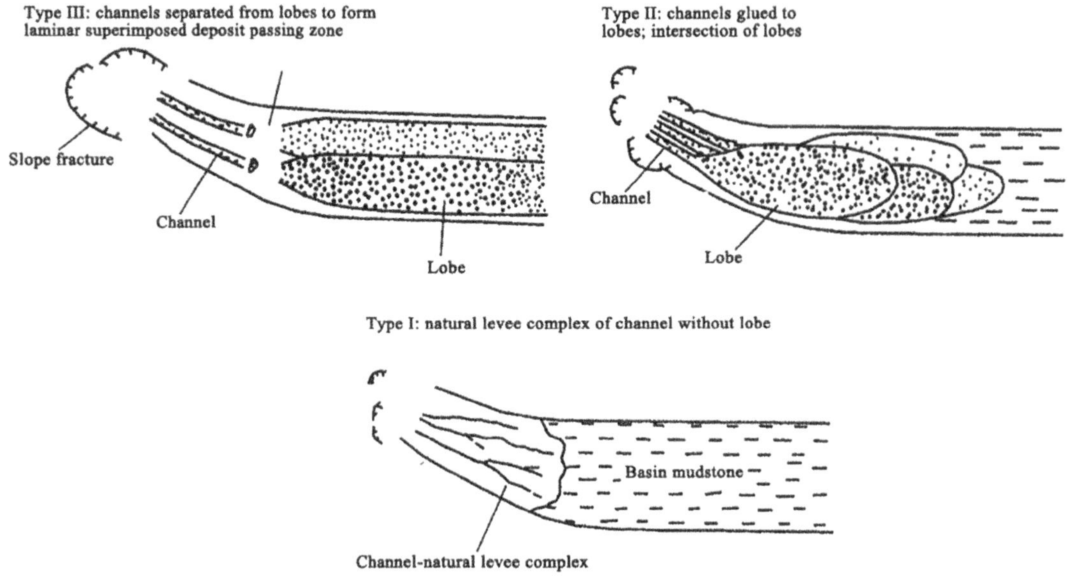

Fig. 13.22 Different types of lobe (after Mutti 1985)

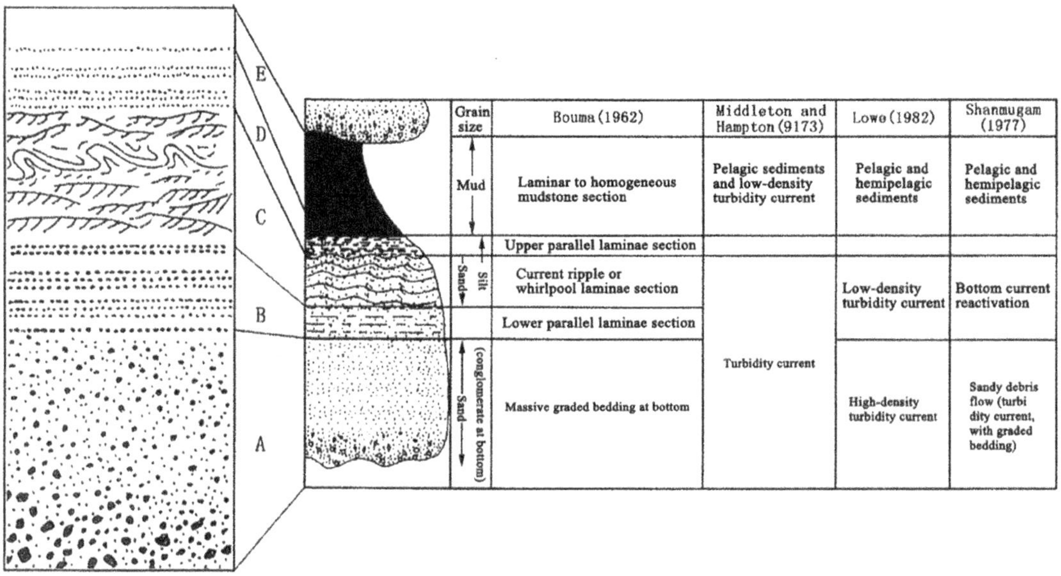

Fig. 13.23 Composition and interpretation of its flow pattern of the Bouma sequence (after Bouma 1962; Middleton and Hampton 1976)

13.4.1 Section A (Bottom Graded Bedding)

This is generally composed of gravel sediments, with gravels developed near the bottom. There is massive or graded bedding with apparently fining upward granularity, in the form of normal grading, reflecting a gradual decrease in turbidity current energy. Reverse-graded bedding several centimeters in thickness may be found at the bottom of thick graded bedding, but it changes rapidly to normal graded bedding. The bottom

(a)

(b)

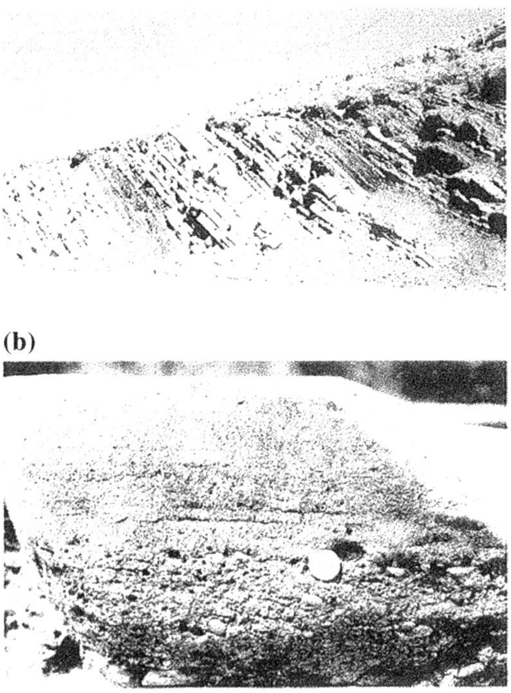

Fig. 13.24 Outcrop of the Bouma sequence. **a** Complex profile of multi-period Bouma sequence. **b** Bouma sequence formed by effusive flow

surface is scoured with filling structures and various mold structures such as flute cast and gutter cast. Section A, which represents the product of graded suspended sediments, is thicker than the other sections (Fig. 13.24).

13.4.2 Section B (Lower Parallel Laminae)

Muddy section B, which is composed of medium-fine grained sandy substances, is thinner than section A. Gradual change relationships exist between sections A and B. It has parallel bedding, without obvious grain size grading. In addition to the obvious grain size change, laminae usually result from the directional distribution of schistose charcoal and elongated debris, and parting lineations can be found when the bedding is uncovered. In most cases, due to unclear bedding, it is difficult to estimate the thickness of section B accurately in the field. If section B is superimposed over section A, the two are continuously transitional; if section B is below A, there is an abrupt relationship between section B and the underlying Bouma unit, between which lies a scour surface. Hence, various mold structures can be found on the bottom bedding of section B.

13.4.3 Section C (Current Ripple Laminae)

This is dominated by silt, with fine sand and mud, including small current ripple cross bedding and climbing ripple cross bedding. Argillaceous laminae, usually convolute bedding, mudstone-tearing clastics, and slump deformation bedding can be found between the sets of ripples, resulting from the recombination action of current transformation and gravity sliding. Owing to the continuous transition between sections C and B, if section C is in abrupt contact with the underlying Bouma unit, a scour surface is present therebetween in addition to various bottom mold structures. With regard to the genesis of various beddings in this section, most scholars believe that turbidity current is converted into hypopycnal flow according to the traction flow mechanism after deposition in sections A and B.

13.4.4 Section D (Upper Parallel Laminae)

This belongs to muddy silt and silty mud deposition, with intermittent horizontal laminae. If section D is overlaid above section C, the two are transited continuously; if it is presented solely, there is a clear boundary between section D and the underlying Bouma unit. Section D is formed by thin boundary layer flow, generally with small thickness.

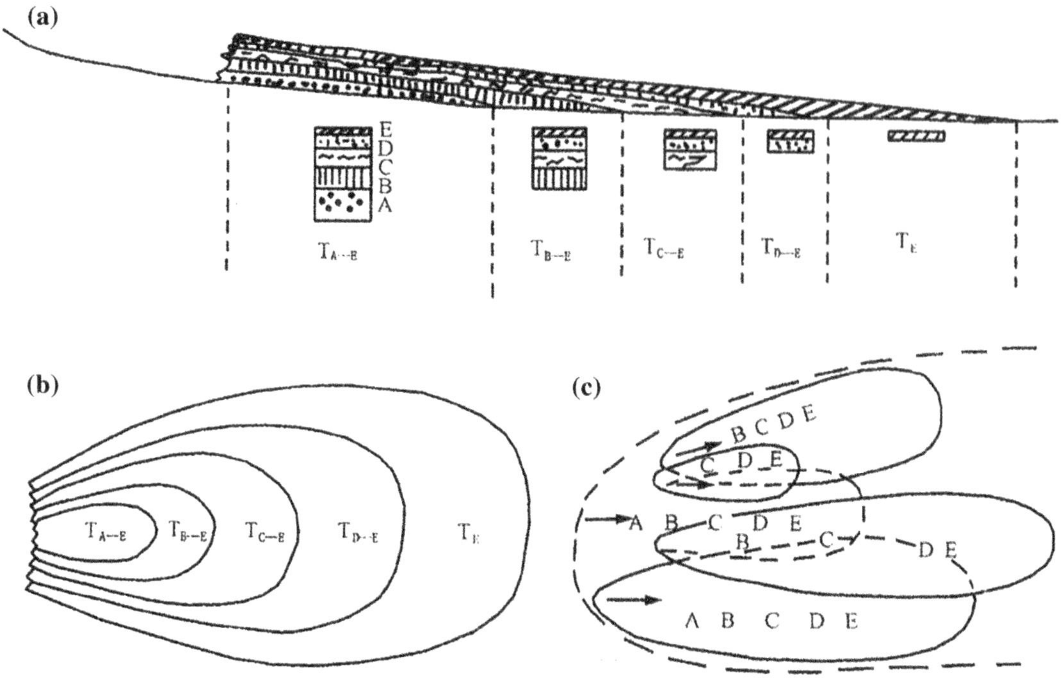

Fig. 13.25 Sketch map of the formation and distribution characteristics of turbidite sequences (after Bouma 1975). **a** Longitudinal section. **b** Plane distribution. **c** Mutual superimposition relationship between turbidity current tongue bodies formed multiple times

13.4.5 Section E (Mudstone Section)

As a pelagic muddy deposit, it is composed of shale or marlstone, biolithite limestone, and mudstone, containing semi-deep water, deep-water biologic fossils, and bioturbation structures. Its slight horizontal bedding is in abrupt or gradual contact with the overlying strata. Actually, section E is not of the turbidity current deposition type, but it acts as an important mark for identifying deep-water turbidity current deposition.

This sequence is the ideal "turbidite facies model" integrated based on many profiles by Bouma. In fact, the complete Bouma sequence is rarely seen, and Bouma never pointed out that the eponymous sequence can be seen only in very thick flysch formations. It is generally incomplete. Sections A and E are easily lost, and only sections BCDE and CD or AE, BCE, and DE can occur. Bouma speculated that each section of turbidite has a tongue-shaped (lobate) distribution (Fig. 13.25). Compared with the lower coarser section, the upper finer section has a larger distribution area. Since this section is eroded and scoured by turbidity currents again, or the second turbidity current occurs shortly after the first one, the front of the second turbidity current is deposited before the tail end of the first one or it is located at the tip of the submarine fan. Hence, only finer materials are deposited in the upper section. That is to say, the maturity level of a turbidite sequence is subject to the frequency and strength of the turbidity current. As a result, sequences that lack a bottom section, have a truncated top section, or lack both sections, are formed.

According to Mutti (1992), the original fluid composition affects the degree of development of the Bouma sequence, while having a certain controlling effect on the geometric shape of the turbidity current (Figs. 13.26 and 13.27). The forward transport distance of a turbidity current with a large coarse grain content is short. Its Bouma sequence is relatively developed, and the sedimentary body has a pronounced lenticular shape. The depositions of coarse- and fine-grained materials separate gradually with increasing fine-grained material content, which results in the formation of a coarse-grained turbidity accretion body in the direction of closing toward the provenance, namely, section A or B or both of the Bouma sequences, and a fine-grained turbidity accretion body in the direction of closing toward the distal provenance, namely, section C, D, or E of the Bouma sequence. The transport distance of the turbidity current is long, the Bouma sequence is not well developed with the development of gradually the upper section D or E, and the turbidity accretion body is mostly tabulate when the content of fine-grained materials is high and coarse-grained materials are basically not developed.

The Bouma sequence can be used for interpreting the underlying hydrodynamics in addition to serving as an indicator for the identification of typical turbidites. A complete Bouma sequence represents a gradual decrease in the turbidity current sequence from the upper flow regime ($F_r > 1$) to the low flow regime ($F_r < 1$) from the bottom to the top. Since the flow velocity of a turbidity current is proportional to the square root of the density difference, and the density of the lower part is large with high flow velocity, coarse-grained materials are concentrated in the bottom front, forming coarse-grained section A sedimentation. Section A has no any other sedimentary structures formed by deposition, except for graded bedding, which shows that huge quantities of the suspended particles sink quickly. When the flow velocity of section B decreases during deposition, section B is still in torrent and supercritical flow, and the particles are dragged, forming parallel bedding. Therefore, sections A and B belong to the upper flow regime; section C is a

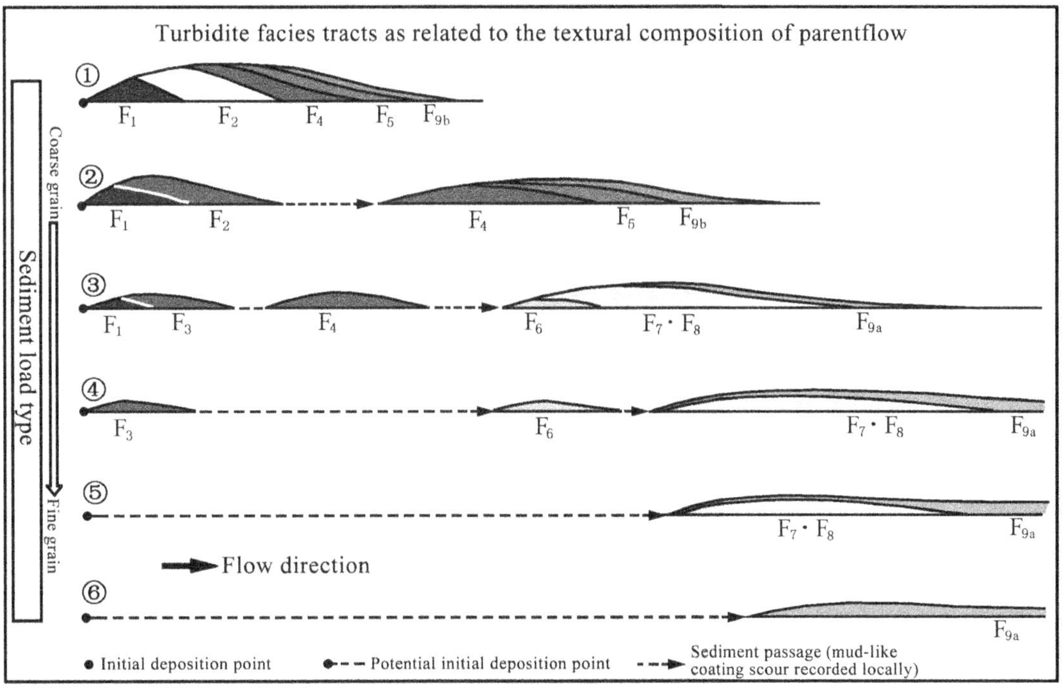

Fig. 13.26 Different developmental degree of the Bouma sequence (after Mutti 1992)

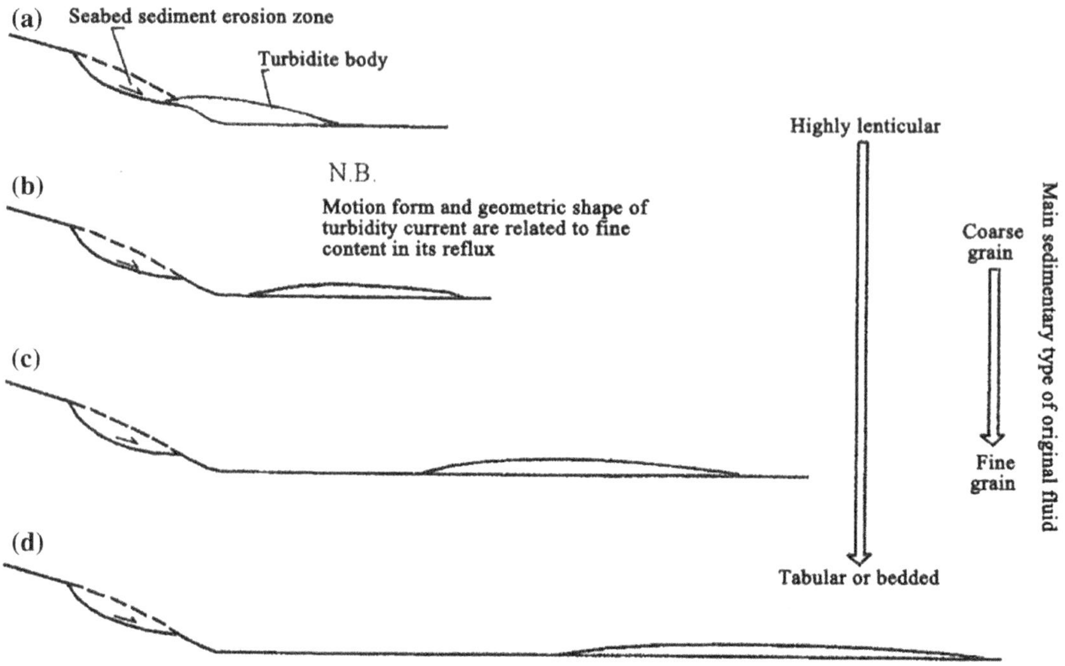

Fig. 13.27 Different geometric forms of controlled structural factors of turbidite (after Mutti 1992)

ripple bedform with a low flow regime; and sections D and E are suspended sediments, which cannot be formed into a bedform, thus reflecting the tail fine-grained sedimentary recession of a turbidity current. If a sequence is composed of dozens of turbidite sections, the associated turbidity current may have high flow velocity and may be close to the provenance area during deposition, starting mainly from section A. On the contrary, if all turbidite layers begin from section B or C, the turbidity current may have low flow velocity and may be far away from the provenance area during deposition. R. G. Walker (1966) proposed the following proximal index (ABC index):

$$p = A + \frac{1}{2}B$$

A and B in the above formula are the percentages of turbidite layers beginning from sections A and B, respectively. If all layers begin from section A, P equals 100%, that is, proximal facies. If all layers begin from section C, P equals 0, that is, distal facies.

13.5 Gravity Flow Sedimentary Facies and Facies Model

Studies on gravity flow sedimentary facies and facies models commenced in the 1950s, developed in the 1960s–1970s, and perfected in the 1980s. A series of turbidite facies sequences and models of marine and lacustrine facies, mainly the flysch facies model, fan facies model, and trough facies model, have been established successively along with deeper research work.

13.5.1 Flysch Facies Model

Flysch facies, in a narrow sense, refers to the turbidite series deposited in the marine facies and deep water environment of geosyncline. Broadly, it contains a turbidite series of clastic rock and carbonate rock in the deep water environment of a fault lacustrine basin and is characterized by the Bouma sequence (Fig. 13.28). The ideal longitudinal section of the turbidity accretion layer is formed by gravity flow. The thickness and grain size of turbidite change regularly along the direction of gravity flow; total thickness of the layer is more than 1 m, length is 15–20 km, and the layer is lens shaped. According to thc degree of perfection of the Bouma sequence, as well as the vertical frequency and plane distribution of SPA-E, B-E, C-E, D-E, and other sequences (Fig. 13.29), flysch facies can be divided into proximal and distal turbidite facies (R.G. Walker 1967). Proximal turbidite facies is characterized by the development of the A-E and B-E sequences, however the distal turbidite facies is characterized by the development of the C-E and D-E sequences (Table 13.3). The layer thickness, granularity scour and wash out, sand-mud ratio, grading, degree of channel development, and Bouma sequence appearance change regularly from proximal to distal turbidite.

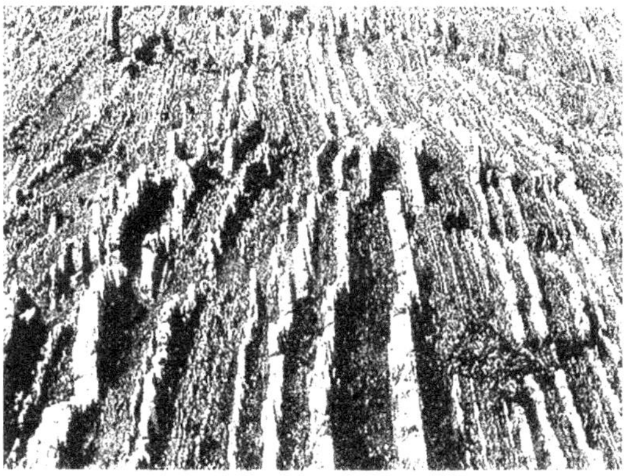

Fig. 13.28 Typical flysch sequences

13.5.2 Fan Facies Model

There exist large differences between marine and lacustrine basins, for example, marine basin accommodation can be regarded as infinite, and it is mainly affected by sea level eustacy, with the influence of other geological factors being small. In contrast, lacustrine basin accommodation is affected by tectonic subsidence, climate, and

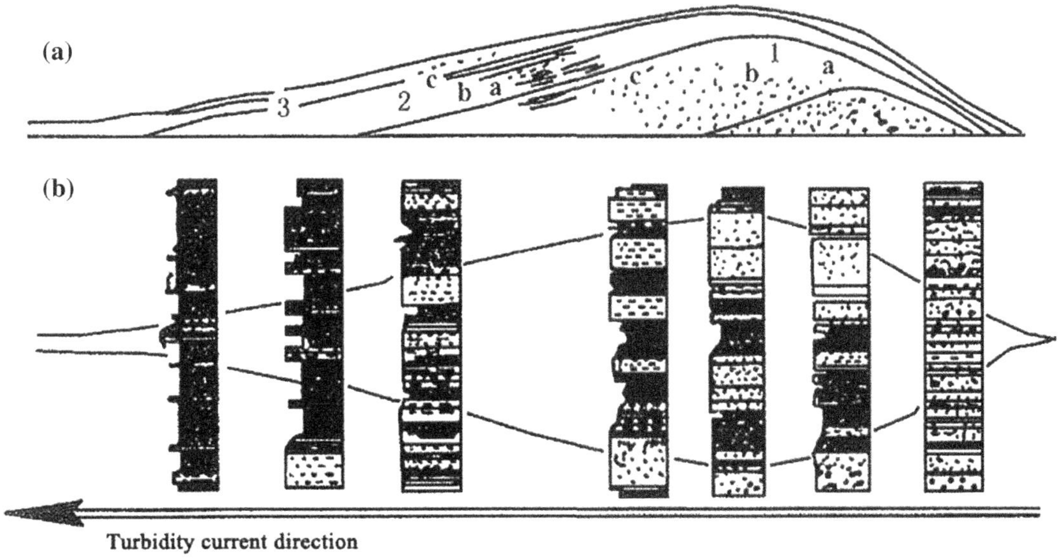

Fig. 13.29 Sequence and facies model of calc turbidites in Germany and Greece (according to Bouma 1962). **a** Ideal longitudinal section of turbidity accretion layer formed by gravity flow. **b** General characteristics of rock stratum in actual section and comparison with ideal layer (**a**)

Table 13.3 Marks of proximal and distal turbidites (according to Flugel 1982)

Sedimentary characteristics	Proximal	Distal	Sedimentary characteristics	Proximal	Distal
Average thickness	Very thick	Thin, often very thin	Geometric shape of rock stratum	Lens with irregular thickness	Regular parallel bedding
Average particle diameter	Large (psammite and psephite)	Small (silty clast and psammite)	Complete Bouma sequence	Common	Rare
Grain sequence	Poor grading or no grading	Almost always graded	Laminae and ripple bedding	Uncommon, limited to thicker layer	Rather common
Reverse grading	Common	None	Upper micritic limestone	Thin or missing	Well developed, usually well preserved
Coarse-tail grading	Common	None	Clastic rock/silty clast ratio	High	Low
Clastic zone bottom	Usually abrupt	Abrupt	Matrix in clastic zone	Sparite cement	Microlite
Clastic zone top	Usually abrupt	Gradually fining sediments	Sedimentary form	Condensing turbidite layers, restricted by sedimentation background	Thick "normal" sediments
Scour, gully, chute and rebuilt gravel at the bottom	Common		Formation-complex association	Slump structure, slump turbidite, conglomerate, and rare flint	Rare coarse fragment, often with flint
Tool mark	Rare				

rainfall. Therefore, here, submarine fans and sublacustrine fans are described separately.

13.5.2.1 Submarine Fan Facies Model

According to modern submarine surveys, the abyssal floor outside the lower submarine canyon in most continental slopes has a huge fan-shaped sedimentary body, usually called the submarine fan. A submarine fan is mainly the product of mud-sand redeposition owing to turbidity currents, thus it is also called a turbidite fan, the shape of which is somewhat similar to that of alluvial fans on the continent. In the case of submarine fans, the turbidity current channel located on the slope acts as the provenance channel. The fan surface can be divided into three main sub-environments: channel, levee, and interchannel. The upper, middle, and outer fans can be distinguished when there is slight tilting outward from the outer fan on the longitudinal section. R. G. Walker proposed the plane distribution model of submarine fan sedimentary facies, which can represent the general characteristics of submarine fans, on the basis of research on modern and ancient abyssal fan sedimentation (Fig. 13.30).

1. **Feeder channel**

The main function of the feeder channel or submarine canyon is to transport grit to the fan, which may also be blocked by coarse-grained materials (slump block, debris flow, gravel, or other coarse-grained materials supplied by

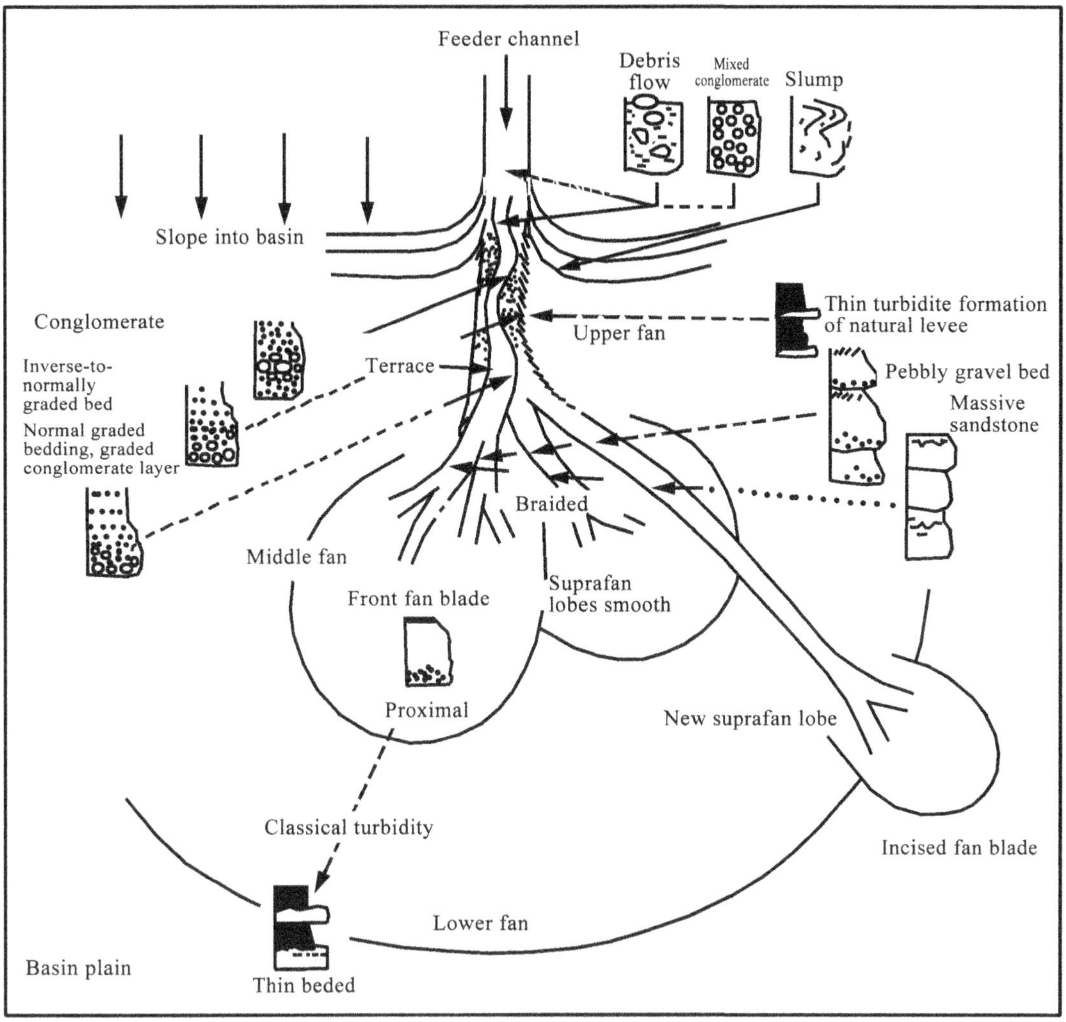

Fig. 13.30 Submarine fan facies model (after Walker 1978)

provenance) or very fine materials (mud, and mudstone). The latter case occurs usually when the original provenance of the fan is cut off due to a relative rise in the sea level.

2. **Inner fan subfacies**

This is located at the outlets of continental slopes and canyons. Silty mudstone, slope channel sand, grit, slump, and crumpled sediments are developed on the slope. At the foot of the slope, a slump layer, disordered layer debris flow, and mixed conglomerate sediments are developed. Slump and original layer massive debris flow deposition occur successively along the current downward. In the channel embankment or terrace edge, turbidite with the C-E sequence can be developed owing to the overflow effect.

Sediment distribution, particularly that of conglomerate, the flow of which is controlled more strictly by the channel, is governed by the terrain. Channel width and depth are different for different terrains. The depth can reach

100–150 m, with a width of 2–3 km. Channel migration and accretion can produce wider conglomeratic distribution. Grain flow and turbidity current deposition can be seen in the channel, particularly at the tail end of main channel of the inner fan.

3. **Middle fan subfacies**

This is located outside the inner fan and inside the outer fan, and is shown usually as a superimposed ligule. Its prominent topographical features include development of braided distributary channels in braided channels or river valleys, abundance of pebbly sandstone (or gravelly sandstone) and massive sandstone, occasional grain flow and fluidized flow deposition, and rare or absent mudstone interbeds. The most active depositions include turbidite of the proximal A-E and B-E sequences in big and small channels of the superimposed ligule. The width of the braided channel is generally 300–400 m, and its depth is 10 m at most. Owing to the migration and accretion of braided rivers on the fan surface, pebbly and massive sandstones deposited by grain flow occur continuously, resulting in the formation of a thick petroleum reservoir with very good porosity and permeability. Shanmugam (1999) deemed that this part of pebbly massive sandstone belongs to sandy debris flow deposit, which not only confirms the existence of deep-water debris flow but is also recognized widely.

4. **Outer fan subfacies and basin plain facies**

The outer fan subfacies, which is connected with the non-channel part of the middle fan, features flat terrain and wide sediment distribution with thin layers. The deposition is typically turbidite and deep-water clay rock with the C-E and D-E sequences. Gravity flow deposits on the abyssal plain are characterized by distal typical turbidite in the broad area, in addition to thickening by the filling of local parts due to its low density bottom current feature for filling low-lying areas (Fig. 13.31). Its thickness is stable, and a few thin silt layers can laterally track dozens to hundreds of kilometers.

5. **Channel filling model**

Each part has a channel during the fan development process. The inner fan channel is wide and deep, the outer fan channel is narrow and shallow, and the middle fan channel has intermediate width and depth (Fig. 13.32). There exist larger differences in terms of the channel filling materials in different parts of the channel. Sand content is the highest in the inner fan; mud content increases while sand content decreases in the front region of the current.

6. **Advancing facies sequence of submarine fan**

Submarine fans generally present the thickening and coarsening sequence from the bottom to the top. If the supply source of the fan tends to be interrupted gradually or has transgression, the thinning and fining upward sequence may be presented. Mutti and Lucchi (1972) proposed a sedimentary facies model for submarine fans (Fig. 13.33). According to their model, the outer fan and inner fan coarsen upward, while the middle fan fines upward because of the development of distributary channels.

Walker (1978) proposed the ideal model of submarine fans by further refining the above model on the basis of previous research (Fig. 13.34). In the figure, CU (coarsen upwards) refers to the thickening and coarsening upward cycle and FU (fining upwards) refers to the fining and thinning upward cycle.

13.5.2.2 Facies Model of Sublacustrine Fan

Research results show that the lacustrine facies and fan facies turbidites are rather developed in the Paleogene of Bohai Bay, with lithology comparable to that in the submarine fan facies model proposed by Walker (1978 and 1982). The vertical sequences present the thickening and coarsening upward sequences with advancing

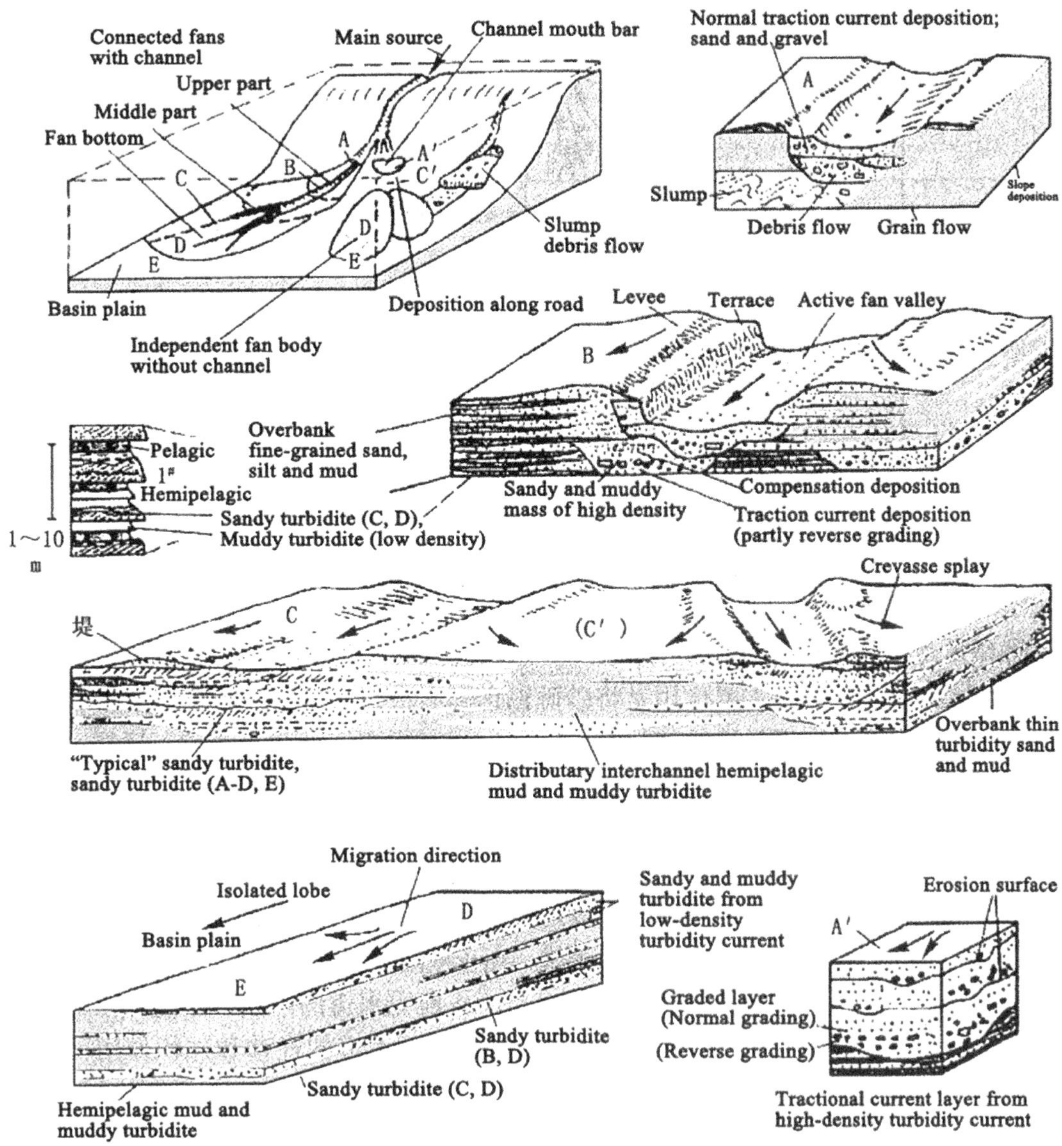

Fig. 13.31 Phase models of deep sea fan (after Mutti and Ricci Lucchi 1972; Walker 1978; Shanmugam and Moiola 1985; from Einsele 2000). Please pay attention to the differences between connected fans with channels and the isolated fans without channels; isolated lobes can better represent the typical turbidity accretion sequence

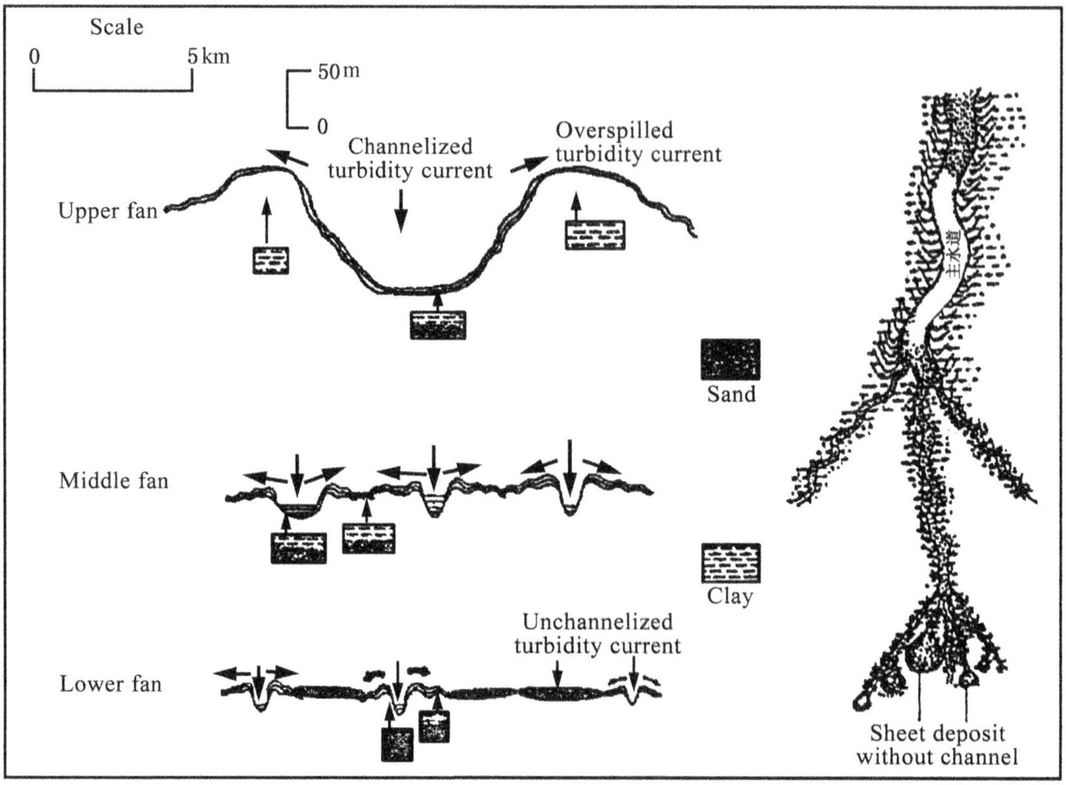

Fig. 13.32 Morphology and filling of channel in submarine fan (after Kolla and Coumes 1987)

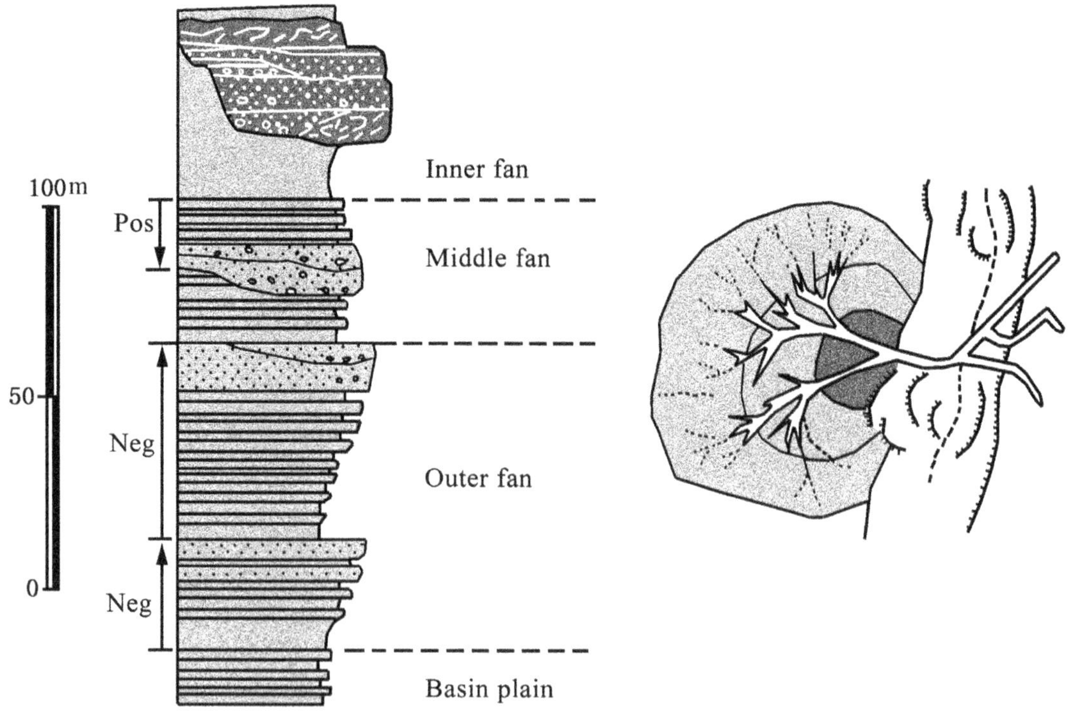

Fig. 13.33 Plane and vertical phase model of submarine fan (after Mutti and Ricci Lucchi 1972)

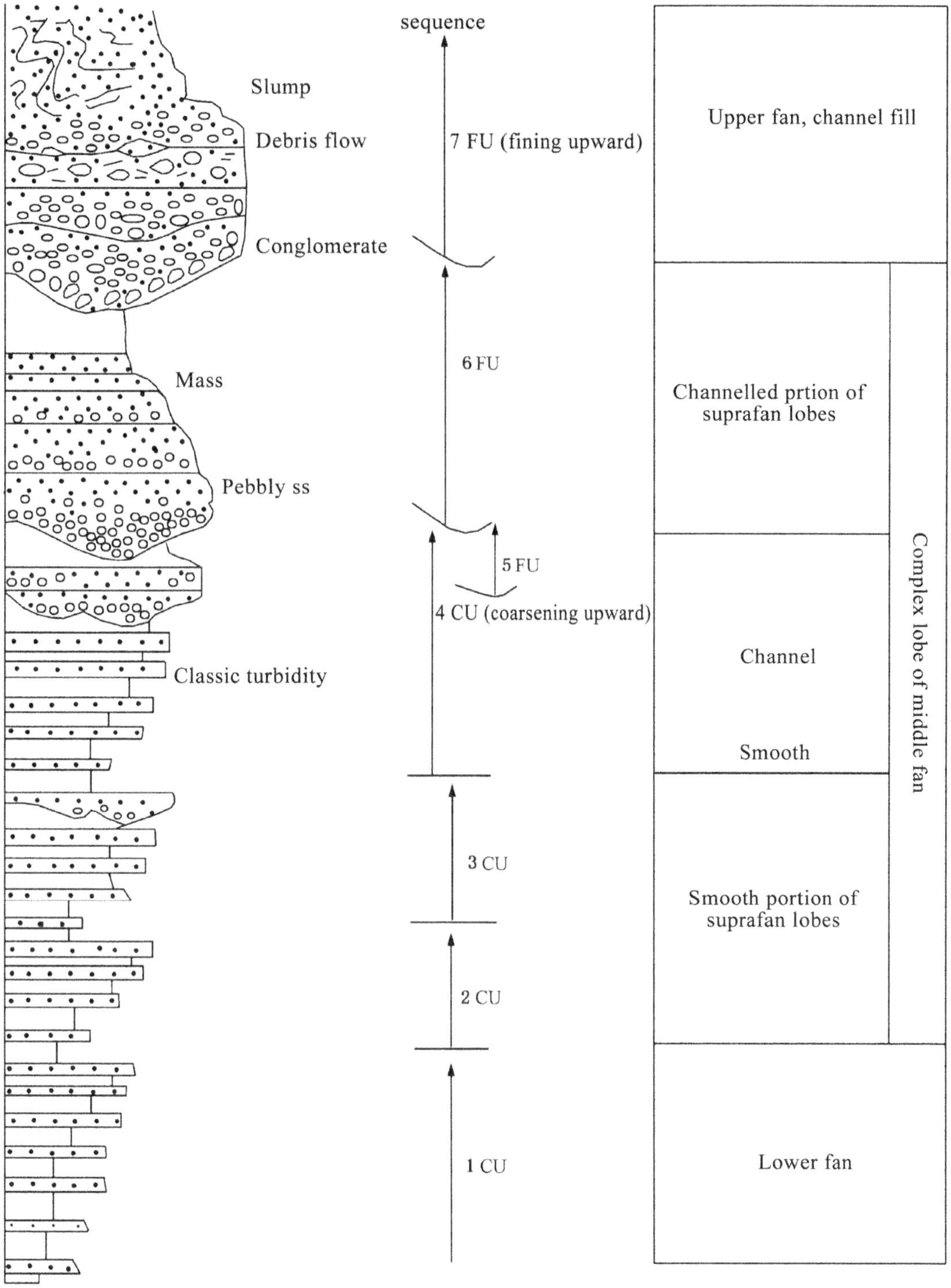

Fig. 13.34 Submarine fan facies models (after Walker 1978)

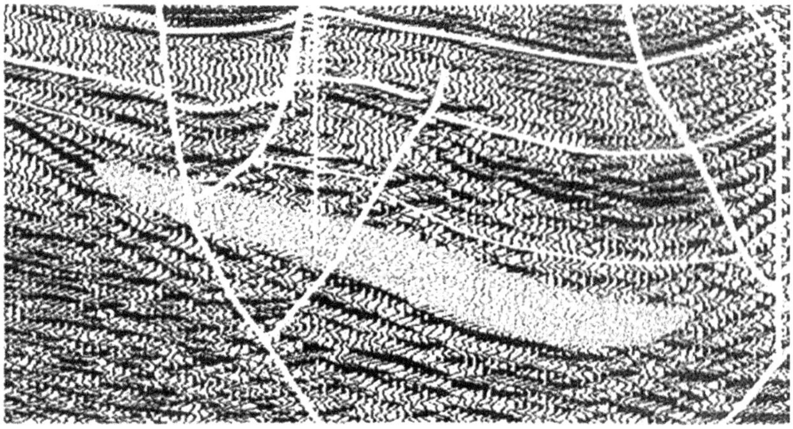

Fig. 13.35 Seismic reflection profile of sublacustrine fan in Oubei-Dawan areas

complex superimposition. Several coarsening upward sequences roughly reflect the superimposition of lobes, which feature the interbedding of fan facies sand bodies, conglomeratic, and deep-water mud shale. Each lobe plane is shown as being fan shaped, of which the cross section is characterized by a flat top and convex bottom, and the longitudinal section or the section in the irradiation direction is wedge shaped. The lithofacies belt from the root to fan edge is of the feeder channel-upper fan (or inner fan)-middle fan-outer fan (or fan edge)-basin plain (possibly with incised channel) types. The corresponding rock types are particle-supported or matrix-supported conglomerate, conglomeratic-pebbly sandstone, and massive sandstone-typical turbidite. The overall change in trend is that channel turbidite decreases and typical turbidite increases, which is a continuous change process.

Sublacustrine fans can be often seen in continental lacustrine basins in China. For example, the sublacustrine fan in the lower sub-member of the S3 member in the Oubei-Dawan area of the middle section of the eastern depression in the Liaohe Basin (Fig. 13.35) presents the onlap reflection structure and a structure with strong amplitude, low frequency, and medium continuity in terms of seismic reflection. This is because the water body is deeper, gradient is steeper, and alluvial fan and fan delta deposition provide adequate source supply during the fault depression expansion period, resulting in the formation of turbidity current deposits through slump.

13.5.2.3 Facies Model of Deep-Water Fan

The two abovementioned facies models mainly consider the sedimentary environment of fan facies development, however the specific classification standard is not clear. Reading and Richards (1994) divided deep-water fans into three categories and 12 types based on provenance and inner grain size structure, namely, point source submarine (subaqueous) fans, multiple-source subaqueous slope fans, and line source subaqueous fan skirts. On the basis of the type and composition of sediment, these three categories are subdivided into 12 types: point source subaqueous fan rich in mud, sand-mud, sand, and gravel; multiple-source subaqueous slope fan rich in mud, sand-mud, sand, and gravel; and line source subaqueous fan skirt rich in mud, sand-mud, sand, and gravel. These 12 types correspond to different facies models (Figs. 13.36, 13.37 and 13.38). G. Einsde (2002) accordingly summarized the classification standards of deep-water fans and the main characteristics of each fan (Fig. 13.39).

13.5.2.4 Mudflow Gully Model

Mudflow deposit, which has been researched on both land and underwater, is a type of relatively

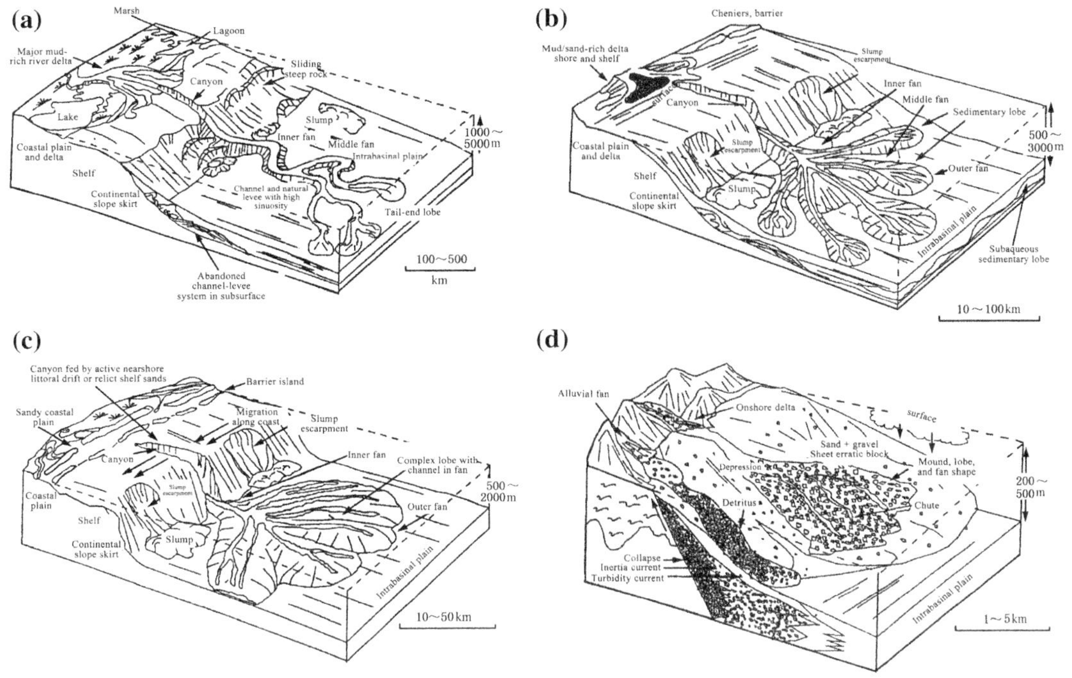

Fig. 13.36 Various submarine (underwater) fan patterns of point provenance (after Reading and Richards 1994). **a** Mud-rich type. **b** Sand-mud type. **c** Sand-rich type. **d** Gravel-rich type

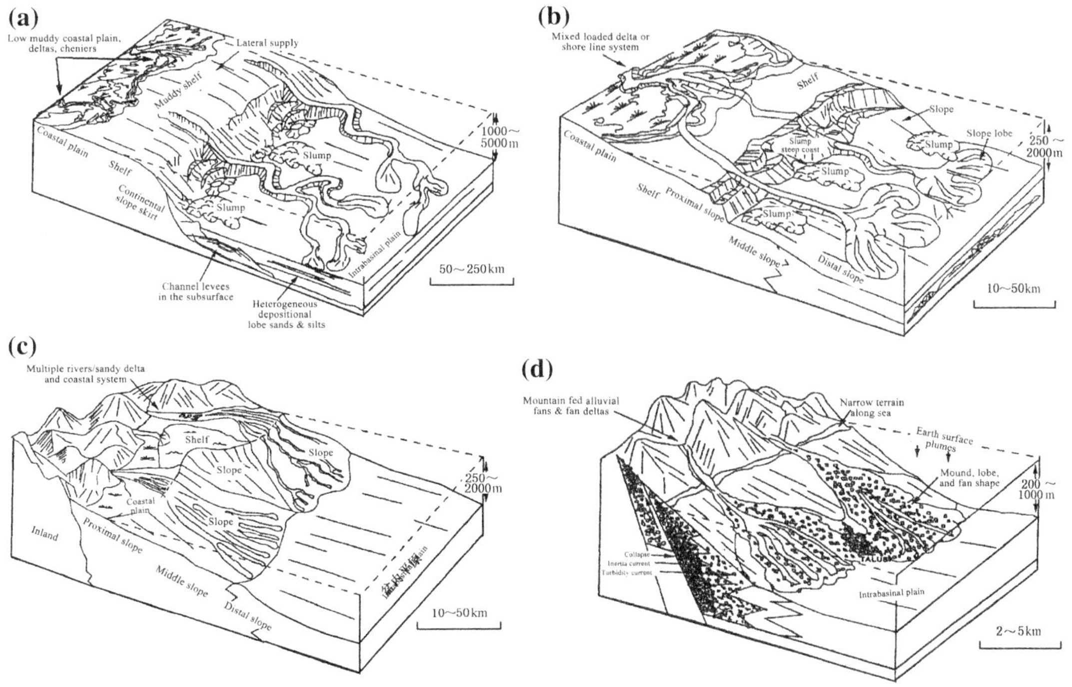

Fig. 13.37 Various submarine (underwater) fan patterns of multi-provenance (after Reading and Richards 1994). **a** Mud-rich type. **b** Sand-mud type. **c** Sand-rich type. **d** Gravel-rich type

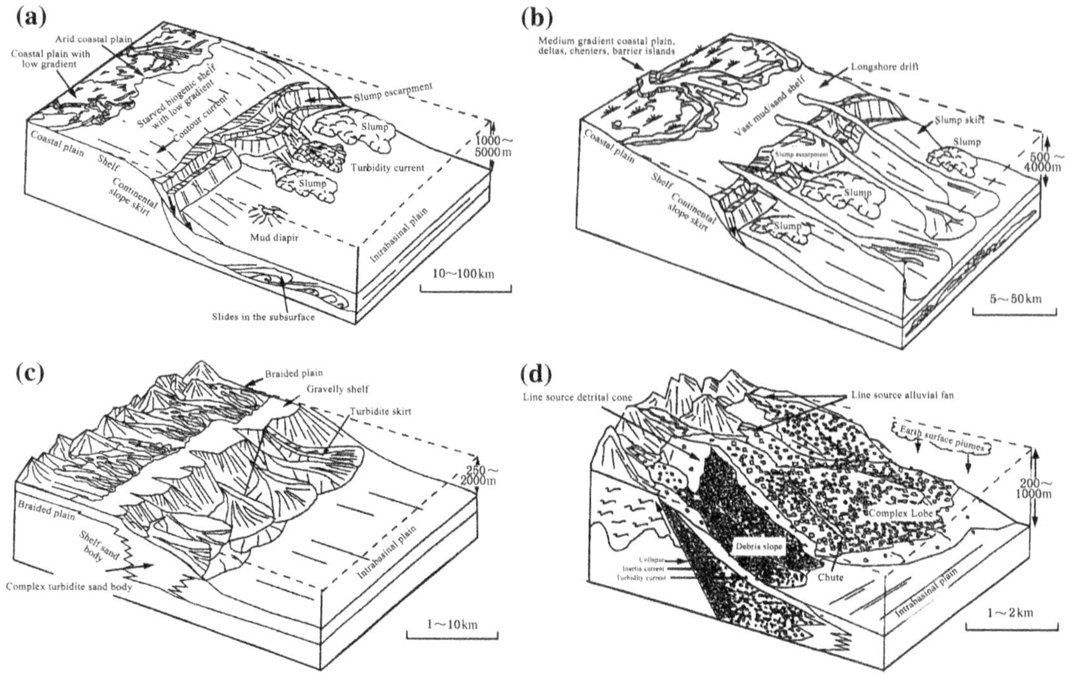

Fig. 13.38 Various submarine (underwater) fan patterns of line-provenance (after Reading and Richards 1994). **a** Mud-rich type. **b** Sand-mud type. **c** Sand-rich type. **d** Gravel-rich type

special gravity flow deposit. Some of its formation features are similar to that of debris flow, however the compositions of mudflow and debris deposits are completely different. Submarine mud flow is generally formed because of the prolonged absence of a coarse clastic sediment supply in the early stage, which means that the basin is in starvation and thicker sediments are deposited. Later tectonic movement (such as diapir) is strong in areas with larger gradient and poor stability, causing the sediments in those areas to slide and erode the plastic seabed to form a mudflow gully; moreover, a mud flow lobe or fan is formed in its front area.

13.5.3 Trough Facies Model

13.5.3.1 Oceanic Trough Facies Model

Oceanic trough turbidity current has been reported before, for example, the turbidite in the Martinsburg Formation of the Ordovician in the Appalachian Mountains in Middle America, turbidites of different ages in the Cordillera Mountain edge zone on the west coast of America, turbidites of different ages in the Tethys Sea in the Alps-Himalayas traversing through Europe and Asia. What is clearer and provides good results in petroleum exploration is the research result pertaining to the oceanic trough turbidite sandstone in the Ventura Basin in America (Xu Jinghua 1980). There are four rock types in the Pliocene-Pleistocene series of the Ventura Basin: mudstone facies, conglomerate facies, graded sandstone facies, and thin sandstone facies. They were formed in the basin slope, submarine canyon or fan, oceanic trough, and basin flank or continental rise environments, respectively. The author particularly emphasizes that oceanic trough-graded sandstone facies is formed in the bend of a submarine canyon or a submarine fan turbidity current, which are caused

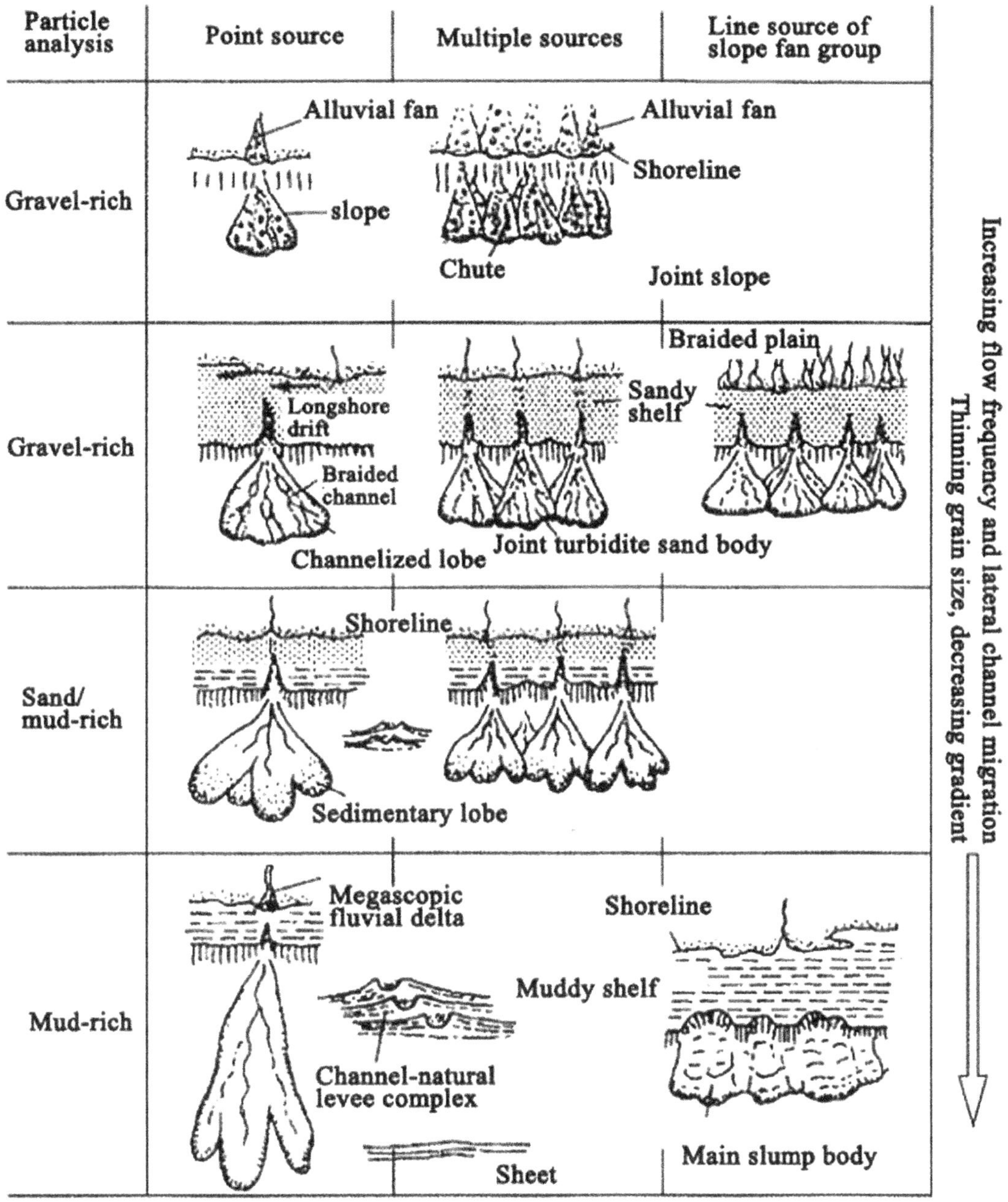

Fig. 13.39 Classification criteria and characteristics of deep water fans (after Reading and Richards 1994; G. Einsele 2002)

Table 13.4 Comparison of the main sedimentary features of single-source submarine (subaqueous) fans

Supply type and main sediment grain size		Mud-rich	Sand-mud	Sand-rich	Gravel-rich
Scale		Big	Large-medium	Medium	Small
Gradient (m/km)		Low 0.20–18	Low-medium 2.5–18	Medium 2.5–36	High 20–250
Radius or length (km)		Elongated 100–300	Lobe 10–450	Radial pattern or lobe 10–100	Radial pattern 1–50
Provenance	Scale	Big	Medium	Medium-small	Small
	Slope	Low	Medium	Medium	High
	Distance	Far	Medium	Near	Near
Supply system		Large-scale fluvial delta rich in mud	Large-scale, mixed-load fluvial delta or downdip canyon	Shelf slump or shelf canyon	Fan delta or alluvial cone
Supply mechanism		Low-density turbidity current and contour current caused by intermittent slump or slump	Mainly high-density or low-density turbidity current	Redeposition or direct deposition of shelf debris; low turbidity current	Frequent mass flow, slump, and turbidity current caused by river
Fluid scale		Very big	Medium	Medium-small	Very small
Channel system		Big, stable channel with stable dyke, from meandering river to straight river	Medium scale; from meandering river to braided river system, lateral migration due to dyke	From braided river to intermittent river or groove with low sinuosity; quick lateral migration	Braided river, miniature intermittent groove
Distal slope fan deposition		Thin, sand-mud interbed formed by sheet flow; thin clastic sheet formed by coarse-grained intermittent flow	Sand-mud interbed lobe formed by mixed load of turbidity current	Low-fluctuation lobe or sheet sand formed by turbidity current rich in sand	Laminar turbidite formed by thin low-density turbidity current
Main basin plain deposit		Turbidity current deposits > offshore deposits	Offshore deposits > turbidity current deposits	Offshore deposits	Offshore deposits

by longitudinal transportation and deposition along the long axis of the basin.

These deep-water fans differ vastly in terms of sedimentary features, strata attributes, and reservoir attributes in different provenance systems and environments (Tables 13.4 13.5, 13.6, 13.7, 13.8 and 13.9). The basic characteristics of each type of deep-water fan are listed in the above-mentioned tables, thus they are not described here.

The gravel deposits of submarine braided channel with terrace in the Cap-Enrage formation of the Cambrian-Ordovician in Quebec, Canada, which were determined by Heimmd and Walker (1982) constitute the most convincing example. These deposits are composed of pebbly and massive sandstone with a thickness of about 270 m. The recovered channel, with a depth of about 300 m and width of about 10 m, stretches

Table 13.5 Comparison between the main underground attributes and reservoir attributes of single-source submarine (subaqueous) fans

Supply type and main sediment grain size		Mud-rich	Sand-mud	Sand-rich	Gravel-rich
Main architectural element	Proximal	Channel embankment	Channel embankment	Channel	Wedge
	Distal	Distal sheet sand	Lobe	Lobe with channel	Distal sheet sand
Seismic configuration		Channel embankment; parallel refection	Channel embankment and crevasse splay mound	Constructive and low-fluctuation crevasse splay mound	Wedge
Sandstone percentage (%)		≤30	30–70	≥70	Changeable, generally at 5–50 (conglomerate >50%)
Geometric shape of sand body		Sand and mud from different periods filled in large-scale lenticular channel; highly heterogeneous sand-mud interbedded distal fan	Lenticular channel, mainly filled with sand or mud; sand-mud interbed or intercalation formed by downdip lobe	Wide sheet low-fluctuation lobate sand body with interior part shown as channel sand body unit	Irregularly connected conglomeratic; clastic rocks and breccia deposited mainly in proximal areas; sandstone deposited mainly in middle-distal areas
Turbidite facies		A, B, E, F	C, D	B, C	A, B, F
Reservoir heterogeneity		High	High-medium	Low	High
Sand body connectedness	Vertical direction	Poor	Medium	Good	Good
	Lateral direction	Poor	Poor	Good	Poor
Trap type		Stratigraphic trap	Stratigraphic trap	Structural trap	Structural trap
Potential production layer		① Stratigraphic trap of outer fan; ② Stratigraphic trap formed by sedimentation of natural levee and crevasse splay	① Stratigraphic trap formed by sedimentation of channel and lobe; ② sandstone trap formed by slope collapse and fluid denudation	① Up-dip pinch-out channel sandstone of inner fan; ② stratigraphic trap formed by channel lobe of middle fan	① Structural trap formed by fan closing toward downthrown side; ② structural trap formed by medium and distal sand
Main exploration risks		Existence, range, and recognition of petroleum reservoir and trap	Existence, range, and recognition of petroleum reservoir and trap; integrally sealed	Structural conditions required for trap; integrally sealed	Reservoir property and integrally sealed
Significance and position in the cycle of sea level		Important; lowstand	Potentially important; highstand and lowstand	Not important	Not important

Table 13.6 Comparison of the main sedimentary features of multiple-source subaqueous slope fans

Supply Type and Main Sediment Grain Size		Mud-rich	Mud-rich	Sand-rich	Gravel-rich
Scale		Medium	Medium	Medium	Small
Gradient (m/km)		Medium 2.5–25	Medium 7–35	Medium > 35	High 20–250
Radius or length (km)		Lobe 50–200	Lobe 5–75	Stripe 1–50	Stripe 1–10
Provenance	Scale	Medium	Medium-small	Small	Very small
	Slope	Low	Medium	Medium	High
	Distance	Far	Medium	Near	Near
Supply system		Large-scale fluvial delta rich in mud	Mixed-load delta, linear shoreline	Sand-rich clastic shoreline or shelf	Alluvial fan/braided river plain/fan delta
Supply mechanism		Low-density turbidity current and contour current caused by intermittent slump or slump	High-density or low-density turbidity current	Redeposition or direct deposition of shelf debris; low turbidity current	Frequent mass flow, slump, and turbidity current of river causes
Fluid scale		Big	Medium	Medium-small	Small
Channel system		Medium scale; channel-embankment complex system	Multi-period, embanked channel, transformed from meandering river to straight river	Multi-period, lateral migration, transformed from braided to low sinuosity channel	Small-scale intermittent groove
Distal slope fan deposition		Sand-mud interbed formed by thin sheet flow	Sand-mud interbedded lobe formed by mixed load of turbidity current	Low-fluctuation lobe or sheet sand formed by turbidity current rich in sand	Laminar distal turbidite formed by low-density turbidity current
Main basin plain deposit		Turbidity current deposits > offshore deposits	Offshore deposits > turbidity current deposits	Offshore deposits	Offshore deposits

Table 13.7 Main underground attributes and reservoir attributes of multiple-source subaqueous slope fans

Supply type and main sediment grain size		Mud-rich	Sand-mud	Sand-rich	Gravel-rich
Main architectural element	Proximal	Channel embankment	Channel embankment	Channel	Wedge
	Distal	Distal sheet sand	Lobe	Lobe with channel	Distal sheet sand
Seismic configuration		Channel embankment and distal parallel refection	Channel embankment and crevasse splay mound	Constructive and low-fluctuation crevasse splay mound	Wedge
Sandstone percentage (%)		≤ 30	30–70	≥ 70	Changeable, generally 5–50 (conglomerate >50%)
Geometric shape of sand body		Medium-scale sand body formed by main channel; sand body generally separated in updip and downdip directions	Deflected accumulation, lenticular channel silty-fine sand, natural levee deflected downward to accumulation body, forming sand shale	Wide sheet low-fluctuation lobate sand body, interior part shown as channel sand body unit	Irregularly connected conglomeratic; clastic rocks and breccia deposited mainly in proximal areas; sandstone deposited mainly in middle-distal areas
Turbidite facies		A, B, E, F	C, D	B, C	A, B, E, F
Reservoir heterogeneity		High	High-medium	Low	High
Sand body connectedness	Vertical direction	Poor-medium	Medium	Good	Good
	Lateral direction	Medium	Medium	Good	Medium
Trap type		Stratigraphic trap	Stratigraphic trap	Structural trap	Structural trap
Potential production layer		① Stratigraphic trap of outer fan; ② stratigraphic trap formed by channel embankment and crevasse splay	① Structure-strata complex trap in proximal slope channel; ② dip pinch-out channel or slope fan	Structure-strata complex trap formed by slope fan	① Structural trap formed by conglomeratic fan closing toward downthrown side; ② structural trap formed by medium and distal sand
Main exploration risks		Existence, range, and recognition of reservoir property and trap	Reservoir recognition, description, and property; integrally sealed	Structural conditions required for trap; integrally sealed	Reservoir property and integrally sealed
Significance and position in sea level cycle		Important; lowstand	Potentially important; highstand and lowstand	Not important; highstand and lowstand (mainly flood event)	Important; lowstand and highstand

Table 13.8 Main sedimentary features of line-source subaqueous fans skirts

Supply type and main sediment grain size		Mud-rich	Sand-mud	Sand-rich	Gravel-rich
Scale		Big	Medium	Small	Very small
Gradient (m/km)		Medium-high 45–150	Medium-high 45–150	Medium-high 45–150	Very high 20–500
Radius or length (km)		Strip 10–100	Strip 10–100	Strip 1–10	Strip 1–5
Provenance	Scale	Big	Medium	Small	Very small
	Slope	Low	Medium	Medium	High
	Distance	Medium	Medium	Near	Near
Supply system		Wide, starved shelf	Residual shelf or narrow shelf	Narrow shelf	Braided plain
Supply mechanism		Slump and related low-density turbidity current, contour current	High-density or low-density turbidity current complex	Low turbidity current caused by shelf debris collapse	Frequent mass flow, slump, and turbidity current caused by river
Fluid scale		Big	Medium	Small	Small
Channel system		None; mainly slump groove	Multi-period, transformed from straight river to meandering river with embankment	Multi-period groove or no well-developed intermittent channel	Small scale, intermittent groove
Distal slope fan deposition		Slump and debris flow, large-scale muddy flow	Small sandy turbidity current, slump	Intermittent sand-rich turbidity current	Low-density turbidity current deposited with large-scale mudstone, occasionally sandstone
Main basin plain deposit		High-density mud flow	Offshore deposits	Offshore deposits	Offshore deposits

Table 13.9 Main underground attributes and reservoir attributes of line-source subaqueous fan skirts

Supply type and main sediment grain size		Mud-rich	Sand-mud	Sand-rich	Gravel-rich
Main architectural element	Proximal	Slump	Sliding	Channel	Wedge
	Distal	Sliding	Branch lobe	Slump	Debris flow
Seismic configuration		Mixed accumulation mound	Mixed accumulation mound	Mound	Wedge
Sandstone percentage (%)		Large change range, generally between 0–20	Large change range	Large change range	Large change range
Geometric shape of sand body		Limited fan development; generally restricted by sliding and lowstand slump	Lateral extension Separated with mudstone	Lobate sand body with mutually connected channel units	Lateral extension; distal limited fan
Turbidite facies		F	D, F	D, E	A, B, F
Reservoir heterogeneity		High	High-medium	Low	High
Sand body connectedness	Vertical direction	Poor	Medium	Very good	Good
	Lateral direction	Poor	Medium	Very good	Very good
Trap type		Stratigraphic trap	Stratigraphic trap	Structural trap	Structural trap
Potential production layer		Stratigraphic trap formed by branch slump due to active or residual delta in updip direction	Stratigraphic trap formed by branch slump due to active or residual delta in updip direction	Structure-strata complex trap containing slope fan skirt turbidite sand	Structural trap formed by conglomeratic fan closing toward downthrown side
Main exploration risks		Existence, range, and recognition of petroleum reservoir and trap; integrally sealed	Existence, range, and recognition of petroleum reservoir and trap; integrally sealed	Structural conditions required for trap; integrally sealed	Reservoir property and integrally sealed
Significance and position in sea level cycle		Very important; highstand and transgressive period	Less important; transgressive and regressive periods	Less important; regressive period	Less important; regressive period

along the trough direction parallel to the foot of the continental slope.

There are eight types of lithofacies: ① cobble conglomerate; ② granule conglomerate and pebbly sandstone with graded bedding; ③ granule conglomerate and pebbly sandstone with obvious grading; ④ graded granule conglomerate, pebbly sandstone, and sandstone with liquid overflow; ⑤ non-graded cross-bedding granule conglomerate, pebbly sandstone, and sandstone; ⑥ pebbly sandstone and sandstone lacking structure; ⑦ sand and silty turbidite; and ⑧ deep-water shale. They also ascribed these eight lithofacies to three facies associations, namely, coarse-grained channel, coarse sandstone with superimposed scour, and non-channel deposit. Structural factors can cause the migration, filling, and abandonment of channels, resulting in the formation of thickening and coarsening sequence, thinning and fining sequence, and other complex sequences, respectively.

13.5.3.2 Lake Trough Facies Model

Lake trough or lake ditch features channel-type gravity flow deposit.

(1) Facies markers: dark mudstone, shale mixed with pebbly sandstone, massive sandstone, particle-supported conglomerate, and matrix-supported conglomerate. Slump deformation structure is well developed, and seen sometimes with typical turbidite. Deep water ostracoda fossils and reticular ichnofossils can be seen commonly in the dark mud shale. The grain size probability graph of sandstone and the C-M figure show dominant graded suspension and suspension population sedimentation.
(2) Facies sequence: A thickening and coarsening upward sequence is mostly developed, reflecting the gradual strength of gravity flow and the superimposition of multiple events, sometimes due to channel migration.
(3) Facies model: The gravity flow deposit of lake troughs can be divided into two subfacies, namely, channel and overflow. Channel can be subdivided into channel axis and point bar, and overflow can be subdivided into proximal overflow and distal overflow. The channel subfacies is characterized by pebbly sandstone, and massive and parallel bedding sandstone, and the overflow microfacies features typical turbidite deposition.
(4) Petroleum geological significance: The plane of a trough turbidite body has an inhomogeneous striped shape. Its profile is presented as a conglomeratic body with lenticular distribution, which may be favorable for hydrocarbon reservoir formation. For example, the western Huzhuangji Oilfield in the Dongpu Depression and the Wendong and Qiaokou Gas Fields in the central uplift zone are oil & gas storage reservoirs with trough gravity flow channel turbidite. Favorable oil & gas reservoirs are subject to the lithology and lithofacies of turbidite.

13.6 Deep-Sea Contourites

Contour current sediments are distributed widely in modern deep seas, occupying a certain proportion. Contour current is mainly ocean underflow driven by thermohaline circulation, which generally flows along the continental slope. Its scale is even equivalent to that of a few submarine fans, and it can transport a large amount of fine-grained sediments to form drifts. Contour current sediments usually co-exist with pelagic and hemipelagic deposits, so they are relatively difficult to identify in geological records. However, great improvements have been made in the observation and research (Stow et al. 2002) of modern ocean floor contourites in the last decade. On the one hand, detailed studies on the sedimentary features, identification marks, and differences with turbidite and pelagic deposits of contour current have been conducted, and the characterized sequence of contourite drift has been recognized by lithofacies classification. On the other hand, the distribution law of contourites and features of contourite drift have been researched fruitfully. Many megascopic contourite drifts have been found in modern oceanic deep-water environments, and their scales can be compared to that of abyssal turbidite fans. Moreover, research on large-scale contourites in ancient strata, that is, study and research of ancient contourite drifts, has progressed considerably in the last three decades. Bern et al. (1976) studied the sedimentary prismoid formed by the craton Cretaceous contour current in Arabia. The prismoid is still located in the Mediterranean seabed. Contourite drift is found successively in the carbonate rock of Jiuxi in the lower Ordovician in China (Duan Taizhong et al. 1990) and in the carbonate rock of Pingliang in the middle Ordovician in Gansu (Gao et al. 1995). The discovery of and research on contourite drifts of ancient strata has opened up a new field in petroleum exploration.

13.6.1 Concept and Definition

Hollister et al. (1963) proposed the concept of contour current and defined it as a thermohaline circulation formed by the Earth's rotation; this circulation, parallel to the submarine contour line, flows at a stable and low speed (5–20 cm/s), mainly presented in continental rise areas. Shanmugam (2000) elaborated upon this definition. In modern ocean basins (such as the Atlantic), thermohaline circulation formed by the Earth's rotation in the deep part or the bottom of a water body is nearly parallel to the submarine contour line (i.e., along the continental slope strike), and it is called "contour current." However, not all bottom currents flow along the contour line. The term "contour current" refers only to the fluids whose current direction is parallel to the contour line, and the other fluids are called "bottom current." Broadly, "bottom current" simply refers to the fluid on seabed that causes sediment transportation and deposition, and it has many types based on genesis. Bottom return current formed by a vortex that reaches the seabed due to circular motion driven by storms, such as (submarine) circulation in the Gulf of Mexico and the gulf stream in the Northern Atlantic Ocean. Bottom currents formed by tides and internal waves can move upward or downward along the slope, particularly in submarine canyons.

Stow et al. (2002) defined the concepts related to contour currents as follows: ① Contourites refer to the sediments deposited by or greatly affected by bottom current activity along the slope (strike of the continental slope). Countourites constitute a contour current facies with broad connotation and comprise muddy-gravel lag facies and various sediments determined by the sedimentary supply system. Many transitional facies exist due to the mutual effects of various mechanisms in deep sea. ② Drift is a general term referring to sedimentary aggregation with an unclear boundary and a unique appearance. Its sedimentary process is controlled by the ocean current. It is not limited to bottom current deposition. Contourite drift is a specific term that refers to the sedimentary rock formed by contour currents with the drift phenomenon. Contourite drift can appear in various environments located at different depths in the ocean, including all deep-water environments (more than 2000 m) and medium deep-water environments (300–2000 m). The sedimentary bodies formed on the uppermost part of the shallow outer shelf or continental slope are not narrowly regarded as contourite drift, however they can be viewed as shallow drift.

According to modern oceanographic surveys, the flow velocity of submarine contour currents is generally 5–20 cm/s, but locally, the velocity can be 50 km/s or higher. Thus, contour current is an particularly important and special geologic agent that not only can erode the seabed but also can transport sediments to form contourites or contourite drift.

The sedimentation of contour currents is generally slow, the deposition rate is low, and the change is relatively large. The deposition rate statistics of modern contourite areas in the Mid-Atlantic show that the deposition rate is 0.6–20 cm/ka, and usually 2–12 cm/ka. The deposition rate of contour currents is related to their flow velocity, source supply, submarine geomorphy, climate change, sea level change, etc.

13.6.2 Classification and Characteristics

Currently, there is little research on the classification of contourite drift (also called "contourite"), particularly on its genesis classification. Two main types of classification can be seen in the literature. The first one is by material composition and grain size; the second is on the basis of geometric shape and formation environment.

13.6.2.1 Clastic Composition and Grain Size Classification

Presently, in China and elsewhere in the world, contourite drift is divided into four basic types according to grain size (Table 13.10), namely, muddy contourite drift, silty contourite drift, sandy contourite drift, and gravelly contourite drift, in addition to several transitional types. Contourite drift is composed of terrigenous debris and carbonate materials.

13.6.2.2 Classification by Geometric Shape and Formation Environment

According to the plane geometrical characteristics of contourite drift (Faugères et al. 1984), it can be divided into four types: ① sheet (including deep sea sheet and adhesive sheet); ② long strip mound (including detached and two types of isolated drifts); ③ drift associated with channel (including patch-like drift and contour current fan); and ④ confined drift (Table 13.11).

The emergence of different types of contourite drifts and geometric shapes depend mainly on the following four related factors: ① geomorphic background or submarine topography; ② fluid velocity and variability; ③ total amount and type of sediment supply; and ④ duration of the underset process The geometric shape of drift can change under the action of continental slopes and the interaction effect of other fluids (Stow and Holbroock 1984; McCave and Tucholke 1986; Faugères et al. 1984) (Fig. 13.40).

According to the appearance and formation environments of contourite drift, Stow et al. (2002) divided it into 6 types: ① sheet drift; ② elongated drift mound body; ③ drift associated with channel; ④ confined drift; ⑤ drift transformed by turbidity current depositional system; and ⑥ filling drift. Of these types, elongated contourite drift mound body is the most important. On the plane, it is shown as a long strip or an elongated shape, but in the cross section, it is shown as a mound with a length of tens to hundreds of kilometers; width of tens of kilometers, which is higher than the surrounding seabed by more than 0.1–1 km; and the thickness

Table 13.10 Classification and characteristics of contourite drift

Classification	Features	Fabric	Sorting	Sedimentary structure
Terrigenous debris	Muddy contourite drift	Silty clay, sand-containing rate of 10–15%	Poor	No obvious tectonic characteristics, developed with lebensspur and bioturbation
	Silty contourite drift	40% of sand, less than 10% of clay	Fine	Common bioturbation and ripples bedding
	Sandy contourite drift	Fine-medium sand (>40%)	Medium-better	Parallel bedding Cross bedding
	Gravelly contourite drift	Gravel	Poor	Irregular
Biologic debris	Biodetritus contourite drift	Dominated by biodetritus contourite drift, muddy silt supported	Poor	None

Table 13.11 Geometric appearance and classification of contourite drift (according to Faugères et al. 1999)

Drift type	Subclassification	Scale (km^2)	Example
Contour current sheet drift	Deep sea sheet	105–106	Gloria drift in Argentine Basin
	Continental slope adhesive sheet	103~104	Campos continental margin in Gulf of Suez
	Continental slope patch sheet	$<10^3$	
Elongated hummocky drift	Detached drift	103–105	Eirik drift; Blake drift
	Isolated drift	10^3–10^4	Feni drift; Faro drift
Drift related to the channel	Patch-like drift/Drift associated with channel (trench)	10–104	Northeast Rockall trough
	Contour current fan	10^3–10^5	Vemma waterway outlet
Confined drift		10^3–10^5	Sumba drift, E Chatham continental rise
Induced drift-turbidity current system	Stretched turbidity current/contour current embankment	10^3–10^4	Columbia dyke (Brazil Basin), Hikurangi fan drift (New Zealand)
	Etched turbidite body	10^3–10^4	Southeast Weddell Sea
	Interactive turbidite-contourite drift	Probably extremely vast	Hatterasl continental rise

of the accumulation can be more than 2 km locally.

13.6.3 Vertical Sedimentary Sequence and Depositional Model

13.6.3.1 Vertical Sedimentary Sequence

The vertical sedimentary sequence of contourite drift facies shows no regularity, unlike that of turbidite (Faugères et al. 1984), and its most outstanding feature is a composite sequence composed of inverse grading coarsening upward and normal grading fining upward. Its thickness is 10–100 cm. The rhythm can occur alone or appear simultaneously to form an inverse-normal grading unit. From the bottom to the top, changes in the facies in sequence include homogeneous mudstone facies, mudstone with discontinuous silt lens facies, sandy siltstone facies, mudstone with discontinuous silt lens facies, siltstone facies, and homogeneous mudstone facies (Fig. 13.41a). The above sequences often have large changes in thickness, integrity, and comprehensiveness, and they may not be totally symmetric. Sometimes, they can be partly or completely asymmetric. A complete inverse grading-normal graded bedding reflects the change situations in the flow velocity of contour current at the same place, which increases gradually at first and then begins to decrease after reaching the maximum value. The average flow velocity variation is 5–25 cm/s, and the duration of the entire sequence is about 1000–30,000 a.

Regarding the contourite drift found in the lower Ordovician of the Jiuxi Area in the northern Hunan Province, China (Fig. 13.41b) and sequences similar to the vertical sequence described by Faugères et al. (Fig. 13.41a), the contact relationships of each section in the sequence can be transition, abrupt, and erosion. The thickness of a single sequence is 10–200 cm, and is usually 30–80 cm. In addition, both complete and incomplete sequence are relatively common. In addition to the above typical sequence, there are some special types, for instance, the contourite drift in carbonate rock in the middle Ordovician of Pingliang City, Gansu

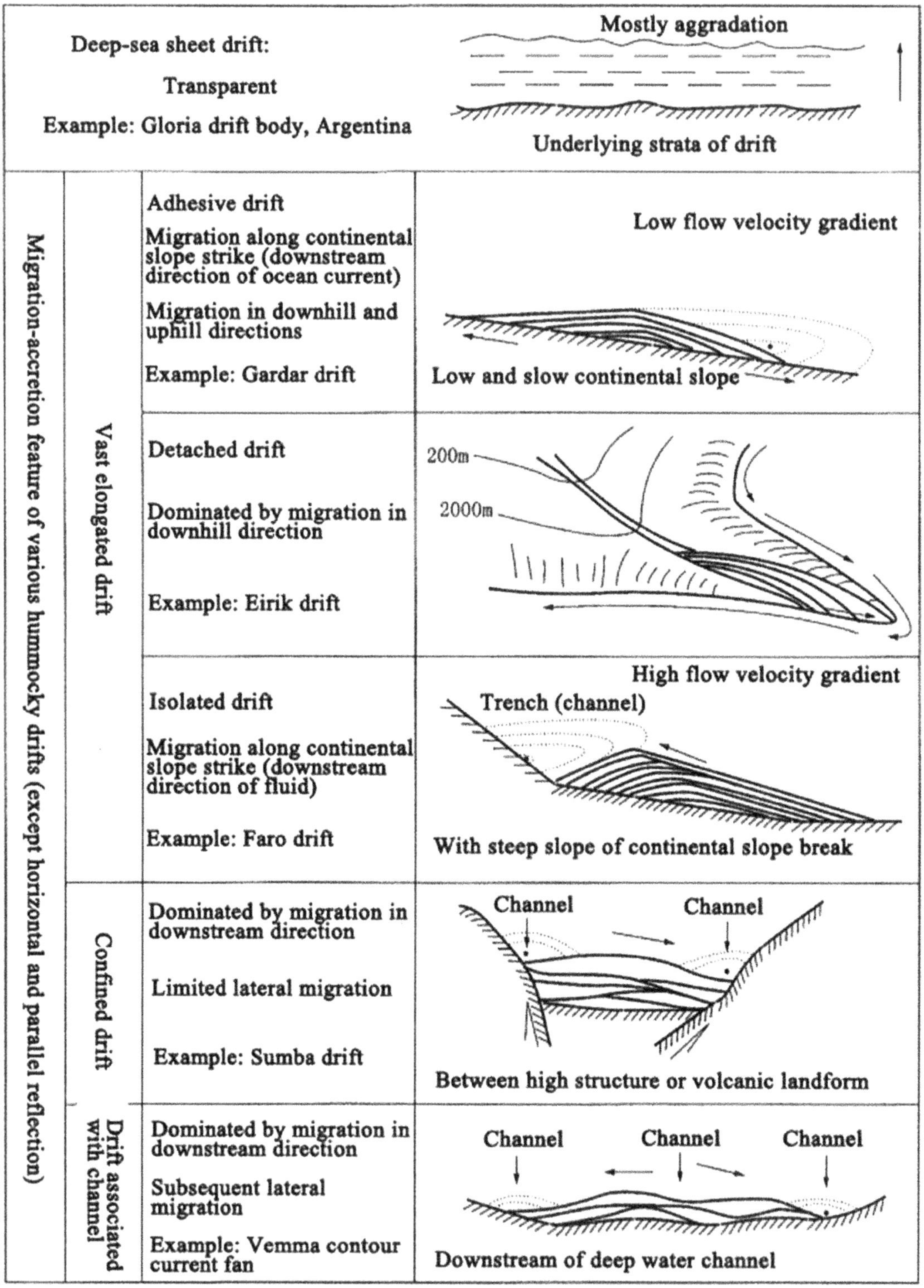

Fig. 13.40 Geometric type and migration-plus addition type schematic diagram of drift rock masses of different isobathic condition (after McCave and Tucholke 1986; Faugères et al. 1993). Double-headed arrow represents migration-accretion tendency

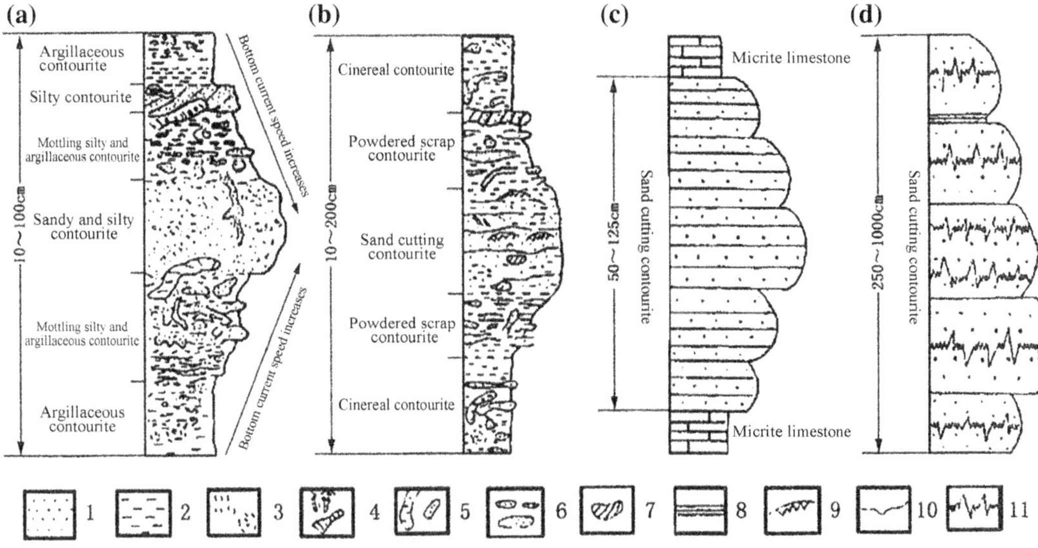

Fig. 13.41 Vertical sequence of the equal deep drift accretion rock is not as regular as turbidite. Its most prominent feature is to become a reverse graded and a coarse to thin normal graded composite sequence. This typical situation can be seen in the Jiuxi Ordovician of Taoyuan, north of Hunan, China (**a**) and equal deep drift accretion rock in the Middle Ordovician carbonate rocks, Pingliang, Gansu (**b**) (after Faugeres J. C, Gonthier E, Stow D. A, 1984. 1—Sand and silt (sand cutting and powdered scrap); 2—Mud (limestone); 3—Shell fragments and biodetritus; 4—Bioturbation; 5—Burrow; 6—Discontinuous silt lenses (or silt-sized strip); 7—Plaster mottling; 8—Dark shale; 9—Masking aperture; 10—Erosion surface; and 11—Suture line

Province, China (Gao et al. 1995), which are composed of a single sand cutting and other contourite drifts (Fig. 13.41c, d). This type of sequence is mainly superimposed by sand cutting and other contourite drift from the medium to the thick layer, where in each single layer, the particle size changes from fine at the top and bottom to coarse in the middle. However, the entire sequence is shown as a fine-coarse-fine cycle overall. Actually, this is a complex sequence. The above sequence feature is widely different from those of turbidite and tempestite, and their hydrodynamic characteristics are different as well.

The vertical sequences of typical turbidite and tempestite represent the sedimentation of a transient event, but that of contourite drift reflects the long-period variation of flow intensity or an activity cycle from weak to strong to weak.

13.6.3.2 Depositional Model

The history of research on contourite drift and limitations on the research on contourite drift in the modern Atlantic ocean is short. Yet, multiple depositional models of contourite drift are available in the literature, for instance, the depositional model of the Faro contourite drift in the east margin of the North Atlantic Ocean, depositional model of the cretaceous contourite drift in the Arabic craton continental margin, and depositional model of the carbonatite contourite drift in the lower Ordovician series of Jiuxi in Northern Hunan province, China.

Contourites are developed mainly in sea level rise periods. Research of the analysis data of carbon oxygen isotopes and micro granularity show that the ice age–interglacial transitional period, or sea level rise period, may be the strongest active period in terms of the underlying circulation. In the low sea level period, gravity flow sediments are dominant, leading to difficulty in the formation or preservation of contourites. With the rise of sea level, the provenance area moves away gradually from the sedimentary basin, injection of coarse clastic substances decreases, and gravity flow activity weakens, thus contourites can be developed. In the high sea level period, sediment supply is low, which retards contourites development. Therefore, contourites can be considered, to a certain extent, the featured sedimentation type of transgressive systems tract. According to the fluctuation and developmental status of sea level, contourite drift formation can be divided into three different stages: initiation, formation, and recession.

(1) Initiation stage: This period is the low sea level period when the depositional district is close to the provenance area, and on the continental slope or slope dominated by gravity flow action (primarily, the turbidity current) moving downward, resulting in the full development of gravity flow sediments and weak contour current activity. A few contourite drifts start forming, but they are too small to form obvious geometric shapes.
(2) Formation stage: At the time of sea level rise, the provenance area is far away from the depositional district, downward gravity flow along the slope is weak, and the granularity of the sediments becomes finer. However, contour current activity is strengthened, so gravity flow sediments are transformed into contour current sediments and accumulated largely in the continental rise or basin margin to form a sediment body with a certain geometric shape scale, that is, contourite drift. Therefore, contourite drift formation occurs mainly in the sea level rise period. The feature of this stage is that coarse-grained contourite drift, particularly sand-class contourite drift, is largely developed.
(3) Recession stage: When the sea level reaches a certain height, or high sea level, the provenance area is far away from the depositional district. Although gravity current action is extremely weak or not developed, the contourites decline gradually because sediment supply is insufficient as a result of the lack of a significant material source. The main feature is that coarse-grained contourite drift decreases obviously, autochthonous deposit increases obviously, and is finally replaced by autochthonous vertically falling sediments (vertical accretion).

13.6.4 Sedimentary Characteristics and Identification Marks

13.6.4.1 Sedimentary Characteristics of Contour Current

(1) Depositional environment: Contourites are associated with deep-water autochthonous deposits and mixed set of deep-water autochthonous deposits, which are shown mainly as irregular thin stratified structures and lenticular occurrences, with a single layer having a thickness of several centimeters overall and tens of centimeters locally. They can be distributed in the environment at any water depth, but mostly in continental rise areas, and can occur in deep-water basins.
(2) Material composition: Contourites have both siliceous debris and carbonate rock, and their sedimentation type is mainly terrigenous clastic rock and carbonate rock (including biodetritus contourite), along with a little volcaniclastic rock.
(3) Granularity: The granularity varies considerably from mud level to gravel level, and a series of transitional types composed of a mix of sand, silt, and clay are present. When a strong contour current corrodes the seabed, gravel lag deposits can be formed. The contourites found currently are generally dominated by the mud level and silt level, followed by the sand level, with the fine gravel level appearing occasionally.
(4) Sorting features: Ordinarily medium-good, but very good locally. On the normal probability curve, there are usually 2–3 depositional populations, in which the slope of the saltation population is large.
(5) Hydrodynamic force: Contour current is a type of deep-water tractional current. Therefore, contourites usually have the features of tractional current sedimentation, for instance, an erosion surface formed by

current scour, water stratification (miniature cross bedding and megascopic cross bedding), and fabric selection (for example, directional alignment of elongated particles).

(6) Occurrence: The stream direction of the contourites formed on the continental slope and continental rise are usually parallel to the slope strike, for instance, the directional alignment of elongated particles is parallel to the slope strike, and the thin layer in cross bedding is generally parallel to the slope strike. This is unlike the features of turbidity current deposit, and internal wave and internal-tide deposits.

(7) Sedimentary structure: Contourites are usually strong bioturbation structures, thus the original sedimentary structure is not well reserved. This is mainly due to the low flow rate of contour currents, hence the sedimentation is relatively slow.

(8) Vertical sequence: Contourites generally have a unique vertical sedimentary sequence, or fine-coarse-fine inverse-to-normal grading of vertical granularity because the flow intensity of contour currents varies periodically.

13.6.4.2 Identification Marks for Contourite Core

Many scholars have put forth their own views on the identification marks for contourite cores. Shanmugam (2000) believes that "a sand body transformed by bottom current can be identified by applying the main primary sedimentary structure," and the sedimentary structure of a tractional current is deemed as the only reliable identification mark of underflow re-transformational sand. For underflow recomposed sand, the vertical facies model under any standard has its limitations, however the features of core and outcrop are beneficial for identifying underflow re-transformational sand (Fig. 13.42).

(1) Fine sand and silt are absolutely dominant;
(2) Sand in the form of thin layers to laminae in deep water mudstone (usually less than 5 cm);
(3) Rhythmic sand and mud layers;
(4) Large numbers of sand layers (50 layers per meter of core or more);
(5) Abrupt (no erosion) contact at the top; abrupt to gradual contact at the bottom;
(6) Inner erosion surface;
(7) Sand with good sorting and low matrix content (pure sand);
(8) Different scales of inverse-graded (coarsening upward) bedding;
(9) Horizontal bedding and low-angle cross bedding;
(10) Cross bedding (although Mutti (1997) suggested that the cross bedding of sediments in an abyssal canyon mouth is formed by the passing of turbidity current, and the flow conditions of stable equilibrium required to form cross bedding are unimaginable given the episodic turbidity current events occurring in a canyon mouth);
(11) Lenticular bedding and starved wavy bedding (in core scale);
(12) Current ripple with original wave crest or erosion wave crest, of which some (symmetric) curve feature ripples can represent wave ripple;
(13) Mud-offshoots;
(14) Double mud layer (tide);
(15) S-shaped cross bedding (tide);
(16) Flaser bedding; and
(17) A sand layer with a drag structure occurs in independent units but is not taken as part of a certain vertical structure, for instance, the Bouma sequence with graded boundary at the bottom. The lack of Bouma sequence in sediments is a mark of underflow products.

However, there is no marker of underflow re-transformation facies among the markers listed above. Notwithstanding, opportunities for identification of underflow re-transformation facies can be strengthened through comprehensive judgment in combination with an understanding of the regional sedimentary environment.

13.6.4.3 Seismic Identification Mark of Contourites

A significant erosion surface is the common characteristic of bottom current sedimentation, thus contourites usually have an obvious inner reflection surface. Although not all erosion

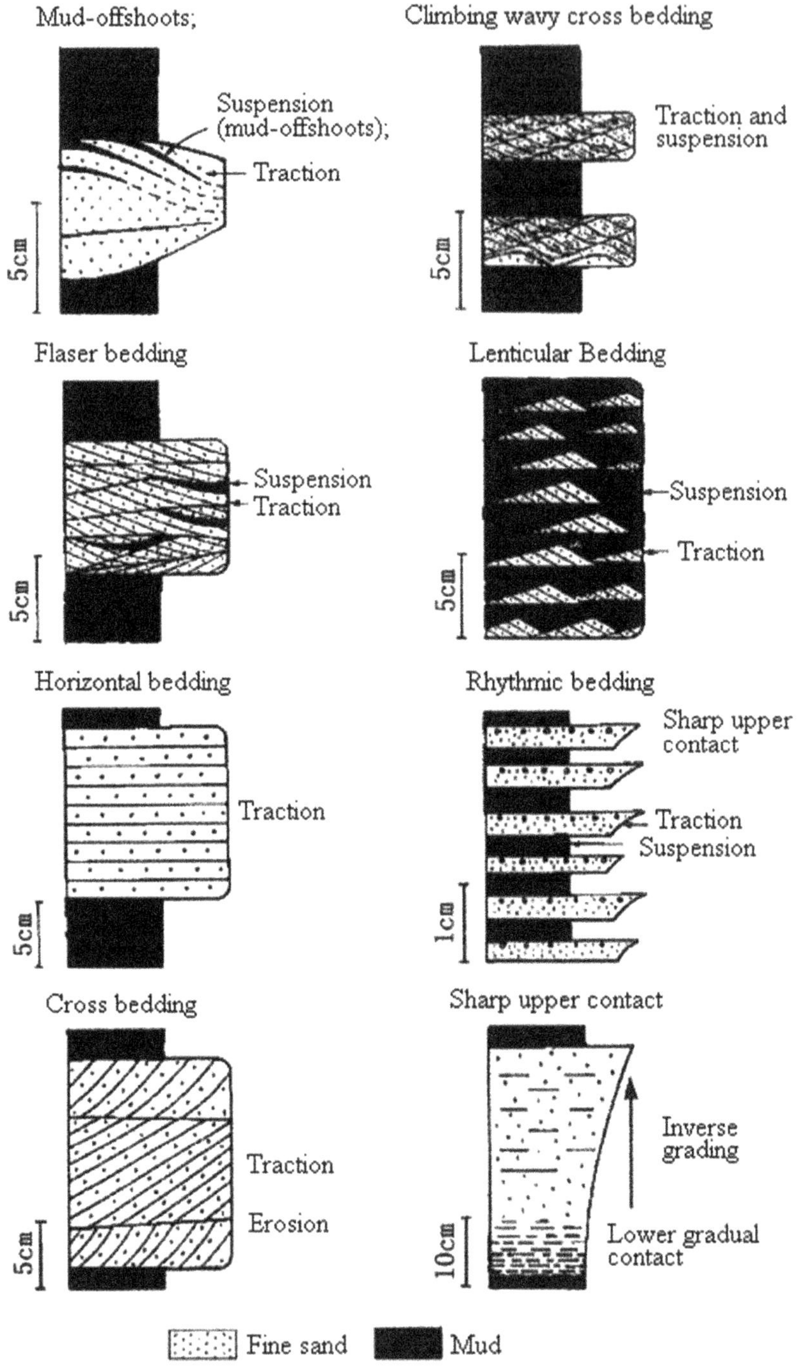

Fig. 13.42 Shanmugam believes that the sand body can be identified with the transformation of the bottom of the main primary sedimentary structure, the drainage sedimentary structure is believed to be the only reliable sign of recognition and then transform the sand bottom (after Shanmugam 2000)

surfaces result in strong bottom current activity (particularly, when their distribution ranges are limited geographically), they are very obvious for contourite drifts.

Wide unconformity is a feature of contourite drift, and it can trace entire sedimentary bodies, leaving a continuous strong reflection as its mark. This unconformity reflects not only the erosion effect caused by an increase in the strength of bottom current, but also significant changes in components or particle diameter in relation to the fluid system. The sedimentary (seismic) unit in the contourite drift body is mainly shown as pisolitic, with a convex shape, however it is not parallel to the erosion surface generated by the preceding erosion event (Fig. 13.43).

The superimposition of sedimentation units reveals the overall migration of sedimentary bodies, and indications of such migration differ for different types of drift bodies. It is mostly shown as foreset downstream or inclined foreset, and it features downlap (toplap) or sigmoid foreset reflection.

Under some circumstances, the seismic reflection image of contourite drift can easily be confused with that of turbidity channel-natural levee systems, and their main difference is that under the action of bottom current, the entire drift body migrates against the slope surface upward or along the contour line of the slope surface, not downward along the slope face.

The seismic facies of contourites are various. In most cases, they have a low-medium amplitude reflection, which is continuous, discontinuous, or even chaotic. In some cases, they have an irregular wavy reflection and regular migration type of sediment wavy reflection, usually with good stratification, which can feature horizontal or parallel high-amplitude reflection (Fig. 13.44).

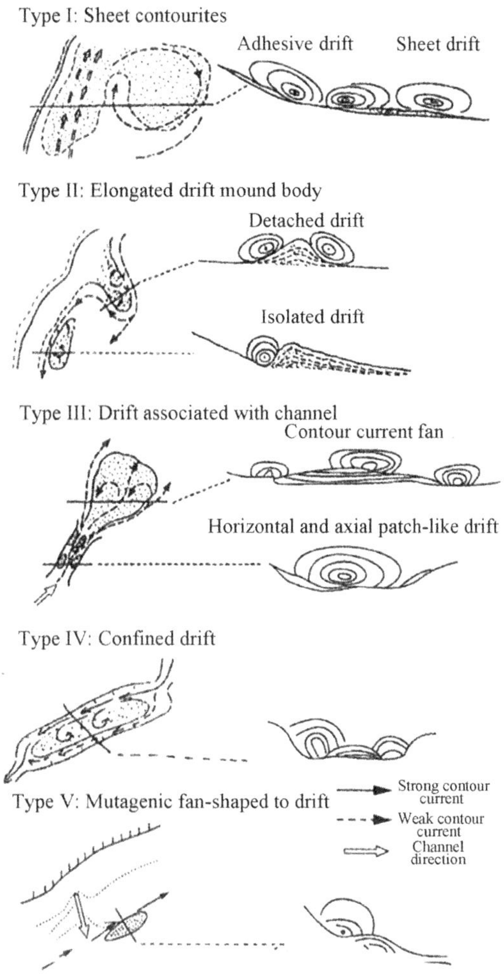

Fig. 13.43 Equal deep (bottom flow) drift rock mass is characterized by generalized unconformity, which can be tracked throughout the sedimentary body, with continuous strong emission (after Stow D. A. et al. 2002)

13.6.5 Interrelationships Between Contour Current and Deep-Water Turbidity Deposit

Both contour current and isolated drift have channels and grooves with long striped berms, or the drift is developed mainly along one flank. On the surface, their geometries and seismic responses can be compared to those of submarine fans (subaqueous fan), slope fan skirts, and turbidity channel-natural levee systems on the basin plain. In addition, the above two seismic facies systems have certain similarities, for instance, strong chaotic seismic reflection representing channels or grooves and parallel-wavy reflection representing natural levees or berms.

However, the contour current and turbidity current systems can be distinguished by their flow directions and the progradation directions in relation to continental margin dipping. In general,

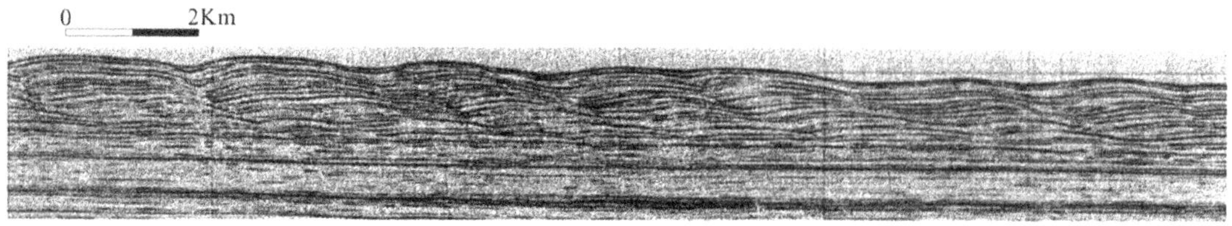

Fig. 13.44 Seismic response characteristics of deep flow deposits in the northern continental slope of the South China Sea (according to Yu Xinghe 2005)

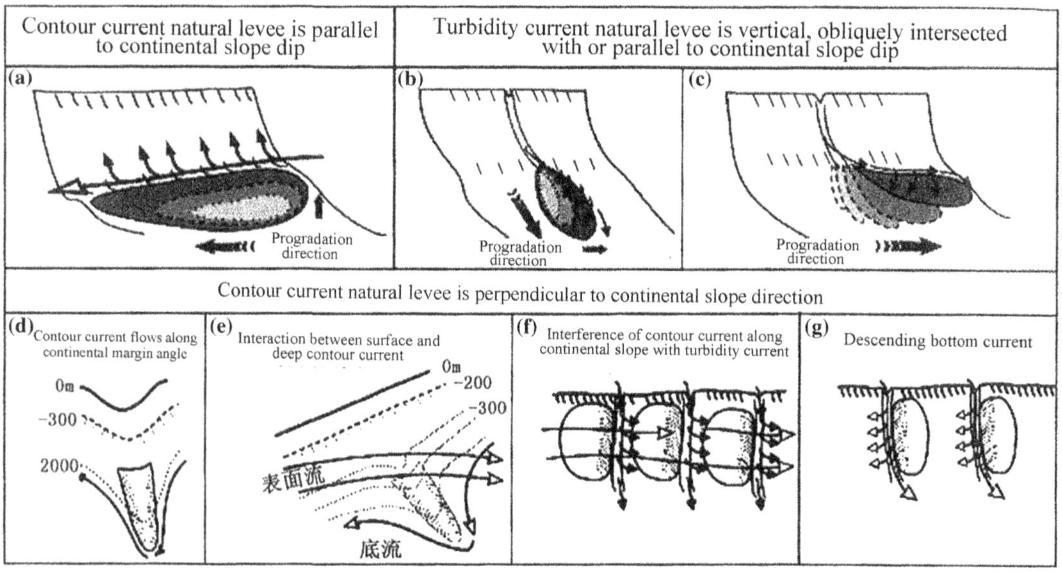

Fig. 13.45 A summary of the relation between the contour current, turbidite natural levee and ocean current direction (after J. C. Faugeres et al. 1999). **a** General flow direction of contourite drift; **b** general direction of turbidite natural levee; **c** possible development direction of turbidite natural levee; and **d**–**g** different developmental stages in which contour current natural levee becomes perpendicular to flow direction of the lobe

the migration for each type of deep sea sedimentary body is controlled by the following four elements: water flow direction, Coriolis effect, landscape environment, and its interrelation with other flows.

First, for any fluid, downstream accretion or migration usually occurs for the turbidity current natural levee or contour current drift. What is representative is that turbidity current flows down the slope, while bottom current flows along the slope (Fig. 13.45a, b).

Second, according to the actions of different types of fluids in the process of deposition-transportation, the lateral migration of natural levees or movement of drift is different from the Coriolis effect. In the Northern hemisphere (Fig. 13.45b), for a turbidity current system, if the water flow is downward along the continental slope, the turbidity current direction is deflected toward the right. When they reach the continental rise or an area where the channel is largely unloaded, they will flow over along the right side of the channel and construct an obvious natural levee on the right side. With continuous sedimentary unloading on the right side, the channel and natural levee will move toward the left side gradually to create a channel-natural levee system under oblique crossing with slope inclination

(Fig. 13.45c). With regard to the bottom current, if water flows along the slope, the Coriolis effect will cause the water flow to deflect toward the right side to effectively limit water flow on the continental slope. Because water flow is strongly limited, erosion will occur on the continental slope to form a channel. The deposition of drift and construction processes tend to occur in the low-velocity zone on the right side of the water flow, and the drift moves upward against the slope and forms a foreset along the slope surface.

Third, submarine topography greatly influences the flow of two fluids. The tendency of the continental margin or the gradient of the continental slope can cause or stop deposition, construction of the bottom current drift, and other changes, for instance, a hummocky drift with obvious progradation features turns into an adhesive continental slope drift with no obvious foreset but obvious accretion. The interaction of turbidity current with other fluids erodes the bottom near the seamount, resulting in deposition.

Finally, the interaction of fluids controls the deposition process and influences its migration type to lead to the generation of different sedimentary characteristics, vertical sequences, and biological differences (Table 13.12).

Both the extension and progradation direction of contour current on the continental margin or in the basin can be changed, depending on landform (for instance, topographic slope and flatness), fluid system, and the interaction between Coriolis force and intensity. Generally, the long axis (extension direction) is parallel or sub-parallel to the continental margin strike (Fig. 13.45a), but progradation can cause the extension direction of a part of the drift to be nearly perpendicular to the continental margin (Fig. 13.45d, e), thus leading to the following situations: ① progradation direction varies with a change in continental margin dip (Eirik drift, Fig. 13.45a, d), and ② surface flow and bottom current interact with each other (Fig. 13.45e, b) (Cape Hatteras and Blake-Bahama drift, McCave and Tucholke 1986). The turbidity channel and the canyon transecting the continental margin along the slope are affected strongly by contour bottom current, and this effect can make part of the turbidity current and the contour current develop an elongated symmetric levee (Weber et al. 1994; Rebesco et al. 1996). The erosion effect of the declining bottom current on the channel under the direction of the continental slope was proposed for the Chatham Rise (Barnes 1992, 1994) and the Cadiz Bay (Fig. 13.45g, Faugères et al. 1985; Nelson et al. 1993) in the east of New Zealand, wherein the "natural levee" is an elongated contour current drift.

Elongated hummocky drift and natural levees with turbidity accretion are hard to distinguish in following cases: ① the above-mentioned drift extends down along the continental slope; ② turbidity channel–natural levee system is developed downward along the continental slope (Fig. 13.45b), but it can also migrate along the continental slope under the Coriolis force or tectonism (Fig. 13.45c); ③ the two have a similar elongated hummocky shape; and ④ there is a mixed drift natural levee.

In the literature, the model (Fig. 13.46) of interaction between turbidity current down the continental slope and contour current along the continental slope can be found frequently, particularly for the eastern continental margin in North America (Locker and Laine 1992; Shanmugam et al. 1993), South Atlantic (Masse et al. 1998), Antarctic Ocean (Kuvaas and Leitchenkov 1992; Larter and Cunningham 1993; Weber et al. 1994; Pudsey and Howe 1998), and New Zealand continental margin (Carter and McCave 1994).

13.6.6 Petroleum Geology Significance of Contourite Drift

Because the silt-level and sand-level sediments in contourite drift are elutriated repeatedly by contour current, the texture maturity of these sediments is much higher than that of turbidite. Primary porosity is developed, together with

Table 13.12 Comparison of characteristics of contourite drift and turbidite

Features		Contourite drift	Turbidite
Fluid property		Deep-water tractional current	Sediment gravity flow
Grain size characteristics	Sorting	Good-very good	Medium-poor sorting
	Probability curve	With 2–3 populations, large slope of saltation population	Only one population, small slope; Parallel to C = M base line on C-M map
Thickness and scale		10–200 cm, usually 30–80 cm	Usually 10–100 cm
Primary sedimentary structure	Grain sequence	Normal and inverse graded bed sequence, clear top and bottom contact	Commonly normal graded bed sequence, clear bottom contact, unclear top contact
	Crossing laminae	Common, shown due to concentration of heavy minerals	Rare, shown due to grain size variation
	Parallel laminae	Existing in entire layer, shown due to concentration of heavy minerals or foraminifera shells	Only seen at the top due to granularity and composition
	Massive bedding	Scarce	Particularly common at the bottom of a rock stratum
	Bioturbation	Developed in the entire sequence	Absent or exists at the top
Grain fabric		Grains generally with preferred orientation	No preferred grain orientation
Silt level component	Matrix (<2 μm)	0–5%	10–30%
	Microfossils	Rare, worn, or broken	Commonly seen in rock stratum, well preserved
	Biological debris	Rare, worn, or broken	Commonly seen in rock stratum, well sorted and preserved
Rock type		Detritus greywacke, arkose, and quartz sandstone	Greywacke, detritus greywacke

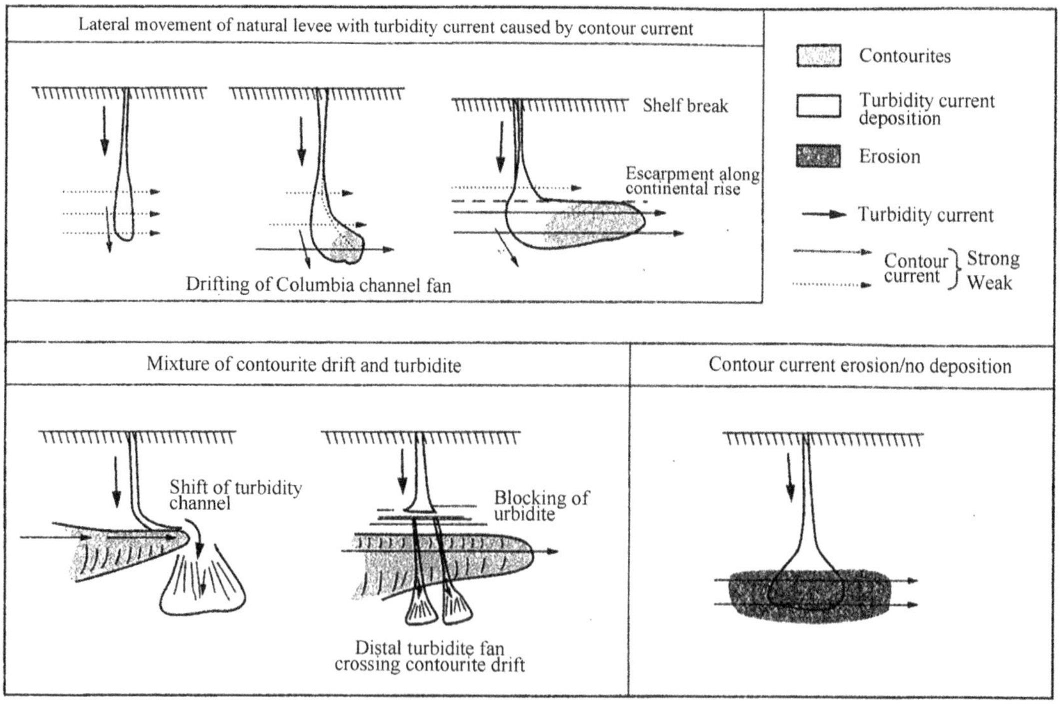

Fig. 13.46 The model of interaction between deep slope and deep slope flow (after J. C. Faugeres et al. 1999)

large-scale sedimentation and interbedding sedimentation with shale and mudstone of deep-water autochthonous deposit. Thus, it possesses good source-reservoir-cap association features, making it a good prospect for oil & gas exploration.

References

Bouma, A.H. 1962. *Sedimentology of some flysch deposits*, 168p. Amsterdam: Elsevier.

Coleman, J.M. 1976. Deltas: Processes of deposition and models for exploration, 102p. Champaign: Continuing Education Publ Comp., Inc.

Einsele, G. 2000. *Sedimentary basins-evolution, facies, and sediment budget*, 1–792. Second: Completely Revised and Enlarged Edition Springer-Verlag.

Faugères, J.-C., E. Gonthier, and D.A.V. Stow. 1984. Contourite drift molded by deep Mediterranean outflow. *Geology* 12 (5): 296–300.

Feng, Zengzhao, Yinghua Wang, Huanjie Liu, Qingan Sha, Wang Defa, et al. 1994. *Sedimentology in China*. Beijing: Petroleum Industry Press.

Galloway, Duncan D., Eric EJ Wolanski, and Brian BA King. (1996). *Modeling eddy formations in coastal waters: a comparison between model capabilities*. American Society of Civil Engineers.

Gao, Zhenzhong, Shunshe Luo, Youbin He, et al. 1995. Middle Ordovician contourites in western margin of Ordos. *Acta Sedimentologica Sinica* 13 (4): 16–25.

Gao, Z.Z., and T.Z. Duan. (1990). Sedimentary facies model of Marine facies in south China. *Journal of sedimentary journal*, (2): 9–21.

George, T.N. (1956). Sedimentary environments of organic reefs. *Science Progress (1933-)*, 44(175): 415–434.

He, Y.B., and S.S Luo. (1997). The recent advances and prospects of the study on deep-water tractive current deposits. *Earth science progress*, 12(3): 247–252.

Heezen, B.C., D.B. Ericson and M. Ewing. 1954. Further evidence for a turbidity current following the 1929 Grand Banks earthquake. *Deep Sea Research* 193–202.

Heezen, B.C., M. Tharp, and M. Ewing. (1959). *The Floors of the Oceans I. The North Atlantic*. Geological Society of America.

Hollister, and M. Walter. (1963). The mission for a manned expedition to mars. *Massachusetts Institute of Technology*, 17(6): 759–65.

Klein, G.V. 1977. Tidal circulation model for deposition of clastic sediments in epeiric and mioclinal shelf sea. *Sedimentary Geology* 18: 1–12.

Kolla, V., and F. Coumes. (1987). Morphology, internal structure, seismic stratigraphy, and sedimentation of indus fan. *AAPG Bulletin*, 71(6): 650–677.

Kruit, Cornelis. (1955). *Sediments of the Rhone delta: grain size and microfauna*. (Doctoral dissertation).

Kuenen, P.H., and C.I. Migiorini. 1950. Turbidity currents as a cause of graded bedding. *Journal Geology* 58: 41–127.

Kuvaas, B., and G. Leitchenkov. (1992). Glaciomarine turbidite and current controlled deposits in prydz bay, antarctica. *Marine Geology*, 108(3–4): 365–381.

Larter, R.D., and A.P. Cunningham. (1993). The depositional pattern and distribution of glacial-interglacial sequences on the antarctic peninsula pacific margin. mar geol. *Marine Geology*, 109(3–4), 203–219.

Locker, S.D., and E.P. Laine. (1992). Paleogene-neogene depositional history of the middle u.s. atlantic continental rise: mixed turbidite and contourite depositional systems. *Marine Geology*, 103(1–3): 137–164.

Lowe, D.R. (1982). Sediment gravity flows: ii. depositional models with special reference to the deposits of high-density turbidity currents. *Journal of Sedimentary Research*, 52(6): 343–61.

Liu, Baojun, and Yunfu Zeng, as chief editor. 1985. *Lithofacies paleogeography basis and working method*. Beijing: Geological Publishing House.

Middleton, G.V. (1973). Sediment gravity flows:mechanics of flow and deposition. *Turbidites and deep-water sedimentation*, 1–38.

Middleton, G.V., and M.A. Hampton. 1976. Subaqueous sediment transport and deposition by sediment gravity flows. In *Marine sediment transport and environmental management*, ed. D.J. Stanley, and D.J.P. Swift, 197–218. New York: Wiley.

Mutti, E. (1977). Distinctive thin-bedded turbidite facies and related depositional environments in the eocene hecho group (south-central pyrenees, spain). *Sedimentology*, 24(1): 107–131.

Mutti, E. 1985. Turbidite systems and their relations (sic) to depositional sequences, 65–93. In Provenance of arenites, ed. G.G. Zuffa, 379p. Boston: D. Reidel Publishing company.

Mutti, E., and F.R. Lucchi. 1972. Le torbiditi dell Appennino settrentrionale: introduzione all analysis di facies. *Società Geologica Italiana, Mem.* 11: 161–199.

Normark, W.R. (1970). Channel piracy on monterey deep-sea fan. *Deep Sea Research & Oceanographic Abstracts*, 17(5): 837–846.

Pudsey, C.J., and J.A. Howe. (1998). Quaternary history of the antarctic circumpolar current: evidence from the scotia sea. *Marine Geology*, 148(1–2): 83–112.

Reading, H.G., and M. Richards. 1994. Turbidite systems in deep-water basin margins classified by grain size and feeder system. *The American Association of Petroleum Geologists* 792–821.

Shanmugam, G. 2003. Deep-marine tidal bottom currents and their reworked sands in modern and ancient submarine canyons. *Marine and Petroleum Geology* 20: 471–491.

Shanmugam, G. and R.J. Moiola. (1985). Submarine fan models: Problems and solutions. In *Submarine fans and related turbidite systems*, eds. A.H. Bouma, et al., 29–34. New York: Springer Verlag.

Shanmugam, T.D. Spalding, and D.H. Rofheart. (1995). Deep-marine bottom-current reworked sand (Pliocene-Pleistocene) Ewing Bank 826 Field, Gulf of Mexico. *Turbidites and associted deep-water facies*, 25–54.

Stow, D.A.V., and M. Mayall. (2000). Deep-water sedimentary systems: new models for the 21st century. *Marine & Petroleum Geology*, 17(2): 125–135.

Stow, D.A.V., J.C. Faugeres, J.A. Howe, et al. 2002. Bottom currents, contourites and deep-sea sediment drifts; current state-of-the-art. In *Deep-water contourite systems: Modern drifts and ancient Selres, seismic and sedimentary characteristics*, ed. D.A.V. Stow, C.J. Pudsey, J.A. Howe et al. Vol. 22, 7–20. Geological Society, London, Memoirs.

Tarbuck, and J. Edward. (1984). The earth. C.E. Merrill Pub. Co.

Walker, R.G. 1978. Deep-water sandstone facies and ancient submarine fans: Models for exploration for stratigraphic traps. *AAPG* 62: 932–966.

Walker, and L. Cherry. (1992). *The volcanic history and geochemical evolution of the Hveragerði Region, S. W. Iceland*. (Doctoral dissertation, Durham University).

Walker, R.G. (1966). Shale grit and grindslow shales: transition from turbidite to shallow water sediments in the upper carboniferous of northern england. *Journal of Sedimentary Research*, 36(1): 90–114.

Appendix

X. Yu et al., *Clastic Hydrocarbon Reservoir Sedimentology*, Advances in Oil and Gas Exploration & Production, https://doi.org/10.1007/978-3-319-70335-0

Atlas I Typical sedimentary structure (by Yu Xinghe)

A. Horizontal bedding

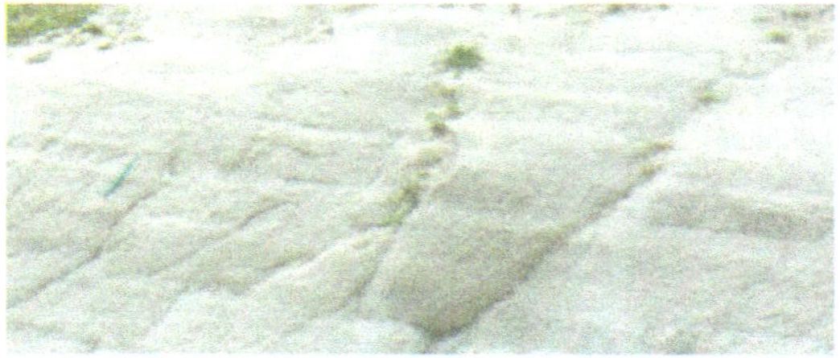

B. Parallel bedding—Jurassic system (photographed from Jinhuagong profile in Datong, Shanxi Province, 1997)

C. Megascopic wedge cross bedding-Jurassic system (photographed from Jinhuagong profile in Datong, Shanxi Province, 2005)

C. Megascopic trough (exocentric) cross bedding-Jurassic system (photographed from Jinhuagong profile in Datong, Shanxi Province, 2003)

E. Megascopic trough (concentric, with same thickness) cross bedding-Jurassic system (photographed in Calgary, Canada, 2000)

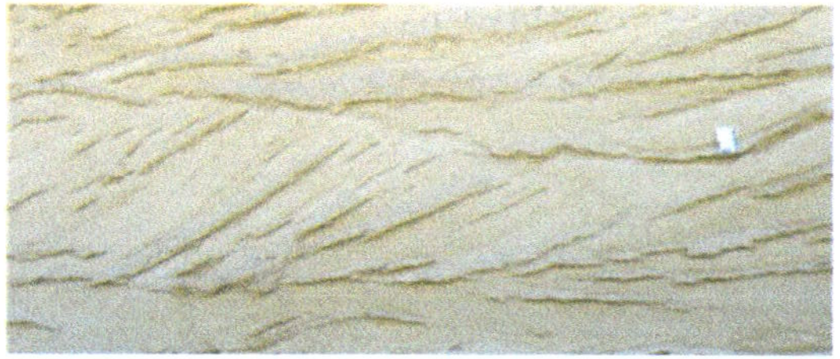

F. Megascopic trough (exocentric, with different thickness) cross bedding-Jurassic system (photographed in Urho Youshashan, Xinjiang, 2006)

Atlas II Various Planar Cross Beddings (by Yu Xinghe)

A. Single-group highly incised planar cross bedding-modern sedimentation (photographed in Muhua River of Daihai Lake, Inner Mongolia, 1995)

B. Single-group highly truncated planar cross bedding-Jurassic system (photographed in Jinhuagong profile in Datong, Shanxi Province, 2002)

C. Single-group low-angle-incised planar cross bedding—modern sedimentation (photographed in the upstream of Buliang River of Daihai Lake in Inner Mongolia, 1993)

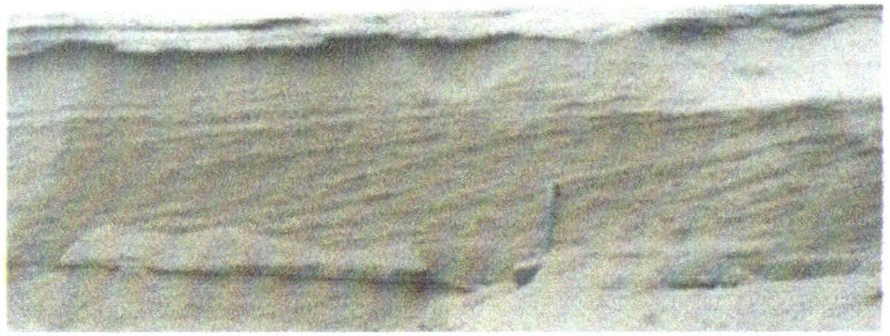

D. Single-group low-angle-truncated planar cross bedding—modern sedimentation (photographed in Luanhe River in Hebei Province, 1997)

E. Multiple-group high-angle-incised planar cross bedding—modern sedimentation (photographed in Luanhe River in Hebei Province, 1997)

F. Multiple-group high-angle-truncated planar cross bedding—modern sedimentation (photographed in Luanhe River in Hebei Province, 1997)

Atlas III Typical sedimentary structures (by Yu Xinghe)

A. Planar cross bedding—modern sedimentation (photographed in Yuanzigou Delta front of Daihai Lake in Inner Mongolia, 1990)

B. Herringbone cross bedding—Sinian system tidal flat sedimentation (photographed in Xiaweidian Village, Mentougou District, Beijing, 2006)

C. Swash cross bedding—modern sedimentation (photographed on seabeach of Xiachuan Island in Guangdong Province, 2003)

D. Hummocky cross bedding—Silurian system (photographed in Yuanbaoshan, Xinjiang, 2003)

E. Current ripples and convolute bedding—modern sedimentation (photographed in Daihai Lake, Inner Mongolia, 1990)

F. Current climbing ripple cross bedding—modern sedimentation (photographed in Hailar River in Inner Mongolia, 2007)

Atlas IV Typical sedimentary structures (by Yu Xinghe)

A. Wavy ripple cross bedding—modern sedimentation (photographed in Daihai Lake, Inner Mongolia, 1990)

B. Wind dune cross bedding—US (quoted from M. Falcon 1998)

C. Complex bedding—modern sedimentation (photographed in Daihai lake, Inner Mongolia, 1990)

D. Graded bedding—modern sedimentation (photographed in Yuanzigou alluvial fan of Daihai Lake in Inner Mongolia, 1990)

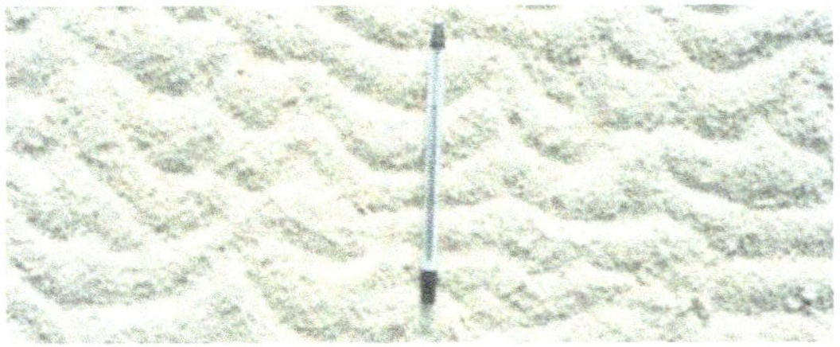

E. Wave ripple—modern sedimentation (photographed on Changli Coast, 1997)

F. Current ripple—modern sedimentation (photographed in Luanhe River, Hebei Province, 1997)

Atlas V Typical sedimentary structures (by Yu Xinghe)

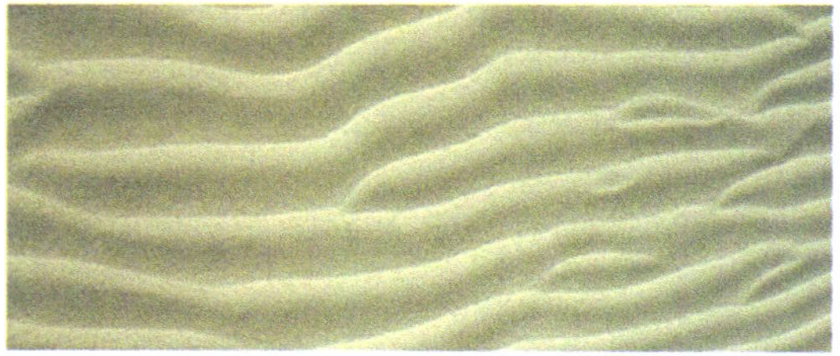

A. Wind ripple—modern sedimentation (photographed in Xinjiang, 2006)

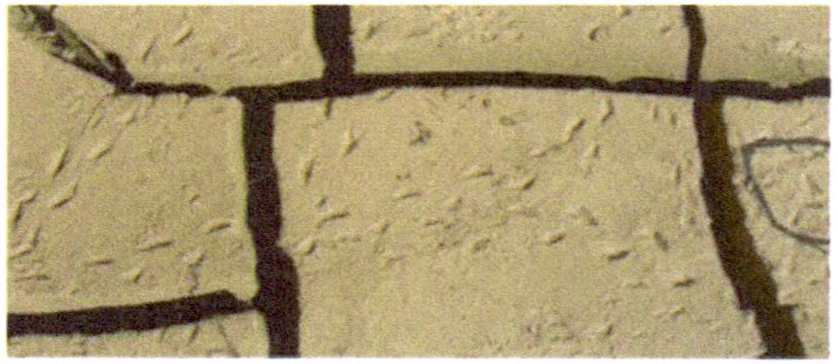

B. Suncrack—modern sedimentation (photographed upstream of the Buliang River of Daihai Lake in Inner Mongolia, 2007)

C. Convolute bedding—Jurassic system (photographed in the north margin of Dzungaria, 2006)

D. Deformation bedding—modern sedimentation (photographed in Hailar River, Inner Mongolia, 2007)

E. Slump deformation bedding—modern sedimentation (photographed in Yuanzigou Delta front of Daihai Lake in Inner Mongolia, 1990)

D. Lebensspur (drilling trace)—modern sedimentation (photographed in Hailar River in Inner Mongolia, 2007)

Atlas VI Various typical sedimentary phenomena of cores (by Yu Xinghe)

A. Parallel bedding
B. Incised planar cross bedding
C. Truncated planar cross bedding
D. Herringbone cross bedding
E. Trough cross bedding (exocentric)
F. Trough cross bedding (concentric)
Complex bedding
G. Lenticular bedding
H. Wavy bedding
I. Graded bedding
J. Slump deformation bedding

Atlas VII Various typical sedimentary phenomena of cores (by Yu Xinghe)

A. Sliding deformation (convolute bedding)
B. Crumpled deformation

C. Liquefaction deformation bedding
D. Bioturbation

E. Vertical and oblique crossing burrow
F. Horizontal burrow

G. Animal debris
H. Individual animal

I. Plant stem
J. Plant debris

Atlas VIII Various typical sedimentary phenomena of cores (by Yu Xinghe)

A. Well depth: 2781.75–2781.95 m, Position: Ezh
Description:
Light-gray middle-layered fine-grained detritus arkose, many visible horizontal burrows (Ophiomorpha), obvious backfill structure, which indicates relatively deep water body with obvious coarsening upward feature, medium-to-fine sorting, and Sm lithofacies

B. Well depth: 2764.75–2765.0 m, Position: Ezh
Description:
Light-gray middle-layered coarse sandstone, coarsening upward feature at the bottom and fining upward feature at the top, with the top obviously coarser than the bottom, scour surface possibly visible, obvious planar cross bedding, and laminae of grain size.

C. Well depth: 2753.75–2754.0 m, Position: Ezh
Description:
Light-gray middle-layered quartz-fine sandstone, dominated by medium sand, with obvious coarsening upward feature. Visible multiple low-angle-incised planar cross beddings, about 2 cm of muddy laminae at the bottom of rhythm, oblique burrow, obvious backfill structure, and a large amount of pyrite

Atlas IX Core description

A

B

Microscopic features: Depth: 2749.30 m Sample number: 419

Fabric feature: Grain size (mm): maximum 0.90/mainly between 0.25-0.50; Cementation type: hole; Sorting: medium; Psephicity: subangular to sub-rounded; Quartz enlargement: common

Detrital component (%)	Polycrystalline quartz	Monocrystal quartz	Feldspar: Anorthosite	Feldspar: Potassium feldspar	Igneous rock: Pelelith	Igneous rock: Granite	Metamorphic rock: Phyllite	Metamorphic rock: Schist	Metamorphic rock: Mylonite	Sedimentary rock: Mudstone	Sedimentary rock: Limestone	Others: Zoisite	Others: Mica	Total detritus	Face hole (%)
	2	92	<1		3	—	—	1	1	—	—	—	1	6	10

Interstitial material (%)	Matrix: Mud	Matrix: Kaolinite	Cement: Calcite	Cement: Ferrocalcite	Cement: Siliceous	Cement: Ferrodolomite	Cement: Ferruginous	Cement: Pyrite	Total interstitial material (%)	Pore type: Intergranular pore	Pore type: Intergranular dissolved pore	Pore type: Intragranular dissolved pore	Pore type: Particle dissolved pore	Pore type: Intercrystalline pore	Pore type: Bioclastic pore in biodetritus	Pore type: Mold pore
	<1	4	—	—	—	—	1	2	7	3.5	6	—		0.5		—

Description:

A. 40x, plane-polarized light (2747.28 m, sample number: 411)
Porous development, good local connectedness

B. 100x, plane-polarized light (2749.30 m, sample number: 419)
Visible quartz enlargement, intergranular filling by kaolinite and mud; visible corrosion in detritus grains

Feature description of Sample 419:
Content of medium sand is about 60%, fine sand 26%, and the rest is coarse sand. Clastic particles are mostly close to axis shape, and few of them are elongated, without directional alignment. Grains are in point contact, or line contact locally, with kaolinite, siliceous, and ferruginous (pyrite) cementation, as well as relatively developed pores, but the distribution is not even, so local connectedness is good.

Atlas X Casting slice

A

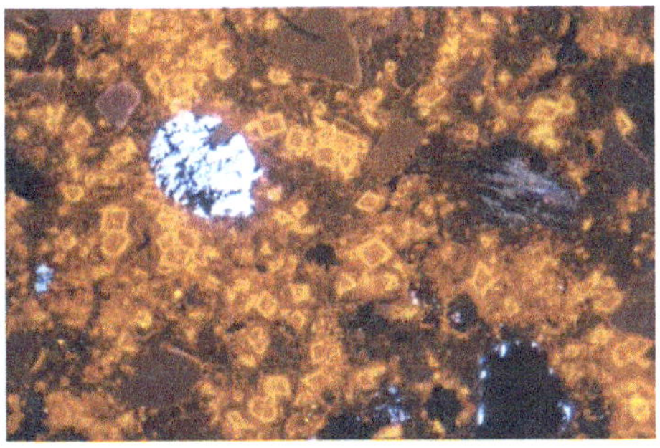

B

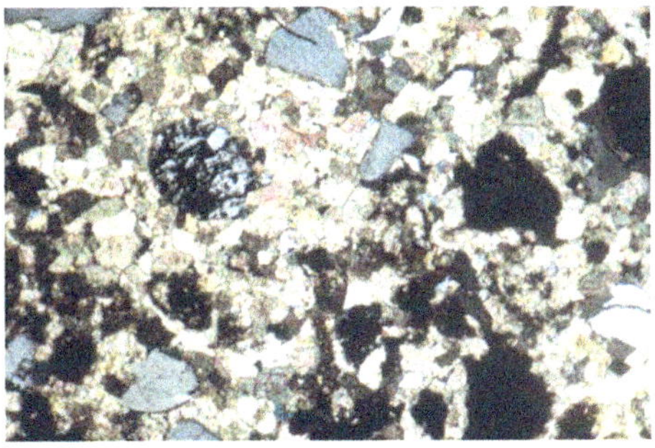

C

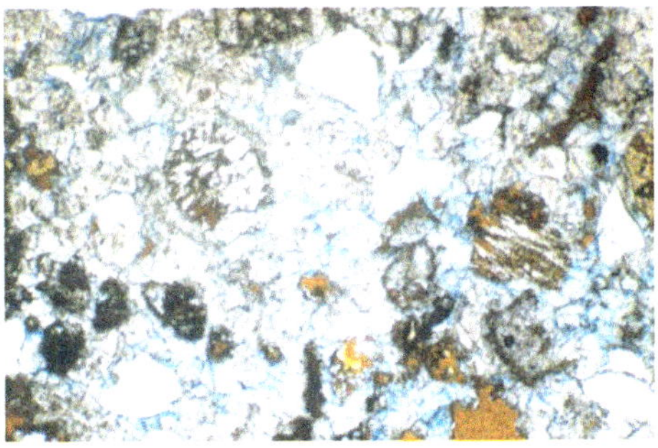

Depth: 2780.3 m Sample number: 569

A. 40x, cathode luminescence
Dolomite has a zone structure, and is euhedral and subhedral. Its core emits red light, and its ring emits orange light; ferruginous dolomite emits orange-red light. It can be seen as superficial ooid with potassium feldspar as the core. The concentric layer is dolomited, seen as white blue feldspar and blue ashes. After dolomite and ferruginous dolomite are filled, feldspar corrodes. Quartz emits dark-brown, dark-purple light, and there is no enlargement

B. 40x, cross-polarized light
At the development site of ferrodolomite and dolomite

C. 40x, plane-polarized light
At the development site of ferrodolomite and dolomite

Atlas XI Cathode luminescence

Atlas XII Scanning electron micrographs

A. XH-18 (2353.58 m, sample number: 167), intergranular kaolinite cement

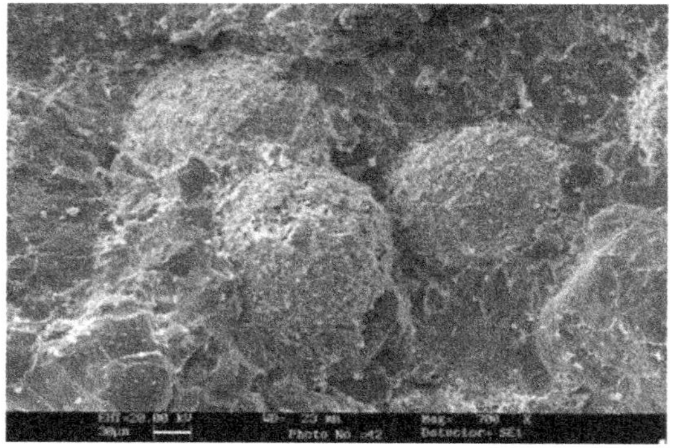

B. XH-22 (2368.28 m, sample number: 207), ooid ($CaCO_3$)

C. XH-35 (2532.26 m, sample number: 223), pocket on quartz surface filled with dolomite and ferrodolomite quartz filled with kaolinite

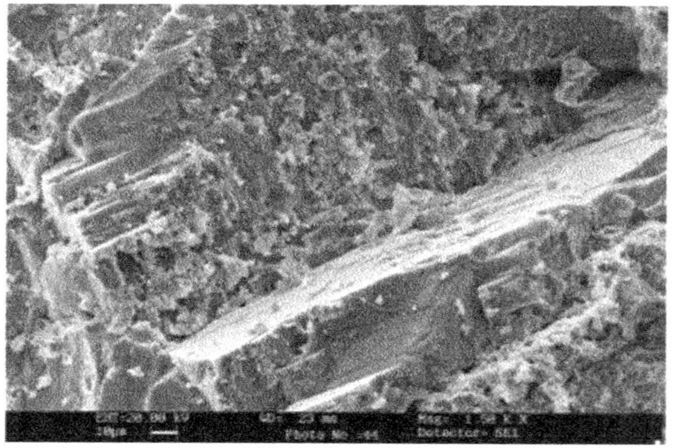

D. XH-48 (2568.39 m, sample number: 258), corroded potassium feldspar surface

Bibliography

Alien, J.R.L. 1964. Studies in fluviatile sedimentation. Six systems from the lower Old Red Sandstone. Anglo-Welsh Basin. *Sedimentology* 3: 163–198.

Allen, J.R.L. 1965. A review of the origin and characteristics of recent alluvial sediments. *Sedimentology* 5.

Allen, J.R.L. 1968. *Current ripples-their relation to patterns of water and sediment.* Amsterdam: North Holland.

Allen, J.R.L. 1983. Studies in fluviatile sedimentation: Bars, bar-complexes and sandstone sheets (low sinuosity braided streams) in the Brownstones (L. Dexonian), Welsh Borders. *Sedimentary Geology* 33: 237–293.

Allen, P.A. 1990. *Basin analysis.* Oxford: Blackwell.

Amaefule, J., M. Altunbay, D. Tiab, D. Kersey, and D. Keelan. 1993. Enhanced reservoir description: Using core and log data to identify hydraulic flow units and predict permeability in uncored intervals/wells. SPE paper 26436, 205–220.

Anatole, Dolgoff. 1998. *Essentials of physical geology.* Boston, New York: Hougton Mifflin Company.

Atlas of Typical Seismic Sections in China. 1987. Beijing: Petroleum Industry Press.

Ayton, C.E. 1977. Seismic stratigraphy—Applications to hydrocarbon exploration, vol. 26, 516pp. Tulsa: American Association of Petroleum Geologists.

Barbara, W.M., J.S. Brian, and C.P. Stephen. 1996. *Environmental geology.* Wiley.

Barwis, J.H, and M.O. Hayes. 1978. Regional patterns of barrier Island and tidal delta deposition applied to hydrocarbon exploration. In *Stratigraphic concepts in hydrocarbon exploration.*

Bathurst, R.G.C. 1971. *Carbonate sedimentary and their diagenesis.* Amsterdam: Elsevier.

Beijing Graduate School, Wuhan College of Geology, Research Institute of Petroleum Exploration and Development, and Dagang Oilfield. 1987. *Sedimentary facies and sedimentary environments of tertiary in the Huanghua depression.* Beijing: Geological Publishing House.

Berg, O.R. 1982. Seismic detection and evaluation of delta and turbidite sequences: Their application to exploration for the subtle trap. *Bulletin of the American Association of Petroleum Geologists* 66: 1271–1288.

Bluck, B.J. 1967. Deposition of some upper Old Red Sandstone conglomerates in the Clyde area: A study in the significance of bedding. *Scottish Journal of Geology* 3 (2): 139–167.

Bluck, B.J. 1974. Structure and directional properties of some valley sandur deposits in southern Island. *Sedimentology* 21: 533–554.

Boersma, J.R., and J.H.J. Terwindt. 1981. Neap-spring tide sequences of intertidal shoal deposits in a mesotidal estuary. *Sedimentology* 28: 151–170.

Boothroyd, J.C. 1972. Coarse grained sedimentation on a Braided Outwash Fan, Northeast Gulf of Alaska, 127. Tech, No. 6, C. R. D. Coastal Research Division, University of South Carolina, Columbia.

Boothroyd, J.C., and G.M. Ashley. 1975. Process, bar morphology and sedimentary structures on braided out-wash fans, North-eastern Gulf of Alaska. In *Glaciofluvial and glaciolacustrine sedimentation*, vol. 23, 193–222. Tulsa: SPEC. Publ. Soc. econ. Paleont. Miner.

Brenchlev, P.J., and B.P. William. 1985. *Sedimentology recent development and applied aspects.* Oxford: Blackwell Scientific Publication.

Brice, J.C. 1982. Stream channel stability assessment. Available from the National Technical Information Service. Springfield VA 22161 as PB 83-118190. Report, 1982.

Bridge, J.S., and J.A. Diemer. 1983. Quantitative interpretation of an evolving ancient river system. *Sedimentology* 30: 55, 599–623.

Buch, D.A. 1974. Stratigraphic traps in sandstones exploration techniques, *AAPG Mem* 21.

Chamberlin, T.C., and R.D. Salisbury. 1909. *A college text-book of geology*. H. Holt and company.

Charles, C.P., and D. McGeary. 1996. *Physical geology with interactive plate tectonics*. Times Mirror Higher Education Group, Inc.

Chen, Jianqiang, Hongrui Zhou, and Xunlian Wang (Compiled). 1998. *Sedimentology and palaeogeography*. Beijing: China University of Geosciences Press.

Chen, Jingshan, and Changming Chen (Trans.). 1981. *Delta sedimentary and oil & gas exploration*. Beijing: Petroleum Industry Press.

Chen, Lihua, and Zaixing Jiang. 1993. *Reservoir experimental testing techniques*, 220–232. Beijing: Petroleum Industry Press.

Chen, Shaozhou, et al. 1982. China tertiary marine-continental transitional facies. *Oil & Gas Geology* 3 (4).

Chilingarian, G.V, R.W. Mannon, and H. Rieke. 1972. Oil and gas production from carbonate rocks, 408pp. Amsterdam: Elsevier Pub. Co.

China Typical Log Interpretation Atlas. 1987. Beijing: Petroleum Industry Press.

Chum, R.M.J., P.R. Vail, and J.B. Sangree. 1977. Seismic stratigraphy and global changes of sea level, Part 6: Stratigraphic interpretation of seismic reflection patterns in depositional sequences. In *Seismic stratigraphy—Applications to hydrocarbon exploration*, vol. 26, ed. C.E. Payton, 117–133. Tulsa: Mem. Am. Ass. Petrol. Geol.

Coleman, J.M. 1982. *Deltas processes models of deposition for exploration*. Boston: International Human Resources Development Corporation.

Coleman, J.M., S.M. Gagliano, and J.E. Webb. 1964. Minor sedimentary structures in a prograding distributary. *Marine Geology* 1: 240–258.

Collinson, J.D., and J. Lewin. 1983. *Modern and ancient fluvial systems*, vol. 6. Int Assoc Sedimental Spec Publ.

Compile group of Sedimentary Structure and Environmental Interpretation. 1984. *Sedimentary structure and environmental interpretation*. Beijing: Science Press.

Cooperation Group of Petroleum Logging Information (Compiled). 1998. *The applications of new well logging*. Beijing: Petroleum Industry Press.

Corllinson, J.D., and D.B. Thomposon. 1982. *Sedimentary structures*. London: George Allen & Unwin Ltd.

Cross, T.A., and M.A. Lessenger. 1998. Sediment volume partitioning: Rationale for stratigraphic model evaluation and high-resolution stratigraphic correlation. In *Sequence stratigraphy concepts and application*, vol. 8, ed. F.M. Gradstein, K.O. Sandvik, and N.J. Milton, 171–195. NPF Special Publication.

Crossey, L.J., B.R. Frost, and R.C. Surdam. 1984. Secondary porosity in laumontite bearing sandstones, clastic diagenesis. *AAPG Memoir* 37: 225–238.

Curtis, D.M. 1970. Miocene deltaic sedimentation, Louisiana Gulf Coast. In *Deltaic sedimentation modern and ancient*, vol. 15, ed. J.P. Morgan, 293–308. Soc. Econ. Paleontologists Mineralogists Spec. Publ.

Davis, R.A. 1992. *Depositional systems: An introduction to sedimentology and stratigraphy*. Englewood Cliffs: Prentice-Hall, Inc.

Demaison, G, and B.J. Huizinga. 1991. Genetic classification of petroleum system. *AAPG Bulletin*.

Deng, Hongwen, and K. Qian. 1993. *Sedimentary geochemistry and environment analysis*, 1–63. Lanzhou: Gansu Science and Technology Press.

Dickinson, W.R. 1974. Plate tectonics and sedimentation. In *Tectonics and sedimentation*. S. E. P. M. Spec. Publ. 22.

Dott, R.H. 1988. *Evolution of the earth*. New York: McGraw Hill.

Duan, Taizhong, Jianhua Guo, and Zhenzhong Gao, et al. 1990. Lower ordovician series carbonate contourite mound of ancient continent margin in Jiuxi of Northern Hunan in South China. *Acta Geologica Sinica* 64 (2): 131–141.

Elliott, T. 1986. Siliciclastic shorelines. In *Sedimentary environments and facies*, ed. H. G. Reading, 155–188. Oxford: Blackwell.

Emery, K.O. 1952. Continental shelf sediments of Southern California. *Geological Society of America Bulletin* 63.

Emery, K.O. 1968. Relict sediments on continental shelves of the world. *AAPG* 52.

Emiliano, Mutn. 1992. *Turbidite sandstones*. Insituto di Geologia University di Parma.

Enderby, A.L. (US), as chief editor. 1981. *Deep-sea sediments* (Physical Mechanical Properties). Beijing: Ocean Press.

Ethridge, F.G. 1985. Modern alluvial fans and fan deltas. In *Recognition of fluvial systems and their resource potential*, vol. 19, ed. R.M. Flores, F.G. Ethridge, A.D. Miall, W.E. Galloway, and T.D. Fouch, 101–126. Soc Econ Paleontol Mineralog Short Course.

Ethridge, F.G., and W.A. Wescott. 1984. Tectonic setting, recognition and hydrocarbon reservoir potential of fan-delta deposits. *Sedimentology of Gravels and Conglomerates*.

Faugeres, Jean Claude, and Dorrik A.V. Stow, Patrice Imbert, et al. 1999. Seismic features diagnostic of contourite drifts. *Marine Geology* 162: 1–38.

Feng, Zengzhao as chief editor. 1993. *Petrology of the sedimentary rocks*. Beijing: Petroleum Industry Press.

Fisher, R.V. 1979. Models for pyroclastic surges and pyroclastic flows. *Journal of Volcanology and Geothermal Research* 6: 305–318.

Fisher, W.L., and L.F. Brown. 1972. Clastic depositional system a genetic approach to facies analysis. Annotated outline and bibliography, Texas University, Austin. *Bureau of Economic Geology*.

Fisher, W.L., et al. 1967. *Delta systems in the exploration for oil and gas*. Austin: Bur. Ecoa University of Texas.

Fisk, H.N. 1960. Recent Mississippi river sedimentation and peat accumulation In *4th Congress Pour Iavancement des' etudes de stratigraphic et de geologic ducarbonifere, Heerlen, Compt Rendu*, ed. Ernest Van Aelst.

Fisk, H.N., E. McFarlan, et al. 1954. Sedimentary frame-work of the modern Mississippi delta. *Journal of Sedimentary Petrology* 24: 76–99.

Foster, J.E. 2004. Observation of quartz cathode-luminescence in a low pressure plasma discharge.

Fraser, H.J. 1935. Experimental study of the porosity and permeability of clastic sediments. *The Journal of Geology* 43: 910–1010.

Friedman, G.M. 1985. The problems of submarine cement in classifying reefrock: An experience in frustration. In *Carbonate cements*, vol. 36, ed. N. Schneiderman and P. M. Harris, 121–177. Tulsa: Spec. Publ. Soc. Econ. Paleont. Miner.

Galloway, W.E. 1985. Meandering streams-modern and ancient. In *Recognition of fluvial systems and their resource potential*, vol. 19, ed. R.M. Flores, F.G. Ethridge, A.D. Miall, W.E. Galloway, and T.D. Fouch, 145–166. SEPM Short Course.

Galloway, W.E. 1998. Siliciclastic slope and base-of slope depositional systems: Component facies: Strati-graphic architecture, and classification. *AAPG Bulletin* 82 (4): 569–595.

Galloway, W.E., and D.K. Hobday. 1989. *Application of terrigenous detrital depositional system in petroleum, coal and oil exploration*. Beijing: Petroleum Industry Press.

Gao, Yanxin. 1988. *Fan delta sedimentation in fault lacustrine basin in Eastern China*. Beijing: Petroleum Industry Press.

Gao, Zhenzhong, and K.A. Enksson. 1993. Ordovician internal-tide deposit in seabed channel in Sue Finncannon of Appalachian Mountains of US. *Acta Sedimentologica Sinica* 11 (1): 12–22.
Gao, Zhenzhong, and Youbin He. 2002. A new field of petroleum exploration in China: Deep-water traction current deposits. *Marine Origin Petroleum Geology* 7 (4): 1–7.
Gao, Zhenzhong, Youbin He, Chengxin Liu, et al. 2006. History, status and prospect of study on deep-water traction current deposits. *Journal of Palaeogeography* 8 (3): 331–337.
Gao, Zhenzhong, Youbin He, Shunshe Luo, et al. 1996. *Deep-water traction current deposits—A study of internal tide, internal wave, contour currents and their deposits*. Beijing: Science Press.
Gawthorpe, R.L, and A. Colella. 1990. Tectonic control on coarse-grained delta depositional systems in rift basins. In *Recent and ancient nonmarine depositional environments: Models for exploration*, vol. 31, ed. A. Colella and D.B. Prior, 127–155. Soc. Econ. Paleontol. Mineral. Spec. Publ.
Geological Research Institute, Dagang Oil Field, et al. 1985. *Lihe River alluvial fan-delta depositional system*. Beijing: Geological Publishing House.
Ginsburg, Klein R.N. 1975. *Tidal deposits: A casebook of recent examples*, vol. 428. Berlin: Springer.
Graham, R. Thompsom, and Turk Jonathan. 1998. *Introduction to physical geology*. Saunders College Publishing in the United States of America.
Guan, Shicong, et al. 1980. *Mesozoic-cenozoic continental deposit, structure and oil gas in Eastern China*. Collected Papers of Petroleum Geology. Beijing: Geological Publishing House.
Guo, Chengxian. 2000. The thirty year study of deep water allogene deposits in China. *Journal of Palaeogeography* 2 (1): 1–9.
Harms, J.C., et al. 1975a. *Depositional environments as interpreted from primary sedimentary structures and stratification sequence*. S. E. P. M. short course No. 2 Dallas Texas.
Harms, J.C., et al. 1975b. *Depositional environments as interpreted from primary sedimentary structures and stratification sequence*. Texas: Society of Sedimentary Geology
Harms, J.C., J.B. Southard, and R.G. Walker. 1982. *Structures and sequences in clastic rock*. Lecture notes for short course No. 9. Sponsored by the S. E. P. M. Tulsa Oklahoma.
Hayes, M. 1980. General morphology and sediment patterns in tidal inlets. *Sedimentary Geology* 26: 139–156.
He, Dengchun, et al. 1987. *Logging stratigraphic analysis and evaluation of oil and gas*. Beijing: Petroleum Industry Press.
He, Jingyu, and Xianghua Meng. 1987. *Sedimentary rock, sedimentary facies model and construction*. Beijing: Geological Publishing House.
He, Youbin, Zhenzhong Gao, and Xingyang Zhang. 2002. Middle-upper ordovician internal-wave and internal-tide deposits and their significance of petroleum geology in the Central Tarim Basin. *Marine Origin Petroleum Geology* 7 (4): 33–40.
Hearn, C.L., W.J. Ebanks, R.S. Tye, et al. 1984. Geological factors influencing reservoir performance of the Hartzog Draw filed. *Wyoming. Journal of Petroleum Technology* 36 (7).
Hillier, S., M.J. Wilson, and R.G. Merriman. 2006. Clay mineralogy of the Old Red Sandstone and Devonian sedimentary rocks of Wales. *Scotland and England Clay Minerals* 41: 433–471.
Hong, Qingyu. 1992. *Geology of gravity flow of sediments*. Chengdu: Chengdu Science and Technology University Press.
Hooke, R.L. 1969. Processes on arid-region alluvial fans. *The Journal of Geology* 75: 438–460.
Houseknecht, D.W. 1987. Assessing the relative importance of compaction processes and cementation to reduction of porosity in sandstones. *AAPG Bulletin* 71: 633–642.

Houseknecht, D.W. 1988. Intergranular pressure solution in four quartzose sandstones. *Journal of Sedimentary Research* 58: 228–246.

Hunt, J.M. 1990. Generation and migration of petroleum from abnormal pressured fluid compartments. *AAPG.*

Hutcheon, I, and H. Abercrombie. 1990. Carbon dioxide in clastic rocks and silicate hydrolysis. *Geology* IB: 541–544.

Illing, L.V. 1954. Bahamian calcareous sands. *AAPG* 38: 1–95.

Ji, Youliang, and Shiqi Zhang. 1998. *Principles of sequence stratigraphy and sequence genetic mechanism mode.* Beijing: Geological Publishing House.

John, G.M., et al. 1988. Fan deltas and braid deltas: Conceptual problems. In *Fan deltas*, ed. W. Nemec and R.J. Steel. Berlin: Springer.

Kang, Yongshang, and Jie Wang. 1999. Fluid dynamic systems and the formation of oil and gas pools. *Acta Petrolei Sinica* 20 (1): 30–33.

Kang, Yongshang, and Wenkuang Wu. 1999. Methodological system of fluid analysis in petroliferous basins: With a discussion on several research subjects for which studies should be enhanced 45 (4): 151–157.

Kang, Yongshang, and Xiongqi Pang. 1998. Principles of oil and gas formation fluid dynamic system analysis and their applications. *Acta Sedimentologica Sinica* 16 (3): 80–84.

Keliing, G., and G.T. George. 1971. Upper Carboniferous sedimentation in the Prmbrokeshire coalfield. In *Geological excursions in South Wales and the Forest of Dean*, ed. D.A. Bassett and M.G. Bassett, 240–259. Cardiff: Geol. Ass. South Wales Group.

Kendall, A.C. 1983. Subaqueous evaporates. In *Facies models*, ed. R.G. Walker, 159–174. Geosci Canada Reprint Ser 1.

Kenneth, J.H., K. Kerry, and W.V. James. 1979. Resedimented facies in Ventura Basin, California and model of longitudinal transport of turbidity currents. *AAPG* 64 (7): 1034–1051.

Klein, G.V. 1972. A sedimentary model for determining Plaeotidal range. *Geological Society of America Bulletin* 82: 2585–2592.

Klein, G.V. 1975. Depositional facies of leg 30, deep sea drilling project sediment cores. In *Initial reports of the deep sea drilling project*, vol. 30, 423–442. Washington: U. S. Government Printing Office.

Kolia, V., and F. Coumes. 1987. Morphology, internal structure, seismic stratigraphy, and sedimentation of Indus fan. *APPG* 71: 650–677.

Krebs, W. 1972. Facies and development of the Meggen Reef (Devonian, West Germany). *Geologische Rundschau* 61 (2): 647–671.

Kruit, C., J. Brouwer, and G. Knox. 1975. Schollnberger W&Vliet Van a. Vne excursion cones.

Krynine, P.D. 1949. Mineralogical research on oil reservoirs. *Mineral Industries* (University Park) 18 (4): 1–3.

Lane, Z.W. 1957. *A study of the shape of channels formed by natural streams flowing in erodible material*, vol. 9, 106. U. S. Armycops of Engineers Missouri River Division, Omaba, Nebraska, Sedimentseries.

Li, Jianfeng, and Xinghe Yu. 1996. *Geological modeling and computer simulation of clastic rock reservoir.* Beijing: Geological Publishing House.

Li, Maolin, and Wen Qing Li. 1979. *Geological basis for oil and gas field development.* Beijing: Petroleum Industry Press.

Li, Pingri, et al. 1987. *Hanjiang delta.* Beijing: Ocean Press.

Li, Sitian. 1988. *Regularity of fault basin and coal accumulation.* Beijing: Geological Publishing House.

Li, Sitian, Hua Wang, Yang Quan Jiao, et al. 2005. *Sedimentary basins analysis basis and application*, 2nd ed. Beijing: Higher Education Press.

Li, Yanfang, and Wenguang Wang, as chief editor. 1998. *Sedimentary rock and sedimentary facies.* Beijing: Petroleum Industry Press.

Li, Zhoubo (Trans.). 1991. Schlumberger Limited (US). *Log interpretation principles/applications*. Beijing: Petroleum Industry Press.
Liu, Jingkui. 1983. Discussion on architectural feature of conglomerate reservoir and reservoir evaluation. *Petroleum Exploration and Development* (2).
Liu, Menghui, et al. 1990. *Translation collection of second international reservoir characterization technical seminar*. Dongying: China University of Petroleum Press.
Liu, Qun, and Yuhua Chen. 1987. *Salt sediments of terrigenous clastic-chemogenic type in meso-cenozoic of China*. Beijing: Beijing Science and Technology Publishing House.
Liu, Zerong, Quanlin Xin, et al. 1993. *Reservoir description method principle*. Beijing: Petroleum Industry Press.
Loch, R.J. 2010. Bedload transport of mud as pedogenic aggregates in modern and ancient rivers[J]. *Sedimentology* 36 (2): 291–306.
Lu, Kezheng (Chief editor). 2006. *Petroliferous basin analysis*. Beijing: China University of Petroleum Press.
Luo, Minggao (Compiled). 1998. *Quantitative reservoir geology*. Beijing: Geological Publishing House.
MacGowan, D., and R.C. Surdam. 1988. Difunctional carboxylic acid anions in oil-field water. *Organic Geochemistry* 12: 245–259.
Matt, A., and M.E. Tucker (Chief editor). 1984. *Modern and ancient lake sediments*. Beijing: Geological Publishing House.
Mckee, E.D., E.J. Crosby, and H.L. Berryhill, Jr. 1967. Flood deposits, Bijoll Greek, Colorado, June 1965. *Journal of Sedimentary Petrology* 37: 829–851.
Miall, A.D. 1977. A review of the braided-river depositional environment. *Earth-Science Reviews* 13: 1–62.
Miall, A.D. 1984. *Principles of sedimentary basin analysis*. New York: Springer.
Miall, A.D. 1990. *Principles of sedimentary basin analysis*. New York Inc.: Springer.
Miall, A.D., and M.R. Gibling. 1978. The Situro-Devonian clastic wedge of somerset island, Arctic Canada, and some regional paleogeographic implication. *Sedimentary Geology* 21.
Mitchell, A.H.G., and H.G. Reading. 1986. Sedimentation and tectonics. In *Sedimentary environments and facies*, 2nd ed, ed. H.G. Reading, 471–519. Oxford: Blackwell.
Morgan, J.P. 1951. Report on the mudlumps at the mouths of the Mississippi River, Partal, the occurrence and origin of the mudlumps at the mouths of the Mississippi River: New Orleans District, U. S. Army corps of Engr, Unpub, Rept, 127pp.
Morgan, J.P. 1970. *Deltaic sedimentation modern and ancient*, vol. 15, 312. Society of Economic Paleontologists and Mineralogists, Spec. Publ.
Mutti, E., and G. Davoli. 1992. *Turbidite sandstones*. Agip, Istituto di geologia, University di Parma.
MvCammon, R.B. 1975. On the efficiency of systematic point-sampling in mapping facies. *Journal of Sedimentary Petrology* 45.
Nelson, C.H., and A. Maldonado. 1988. Factors controlling depositional patterns of Ebro turbidite systems, Mediterranean Sea. *The American Association of Petroleum Geologists* 698–716.
Nemec, W., and R.J. Steel. 1984. Alluvial and costal conglomerates: Their significant features and some comments on gravelly mass-flow deposits in sedimentology of gravels and conglomerates.
Nemec, W., and R.J. Steel. 1988. What is a fan delta and how do we recognize it? In *Fan deltas: Sedimentology and tectonic settings*, ed. W. Nemec and R.J. Steel, 3–13.
Nilsen, T.H. 1981. Alluvial fan deposits. In *Sandstones depositional environments*, 49–85.
Noel, P.J., and A.C. Kendall. 1992. *Introduction to carbonate and evaporite facies models (in Facies models; response to sea level change)*, 265–275. Geological Association of Canada, St. Johns, NL, Canada.
Passega, R. 1977. Significance of CM diagrams of sediments by suspensions. *Sedimentology* 24.

Payton, C.E. 1977. Seismic stratigraphy-application to hydrocarbon exploration. *American Association of Petroleum Geologists Memoir* 26

Payton, C.E. 1980. *Seismic stratigraphy application to hydrocarbon exploration*. Huaida Xu (Trans.). Beijing: Petroleum Industry Press.

Petroleum Geology Institute of Chinese Petroleum Society (Trans. and Ed.). 1986. *Study of turbidite and fan delta overseas*. Beijing: Petroleum Industry Press.

Pettijohn, F.J., P.E. Potter, and R. Siecer. 1987. *Sand and sandstone*, 2nd ed, 553pp. New York, Heidelberg: Springer.

Plummer, C.C., and D. McGeary. 1996. *Physical geology with interactive plate tectonics*. Dubuque in United States: Wm. C Brown Publishers.

Qian, Liying. 1990. Fan delta and braided delta-two different types of coarse delta. *Sedimentary Geology and Tethyan Geology* (5): 60–67.

Qiu, Longwei, Zaixing Jiang, et al. 2001. Alkaline diagenesis and its effect on reservoirs in Biyang Sag. *Science in China (Series D)* 31 (9): 752–758.

Qiu, Yinan, Shuhao Xue, and Fengxiang Ying. 1991. *China continental facies petroleum reservoir*. Beijing: Petroleum Industry Press.

Qiu, Yinan, and Ziqi Chen. 1996. *Reservoir description*. Beijing: Petroleum Industry Press.

Reading, H.G. 1978. *Sedimentary environments and facies*. Oxford: Blackwell.

Reading, H.G. 1986. *Sedimentary environments and facies*, 2nd ed. Oxford: Blackwell Scientific Publications.

Reading, H.G. 1996. *Sedimentary environments: Processes, facies and stratigraphy*, 3rd ed. Blackwell Scientific Publications.

Reading, H.G. (Chief editor). 1986. *Sedimentary environments and facies*. Beijing: Science Press.

Reinecke, H.E., and I.B. Singh. 1979. *Depositional sedimentary environments with reference to terrigenous clastics*. Changming Chen and Jiliang Li (Trans.). Beijing: Petroleum Industry Press.

Ren, Mingda, and Nailiang Wang. 1981. *Introduction to the modern sedimentary environment*. Beijing: Science Press.

Russell, R.J. 1967. *River plains and sea coasts*, vol. 173. Berkeley and Los Angeles: University of California Press.

Rust, B.R. 1978. A classification of alluvial channel systems. In *Fluvial sedimentology*, ed. A.D. Miall, 187–198. Canadian Society of Petroleum Geologists Memoir.

Rust, B.R, and A.S. Legun. 1983. *Modern anastomosing-fluvial deposits in arid central Australia, and a Carboniferous analogue in New Brunswick, Canada*, vol. 6, 385–392. Collinson Int Assoc Sedimental Spec Publ.

Sangree, J.B., P.R. Vail, and R.M Sneider. 1988. Evolution of facies interpretation of the shelfslope: Application of the new eustatic framework to the Gulf of Mexico. In *Offshore Technology Conference Proceeding*, vol. 5695, 133–144.

Schlumberger, L. 1970. *Fundamentals of dipmeter interpretation*, vol. 145. New York.

Schmidt, V., and D.A. McDonald. 1979a. Texture and recognition of secondary porosity in sandstones. *SEPM*. SpecPub. No 26: 209–225.

Schmidt, V., and D.A. McDonald. 1979b. The role of secondary porosity in the course of sandstone diagenesis (in Aspects of diagenesis). *Special Publication-Society of Economic Paleontologists and Mineralogists?* (26): 175–207.

Schumm, S.A. 1963. Sinuosity of alluvial rivers on the Great Plains. *Bulletin of the Geological Society of America* 74: 1089–1100.

Schumm, S.A. 1981. Evolution and response of the fluvial system, sedimentological implications. In *Recent and ancient nonmarine depositional systems: Models for exploration*, vol. 31, ed. F.G. Ethridge and R. M. Flores, 19–29. Soc Econ Paleontol Mineralog Spec Publ.

Scott, A.J, and J.H. McGowen. 1969. Deltas systems in the exploration for oil and gas: A research colloquium: University of Texas (Austin). *Bureau of Economic Geology* 178.

Serra, O. 1992. *Geological interpretation of logging data*. Beijing: Petroleum Industry Press.

Shanmugam, G., R.B. Bloch, R.B. Damuth, et al. 1997. Basin-floor fans in the North Sea: Sequence stratigraphic models vs. sedimentary facies: Reply. *AAPG Bulletin* 81: 662–672.

Shen, Xichang, and Guo Buying (Compiled). 1993. *Marine geology*. Wuhan: China University of Geosciences Press.

Sheriff, R.E. 1980. *Seismic stratigraphy*. Boston, MA: International Human Resources Development Corporation.

Shi, Hesheng, Wenxiang Li, Xinghe Yu, Xiaoping Zou, et al. 1999. A study on sedimentary facies of sandstone oil fields in Eastern Pearl River Mouth Basin. *China Offshore Oil and Gas (Geology)* 13 (3): 181–188.

Shi, Hesheng, Wenxiang Li, Xinghe Yu, Xiaoping Zou, et al. 2000. The application of sequence stratigraphy in developing sandstone oilfields in Eastern Pearl River Mouth Basin. *China Offshore Oil and Gas (Geology)* 14 (1): 15–20.

Smith, D.G. 1983. Anastomosed fluvial deposits, modern examples from Western Canada. In *Modem and ancient fluvial systems*, vol. 6, ed. J.D. Collinson and J. Lewin, 155–168. Int Assoc Sedimental Spec Publ.

Smith, D.G., and P.E. Putnam. 1980. Anastomosed river deposits, modern and ancient examples in Alerta, Canada. *Canadian Journal of Earth Sciences*.

Smith, N.D. 1974. Sedimentology and bar formation in the upper Kicking Horse River, a braided river. *Journal of Sedimentary Petrology* 42: 624–634.

Sprearing, D.R. 1974. Summary sheets of sedimentary deposits with bibliographies, Mc-8. *Geological Society of America*, 7 sheets, Boulder.

Stow, D.A.V. 1985. Deep Sea clastic: Where are we and where are we going? In *Sedimentology recent developments and applied aspects*, ed. P.J. Brenchley and Williams, 67–93, 13 pp. Oxford: Blackwell Scientific Pub.

Stow, D.A.V., and M. Johansson. 2000. Deep-water massive sands: Nature, origin and hydrocarbon implications. *Marine and Petroleum Geology* 17: 145–174.

Surdam, R.C., L.J. Crossey, E.S. Hagen, et al. 1989. Heasler. Organic-inorganic and sandstone diagenesis. *AAPG Bulletin* 73: 1–23.

Tankard, A.J., and D.K. Hobday. 1977. Tide dominated back-barrier sedimentation, early Ordovician cape Basin, Cape Peninsula, South Africa. *Sedimentary Geology* IB: 135–159.

Tankard, A.J., M.P.A. Jackson, K.A. Eriksson, et al. 1982. *Crustal evolution of southern Africa; 3.8 billon years of earth history*. New York: Springer.

Thompson, G.R., and T. Jonathan. 1991. *Modem physical geology*. Saunder College Publishing, A Division of Holt, Rinehart and Winston, Inc.

Tian, Zaiyi. 1983. Potential evaluation on China petroleum geological structure and oil gas. *Acta Petrolei Sinica* 4 (1): 1–10.

Van Hinte, J.E. 1987. Geohistory analysis-application of micropaleontology in exploration geology. *AAPG* 62.

Walker, R.G. 1984a. *Facies models*, 2nd ed. Toronto: Geological Ass. of Canada Business and Economic Service Ltd.

Walker, R.G. 1984b. Shelf and shallow marine sands. In *Facies models*, ed. R.G. Walker, 141–170. Geosci. Can. Reprint Ser., 1. Geol. Soc. Can. Waterloo, Ontario.

Walker, R.G., and D.J. Cant. 1979. Facies models 3. Sandy fluvial systems. In *Facies models*: Geosci. Can. Reprint Series, vol. 1, ed. R.G. Walker, 23–31.

Walker, R.G., and N.P. James. 1992. Facies models: Response to sea level change. *Geological Association of Canada.*
Walker, T.R. 1976. Diagenetic origin of continental red beds. In *The continental Permian in Central, West, and South Europe, Reidel, Dordrecht*, ed. H. Falke, 240–282.
Wang, Defa, et al. 1983. Depositional features and oil and gas distribution in some of the faulted basins in North China. *Acta Petrolei Sinica* 14 (1).
Wang, Hengjian, et al. 1983. A study of fresh water carbonates in the cretaceous Qiangshankou and Nenjiang formations in Western Songliao Basin. *Petroleum Exploration and Development (Geology).*
Wang, Liangchen, and Jinliang Zhang. 1996. *Sedimentary environments and sedimentary facies.* Beijing: Petroleum Industry Press.
Wang, Puju, and Wanzhu Liu (Compiled). 2001. *Depositional events: Introduction, example, application.* Changchun: Jilin Science & Technology Press.
Wang, Shouqing. 1987. A heterogeneous model describing the heterogeneity of the production reservoir in Shuanghe oil field. *Petroleum Exploration and Development* (1).
Wang, Yuncheng, and Zhetan Luo. 1986. *Pore structure of oil and gas reservoir.* Beijing: Science Press.
Wardlaw, N.C., and R.P. Tayler. 1976. Mercury capillary pressure curves and the interpretation of pore structure and capillary behavior in reservoir rocks. *BCPG* 24 (2): 225–262.
Weber, K.J. 1982. Influence of common sedimentary structure on fluid flow in reservoir model. *JPT* 665–672.
Weber, K.J., and L.C. VanGeuns. 1990. Framework for constructing clastic reservoir simulation models. *JPT* (October).
Weeks, L.G. 1948. Paleogeography of South America. *America, Bulletin, Geological Society* 59: 249–282.
Wilgus, C.K. 1993. *Principle of sequence stratigraphy.* Beijing: Petroleum Industry Press.
Williams, P.F., and B.R. Rust. 1969. The sedimentology of a braided river. *Sedimentary Petrol* 39: 649–679.
Wilson, J.G. 1975. *Carbonate facies in geologic history.* New York: Springer.
Wright, A.E., and F. Moseley. 1975. Ice Ages: Ancient and modern. *The Journal of Geology.* Spec. Issue 6: 320pp.
Wu, Chongjun. 1981. The division of lake sedimentary facies. *Petroleum Exploration and Development* (2).
Wu, Chongjun. 1982. Sandstone oil-gas oilfield geomorphology. *Developments in Petroleum Science* 4.
Wu, Chongjun. 1983. Structural lake deltas and oil-gas distribution. *Acta Sedimentologica Sinica* (1).
Wu, Shenghe, and Qihua Xiong. 1998. *Reservoir geology.* Beijing: Petroleum Industry Press.
Wu, Yuanyan, and Bijue Chen. 1996. *Industrial geology of petroleum.* Beijing: Petroleum Industry Press.
Xu, Huaida, Shifeng Wang, and Kaiyuan Chen. 1990. *The basic interpretation of seismic stratigraphy.* Wuhan: China University of Geosciences Press.
Ye, Lianjun. 1960. Comment on classification principle and method of depositional facies and sedimentary structure. *Geological Science* (6).
Ying, Fengxiang, Dongbo He, Yumei Long, et al. 2003. The division of diagenetic stages in clastic rocks. People's Republic of China Natural Gas Industry Standards.
Yu, Xinghe, Daojian Zhang, et al. 1999. Depositional facies model of deep Shahejie formation in eastern and western depressions in Liaohe oilfield. *Journal of Palaeogeography* 1 (3).
Yu, Xinghe, and Defa Wang. 1997. The architectural elements of the deltaic system in the terrestrial faulted basin and the significance of its reservoir geological model. *Geological Review* 43 (3).
Yu, Xinghe, Defa Wang, and Junmao Zheng. 1992. Lithofacies association types or

sequences and depositional system of Permian sandstones in North China. *Acta Sedimentologica Sinica* 10 (1): 27–34.

Yu, Xinghe, and Defa Wang, et al. 1994. Braided River delta sandbody characteristics and sandbody distribution model-modern delta sedimentary study in Daihai Lake of Inner Mongolia. *Acta Sedimentologica Sinica* 15 (1): 27–36.

Yu, Xinghe, and Hui Jiang, et al. 2007. Depositional filling models and controlling factors on mesozoic and cenozoic fault basins of terrestrial facies in Eastern China—A case study of Dongying Sag of Jiyang depression. *Lithologic Reservoirs* 19 (1): 39–45.

Yu, Xinghe, and Jianfeng Li. 1995. Challenges and trends in the study of hydrocarbon reservoir. *Earth Science Frontiers* 2 (3–4).

Yu, Xinghe, Junmao Zheng, et al. 1997. The establishment of integrated model on structure, deposition and diagenesis. *Acta Sedimentologica Sinica* 15 (3).

Yu, Xinghe, Xingxiang Ma, Longxin Mu, et al. 2004. *Braided River reservoir geologic mode and analysis of hierarchy interfaces*. Beijing: Petroleum Industry Press.

Yu, Xinghe, and Xiujuan Zheng. 2004. Review of the development and prospect of sedimentology. *Advance in Earth Sciences* 19 (2): 173–182.

Yu, Xinghe, and Yongjiao Chen. 2004. Eight major depositional processes of clastic rocks and their petroleum reservoir characterization. *Petroleum Geology and Experiment* 26 (6): 517–524.

Yu, Xinghe, Zhijie Zhang, Xin Su, et al. 2004. Conditions for sedimentation of gas hydrate and its distribution in South China Sea. *Earth Science Frontiers* 11 (1): 311–315.

Zhang, Xingyang. 2000. Bedform unit assemblages of deep-water traction-current origin. *Journal of Palaeogeography* 2 (2): 28–35.

Zhang, Xingyang, Youbin He, Shunshe Luo, et al. 2002. Deep-water sediment waves formed by internal waves. *Journal of Palaeogeography* 4 (1): 83–88.

Zhang, Xueru, Zhongren Niu, et al. 1995. *Liaohe River fault lacustrine basin clastic rocks development reservoir research*. Beijing: Petroleum Industry Press.

Zhao, Chenglin. 2001. *Principles of sedimentology*. Beijing: Petroleum Industry Press.

Zhao, Chenglin, et al. 1992. *Depositional systems and diagenesis of lower tertiary clastic rocks in Dongpu Sag*. Beijing: Petroleum Industry Press.

Zhu, Guohua. 1985a. Formation of lomonitic sand bodies with secondary porosity and their relationship with hydrocarbons. *Acta Sedimentologica Sinica* 6 (1): 1–8.

Zhu, Guohua. 1985b. Formation of low permeability sandbodies and secondary pore sandbodies in the upper Permian Yanchang series of South-western Shan-Gan-Ning basin. *Acta Sedimentologica Sinica* 3 (2).

Zhu, Yongan, et al. 1983. Sedimentary features of the Shuanghe subaqueous alluvial fan in the Miyang depression. *Acta Sedimentologica Sinica* 4 (1).

Zimkernagel, M.U. 1978. Cathodoluminescence of quartz and its application to sandstone petrology[J]. Contributions to Sedimentary Geology.

The manufacturer's authorised representative in the EU is Springer
Nature Customer Service Centre GmbH, Europaplatz 3, 69115 Heidelberg,
Germany. If you have any concerns regarding our products, please
contact ProductSafety@springernature.com

Printed and bound by CPI Group (UK) Ltd, Croydon, CR0 4YY
30/09/2026
02225587-0013